10 0670205 7
WITH
FROM THE LIBRARY

The Diatoms

Applications for the Environmental and Earth Sciences
Second Edition

This much revised and expanded edition provides a valuable and detailed summary of the many uses of diatoms in a wide range of applications in the environmental and earth sciences. Particular emphasis is placed on the use of diatoms in analyzing ecological problems related to climate change, acidification, eutrophication, and other pollution issues. The chapters are divided into sections for easy reference, with separate sections covering indicators in different aquatic environments. A final section explores diatom use in other fields of study such as forensics, oil and gas exploration, nanotechnology, and archeology. Sixteen new chapters have been added since the First Edition, including introductory chapters on diatom biology and the numerical approaches used by diatomists. The extensive glossary has also been expanded and now includes over 1000 detailed entries, which will help non-specialists to use the book effectively.

JOHN P. SMOL is a Professor in the Department of Biology at Queen's University (Ontario, Canada), with a cross appointment at the School of Environmental Studies. He is also co-director of the Paleoecological Environmental Assessment and Research Lab (PEARL). Since 1990, he has won over 25 research awards and fellowships, including the 2004 NSERC *Herzberg Canada Gold Medal* as Canada's top scientist or engineer.

EUGENE F. STOERMER is a past-President of the Phycological Society of America and the International Society for Diatom Research. He has worked at the University of Michigan (Ann Arbor, USA) since 1965, where he is currently Professor Emeritus in the School of Natural Resources and Environment. He directed the "phyto-lab", which undertook a wide variety of research topics, specializing in diatom systematics and ecology.

Edited by

John P. Smol
Eugene F. Stoermer

The Diatoms

Applications for the Environmental and Earth Sciences

Second edition

CAMBRIDGE UNIVERSITY PRESS
Cambridge, New York, Melbourne, Madrid, Cape Town, Singapore,
São Paulo, Delhi, Dubai, Tokyo, Mexico City

Cambridge University Press
The Edinburgh Building, Cambridge CB2 8RU, UK

Published in the United States of America by Cambridge University Press, New York

www.cambridge.org
Information on this title: www.cambridge.org/9780521509961

First published 1999
Second Edition 2010

Printed in the United Kingdom at the University Press, Cambridge

A catalog record for this publication is available from the British Library

Library of Congress Cataloging in Publication data
The diatoms : applications for the environmental and earth sciences / edited by
John Smol, Eugene Stoermer. – 2nd ed.
p. cm.
Includes index.
ISBN 978-0-521-50996-1 (hardback)
1. Diatoms. 2. Diatoms – Ecology. 3. Plant indicators. I. Smol, J. P. (John P.)
II. Stoermer, Eugene F., 1934– III. Title.
QK569.D54D536 2010
579.8′5 – dc22 2010022107

ISBN 978-0-521-50996-1 Hardback

This book is dedicated to the memory of our friend and colleague, Dr. John Platt Bradbury (1935–2005), a leader in the application of diatoms to the study of earth and environmental issues.

Dr. John Platt Bradbury, sampling peat in Tierra del Fuego (photo by Vera Markgraf).

Contents

Contributors

Leanne K. Armand
Quantitative Marine Science (CSIRO-University of Tasmania) and the Antarctic Climate and Ecosystems Cooperative Research Centre
Hobart, Tasmania, Australia
Current address:
Climate Futures at Macquarie – Department of Biological Sciences
Macquarie University
North Ryde, NSW, 2109
Australia

Richard W. Battarbee
Environmental Change Research Centre (ECRC)
Department of Geography, UCL
Gower Street
London WC1E 6BT
United Kingdom

Helen Bennion
Environmental Change Research Centre (ECRC)
Department of Geography, UCL
Gower Street
London, WC1E 6BT
United Kingdom

Christian Bigler
Department of Ecology and Environmental Science
Umeå University
SE-90187 Umeå
Sweden

H. John B. Birks
Department of Biology
University of Bergen
Thormøhlensgate 53A

N-5007 Bergen
Norway
and
Environmental Change Research Centre (ECRC)
Department of Geography, UCL
Gower Street
London, WC1E 6BT
United Kingdom

Nigel G. Cameron
Environmental Change Research Centre (ECRC)
Department of Geography, UCL
Gower Street
London WC1E 6BT
United Kingdom

Hunter J. Carrick
The Pennsylvania State University
School of Forest Resources
434 Forest Resources Building
University Park
PA 16802
USA

Donald F. Charles
Patrick Center for Environmental Research
The Academy of Natural Sciences
1900 Benjamin Franklin Parkway
Philadelphia, PA 19103-1195
USA

Sherri Cooper
Bryn Athyn College
Bryn Athyn, PA 19009-0717
USA

Xavier Crosta
UMR-CNRS 5805 EPOC
Université Bordeaux 1
Avenue des Facultés,
33405 Talence Cedex
France

Brian F. Cumming
Paleoecological Environmental and Assessment and Research Laboratory (PEARL)
Department of Biology
Queen's University
Kingston, Ontario, K7L 3N6
Canada

Christina L. De La Rocha
Institut Universitaire Européen de la Mer (IUEM)
Université de Bretagne Occidentale (UBO)
Technopôle Brest-Iroise, Place Nicolas Copernic
29280 Plouzané
France

Gregory J. Doucette
NOAA/National Ocean Service
219 Fort Johnson Road
Charleston, SC 29412
USA

Marianne S. V. Douglas
Department of Earth and Atmospheric Sciences,
University of Alberta
Edmonton, Alberta T6G 2E3
Canada

Mark B. Edlund
St. Croix Watershed Research Station
Science Museum of Minnesota
16910 152nd St. North
Marine on St. Croix, Minnesota 55047
USA

Sherilyn C. Fritz
Department of Geosciences and School of Biological Sciences
University of Nebraska – Lincoln
Lincoln, NE 68588-0340
USA

Evelyn Gaiser
Department of Biological Sciences and Southeast Environmental Research Center
Florida International University
Miami, FL 33199
USA

Françoise Gasse
CEREGE, UMR 6635
Marseille University-CNRS-IRD
BP 80
13545 Aix-en-Provence cedex 4
France

Andrey Yu. Gladenkov
Geological Institute
Russian Academy of Sciences
Pyzhevskii per., 7
Moscow 119017
Russia

Richard Gordon
Department of Radiology
University of Manitoba
Winnipeg, Manitoba R3A 1R9
Canada

Irene Gregory-Eaves
Department of Biology
McGill University
1205 Dr. Penfield
Montreal, Québec H3A 1B1
Canada

Roland I. Hall
Department of Biology,
University of Waterloo,
200 University Avenue West,
Waterloo, Ontario N2L 3G1
Canada

Margaret A. Harper
School of Geography, Environment and Earth Sciences
Victoria University of Wellington
PO Box 600 Wellington
New Zealand 6140

David M. Harwood
Department of Geosciences
University of Nebraska-Lincoln
Lincoln, NE 68588-0340
USA

Dominic A. Hodgson
British Antarctic Survey
High Cross
Madingley Road
Cambridge, CB3 0ET
United Kingdom

Benjamin P. Horton
Department of Earth and Environmental Science
University of Pennsylvania
Philadelphia, PA 19104
USA

Jeffrey R. Johansen
Department of Biology
John Carroll University
20700 North Park Blvd.
University Heights
OH 44118
USA

Gareth D. Jones
3 Meadowlands Drive
Westhill, Aberdeenshire AB32 6EJ
United Kingdom

Richard W. Jordan
Department of Earth & Environmental Sciences
Faculty of Science
Yamagata University
Yamagata, 990-8560
Japan

Steve Juggins
School of Geography, Politics and Sociology
Newcastle University
Newcastle upon Tyne NE1 7RU
United Kingdom

Matthew L. Julius
Aquatic Toxicology Laboratory
Department of Biological Sciences
St. Cloud State University
St. Cloud, MN 56301
USA

Irena Kaczmarska
Department of Biology
Mount Allison University
63B York St.
Sackville, NB E4L 1G7
Canada

Bronwyn E. Keatley
Department of Biology
McGill University
1205 Dr. Penfield
Montreal, Québec H3A 1B1
Canada
and
Department of Natural Resource Sciences
McGill University
21,111 Lakeshore Road
Ste. Anne de Bellevue, Québec H9X 3V9
Canada

Galina Khursevich
M. Tank State Pedagogical University
Department of Botany
18 Sovetskaya Street
Minsk 220809
Republic of Belarus

Cathy Kilroy
National Institute of Water and Atmospheric Research
PO Box 8602
Christchurch
New Zealand

William N. Krebs
Petronas Carigali SDN. BHD.
Petronas Twin Towers
Kuala Lumpur, Malaysia
Current address:
2506 Plumfield Lane
Katy
Texas 77450
USA

Kathleen R. Laird
Paleoecological Environmental Assessment and Research Lab (PEARL)
Department of Biology
Queen's University
Kingston, Ontario K7L 3N6
Canada

Melanie J. Leng
NERC Isotope Geosciences Laboratory
British Geological Survey
Keyworth
Nottingham NG12 5GG
United Kingdom
and
School of Geography
University of Nottingham
Nottingham NG7 2RD
United Kingdom

Amy Leventer
Department of Geology
Colgate University
Hamilton, NY 13346
USA

Aude Leynaert
UMR CNRS 6539
Institut Universitaire Européen de la Mer (IUEM)
Technôpole Brest-Iroise, Place Nicholas Copernic
29280 Plouzané
France

Christopher S. Lobban
Division of Natural Sciences
University of Guam
Mangilao
Guam 96923
USA

André F. Lotter
Institute of Environmental Biology
Laboratory of Palaeobotany and Palynology
Utrecht University
Budapestlaan 4
3584 CD Utrecht
The Netherlands

Anson W. Mackay
Environmental Change Research Centre (ECRC)
Department of Geography, UCL
Gower Street
London WC1E 6BT
United Kingdom

Robert M. McKay
Antarctic Research Centre
Victoria University of Wellington
PO Box 600 Wellington
New Zealand 6140

Diane M. McKnight
INSTAAR
1560 30th Street
University of Colorado
Boulder CO 80303
USA

Yangdong Pan
Environmental Sciences and Management
Portland State University
Portland, Oregon 97207
USA

Anthony J. Peabody
Metropolitan Forensic Science Laboratory
109 Lambeth Rd,
London SE1 7LP
United Kingdom

Reinhard Pienitz
Aquatic Paleoecology Laboratory
Centre for Northern Studies
2405 rue de la Terrasse
Université Laval
Québec (Québec) G1V 0A6
Canada

Jennifer Pike
School of Earth and Ocean Sciences
Cardiff University
Main Building, Park Place
Cardiff, CF10 3YE
United Kingdom

Olivier Ragueneau
UMR CNRS 6539
Institut Universitaire Européen de la Mer (IUEM)
Technôpole Brest-Iroise, Place Nicholas Copernic
29280 Plouzané
France

Euan D. Reavie
Center for Water and the Environment
Natural Resources Research Institute
University of Minnesota Duluth
1900 East Camp Street
Ely, Minnesota 55731
USA

Ingemar Renberg
Department of Ecology and Environmental Science
Umeå University
SE-90187 Umeå
Sweden

Oscar E. Romero
Instituto Andaluz de Ciencias de la Tierra (IACT-CSIC)
Facultad de Ciencias, Universidad de Granada
Campus Fuentenueva
18002 Granada
Spain

Kathleen Rühland
Paleoecological Environmental Assessment and Research Laboratory (PEARL)
Department of Biology
Queen's University
Kingston, Ontario K7L 3N6
Canada

Yuki Sawai
Geological Survey of Japan
National Institute of Advanced Industrial Science and Technology
Site C7 1-1-1 Higashi
Tsukuba, Ibaraki, 305-8567
Japan

Carl D. Sayer
Environmental Change Research Centre (ECRC)
Department of Geography, UCL
Gower Street
London WC1E 6BT
United Kingdom

Roland Schmidt
Institute for Limnology
Austrian Academy of Sciences
Mondseestrasse 9
A-5310 Mondsee
Austria

John P. Smol
Paleoecological Environmental Assessment and Research Lab (PEARL)
Department of Biology
Queen's University
Kingston, Ontario K7L 3N6
Canada

Pauline Snoeijs
Department of Systems Ecology
Stockholm University
SE-10691 Stockholm
Sweden

Sarah A. Spaulding
US Geological Survey
INSTAAR, 1560 30th Street
University of Colorado

Boulder CO 80303
USA

Lee Stanish
INSTAAR
1560 30th Street
University of Colorado
Boulder CO 80303
USA

R. Jan Stevenson
Center for Water Sciences
Department of Zoology
Michigan State University
East Lansing, Michigan 48824
USA

Catherine E. Stickley
Department of Geology
University of Tromsø
N-9037 Tromsø
Norway
and
Norwegian Polar Institute
Polar Environmental Centre
N-9296 Tromsø
Norway

Eugene F. Stoermer
4392 Dexter Road
Ann Arbor, MI 48103
USA

Jeffery R. Stone
Department of Geosciences
University of Nebraska – Lincoln
Lincoln, Nebraska 68588-0340
USA

Michael J. Sullivan
St. Andrews's North Campus
370 Old Agency Road
Ridgeland, Mississippi 39157
USA

George E. A. Swann
NERC Isotope Geosciences Laboratory
British Geological Survey
Keyworth
Nottingham NG12 5GG
United Kingdom

Edward C. Theriot
Texas Natural Science Center
The University of Texas at Austin
Austin, TX 78705
USA

John Tibby
Geographical and Environmental Studies
The University of Adelaide
Adelaide SA 5005
Australia

Rosa Trobajo
IRTA-Aquatic Ecosystems
Crta de Poble Nou, Km 5.5
PO Box 200
E- 43540 Sant Carles de la Ràpita
Tarragona, Catalonia
Spain

Herman van Dam
Consultancy for Water and Nature
P.O. Box 37777
1030 BJ Amsterdam
The Netherlands

Bart Van de Vijver
National Botanic Garden of Belgium
Department of Cryptogamy (Bryophyta & Thallophyta)
Domein van Bouchout
B-1860 Meise
Belgium

Elie Verleyen
Ghent University
Protistology and Aquatic Ecology
Krijgslaan 281 S8, B-9000
Gent
Belgium

Maria Célia Villac
Department of Biology
Mount Allison University
63B York St.
Sackville, NB E4L 1G7
Canada

Anna Wachnicka
Southeast Environmental Research Center
Florida International University
University Park OE 148
11200 SW 8th Street
Miami, FL 33199
USA

Kaarina Weckström
Geological Survey of Denmark and Greenland (GEUS)
Department of Marine Geology and Glaciology
Øster Voldgade 10
DK-1350 Copenhagen K
Denmark

Julie A. Wolin
Department of Biological, Geological and
Environmental Sciences
Cleveland State University
Cleveland, Ohio 44115
USA

Preface

> "If there is magic on this planet, it is contained in water." (Loran Eiseley, *The Immense Journey*, 1957)

Diatoms are being used increasingly in a wide range of applications, and the number of diatomists and their publications continues to increase rapidly. Although several books have dealt with various aspects of diatom biology, ecology, and taxonomy, the first edition of this volume, published over a decade ago, was the first to summarize the many applications and uses of diatoms. However, many new and exciting papers have been published in the intervening years. This, coupled with the fact that research on environmental and earth science applications of diatoms has continued at a frenetic pace, prompted us to undertake a major revision of our first edition.

Our overall goal was to collate a series of review chapters that would cover most of the key applications and uses of diatoms in the environmental and earth sciences. Due to space limitations, we could not include all types of applications, but we hope to have covered the main ones. Moreover, many of the chapters could easily have been double in size, and in fact several chapters could have been expanded to the size of books. Nonetheless, we hope material has been reviewed in sufficient breadth and detail to make this a valuable reference book for a wide spectrum of scientists, managers, and other users. In addition, we hope that researchers who occasionally use diatoms in their work, or at least read about how diatoms are being used by their colleagues (e.g., archeologists, forensic scientists, climatologists, engineers, etc.), will also find the book useful.

Compared to our 1999 volume, the current edition is significantly expanded and revised. It includes 16 new chapters, and an extensive revision and expansion of the original chapters, including the addition of many new co-authors.

The volume is broadly divided into six parts. Following our brief prologue, there is an introduction to the biology of diatoms and a chapter on numerical approaches commonly used by diatomists. Part II contains eight chapters that review how diatoms can be used as indicators of environmental change in flowing waters and lakes. Part III summarizes work completed on diatoms from cold and extreme environments, such as subarctic and alpine regions, Antarctica and the High Arctic. These ecosystems are often considered to be especially sensitive bellwethers of environmental change. Part IV contains nine chapters dealing with diatoms in marine and estuarine environments. The final part (Part V) summarizes most other applications (e.g., using diatoms as indicators in subaerial, wetland and peatland environments, and other applications such as tracking past wildlife populations, archeology, oil exploration and correlation, forensic studies, toxic effects, atmospheric transport, invasive species, diatomites, isotopes, and nanotechnology). We conclude with a short epilogue (Part VI), followed by a much expanded glossary (including acronyms and abbreviations) and an index.

Many individuals have helped with the preparation of this volume. We are especially grateful to Martin Griffiths (Cambridge University Press), who shepherded this project throughout much of its development. As always, the reviewers provided excellent suggestions for improving the chapters. We are also grateful to our colleagues at Queen's University and the University of Michigan, and elsewhere, who helped in many ways to bring this volume to its completion. And, of course, we thank the authors.

These are exciting times for diatom-based research. We hope that the following chapters effectively summarize how these powerful approaches can be used by a diverse group of users.

John P. Smol
Kingston, Ontario, Canada
Eugene F. Stoermer
Ann Arbor, Michigan, USA

Part I

Introduction

1 Applications and uses of diatoms: prologue

JOHN P. SMOL AND EUGENE F. STOERMER

This book is about the uses of diatoms (Class Bacillariophyceae), a group of microscopic algae abundant in almost all aquatic habitats. There is no accurate count of the number of diatom species; however, estimates on the order of 10^4 are often given (Guillard & Kilham, 1977), although Mann & Droop (1996) point out that this number would be raised to at least 10^5 by application of modern species concepts. Diatoms are characterized by a number of features, but are most easily recognized by their siliceous (opaline) cell walls, composed of two valves, that, together with the girdle bands, form a frustule (Figure 1.1). The size, shape, and sculpturing of diatom cell walls are taxonomically diagnostic. Moreover, because of their siliceous composition, they are often very well preserved in fossil deposits and have a number of industrial uses.

The main focus of this book is not the biology and taxonomy of diatoms, although Julius & Theriot (this volume) provide a "primer" on this subject, and a number of chapters touch on these topics. Other books (e.g. Round *et al.*, 1990) and the review articles and books cited in the following chapters, provide introductions to the biology, ecology, and taxonomy of diatoms. Instead, our focus is on the applications and uses of diatoms to the environmental and earth sciences. Although this book contains chapters on practical uses, such as uses of fossilized diatom remains in industry, oil exploration, and forensic applications, most of the book deals with using these indicators to decipher the effects of long-term ecological perturbations, such as climatic change, lake acidification, and eutrophication. As many others have pointed out, diatoms are almost ideal biological monitors. There are a very large number of ecologically sensitive species, which are abundant in nearly all habitats where water is at least occasionally present. Importantly, diatom valves are typically well preserved in the sediments of most lakes and many areas of the oceans, as well as in other environments.

The Diatoms: Applications for the Environmental and Earth Sciences, 2nd Edition, eds. John P. Smol and Eugene F. Stoermer. Published by Cambridge University Press.

Precisely when and how people first began to use the occurrence and abundance of diatom populations directly, and to sense environmental conditions and trends, is probably lost in the mists of antiquity. It is known that diatomites were used as a palliative food substitute during times of starvation (Taylor, 1929), and Bailey's notes attached to the type collection of *Gomphoneis herculeana* (Ehrenb.) Cleve (Stoermer & Ladewski, 1982) indicate that masses of this species were used by native Americans for some medicinal purpose, especially by women (J. W. Bailey, unpublished notes associated with the type gathering of *G. herculeana*, housed in the Humboldt Museum, Berlin). It is interesting to speculate how early peoples may have used the gross appearance of certain algal masses as indications of suitable water quality (or contra-indications of water suitability!), or the presence of desirable and harvestable fish or invertebrate communities.

However, two great differences separate human understanding of higher plants and their parallel understanding of algae, particularly diatoms. The first is direct utility. Anyone can quite quickly grasp the difference between having potatoes and not having potatoes. It is somewhat more difficult to establish the consequences of, for example, *Cyclotella americana* Fricke being extirpated from Lake Erie (Stoermer *et al.*, 1996).

The second is perception. At this point in history, nearly any person living in temperate latitudes can correctly identify a potato. Some people whose existence has long been associated with potato culture can provide a wealth of information, even if they lack extended formal education. Almost any university will have individuals who have knowledge of aspects of potato biology or, at a minimum, know where this rich store of information may be obtained. Of course, knowledge is never perfect, and much research remains to

Figure 1.1 Scanning electron micrographs of some representative diatoms: (a) *Hyalodiscus*; (b) *Diploneis*; (c) = *Surirella*; and (d) *Stephanodiscus*. Micrographs a, b, and c courtesy of I. Kaczmarska and J. Ehrman; micrograph d courtesy of M. B. Edlund.

be done before our understanding of potatoes approaches completeness.

Diatoms occupy a place near the opposite end of the spectrum of understanding. Early peoples could not sense individual diatoms, and their only knowledge of this fraction of the world's biota came from mass occurrences of either living (e.g. biofilms) or fossil (diatomites) diatoms. Even in today's world, it is difficult to clearly and directly associate diatoms with the perceived values of the majority of the world's population. The consequences of this history are that the impetus to study diatoms was not great. Hence, many questions concerning basic diatom biology remain to be addressed. Indeed, it is still rather rare to encounter individuals deeply knowledgeable about diatoms even amongst university faculties. This, however, is changing rapidly.

What is quite clear is that people began to compile and speculate upon the relationships between the occurrence of certain diatoms and other things which were useful to know almost as soon as optical microscopes were developed. In retrospect, some of the theories developed from these early observations and studies may appear rather quaint in the light of current knowledge. For instance, Ehrenberg (see Jahn, 1995) thought that diatoms were animal-like organisms, and his interpretation of their cytology and internal structure was quite different from our modern understanding. Further, his interpretations of the origins of airborne diatoms (he thought they were directly associated with volcanoes) seem rather outlandish today. On the other hand, Ehrenberg did make phytogeographic inferences which are only now being rediscovered.

As will be pointed out in chapters following, knowledge about diatoms can help us know about the presence of petroleum, if and where a deceased person drowned, when storms over the Sahara and Sub-Saharan Africa were of sufficient strength to transport freshwater diatom remains to the mid-Atlantic, and indeed to the most remote areas of Greenland, as well as many other applications and uses. As will be reflected in the depth of presentation in these chapters, diatoms provide perhaps the best biological index of annual to millennial changes in Earth's biogeochemistry. As it becomes increasingly evident that human actions are exercising ever greater control over the conditions and processes that allow for our existence (Crutzen & Stoermer, 2000), fully exercising all the tools which may serve to infer the direction and magnitude of change, and indeed the limits of change, becomes increasingly imperative. This need has fueled a considerable increase in the number of studies that deal with diatoms, particularly as applied to the problems alluded to above.

The primary motivation for this much-revised and expanded second edition is to compile this rapidly accumulating and scattered information into a form readily accessible to interested readers. A perusal of the literature will show that the authors of the different chapters are amongst the world leaders in research on the topics addressed.

The perceptive reader will also note that, despite their great utility, the store of fundamental information concerning diatoms is not as great as might logically be expected for a group of organisms that constitute a significant fraction of Earth's biomass. For example, readers will find few references to direct experimental physiological studies of the species discussed. Sadly, there are still only a few studies to cite, and practically none of those available was conducted on the most ecologically sensitive freshwater species. Readers may also note that there are some differences of opinion concerning taxonomic limits, even of common taxa, and that naming conventions are presently in a state of flux. These uncertainties are real, and devolve from the history of diatom studies.

As already mentioned, the study of diatoms started relatively late, compared with most groups of macroorganisms. Diatoms have only been studied in any organized fashion for about 200 years, and the period of effective study has only been about 150 years. It is also true that the history of study has been quite uneven. After the development of fully corrected optical microscopes, the study of microorganisms in general, and diatoms in particular, attracted immense interest and the attention of a number of prodigiously energetic and productive workers. This grand period of exploration and description produced a very substantial, but poorly assimilated, literature. Diatomists who worked toward the end of this grand period of growth produced remarkably advanced insights into cytology and similarly advanced theories of biological evolution (Mereschkowsky, 1903). This, and the fact that sophisticated and expensive optical equipment is required for their study, gained diatoms the reputation of a difficult group of organisms to study effectively. Partially for this reason, basic diatom studies entered a period of relative decline beginning c. 1900, although a rich, if somewhat eclectic, amateur tradition flourished, especially in England and North America. The area that remained most active was ecology. As Pickett-Heaps *et al.* (1984) have pointed out, Robert Lauterborn, an exceptionally talented biologist who was well known for his studies of diatom cytology, could also be appropriately cited as one of the founders of aquatic ecology.

The people who followed often did not command the degree of broad recognition enjoyed by their predecessors, and many of them operated at the margin of the academic world. As examples, Friedrich Hustedt, perhaps the best known diatomist in the period from 1900 to 1960 (Behre, 1970), supported himself and his family as a high-school teacher for much of his career. B. J. Cholnoky (Archibald, 1973) was caught up in the vicissitudes of the Second World War, and produced his greatest works on diatom autecology, including his large summary work (Cholnoky, 1968), after he became an employee of the South African Water Resources Institute. Although many workers of this era produced notable contributions, they were peripheral to the main thrusts of academic ecological thought and theory, particularly in North America. Although this continent had numerous individuals who were interested in diatoms, and published on the group, most of them were either interested amateurs or isolated specialists working in museums or other non-university institutions. For example, when one of us (E. F. S.) decided to undertake advanced degree work on diatoms in the late 1950s, there was no university in the United States with a faculty member specializing in the study of freshwater diatoms.

One of the most unfortunate aspects of separation of diatom studies from the general course of botanical research was substantial separation from the blossoming of new ideas. The few published general works on diatoms had a curious "dated" quality, and relatively little new understanding, except for descriptions of new species. The main impetus that kept this small branch of botanical science alive was applied ecology, and this was the area that, in our opinion, produced the most interesting new contributions.

The above situation began to change in the late 1950s, partially as a result of the general expansion of scientific research in the post-Sputnik era, and partially as the result of technological advances, particularly in the area of electronics. The general availability of electron microscopes opened new orders of magnitude in resolution of diatom structure, which made it obvious that many of the older, radically condensed, classification schemes were untenable. This released a virtual flood of new, rediscovered, and reinterpreted entities (Round *et al.*, 1990), which continues to grow today. At the same time, the general availability of high-speed digital computers made it possible to employ multivariate statistical techniques ideally suited to objective analysis of modern diatom communities and those contained in sediments (Birks, this volume).

The history of ecological studies centered on diatoms can be roughly categorized as consisting of three eras. The first is what we might term the "era of exploration." During this period (c. 1830–1900), most research focused on diatoms as objects of study. Work during this period was largely descriptive, be it the topic of the description of new taxa, discovery of their life cycles and basic physiology, or observations of their geographic and temporal distributions. One of the hallmarks of this tradition was the "indicator species concept." Of course, the age of exploration is not over for diatoms. New taxa have been described at a rate of about 400 per year over the past four decades, and this rate appears to be accelerating in recent years. Basic information concerning cytology and physiology of some taxa continues to accumulate, although at a lesser rate than we might desire.

The second era of ecological studies can be termed the "era of systematization" (c. 1900–1970). During this period, many researchers attempted to reduce the rich mosaic of information and inference concerning diatoms to more manageable dimensions. The outgrowths of these efforts were the so-called systems and spectra (e.g. halobion, saprobion, pH, temperature, etc.). Such devices are still employed, and sometimes modified and improved. Indeed, there are occasional calls for simple indices as a means of conveying information more clearly to managers and the public.

We would categorize the current era of ecological studies focused on diatoms as the "age of objectification." Given the computational tools now generally available, it is possible to determine more accurately which variables affect diatom occurrence and growth and, more importantly, do so quantitatively, reproducibly, and with measurable precision.

(a)

Figure 1.2 Some of the diverse ways that diatoms are collected for use in environmental and earth science applications. (a) Collecting a sediment core from Lake of the Woods. Photograph by K. Rühland. (b) Epilithic diatoms being sampled from a rock substrate from a High Arctic pond. Photograph by J. P. Smol. (c) Scuba divers collecting marine diatoms near Guam. Photograph by M. Schefter.

(b)

(c)

Figure 1.2 *(cont.)*

Thus, applied studies based on diatoms have been raised from a little-understood art practised by a few extreme specialists, to a tool that more closely meets the general expectations of science and the users of this work, such as environmental managers.

The result is that we now live in interesting times. Diatoms have proven to be extremely powerful indicators with which to explore and interpret many ecological and practical problems. They are used in a variety of settings, using different approaches (Figure 1.2). The continuing flood of new information will, without doubt, make the available tools of applied ecology even sharper. It is also apparent that the maturation of this area of science will provide additional challenges. Gone are the comfortable days when it was possible to learn the characteristics of most freshwater genera in a few days and become familiar with the available literature in a few months. Although we might sometimes wish for the return of simpler days, it is clear that this field of study is rapidly expanding, and it is our conjecture that we are on the threshold of even larger changes. The motivation for producing this updated volume is to summarize recent accomplishments and, thus, perhaps make the next step easier.

References

Archibald, R. E. M. (1973). Obituary: Dr. B. J. Cholnoky (1899–1972). *Revue Algologique*, N. S., **11**, 1–2.

Behre, K. (1970). Friedrich Hustedt's Leben und Werke. *Nova Hedwigia, Beiheft*, **31**, 11–22.

Cholnoky, B. J. (1968). *Die Ökologie der Diatomeen in Binnengewässern*. Verlag von J. Cramer: Lehre.

Crutzen, P. J. & Stoermer, E. F. (2000). The Anthropocene. *The International Geosphere-Biosphere Programme (IGBP) Global Change Newsletter*. **41**: 17–18.

Guillard, R. R. L. & Kilham, P. (1977). The ecology of marine planktonic diatoms. In *The Biology of Diatoms. Botanical Monographs*, vol. 13, ed. D. Werner, Oxford: Blackwell Scientific Publications, pp. 372–469.

Jahn, R. (1995). C. G. Ehrenberg's concept of the diatoms. *Archiv für Protistenkunde*, **146**, 109–16.

Mann, D. G. & Droop, J. M. (1996). Biodiversity, biogeography and conservation of diatoms. *Hydrobiologia*, **336**, 19–32.

Mereschkowsky, C. (1903). Nouvelles recherches sur la structure et la division des Diatomèes. *Bulletin Societé Impèriale des Naturalistes de Moscou*, **17**, 149–72.

Pickett-Heaps, J. D., Schmid, A.-M., & Tippett, D. H. (1984). Cell division in diatoms: a translation of part of Robert Lauterborn's treatise of 1896 with some modern confirmatory observations. *Protoplasma*, **120**, 132–54.

Round, F. E., Crawford, R. M., & Mann, D. G. (1990). *The Diatoms: Biology and Morphology of the Genera*. Cambridge: Cambridge University Press.

Stoermer, E. F. & Ladewski, T. B. (1982). Quantitative analysis of shape variation in type and modern populations of Gomphoneis herculeana. *Nova Hedwigia, Beiheft*, **73**, 347–86.

Stoermer, E. F., Emmert, G., Julius, M. L., & Schelske, C. L. (1996). Paleolimnologic evidence of rapid recent change in Lake Erie's trophic status. *Canadian Journal of Fisheries and Aquatic Sciences*, **53**, 1451–8.

Taylor, F. B. (1929). *Notes on Diatoms*. Bournemouth: Guardian Press.

2 The diatoms: a primer

MATTHEW L. JULIUS AND
EDWARD C. THERIOT

2.1 Introduction

Diatoms have long been lauded for their use as powerful and reliable environmental indicators (Cholnoky, 1968; Lowe, 1974). This utility can be attributed to their high abundance and species diversity, which are distributed among most aquatic environments. Additionally, their remains are highly durable and well preserved in accumulated sediments. Often, scientists exploiting the group simply as environmental proxies give little thought as to how and why the species diversity exists in these environments. This may be a by-product of how diatoms are collected and identified. Diatoms are most often recognized by the presence of a siliceous cell wall, the frustule. This structure varies considerably in shape and architecture among species (Figure 2.1) and virtually all taxonomic diagnosis of taxa is based upon frustular morphology. To properly observe diatom frustules for taxonomic identification, living and sedimentary collections are typically subjected to various "cleaning" techniques designed to remove all organic materials (e.g. Battarbee *et al.*, 2001; Blanco *et al.*, 2008), allowing unobstructed observation of the frustule in the microscope. This frequent observation of inorganic components of the cell without reference to the organic features allows observers to "forget" that the specimens seen in the microscope represent individual organisms competing in the selective environments driven by biotic and abiotic ecological pressures. The abundance and taxonomic diversity can be attributed to the extraordinary success of diatoms in the competitive ecological arena.

The casual observer frequently regards diatoms, like most protists, as primitive ancestral lineages to multicellular organisms. While some protists may fit this description, diatoms do not. Diatoms are a relatively recent evolutionary group with the common ancestor's origin considered to be between 200 and 190 million years before present (Rothpletz, 1896, 1900; Medlin *et al.*, 1997). As a point of reference, the origin of this "first diatom" is approximately 60 to 70 million years younger than the specialized teeth found in mammals, including those in the reader's mouth (Shubin, 2008). Dates for the origin of the diatom common ancestor are bracketed by molecular clock estimates (Sorhannus, 2007) and the oldest stratigraphic observation (Rothpletz, 1896, 1900). Both of these estimates are inherently biased. The temporal proximity of each estimate to one another does, however, suggest a certain degree of accuracy, given the complimentary nature of the biases. Molecular estimates represent an attempt to identify the absolute moment two populations diverged from one another. The oldest stratigraphic observation represents a period where fossil remains were sufficiently abundant to allow discovery. Given the expected disparity between the moment two populations diverged, and the time it would take divergent populations to develop sufficient numbers allowing paleontological discovery, the 10 million year gap between the two estimates does not appear to be overly large in context of other estimates in this temporal range.

Discussion of when diatoms originated begs the question: what did they originate from? Diatoms share ancestry with heterokonts. Heterokonts (or stramenopiles) are a group of protists with unequal flagella (Leedale, 1974; Hoek, 1978) that includes both chloroplast-bearing and non-chloroplast-bearing representatives (Patterson, 1989) whose common ancestor is thought to have arisen ~725 million years before present (Bhattacharya & Medlin, 2004). The group contains an array of morphologically diverse groups including giant kelps (>60 m) at the large end of the size spectrum and the Bolidomonads and Pelagomonads (1–2 μm) at the small end of the size spectrum (North, 1994; Andersen *et al.*, 1993; Guillou *et al.*, 1999). The heterokonts may be part of the larger "chromalveolate" evolutionary group, which includes

The Diatoms: Applications for the Environmental and Earth Sciences, 2nd Edition, eds. John P. Smol and Eugene F. Stoermer. Published by Cambridge University Press.

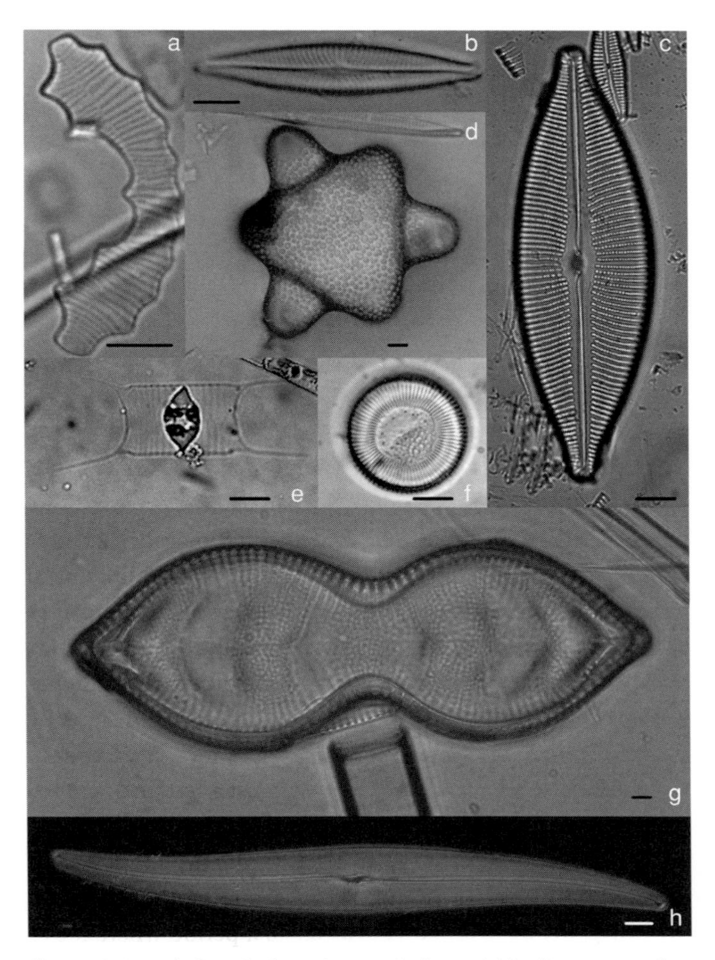

Figure 2.1 Variations in frustule morphology within diatoms species. (a) *Amphicampa mirabilis*, (b) *Navicula cryptocephala*, (c) *Cymbella inaequalis*, (d) *Hydrosera whampoensis*, (e) *Acanthoceras magdeburgensis*, (f) *Cyclotella striata*, (g) *Cymatopleura solea*, (h) *Gyrosigma acuminatum*. Scale bars equal 5 μm.

cryptophytes, dinoflagellates, ciliates, apicomplexans, and haptophytes (Yoon *et al.*, 2002; Cavalier-Smith, 2003; Harper & Keeling, 2003; Ryall *et al.*, 2003; Bachvaroff *et al.*, 2005; Harper *et al.*, 2005). This proposed relationship is controversial and highly debated (e.g., Falkowski *et al.*, 2004; Grzebyk *et al.*, 2004; Keeling *et al.*, 2004; Bachvaroff *et al.*, 2005; Bodyl, 2005). Within heterokonts, individual groups are well established and easily diagnosable, but the relationship between these groups has yet to be definitively identified (Saunders *et al.*, 1995; Sorhannus, 2001; Goertzen & Theriot, 2003). Molecular techniques utilizing multiple data sets have identified the bolidophytes (Goertzen & Theriot, 2003) as the heterokont most closely related to diatoms. The bolidophytes are a group of marine unicellular flagellates that were unknown to science prior to the late 1990s (Guillou *et al.*, 1999). This relatively recent discovery of the diatoms sister group reflects how much discovery and description-level science remains uncompleted in heterokont biology.

This statement about heterokonts easily extends to diatoms. Once an understanding of origin is achieved, an appreciation should be given to the speed of diversification. Currently, >24,000 diatom species have valid scientific names (Fourtanier & Kociolek, 2009a, b). Many of these have only been illustrated in the literature with light microscopy, and few have yet been the subject of any other genetic, ecological, or physiological study. Mann and Droop (1996) conservatively estimated that there are 200,000 diatom species. If these numbers are taken at face value, 12% of the diatom flora is currently described. This means the modern diatom taxonomic community has a majority of the 24,000 described species to observe in the electron microscope, and an additional 176,000 species to describe. In addition, completing a phylogeny for these 200,000 species should also be an objective. Julius (2007a) demonstrated that the rate of species description in diatoms is approximately 183–185 per year and that this rate has remained constant for nearly a century. At this pace it will take approximately 951 years to describe diatom species completely!

With this in mind, it is easy to understand why the diatom systematics community is still grappling with the collection of detailed ultrastuctural information for most species, description of species, and the proper way to analyze these data. Many modern diatom systematic studies deal with taxa at the generic or higher level, avoiding unresolved issues concerning ultrastructure and species concepts. Several recent studies suggest diatom diversity is much greater than previously imagined (Theriot & Stoermer, 1984; Bourne *et al.*, 1992), causing researchers to suggest continued emphasis on species description is most essential in developing phylogenetic hypotheses (Kociolek, 1997; Lange-Bertalot, 1997; Mann, 1997; Round, 1997). In many instances, researchers also continue to argue about what classes of data should be emphasized, valve morphology or cytological features, in classification systems (Round, 1996) without regard to any sense of evolution.

Systematic studies in the twenty-first century must incorporate all types of character information in some sort of an analysis emphasizing the similarity between evolved features (cladistic analysis is currently the most prominent system). This character information must be presented for individual species, not broad generic groups. This requires considerable additional descriptive work. Gradually, a classification more reflective of evolutionary history will develop. Simply put, there is a great deal of work to be done. We are gradually developing a more structured way of handling the problem and, with luck, progress will be made. One indication of this activity is the distinctly non-linear trend seen in the rate of generic description over the last two decades (Fourtanier & Kociolek, 1999). Generic descriptions increased at an exponential rate during this time, contrasting with the linear rate seen in species. This may indicate that existing taxa are being placed into newly created higher taxonomic categories and a greater interest is being taken in the relationship between one species and another.

2.2 Classification

Modern systematics strives to achieve natural, or monophyletic, groups when designating categories above the species level. These natural groups contain an ancestral lineage and all of its descendants (monophyly). Many taxonomic groups were established prior to the acceptance of monophyly as a goal. Taxonomic designations for diatoms are no exception, and researchers have only recently begun attempts to test and adjust diatom taxonomic schemes to reflect monophyletic groupings. Not all individuals establishing genera and other categories for diatoms view monophyly as a goal, despite its widespread acceptance elsewhere in biology, and proceed in their endeavors in an evolutionary free context (Williams & Kociolek, 2007). Individuals utilizing taxonomic schemes for diatoms should be aware of the unstable status of many higher taxonomic categories (genus and above). Species are frequently moved in and out of categories and new categories are continually being established. This process, hopefully, reflects the gradual transition to a monophyletic taxonomic system and the overwhelming level of species description remaining incomplete. Individuals utilizing diatoms as indicator species often find this fluctuation in higher taxonomic categories frustrating. To circumvent this taxonomic instability, identifications should be made to the species level whenever possible, because species names can always be referenced back to the original population described no matter how many times a name is modified nomenclaturally.

Diatoms are traditionally classified as one of two biological orders, the Centrales (informally referred to as centrics) and the Pennales (or pennates). Diagnostic features cited supporting the two classes typically include (1) valve formation developing radially around a "point" in centrics, contrasted by deposition originating along a "plane" in pennates and (2) oogamous sex with relatively small motile flagella bearing sperm and a large non-motile egg in the centrics, contrasted by isogamous sex with ameboid gametes in the pennates. These features are not distinctly distributed among bilaterally and radially symmetric morphologies on the diatom evolutionary tree, but are instead distributed along a gradient moving from basal radially symmetric groups to more recently diverged bilaterally symmetric groups.

Simonsen (1979) was the first to discuss a phylogeny for all diatoms in the context of a taxonomic system. While Simonsen presented an evolutionary tree for diatom families, he was reluctant to deconstruct class and order designations in a manner reflecting monophyly. Most notable is the presentation of centric diatoms as distinctly non-monophyletic while maintaining the traditional taxonomic category for the group. Round *et al.* (1990) presented a taxonomic system for genera and higher-level groups. This work treated the diatoms as a division with three classes consisting of the radially symmetric taxa, the araphid pennate taxa, and raphid pennate taxa, suggesting that evolution was along this line and that the centric diatoms preceded pennates. This text remains the most recent comprehensive coverage for diatoms, but the classification system was not developed in an evolutionary context and many of the taxonomic designations are being reconsidered and modified.

A comprehensive evolutionary tree for the diatoms is currently a popular research topic (summarized in Alverson *et al.*, 2006 and Theriot *et al.*, 2009). While molecular systematics has advanced rapidly in other areas of biology, only the small subunit of the nuclearly encoded ribosomal rDNA gene (SSU) has been used for comprehensive analyses of diatom phylogeny (other genes and morphology have generally just been employed selectively at the ordinal level or below). Trees produced using the SSU molecule have uniformly obtained a grade of multiple lineages of centric diatoms generally with radial symmetry of valve elements, then a series of lineages of centrics with generally bipolar or multipolar symmetry, then a series of araphid taxa and, finally, the clade of raphe-bearing pennates. The only exceptions to this are when only a few diatoms were included in the analysis or the analytical techniques used were improperly applied (Theriot *et al.*, 2009).

Goertzen and Theriot (2003) noted the effect of taxon sampling on topologies generated in phylogenetic analyses of heterokont taxa. Diatom diversity presents a challenge, in this context, to attempts at reconstructing phylogenies for the group using molecular data. All attempts to date have sampled $<0.01\%$ of the described diatom flora (and obviously have not included extinct taxa). A very small amount of morphological data has been formally analyzed and it supports the trees derived from the SSU gene (Theriot *et al.*, 2009). Again, however, results are weakly supported and the analyses did not include extinct taxa. What can be conservatively concluded is that no investigation has produced a comprehensive and robust phylogenetic hypothesis for the group. Nevertheless, there seems to be mounting evidence that only raphid pennates are monophyletic among the major diatom taxa. Readers should understand that, like the terms "invertebrate" or "fish", the words "centric" and "araphid pennate" are simply terms of convenience and do not reflect independent phylogenetic groups.

While a comprehensive diatom phylogeny is desirable, it is not a requirement for utilizing diatom species for environmental reconstruction and other applications. Reliable species identification is, however, essential. Two text collections represent the standards for freshwater species identification with the light microscope. Patrick and Reimer (1966, 1975) are English guides to the North American flora. Taxa are illustrated with line drawings. Unfortunately, the work was never completed and contains only a partial coverage of pennate diatoms. Some major pennate groups (e.g. keel-bearing genera such as *Nitzschia* and the order Surirellales) and centric diatoms were not treated. Volume 1 (Patrick & Reimer, 1966) is still available for purchase, but volume 2 (Patrick & Reimer, 1975) has become difficult to acquire. Krammer and Lange-Bertalot (1986, 1988, 1991a, 1991b) are German language guides to the central European flora. Taxa are illustrated with photographs. The texts have been used to identify species from around the world and represent a reasonably comprehensive coverage of species. All four volumes are currently available for purchase.

2.3 The diatom cell

The diatom cell has the same general pattern as many plastid-bearing protists. It has an outer cell wall lined internally with a plasma membrane containing the cytoplasm and a collection of organelles. The cell wall, as previously mentioned, is distinct and is composed of silicon dioxide (SiO_2). Generally, the cell wall consists of two halves, termed valves. The two valves have slightly different sizes. The larger valve is termed the epitheca and the smaller the hypotheca. The two valves fit together to form a box. Between the valves are a series of siliceous bands or belts termed girdle bands. The valves plus all girdle bands comprise a complete cell wall that is termed a frustule (Figure 2.2).

The frustule has many openings allowing the cell's organic component to make contact with its surrounding environment. The openings can range from simple pores to specialized structures of extremely complex micro-architecture. The cell's plasma membrane, like other eukaryotes, is rich with proteinaceous receptors (Scherer *et al.*, 2007). Little work has been done

Figure 2.2 Complete *Stephanodiscus minutulus* frustule. Both valves and girdle bands comprising the "pill box" architecture are visible. Scale bar equals 5 μm.

concerning these receptors (Julius *et al.*, 2007), but basic physical properties of extra cellular material emanating from the membrane can be altered quickly in response to environmental challenges (Higgins *et al.*, 2003). Adhesion responses have been tied to specific plasma membrane receptors (Almqvist *et al.*, 2004).

The cell interior (Figure 2.3) typically has a central cytoplasmic bridge containing the nucleus. This is bounded by a large vacuole, which can account for up to 70% of the total cell biovolume (Sicko-Goad *et al.*, 1984). The vacuole is frequently associated with laterally positioned chloroplasts, which may deposit photosynthetic products in the vacuole. The silica deposition vesicles (SDVs) are closely associated with the golgi, and in some species mitochondria may be found in close proximity to both the SDVs and golgi. Little is understood about diatom physiology and most efforts focus on frustule formation or plastid function.

2.3.1 Reproduction

Diatoms, like many other unicells, reproduce primarily via asexual mitotic divisions with relatively rare instances of sexual reproduction. Cells are typically diploid with chromosome numbers varying considerably across species, having as few as 4 and as many as 68 (Kociolek & Stoermer, 1989). Some species exhibit high growth rates and may divide once per day in optimal environments (Rivkin, 1986). The frustule is involved in the divisional process to varying degrees with some taxa,

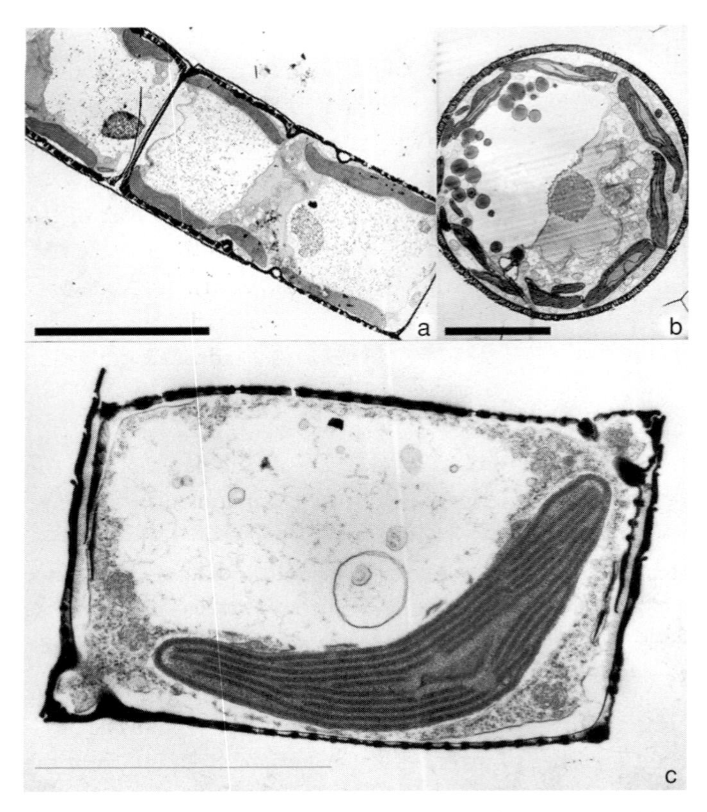

Figure 2.3 Transmission electron micrographs of *Aulacoseira ambigua* and *Nitzschia* sp. (a) *A. ambigua*, transmission electron micrograph (TEM) cross section through region of central cytoplasmic bridge. (b) *A. ambigua*, TEM longitudinal section showing fine structure of frustule and arrangements of cytoplasmic components. (c) *Nitzschia* sp. cross section: the hypothecal valve can be seen resting inside the epithecal valve. Scale bars equal 10 μm. Photographs courtesy Linda Goad.

specifically Thalassiosiraceae and Coscinodiscaceae, possessing dedicated structures for positioning and manipulating the nucleus (Schmid & Volcani, 1983; Chepurnov *et al.*, 2006). In most species, the arrangement of the epitheca and hypotheca influence mitotic products. Both the epitheca and hypotheca of the mother cells will become epitheca in the next generation, one for each daughter cell. The mitotic products result in one daughter cell equal in size to the mother cell and one daughter cell smaller than the mother cell. Repeated division results in a gradual decrease in population mean cell size over time. MacDonald (1869) and Pfitzer (1869, 1871) described this process over a century ago.

Cell size is restored via sexual reproduction, with a syngamy event producing a single maximally sized cell. Triggers for sexual reproduction are not well understood, but are generally thought to include both environmental factors and size cues (Mann, 1993). A collection of genes that are expressed at the onset of sexual reproduction has been identified (Armbrust, 1999). The three genes identified in this collection were named Sig 1, Sig 2, and Sig 3 or sexually induced genes 1–3. Armbrust (1999) found that the genes encoded for a protein are similar in structure to an animal epidermal growth protein potentially involved in sperm and egg recognition. Sexual events are typically observed in cells approximately one third of the maximum cell size. A single cell cycle, the process of cell size reduction and restoration via sex, can take place over decades (Julius *et al.*, 1998). These factors must be accounted for when making decisions about the morphological variability of a population over time. Size ranges contained within a species category should have the largest individuals being at least three times the size of the smallest individuals. It may only be possible to observe this range in size over several seasons or numerous sediment intervals, depending upon where the population is in the size reduction/sexual cycle when observed.

Two basic forms of sexual reproduction are presented in reviews dealing with the subject (Mann, 1993; Edlund & Stoermer, 1997): oogamy in centric diatoms and isogamy in pennate diatoms. As mentioned in the introduction, these features are not distributed between radially and bilaterally symmetric diatoms in a strictly categorical fashion. Variety also exists and numerous modifications in sexual reproduction are known in the gradation ancestral to more recently diverged diatom lineages. Mann (1993) illustrates a transitional form of oogamy in an early pennate lineage in which flagella are absent in "sperm" but the gamete migrates toward a large egg to complete syngamy. All sexual events in diatoms result in auxospore production. The auxospore is a covering composed of siliceous scales or bands, which surrounds a newly formed zygote. The zygote and the auxospore expand together, ranging in shape from spheres to cigar-shaped structures. Within the auxospore a maximally sized frustule is formed, called the initial valve. Initial valves can appear slightly irregular in form as they are sometimes formed directly against the auxospore and resemble the inflated shape of this structure. These irregularities diminish as mitotic products are produced. Some taxa are capable of producing auxospores without sexual reproduction and will undergo automixis in which gametes form within a single individual and fuse with one another giving rise to an auxospore (Geitler, 1973). Parthenogenesis has also been described in some taxa (Geitler, 1982). These non-sexual mechanisms for regenerating size appear to be adaptations existing in addition to standard sexual processes producing auxospores and initial cells.

2.3.2 Resting cells

The ability to produce resting cells and spores is thought to have arisen early in diatoms (Round & Sims, 1981; Harwood & Nikolaev, 1995). This speculation is unconfirmed, but resting cell and spore production is a known component of many diatoms since the Cretaceous (Harwood & Nikolaev, 1995). Resting spores are regarded as an ancestral structure (Hargraves & French, 1983), and they are given credit as one reason the diatoms were able to survive the Cretaceous extinction (Kitchell *et al.*, 1986; Harwood, 1988). Resting spores produced through asexual division generally differ morphologically from vegetative valves and generally possess two distinctive, heavily silicified valves. The spores are rich in storage products and have faster sinking rates and higher chlorophyll content than vegetative valves (Hargraves & French, 1983). Resting spores are viable for a relatively short time (<3 years); in contrast, resting cells are viable for long periods when in anoxic sediments (at least 20–30 years; Sicko-Goad *et al.*, 1986). Resting cells are morphologically indistinguishable from vegetative valves, differing only in the dark, highly condensed cytoplasm of the cell. Resting cell formation has largely replaced resting spore production in freshwater diatoms (Sicko-Goad *et al.*, 1986). Resting cells are physiologically similar to resting spores, having high quantities of storage products.

Resting cells, however, do not undergo a mitotic division during formation. Instead, they form from a vegetative cell by condensing the chloroplast and accumulating storage products (Sicko-Goad *et al.*, 1986). This fact suggests that although both resting spores and resting cells are very effective survival strategies, the resting cells are more physiologically efficient, bypassing the cellular expense (time and energy) of an additional mitotic division.

Both of the above resting structures are triggered into formation or encystment by very similar environmental cues. These include the availability of various nutrients (N, P, Si), temperature, light intensity, and photoperiod (Round *et al.*, 1990). Once formation has occurred, both resting spores and resting cells sink, to the pycnocline in many marine species (Hargraves & French, 1983) and into the sediments for many freshwater taxa. Eventually the resting structures will be resuspended into the water column during favorable environmental conditions, when the above environmental triggers will cause vegetative cell production.

2.3.3 Silica and frustule formation

Diatoms are the primary organisms involved in the mediation of silica in modern oceans (Conley, 2002) and are thought to have occupied this role since the Cenozoic (Martin-Jezequel *et al.*, 2000; Katz *et al.*, 2004). They also achieved dominance over other phytoplankton during this same period. Diatom dominance in the marine and freshwater fossil record since the Cenozoic is often attributed to geochemical coupling between terrestrial grasslands and aquatic ecosystems as part of the global silicon cycle (Kidder & Gierlowski-Kordesch, 2005; Falkowski *et al.*, 2004; Kidder & Erwin, 2001). Proponents of this hypothesis argue that this process makes silica readily available to diatoms, providing a competitive advantage allowing displacement of other photosynthetic protists. This hypothesis is not corroborated by the current data set available for fossil diatoms (Rabosky & Sorhannus, 2009), and development and dominance of diatoms over the last 65 millions years is a more complex story than previously postulated.

While the mechanisms and extent of diatom dominance since the Cenozoic remains an open question, the immediate availability of silica in aquatic systems is an important factor in determining the locally dominant primary producer (Schelske & Stoermer, 1971; Conley, 1988). Excess silica in combination with excess nitrogen frequently results in diatom-dominated environments. Generalizations can also be made about silica utilization in marine and freshwater species. Marine taxa typically contain ten times less silica than freshwater taxa (Conley *et al.*, 1989). This reflects a dichotomy in the availability of silica in freshwater and marine systems, in that silica is far less abundant in marine environments (Conley, 2002). Despite this, silica availability does not appear to be a major factor driving selection of silica transport genes in modern marine and freshwater species (Alverson, 2007; Alverson *et al.*, 2007).

Diatom cell wall formation is the subject of considerable interest within the nanotechnology community (see Gordon, this volume). The structure is unique to diatoms and is composed of opaline silica (SiO_2) deposited via a specialized organelle, the silica deposition vesicle. Within this organelle, silica is sequestered and deposited along a proteinaceous template during cytokinesis following the initiation of mitosis. In older lineages valve formation is initiated during anaphase/telophase; and in more recent pennate diatom lineages the process begins after cleavage (Schmid, 1994). Frustule formation involves an unusual interaction between organic and inorganic molecules. Wetherbee (2002) provides a summary of these interactions. Organic molecules regulating silica deposition in diatoms are proteins called silafins (Kröger *et al.*, 1999), and a small collection of silica transporter genes (SITs) is responsible for silica uptake in the cell (Hildebrand *et al.*,

1998). Silica available for valve formation in aquatic systems is in the form of silicic acid ($Si(OH)_4$). Acquisition of this material is synchronized with the cell cycle, providing adequate raw material for valve formation during cytokinesis, with peak silicic acid intake from the environment occurring during the G2 and M phases of the cell cycle (Sullivan, 1977; Brzezinski *et al.*, 1990; Brzezinski, 1992).

An understanding of the valve formation process is still developing and is best understood from microscopic observations of the process. Molecular-level dynamics of the process are the subject of intense study, and information concerning this process should develop rapidly. Gordon and Drum (1994) present a preliminary review of the chemical and physical parameters accounting for certain siliceous structures, and Schmid (1994) performed the most extensive studies of valve morphogenesis via light microscopic observation and provides a summary of nearly two decades of her work on the subject.

Following nuclear division, the valves of the mother cell become the new epithecal valves of daughter cells. Between these mother-cell valves each hypothecal valve is formed inside an individual vesicle, called a silica deposition vesicle. The membrane of this vesicle is called the silicalemma. The silicalemma controls the release (flow and position) of silica from the SDVs. In all species of diatoms investigated, a structure known as the cleavage plasmalemma adheres to the silicalemma during expansion and shaping of the new valve. The pattern of siliceous structures (pore fields, chambers, the raphe, or strutted processes) is produced as a result of the cell's ability to localize silica and molding organelles to particular regions of the cleavage plasmalemma. The morphogenic information for a new valve is carried from the plasmalemma to the silicalemma. The mechanics and regulation of this information transfer remain an open question.

As with all proteins, silafins are directly coded by nuclear DNA, but function and shape can be modified by environmental conditions. This undoubtedly causes a certain degree of environmentally regulated plasticity, but evolutionary processes at the genetic level inherently regulate expression in valve morphology. Prior work has documented environmentally induced variation in valve morphology. Julius *et al.* (1997) found that specific features commonly used in identifying radially symmetric diatoms may be highly prone to environmental variation and encouraged the use of features passing tests for evolutionary homology when diagnosing species. It is commonly suggested (Round *et al.*, 1990) that researchers look for "janus valves" (cells in which the frustule has two distinctly different valve types in species for which the two valves normally appear identical (MacBride & Edgar, 1998)) if they are concerned about differing environmental forms. Stoermer (1967) identified this phenomenon in field collections of the pennate diatom *Mastogloia*. Modifications in levels of total dissolved solids apparently caused two distinct morphotypes to appear. Tuchman *et al.* (1984) noted similar modifications in the valve structure of the centric diatom *Cyclotella meneghiniana* with increased salinity under experimental conditions. Other researchers have observed similar taxonomic forms occurring over an environmental gradient and suggested that the forms represent a single highly plastic species. This variation should encourage caution when making taxonomic diagnoses based upon valve morphology. The feature is absolutely the product of evolutionary processes, but is variable like morphological features in all organisms. Conclusions should be reached via observation of multiple valves from repeated collections of a specific locality.

2.3.4 Plastids

The diatom plastid, like that of all other photosynthetic eukaryotes, originated as the result of an endosymbiotic event involving an as-yet unidentified eukaryotic protist or a photosynthetic bacterium (Sanchez-Puerta & Delwiche, 2008). The engulfed bacterium likely shared an ancestry with the cyanobacteria. In addition to this original endosymbiotic event, additional kleptoplastic events involving only eukaryotes have also occurred. One of these events involving a non-plastid bearing eukaryote and a red alga presumably resulted in the plastid-bearing cell giving rise to organisms with plastids containing chlorophyll *a* and *c* (Keeling, 2004) (Figure 2.4). The diatom plastid arose from this chlorophyll *a*- and *c*-bearing cell through an additional endosymbiotic event (Sanchez-Puerta & Delwiche, 2008).

Plastid acquisition history in diatoms encourages caution when making generalizations about diatoms and other photosynthetic organisms. Consideration must be given to the plastid's history as a cyanobacterial cell, the time it spent within the red algal lineage, and events since its removal from that lineage. To do this, it is best to think of the plastid's evolutionary history separately from the "host" organism (Figure 2.5). Photosynthesis in diatoms mechanistically resembles the process in cyanobacteria and other photosynthetic eukaryotes, in that it is chlorophyll *a* centered, non-cyclic, and involves the same number of proteins in the reactions. Diatoms can be distinguished from other non-heterokont algae through the possession of (a) chlorophyll *a*, *c*, fucoxanthin, and β-carotene

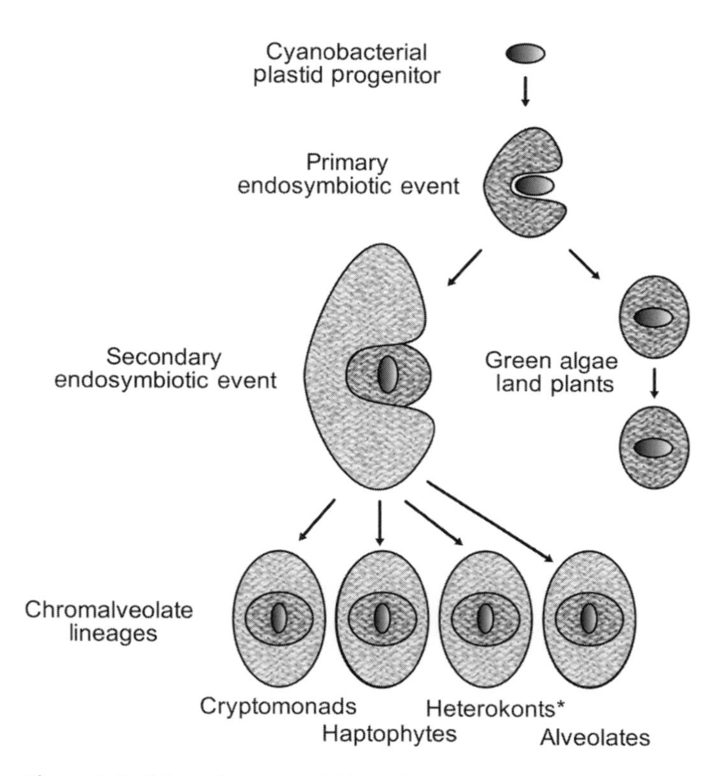

Figure 2.4 Chloroplast acquisition by photosynthetic protists. Endosymbiotic events from acquisition of original cyanobacterial endosymbiont through events leading to modern photosynthetic protist lineages. The asterisk indicates a group containing diatoms.

as pigments, (b) four plastid membranes with the outer membrane functionally bound to the golgi apparatus, and (c) the utilization of chrysolaminarin as a photosynthetic product.

Pigments in diatom plastids are bound together by nuclear encoded proteins homologous to those in other eukaryotic phototrophs (Owens, 1986; Bhaya & Grossman, 1993; Apt *et al.*, 1994). Specific proteins found in diatoms are termed the fucoxanthin, chlorophyll *a*/*c* binding proteins (FCPs). These proteins function on the thylakoid membranes and channel light energy to chlorophyll *a* (Owens, 1986). At least two FCP gene clusters exist with multiple FCP gene copies contained within each cluster (Bhaya & Grossman, 1993). These proteins are nuclear encoded, and protein transport to the plastids differs from that in other photosynthetic eukaryotes with less than four plastid membranes (Bhaya & Grossman, 1991).

Plastid number and morphology varies among different diatom evolutionary lineages. Non-pennate diatoms and basal pennate lineages generally have numerous discoid chloroplasts. Chloroplast number in these diatom lineages varies with environmental cues. Modifications in chloroplast number associated with such cues can be tied to variations in FCP gene activity (Leblanc *et al.*, 1999). More recently diverged diatom lineages can have fixed numbers of morphologically distinct plastids. These attributes have been used as an identification

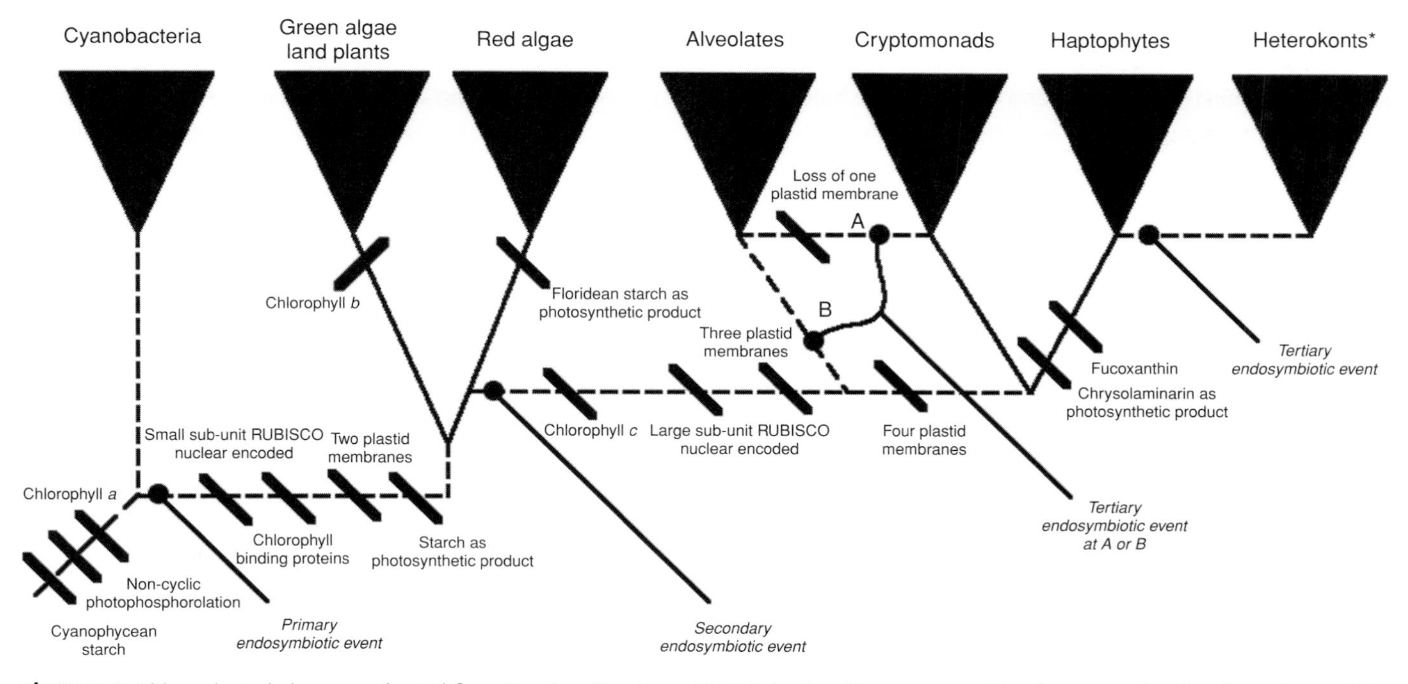

Figure 2.5 Chloroplast phylogeny, adapted from Sanchez-Puerta and Delwiche (2008), as proposed by the chromalveolate hypothesis. Solid lines in the tree represent phylogenetic relation at the chloroplast and host level. Dashed lines represent phylogenetic relation for only the chloroplast. Hash marks on the tree represent acquisition of specific features. The asterisk indicates a group containing diatoms.

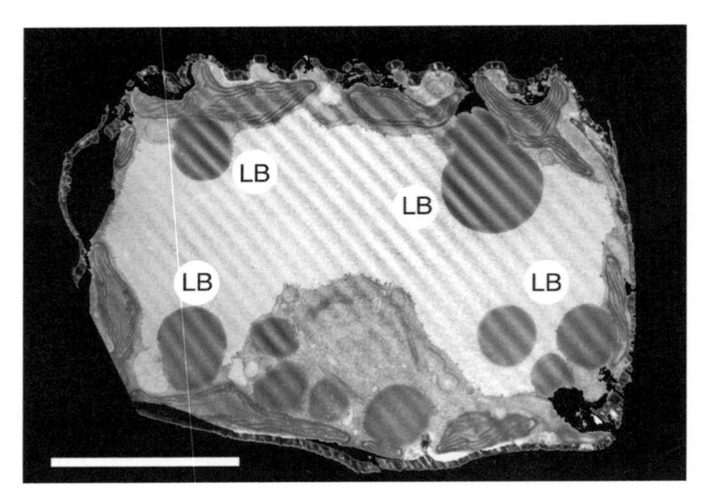

Figure 2.6 Transmisssion electron micrograph of *Cyclotella meneghiniana* in cross section. LB denotes lipid bodies in the vacuole, which are often located near chloroplasts. Scale bar equals 5 μm.

tool for distinguishing living specimens from one another (Cox, 1986).

Chloroplasts directly produce chrysolaminarin, a biological polymer composed of β 1–3 glucose molecules. This is used directly as a food source by the diatoms, but is transformed and stored as lipid in the cell's vacuole. Diatoms typically have one or two large vacuoles which can account for >60% of the cells volume (Sicko-Goad *et al.*, 1977). Stored lipid can occupy a major portion of the vacuole (Figure 2.6) with lipid accounting for a large percentage of cell dry weight. Modifications in chloroplast number may be made in times of environmental stress to maintain lipid reserves, and chlorophyll to lipid ratios have been successfully used as a metric for cell health (Julius *et al.*, 2007).

2.4 Ecology and growth habits

The production of high-energy lipid reserves make diatoms a highly desirable food source for many aquatic heterotrophs (Volkman *et al.*, 1989; Ahlgren *et al.*, 1990). Lipids, especially those from diatoms, play a critical role in the aquatic food web providing essential fatty acids (EFAs) to higher trophic levels (Arts *et al.*, 2009). While animals are predominately associated with these heterotrophic consumers, members of other evolutionary groups also graze on diatom taxa. Canter and Lund (1948, 1951, 1953), for example, detail the relationship between fungal chytrids and diatom populations. This description is frequently used as a model for describing predator–prey dynamics in ecology texts. While discussions of microalgae as a food source for animals frequently focus on members of the invertebrate community (Bott, 1996), spectacular examples exist in among vertebrate grazers. Julius *et al.* (2005), for example, detail the behavior and feeding mechanics of a gobid fish adapted for grazing on planktonic diatoms as a larvae and stalk-forming benthic diatoms as an adult.

The abundance and diversity of diatom grazers directly reflects diatom species abundance and diversity. Diatom species are found in a wide array of nutrient, pH, salinity, and temperature regimes. The many combinations of these chemical and physical parameters provide the diversity of niches required to sustain the thousands of known species. Functionally the diatoms can be described by growth habit, with the understanding that certain strategies likely evolved multiple times in distantly related diatom species.

Species can be easily found with adaptations for planktonic, tychoplanktonic, benthic, and subaerial existences. Taxa found in each of these environments can occur as unicells or colonially. The widest diversity in growth habits is found in benthic taxa. Adaptations can be found for growth on a variety of substrates including sediments, rocks, photosynthetic macrophytes, and animals (Figure 2.7). Other species remain motile and rarely adhere to substrates. Substrate availability (Townsend & Gell, 2005) and flow (Tuji, 2000) can dramatically restructure diatom communities. Substrate disturbance frequency and magnitude is also a major consideration (Tuji, 2000), as many taxa dominate in disturbed rather than stable systems or vice versa (see Stevenson *et al.*, this volume). Variations in growth forms directly impact the availability of cells to grazers. Alterations in physical parameters can alter dominant diatom growth types making taxa unavailable to grazers (Julius, 2007b) with subtle changes in physical parameters. Variations in growth habit can also cause selective collection of some species over others when sampling benthos. Factors influencing growth habit can also dramatically impact the quantity of biomass in a benthic system. Peak biomass typically occurs in low to moderate flow settings with rock substrates where diatom taxa can readily grow on attached green algae (Biggs, 1996).

Growth habit often may or may not reflect shared evolutionary history. Barriers encountered by taxa requiring adaptations for growth may, in some instances, have provided an advantage to species allowing them to radiate and colonize many similar habitats. In other cases, chemical barriers may have presented a greater challenge, and adaptation to nutrient regimes many have provided selective advantage, allowing variations in growth forms to evolve multiple times. Monoraphid diatoms, for example, are now known to have multiple evolutionary origins (Kingston, 2003) and

(a) (b)

Figure 2.7 Some growth variations in attached benthic diatoms: (a) the tuft-forming diatom *Synedra acus*; (b) the stalk-forming diatom *Rhoicosphenia curvata*. Scale bars equal 200 μm for A and 10 μm for B.

taxa adapted for disparate nutrient conditions achieve similar growth solutions for attachment to substrates. Convergent evolution has also been described for stalk-forming clavate-shaped diatoms (Kociolek & Stoermer, 1988). Caution should be taken when taxonomic classification is based upon growth habit without reference to other features useful for diagnosis.

Variations in habitat provide opportunity for speciation and divergence of diatom populations. This can occur rapidly through seemingly minor environmental alterations. Theriot *et al.* (2006) documented species divergence of two *Stephanodiscus* species in Yellowstone Lake (Wyoming, USA) occurring in fewer than 10,000 years. The number of possible habitat combinations and abundance of diatoms mean new species discovery is possible in any ecological endeavor involving diatoms. Investigators should not hesitate to label unknown taxa as unknown, rather than forcing a specific entity into an existing species category (see Stoermer, 2001). Materials used in these investigations should be archived and made available for investigation and classification.

2.5 Sample preparation and material preservation

Diatom studies are somewhat of an enigma. The diversity and ecological dominance in aquatic systems demonstrate their importance and utility in ecological investigations. At the same time, the lack of understanding in virtually all aspects of the group's biology is, at times, overwhelming. This contrast should not intimidate investigators interested in utilizing diatoms in research pursuits. It need only encourage investigators to make the best estimates possible and voucher samples used to make those estimates. Diatoms are inherently suited for vouchering. Appropriate preparation and curation of samples will allow vouchered material to be available to future generations interested in reexamination in light of new discoveries and an expanded knowledge base.

When a sample is collected, two considerations must be given to the collection in context of archival preservation. First, some portion of the material has to be processed in a manner resulting in the removal of organic materials, with only siliceous components remaining (Battarbee *et al.*, 2001). Second, the expanding techniques and power of molecular data sets also encourage the preservation of organic components. Protocols exist for each sample preparation method (Blanco *et al.*, 2008).

A variety of techniques exist for removal of organic material allowing observation of the frustule (Battarbee *et al.*, 2001). These methods vary in time, expense, and the generation of dangerous by-products. Investigators must select a method producing the best product within constraints of specific laboratory settings. Until recently, frustules were most often cleaned with concentrated nitric acid (HNO_3) augmented with potassium dichromate (KCr_2O_7) as a catalyst. The chromium disassociated in this process has made the technique less popular. Similar results can be achieved via extended exposure (multiple hours) to heated nitric acid. Diatom acid slurries must be repeatedly rinsed with distilled water. In this process, water is added to a sample being cleaned and diatoms are allowed to settle. Liquids are decanted and distilled water is again added and diatoms are allowed to settle. This process continues until a neutral pH is achieved. Forcing diatom sinking via centrifugation can accelerate the process. Caution must be taken prior to application of nitric acid. If calcium carbonate ($CaCO_3$) or ethanol (C_2H_6O) is present in samples nitric acid addition may cause a violent reaction. Calcium carbonate can be eliminated via pretreatment with hydrochloric acid (HCl). In the case of ethanol, nitric acid must be avoided completely.

Once samples are cleaned, a portion of the siliceous remnants should be archived for electron microscope and other types of observation. Cleaned material should be dried before archiving to avoid individual frustules dissolving and reprecipitating due to Ostwald ripening (a phenomenon in which larger particles are thermodynamically favored over many small particles). To prepare a light microscope slide, materials should be mounted in a highly refractive material, ideally one with a refractive index of ~1.7. Two products are currently commonly used for these preparations, Naphrax™ and Meltmount™. Cleaned material is dried onto cover slips and then adhered to a microscope slide with one of these high-refractive-index materials. Observation of slides prepared is typically conducted at 1000× with a high-resolution objective.

When preserving organic materials for future molecular investigations formalin-based preservatives should be avoided in favor of ethanol. While pigments will be extracted into the preservation media, nucleic acids will not be damaged. If organics need to be preserved in a manner leaving pigments intact, a separate preparation should be made.

The degree to which samples should be processed and archived is at the discretion of the investigator. At a minimum, materials should be prepared to accomplish the project for which collections were made. Maximally, materials should be archived in light of potential future investigations. Materials archived in a manner allowing access to future investigators lend themselves to a more rapid advancement in an understanding of diatom biology.

2.6 Summary

Diatoms diverged from other "golden" algae relatively recently and represent a unique evolutionary lineage less than 200 million years old. During this time, the group diversified to a degree rivaled only by the insects and contains at least 200,000 species. Diatom researchers have examined less than 15% of this estimated diversity, and many gaps exist in the current understanding of diatom biology. Diatoms can be distinguished from other unicellular organisms by the presence of a siliceous cell wall, which typically consists of two halves fitting together to form a box containing the organic components of the cell. Cellular material within the box, known as a frustule, is similar in composition to other photosynthetic protists. Diatoms are members of a larger evolutionary group known as the heterokonts. Like many heterokonts, diatoms possess chlorophyll *a* and *c* and convert photosynthetic products into high-energy lipids. These lipids contain fatty acids required for animal metabolism. Diatoms are the most abundant group of heterokonts containing these lipid storage products, making them the preferred food source of many animals and aquatic fungi. The high species diversity and critical role in the aquatic food web make diatoms an ideal choice for environmental investigations; additionally the diatom frustule preserves well in sediments providing a data set ideal for paleoecological-oriented studies. Historically diatom investigations have focused on descriptive aspects of biology, but recent technological developments, specifically molecular tools, have allowed process-oriented questions to be explored. These activities are rapidly expanding the diatom knowledgebase and prior descriptive efforts are beginning to be understood at a genetic and metabolic level. This understanding should continue to expand at an unprecedented rate during the next few decades.

References

Ahlgren, G., Lundstedt, L., Brett, M., & Forsberg, C. (1990). Lipid composition and food quality of some freshwater phytoplankton for cladoceran zooplankters. *Journal of Plankton Research*, **12**, 809–18.

Almqvist, N., Bhatia, R., Primbs, G., *et al.* (2004). Elasticity and adhesion force mapping reveals real-time clustering of growth factor receptors and associated changes in local cellular rheological properties. *Biophysical Journal*, **86**, 1753–62.

Alverson, A. J. (2007). Strong purifying selection in the silicon transporters of marine and freshwater diatoms, *Limnology and Oceanography*, **52**, 1420–9

Alverson, A. J., Cannone, J. J., Gutell, R. R., & Theriot, E. C. (2006). The evolution of elongate cell shape in diatoms. *Journal of Phycology*, **42**, 655–68.

Alverson, A. J., Jansen, R. K., and Theriot, E. C. (2007). Bridging the Rubicon: phylogenetic analysis reveals repeated colonizations of marine and fresh waters by thalassiosiroid diatoms. *Molecular Phylogenetics and Evolution*, **45**, 193–210.

Andersen, R. A., Saunders, G. W., Paskind, M. P., & Sexton, J. P. (1993). Ultrastructure and 18S rRNA gene sequence for *Pelagomonas calceolata* gen. et sp. nov. and the description of a new algal class, the Pelagophyceae classis nov. *Journal of Phycology*, **29**, 701–15.

Apt, K. K., Bhaya, D., & Grossman, A. R. (1994) Characterization of the genes encoding the light-harvesting proteins in diatoms: the biogenesis of the fucocanthin chlorophyll a/c protein complex, *Journal of Applied Phycology* **6**, 225–30.

Armbrust, E. V. (1999). Identification of a new gene family expressed during the onset of sexual reproduction in the centric diatom Thalassiosira weissflogii. *Applied Environmental Microbiology*, **65**, 121–8.

Arts, M. T., Brett, M. T., & Kainz, M. J. (2009). *Lipids in Aquatic Ecosystems*. New York: Springer.

Bachvaroff, T. R., Sanchez-Puerta, M. V., & Delwiche, C. F. (2005). Chlorophyll c-containing plastid relationships based on analyses of a multigene data set with all four chromalveolate lineages. *Molecular Biology Evolution*, **22**, 1772–82.

Battarbee, R. W., Carvalho, L., Jones, V. J., *et al.* (2001). Diatom analysis. In *Tracking Environmental Change using Lake Sediments*, ed. J. P. Smol, H. J. B. Birks and W. M. Last, Dordrecht: Kluwer Academic Publishers, pp. 155–202.

Bhattacharya, D. and Medlin, L. K. (2004). Dating and algal origin using molecular clock methods. *Protist*, **155**, 9–10.

Bhaya, D. & Grossmann (1991). A new route for targeting proteins into plastids; evidence from diatoms. *Molecular and General Genetics*, **229**, 400–4.

Bhaya, D. and Grossman, A. R. (1993). Characterization of gene clusters encoding the fucoxanthin chlorophyll proteins of the diatom Phaeodactylum tricornutum. *Nucleic Acids Research*, **21**, 4458–66.

Biggs, B. J. F. (1996). Patterns in benthic algae of streams. In *Algal Ecology: Freshwater Benthic Ecosystems*, ed. R. J. Stevenson, M. L. Bothwell and R. L. Lowe, San Diego, CA: Academic Press, pp. 31–56.

Blanco, S., Alvarez, I. & Cejudo, C. (2008). A test on different aspects of diatom processing techniques. *Journal of Applied Phycology*, **20**, 445–50.

Bodyl, A. (2005). Do plastid-related characters support the chromalveolate hypothesis? *Journal of Phycology*, **41**, 712–19.

Bott, T. L. (1996). Algae in microscopic food webs. In *Algal Ecology: Freshwater Benthic Ecosystems*, ed. R. J. Stevenson, M. L. Bothwell and R. L. Lowe, San Diego, CA: Academic Press, pp. 574–607.

Bourne, C. M., Palmer, J. D., & Stoermer, E. F. (1992). Organization of the chloroplast genome of the freshwater centric diatom *Cyclotella meneghiniana*. *Journal of Phycology*, **28**, 347–55.

Brzezinski, M. A. (1992). Cell-cycle effects on the kinetics of silicic acid uptake and resource competition among diatoms. *Journal of Plankton Research*, **14**, 1511–39.

Brzezinski, M. A., Olson, R. J., & Chisholm, S. W. (1990). Silicon availability and cell-cycle progression in marine diatoms. *Marine Ecological Progress Series*, **67**, 83–96.

Canter, H. M. & Lund, J. W. G. (1948). Studies on plankton parasites. I. Fluctuations in the numbers of *Asterionella formosa* Hass. in relation to fungal epidemics. *New Phytologist*, **47**, 238–61.

Canter, H. M. & Lund, J. W. G. (1951). Studies on plankton parasites. III. Examples of the interaction between parasitism and other factors determining the growth of diatoms. *Annals of Botany*, **15**, 359–71.

Canter, H. M. & Lund, J. W. G. (1953). Studies on plankton parasites. II. The parasitism of diatoms with special reference to lakes in the English Lake District. *Transactions of the British Mycological Society*, **36**, 13–37.

Cavalier-Smith, T. (2003) Protist phylogeny and the high-level classification of Protozoa. *European Journal of Phycology*, **39**, 338–48.

Chepurnov, V. A., Mann, D. G., Dassow, P., *et al.* (2006). Oogamous reproduction, with two-step auxosporulation, in the centric diatom *Thalassiosira puntigera* (Bacillariophyta). *Journal of Phycology*, **42**, 845–58.

Cholnoky, B. J. (1968). *Die Ökologie der Diatomeen in Binnengewässern*. Lehere: Verlag von J. Cramer.

Conley, D. J. (1988). Biogenic silica as an estimate of siliceous microfossil abundance in Great Lakes sediments. *Biogeochemistry*, **6**, 161–79.

Conley, D. J. (2002). Terrestrial ecosystems and the global biogeochemical silica cycle. *Global Biogeochemical Cycles*, **16**, 1121.

Conley, D. J., Kilham, S. S., & Theriot, E. C. (1989). Differences in silica content between marine and freshwater diatoms. *Limnology and Oceanography*, **3**, 205–13.

Cox, E. J. (1996). *Identification of freshwater diatoms from live material*. London: Chapman & Hall.

Edlund, M. B. & Stoermer, E. F. (1997). Ecological, evolutionary, and systematic significance of diatom life histories. *Journal of Phycology* **33**, 897–918.

Falkowski, P. G., Katz, M. E., Knoll, A. H., *et al.* (2004). The evolution of modern eukaryotic phytoplankton. *Science*, **305**, 354–60.

Fourtanier, E. & Kociolek, J. P. (1999). Catalogue of the diatom genera. *Diatom Research*, **14**, 1–190.

Fourtanier, E. & Kociolek J. P. (2009a). Catalogue of Diatom names, part II: Abas through Bruniopsis. *Occasional Papers of the California Academy of Sciences*, **156**(1).

Fourtanier, E. & Kociolek J. P. (2009b). Catalogue of diatom names, part I: introduction and bibliography. *Occasional Papers of the California Academy of Sciences*, **156**(2).

Geitler, L. (1973). Auxosporenbildung und Systematik bei pennaten Diatomeen und die Cytologie von *Cocconeis*-Sippen. *Österreichische Botanische Zeitschrift*, **122**, 299–321.

Geitler, L. (1982). Die infraspeczifischen Sippen von *Cocconeis placentula* des Lunzer Seebachs. *Archiv für Hydrobiologie. Supplement*, **63**, 1–11.

Goertzen, L. R. & Theriot, E. C. (2003). Effect of taxon sampling, character weighting, and combined data on the interpretation of relationships among the heterokont algae. *Journal of Phycology*, **39**, 423–39.

Gordon, R. & Drum, R. (1994). The chemical basis of diatom morphogenesis. *International Review of Cytology*, **150**, 243–372.

Grzebyk, D., Katz, M. E., Knoll, A. H., *et al.* (2004). Response to comment on "The evolution of modern eukaryotic phytoplankton." *Science*, **306**, 2191.

Guillou, L., Chrétiennot-Dinet, M. J., Medlin, L. K., *et al.* (1999). Bolidomonas: a new genus with two species belonging to a new algal class, Bolidophyceae (Heterokonta). *Journal of Phycology*, **35**, 368–81.

Hargraves, P. E. & French, F. W. (1983). Diatom resting spores: significance and strategies. In *Survival Strategies of the Algae*, ed. G. Fryxell, Cambridge: Cambridge University Press, pp. 49–68.

Harper, J. T. & Keeling, P. J. (2003). Nucleus-encoded, plastid-targeted glyceraldehyde-3-phosphate dehydrogenase (GAPDH) indicates a single origin for chromalveolate plastids. *Molecular Biology and Evolution*, **20**, 1730–5.

Harper, J. T., Waanders, E., & Keeling, P. J. (2005). On the monophyly of chromalveolates using a six-protein phylogeny of eukaryotes. *International Journal of Systematic and Evolutionary Microbiology*, **55**, 487–96.

Harwood, D. M. (1988). Upper Cretaceous and lower Paleocene diatom and silicoflagellate biostratigraphy from Seymour Island, eastern Antarctic Peninsula. In *Seymour Island Geology and Paleontology*, ed. R. M. Feldman and M. O. Woodburne, Geological Society of America Memoir, vol. 169, pp. 55–129.

Harwood, D. M. & Nikolaev, V. A. (1995). Cretaceous diatoms: morphology, taxonomy, biostratigraphy. In *Siliceous Microfossils*, ed. C. E. Blome, P. M. Whalen, & K. M. Reed, Paleontological Society Short Course 8, Knoxville, TN: The Paleontology Society, pp. 81–106.

Higgins, M. J., Molino, P., Mulvaney, P., & Wetherbee, R. (2003). The structure and nanomechanical properties of the adhesive mucilage that mediates diatom–substratum adhesion and motility. *Journal of Phycology*, **39**, 1181–93.

Hildebrand, M. K., Dahlin, K., & Volcani, B. E. (1998). Characterization of a silicon transporter gene family in *Cylindrotheca fusiformis*: Sequences, expression analysis, and identification of homologs in other diatoms, *Molecular and General Genetics*, **260**, 480–6.

Hoek, C. Van Den. (1978). *Algen: Einführung in die Phycologie*. Stuttgart: G. Thieme Verlag.

Julius, M. L. (2007a). Perspectives on the evolution and diversification of the diatoms. In *Pond Scum to Carbon Sink: Geological and Environmental Applications of the Diatoms*, ed. S. Starratt, Paleontological Society Short Course 13, Knoxville, TN: Paleontological Society, pp. 1–13.

Julius, M. L. (2007b). Why sweat the small stuff: the role of microalgae in sustaining Hawaiian ecosystem integrity. *Bishop Museum Bulletin in Cultural and Environmental Studies*, **3**, 183–93.

Julius, M. L., Blob, R., & Schoenfuss, H. L. (2005). The survival of Sicyopterus stimpsoni, an endemic amphidromous Hawaiian gobiid fish, relies on the hydrological cycles of streams: evidence from changes in algal composition of diet through growth stages. *Aquatic Ecology*, **39**, 473–84.

Julius, M. L., Estabrook, G. F., Edlund, M. B., & Stoermer, E. F. (1997). Recognition of taxonomically significant clusters near the species level, using computationally intensive methods, with examples from the Stephanodiscus niagarae species complex (Bacillariophyceae). *Journal of Phycology*, **33**, 1049–54.

Julius, M. L., Stepanek, J., Tedrow, O., Gamble, C., & Schoenfess, H. L. (2007). Estrogen-receptor independent effects of two ubiquitous environmental estrogens on *Melosira varians* Agardh, a common component of the aquatic primary production community. *Aquatic Toxicology*, **85**, 19–27.

Julius, M. L., Stoermer, E. F., Taylor, C. M., & Schelske, C. L. (1998). Local extinction of *Stephanodiscus niagarae* Ehrenb. (Bacillariophyta) in the recent limnological record of Lake Ontario. *Journal of Phycology*, **34**, 766–71.

Katz, M. E., Finkel, Z. V., Grzebyk, D., Knoll, A. H., & Falkowski, P. G. (2004). Evolutionary trajectories and biogeochemical impacts of marine eukaryotic phytoplankton. *Annual Review of Ecology and Systematics*, **35**, 523–56.

Keeling, P. J. (2004). Diversity and evolutionary history of plastids and their hosts. *American Journal of Botany*, **91**, 1481–93.

Keeling, P. J., Archibald, J. M., Fast, N. M., and Palmer, J. D. (2004). Comment on "The evolution of modern eukaryotic phytoplankton," *Science*, **306**, 2191.

Kidder, D. L. & Erwin, D. H. (2001). Secular distribution of biogenic silica through the phanerozoic: comparison of silica-replaced fossils and bedded cherts at the series level. *Journal of Geology*, **109**, 509–22.

Kidder, D. L. & Gierlowski-Kordesch, E. H. (2005). Impact of grassland radiation on the nonmarine silica cycle and Miocene diatomite. *Palaios*, **20**, 198–206.

Kingston, J. C. (2003). Araphid and monorhaphid diatoms. In *Freshwater Algae of North America: Classification and Ecology*, ed. J. D. Wehr and R. G. Sheath, San Diego, CA: Academic Press, pp. 595–636.

Kitchell, J. A., Clark, D. L., and Gombos, A. M. Jr. (1986). Biological selectivity of extinction: a link between background and mass extinction. *Palios*, **1**, 504–11.

Kociolek, J. P. (1997). Historical constraints, species concepts and the search for a natural classification of diatoms. *Diatom*, **13**, 3–8.

Kociolek, J. P. & Stoermer, E. F. (1988). A preliminary investigation of the phylogenetic relationships of the freshwater, apical pore field-bearing cymbelloid and gomphonemoid diatoms (Bacillariophyceae). *Journal of Phycology*, **24**, 377–85.

Kociolek, J. P. & Stoermer, E. F. (1989). Chromosome numbers in diatoms: a review. *Diatom Research*, **4**, 47–54.

Krammer, K. & Lange-Bertalot, H. (1986). Bacillariophyceae, 1 Teil, Naviculaceae. In *Süsswasserflora von Mitteleuropa, Band 2*, ed. H. Ettl, J. Gerloff, H. Heynig, and D. Mollenhauer, Stuttgart: Gustav Fischer.

Krammer, K. & Lange-Bertalot, H. (1988). Bacillariophyceae, 2 Teil, Bacillariaceae, Epithemiaceae, Surirellaceae. In *Süsswasserflora von Mitteleuropa, Band 2*, ed. H. Ettl, J. Gerloff, H. Heynig, and D. Mollenhauer. Jena: Gustav Fischer.

Krammer, K. & Lange-Bertalot, H. (1991a). Bacillariophyceae, 3 Teil, Centrales, Fragilariaceae, Eunotiaceae. In *Süsswasserflora von Mitteleuropa, Band 2*, ed. H. Ettl, J. Gerloff, H. Heynig, and D. Mollenhauer. Jena, Stuttgart: Gustav Fischer.

Krammer, K. & Lange-Bertalot, H. (1991b). Bacillariophyceae, 4 Teil, Achnanthaceae, Kritische Erganzungen zu Navicula (Lineolatae) und Gomphonema, In *Süsswasserflora von Mitteleuropa, Band 2*, ed. H. Ettl, J. Gerloff, H. Heynig and D. Mollenhauer. Jena, Stuttgart: Gustav Fischer.

Kröger, N., Deutzmann, R., & Sumper, M. (1999). Polycationic peptides from diatom biosilica that direct silica nanosphere formation. *Science*, **286**, 1129–32.

Lange-Bertalot, H. (1997). As a practical diatomist, how does one deal with the flood of new names? *Diatom*, **13**, 9–12.
Leblanc, C., Falciatore, A., & Bowler, C. (1999). Semi-quantitative RT-PCR analysis of photoregulated gene expression in marine diatoms. *Plant Molecular Biology*, **40**, 1031–44.
Leedale, G. F. (1974). How many are the kingdoms of organisms? *Taxon*, **23**, 261–70.
Lowe, R. L. (1974). Environmental requirements and pollution tolerance of freshwater diatoms. EPA-670/4-74-005. Cincinnati, OH: US Environmental Protection Agency.
MacDonald, J. D. (1869). On the structure of the diatomaceous frustule, and its genetic cycle. *Annual Magazine of Natural History*, **4**, 1–8.
Mann, D. G. (1993). Patterns of sexual reproduction in diatoms. *Hydrobiologia*, **269/270**, 11–20.
Mann, D. G. (1997). Shifting sands: the use of the lower taxonomic ranks in diatoms. *Diatom*, **13**, 13–17.
Mann, D. G. & Droop, S. J. M. (1996). Biodiversity, biogeography and conservation of diatoms. In *Biogeography of Freshwater Algae: Proceedings of the Workshop on Biogeography of Freshwater Algae, Developments in Hydrobiology 118*, ed. J. Kristiansen, Dordecht: Kluwer Academic Publishers, pp. 19–32.
Martin-Jezequel, V., Hildebrand, M., and Brzezinski, M. A. (2000). Silicon metabolism in diatoms: implications for growth. *Journal of Phycology*, **36**, 821–40.
MacBride, S. A. & Edgar, R. K. (1988) Janus cells unveiled: frustular morphometric variability in *Gomphonema angustatum*. *Diatom Research*, **13**, 293–310.
Medlin, L. K., Kooistra, W. H. C. F., Gersonde, R., Sims, P. A., & Wellbrock, U. (1997). Is the origin of the diatoms related to the end-Permian mass extinction? *Nova Hedwigia*, **65**, 1–11.
North, W. J. (1994). Review of Macrocystis biology. In *Biology of Economic Algae*, ed. I. Akatsura, The Hague: SPB Academic Publishing, pp. 447–528.
Owens T. G. (1986). Light-harvesting function in the diatom Phaeodactylum tricornutum. *Plant Physiology*, **80**, 732–8
Patrick, R. & Reimer, C. W. (1966). The diatoms of the United States I. Academy of Natural Sciences Philadelphia, Monograph no. 13.
Patrick, R. & Reimer, C. W. (1975). The diatoms of the United States II, part 1. Academy of Natural Sciences Philadelphia, Monograph no. 13.
Patterson, D. J. (1989). Stramenopiles: chromophytes from a protistan perspective. In *The Chromophyte Algae: Problems and Perspectives*, ed. J. C. Green, B. S. C. Leadbeater and W. L. Diver. Systematics Association Special Volume No. 38, Oxford: Clarendon Press, pp. 357–79.
Pfitzer, E. (1869). Über den Bau und Zellteilung der Diatomeen, *Botanische Zeitung*, **27**, 774–6.
Pfitzer, E. (1871). Üntersuchungen über Bau und Entwicklung der Bacillariaceen (Diatomeen). In *Botanische Abhandlungen aus dem Gebiet der Mophologie und Physiologie*, ed. J. Hanstein, Bonn: Adolph Marcus Publishing, pp. 1–189.
Rabosky, D. L. & Sorhannus, U. (2009). Diversity dynamics of marine planktonic diatoms across the Cenozoic. *Nature*, **457**, 183–6.
Rivkin, R. B. (1986). Radioisotopic method for measuring cell division rates of individual species of diatoms from natural populations. *Applied Environmental Microbiology*, **51**, 769–75.
Rothpletz, A. (1896). Über die Flysch-Fucoiden und einige andere fossile Algae, sowie über laisische, Diatomeen führende Hornschwämme. *Zeitschrift der Deutschen Geologischen Gesellschaft*, **4**, 854–914.
Rothpletz, A. (1900). Über einen neuen jurassichen Hornschwämme und die darin eingeschlossenen Diatomeen. *Zeitschrift der Deutschen Geologischen Gesellschaft*, **52**, 154–60.
Round, F. E. (1996). What characters define diatom genera, species and infraspecific taxa? *Diatom Research*, **11**, 203–18.
Round, F. E. (1997). Genera, species and varieties B are problems real or imagined? *Diatom*, **13**, 25–9.
Round, F. E., Crawford, D. M., & Mann, D. G. (1990). *The Diatoms: Biology and Morphology of the Genera*. Cambridge: Cambridge University Press.
Round, F. E. & Sims, P. A. (1981). The distribution of diatom genera in marine and freshwater environments and some evolutionary considerations. In *Proceedings of the Sixth Symposium of Recent and Fossil Diatoms*, ed. R. Ross, Königstein: Koeltz Scientific Books, pp. 301–20.
Ryall, K., Harper, J. T., & Keeling, P. J. (2003). Plastid-derived Type II fatty acid biosynthetic enzymes in chromists. *Gene*, **313**, 139–48.
Sanchez-Puerta, M. V. & Delwiche, C. F. (2008). Minireview: a hypothesis for plastid evolution in chromalveolates. *Journal of Phycology*, **44**, 1097–107.
Saunders, G. W., Potter, D., Paskind, M. P., & Andersen, R. A. (1995). Cladistic analyses of combined traditional and molecular data sets reveal an algal lineage. *Proceedings of the National Academy of Sciences of the USA*, **92**, 244–8.
Schelske, C. L. & Stoermer E. F. (1971). Eutrophication, silica depletion, and predicted changes in algal quality in Lake Michigan. *Science*, **173**, 423–4.
Scherer, C., Wiltshirea, K., & Bickmeyer, U. (2007). Inhibition of multidrug resistance transporters in the diatom Thalassiosira rotula facilitates dye staining. *Plant Physiology and Biochemistry*, **46**, 100–3.
Schmid, A. M. (1994). Aspects of morphogenesis and function of diatom cell walls with implications for taxonomy. *Protoplasma*, **181**, 43–60.
Schmid, A. M. & Volcani, B. E. (1983). Wall morphogenesis in Cosinodiscus wailesii Gran and Angst. I. Valve morphology and development of its architecture. *Jorunal of Phycology*, **19**, 387–402
Shubin, N. (2008). *Your Inner Fish: A Journey into the 3.5-Billion-Year History of the Human Body*. New York: Pantheon Publishers.
Sicko-Goad, L., Schelske, C. L., & Stoermer, E. F. (1984). Estimation of carbon and silica content of diatoms from natural assemblages using morphometric techniques. *Limnology and Oceanography*, **29**, 1170–8.
Sicko-Goad, L., Stoermer, E. F., & Fahnenstiel, G. (1986). Rejuvenation of Melosira granulata (Bacillariophyceae) resting cells from anoxic sediments of Douglas Lake, Michigan. I. Light and 14C uptake. *Journal of Phycology*, **22**, 22–8.

Sicko-Goad, L., Stoermer, E. F., & Ladewski, B. G. (1977). A morphometric method for correcting phytoplankton cell volume estimates. *Protoplasma*, **93**, 147–63.

Simonsen, R. (1979). The diatom system: ideas on phylogeny. *Bacillaria*, **2**, 9–71.

Sorhannus, U. (2001). A "total evidence" analysis of the phylogenetic relationships among the photosynthetic stramenopiles. *Cladistics*, **17**, 227–41.

Sorhannus, U. (2007). A nuclear-encoded small-subunit ribosomal RNA timescale for diatom evolution. *Marine Micropaleontology*, **65**, 1–12.

Stoermer, E. F. (1967). Polymorphism in Mastogloia. *Journal of Phycology*, **3**, 73–7.

Stoermer, E. F. (2001). Diatom taxonomy for paleolimnologists. *Journal Paleolimnology*, **25**, 393–398.

Sullivan, C. W. (1977). Diatom mineralization of silicic acid. II. Regulation of $Si(OH)_4$ transport rates during the cell cycle of *Navicula pelliculosa*. *Journal of Phycology*, **13**, 86–91.

Townsend, S. A. & Gell, P. A. (2005) The role of substrate type on benthic diatom assemblages in the Daly and Roper rivers of the Australian wet/dry tropics. *Hydrobiologia*, **548**, 101–15.

Theriot, E. C., Cannone, J. J., Gutell, R. R., & Alverson, A. J. (2009). The limits of nuclear-encoded SSU rDNA for resolving the diatom phylogeny. *European Journal of Phycology*, **44**, 277–90.

Theriot, E. C., Sherilyn, C. F., Whitlock, C., & Conley, D. J. (2006). Late Quaternary rapid morphological evolution of an endemic diatom in Yellowstone Lake, Wyoming. *Paleobiology*, **32**, 38–54.

Theriot, E. C. & Stoermer, E. F. (1984). Principal component analysis of *Stephanodiscus*: observations on two new species from the *Stephanodiscus niagarae* complex. *Bacillaria*, **7**, 37–58.

Tuchman, M. L., Theriot, E. C., & Stoermer, E. F. (1984). Effects of low level salinity concentrations on the growth of *Cyclotella meneghiniana* Kütz. *Archiv für Protistenkunde*, **128**, 319–26.

Tuji, A. (2000). Observation of developmental processes in loosely attached diatom (Bacillariophyceae) communities. *Phycological Research*, **48**, 75–84.

Volkman, J. K., Jeffery, S. W., Nichols, P. D., Rogers, G. I., & Garland, C. D. (1989). Fatty acid and lipid composition of 10 species of microalgae used in mariculture. *Journal of Experimental Marine Biology and Ecology*, **128**, 219–40.

Yoon, H. S., Hackett, J. D., Pinto, G., & Bhattacharya, D. (2002). The single, ancient origin of chromist plastids. *Proceedings of the National Academy of Science, USA*, **99**, 1507–12.

Wetherbee, R. (2002). The diatom glasshouse. *Science*, **298**, 13–14.

Williams, D. M. & Kociolek, J. P. (2007). Pursuit of a natural classification of diatoms: history, monophyly and the rejection of paraphyletic taxa. *European Journal of Phycology*, **42**, 313–19.

3

Numerical methods for the analysis of diatom assemblage data

H. JOHN B. BIRKS

3.1 Introduction

Research involving diatom assemblages, both modern and fossil, has expanded enormously in recent decades. Because many of the questions asked in such research are quantitative in character (e.g. what was the lake-water pH at AD 1850, what are the major environmental gradients determining the modern diatom assemblages in a set of lakes on the Isle of Skye), there has been a similar development and application of numerical methods appropriate for the quantitative analysis of diatom assemblage data.

Despite diatom ecology and paleoecology being over 100 years old, the relevant statistical methods for assessing the inherent uncertainties associated with diatom counts were only relatively recently developed (in the context of pollen counting) by Mosimann (1965). The application of multivariate data analytical techniques such as cluster analysis, principal components analysis, and correspondence analysis to diatom assemblage data began in the early 1970s. With the upsurge of interest in the mid 1980s in diatom ecology and paleoecology in response to research on the causes of surface-water acidification in Europe and North America, the development and application of data analytical techniques such as canonical correspondence analysis (ter Braak, 1985, 1986) and weighted-averaging regression and calibration (ter Braak & van Dam, 1989) in diatom research expanded greatly (Birks, 1998). Such techniques are now widely used items in the diatomist's tool-kit (Smol *et al.*, 2011 and chapters in this volume).

Numerical methods are useful tools in many stages in the study of modern and fossil diatom assemblages. During **data collection** they can be useful in morphometric studies and in identification. In **data assessment**, statistical techniques are essential in estimating the inherent errors associated with the resulting diatom counts. A range of numerical techniques can be used to detect and **summarize** the major underlying patterns in diatom assemblages. For single stratigraphical sequences or spatial data sets of modern surface-samples, ordination procedures can help to summarize temporal, spatial, and/or environmental trends in such data sets. Numerical methods are essential for quantitative **data analysis** in terms of establishing directly diatom–environment relationships in modern spatial data sets, estimating diatom taxonomic richness, quantifying rates of change in diatom stratigraphical data, detecting temporal patterns such as periodicities in diatom stratigraphical data and resulting environmental reconstructions, and quantitatively reconstructing aspects of past environments such as lake-water pH, total phosphorus content, or specific conductivity from fossil diatom assemblages. Statistical techniques can greatly assist in the final stage of **data interpretation**, namely the testing of competing hypotheses about underlying causative factors such as climate change and human activity that may have influenced fossil diatom assemblage changes in the past or may be influencing modern diatom assemblages today.

The structure of this chapter is as follows. After summarizing the different types of diatom assemblage data and their numerical properties, I consider numerical techniques that can assist in data collection and data assessment. I then consider appropriate numerical techniques for data summarization where basic patterns within a diatom data set are detected and for data analysis where particular numerical characteristics are estimated from diatom assemblage data. Data interpretation in terms of possible causative factors is then discussed. I conclude with some ideas on future challenges.

This chapter draws extensively on reviews by Birks and Gordon (1985), Birks (1985, 1987, 1992, 1995, 1998, 2008), ter Braak (1995, 1996), and Birks and Seppä (2004), and on several chapters in the forthcoming volume in the *Developments in Paleoenvironmental Research* series on *Tracking Environmental*

The Diatoms: Applications for the Environmental and Earth Sciences, 2nd Edition, eds. John P. Smol and Eugene F. Stoermer. Published by Cambridge University Press.

Change Using Lake Sediments, Volume 5: Data Handling and Numerical Techniques (Birks *et al.*, 2011).

No attempt is made in this chapter to present the underlying mathematics behind the different numerical methods. For that, the reader is referred to Birks and Gordon (1985), Jongman *et al.* (1987), Birks (1995), ter Braak (1995, 1996), Legendre and Legendre (1998), Gordon (1999), Reyment and Savazzi (1999), Davis (2002), Hammer and Harper (2006), Manly (2007), and Zuur *et al.* (2007). Instead, the emphasis is on the ecological assumptions of the numerical methods discussed and on the role of different numerical techniques in diatom research; and on the reader acquiring a feel for "numerical thinking" in her/his diatom research, rather than providing a detailed account of "numerical calculations." Similarly, no attempt is made to review the now vast literature where numerical techniques have been used in diatom research.

Many of the numerical methods discussed here are widely used in other branches of paleolimnology (Birks, 1998) and in Quaternary pollen analysis (Birks & Gordon, 1985; Birks, 1987; 1992; Birks & Seppä, 2004).

3.2 Diatom assemblage data

Diatom assemblage data consist of counts of different diatom taxa present in one or more samples. The samples can be from one lake-sediment core collected from the study site (fossil, temporal, or stratigraphical data) or from one age at one or several sites (spatial, geographical data). The most common type of spatial data consists of diatom counts from the uppermost, surficial lake sediments (modern surface samples). Other types of spatial data include diatom counts from particular habitat types within or between lakes that support plankton, epilithon, epiphyton, epipsammon, and epipelon assemblages (Battarbee *et al.*, 2001a). Stratigraphical and spatial diatom data usually contain counts of many (100–500) different diatom taxa in a large number of samples (50–500). Such data are highly multivariate, containing many variables (diatom taxa), and many objects (sediment or habitat samples). The number of taxa usually exceeds the number of samples, in contrast to, for example, pollen analysis where the number of samples is usually larger than the number of pollen and spore types.

Diatom assemblage data are almost always expressed as percentages or proportions of some calculation or base sum (e.g. total diatom valves), thereby transforming the counts into relative proportional or percentage compositional data. The sample sum itself is relatively uninformative in many numerical analyses. It is largely determined by the initial research questions and sampling design, namely the number of valves to be counted per sample (Battarbee *et al.*, 2001a). The numerical properties of relative diatom data are that the data are "closed" and add up to 100%, that there may be many zero values (absences, with up to 75% of the elements in the data matrix containing zero values), and that there are many diatom taxa (usually more than the number of samples). These numerical properties necessitate the use of particular numerical methods. Such closed compositional data require special methods for rigorous statistical analysis. Aitchinson (1986) and Reyment and Savazzi (1999) discuss statistical approaches to analyzing closed data using log ratios. The very large number of zero values in diatom data causes problems, however, in the use of log ratios. ter Braak and Šmilauer (2002) discuss in detail the problems of analyzing compositional data containing many zero values and suggest robust approximations via correspondence analysis.

Diatom counts are more rarely expressed as diatom concentrations (valves per unit volume: valves cm^{-3}, or dry weight of sediment: valves g^{-1}). If the sediment matrix accumulation rate (net thickness of sediment accumulation rate per unit time: $cm\ yr^{-1}$) can be estimated by, for example, age–depth modeling (Blaauw & Heegaard, 2011), diatom accumulation rates ('influx': valves $cm^{-2}\ yr^{-1}$) can be estimated. Such "absolute" data are important in, for example, the interpretation of fossil diatom assemblages from systems where silica may become limiting to diatom growth such as the Great Lakes of North America (Schelske & Stoermer, 1971; Schelske *et al.*, 1983; Stoermer *et al.*, 1985). As diatoms vary considerably in size, diatom accumulation rates may not, however, be very relevant in terms of paleoproductivity (Battarbee *et al.*, 2001a), unless some transformation from valve number to biovolume or biomass is applied. Mean diatom volumes can be estimated and used to transform valve numbers to biovolumes for individual taxa, and then the totals for individual taxa summed, corrected for the sediment matrix accumulation rate, and expressed as diatom biovolume ($\mu m^3\ cm^{-2}\ yr^{-1}$) (Anderson *et al.*, 1995; Battarbee *et al.*, 2001a).

Such "absolute" data are harder to obtain than relative counts. Although their numerical properties are less complex than relative abundance data, accumulation rates and biovolume estimates have a high inherent variance and some variance-stabilizing transformation is usually essential prior to any numerical analysis (Birks & Gordon, 1985).

3.3 Data collection and data assessment

Numerical methods can help in some aspects of the taxonomy and identification of diatom taxa and hence in basic data collection. Statistical techniques are essential in the assessment of the statistical uncertainties associated with diatom counting, namely error estimation.

3.3.1 Identification

Diatom identification is almost entirely done by comparison of the modern or fossil diatom with descriptions and illustrations in floras for specific geographical areas or particular ecological systems such as acidic lakes (Battarbee *et al.*, 2001a). Floras in an electronic format are increasingly becoming available that can be accessed either by CD-ROM (e.g. Kelly & Telford, 2007) or through the internet (Battarbee *et al.*, 2001b). The development of easy-to-use electronic monographs and updated floras is an important contribution in this time of rapidly declining basic taxonomic training and expertise (Stoermer, 2001).

Attempts have been made at automatic diatom identification by the ADIAC project (du Buf *et al.*, 1999; du Buf & Bayer, 2002) involving size and shape analysis of digital images. Automatic identification is a major challenge not only in diatom research but in other branches of paleoecology involving microfossils (e.g. pollen and spores). With ever-increasing computing power and improved and faster image-capture and image-analytical techniques, major advances can be expected in the coming years (Flenley, 2003).

Numerical techniques for morphometric analysis, such as Fourier analysis of diatom valve outlines to provide robust shape descriptors for subsequent multivariate analysis, have been used in diatom identification and taxonomy of critical taxonomic groups (e.g. Mou & Stoermer, 1992; Julius *et al.*, 1997; Pappas *et al.*, 2001; Pappas & Stoermer, 2001; Kingston & Pappas, 2009). Claude (2008) provides a thorough and up-to-date introduction to the wide range of quantitative morphometric techniques such as Fourier analysis, landmark configuration, and thin-plate splines available for rigorous morphometric analysis. These specialized but numerically robust methods are increasingly used for morphometric analysis of fossil and living organisms (Birks, 1985; Bookstein *et al.*, 1985; Reyment, 1991; Hammer & Harper, 2006). These multivariate techniques and methods such as principal components analysis, canonical correspondence analysis, and redundancy analysis have also been used in the study of diatom morphometric phenotypic variation in relation to contemporary environmental factors (Stoermer & Ladewski, 1982; Theriot & Stoermer, 1984; Theriot, 1987; Theriot *et al.*, 1988; Pappas & Stoermer, 1995; Hausmann & Lotter, 2001). These multivariate techniques are described in later sections on data summarization and data analysis. Besides detecting morphometric variation in relation to modern environmental variables, the same techniques can potentially be useful in detecting and summarizing temporal patterns of morphological change in, for example, the Yellowstone Lake endemic *Stephanodiscus yellowstonensis* Theriot & Stoermer (Theriot *et al.*, 2006). With increasing interest in contemporary diatom biogeography (Bouchard *et al.*, 2004; Vanormelingen *et al.*, 2008), there is the need to resolve any possible biogeographical components inherent in diatom taxonomy (Stoermer *et al.*, 1986; Tyler, 1996). Numerical morphometric techniques may play a critical role in this demanding research.

Although outside the scope of this chapter, it is important to note that modern cladistic analysis and phylogenetic analysis involve sophisticated and computer-intensive numerical methods. Paradis (2006) provides an excellent introduction to these methods and the computational challenges that arise with phylogenetic data and the estimation of phylogenies. Early examples of the application of these and related techniques in diatom research include Williams (1985) and Kociolek and Stoermer (1986, 1988).

3.3.2 Error estimation

All diatom counts are sample counts of the total amount of diatoms in the sample of interest, whether it be a sediment sample, an epilithon sample, or an epiphyton sample. The resulting count is a sample, hopefully an unbiased sample count, of the total amount of diatoms present in the sample of interest. As in all sampling, there are statistical uncertainties associated with any sample count. The larger the diatom count, the smaller the uncertainties become (Battarbee *et al.*, 2001a). As larger counts require more time to obtain, there is an inevitable trade-off between time spent counting and the acceptable level of uncertainty. It is therefore important (but very rarely done) to estimate and display the uncertainty associated with all diatom counts.

Statistical methods for estimating counting errors and confidence intervals associated with relative microfossil data were developed by Mosimann (1965) for pollen data based on the multinomial and negative multinomial distributions (Birks & Gordon, 1985; Maher *et al.*, 2011). Errors depend on the size of the count and on the proportion in the count sum of the diatom taxon of interest. The degree of uncertainty decreases with the size of the count (Maher *et al.*, 2011). Statistical methods for estimating errors and confidence intervals associated with diatom concentrations are also available (Maher *et al.*, 2011), if the diatom concentrations have been estimated by adding a known number of "exotic" plastic microspheres to a known volume of sediment (Battarbee *et al.*, 2001a). Estimating the total error requires careful propagation of the errors associated with the various stages involved in deriving concentration values. The total error depends not only on estimating and combining these errors but also on the number of diatoms counted. Estimating errors for diatom or biovolume accumulation rates is similar to estimating concentration errors (Bennett, 1994), but with the additional complexity of incorporating the uncertainties associated with estimating sediment accumulation rates based on radiometric dates (e.g. ^{210}Pb, ^{14}C) and, in the case of radiocarbon dates, the calibration of radiocarbon ages into calibrated ages (Heegaard *et al.*, 2005; Blaauw & Heegaard, 2011). Additional errors will arise in the conversion of diatom accumulation rates to biovolume accumulation rates. The total error associated with diatom accumulation rates may be 20–40% or more of the estimated values. It is therefore essential to derive and display reliable estimates of the total errors before attempting interpretation of observed changes in diatom or biovolume accumulation rates (Bennett, 1994; Maher *et al.*, 2011).

In cases where research projects involve more than one diatomist or more than one laboratory, inter-analyst or inter-laboratory comparisons become important. Such comparisons have, however, rarely been conducted. Munro *et al.* (1990) present the results of an inter-laboratory comparison involving four different diatom analysts. They emphasize the importance of harmonizing nomenclature and of establishing clear and well-documented criteria for the identification of critical taxa (Battarbee *et al.*, 2001b). Kreiser and Battarbee (1988) discuss this in more detail and recommend the use of ordination methods (see below) for comparing the results of inter-analyst comparisons of fossil diatom samples. Wolfe (1997) made an inter-laboratory comparison where 15 different laboratories estimated the total diatom concentration in a homogenized uniform lake-sediment sample. The results of the comparison highlight the need for harmonizing and improving techniques currently used for estimating total diatom concentrations. Besse-Lototskaya *et al.* (2006) present the results of an inter-laboratory assessment of uncertainties in diatom taxonomy and counting and substrate type in using periphyton diatoms for environmental monitoring in European rivers. This study reveals important differences in taxonomy and counting between analysts (see also Kelly *et al.*, 2002).

An important but frequently ignored aid in all subsequent numerical analysis of diatom assemblage data is to store the basic data and associated sample, site, environmental, and chronological information in a carefully designed relational data base. Juggins (1996) describes the various questions and stages in the design and implementation of such a data base in a large multiproxy collaborative research project involving many biological proxies, including diatom assemblages.

3.4 Data summarization

Data summarization involves detecting the major patterns of variation within a single diatom assemblage data set (temporal or spatial data) and the patterns of similarity and difference between two or more stratigraphical sequences or geographical data sets. An important preliminary step before data summarization, data analysis, or data interpretation is exploratory data analysis.

3.4.1 Exploratory data analysis

Exploratory data analysis (EDA) is curiously given little attention in the numerical analysis of diatom assemblage data. It involves summarizing data sets in terms of measures of location or "typical value" such as means, medians, trimmed means, and geometric means; measures of dispersion such as range, quartiles, variance, standard deviation, coefficient of variation, and standard error of the mean; and measures of skewness and kurtosis (see Sokal & Rohlf, 1995; Fox, 2002 for details of these basic tools). Simple graphical tools like histograms, kernel density estimation plots, quantile–quantile plots, and box-whisker plots can help in the critical decision about data transformation (Fox, 2002), particularly of environmental data such as lake-water chemical measurements in modern diatom-chemistry data sets.

For data consisting of two or three variables only, simple scatter plots and matrices of scatter plots can be useful in helping to assess relationships between variables. For multivariate data, there are several useful graphical tools (Everitt, 1978), but with diatom data sets with 100–300 taxa these graphical tools have limited value and methods of clustering, partitioning, and ordination (Legendre & Legendre, 1998; Legendre & Birks, 2011a, 2011b) are generally more useful and the results easier to interpret than more strictly EDA graphical tools. EDA provides means of identifying potential outlying or "rogue" samples – samples that are, in some way, inconsistent with the rest of the samples in the data set. A sample can be an outlier for various reasons, such as the value of a diatom taxon in a sample lies well outside the expected range. Exploratory data analysis provides powerful means of detecting outliers using measures of leverage (the potential for influence resulting from unusual values) and of influence (an object or variable is influential if its deletion substantially changes the results) (Fox, 2002). After detecting potential outliers, the diatomist then has the challenge of trying to ascertain why the sample or variable is an outlier – do the unusual values result from incorrect determinations or measurements, incorrect data entry, transcription or recording errors, or unusual site features? It is important to realize that the concept of "outlier" is model dependent, so a sample may appear as an outlier in, say, a linear regression but not be an outlier in, say, a principal components analysis. Good practice is not to delete a sample or a variable until there is very strong evidence that they are really outliers and have significant influence on the results. Birks *et al.* (1990a), Jones and Juggins (1995), and Lotter *et al.* (1997) illustrate different approaches to detecting outliers in modern diatom-environment data sets.

Graphical techniques and effective data display are key aspects of modern EDA (Chambers *et al.*, 1983; Tufte, 1983; Hewitt, 1992; Cleveland, 1993, 1994). As Cleveland (1994) notes "graphs allow us to explore data to see overall patterns and to see the detailed behavior; no other approach can compete in revealing the structure of data so thoroughly." In a penetrating graphical analysis of a standard hunting-spider ecological data set that is widely used in ordination analyses, Warton (2008) shows how this data set has been misinterpreted in several influential methodological papers because the data were interpreted from ordination plots **alone**, with no consideration of simple graphical plots of the basic raw data.

3.4.2 Clustering, partitioning, and ordination approaches

Clustering, partitioning, and ordination approaches overlap with EDA but attempt to detect clusters, groups, gradients, and patterns of variation when the diatom data set is considered as multivariate data rather than as univariate or bivariate data sets. These approaches provide an essential stage in data summarization as they provide low-dimensional representations or groupings of samples that can provide a convenient basis for description, discussion, hypothesis generation, and interpretation. The low-dimensional representation or groupings are selected to fulfill predefined mathematical criteria. For example the selected dimensions and associated axes capture as much of the total variation in the data set with the constraint that all the axes are uncorrelated to each other, or the groups are delimited to maximize the between-group variation relative to the total variation in the data set.

3.4.2.1 Clustering and partitioning approaches Standard techniques for clustering and partitioning (e.g. hierarchical cluster analysis, *k*-means partitioning, two-way indicator species analysis) do not take account of the associated stratigraphical, geographical, or environmental features of a diatom data set. Hence, one may discard potentially important information when these methods are applied to diatom assemblage data, with the result that samples with similar diatom composition may be grouped together even though the samples are far apart stratigraphically or geographically.

In the case of the zonation or partitioning of a stratigraphical diatom data set, it is important to use clustering or partitioning methods that take account of the stratigraphical nature of the data, so-called constrained methods (Birks & Gordon, 1985; Gordon, 1999; Birks, 2011). A numerically derived zonation is implemented by a clustering or partitioning method but with the stratigraphical constraint that similar samples are only grouped together if they are stratigraphically adjacent. The end result is a series of site or local diatom-assemblage zones. The question immediately arises as to how many zones should be distinguished and which zones are "significant" in a statistical sense. Bennett (1996) has shown how the significance of zones can be assessed against a model of random partitioning of the stratigraphical sequence using the simple broken-stick model (Jolliffe, 1986; Jackson, 1993). The broken-stick model provides an indication of the number of significant partitions. The significant diatom zones are simply those partitions that account for a portion of the total variation greater than would be expected from the broken-stick model for the same number of segments (Bennett, 1996). Lotter (1998) has applied this approach to a diatom stratigraphy with an annual resolution from Baldeggersee (Switzerland) to delimit seven statistically significant partitions in a 110-year-long sequence. The broken-stick approach is directly applicable to divisive zonation procedures such as optimal divisive partitioning and binary divisive techniques (Birks & Gordon, 1985; Gordon, 1999). It can also be applied with care to agglomerative clustering procedures such as the widely used CONISS (constrained incremental sum-of-squares clustering (Grimm, 1987)) zonation procedure available in the Tilia-Graph/TGView software. Bennett's (1996) broken-stick approach puts the identification and delimitation of partitions in diatom stratigraphical sequences on a rigorous and consistent basis. Of the many constrained zonation techniques currently available, optimal divisive partitioning using the sum-of-squares criterion with square-root transformed diatom percentage values and implemented by a dynamic programming algorithm (Birks & Gordon, 1985), followed by a comparison of the variation of the zones with the variation expected under the broken-stick model, is consistently the most robust and hence most useful zonation procedure for complex multivariate stratigraphical diatom data. In the zonation of high-resolution diatom stratigraphical data, the application of this approach involving the broken-stick model may result in a large number (>20) of seemingly significant partitions (e.g. Laird *et al.*, 2003). Such a large number of zones can defeat the major purpose of zonation, namely as a tool in data summarization. High-resolution data often have a high inherent sample-to-sample variance (Birks, 1998). An effective and more useful summarization of such data may be achieved by using ordination methods, which, by design, concentrate the "signal" in a data set into the first few ordination axes and relegate the "noise" to later axes (Gauch, 1982).

In the analysis of modern diatom data sets, it can be useful to cluster or partition the data into a smaller number of groups of samples with similar diatom composition or environmental characteristics (Legendre & Birks, 2011a). There are many potential uses of the results from such analyses, for example detecting groups of samples with similar diatom composition or with similar environmental features, detecting indicator species for the groups, relating biologically based groups to environmental variables, and assessing similarities between fossil diatom samples and groups of modern diatom samples from known environmental settings. Besides imposing one-dimensional temporal or stratigraphical constraints, two-dimensional (geographical coordinates) constraints (Legendre & Legendre, 1998; Legendre & Birks, 2011a) can be imposed to detect spatially contiguous groups of samples with similar diatom composition and/or environmental characteristics. The question of whether to apply such constraints depends very much on the research problem of interest – do similar modern diatom assemblages all occur together geographically, or how can modern diatom assemblages be partitioned into coherent geographical areas of similar diatom floras? In the first question, unconstrained analyses would be appropriate; in the second case, spatially constrained analyses would be required. The results of such clustering or partitioning should be displayed as maps of the group membership of samples or on low-dimensional ordination plots of the sample scores on the ordination axes. Clustering and partitioning techniques are surprisingly little used in the numerical analysis of diatom assemblage data, except in the partitioning of stratigraphical sequences into assemblage zones.

3.4.2.2 Ordination approaches In contrast to cluster and partitioning approaches, ordination techniques or indirect gradient analysis are widely used in diatom research (Birks, 1998; Legendre & Birks, 2011b) to summarize patterns of variation in complex modern or stratigraphical diatom data sets and to provide convenient low-dimensional representations of such data.

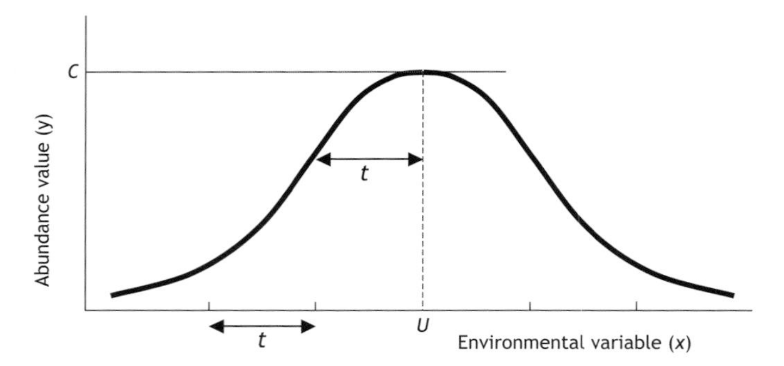

Figure 3.1 The Gaussian response curve for the abundance (y) of a diatom taxon in relation to an environmental gradient (x). The estimated optimum for the species is shown as U, its tolerance is labeled t, and the maximum height of the response curve is designated C (ter Braak, 1987b).

The major ordination or dimension-reduction techniques for data summarization, as distinct from constrained ordination techniques used in data analysis where particular numerical statistics are estimated from diatom data (see below), are principal components analysis (PCA), correspondence analysis (CA), detrended correspondence analysis (DCA), principal coordinates analysis (PCoA) (= metric or classical multidimensional scaling), and non-metric multidimensional scaling (NMDS). All these ordination techniques can provide a low-dimensional representation of the samples and the variables (indirectly in PCoA and NMDS) in the diatom data set so that the points in the low-dimensional plot (usually two-dimensional) that are close together correspond to samples that are similar in their diatom composition, and points that are apart correspond to samples that are dissimilar in their diatom composition. There is, however, a second more ambitious purpose of an ordination (ter Braak, 1987a), namely to detect the underlying latent structure in the data: in other words the occurrence and/or abundances of all the species in the data set are assumed to be determined by a few unknown environmental variables (latent variables) according to a simple species–environment response model (ter Braak & Prentice, 1988). If the aim of an ordination is to detect the underlying structure in the data, we need to assume a species–response model (e.g. linear or unimodal) *a priori*. The ordination problem is thus to construct the single "hypothetical environmental variable" that gives the best fit to the species data under the assumption of a linear or unimodal response model. Principal components analysis relates to a linear response model in which the abundance of any species increases or decreases with the value of each latent environmental variable. Corresondence analysis and its DCA relative are related to a unimodal response model (Figure 3.1) in which any species occurs in a limited range of values for each of the latent variables (ter Braak, 1987b). Principal coordinates analysis and NMDS are solely dimension-reduction techniques and are not guaranteed to extract the latent variables that give the best fits to the species data because they make no assumptions about species–response models. They only involve dissimilarities (PCoA) or rank dissimilarities (NMDS) between all pairs of samples. Detailed accounts of PCA, CA, DCA, PCoA, and NMDS are given by ter Braak (1987a; 1996), Legendre and Legendre (1998), and Lepš and Šmilauer (2003).

As it is generally more useful ecologically to extract the latent variables, and thus to use PCA, CA, or DCA, the question immediately arises of which response model and hence which methods are appropriate for a given data set. ter Braak and Prentice (1988) suggest that the length of the first gradient of variation in multivariate biological data, such as diatom assemblage data, as estimated by DCA (in standard deviation (SD) units of compositional turnover) is a useful guide as to whether species responses are primarily monotonic (gradient length <2 SD) or primarily unimodal (>3 SD). My experience with DCA with many diatom percentage data sets is that turnover estimation should be based on square-root transformed data and without down-weighting of rare species (ter Braak & Šmilauer, 2002). Which method should be used if the gradient length is 2.4 SD? My experience is that CA is remarkably robust with percentage data and it can be safely used for data with gradient lengths as low as 1.5 SD, whereas PCA is more erratic for data with gradient lengths >2.5 SD. If the final ordination is based on PCA, square-root transformed data and a covariance matrix between species are reliable choices, whereas if the final ordination is based on CA or DCA, square-root transformed data and down-weighting of rare species are appropriate. In the construction of ordination plots for PCA or CA, care should be given to the choice of axis scaling. The choice of scaling focused on inter-sample distances is generally appropriate when the primary interest is on the configuration of the samples, whereas the scaling focused on inter-species distance is generally appropriate when the interest is on the configuration of the species. Symmetric scaling (Gabriel, 2002) is often a very good compromise.

In general, in all uses of the unimodal-based methods of CA, DCA, and canonical correspondence analysis (CCA), down-weighting of rare species should be applied because individual samples with rare species may distort the results of the analyses (ter Braak & Šmilauer, 2002). In estimating gradient length or compositional turnover along the first DCA axis or along the first constrained axis (e.g. pH or age) in detrended canonical correspondence analysis (DCCA), experience shows that **no** down-weighting of rare species is desirable as it usually leads to more robust estimates of compositional turnover.

Ordination results of analyzing diatom stratigraphical data can usefully be plotted in a stratigraphical context with axis 1 sample scores plotted against depth or age, axis 2 sample scores plotted in the same way, etc. The appropriate number of ordination axes to retain and plot can be assessed by comparison with the broken-stick model (Jolliffe, 1986; Jackson, 1993; Legendre & Legendre, 1998). Such stratigraphical ordination plots provide a useful summary of the major patterns of variation and the nature of changes in the stratigraphical sequence. These plots allow the detection of trends and abrupt changes within the sequence which may be obscured in zonation where the primary aim is partitioning the data sequence into "homogeneous" units or zones. Careful examination of the species (variable) scores or loadings on the ordination axes can reveal which diatom species are most influential to the sample scores for a given ordination axis, thereby providing an ecological interpretation of the observed patterns in the latent variables (= ordination axes) (see Bradshaw *et al.* (2005) for a detailed example of using DCA to summarize and compare patterns in stratigraphical diatom and pollen data).

For summarizing biostratigraphical data, DCA is my preferred ordination technique because the sample scores are scaled in standard deviation units of compositional change or turnover (Hill & Gauch, 1980). It is thus possible to obtain a graphical summary of the magnitude of diatom compositional change within the stratigraphical sequence. Other ordination techniques such as PCA and CA lack this ecologically attractive and useful scaling of sample scores. Using PCoA and NMDS is not really appropriate for stratigraphical plotting as their axes are not selected to capture the latent structure of the data.

Ordination of spatial modern diatom data sets can usefully be done with PCA, CA, or DCA depending on gradient length (PCA or CA) and research questions (CA or DCA). However, many modern diatom data sets also include environmental variables such as lake-water chemical data and, in these cases, constrained ordination or direct gradient analytical techniques are more useful. These are discussed below under data dnalysis.

3.4.2.3 Two or more stratigraphical sequences When two or more paleoecological variables (e.g. diatoms, pollen, chironomids) have been studied in the same stratigraphical sequence, numerical zonations based on each set of variables and a comparison of the resulting partitions can help to identify common and unique changes in different variables (Birks & Gordon,

1985). Separate ordinations (PCA, CA, or DCA) of the different data sets can help to summarize the major patterns within each data set, and these patterns can be compared by, for example, oscillation logs (Birks, 1987). In cases where one data set can be considered as representing "response" variables (e.g. diatoms) and another data set can be regarded as reflecting potential "predictor" or explanatory variables (e.g. pollen, independent climate estimates (e.g. St Jacques *et al.*, 2009)), constrained ordinations such as redundancy analysis (RDA = constrained PCA), canonical correspondence analysis (CCA = constrained CA), or detrended CCA (DCCA = constrained DCA) and associated Monte Carlo permutation tests (ter Braak & Šmilauer, 2002) can be used to assess the statistical relationship between the two data sets (Haberle *et al.*, 2006). This approach is discussed further in the data interpretation section.

3.4.2.4 Other data summarization techniques An important part of data summarization is clear display of the data. For fossil diatom assemblages with 50 or more species, it is important to plot their relative frequencies in basic stratigraphical plots in a way that displays the major patterns of variation within the data as a whole. A very simple but effective way is to calculate the weighted average or "optimum" of each species for age or depth and to re-order the species in order of their optima, with species having high optima for age or depth being plotted first at the bottom left of the stratigraphical plot and with species having low optima being plotted at the top right of the plot (Janssen & Birks, 1994). Alternatively, species can be ordered on the basis of their modern weighted-average optima for a particular environmental variable (e.g. lake-water pH – see the section on data analysis below) or on the basis of the species scores on the first DCA or CA axis.

Non-parametric regression techniques such as locally weighted scatterplot-smoothing (LOWESS or LOESS) regression provide useful graphical tools for highlighting the "signal" or major patterns in stratigraphical sequences of individual species in stratigraphical plots of sample scores on ordination axes, and of time series of reconstructed environmental variables. The LOESS technique (Cleveland, 1979; 1993; 1994) can be used to model the relationship between a response or dependent variable (e.g. abundances of *Tabellaria binalis* (Ehrenberg) Grunow) and an independent variable (e.g. age or depth) when no single functional form such as a linear or quadratic model is appropriate or assumed. The technique provides a graphical summary that helps to assess the relationship and to detect the major patterns of change within "noisy" data. Unlike conventional regression modeling, LOESS fits a series of locally weighted regression curves for different values of the independent variable, in each case using data points weighted by their distances to the values of interest in the independent variable. A LOESS curve is a non-parametric regression estimate because it does not assume a particular parametric form (e.g. quadratic) for the regression (Cleveland & Devlin, 1988). It is conceptually similar to "running means" except that LOESS takes into account the uneven spacing of the independent time or depth variable. In LOESS fitting, the degree of smoothing or span can be varied and lies between 0 and 1. As the span is increased, the fitted curve becomes smoother. Choosing the appropriate span requires some judgment for each data set. The goal is generally to make the span as large as possible and thus to make the fitted curve as smooth as possible without distorting the underlying pattern or "signal" in the data (Cleveland, 1994). The LOESS technique can also be used as a scatter-plot smoother when, for example, the abundance of *Eunotia exigua* Brébisson ex Kützing Rabenhorst (dependent variable) in a series of modern samples is plotted in relation to lake-water pH (independent variable).

3.5 Data analysis

Data analysis is used here specifically for specialized techniques that estimate particular numerical characteristics from modern or fossil diatom assemblage data. Examples include taxonomic richness, rates of change, modern diatom assemblage–environment relationships, species optima and tolerances for contemporary environmental variables, inferred past environmental variables, and periodicities and power spectra in diatom stratigraphical data.

3.5.1 Taxonomic richness

Rarefaction analysis (Birks & Line, 1992) estimates the taxonomic richness within and between stratigraphical diatom sequences and within and between modern diatom data sets. It estimates how many species would have been found if all the diatom counts had been the same size. The actual minimum count among the data set(s) of interest is generally used as the base value. Examples of the use of rarefaction analysis in diatom research include Anderson *et al.* (1996) and Lotter (1998), whereas Weckström and Korhola (2001), Rusak *et al.* (2004), Heegaard *et al.* (2006), Telford *et al.* (2006), and Vyverman *et al.* (2007), highlight the potential of exploring rigorously richness patterns of diatom assemblages in space and time. The pioneering approach of Patrick and colleagues (Patrick, 1949; Patrick *et al.*, 1954; Patrick & Strawbridge, 1963) used diatom assemblage structure to assess richness and diversity of diatom communities. Pappas and Stoermer (1996) also exploited aspects of diatom assemblage structure to evaluate counting accuracy. Much work is needed in estimating and interpreting taxonomic richness and biodiversity of diatom assemblage data.

3.5.2 Rate-of-change analysis

Rate-of-change analysis (Jacobson & Grimm, 1986; Grimm & Jacobson, 1992) estimates the amount of compositional change per unit time in stratigraphical data. It is estimated by calculating the multivariate dissimilarity (e.g. Euclidean distance, chord distance (= Euclidean distance of square-root transformed percentage data)) between stratigraphically adjacent samples and by dividing the dissimilarity by the estimated age interval between the sample pairs. An alternative approach is to interpolate the stratigraphical data to constant time intervals, calculate the dissimilarity, and divide by a constant time interval. The data can be smoothed prior to interpolation. Although the basic idea is attractive, in practice there are several problems (Lotter *et al.*, 1992; Laird *et al.*, 1998). Rate-of-change analysis is critically dependent on the time standardization unit used to standardize the estimated dissimilarity between adjacent stratigraphical samples. As radiocarbon years do not equal calendar years, a carefully calibrated timescale or an independent absolute chronology (e.g. from laminated sediments) is essential for reliable rate-of-change estimation. Rate-of-change analysis of fine-resolution data from annually laminated sediments (e.g. Lotter *et al.*, 1995) is often dominated by the high sample-to-sample variability in composition that commonly occurs in fine-resolution data. This inherent variability is reduced in coarser resolution studies as stratigraphical data from such studies are effectively time averaged and temporally smoothed. One approach is to summarize the stratigraphical data as the first few statistically significant ordination axes or latent variables and to use these composite summarizing variables only in the rate-of-change analysis (Jacobson & Grimm, 1986). The rationale here is that, as discussed above, ordinations like PCA, CA, or DCA generally represent the "signal" in complex data such as diatom assemblage data as the first few ordination axes and relegate the "noise" to later, non-significant ordination axes (Gauch, 1982; Birks, 1998). Examples of rate-of-change analysis with diatom data include Lotter *et al.* (1995), Laird *et al.* (1998), Birks (1997), and Birks *et al.* (2000).

3.5.3 Constrained ordination methods

Prior to the development by Cajo ter Braak of constrained ordination methods such as canonical correspondence analysis (ter Braak, 1986), the classical approach to assessing what environmental variables might influence modern diatom assemblages was to do a PCA or CA of the diatom data and then relate the resulting ordination axes to particular environmental variables by correlation or regression analysis. What ter Braak did was to combine the ordination stage and the regression stage into one technique that finds a weighted sum of the environmental variables that fits the diatom species data best statistically, namely that gives the maximum regression sum of squares. The resulting ordination plots show the patterns of variation in the species and the samples and the main relationships between the species and the environmental variables. The ordination axes are constrained to be linear combinations of the environmental variables. The constrained version of PCA that assumes linear species responses is redundancy analysis (RDA) and the constrained version of CA that assumes unimodal species responses is canonical correspondence analysis (CCA). In addition, the constrained version of DCA is detrended canonical correspondence analysis (DCCA). Details of the mathematics and how to use RDA, CCA, and DCCA are presented in ter Braak, 1986, ter Braak and Prentice (1988), ter Braak and Verdonschot (1995), Legendre and Legendre (1998), ter Braak and Šmilauer (2002), and Lepš and Šmilauer (2003).

The commonest way that RDA or CCA are used in the analysis of modern diatom assemblage data such as diatom counts from surface sediments from a set of lakes with associated environmental data such as lake-water pH, specific conductivity, total nitrogen, etc. is to analyze the two data sets together to establish how well, in a statistical sense, the environmental variables explain the modern diatom assemblages. This is done, if we are using CCA, by comparing the eigenvalues of an unconstrained CA or DCA with the eigenvalues of the CCA which are measures of how well the species scores or optima are dispersed along the axes. The species–environment correlations and the percentage variation of the species data should also be compared between the CA (or DCA) and the CCA. The eigenvalues in the CCA will always be lower than the unconstrained CA or DCA eigenvalues because of the constraint that the axes must be linear combinations of the environmental variables, whereas the species–environment correlations will normally be higher in CCA. If the environmental variables are important in explaining the diatom data and are highly correlated to the latent variables in the CA or DCA, then the eigenvalues of the CA or DCA and the CCA should be fairly similar. The sum of all the constrained eigenvalues (species–environmental relationships) should be compared with the sum of all the unconstrained eigenvalues ("inertia") to assess how much of the total variation in the diatom data is explained by the environmental data. Experience shows that a CCA eigenvalue >0.3 suggests a strong gradient, and an eigenvalue >0.4 suggests a good niche separation of species along the overall environmental gradient. A similar argument applies to PCA and RDA except that the eigenvalues quantify the sum of squares of fitted values to the latent variables (PCA) and to the linear combination of environmental variables (RDA). RDA, CCA, and DCCA are multivariate direct gradient analytical techniques where all species and environmental variables are analyzed together, in contrast to simple direct gradient analyses (= regression analysis) where one species and one or more environmental variable(s) are analyzed, and multivariate indirect gradient analyses where only the species data are analyzed as in PCA, CA, or DCA.

As CCA, RDA, and DCCA are, in reality, multivariate regressions of complex diatom species response data in relation to many environmental predictors, they are statistical modeling techniques as well as graphical data-summarization tools. As a basic model is fitted by RDA, CCA, or DCCA (species respond to the environmental predictors), it is essential to evaluate the statistical significance of the model, namely to do a confirmatory or hypothesis-testing data analysis. This is done by Monte Carlo permutation tests. The environmental data are permuted many times (usually 99 or 999 times) and test statistics for the first canonical axis and for all canonical axes are computed for the observed and the permuted data. If the diatom species are statistically related to the environmental variables, the statistic for the observed data should be larger than most (e.g. 95%) of the statistics calculated for permuted data. If the observed statistics are in the top 5%, one can conclude that the species response is statistically significant to the environmental variable considered.

In addition to the question of statistical significance of the species–environment relationships, a critically important part of RDA, CCA, and DCCA is the graphical presentation of the results (ter Braak, 1994; ter Braak & Verdonschot, 1995). Given that there are ten possible data tables that can be plotted, it is important to decide which aspects of the results are most important in a particular research project. In most diatom studies using CCA, the most relevant are the species scores, the correlations of the quantitative environmental variables (e.g. pH) with the ordination axes, the centroids of nominal environmental variables (e.g. presence or absence of fringing bog around the lakes sampled), and the sample (lake) scores. When the emphasis is on the species–environment relationships, a species-conditional plot or correlation biplot is appropriate where the species are represented by the niche centre along each axis. If the emphasis is on the sample scores, a distance biplot, possibly with Hill's scaling so that distances between the samples are compositional turnover distances, is most useful. Ter Braak and Verdonschot (1995) describe in detail the options in the construction of CCA plots and provide a guide as to how to interpret such plots in terms of the centroid, distance, and biplot principles. ter Braak (1994) discusses these topics in the context of RDA. One aspect in CCA and RDA that can cause some confusion is that there are two sets of sample scores – one is the linear combination scores that are predicted or "fitted" values of the multiple regression with the constraining environmental variables, the other is the weighted averages of the species scores. In general the linear combination scores should be used, except when all the environmental predictor variables are presence/absence variables.

A useful tool in diatom studies where there are both modern diatom and environmental data and fossil diatom data is to do an RDA or CCA of the modern data and to treat the fossil data as "passive" or supplementary samples (ter Braak & Šmilauer, 2002). The fossil samples are positioned on the basis of the compositional similarities with the modern samples that are, in turn, positioned in relation to the modern environmental variables. It is thus possible to trace the fossil diatom sequence through modern environmental space (e.g. Birks *et al.*, 1990b), if the fossil samples are well fitted, namely have a low squared residual distance from the plane formed by the first two ordination axes.

The RDA or CCA techniques can be used to quantify the explanatory power of single environmental variables on the modern diatom assemblages by a series of constrained analyses with each environmental variable as the sole constraint (e.g. Lotter *et al.*, 1997). The DCCA technique can also be used to estimate the amount of diatom compositional turnover along specific environmental gradients (e.g. lake-water pH). Such estimates are a guide to the type of regression procedure that should be used in the development of transfer functions from modern calibration sets for quantitative environmental reconstructions (Birks, 1995; 1998). In these analyses, DCCA should be based on detrending by segments, non-linear rescaling of the axes, and **no** down-weighting of rare species. The same DCCA approach can be used to estimate compositional turnover in fossil diatom assemblages over a specific time period at many sites (Smol *et al.*, 2005; Birks, 2007) or in different groups of organisms over a specific time period at one or more sites (Birks & Birks, 2008), thereby providing a common basis for comparing the amount of compositional change between sites, between organisms, or between different time periods.

In addition to RDA, CCA, and DCCA, there are related partial techniques where the effects of particular variables or "nuisance" variables (e.g. different sampling dates) are factored out statistically as in partial regression analysis (ter Braak & Prentice, 1988). Partial RDA or partial CCA provide means of partitioning the total variation in diatom assemblages into fractions explained by, for example, lake characteristics (size, depth, etc.) and lake-water chemistry, and the covariation between lake morphometry and water chemistry, plus the unexplained fraction. This approach, pioneered by Borcard *et al.* (1992) has been expanded to three or more sets of predictor variables (e.g. Jones & Juggins, 1995) and has recently been put on a more rigorous statistical basis to obtain unbiased estimates of the fractions (Peres-Neto *et al.*, 2006). Although rarely used in diatom research, there is also partial PCA, CA, or DCA where the effects of particular variables can be factored out prior to an indirect ordination (ter Braak & Prentice, 1988).

Ordinations of fossil diatom stratigraphical data, constrained by sample depth or age, using RDA, CCA, or DCCA, can be valuable in identifying the major patterns of variation, the nature of the changes, and trends in the sequence (Birks, 1987). For some research questions, it may not be useful to impose the stratigraphical constraint and to use PCA, CA, or DCA (see above) and simply detect the major patterns of variation irrespective of the ordering of the objects. In other research problems, it may be relevant to partial out, as covariables, the effects of variables not of primary interest, such as sample age or depth in partial PCA, CA, DCA, RDA, CCA, or DCCA (e.g. Odgaard, 1994; Bradshaw *et al.*, 2005).

A basic principle of statistical modeling is the minimal adequate model, a realization of the principle of parsimony in statistics and of Ockham's principle of simplicity in science (Ockham's razor). Various aids in RDA, CCA, or DCCA such as variance inflation factors and forward selection and associated permutation tests (ter Braak & Šmilauer, 2002) can help to derive a minimal adequate model with the smallest number of significant environmental variables that explains, in a statistical sense, the diatom data about as well as the full set of explanatory environmental variables. Oksanen *et al.* (2008) have developed experimental "unfounded and untested" statistics in CCA and RDA that resemble deviance and the Akaike Information Criterion (AIC) (Godínez-Domínguez & Freire, 2003) used in the building and selection of statistical regression models. These new statistics can help model building and selection in CCA and RDA. If the investigator is lucky there may only be one minimal adequate model but it is important to assume that there may not be only one such model.

Important recent developments in the area of constrained ordinations are constrained methods that use sample–sample proximity or distance measures other that those implicit in RDA/PCA (Euclidean distances) and CCA/CA (chi-squared distance). Distance-based RDA (Legendre & Anderson, 1999), canonical analysis of principal coordinates (CAP) (Anderson & Willis, 2003), and non-linear CAP (Millar *et al.*, 2005) use proximity measures such as the Bray and Curtis, Jaccard, or Gower coefficients but in a constrained ordination framework. These developments considerably widen the choice of proximity measures and also allow (non-linear CAP) for non-linear relationships between diatom assemblages and environmental variables. They warrant use in several areas of diatom research where aspects of both assemblage composition and species abundance need to be considered simultaneously and where the environmental variables show non-linear relationships, for example with increasing distance from a pollution source.

Constrained ordination techniques like RDA are also very useful with field or laboratory experimental data of diatoms (e.g. Pappas & Stoermer, 1995) because all species can be analyzed, interactions between variables can be included, and statistical significance can be assessed by permutation tests, thereby relaxing some of the demanding assumptions of classical multivariate analysis of variance. Several permutation tests are available in the CANOCO software (ter Braak & Šmilauer, 2002) to take account of particular sampling designs, e.g. time series, line-transects, rectangular grids, repeated measures, and split-block with whole-plots containing split-plots. Whole-plots, split-plots, or both can be permuted. Within plots, samples can represent time series, line-transects, spatial grids, or be freely exchangeable. Thus multivariate data from different designs can be analyzed and tested (Lepš & Šmilauer, 2003). A further development relevant in experimental diatom studies involving a repeated-measure design is the principal response curve (PRC) technique (van der Brink & ter Braak, 1998) developed in ecotoxicology. This technique focuses on time-dependent treatment effects by adjusting for changes across time in the control and provides a succinct and elegant way of displaying and testing (by Monte Carlo permutation tests) effects across time.

3.5.4 Estimating species optima and tolerances

In some diatom studies involving modern data sets from different geographical areas or from different habitats (e.g. Cameron

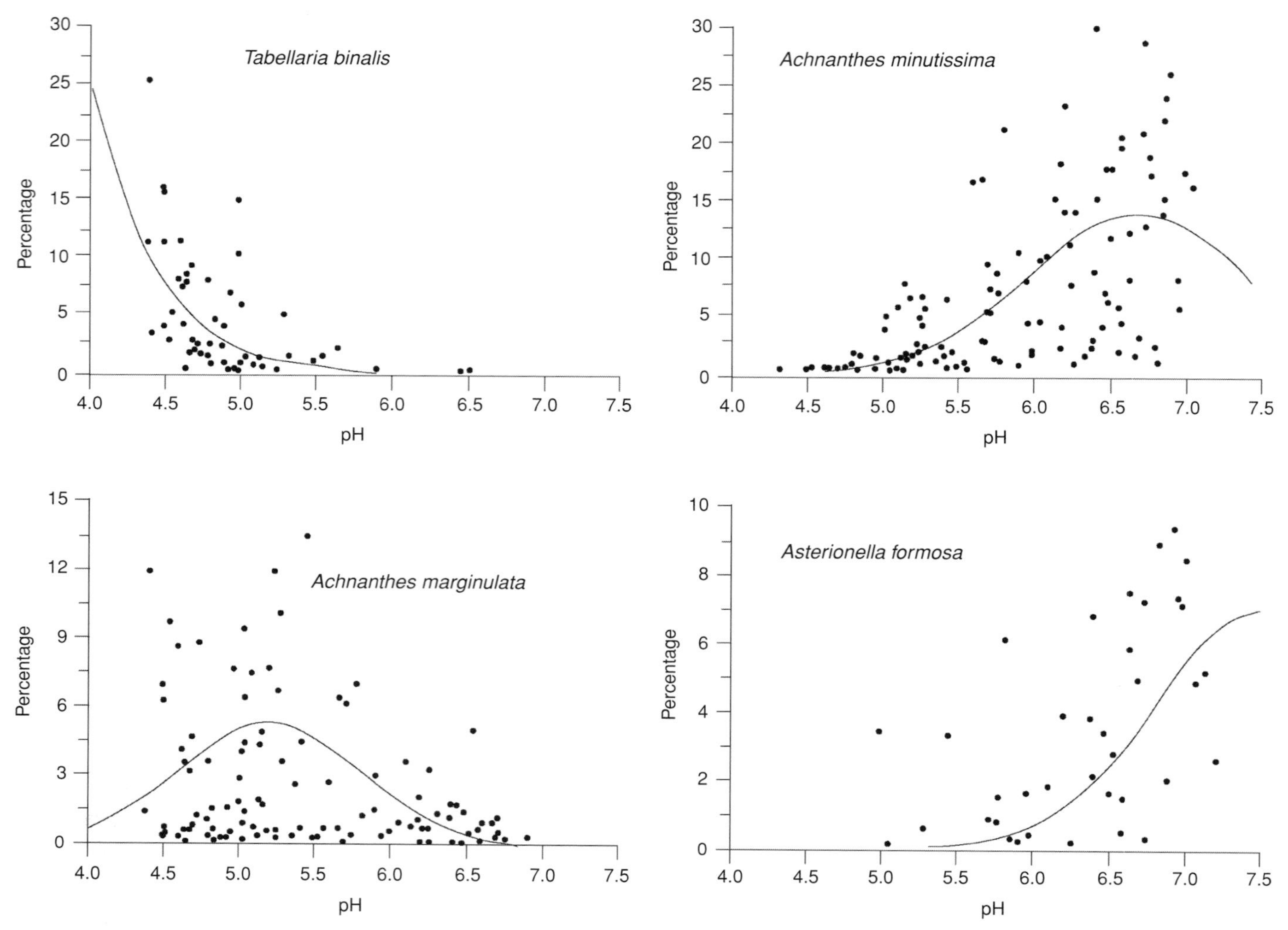

Figure 3.2 Fitted Gaussian response curves for four diatom species in relation to pH. Data are from the SWAP data set of 167 lakes in England, Wales, Scotland, Norway, and Sweden (Stevenson *et al.*, 1991). Diatom nomenclature follows Stevenson *et al.* (1991).

et al., 1999), it is useful to compare the realized niche of the same diatom species in different areas or habitats (e.g. Stoermer & Ladewski, 1976). Such comparisons require statistical modeling of the diatom responses in relation to environmental variables of interest. There are many types of ecological response curves. A compromise is necessary between ecological realism and simplicity (ter Braak & van Dam, 1989); the Gaussian response model with symmetric unimodal curves is a suitable compromise (Figure 3.1) (ter Braak, 1996). The Gaussian logit model is usually applied to presence/absence data (ter Braak & Looman, 1986) but it can be used as a quasi-likelihood model for proportions (Figure 3.2) and as an approximation to the more complex multinomial logit model (ter Braak & van Dam, 1989). Using Gaussian logit regression (GLR) (binomial error structure), one can estimate the optimum, tolerance, and height of its curve peak from the regression coefficients (Birks *et al.*, 1990a), along with the approximate 95% confidence intervals for the estimated optimum and the standard error of the estimated tolerance for each species in relation to the environmental variable of interest (ter Braak & Looman, 1986). For each species, the significance of the Gaussian logit model can be tested against the simpler linear-logit (sigmoidal) model, and the linear-logit model can be tested against the null model that the species has no statistically significant relationship to the environmental variable of interest. For species with estimated optima clearly outside the range of sampled environmental variables (see Figure 3.3), and with a significant linear logit model, the optimum can be assumed (as a minimal estimate) to be the lowest value for the environmental variable sampled for decreasing linear logit curves and the highest environmental value sampled for increasing linear logit curves (Birks *et al.*, 1990a). Tolerances are defined

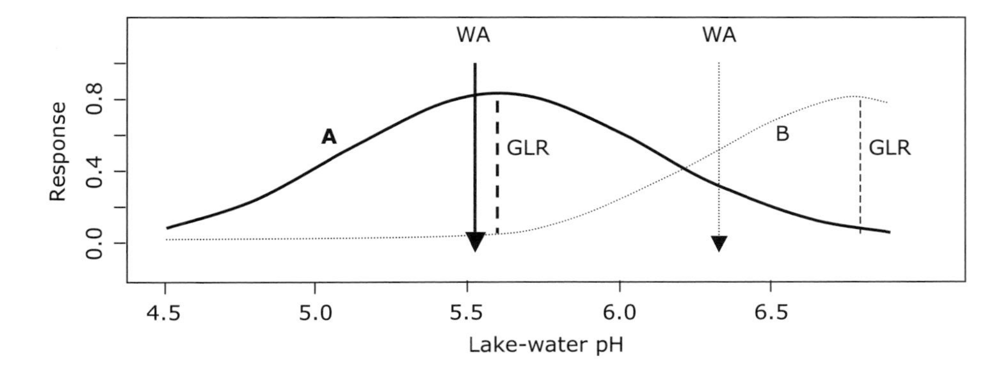

Figure 3.3 Comparison of estimates of pH optima for two diatom species by Gaussian logit regression (GLR) and simple weighted-averaging regression (WA). Note how close the estimates are for species A whose range is well covered by the pH gradient, whereas the WA optima for species B is grossly underestimated because of the truncation of species B's range by the sampled gradient. The GLR estimate is more reliable because GLR allows some extrapolation under the assumption of a Gaussian response model. (Based on a diagram in Oksanen (2004); see http://cc.oulu.fi/~jarioksa/opetus/plancom/commind.pdf.)

only for taxa with unimodal response curves. The GLR optima and tolerances can be compared between species and between regions because their estimation is not dependent on the distribution of lakes along the environment gradient of interest, in contrast to optima and tolerances estimated by weighted-averaging (WA) regression (ter Braak & Looman, 1986; ter Braak & Prentice, 1988) that are influenced by lake distribution (Figure 3.3).

Huisman *et al.* (1993) (see Oksanen & Minchin, 2002) have developed a hierarchical set of taxon response models consisting of a skewed unimodal response model, a symmetric Gaussian unimodal model, a plateau response model, a monotonically increasing or decreasing sigmoidal response model, and a null model of no relationship to the environmental variable. The simplest statistically significant response model for each species is found by fitting the most complex model first and progressively removing parameters from the regression model. This modeling approach allows the investigator to assess how many species of diatom show skewed, symmetric, sigmoidal, and null responses or fits to a particular environmental variable (e.g. Lotter *et al.*, 1997).

More flexible and less restrictive methods for modeling species response are available through generalized additive modeling that assumes no underlying response model and allows the response to be derived from the available data alone (Yee & Mitchell, 1991; Heegaard, 2002). Šmilauer and Birks (1995) present a preliminary analysis of diatom–environment responses using generalized additive models and Yuan (2004) illustrates the use of these models in deriving tolerance classifications for aquatic biota.

3.5.5 Reconstruction of environmental variables

The use of fossil diatom assemblages as the basis for quantitatively reconstructing past environmental conditions (e.g lake-water pH) is one of the important applications of diatoms to the environmental sciences (Smol, 2008). Attempts to derive quantitative estimates of environmental change from diatoms have a relatively long history in diatom research (Battarbee *et al.*, 2001a). Although numerical techniques had been developed in the 1970s to reconstruct sea-surface temperatures and salinity from fossil assemblages of foraminifera and to reconstruct climate from fossil pollen assemblages (see Birks, 1995, 2003 for reviews), it was not until the "acid-rain controversy" in the late 1980s and the coincidence of the Surface Waters Acidification Project (SWAP) and the Paleoecological Investigation of Recent Lake Acidification (PIRLA) project with Cajo ter Braak's seminal work on unimodal models and WA in his doctoral thesis (ter Braak, 1987b) that ecologically realistic and numerically robust methods were developed to reconstruct environmental variables from diatom assemblages (ter Braak & van Dam, 1989; Birks *et al.*, 1990a).

The basic idea of quantitative environmental reconstructions involves transfer functions estimated from modern diatom assemblages with known environmental variables. If we want to reconstruct lake-water pH from fossil diatom assemblages consisting of m taxa preserved in the sediment samples, the abundances of the same m taxa today are modeled numerically in relation to pH. This involves building up a modern "training set" or "calibration set" of diatom assemblages containing m taxa in surface sediments from a large number (30–200) of lakes with associated modern pH values. The modern relationships between the diatom assemblages and pH are modeled numerically and the resulting model or "transfer function" is used to transform the fossil diatom assemblages into quantitative estimates of past lake-water pH.

There are now many numerical techniques for deriving transfer functions (Birks, 1995; ter Braak, 1995; Birks, 1998, 2003; Telford *et al.*, 2004; Guiot & de Vernal, 2007; Juggins & Birks, 2011). Some have a stronger theoretical basis, either statistically, ecologically, or both, than others. Some (e.g. simple two-way WA regression and calibration and its closely related technique of weighted-averaging partial least squares (WA-PLS) regression and calibration and the more theoretically rigorous GLR and maximum likelihood (ML) calibration) fulfill all the basic requirements for quantitative reconstructions, perform consistently well with a range of data, do not involve an excessive number of parameters to be estimated and fitted, involve "global" rather than "local" parametric estimation from the modern data, and are thus relatively robust statistically and computationally economical.

3.5.2.1 Basic approaches and assumptions There are four basic considerations when estimating transfer functions. First, there is the choice of inverse or classical regression approaches. The classical approach is of the general form:

$$\mathbf{Y} = f(\mathbf{X}) + \text{error}$$

where **Y** are species responses that are modeled as a function of the environment **X** with some error. The function f() is estimated by linear, non-linear, and/or multivariate regression from the modern training set. Estimated f() is then "inverted" to infer the past environment from $\mathbf{Y}_f$, the fossil diatom assemblage. "Inversion" involves finding the past environmental variable that maximizes the likelihood of observing the fossil assemblage in that environment. If function f() is non-linear, which it almost always is, non-linear optimization procedures are required. Alternatively there is the simpler inverse approach of

$$\mathbf{X} = g(\mathbf{Y}) + \text{error}$$

where the difficult inversion step is avoided by estimating directly the function of g() from the training set by inverse regression of **X** on **Y**. The inferred past environment ($\mathbf{X}_f$) given a fossil assemblage ($\mathbf{Y}_f$) is simply

$$X_f = g(\mathbf{Y}_f).$$

As ter Braak (1995) discusses, statisticians have debated the relative merits of the classical and inverse approaches. In the few comparisons of these two major approaches (ter Braak *et al.*, 1993; Birks, 1995; ter Braak, 1995; Telford & Birks, 2005), inverse models (WA or WA-PLS) nearly always perform as well as classical methods (ML). Inverse models appear to perform best if the fossil assemblages are similar in composition to samples in the central part of the modern data, whereas classical models may be better at the extremes and with some extrapolation, as in "no-analog" situations.

Second, there is the assumed species–response model. Many transfer-function methods assume a linear or a unimodal response whereas others (e.g. modern analog techniques, artificial neural networks) do not assume any response model. It is a general law of nature that species–environment relationships are usually non-linear and species abundance is often a unimodal function of the environmental variable (Figure 3.1). Each species grows best at a particular optimum value of the environmental variable and cannot survive where the value of that variable is too low or too high (ter Braak, 1996). Thus all species tend to occur over a characteristic but limited environmental range and within this range tend to be most abundant near their environmental optimum.

Third, there is the question of dimensionality. Should all species be considered individually (full dimensionality) or should some axes or components of species abundance (analogous to principal components) (reduced dimensionality) be used?

Fourth, there is the estimation procedure to consider. Should a global estimation procedure be used that estimates parametric functions across the complete training set and thus allows some extrapolation or should a local estimation procedure be adopted that estimates non-parametric local functions which do not allow any extrapolation?

There are strong theoretical and empirical reasons for preferring methods with an assumed species response model, full dimensionality, and global parametric estimation. These reasons are:

(1) It is possible to test statistically if a species has a statistically significant relation to a particular environmental variable by GLR (see above).
(2) It is possible to develop "artificial" simulated data with realistic assumptions for numerical experiments (e.g. ter Braak *et al.*, 1993; ter Braak, 1995).
(3) Such methods have clear and testable assumptions and are less of a "black box" than, for example, artificial neural networks (Racca *et al.*, 2001, 2003; Telford *et al.*, 2004; Telford & Birks, 2005).

(4) It is possible to develop model evaluation or diagnostic procedures analogous to regression diagnostics in statistical modeling (Birks, 1995, 1998).
(5) As such methods have a statistical basis, it is possible to adopt the well-established principles of statistical model selection and testing (e.g. minimal adequate model, maximum numbers of degrees of freedom), thereby minimizing "ad hoc" aspects of some transfer function methods (Birks, 1995).

In this section, I will only consider WA and WA-PLS in detail as they currently represent simple, robust, and widely used approaches for quantitative reconstructions. ter Braak *et al.* (1993) concluded "until such time that such sophisticated methods mature and demonstrate their power for species–environment calibration, WA-PLS is recommended as a simple and robust alternative." I will also briefly discuss ML now that some programming problems (Birks, 2001) have been overcome and computing power has greatly increased in recent years. Maximim likelihood is now a feasible and attractive competitor to WA and WA-PLS with many excellent properties (Telford & Birks, 2005; Ginn *et al.*, 2007).

Although I have listed the major assumptions in quantitative paleoenvironmental reconstructions before (Birks *et al.*, 1990a; Birks, 1995; 2003), I repeat them here not only because they are important but also because there is now an additional critical assumption. The six major assumptions are:

(1) The *m* taxa in the training set (**Y**) are systematically related to the environment (**X**) in which they live.
(2) The environmental variable ($\mathbf{X_f}$) to be reconstructed is, or is linearly related to, an ecologically important determinant in the system of interest.
(3) The *m* taxa in the training set (**Y**) are the same biological entities as in the fossil data ($\mathbf{Y_f}$) and their ecological responses to **X** have not changed over the time represented by the fossil assemblages. Contemporary spatial patterns of species abundances in relation to **X** can thus be used to reconstruct change in **X** through time.
(4) The mathematical models used adequately model the species responses to **X** and yield transfer functions with sufficient predictive power to allow accurate and unbiased reconstructions of **X**.
(5) Other environmental variables than the one of interest (**X**) have negligible influence on **Y**, or their joint distribution with **X** in the past is the same as today.

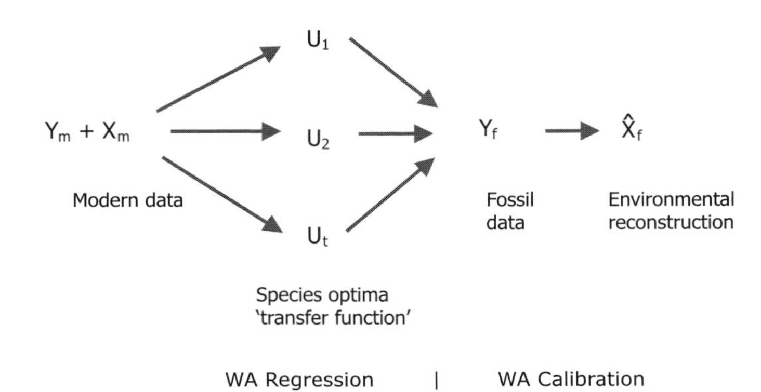

Figure 3.4 Schematic representation of two-way weighted-averaging (WA) regression and calibration to reconstruct environmental variable $\hat{X}_f$ from fossil diatom assemblage data Y_f. Y_m and X_m are modern diatom assemblage and associated environmental data, respectively, and $U_1, U_2, \ldots, U_t$ are the WA estimates of the optima for species 1, 2, . . . , t.

(6) In model evaluation by some form of cross-validation, the test data are statistically independent of the training set (Telford & Birks, 2005, 2009).

3.5.5.2 Two-way weighted averaging The basic idea behind WA (ter Braak, 1996) is that in a lake with a particular environmental variable (e.g. pH), species with their optima for pH close to the lake's pH value will tend to be the most abundant species present if the species shows a unimodal relationship with pH. A simple and ecologically reasonable estimate of a species' optimum for pH is thus the average of all the pH values for lakes at which the species occurs, weighted by the species relative abundances. The estimated optimum is the weighted average of pH. Species absences have no weight. The species' tolerance (Figure 3.1) can be estimated as the weighted standard deviation of pH. An estimate of a lake's pH value is the weighted average of the optima for pH for all the species present. Species with a narrow tolerance for pH can, if required, be given greater weight than species with a wide tolerance. The underlying theory of WA and the conditions under which it approximates Gaussian logit regression and calibration are discussed by ter Braak and Barendregt (1986), ter Braak and Looman (1986), and ter Braak (1996). Simple two-way WA is summarized in Figure 3.4.

Because the computations involved in WA are simple and fast, computer-intensive bootstrapping can be used to estimate the root-mean-square error of prediction (RMSEP) for inferred values of pH for all modern samples, the whole training set, and for individual fossil samples (Birks *et al.*, 1990a). The basic idea of bootstrap error estimation is to do many bootstrap cycles, say

1000. In each, a subset of modern samples is selected randomly but with replacement from the training set to form a bootstrap training set of the same size as the original training set. As sampling is with replacement, some samples may be selected more than once in a cycle. Any modern samples not selected form a bootstrap *test* set for that cycle. Weighted averaging is then used with the bootstrap training set to infer pH for the modern samples (all with known measured modern values) in the bootstrap *test* set. In each cycle, WA is also used to infer pH for each fossil sample. The standard deviation of the inferred pH values for both modern and fossil samples is calculated. This is one component of the overall prediction error, namely the estimation error for the species parameters. The second component, due to variations in species abundance at a given pH value, is estimated from the training set by the root mean square of the difference between observed values of pH and the mean bootstrap of pH when the modern sample is in the bootstrap *test* set. The first component varies from one fossil sample to another depending on the composition of the diatom assemblage, whereas the second component is constant for all fossil samples. The estimated RMSEP for a fossil sample is the square root of the sum of squares for these two components (Birks *et al.*, 1990a).

WA has gained considerable popularity in diatom paleoecology for several reasons:

(1) It combines ecological realism with mathematical and computational simplicity, rigorous underlying theory, and good empirical power.
(2) It does not assume linear species–environment responses, it is relatively insensitive to outliers, and it is not hindered by the high correlations between species or by the large number of species in diatom training sets.
(3) It performs well in no-analog situations (ter Braak *et al.*, 1993). Such situations are not uncommon in diatom paleoecology. In such situations, pH inferences are based on the WA of the optima of species in common between the modern and fossil assemblages. As long as there are reliable estimates of the optima for the fossil species of high numerical importance, WA reconstructions are often relatively realistic. Weighted averaging is thus a global or multivariate indicator species approach rather than a strict analog-matching procedure (Birks, 2003).
(4) WA performs best with noisy, species-rich compositional data with many species absent from many samples and extending over a relatively long environmental gradient (>2–3 standard deviations in a DCCA).

Weighted averaging does, however, have two important limitations (ter Braak & Juggins, 1993):

(1) WA is sensitive to the distribution of lakes within the training set along the environmental gradient of interest (ter Braak & Looman, 1986), although with large data sets (>400 lakes), WA has been shown to be quite robust to this distributional requirement (Ginn *et al.*, 2007).
(2) WA disregards residual correlations in the modern diatom data, namely correlations that remain in the diatom data after fitting the environmental variables of interest that result from environmental variables not considered directly in WA. The incorporation of partial least squares (PLS) regression into WA (ter Braak & Juggins, 1993) helps to overcome this by utilizing residual correlations to improve the estimates of species optima.

3.5.5.3 Weighted averaging partial least-squares regression (WA-PLS) The key feature of linear PLS regression is that components of predictor variables (in this inverse type of model, the predictors **Y** are diatom species) are selected not to maximize the variance of each component within **Y**, as in PCA, but to maximize the covariance between components that are linear combinations of the variables within **Y** and **X**. In the unimodal equivalent, WA-PLS, components are selected to maximize the covariance between **X** and the vector of weighted averages of **Y**. Subsequent components are chosen to maximize the same criterion but with the restriction that they are orthogonal and hence uncorrelated to earlier components (ter Braak *et al.*, 1993). In WA-PLS, components two and more are obtained as weighted averages of the residuals for the environmental variable **X**, in other words the regression residuals of **X** on the components extracted to date are used as new sample scores in the basic two-way WA algorithm. A joint estimate of **X** is, in PLS, a linear combination of the WA-PLS components, each of which is a weighted average of the species scores, hence the name WA-PLS. The final transfer function is a weighted average of updated "optima," but in contrast to WA, the optima are updated by considering residual correlations in the diatom data **Y**. In practice, species abundant in samples with large residuals are most likely to have updated optima (ter Braak & Juggins, 1993). Figure 3.5 summarizes WA-PLS in terms of its various stages.

Although WA-PLS involves a model of reduced dimensionality in contrast to WA, it usually produces transfer functions with lower RMSEP and lower mean or maximum bias than WA (ter Braak & Juggins, 1993). There are two reasons for this.

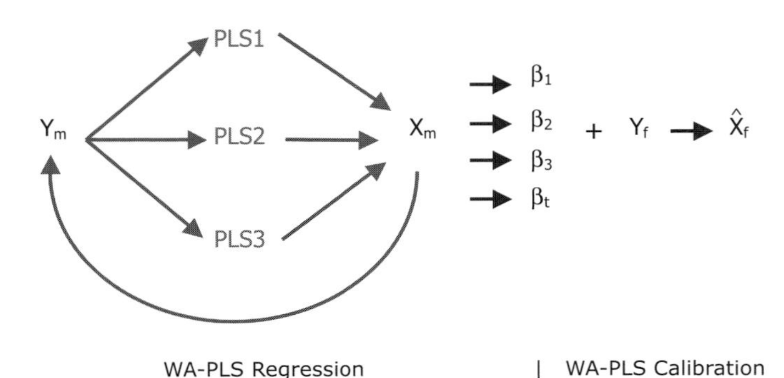

Figure 3.5 Schematic representation of WA-PLS regression and calibration. Partial least squares components are selected to maximize the covariance between the species weighted averages of Y_m with the environmental variable X_m, to derive a set of regression coefficients $\beta_1, \beta_2, \beta_3, \ldots, \beta_t$ that are then applied to fossil diatom assemblage data Y_f to produce an estimate of the past environment $\hat{X}_f$.

(1) Weighted-averaging-based models suffer from "edge effect" problems that result in the inevitable overestimation of species optima at the low end of the environmental gradient of interest and underestimation at the high end of the gradient (Figure 3.3). As a result there is a systematic bias in the inferred values and in the residuals. WA-PLS implicitly involves a weighted inverse deshrinking regression that pulls the inferred values towards the training-set mean. WA-PLS uses patterns in the residuals to update the transfer function, thereby reducing errors and patterns in the systematic bias.

(2) In real life, there are often additional environmental variables that influence the diatom assemblages. Weighted averaging ignores structure resulting from these variables and assumes that environmental variables other than the one of interest have negligible influence. WA-PLS uses this additional structure to improve estimates of the species "optima" in the final transfer function. For optimal performance, a further assumption is that the joint distribution of these environmental variables in the past should be the same as in the modern data (ter Braak & Juggins, 1993).

The main disadvantage of WA-PLS compared to WA is that great care is need in model selection to avoid model over-fitting. As more components are added, the WA-PLS model seems to fit the data better as the root-mean-square error (RMSE) decreases and becomes 0 when the number of components equals the number of samples. The RMSE is an "apparent" statistic of no predictive value as it is based on the training set alone. An independent test set is needed to evaluate different models as the optimal model is the model giving the lowest RMSEP for the test set (Birks, 1998). In real-life diatom research, there are usually no independent test sets and model evaluation is based on cross-validation to derive approximate estimates of RMSEP. In leave-one-out cross-validation, the simplest form of cross-validation, the WA-PLS models for $1, \ldots, p$ components (where p (usually 6–10) is less than n, the number of samples) is applied $n-1$ times using a training set of size $n-1$. In each of the $n-1$ models, one sample is left out and the transfer function based on the $n-1$ modern samples is applied to the one test sample omitted from the training set, giving a predicted value of the environmental variable for that sample. This is subtracted from the known value of the environmental variable x for that sample to give a corresponding prediction error. Thus in each model, individual samples act in turn as a test set, each of size 1, the prediction errors are accumulated to form a leave-one-out RMSEP. The final WA-PLS model to use in reconstructions is selected on the basis of low RMSEP, low mean and maximum bias (ter Braak & Juggins, 1993; Birks, 1998), and a small number of "useful" components (a useful component gives an RMSEP reduction of >5% of the one-component WA-PLS model; Birks, 1998). A one-component WA-PLS is identical to a simple WA model that uses an inverse deshrinking regression to compensate for taking averages twice (Birks *et al.*, 1990a; Birks, 1995). Sample-specific errors of prediction for fossil assemblages can be estimated by cross-validation or Monte Carlo simulation (Birks, 1995, 1998). With improved computer power, WA-PLS model selection can also be based on bootstrapping or ten-fold cross-validation and sample-specific errors estimated by bootstrapping (Juggins, 2005). In some cases the arbitrary 5% RMSEP threshold does not always guard against selecting an over-fitted model (Juggins & Birks, 2011). A simple solution to avoid over-fitting is to use van der Voet's (1994) randomization t-test to test the equality of predictions from two models (see Racca *et al.* (2001) for an application to diatom-pH transfer functions).

Like WA, WA-PLS performs surprisingly well and considerably better than modern analog procedures when the fossil assemblages are no-analogs and are dissimilar to the modern samples (ter Braak *et al.*, 1993; ter Braak, 1995). For very strong extrapolation beyond the training set, WA often performs better than WA-PLS.

3.5.5.4 Gaussian logit regression and maximum likelihood calibration (ML) In this classical approach, the responses of all the diatom species in the training set to the environmental

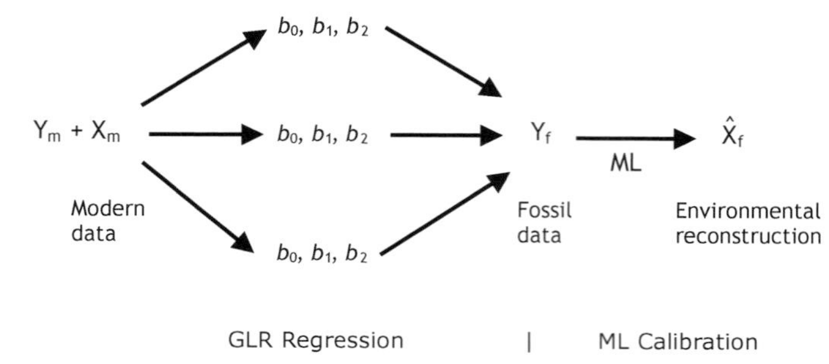

Figure 3.6 Schematic representation of GLR and ML calibration in the reconstruction of environmental variable $\hat{X}_f$. Y_m = modern diatom assemblage, X_m = modern environmental data, Y_f = fossil diatom assemblage data, and b_0, b_1, b_2 are estimated GLR regression coefficients for all species for the equation

$$\log\left(\frac{p}{1-p}\right) = b_0 + b_1 x + b_2 x^2$$

where p = proportion of the species of interest and x = environment variable in $\hat{X}_m$.

variable of interest are modeled by GLR (see above). These responses and their assumed error structure are then used in ML calibration to find the environmental variable with the highest probability of producing the observed diatom assemblages for each sample in both the training set and the fossil data sets (see Figure 3.6) (ter Braak & van Dam, 1989; Birks *et al.*, 1990a). With improved computer programming (Birks, 2001; Juggins, 2005) and greater computing power, GLR and ML can now be done with cross-validation or bootstrapping to derive realistic estimates of RMSEP and mean and maximum bias (Juggins, 2005). With several diatom data sets, ML appears to perform as well or marginally better than WA or WA-PLS models. The advantages of ML are that it is a classical, global, fully dimensional, and unimodal-based model, whereas WA is an inverse model and WA-PLS is a reduced dimensionality model. Yuan (2007a; 2007b) has extended the ML approach to infer more than one environmental variable simultaneously.

3.5.5.5 Spatial autocorrelation and diatom-environment reconstructions The estimation of the predictive power and performance of a training set in terms of RMSEP, coefficient of determination, mean and maximum bias, etc. by cross-validation assumes that the test set (one or many samples) is statistically independent of the training set. Cross-validation in the presence of spatial autocorrelation seriously violates this assumption (Telford & Birks, 2005; 2009). Positive spatial autocorrelation is the tendency of sites close to each other to resemble one another more than randomly selected sites. Using a large foraminiferal training set from the North Atlantic, Telford and Birks (2005) used cross-validation with a test set 10% the size of the full data set to compare the RMSEP in ML, WA, WA-PLS, modern analog technique (MAT), and artificial neural networks (ANN). They found the lowest RMSEP for sea-surface temperature (SST) in the MAT, ANN, and WA-PLS models. If, however, they used the South Atlantic data as a test set, where there can be no spatial autocorrelation with the North Atlantic training set, the lowest RMSEP are for GLR and WA. The authors interpret these findings as a result of spatial autocorrelation resulting in the apparently superior performance of ANN and MAT, both of which involve local non-parametric estimation, when applied to a test set within the same geographical area as the training set. In contrast, WA and ML only model the variance in the foraminiferal assemblages that is correlated with SST. They are therefore robust to the spatial structure in the data and they involve global parametric estimation. The models therefore perform best with a spatially independent test set, namely the South Atlantic. How widespread is the spatial autocorrelation problem? Is it important for diatom training sets?

Telford and Birks (2009) examined several training sets, including a large diatom-pH data set from the northeastern United States (Dixit *et al.*, 1999). They devised a simple test to detect spatial autocorrelation by (1) deleting samples at random and deriving a transfer function and its performance statistics and (2) deleting sites geographically close to the test sample and deriving a new transfer function and its performance statistics based on the remaining samples. If strong spatial autocorrelation is present, deleting geographically close sites will preferentially delete the environmentally closest sites. With autocorrelated data these will bias the apparent "good" performance of the transfer function, and their deletion should drastically decrease the performance of the transfer function. In contrast, random deletion should have much less effect on the performance of the transfer function. This test showed no difference in transfer-function performance between the randomly deleted and neighbour-deleted diatom data sets, suggesting that spatial autocorrelation is not a problem in this diatom-pH data set.

Telford and Birks (2009) also developed a method for cross-validating a transfer function in the presence of spatial autocorrelation. The method is based on *h*-block cross-validation where a test sample is deleted from the training set along with *h* observations within a certain radius of the test sample. Not surprisingly, there is no difference in the RMSEP between conventional leave-one-out cross-validation and *h*-block cross-validation with the diatom-pH data set as there is no spatial autocorrelation in these pH data. On the other hand there is a large increase in RMSEP in *h*-block cross-validation for planktonic and benthic foraminiferal training sets and a pollen-climate training set, where there is high spatial autocorrelation in the SST, salinity, or climate data.

3.5.5.6 Evaluation and validation of environmental reconstructions This is a topic that has received surprisingly little attention (Birks, 1995; 1998; 2003). It is very important as all transfer function procedures will produce a reconstruction; but an obvious and critically important question is how reliable is the reconstruction?

The most powerful validation is to compare reconstructions, at least for the recent past, against recorded historical records (e.g. Fritz, 1990; Laird *et al.*, 1996; Lotter, 1998). An alternative approach compares reconstructions with independent paleoenvironmental data. Such comparisons are part of the importance of multi-proxy approaches (Birks & Birks, 2006). Without historical validation or independent paleoenvironmental data, evaluation must be indirectly using numerical criteria (Birks, 1995; 1998). These include:

(1) Sample-specific RMSEP for individual samples in the stratigraphical sequence (Birks *et al.*, 1990a).
(2) "Goodness-of-fit" statistics assessed by fitting fossil samples passively onto a CCA axis only constrained by the environmental variable being reconstructed for the modern training set, and evaluating how well the individual samples fit onto this axis in terms of their squared residual distances (Birks *et al.*, 1990a).
(3) Analog measures for each individual fossil sample in comparison with the training set (Simpson, 2007). A reconstructed variable is likely to be more reliable if the fossil sample has modern analogs within the training set (ter Braak, 1995).
(4) The percentages of the total fossil diatom assemblage that consist of species (a) that are not represented (e.g. less than 10 occurrences) in the training set and hence whose transfer function parameters (WA optima, WA-PLS coefficients, GLR optima and tolerances) are poorly estimated and have high associated standard errors in cross-validation (Birks, 1998).

Bigler *et al.* (2002) and Birks *et al.* (1990a) illustrate the use of these evaluation criteria with fossil diatom data.

3.5.5.7 Conclusions The major conclusion is that quantitative environmental reconstructions of variables such as lake-water pH, total phosphorus, dissolved organic carbon, specific conductivity, and salinity are possible from diatom fossil assemblages but we are probably near the resolution of current data and methods. The whole transfer function approach is dependent on modern and fossil data of high taxonomic and analytical quality. The acquisition of such data is time-consuming. Despite considerable advances in transfer function methodology and in developing diatom-environment training sets, our abilities to interpret and compare time series of paleoenvironmental reconstructions have hardly developed beyond their visual comparisons (Birks, 2003).

3.5.6 Temporal-series analysis

Diatom stratigraphical data represent temporal series of the changing percentages or accumulation rates of different species or can be transformed via transfer functions into temporal series of reconstructed environmental variables at selected times in the past. In many time series analytical methods, the term "time series" is reserved for partial realizations of a discrete-time stochastic process, namely that the observations are made repeatedly on the same random variable at equal time spacings (Dutilleul *et al.*, 2011). There are clearly difficulties in meeting this requirement in many diatom studies unless the observations are from annually laminated sediments. Dutilleul *et al.* (2011) propose that diatom stratigraphical data sets comprise "temporal series" rather than time series in the strict statistical sense.

Prior to any statistical analysis, it is usually useful to perform exploratory data analysis on the individual species or environmental variables within the temporal series, such as simple tests for trends within each variable and estimating the statistical significance of the trends by randomization tests (Manly, 2007). The LOESS smoothers (see above) are useful graphical tools for highlighting the "signal" or major patterns in individual temporal series. An alternative approach to LOESS smoothing that combines graphical display, hypothesis-testing, and temporal-series analysis is provided by SiZer (significance of zero crossings of the derivative) (Chaudhuri & Marron, 1999; Holmström & Erästö, 2002; Sonderegger *et al.*, 2009). The SiZer approach was introduced into diatom paleoecology by Korhola *et al.* (2000). This approach asks which features in a series of smoothers are real and which are sampling artifacts. It fits trends and curves within a temporal series that are statistically significant. It uses a whole family of smooth curves fitted to the temporal series, each based on various smoothing window sizes and provides information about the underlying curve at different levels of detail. The features detected typically depend on the level of detail for which the temporal series is considered. The SiZer technique is a useful tool that can help to detect signal in temporal series and to test if particular changes are statistically significant or not.

There are two main approaches to time series analysis (Dutilleul *et al.*, 2011). First, there is the time-domain approach that is based on the concept of temporal autocorrelation, namely the correlation between samples in the same sequence that are *k* time intervals apart. The autocorrelation coefficient is a measure of similarity between samples separated by different time intervals and it can be plotted as a correlogram to assess the behavior (e.g. periodicities) of the diatom species of interest over time. Time-series of two different species can be compared by the cross-correlation coefficient to detect patterns of temporal variation and relationships between species (Legendre & Legendre, 1998; Davis, 2002).

The second approach involves the frequency domain and focuses on bands of frequency or wavelength over which the variance of a time series is concentrated. It estimates the proportion of the variance that can be attributed to each of a continuous series of frequencies. The power spectrum of a time series can help detect periodicities within the data and the main tools are spectral density functions, cross-spectral functions, and the periodogram. Birks (1998) discusses various examples of Fourier (amplitude) and power (variance) spectral analysis in paleolimnology.

Conventional time series analysis makes stringent assumptions of the data, namely that the inter-sample time intervals are constant and that the data are stationary and thus there are no trends in means or variances in the time series. In the absence of equally spaced samples, the usual procedure is to interpolate samples to equal time intervals. This is equivalent to a low-pass filter and may result in an underestimation of the high-frequency component in the spectrum. Thus the estimated spectrum of an interpolated time series becomes too "red" compared to the time spectrum (Schulz & Mudelsee, 2002). Techniques have been developed for spectral and cross-spectral analysis for unevenly spaced time series (Schulz & Stattegger, 1997) but they do not appear to have been applied to diatom stratigraphical data. Dutilleul *et al.* (2011) provide a critique of these techniques.

A limitation of power-spectral analysis is that it provides an integrated estimate of variance for the entire time series. This can be overcome by wavelet power-spectral analysis (Torrence & Compo, 1999) that identifies the dominant frequencies in different variables and displays how these frequencies have varied through the time series (see Brown *et al.* (2005) for a detailed paleoecological example). Witt and Schumann (2005) have developed an approach for deriving wavelets for unevenly sampled time series.

The rigorous analysis of diatom-based temporal series remains a major challenge in quantitative paleoecology and new methods are being developed as a result of close collaboration between applied statisticians and paleolimnologists. For example, Dutilleul *et al.* (2011) illustrate the use of the autocorrelogram using distance classes and a novel frequency-domain technique, so-called multifrequential periodogram analysis, to analyze four different diatom-based temporal series, all of which consist of unequally spaced observations.

3.6 Data interpretation

3.6.1 Introduction

Diatom assemblages are a complex reflection of the limnological conditions at the site(s) sampled. These conditions can be influenced by many factors, such as biotic interactions, light, turbulence, resource limitation, water chemistry, and season duration. These factors can themselves be influenced by factors such as climate change, human activity, volcanic eruptions, catchment-soil development and erosion, and catchment vegetation. Interpretation of diatom data involves testing hypotheses about possible "forcing factors" that may have influenced changes in diatom assemblages.

Birks (1998) discusses the importance of hypothesis testing in paleolimnology. In the analytical phase (*sensu* Ball, 1975) of any primarily non-experimental science, falsifiable working hypotheses must be proposed to explain the observed patterns. As alternative explanations for the observed patterns are nearly always possible, there is the need for considering multiple working hypotheses (Chamberlain, 1965). Despite the obvious importance of hypothesis testing, there has been relatively little such testing in diatom research, except in the statistical evaluation of modern diatom calibration data sets. Birks (1998) proposed that there are two main approaches to testing hypotheses in diatom paleoecology – the direct approach and the indirect approach.

The direct hypothesis-testing approach (Birks, 1998) involves collecting diatom data from sites in contrasting

Table 3.1 List of relevant numerical methods for data summarization (DS), data analysis (DA), and data interpretation (DI) in relation to the assumed diatom species response model to the environmental variable(s) of interest (linear or unimodal).

	Response model	
Problem and use	Linear	Unimodal
Regression (DA, DI)	Linear regression Multiple linear regression	Gaussian logit regression (GLR) Multiple Gaussian logit regression Weighted averaging (WA) of sample scores
Calibration (DA)	Linear calibration "inverse regression"	Maximum likelihood (ML) calibration using GLR coefficients WA of species scores and simple two-way WA
	Partial least squares (PLS)	WA-PLS (WAPLS)
Ordination (DS, DA)	Principal components analysis (PCA)	Correspondence analysis (CA) and detrended CA (DCA)
Constrained ordination (= reduced rank regression) (DS, DA, DI)	Redundancy analysis (RDA)	Canonical correspondence analysis (CCA)
Partial ordination (DS, DA)	Partial PCA	Partial CA
Partial constrained ordination (DS, DA, DI)	Partial RDA	Partial CCA

settings (e.g. geology, land-use, vegetation) to test a specific hypothesis, what Deevey (1969) suggested as "coaxing history to perform experiments." Diatom stratigraphical studies within the Surface Waters Acidification Project that was designed to test specific causes of recent lake-acidification (Battarbee, 1990; Renberg & Battarbee, 1990) are elegant examples of direct hypothesis testing (e.g. Anderson & Korsman, 1990; Birks *et al.*, 1990c; Korsman *et al.*, 1994). Similarly the impact of clear-cutting was evaluated in British Columbia using control and reference systems before and following forestry activities (Laird & Cumming, 2001; Laird *et al.*, 2001).

In this section I will discuss indirect hypothesis testing using statistical techniques (Birks, 1998) where results from some form of regression modeling are used to test alternative hypotheses. I will also outline the value of synthesizing diatom stratigraphical data from many sites for hypothesis testing.

3.6.2 Indirect hypothesis testing

In this approach, results from statistical tests involving regression modeling are used to evaluate the probability of specific null hypotheses. The effects of other, possibly confounding variables can be allowed for statistically in the regression model by partialling out their effects as covariables.

The basic statistical tests all involve regression analysis (Birks, 1997) based on an underlying linear or unimodal response model (ter Braak & Prentice, 1988). The appropriate technique depends on the numbers of response variables and the numbers of predictor or explanatory variables (Table 3.1). With one response variable and one predictor, simple linear or GLRs are appropriate. With one response model and many predictors, multiple linear regression or multiple GLRs are relevant. With more than one response variable and one or more predictors, the appropriate techniques are RDA or CCA. There are, as discussed above, partial versions of all these techniques where the effects of covariables can be adjusted for and partialled out statistically. Partial techniques provide a powerful means of testing competing hypotheses, as the effects of particular predictor variables can be partialled out and the relative importance of other predictors assessed statistically (e.g. Lotter & Birks, 1993; Lotter *et al.*, 1995; Lotter & Birks, 2003; Bradshaw *et al.*, 2005).

From a numerical viewpoint, the critical problem is how to evaluate the statistical significance of the fitted regression model, possibly with covariables, given the complex numerical properties of diatom stratigraphical data (closed percentage data, many zero values, many variables, temporally ordered samples). It is not possible to rely on classical statistical tests given the nature of the data. Instead distribution-free Monte Carlo permutation tests (e.g. ter Braak & Šmilauer, 2002; Lepš & Šmilauer, 2003; Manly, 2007) are used. In these an empirical distribution of the test statistic of relevance is derived by repeated permutations of the predictor variables or the regression residuals (ter Braak & Šmilauer, 2002), and a comparison made between the observed test statistic and, say, 999 values of the same statistic based on permuted data to derive a Monte Carlo test of significance and an exact probability for the observed statistic. Such Monte Carlo tests are distribution-free as they do not assume normality of the error distribution. They do, however, require independence or exchangeability of samples. The validity of any permutation test rests on the appropriateness of the type of permutation test used (ter Braak & Šmilauer, 2002). Time-ordered stratigraphical data require a special permutation type. Such restricted permutation tests provide a non-parametric means of overcoming the problems of performing statistical tests in the presence of temporal autocorrelation in diatom stratigraphical data. Examples of this type of hypothesis-testing approach include Lotter and Birks (1993), Renberg *et al.* (1993), Birks and Lotter (1994), Korsman *et al.* (1994), Lotter *et al.* (1995), Anderson *et al.* (1995; 1996), Korsman and Segerström (1998), and Haberle *et al.* (2006). Lotter and Anderson (2011) review this approach in detail.

In their analysis of what factors may have driven the observed changes in diatom stratigraphy at Dallund Sø, a small eutrophic lake on the island of Funnen (Denmark), Bradshaw *et al.* (2005) argued that over the last 7000 years different external factors may have been important at different times. They therefore did a series of partial CCAs with diatom taxa as the response variables, sample age partialled out as a covariable, and a wide range of potential predictor variables (e.g. pollen taxa, loss-on-ignition, plant macrofossils). The stratigraphical data were divided into a series of segments (samples 1–20, 2–21, 3–22, etc.) and a partial CCA done on each segment. The statistical significance of each partial CCA was assessed by Monte Carlo permutation tests. The results showed that early in the lake's history, diatom changes were influenced by external catchment processes and terrestrial vegetation, after about 2600 BC by changing water depth and land-use changes in the catchment, and in the last 2500 years by nutrient enrichment as a result of retting *Cannabis sativa* L. (hemp) and also changes in water depth and water clarity. This study highlights the importance of fitting "local" regression models to test hypotheses specific to particular times rather than fitting one "global" regression model to the entire sequence.

The CCA, RDA techniques, and their partial relatives, are widely used to assess the statistical importance of environmental factors in determining the composition and abundance of diatom assemblages in modern calibration data sets and in testing specific hypotheses relating to modern calibration data sets. Partitioning the variation (see above) can help in assessing the relative importance of different sets of environmental or other explanatory variables in explaining the variation in modern diatom assemblages. Variation partitioning can also be used with stratigraphical diatom data to try to partition the variation into different independent unique components, their covariance terms, and the unexplained component (e.g. Lotter & Birks, 1993; Birks & Lotter, 1994; Lotter *et al.*, 1995; Lotter & Birks, 1997; Lotter, 1998; Hall *et al.*, 1999). As the statistical importance of the various components can be assessed by Monte Carlo permutation tests, the relative importance of competing hypotheses can be evaluated and quantified.

A summary of the major numerical techniques used in the summarization, analysis, and interpretation of modern and fossil diatom assemblage data is given in Table 3.1. The major distinction between the methods is the assumed species response model (linear or unimodal) (ter Braak & Prentice, 1988).

3.6.3 Data synthesis

The interpretation of the observed diatom stratigraphy at one site can be fraught with difficulties – which changes are unique to the site, which changes may be more general, etc? Assembling together diatom stratigraphical data from many sites within a geographical area can aid interpretation. Because of the inevitable differences between diatom analysts in their identification and nomenclature and the lack of analytical quality control and taxonomic harmonization (e.g. Munro *et al.*, 1990), can anything useful be done numerically with diatom stratigraphical data from many sites?

For example, Smol *et al.* (2005) made such a synthesis of diatom stratigraphies from 42 lakes and ponds in the Arctic to test the hypothesis that there had been major stratigraphical changes in the last 150 years and to evaluate whether the magnitude of these changes varied geographically. The question is how to quantify in a consistent way the amount of compositional change in each profile without taxonomic harmonization between, for example, the Canadian Arctic, Svalbard, and Arctic Finland. Smol *et al.* (2005) used detrended CCA (DCCA) with sample age as the sole constraint, detrending by segments, and non-linear rescaling of the axes to estimate the total amount of compositional change or turnover at each site. They then compared these estimates with turnover in the last 150 years in temperate lakes not impacted by acidification, eutrophication, or other processes. They were thus able to establish a median reference turnover value and could show that 83% of sites above the Arctic Circle show greater turnover than the reference level, whereas south of the Arctic Circle turnover at only 59% of the sites exceed the reference value.

The value of synthesizing diatom-stratigraphical data from many sites in a meta-analysis is shown not only by Smol *et al.* (2005) but also in the extensive synthesis of recent changes in planktonic diatoms in over 200 lakes in North America and Europe by Rühland *et al.* (2008).

3.7 Future challenges

Considerable advances have been made in the numerical analysis of modern and fossil diatom data since the pioneering attempts in the early 1970s, particularly in the descriptive and narrative phases (*sensu* Ball, 1975) of diatom analytical studies. In the descriptive phase, basic patterns are detected, assessed, described, and classified and the relevant numerical techniques for data collection, assessment, and summarization are all valuable tools in the descriptive phase. The various numerical approaches to data analysis discussed above contribute to the narrative phase where inductively based explanations, generalizations, and reconstructions are derived from the observed patterns. Hypothesis-testing techniques such as CCA, RDA, and their partial relatives and associated Monte Carlo permutation tests provide one approach to data interpretation and to the analytical phase where testable and falsifiable hypotheses are proposed, evaluated, tested, and rejected.

What are the future challenges? First, despite considerable advances in transfer function methodology (Holden *et al.*, 2008; Hübener *et al.*, 2008) and in developing diatom-environment training sets, our abilities to interpret and compare temporal series of environmental reconstructions has hardly developed beyond visual comparisons. There is an urgent need for developing robust approaches for comparing temporal series. A related problem is how to interpret sample-specific errors in environmental reconstructions in the presence of the strong temporal autocorrelation that is a basic property of stratigraphical data and derived environmental reconstructions.

Second, many of our current techniques for data analysis and data interpretation assume a symmetric Gaussian unimodal species response to environmental variables that can be approximated by simple weighted averaging. With the great increase in computing power, Yee (2004) has returned to ter Braak's (1986) idea of constrained Gaussian ordination (CGO) with estimation by ML rather than by the simple WA algorithm used in CCA. Quadratic reduced-rank vector-based generalized linear models (GLMs) implement CGO but assume symmetric unimodal species responses. Yee (2006) has extended this approach to reduced-rank vector-based generalized additive models where *no* response model is assumed. Instead the data determine the response model as in generalized additive models (GAMs). These techniques are still under development as they are "fragile with dirty data" (Yee, 2006). Further developments will give diatom researchers powerful ordination and constrained ordination techniques within the GLM/GAM theoretical framework incorporating a mixture of linear, quadratic, and smooth responses. This general approach can, in theory, be extended to regression or prediction and to calibration and environmental reconstructions (Yee, 2006).

Third, more work is need to improve and quantify comparisons and interpretations of diatom stratigraphical data and to test competing hypotheses about possible causative factors in influencing diatom assemblages. The major problem is deriving, for statistical analysis, data that reflect potential forcing factors that are independent of the diatom response data. This is a major challenge in the interpretation of diatom stratigraphical data. Multidisciplinary paleoecological investigations are resulting in major advances because a multiproxy approach can help to test competing hypotheses (Lotter & Birks, 2003; Bradshaw *et al.*, 2005; Birks & Birks, 2006). However, the rigorous statistical analysis of multiproxy data is still in its infancy.

Fourth, there is a wealth of ecological and biogeographical information hidden in the many modern diatom-environmental data sets that have been developed in the recent 20 years. The studies by Tyler (1996), Weckström and Korhola (2001), Bouchard *et al.* (2004), Telford *et al.* (2006), Vyverman *et al.* (2007), and Vanormelingen *et al.* (2008) show how such data sets can be used to address critical questions in biogeography, biodiversity studies, and metacommunity ecology. There is very great scope for further such studies using appropriate statistical numerical techniques for data summarization and data analysis.

Fifth, ecologists and biogeographers, using basic but robust regression and variation partitioning techniques and large continental-scale data sets, are beginning to explore novel questions concerning population density, body size, and species distribution in diatom assemblages (Passy, 2007, 2008a, 2009), diatom biomass and species richness relationships (Potapova & Charles, 2002; Passy & Legendre, 2006a), diatom diversity gradients in relation to resources (Passy, 2008b) and disturbance (Cardinale *et al.*, 2006), power-law relationships among hierarchical taxonomic categories in algae (Passy & Legendre, 2006b), diatom valve-size variations along gradients (Soinenen

& Kokocinski, 2006), aspects of assembly rules, persistence, local abundances, regional occupancy, community structure, and biodiversity assessment in diatom assemblages (Soinenen & Eloranta, 2004; Soinenen *et al.*, 2004; Heino & Soinenen, 2005; Soinenen & Heino, 2005; Heino *et al.*, 2009; Soinenen *et al.*, 2009), and the rate and scale of diatom dispersal in relation to global abundance (Finlay *et al.*, 2002). These studies highlight the potential of diatom assemblages as "model" systems in different aspects of community ecology, metapopulation and metacommunity biology, and macroecology.

Sixth, with increasing interest in the compilation and analysis of huge diatom data sets (over 1000 samples) and the necessity to detect the major patterns of variables within such data sets, to predict responses to future environmental change, and, at the same time, to summarize them as simple groups, classification and regression trees (CART) and associated decision trees have considerable potential in these research areas. De'ath and Fabricus (2000) and Fielding (2007) provide clear introductions to these robust techniques. De'ath (2002, 2007), Prasad *et al.* (2006), Cutler *et al.* (2007), and Elith *et al.* (2008) discuss recent developments in CART in terms of bagging trees, boosted trees, random forests, and multivariate adaptive regression splines in the context of ecological studies. They have considerable potential in diatom research, particularly in terms of data summarization and future predictions. Peters *et al.* (2007) illustrate the use of random forests in ecohydrological distribution-modeling and prediction of vegetation types in Belgian wet habitats and Pelánková *et al.* (2008) document the use of CART to explore relationships between modern pollen assemblages and contemporary climate variables, vegetation types, and landscape features in southern Siberia. These techniques have considerable potential in the analysis of diatom assemblage data sets (see Legendre & Birks, 2011a for a simple example with a modern diatom data set).

Seventh, diatom researchers should try to keep up with the many important advances being made in applied statistics that are of potential relevance in the analysis of diatom assemblage data. Important new and applicable approaches include generalized linear mixed models (Bolker *et al.*, 2009; Zuur *et al.*, 2009), quantile regression (Cade & Noon, 2003; Cade *et al.*, 2005; Schröder *et al.*, 2005; Hao & Naiman, 2007; Kelly *et al.*, 2008), non-parametric additive modeling of temporal series (Ferguson *et al.*, 2008), Bayesian approaches (Ellison, 2004; Clark, 2005; McCarthy, 2007), and related hierarchical Bayesian inference methods (Clark & Gelfand, 2006a, 2006b). As these methods and approaches are increasingly complex and computationally demanding, close collaboration between diatom researchers and applied statisticians is even more important than it was in the past.

Given the great advances in the last 20 years in the numerical analysis of diatom assemblage data, the central role that diatoms play in paleolimnology and environmental sciences, and the many future challenges that have arisen as a result of previous work, the next 20 years promise to be as exciting and as intellectually stimulating as the previous 20 years and will no doubt lead to major advances in diatom research.

3.8 Summary

Numerical methods are useful tools in many stages in the study of modern and fossil diatom assemblages. During data collection, numerical and computer-based procedures can be useful in morphometric studies and in identification. In data assessment, statistical techniques are essential in estimating the inherent errors associated with the resulting diatom counts. A range of numerical techniques can be used to detect and summarize the major underlying patterns in diatom assemblages. For single stratigraphical sequences or a modern diatom-environment data set, ordination procedures such as correspondence analysis can summarize temporal or spatial patterns in such data sets. Numerical methods are essential for quantitative data analysis in terms of establishing modern diatom–environment relationships by, for example, canonical correspondence analysis, estimating diatom taxonomic richness, quantifying rates of change in diatom stratigraphical data, and reconstructing quantitatively aspects of the past environment such as lake-water pH from fossil diatom assemblages using modern transfer functions and techniques like simple two-way weighted averaging. Statistical techniques are essential in the final stage of data interpretation where competing hypotheses about underlying causative factors such as human activity or climate change that may have influenced fossil diatom assemblages can be tested. The chapter concludes with ideas on future challenges.

Acknowledgments

I am indebted to John Smol and Gene Stoermer for their invitation to write this chapter. I am very grateful to the many diatomists I have collaborated with in the numerical analysis of their hard-earned data, in particular Rick Battarbee, Christian Bigler, Emily Bradshaw, Nigel Cameron, Brian Cumming, Roger Flower, Roland Hall, Will Hobbs, Viv Jones, the late John Kingston, Tom Korsman, Andy Lotter, Reinhard Pienitz, Ingemar Renberg, John Smol, and Alex Wolfe. I am also indebted to Cajo ter Braak, Allan Gordon, Steve Juggins, Petr Šmilauer, and Richard Telford for much stimulating statistical collaboration. I am grateful to Brian Cumming, John Smol, and Gene Stoermer for their comments on an earlier draft of this chapter. I am particularly grateful to Cathy Jenks for invaluable help in preparing this manuscript in such a short time.

References

Aitchinson, J. (1986). *The Statistical Analysis of Compositional Data*, Chapman and Hall, London.

Anderson, M. J. & Willis, T. J. (2003). Canonical analysis of principal co-ordinates: a useful method of constrained ordination for ecology. *Ecology*, **84**, 511–25.

Anderson, N. J. & Korsman, T. (1990). Land-use change and lake acidification: Iron Age de-settlement in northern Sweden as a pre-industrial analogue. *Philosophical Transactions of the Royal Society of London B*, **327**, 373–6.

Anderson, N. J., Odgaard, B. V., Segerström, U. & Renberg, I. (1996). Climate–lake interactions recorded in varved sediments from a Swedish boreal forest lake. *Global Change Biology*, **2**, 399–405.

Anderson, N. J., Renberg, I., & Segerstrom, U. (1995). Diatom production response to the development of early agriculture in a boreal forest lake-catchment (Kassjön, northern Sweden). *Journal of Ecology*, **88**, 809–22.

Ball, I. R. (1975). Nature and formulation of biogeographic hypotheses. *Systematic Zoology*, **24**, 407–30.

Battarbee, R. W. (1990). The causes of lake acidification, with special reference to the role of acidification. *Philosophical Transactions of the Royal Society of London B*, **327**, 339–47.

Battarbee, R. W., Jones, V. J., Flower, R. J., *et al.* (2001a). Diatoms. In *Tracking Environmental Change Using Lake Sediments, Volume 3: Terrestrial, Algal, and Siliceous Indicators*, ed. J. P. Smol, H. J. B. Birks, & W. M. Last, Kluwer Academic Publishers: Dordrecht, pp. 155–202.

Battarbee, R. W., Juggins, S., Gasse, F., *et al.* (2001b). European Diatom Database (EDDI). An information system for palaeoenvironmental reconstruction. Environmental Change Research Centre, University College London, pp. 210.

Bennett, K. D. (1994). Confidence intervals for age estimates and deposition times in late-Quaternary sediment sequences. *The Holocene*, **4**, 337–48.

Bennett, K. D. (1996). Determination of the number of zones in a biostratigraphical sequence. *New Phytologist*, **132**, 155–70.

Besse-Lototskaya, A., Verdonschot, P. F. M., & Sinkeldam, J. A. (2006). Uncertainty in diatom assessment: sampling, identification and counting varition. *Hydrobiologia*, **566**, 247–60.

Bigler, C., Larocque, I., Peglar, S. M., Birks, H. J. B., & Hall, R. I. (2002). Quantitative multiproxy assessment of long-term patterns of Holocene environmental change from a small lake near Abisko, northern Sweden. *Holocene*, **12**, 481–96.

Birks, H. H., Battarbee, R. W., & Birks, H. J. B. (2000). The development of the aquatic ecosystem at Krakenes Lake, western Norway, during the late glacial and early Holocene – a synthesis. *Journal of Paleolimnology*, **23**, 91–114.

Birks, H. H. & Birks, H. J. B. (2006). Multi-proxy studies in palaeolimnology. *Vegetation History and Archaeobotany*, **15**, 235–51.

Birks, H. J. B. (1985). Recent and possible future mathematical developments in quantitative paleoecology. *Palaeogeography Palaeoclimatology Palaeoecology*, **50**, 107–47.

Birks, H. J. B. (1987). Multivariate-analysis of stratigraphic data in geology – a review. *Chemometrics and Intelligent Laboratory Systems*, **2**, 109–26.

Birks, H. J. B. (1992). Some reflections on the application of numerical methods in Quaternary palaeoecology. *Publications of the Kareliàn Institute, University of Joensuu*, **102**, 7–20.

Birks, H. J. B. (1995). Quantitative palaeoenvironmental reconstructions. In *Statistical Modelling of Quaternary Science Data. Technical Guide* 5, ed. D. Maddy & J. S. Brew, Cambridge: Quaternary Research Association pp. 161–254.

Birks, H. J. B. (1997). Environmental change in Britain – a long-term palaeoecological perspective. In *Britain's Natural Environment: a State of the Nation Review*, ed. A. W. Mackay & J. Murlis, London: Ensis Publications, pp. 23–8.

Birks, H. J. B. (1998). Numerical tools in palaeolimnology – progress, potentialities, and problems. *Journal of Paleolimnology*, **20**, 307–32.

Birks, H. J. B. (2001). Maximum likelihood environmental calibration and the computer program WACALIB – a correction. *Journal of Paleolimnology*, **25**, 111–15.

Birks, H. J. B. (2003). Quantitative palaeoenvironmental reconstructions from Holocene biological data. *Global Change in the Holocene*, ed. A. W. Mackay, R. W. Battarbee, H. J. B. Birks & F. Oldfield, London: Arnold, pp. 342–57.

Birks, H. J. B. (2007). Estimating the amount of compositional change in late-Quaternary pollen-stratigraphical data. *Vegetation History and Archaeobotany*, **16**, 197–202.

Birks, H. J. B. (2008). Ordination – an ever-expanding tool-kit for ecologists? *Bulletin of the British Ecological Society*, **39**, 31–3.

Birks, H. J. B. (2011). Stratigraphical data analysis. In *Tracking Environmental Change Using Lake Sediments, Volume 5: Data Handling and Numerical Techniques*, eds. H. J. B. Birks, A. F. Lotter, S. Juggins & J. P. Smol, Dordrecht: Springer (in press).

Birks, H. J. B., Berge, F., Boyle, J. F. & Cumming, B. F. (1990c). A palaeoecological test of the land use hypothesis for recent lake acidification in south west Norway using hill top lakes. *Journal of Paleolimnology*, **4**, 69–85.

Birks, H. J. B. & Birks, H. H. (2008). Biological responses to rapid climate changes at the Younger Dryas–Holocene transition at Kråkenes, western Norway. *The Holocene*, **18**, 19–30.

Birks, H. J. B. & Gordon, A. D. (1985). *Numerical Methods in Quaternary Pollen Analysis*, London: Academic Press.

Birks, H. J. B., Juggins, S. & Line, J. M. (1990b). Lake surface-water chemistry reconstructions from palaeolimnological data. In *The*

Surface Waters Acidification Programme, ed. B. J. Mason, Cambridge: Cambridge University Press, pp. 301–13.

Birks, H. J. B. & Line, J. M. (1992). The use of rarefraction analysis for estimating palynological richness from Quaternary pollen-analytical data. *The Holocene*, **2**, 1–10.

Birks, H. J. B., Line, J. M., Juggins, S., Stevenson, A. C., & ter Braak, C. J. F. (1990a). Diatoms and pH reconstruction. *Philosophical Transactions of the Royal Society of London Series B – Biological Sciences*, **327**, 263–78.

Birks, H. J. B. & Lotter, A. F. (1994). The impact of the Laacher See Volcano (11000 yr B.P.) on terrestrial vegetation and diatoms. *Journal of Paleolimnology*, **11**, 313–22.

Birks, H. J. B., Lotter, A. F., Juggins, S., & Smol, J. P. (eds.) (2011). *Tracking Environmental Change Using Lake Sediments, Volume 5: Data Handling and Numerical Techniques*, Dordrecht: Springer (in press).

Birks, H. J. B. & Seppä, H. (2004). Pollen-based reconstructions of late-Quaternary climate in Europe – progress, problems, and pitfalls. *Acta Palaeobotanica*, **44**, 317–34.

Blaauw, M. & Heegaard, E. (2011). Estimation of age–depth relationships. In *Tracking Environmental Change Using Lake Sediments, Volume 5: Data Handling and Numerical Techniques*, ed. H. J. B. Birks, A. F. Lotter, S. Juggins, & J. P. Smol, Dordrecht: Springer (in press).

Bolker, B. M., Brooks, M. E., Clark, C. J., *et al.* (2009). Generalized liner mixed models: a practical guide for ecology and evolution. *Trends in Ecology and Evolution*, **24**, 127–35.

Bookstein, F., Chernoff, B., Elder, R., *et al.* (1985). *Morphometrics in Evolutionary Biology*, Philadelphia: Academy of National Sciences of Philadelphia Special Publication.

Borcard, D., Legendre, P., & Drapeau, P. (1992). Partialling out the spatial component of ecological variation. *Ecology*, **73**, 1045–55.

Bouchard, G., Gajewski, K., & Hamilton, P. B. (2004). Freshwater diatom biogeography in the Canadian Arctic archipelago. *Journal of Biogeography*, **31**, 1955–73.

Bradshaw, E. G., Rasmussen, P., & Odgaard, B. V. (2005). Mid- to late-Holocene land-use change and lake development at Dallund Sø, Denmark: synthesis of multiproxy data, linking land and lake. *The Holocene*, **15**, 1152–1162.

Brown, K. J., Clark, J. S., Grimm, E. C., *et al.* (2005). Fire cycles in North American interior grasslands and their relation to prairie drought. *Proceedings of the National Academy of Sciences of the USA*, **102**, 8865–70.

Cade, B. S. & Noon, B. (2003). A gentle introduction to quantile regression for ecologists. *Frontiers in Ecology and the Environment*, **1**, 412–20.

Cade, B. S., Noon, B. R., & Flather, C. H. (2005). Quantile regression reveals hidden bias and uncertainty in habitat models. *Ecology*, **86**, 786–800.

Cameron, N. G., Birks, H. J. B., Jones, V. J., *et al.* (1999). Surface-sediment and epilithic diatom pH calibration sets for remote European mountain lakes (AL:PE Project) and their comparison with the Surface Waters Acidification Programme (SWAP) calibration set. *Journal of Paleolimnology*, **22**, 291–317.

Cardinale, B. J., Hillebrand, H., & Charles, D. F. (2006). Geographic patterns of diversity in streams are predicted by a multivariate model of disturbance and productivity. *Journal of Ecology*, **94**, 609–18.

Chamberlain, T. C. (1965). The method of multiple working hypotheses. *Science*, **148**, 754–9.

Chambers, J. M., Cleveland, W. S., Kleiner, B., & Tukey, P. A. (1983). *Graphical Methods for Data Analysis*, Monterey, CA: Wadsworth.

Chaudhuri, P. & Marron, J. S. (1999). SiZer for exploration of structures in curves. *Journal of the American Statistical Association*, **94**, 807–23.

Clark, J. S. (2005). Why environmental scientists are becoming Bayesians. *Ecology Letters*, **8**, 2–14.

Clark, J. S. & Gelfand, A. E. (2006a). A future for models and data in environmental science. *Trends in Ecology and Evolution*, **21**, 375–380.

Clark, J. S. & Gelfand, A. E. (eds.) (2006b). *Hierarchical Modelling for the Environmental Sciences*, Oxford: Oxford University Press.

Claude, J. (2008). *Morphometrics with R*, New York, NY: Springer.

Cleveland, W. A. (1979). Robust locally weighted regression and smoothing scatterplots. *Journal of the American Statistical Association*, **74**, 829–36.

Cleveland, W. A. & Devlin, S. J. (1988). Locally weighted regression: an approach to regression analysis by local fitting. *Journal of the American Statistical Association*, **83**, 596–610.

Cleveland, W. S. (1993). *Visualizing Data*, Murray Hill, NJ: AT&T Bell Laboratories.

Cleveland, W. S. (1994). *The Elements of Graphing Data*, Murray Hill, NJ: AT&T Bell Laboratories.

Cutler, D. R., Edwards, T. C., Beard, K. H., *et al.* (2007). Random forests for classification in ecology. *Ecology*, **88**, 2783–92.

Davis, J. C. (2002). *Statistics and Data Analysis in Geology*, New York, NY: Wiley.

De'ath, G. (2002). Multivariate regression trees: a new technique for modeling species–environment relationships. *Ecology*, **83**, 1105–17.

De'ath, G. (2007). Boosted trees for ecological modeling and prediction. *Ecology*, **88**, 243–51.

De'ath, G. & Fabricus, K. E. (2000). Classification and regression trees: a powerful yet simple technique for ecological data analysis. *Ecology*, **81**, 3178–92.

Deevey, E. S. (1969). Coaxing history to conduct experiments. *BioScience*, **19**, 40–3.

Dixit, S. S., Smol, J. P., Charles, D. F., *et al.* (1999). Assessing water quality changes in the lakes of the northeastern United States using sediment diatoms. *Canadian Journal of Fisheries and Aquatic Science*, **56**, 131–52.

du Buf, J. M. H. & Bayer, M. M. (2002). *Automatic Diatom Identification*, Singapore: World Scientific Publishing.

du Buf, J. M. H., Bayer, M. M., Droop, S. J. M., *et al.* (1999). Diatom identification: a double challenge called ADIAC. Proceedings of the 10th International Conference on Image Analysis and Processing, Venice, Italy.

Dutilleul, P., Cumming, B. F., & Lontoc-Roy, M. (2011). Autocorrelogram and periodogram analyses of palaeolimnological temporal series from lakes in central and western North America to assess shifts in drought conditions. In *Tracking Environmental Change Using Lake Sediments, Volume 5: Data Handling and Numerical Techniques*, ed. H. J. B. Birks, A. F. Lotter, S. Juggins, & J. P. Smol, Dordrecht: Springer (in press).

Elith, J., Leathwick, J. R., & Hastie, T. (2008). A working guide to boosted regression trees. *Journal of Animal Ecology*, **77**, 802–13.

Ellison, A. M. (2004). Bayesian inference in ecology. *Ecology Letters*, **7**, 509–20.

Everitt, B. (1978). *Graphical Techniques for Multivariate Data*, London: Heinemann.

Ferguson, C. A., Carvalho, L., Scott, E. M., Bowman, A. W., & Kirika, A. (2008). Assessing ecological responses to environmental change using statistical models. *Journal of Applied Ecology*, **45**, 193–203.

Fielding, A. H. (2007). *Cluster and Classification Techniques for the Biosciences*, Cambridge: Cambridge University Press.

Finlay, B. J., Monaghan, E. B., & Maberly, S. C. (2002). Hypothesis: the rate and scale of dispersal of freshwater diatom species is a function of their global abundance. *Protist*, **153**, 261–73.

Flenley, J. (2003). Some prospects for lake sediment analysis in the 21st century. *Quaternary International*, **105**, 77–80.

Fox, J. (2002). *An R and S-Plus Companion to Applied Regression*, Thousand Oaks, CA: Sage Publishers.

Fritz, S. C. (1990). Twentieth-century salinity and water-level fluctuations in Devil's Lake, North Dakota: test of a diatom-based transfer function. *Limnology and Oceanography*, **35**, 1171–81.

Gabriel, K. R. (2002). Goodness of fit of biplots and correspondence analysis. *Biometrika*, **89**, 423–36.

Gauch, H. G. (1982). Noise reduction by eigenvector ordination. *Ecology*, **63**, 1643–9.

Ginn, B. K., Cumming, B. F., & Smol, J. P. (2007). Diatom-based environmental inferences and model comparisons from 494 northeastern North American lakes. *Journal of Phycology*, **43**, 647–61.

Godínez-Domínguez, E. & Freire, F. (2003). Information-theoretic approach for selection of spatial and temporal models of community organization. *Marine Ecology Progress Series*, **253**, 17–24.

Gordon, A. D. (1999). *Classification*, Boca Raton, FL: Chapman & Hall/CRC Press.

Grimm, E. C. (1987). CONISS – a FORTRAN-77 program for stratigraphically constrained cluster-analysis by the method of incremental sum of squares. *Computers & Geosciences*, **13**, 13–35.

Grimm, E. C. & Jacobson, G. L. (1992). Fossil-pollen evidence for abrupt climate changes during the past 18 000 years in eastern North America. *Climate Dynamics*, **6**, 179–84.

Guiot, J. & de Vernal, A. (2007). Transfer functions: methods for quantitative paleoceanography based on microfossils. In *Proxies in Late Cenozoic Paleoceanography*, ed. C. Hillaire-Marcel & A. de Vernal, Amstedam: Elsevier, pp. 523–63.

Haberle, S. G., Tibby, J., Dimitriadis, S., & Heijnis, H. (2006). The impact of European occupation on terrestrial and aquatic dynamics in an Australian tropical rain forest. *Journal of Ecology*, **94**, 987–1102

Hall, R. I., Leavitt, P. R., Quinlan, R., Dixit, A. S., & Smol, J. P. (1999). Effects of agriculture, urbanization, and climate on water quality in the northern Great Plains. *Limnology and Oceanography*, **44**, 736–56.

Hammer, Ø. & Harper, D. A. T. (2006). *Paleontological Data Analysis*, Oxford: Blackwell.

Hao, L. & Naiman, D. Q. (2007). *Quantile Regression*, Thousand Oaks, CA: Sage Publishers.

Hausmann, S. & Lotter, A. F. (2001). Morphological variation within the diatom taxon Cyclotella comensis and its importance for quantitative temperature reconstructions. *Freshwater Biology*, **46**, 1323–33.

Heegaard, E. (2002). The outer border and central border for species environmental relationships estimated by non-parametric generalised additive models. *Ecological Modelling*, **157**, 131–9.

Heegaard, E., Birks, H. J. B., & Telford, R. J. (2005). Relationships between calibrated ages and depth in stratigraphical sequences: an estimation procedure by mixed-effect regression. *The Holocene*, **15**, 612–18.

Heegaard, E., Lotter, A. F., & Birks, H. J. B. (2006). Aquatic biota and the detection of climate change: are there consistent aquatic ecotones? *Journal of Paleolimnology*, **35**, 507–18.

Heino, J., Ilmonen, J., Kotanen, J., *et al.* (2009). Surveying biodiversity in protected and managed areas: algae, macrophytes and macroinvertebrates in boreal forest streams. *Ecological Indicators*, **9**, 1179–87.

Heino, J. & Soininen, J. (2005). Assembly rules and community models for unicellular organisms: patterns in diatoms of boreal streams. *Freshwater Biology*, **50**, 567–77.

Hewitt, C. N. (ed.) (1992). *Methods of Environmental Data Analysis*, London: Elsevier.

Hill, M. O. & Gauch, H. G. (1980). Detrended correspondence analysis, an improved ordination technique. *Vegetatio*, **42**, 47–58.

Holden, P. B., Mackay, A. M., & Simpson, G. L. (2008). Bayesian palaeoenvironmental transfer function model for acidified lakes. *Journal of Paleolimnology*, **39**, 551–66.

Holmström, L. & Erästö, P. (2002). Making inferences about past environmental change using smoothing in multiple timescales. *Computational Statistics and Data Analysis*, **41**, 289–309.

Hübener, T., Dressler, M., Schwarz, A., Langner, K., & Adler, S. (2008). Dynamic adjustment of training set ("moving-window" reconstruction) by using transfer functions in paleolimnology – a new approach. *Journal of Paleolimnology*, **40**, 79–95.

Huisman, J., Olff, H. & Fresco, L. F. M. (1993). A hierarchical set of models for species response models. *Journal of Vegetation Science*, **4**, 37–46.

Jackson, D. A. (1993). Stopping rules in principal components analysis: a comparison of heuristical and statistical approaches. *Ecology*, **74**, 2204–14.

Jacobson, G. L. & Grimm, E. C. (1986). A numerical-analysis of Holocene forest and prairie vegetation in central Minnesota. *Ecology*, **67**, 958–66.

Janssen, C. R. & Birks, H. J. B. (1994). Recurrent Groups of Pollen Types in Time. *Review of Palaeobotany and Palynology*, **82**, 165–73.

Jolliffe, I. T. (1986). *Principal Components Analysis*, New York, NY: Springer.

Jones, V. J. & Juggins, S. (1995). The construction of a diatom-based chlorophyll a transfer function and its application at three lakes on Signy Islands (Maritime Antarctic) subject to differing degrees of nutrient enrichment. *Freshwater Biology*, **34**, 433–45.

Jongman, R. H. G., ter Braak, C. J. F., & van Tongren, O. F. R. (1987). *Data Analysis in Community and Landscape Ecology*, Wageningen: Pudoc. (Reissued in 1995 by Cambridge University Press, Cambridge).

Juggins, S. (1996). The PALICLAS database. *Memorie dell'Instituto Italiano di idrobiologia*, **55**, 321–8.

Juggins, S. (2005). *C2 User Guide*. Software for ecological and palaeoecological data analysis and visualisation, University of Newcastle, Newcastle-upon-Tyne.

Juggins, S. & Birks, H. J. B. (2011). Quantitative environmental reconstructions from biostratigraphical data. In *Tracking Environmental Change Using Lake Sediments, Volume 5: Data Handling and Numerical Techniques*, ed. H. J. B. Birks, A. F. Lotter, S. Juggins & J. P. Smol, Dordrecht: Springer (in press).

Julius, M. L., Estabrook, G. F., Edlund, M. B., & Stoermer, E. F. (1997). Recognition of taxonomically significant clusters near the species level, using computationally intense methods, with examples from the Stephanodiscus niagarae complex. *Journal of Phycology*, **33**, 1049–54.

Kelly, M. G., Bayer, M. M., Hürlmann, J., & Telford, R. J. (2002). Human error and quality assurance in diatom analysis. In *Automatic Diatom Identification*, ed. H. du Buf & M. M. Bayer, Singapore: World Scientific Publishing, pp. 75–91.

Kelly, M., Juggins, S., Guthrie, R., *et al.* (2008). Assessment of ecological status in UK rivers using diatoms. *Freshwater Biology*, **53**, 403–22.

Kelly, M. & Telford, R. J. (2007). Common freshwater diatoms of Britain and Ireland: an interactive identification key (CD ROM). Environment Agency, UK.

Kingston, J. C. & Pappas, J. L. (2009). Quantitative shape analysis as a diagnostic and prescriptive tool in determining Fragilarioaforma (Bacillariophyta) taxon status. *Nova Hedwigia*, **135**, 103–19.

Kociolek, J. P. & Stoermer, E. F. (1986). Phylogenetic relationships and classification of monoraphid diatoms based on phenetic and cladistic methodologies. *Phycologia*, **25**, 297–303.

Kociolek, J. P. & Stoermer, E. F. (1988). A preliminary investigation of the phylogenetic relationships among the freshwater, apical pore field-bearing cymbelloid and gomphonemoid diatoms (Bacillariophyceae). *Journal of Phycology*, **24**, 377–85.

Korhola, A., Weckström, J., Holmström, L., & Erästö, P. (2000). A quantitative Holocene climatic record from diatoms in northern Fennoscandia. *Quaternary Research*, **54**, 284–94.

Korsman, T., Renberg, I., & Anderson, N. J. (1994). A palaeolimnological test of the influence of Norway spruce (Picea abies) immigration on lake-water acidity. *The Holocene*, **4**, 132–40.

Korsman, T. & Segerström, U. (1998). Forest fire and lake-water acidity in a northern Swedish boreal area: Holocene changes in lake-water quality at Makkassjon. *Journal of Ecology*, **86**, 113–24.

Kreiser, A. M. & Battarbee, R. W. (1988). Analytical quality control (AQC) in diatom analysis. *Proceedings of Nordic Diatomist Meeting, Stockholm, June 10–12, 1987*, ed. U. Miller & A. M. Robertsson, University of Stockholm, Dept Quaternary Research Report, pp. 41–4.

Laird, K. R. & Cumming, B. F. (2001). A regional paleolimnological assessment of the impact of clear-cutting on lakes from the central interior of British Columbia. *Canadian Journal of Fisheries and Aquatic Science*, **58**, 492–505.

Laird, K. R., Cumming, B. F., & Nordin, R. (2001). A regional paleolimnological assessment of the impact of clear-cutting on lakes from the west coast of Vancouver Island, British Columbia. *Canadian Journal of Fisheries and Aquatic Science*, **58**, 479–91.

Laird, K. R., Cumming, B. F., Wunsum, S., *et al.* (2003). Large-scale shifts in moisture regimes from lake records across the Northern Plains of North America during the past two millennia. *Proceedings of the National Academy of Sciences of the USA*, **100**, 2483–8.

Laird, K. R., Fritz, S. C., Cumming, B. F., & Grimm, E. C. (1998). Early-Holocene limnological and climate variability in the Northern Great Plains. *The Holocene*, **8**, 275–85.

Laird, K. R., Fritz, S. C., Maasch, K. A., & Cumming, B. F. (1996). Greater drought intensity and frequency before AD 1200 in the Northern Great Plains, USA. *Nature*, **384**, 552–4.

Legendre, P. & Anderson, M. J. (1999). Distance-based redundancy analysis: testing multispecies responses in multifactorial ecological experiments. *Ecological Monographs*, **69**, 1–24.

Legendre, P. & Birks, H. J. B. (2011a). Clustering and partitioning. In *Tracking Environmental Change Using Lake Sediments, Volume 5: Data Handling and Numerical Techniques*, ed. H. J. B. Birks, A. F. Lotter, S. Juggins, & J. P. Smol, Dordrecht: Springer (in press).

Legendre, P. & Birks, H. J. B. (2011b). From classical to canonical ordination. In *Tracking Environmental Change Using Lake Sediments, Volume 5: Data Handling and Numerical Techniques*, ed. H. J. B. Birks, A. F. Lotter, S. Juggins, & J. P. Smol, Dordrecht: Springer (in press).

Legendre, P. & Legendre, L. (1998). *Numerical Ecology*, Amsterdam: Elsevier.

Lepš, J. & Šmilauer, P. (2003). *Multivariate Analysis of Ecological Data using CANOCO*, Cambridge: Cambridge University Press.

Lotter, A. F. (1998). The recent eutrophication of Baldeggersee (Switzerland) as assessed by fossil diatom assemblages. *Holocene*, **8**, 395–405.

Lotter, A. F., Ammann, B., & Sturm, M. (1992). Rates of change and chronological problems during the late-glacial period. *Climate Dynamics*, **6**, 233–9.

Lotter, A. F. & Anderson, N. J. (2011). Limnological responses to environmental changes at inter-annual to decadal time-scales. In *Tracking Environmental Change Using Lake Sediments, Volume 5: Data Handling and Numerical Techniques*, ed. H. J. B. Birks, A. F. Lotter, S. Juggins & J. P. Smol, Dordrecht: Springer (in press).

Lotter, A. F. & Birks, H. J. B. (1993). The impact of the Laacher See tephra on terrestrial and aquatic ecosystems in the Black-Forest, Southern Germany. *Journal of Quaternary Science*, **8**, 263–76.

Lotter, A. F. & Birks, H. J. B. (1997). The separation of the influence of nutrients and climate on the varve time series of Baldeggersee, Switzerland. *Aquatic Sciences*, **59**, 362–75.

Lotter, A. F. & Birks, H. J. B. (2003). The Holocene palaeolimnology of Sagistalsee and its environmental history – a synthesis. *Journal of Paleolimnology*, **30**, 333–42.

Lotter, A. F., Birks, H. J. B., Hofmann, W., & Marchetto, A. (1997). Modern diatom, cladocera, chironomid, and chrysophyte cyst assemblages as quantitative indicators for the reconstruction of past environmental conditions in the Alps. 1. Climate. *Journal of Paleolimnology*, **18**, 395–420.

Lotter, A. F., Birks, H. J. B., & Zolitschka, B. (1995). Late-Glacial pollen and diatom changes in response to two different environmental perturbations – volcanic-eruption and Younger Dryas cooling. *Journal of Paleolimnology*, **14**, 23–47.

Maher, L. J., Heiri, O., & Lotter, A. F. (2011). Assessment of uncertainties associated with palaeolimnological laboratory methods and microfossil analysis. In *Tracking Environmental Change Using Lake Sediments, Volume 5: Data Handling and Numerical Techniques*, ed. H. J. B. Birks, A. F. Lotter, S. Juggins, & J. P. Smol, Dordrecht: Springer (in press).

Manly, B. J. F. (2007). *Randomization, Bootstrap, and Monte Carlo Methods in Biology*, London: Chapman & Hall/CRC Press.

McCarthy, M. A. (2007). *Bayesian Methods in Ecology*, Cambridge: Cambridge University Press.

Millar, R. B., Anderson, M. J., & Zunun, G. (2005). Fitting nonlinear environmental gradients to community data: a general distance-based approach. *Ecology*, **86**, 2245–51.

Mosimann, J. E. (1965). Statistical methods for the pollen analyst: multinomial and negative multinomial techniques. In *Handbook of Paleontological Techniques*, ed. B. Kummel & D. Raup, San Francisco, CA: W.H. Freeman.

Mou, D. & Stoermer, E. F. (1992). Separating Tabellaria (Bacillariophyceae) shape groups based on Fourier descriptors. *Journal of Phycology*, **28**, 386–95.

Munro, M. A. R., Kreiser, A. M., Battarbee, R. W., *et al.* (1990). Diatom quality control and data handling. *Philosophical Transactions of the Royal Society of London Series B – Biological Sciences*, **327**, 257–61.

Odgaard, B. V. (1994). The Holocene vegetation history of northern West Jutland, Denmark. *Opera Botanica*, **123**, 1–171.

Oksanen, J. (2004). Multivariate analysis in ecology – lecture notes. See http://cc.oulu.fi/~jarioksa/opetus/metodi/index.html (accessed 25 January 2010).

Oksanen, J., Kindt, R., Legendre, P., *et al.* (2008). VEGAN: community ecology package. R package version 1.13–1. http://cran.r-project.org/, http://vegan.r-forge.r-project.org/.

Oksanen, J. & Minchin, P. R. (2002). Continuum theory revisited: what shape are species responses along ecological gradients? *Ecological Modelling*, **157**, 119–29.

Pappas, J. L., Fowler, G. W., & Stoermer, E. F. (2001). Calculating shape descriptors from Fourier analysis: shape analysis of *Asterionella* (Heterokontophyta, Bacillariophyceae). *Phycologia*, **40**, 440–56.

Pappas, J. L. & Stoermer, E. F. (1995). Multidimensional analysis of diatom morphologic and morphometric phenotypic variation and relation to niche. *Ecoscience*, **2**, 357–67.

Pappas, J. L. & Stoermer, E. F. (1996). Formulation of a method to count number of individuals representative of number of species in algal communities. *Journal of Phycology*, **32**, 693–6.

Pappas, J. L. & Stoermer, E. F. (2001). Fourier shape analysis and fuzzy measure shape group differentiation of Great Lakes *Asterionella* Hassall (Heterokontophyta, Bacillariophyceae). In *Proceedings of the Sixteenth International Diatom Symposium*, ed. A. Economou-Amilli, Athens: Amvrosiou Press, pp. 485–501.

Paradis, E. (2006). *Analysis of Phylogenetics and Evolution with R*, New York, NY: Springer.

Passy, S. I. (2007). Differential cell size optimization strategies produce distinct diatom richness–body size relationships in stream benthos and plankton. *Journal of Ecology*, **95**, 745–54.

Passy, S. I. (2008a). Species size and distribution jointly and differentially determine diatom densities in US streams. *Ecology*, **89**, 475–84.

Passy, S. I. (2008b). Continental diatom biodiversity in stream benthos declines as more nutrients become limiting. *Proceedings of the National Academy of Sciences of the USA*, **105**, 9663–7.

Passy, S. I. (2009). The relationship between local and regional diatom richness is mediated by the local and regional environment. *Global Ecology and Biogeography*, **18**, 383–91.

Passy, S. I. & Legendre, P. (2006a). Are algal communities driven toward maximum biomass? *Proceedings of the Royal Society B*, **273**, 2667–74.

Passy, S. I. & Legendre, P. (2006b). Power-law relationships among hierarchical taxonomic categories in algae reveal a new paradox of the plankton. *Global Ecology and Biogeography*, **15**, 528–35.

Patrick, R. (1949). A proposed biological measure of stream conditions based on a survey of the Conestaga Basin, Lancaster County, Pennsylvania. *Proceedings of the National Academy of Sciences of the USA*, **101**, 277–341.

Patrick, R., Hohn, M. H., & Wallace, J. H. (1954). A new method for determining the pattern of the diatom flora. *Notulae Naturae*, **259**, 1–12.

Patrick, R. & Strawbridge, D. (1963). Variation in the structure of natural diatom communities. *American Naturalist*, **97**, 51–7.

Pelánková, B., Kunes, P., Chytry, M., *et al.* (2008). The relationships of modern pollen spectra to vegetation and climate along a steppe–forest–tundra transition in southern Siberia, explored by decision trees. *The Holocene*, **18**, 1259–71.

Peres-Neto, P. R., Legendre, P., Dray, S., & Borcard, D. (2006). Variation partitioning of species data matrices: estimation and comparison of fractions. *Ecology*, **87**, 2614–25.

Peters, J., De Baets, B., Verhoest, N. E. C., *et al.* (2007). Random forests as a tool for ecohydrological distribution modelling. *Ecological Modelling*, **207**, 304–18.

Potapova, M. G. & Charles, D. F. (2002). Benthic diatoms in USA rivers: distribution along spatial and environmental gradients. *Journal of Biogeography*, **29**, 167–87.

Prasad, A. M., Iverson, L. R., & Liaw, A. (2006). Newer classification and regression tree techniques: bagging and random forests for ecological predictions. *Ecosystems*, **9**, 181–99.

Racca, J. M. J., Philibert, A., Racca, R., & Prairie, Y. T. (2001). A comparison between diatom-based pH inference models using artificial neural networks (ANN), weighted averaging (WA) and weighted averaging partial least squares (WA-PLS) regressions. *Journal of Paleolimnology*, **26**, 411–22.

Racca, J. M. J., Wild, M., Birks, H. J. B., & Prairie, Y. T. (2003). Separating wheat from chaff: diatom taxon selection using an artificial neural network pruning algorithm. *Journal of Paleolimnology*, **29**, 123–33.

Renberg, I. & Battarbee, R. W. (1990). The SWAP Palaeolimnology Programme: a synthesis. In *The Surface Waters Acidification Programme*, ed. B. J. Mason, Cambridge: Cambridge University Press, pp. 281–300.

Renberg, I., Korsman, T., & Birks, H. J. B. (1993). Prehistoric increases in the pH of acid-sensitive Swedish lakes caused by land-use changes. *Nature*, **362**, 824–7.

Reyment, R. A. (1991). *Multidimensional Palaeobiology*, Oxford: Pergamon Press.

Reyment, R. A. & Savazzi, E. (1999). *Aspects of Multivariate Statistical Analyses in Geology*, Amsterdam: Elsevier.

Rühland, K., Paterson, A. M., & Smol, J. P. (2008). Hemispheric-scale patterns of climate-related shifts in planktonic diatoms from North American and European lakes. *Global Change Biology*, **14**, 2470–5.

Rusak, J. A., Leavitt, P. R., & McGowan, S. (2004). Millennial-scale relationships of diatom species richness and production in two prairie lakes. *Limnology and Oceanography*, **49**, 1290–9.

Schelske, C. L. & Stoermer, E. F. (1971). Eutrophication, silica depletion, and predicted changes in algal quality in Lake Michigan. *Science*, **173**, 423–4.

Schelske, C. L., Stoermer, E. F., Conley, D. J., Robbins, J. A., & Glover, R. M. (1983). Early eutrophication of the lower Great Lakes: new evidence from biogenic silica in the sediments. *Science*, **222**, 320–2.

Schröder, H. K., Andersen, H. E., & Kiehl, K. (2005). Rejecting the mean: estimating the response of fen plant species to environmental factors by non-linear quantile regression. *Journal of Vegetation Science*, **16**, 373–82.

Schulz, M. & Mudelsee, M. (2002). REDFIT: estimating red-noise spectra directly from unevenly spaced paleoclimatic time series. *Computations and Geoscience*, **28**, 421–6.

Schulz, M. & Stattegger, K. (1997). SPECTRUM: spectral analysis of unevenly spaced paleoclimatic time series. *Computations and Geoscience*, **23**, 929–945.

Simpson, G. L. (2007). Analogue methods in palaeoecology: using the analogue package. *Journal of Statistical Software*, **22**, 1–29.

Šmilauer, P. & Birks, H. J. B. (1995). The use of generalised additive models in the description of diatom–environment response surfaces. *Geological Survey of Denmark (DGU) Service Report*, **7**, 42–7.

Smol, J. P. (2008). *Pollution of Lakes and Rivers*, Oxford: Blackwell.

Smol, J. P., Birks, H. J. B., Lotter, A. F., & Juggins, S. (2011). The march towards the quantitative analysis of palaeolimnological data. In *Tracking Environmental Change Using Lake Sediments, Volume 5: Data Handling and Numerical Techniques*, ed. H. J. B. Birks, A. F. Lotter, S. Juggins & J. P. Smol, Dordrecht: Springer (in press).

Smol, J. P., Wolfe, A. P., Birks, H. J. B., *et al.* (2005). Climate-driven regime shifts in the biological communities of arctic lakes. *Proceedings of the National Academy of Sciences of the USA*, **102**, 4397–402.

Soininen, J. & Eloranta, P. (2004). Seasonal persistence and stability of diatom communities in rivers: are there habitat specific differences? *European Journal of Phycology*, **39**, 153–60.

Soininen, J. & Heino, J. (2005). Relationships between local population persistence, local abundance and regional occupancy of species: distribution patterns of diatoms in boreal streams. *Journal of Biogeography*, **32**, 1971–8.

Soininen, J., Heino, J., Kokocinski, M., & Muotka, T. (2009). Local–regional diversity relationship varies with spatial scale in lotic diatoms. *Journal of Biogeography*, **36**, 720–7.

Soininen, J. & Kokocinski, M. (2006). Regional diatom body size distributions in streams: does size vary along environmental, spatial and diversity gradients? *Ecoscience*, **13**, 271–4.

Soininen, J., Paavola, R., & Muotka, T. (2004). Benthic diatom communities in boreal streams: community structure in relation to environmental and spatial gradients. *Ecography*, **27**, 330–42.

Sokal, R. R. & Rohlf, F. J. (1995). *Biometry – The Principles and Practice of Statistics in Biological Research*, New York, NY: W.H. Freeman.

Sonderegger, D. L., Wang, H., Clements, W. H., & Noon, B. R. (2009). Using SiZer to detect thresholds in ecological data. *Frontiers in Ecology and the Environment*, **7**, 190–5.

St Jacques, J.-M., Cumming, B. F., & Smol, J. P. (2009). A 900-year diatom and chrysophyte record of spring mixing and summer stratification from varved Lake Mina, west-central Minnesota, USA. *The Holocene*, **19**, 537–47.

Stevenson, A. C., Juggins, S., Birks, H. J. B., *et al.* (1991). *The Surface Waters Acidification Project Palaeolimnology Programme: Modern Diatom/Lake-Water Chemistry Data-Set*, London: ENSIS Publishing.

Stoermer, E. F. (2001). Diatom taxonomy for paleolimnologists. *Journal of Paleolimnology*, **25**, 393–8.

Stoermer, E. F. & Ladewski, T. B. (1976). Apparent optimal temperatures for the occurrence of some common phytoplankton species in southern Lake Michigan. University of Michigan, Great Lakes Research Division, Publication 18.

Stoermer, E. F. & Ladewski, T. B. (1982). Quantitative analysis of shape variation in type and modern populations of Gomphoneis herculeana. *Nova Hedwigia*, **73**, 347–86.

Stoermer, E. F., Qi, Y.-Z., & Ladewski, T. B. (1986). A quantitative investigation of shape variation in *Didymosphenia (Lyngb.)* M Schmidt. *Phycologia*, **25**, 494–502.

Stoermer, E. F., Wolin, J. A., Schelske, C. L., & Conley, D. J. (1985). An assessment of ecological changes during the recent history of Lake Ontario based on siliceous microfossils preserved in the sediments. *Journal of Phycology*, **21**, 257–76.

Telford, R. J., Andersson, C., Birks, H. J. B., & Juggins, S. (2004). Biases in the estimation of transfer function prediction errors. *Paleoceanography*, **19**, PA4014, doi: 10.1029/2004PA001072.

Telford, R. J. & Birks, H. J. B. (2005). The secret assumption of transfer functions: problems with spatial autocorrelation in evaluating model performance. *Quaternary Science Reviews*, **24**, 2173–9.

Telford, R. J. & Birks, H. J. B. (2009). Design and evaluation of transfer functions in spatially structured environments. *Quaternary Science Reviews*, **28**, 1309–16.

Telford, R. J., Vandvik, V., & Birks, H. J. B. (2006). Dispersal limitations matter for microbial morphospecies. *Science*, **312**, 1015.

ter Braak, C. J. F. (1985). Correspondence analysis of incidence and abundance data: properties in terms of a unimodal response model. *Biometrics*, **41**, 859–73.

ter Braak, C. J. F. (1986). Canonical correspondence analysis: a new eigenvector technique for multivariate direct gradient analysis. *Ecology*, **67**, 1167–79.

ter Braak, C. J. F. (1987a). Ordination. In *Data Analysis in Community and Landscape Ecology*, ed. R. H. G. Jongman, C. J. F. ter Braak, & O. F. R. van Tongren, Wageningen: Pudoc, pp. 91–173.

ter Braak, C. J. F. (1987b). Unimodal models to relate species to environment. Unpublished Ph.D. thesis, University of Wageningen.

ter Braak, C. J. F. (1994). Canonical community ordination. Part I: basic theory and linear methods. *Ecoscience*, **1**, 127–40.

ter Braak, C. J. F. (1995). Non-linear methods for multivariate statistical calibration and their use in palaeoecology: a comparison of inverse (k-nearest neighbours, partial least squares and weighted averaging partial least squares) and classical approaches. *Chemometrics and Intelligent Laboratory Systems*, **28**, 165–80.

ter Braak, C. J. F. (1996). *Unimodal Models to Relate Species to Environment*, Wageningen: DLO-Agricultural Mathematics Group.

ter Braak, C. J. F. & Barendregt, L. G. (1986). Weighted averaging of species indicator values: its efficiency in environmental calibration. *Mathematical Biosciences*, **78**, 57–72.

ter Braak, C. J. F. & Juggins, S. (1993). Weighted averaging partial least-squares regression (WA-PLS) – an improved method for reconstructing environmental variables from species assemblages. *Hydrobiologia*, **269/270**, 485–502.

ter Braak, C. J. F., Juggins, S., Birks, H. J. B., & van der Voet, H. (1993). Weighted averaging partial least squares regression (WA-PLS): definition and comparison with other methods for species–environment calibration. In *Multivariate Environmental Statistics*, ed. G. P. Patil & C. R. Rao), Amsterdam: Elsevier, pp. 529–560.

ter Braak, C. J. F. & Looman, C. W. N. (1986). Weighted averaging, logit regression and the Gaussian response model. *Vegetatio*, **65**, 3–11.

ter Braak, C. J. F. & Prentice, I. C. (1988). A theory of gradient analysis. *Advances in Ecological Research*, **18**, 271–317.

ter Braak, C. J. F. & Šmilauer, P. (2002). *CANOCO Reference Manual and CanoDraw for Windows User's Guide: Software for Canonical Community Ordination (version 4.5)*, Ithaca, NY: Microcomputer Power.

ter Braak, C. J. F. & van Dam, H. (1989). Inferring pH from diatoms – a comparison of old and new calibration methods. *Hydrobiologia*, **178**, 209–23.

ter Braak, C. J. F. & Verdonschot, P. F. M. (1995). Canonical correspondence analysis and related multivariate methods in aquatic ecology. *Aquatic Sciences*, **57**, 255–89.

Theriot, E. C. (1987). Principal component analysis and taxonomic interpretation of environmentally related variation in silification in *Stephanodiscus* (Bacillariophyceae). *British Phycological Journal*, **22**, 359–73.

Theriot, E. C., Fritz, S. C., Whitlock, C., & Conley, D. J. (2006). Late Quaternary rapid morphological evolution of an endemic diatom in Yellowstone Lake, Wyoming. *Paleobiology*, **32**, 38–54.

Theriot, E. C., Håkansson, H., & Stoermer, E. F. (1988). Morphometric analysis of *Stephanodiscus alpinus* (Bacillariophyceae) and its morphology as an indicator of lake trophic status. *Phycologia*, **27**, 485–93.

Theriot, E. C. & Stoermer, E. F. (1984). Principal component analysis of character variation in *Stephanodiscus niagarae* Ehrenb.: morphological variation related to lake trophic status. *In Proceedings of the Seventh Diatom Symposium* 1982, ed. D. G. Mann, Königstein: Koeltz Scientific Publishers, pp. 91–111.

Torrence, C. & Compo, G. P. (1999). A practical guide to wavelet analysis. *Bulletin of the American Meteorological Society*, **79**, 61–78.

Tufte, E. R. (1983). *The Visual Display of Quantitative Information*, Cheshire, CT: Graphics Press.

Tyler, P. A. (1996). Endemism in freshwater algae. *Hydrobiologia*, **336**, 127–35.

van der Brink, P. J. & ter Braak, C. J. F. (1998). Multivariate analysis of stress in experimental ecosystems by principal response curves and similarity analysis. *Aquatic Ecology*, **32**, 163–78.

van der Voet, H. (1994). Comparing the predictive accuracy of models using a simple randomization test. *Chemometrics and Intelligent Laboratory Systems*, **25**, 313–23.

Vanormelingen, P., Verleyen, E., & Vyverman, W. (2008). The diversity and distribution of diatoms: from cosmopolitanism to narrow endemism. *Biodiversity and Conservation*, **17**, 393–405.

Vyverman, W., Verleyen, E., Sabbe, K., *et al.* (2007). Historical processes constrain patterns in global diatom diversity. *Ecology*, **88**, 1924–31.

Warton, D. I. (2008). Raw data graphing: an informative but under-utilized tool for the analysis of multivariate abundances. *Australian Ecology*, **33**, 290–300.

Weckström, J. & Korhola, A. (2001). Patterns in the distribution, composition and diversity of diatom assemblages in relation to ecoclimatic factors in Arctic Lapland. *Journal of Biogeography*, **28**, 31–45.

Williams, D. M. (1985). Morphology, taxonomy and inter-relationships of the ribbed araphid diatoms from the genera *Diatoma* and *Meridion* (Diatomaceae: Bacillariophyta). *Bibliotheca Diatomologica*, **8**, 1–228.

Witt, A. & Schumann, A. Y. (2005). Holocene climate variability on millennial scales recorded in Greenland ice cores. *Non-linear Processes in Geophysics*, **12**, 345–52.

Wolfe, A. P. (1997). On diatom concentrations in lake sediments: results from an inter-laboratory comparison and other tests performed on a uniform sample. *Journal of Paleolimnology*, **18**, 261–8.

Yee, T. W. (2004). A new technique for maximum-likelihood canonical Gaussian ordination. *Ecological Monographs*, **74**, 685–701.

Yee, T. W. (2006). Constrained additive ordination. *Ecology*, **97**, 203–13.

Yee, T. W. & Mitchell, N. D. (1991). Generalized additive models in plant ecology. *Journal of Vegetation Science*, **2**, 587–602.

Yuan, L. L. (2004). Assigning macroinvertebrate tolerance classification using generalised additive models. *Freshwater Biology*, **49**, 662–77.

Yuan, L. R. (2007a). Maximum likelihood method for predicting environmental conditions from assemblage composition: the R package bio.infer. *Journal of Statistical Software*, **22**, 1–20.

Yuan, L. R. (2007b). Using biological assemblage composition to infer the values of covarying environmental factors. *Freshwater Biology*, **52**, 1159–75.

Zuur, A. F., Ieno, E. N., & Smith, G. M. (2007). *Analyzing Ecological Data*, New York, NY: Springer.

Zuur, A. F., Ieno, E. N., Walker, N. J., Savelier, A. A., & Smith, G. M. (2009). *Mixed Effect Models and Extensions in Ecology with R*, New York, NY: Springer.

Part II

Diatoms as indicators of environmental change in flowing waters and lakes

4 Assessing environmental conditions in rivers and streams with diatoms

R. JAN STEVENSON, YANGDONG PAN, AND HERMAN VAN DAM

4.1 Introduction

Assessments of environmental conditions in rivers and streams using diatoms have a long history in which two basic conceptual approaches emerged. First, based on the work of Kolkwitz and Marsson (1908), autecological indices were developed to infer levels of pollution based on the species composition of assemblages and the ecological preferences and tolerances of taxa (e.g. Butcher, 1947; Fjerdingstad, 1950; Zelinka and Marvan, 1961; Lowe 1974; Lange-Bertalot, 1979). Second, Patrick's early monitoring studies (Patrick, 1949; Patrick *et al.*, 1954; Patrick and Strawbridge, 1963) relied primarily on diatom diversity as a general indicator of river health (i.e. ecological integrity), because species composition of assemblages varied seasonally and species diversity varied less. The conceptual differences in these two approaches really address two different goals for environmental assessments, one inferring pollution levels and the other determining biodiversity, a more valued ecological attribute (Stevenson, 2006). Thus, the concepts and tools for assessing ecosystem health and diagnosing causes of impairment in aquatic habitats, particularly rivers and streams, were established and developed between ~50 and 100 years ago.

Today, diatoms are being used to assess ecological conditions in streams and rivers around the world (Asai, 1996; Kelly *et al.*, 1998; Wu, 1999; Lobo *et al.*, 2004; Wang *et al.*, 2005; Chessman *et al.*, 2007; Taylor *et al.*, 2007; Porter *et al.*, 2008). They have become valuable elements in large-scale national and international assessment programs of the United States and Europe (e.g. Kelly *et al.* 2009a). Keeping up with the large number of papers being published in locally, nationally, and internationally recognized journals has become a challenge, but this review shows the emergence of a great diversity of methods and findings, identifying regionally refined tools, and organizing the application of these tools in scientifically sound protocols for solving environmental problems. Although initially overwhelming, the diversity of methods and findings for diatom assessments produced over the last ~100 years can be organized by relating them to frameworks for environmental assessment and management. Using the correct diatom assessment tools for the correct reason during an assessment is critical for scientifically rigorous and effective support of environmental management.

The Diatoms: Applications for the Environmental and Earth Sciences, 2nd Edition, eds. John P. Smol and Eugene F. Stoermer. Published by Cambridge University Press.

The many advances in using diatoms and other algae for monitoring stream and river quality have been reviewed by Patrick (1973) and, more recently, by Stevenson and Lowe (1986), Rott (1991), Round (1991), Whitton *et al.* (1991), Coste *et al.* (1991), Whitton and Kelly (1995), Rosen (1995), Lowe and Pan (1996), Ector *et al.* (2004), and Stevenson and Smol (2002). In this chapter, we update our last review completed for the first edition of this book (Stevenson and Pan, 1999). First, we provide the foundation for how diatoms can be used in ecological assessments of rivers and streams and why they should be used. Second, we review the many characteristics of diatom assemblages that could be used in assessments and the methods used for sampling, sample analysis, and calculating indicators. Finally, we provide examples and discuss how a set of diatom indicators can be used in ecological assessments with different purposes for each indicator, such as establishing reference conditions and numeric water-quality criteria, assessing physical and chemical as well as biological condition, and diagnosing stressors of ecosystem services.

4.2 Rationale for using diatoms

Two fundamental questions need to be answered in almost all ecological assessments: "Is there a problem?" and "What is causing the problem?" (Stevenson *et al.*, 2004a). Understanding the meaning of these questions, and how they will be asked

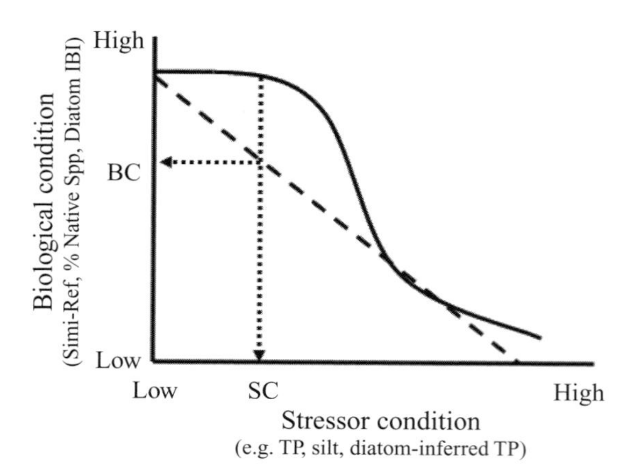

Figure 4.1 A scheme for using a suite of indicators for stressor and biocriteria development. The solid line represents a valued ecological attribute, such as the percentage of individuals in native species in the habitat, having a non-linear response to a stressor or human-disturbance gradient. The dashed line represents an indicator with linear response to a stressor or human-disturbance gradient. After identifying indicators with linear and non-linear responses, stressor criteria can be proposed for the highest stressor level (vertical dashed line) that has a high probability of protecting the valued ecological attribute with the non-linear response. The biological criterion (horizontal dashed line) can be proposed for an indicator with the linear response at a level predicted to occur when the stressor is at its criterion level.

and answered by government agencies or other scientists, is important for determining how diatoms can help answer both of these questions. Problems can usually be defined relative to the uses of water bodies and their ecosystem services. "Use" is a term that is defined in regulations developed in the United States of America (USA) associated with implementation of the US Clean Water Act (US CWA) and is one part of water quality standards. Examples of uses are aquatic-life use, drinking-water supply, and recreation. The other parts of water-quality standards in the USA are water-quality criteria, which are established to protect uses of waters, and an antidegradation policy that protects future degradation of waters. Water-quality criteria can be narrative or numeric descriptions of the physical, chemical, and biological condition of waters that define whether a use is being supported or not. The variables used for water-quality criteria are indicators of either uses or the contaminants and habitat alterations (i.e. "stressors") that degrade or threaten uses (Figure 4.1). Throughout this chapter, we will emphasize that slight differences in how diatom indicators are developed and applied determines whether indicators should be used to assess valued ecological attributes (uses and ecosystem services) or stressors (Stevenson, 2006).

Protection of the biota that we would expect in minimally disturbed waters is a common policy of governments around the world. In the USA, many state governments refer to this as "aquatic life use support" and measure it with indicators of biological condition (Davies and Jackson, 2006). Waters with high levels of biological condition have biological integrity, which is a defined goal of the US CWA. Biological condition is usually measured as some deviation from natural or minimally disturbed condition (Davies and Jackson, 2006; Stoddard et al. 2006). In the European Union (EU), all water bodies are supposed to achieve "good ecological status" by 2015 according to the Water Framework Directive (WFD: European Union, 2000). "Good ecological status" is defined in the EU WFD as biota consistent with only slight alterations from that expected in the absence of human impacts (reference conditions), which is the same as the concept of high biological condition in the USA. Although we usually measure structure, such as species composition of assemblages, both the US and EU concepts of reference conditions also include the function of aquatic ecosystems, such as productivity and nutrient uptake (Karr and Dudley, 1981).

One of the main reasons for convergence of management goals on "good ecological status" or "biological integrity" is that biodiversity is a valued ecological attribute, thus an ecosystem service. In addition, waters with high biological condition are assumed to support other ecosystem services, such as safe and aesthetically pleasing drinking water and recreational uses. In addition, rivers and streams are complex ecosystems in which many environmental factors vary on different spatial and temporal scales. Even though as many variables as possible should be measured to determine environmental conditions in a habitat (Barbour et al., 1999; Norris and Norris, 1995), full assessment of all the potentially important variables is impractical. Biological indicators respond to altered physical and chemical conditions that may not have been measured and thereby reflect these altered conditions in assessments. Using a suite of biological indicators, based on organisms living from one day to several years, provides an integrated assessment of environmental conditions in streams and rivers which are spatially and temporally highly variable.

Using diatoms as indicators of environmental conditions in rivers and streams is important for three basic reasons: their importance in ecosystems, their utility as indicators of environmental conditions, and their ease of use. Diatom importance in river and stream ecosystems is based on their fundamental role in food webs (e.g., Mayer and Likens, 1987; see review by Lamberti, 1996) and linkage in biogeochemical cycles (e.g.

Newbold *et al.*, 1981; Kim *et al.*, 1990; Mulholland, 1996). As one of the most species-rich components of river and stream communities, diatoms are important elements of biodiversity and genetic resources in rivers and streams (Patrick, 1961). In addition, diatoms are the source of many nuisance algal problems, such as taste and odor impairment of drinking water, reducing water clarity, and clogging water filters (e.g. Palmer, 1962).

Diatoms are valuable indicators of environmental conditions in rivers and streams, because they respond directly and sensitively to many physical, chemical, and biological changes in river and stream ecosystems, such as temperature (Squires *et al.*, 1979; Descy and Mouvet, 1984), nutrient concentrations (Pan *et al.*, 1996; Kelly, 1998), and herbivory (Steinman *et al.*, 1987a; McCormick and Stevenson, 1989). The species-specific sensitivity of diatom physiology to many habitat conditions is manifested in the great variability in biomass and species composition of diatom assemblages in rivers and streams (e.g. Patrick, 1961). This great variability is the result of complex interactions among a variety of habitat conditions that differentially affect physiological performance of diatom species, and thereby, diatom assemblage composition. Stevenson (1997) organizes these factors into a hierarchical framework in which higher-level factors (e.g. climate and geology) can restrict effects of low-level factors. Low-level, proximate factors, such as resources (e.g. light, nitrogen (N), phosphorus (P)) and stressors (e.g. pH, temperature, toxic substances), directly affect diatoms. At higher spatial and temporal levels, effects of resources and stressors on diatom assemblages can be constrained by climate, geology, and land use (Biggs, 1995; Stevenson, 1997). The sensitivity of diatoms to so many habitat conditions can make them highly valuable indicators, particularly if effects of specific factors can be distinguished.

Diatoms occur in relatively diverse assemblages, and most species, especially common species, are relatively easily distinguished when compared to other algae and invertebrates that also have diverse assemblages. Diatoms are readily distinguished to species and subspecies levels based on unique morphological features, whereas many other algal classes have more than one stage in a life cycle and some of those stages are either highly variable ontogenically (e.g. blue-green algae), cannot be distinguished without special reproductive structures (e.g. Zygnematales), or cannot be distinguished without culturing (many unicellular green algae). Diverse assemblages provide more statistical power in inference models (see Birks, this volume). Identification to species level improves precision and accuracy of indicators that could arise from autecological variability within genera. Diatoms are relatively similar in size compared to the variability among all groups of algae, even though the size of diatoms does vary by several orders of magnitude; therefore diatom assemblage characterizations accounting for cell size (biovolume and relative biovolume) are probably not as necessary as when using all groups of algae together.

Diatoms can be found in almost all aquatic habitats, so that the same group of organisms can be used for comparison of streams, lakes, wetlands, oceans, estuaries, and even some ephemeral aquatic habitats. Diatoms can be found on substrata in streams, even when the stream is dry, so they can be sampled at most times of the year. If undisturbed sediments can be found in lotic ecosystems, such as in reservoirs, deltas, wetlands, and floodplains where rivers and streams drain, diatom frustules are preserved in sediments and record habitat history (Amoros and van Urk, 1989; Cremer *et al.*, 2004; Gell *et al.* 2005). Historical conditions in streams and rivers have also been assessed by using museum collections of diatoms on macrophytes and in fish (van Dam and Mertens, 1993; Rosati *et al.*, 2003; Yallop *et al.*, 2006). Diatoms have shorter generation times than fish and macroinvertebrates. Therefore we assume they reproduce and respond rapidly to environmental change, thereby providing early warning indicators of both pollution increases and habitat restoration success. The combined costs of sampling and sample assay are relatively low when compared to other organisms. Samples can be archived easily for long periods of time for future analysis and long-term records.

Thus, the study of diatoms has become an important element of monitoring and assessment programs in countries around the world. In the EU, each member state has established methods for assessing ecological status for a range of water-quality elements, including phytobenthos in running waters. Diatoms are used in water-quality assessments in many states in the USA and were included in the National Rivers and Streams Assessment being conducted from 2008–2011 by the US Environmental Protection Agency (EPA). Diatoms provide as precise an assessment, as macrophytes, invertebrates, and fish and actually are more sensitive to some stressors than other organisms (Griffith *et al.*, 2005; Johnson *et al.*, 2006; Chessman *et al.*, 2007; Carlisle *et al.*, 2008). Hering *et al.* (2006) state that diatoms are most sensitive to changes in nutrient and organic-matter concentrations, macrophytes are sensitive to changes in hydromorphology, and macroinvertebreates and fish are sensitive to changes in hydromorphology at larger scales of the landscape.

4.3 Assemblages and characteristics used in assessment

4.3.1 Periphyton, plankton, and sampling

Both planktonic and benthic diatoms (i.e. phytobenthos or periphyton) can be used in assessments of rivers and streams. The advantages of sampling plankton and periphyton vary with size of the river and objective of the research. Plankton should usually be sampled in large rivers and periphyton should be sampled in shallow streams, where each, respectively, is the most important source of primary production (Vannote *et al.*, 1980). However, periphyton sampling could be more appropriate than phytoplankton in large rivers if assessing point sources of pollution and if high spatial resolution in water-quality assessment are objectives. If greater spatial integration is desired, then plankton sampling may be an appropriate approach even in small streams. Suspended algae originate from benthic algae in small rivers and streams and are transported downstream (Swanson and Bachman, 1976; Müller-Haeckel and Håkansson, 1978; Stevenson and Peterson, 1991). Therefore, plankton may provide a good spatially integrated sample of benthic algae in a stream. Further studies of the value of sampling plankton in small rivers and streams are warranted for assessment of watershed conditions.

Benthic algae on natural substrata and plankton should be sampled in stream assessments whenever objectives call for accurate assessment of ecosystem components (Aloi, 1990; Cattaneo and Amireault, 1992) or when travel costs to sites are high. Artificial substratum sampling is expensive because it requires two separate trips to the field, and because artificial substrata are highly susceptible to vandalism and damage from floods. One problem with periphyton samples on natural substrata is that they can be highly variable. Targeted habitat sampling, such as rocks and hard surfaces, are recommended in the EU and some US programs (Kelly *et al.*, 1998; Moulton *et al.*, 2002; CEN, 2003, 2004). However, other US programs (Weilhoefer and Pan, 2007; Peck *et al.*, 2006) randomly sample whatever substratum occurs throughout a defined section of the stream to provide an accurate characterization of assemblages in that section (i.e. reach). Composite sampling approaches have been used to reduce within-habitat variability when sampling stream and river periphyton. Composite samples are collected by sampling periphyton on rocks at random locations along three or more random transects in a habitat and combining the samples into one composite (e.g. Weilhoefer and Pan, 2007).

Using artificial substrata is a valuable approach when objectives call for precise assessments in streams with highly variable habitat conditions, or when natural substrata are unsuitable for sampling. The latter may be the case in deep, channelized, or silty habitats. Benthic algal communities on artificial substrata are commonly different than those on natural substrata (e.g. Tuchman and Stevenson, 1980). However, when the ecology of the natural habitat is simulated, benthic diatom assemblages developing on artificial substrata can be similar to assemblages on natural substrata (see review by Cattaneo and Amireault, 1992). Cattaneo and Amireault recommend cautious use of artificial substrata, because algal quantity often differs and non-diatom algae are underrepresented on artificial substrata. When detecting change in water quality is a higher priority than assessing effects of water quality on natural assemblages of periphyton in that habitat, then the advantages of the high precision and sensitivity of diatoms on artificial substrata for assessing the physical and chemical conditions in the water may outweigh the disadvantages of questionable simulation of natural communities.

4.3.2 An organizational framework for assemblage attributes

To help organize the tools for the assessment toolbox, we have classified assemblage attributes based on how they are developed, because that affects how they should be applied. Diatom assemblage characteristics are typically used in conjunction with the characteristics of entire periphyton or plankton assemblages, thereby accounting for changes in other algae and microorganisms that occur in benthic and planktonic samples. These characteristics occur in two categories, structural and functional (Table 4.1). Taxonomic composition and biomass (measured as cell density or biovolume) are the only diatom assemblage characteristics that can be distinguished from other algae and microbes in periphyton and plankton samples, which can be done with microscopic examination, identification, and counting cells in samples. It is worthwhile to note that chlorophyll *a* (chl *a*) concentrations, ash-free dry mass (AFDM), chemical composition, and functional characteristics of diatom assemblages cannot be distinguished from other algae, bacteria, and fungi in a periphyton or plankton sample. Little is known about the accuracy of diatom biomass assessments with chl *c*, so measurements of diatom cell numbers or biovolumes are probably the most reliable estimates of diatom biomass.

The most commonly used diatom indicators are those using taxonomic composition of assemblages. There is a great diversity of these indices based on how they are calculated and how

Table 4.1 Characteristics of algal assemblages that could be used to assess the ecological (biological and stressor) condition of a habitat. Attribute type indicates the kind of analysis, whether attributes are structural or functional characteristics, and whether they are usually (not always) indicators of biological or stressor condition. For indicators and citations, we've provided early and recent examples, but have not attempted to be encyclopedic. (Abbreviations: SPI, Species Pollution Index; GDI, Indice Générique Diatomique; DAI, Diatom Autecological Index; TDI, Trophic Diatom Index)

Attribute type	Indicator	Citations
Biomass (structural characteristic, biological condition)	AFDM, chl *a* and other pigments, cell densities, cell biovolumes, other elements that are most common in microbial biomass (N or P)	APHA, 1998; Dodds *et al.* 1997; Stevenson *et al.* 2006
Diversity (structural characteristics, biological condition)	Composite diversity, total species or generic richness, richness of sensitive native taxa, species evenness, percent dominant taxon	Stevenson *et al.*, 1984b, 2008a
Taxonomic composition (structural characteristic, biological condition)	Similarity to reference condition, relative abundances of individuals in genera and species, pigment ratios	Common throughout the literature, e.g. Schoemann, 1976; Lange-Bertalot, 1979; Stevenson and Bahls, 1999; Passy and Bode 2004; Wang *et al.*, 2005; Lavoie *et al.*, 2006; Stevenson *et al.*, 2008a
Guild indicators (structural characteristics, biological condition)	Percent of diatom individuals or taxa that are stalked, adnate, motile, native (reference), non-native (non-reference), pollution sensitive, pollution tolerant or are adapted to low P, high P, low pH, high pH, etc.	Palmer, 1969; Lange-Bertalot, 1979; Wang *et al.*, 2005; Stevenson *et al.* 2008a, 2008b
Autecological indices (structural characteristic, stressor condition)	Pollution Tolerance Index, SPI, GDI, DAIpo, DAI-pH, DAI-TP, TDI, human disturbance, conductivity, pH, trophic status, saprobity (organic pollution), dissolved oxygen, nitrogen, phosphorus, sediments	Lange-Bertalot, 1979; Rumeau and Coste, 1988; Watanabe *et al.*, 1986; Prygiel and Coste, 1993; van Dam *et al.*, 1994; Kelly *et al.*, 1995; Pan *et al.*, 1996; Walley *et al.*, 2001; Gómez and Licursi, 2001; Rott *et al.*, 2003; Potapova and Charles, 2003, 2007; Lobo *et al.*, 2004; Kovács *et al.*, 2006; Stevenson *et al.*, 2008a, 2008b
Morphology (structural characteristic, biological & stressor condition)	Larger cells with UV effects, deformed frustules with metals	Bothwell *et al.*, 1993; McFarland *et al.*, 1997; Falasco *et al.*, 2009
Chemical ratios (structural characteristic, biological & stressor condition)	chl *a*:AFDM, chl *a*:phaeophytin, N:P, N:C, P:C, heavy metals:AFDM	Weber, 1973; Peterson and Stevenson, 1992; Humphrey and Stevenson, 1992; Biggs, 1995
Growth and dispersal rates (functional characteristic, biological condition)	Reproduction rate, growth rate, accrual rate, immigration rate, emigration rate	Müller-Haeckel and Hakansson, 1978; Stevenson, 1983; Biggs, 1990
Metabolic rates (functional characteristic, biological condition)	Photosynthetic rate, respiration rate, phosphatase activity	Blanck, 1985; Mulholland and Rosemond, 1992; Hill *et al.*, 1997

they should be applied. Often, two pieces of information are needed to calculate diatom indicators that use taxonomic composition of assemblages, the relative abundance of taxa in the sample (i.e. taxonomic composition) and the environmental preferences and tolerances of taxa (autecological attributes of taxa or taxa traits). These calculations are usually done at the species level, but they can be done at higher levels of taxonomic organization as well.

Taxa autecologies or traits (*sensu* Stevenson *et al.* 2008b) were historically determined by non-quantitative rankings based on the types of habitats in which taxa had been collected in the past. The compilations of literature with indicator values for diatoms by Lowe (1974) and van Dam *et al.* (1994) are two examples. The weighted-averaging method, developed and refined from 1955 to 1989 (Pantle and Buck, 1955; Zelinka and Marvan, 1961; ter Braak and van Dam, 1989), is a more quantitative method for determining species environmental optima and the tolerance of taxa for environmental variability. Morphological traits, such as stalked or adnate growth forms and raphe structures (indicating ability to move on substrata and through a periphyton matrix and sediments), are attributes that can be used to group taxa and start identifying guilds of taxa, i.e. groups with similar functions in ecosystems (see Wang *et al.* 2005). Physiological guilds of species can also be defined with low or high trait values, such as: low and high nutrient taxa, native and non-native taxa, taxa sensitive and tolerant to pollution, or low pH and high pH taxa (e.g. van Dam *et al.*, 1994; Stevenson *et al.*, 2008a). Guilds can also be identified using indicator species analysis (Dufrêne and Legendre, 1997), which allows selection of species that are more consistently found in one type of habitat versus another. Alternatively, we can assume that taxa, such as genera, families, orders, and classes have greater physiological similarity within taxa than across taxa. Thereby, percent *Nitzschia*, araphids, and Centrales become indicators based on the assumption that their physiologies or performance in ecosystems differ from other taxa. Testing guild indicators can be as simple as a regression analysis (e.g. Stevenson *et al.*, 2008b). Alternatively, Potapova and Charles (2007) use indicator species analysis to identify taxa characteristically found in high and low total phosphorus (TP) or total nitrogen (TN) conditions.

Usually, diatom indicators are calculated based on a weighted-averaging model using relative abundances of all taxa in a sample from a site and the environmental optima or autecological values for the taxa. However, they can be calculated many other ways. One major distinction is using subsets of all taxa in indicators (such as guilds) versus all taxa. The coarsest resolution for indicators using multiple species would be the simple average of species traits (environmental optima or autecological values) for all species in the sample; thus all species are weighted equally rather than by their relative abundances (Stevenson *et al.*, 2008b). Other calculation methods include the number of taxa in one specific guild, the proportion of all taxa in samples from one specific guild, or the proportion of all individuals in samples from one specific guild (Wang *et al.*, 2005).

Diatom indicators based on taxonomic characteristics of assemblages and taxa traits have been used to assess both the biological condition of rivers and streams and causes of their degradation, thus physical, chemical, and biological condition. Traditionally, diatom assemblages have been used to assess the chemical condition, using saprobic, pH, nutrient, and other stressor indicator systems (e.g. van Dam *et al.*, 1994; Kelly and Whitton, 1995; Potapova and Charles, 2003; Walley *et al.*, 2001). More recently, diatom research has also emphasized assessment of the biological condition of an ecosystem, i.e. the similarity of an assessed assemblage to assemblages in reference ecosystems (Passy and Bode, 2004; Wang *et al.*, 2005; Lavoie *et al.*, 2006; Stevenson, 2006).

There is a fine distinction between how these indicators are calculated, but there should be a substantial difference in how they are applied. This distinction is discussed extensively in Stevenson (2006). The basic point is we should distinguish conceptually between indicators of physical, chemical, and biological condition because they are used differently in management. They are used differently because of how we manage ecosystems; we want to manage ecosystems to minimize physical alteration and chemical pollution of habitats to maximize the biological condition (Karr and Dudley, 1981; Karr and Chu 1997; Davies and Jackson, 2006) or some other valued ecological attributes (Stevenson *et al.*, 2004a). Therefore, indicators such as similarity of species composition to reference condition, percent native and non-native taxa (i.e. reference or non-reference taxa), percent diatoms, percent *Achnanthidium* A. G. Agardh, and percent *Cymbella* Kutzing and *Encyonema* Kutzing are indicators of biological condition because they do not specifically infer abiotic conditions and they do reflect elements of biological condition (i.e. "the natural balance of flora and fauna . . . ", Karr and Dudley, 1981). Inferred TP, pH, and specific conductivity are indicators that reflect shifts in diatom species composition from reference condition, but they more specifically address abiotic conditions. Inferred abiotic conditions are not easily interpretable biological attributes. Indicators such as percent low P species or percent motile species are more easily interpreted and could be considered valued attributes if they were thought of as elements of biological condition, even though they are stressor-specific guilds.

4.3.3 Biomass assay

Periphyton and phytoplankton biomass can be estimated with assays of dry mass (DM), AFDM, chl *a*, cell densities, cell biovolumes, and cell surface area (Aloi, 1990; APHA, 1998). All these estimates have some bias in their measurement of algal biomass. Dry mass varies with the amount of inorganic as well as organic matter in samples. The AFDM varies with the amount of detritus as well as the amount of bacteria, fungi, microinvertebrates, and algae in the sample. Chl *a*:algal carbon (C) ratios can vary with light and N availability (Rosen and Lowe, 1984). Chl *c* concentration in habitats could be a good indicator of diatom biomass in a habitat (APHA, 1998). Cell density:algal C ratios vary with cell size and shape. Even cell volume:algal C ratios vary among species, particularly among some classes of algae, because vacuole size in algae varies (Sicko-Goad *et al.*, 1977). Cell surface area may be a valuable estimate of algal biomass because most cytoplasm is adjacent to the cell wall. Elemental and chemical mass per unit area (other than chl *a*, such as μg P cm^{-2}, μg N cm^{-2}) could also be used to assess algal biomass, but many of the chemical methods of assessing biomass have not been studied extensively.

We recommend using as many indicators of algal biomass as possible. We typically do not restrict our assays of algal biomass to diatom density and biovolume. We usually assess chl *a* and AFDM of samples and count and identify all algae to the lowest possible taxonomic level in Palmer counting cells or wet mounts to determine algal cell density and biovolume. More recently we have employed rapid periphyton surveys to visually assess biomass of microalgae and macroalgae with substantial success (Stevenson and Bahls, 1999; Stevenson *et al.*, 2006). In ecological studies when distinguishing live and dead cells is important (e.g. experiments), diatoms are counted in syrup (Stevenson, 1984a) or high refractive-index media using vapor substitution (Stevenson and Stoermer, 1981; Crumpton, 1987). In large-scale ecological surveys, when distinguishing live and dead diatoms has not been shown to be important, we count acid-cleaned diatoms using a highly refractive mounting medium such as Naphrax® to ensure the best taxonomic assessments. Conceptually, counting dead diatoms that may have drifted into an area or persisted from the past should only increase the spatial and temporal scale of an ecological assessment. Gillett *et al.* (2009) show that indicators performed similarly if they were based on diatom frustules with protoplasm or all diatoms in acid-cleaned material.

Periphyton and phytoplankton biomass is highly variable in streams and rivers, and periphyton biomass in particular has been criticized as a reliable indicator of water quality (Whitton and Kelly, 1995; Leland, 1995). According to theories of community adaptation to stress (Stevenson 1997), biomass should be less sensitive than species composition to environmental stress, because communities can adapt to environmental stress by changing species composition.

Another problem with using algal biomass as an indicator of nutrient enrichment and toxicity is that low biomass may be the result of a recent natural physical and biotic disturbance (e.g. Tett *et al.*, 1978; Steinman *et al.*, 1987) or toxicity (e.g. Gale *et al.*, 1979). A more reliable indicator of environmental impacts on algal and diatom biomass in a habitat may be the peak biomass that can accumulate in a river or stream after a disturbance (Biggs, 1996; Stevenson, 1996). Peak biomass is the maximum biomass in the phytoplankton or the periphyton that accumulates after a disturbance. These maxima develop during low discharge periods, usually seasonally, for both phytoplankton and periphyton and, theoretically, should be highly correlated to nutrient and light availability in a system. Clear relationships have been shown between phosphate and nitrate concentration and peak biomass of periphytic diatoms in experimental systems (Bothwell, 1989; Rier and Stevenson, 2006). Peak biomass is also a valuable parameter because it indicates the potential for nuisance-levels of algal biomass accumulation. In practical application, we seldom sample at the period of peak biomass; however adjusting the timing of sampling to get as close as possible to peak biomass state should reduce variability in biomass relationships with nutrient concentrations. For river phytoplankton, consider use of remote sensing from satellites to assess algal biomass at large spatial scales.

More recent field studies have shown clear relationships between benthic algal biomass and nutrient concentrations in streams by either accounting for temporal variability with multiple measurements during a season (Stevenson *et al.*, 2006), time since last disturbance (Biggs, 2000), or by high sample size (Dodds *et al.*, 1997). In addition, Stevenson *et al.* (2006) show that response of biomass to nutrients is much more related to increases in filamentous green algae than diatoms; the response of diatom biomass is saturated at relatively low phosphorus concentrations. These relationships were characterized by fairly high variability, but also by a high magnitude of effect.

4.3.4 Diversity

Many indices have been developed to characterize the number of species in a sample (species richness), the evenness of species abundances, and composite diversity. Composite diversity is represented in indices that respond to changes in both richness and evenness (e.g. Shannon, 1948; Simpson, 1949). High correlation between all of these indices has been observed (Archibald, 1972). This is probably because composite diversity and species richness measurements are highly dependent on evenness of species abundances in short counts (e.g. 600 valves; Stevenson and Lowe, 1986).

Assessment of species richness in a habitat is particularly problematic because species numbers are highly correlated to species evenness in counts when a predetermined and low number of diatoms is counted (e.g. 600 valves). Better assessments of species richness can be determined by developing the relationship between species numbers observed and the number of cells that have been counted. Species richness can be defined as the number of species in a count when no new species are observed with a specified additional counting effort.

Alternatively, non-linear regression can be used to determine the asymptote of the relationship between number of species observed and number of cells counted – the asymptote is an estimate of the number of species in the sample. The precision of the asymptote, and thus the estimate of species richness, will be reported by most statistical programs. Stratified counting efforts can be employed to assess different community parameters, such as relative abundance of dominant taxa and species richness. This is done by identifying and counting all valves until a pre-specified number of valves is enumerated to determine relative abundance of the dominants; then species richness is estimated by continuing to scan the sample, counting valves, identifying only the new species, and stopping the count when no new species are identified after a fixed counting effort, such as 100 or 200 valves. When the budget permits, species richness could be determined with long diatom counts (3000–8000 valves) and estimation using the assumption that the number of species in different density categories fit a lognormal curve (Patrick *et al.*, 1954). Assessments of species evenness do vary with the evenness parameter used (e.g. Hurlbert's versus Alatalo's evenness, Hurlbert, 1971; Alatalo, 1981), but the utility of differential sensitivity of these characteristics has not be extensively investigated.

The best use of diversity-related indices in river and stream assessments is probably as an indicator of changes in species composition when comparing impacted and reference assemblages (Stevenson, 1984b; Jüttner *et al.*, 1996). Some investigators have found that diversity decreases with pollution (e.g. Rott and Pfister, 1988; Sonneman *et al.* 2001), that diversity can increase with pollution (e.g. Archibald, 1972; van Dam, 1982; Stevenson *et al.* 2008a), and that diversity changes differently depending upon the type of pollution (Jüttner *et al.*, 1996). Patrick (1973) hypothesized ambiguity in diversity assessments of pollution when using composite diversity indices because of differing effects of pollutants on species richness and evenness. She predicted that some pollutants (e.g. organic pollution) would differentially stimulate growth of some species and thereby decrease evenness of species abundances. The author also predicted that toxic pollution could increase evenness and that severe pollution could decrease species numbers (Patrick, 1973). Therefore, depending upon the kind and severity of pollution, human alteration of river and stream conditions could decrease or increase the diversity that was characterized with composite indices that incorporate both the richness and evenness elements of diversity.

More recently Stevenson *et al.* (2008a) constrained diversity to just the sensitive taxa found in reference sites. Even though the total number of species in 600 valve counts increased over the TP gradient in this set of streams, the number of low-P native taxa decreased. After taking into account concerns about underestimates of taxa numbers in samples when using 600 valve counts and effects of evenness on taxa numbers observed in 600 valve counts, it is difficult to draw firm conclusions about effects of enrichment on diatom diversity in streams. However, we know that low-nutrient streams are dominated by a few small species of *Achnanthidium* (Kawecka, 1993; Stevenson *et al.*, 2008a) and that increases in nutrient concentrations may stimulate growth rates of other taxa more than the small *Achnanthidium* (Manoylov and Stevenson, 2006). Therefore, evenness of species densities likely increases with modest levels of nutrient enrichment causing an apparent increase in species numbers in samples when based on 600 valve counts. Key unanswered questions include, "How do we define and detect species extinctions?" and "Under what conditions do we actually have local, regional, and global extinctions?"

4.3.5 Taxonomic similarity

Changes in species composition tend to be the most sensitive responses of diatoms and other microbes to environmental change (van Dam, 1982; Niederlehner and Cairns, 1994). However, the temporal scale of the observation is important. In the very short term of a bioassay, algal metabolism responds sensitively to environmental stress (Blanck 1985). Quickly, however, communities can adapt to many environmental stresses by changing species composition and, thereby, may achieve biomass and metabolic rates like those in unimpacted areas (Stevenson, 1997). Diatom assemblages in most field situations have had this time to adapt to moderate environmental stresses by changing species composition. Therefore, in most field sampling situations, when stresses have existed long enough for immigration of new species and accrual of rare taxa that are stress-tolerant, species composition should be more sensitive to changes in environmental conditions than changes in biomass or metabolic rates (e.g. Schindler, 1990).

Ordination, clustering, and community similarity indices are three approaches to assess variation in species composition among communities. Ordination (correspondence analysis, detrended correspondence analysis, non-metric multidimensional scaling) is typically used to assess the multidimensional pattern in the relationships between assemblages based on species composition. Species and sample scores are related to ordination axes and can be used to determine which species were most important in groups of samples. Environmental conditions can also be related to the ordination axes by using canonical correspondence analysis and detrended canonical correspondence analysis (ter Braak and Šmilauer, 2002). Ordination and clustering can be used to show which assemblages are the most different from other assemblages, which may be caused by anthropogenic impacts (e.g. Chessman, 1986; Stevenson and White, 1995).

Community dissimilarity or similarity indices (see reviews in Wolda, 1981 and Pielou, 1984) can be used to test specific hypotheses about correlations between changes in species composition and the environment (Cairns and Kaesler, 1969; Moore and McIntire, 1977; Peterson and Stevenson, 1989, 1992; Kelly, 2001; Passy and Bode, 2004; Stevenson *et al.* 2008a). Similarity of species composition to reference condition is one of the key metrics of biological condition recommended in the USEPA's Rapid Bioassessment Protocols (Stevenson and Bahls, 1999). For example, Stevenson *et al.* (2008a) observed thresholds in community similarity to reference condition along TP gradients. This metric provides a simple, easy-to-understand description of the percent change in species relative abundances between assemblages at assessed and reference sites. Lavoie *et al.* (2006) used the dominant axes in correspondence analysis to measure dissimilarity among communities along a human disturbance gradient, but the interpretation of this metric can be challenging for non-scientists. Cluster analysis (e.g. TWINSPAN, Hill, 1979) groups assemblages based on the similarity in species composition between assemblages (Leland, 1995). Community dissimilarity indices can also be used to distinguish among groups of assemblages by testing the hypothesis that dissimilarity among assemblages within a group is significantly less than between groups. Discriminant analysis can also be used to determine whether species composition of groups of assemblages differ significantly between clusters (ter Braak, 1986; Peterson and Stevenson, 1989). Multivariate analyses of dissimilarity among groups of assemblages is available in many modern statistical packages (e.g. McCune and Grace, 2002; Oksanen, 2004; Clarke and Gorley, 2006).

Taxonomic similarity among assemblages can also be evaluated with subsets of the data, such as percent Centrales, percent araphids, and percent *Nitzschia* Hass. Wang *et al.* (2005) tested the proportion of species and individuals in assemblages that were within one genus as potential metrics of biological condition. Many genera responded predictably to one environmental gradient or another. Percent *Cymbella (sensu lato)* individuals was an example of a taxon that has been adopted for use as a metric in some assessments (KDOW 2008). The transferability of genus level metrics from one study to another may be problematic because of the variability in species that may occur across seasons or regions. Although these metrics do not explicitly involve morphological or physiological traits, they do implicitly; they will only be transferable if the function of the diatoms in ecosystems is genetically regulated by traits that vary less within genera than between genera. Thus they are similar to the following group of indicators, the guild indicators.

4.3.6 Guild indicators

Changes in relative abundance of indicator taxa or guilds are another type of measure of biological or stressor condition based on taxonomic composition, but they are also based on taxa traits. We will refer to this type of indicator as a guild indicator because these measures are based on subsets of the all taxa having similar physiologies and functions in the ecosystem. They differ from the weighted-average style indices because the latter use all taxa in the indicator calculation. In streams and rivers, the ratio between centrics and pennates is not a commonly used indicator, but centric and pennate diatoms are examples of indicator taxa at the class level of taxonomy (Brugam and Patterson, 1983). This ratio could be a valuable indicator because these classes of diatoms have different functions in streams and rivers, as centrics are primarily adapted to planktonic habitats and most pennates are benthic. Therefore a centric to pennate ratio could indicate both biological condition and hydrologic alteration by impoundments (a stressor in the environment) (Brugam *et al.* 1998).

Many guilds are genera or groups of genera, and most genus level indicators are at least implicitly based on the indicator taxon or guild concept. For example, the number of *Nitzschia (sensu lato)* taxa and relative abundance of *Nitzschia* taxa or individuals in samples are examples of guild indicators in which *Nitzschia* is the indicator taxon as well as the guild. This is because *Nitzschia* is often assumed to be a group of taxa that tend to be tolerant to a variety of pollutants (e.g. Wang *et al.*, 2005). The percentage of *Eunotia* Ehr. in a habitat can be used to infer the pH of habitats, particularly the relative pH of two habitats when diatom assemblages are compared. The dominance of a few small *Achnanthidium* taxa in low nutrient streams, because they grow faster than other taxa in low nutrients (Manoylov and Stevenson 2006), underpins the value of two indices: the ratio between percent *Achnanthes* Bory, *Cocconeis* Ehr. plus *Cymbella* versus percent *Cyclotella* (Kütz) Bréb., *Melosira* C. A. Agardh plus *Nitzschia* (Wu, 1999; note here *Achnanthes*, *Cocconeis*, *Cyclotella*, and *Melosira* were *sensu lato*, and *Achnanthes sensu lato* and *Achnanthidium* are the same in most freshwater cases); and the ratio between percent *Achnanthidium* versus percent *Achnanthidium* plus *Navicula* Bory (Wang *et al.*, 2005).

Diatoms with the morphological capability of moving through fine sediments (e.g. *Cylindrotheca* Rab., *Gyrosigma* Hass., *Navicula*, *Nitzschia*, and *Surirella* Turp. in Wang *et al.*, 2005 and modified from Bahls, 1993) represent another guild in the community. Motile diatoms reflect an important function in habitats where fine sediments occur naturally, and therefore also reflect an element of biological condition. Motile diatoms can also be used as an indicator of excess siltation, one of the most widespread stressor conditions in streams and rivers. *Epithemia* Kütz. and *Rhopalodia* Müll. represent another guild of diatoms because they can have N-fixing blue-green endosymbionts (DeYoe *et al.*, 1992), and they respond when N occurs in relatively low supply compared to P in streams (Peterson and Grimm, 1992; Kelly, 2003); thus the percentage of these two genera in assemblages could be used as an indicator of low N concentrations or low N:P ratios. Growth forms could also be defined as a guild and tested as indicators of ecological condition (Wang *et al.*, 2005). In general, genus level metrics are not as precise as species level indicators, but they are valuable when only genus level taxonomy is available (Coste *et al.*, 1991; Prygiel and Coste, 1993; Kelly *et al.*, 1995; Hill *et al.*, 2001). In addition, guild indicators that usually use less than half the taxa in an assemblage are usually not quite as precise and transferable as weighted-average autecological indices that use all taxa. However, guild indicators can be more clearly interpreted for some applications than indicators using all taxa.

4.3.7 Weighted average autecological indices

Weighted average autecological indices are defined here as indices using traits of species. Historically, traits of species were defined categorically on an ordinal scale (e.g. Lowe 1974, van Dam *et al.* 1994, Stevenson *et al.* 2008a). Much trait information in the literature (e.g., Hustedt, 1957; Cholnoky, 1968; Slàdecek, 1973; Lowe, 1974; Descy, 1979; Lange-Bertalot, 1979; Beaver, 1981; Fabri and Leclercq, 1984; Steinberg and Schiefele, 1988; van Dam *et al.*, 1994; Gómez and Licursi, 2001; Rott *et al.*, 2003; Lobo *et al.*, 2004; and see reviews in Lowe 1974; van Dam *et al.*, 1994; Kelly and Whitton, 1995; Stevenson *et al.* 2008b) uses an ordinal scale for the species traits with six or less ranks for a specific environmental stressor (pH, nutrient requirements, temperature, salinity, organic pollution, etc.). See van Dam *et al.* (1994) for additional references. Remarkably precise autecological indices of environmental conditions can be calculated with the ordinally ranked traits using this great wealth of autecological information in the literature and a simple formula (Pantel and Buck, 1955; Zelinka and Marvan, 1961):

$$\sum p_i \Theta_i / \sum p_i$$

in which p_i is the proportional abundance of the ith taxon for which autecological information is known (p_i, for $i = 1, 2, \ldots S$ species) and Θ_i is the autecological rank of a species for a specific stressor. These autecological indices tend to work across broad geographic ranges when more accurate information is not known about diatom traits in the region studied (Kelly *et al.*, 2009a; Porter *et al.*, 2008; Lavoie *et al.*, 2009; Stevenson *et al.*, 2009).

More recently, weighted-average optima for taxa have been developed extensively for many taxa, for many environmental variables, and in many regions (Table 4.1). Optima and sometimes tolerances have been developed in many regions for nutrient concentrations (Bennion *et al.*, 1996; Pan *et al.*, 1996; Schönfelder *et al*, 2002; Soininen and Niemelä, 2002; Potapova and Charles, 2007; Ponander *et al.* 2007). The tolerance of species to environmental variability can be used to downweight species influence on stressor inference models in a slightly modified version of the Zelinka and Marvan equation:

$$\sum p_i \upsilon_i t_i / \sum p_i t_i$$

in which υ_i is the optimum environmental condition for a taxon and t_i is the tolerance of species to variation in environmental conditions (ter Braak and van Dam, 1989). Optima have also been developed for pH and conductivity gradients (Pan *et al.*, 1996; Potapova and Charles 2003). Stevenson *et al.* (2008b) developed optima for species performance along a human disturbance gradient for streams in the western USA. In this paper, they also document the importance of linking causality to indicators in regions when many environmental factors covary. Caution should be exercised when developing weighted-average models for stressors in large-scale data sets when environmental factors covary. Resulting inference models may not be robust over time and may not have a causal foundation, i.e. truly respond to the correlated physicochemical variable. Weighted-average optima have the advantage over ordinally ranked traits because they are continuous variables and may provide more precision than autecological indices that use categorical traits or guild indicators using subsets of species (Stevenson *et al.* 2008b).

Diatom-based autecological indices can be particularly valuable in stream and river assessments because one-time assay of species composition of diatom assemblages in streams could provide better characterizations of physical and chemical

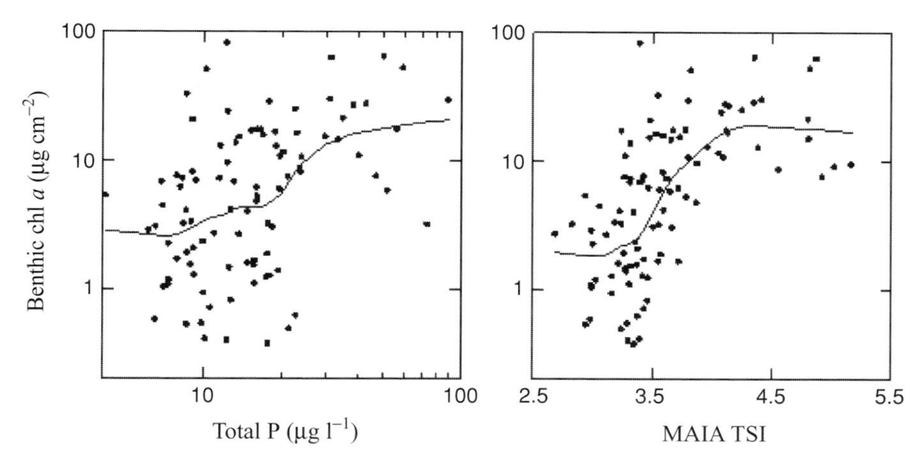

Figure 4.2 Refining stressor–response relationships with diatom indicators (Stevenson, unpublished data). The relationship between benthic algal biomass (chlorophyll *a*, $\mu g\ cm^{-2}$) and either total phosphorus concentrations or a trophic status index (Stevenson *et al.* 2006) in streams of Michigan and Kentucky, USA. Data for the same streams are plotted in both figures. The lines through the data show statistically significant linear and non-linear trends. The Mid-Atlantic Integrated Assessment Trophic Status Index (MAIA TSI) was developed in Stevenson *et al.* (2008a).

conditions than one-time measurement of those conditions. Precise characterization of environmental conditions in rivers and streams is difficult because of the high variability in discharge, water chemistry, temperature, and light availability associated with weather-related events. Charles (1985) showed that diatom-inferred pH was a better characterization of mean annual pH than one-time sampling of pH for lakes. The RMSE for a weighted-average index of TP based on diatom species composition in streams of the Mid-Atlantic Highlands (USA) was 0.32 log(TP $\mu g\ l^{-1}$) (Pan *et al.*, 1996) and was substantially less than the range in TP concentration that is commonly observed in streams. For example, TP ranged from 1.0–4.0 log(TP $\mu g\ l^{-1}$) over an eight week period in Kentucky streams (Stevenson, unpublished data). Although additional results from this study are not yet published, they provide evidence that diatom-inferred TP concentrations can more accurately assess bioavailable phosphorus than even repeated measurements of TP. Stevenson (unpublished data) shows that the conditions in which *Cladophora* accumulates across ecoregions can be better predicted with diatom-inferred TP than average TP measured over a two-month period (Figure 4.2). These results show how diatom-inference models for TP can aid refinement of stressor–response relationships, development of nutrient criteria to prevent nuisance algal blooms, and complement measured TP in use as nutrient criteria (see later discussion).

Variations on the weighted-average modeling approach have also been explored and provide some improvements in performance under specific circumstances. One of the most useful is weighted-average partial least squares (WA-PLS) technique, which builds a complex regression model that predicts environmental conditions based on species composition and regression coefficients for each species (ter Braak and Juggins, 1993). Indeed, Stevenson and Pinowska (2007) found WA-PLS to be a valuable indicator development method for nutrient models in Florida Springs, where variability in species composition was likely less than in broad-scale regional studies. In the northern Piedmont area of the eastern USA, Potapova *et al.* (2004) also document the utility of WA-PLS models for nutrients. In addition, Potapova *et al.* revisit the concept used by Zelinka and Marvan (1961) in which multiple indicator values are used per species and indicator system, such that each species is assigned a probability of being observed in a specific level of ecological condition, e.g. five successively higher TP ranges along a TP gradient (Potapova *et al.*, 2004). The authors conclude that despite the non-symmetrical distributions of many diatom taxa along the TP gradient, indicators perform just as well if they are based on the simplifying assumption that species have a symmetrical distribution along environmental gradients.

4.3.8 Sources of error

Considerable debate exists over sources of error in diatom assessments and how to reduce them. What habitats should be sampled and how many samples should be collected at a site? How many diatoms should be counted? How important is the level of taxonomic resolution and variability in taxonomy among operators/counters/technicians? The answer to these questions again depends on the objectives and budget of the project and indicator metrics.

First, the sampling substrata should be standardized if possible. Although some studies show that sampling multiple habitats and differing habitats among streams has relatively little effect on indicators and assessments of ecological conditions (Winter and Duthie, 2000; Potapova and Charles, 2005; Weilhoefer and Pan, 2007), others show or recommend that substrata sampled should be standardized as much as possible (Kelly *et al.*, 1998; Besse-Lototskaya *et al.*, 2006). Given that all streams do not have the preferred hard substrata for sampling, one reasonable approach is to separate stream types in analyses to have different *a-priori* defined reference conditions for low-gradient streams, in which there are few "hard" substrata, and higher-gradient streams with hard substrata.

Another sampling consideration is the number of samples that should be collected at a site. The answer to this issue depends upon the objectives of the study and definition of "samples." First, almost all protocols call for collecting multiple samples at a site during each visit and combining those samples into one or more composite samples that will be taken to the laboratory for separate analysis. For example, the USEPA protocols (Lazorchak *et al.*, 1998; Peck *et al.*, 2006; Stevenson *et al.*, 2008b) require sampling of substrata at randomly selected locations along each of nine transects at a site and combining those samples into a composite sample for analysis in the laboratory. It is assumed that this form of composite sampling reduces variation in assessments of a site. Taking one sample from a site is routine for determining the status of water bodies in a region when many water bodies are sampled or for development of relationships that can be used for development of water-quality criteria. If multiple composite samples were collected from a site, then only one of the composite samples could be used in statistics for regional assessments or developing relationships between ecological condition and stressors because replicate samples from an site are not statistically independent (Hurlbert, 1984). Alternatively advanced statistical methods can be used to account for lack of statistical independence for repeated measurements from a site.

However, if the objective of the study were to assess one specific site and compare it to standards, then multiple composite samples should be collected from one site to establish the central tendency and variation in assessed conditions. Kelly *et al.* (2009b) show that increased replication of samples, in this case taken at least six times over multiple years, reduces risk of misclassification of sites. They show that assessed conditions are most variable at intermediate levels of ecological condition. They recommend up to six samples at separate times for assessments of sites, with number of samples increasing if assessed conditions are close to those established as water-quality criteria. These considerations are particularly applicable to situations where individual sites are suspected of failing minimum standards, are candidates for restoration, or any situation when assessment of a specific site is needed.

Laboratory methods, as well as field-sampling methods, should be considered to minimize errors in assessments. The recommended number of valves to count in a sample ranges from 200 to tens of thousands, depending upon objectives. If an objective of the study is to determine the number and identity of most of the species in an assemblage, then 3000–8000 or even more cells may need to be counted (Patrick *et al.*, 1954). Alternatively, if a precise characterization of just the dominant taxa is necessary, then between 500 and 1000 diatoms should be counted, depending upon the number and evenness of species in the community (Pappas and Stoermer, 1996). The European Standard (CEN 2004) for the WFD has to be followed for identification and enumeration of diatoms. A typical count size is 300 to 500 valves, although lower or higher numbers may be appropriate for some purposes. Prygiel *et al.* (2002) show little effect for calculating the biological diatom index if more than 300 frustules or valves are counted. Routinely, 600 valves of diatoms are counted in the large-scale national programs in the USA (Charles *et al.*, 2002; Stevenson *et al.*, 2008b). One rationale for that count size is that 30–50 valves may need to be counted to estimate precisely the relative abundance of a taxon (Alverson *et al.*, 2003). Probably the best advice is to err on the high side of counting unless specific indices and tests of those indices have been calculated. Ultimately, we should be counting each sample until a desired level of precision in metrics is achieved. This varies with diversity and other factors, but will be easily incorporated into assessment protocols when software is developed so that analysts enter counts directly into the computer and have instantaneous updates of metric values as they count.

With species-level indicators, inter-analyst error in diatom identifications is an important source of variation in assessments (Prygiel *et al.*, 2002; Besse-Lototskya *et al.* 2006). Prygiel *et al.* (2002) found that misidentifications of small *Achnanthes* and *Cocconeis placentula* varieties were major sources of error for the Indice Biologique Diatomées (IBD) scale, and they recommend intercomparison exercises and internet exchange of materials to improve inter-analyst errors in taxonomy. Kahlert *et al.* (2009) show in a EU calibration test that the main sources of error were wrong calibration scales, overlooking small taxa (especially small *Navicula*), misidentifications (e.g. *Eunotia rhomboidea* was mistaken for *E. incisa*), and unclear separation between certain taxa in the identification literature. In addition, Kahlert *et al.* (2009) show harmonization (frequent communication about taxonomy among analysts) is even more important than many years of experience in getting reliable monitoring results. Harmonization efforts with annual meetings, frequent internet communication, and a standardized taxa list have been important elements of taxonomic consistency in diatom programs in the USA, but results of these efforts have not been evaluated. In the USA, a standardized list of taxa names, with literature citations, and a type image was established for the US Geological Survey (USGS) National Water Quality Assessment (NAWQA) program. This protocol with a standardized list and taxa references is being followed in the National Lakes Survey and National Rivers and Streams Assessment sponsored by the USEPA.

One possible solution for the problem with taxonomic consistency is to use genus-level indicators. Although such indicators have significant value (Kelly *et al.*, 1995; Hill *et al.*, 2001; Wunsam *et al.*, 2002), they do not work as reliably in different regions and across larger geographic scales as species-level indicators (Chessman *et al.*, 1999; see early discussion of genus level indicators under headings for taxonomic and guild indicators).

Regionally varying factors and cryptic species can also be sources of error in all types of diatom indices. Potapova and Charles (2002, 2007) show that refining diatom optima for different regions of the USA improves performance of diatom indicators. Stevenson *et al.* (2008b), however, show that diatom indices are most precise when using data from multiple regions; but this may be due to increasing the length of environmental gradients with multiple regions in analyses. Regional variation in population traits, biogeography of taxa, interactions among populations present, or physical and chemical determinants of diatom ecology could affect transferability of metrics from one region to another. Biogeographic patterns in species, endemic species, and cryptic species indicate global distribution of populations is not sufficiently great to homogenize populations. Therefore, species traits may vary geographically. In addition, variation in species within genera could affect genus and guild indicator performance in the same way that variation in population genetics among regions affects species-level indicators. We will discuss ways to account for regional and between-stream variation in reference condition in a later section on defining reference condition. Thus, testing indicators in the regions that they will be used is important for evaluating their precision and accuracy, i.e. "Can they detect differences?" and "Are they detecting the differences we think they are?"

4.3.9 Robust indicators

Given all these sources of error, why are diatom indicators so robust? Why can we use them in multiple regions and incorporate results of many technicians in an assessment? It is generally agreed that regional refinements of indicators can improve indicator performance, and greater taxonomic consistency among analysts will improve assessments. But diatom indicators seem to be greatly transferable in space and time given the natural, sampling, and inter-analyst sources of variability. For example, the indicator values of van Dam *et al.* (1994) have been used in ecological assessments and indicator refinements in the USA (e.g. Porter *et al.*, 2008; Lavoie *et al.*, 2009; Stevenson *et al.*, 2009). Surely there was great variability in sampling and inter-analyst taxonomy between the sources of the indicator values in van Dam *et al.* (1994) and the cited projects. Pan (personal communication) has referred to this robustness of diatom indicators to sources of error as diatom-indicator magic.

Since this property of diatom indicators has not been evaluated in detail, we decided to postulate that the large number of taxa in an assemblage, unbiased error, and the Law of Large Numbers might explain why some types of diatom indicators are so robust. The Law of Large Numbers describes the stable behavior of the sampling mean of a random variable; such that repeated sampling from the population of a random variable with a finite expected mean will produce an estimated mean that approaches and stays close to the finite expected mean, when the sample size is large. Just as with flipping a coin repeatedly, the more times you flip the coin the closer you get to having 50% heads and 50% tails. The more diatoms you count, the closer you get to a stable mean of their relative abundances (Alverson *et al.*, 2003).

Let us start with a simple example to illustrate the concept, an assemblage in which the relative abundances of all species are equal. If we repeatedly sample species traits from the sample, then our estimate of the average trait value in the assemblage increases with the number of species observed from the assemblage. In addition, errors due to assigning traits to individuals or species are probably unbiased when they are related to biogeography, sampling, and inter-analyst taxonomic errors. Thus, with increasing error due to these sources, our assessments become less precise, but probably not less accurate. In addition, longer counts with more taxa and individuals should be more robust than shorter counts.

The manifestation of these properties for diatom metrics can be observed in the analyses conducted by Kahlert *et al.* (2009) during their analysis of taxonomic inconsistency among analysts in northern Europe. Even though there was considerable variation in taxonomic identifications of some taxa among analysts, these errors seemed to be muted after calculation of the metrics. In Figures 1 and 2 of Kahlert et al. (2009), there is clear distinction between values of the Indice de Polluo-sensibilité Spécifique (Coste, 1982) and Acidity Index for Diatoms (Andrén and Jarlman, 2008) metrics among samples, despite the variation in taxonomic identifications of the analysts.

Our experience indicates that the robustness of diatom indicators seems to be greatest for weighted-average metrics. Indicators derived from cluster analyses and ordinations are more likely to be sensitive to taxonomic inconsistencies, because they do not involve an average of traits. The properties of diatom indicators should be studied more thoroughly so that we can understand how to improve their performance and which are more sensitive than others to different types of errors. Without question, we should strive to improve taxonomic consistency and accuracy of analysts, which will improve performance of all types of indicators. In addition, we should develop better traits for species as well as an understanding of the evolution of species, their adaptation to environments, and function in ecosystems. The collections of thousands of samples with complementary environmental information in assessment programs around the world provide an unprecedented opportunity for advancing these topics.

4.3.10 Morphological characteristics

Little research has been conducted to evaluate the effects of stressors on diatom size, striae density, shape, and other morphological characteristics. Sexual reproduction and auxospore formation in high-density periphyton assemblages after substantial colonization was hypothesized to be related to lower nutrient availability, in this case resulting from nutrient depletion developing during colonization and return of stream to baseflow conditions (Stevenson, 1990). Ultraviolet (UV) radiation may cause an increase in cell size and abundance of stalked diatoms (Bothwell et al., 1993). Aberrant diatom shape, such as indentations and unusual bending in frustules, has been shown to be related to heavy-metal stress and many other stressors in streams (McFarland et al., 1997). Falasco et al. (2009) provide an up-to-date review of the potential causes of different teratological forms for different species and specific stressors. Cells size, striae density, and shape of diatoms may also respond to environmental conditions. More research is justified to pursue this potentially sensitive set of metrics that could assist assessments of stressor as well as biological condition.

4.3.11 Chemical characteristics

Sediments and periphyton are important sinks for nutrients as well as many toxic inorganic and organic chemicals (Kelly and Whitton, 1989; Genter, 1996; Hoagland et al., 1996). In addition to their potential as indicators of biomass, chemical characteristics of periphyton may provide valuable indications of the environmental conditions that affect periphytic diatoms. The TN and TP of periphyton communities have been used by Biggs (1995) to infer nutrient limitation and eutrophication in habitats (see also Humphrey and Stevenson, 1992). Kelly and Whitton (1989) demonstrate the accumulation of heavy metals in periphyton. Similarly, assays of particulate chemicals of the water column may provide insight into the chemical environment of phytoplankton that is not evident from assay of dissolved chemicals alone. For example, TP is used routinely to assess trophic status of lakes (Vollenweider and Kerekes, 1981). Thus chemical characteristics of assemblages could be used to assess biological and stressor condition of assemblages, but these indicators have not been fully developed and tested at broad scales.

4.3.12 Functional characteristics

Functional characteristics of diatom-dominated assemblages, such as photosynthesis and respiration rates, nutrient uptake rate and spiralling, phosphatase activity, and growth rate, have been used as indicators of environmental conditions in streams and rivers. Phosphatase activity is a valuable indicator of P limitation (Healey and Henzel, 1979; Mulholland and Rosemond, 1992). Photosynthesis and respiration can be used as measures of community productivity and health, but these assays are not commonly used in field surveys. Hill et al. (1997) use the response of periphyton respiration rate to experimentally manipulated stressors as an indicator of those stressors in the habitat. Since assemblages can adapt to environmental stressors by changing species composition and maintaining functional ecological integrity (Stevenson, 1997), Hill et al. (1997) predict that respiration rates of assemblages will not be inhibited by exposure to that stressor. Based on regional-scale patterns in phosphatase and respiration activity varying with TP concentration and water chemistry (Hill et al., 1998; Stevenson et al. 2008a), metabolic indicators could be valuable in large-scale assessments of streams and rivers.

Growth rate has been used as an indicator of algal biomass production and can be assessed at population as well as assemblage levels (Stevenson, 1996). Schoeman (1976) and Biggs (1990) used growth rates as an indicator of nutrient limitation in habitats by resampling habitats after a short time

(3–7 days). Assessment of differing responses of species growth rates to environmental conditions may enhance the simple characterization of the autecology of species based on changes in their relative abundance. The main reason is that abundance is a function both of immigration and reproduction, as well as other processes (Stevenson, 1996). Reproduction is probably much more directly responsive to local conditions in a stream or river than immigration rates. However, assessing species growth rates in the field for assessments is impractical in most cases because of the need for multiple visits and perhaps use of artificial substrata. Assessing species growth rates in experiments and relating the data to field traits can be valuable. Manoylov and Stevenson (2006) link species responses to nutrients in experiments to the relative abundances in large-scale surveys, which helps to establish cause–effect relationships.

Thus, it is possible to measure ecosystem function in assessments of ecological condition, develop relationships between function and stressors, and thereby establish criteria for management. However, functional attributes are rarely measured in water-body assessments because the time needed for functional measures is greater than collecting materials in the field and later analysis of structural attributes in the laboratory. In addition, variability in ecological relationships is usually greater with functional attributes than many structural attributes, such as metrics based on species composition (Stevenson *et al.*, 2008a). Assessments of ecosystem function could be improved by better controlling for algal biomass. Metabolic rates are affected greatly by algal biomass (μg chl *a* cm^{-2}), whether metabolic rates are normalized based on area or biomass. Area-specific metabolic rates (e.g. mg C m^{-2} h^{-1}) increase with algal biomass, but biomass-specific metabolic rates (mg C μg^{-1} chl *a* h^{-1}) decrease with algal biomass (Stevenson, 1990; Hill and Boston, 1991). Future research should quantify the biomass effect on metabolism so that functional attributes of algal assemblages can be assessed.

Structural attributes can also be used to infer function. Numbers and biomass of functional groups of algae, invertebrates, and fish indicate the function of algae, grazing, and predation. Kelly *et al.* (2008) related shifts in diatom species composition to inferred changes in periphyton function. Trophic-state indices indicate nutrient conditions and could more specifically be related to algal biomass, productivity, and nutrient uptake. Future research should more directly address relationships between structure and function (similar to pattern and process) to confirm the often-implied assumption that water bodies with similar species composition and biomass have similar function, and to develop predictive models of ecosystem function.

4.3.13 Multimetric indices

Managing stream and river ecosystems calls for an assessment of integrity of the ecosystem and a diagnosis of causes of degradation (Figure 4.1). Indices of biotic integrity (IBI) of aquatic invertebrates and fish are being widely used to characterize streams (Karr, 1981; Hilsenhoff, 1988; Plafkin *et al.*, 1989; Lenat, 1993). These indices are often called multimetric indices of biological condition (IBC), because we are really assessing condition; integrity refers to high levels of biological condition. More recently, these multimetric indices (MMIs) have been developed for diatoms (Hill *et al.*, 2000; Fore and Grafe, 2002; Wang *et al.*, 2005; Kelly *et al.*, 2009a). The value of multimetric indices is they tend to be more precise than univariate metrics and they tend to be more linear (Fore *et al.*, 1994). In addition, they help provide a summary index and simplify communication of results. Thus, these indices are probably more valuable for assessing biological condition than non-linear univariate indicators because they respond sensitively to environmental change all along the human disturbance gradient (Figure 4.1).

Multimetric indices for diatoms can be constructed many ways, depending upon their goals. The basic steps for developing MMIs have been outlined by Plafkin *et al.* (1989) and more recently by Barbour *et al.* (1999), Wang *et al.* (2005), and Stoddard *et al.* (2008). First, the goal for the multimetric index should be established, such as an index for biological condition, stressor condition, or overall ecological condition (which would include biological and stressor conditions). A list of possible metrics should be developed based on the goal. Then the metrics should be classified for different elements of biological or stressor condition. For biological condition, examples of indicator classes are diversity, similarity to reference condition, sensitive and tolerant taxa, functional group, habitat, or growth form (Wang *et al.*, 2005; Stoddard *et al.*, 2008). For stressor condition, metrics should be considered for nutrient enrichment, siltation, acidification/alkalization, toxic substances, and hydrologic alteration. Both biological- and stressor-condition metrics would be included in an MMI for ecological condition. Metrics should then be adjusted for effects of natural variability and tested for having adequate range, reproducibility, and responsiveness (Stoddard *et al.*, 2008). Then the metrics with the highest responsiveness should be selected from each category in an iterative process that minimizes correlations with metrics from other categories. Metrics need to be rescored so they are all on the same scale, e.g. 0–1 or 0–100. Four to ten metrics are commonly used in MMIs for periphyton and diatoms (e.g. Hill *et al.*, 2000; Schaumburg *et al.*, 2004; Wang *et al.*, 2005; KDOW, 2008). Most assessment processes use the average of metrics in MMIs to evaluate the condition, but other decision processes could apply rules with joint criteria involving individual assessment of metrics with summarization in MMIs (see discussion in Wang *et al.*, 2005). An example of a rule with joint criteria is that all metrics must pass individual tests or a site fails to meet the criteria.

4.4 Assessment of biological condition and diagnosis of environmental stressors

4.4.1 Assessing ecological condition

Here we define ecological condition to include physical, chemical, and biological condition, and ecological integrity to be a high level of ecological condition. As emphasized previously, with diatoms we can measure biological condition of diatoms and infer chemical and physical condition, e.g. pH, nutrient concentrations, conductivity, and temperature. Ecological condition can be assessed as the absolute value of an attribute or as a deviation from an expected condition. For example, we could describe the number of pollution-sensitive species in a habitat, or we could describe the ratio between the observed and expected number of pollution-sensitive species in a habitat. Simple use of the number without comparison to an expected condition limits interpretation of the meaning. For example, when expected condition is a reference condition, then deviation from the reference condition can more clearly describe the effects of human activities. When the expected conditions are indicator values used as ecological status boundaries or water-quality criteria in the EU and USA, respectively, then they are triggers for management actions such as development and implementation of restoration plans.

4.4.1.1 Reference conditions and assessment Minimally disturbed conditions have been defined using two basic methods, frequency distributions and predictive models. Assuming that we have an indicator that is highest in reference conditions (e.g. number of pollution-sensitive species), the 25th percentile of a frequency distribution of all sites or the 75th percentile of a frequency distribution of reference sites has been used to define the lower boundary of reference conditions (Figure 4.3). If the indicator (e.g. percent of all individuals that are in pollution-tolerant species or diatom-inferred total phosphorus concentration) is lowest in reference conditions, the 25th percentile of a frequency distribution of all sites or the 75th percentile of a frequency distribution of reference sites has been used

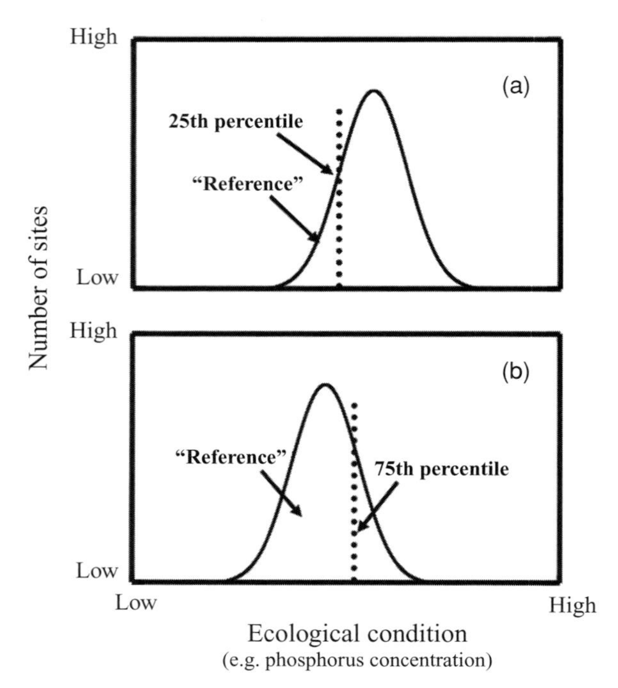

Figure 4.3 The frequency-distribution approach for establishing environmental criteria (Barbour *et al.*, 1999). The number of sites with different levels of an ecological attribute are plotted. In this case, the ecological attribute is a negative attribute, such as algal biomass or a pollutant; therefore sites with low levels of the ecological attribute characterize reference condition. Often, quartiles of frequency distributions are used to establish management targets. (A) The lower quartiles of frequency distributions that include all sites are assumed to provide reasonable restoration targets in a region in which substantial alteration of the landscape has occurred, but these quartiles may be overly protective in regions in which little alteration of the landscape has occurred. The upper quartile of frequency distributions using reference sites allows for some variation in measurement and actual expected value of expected condition; it is assumed to provide a balance of type I and II statistical errors for protecting the mean or median of reference conditions.

to define the upper-boundary of reference conditions. Several problems have been noted using the frequency-distribution method for establishing criteria for reference condition, such as arbitrary selection of percentiles and lack of established relationships between measures of ecological condition and human activities. Predictive models and non-linear relationships help resolve the latter problems.

Predictive models are used to define reference condition by relating measures of ecological condition to indicators of human alterations of watersheds that produce contamination or habitat alterations (Figure 4.4). For example, the minimally disturbed reference condition of TP can be predicted by relating TP concentrations to the percentage of agricultural and urban land use in watersheds (Dodds and Oakes 2004; Stevenson *et al.*,

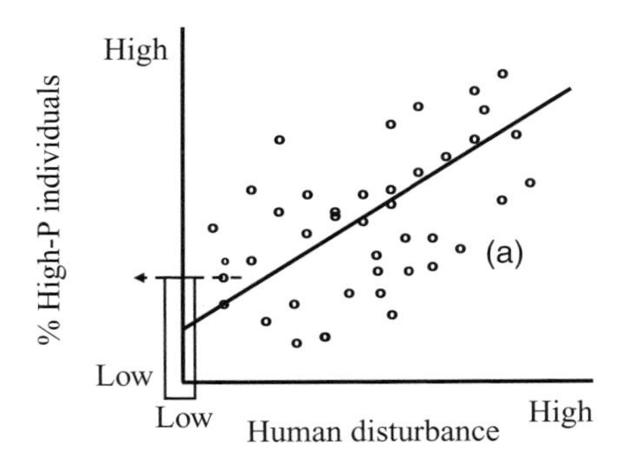

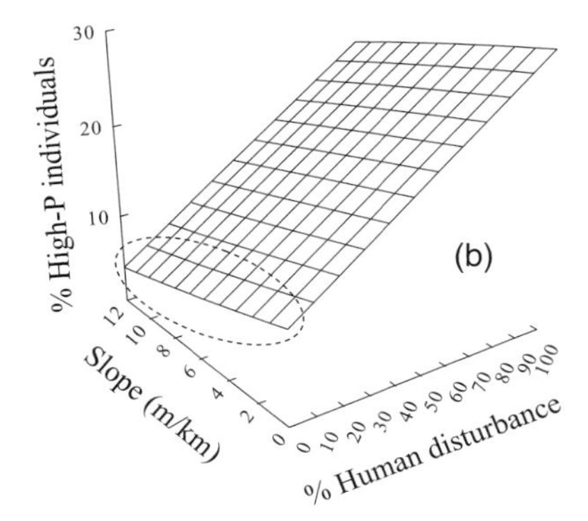

Figure 4.4 The predictive model approach (a & b) for characterizing reference condition. Using relationships between responses (e.g. percent high-P individuals) and indicators of human activities (e.g. percentages of urban and agricultural land use in watersheds) to determine natural or best attainable conditions. (a) Prediction of natural condition (*y*-intercept) using the regression relationship between biological condition and human activities assuming little natural variation in ecological condition. The arrow indicates the upper 75th percentile of the estimate of the *y*-intercept; i.e. when human disturbance is equal to zero. (b) Predictive model of natural condition (indicated by dashed oval) using a regression relationship between biological condition and human activities assuming considerable variation due to other natural factors (e.g. stream slope). Natural condition is calculated for a site with a model, measures of natural factors at assessed sites, and setting indicators of human activities to zero or an acceptably low value.

2008a). Similarly, the biologically inferred abiotic condition of minimally disturbed reference conditions can be determined by relating diatom indicators to the percentage of agricultural and urban land use in watersheds and accounting for effects of natural variation among streams on expected condition (Stevenson *et al.*, 2009). The strength of this predictive modeling approach is that it enables modeling of natural, minimally disturbed, or even best attainable conditions without having to have a large number of reference sites, and it can account for natural variability among sites.

Reference condition can be explicitly or implicitly incorporated into indicators of ecological condition. In the EU, the WFD calls for determination of the ecological quality ratio (EQR), which is the ratio between the observed status and expected status, where expected status is a minimally disturbed reference condition. Therefore the EQR explicitly measures deviation between the observed and expected condition. The EQR varies between 0 and 1 for low and high ecological quality, respectively. States in the USA commonly use indicators or multimetric indices of biological condition in characterizations of biological condition without explicitly incorporating the reference condition in the calculation; therefore interpreting the magnitude of deviation from reference condition requires an explicit statement of reference condition or only relative condition can be interpreted by comparison to other sites.

4.4.1.2 Ecological criteria and assessment There is an important distinction between reference condition and the management goals for waters, because the natural or minimally disturbed condition is not often a practical management goal in many landscapes, such as watersheds with extensive alteration by farming and urban activities. In the US CWA, biological integrity is the ultimate goal for US waters, but "the protection and propagation of fish, shellfish, and wildlife" is considered an interim goal, even though it falls short of biological integrity in most definitions used by states. This interim goal recognizes the need for a practical, but acceptable goal for protecting waters in the USA. The EU WFD calls for all surface waters to have "good ecological status," which is defined as "having biota consistent with only slight alterations from that expected in that absence of human impacts" (Kelly *et al.*, 2008). Thus, in many cases, we need more than one level of ecological condition for setting appropriate management goals for the diversity of our waters (Davies and Jackson, 2006).

As a result, multiple management goals are needed for a more flexible and arguably more protective approach that enables different goals for different water bodies. In the USA, this approach has been referred to as tiered aquatic life "uses" (Davies and Jackson, 2006), and in the EU this is manifested in the different ecological-status categories (high, good, moderate, poor, and bad) and acceptability of the two higher goals (e.g. Kelly *et al.*, 2008). The lower bounds of indicator values for each category of ecological condition become the targets for restoration and protection. These lower bounds are water-quality criteria in the USA, and they have similar meaning in the EU. Here we will refer to them as ecological criteria.

Several approaches have been used to establish ecological criteria. When there is just one acceptable use and associated criterion, then the frequency-distribution approach is often satisfactory. When there is more than one level, governments have chosen simply to divide the range of conditions into a pre-specified number of categories. However, a more scientifically defensible approach is to relate loss of ecological condition to stressors and human activities, or what has been referred to as the pressure gradient. The challenge with this approach is to decide what specific levels of ecological condition should be chosen as criteria (or the boundaries of the ecological-status categories), especially when responses are linear.

Non-linear relationships between valued ecological attributes and stressors or measures of human activities are particularly valuable for establishing criteria for reference conditions, especially if they demonstrate thresholds (Muradian, 2001; Stevenson *et al.*, 2004a, 2008a; Figure 4.1). Thresholds help justify where to establish both stressor and biological criteria. With this approach, we first identify an indicator of a valued ecological attribute that non-linearly responds to increasing stressor levels, and then set a stressor criterion at a level of stressor that provides a margin of safety for protection of the valued attribute. Then we identify an indicator of valued attributes that responds linearly to increasing stressor levels, because it provides a sensitive and consistent response to changing stressor levels over the entire range of stressor conditions. Thus, a linear indicator provides an early warning of risk to a valued attribute that has a threshold response as stressor levels increase. Finally, the biological criterion for delineating reference condition can be established at a level that corresponds to the stressor criterion. If assessed biological condition decreases below the biological criterion, then the risk of unacceptable degradation of valued attributes is too high.

Other analytical approaches could be used to establish biological criteria or boundaries for defining ecological status. Kelly *et al.* (2008) relate the relative abundance of nutrient-sensitive and nutrient-tolerant taxa to an EQR, based on a trophic diatom index. They use the point where nutrient tolerant-species become more abundant than nutrient-sensitive species as a benchmark for the boundary between moderate and good ecological status. Kelly *et al.* argue that the benchmark for good status represents a point below which there is a shift in periphyton functioning; therefore it represents an objective and defensible criterion. Variations on this approach could evaluate metric thresholds along MMI-defined gradients, in which the MMI is a proxy for the human-disturbance gradient. The values of MMIs where thresholds in various metrics occur could be used as benchmarks for criteria for the tiered (successively higher) ecological-status categories (Figure 4.1).

The above methods for assessing ecological condition provide a set of diatom indicators that could be used complementarily in environmental assessments. Diatom indicators of valued ecological attributes with non-linear responses to stressors can be used to establish stressor criteria and ecological-status boundaries. Diatom indicators of biological condition with linear responses can be used to assess this condition with higher precision than indicators with non-linear responses. And finally, diatom indicators inferring stressor conditions could be used to complement measurement of stressors, to provide more precise and potentially accurate assessments of stressor conditions than simply measuring stressor conditions (e.g. water chemistry) on one sampling day (Stevenson 2006, unpublished data). Because of the high spatial and temporal variability in nutrient concentrations in streams and rivers, diatom indicators inferring nutrients could be very valuable as nutrient criteria in a multimetric index that includes actual measurements of nutrient concentrations.

4.4.2 Refining assessments by accounting for natural variability

Due to the great natural variability in ecological conditions of minimally disturbed streams and rivers, refining definitions of reference condition can increase precision in assessments (Figure 4.5). For example, streams in two geologically different settings could naturally have different pH and conductivity levels and therefore, naturally different diatom species occurring in them. Accounting for natural differences among ecosystems provides a more precise definition of reference condition, and consequently a more precise assessment of deviation from reference condition (Hawkins *et al.* 2000a). This concept can be integrated into all three assessment approaches described above.

Over the last 10 years, many conceptual and statistical approaches have been developed and evaluated for refining definitions of reference condition and incorporating them into assessments. Some of the earliest approaches for refining definitions of reference condition were *a-priori* separations of sites into groups by regions and stream size. Thus, before data were collected and analyzed, we assumed *a priori* that one or more naturally occurring ecological factors could be used to group sites. Regional variation in determinants of stream condition underpin the ecoregion approach, in which we assume that regional variation in geologic, climatic, and biogeographic factors explain significant amounts of variation in causal pathways and that they regulate diatom species composition (Biggs, 1995; Stevenson, 1997; Biggs *et al.*, 1998; Soininen, 2007). The

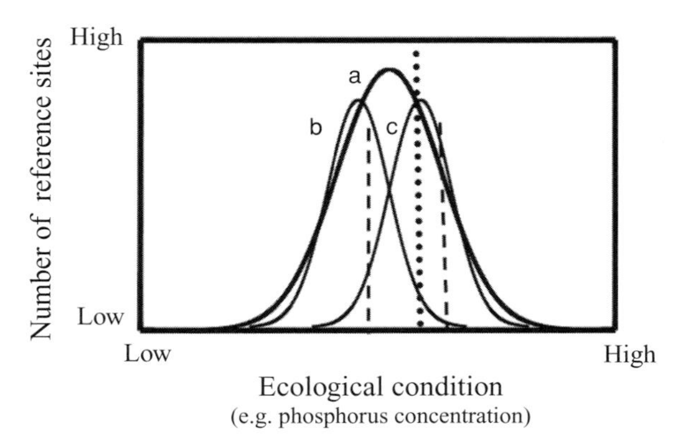

Figure 4.5 Refined frequency-distribution approach for defining reference condition. Separate distributions are developed for two or more classes of sites (e.g. b and c), where classes are based on natural variation in ecological condition or regional differences in management goals that may be based on the extent of human activities in watersheds. The vertical lines indicate the 75th percentiles of reference condition, with the longer dotted line for the aggregated distribution a and shorter left and right dashed lines for distributions b and c, respectively. Different criteria are established for each class of sites. The refined definition of reference condition based on natural factors provides more appropriate protection for different classes of sites, such that a higher level of protection is possible for one class of sites without being overly protective for the other class of sites.

ecoregion approach for refining the definition of reference condition has been used extensively in the USA for stream assessments. The approach has been criticized because relatively little variation in diatom species composition is explained by ecoregions (Pan *et al.* 2000) compared to the many variables which can vary naturally within ecoregions, such as stream pH, size, slope, substratum, watershed geology, and wetlands proportions of watersheds (Lavoie *et al.*, 2006; Cao *et al.*, 2007). However, Fore and Grafe (2002) successfully aggregated ecoregions to develop different metric expectations in different stream typologies based on altitude and regional land use.

More recently, refined characterizations of reference condition have been determined by using so-called *post-priori* approaches. These approaches are called *post priori* because after data are collected to develop indicators, they are analyzed to develop models for refining definitions of reference condition. After these indicators and models are developed, they are applied in assessments. The analytical methods vary greatly among all these approaches, which present a challenge to group and describe. One way to group methods is by establishing whether the methods call for separating reference sites from all sites in developing the model, which can limit usefulness when the numbers of sites in regions are small.

Kelly *et al.* (2008) developed site-specific expectations for indicator values by using regression analysis to determine effects of natural factors on Trophic Diatom Index (TDI) values across the United Kingdom. In this case, stream alkalinity was assumed to be independent of human activities in watersheds and the alkalinity explained the greatest variation in TDI among sites. The biological condition at a site can then be determined by comparing the observed TDI with the the expected TDI based on measured alkalinity at a site and the alkalinity model for TDI (Kelly *et al.*, 2008). Stevenson *et al.* (2009) used multiple regression to predict natural values of three trophic indices in Michigan streams, in which both total land use (percent agriculture and urban land use) and either wetland or geological conditions in watersheds were important. Natural factors explained almost as much variation in the trophic indices as total land use. Again, site-specific assessments were calculated as the difference between the observed and predicted trophic index values at a site, based on the wetland and geological conditions in watersheds and with land use set to zero. Cao *et al.* (2007) also developed an approach using site-specific predictions of the reference condition, but they used classification and regression tree (CART) and only reference sites to identify natural factors regulating metric values and to develop the predictive model of metrics in the reference condition.

As with the Cao *et al.* (2007) example, we can refine our expectations for species composition and metric values at a site by developing a better understanding of the natural factors that explain variation in diatom species composition using only reference sites. These models are like those used in the River Invertebrate Prediction and Classification System (RIVPACS, *sensu* Moss *et al.* 1987; Hawkins *et al.* 2000b) and they use cluster analysis to identify groups of sites with similar species composition; then discriminant function analysis is used to identify distinguishing natural factors among the groups of sites and to build models for assigning sites with a predicted condition (Chessman *et al.*, 1999; Cao *et al.*, 2007; Carlisle *et al.*, 2008). The RIVPACS models usually assess the observed number of species typical of reference sites compared to the expected number at a site with similar typology. The value of RIVPACS models in the assessment of the proportion of reference taxa at an assessed site, which is a direct measure of the biodiversity, a valued element of biological condition and it could be applied consistently in all assessments (Hawkins *et al.*, 2000b); however the hump-shaped relationships between species numbers in counts and environmental gradients can cause problems with this metric (Stevenson *et al.*, 2008a). A similar approach is being developed using self-organizing maps to group reference sites with species composition; then discriminant function analysis and similar methods (such as multilayer perception) can be used to predict what an assemblage should be in the absence of anthropogenic impacts (Tison *et al.*, 2007). In a very simple version of this approach, Grenier *et al.* (2006) found that the natural conditions of sites in the St. Lawrence basin of Quebec, Canada, were best explained by differences in pH; then Lavoie *et al.* (2006) developed separate metric expectations in streams with low and high pH.

4.4.3 Stressors condition and diagnosis

Many diatom indicators that have been developed in the last century really infer the stressor conditions in a habitat, rather than assess the biological condition (Stevenson 2006). Thus, diatom indicators of stressor condition can be used to complement actual measures of stressor condition. This has been particularly important in paleoecology and such efforts as inferring pH, trophic status, and climate change in lakes for which diatoms are especially valuable proxies of past conditions that can not be measured directly (Charles *et al.* 1990, Dixit *et al.* 1999, Smol *et al.* 2005, and several chapters in this volume). Diatom indicators are also valuable in streams and rivers because of the great temporal variability in some variables, such as nutrient concentrations. Stevenson (2006) summarized results of unpublished data showing how diatom indicators of TP concentration can be more precise and accurate indicators of TP availability than measured TP concentrations (Figure 4.2).

Diatom indicators of nutrient concentrations, dissolved oxygen, organic matter, pH, conductivity, and sediments have been developed for streams (Table 4.1). The diatom indicators for nutrients, dissolved oxygen, and pH should be particularly useful because these factors vary so much on diurnal and weekly timescales due to weather and diurnal variation in metabolic processes. However, diatom indicators of stressor conditions should be rigorously tested to confirm their causal linkage to the stressor that they are designed to indicate. Stevenson *et al.* (2008b) illustrated this problem showing the challenge of identifying causal versus correlational relationships for diatom indicators of stressors in large regional surveys. Despite reasonable levels of precision in indicators of pH, conductivity, fine sediments, and embeddedness, these diatom indicators were actually more precisely related to percent disturbance in watersheds than the stressors for which the weighted-average models were developed. Covariation among environmental factors can be a significant problem for indicator development in large regional projects. Subdividing large data sets may help reduce covariation and the development of more causally related indicators.

Diatom indicators of stressors can be used in stressor diagnosis in a number of ways. First, diatom indicators of stressors could be used to refine stressor–response relationships, such as responses of algal biomass to TP (Stevenson 2006). By refining stressor–response relationships, they can help resolve threshold relationships and establish expected conditions or criteria for stressors (Figure 4.2). Second, they could be used for the development of criteria for expected stressor conditions. In addition to helping establish the specific level of a stressor that should be designated as the water-quality criterion, the inferred stressor condition by the diatom indicator, as well as the analytically measured stressor condition, could be used as water-quality criteria. For example, we could use a diatom indicator of TP concentration as well as measured TP concentrations to determine whether TP criteria were being violated or not. Using multiple lines of evidence, or multiple indicators, usually increases the precision and accuracy of assessments. Finally, we need to assess stressor conditions in the habitat for stressor diagnosis. Use of the diatom-inferred stressor condition alone or in combination as a multimetric index with an actual measured condition could improve the accuracy and precision of stressor assessments.

Stressor diagnosis is critical for both protection and restoration of ecosystems (Stevenson *et al.*, 2004b). Beyers (1998) amended and refined the postulates of Hill (1965) to list a set of criteria that should be considered when trying to infer the cause of an un-replicated environmental impact. These criteria are: extent of alterations (strength), observation by other investigators (consistency), unique effects specific to the stressor (specificity), exposure to the stressor prior to the observed effect (temporality), the relationship between the magnitude of the stressor and the effect (stressor–response relationship), plausibility of causal linkage, experimental evidence, transferability of stressor/effects elsewhere, causal hypothesis consistent with existing knowledge, and exposure sufficient for the effect. Thus, proper identification of the contaminant and habitat alterations that are causing problems can be facilitated with diatom indicators of stressors. First, if stressor levels are greater than stressor criteria, then those stressors should be targeted for remediation. Second, ratios and differences between stressor levels and stressor criteria can be used to rank the importance of different stressors. Ratios between stressor levels and criteria follow the concept of toxic units in the toxicological literature (Stevenson *et al.*, 2004b). Differences between stressor levels and criteria have been related to the sustainability and restorability of habitats (Stevenson, 1998). Third, and more recently, the probability of losing valued attributes has been related to stressor levels in a risk-based approach (Paul and McDonald, 2005; Rollins and Stevenson, unpublished manuscript). Thus, diatom indicators of stressors can be used in stressor diagnosis in the same way as actual measurements of the physical, chemical, and biological alterations of habitats by humans.

4.5 Concluding remarks

Developing approaches and indices for environmental assessment is an interactive process between scientists and policy makers. Interactions should focus on furthering scientists' understanding of policy issues and environmental problems and on helping policy makers translate their goals into testable hypotheses and practical approaches for environmental assessment and problem solving. Priorities of some policy makers have been directed toward understanding relationships between land use, physical and chemical changes in streams and rivers, and ecological responses. Land-use planning and zoning are important strategies for slowing environmental degradation in areas under development pressure from urban and suburban sprawl. Assessing watershed-scale changes in stream and river conditions could be valuable for inferring the land-use effects and the geological and climatic factors that make watersheds sensitive or tolerant to land-use changes (Robinson *et al.*, 1995; Richards *et al.*, 1996; Kutka and Richards, 1996).

Ecological theory should become a more important foundation for the environmental assessments and the indices used in environmental assessments. For example, Ruth Patrick's pioneering work on species diversity as an indicator of water quality was well founded in the ecological theory that was being explored at the time (Patrick *et al.*, 1954; Patrick and Strawbridge, 1963). By placing research in an ecological context and testing a broader ecological theory, such as Odum's predictions for stressed ecosystems (Odum *et al.*, 1979; Odum, 1985), as well as specific diatom-based hypotheses, results of our research become more transferable to assessments with other organisms and to assessments of other habitats with diatoms.

Diatoms are valuable indicators of biological condition and the environmental factors that impair rivers and streams. Public concern often focuses on the biodiversity of other organisms, but partly because they do not appreciate the diversity, beauty, and ecology of algae, particularly diatoms. Greater efforts should be made to inform the public and develop their appreciation for diatoms in aquatic ecosystems so they know what valued attributes are at risk. In addition, development of diatom indicators of ecosystem function and services provides a direct linkage to valued ecological attributes. Relating diatom assessments to questions being asked by stakeholders or to questions that they should be asking is essential for sustainable diatom monitoring programs. Great investments are being made in diatom assessment programs, on the scale of millions of dollars per year. They require our accountability and creativity to meet the needs of the stakeholders.

4.6 Summary

Diatoms have a long history of use in assessing the ecological integrity of streams. Diatom assemblages respond rapidly and sensitively to environmental change and provide highly informative assessments of the biotic integrity of streams and rivers and causes of ecosystem impairment. Periphytic diatoms from natural and artificial substrata are usually sampled from streams and small rivers, but plankton provides valuable assessments of conditions in large rivers. Structural and functional characteristics of diatom communities can be used in bioassessments, but relative abundances of diatom genera and species are usually used as the most valuable characteristics of diatom assemblages for bioassessment. Using these characteristics, multimetric indices of biotic and ecological condition have been developed that enable use of diatom assemblages in risk assessment and management of stream and river ecosystems.

Exciting new approaches for indicator refinements and applications provide opportunities for research at the cutting edge of applied ecological science. With near-natural condition as a management target for many governments, the refining of indicators to account for natural variability and the differing response to stressors in different types of streams and rivers present challenges for future study. The refining of indicators is also dependent upon sound analytical technique, including better knowledge of diatom taxonomy and communication of that taxonomy. Understanding threshold responses in diatom assemblages along environmental gradients is valuable for justifying specific management targets. Thus, a more complete understanding of the taxonomy and ecology of diatoms remains at the foundation of sound science for environmental assessments and advancements in that science.

Acknowledgments

We thank John Smol and Eugene Stoermer for providing the opportunity to write this paper and present our concepts of using diatoms to assess rivers and streams. We thank our colleagues and students for the many rigorous and earnest debates about algal ecology and ecological assessment. In particular, we thank Martyn Kelly for his insightful contributions while reviewing this chapter. Research enabling this synthesis was supported by grants and contracts from the US Environmental Protection Agency. We dedicate this chapter to Dr. Charles Reimer for his inspiration and teaching.

References

Alatalo, R. V. (1981). Problems in the measurement of evenness in ecology. *Oikos*, **37**, 199–204.

Aloi, J. E. (1990). A critical review of recent freshwater periphyton methods. *Canadian Journal of Fisheries and Aquatic Sciences*, **47**, 656–70.

Alverson, A. J., Manoylov, K. M., and Stevenson, R. J. (2003). Laboratory sources of error for algal community attributes during sample preparation and counting. *Journal of Applied Phycology*, **15**, 357–69.

American Public Health Association (APHA) (1998). *Standard Methods for the Evaluation of Water and Wastewater*, 20th edition, American Public Health Association, Washington, DC.

Amoros, C. and Van Urk, G. (1989). Palaeoecological analyses of large rivers: some principles and methods. In *Historical Change of Large Alluvial Rivers: Western Europe*, ed. G. E. Petts, H. Möller and A. L. Roux, Chichester: John Wiley & Sons, pp. 143–65.

Andrén, C. & Jarlman, A. (2008). Benthic diatoms as indicators of acidity in streams. *Fundamental and Applied Limnology*, **173**, 237–53.

Archibald, R. E. M. (1972). Diversity in some South African diatom assemblages and its relation to water quality. *Water Research*, **6**, 1229–38.

Asai, K. (1996). Statistical classification of epilithic diatom species into three ecological groups relating to organic water pollution. (1) Method with coexistence index. *Diatom*, **10**, 13–34.

Bahls, L. L. (1993). Periphyton Bioassessment Methods for Montana Streams. Water Quality Bureau, Department of Health and Environmental Sciences, Helena, MT.

Barbour, M. T., Gerritsen, J., Snyder, B. D., and Stribling, J. B. (1999). Revision to Rapid Bioassessment Protocols for Use in Streams and Rivers: Periphyton, Benthic Macroinvertebrates, and Fish. EPA 841-D-97–002, Washington, DC: United States Environmental Protection Agency.

Beaver, J. (1981). Apparent Ecological Characteristics of Some Common Freshwater Diatoms. Rexdale, Ontario: Ontario Ministry of the Environment.

Bennion, H., Juggins, S., and Anderson, N. J. (1996). Predicting epilimnetic phosphorus concentrations using an improved diatom-based transfer function and its application to lake eutrophication management. *Environmental Science & Technology*, **30**, 2004–7.

Besse-Lototskaya, A., Verdonschot, P. F. M., and Sinkeldam, J. A. (2006). Uncertainty in diatom assessment: sampling, identification and counting variation. *Hydrobiologia*, **566**, 247–60.

Beyers, D. W. (1998). Causal inference in environmental impact studies. *Journal of the North American Benthological Society*, **17**, 367–73.

Biggs, B. J. F. (1990). Use of relative specific growth rates of periphytic diatoms to assess enrichment of a stream. *New Zealand Journal of Marine and Freshwater Research*, **24**, 9–18.

Biggs, B. J. F. (1995). The contribution of flood disturbance, catchment geology and land use to the habitat template of periphyton in stream ecosystems. *Freshwater Biology*, **33**, 419–38.

Biggs, B. J. F. (1996). Patterns of benthic algae in streams. In *Algal Ecology: Freshwater Benthic Ecosystems*, ed. R. J. Stevenson, M. Bothwell, and R. L. Lowe, San Diego, CA: Academic Press, pp. 31–55.

Biggs, B. J. F. (2000). Eutrophication of streams and rivers: dissolved nutrient- chlorophyll relationships for benthic algae. *Journal of the North American Benthological Society*, **19**, 17–31.

Biggs, B. J. F., Stevenson, R. J. & Lowe, R. L. (1998). A habitat matrix conceptual model for stream periphyton. *Archiv für Hydrobiologie*, **143**, 21–56.

Blanck, H. (1985). A simple, community level, ecotoxicological test system using samples of periphyton. *Hydrobiologia*, **124**, 251–61.

Bothwell, M. L. (1989). Phosphorus-limited growth dynamics of lotic periphytic diatom communities: areal biomass and cellular growth rate responses. *Canadian Journal of Fisheries and Aquatic Sciences*, **46**, 1293–301.

Bothwell, M. L., Sherbot, D., Roberge, A. C., and Daley, R. J. (1993). Influence of natural ultraviolet radiation on lotic periphytic diatom community growth, biomass accrual, and species composition: short-term versus long-term effects. *Journal of Phycology*, **29**, 24–35.

Brugam, R., McKeever, K., and Kolesa, L. (1998). A diatom-inferred water depth reconstruction for Upper Peninsula, Michigan lake. *Journal of Paleolimnology*, **20**, 267–76.

Brugam, R. & Patterson, C. (1983). The A/C (Araphidineae/Centrales) ratio in high and low alkalinity lakes in eastern Minnesota. *Freshwater Biology*, **13**, 47–55.

Butcher, R. W. (1947). Studies in the ecology of rivers. IV. The algae of organically enriched water. *Journal of Ecology*, **35**, 186–91.

Cairns, J. Jr. and Kaesler, R. L. (1969). Cluster analysis of Potomac River survey stations based on protozoan presence–absence data. *Hydrobiologia*, **34**, 414–32.

Cao, Y., Hawkins, C. P., Olson, J., and Kosterman, M. A. (2007). Modeling natural environmental gradients improves the accuracy and precision of diatom-based indicators. *Journal of North American Benthological Society*, **26**, 566–84.

Carlisle, D. M., Hawkins, C. P., Meador, M. R., Potapova, M., and Falcone, J. (2008). Biological assessments of Appalachian streams based on predictive models for fish, macroinvertebrate, and diatom assemblages. *Journal of North American Benthological Society*, **27**, 16–37.

Cattaneo, A. & Amireault, M. C. (1992). How artificial are artificial substrata for periphyton? *Journal of the North American Benthological Society*, **11**, 244–56.

CEN (Comité Européen de Normalisation) (2003). Water quality – guidance standard for the routine sampling and pretreatment of benthic diatoms from rivers. European Standard EN 13946, Brussels.

CEN (2004). Water quality – guidance standard for the identification, enumeration and interpretation of benthic diatom samples from running waters. European Standard EN 14407.

Charles, D. F. (1985). Relationships between surface sediment diatom assemblages and lake-water characteristics in Adirondack lakes. *Ecology*, **66**, 994–1011.

Charles, D. F., Binford, M. W., Furlong, E. T., *et al.* (1990). Paleoecological investigation of recent lake acidification in the Adirondack Mountains, N. Y. *Journal of Paleolimnology*, **3**, 195–241.

Charles, D. F., Knowles, C., and Davis, R.S. (eds.) (2002). Protocols for the analysis of algal samples collected as part of the U.S. Geological Survey National Water-Quality Assessment Program. Report No. 02–06, Patrick Center for Environmental Research, The Academy of Natural Sciences, Philadelphia, PA. See http://diatom.acnatsci.org/nawqa/pdfs/ProtocolPublication.pdf

Chessman, B. C. (1986). Diatom flora of an Australian river system: spatial patterns and environmental relationships. *Freshwater Biology*, **16**, 805–19.

Chessman, B., Growns, I., Currey, J., and Plunkett-Cole, N. (1999). Predicting diatom communities at the genus level for the rapid biological assessment of rivers. *Freshwater Biology*, **41**, 317–31.

Chessman, B. C., Bate, N., Gell, P. A., and Newall, P. (2007). A diatom species index for bioassessment of Australian rivers. *Marine and Freshwater Research*, **58**, 542–57.

Cholnoky, B. J. (1968). *Ökologie der Diatomeen in Binnengewässern*. Lehre: Cramer.

Clarke, K.R. & Gorley, R.N. (2006). PRIMER v6: Users Manual/Tutorial. PRIMER-E, Plymouth.

Coste, M. (1982). *Etude des méthodes biologiques d'appréciation quantitative de la qualité des eaux*. Lyon: CEMAGREF Division Qualité des Eaux, Agence de L'eau Rhône-Méditerrané Corse.

Coste, M., Bosca, C., and Dauta, A. (1991). Use of algae for monitoring rivers in France. In *Use of Algae for Monitoring Rivers*, ed. B. A. Whitton, E. Rott, and G. Friedrich, Innsbruck: Universität Innsbruck, pp. 75–88.

Cremer, H., Buijse, A., Lotter, A., Oosterberg, W., and Staras, M. (2004). The palaeolimnological potential of diatom assemblages in floodplain lakes of the Danube Delta, Romania: a pilot study. *Hydrobiologia*, **513**, 7–26.

Crumpton, W. G. (1987). A simple and reliable method for making permanent mounts of phytoplankton for light and fluorescence microscopy. *Limnology and Oceanography*, **32**, 1154–9.

Davies, S. P. & Jackson, S. K. (2006). The biological condition gradient: a descriptive model for interpreting change in aquatic ecosystems. *Ecological Applications*, **16**, 1251–66.

Descy, J. P. (1979). A new approach to water quality estimation using diatoms. *Nova Hedwigia*, **64**, 305–23.

Descy, J. P. & Mouvet, C. (1984). Impact of the Tihange nuclear power plant on the periphyton and the phytoplankton of the Meuse River (Belgium). *Hydrobiologia*, **119**, 119–28.

DeYoe, H. R., Lowe, R. L., and Marks, J. C. (1992). The effect of nitrogen and phosphorus on the endosymbiot load of *Rhopalodia gibba* and *Epithemia turgida* (Bacillariophyceae). *Journal of Phycology*, **23**, 773–7.

Dixit, S. S., Smol, J. P., Charles, D. F., *et al.* (1999). Assessing water quality changes in the lakes of the northeastern United States using sediment diatoms. *Canadian Journal of Fisheries and Aquatic Sciences*, **56**, 131–52.

Dodds, W. K. & Oakes, R. M. (2004). A technique for establishing reference nutrient concentrations across watersheds affected by humans. *Limnology and Oceanography: Methods*, **2**, 331–41.

Dodds, W. K., Smith, V. H., and Zander, B. (1997). Developing nutrient targets to control benthic chlorophyll levels in streams: a case study of the Clark Fork River. *Water Research*, **31**, 1738–50.

Dufrêne, M. & Legendre, P. (1997). Species assemblages and indicator species: the need for a flexible asymmetrical approach. *Ecological Monographs*, **67**, 345–66.

Ector, L., Kingston, J. C., Charles, D. F., *et al.* (2004). Workshop report: freshwater diatoms and their role as ecological indicators. In *Proceedings of the Seventeenth International Diatom Symposium 2002*, ed. M. Poulin, Bristol: Biopress Ltd, pp. 469–80.

European Union (EU) (2000). Directive 2000/60/EC of the European Parliament and of the Council of 23 October 2000 establishing a framework for Community action in the field of water policy. The European Parliament and the Council of the European Union. *Official Journal of the European Communities*, **L 327/1**, 1–72.

Fabri, R. & Leclercq, L. (1984). *Etude écologique des riviéres du nord du massif Ardennais (Belgique): flore et végétation de diatomées et physico-chimie des eaux*. 1. Robertville: Station scientifique des Hautes Fagnes.

Falasco, E., Bona, F., Badino, G., Hoffmann, L., and Ector, L. (2009). Diatom teratological forms and environmental alterations: a review. *Hydrobiologia*, **623**, 1–35

Fjerdingstad, E. (1950). The microflora of the River Molleaa with special reference to the relation of benthic algae to pollution. *Folia Limnologica Scandanavica*, **5**, 1–123.

Fore, L. & Grafe, C. (2002). Using diatoms to assess the biological condition of large rivers in Idaho (U.S.A.). *Freshwater Biology*, **47**, 2015–37.

Fore, L. S., Karr, J. R., and Conquest, L. L. (1994). Statistical properties of an index of biotic integrity used to evaluate water resources. *Canadian Journal of Fisheries and Aquatic Sciences*, **51**, 1077–87.

Gale, W. F., Gurzynski, A. J., and Lowe, R. L. (1979). Colonization and standing crops of epilithic algae in the Susquehanna River, Pennsylvania. *Journal of Phycology*, **15**, 117–23.

Gell, P., Bulpin, S., Wallbrink, P., Hancock, G., and Bickford, S. (2005). Tareena Billagong – a palaeolimnological history of an ever-changing wetland, Chowilla Floodplain, lower Murray-Darling Basin, Australia. *Marine and Freshwater Research*, **56**, 441–56.

Genter, R. B. (1996). Ecotoxicology of inorganic chemical stress to algae. In *Algal Ecology: Freshwater Benthic Ecosystems*. ed. R. J. Stevenson, M. Bothwell and R. L. Lowe, San Diego, CA: Academic Press, pp. 403–68.

Gillett, N., Pan, Y. & Parker, C. (2009). Should only live diatoms be used in the bioassessment of small mountain streams? *Hydrobiologia*, **620**, 135–47.

Gómez, N. & Licursi, M. (2001). The Pampean Diatom Index (DPI) for assessment of rivers and streams in Argentina. *Aquatic Ecology*, **35**, 173–81.

Grenier, M., Campeau, S., Lavoie, I., Park, Y.-S. & Lek, S. (2006). Diatom reference communities in Québec (Canada) streams based on Kohonen self-organized maps and multivariate analyses. *Canadian Journal of Fisheries and Aquatic Sciences*, **63**, 2087–106.

Griffith, M. B., Hill, B. H., McCormick, F. H., *et al.* (2005). Comparative application of indices of biotic integrity based on periphyton, macroinvertebrates, and fish to southern Rocky Mountain streams. *Ecological Indicators*, **5**, 117–36.

Hawkins, C. P., Norris, R. H., Gerritsen, J., *et al.* (2000a). Evaluation of the use of landscape classifications for the prediction of freshwater biota: synthesis and recommendations. *Journal of the North American Benthological Society*, **19**, 541–56.

Hawkins, C. P., Norris, R. H., Hogue, J. N. & Feminella, J. W. (2000b). Development and evaluation of predictive models for measuring the biological integrity of streams. *Ecological Applications*, **10**(5), 1456–77.

Healey, F. P. and Henzel, L. L. (1979). Fluorometric measurement of alkaline phosphatase activity in algae. *Freshwater Biology*, **9**, 429–39.

Hering, D., Johnson, R. K., Kramm, S., *et al.* (2006). Assessment of European streams with diatoms, macrophytes, macroinvertebrates and fish: a comparative metric-based analysis of organism response due to stress. *Freshwater Biology*, **51**, 1757–85.

Hill, A. B. (1965). The environment and disease: association or causation? *Proceedings of the Royal Society of Medicine*, **58**, 295–300.

Hill, B. H., Herlihy, A. T., Kaufmann, P. R. & Sinsabaugh, R. L. (1998). Sediment microbial respiration in a synoptic survey of mid-Atlantic region streams. *Freshwater Biology*, **39**, 493–501.

Hill, B., Herlihy, A., Kaufmann, P., *et al.* (2000). Use of periphyton assemblage data as an index of biotic integrity. *Journal of North American Benthological Society*, **19**, 50–67.

Hill, B. H., Lazorchak, J. M., McCormick, F. H., and Willingham, W. T. (1997). The effects of elevated metals on benthic community metabolism in a Rocky Mountain stream. *Environmental Pollution*, **95**, 183–90.

Hill, B. H., Stevenson, R. J., Pan, Y., *et al.* (2001). Correlations of stream diatoms with their environment: a comparison of genus-level and species-level identifications. *Journal of the North American Benthological Society*, **20**, 299–310.

Hill, M. O. (1979). TWINSPAN-A FORTRAN Program for Detrended Correspondence Analysis and Reciprocal Averaging. Cornell University, Ithaca, NY.

Hill, W. R. & Boston, H. L. (1991). Community development alters photosynthesis-irradiance relations in stream periphyton. *Limnology and Oceanography*, **36**, 1375–89.

Hilsenhoff, W. L. (1988). Rapid field assessment of organic pollution with a family level biotic index. *Journal of the North American Benthological Society*, **7**, 65–8.

Hoagland, K. D., Carder, J. P. & Spawn, R. L. (1996). Effects of organic toxic substances. In *Algal Ecology: Freshwater Benthic Ecosystems*, ed. R. J. Stevenson, M. Bothwell and R. L. Lowe, San Diego, CA, Academic Press, pp. 469–97.

Humphrey, K. P. & Stevenson, R. J. (1992). Responses of benthic algae to pulses in current and nutrients during simulations of subscouring spates. *Journal of the North American Benthological Society*, **11**, 37–48.

Hurlbert, S. H. (1971). The nonconcept of species diversity: a critique and alternative parameters. *Ecology*, **52**, 577–86.

Hurlbert, S. J. (1984). Pseudoreplication and design of ecological field experiments. *Ecological Monographs*, **54**, 187–211.

Hustedt, F. (1957). Die Diatomeenflora des Flusssystems der Weser im Gebiet der Hansestadt Bremen. *Bremen: Abhandlungen Naturwissenschaftlichen Verein*, **34**(3), 181–440.

Johnson, R. K., Hering, D., Furse, M. T., and Clarke, R. T. (2006). Detection of ecological change using multiple organism groups: metrics and uncertainty. *Hydrobiologia*, **566**, 115–37.

Jüttner, I., Rothfritz, H., and Omerod, S. J. (1996). Diatoms as indicators of river water quality in the Nepalese Middle Hills with consideration of the effects of habitat-specific sampling. *Freshwater Biology*, **36**, 475–86.

Kahlert, M., Albert, R.-L., Anttila, E.-L., *et al.* (2009). Harmonization is more important than experience – results of the first Nordic–Baltic diatom intercalibration exercise 2007 (stream monitoring). *Journal of Applied Phycology*, **21**, 471–82.

Karr, J. R. (1981). Assessment of biotic integrity using fish communities. *Fisheries*, **6**, 21–7.

Karr, J. R. & Chu, E. W. (1997). Biological Assessment: Using Multimetric Indexes Effectively. United States Environmental Protection Agency, Washington, D. C.

Karr, J. R. & Dudley, D. R. (1981). Ecological perspective on water quality goals. *Environmental Management*, **5**, 55–68.

Kawecka, B. (1993). Ecological characteristics of sessile algal communities in streams flowing from the Tatra Mountains in the area of Zakopane (southern Poland) with special consideration of their requirements with regard to nutrients. *Acta Hydrobiologica*, **35**, 295–306.

KDOW (Kentucky Division of Water) (2008). Standard Methods for Assessing Biological Integrity of Surface Waters in Kentucky. Lexington, KY: Kentucky Division of Water.

Kelly, M., Bennett, C., Coste, M., *et al.* (2009a). A comparison of national approaches to setting ecological status boundaries in phytobenthos assessment for the European Water Framework Directive: results of an intercalibration exercise. *Hydrobiologia*, **621**, 169–82.

Kelly, M., Bennion, H. Burgess, A., *et al.* (2009b). Uncertainty in ecological status assessments of lakes and rivers using diatoms. *Hydrobiologia*, **633**, 5–15.

Kelly, M., Juggins, S., Guthrie, R., *et al.* (2008). Assessment of ecological status in U.K. rivers using diatoms. *Freshwater Biology*, **53**, 403–22.

Kelly, M. & Whitton, B. A. (1995). The trophic diatom index: a new index for monitoring eutrophication in rivers. *Journal of Applied Ecology*, **7**, 433–44.

Kelly, M. G. (1998). Use of the trophic diatom index to monitor eutrophication in rivers. *Water Research*, **32**, 236–42.

Kelly, M. G. (2001). Use of similarity measures for quality control of benthic diatom samples. *Water Research*, **35**, 2784–88.

Kelly, M. G. (2003). Short term dynamics of diatoms in an upland stream and implications for monitoring eutrophication. *Environmental Pollution*, **125**, 117–22.

Kelly, M. G., Cazaubon, A., Coring, E., *et al.* (1998). Recommendations for the routine sampling of diatoms for water quality assessments in Europe. *Journal of Applied Phycology*, **10**, 215–24.

Kelly, M. G., Penny, C. J., and Whitton, B. A. (1995). Comparative performance of benthic diatom indices used to assess river water quality. *Hydrobiologia*, **302**, 179–88.

Kelly, M. G. & Whitton, B. A. (1989). Interspecific differences in Zn, Cd and Pb accumulation by freshwater algae and bryophytes. *Hydrobiologia*, **175**, 1–11.

Kim, B. K., Jackman, A. P., and Triska, R. J. (1990). Modeling transient storage and nitrate uptake kinetics in a flume containing a natural periphyton community. *Water Resources Research*, **26**, 505–15.

Kolkwitz, R. & Marsson, M. (1908). Ökologie der pflanzliche Saprobien. *Berichte der Deutsche Botanische Gesellschaften*, **26**, 505–19.

Kovács, C. M., Kahlert, M., and Padisák, J. (2006). Benthic diatom communities along pH and TP gradients in Hungarian and Swedish streams. *Journal of Applied Phycology*, **18**, 105–17.

Kutka, F. J. & Richards, C. (1996). Relating diatom assemblage structure to stream habitat. *Journal of the North American Benthological Society*, **15**, 469–80.

Lamberti, G. A. (1996). The role of periphyton in benthic food webs. In *Algal Ecology: Freshwater Benthic Ecosystems*, ed. R. J. Stevenson, M. Bothwell and R. L. Lowe, San Diego, CA: Academic Press, pp. 533–72.

Lange-Bertalot, H. (1979). Pollution tolerance of diatoms as a criterion for water quality estimation. *Nova Hedwigia* **64**, 285–304.

Lavoie, I., Campeau, S., Grenier, M., and Dillon, P. J. (2006). A diatom-based index for the biological assessment of eastern Canadian rivers: an application of correspondence analysis (CA). *Canadian Journal of Fisheries and Aquatic Sciences*, **8**, 1793–811.

Lavoie, I., Hamilton, P. B., Wang, Y. K., Dillon, P. J., and Campeau, S. (2009). A comparison of stream bioassessment in Québec (Canada) using six European and North American diatom-based indices. *Nova Hedwigia*, **135**, 37–56.

Lazorchak, J. M., Klemm, D. J., and Peck, D. V. (1998). Environmental Monitoring and Assessment Program – Surface Waters: field operations and methods for measuring the ecological condition of wadeable streams. EPA-620-R-94–004F, Washington, DC: United States Environmental Protection Agency.

Leland, H. V. (1995). Distribution of phytobenthos in the Yakima River basin, Washington, in relation to geology, land use and other environmental factors. *Canadian Journal of Fisheries and Aquatic Sciences*, **52**, 1108–29.

Lenat, D. R. (1993). A biotic index for the southeastern United States: derivation and list of tolerance values, with criteria for assigning water quality ratings. *Journal of the North American Benthological Society*, **12**, 279–90.

Lobo, E. A., Callegaro, V. L. M., Hermany, G., Gómez, N., and Ector, L. (2004). Review of the use of microalgae in South America for monitoring rivers, with special reference to diatoms. *Vie Milieu*, **54**, 105–14.

Lowe, R. L. (1974). Environmental Requirements and Pollution Tolerance of Freshwater Diatoms. US Environmental Protection Agency, EPA-670/4–74-005, Cincinnati, OH.

Lowe, R. L. & Pan, Y. (1996). Benthic algal communities and biological monitors. In *Algal Ecology: Freshwater Benthic Ecosystems*, ed. R. J. Stevenson, M. Bothwell and R. L. Lowe, San Diego, CA: Academic Press, pp. 705–39.

Manoylov, K. M. & Stevenson, R. J. (2006). Density-dependent algal growth along N and P nutrient gradients in artificial streams. In *Advances in Phycological Studies*, ed. N. Ognjanova-Rumenova and K. Manoylov, Moscow: Pensoft Publishers, pp. 333–52.

Mayer, M. S. and Likens, G. E. (1987). The importance of algae in a shaded headwater stream as food for an abundant caddisfly (Trichoptera). *Journal of the North American Benthological Society*, **6**, 262–9.

McCormick, P. V. & Stevenson, R. J. (1989). Effects of snail grazing on benthic algal community structure in different nutrient environments. *Journal of the North American Benthological Society*, **82**, 162–72.

McCune, B., and Grace, J. B. (2002). Analysis of ecological communities. MjM software design, Gleneden Beach, OR.

McFarland, B. H., Hill, B. H., and Willingham, W. T. (1997). Abnormal *Fragilaria* spp. (Bacillariophyceae) in streams impacted by mine drainage. *Journal of Freshwater Ecology*, **12**, 141–9.

Moore, W. W. and McIntire, C. D. (1977). Spatial and seasonal distribution of littoral diatoms in Yaquina Estuary, Oregon (U.S.A.). *Botanica Marina*, **20**, 99–109.

Moss, D., Furse, M. T., Wright, J. F., and Armitage, P. D. (1987). The prediction of the macro-invertebrate fauna of unpolluted running-water sites in Great Britain using environmental data. *Freshwater Biology*, **17**, 41–52.

Moulton, S. R., Kennen, J. G., Goldstein, R. M., and Hambrook, J. A. (2002). Revised protocols for sampling algal, invertebrate, and fish communities as part of the National Water-Quality Assessment Program. Open-file Report 02–150, United States Geological Survey, Reston, VA.

Mulholland, P. J. (1996). Role of nutrient cycling in streams. In *Algal Ecology: Freshwater Benthic Ecosystems*, ed. R. J. Stevenson, M. Bothwell, and R. L. Lowe, San Diego, CA: Academic Press, pp. 609–39.

Mulholland, P. J. & Rosemond, A. D. (1992). Periphyton response to longitudinal nutrient depletion in a woodland stream: evidence of upstream-downstream linkage. *Journal of the North American Benthological Society*, **11**, 405–19.

Müller-Haeckel, A. and Håkansson, H. 1978. The diatom-flora of a small stream near Abisko (Swedish Lapland) and its annual periodicity, judged by drift and colonization. *Archiv für Hydrobiologie*, **84**, 199–217.

Muradian, R. (2001). Ecological thresholds: a survey. *Ecological Economics*, **38**, 7–24.

Newbold, J. D., Elwood, J. W., O'Neill, R. V., and Van Winkle, W. (1981). Measuring nutrient spiralling in streams. *Canadian Journal of Fisheries and Aquatic Sciences*, **38**, 680–3.

Niederlehner, B. R. & Cairns, J. C., Jr. (1994). Consistency and sensitivity of community level endpoints n microcosm tests. *Journal of Aquatic Ecosystem Health*, **3**, 93–9.

Norris, R. H. and Norris, K. R. (1995). The need for biological assessment of water quality: Australian perspective. *Australian Journal of Ecology*, **20**, 1–6.

Odum, E. P. (1985). Trends expected in stressed ecosystems. *BioScience*, **35**, 412–22.

Odum, E. P., Finn, J. T., and Franz, E. H. (1979). Perturbation theory and the subsidy-stress gradient. *BioScience*, **29**, 349–52.

Oksanen, J. (2004). 'Vegan' Community Ecology Package: ordination methods and other functions for community and vegetation ecologists. University of Oulu, Oulu, Finland.

Palmer, C. M. (1962). *Algae in Water Supplies.*, Washington, DC: US Department of Health, Education and Welfare.

Palmer, C. M. (1969). A composite rating of algae tolerating organic pollution. *Journal of Phycology*, **5**, 78–82.

Pan, Y. D., Stevenson, R. J., Hill, B. H., Herlihy, A. T., and Collins, G. B. (1996). Using diatoms as indicators of ecological conditions in lotic systems: a regional assessment. *Journal of the North American Benthological Society*, **15**, 481–95.

Pan, Y. D., Stevenson, R. J., Hill, B. H., and Herlihy, A. T. (2000). Ecoregions and benthic diatom assemblages in Mid-Atlantic Highlands streams, USA. *Journal of the North American Benthological Society*, **19**, 518–40.

Pantle, R. & Buck, H. (1955). Die biologische Überwachung der Gewässer und die Darstellung der Ergebnisse. *Gas- und Wasserfach*, **96**, 604.

Pappas, J. L. & Stoermer, E. F. (1996). Quantitative methods for determining a representative algal count. *Journal of Phycology*, **32**, 693–6.

Passy, S. I. & Bode, R. W. (2004). Diatom model affinity (DMA), a new index for water quality assessment. *Hydrobiologia*, **524**, 241–51.

Patrick, R. (1949). A proposed biological measure of stream conditions based on a survey of the Conestoga Basin, Lancaster County, Pennsylvania. *Proceedings of the Academy of Natural Sciences of Philadelphia*, **101**, 277–341.

Patrick, R. (1961). A study of the numbers and kinds of species found in rivers of the Eastern United States. *Proceedings of the Academy of Natural Sciences of Philadelphia*, **113**, 215–58.

Patrick, R. (1973). Use of algae, especially diatoms, in the assessment of water quality. In *Biological Methods for the Assessment of Water Quality*, ASTM STP 528, Philadelphia, PA: American Society for Testing and Materials, pp. 76–95.

Patrick, R., Hohn, M. H., and Wallace, J. H. (1954). A new method for determining the pattern of the diatom flora. *Notulae Naturae*, No. 259.

Patrick, R. and Strawbridge, D. (1963). Variation in the structure of natural diatom communities. *The American Naturalist*, **97**, 51–7.

Paul, J. F. & McDonald, M. E. (2005). Development of empirical, geographically specific water quality criteria: a conditional probability analysis Approach. *Journal of the American Water Resources Association*, **41**, 1211–23.

Peck, D. V., Averill, D. K., Herlihy, A. T., *et al.* (2006). Environmental Monitoring and Assessment Program – Surface Waters Western Pilot Study: field operations manual for non-wadeable rivers and streams. EPA 620/R-06/003. Washington, DC: Office of Research and Development, US Environmental Protection Agency.

Peterson, C. G. and Grimm, N. B. (1992). Temporal variation in enrichment effects during periphyton succession in a nitrogen-limited desert stream ecosystem. *Journal of the North American Benthological Society*, **11**, 20–36.

Peterson, C. G. & Stevenson, R. J. (1989). Seasonality in river phytoplankton: multivariate analyses of data from the Ohio River and six Kentucky tributaries. *Hydrobiologia*, **182**, 99–114.

Peterson, C. G. and Stevenson, R. J. (1992). Resistance and recovery of lotic algal communities: importance of disturbance timing, disturbance history, and current. *Ecology*, **73**, 1445–61.

Pielou, E. C. (1984). *The Interpretation of Ecological Data, A Primer on Classification and Ordination*. New York, NY: John Wiley & Sons.

Plafkin, J. L., Barbour, M. T., Porter, K. D., Gross, S. K., and Hughes, R. M. (1989). Rapid bioassessment protocols for use in streams and rivers: benthic macroinvertebrates and fish. EPA/444/4–89-001, Washington, DC: US EPA Office of Water.

Ponander, K. C., Charles, D. F., and Belton, T. J. (2007). Diatom based TP and TN inference models and indices for monitoring nutrient enrichment of New Jersey streams. *Ecological Indicators*, **7**, 79–93.

Porter, S. D., Mueller, D. K., Spahr, N. E., Munn, M. D., & Dubrovsky, N. M. (2008). Efficacy of algal metrics for assessing nutrient and organic enrichment in flowing waters. *Freshwater Biology*, **53**, 1036–54.

Potapova, M. G. & Charles, D. F. (2002). Benthic diatoms in USA rivers: distributions along spatial and environmental gradients. *Journal of Biogeography*, **29**, 167–87.

Potapova, M. G. & Charles, D. F. (2003). Distribution of benthic diatoms in U.S. rivers in relation to conductivity and ionic composition. *Freshwater Biology*, **48**, 1311–28.

Potapova, M. G. & Charles, D. F. (2005). Choice of substrate in algae-based water-quality assessment. *Journal of North American Benthological Society*, **24**, 415–27.

Potapova, M. G. & Charles, D. F. (2007). Diatom metrics for monitoring eutrophication in rivers of the United States. *Ecological Indicators*, **7**, 48–70.

Potapova, M. G., Charles, D. F., Ponader, K. C., and Winter, D. M. (2004). Quantifying species indicator values for trophic diatom indices: a comparison of approaches. *Hydrobiologia*, **517**, 25–41.

Prygiel, J. and Coste, M. (1993). The assessment of water quality in the Artois-Picardie water basin (France) by the use of diatom indices. *Hydrobiologia*, **269/270**, 343–9.

Prygiel, J., Carpentier, P., Almeida, S., *et al.* (2002). Determination of the Biological Diatom Index (IBD NF T 90–354). Results of an intercomparison excercise. *Journal of Applied Phycology*, **14**, 27–39.

Richards, C., Johnson, L. B., and Host, G. E. (1996). Landscape-scale influences on stream habitats and biota. *Canadian Journal of Fisheries and Aquatic Sciences*, **53** (Supplement 1), 295–311.

Rier, S. T. & Stevenson, R. J. (2006). Response of periphytic algae to gradients in nitrogen and phosphorus in streamside mesocosms. *Hydrobiologia*, **561**, 131–47.

Robinson, C. T., Rushforth, S. R., and Liepa, R. A. (1995). Relationship of land use to diatom assemblages of small streams in Latvia. In *A Century of Diatom Research in North America: A Tribute to the Distinguished Careers of Charles W. Reimer and Ruth Patrick*, ed. J. P. Kociolek and M. J. Sullivan, Champaign, IL: Koeltz Scientific Books, pp. 47–59.

Rosati, T. C., Johansen, J. R., and Coburn, M. M. (2003). Cyprinid fishes as samplers of benthic diatom communities in freshwater streams of varying water quality. *Canadian Journal of Fisheries and Aquatic Sciences*, **60**, 117–25.

Rosen, B. H. (1995). Use of periphyton in the development of biocriteria. In *Biological Assessment and Criteria: Tools for Water Resource Planning and Decision Making*, ed. W. S. Davis and T. P. Simon, Boca Raton, FL: Lewis Publishers, pp. 209–15.

Rosen, B. H. & Lowe, R. L. (1984). Physiological and ultrastructural responses of *Cyclotella meneghiniana* (Bacillariophyta) to light intensity and nutrient limitation. *Journal of Phycology*, **20**, 173–93.

Rott, E. (1991). Methodological aspects and perspectives in the use of periphyton for monitoring and protecting rivers. In *Use of Algae for Monitoring Rivers*, ed. B. A. Whitton, E. Rott and G. Friedrich, Innsbruck: Universität Innsbruck, pp. 9–16.

Rott, E. and Pfister, P. (1988). Natural epilithic algal communities in fast-flowing mountain streams and rivers and some man-induced changes. *Verhandlungen Internationale Vereinigung für Theoretische und angewandte Limnologie*, **23**, 1320–4.

Rott, E., Pipp, E., and Pfister, P. (2003). Diatom methods developed for river quality assessment in Austria and a cross-check against numerical trophic indication methods used in Europe. *Algological Studies*, **110**, 91–115.

Round, F. E. (1991). Diatoms in river water-monitoring studies. *Journal of Applied Phycology*, **3**, 129–45.

Rumeau, A., and Coste, M. (1988). Initiation à la systématique des diatomées d'eau douce pour l'utilisation pratique d'un indice diatomique générique. *Bulletin Francais de la Pêche et de la Pisciculture*, **309**, 1–69.

Schaumburg, J., Schranz, C., Hofmann, G., *et al.* (2004). Macrophytes and phytobenthos as indicators of ecological status in German lakes – a contribution to the implementation of the Water Framework Directive. *Limnologica*, **34**, 302–14.

Schindler, D. W. (1990). Experimental perturbations of whole lakes as tests of hypotheses concerning ecosystem structure and function. *Oikos*, **57**, 25–41.

Schoeman, F. R. (1976). Diatom indicator groups in the assessment of water quality in the Jukskei-Crocodile River System (Transvaal, Republic of South Africa). *Journal of the Limnological Society of South Africa*, **2**, 21–4.

Schönfelder, I., Gelbrecht, J., Schönfelder, J. & Steinberg, C.E.W. (2002). Relationships between littoral diatoms and their chemical environment in northeastern German lakes and rivers. *Journal of Phycology*, **38**, 66–82.

Shannon, C. F. (1948). A mathematical theory of communication. *Bell Systems Technical Journal*, **27**, 37–42.

Sicko-Goad, L., Stoermer, E. F., and Ladewski, B. G. (1977). A morphometric method for correcting phytoplankton cell volume estimates. *Protoplasma*, **93**, 147–63.

Simpson, E. H. (1949). Measurement of diversity. *Nature*, **163**, 688.

Slàdecek, V. (1973). System of water quality from the biological point of view. *Archiv für Hydrobiologie und Ergebnisse Limnologie*, **7**, 1–218.

Smol, J. P., Wolfe, A. P., Birks, H. J. B., *et al.* (2005). Climate-driven regime shifts in the biological communities of arctic lakes. *Proceedings of the National Academy of Sciences of the USA*, **102**, 4397–402.

Soininen, J. (2007). Environmental and spatial control of freshwater diatoms – a review. *Diatom Research*, **22**, 473–90.

Soininen, J. & Niemelä, P. (2002). Inferring the phosphorus levels of rivers from benthic diatoms using weighted averaging. *Archiv für Hydrobiologie*, **154**, 1–18.

Sonneman, J. A., Walsh, C. J., Breen, P. F., and Sharpe, A. K. (2001). Effects of urbanization on streams of the Melbourn region, Victoria, Australia. II. Benthic diatom communities. *Freshwater Biology*, **46**, 553–65.

Squires, L. E., Rushforth, S. R., and Brotherson, J. D. (1979). Algal response to a thermal effluent: study of a power station on the Provo River, Utah, USA. *Hydrobiologia*, **63**, 17–32.

Steinberg, C. and Schiefele, S. (1988). Indication of trophy and pollution in running waters. *Zeitschrift für Wasser-Abwasser Forschung*, **21**, 227–34.

Steinman, A. D., McIntire, C. D., Gregory, S. V., Lamberti, G. V., and Ashkenas, L. (1987). Effect of herbivore type and density on taxonomic structure and physiognomy of algal assemblages in laboratory streams. *Canadian Journal of Fisheries and Aquatic Sciences* **44**, 1640–8.

Stevenson, R. J. (1983). Effects of current and conditions simulating autogenically changing microhabitats on benthic algal immigration. *Ecology*, **64**, 1514–24.

Stevenson, R. J. (1984a). Procedures for mounting algae in a syrup medium. *Transactions of the American Microscopical Society*, **103**, 320–1.

Stevenson, R. J. (1984b). Epilithic and epipelic diatoms in the Sandusky River, with emphasis on species diversity and water quality. *Hydrobiologia*, **114**, 161–75.

Stevenson, R. J. (1990). Benthic algal community dynamics in a stream during and after a spate. *Journal of the North American Benthological Society*, **9**, 277–88.

Stevenson, R. J. (1996). An introduction to algal ecology in freshwater benthic habitats. In *Algal Ecology: Freshwater Benthic Ecosystems*, ed. R. J. Stevenson, M. Bothwell, and R. L. Lowe, San Diego, CA: Academic Press, pp. 3–30.

Stevenson, R. J. (1997). Scale-dependent causal frameworks and the consequences of benthic algal heterogeneity. *Journal of the North American Benthological Society*, **16**, 248–62.

Stevenson, R. J. (1998). Diatom indicators of stream and wetland stressors in a risk management framework. *Environmental Monitoring and Assessment*, **51**, 107–18.

Stevenson, R. J. (2006). Refining diatom indicators for valued ecological attributes and development of water quality criteria. In *Advances in Phycological Studies*, ed. N. Ognjanova-Rumenova and K. Manoylov, Moscow: Pensoft Publishers, pp. 365–83.

Stevenson, R. J. & Bahls, L. L. (1999). Periphyton protocols. In *Rapid Bioassessment Protocols for Use in Wadeable Streams and Rivers: Periphyton, Benthic Macroinvertebrates, and Fish*, 2nd edn., ed. M. T. Barbour, J. Gerritsen, and B. D. Snyder, EPA 841-B-99-002. United States Environmental Protection Agency, Washington, DC, pp. 6–1 to 6–22.

Stevenson, R. J., Bailey, B. C., Harass, M. C., *et al.* (2004a). Designing data collection for ecological assessments. In *Ecological Assessment of Aquatic Resources: Linking Science to Decision-Making*, ed. M. T. Barbour, S. B. Norton, H. R. Preston, and K. W. Thornton, Pensacola, FL: Society of Environmental Toxicology and Chemistry, pp. 55–84.

Stevenson, R. J., Bailey, B. C., Harass, M. C., *et al.* (2004b). Interpreting results of ecological assessments. In *Ecological Assessment of Aquatic Resources: Linking Science to Decision-Making*, ed. M. T. Barbour, S. B. Norton, H. R. Preston, and K. W. Thornton, Pensacola, FL: Society of Environmental Toxicology and Chemistry, pp. 85–111.

Stevenson, R. J., Bothwell, M., and Lowe, R. L. (1996). *Algal Ecology: Freshwater Benthic Ecosystems*. San Diego, CA: Academic Press.

Stevenson, R. J., Hill, B. E., Herlihy, A. T., Yuan, L. L., and Norton, S. B. (2008a). Algae–P relationships, thresholds, and frequency distributions guide nutrient criterion development. *Journal of the North American Benthological Society*, **27**, 259–75.

Stevenson, R. J. & Lowe, R. L. (1986). Sampling and interpretation of algal patterns for water quality assessment. In *Rationale for Sampling and Interpretation of Ecological Data in the Assessment of Freshwater Ecosystems*, ASTM STP 894, American Society for Testing and Materials Publication, Philadelphia, PA, pp. 118–49.

Stevenson, R. J. & Pan, Y. (1999). Assessing environmental conditions in rivers and streams with diatoms. In *The Diatoms: Applications for the Environmental and Earth Sciences*, ed. E. F. Stoermer and J. P. Smol, Cambridge: Cambridge University Press, pp. 11–40.

Stevenson, R. J., Pan, Y., Manoylov, K., *et al.* (2008b). Development of diatom indicators of ecological conditions for streams of the western United States. *Journal of the North American Benthological Society*, **27**, 1000–16.

Stevenson, R. J. and Peterson, C. G. (1991). Emigration and immigration can be important determinants of benthic diatom assemblages in streams. *Freshwater Biology*, **6**, 295–306.

Stevenson, R. J. & Pinowska, A. (2007). Diatom indicators of ecological conditions in Florida springs. Report to Florida Department of Environmental Protection, Tallahassee, FL.

Stevenson, R. J., Rier, S. T., Riseng, C. M., Schultz, R. E., and Wiley, M. J. (2006). Comparing effects of nutrients on algal biomass in streams in 2 regions with different disturbance regimes and with applications for developing nutrient criteria. *Hydrobiologia*, **561**, 149–65.

Stevenson, R. J. & Smol, J. P. (2002). Use of algae in environmental assessments. In *Freshwater Algae in North America: Classification and Ecology*, ed. J. D. Wehr and R. G. Sheath, San Diego, CA: Academic Press, pp. 775–804

Stevenson, R. J. and Stoermer, E. F. (1981). Quantitative differences between benthic algal communities along a depth gradient in Lake Michigan. *Journal of Phycology*, **17**, 29–36.

Stevenson, R. J. and White, K. D. (1995). A comparison of natural and human determinants of phytoplankton communities in the Kentucky River basin, USA. *Hydrobiologia*, **297**, 201–16.

Stevenson, R. J., Novoveska, L., Riseng, C. M., and Wiley, M. J. (2009). Comparing responses of diatom species composition to natural and anthropogenic factors in streams of glaciated ecoregions. *Nova Hedwigia*, **135**, 1–13.

Stoddard, J. L., Herlihy, A. T., Peck, D. V., *et al.* (2008). A process for creating multimetric indices for large-scale aquatic surveys. *Journal of the North American Benthological Society*, **27**, 878–91.

Stoddard, J. L., Larsen, D. P., Hawkins, C. P., Johnson, R. K., and Norris, R. H. (2006). Setting expectations for the ecological condition of streams: the concept of reference condition. *Ecological Applications*, **16**, 1267–76.

Swanson, C. D. and Bachmann, R. W. (1976). A model of algal exports in some Iowa streams. *Ecology*, **57**, 1076–80.

Taylor, J. C., Prygiel, J,. Vosloo, A., Rey, P. A., d. l., and Rensburg, L., v. (2007). Can diatom-based pollution indices be used for biomonitoring in South Africa? A case study of the Crocodile West and Marico water management area. *Hydrobiologia*, **592**, 455–64.

ter Braak, C. J. F. (1986). Interpreting a hierarchical classification with simple discriminate functions: an ecological example. In *Data Analysis and Informatics*, ed. E. Diday, Amsterdam: North Holland, pp. 11–21.

ter Braak, C. J. F. & Šmilauer, P. (2002). CANOCO Reference manual and CanDraw for Windows User's Guide. Software for Canonical Community Ordination (version 4.5). Biometris, Wageningen and České Budějovice.

ter Braak, C. J. F. & Juggins, S. (1993). Weighted averaging partial least squares regression (WA-PLS): an improved method for reconstructing environmental variables from species assemblages. *Hydrobiologia*, **269/270**, 485–502.

ter Braak, C. J. F. & van Dam, H. (1989). Inferring pH from diatoms: a comparison of old and new calibration methods. *Hydrobiologia* **178**, 209–23.

Tett, P., Gallegos, C., Kelly, M. G., Hornberger, G. M., and Cosby, B. J. (1978). Relationships among substrate, flow, and benthic microalgal pigment density in the Mechums River, Virginia. *Limnology and Oceanography*, **23**, 785–97.

Tison, J., Park, Y.-S., Coste, M., *et al.* (2007). Predicting diatom reference communities at the French hydrosystem scale: a first step towards the definition of the good ecological status. *Ecological Modelling*, **203**, 99–108.

Tuchman, M. & Stevenson, R. J. (1980). Comparison of clay tile, sterilized rock, and natural substrate diatom communities in a small stream in southeastern Michigan, U.S.A. *Hydrobiologia*, **75**, 73–9.

van Dam, H. (1982). On the use of measures of structure and diversity in applied diatom ecology. *Nova Hedwigia*, **73**, 97–115.

van Dam, H. & Mertens, A. (1993). Diatoms on herbarium macrophytes as indicators for water quality. *Hydrobiologia*, **269–270**: 437–45.

van Dam, H., Mertens, A. & Sinkeldam, J. (1994). A coded checklist and ecological indicator values of freshwater diatoms from the Netherlands. *Netherlands Journal of Aquatic Ecology*, **28**, 117–33.

Vannote, R. L., Minshall, G. W., Cummins, K. W., Sedell, J. R., and Cushing, C. E. (1980). The River Continuum Concept. *Canadian Journal of Fisheries and Aquatic Sciences*, **37**, 130–7.

Vollenweider, R. A. & Kerekes, J. J. (1981). Background and summary results of the OECD cooperative program on eutrophication. In *Restoration of Inland Lakes and Waters*, Washington, DC: US Environmental Protection Agency, pp. 25–36.

Walley, W. J., Grbović, J., and Džeroski, S. (2001). A reappraisal of saprobic values and indicator weights based on Slovenian river quality data. *Water Research*, **35**, 4285–92.

Wang Y. K., Stevenson R. J., and Metzmeier L. (2005). Development and evaluation of a diatom-based index of Biotic Integrity for the Interior Plateau Ecoregion, USA. *Journal of the North American Benthological Society*, **24** (4), 990–1008.

Watanabe, T., Asai, K., Houki, A., Tanaka, S., and Hizuka, T. (1986). Saprophilous and eurysaprobic diatom taxa to organic water pollution and diatom assemblage index (DAIpo). *Diatom*, **2**, 23–73.

Weber, C. I. (1973). Recent developments in the measurement of the response of plankton and periphyton to changes in their environment. In *Bioassay Techniques and Environmental Chemistry*, ed. G. Glass, Ann Arbor, MI: Ann Arbor Science Publishers, pp. 119–38.

Weilhoefer, C. L. & Pan, Y. (2007). A comparison of periphyton assemblages generated by two sampling protocols. *Journal of North American Benthological Society*, **26**, 308–18.

Whitton, B. A. & Kelly, M. G. (1995). Use of algae and other plants for monitoring rivers. *Australian Journal of Ecology*, **20**, 45–56.

Whitton, B. A., Rott, E., and Friedrich, G. (ed.) (1991). *Use of Algae for Monitoring Rivers*, Innsbruck: Universität Innsbruck.

Winter, J. G. & Duthie, H. C. (2000). Epilithic diatoms as indicators of stream total N and total P concentration. *Journal of the North American Benthological Society*, **19**, 32–49.

Wolda, H. (1981). Similarity indices, sample size and diversity. *Oecologia*, **50**, 296–302.

Wu, J. T. (1999). A generic index of diatom assemblages as bioindicator of pollution in the Keelung River of Taiwan. *Hydrobiologia*, **397**, 79–87.

Wunsam, S., Cattaneo, A., and Bourassa, N. (2002). Comparing diatom species, genera and size in biomonitoring: a case study from streams in the Laurentians (Québec, Canada). *Freshwater Biology*, **47**, 325–40.

Yallop, M., Hirst, H., Kelly, M., *et al.* (2006). Validation of ecological status concepts in UK rivers using historic diatom samples. *Aquatic Botany* **90**, 289–95.

Zelinka, M. & Marvan, P. (1961). Zur Präzisierung der biologischen Klassifikation des Reinheit fliessender Gewässer. *Archiv für Hydrobiologie*, **57**, 389–407.

5 Diatoms as indicators of long-term environmental change in rivers, fluvial lakes, and impoundments

EUAN D. REAVIE AND MARK B. EDLUND

5.1 Introduction

Natural succession of rivers, such as sediment filling, migration, and the development of abandoned river channels into terrestrial systems, tends to occur over geological timescales. In recent centuries rivers have been used as water supplies for domestic, agricultural, and industrial activities, and for navigation, fisheries and water power. These anthropogenic activities have accelerated changes in river systems through pollution, habitat destruction, non-native species introductions, hydrologic manipulation and other physical disturbances. In addition to the obvious physical impairments, human activities have had numerous deleterious impacts on water quality and biotic communities inhabiting rivers (Smol, 2008).

The anthropogenic nutrient and particulate loads carried by rivers have markedly increased over the past few centuries, causing an overall increase in organic matter flux. However, one of the most significant manipulations of rivers has been damming, resulting in impounded aquatic systems that plainly contrast their previous conditions. Dam construction has reduced organic flux in many regions (Meade *et al.*, 1990), and it is estimated that seven times the natural volume of rivers is stored in the world's reservoirs (Vörösmarty *et al.*, 1997). Dams have also changed the global silica cycle by storing large amounts of biogenic silica in reservoir deposits and preventing its delivery to oceans (Humborg *et al.*, 2000). Enhanced diatom productivity fueled by excess nutrients in the world's rivers has further increased the trapping efficiency of silica within impoundments (Triplett *et al.*, 2008). If attempts are made to understand changing conditions in rivers or achieve ecological management, conceptualization of long-term conditions is essential (Reid & Ogden, 2006). Because diatoms are sensitive to a wide range of conditions, they are particularly suited to applications in rivers that can be subject to complex physical, chemical and biological shifts (see Stevenson *et al.*, this volume). In rare cases, long-term monitoring data may be used to reconstruct the ecological history of lotic systems. For instance, Berge (1976) investigated acidification through the comparison of old (1949) and recent (1975) diatom samples collected from Norwegian streams. More recently, Van Dam & Mertens (1995) used water quality and benthic diatom collections from 1974, 1981, and 1990 to provide a 16-year acidification trend for streams in the Netherlands. They noted that many of the historical environmental data were inadequate to support their results, which is an expected consequence when older monitoring data are sought for aquatic management (Smol, 2008). In the 1980s, river researchers started to explore long-term paleoecological analyses of diatoms in hopes that, like many similar, highly successful applications in lakes, they could provide baseline data and clear indications of natural and artificial ecological transitions. As will be discussed in this chapter, trial-and-error was the norm for these early studies, but significant advancements in diatom applications in river paleoecology have been achieved.

The use of diatoms as monitoring tools in river systems is treated in Stevenson *et al.* (this volume). Indeed, numerous studies have applied diatoms in lotic settings, largely focusing on modern diatom ecology and littoral habitats, with the primary objective of monitoring environmental quality. In this chapter we aim to illustrate how diatom tools are used to reveal long-term changes in rivers (including former and recurrent river systems), particularly through the analysis of diatom remains in sediments. In many cases it will be clear that diatom indicators developed from assessments of modern materials can be used

The Diatoms: Applications for the Environmental and Earth Sciences, 2nd Edition, eds. John P. Smol and Eugene F. Stoermer. Published by Cambridge University Press.

in paleoecological contexts, demonstrating that uniformitarian principles are critical to understanding river ecology. Diatom paleolimnological applications are also given treatment in several other chapters in this volume, but river ecosystems necessitate very different considerations than similar studies on more physically stable systems such as lakes.

While they may seem obvious from a limnological standpoint, the differences between river and lake systems are worth discussing in terms of diatoms and paleolimnology. In contrast to lakes, obtaining suitable stratigraphically dated sediment cores from rivers can be problematic. Acquiring continuous lotic sedimentary records can be difficult because of constant or frequent reworking of riverine material, and so it is often necessary to collect material from more stable environments such as fluvial lakes, former channels (e.g. oxbows), bays, lagoons, deltas, backwaters, and floodplains. Although sites near river channels might be considered to have the most unreliable sedimentary records due to frequent scouring, such sites may also be used if they are very carefully selected.

Early paleoecological investigations in rivers were performed on the Rhône and Rhine rivers in the 1980s (Bravard *et al.*, 1986; Klink, 1989). These studies focused, respectively, on the use of cladoceran and insect remains in the sediments of abandoned channels to formulate objectives for mitigating and predicting the impacts of massive civil engineering projects in the rivers. These data from the Netherlands illustrated that paleolimnological information was not limited to lake systems, and it was not surprising that down-core diatom studies in rivers were soon to follow. The sections that follow describe case studies using diatoms in paleoecological assessments in extant and former rivers. In a relative sense, river paleolimnology is still in its infancy, but, using the literature to date, we summarize methods and recommendations from scientists who have overcome the often daunting problems associated with lotic systems.

Some ambiguity surrounds the definition of this chapter as one that deals with diatoms in river systems. Most lakes are part of a flow-through network of some kind, with the possible exception of certain lake types such as headwaters and perched lakes. Hence, some limnologists might dispute whether a particular lake is a closed or fluvial lake. Take the Rideau Canal system (Canada) as an example: a series of lakes, rivers, and locks connecting the Ottawa River to Lake Ontario (Legget, 1975). A casual limnological assessment of many of the Rideau Canal lakes would conclude that they are typical temperate, freshwater lakes with few or no detectable river characteristics. But, because the majority of these lakes have major inflows and outflows, one might define them as fluvial. Christie & Smol (1996) performed a diatom-paleolimnological analysis on one of these lakes, and such a study could well apply to this chapter as well. Although studies on lotic systems have a unique set of concerns and strategies, it is worth noting that diatom-based methods used in lakes, rivers, and other aquatic systems are dictated by the unique properties of each system.

5.2 Are the fluvial diatoms allochthonous or autochthonous?

Several researchers have touted the value of nearshore diatoms as indicators of river condition using periphytic assemblages collected from rocks (e.g. Lavoie *et al.*, 2008), plants (e.g. Reavie & Smol, 1998a), sediments (e.g. Tibby, 2004), and artificial substrates (e.g. Hoagland *et al.*, 1982). These researchers have the advantage of knowing that the diatom species they encounter probably lived at or near their sample locations. These locally derived species, termed autochthonous, make ideal indicator organisms because they can be calibrated to the water quality and other environmental conditions present at their respective sample locations. Furthermore, if there is sample contamination from diatoms that have been carried from elsewhere (i.e. allochthonous taxa), one may make assumptions to discern living from dead diatoms; for instance by examining whether the cells contain cytoplasmic contents, or if they are old, empty frustules that may have been carried great distances. Identifying the source of sedimented diatoms from rivers and impoundments can be problematic as cell contents often degrade rapidly; new frustules may look identical to frustules that have been resuspended in the river for several years. One might assume that highly fragmented or dissolved valves are allochthonous, but that assumption has been shown to be incorrect at times (Beyens & Denys, 1982), and such a consideration is probably of little use in most studies. It has been suggested that planktonic diatoms in sediments are by definition allochthonous and so should not be used in paleolimnological studies (Simonsen, 1969), but as summarized in case studies below, phytoplanktonic diatoms can provide critical information on large spatial scales in rivers. Until more advanced methods are developed to identify allochthonous diatom remains in river sediments, researchers should assume that at least some of the ecological interpretation that arises from these assemblages represent conditions up-gradient from the sample location and that a subjective assessment of the allochthonous component of an assemblage may be necessary.

5.3 Dealing with uncertain temporal profiles

Sediment dating usually has little to do with diatoms, but this chapter would not be complete without some background on how researchers have obtained suitable temporal profiles for paleoecological analyses in river systems. Even with detailed hydrological and bathymetric data, any paleolimnological investigation on a river demands an exploratory period to find suitable core sites. Good planning (e.g. historical aerial photos), determination, and fastidious groundwork appear to have prevailed in such studies to date, as we are unaware of river paleolimnology studies that had to be abandoned due to "uncorable" conditions. Unlike lake studies, where the single deepest point in a basin is typically suited to an excellent temporal record (Charles *et al.*, 1991), the shrewd river paleolimnologist will have time, effort and budget set aside to identify, and discard if necessary, sediment cores from areas with poor temporal records. Carignan *et al.* (1994) performed isotopic dating on several cores collected from fluvial lakes throughout the St. Lawrence River. Several of these cores had monotonous or erratic isotope profiles, indicating that they had been homogenized, scoured, or otherwise physically disturbed. However, select cores had strong exponential isotope decay profiles, and additional analyses allowed for exceptional resolution in nutrient, trace-metal, and organic contaminant trends for the previous 50 years.

Care must be exercised in dating, as isotopic supplies to rivers are rarely dominated by atmospheric sources, and with land-use changes the delivery and sources of radioisotopes likely have changed over time (Gell *et al.*, 2005a), in direct violation of most common dating models (e.g. constant rate of supply, CRS; Triplett *et al.*, 2009). Researchers are encouraged to use all available dating tools and test multiple cores when analyzing riverine sediments. A combination of ^{210}Pb, ^{137}Cs, accelerator mass spectrometry (AMS) ^{14}C, pollen, and magnetic susceptibility was critical for dating and correlating sediment cores from the natural impoundments of Lake Pepin and Lake St. Croix in the Upper Mississippi River basin (Engstrom *et al.*, 2009; Triplett *et al.*, 2009). Only with well-dated cores were diatom studies in the St. Lawrence and Mississippi basin performed, as described in case studies below.

5.4 Natural fluvio-lacustrine systems

Retrospective diatom applications in fluvial lakes have become more common in the last decade. A fluvial lake is somewhat arbitrarily described as an open-water widening of a river with a surface area large enough that it may be considered a lake, but with river-like flow characteristics. Because of their lake-like properties, fluvial lakes are valuable sources of sedimentary records that may otherwise be unavailable in the high-flow regions of a river. Some fluvial lakes are naturally impounded lakes formed by processes such as tributary deltas, dune blockage, landslides, and lava flows. Unlike artificial impoundments, which largely reflect human interventions, natural impoundments allow for investigations of processes in rivers that have occurred over longer geologic timescales. The same principles are involved when applying diatoms in investigations of natural impoundments or fluvial lakes and we describe these applications in the context of several case studies on large transboundary rivers.

5.4.1 St. Lawrence River case study

As an economically important waterway connecting the Atlantic Ocean to the Laurentian Great Lakes, the St. Lawrence River, which is also an international boundary between Canada and the United States, has been subjected to considerable anthropogenic disturbance. Rapid agricultural expansion in the late nineteenth and early twentieth centuries increased nutrient loads. Canalization and rapid shoreline industrialization occurred in the early twentieth century, and in the 1950s the St. Lawrence Seaway was constructed, resulting in dramatic modifications to the hydrologic regime of the river. The Seaway is a series of 15 shipping locks which reduced the flow rate and amplitude of water-level fluctuations. Not surprisingly, these hydrologic alterations resulted in significant changes to sedimentary regimes and nutrient cycles, and provided a highly favorable environment for aquatic macrophytes.

As mentioned previously, significant isotope work identified several sites in the St. Lawrence River containing conformable sedimentary profiles (Carignan *et al.*, 1994). Many investigations followed these findings, including a diatom-based paleolimnological study on several cores from the river's fluvial lakes (Reavie *et al.*, 1998). In particular, a core from Lake Saint-François, a fluvial lake between the cities of Cornwall and Montréal (Canada), contained strong shifts in diatom assemblages related to industrialization and other human activities (Figure 5.1). Simultaneously, studies were being performed on the modern diatom assemblages of the St. Lawrence River to determine the chemical and habitat characteristics of the diatom species (O'Connell *et al.*, 1997; Reavie & Smol, 1997, 1998a, b). One of these studies (Reavie & Smol, 1997) used modern diatom assemblages collected from rocks, macrophytes, and the macroalga *Cladophora* Kütz. to develop a diatom-based

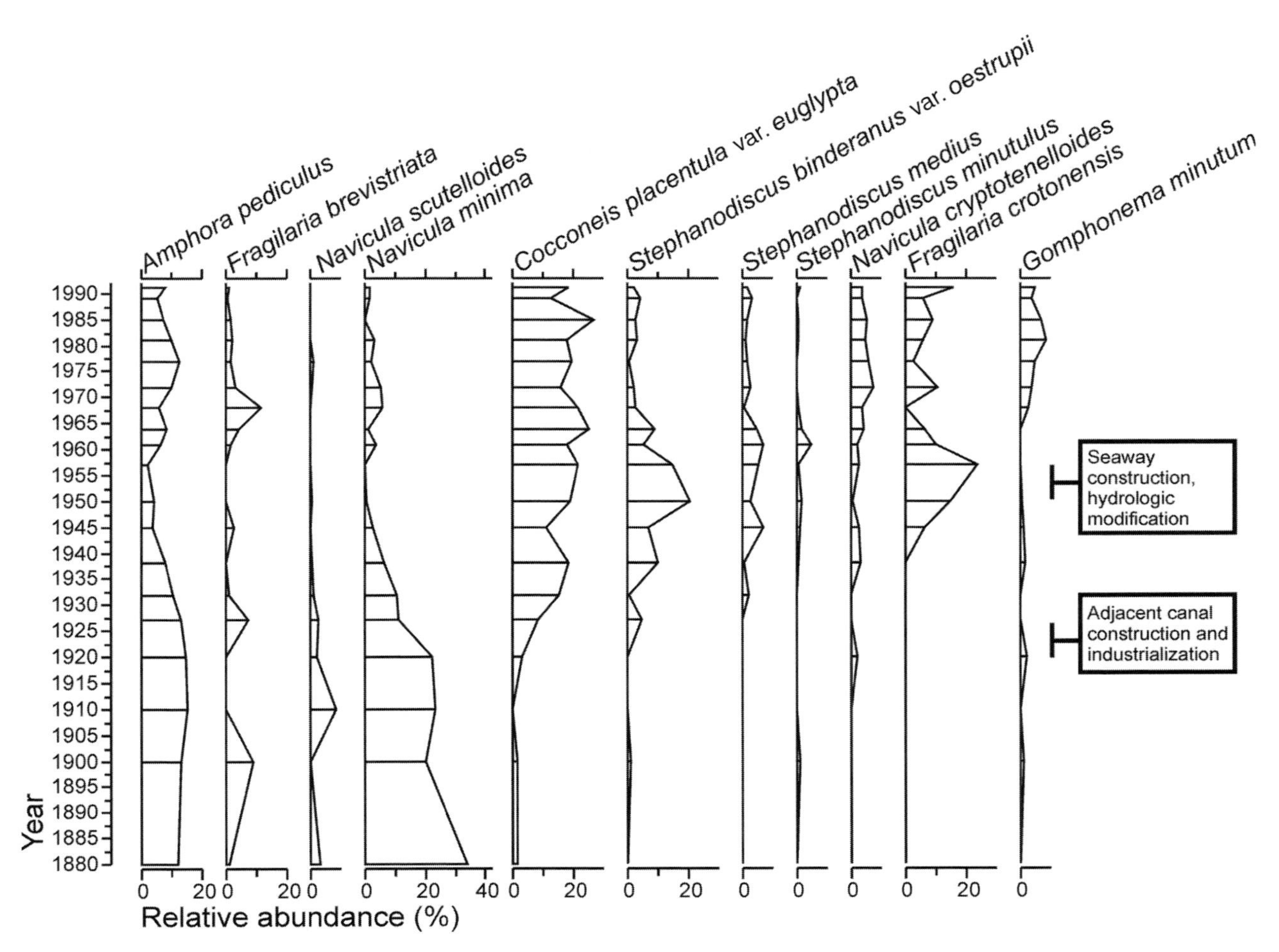

Figure 5.1 Diatom microfossil profiles of common species in a core from Lake Saint-François, St. Lawrence River. Major periods of local development are indicated on the right. (Modified from Reavie *et al.* (1998).)

habitat model. As it is generally known which taxa are planktonic, a robust habitat inference model was developed that was subsequently used to infer shifts in littoral habitats based on diatom assemblages in sediment cores.

The core profile from Lake Saint-François (Figure 5.1) indicated a notable shift from benthic species (e.g. *Navicula minima* Grunow) in the late twentieth century to epiphytic taxa such as *Cocconeis placentula* var. *euglypta* Ehrenb., a taxon shown to dominate the modern assemblages living on macrophytes and *Cladophora*. These shifts were largely attributed to hydrologic modifications that resulted in stabilization of littoral areas and an augmented standing crop of macrophytes. The mid 1900s also experienced an increase in planktonic taxa that require high levels of nutrients such as *Stephanodiscus* Ehrenb. and *Fragilaria crotonensis* Kitton, reflecting the increased nutrient load and stagnation of flow. While this study provided a clear ecological history of the river, it also showed that diatom-based paleolimnological studies can be a valuable addition to monitoring programs in large rivers.

5.4.2 The Upper Mississippi River – Lakes Pepin and St. Croix case study

The Mississippi River drains 40% of the contiguous United States and is the largest and most economically important river system in North America. The Mississippi River, with its culturally enhanced burden of phosphorus, nitrogen, and sediment, has been linked to coastal eutrophication, in particular the annual formation of the Gulf of Mexico "Dead Zone" (Rabalais *et al.*, 2003). Modern nutrient exports to the Gulf are significantly derived from the Upper Mississippi River (Alexander *et al.*, 2000; Rabalais *et al.*, 2002); however, only recently have retrospective studies addressed the role of the Upper Mississippi and its tributaries in regulating land-use impacts and mass transport from the continental to coastal environments.

Two natural impoundments in the Upper Mississippi basin, Lake Pepin on the Mississippi River and Lake St. Croix on the St. Croix River, have been the focus of intense paleoecological

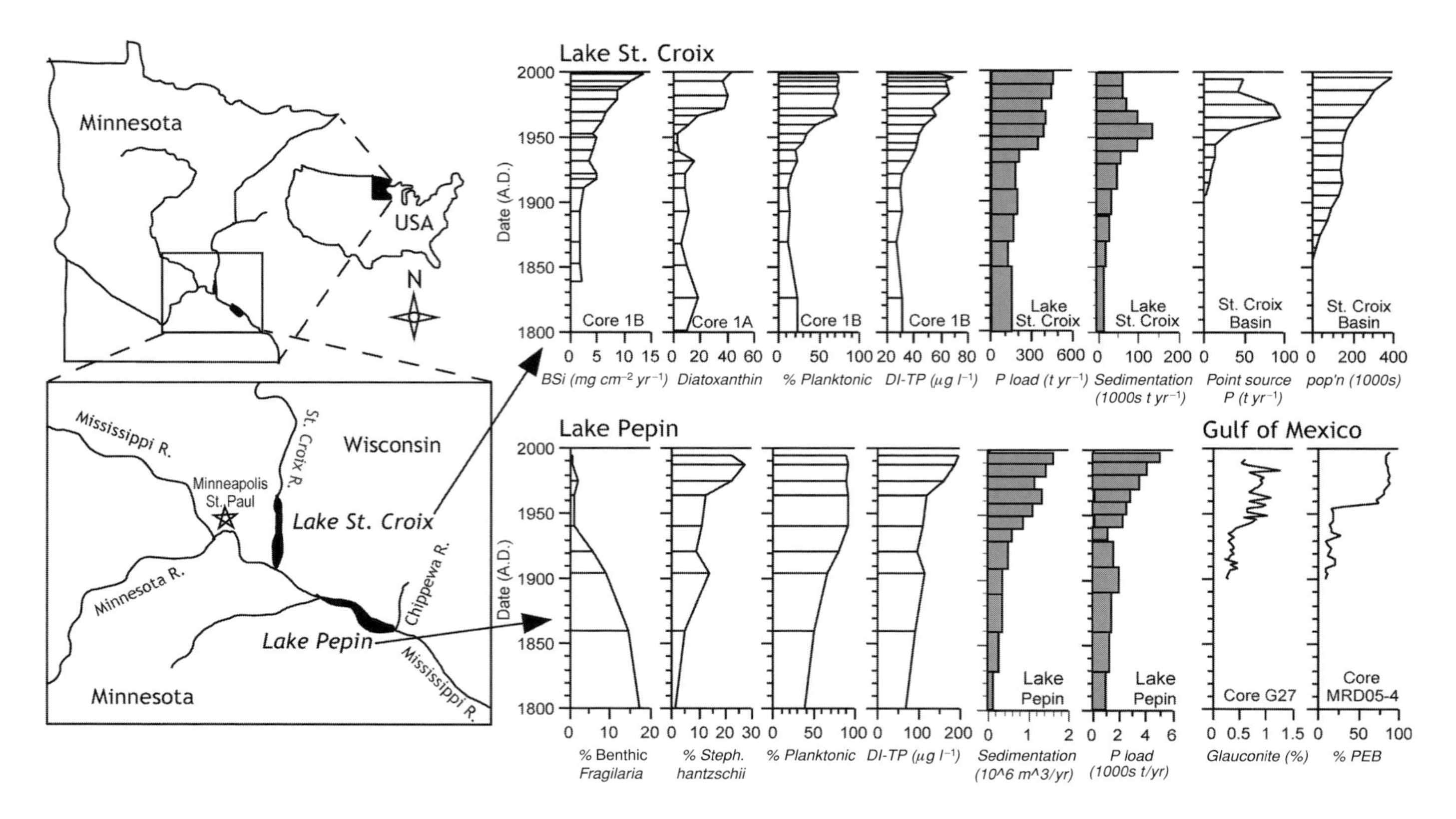

Figure 5.2 Diatom and biogeochemical profiles in sediment cores and historical reconstructions from Lake St. Croix and Lake Pepin (Upper Mississippi River basin) and the Gulf of Mexico. Lakes St. Croix and Pepin proxies include biogenic silica (BSi, mg cm^{-2} yr^{-1}), diatoxanthin (nmol g^{-1} organic matter), percent diatom groups, diatom-inferred total phosphorus (DI-TP, µg L^{-1}, ppb), total phosphorus load to each lake (TP load, tonnes yr^{-1}), total sedimentation (tonnes yr^{-1} or m^3 yr^{-1}), basin population, and point-source phosphorus loadings (tonnes yr^{-1}). Lake St. Croix proxies adapted and modified from Edlund *et al.* (2009a, b) and Triplett *et al.* (2009); Lake Pepin proxies adapted and modified from Engstrom *et al.* (2009). Gulf of Mexico proxies that are indicative of anoxia include percent glauconite on coarse grains (modified from Rabalais *et al.*, 2007) and PEB index (% low-oxygen-tolerant benthic foraminifers; modified from Osterman *et al.*, 2008). Portions of figure previously published in the *Journal of Paleolimnology* are used with kind permission of Springer Science and Business Media.

investigation (Figure 5.2). The lakes formed approximately 10.3 ka after drainage from proglacial lakes Agassiz and Duluth receded. Deltaic fans from the Chippewa, Vermillion and Cannon rivers were deposited to form a series of riverine lakes in the Mississippi River valley, the former creating Lake Pepin (Blumentritt *et al.*, 2009). Lake St. Croix is a narrow, 37 km long lake with four major basins (10–22 m deep) along the Minnesota–Wisconsin border. The riverine character of the lakes is obvious in their short residence times (days to weeks), but both lakes retain full post-glacial sediment records that were used to reconstruct ecological change and historical phosphorus and sediment loads using a combination of diatom-inferred phosphorus reconstructions and whole-lake mass balance techniques (Eyster-Smith *et al.*, 1991; Blumentritt *et al.*, 2009).

In both natural impoundments, the diatoms indicated dramatic ecological changes from clear-water benthic forms to planktonic dominance in the last 200 years (Figure 5.2; Edlund *et al.*, 2009a; Engstrom *et al.*, 2009). Diatom-inferred total phosphorus increased in both impoundments following Euro-American settlement, with especially large increases after AD 1950 (Edlund *et al.*, 2009a; Engstrom *et al.*, 2009). Historical phosphorus mass balances indicated that phosphorus loading to each impoundment had also increased rapidly after World War II in response to growing populations and increased point- and non-point-source loadings (Figure 5.2; Edlund *et al.*, 2009b; Engstrom *et al.*, 2009; Triplett *et al.*, 2009). Whole-basin sedimentation patterns differed between the lakes. Sedimentation rates in Lake Pepin continue to increase (Engstrom *et al.*, 2009), whereas sedimentation rates peaked in Lake St. Croix during the 1960s, but remain threefold higher than background levels (Triplett *et al.*, 2009).

Results of these paleolimnological analyses have been instrumental in determining that both rivers suffer from nutrient impairment and that the Mississippi is further impaired for turbidity. Background nutrient and sedimentation rates have guided development of nutrient and sedimentation targets as federal and interstate agencies institute remediation policies (Edlund *et al.*, 2009b). These studies clearly establish a historical linkage among increased nutrient and sediment loads, their ecological impacts, and mass transport from the Upper Mississippi drainage with coastal eutrophication and anoxia in the Gulf of Mexico (Figure 5.2; Rabalais *et al.*, 2007; Osterman *et al.*, 2008; Edlund *et al.*, 2009a; Engstrom *et al.*, 2009).

5.5 Backwaters and floodplain lakes

Geomorphologic evolution of a river is a dynamic process driven by downcutting, erosion, and sediment deposition that results in channel movement, channel abandonment, and creation of floodplain features such as wetlands, meanders, and oxbow lakes. Floodplain water bodies have been subject to paleoecological analysis to address research questions on geological and human impact timescales. On geologic scales, Gaiser *et al.* (2001) investigated an oxbow wetland near the Savannah River (South Carolina, United States) to reconstruct the hydrologic environment over the last 5200 years. Diatoms recorded an initial period of mid-Holocene lake-like conditions from 4600–3800 BP. A shift to aerophilic taxa followed when the wetland became a temporary pond during the late Holocene that was subject to periodic drying until a nearby reservoir constructed in 1985 permanently flooded the wetland.

Schönfelder & Steinberg (2002) studied a 4000-year record of long-term human impact using sediments from paleomeanders and oxbow lakes in Germany's rivers Havel and Spree. Diatoms were used to reconstruct historical nutrient dynamics and baseline conditions and showed that both rivers had long been moderately eutrophic systems and that increased intensity of land use (e.g. deforestation) raised nutrient levels especially in the last 800–1000 years. More intense human impacts were noted in post 1800s diatom assemblages.

Southeast Australia's river systems (Murray, Yarra, Darling) have numerous wetland and billabong habitats in their floodplains. To aid interpretation of paleoecological studies in billabongs, Reid & Ogden (2009) recently developed a powerful diatom-based model by relating surface-sediment diatom assemblages to nutrients and adjacent human activities such as farming. Numerous paleolimnological studies in Australian rivers have evaluated changes on human and geological timescales. For instance, Thoms *et al.* (1999) and Reid *et al.* (2007) were able to track declines in submerged macrophytes through the decline of periphytic diatoms in sediment profiles, indicating that since European settlement turbidity had increased due to anthropogenic soil erosion. Thoms *et al.* (1999) further used diatoms and other biological and geochemical proxies on Murray River billabongs to set benchmark conditions and show that ecological changes began with early post-European settlement and before flow regulation. Post-European impacts were also identified in a billabong in the Yarra River floodplain by using diatom fossils to track significant erosion and nutrient enrichment (Leahy *et al.* 2005). In particular, a shift from an assemblage dominated by *Cyclotella stelligera* (Cleve & Grunow) Van Heurck to a diverse assemblage of epipelic and aerophilous species marked the transition from low-nutrient, pre-settlement conditions to a post-settlement period of significant soil erosion. Gell *et al.* (2005a) used diatom and other records to identify post-European changes in salinity, pH, turbidity, and nutrients, especially along the regulated river systems. Critical to the interpretation of sediment records were cautious approaches to dating and an understanding of connectivity between floodplain habitats and the main rivers.

Billabongs have also been the subject of long-term climate studies. A ~5000 year diatom-based paleolimnological analysis of Tareena Billabong (Australia) indicated that it was naturally a freshwater system, and several thousand years of inferred shifts in sedimentation and turbidity reflected climate-related changes in river influence (Gell *et al.* 2005b). In order to determine if recent pH changes in the Goulburn River were within the natural range for the river, Tibby *et al.* (2003) analyzed the diatoms from a sediment core collected from a fluvial billabong. Using a diatom-based pH model, diatom-inferred pH indicated an over 3000-year period of stable pH, and that ion concentrations shifted outside the natural range shortly following European arrival in the area.

5.6 Artificial impoundments

Artificial impoundments change the characteristics of water bodies from those of rivers to lakes, impacting physical, chemical, and biological properties (Baxter, 1977; Friedl & Wüest, 2002). Damming typically increases residence time of water up-gradient of the dam, resulting in temperature increases, stratification changes, materials retention, and an increase in primary production in the newly formed reservoir. Further, tailwater reaches below dams are also modified, sometimes reduced to lower-grade rivers or streams, and in cases of hydroelectric dams, these down-gradient reaches can be subject to discharge fluctuations resulting from dam operations. Impoundments and tailwaters also serve as primary sites for introduction and establishment of invasive species including the diatom *Didymosphenia geminata* (Lyngb.) M. Schmidt (see Spaulding *et al.*, this volume). In this section we summarize selected case studies of diatom applications for evaluating the impacts of artificial impoundment.

A natural aging process occurs in lakes due to nutrient and sediment loading and gradual filling of a basin. With rare exceptions (Melcher & Sebetich, 1990), reservoirs tend to fill up with sediment more rapidly due to more efficient trapping of suspended particles and nutrients in these artificial lacustrine basins (Benson, 1982). When a river is dammed and a new lake is formed, lotic benthos typically perish and are replaced by plankton. For instance, Zeng *et al.* (2007) observed significant increases in planktonic diatoms and other algal taxa in reservoirs following dam construction in the Yangtze River (China), and Wehr & Thorp (1997) noticed reductions in benthic species up-gradient of navigation dams in Ohio rivers. Diatom-paleolimnological analyses from artificial impoundments have mainly focused on reconstructing ecological shifts since damming. Hall *et al.* (1999) used diatom microfossils and supporting algal pigment analyses to describe the long-term impacts of impoundment in two southern Saskatchewan (Canada) impoundments. Sedimentary profiles from one reservoir had high concentrations of eutrophic diatoms (e.g. *Stephanodiscus* species) after damming in 1967, followed by oligotrophication denoted by increases in lower-nutrient taxa such as *Tabellaria* Ehrenb. ex Kütz. Hall *et al.* emphasized that, contrary to what is typically assumed, reservoir ontogeny does not always lead to eutrophication due to the multitude of environmental factors that must be considered for each system.

Unlike a physically stabilized reservoir, the downstream portion often has non-uniform currents, higher turbidity, and increased potential for bank erosion (Ward & Stanford, 1979). Ecological changes are particularly common downstream of hydroelectric dams, where short-term fluctuations in flow are the typical result of operation. Although not as common, studies of diatoms to infer ecological changes downstream of dams, weirs, and other storages indicate that r-selected species dominate due to frequent physical disturbance of the aquatic system (Silvester & Sleigh, 1985). In a case from impounded rivers in Colorado (Zimmerman & Ward, 1984), it was noted that, compared to unregulated rivers, diatoms were extirpated in favor of filamentous green algae at sample locations below impoundments. Further complexity of diatom colonization below a hydroelectric dam in the Colorado River was explored by Peterson (1986), who observed that variations in water shear stresses resulted in variations in the attached diatom communities, a result of colonization ability and immigration rates among the diatom taxa. For instance, the epiphyte *Cocconeis pediculus* Ehrenb. was identified as a likely "poor immigrant" in high shear habitats. Such findings are important to river monitoring programs that rely on understanding diatom community dynamics and species composition.

Significant multidisciplinary work has been performed on the impacts of the Iron Gate I dam and reservoir on nutrient flow through the Danube River (Serbia, Romania). The Iron Gate I dam, built in 1972, was expected to decrease nutrient and sediment loads downstream to the Black Sea. Unlike the majority of diatom-based paleoecological studies, Teodoru *et al.* (2006) used biogenic silica from sediment cores to track the change in diatoms following dam installation. Dissolved silica is taken up by diatoms to create biogenic silica (BSi) as the primary component of their frustules. Downcore trends in BSi can be used to track historic changes in diatom production (Schelske 1999). Using BSi from sediment cores collected from the Iron Gate I reservoir, Teodoru *et al.* (2006) noted an increasing trend over the previous 20 years, suggesting enhanced diatom growth. However, despite previous assumptions to the contrary, the inferred environmental impacts of Iron Gate I were minor compared to changes occurring due to smaller dams in the Danube's headwaters and coastal pollution. Although the Iron Gate I impoundment is the largest in the Danube, it was not playing a dominant role in decreasing silica loads downstream. The work of Teodoru *et al.* not only demonstrated the power of diatom studies in reservoirs, but further that these studies can put environmental impacts in a holistic, quantitative perspective for a large river.

Finally, dam removal has been suggested as a way to restore the ecological integrity of a river, and diatom-based monitoring data have been used to track the expected recovery following dam decommissioning. Thomson *et al.* (2005) studied the upstream and downstream algal assemblages before, during, and after the removal of a small Pennsylvania dam. Abundance and taxonomic composition of diatoms and other algae were evaluated. Although they acknowledged that every dam removal situation would be different, the authors found that diatom abundance and species richness declined dramatically following removal due to the sudden massive flux of sediments downstream. However, they also suggested that such impacts as indicated by the diatom communities are likely to be short-lived, and so dam removal would have a net ecological benefit. To our knowledge, no paleoecological application has used diatoms to track recovery following decommissioning of a dam.

5.7 Terminal lagoons, deltas, and estuaries

While identifying depositional basins for river paleolimnology is a challenge, river mouths are an ideal place to look for diagnostic diatom-based records. Deltas, riverine wetlands and receiving lagoons are likely to contain microfossil material that may have been carried great distances by the upgradient river, and so those microfossils may be used to provide an integrated picture of the river's ecological condition. To minimize topical overlap, we refer the reader to Cooper *et al.* (this volume) for a discussion of diatom applications in estuarine river mouths, with particular attention to work performed on Chesapeake Bay (Cooper, 1995). Further, refer to Trobajo and Sullivan (this volume) for a summary of the ways that diatoms are used to reconstruct the histories of coastal environments.

One modern diatom-based model was designed with the intent of using the model equations to infer environmental conditions in the Mackenzie Delta lakes (Hay *et al.*, 1997). Delta floodplain lakes allow for paleolimnological investigations of watersheds due to close interaction between floodplain lakes and rivers. In the case of Hay *et al.* (1997), a training set of floodplain lakes was used to develop a diatom model that can reconstruct the frequency and magnitude of flooding and other river influences on these lakes. The frequency and duration of lake flooding by a river have strong control over the diatom community through hydrological and chemical shifts. For instance, frequently flooded lakes tend to be turbid and so favor tolerant phytoplanktonic diatom species (e.g. *Asterionella* Hassall and *Aulacoseira* Thwaites), whereas infrequently flooded lakes are typically transparent in the summer and support dense macrophyte populations; so those lakes tend to contain rich assemblages of benthic and epiphytic diatom species (e.g. *Navicula* Bory). Further, Hay *et al.* identified a relationship between the structure of diatom assemblages and methane (CH_4), a variable which reflects decomposition of organic material, especially in non-flushed lakes with high macrophyte densities. The authors acknowledged that diatoms were unlikely to be directly responding to methane concentrations, but that the methane variable captured diatom community changes that are associated with flooding. A regression model was developed to infer sub-ice winter methane concentrations from sedimentary diatom assemblages.

The Hay *et al.* (1997) model was applied to down-core diatom assemblages in eight Mackenzie Delta lakes (Michelutti *et al.*, 2001). Not surprisingly, the lake with the most common riverine influence had high densities of planktonic species and very low diatom-inferred methane (DI-CH_4), whereas lakes with rare or no connections to rivers had higher modern DI-CH_4. The Michelutti *et al.* (2001) sediment cores each represented a few hundred years of accumulated material, but an important future application of this model may be to generate regional climate proxy data that can provide a record to evaluate warming scenarios in the Arctic. The importance of such work is underscored by more recent studies on the Slave River Delta, which flows into Great Slave Lake, northern Canada (Sokal *et al.*, 2008), and on the Peace-Athabasca Delta (PAD), which connects to the Slave River (Wolfe *et al.*, 2005). Wolfe *et al.* (2005) evaluated diatom profiles from a small lake in the PAD to explore the affects of flow regulation and climatic variability on lake ecology. In concord with several ancillary indicators (dendrochronology, magnetic susceptibility, isotopes, organic content, plant and invertebrate macrofossils), they were able to reconstruct riverine influences based on past accumulations of *Fragilaria pinnata* Ehrenb., a diatom known to be carried in turbid river water. Clear trends related to dry and wet conditions were reconstructed for the last 300 years.

Terminal floodplain lakes in Australia have also been subjects of diatom paleoecological studies. Fluin *et al.* (2007) used diatoms to explore the degree of historical connectivity between two adjacent terminal lagoons of the Murray River. Comparison of the relative numbers of riverine diatoms in sedimentary profiles revealed that the two systems evolved independently due to geomorphic separation. MacGregor *et al.* (2005) used diatom records from terminal basins of the Lower Snowy River to describe the Holocene record. For instance, diatom-inferred salinity indicated that Lake Curlip was a saline system (~7000 BP) that became brackish as sea levels dropped over subsequent millennia and the diatom record reflected greater riverine influence. As is typical in similar systems in Australia, human intervention in the forms of land clearance and hydrological manipulation resulted in dramatic changes to water quality in the lake, as indicated by a shift to a brackish and nutrient-tolerant diatom flora over the last century.

5.8 Flowing rivers

Of all paleoecological river studies, those based on diatom remains collected at or near a river channel might be assumed to be the most problematic, owing to the most physically harsh conditions for sedimentary regimes. Nonetheless, select studies have been successful in reconstructing historical conditions in these high-flow systems.

5.8.1 Zippel Bay case study

In a recent paleolimnological study in northern Minnesota of a fluvial bay, diatom remains were used to describe long-term eutrophication and erosion, the resulting dissolved-oxygen declines, and other threats to biological integrity (Reavie & Baratono, 2007). The impacted area comprised two primary tributaries flowing into a protected fluvial bay; sediment cores were collected from these three components. Cores from the tributaries were collected approximately 100 m perpendicular to the shipping channel in order to minimize disturbance from boating activities and scouring. The bay core was collected from the flat, deep basin.

Despite high flow rates in the tributaries, ^{210}Pb dating analyses indicated that the cores were suitable for retrospective analyses. These tributary profiles were particularly valuable because they allowed diatom assemblages and their ecological interpretations to be linked with similar trends in the downstream bay core, enabling isolation of the relative impacts of human activities in each tributary's respective catchment. Further, by analyzing the three profiles relative amounts of allochthonous and autochthonous diatoms in the bay core could be estimated.

5.8.2 Thames estuary case study

Juggins (1992) undertook a large project aimed at deriving paleosalinity estimates for the Thames estuary in central London, UK. The Thames has been subject to hydrologic modification since the fourteenth century. Embankments in particular have acted to straighten and smooth the banks of the estuary, resulting in an increase in tidal amplitude. Diatoms from 21 archeological samples, dating from the first to the seventeenth century, were collected from three sites. In contrast to higher-resolution lotic paleoecological studies, this work was less prone to unpredictable temporal shifts in the diatom assemblages simply because of the long temporal record and relatively small number of samples that were each likely to contain diatom remains that had been integrated from several years of deposition.

Diatoms from the Thames archeological sites were used to track eutrophication, indicated largely by the appearance of *Stephanodiscus* species, as far back as the medieval period. Paleosalinity estimates were generated from archeological assemblages using a weighted-averaging transfer function. The mix of freshwater and marine forms in historic sediments indicated, contrary to general beliefs, that the Thames was tidal in central London during the Roman period. Greater detail on this project is provided in Juggins & Cameron (this volume).

5.9 Summary

Diatoms are known to be excellent indicators of environmental conditions in rivers. However, their use in paleoecological applications of lotic ecosystems is more recent due to slower development of techniques to deal with the possibility of inconsistent temporal profiles in sediment cores. In many cases, successful applications in rivers have involved selecting sites with lake-like behavior such as fluvial lakes. Studies have proven that a researcher with patience and a sufficient budget for exploratory work can successfully apply diatom indicators down-core in more challenging systems. This move by scientists to tackle problems in rivers is due to an increasing recognition of the environmental losses resulting from human impacts, such as hydrologic manipulation, pollution, and eutrophication. Furthermore, the impacts of river regulation have been dramatic, and diatom indicators for impounded rivers and their reservoirs are needed to track environmental impacts in these artificial systems.

As in the Thames and Canadian delta work described in this chapter, the application of riverine diatoms in paleolimnology needs not be performed on an extant river. Instead, one might use the riverine characteristics of diatom taxa to interpret the distant past of an aquatic, or even terrestrial, system that was once influenced by flowing water. We anticipate that as socioeconomic concerns grow due to climate changes, such studies will play an increasing role in defining background and human-induced changes.

Information on riverine diatom species descriptions and ecological characterizations is constantly being gathered (see Stevenson *et al.*, this volume). These important data will support ongoing efforts to describe long-term trends in rivers using sedimentary diatom assemblages. Current applications are based on fairly sparse available literature. The most pressing needs are to continue taxonomic and ecological characterization of diatom indicators and refinement of paleoecological methods that are suited to river applications.

References

Alexander, R. B., Smith, R. A., & Schwarz, G. E. (2000). Effect of stream channel size on the delivery of nitrogen to the Gulf of Mexico. *Nature*, **403**, 758–61.

Baxter, R. M. (1977). Environmental effects of dams and impoundments. *Annual Review of Ecological Systems*, **8**, 255–83.

Benson, N. G. (1982). Some observations on the ecology and fish of reservoirs in the United States. *Canadian Water Resources Journal*, **7**, 2–25.

Berge, F. (1976). *Kiselalger og pH I noen elver og insjoer I Agder og Telemark. En Sammenlikning mellom arene 1949 og 1975*. SNSF – prosjektet IR 18/76.

Beyens, L. & Denys, L. (1982). Problems in diatom analysis of deposits: allochthonous valves and fragmentation. *Geologie en Mijnbouw*, **61**, 159–62.

Blumentritt, D. J., Wright, H. E. Jr., & Stefanova, V. (2009). Formation and early history of Lakes Pepin and St. Croix of the upper Mississippi River. *Journal of Paleolimnology*. **41**, 545–62.

Bravard, J., Amoros, C., & Patou, G. (1986). Impact of civil engineering works on the successions of communities in a fluvial system. *Oikos*, **47**, 92–111.

Carignan, R., Lorrain, S., & Lum, K. R. (1994). A 50-year record of pollution by nutrients, trace metals and organic chemicals in the St. Lawrence River. *Canadian Journal of Fisheries and Aquatic Sciences*, **51**, 1088–100.

Charles, D. F., Dixit, S. S., Cumming, B. F., & Smol, J. P. (1991). Variability in diatom and chrysophyte assemblages and inferred pH: paleolimnological studies of Big Moose L., N.Y. *Journal of Paleolimnology*, **5**, 267–84.

Christie C. E. & Smol, J. P. (1996). Limnological effects of 19th century canal construction and other disturbances on the trophic state history of Upper Rideau Lake, Ontario. *Lake and Reservoir Management*, **12**, 78–90.

Cooper, S. R. (1995). Diatoms in sediment cores from the mesohaline Chesapeake Bay, USA. *Diatom Research*, **10**, 39–89.

Edlund M. B., Engstrom D. R., Triplett L. D., Lafrancois, B. M., & Leavitt, P. R. (2009a). Twentieth-century eutrophication of the St. Croix River (Minnesota–Wisconsin, USA) reconstructed from the sediments of its natural impoundment. *Journal of Paleolimnology*, **41**, 641–57.

Edlund M. B., Triplett L. D., Tomasek, M., & Bartilson, K. (2009b). From paleo to policy: partitioning the historical point and non-point phosphorus loads to the St. Croix River, USA. *Journal of Paleolimnology*. **41**, 679–89.

Engstrom, D. R., Almendinger, J. E., & Wolin, J. A. (2009). Historical changes in sediment and phosphorus loading to the upper Mississippi River: mass-balance reconstructions from the sediments of Lake Pepin. *Journal of Paleolimnology*, **41**, 563–88.

Eyster-Smith, N. M., Wright, H. E. Jr., & Cushing, E. J. (1991). Pollen studies at Lake St. Croix, a river lake on the Minnesota/Wisconsin border, USA. *The Holocene*, **1**, 102–11.

Fluin, J., Gell, P., Haynes, D., Tibby, J., & Hancock, G. (2007). Palaeolimnological evidence for the independent evolution of neighbouring terminal lakes, the Murray Darling Basin, Australia. *Hydrobiologia*, **591**, 117–34.

Friedl, G. & Wüest, A. (2002). Disrupting biogeochemical cycles – consequences of damming. *Aquatic Science*, **64**, 55–65.

Gaiser, E. E., Taylor, B. E., & Brooks, M. J. (2001). Establishment of wetlands on the southeastern Atlantic Coastal Plain: paleolimnological evidence of a mid-Holocene hydrologic transition from a South Carolina Pond. *Journal of Paleolimnology*, **26**, 373–91.

Gell, P., Tibby, J., Fluin, J., *et al.* (2005a). Accessing limnological change and variability using fossil diatom assemblages, south-east Australia. *River Research and Applications*, **21**, 257–69.

Gell, P. A., Bulpin, S., Wallbrink, P., Hancock, G., & Bickford, S. (2005b). Tareena Billabong – a palaeolimnological history of an ever-changing wetland, Chowilla Floodplain, lower Murray–Darling basin, Australia. *Marine and Freshwater Research*, **56**, 441–56.

Hall, R. I., Leavitt, P. R., Dixit, A. S., Quinlan, R., & Smol, J. P. (1999). Limnological succession in reservoirs: a paleolimnological comparison of two methods of reservoir formation. *Canadian Journal of Fisheries and Aquatic Sciences*, **56**, 1109–21.

Hay, M. B., Smol, J. P., Pipke, K. J., & Lesack, L. F. W. (1997). A diatom-based paleohydrological model for the Mackenzie Delta, Northwest Territories, Canada. *Arctic and Alpine Research*, **29**, 430–44.

Hoagland, K. D., Roemer, S. C., & Rosowski, J. R. (1982). Colonization and community structure of two periphyton assemblages, with emphasis on the diatoms (Bacillariophyceae). *American Journal of Botany*, **69**, 188–213.

Humborg, C., Conley, D. J., Rahm, L., Wulff, F., Cociasu, A., & Ittekkot, V. (2000). Silicon retention in river basins: far-reaching effects on biogeochemistry and aquatic food webs in coastal marine environments. *Ambio*, **29**, 45–50.

Juggins, S. (1992). *Diatoms in the Thames Estuary, England: Ecology, Palaeoecology, and Salinity Transfer Function*. Bibliotheca Diatomologica Band 25, Stuttgart: E. Schweizerbart.

Klink, A. (1989). The Lower Rhine: palaeoecological analysis. In *Historical Change of Large Alluvial Rivers (Western Europe)*, ed. G. E. Petts, H. Möller & A. L. Roux, Chichester, UK: John Wiley & Sons, pp. 183–201.

Lavoie, I., Campeau, S., Darchambeau, F., Cabana, G., & Dillon, P. J. (2008). Are diatoms good integrators of temporal variability in stream water quality? *Freshwater Biology*, **53**, 827–41.

Leahy, P. J., Kershaw, A. P., Tibby, J., & Heinjis, H. (2005). A palaeoecological reconstruction of the impact of European settlement on Bolin Billabong, Yarra River floodplain, Australia. *River Research and Applications*, **21**, 131–149.

Legget, R. (1975). *Rideau Waterway*. Toronto: University of Toronto Press.

MacGregor, A. J., Gell, P. A., Wallbrink, P. J., & Hancock, G. (2005). Natural and post-European settlement variability in water quality of the Lower Snowy River floodplain, eastern Victoria, Australia. *River Research and Applications*, **21**, 201–13.

Meade, R. H., Yuzyk, T. R., & Day, T. J. (1990). Movement and storage of sediment in rivers of the United States and Canada. In *Surface Water Hydrology*. Boulder, CO: Geological Society of America, pp. 255–80.

Melcher, J. & Sebetich, M. J. (1990). Primary productivity and plankton communities in a two-reservoir series. *Water Resources Bulletin*, **26**, 949–58.

Michelutti, N., Hay, M. B., Marsh, P., Lesack, L., & Smol, J. P. (2001). Diatom changes in lake sediments from the Mackenzie Delta, N.W.T., Canada: paleohydrological applications. *Arctic, Antarctic, and Alpine Research*, **33**, 1–12.

O'Connell, J. M., Reavie, E. D., & Smol, J. P. (1997). Assessment of water quality using epiphytic diatom assemblages on Cladophora from the St. Lawrence River. *Diatom Research*, **12**, 55–70.

Osterman, L. E., Poore, R. Z., & Swarzenski, P. W. (2008). The last 1000 years of natural and anthropogenic low-oxygen bottom-water on the Louisiana shelf, Gulf of Mexico. *Marine Micropaleontology*, **66**, 291–303.

Peterson, C. G. (1986). Effects of discharge reduction on diatom colonization below a large hydroelectric dam. *Journal of the North American Benthological Society*, **5**, 278–89.

Rabalais, N. N., Turner, R. E., & Scavia, D. (2002). Beyond science into policy: Gulf of Mexico hypoxia and the Mississippi River. *BioScience*, **52**, 129–142.

Rabalais, N. N., Turner, R. E., Sen Gupta, B. K., Platon, E., & Parsons, M. L. (2007). Sediments tell the history of eutrophication and hypoxia in the northern Gulf of Mexico. *Ecological Applications*, **17**(5, Suppl.), S129–S143.

Rabalais, N. N., Turner, R. E., & Wiseman, W. J. Jr. (2003). Gulf of Mexico hypoxia, a.k.a. "The Dead Zone." *Annual Review of Ecology and Systematics*, **33**, 235–63.

Reavie, E. D. & Baratono, N. G. (2007). Multi-core investigation of a lotic bay of Lake of the Woods (Minnesota, USA) impacted by cultural development. *Journal of Paleolimnology*, **38**, 137–156.

Reavie, E. D. & Smol, J. P. (1997). Diatom–based model to infer past littoral habitat characteristics in the St. Lawrence River. *Journal of Great Lakes Research*, **23**, 339–348.

Reavie, E. D. & Smol, J. P. (1998a). Diatom epiphytes on macrophytes in the St. Lawrence River (Canada): characterization and relation to environmental conditions. In *Proceedings of the Fourteenth International Diatom Symposium, Tokyo, Japan 1996*, ed. S. Mayama, M. Idei, & I. Koizumi. Bristol: Biopress Ltd., pp. 489–500.

Reavie, E. D. & Smol, J. P. (1998b). Epilithic diatoms from the St. Lawrence River and their relationships to water quality. *Canadian Journal of Botany*, **76**, 251–7.

Reavie, E. D., Smol, J. P., Carignan, R., & Lorrain, S. (1998). Diatom paleolimnology of two fluvial lakes in the St. Lawrence River: a reconstruction of environmental changes during the last century. *Journal of Phycology*, **34**, 446–56.

Reid, M. A. & Ogden, R. W. (2006). Trend, variability or extreme event? The importance of long-term perspectives in river ecology. *River Research and Applications*, **22**, 167–177.

Reid, M. A. & Ogden, R. W. (2009). Factors affecting diatom distribution in floodplain lakes of the southeast Murray Basin, Australia and implications for palaeolimnological studies. *Journal of Paleolimnology*, **41**, 453–70.

Reid, M. A., Sayer, C. D., Kershaw, A. P., & Heijnis, H. (2007). Palaeolimnological evidence for submerged plant loss in a floodplain lake associated with accelerated catchment soil erosion (Murray River, Australia). *Journal of Paleolimnology*, **38**, 191–208.

Schelske, C. L. (1999). Diatoms as mediators of biogeochemical silica depletion in the Laurentian Great Lakes. In *Diatoms: Applications for the Environmental and Earth Sciences*, ed. E. F. Stoermer & J. P. Smol. Cambridge: Cambridge University Press, pp. 73–84.

Schönfelder, I. & Steinberg, C. E. W. (2002). How did the nutrient concentrations change in northeastern German lowland rivers during the last four millennia? – a paleolimnological study of floodplain sediments. *Studia Quaternaria*, **21**, 129–138.

Silvester, N. R. & Sleigh, M. A. (1985). The forces on microorganisms at surfaces in flowing water. *Freshwater Biology*, **15**, 433–48.

Simonsen, R. (1969). Diatoms as indicators in estuarine environments. *Veröffentlichungen des Institut für Meeresforschung*, **11**, 287–91.

Smol, J. P. (2008). *Pollution of Lakes and Rivers, a Paleoenvironmental Perspective*, 2nd edition, Malden, MA: Blackwell Publishing.

Sokal, M. A., Hall, R. I., & Wolfe, B. B. (2008). Relationships between hydrological and limnological conditions in lakes of the Slave River Delta (NWT, Canada) and quantification of their roles on sedimentary diatom assemblages. *Journal of Paleolimnology*, **39**, 533–50.

Teodoru, C., Dimopoulos, A., & Wehrli, B. (2006). Biogenic silica accumulation in the sediments of the Iron Gate I reservoir on the Danube River. *Aquatic Sciences*, **68**, 469–81.

Thoms, M. C., Ogden, R. W., & Reid, M. A. (1999). Establishing the condition of lowland flood-plain rivers: a palaeo-ecological approach. *Freshwater Biology*, **41**, 407–23.

Thomson, J. R., Hart, D. D., Charles, D. F., Nightengale, T. L., & Winter, D. M. (2005). Effects of removal of a small dam on downstream macroinvertebrate and algal assemblages in a Pennsylvania stream. *Journal of the North American Benthological Society*, **24**, 192–207.

Tibby, J. (2004). Development of a diatom-based model for inferring total phosphorus in southeastern Australian water storages. *Journal of Paleolimnology*, **31**, 23–36.

Tibby, J., Reid, M. A., Fluin, J., Hart, B. T., & Kershaw, A. P. (2003). Assessing long-term pH change in an Australian river catchment using monitoring and palaeolimnological data. *Environmental Science & Technology*, **37**, 3250–5.

Triplett, L. D., Engstrom, D. R., Conley, D. J., & Schellhaass, S. M. (2008). Silica fluxes and trapping in two contrasting natural impoundments of the upper Mississippi River. *Biogeochemistry*, **87**, 217–30.

Triplett, L. D., Engstrom, D. R., & Edlund, M. B. (2009). A whole-basin stratigraphic record of sediment and phosphorus loading to the St. Croix River, USA. *Journal of Paleolimnology*, **41**, 659–77.

van Dam, H. & Mertens, A. (1995). Long-term changes of diatoms and chemistry in headwater streams polluted by atmospheric deposition of sulphur and nitrogen compounds. *Freshwater Biology* **34**, 579–600.

Vörösmarty, C. J., Sharma, K. P., Fekete B. M., *et al.* (1997). The storage and aging of continental runoff in large reservoir systems of the world. *Ambio*, **26**, 210–19.

Ward, J. & Stanford, J. (1979). *The Ecology of Regulated Streams*. New York, NY: Plenum Press.

Wehr, J. D., & Thorp, J. H. (1997). Effects of navigation dams, tributaries, and littoral zones on phytoplankton of the Ohio River. *Canadian Journal of Fisheries and Aquatic Sciences*, **54**, 378–95.

Wolfe, B. B., Karst-Riddoch, T. L., Vardy, S. R., *et al.* (2005). Impacts of climate and river flooding on the hydro-ecology of a floodplain basin, Peace-Athabasca Delta, Canada since A.D. 1700. *Quaternary Research*, **64**, 147–62.

Zeng, H., Song, L., Yu, Z., & Chen, H. (2007). Post-impoundment biomass and composition of phytoplankton in the Yangtze River. *International Review of Hydrobiology*, **92**, 267–80.

Zimmerman H. J. & Ward, J. V. (1984). A survey of regulated streams in the Rocky Mountains of Colorado, USA. In *Regulated Rivers*, ed. A. Lillehammer & S. J. Saltveit. Oslo, Norway: Universitetsforlaget AS, pp. 251–62.

6 Diatoms as indicators of surface-water acidity

RICHARD W. BATTARBEE, DONALD F. CHARLES, CHRISTIAN BIGLER, BRIAN F. CUMMING, AND INGEMAR RENBERG

6.1 Introduction

Lake acidification became an environmental issue of international significance in the late 1960s and early 1970s when Scandinavian scientists claimed that "acid rain" was the principal reason why fish populations had declined dramatically in Swedish and Norwegian lakes (Odén, 1968; Jensen & Snekvik, 1972; Almer *et al.*, 1974). Similar claims were being made at about the same time in Canada (Beamish & Harvey, 1972). However, these claims were not immediately accepted by all scientists. It was argued by some that acidification was due to natural factors or to changes in catchment land use and management (Rosenqvist 1977, 1978; Krug & Frink, 1983; Pennington, 1984).

In the scientific debate that followed, diatom analysis played a pivotal role. It enabled the timing and extent of lake acidification to be reconstructed (Charles *et al.*, 1989; Battarbee *et al.*, 1990; Dixit *et al.*, 1992a) and allowed the various competing hypotheses concerning the causes of lake acidification to be evaluated (Battarbee *et al.*, 1985; Battarbee & Charles, 1994; Emmett *et al.*, 1994). However, diatoms had been recognized and used as indicators of water pH well before the beginning of this controversy. The acid rain issue served to highlight the importance of diatoms and stimulated the advance of more robust and sophisticated techniques, especially the development of transfer functions for reconstructing lake-water pH and related hydrochemical variables.

This chapter outlines the history of diatoms as pH indicators, and describes how diatoms are currently used in studies of acid and acidified waters. It then describes how diatom-based paleolimnological methods have been used to trace the pH and acidification history of lakes and how diatoms are being used to monitor acidity trends in streams and lakes. It draws on a wide literature and on a range of previous reviews (e.g. Charles & Norton, 1986; Battarbee & Charles, 1986; Charles *et al.*, 1989, 1994; Charles and Smol, 1994; Battarbee, 1991).

The Diatoms: Applications for the Environmental and Earth Sciences, 2nd Edition, eds. John P. Smol and Eugene F. Stoermer. Published by Cambridge University Press.

6.2 Diatoms as indicators of pH in acid waters

Modern measurements of stream- and lake-water chemistry show that acid waters occur in many parts of the world, but almost always in regions where catchment soils have low concentrations of base cations. Some acid waters are very clear, others, with peaty catchments, can be colored containing high concentrations (>10 mg l^{-1}) of dissolved organic carbon (DOC).

Almost all these waters are unproductive, but, except in the most extreme of cases, they are capable of supporting highly diverse, characteristic diatom floras. In streams, the epilithic community is usually the most common and diverse. In lakes there are more habitats. The epilithon and epiphyton have many taxa in common and have floras that can be similar to stream floras. For those lakes with sandy shorelines distinctive epipsammic communities occur, and most lakes support epipelic communities, but these are the least easy to sample and characterize and are therefore least known. In contrast to the benthos, diatom plankton in acid lakes is often poorly developed and some acid and acidified lakes completely lack planktonic diatoms.

Most early studies of diatoms in acid waters have focused on standing waters, principally because of the use made of diatoms in paleolimnological research, but there has been increasing attention given to running waters, principally as biological indicators used to characterize acid streams and to monitor changes in water quality (e.g. Round, 1991; Steinberg & Putz, 1991; van Dam & Mertens, 1995; Coring, 1996; Lancaster *et al.*, 1996; Planas, 1996; Kahlert & Andrén, 2005; Persson *et al.*, 2007; Stevenson *et al.*, this volume).

The strong relationship between diatom distribution and pH has been recognized for many decades, and the relationship extends throughout the pH range (e.g. Watanabe & Asai, 1999). The earliest classification of diatoms according to pH was presented by Hustedt in his monograph on the diatoms of Java, Bali, and Sumatra (Hustedt, 1937–1939). He studied over 650 samples from a wide variety of habitats and from his data argued that diatoms have different pH "preferences," and could be classified into the following groups:

1. alkalibiontic: occurring at pH values > 7
2. alkaliphilous: occurring at pH ~ 7 with widest distribution at pH > 7
3. indifferent: equal occurrences on each side of pH 7
4. acidophilous: occurring at pH ~ 7 with widest distribution at pH < 7
5. acidobiontic: occurring at pH values <7, with the optimum distribution at pH $= 5.5$

This classification has been immensely influential. It was adopted by most later diatomists and it became common practice for diatom floras to use the Hustedt terminology and concepts (Cleve-Euler, 1951–1955, Foged, 1977).

Hustedt's classification also became the basis of early attempts to reconstruct pH from diatom assemblages in lake sediment cores. The first was by Nygaard (1956). He developed a number of indices for deriving pH, the most used of which, index α, uses the ratio of the abundance of acidophilous and acidobiontic taxa to the abundance of alkaliphilous and alkalibiontic taxa in a sample, with the acidobiontic and alkalibiontic taxa being weighted by a factor of 5.

This basic approach was further developed by Meriläinen (1967), Renberg & Hellberg (1982), Charles (1985), Davis & Anderson (1985), Flower (1986), Baron *et al.* (1986), Arzet *et al.* (1986a, b), Charles & Smol (1988), Dixit *et al.* (1988), Steinberg *et al.* (1988) and Whiting *et al.* (1989) using a variety of simple and multiple linear regression techniques. These methods have been extensively reviewed (Battarbee, 1984; Battarbee & Charles, 1987; Davis, 1987).

Despite the demonstrable success of using the above approach, it was recognized that it had many weaknesses with several unjustifiable assumptions. The main problem, as with all classification systems, is the difficulty of assigning a taxon unambiguously to an individual class. The diatom literature has many examples of authors placing the same taxon into different classes, and pH reconstruction models can be shown to be extremely sensitive to mis-allocation of this kind (Oehlert, 1988).

Moreover, a number of invalid ecological and statistical assumptions associated with the linear regression approach to pH reconstruction have been pointed out by Birks (1987; this volume). In particular linear regression requires that species have a linear or monotonic relationship to pH. Whilst this is sometimes the case along short gradients, examination of species' distribution along longer gradients shows, as would be predicted from ecological theory, that species' responses are non-linear and predominantly unimodal, with a species' abundance rising to a maximum at the center of its pH range. Consequently, contemporary work centers on the use of unimodal response functions, whereby the pH optima of individual taxa are derived directly from modern observations of diatom distributions from sites situated along a pH gradient. Data sets generated by such observations are termed "training sets" (cf. Birks, this volume), and Charles (1990a) provides a useful guide to their compilation.

Whilst it would seem preferable, on ecological grounds, to use samples of living diatoms collected from individual benthic and planktonic habitats, paleolimnologists wishing to reconstruct past pH values from the diatom analysis of lake sediment cores have preferred to base their training sets on diatoms in surface-sediment samples (usually the top 0.5 or 1 cm levels). Surface-sediment samples have the advantage that they contain, or are assumed to contain, an integrated sample of diatoms representative of all habitats in the lake (cf. DeNicola, 1986; Cameron, 1995) in time and space. This assemblage is also the most analogous, in terms of sedimentary environment, to the core material used in pH reconstruction. The disadvantage of such samples, on the other hand, is that they may contain contaminants from older reworked sediment and the age span of the uppermost sample is usually unknown and varies both within and between lakes.

The earliest pH training sets were relatively small, often between 30 and 40 sites (e.g. Charles, 1985; Flower, 1986). Whilst such data sets perform quite well with older methods of pH reconstruction (e.g. Renberg & Hellberg, 1982), where pH classes based on literature reports are used, they are too small for methods that require pH optima to be calculated from contemporary measurements and from newly collected diatom samples. In particular, more sites are needed across longer pH gradients both to avoid artificial truncation of species' ranges and to provide more reliable estimates of pH optima, especially for uncommon taxa.

Larger training sets can be built most rapidly by combining smaller ones. However, where this involves more than one laboratory, problems of taxonomic consistency between laboratories becomes important. This is a concern, not only in ensuring correct estimates of optima, but also in allowing the correct application of transfer functions to sediment core assemblages (Birks, 1994). This issue has been addressed in both the Surface Water Acidification Project (SWAP) (Battarbee et al., 1990) and the Paleoecological Investigation of Recent Lake Acidification (PIRLA) project (Charles & Whitehead, 1986; Kingston et al., 1992a). In Europe it culminated in the European Diatom Database (EDDI) project (Battarbee et al., 2001) where regional training sets from all over Europe were combined into a single continental-scale training set. In each example differences in conventions between laboratories were resolved following taxonomic workshops, slide exchanges, the circulation of agreed nomenclature (e.g. Williams et al., 1988), taxonomic protocols (Stevenson et al., 1991) and taxonomic revisions (e.g. Flower & Battarbee, 1985; Camburn & Kingston, 1986). The EDDI project also developed a website with light microscope images to illustrate the taxonomic decisions taken in the harmonization process (http://craticula.ncl.ac.uk/Eddi/jsp/index.jsp).

In North America, the PIRLA project (Charles & Whitehead, 1986) produced a diatom iconograph (Camburn et al., 1986, Camburn & Charles, 2000) to aid in taxonomic identification and consistency between labs. The Diatom Paleolimnology Data Cooperative (DPDC) was subsequently created to store and allow easy access to diatom calibration and stratigraphic data sets, including many from acidification studies. Over 30 data sets can be downloaded from the DPDC website and combined to create larger data bases. Similar to the European approach, the groundwork laid by the iconograph and the DPDC allowed for the development of a 494-lake northeastern North America (NENA) diatom calibration set (Ginn et al., 2007a).

The availability of such training sets, which define relationships between diatom taxa and measured lake-water pH, is a prerequisite for the generation of transfer functions needed to reconstruct pH from sediment cores. Statistical methods developed in the 1980s (ter Braak & van Dam, 1989; Birks et al., 1990a; Birks, this volume) are still used today, but with some modifications (e.g. ter Braak & Juggins, 1993; Telford et al., 2006) and increasing realism (e.g. Cameron et al., 1999). Telford et al. (2006) assessed the proportion of more abundant diatom taxa that had a significant relationship to pH in a data set of 239 lakes from the northeastern USA (Dixit et al. 1999). Not surprisingly, of the 276 taxa examined, nearly 80% exhibited a statistically significant relationship to lake-water pH (Telford et al., 2006). Similar relationships using regression approaches have been found in other diatom data sets. For example, Denys (2006) showed that over 90% of 286 most abundant diatom taxa from 186 freshwater Belgium ponds were significantly related to lake-water pH, and Cameron et al. (1999) showed that ~ 90% of the more abundant diatom taxa had significant relationships to pH in a data set of almost 300 lakes. These results stand in strong contrast to a null-modeling approach that was used by Pither & Aarssen (2005) who concluded that only a small percentage (29%) of the 401 species examined in the data set of Dixit et al. (1999) exhibited a narrower pH breadth than expected by chance, and concluded that the majority of diatom taxa are pH generalists (cf. Pither & Aarssen, 2005, 2006; Telford et al., 2006). The discrepancies rest in the large type-II errors for rare taxa in the null-modeling approach, which is further confounded by the non-uniform distribution of the lakes across the pH gradient (Telford et al., 2006). It is clear from these studies that diatom taxa do exhibit a strong relationship to lake-water pH and that diatom-based pH reconstructions are robust.

6.3 Application of diatom analysis to questions of changing surface-water acidity

Surface-water acidification can be caused in many ways relating either to changes in the characteristics of stream and lake catchments or to the incidence of acid pollution. The main processes are long-term soil acidification over the post-glacial period, changes in the base-cation status of soils associated with changes in catchment characteristics and land use, and an increase in sulfur (S) and nitrogen (N) deposition from combustion of fossil fuels. More locally, acidification can also be caused by acid-mine drainage, by the direct discharge of industrial effluents, and by the disturbance and drainage of sulfide-rich soils.

6.3.1 Natural acidification processes

6.3.1.1 Long-term soil acidification The tendency for certain lakes with catchments on base-poor or slow-weathering bedrock to become gradually more acid during the post-glacial time period has been recognized for many decades (Lundqvist, 1924; Iversen, 1958). Long-term acidification has been reported from many countries (Charles et al., 1989), and has been ascribed to base-cation leaching and paludification of catchment soils (Pennington, 1984).

Round's study of Kentmere in the English Lake District (Round, 1957) showed, from diatom analysis, that the beginning of acidification was coincident with the Boreal-Atlantic transition (about 7500 BP), a time when upland blanket peat was beginning to form in the surrounding hills.

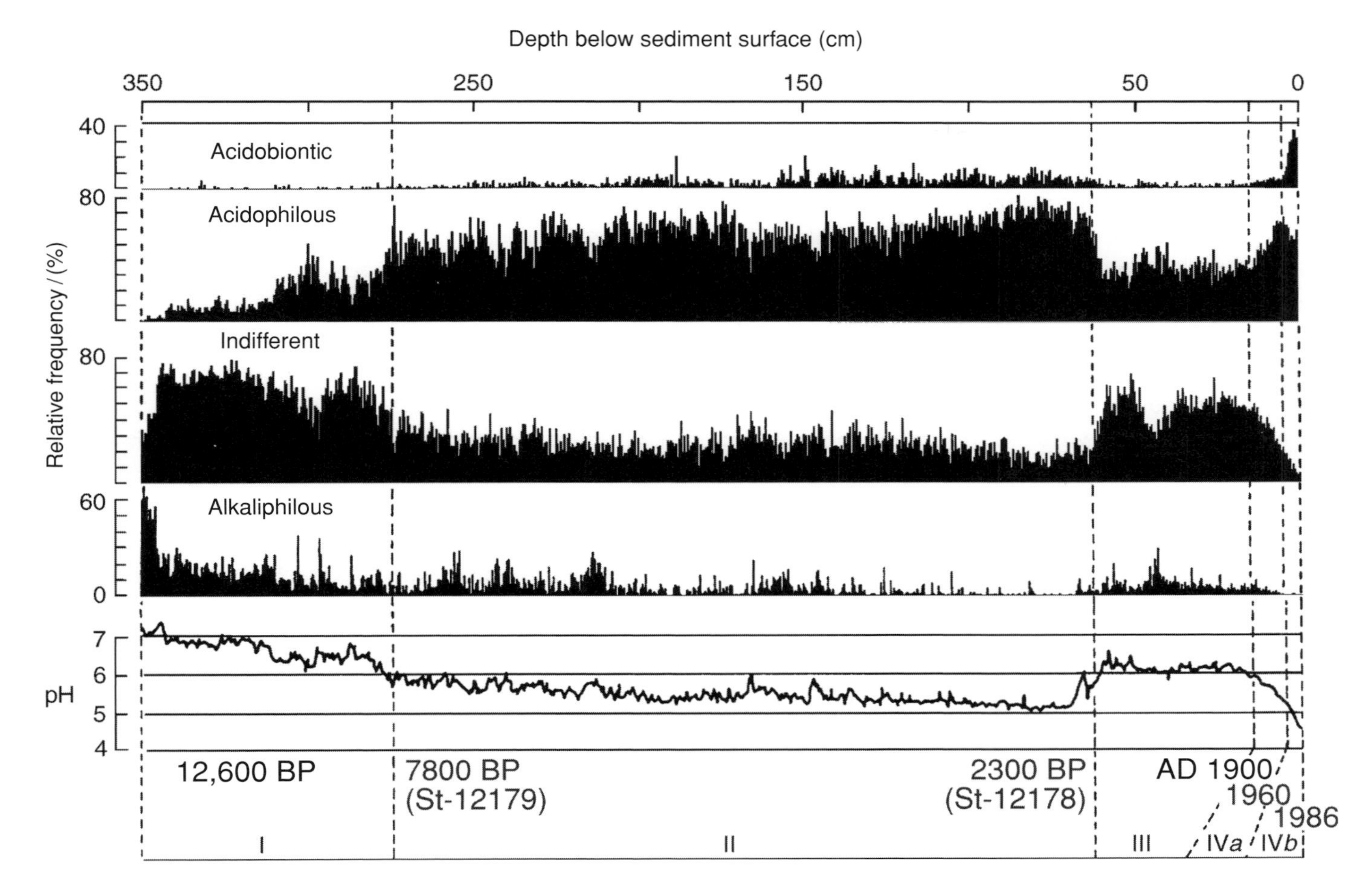

Figure 6.1 Hustedt pH categories, diatom-inferred pH values (from weighted averaging, WA), calibrated radiocarbon dates, ^{210}Pb dates and pH periods in the history of Lilla Öresjön, southwest Sweden (from Renberg 1990).

The change was characterized by a replacement of a predominantly alkaliphilous *Epithemia–Fragilaria–Cyclotella* assemblage by a more acidophilous *Eunotia–Achnanthes–Cymbella–Anomoeoneis* (= *Brachysira*)–*Gomphonema* assemblage. Further early studies of this kind, using diatoms and identifying long-term acidifying trends in lakes, have been carried out, especially in northwest Europe, including Scotland (Alhonen, 1968; Pennington *et al.*, 1972), northwest England (Round, 1961; Haworth, 1969; Evans, 1970), north Wales (Crabtree, 1969; Evans & Walker, 1977), Sweden (Renberg, 1976, 1978), and Finland (Huttunen *et al.*, 1978).

Following the emergence of interest in acid deposition and its effects, there was an increase in long-term acidification studies (Renberg & Hellberg, 1982; Whitehead *et al.*, 1986, 1989; Jones *et al.*, 1986, 1989; Winkler, 1988; Renberg, 1990; Ford, 1990; Korhola & Tikkanen, 1991; Korsman, 1999; Bradshaw *et al.*, 2000; Bigler *et al.*, 2002; Solovieva & Jones, 2002). There were a number of reasons for this interest: (i) to test the hypothesis that present low values of pH in many upland waters were not due to acid deposition but to long-term natural processes (cf. Pennington, 1984; Jones *et al.*, 1986); (ii) to show that long-term acidification could reduce catchment alkalinity and make lakes more vulnerable to the impact of acid deposition (Battarbee, 1990); and (iii) to demonstrate the difference between the nature of recent change in comparison to the rate and amplitude of earlier natural changes (Jones *et al.*, 1986; Winkler, 1988; Renberg *et al.*, 1990; Ford, 1990; Birks *et al.*, 1990b). There is now renewed interest in diatom records from early Holocene sediments of upland lakes following Boyle's suggestion that the cause of acidification might be due to mineral depletion, specifically through weathering loss of apatite from catchment soils (Boyle, 2007). Engstrom *et al.* (2000) have demonstrated how a comparison of newly formed lakes of different ages created by glacial retreat in the subarctic can be used to test hypotheses about diatom responses to early-Holocene lake evolution.

Although some present-day lakes have been acid (pH < 5.6) for the entire Holocene (Jones *et al.*, 1989; Winkler

1988), most acid lakes studied so far were less acid in their initial stages. In the case of Lilla Öresjön (Figure 6.1) and other lakes in southern Sweden, Renberg (1990) and Renberg *et al.* (1993a, b) recognized a natural long-term acidification period following deglaciation around 12,000 BP to about 2000 BP during which period pH decreased from about 7 to 5.5 resulting from soil acidification. At Loch Sionascaig in Scotland, and Devoke Water, in England (Atkinson & Haworth, 1990), the basal diatom assemblages indicate pH values of 7 and above. At both sites pH declined during the early post-glacial period, stabilizing at about pH 6.5 before 8500 BP in Loch Sionascaig, and by about 5000 BP in Devoke Water. Unlike Lilla Öresjön, neither of these two lakes continued to acidify and pH at both sites remained constant at about 6.2–6.6 until the twentieth century. Further examples of early acidification are Pieni Majaslampi (Korhola & Tikkanen, 1991) and Njargajavri (Sarmaja-Korjonen *et al.*, 2006), both in Finland. In Pieni Majaslampi the earliest lake phase was acid (pH about 5.5 to 6.0) and fell rapidly within the first 2000 years of lake history to pH < 5.0. This is a rare example of acidification to such extremely low pH values as a result of natural processes. In Njargajavri, the decline is from approximately pH 7.5 to 5.5, as a result of base depletion and changes in catchment land cover.

In North America, Whitehead *et al.* (1986, 1989) studied the post-glacial acidification history of three sites along an elevational gradient in the Adirondack Mountains (NY). They argued that the rate of acidification was most rapid at the highest site and slowest at the lowest site, corresponding to catchment soil depth, slope, and amount of rainfall. Ford (1990) demonstrated that the pH of Cone Pond in New Hampshire had dropped to <5.5 by 6500 BP and to around pH 5.0 by as early as 2000 BP, whereas nearby S. King Pond, which has thicker glacial tills and some limestone in its catchment, did not undergo acidification. In contrast, Duck Pond, Massachusetts, a kettle lake located in an outwash plain composed of crystalline sands, has been very acid, with a pH *c.* 5.3 throughout its 12,000 year history (Winkler, 1988).

6.3.1.2 Post-glacial changes in catchment vegetation

Whilst the dominant time-dependent acidifying process during the post-glacial period is thought to be the mineral depletion of catchment soils, the rate and extent of acidification is also moderated by vegetation changes, and the impact of vegetation change on soils. In some cases the vegetation change may be driven by changing soil nutrient and base status, but in other cases climate change, catchment disturbance (e.g. by fire or wind, or early human interference) may be more important (see below). What is unclear at present is the extent to which studies of diatom assemblages in lake sediments can be used to discriminate between these different acidifying processes and identify the specific role of vegetation change.

For example, in the study of the Round Loch of Glenhead, originally designed to assess the extent to which mid-post-glacial catchment paludification and the spread of blanket mires could be the cause of lake acidification, diatom changes that were coincident with changes in the vegetation and soils did not indicate any significant depression in lake-water pH (Jones *et al.*, 1986). Changes in DOC may have been more important (Jones *et al.*, 1986; Birks *et al.*, 1990b).

In similar studies, Renberg *et al.* (1990) and Korsman *et al.* (1994) assessed the potentially acidifying influence of the colonization of lake catchments by spruce by examining the diatom response to the natural immigration of spruce about 3000 years ago into northern and central Sweden. Eight lakes were selected that were acidified or sensitive to acidification at the present time and that had minimal amounts of peatland in their catchments. Pollen analysis was used to identify levels in the sediment cores that corresponded to the arrival of spruce in the region and diatom analysis to identify any consequent acidification. The results showed that despite a major shift in vegetational composition towards spruce no change in lake-water pH was indicated by diatoms.

In contrast, in the study of North Pond, Massachusetts, a long-term decrease in diatom-inferred pH occurred as hemlock (*Tsuga*) gradually expanded between 7500 and 3500 BP (Huvane & Whitehead, 1996). In Nygaard's classic study of Lake Gribsø in Denmark, early acidification was shown to have occurred at around AD 400 coincident with the spread of beech forest (Nygaard, 1956), and, in a Finnish study, the rapid expansion of the acidobiontic diatom taxa *Tabellaria binalis* (Ehrenb.) Grun. and *Semiorbis hemicyclus* (Ehrenb.) Patr., coincided with the appearance and increase in the pollen of spruce (*Picea*) from around 3700 BP (Salomaa & Alhonen, 1983).

Changes in catchment vegetation do not always lead towards acidification. There are examples where both decreased and increased acidity in lake water has occurred. Decreases in acidity are usually associated with transient catchment disturbances such as wind throw of forest trees or fire (Rhodes & Davis, 1995; Korhola *et al.*, 1996; Huvane & Whitehead, 1996; Korsman & Segerström, 1998). Huvane and Whitehead (1996) showed that a sudden drop in the pollen frequency for *Tsuga* in North Pond, Massachusetts, coincided with an increase in charcoal particles and a marked peak of the planktonic diatoms *Asterionella formosa* Hassall, *Cyclotella comta* (Ehrenb.) Kütz. and

Synedra radians Kütz. It seems probable that this was the result of a fire in the catchment causing the rapid release of nutrients and base cations to the lake as organic matter was burnt. Likewise in Makkassjön, a small lake located in northern Sweden, long-term changes in lake-water pH occurred simultaneously with a change from deciduous to coniferous forest, but with an increasing proportion of coniferous trees in the lake catchment, forest fires became more frequent. As a consequence, fire, as indicated by the charcoal record in the sediment core, became a significant driver of lake alkalinity during the conifer period (Korsman & Segerström, 1998).

6.3.2 Human modification of catchment soils and vegetation

Diatom analysis has been used frequently to assess the extent to which human modification of catchment soils and vegetation has changed the acidity of lakes. There are some cases where there is clear evidence of an increase in pH following land-use change, cases where there has been no impact, and, in special circumstances involving modern forestry, cases where land-use change has exacerbated the acidifying impact of acid rain.

With the exception of natural fires described above (e.g. Huvane & Whitehead, 1996; Virkanen *et al.* 1997), there is little or no evidence that poorly buffered lakes have experienced an increase in pH during the Holocene that can be ascribed to natural causes. The only sustained increases in pH for such lakes appears to be associated with the impact of early agriculture, probably also involving fire (Renberg *et al.*, 1990). Following the observation (Figure 6.1) of increased pH of Lilla Öresjön at about 2300 BP, Renberg *et al.* (1993b) carried out pH reconstruction for a further 12 lakes in Southern Sweden. In almost all cases they demonstrated that increases in diatom-inferred pH had occurred, in some cases by over 1 pH unit, and they showed from pollen and charcoal analysis that the increases were associated with evidence for agriculture dating between 2300 BP and 1000 BP. Elevated pH was maintained, probably due to continued burning, preventing the accumulation of acid humus in catchment soils, until the abandonment of agriculture and subsequent reafforestation in the nineteenth century. It was also suggested by Bindler *et al.* (2002) that deposition of pre-industrial atmospheric pollution contributed to a higher pH of the water. Suggested mechanisms are that sulfur deposition from mining and metal industries in Europe in the period AD 1200–1800, which was larger than usually assumed, caused enhanced cation exchange in catchment soils; and also that nitrogen and trace elements could have contributed to the fertilization of these very oligotrophic lakes. The more recent acidification of many of these lakes, although mainly due to atmospheric acid deposition, was also in part due to a re-acidification process following the decline in agriculture. A similar conclusion was reached by Davis *et al.* (1994), who studied the acidification history of 12 lakes in New England. In this case, catchment disturbance associated with logging caused an increase in lake-water pH. As in Sweden acidification of these lakes in the last century was ascribed first to re-acidification following the regrowth of conifers and secondly, as pH values fell below pre-logging levels, to acid deposition.

In the SWAP Palaeolimnology Programme (Battarbee & Renberg, 1990), studies were designed specifically to test the hypothesis that land-use change was an alternative cause of surface-water acidification to acid deposition. Sites, such as Lilla Öresjön, where both catchment land-use change and acid deposition could be acting together were complemented by sites where either land-use change or acid deposition could be eliminated as a potential cause (see below).

For this latter approach it was necessary to choose sites situated in sensitive areas with low levels of acid deposition at the present time, or to use historical analogues. Anderson & Korsman (1990) argued that a good historical analog for modern land-use changes is provided by the effects of depopulation of farms and villages in Hälsingland, northern Sweden during Iron Age times, an event well established by archeologists. Following abandonment of the area from *c.* AD 500, re-settlement did not occur before the Middle Ages (*c.* AD 1100). Although evidence for regeneration of forest vegetation during this period can be identified in pollen diagrams from lakes in this region, there is no evidence from diatom analysis for lake acidification at the two sites studied, Sjösjön and Lill Målsjön (Anderson & Korsman, 1990). Similarly, Jones *et al.* (1986) showed that there was no evidence for pH decline at the Round Loch of Glenhead in Scotland at the time when pollen evidence indicated deforestation and an increase in catchment peat cover during Neolithic- and Bronze-Age times.

An exception to the proposition that land-use change is unimportant in causing surface-water acidification is the apparent increased acidity of lakes and streams in the United Kingdom associated with modern forestry plantations. Comparisons between adjacent afforested and moorland streams in the uplands of the UK have shown that afforested streams have lower pH, higher aluminum and sulfate concentrations, and poorer fish populations than moorland streams (Harriman & Morrison, 1982). These differences have been mainly attributed either to the direct effect of forest growth (Nilsson *et al.*, 1982) or to the indirect effect of the forest canopy enhancing ("scavenging") dry and occult sulfur deposition (e.g. Unsworth 1984), or to a combination of these effects. In an attempt to separate these factors in time and space Kreiser *et al.* (1990) devised a paleolimnological study comparing the acidification histories of afforested (Loch Chon) and non-afforested (Loch Tinker) sites in the Trossachs region of Scotland, an area of high sulfur deposition, with afforested (Loch Doilet) and non-afforested (Lochan Dubh) sites in the Morvern region, an area of relatively low sulfur deposition. All four sites showed evidence of acidification over the last century or so and all show contamination over this time period by trace metals and carbonaceous particles, although concentrations of these contaminants in the northwest region of Scotland are very low (Kreiser *et al.*, 1990).

6.3.3 The influence of strong acids

Unless significant quantities of organic acids are present, bicarbonate alkalinity generation in stream and lake catchments is usually sufficient to maintain the pH of naturally acid, or naturally acidified, waters above pH 5.0–5.3. Unusually acid waters (pH < 5.0), consequently, only occur where there are natural sources of strong mineral acids (Renberg, 1986; Yoshitake & Fukushima, 1995) or where sites are polluted either by acid-mine drainage (Fritz & Carlson, 1982; Brugam & Lusk, 1986), the direct discharge of acidifying industrial effluents (van Dam & Mertens, 1990), or by acid deposition associated with fossil-fuel combustion.

6.3.3.1 Natural sources of sulfur Unusually acid waters can occur where lake waters are influenced directly by sulfur, as in the case of some volcanic lakes (Yoshitake & Fukushima, 1995), or where catchment soils are sulfur rich (Renberg, 1986). In a study of six volcanic lakes in Japan with pH ranging from 2.2 to 6.2, Yoshitake & Fukushima (1995) observed high standing crops of diatoms even at pH 2.2 to 3.0. At these extremely low values, *Pinnularia braunii* (Grun.) Cleve was the dominant taxon. Other common diatoms included *Eunotia exigua* (Bréb. ex Kütz.) Rabenh., *Aulacoseira distans* (Ehrenb.) Simonsen, and *Anomoeoneis brachysira* (= *Brachysira brebissonii* R. Ross in Hartley). Renberg (1986) describes an example of a lake in Northern Sweden recently isolated from Bothnian Bay (Baltic Sea) with a modern pH of 3. Diatom analysis of a core from this lake showed that shortly after isolation the lake was dominated by a largely circumneutral diatom flora of *Fragilaria* spp. and *Aulacoseira* spp. However, a rapid change in the flora took place with these taxa being completely replaced, the sole dominant becoming *E. exigua*. Renberg ascribes the acidification of the lake to the exposure and oxidation of sulfides that had previously been deposited in the Baltic Sea, exacerbated by drainage activities that took place in the catchment to enhance cultivation of the soils.

6.3.3.2 Acid-mine drainage (AMD) and industrial waste-water discharge Drainage from mines, especially abandoned mines, can cause extreme surface-water acidification through bacterial oxidation of pyrite (FeS_2) to sulfuric acid. Values of pH as low as 2 can occur, although most diatom species cannot survive pH values below 3.5 (DeNicola, 2000). *E. exigua* in particular is often abundant (Hargreaves *et al.*, 1975) along with *Nitzschia* and *Pinnularia* spp. (DeNicola, 2000).

In order to assess whether acid-mine waters progressively neutralize over time, Brugam & Lusk (1986) studied 48 mine drainage lakes in the US midwest and carried out diatom analysis on sediment cores from 20 of them. Six of the lakes were initially acid, characterized by species such as *Pinnularia biceps* Greg., but became alkaline during their post-mining history, indicated by dramatic taxonomic shifts to strongly alkaliphilous taxa such as *Rhopalodia gibba* (Ehrenb.) O. Müll. and *Navicula halophila* (Grun. ex Van Heurck). These floristic changes represent changes in water pH of 4 or 5 pH units and are far more striking than changes associated with effects of acid deposition where changes of 1 pH unit or less are more usual. Recovery studies (e.g. Fritz & Carlson, 1982) document the typical transition from assemblages dominated by very few benthic taxa when pH is below 4.5 to increasing numbers of epiphytic and planktonic species as pH increases.

Diatoms are used increasingly to assess AMD damage to streams and to determine the effectiveness of reclamation efforts. Studies of rivers influenced by AMD in the USA, New Zealand, and Europe, among other regions, show strong relationships between acidity-related chemistry and taxonomic composition (Verb & Vis, 2000, 2005; Sabater *et al.*, 2003; Bray *et al.*, 2008; Luís *et al.*, 2009). Concentrations of metals are particularly important. These relationships were defined primarily using multivariate analyses and metrics and were used to assign sites to AMD categories (e.g. unaffected, acidified, recovering, recovered). Species composition, taxon richness, biomass and productivity can all be affected by AMD, but in different ways depending on combinations of pH, metal concentration, physical stress (e.g. metal oxide precipitate), nutrients, and grazing pressure. If pH is sufficiently low and metals or metal oxides are high, the number of species and biomass both decline due to direct physiological effects. But if chemistry and physical effects are moderate, enough to suppress grazers but not algae, then biomass can increase significantly and have a large influence on the organic matter budget of a stream (Planas *et al.*, 1989; Niyogi *et al.*, 2002; Sabater *et al.*, 2003).

Perhaps the best known example of acidification from industrial discharges is the case of Lake Orta in northern Italy where nitrification processes following the release of ammonia from a cupro-ammonia rayon factory since 1926 caused pH levels to fall to *c.* pH 4 (Mosello *et al.*, 1986). Liming began in 1989 and the lake pH has increased to about 5.2. Van Dam & Mertens (1990) collected epilithic samples from the lake and showed that the present flora is dominated by acid and metal-tolerant taxa, especially *E. exigua* and *Pinnularia subcapitata* var. *hilseana* (Janisch ex Rabenh.) O. Müll. Examination of earlier descriptions of diatom samples from the lake in 1884 and 1915 showed that these taxa were absent and that the lake was dominated by circumneutral and alkaliphilous taxa.

6.3.4 Acid deposition

6.3.4.1 Diatom evidence for recent surface-water acidification As already pointed out it was the acid-rain issue that stimulated an upsurge in interest in diatoms as indicators of water acidity. In this context the earliest use of diatom analysis in Europe was by Miller (1973), Berge (1976, 1979, and Davis & Berge, 1980), van Dam & Kooyman-van Blokland (1978), van Dam *et al.* (1981), Renberg and Hellberg (1982), and Flower & Battarbee (1983). In North America similar studies were underway most notably by Del Prete & Schofield (1981) and Charles (1985) in the Adirondacks, by Davis *et al.* (1983) in New England, and Dixit (1986) and Dixit *et al.* (1987) in Canada. Whilst most of these studies focused on sediment cores, Berge (1976) and van Dam *et al.* (1981) demonstrated the usefulness of comparing old diatom samples from herbarium collections with present-day samples collected from the same locations. In the study by van Dam *et al.* (1981), old diatom samples, collected in 1920 from 16 moorland ponds, were used to infer the past range of pH in the pools. They estimated that pH amongst the 16 ponds ranged between 4 and 6 in the 1920s, but this had narrowed to a range from pH 3.7 to 4.6 by 1980 owing to acidification. Floristically, there had been a major increase in the abundance of *E. exigua* in the humic-poor pools, and decreases in a range of taxa including *Eunotica veneris* (= *incisa* W. Sm. ex Greg.), *Fragilaria virescens* Ralfs and *Frustulia rhomboides* var. *saxonica* (Rabenh.) De Toni. These sites have subsequently been resurveyed to assess the response of diatom assemblages to acid deposition reduction (see below).

Early studies of sediment cores cited above did not have the benefit of using fully developed transfer functions, but the various techniques of pH reconstruction that were used showed beyond doubt that rapid, recent acidification had taken place in many areas of North America and Europe that were both sensitive to acidification and that were receiving significant amounts of acid, especially sulfur, deposition. In particular it was clear that planktonic floras of *Cyclotella* taxa were disappearing, and that acid-tolerant taxa such as *Tabellaria binalis*, *T. quadriseptata* Knudson, *Navicula subtilissima* Cleve, and *Eunotia incisa* were becoming more abundant. Moreover, inspection of the dates of these changes (Battarbee, 1984) showed that the onset of acidification at these lakes approximately coincided with the history of fossil-fuel combustion in the countries concerned, with the earliest acidification taking place in the UK (mid-nineteenth century). In North America and Scandinavia, the same changes did not occur until the twentieth century. A more recent compilation of acidification dates for European lakes has confirmed that the initial acidification of almost all sites post-dates 1850 (Battarbee *et al.*, in press).

A typical acidification sequence can be illustrated at the Round Loch of Glenhead in Scotland (Figure 6.2) that has been intensively studied (Flower & Battarbee, 1983; Flower *et al.*, 1987; Jones *et al.*, 1989). The lake and its catchment are situated entirely on granitic bedrock and the catchment has mainly peaty soils and a moorland vegetation that is grazed by sheep.

The Round Loch of Glenhead sediment core was dated by the ^{210}Pb method (Appleby *et al.*, 1986) and the proportion of the main diatom species at successive sediment levels were calculated. The diagram (Figure 6.2) shows that there had been little change in the diatom flora of this lake until the mid-nineteenth century when diatoms characteristic of circumneutral water such as *Brachysira vitrea* (Grun.) R. Ross began to decline and be replaced by more acidophilous species. Acidification continued through the twentieth century and by the early 1980s the diatom flora of the lake was dominated only by acid-tolerant taxa such as *T. quadriseptata* and *T. binalis*. The reconstructed pH curve at 40 cm depth (about AD 1700) in the core (Figure 6.2) shows a stable pH of about 5.4 until about 1850, followed by a reduction of almost 1 pH unit in the following ~130 years to almost the top of the core, taken in 1985.

Despite the excellent correspondence between the incidence of acid deposition and this kind of evidence for the timing and extent of surface-water acidification, there was far from universal acceptance that the acute decline in fish populations in many upland lakes was caused by acid deposition.

Consequently, many diatom studies that were carried out to address the acid-rain problem focused on research designs to test alternative hypotheses, especially those outlined above

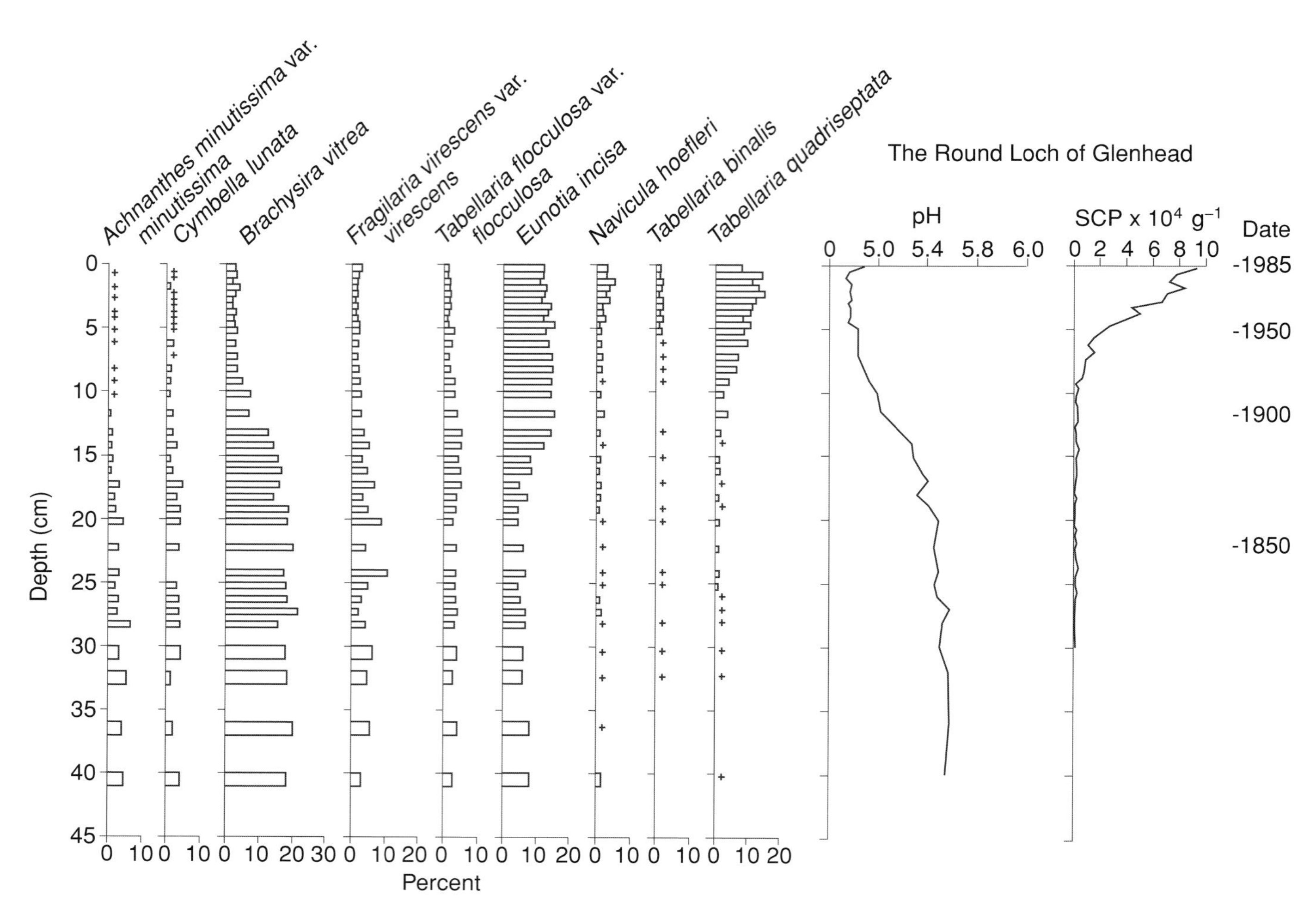

Figure 6.2 Summary diatom diagram from the Round Loch of Glenhead with ^{210}Pb dates, diatom-inferred pH and spheroidal carbonaceous particles (SCPs) . (Modified from Jones *et al.* 1989.)

relating to catchment influences. In the case of the Round Loch of Glenhead, and similar lakes in the UK, it was possible to argue that there were no alternative explanations, as catchment burning and grazing intensity had not decreased over this period, and there was no afforestation of the catchment. Moreover, whilst base-cation leaching was an important acidifying process in some catchments, especially in the early phases of lake development, this was a very slow process, and for most lakes the rate of base-cation loss is balanced by primary mineral weathering.

Perhaps the most conclusive evidence for the impact of acid rain comes from the diatom analysis of remote lakes perched on hilltops, where only changes in the acidity of precipitation can account for lake acidification. Many such sites occur in the Cairngorm Mountains of Scotland. The high corrie lakes in this area have steep-sided boulder-strewn catchments with little soil and sparse vegetation. Despite their isolation and lack of catchment change, these sites are all acidified, and their sediments contain high levels of atmospheric contaminants (Battarbee *et al.*, 1988; Jones *et al.*, 1993). Similar results were obtained in Norway, where two hill-top lakes, Holetjörn and Ljosvatn, in Vest-Agder, southwest Norway, were studied (Birks *et al.*, 1990c). Although the sites were naturally acid, both showed a rapid drop in reconstructed pH values in the early twentieth century, coincident with increasing contamination of the lake by trace metals and fossil-fuel derived carbonaceous particles.

Evidence that acid deposition could not only cause lake acidification, but was the primary reason for it, was established when the distribution of acidified sites was examined and compared to the regional pattern of sulfur deposition, both in Europe (e.g. Renberg & Hellberg, 1982; Davis *et al.*, 1983; Arzet *et al.* 1986a, b; Flower *et al.*, 1987; Battarbee *et al.*, 1988; van Dam, 1988) and in North America (Charles *et al.*, 1989; Davis *et al.*, 1990, 1994b; Sweets *et al.*, 1990; Dixit *et al.*, 1992b, c; Cumming *et al.*, 1992, 1994; Ginn *et al.*, 2007b, c). Strongly acidified sites

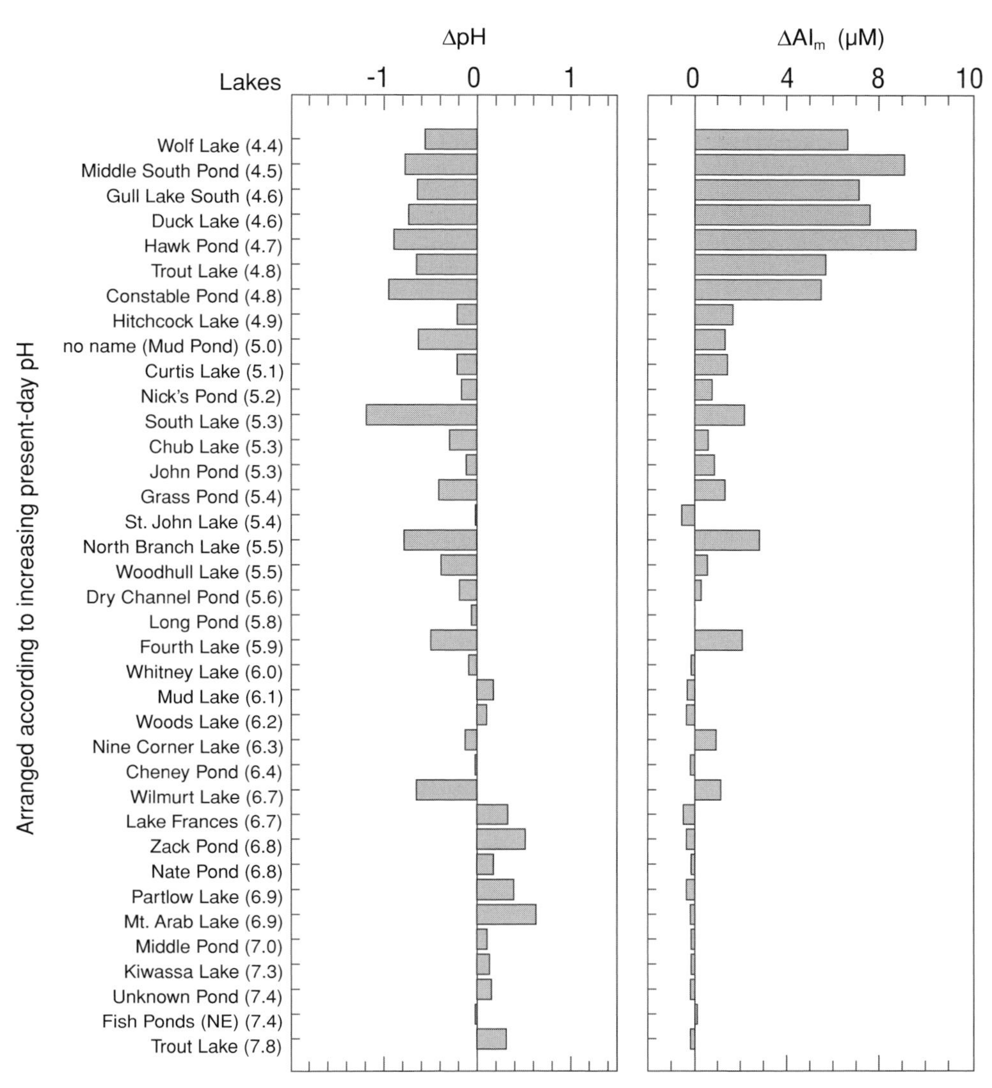

Figure 6.3 Diatom-based estimates of historical change (pre-industrial to 1987) in pH and total monomeric aluminum from 37 randomly selected Adirondack lakes. The estimates are presented as the differences in diatom-inferred chemistry between the top (0–1 cm) and bottom (>20 cm, usually >25 cm) sediment core intervals. The lakes are arranged according to increasing present-day measured pH. See Cumming *et al.* (1994) for further details.

are only found in areas of high sulfur deposition. In areas of very low sulfur deposition acidification on a regional scale only occurs in the most sensitive areas (Battarbee *et al.*, 1988; Whiting *et al.*, 1989; Holmes *et al.*, 1989; Charles, 1990b; Cumming *et al.*, 1994; Ginn *et al.* 2007c) and, at the most remote, sensitive sites show no evidence for recent acidification (e.g. Korsman, 1999; Cameron *et al.*, 2002). Such a dose–response relationship has been demonstrated especially in the case of the UK where Battarbee (1990) was able to quantify the relationship and show how the incidence of lake acidification could be predicted using an empirical model relating sulfur deposition at a site to lake-water calcium concentration (Battarbee *et al.*, 1996).

For some regions of North America, studies of lake acidification have been carried out by comparing the diatom-inferred pH of samples taken from the bottoms (pre-acidification) and tops (modern) of sediment cores (e.g. Cumming *et al.*, 1992; Dixit & Smol, 1994; Ginn *et al.*, 2007b). This approach allows many lakes to be studied in a relatively short period of time, and by selecting lakes on a statistical basis, population estimates of the extent of acidification have been made (Sullivan *et al.*, 1990; Cumming *et al.*, 1992). These studies showed that the currently most acid lakes acidified the most since pre-industrial times and that lakes with higher buffering capacities did not acidify (e.g. Figure 6.3). Similar conclusions were reached by Weckström *et al.* (2003) for sites in the

Kola Peninsula that were close to and more distant from metal smelters, and by Michelutti *et al.* (2001) in the Norilsk region of Russia.

Acidification of streams, especially episodic acidification following snowmelt or rainstorms, affects many systems in acid-sensitive areas. But until recently, diatoms were not often used to study biological responses. Several experimental studies have shown that stream diatoms respond to pH-related factors and that forms, concentrations, and interactions of aluminum and DOC in particular can have a large influence on diatom species composition and biomass (Planas *et al.*, 1989; Planas, 1996; Genter, 1995; Gensemer and Playle, 1999). These findings are corroborated by field studies (e.g. Round, 1991; Kovács *et al.*, 2006; Passy, 2006), and there has been at least one transplant study that shows species composition can change relatively rapidly, within three to nine days (Hirst *et al.* 2004). Several existing methods work reasonably well to characterize pH-related conditions, including the use of pH categories (Kahlert & Andrén, 2005), multivariate analysis (e.g. Lancaster *et al.*, 1996), and inference models (Steinberg & Putz, 1991; Kahlert & Andrén, 2005; Kovács *et al.*, 2006). Methods to assess episodic conditions are particularly difficult because of rapid temporal changes in pH and relationships between related chemistry (Coring, 1996). Passy (2006) studied episodic acidification of two streams in the Adirondack Mountains, USA. One drains a wetland and has chronic low pH and the highest DOC. The other has no wetland, and greater episodic declines in pH and increases in aluminum due to pulses of nitrate. The stream draining the wetland had greater diatom taxonomic diversity and species composition was explained well by environmental factors. The clearer, more episodic stream had a less diverse and variable assemblage, and temporal factors (e.g. day of year) explained a larger component of variability. MacDougal *et al.* (2008) characterized the most stressful episodic conditions in five Pennsylvania, USA, streams (pH 5–6.5) as those having the highest total dissolved aluminum as calculated by a model. This measure correlated with algal biovolume and non-linearly with the abundance of *Eunotica exigua*, but not with other species or species richness. One of the most successful approaches for dealing with temporal variation of pH is the Diatom Assemblage Type Analysis (DATA) described by Coring (1996). This system defines five acidification categories (e.g. type 3: periodically acid streams – pH normally < 6.5, minima < 5.5). It provides verbal descriptions of each category, listing combinations of specific taxa and expected range of abundances, and a key to classify a sample into one of the categories. This system performed better than the inference model (Index B) approach for streams in Germany (Coring, 1996) and Sweden (Kahlert & Andrén, 2005).

6.3.4.2 Use of diatoms in monitoring recovery from the effects of acid deposition Mitigating the effects of surface-water acidification caused by acid rain or other pollution sources requires either increasing the base status of the water body (by liming) or reducing acid inputs, e.g. by controlling emissions of acidifying sulfur and nitrogen compounds from fossil-fueled industrial plants. Whilst liming has been carried out systematically for decades in Sweden (Appelberg & Svenson, 2001; Guhrén *et al.*, 2007), and more recently and occasionally in other countries (Sandøy & Langåker 2001; Kwandrans, 2007), long-term management of acidification requires reduction of acid emissions at the source (e.g. Convention on Long-range Transboundary Air Pollution, 1979; US Clean Water Act, 1972; European Water Framework Directive, 2000).

As a consequence of the dominant role played by diatom analysis in acid-rain studies, diatoms are now recognized as the premier biological indicators of surface-water acidity and are being used throughout Europe and North America for water-quality monitoring to assess the response of lakes and streams to the major reduction in sulfur emissions that has taken place over recent decades.

Europe

Diatom response to liming Liming has been used as an acidification mitigation strategy in Sweden more than in any other country. The Swedish liming program was initiated in 1976 and in 1989 a monitoring program (Integrated Studies of the Effects of Liming in Acidified Waters, or ISELAW) was designed to assess the long-term ecological effects of liming, the extent of recovery to a pre-acidification state, and detection of possible collateral effects of lime treatment (Appelberg & Svenson, 2001). In the project, sediment records from 12 lakes were analyzed. In eight lakes with complete post-glacial sediment records, natural pH values (prior to human disturbance) as inferred by diatoms indicated pH values between 5.3 and 6.5. The early agricultural activities, as inferred by pollen, started 1000 to 2000 years ago, and pH increased as a result of changing land use in six of the lakes. Overall, five of the investigated lakes have acidified during recent decades, and in all 12 lakes some effects of liming were detected within the diatom communities. The lakes show different responses to liming, including a return to pre-acidification communities or a shift to a state previously not recorded in the lake's histories (Guhrén *et al.*, 2007).

Liming has not been used as a general means of mitigating surface-water acidification in the UK, but several sites have been limed for experimental or demonstration purposes. The most thoroughly studied site was Loch Fleet in Galloway, Scotland (Howells & Dalziel, 1991). The catchment of this lake was limed in 1986 and Cameron (1995) was able to sample the benthic diatom flora and take sediment cores from the lake prior to liming, enabling an assessment of the response of the flora to a more or less instantaneous increase in pH. He used sediment traps and repeated sediment coring to assess how quickly and how faithfully floristic changes were recorded in the sediment. It was shown from diatom analysis of a long sediment core from the lake that the liming gave rise to planktonic diatom communities that had not previously occurred in the lake and was therefore not a technique to restore the lake's natural pre-acidification flora and fauna.

Diatom response to a reduction in acid deposition Emissions of acidifying gases, especially sulfur dioxide, from power stations and metal smelters has fallen dramatically over the last 20 years and there have been equally significant reductions in sulfur deposition and in non-marine sulfate concentration in streams and lakes across the continent (Skjelvåle *et al.*, 2005, 2007). There have been many studies using diatoms to assess the response of surface waters to a reduction in acid deposition in Europe (Jüttner *et al.*, 1997; Sienkiewicz *et al.*, 2006; Kwandrans, 2007) and in some countries, principally Sweden, the UK, and the Netherlands, diatoms have been included in national acidification monitoring programs.

In Sweden, diatoms have been used to assess the ecological responses of lakes to emission reduction both from local smelters and from long-range transported air pollutants. One of the most important local sources of sulfur dioxide emissions in Sweden was the Falun copper mine. After running for about one millennium and with a peak in the seventeenth century it was closed in 1990. A diatom study of fourteen lakes in the vicinity showed that pH had decreased at eight lakes by 0.4 to 0.8 pH units (Ek & Renberg, 2001) with lakes with high total organic carbon (>9 mg l^{-1}) having acidified less than those with lower total organic carbon (<7 mg l^{-1}), highlighting the important role of the buffering capacity of organic compounds (Ek & Korsman, 2001). None of the acidified lakes, however, showed any signs of recovery, possibly because large amounts of sulfur still remain in the catchment soils or because afforestation in the area over the last 100 years is playing a role in delaying the recovery process (Ek & Renberg, 2001; Ek *et al.*, 2001).

In other Swedish lakes that have been severely acidified by long-range air pollutants, sulfur deposition has decreased substantially and water chemistry has improved. However, diatom communities do not indicate a general pH recovery (Ek *et al.*, 1995). An exception is Lake Örvattnet where analysis of the uppermost sediments showed a decrease in the relative abundance of *Tabellaria binalis*, indicating a slight rise in pH, dating from the 1970s. Although there was no clear increase in measured water pH during this time, acid episodes in spring have become less severe and there has been a decrease in sulfate deposition by between 30–40%. In this case, diatom analysis of the core was a more sensitive and integrative measure of recovery than water chemistry monitoring (Ek *et al.*, 1995).

In the United Kingdom an acid waters monitoring network of 22 sites was established in 1988 (Patrick *et al.*, 1995; 1996). This network includes sensitive streams and lakes in areas of high and low acid deposition, and at each site diatoms, in addition to other key chemical and biological indicators, are monitored. Epilithic diatoms are collected once per year from both lake and stream sites and sediment trap samples are collected annually from the lakes. Chemical data from almost all sites show that sulfate and labile aluminum levels have decreased and that there has been a general increase in alkalinity. However, calcium levels have decreased, DOC has increased at all sites, and nitrate has changed little. As a result there have been only modest increases in pH at the most acidified sites and only small, but clear, changes in epilithic diatom floras. At the Round Loch of Glenhead, for example, there has been a decrease in the abundance of the acid-tolerant species *Tabellaria quadriseptata*, a species that was rare in the lake before acidification (see Figure 6.2) and an increase in *Navicula leptostriata* Jorgensen (not shown in Figure 6.2). This latter species appears now to be more abundant than in the past indicating that the reduction in acidity is not simply leading to a reversal in assemblage composition but to new assemblages. This response can be observed at several other sites in the network prompting the hypotheses either that diatom recovery is following a hysteretic trajectory or that other environmental factors are at play, possibly climate change.

Diatoms have also been used to assess the extent of recovery from acidification in the Netherlands. Van Dam & Mertens (2008) have monitored a series of Dutch moorland pools since 1978 using repeated surveys of the diatom floras every year for three pools and every four years for a further eight pools. They used, as reference conditions, samples from an early study of the same pools conducted by Heimans in 1920 (see

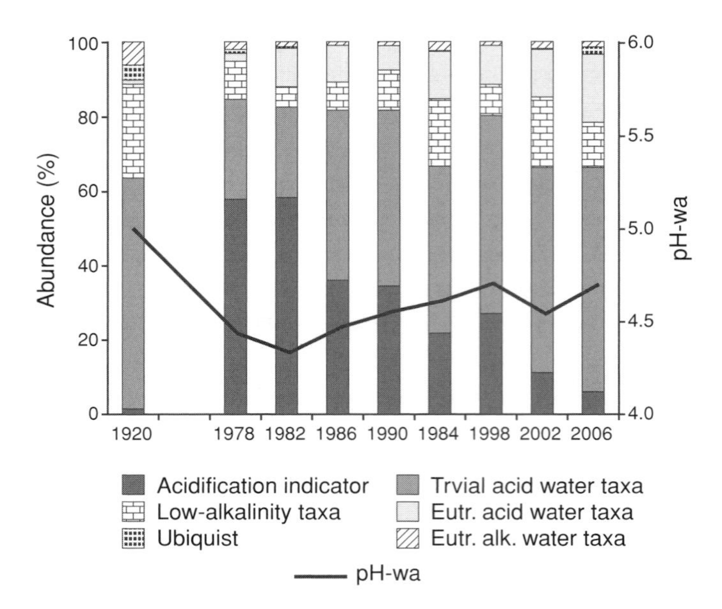

Figure 6.4 Overall changes in the composition of diatoms in 93 samples from 11 Dutch soft-water moorland pools compared to the "reference" collection of ca 1920. The solid line represents changes in the main inferred pH values for all samples, calculated using a weighted-average statistic. (Modified from van Dam & Mertens, 2008.)

above). In order to make the old and new samples comparable they followed the same sampling technique as Heimans by pulling a plankton net along the bottom of the pool and across the littoral macrophyte beds. The authors classified the diatoms into six groups, depending on their indicator value for acid waters, and showed that a striking decline in the proportion of acid-tolerant taxa has occurred since 1978 (Figure 6.4) reflecting a strong decrease in ammonium, sulfate, and nitrate in bulk deposition. However, in comparison to the reference samples, it is evident that the recovery from acidification is only a partial one and that the detailed composition of diatom assemblages at the present day is dissimilar from those in 1920. The changes observed therefore may not only represent a response to a decrease in acid deposition but, as observed above for the UK, be related to other processes, possibly increased nutrient recycling associated with pronounced warming.

North America

In North America, industrial emissions of sulfur dioxide (SO_2) have been reduced by at least 40% since the early 1970s (USEPA, 2000; Niemi, 2005) resulting in a decrease in lake-water sulfate concentrations and in some cases an increase in pH or alkalinity, together with a decrease in monomeric aluminum (Stoddard *et al.*, 1999; Driscoll *et al.*, 2003, 2007; Skjelkvåle *et al.*, 2005). While any increases in pH and alkalinity have been small, a few regions in North America (e.g. Sudbury and Wawa, Ontario) have experienced reductions in SO_2 emissions in excess of 90%. These reductions have resulted in large increases in pH and alkalinity and biological recovery in a number of aquatic ecosystems (e.g. Gunn & Keller, 1992; Locke *et al.*, 1994).

Paleolimnological studies from the Sudbury region in the early 1990s provided convincing evidence that diatom assemblages in sedimentary records can track short-term changes in lake-water pH following reductions in sulfate deposition (Dixit *et al.*, 1992a, b). For example, Baby Lake, located about 1 km southwest of the Coniston Smelter near Sudbury, showed marked changes in diatom species composition associated with acidification, and then recovery, as emissions from the smelter declined (Dixit *et al.*, 1992a). More recently, Tropea (2008) examined diatom and chrysophyte assemblages from sediment cores from three Sudbury-region lakes that increased in lake-water pH in recent decades. Paleolimnological evidence from two of these lakes (Swan Lake, Dixit *et al.*, 1989; Daisy Lake, Dixit *et al.*, 1996) indicate that these lakes reached inferred pH values of < 4 and 5 in the 1970s (Dixit *et al.*, 1989) and the late 1980s (Dixit *et al.*, 1996). The currently measured lake-water pH for Swan Lake and Daisy Lake was 5.5 and 6.6 in 2006 (Tropea, 2008). Measured lake-water pH for the third lake studied was 6.6 in 2006, up from a measured value of 5.0 in 1981 (Tropea, 2008). In cores from all three lakes, Tropea (2008) shows clear shifts from circumneutral diatom assemblages to acid and metal-tolerant taxa in the early to mid 1900s following the onset of open-pit roasting and smelting operations in Sudbury. The impact of these emissions was severe, causing acidification of many lakes (e.g. Dixit *et al.*, 1992c). Following large post-1970 emission declines, and increases in measured lake-water pH, diatom assemblages in these study lakes showed little evidence of a change towards pre-industrial assemblages, despite clear trends towards more neutral assemblages of scaled chrysophytes in all three cores (Tropea, 2008). The lack of recovery in diatom assemblages was related to the high levels of metal contamination in the sediments. Similarly, Greenaway (2009) examined the acidification history of lakes affected by a sintering plant in Wawa (Ontario) that reached peak emissions of over 250,000 gigatonnes in 1975 and that subsequently underwent huge declines from the mid 1970s to the time it closed in 1998 (Rowe, 1999). The local emissions from the sintering plant resulted in acidification of a number of lakes northeast of the site despite the carbonate-rich geology underlying lakes in this region (Somers & Harvey, 1984). Greenaway (2009) examined five lakes in the "fume-kill" area that recovered in lake-water pH from pH values between pH 3 and 4 (Somers & Harvey, 1984) to current pH values between 6.5 and 7.5 (Greenaway, 2009). Pre-disturbance assemblages contained circumneutral to alkaline diatoms (e.g. *Cyclotella stelligera/pseudostelligera*, *Aulacoseira* spp., and *Achnanthes minutissima*), and all five lakes exhibited substantial increases in *Eunotia* and other acid-tolerant taxa, from trace amounts to between ~50 and 70% in post-1970 sediments. The changes in diatom assemblages were concurrent with large increases in sedimentary concentrations of manganese, aresenic and iron associated with the sintering activities. Despite the clear recovery in lake-water pH, diatom changes consistent with recovery were clearly evident in only two lakes (Otter Lake and Little Soulier Lake) where increases in the relative abundance of *Cyclotella* spp., *A. minutissima* and *Brachysira vitrea* (Grun.) R. Ross in Hartley, have occurred in the most recent sediments. Trends of recovery in the other three lakes were less obvious, and the most recent diatom assemblages were still more similar to the acid flora than to the pre-industrial assemblages. Greenaway (2009) also attributed the minimal diatom response to the elevated concentrations of metals in the sediments, but resolution and sediment chronology were likely complicating factors as well. Although changes in diatom assemblages have tracked increasing pH in many recovering Sudbury lakes, the communities may not return to their pre-disturbance condition. Dissimilarities between pre-disturbance and present-day communities result from changes in several factors at both regional and local scales (e.g. Dixit *et al.*, 2002; Tropea, 2008; Greenaway, 2009).

Paleolimnological investigation of changes in diatom assemblages following a ~40% reduction in acid deposition have also been undertaken in regions lacking metal contamination, including the Atlantic region of Canada and the Adirondacks (NY, USA). Atlantic Canada has some of the most acid lakes in North America due to the low buffering provided by the regional bedrock and high amounts of organic acids from wetlands (Clair *et al.*, 2007). Despite measured reductions in sulfate deposition in Atlantic Canada, improvements in lake-water pH and alkalinity have not been observed, although there has been a reduction in the severity of acid pulses in the spring (Clair *et al.*, 2007). Ginn *et al.* (2007c) examined the acidification histories of 14 Nova Scotia lakes including eight lakes in Kejimkujik National Park, a region of relatively high sulfate deposition in the southwestern part of Nova Scotia and six lakes in the Cape Breton Highlands, northern Nova Scotia, a region of much lower sulfate deposition. Not surprisingly, lakes that received the highest sulfate deposition all showed increases in acid diatom taxa and inferred acidification between 1925 and 1940. Lakes with lower *c.* 1850 pH values acidified first (Ginn *et al.*, 2007c). Consistent with the monitoring data, all of these lakes show no or little change in taxonomic composition of diatom assemblages (Ginn *et al.*, 2007c) over the past few decades. Sediment diatom assemblages in the northern lakes were consistent with little impact from acid deposition, with only one exception, Glasgow Lake, a very acid-susceptible lake (Ginn *et al.*, 2007c; Gerber *et al.*, 2008).

Paleolimnological assessments of recovery in the Adirondack region are just beginning. Changes in scaled chrysophyte assemblages (Cumming *et al.*, 1994; Majewski & Cumming, 1999) suggest that some lakes were still acidifying or were in "steady state" up until 1988, and 2004, respectively, when the cores were taken. This result is consistent with measured water-chemistry data that show no increases in pH and acid neutralizing capacity (ANC) in Adirondack lakes until the mid to late 1990s (Driscoll *et al.*, 2003, 2007). Analysis of a 2007 core from Big Moose Lake, the most intensively studied lake in the Adirondacks from a paleolimnological perspective (e.g. Charles *et al.*, 1990, 1991; Cumming *et al.*, 1992, 1994; Majewski & Cumming, 1999), shows a decrease in the relative abundances of the most acid-tolerant taxa (e.g. *Fragilaria acidobiontica* (Charles) M. Williams et Round, *Navicula tenuicephala* Hustedt) with a shift to taxa with slightly higher pH optima (e.g. *Frustulia pseudomagaliesmontana* Camburn et Charles, *Frustulia rhomboides* var. *saxonica* (Rabenh.) De Toni) (Cumming *et al.* unpublished data). These results suggest that diatom assemblages are in the early stages of recovery in this lake.

6.3.5 Reference conditions, restoration targets, and models

Whereas diatom analysis in the 1980s played a major role in providing evidence for the timing and causes of acidification, it is now a principle technique being used in restoration strategies that rely on establishing past pre-acidification reference conditions to define management targets for the future (e.g. Bennion & Battarbee, 2007; Norberg *et al.*, 2008; Renberg *et al.*, 2009; Norberg *et al.*, 2010). This can be achieved either by using the diatom assemblage from old sediments immediately predating the first sign of acidification as a biological reference itself or by using diatom-inferred pH as a hydrochemical reference (cf. Battarbee *et al.*, in press).

In the case of diatom assemblages, the similarity (or dissimilarity) between the diatom composition of reference assemblages and that of modern (core-top) assemblages can be used as a measure of ecological status. It can also be used as a measure of progress towards a restoration objective, especially where the temporal trajectory represented by core assemblages

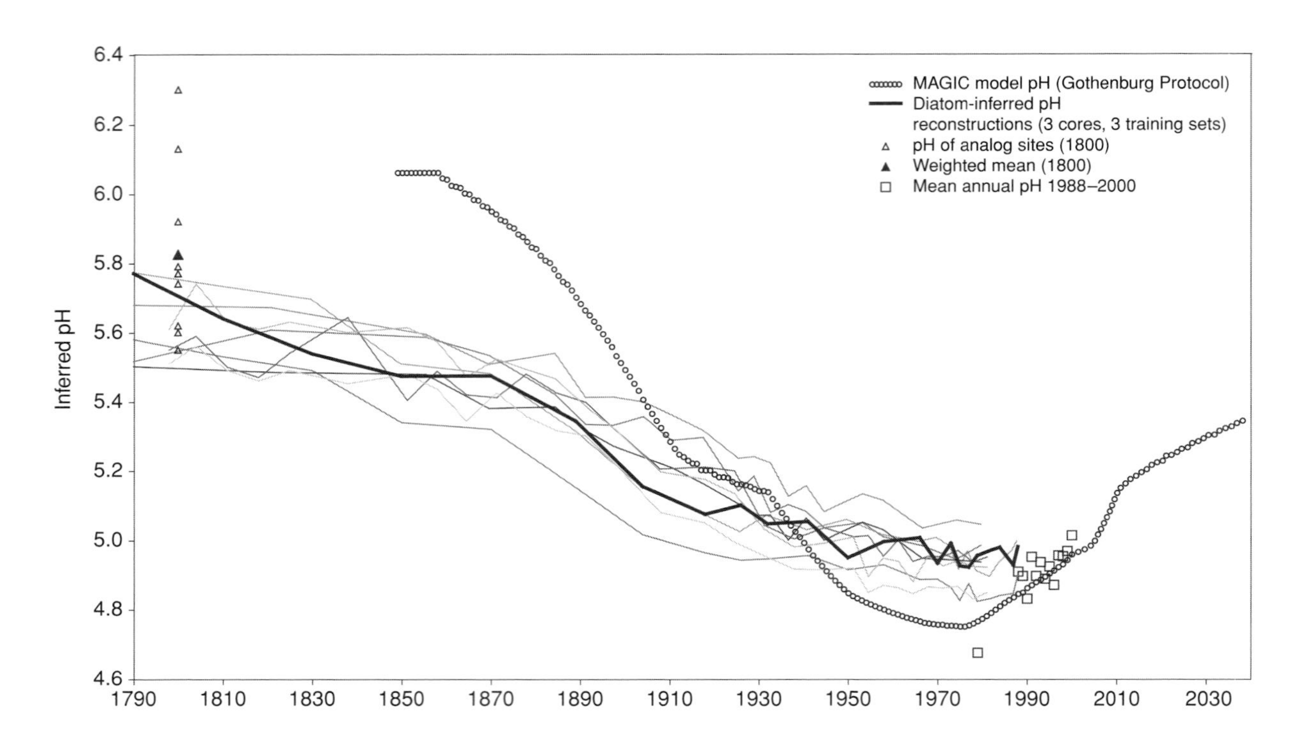

Figure 6.5 Comparison of pH reconstruction outputs and annual measured pH at the Round Loch of Glenhead. Chronology of diatom-inferred pH according to SWAP, UK and EDDI models (fine lines) for ^{210}Pb-dated samples from three sediment cores (RLGH 81, RLGH 3 and K05). The RMSEP of the SWAP, UK and EDDI training sets are 0.38, 0.31, and 0.25 pH units respectively. The modern annual pH values of nine lakes providing the strongest biological analogs for a pre-acidification (*c.* 1800) sediment sample (open triangles) are shown together with the weighted average of these (filled triangle). The MAGIC model pH reconstruction is indicated by open circles and mean annual average pH values for the period 1988–2000 and the year 1979 are shown by open squares. (From Battarbee *et al.*, 2005.)

can be compared to diatom assemblages from other acid-sensitive lakes with analogous floras. Using data from the UK Acid Waters Monitoring Network Battarbee *et al.* (unpublished data) have used this approach to assess the extent of recovery with respect to reference assemblages and whether the diatom response is in the direction of the reference assemblage or deflected from it.

Reference diatom assemblages can also be used to infer reference pH values using a diatom-based pH transfer function. Although the transfer-function technique is well established as described above there has been little work designed to assess the uncertainty of the method, and this becomes an issue of considerable importance if the diatom-reconstructed pH values for reference periods are to be used as targets for restoration by lake managers or are to be used to verify process-based models. To address this issue, Battarbee *et al.* (2008) compared pH inferences from diatom data collected annually in sediment traps from lakes in the UK acid-waters monitoring network with measured water-column pH over the last 18 years using three different training sets. Whilst the trends in pH inferred from the trapped diatoms followed closely the trends in mean measured pH, there was a tendency for the diatom-inferred values to underestimate the measured values, with the pH values generated by the SWAP training set being the most biased. Nevertheless, given the natural variability in interannual pH in acid waters, the results indicated that the approach was robust.

The robustness of the transfer-function approach for pH allows diatom-based pH reconstruction to be used confidently as a method for testing models such as the Model of Acidification of Groundwater in Catchments (MAGIC; Cosby *et al.*, 2001) and there have been several studies comparing diatom reconstructions with MAGIC model pH hindcasts (e.g. Wright *et al.*, 1986; Jenkins *et al.*, 1990; Sullivan *et al.*, 1996; Hinderer *et al.*, 1998; Korsman, 1999; Kram *et al.*, 2001; Stuchlík *et al.*, 2002; Battarbee *et al.*, 2005; Erlandsson *et al.*, 2008). In most cases, the results show reasonably good agreement between model and diatom-based pH, although there are some significant discrepancies at some sites (e.g. Hinderer *et al.*, 1998) and overall there is a tendency for model estimates to yield higher reference pH values than diatom-based estimates, as in the case of the Round Loch of Glenhead (Battarbee *et al.*, 2005, Figure 6.5).

In this case the mismatch was attributed to a weakness in the model with respect to its treatment of DOC, a key variable which is not treated dynamically. There is now increasing evidence that DOC concentrations in acidified lakes and streams are closely related to catchment soil acidity (Monteith *et al.*, 2007) implying that DOC as well as pH was higher in the past under reference conditions.

6.3.6 Climate change, acidity, and diatoms

During the acid-rain debate, there was little concern for the potential influence of climate change on surface-water acidity and the composition of diatom floras. It is now a central focus of research, stemming from the study by Psenner & Schmidt (1992) of diatom-inferred pH from two cores in the Austrian Tyrol and their correspondence to instrumental temperature records over the last ~200 years. Their work showed not only that temperature was probably the principal driver of pH in the period before industrialization, but that excursions of increased pH in the twentieth century could be attributed to climate warming, even though the dominant trend towards decreasing pH was the result of acid deposition. A more striking example of the dependence of pH on climate variability in the absence of acid deposition was presented by Sommaruga-Wögrath *et al.* (1997), who compared diatom records and instrumental temperature records at a higher alkalinity site insensitive to acid deposition. At this site, pH increased towards the present day in line with global warming. Wolfe (2002) has also argued for a strong relationship between climate and pH from his analysis of cores from Baffin Island in the Canadian Arctic. In this case a close correspondence between neoglacial cooling over the last 5000 years and decreasing pH was observed.

As sulfur deposition continues to decrease in North America and Europe, temperature increase is likely to emerge as the strongest driver of lake acidity at many sites, and, at some, pH values may reach levels only previously experienced in the warmer early Holocene (Larsen *et al.*, 2006), probably higher than nineteenth century pre-acidification levels. A key consideration for future research is the extent to which diatom populations in acid waters will respond to the combined effects of decreasing acid deposition and increasing temperature. The potential effects of changing precipitation patterns will also need to be taken into account, as increasing precipitation and storminess predicted for some mid- to high-latitude regions may cause more acid conditions in future through higher stream-flow episodes (Evans *et al.*, 2008). In addition, increased drought frequency may delay the recovery of acidified lakes through drying and oxidation of sulfur-polluted catchment wetlands (Faulkenham *et al.*, 2003).

6.4 Conclusions

Although the use of diatoms as indicators of pH has a long history, it was the emergence of surface-water acidification as a major environmental problem that brought diatoms to prominence. This rapidly accelerated the development of the robust statistical methodology that is currently used not only for pH reconstruction from sediment cores but also for the reconstruction of other hydrochemical variables, such as total phosphorus (Bennion, 1994; Hall & Smol, this volume), salinity (Fritz *et al.*, 1991, and this volume), and other variables.

Traditionally, diatom analysis has been used to show how lakes have developed over the Holocene period in relation to base-cation leaching and catchment paludification, fire, and the impact of early agriculture. In the acid-rain controversy of the 1980s, diatom analysis played a pivotal role in demonstrating how the recent acidifying impacts of acid deposition could be differentiated in space and time from the influence of these largely natural catchment-based processes.

Diatom techniques are now playing important roles in the management of acidified waters. In particular diatom analysis is being used to differentiate between acid and recently acidified surface waters, to determine critical loads of acidity, to evaluate models of acidification, to assess the extent of episodic acidification of streams, and to monitor the response of acidified waters to reduced acid deposition following mitigation policies.

In less-acidified regions, diatom-based pH reconstruction is important in identifying pristine and relatively pristine sites where variations in lake-water pH are the result of natural forces, including climatic variability. The correspondence between pH and temperature demonstrated in the Austrian Alps indicates that diatoms, in the absence of masking by the effects of acid deposition, could be used as proxies for lake temperature. Future work on diatoms as indicators of pH needs to evaluate these observations and assess the way in which variations in lake-water temperature and other climate variables can influence pH in low-alkalinity lakes.

Acknowledgments

The authors wish to thank Herman van Dam for his valuable contributions to this chapter. Don Charles is grateful for support through the Ruth Patrick Chair in Environmental Science and Rick Battarbee acknowledges funding from the EU FP6 project Euro-limpacs (GOCE-CT-2003–505540).

References

Alhonen, P. (1968). On the late-glacial and early post-glacial diatom succession in Loch of Park, Aberdeenshire, Scotland. *Memoranda Societatis Pro Fauna et Flora Fennica*, **44**, 13–20.

Almer, B., Dickson, W., Ekström, C., Hörnström, E., & Miller, U. (1974). Effects of acidification on Swedish lakes. *Ambio*, **3**, 30–6.

Anderson, N. J. & Korsman, T. (1990). Land-use change and lake acidification: Iron-Age de-settlement in northern Sweden as a pre-industrial analogue. *Philosophical Transactions of the Royal Society, London*, **B 327**, 373–6.

Appelberg, M. & Svenson, T. (2001). Long-term ecological effects of liming – the ISELAW programme. *Water, Air and Soil Pollution*, **130**, 1745–50.

Appleby, P. G., Nolan, P. J., Gifford, D. W., *et al.* (1986). ^{210}Pb dating by low background gamma counting. *Hydrobiologia*, **143**, 21–7.

Arzet, K., Krause-Dellin, D., & Steinberg, C. (1986b). Acidification of four lakes in the Federal Republic of Germany as reflected by diatom assemblages, cladoceran remains, and sediment chemistry. In *Diatoms and Lake Acidity*, ed. J. P. Smol, R. W. Battarbee, R.B. Davis, & J. Meriläinen. Dordrecht, The Netherlands: Dr. W. Junk, pp. 227–50.

Arzet, K., Steinberg, C., Psenner, R., & Schulz, N. (1986a). Diatom distribution and diatom inferred pH in the sediment of four alpine lakes. *Hydrobiologia*, **143**, 247–54.

Atkinson, K. M. & Haworth, E. Y. (1990). Devoke Water and Loch Sionascaig: recent environmental changes and the post-glacial overview. *Philosophical Transactions of the Royal Society, London*, **B 327**, 349–55.

Baron, J., Norton, S. A., Beeson, D. R., & Herrmann, R. (1986). Sediment diatom and metal stratigraphy from Rocky Mountain lakes with special reference to atmospheric deposition. *Canadian Journal of Fisheries and Aquatic Sciences*, **43**, 1350–62.

Battarbee, R. W. (1984). Diatom analysis and the acidification of lakes. *Philosophical Transactions of the Royal Society, London*, **B 305**, 451–77.

Battarbee, R. W. (1990). The causes of lake acidification, with special reference to the role of acid deposition. *Philosophical Transactions of the Royal Society, London*, **B 327**, 339–47.

Battarbee, R. W. (1991). Recent palaeolimnology and diatom-based reconstruction. In *Quaternary Landscapes*, ed. L. C. K. Shane & E. J. Cushing. Minneapolis: University of Minnesota Press, pp. 129–74.

Battarbee, R. W., Allott, T. E. H., Juggins, S., *et al.* (1996). An empirical critical loads model for surface water acidification, using a diatom-based palaeolimnological approach. *Ambio*, **25**, 366–9.

Battarbee, R. W., Anderson, N. J., Appleby, P. G., *et al.* (1988). *Lake Acidification in the United Kingdom 1800–1986: Evidence from Analysis of Lake Sediments*. London: Ensis.

Battarbee, R. W. & Charles, D. F. (1986). Diatom-based pH reconstruction studies of acid lakes in Europe and North America: A synthesis. *Water, Air and Soil Pollution*, **31**, 347–54.

Battarbee, R. W. & Charles, D. F. (1987). The use of diatom assemblages in lake sediments as a means of assessing the timing, trends, and causes of lake acidification. *Progress in Physical Geography*, **11**, 552–580.

Battarbee, R. W. & Charles, D. F. (1994). Lake acidification and the role of paleolimnology. In *Acidification of Freshwater Ecosystems: Implications for the Future*, ed. C. Steinberg & R. Wright, Dahlem Workshop Environmental Sciences Research Report 14, Chichester: Wiley, pp. 51–65.

Battarbee, R. W., Flower, R. J., Stevenson, A.C., & Rippey, B. (1985). Lake acidification in Galloway: a palaeoecological test of competing hypotheses. *Nature*, **314**, 350–2.

Battarbee, R. W., Juggins, S., Gasse, F., *et al.* (2001). European Diatom Database (EDDI). An Information System for Palaeoenvironmental Reconstruction. University College London, Environmental Change Research Centre, ECRC Research Report, 81.

Battarbee, R. W., Mason, J., Renberg, I., & Talling, J. F. (eds.) (1990). *Palaeolimnology and Lake Acidification*. London: The Royal Society.

Battarbee, R. W., Monteith, D. T., Juggins, S., *et al.* (2005). Reconstructing pre-acidification pH for an acidified Scottish loch: a comparison of palaeolimnological and modelling approaches. *Environmental Pollution*, **137**, 135–49.

Battarbee, R. W., Monteith, D. T., Juggins, S., *et al.* (2008). Assessing the accuracy of diatom-based transfer functions in defining reference pH conditions for acidified lakes in the United Kingdom. *The Holocene* **8**, 57–67.

Battarbee, R. W., Morley, D., Bennion, H., *et al.* (in press). A palaeolimnological meta-database for assessing the ecological status of lakes. *Journal of Paleolimnology*.

Battarbee, R. W. & Renberg, I. (1990). The Surface Water Acidification Project (SWAP) Palaeolimnology Programme. *Philosophical Transactions of the Royal Society, London*, **B 327**, 227–32.

Beamish, J. & Harvey, H. H. (1972). Acidification of La Cloche Mountain lakes, Ontario, and resulting fish mortalities. *Journal of the Fisheries Research Board of Canada*, **29**, 1131–43.

Bennion, H. (1994). A diatom–phosphorus transfer function for shallow, eutrophic ponds in southeast England. *Hydrobiologia*, **275**, 391–410.

Bennion, H. & Battarbee, R.W. (2007). The European Union Water Framework Directive: opportunities for palaeolimnology. *Journal of Paleolimnology*, **38**, 285–95.

Berge, F. (1976). Kiselalger og pH i noen elver og innsjöer i Agder og Telemark. En sammenlikning mellom årene 1949 og 1975. [Diatoms and pH in some rivers and lakes in Agder and Telemark (Norway): a comparison between the years 1949 and 1975.] Sur Nedbörs Virkning på Skog og Fisk. IR18/76. Aas, Norway: Norwegian Forest Research Institute.

Berge, F. (1979). Kiselalger og pH i noen innsjöer i Agder og Hordaland. [Diatoms and pH in some lakes in the Agder and Hordaland counties, Norway.] Sur Nedbörs Virkning på Skog og Fisk. IR42/79. Aas, Norway: Norwegian Forest Research Institute.

Bigler, C., Larocque, I., Peglar, S. M., Birks, H. J. B., & Hall, R. I. (2002). Quantitative multiproxy assessment of long-term patterns of Holocene environmental change from a small lake near Abisko, northern Sweden. *The Holocene*, **12**, 481–96.

Bindler, R., Korsman, T., Renberg, I., & Högberg, P. (2002). Pre-industrial atmospheric pollution: was it important for the pH of acid-sensitive Swedish lakes? *Ambio*, **6**, 460–5.

Birks, H. J. B. (1987). Methods for pH calibration and reconstruction from palaeolimnological data: procedures, problems, potential techniques. Surface Water Acidification Programme, Mid-term Review Conference, Bergen, Norway, pp. 370–80.

Birks, H. J. B. (1994). The importance of pollen and diatom taxonomic precision in quantative palaeoenvironmental reconstructions. *Review of Palaeobotany and Palynology*, **83**, 107–17.

Birks, H. J. B., Berge, F., Boyle, J. F. & Cumming, B. F. (1990c). A palaeoecological test of the land-use hypothesis for recent lake acidification in South-West Norway using hill-top lakes. *Journal of Paleolimnology*, **4**, 69–85.

Birks, H. J. B., Juggins, S., & Line, J. M. (1990b). Lake surface-water chemistry reconstructions from palaeolimnological data. In *The Surface Waters Acidification Programme*, ed. B.J. Mason. Cambridge: Cambridge University Press, pp. 301–313.

Birks, H. J. B., Line, J. M., Juggins, S., Stevenson, A.C., & ter Braak, C. J. F. (1990a). Diatoms and pH reconstruction. *Philosophical Transactions of the Royal Society, London*, **B 327**, 263–78.

Boyle, J. F. (2007). Loss of apatite caused irreversible early Holocene lake acidification. *The Holocene*, **17**, 539–40.

Bradshaw, E. G., Jones, V. J., Birks, H. J. B., & Birks, H. H. (2000). Diatom responses to late-glacial and early-Holocene environmental changes at Kråkenes, western Norway. *Journal of Paleolimnology*, **23**, 21–34.

Bray, J., Broady, P. A., Niyogi, D. K., & Harding, J. (2008). Periphyton communities in New Zealand streams impacted by acid mine drainage. *Marine and Freshwater Research*, **59**, 1084–91.

Brugam, R. B. & Lusk, M. (1986). Diatom evidence for neutralization in acid surface mine lakes. In *Diatoms and Lake Acidity*, ed. J. P. Smol, R. W. Battarbee, R. B. Davis, & J. Meriläinen. Dordrecht: Dr. W. Junk, pp. 115–29.

Camburn, K. E. & Charles, D.F. (2000). *Diatoms of Low-alkalinity Lakes in the Northeastern United States*. Special Publication 18. Philadelphia, PA: Academy of Natural Sciences of Philadelphia.

Camburn, K. E. & Kingston, J. C. (1986). The genus *Melosira* from soft-water lakes with special reference to northern Michigan, Wisconsin and Minnesota. In *Diatoms and Lake Acidity*, ed. J. P. Smol, R. W. Battarbee, R. B. Davis, & J. Meriläinen. Dordrecht: Dr. W. Junk, pp. 17–34.

Camburn, K. E., Kingston, J. C. & Charles, D. F. (1986). *PIRLA Diatom Iconograph*. PIRLA Unpublished Report Number 3. Bloomington, IN: Indiana University.

Cameron, N. G. (1995). The representation of diatom communities by fossil assemblages in a small acid lake. *Journal of Paleolimnology*, **14**, 185–223.

Cameron, N. G., Birks, H. J. B., Jones, V. J., *et al.* (1999). Surface-sediment and epilithic diatom pH calibration sets for remote European mountain lakes (AL:PE Project) and their comparison with the Surface Waters Acidification Programme (SWAP) calibration set. *Journal of Paleolimnology*, **22**, 291–317.

Cameron, N. G., Schnell, O. A., Rautio, M. L., *et al.* (2002). High-resolution analyses of recent sediments from a Norwegian mountain lake and comparison with instrumental records of climate. *Journal of Paleolimnology*, **28**, 79–93.

Charles, D. F. (1985). Relationships between surface sediment diatom assemblages and lakewater characteristics in Adirondack lakes. *Ecology*, **66**, 994–1011.

Charles, D. F. (1990a). A checklist for describing and documenting diatom and chrysophyte calibration data sets and equations for inferring water chemistry. *Journal of Paleolimnology*, **3**, 175–8.

Charles, D. F. (1990b). Effects of acidic deposition on North American lakes: paleolimnological evidence from diatoms and chrysophytes. *Philosophical Transactions of the Royal Society of London*, **B 327**, 403–12.

Charles, D. F., Battarbee, R. W., Renberg, I. van Dam, H., & Smol, J. P. (1989). Paleoecological analysis of lake acidification trends in North America and Europe using diatoms and chrysophytes. In *Soils, Aquatic Processes, and Lake Acidification*, eds. S. A. Norton, S. E. Lindberg & A. L. Page, Acid Precipitation, vol. 4, New York: Springer-Verlag, pp. 207–276.

Charles, D. F., Binford, M. W., Furlong, E. T., *et al.* (1990). Paleoecological investigation of recent lake acidification in the Adirondack Mountains, N.Y. *Journal of Paleolimnology*, **3**, 195–241.

Charles, D. F., Dixit, S. S., Cumming, B. F., & Smol, J. P. (1991). Variability in diatom and chrysophyte assemblages and inferred pH: paleolimnological studies of Big Moose Lake, New York, (USA). *Journal of Paleolimnology*, **5**, 267–84.

Charles, D. F. & Norton, S. A. (1986). Paleolimnological evidence for trends in atmospheric deposition of acids and metals. In *Acid Deposition: Long-Term Trends*. Washington: National Academy Press, pp. 335–435.

Charles, D. F. & Smol, J. P. (1988). New methods for using diatoms and chrysophytes to infer past pH of low-alkalinity lakes. *Limnology and Oceanography*, **33**, 1451–62.

Charles, D. F. & Smol, J. P. (1994). Long-term chemical changes in lakes: Quantitative inferences from biotic remains in the sediment record. In *Environmental Chemistry of Lakes and Reservoirs*, ed. L. Baker, Washington, DC: American Chemical Society, pp. 3–31.

Charles, D. F., Smol, J. P., & Engstrom, D. R. (1994). Paleolimnological approaches to biological monitoring. In *Biological Monitoring of Aquatic Systems*, ed. L. L. Loeb & A. Spacie, Boca Raton, FL: CRC Press, pp. 233–93.

Charles, D. F. & Whitehead, D. R. (1986). The PIRLA project: Paleoecological investigation of recent lake acidification. *Hydrobiologia*, **143**, 13–20.

Clair, T. A., Dennis, I. F., Scruton, D. A., & Gilliss, M. (2007). Freshwater acidification research in Atlantic Canada: a review of results and predictions for the future. *Environmental Reviews*, **15**, 153–67.

Cleve-Euler, A. (1951–1955). Die Diatomeen von Schweden und Finnland. Kungliga Vetenskapsakademiens Handlingar Series 4, 2(1), 3–163; 4(1), 3–158; 4(5), 3–355; 5(4), 3–231; 3(3), 3–153.

Coring, E. (1996). Use of diatoms for monitoring acidification in small mountain rivers in Germany with special emphasis on "diatom assemblage type analysis" (DATA). In *Use of Algae for Monitoring Rivers II*, ed. B.A. Whitton & E. Rott, Innsbruck: Institut für Botanik, Universität Innsbruck, pp. 7–16.

Cosby, B. J., Ferrier, R. C., Jenkins, A., & Wright, R. F. (2001). Modelling the effects of acid deposition: refinements, adjustments and inclusion of nitrogen dynamics in the MAGIC model. *Hydrology and Earth System Sciences*, **5**, 499–517.

Crabtree, K. (1969). Post-glacial diatom zonation of limnic deposits in north Wales. *Mitteilungen Internationale Vereinigung für Theoretische und Angewandte Limnologie*, **17**, 165–71.

Cumming, B. F., Davey, K. A., Smol, J. P., & Birks, H. J. B. (1994). When did acid-sensitive Adirondack lakes (New York, USA) begin to acidify and are they still acidifying? *Canadian Journal of Fisheries and Aquatic Sciences*, **51**, 1550–68.

Cumming, B. F., Smol, J. P., Kingston, J. C., *et al.* (1992). How much acidification has occurred in Adirondack region lakes (New York, USA) since preindustrial times? *Canadian Journal of Fisheries and Aquatic Sciences*, **49**, 128–41.

Davis, R. B. (1987). Paleolimnological diatom studies of acidification of lakes by acid rain: an application of quaternary science. *Quaternary Science Reviews*, **6**, 147–63.

Davis, R. B. & Anderson, D. S. (1985). Methods of pH calibration of sedimentary diatom remains for reconstructing history of pH in lakes. *Hydrobiologia*, **120**, 69–87.

Davis, R. B., Anderson, D. S., Norton, S. A., & Whiting, M. C. (1994). Acidity of twelve northern New England (U.S.A.) lakes in recent centuries. *Journal of Paleolimnology*, **12**, 103–154.

Davis, R. B., Anderson, D. S., Whiting, M. C., Smol, J. P. & Dixit, S. S. (1990). Alkalinity and pH of three lakes in northern New England, U.S.A. over the past 300 years. *Philosophical Transactions of the Royal Society, London*, **B 327**, 413–21.

Davis, R. B. & Berge, F. (1980). Atmospheric deposition on Norway during the last 300 years as recorded in SNSF lake sediments. II. Diatom stratigraphy and inferred pH. In *Ecological Impact of Acid Precipitation; Proceedings of an International Conference*, ed. D. Drablos & A. Tollan, Oslo-Ås: SNSF, pp. 270–1.

Davis, R. B., Norton, S. A., Hess, C. T., & Brakke, D. F. (1983). Paleolimnological reconstruction of the effects of atmospheric deposition of acids and heavy metals on the chemistry and biology of lakes in New England and Norway. *Hydrobiologia*, **103**, 113–23.

Del Prete, A. & Schofield, C. (1981). The utility of diatom analysis of lake sediments for evaluating acid precipitation effects on dilute lakes. *Archiv für Hydrobiologie*, **91**, 332–40.

DeNicola, D. M. (1986). The representation of living diatom communities in deep-water sedimentary diatom assemblages in two Maine (U.S.A.) lakes. In *Diatoms and Lake Acidity*, ed. J. P. Smol, R. W. Battarbee, R. B. Davis, & J. Meriläinen. Dordrecht: Dr. W. Junk, pp. 73–85.

DeNicola, D. M. (2000). A review of diatoms found in highly acidic environments. *Hydrobiologia*, **433**, 111–22.

Denys, L. (2006). Calibration of littoral diatoms to water chemistry in standing fresh waters (Flanders, lower Belgium): inference models for historical sediment assemblages. *Journal of Paleolimnology*, **35**, 763–87.

Dixit, A. S., Dixit, S. S., & Smol, J. P. (1989). Lake acidification recovery can be monitored using chrysophycean microfossils. *Canadian Journal of Fisheries and Aquatic Sciences*, **46**, 1309–12.

Dixit, A. S., Dixit, S. S., & Smol, J. P. (1992a). Algal microfossils provide high temporal resolution of environmental change. *Water, Air and Soil Pollution*, **62**, 75–87.

Dixit, A. S., Dixit, S. S., & Smol, J. P. (1992b). Long-term trends in lake water pH and metal concentrations inferred from diatoms and chrysophytes in three lakes near Sudbury, Ontario. *Canadian Journal of Fisheries and Aquatic Sciences*, **49**, 17–24.

Dixit, A. S., Dixit, S. S., & Smol, J. P. (1996). Setting restoration goals for an acid and metal-contaminated lake: a paleolimnological study of Daisy Lake (Sudbury, Canada). *Journal of Lake and Reservoir Management*, **12**, 323–30.

Dixit, S. S. (1986). Diatom-inferred pH calibration of lakes near Wawa, Ontario. *Canadian Journal of Botany*, **64**, 1129–33.

Dixit, S. S., Dixit, A. S. & Evans, R. D. (1987). Paleolimnological evidence of recent acidification in two Sudbury (Canada) lakes. *Science of the Total Environment*, **67**, 53–63.

Dixit, S. S., Dixit, A. S., & Evans, R. D. (1988). Sedimentary diatom assemblages and their utility in computing diatom-inferred pH in Sudbury Ontario lakes. *Hydrobiologia*, **169**, 135–48.

Dixit, S. S., Dixit, A. S. & Smol, J. P. (1992c). Assessment of changes in lake water chemistry in Sudbury area lakes since preindustrial times. *Canadian Journal of Fisheries and Aquatic Sciences*, **49**, 8–16.

Dixit, S. S., Dixit, A. S., & Smol, J. P. (2002). Diatom and chrysophyte transfer functions and inferences of post-industrial acidification and recent recovery trends in Killarney lakes (Ontario, Canada). *Journal of Paleolimnology*, **27**, 79–96.

Dixit, S. S. & Smol, J. P. (1994). Diatoms as indicators in the Environmental Monitoring and Assessment Program – Surface Waters (EMAP – SW). *Environmental Monitoring and Assessment*, **31**, 275–306.

Dixit, S. S., Smol, J. P., Charles, D. F., *et al.* (1999). Assessing water quality changes in the lakes of the northeastern United States using sediment diatoms. *Canadian Journal of Fisheries and Aquatic Sciences*, **56**, 131–52.

Driscoll, C. T., Driscoll, K. M., Roy, K. M., & Mitchell, M. J. (2003). Chemical response of lakes in the Adirondack region of New York to declines in acidic deposition. *Environmental Science and Technology*, **37**, 2036–42.

Driscoll, C. T., Driscoll, K. M., Roy, K. M., & Dukett, J. (2007). Changes in the chemistry of lakes in the Adirondack region of New York following declines in acidic deposition. *Applied Geochemistry*, **22**, 1181–8.

Ek, A. S., Grahn, O., Hultberg, H., & Renberg, I. (1995). Recovery from acidification in lake Örvattnet, Sweden. *Water, Air and Soil Pollution*, **85**, 1795–800.

Ek, A. S. & Korsman, T. (2001). A paleolimnological assessment of the effects of post-1970 reductions of sulfur deposition in Sweden. *Canadian Journal of Fisheries and Aquatic Sciences*, **58**, 1692–700.

Ek, A. S., Löfgren, S., Bergholm, J., & Qvarfort, U. (2001). Environmental effects of one thousand years of copper production at Falun, central Sweden. *Ambio*, **30**, 96–103.

Ek, A. S. & Renberg, I. (2001). Heavy metal pollution and lake acidity changes caused by one thousand years of copper mining at Falun, central Sweden. *Journal of Paleolimnology*, **26**, 89–107.

Emmett, B., Charles, D. F., Feger, & K. H. (1994). Group report: can we differentiate between natural and anthropogenic acidification? In *Acidification of Freshwater Ecosystems: Implications for the Future*, ed. C. Steinberg & R. Wright, Dahlem Workshop Environmental Sciences Research Report 14, Chichester: Wiley, pp. 118–40.

Engstrom, D. R., Fritz, S. C., Almendinger, J. E., & Juggins, S. (2000). Chemical and biological trends during lake evolution in recently deglaciated terrain. *Nature*, **408**, 161–6.

Erlandsson M., Bishop K., Fölster J., *et al.* (2008). A comparison of MAGIC and paleolimnological predictions of preindustrial pH for 55 Swedish lakes. *Environmental Science and Technology*, **42**, 43–8.

Evans, C., Reynolds, B., Hinton, C., Hughes, *et al.* (2008). Effects of decreasing acid deposition and climate change on acid extremes in an upland stream. *Hydrology and Earth System Sciences*, **12**, 337–51.

Evans, G. H. (1970). Pollen and diatom analysis of late-Quaternary deposits in the Blelham Basin, north Lancashire. *New Phytologist*, **69**, 821–74.

Evans, G. H. & Walker, R. (1977). The late-Quaternary history of the diatom flora of Llyn Clyd and Llyn Glas, two small oligotrophic high mountain tarns in Snowdonia (Wales). *New Phytologist*, **78**, 221–36.

Faulkenham, S. E., Hall, R. I., Dillon, P. J., & Karst-Riddoch, T. (2003). Effects of drought-induced acidification on diatom communities in acid-sensitive Ontario lakes. *Limnology & Oceanography*, **48**, 1662–73.

Flower, R. J. (1986). The relationship between surface sediment diatom assemblages and pH in 33 Galloway lakes: some regression models for reconstructing pH and their application to sediment cores. *Hydrobiologia*, **143**, 93–103.

Flower, R. J. & Battarbee, R. W. (1983). Diatom evidence for recent acidification of two Scottish lochs. *Nature*, **20**, 130–3.

Flower, R. J. & Battarbee, R. W. (1985). The morphology and biostratigraphy of *Tabellaria quadriseptata* Knudson (Bacillariophyta) in acid waters and lake sediments in Galloway, south-west Scotland. *British Phycological Journal*, **20**, 69–79.

Flower, R. J., Battarbee, R. W., & Appleby, P. G. (1987). The recent palaeolimnology of acid lakes in Galloway, south-west Scotland: diatom analysis, pH trends and the role of afforestation. *Journal of Ecology*, **75**, 797–824.

Foged, N. (1977). *Freshwater Diatoms in Ireland*. Ruggell: A. R. G. Ganter Ford, M. S. Verlag KG. (1990). A 10,000 year history of natural ecosystem acidification. *Ecological Monographs*, **60**, 57–89.

Fritz, S. C. & Carlson, R. E. (1982). Stratigraphic diatom and chemical evidence for acid strip-mine lake recovery. *Water, Air and Soil Pollution*, **17**, 151–63.

Fritz, S. C., Juggins, S., Battarbee, R. W., & Engstrom, D. R. (1991). A diatom-based transfer function for salinity, water-level, and climate reconstruction. *Nature*, **352**, 706–8.

Gensemer, R. W. & Playle, R. C. (1999). The bioavailability and toxicity of aluminum in aquatic environments. *Critical Reviews in Environmental Science and Technology*, **29**, 315–450.

Genter, R. B. (1995). Benthic algal populations respond to aluminum, acid, and aluminum-acid mixtures in artificial streams. *Hydrobiologia*, **306**, 7–19.

Gerber, A. M., Ginn, B. K., Whitfield, C. J., *et al.* (2008). Glasgow Lake: an early-warning sentinel of lake acidification in Cape Breton Highlands National Park (Nova Scotia), Canada. *Hydrobiologia*, **614**, 299–307.

Ginn, B. K., Cumming, B. F., & Smol, J. P. (2007a). Diatom-based environmental inferences and model comparisons from 494 northeastern North American lakes. *Journal of Phycology*, **43**, 647–61.

Ginn, B. K., Cumming, B. F., & Smol, J. P. (2007b). Assessing pH changes since preindustrial times in 51 low-alkalinity lakes from Nova Scotia, Canada. *Canadian Journal of Fisheries and Aquatic Sciences*, **64**, 1043–54.

Ginn, B. K., Cumming, B. F., & Smol, J. P. (2007c). Long-term acidification trends in high- and low-sulphate deposition regions from Nova Scotia, Canada. *Hydrobiologia*, **586**, 261–75.

Greenaway, C. M. (2009). Diatom responses to water-quality improvements in lakes recovering from acidification and metal-contamination near Wawa, Ontario: a paleolimnological perspective. Unpublished M.Sc. Thesis. Queen's University, Kingston, Ontario.

Guhrén, M., Bigler, C., & Renberg, I. (2007). Liming placed in a long-term perspective: a paleolimnological study of 12 lakes in the Swedish liming program. *Journal of Paleolimnology*, **37**, 247–58.

Gunn, J. M. & Keller, W. (1992). Biological recovery of an acid lake after reductions in industrial emissions of sulphur. *Nature*, **345**, 431–3.

Hargreaves, J. W., Lloyd, E. J. H., Whitton, B. A. (1975). Chemistry and vegetation of highly acidic streams. *Freshwater Biology*, **5**, 563–76.

Harriman, R. & Morrison, B. R. S. (1982). The ecology of streams draining forested and non-forested catchments in an area of central Scotland subject to acid precipitation. *Hydrobiologia*, **88**, 251–63.

Haworth, E. Y. (1969). The diatoms of a sediment core from Blea Tarn, Langdale. *Journal of Ecology*, **57**, 429–39.

Hinderer, M., Jüttner, I., Winkler, R., Steinberg, C. E. W., & Kettrup, A. (1998). Comparing trends in lake acidification using hydrochemical modelling and paleolimnology: the case of the Herrenwieser See, Black Forest, Germany. *Science of the Total Environment*, **218**, 113–21.

Hirst, H., Chaud, F., Delabie, C., Jüttner, I., & Ormerod, S. J. (2004). Assessing the short-term response of stream diatoms to acidity using inter-basin transplantations and chemical diffusing substrates. *Freshwater Biology*, **49**, 1072–88.

Holmes, R. W., Whiting, M. C. & Stoddard, J. L. (1989). Changes in diatom-inferred pH and acid neutralizing capacity in a dilute, high elevation, Sierra Nevada lake since A.D. 1825. *Freshwater Biology*, **21**, 295–310.

Howells, G. & Dalziel, T. R. K. (1991). *Restoring Acid Waters: Loch Fleet 1984–1990*. London: Elsevier.

Hustedt, F. (1937–1939). Systematische und ökologische Untersuchungen über die Diatomeen-Flora von Java, Bali, Sumatra. *Archiv für Hydrobiologie* (Suppl.), **15** & **16**.

Huttunen, P., Meriläinen, J., & Tolonen, K. (1978). The history of a small dystrophied forest lake, southern Finland. *Polskie Archiwum Hydrobiologii*, **25**, 189–202.

Huvane, J. K. & Whitehead, D. R. (1996). The paleolimnology of North Pond: watershed-lake interactions. *Journal of Paleolimnology*, **16**, 323–54.

Iversen, J. (1958). The bearing of glacial and interglacial epochs on the formation and extinction of plant taxa I. In *Systematics of Today*, ed. O. Hedberg, Uppsala Universitets årsskrift 6, Uppsala: University of Uppsala, pp. 210–15.

Jenkins, A., Whitehead, P. G., Cosby, B. J., & Birks, H.J.B. (1990). Modelling long-term acidification: a comparison with diatom reconstructions and the implications for reversibility. *Philosophical Transactions of the Royal Society, London*, **B327**, 209–14.

Jensen, K. W. & Snekvik, E. (1972). Low pH levels wipe out salmon and trout populations in southernmost Norway. *Ambio*, **1**, 223–5.

Jones, V. J., Flower, R. J., Appleby, P. G., *et al.* (1993). Palaeolimnological evidence for the acidification and atmospheric contamination of lochs in the Cairngorms and Lochnagar areas of Scotland. *Journal of Ecology*, **81**, 3–24.

Jones, V. J., Stevenson, A. C. & Battarbee, R. W. (1986). Lake acidification and the land-use hypothesis: a mid-post-glacial analogue. *Nature*, **322**, 157–8.

Jones, V. J., Stevenson, A. C., & Battarbee, R. W. (1989). Acidification of lakes in Galloway, south west Scotland: a diatom and pollen study of the post-glacial history of the Round Loch of Glenhead. *Journal of Ecology*, **77**, 1–23.

Jüttner, I., Lintelmann, J., Michalke, B., *et al.* (1997). The acidification of the Herrenwieser See, Black Forest, Germany, before and during industrialisation. *Water Research*, **31**, 1194–206.

Kahlert, M. & Andrén, C. (2005). Benthic diatoms as valuable indicators of acidity. *Verhandlungen der Internationalen Vereinigung für Theoretische und Angewandte Limnologie*, **29**, 635–9.

Kingston, J. C., Cumming, B. F., Uutala, A. J., *et al.* (1992a). Biological quality control and quality assurance: a case study in paleolimnological biomonitoring. In *Ecological Indicators*, ed. D. H. McKenzie, D. E. Hyatt, & V. J. McDonald, New York: Elsevier Applied Science, pp. 1542–3.

Korhola, A. A. & Tikkanen, M. J. (1991). Holocene development and early extreme acidification in a small hilltop lake in southern Finland. *Boreas*, **20**, 333–56.

Korhola, A., Virkanen, J., Tikkanen, M., & Blom, T. (1996). Fire-induced pH rise in a naturally acid hill-top lake, southern Finland: a palaeoecological survey. *Journal of Ecology*, **84**, 257–65.

Korsman, T. (1999). Temporal and spatial trends of lake acidity in northern Sweden. *Journal of Paleolimnology*, **22**, 1–15.

Korsman, T., Renberg, I., & Anderson, N.J. (1994). A palaeolimnological test of the influence of Norway spruce (*Picea abies*) immigration on lake-water acidity. *The Holocene*, **4**, 132–40.

Korsman, T. & Segerström, U. (1998). Forest fire and lake water acidity in a northern Swedish boreal area: Holocene changes in lake-water quality at Makkassjön. *Journal of Ecology*, **86**, 113–24.

Kovacs, C., Kahlert, M., & Padisak, J. (2006). Benthic diatom communities along pH and TP gradients in Hungarian and Swedish streams. *Journal of Applied Phycology*, **18**, 105–17.

Kram, P., Laudon, H., Bishop, K., Rapp, L., & Hruska, J. (2001). MAGIC modeling of long-term lake water and soil chemistry at Abborrträsket, northern Sweden. *Water, Air and Soil Pollution*, **130**, 1301–6.

Kreiser, A. M., Appleby, P. G., Natkanski, J., Rippey, B., & Battarbee, R.W. (1990). Afforestation and lake acidification: a comparison of four sites in Scotland. *Philosophical Transactions of the Royal Society, London*, **B 327**, 377–83.

Krug, E. C. & Frink, C. R. (1983). Acid rain on acid soil: a new perspective. *Science*, **221**, 520–5.

Kwandrans, J. (2007). *Diversity and Ecology of Benthic Diatom Communities in Relation to Acidity, Acidification and Recovery of Lakes and Rivers*, Diatom Monographs, vol. 9, ed. A. Witkowski, ed., Ruggell: A. R. G. Gantner Verlag.

Lancaster, J., Real, M., Juggins, S., *et al.* (1996). Monitoring temporal changes in the biology of acid waters. *Freshwater Biology*, **36**, 179–202.

Larsen, J., Jones, V. J., & Eide, W. (2006). Climatically driven pH changes in two Norwegian alpine lakes. *Journal of Paleolimnology*, **36**, 57–69.

Locke, A., Sprules, G. W., Keller, W., & Pitblado, R. J. (1994). Zooplankton communities and water chemistry of Sudbury area lakes: changes related to pH recovery. *Canadian Journal of Fisheries and Aquatic Sciences*, **51**, 151–60.

Luís, A., Teixeira, P., Almeida, S., *et al.* (2009). Impact of acid mine drainage (AMD) on water quality, stream sediments and periphytic diatom communities in the surrounding streams of Aljustrel mining area (Portugal). *Water Air and Soil Pollution*, **200**, 147–67.

Lundqvist, G. (1924). Utvecklingshistoriska insjöstudier i Syd-sverige. *Sveriges Geologiska Undersökning*, **C 330**, 1–129.

MacDougall, S., Carrick, H. J. & DeWalle, D. R. (2008). Benthic algae in episodically acidified Pennsylvania streams. *Northeastern Naturalist*, **15**, 189–208.

Majewski S.P. & Cumming B. F. (1999). Paleolimnological investigation of the effects of post-1970 reductions of acidic deposition on an acidified Adirondack lake. *Journal of Paleolimnology*, **21**, 207–13.

Meriläinen, J. (1967). The diatom flora and the hydrogen ion concentration of the water. *Annales Botanici Fennici*, **4**, 51–8.

Michelutti, N., Laing, T. E., & Smol, J. P. (2001). Diatom assessment of past environmental changes in lakes located near the Noril'sk (Siberia) smelters. *Water, Air and Soil Pollution*, **125**, 231–41.

Miller, U. (1973). Diatoméundersökning av bottenproppar från Stora Skarsjön, Ljungskile. *Statens Naturvårdsverk Publikationer*, **7**, 43–60.

Monteith, D. T., Stoddard, J. L., Evans, C. D., *et al.* (2007). Dissolved organic carbon trends resulting from changes in atmospheric deposition chemistry. *Nature*, **450**, 537–40.

Mosello, R., Bonacina, C., Carollo, A., Libera, V., & Tartari, G.A. (1986). Acidification due to in-lake ammonia oxidation: an attempt to quantify the proton production in a highly polluted subalpine Italian lake. *Memorie dell'Istituto Italiano di Idrobiologia*, **44**, 47–71.

Niemi, D. (2005). Emissions of pollutants related to acid deposition in North America. In *2004 Canadian Acid Deposition Science Assessment*, Ottawa, ON: Environment Canada, pp. 5–14.

Nilsson, I. S., Miller, H. G., & Miller, J. D. (1982). Forest growth as a possible cause of soil and water acidification: an examination of the concepts. *Oikos*, **39**, 40–9.

Niyogi, D. K., Lewis, W. M., & McKnight, D. M. (2002). Effects of stress from mine drainage on diversity, biomass, and function of primary producers in mountain streams. *Ecosystem*, **5**, 554–67.

Norberg, M., Bigler, C., & Renberg, I. (2008). Monitoring compared with paleolimnology: implications for the definition of reference condition in limed lakes in Sweden. *Environmental Modelling and Assessment*, **146**, 295–308.

Norberg, M., Bigler, C., & Renberg, I. (2010). Comparing pre-industrial and post-limed diatom communities in Swedish lakes, with implications for defining realistic management targets. *Journal of Paleolimnology*, **44**: 233–42.

Nygaard, G. (1956). Ancient and recent flora of diatoms and chrysophyceae in Lake Gribsø. Studies on the humic acid lake Gribsø. *Folia Limnologica Scandinavica*, **8**, 32–94.

Odén, S. (1968). *The Acidification of Air Precipitation and its Consequences in the Natural Environment*, Energy Committee Bulletin, 1, Stockhom: Swedish Natural Sciences Research Council.

Oehlert, G. W. (1988). Interval estimates for diatom inferred lake pH histories. *Canadian Journal of Statistics*, **16**, 51–60.

Passy, S. I. (2006). Diatom community dynamics in streams of chronic and episodic acidification: the roles of environment and time. *Journal of Phycology*, **42**, 312–23.

Patrick, S. T., Monteith, D. T., & Jenkins, A. (eds.) (1995). *UK Acid Waters Monitoring Network: the First Five Years. Analysis and Interpretation of Results April 1988-March 1993*. London: ENSIS.

Patrick, S. T., Battarbee, R. W., & Jenkins, A. (1996). Monitoring acid waters in the U.K.: an overview of the U.K. Acid Waters Monitoring Network and summary of the first interpretative exercise. *Freshwater Biology*, **36**, 131–50.

Pennington, W. (1984). Long-term natural acidification of upland sites in Cumbria: evidence from post-glacial lake sediments. *Freshwater Biological Association Annual Report*, **52**, 28–46.

Pennington, W., Haworth, E. Y., Bonny, A. P., & Lishman, J. P. (1972). Lake sediments in northern Scotland. *Philosophical Transactions of the Royal Society, London*, **B 264**, 191–294.

Persson, J., Nilsson, M., Bigler, C., Brooks, S. J., & Renberg, I. (2007). Near-infrared spectroscopy (NIRS) of epilithic material in streams has a potential for monitoring impact from mining. *Environmental Science and Technology*, **41**, 2874–80.

Pither, J. & Aarssen, L. W. (2005). Environmental specialists: their prevalence and their influence on community – similarity analyses. *Ecology Letters*, **8**, 261–71.

Pither, J. & Aarssen, L. W. (2006). How prevalent are pH-specialist diatoms? A reply to Telford *et al.* (2006) *Ecology Letters*, **9**, E6–E12.

Planas, D. (1996). Acidification effects. In *Algal ecology: Freshwater benthic ecosystems*, (eds. R. J. Stevenson, M. L. Bothwell and R. L. Lowe). Academic Press: San Diego, pp. 497–530.

Planas, D., Lapierre, L., Moreau, G., & Allard, M. (1989). Structural organization and species composition of a lotic periphyton community in response to experimental acidification. *Canadian Journal of Fisheries and Aquatic Sciences*, **46**, 827–35.

Psenner, R. & Schmidt, R. (1992). Climate-driven pH control of remote alpine lakes and effects of acid deposition. *Nature*, **356**, 781–3.

Renberg, I. (1976). Palaeolimnological investigations in Lake Prästsjön. *Early Norrland*, **9**, 113–60.

Renberg, I. (1978). Palaeolimnology and varve counts of the annually laminated sediment of Lake Rudetjärn, northern Sweden. *Early Norrland*, **11**, 63–92.

Renberg, I. (1986). A sedimentary diatom record of severe acidification in Lake Blåmissusjön, N. Sweden, through natural processes. In *Diatoms and Lake Acidity*, ed. J. P. Smol, R. W. Battarbee, R. B. Davis, & J. Meriläinen. Dordrecht: Dr. W. Junk, pp. 213–19.

Renberg, I. (1990). A 12,600 year perspective of the acidification of Lilla Öresjön, southwest Sweden. *Philosophical Transactions of the Royal Society, London*, **B 327**, 357–61.

Renberg, I., Bigler, C., Bindler, R., *et al.* (2009). Environmental history: a piece in the puzzle for establishing plans for environmental management. *Journal of Environmental Management*, **90**, 2794–800.

Renberg, I. & Hellberg, T. (1982). The pH history of lakes in southwestern Sweden, as calculated from the subfossil diatom flora of the sediments. *Ambio*, **11**, 30–3.

Renberg, I., Korsman, T., & Anderson, N. J. (1990). Spruce and surface water acidification: an extended summary. *Philosophical Transactions of the Royal Society, London*, **B 327**, 371–2.

Renberg, I., Korsman, T., & Anderson, N. J. (1993a). A temporal perspective of lake acidification in Sweden. *Ambio*, **22**, 264–71.

Renberg, I., Korsman, T., & Birks, H. J. B. (1993b). Prehistoric increases in the pH of acid-sensitive Swedish lakes caused by land-use changes. *Nature*, **362**, 824–6.

Rhodes, T. E. & Davis, R. B. (1995). Effects of late Holocene forest disturbance and vegetation change on acidic Mud Pond, Maine, USA. *Ecology*, **76**, 734–46.

Rosenqvist, I. T. (1977). *Acid Soil – Acid Water*. Oslo: Ingeniörforlaget.

Rosenqvist, I. T. (1978). Alternative sources for acidification of river water in Norway. *The Science of the Total Environment*, **10**, 39–49.

Round, F. E. (1957). The late-glacial and post-glacial diatom succession in the Kentmere Valley deposit: I. Introduction, methods and flora. *New Phytologist*, **56**, 98–126.

Round, F. E. (1961). Diatoms from Esthwaite. *New Phytologist*, **60**, 98–126.

Round, F. E. (1991). Epilithic diatoms in acid water streams flowing into the reservoir Llyn Brianne. *Diatom Research*, **6**, 137–45.

Rowe, J. M. (1999). *Heart of a Mountain, Soul of a Town – The Story of Algoma Ore and the Town of Wawa*, Altona, Canada: Friesens.

Sabater, S., Buchaca, T., Cambra, J., *et al.* (2003). Structure and function of benthic algal communities in an extremely acid river. *Journal of Phycology*, **39**, 481–9.

Salomaa, R. & Alhonen, P. (1983). Biostratigraphy of Lake Spitaalijärvi: an ultraoligotrophic small lake in Lauhanvuori, western Finland. *Hydrobiologia*, **103**, 295–301.

Sandøy, S. & Langåker, R. M. (2001). Atlantic salmon and acidification in Southern Norway: a disaster in the 20th century, but a hope for the future? *Water, Air and Soil Pollution*, **130**, 1343–8.

Sarmaja-Korjonen, K., Nyman, M., Kultti, S., & Väliranta, M. (2006). Palaeolimnological development of Lake Njargajavri, northern Finnish Lapland, in a changing Holocene climate and environment. *Journal of Paleolimnology*, **35**, 65–81.

Sienkiewicz, E., Gasiorowski, M., & Hercman, H. (2006). Is acid rain impacting the Sudetic lakes? *Science of the Total Environment*, **369**, 139–49.

Skjelkvåle, B. L., Borg, H., Hindar, A., & Wilander, A. (2007). Large-scale patterns of chemical recovery in lakes in Norway and Sweden: importance of seasalt episodes and changes in dissolved organic carbon. *Applied Geochemistry*, **22**, 1174–80.

Skjelkvåle, B. L., Stoddard, J. L., Jeffries, D. S., *et al.* (2005). Regional scale evidence for improvements in surface water chemistry 1990–2001. *Environmental Pollution*, **137**, 165–76.

Solovieva, N. & Jones, V. J. (2002). A multiproxy record of Holocene environmental changes in the central Kola Peninsula, northwest Russia. *Journal of Quaternary Science*, **17**, 303–18.

Somers, K. M. & Harvey, H. H. (1984). Alteration of lake fish communities in response to acid precipitation and heavy metal loading near Wawa, Ontario. *Canadian Journal of Fisheries and Aquatic Sciences*, **41**, 20–9.

Sommaruga-Wögrath, S., Koinig, K. A., Schmidt, R., *et al.* (1997). Temperature effects on the acidity of remote alpine lakes. *Nature*, **387**, 64–7.

Steinberg, C., Hartmann, H., Arzet, K., & Krause-Dellin, D. (1988). Paleoindication of acidification in Kleiner Arbersee (Federal Republic of Germany, Bavarian Forest) by chydorids, chrysophytes, and diatoms. *Journal of Paleolimnology*, **1**, 149–57.

Steinberg, C. & Putz, R. (1991). Epilithic diatoms as bioindicators of stream acidification. *Verhandlungen der Internationalen Vereinigung für Theoretische und Angewandte Limnologie*, **24**, 1877–80.

Stevenson, A. C., Juggins, S., Birks, H. J. B., *et al.* (1991). *The Surface Waters Acidification Project Palaeolimnology Programme: Modern Diatom/Lake-Water Chemistry Data-Set*. London: Ensis Ltd.

Stoddard J., Jeffries, D. S, Lükewille, A., *et al.* (1999). Regional trends in aquatic recovery from acidification in North America and Europe. *Nature*, **401**, 575–8.

Stuchlík, D., P. Appleby, P. Bituscaroník, C. *et al.* (2002). Reconstruction of long-term changes in lake water chemistry, zooplankton and benthos of a small, acidified high-mountain lake: MAGIC modelling and palaeolimnogical analysis. *Water, Air and Soil Pollution: Focus*, **2**, 127–38.

Sullivan, T. J., Charles, D. F., Smol, J. P, *et al.* (1990). Quantification of changes in lakewater chemistry in response to acidic deposition. *Nature*, **345**, 54–8.

Sullivan, T. J., McMartin, B., & Charles, D. F. (1996). Re-examination of the role of landscape change in the acidification of lakes in the Adirondack Mountains, New York. *The Science of the Total Environment*, **183**, 231–48.

Sweets, P. R., Bienert, R. W., Crisman, T. L. & Binford, M. W. (1990). Paleoecological investigations of recent lake acidification in northern Florida. *Journal of Paleolimnology*, **4**, 103–37.

Telford, R. J., Vandvik, V. & Birks, H. J. B. (2006). How many freshwater diatoms are pH specialists? A response to Pither & Aarssen (2005). *Ecology Letters*, **9**, E1–E5.

ter Braak, C. J. F. & Juggins, S. (1993). Weighted averaging partial least squares regression (WA-PLS): an improved method for reconstructing environmental variables from species assemblages. *Hydrobiologia*, **269/270**, 485–502.

ter Braak, C. J. F & van Dam, H. (1989). Inferring pH from diatoms: a comparison of old and new calibration methods. *Hydrobiologia*, **178**, 209–23.

Tropea, A. E. (2008). Assessing biological recovery from acidification and metal contamination in urban lakes from Sudbury, Canada: a paleolimnological approach. Unpublished M.Sc. thesis, Queen's University, Kingston, Ontario, pp. 232.

Unsworth, M. H. (1984). Evaporation from forests in cloud enhances the effect of acid deposition. *Nature*, **312**, 262–4.

USEPA. (2000). National air pollutant emissions trends, 1900–1998. Washington, DC: EPA.

van Dam, H. (1988). Acidification of three moorland pools in The Netherlands by acid precipitation and extreme drought periods over seven decades. *Freshwater Biology*, **20**, 157–76.

van Dam, H. & Kooyman-van Blokland, H. (1978). Man-made changes in some Dutch moorland pools, as reflected by historical and recent data about diatoms and macrophytes. *Internationale Revue gesamten Hydrobiologie*, **63**, 587–607.

van Dam, H. & Mertens, A. (1990). A comparison of recent epilithic diatom assemblages from the industrially acidified and copper polluted Lake Orta (Northern Italy) with old literature data. *Diatom Research*, **5**, 1–13.

van Dam, H. & Mertens, A. (1995). Long-term changes of diatoms and chemistry in headwater streams polluted by atmospheric deposition of sulphur and nitrogen compounds. *Freshwater Biology*, **34**, 579–600.

van Dam, H. & Mertens, A. (2008). Vennen minder zuur maar warmer. *H_2O*, **41**, 36–9.

van Dam, H., Suurmond, G., & ter Braak, C. J. F. (1981). Impact of acidification on diatoms and chemistry of Dutch moorland pools. *Hydrobiologia*, **83**, 425–59.

Verb, R. G. & Vis, M. L. (2000). Comparison of benthic diatom assemblages from streams draining abandoned and reclaimed coal mines and nonimpacted sites. *Journal of the North American Benthological Society*, **19**, 274–88.

Verb, R. G. & Vis, M. L. (2005). Periphyton assemblages as bioindicators of mine-drainage in unglaciated Western Allegheny Plateau lotic systems. *Water, Air and Soil Pollution*, **161**, 227–65.

Virkanen, J., Korhola, A., Tikkanen, M., & Blom, T. (1997). Recent environmental changes in a naturally acidic rocky lake in southern Finland, as reflected in its sediment geochemistry and biostratigraphy. *Journal of Paleolimnology*, **17**, 191–213.

Watanabe, T. & Asai, K. (1999). Diatoms on the pH gradient from 1.0 to 12.5. In *Proceedings of the 14th Diatom Symposium 1996*, ed. S. Mayama, M. Idei, and I. Koizumi, Königstein: Koeltz Scientific Books, pp. 383–412.

Weckström, J., Snyder, J. A., Korhola, A., Laing, T. E., & MacDonald, G. M. (2003). Diatom inferred acidity history of 32 lakes on the Kola Peninsula, Russia. *Water, Air and Soil Pollution*, **149**, 339–61.

Whitehead, D. R., Charles, D. F., Jackson, S. T., Smol, J. P., & Engstrom, D. R. (1989). The developmental history of Adirondack (N.Y.) lakes. *Journal of Paleolimnology*, **2**, 185–206.

Whitehead, D. R., Charles, D. F., Reed, S. E., Jackson, S. T., & Sheehan, M. C. (1986). Late-glacial and holocene acidity changes in Adirondack (NY) lakes. In *Diatoms and Lake Acidity*, ed. J. P. Smol, R. W. Battarbee, R. B. Davis, & J. Meriläinen, Dordrecht: Dr. W. Junk, pp. 251–74.

Whiting, M. C., Whitehead, D. R., Holmes, R. W., & Norton, S. A. (1989). Paleolimnological reconstruction of recent acidity changes in four Sierra Nevada lakes. *Journal of Paleolimnology*, **2**, 285–304.

Williams, D. M., Hartley, B., Ross, R., Munro, M. A. R., Juggins, S., & Battarbee, R. W. (1988). *A Coded Checklist of British Diatoms*. London: ENSIS Publishing.

Winkler, M. G. (1988). Paleolimnology of a Cape Cod Kettle Pond: diatoms and reconstructed pH. *Ecological Monographs*, **58**, 197–214.

Wolfe A. P. (2002). Climate modulates the acidity of Arctic lakes on millennial time scale. *Geology*, **30**, 215–18.

Wright, R. F., Cosby, B. J., Hornberger, G. M., & Galloway, J. N. (1986). Comparison of paleolimnological with MAGIC model reconstructions of water acidification. *Water, Air and Soil Pollution*, **30**, 367–80.

Yoshitake, S. & Fukushima, H. (1995). Distribution of attached diatoms in inorganic acid lakes in Japan. In *Proceedings of the Thirteenth International Diatom Symposium*, ed. D. Marino & M. Montresor, Bristol: Biopress Ltd., pp. 321–33.

7 Diatoms as indicators of lake eutrophication

ROLAND I. HALL AND JOHN P. SMOL

7.1 Introduction

The term eutrophication broadly refers to the enrichment of aquatic systems by inorganic plant nutrients (Mason, 1991; Wetzel, 2001). Lake eutrophication occurs when nutrient supplies, usually phosphorus (P) and nitrogen (N), are elevated over rates that occur in the absence of any system perturbation, and results in increased production. Causes of eutrophication include human (anthropogenic eutrophication) and non-human (natural eutrophication) disturbances. Marked natural eutrophication events are relatively rare and may result from dramatic episodes, such as forest fire (e.g. Hickman *et al.*, 1990), tree die-off (Boucherle *et al.*, 1986; Hall & Smol, 1993; St. Jacques *et al.*, 2000) and prolific returns of spawning salmon to nursery lakes (Gregory-Eaves & Keatley, this volume), to name a few mechanisms. Climatic episodes, such as droughts, may also concentrate lake-water nutrients by increasing contributions of nutrient-rich groundwater (e.g. Webster *et al.*, 1996), or reducing flushing rates and increasing deepwater anoxia leading to elevated internal P loading from sediments to the illuminated surface waters (Brüchmann & Negendank, 2004). Some lakes lie in naturally fertile catchments or receive high natural loads of nutrients from groundwater and are naturally eutrophic (e.g. Hall *et al.*, 1999). In most cases, however, eutrophication is caused by anthropogenic nutrient inputs from domestic and industrial sewage disposal, farming activities, soil erosion, and numerous other activities.

Eutrophication is the most widespread form of lake pollution on a global scale, and has many deleterious effects on aquatic systems (Harper, 1992; Smith *et al.*, 2006). In addition to increasing overall primary production, eutrophication causes considerable changes to biochemical cycles and biological communities (Schelske, 1999). Marked changes occur at all levels in the food web and entire communities can change or die out (Carpenter *et al.*, 1995). For example, changes in the ratio of N:P often results in primary production shifting from primarily diatoms and other smaller edible algae towards larger cyanobacteria that are better competitors for N (Tilman *et al.*, 1986) and more resistant to grazing (Reynolds, 1984). As the light climate changes with increased phytoplankton turbidity, lakes may lose their submerged aquatic macrophyte communities, which in turn affect higher trophic levels (e.g. fish, waterfowl; see Bennion *et al.*, this volume). Decomposition of plant and algal biomass reduces oxygen availability in deep waters, causing declines in available habitat for fish and other biota (e.g. benthic invertebrates) and, under extreme conditions, massive fish kills. A major concern is that lake ecosystems may become unstable and biodiversity may be lost during eutrophication (Carpenter *et al.*, 1995; Cottingham *et al.*, 2000).

Eutrophication is a costly economic problem (Dodds *et al.*, 2009). For example, algal blooms increase water-treatment costs and sometimes cause treatment facilities to malfunction (Vaughn, 1961; Hayes & Greene, 1984). Furthermore, algal blooms (including diatoms) can create taste and odor problems (Mason, 1991), and algal breakdown products may chelate with iron and aluminum to increase metal contamination of drinking water (Hargesheimer & Watson, 1996). Toxins produced by cyanobacteria (e.g. microcystin; Kotak *et al.*, 1995) pose risks to human health, livestock, and wildlife. Diatom blooms are known to clog fishing nets, adversely affecting freshwater fisheries (Weyhenmayer *et al.*, 2008).

In agricultural regions of Europe, Asia, and Africa, lakes may have been disturbed over long timescales (several centuries or millennia) in response to forest clearance and the onset of agriculture (e.g. Fritz, 1989; Verschuren *et al.*, 2002; Bradshaw *et al.*, 2005; Dong *et al.*, 2008). In North America, marked and long-lasting nutrient enrichment has been observed even at

The Diatoms: Applications for the Environmental and Earth Sciences, 2nd Edition, eds. John P. Smol and Eugene F. Stoermer. Published by Cambridge University Press.

lakes where subsistence hunting (Douglas *et al.*, 2004; Douglas & Smol, this volume) and farming (Ekdahl *et al.*, 2007) activities of low densities of indigenous people have occurred. In many regions, however, major human impacts became noticeable more recently (e.g. the last 50–150 years or less; Boyle *et al.*, 1999; Little *et al.*, 2000; Finsinger *et al.*, 2006; Franz *et al.*, 2006).

The long history of human impacts may complicate remediation efforts in at least two ways. First, nutrients build up in lake sediments (Marsden, 1989; Rippey *et al.*, 1997) and catchment soils (Jordan *et al.*, 2001; Jordan *et al.*, 2002; Foy *et al.*, 2003) over time, causing delays in recovery following reduction or stabilization of external nutrient loads due to internal P loading and P saturation of soils (or, desorption of soluble P), respectively. Second, long-term data are unavailable for most lakes over the full timescale of human impacts (usually $<$20 years; Likens, 1989). We rarely undertake extensive studies until "after the fact" (Smol, 2008). Nonetheless, successful preventative and restorative management programs require detailed guidelines as to how much ecological and water-quality change has occurred in order to set realistic target conditions for remediation and to anticipate the degree of improvement that is possible. Increasingly, water-protection legislation is adopting approaches based on reference conditions and state-change to assess water quality, biotic integrity, and ecological health (e.g. Karr & Dudley, 1981; Pollard & Huxham, 1998; Barbour *et al.*, 2000). A striking example is the European Council Water Framework Directive (EC WFD) that was initiated recently (European Union, 2000). As detailed by Bennion & Battarbee (2007) and Bennion *et al.* (this volume), the EC WFD is in sharp contrast with most other water directives that traditionally have placed primary importance on chemical targets. Instead, the EC WFD places emphasis on the ecological structure and function of aquatic ecosystems, with biological elements (fish, invertebrates, macrophytes, phytobenthos, and phytoplankton) at the center of the status assessments, and hydromorphology and physico-chemical criteria play supporting roles. Ecological quality is judged by the degree to which present-day conditions deviate from reference conditions that occur in the absence of anthropogenic influence. Consequently, alternative monitoring and assessment approaches must be developed to determine what the natural productivity and variability was before impact, and to identify when individual lakes began to change and what the causal mechanisms were.

The EC WFD allows for several methods to define reference conditions, including expert judgment, space-for-time substitution, historical data, modelling, and paleolimnology (Pollard & Huxham, 1998; European Union, 2000). Because diatoms are sensitive, abundant bioindicators that can be sampled with relative ease, their use in biomonitoring programs will undoubtedly continue to increase. Moreover, because diatoms are well preserved in most lake sediments, the recent shifts in water-protection policy and legislation, such as the EC WFD, will continue to promote expansion of their use in applied paleolimnological studies to determine reference conditions and quantify the magnitude of ecological change. A combined approach of long-term monitoring and paleolimnological assessments is likely to prove particularly effective (Battarbee, 1999; Bennion & Battarbee, 2007).

In this chapter, we focus on the use of diatom algae for applied eutrophication studies in freshwater lakes. We further focus on paleolimnological applications. Our review is primarily on applications for deep lakes, as other chapters in this book discuss nutrient fluctuations in other aquatic systems; including, shallow lakes (Bennion *et al.*, this volume), fluvial lakes and impoundments (Reavie & Edlund, this volume), wetlands and peatlands (Gaiser & Rühland, this volume), rivers and streams (Stevenson *et al.*, this volume), and estuarine, brackish, and coastal ecosystems (Cooper *et al.*, this volume; Trobajo & Sullivan, this volume; Snoeijs & Weckström, this volume).

7.2 Why diatoms are useful indicators of lake eutrophication

Diatoms are an abundant, diverse, and important component of algal assemblages in freshwater lakes. They comprise a large portion of total algal biomass over a broad spectrum of lake trophic status (Kreis *et al.*, 1985). Because diatoms are a high-quality food source for herbivores, they often play an important role in aquatic food-web structure and function (Round *et al.*, 1990). Thus, responses of diatoms to lake eutrophication can have important consequences for other components of aquatic ecosystems.

While diatoms collectively show a broad range of environmental tolerances along a gradient of lake-nutrient status, individual species have specific habitat and water-chemistry requirements (Patrick & Reimer, 1966; Werner, 1977; Round *et al.*, 1990). In addition, distinct diatom communities live in open waters of lakes (plankton), or primarily in association with plants (epiphyton), rocks (epilithon), sand (epipsammon), or mud (epipelon) in littoral, nearshore habitats. Consequently, diatoms can be used to track shifts from littoral to planktonic algal production as well as declines of macrophyte cover that occur during eutrophication (Bradshaw *et al.*, 2005; McGowan *et al.*, 2005; Yang *et al.*, 2008; see also Julius & Theriot, this volume).

Diatoms are well suited to studies of lake eutrophication, because individual species are sensitive to changes in concentrations, supply rates, and ratios of nutrients (e.g. Si:P; Tilman, 1977; Tilman *et al.*, 1982). Each taxon has an optimum and a tolerance for nutrients that can usually be quantified to varying degrees (e.g. P: Fritz *et al.*, 1993; Hall & Smol, 1992; Bennion, 1994, 1995; Reavie *et al.*, 1995; Bennion *et al.*, 1996; N: Christie & Smol, 1993). This ability to quantify responses of individual taxa to nutrient concentrations has provided diatomists with a powerful tool for quantifying environmental changes that accompany eutrophication and recovery.

Diatom assemblages are typically species-rich. It is common to find over a hundred taxa in a single sample collected from a rock, aquatic plant, or sediment. This diversity of diatoms contains considerable ecological information concerning lake eutrophication. Moreover, the large numbers of taxa provide redundancies of information – an important internal check on data sets that increases confidence of environmental inferences (Dixit *et al.*, 1992).

Diatoms respond rapidly to eutrophication and recovery (e.g. Zeeb *et al.*, 1994; Alefs & Müller, 1999; Anderson *et al.*, 2005). Because diatoms are primarily photoautotrophic organisms, they are directly affected by changes in availability of nutrients and light (Tilman *et al.*, 1982). Their rapid growth and immigration rates and the general lack of physical dispersal barriers ensure there is little lag-time between perturbation and response (Vinebrooke, 1996). Consequently, diatoms are early indicators of environmental change. Because the cell wall is composed of resistant opaline silica (SiO_2), diatom valves are usually well preserved in most samples, including lake sediments. Consequently, by taking sediment cores and analyzing diatom assemblages, it is possible to infer past environmental conditions using paleolimnological techniques. Schelske (1999) reviews how measurements of biogenic silicon contained in sedimentary diatom assemblages can be used to track lake eutrophication.

Below, we present an overview of many of the approaches that employ diatoms to study lake eutrophication, and then illustrate the contributions of diatoms to our knowledge base by presenting applications, or case studies, of the different approaches.

7.3 Approaches

Lake eutrophication operates over a range of spatial and temporal scales. Consequently, scientists and lake managers must rely on a number of different approaches to study the effects of nutrient enrichment and to identify the best possible management strategies. Applied diatom studies generally fall into two broad categories – experiments and field observations. Some studies effectively utilize both approaches (e.g. Tilman *et al.*, 1986; Jeziorski *et al.*, 2008).

7.3.1 Experiments

Experimental approaches have been useful for documenting the importance of resource limitation as a mechanism for determining species composition of algal communities in lakes, and for identifying the specific resource most strongly limiting growth of diatom species (e.g. Tilman *et al.*, 1982). In applied studies, often the first piece of information lake managers need to know is exactly which resource (e.g. which nutrient) most strongly limits nuisance algal growth, in order to devise effective control strategies.

Controlled experiments can be performed in a laboratory or in lakes, and can vary in scale from small bottles and chemostats (e.g. Schelske *et al.*, 1974; Tilman, 1977), to mesocosms (e.g. Schelske & Stoermer, 1971; Lund & Reynolds, 1982), and even to entire lakes (e.g. Schindler, 1977; Yang *et al.*, 1996). In general, larger-scale experiments provide information that is more realistic of aquatic ecosystems, because they may include important ecological processes, such as biogeochemical exchanges with the sediments and the atmosphere, and food-web interactions. Smaller-scale experiments, however, provide more precise experimental control, and are cheaper and easier to perform, but they are oversimplified systems compared to real lakes (Schindler, 1977). Whole-lake experiments provide the most realistic responses of aquatic communities to eutrophication, but the rate of eutrophication is often greatly accelerated above rates at which anthropogenic eutrophication normally operates (e.g., Zeeb *et al.*, 1994).

One important contribution of experimental bioassay research has been the production of detailed physiological data for individual diatom species (e.g. Tilman *et al.*, 1982). By combining physiological data with resource-based competition models, researchers have begun to provide an important bridge between modern limnological studies and paleolimnology (sediment-based studies), because they can provide a mechanistic link to interpret observed species shifts in lakes (e.g. Kilham *et al.*, 1996).

7.3.2 Field observation

Field observation is a widely used approach to gather ecological information concerning distributions of individual diatom species and entire assemblages in lakes. Field-observation studies usually involve lake surveys along environmental gradients to describe correlative relationships between diatom community composition and water chemistry. While these studies do not measure the causal factors to which diatoms respond, they provide information on the net outcome of diatom assemblages to the complex interactions of all important control factors. Consequently, field surveys provide a useful approach for describing apparent nutrient and habitat preferences of diatom species along a gradient of lake productivity.

One example of the above approach is the survey of planktonic diatoms along a nutrient gradient in Lake Michigan by Stoermer & Yang (1970). They related the distributions of diatom species in algae samples collected over ~100 years from polluted and unpolluted sites with measured limnological variables. Using this approach, they were able to identify dominant diatom taxa characteristics of highly oligotrophic open-lake environments, as well as eutrophic small-basin environments. Ecological generalizations provided by this survey have been extremely useful for interpreting past and present lake trophic status in many other regions (Bradbury, 1975).

With the development of the concept of the "training set" approach (Birks *et al.*, 1990b), as well as multivariate (e.g. canonical ordination) and weighted-averaging numerical approaches that quantify species–environment relationships (see Birks, this volume), the use of lake surveys has increased tremendously in recent decades (Charles & Smol, 1994; Charles *et al.*, 1994; Reavie *et al.*, 2006). In particular, large surveys of surface-sediment diatom assemblages along productivity gradients have provided considerable ecological information concerning the relationships between diatoms and nutrients, and they have generated transfer functions that allow scientists to quantify lake-water nutrient concentrations from taxonomic composition of diatom communities (e.g. Hall & Smol, 1992; Bennion *et al.*, 1996; see Birks, this volume).

Long-term monitoring programs (LTMPs) of lakes are increasingly used to assess regional and national water quality, with the aim of providing an early warning of eutrophication and other water-quality problems (Smol, 1995, 2008). Water chemistry has traditionally formed the foundation of most LTMPs, but biological indicators such as diatoms are increasingly incorporated into these programs. For example, diatoms are being incorporated into the US Environmental Protection Agency's Environmental Monitoring and Assessment Program (EMAP) (Whittier & Paulsen, 1992). Ongoing research is actively designing programs and protocols that can meet requirements of the European Union Water Framework Directive (EC WFD) (King *et al.*, 2006; Bennion & Battarbee, 2007), and new programs that assess environmental changes in the coastal zones of the Laurentian Great Lakes (Niemi *et al.*, 2007) and elsewhere (e.g. Rosenberger *et al.*, 2008). Diatoms are increasingly utilized because most LTMPs work on a tiered system, with a few sites monitored intensively (e.g. weekly), others less frequently (e.g. seasonally), and the majority of lakes sampled infrequently (e.g. less than once a year). Because much of the information collected by LTMPs is based on a few "snapshots" of the system, it is important to use biological assemblages such as diatoms. Diatoms integrate environmental conditions over larger spatial and temporal scales compared to a single water-chemistry sample, thus improving the signal-to-noise ratio (Battarbee *et al.*, 2005; King *et al.*, 2006; Reavie *et al.*, 2006). Consequently, the cost-to-benefit ratio is relatively low for LTMPs using diatoms.

7.4 Responses of periphytic diatoms to eutrophication

Diatoms are often an important component of algal communities growing attached to substrata located within the nearshore, photic zone of lakes, known as the periphyton. Most experimental and field studies have focused on planktonic algae, and responses of periphytic diatom communities to eutrophication remain less well understood (Lowe, 1996; King *et al.*, 2006). However, periphyton can dominate primary production in many lakes (Hargrave, 1969; Wetzel, 1964; Wetzel *et al.*, 1972; Anderson 1990a), especially the numerous small, shallow lakes that dominate many landscapes. Consequently, periphytic diatom communities may provide information concerning littoral-zone responses to eutrophication.

The use of periphytic diatoms in applied diatom studies has been hampered in part because mechanisms regulating periphyton have remained poorly understood (Burkhardt, 1996; Lowe, 1996; Bennion *et al.*, this volume). Nutrient availability plays an important role in determining the quantity and distribution of periphyton, but periphyton are potentially subject to more complex, multiple controls than phytoplankton (Fairchild & Sherman, 1992, 1993; Niederhauser & Schanz, 1993). For example, steep resource gradients of space and light availability, water turbulence, and grazing (Lowe & Hunter, 1988) are important factors, but may not respond in a predictable or direct manner to eutrophication. In some lakes, substratum and algal-mat chemistry appears to play a stronger role on periphyton communities than water-column chemistry (Cattaneo, 1987; Bothwell, 1988; Hansson, 1988, 1992), and chemical conditions may be strikingly different between the water column and algal mats (Revsbeck *et al.*, 1983; Revsbeck & Jorgensen, 1986). Consequently, periphyton may respond less directly than phytoplankton to water-column nutrient additions, especially in mesotrophic to eutrophic systems where nutrient limitation of periphyton may not be severe (Hansson, 1992; Fairchild & Sherman, 1992; Turner *et al.*, 1994). Lowe (1996) and Burkhardt (1996) review factors regulating periphytic algal communities, and Bennion *et al.* (this volume) discuss associated problems for use of diatom-P transfer functions in shallow, macrophyte-dominated lakes.

In spite of the complexities identified above, there is recent and increasing recognition that periphytic communities in the littoral zones of lakes may actually respond more rapidly and more sensitively to eutrophication than indicators measured in the pelagic zones of lakes (e.g. water chemistry or phytoplankton) (Kann & Falter, 1989; Lambert *et al.*, 2008). For example, Goldman (1981) was able to relate increased nutrient loading into oligotrophic Lake Tahoe (USA) to increases in benthic algal biomass. Significant increases in periphytic diatoms were also observed in the littoral zone of Lake Taupo (New Zealand) following sewage loading (Hawes & Smith, 1992). Recently, Lambert *et al.* (2008) showed that the biomass of epilithic algae in littoral zones of oligotrophic lakes in central Canada affected by recreational uses responded more to gradients of lakeshore perturbations (dwellings, land clearance) than did phytoplankton in the pelagic zone, likely because littoral algae are closer to the sources of nutrient supply and because there is inertia inherent in raising nutrient concentrations of the pelagic lake volume to sufficient thresholds to affect metrics based on phytoplankton. Although methods are well developed for using diatoms to assess eutrophication of flowing waters (see Stevenson *et al.*, this volume), they remain less developed for lakes. New protocols and research methods employing littoral zone diatoms are, however, being developed at a rapid and increasing rate. This is in recognition of recent findings that the community composition of periphytic diatoms (1) responds rapidly and sensitively to watershed disturbances, (2) can be sampled relatively easily, and (3) has the potential to contribute data that are very useful for biomonitoring programs (e.g. King *et al.*, 2006; Reavie, 2007; Kireta *et al.*, 2007; Reavie *et al.*, 2008). King *et al.* (2006) provide a number of important recommendations for designing sampling protocols that employ littoral diatoms to monitor effectively changes in ecological and water-chemistry conditions. Their recommendations attempt to minimize noise due to inherent within-site variability, successional patterns of periphyton, and substratum specificity, in order to maximize the ability to detect signals of human impacts.

An excellent example of an innovative new monitoring program that employs periphytic diatoms is the Great Lakes Environmental Indicators (GLEI) project. This large project developed and tested indicators of ecological condition for coastal ecosystems along the US side of the five Laurentian Great Lakes and has included measurements of birds, fish, amphibians, aquatic macroinvertebrates, and wetland vegetation, in addition to diatoms (Niemi *et al.*, 2007). A study by Reavie *et al.* (2006) developed and tested several diatom-based transfer functions derived from a training set of coastal surface-sediment and epilithic samples from 115 sites that included five coastal ecosystem types (embayments, high-energy shorelines, coastal wetlands, riverine wetlands, and protected wetlands). Multivariate ordination (canonical correspondence analysis; CCA) identified a strong correlation between composition of the diatom assemblages and total phosphorus (TP) concentration. They were able to develop a diatom–TP transfer function that performs well ($r^2_{jackknife} = 0.65$; RMSEP = 0.26 log (μg l^{-1})). Then they assessed the relative abilities of measured and diatom-inferred lake-water TP concentrations to track the influence of stressor gradients by regressing values against watershed characteristics that included gradients of agriculture, atmospheric deposition, and industrial facilities. This analysis showed that diatom-inferred TP values were better predicted by watershed characteristics than were measured lake-water TP concentrations. On this basis, the authors concluded that diatom communities can integrate water-quality information on coastal conditions better than snapshot water-chemistry measurements, likely because diatoms integrate water-quality conditions over longer temporal periods (e.g. days to months) compared to water-chemistry measurements (e.g. hours to days). They identified that assessment of coastal diatom communities for monitoring of environmental conditions is a cost-effective alternative to snapshot assessment based on water chemistry, and that diatoms have the advantage that samples can be stored indefinitely for reference purposes. Their study provides strong support for the use of diatoms in monitoring programs in the nearshore zones of the Laurentian Great Lakes and elsewhere. We anticipate that nearshore diatom-based biomonitoring programs will increase over the next years to decades as agencies responsible for water-quality management begin to realize these advantages in lakes of all sizes.

7.5 Using diatoms to investigate past eutrophication trends

Water-quality management and lake restoration are increasingly important aspects of environment and natural-resource agencies. The susceptibility of lakes to eutrophication, however, varies with a number of site- and region-specific factors (e.g. suitability of the catchment for agriculture, depth and quality of soil, industrialization and urbanization, geology, climate, lake depth, natural trophic status), and the magnitude and duration of past pollution. The timescales of eutrophication also vary enormously between lakes, from years to millennia (Fritz, 1989; Anderson 1995a, b; Dong *et al.*, 2008). Consequently, the lack of sufficient long-term data for most lakes creates a clear need for approaches which can assess past environmental conditions. For example, scientists and aquatic managers need to know if, when, and how much a lake has changed through time (Smol, 2008). They must also establish the pre-impact conditions (both chemical and ecological) and natural variability in order to set realistic goals for restoration and management decisions (Bennion & Battarbee, 2007). Research that quantifies the responses of a variety of lake types to past nutrient pollution (which have varied in duration, nutrient load, and nutrient combinations) will undoubtedly be very valuable for predicting lake responses to likely future scenarios (Smol, 2008). It is in this area that sedimentary diatom records and paleolimnology can be particularly valuable (Battarbee, 1999; Bennion & Battarbee, 2007; Smol, 2008).

Our ability to infer eutrophication trends quantitatively using diatoms has increased greatly during the past two decades (Hall & Smol, 1992; Bennion *et al.* 1996; Wessels *et al.*, 1999; Bradshaw & Anderson, 2001; Marchetto *et al.*, 2003). Prior to the 1990s, most diatom-based assessments of lake eutrophication were qualitative, based on the ecological interpretation of shifts in the abundance of individual species in sediment cores (e.g. Bradbury, 1975; Battarbee, 1978). Information on ecological preferences of individual indicator taxa were largely gleaned from contemporary phycological surveys that described patterns in diatom distributions among lakes of differing productivity (e.g. Stoermer & Yang, 1970), or anecdotal descriptions in taxonomic books (e.g. Hustedt, 1930; Patrick & Reimer, 1966; Cholnoky, 1968; Krammer & Lange-Bertalot, 1986–1991). Using this approach, investigators were able to infer, albeit qualitatively, anthropogenic eutrophication trajectories in lakes and correlate them with known changes in sewage inputs and land-use activities, such as agriculture, road construction, and urbanization (e.g. Bradbury, 1975; Smol & Dickman, 1981; Engstrom *et al.*, 1985). An historical overview of some of the developments has been summarized by Battarbee *et al.* (2001).

The composition of diatom assemblages in sediment cores is now widely used to quantify past lake-water total phosphorus concentration and other variables related to eutrophication. Hall & Smol (1999) summarize the stages of advancement that have allowed the field to develop from qualitative to quantitative diatom-based estimates, including some early attempts with the use of diatom ratios (e.g. Centrales:Penales ratio (Nygaard, 1949), Araphidineae:Centrales ratio (Stockner, 1971)) and indices based on multiple linear regression models (Agbeti & Dickman, 1989; Whitmore, 1989). Quantitative estimates now typically rely on the use of spatial surveys or training sets of diatom assemblages in surface sediments of a large sample of lakes (typically about 50 or more) situated along a trophic status gradient to develop transfer functions that estimate reliably the optima and tolerances of diatoms to nutrient concentration (or other eutrophication-related variables). These transfer functions, in combination with percent abundance data of diatom taxa in sedimentary assemblages, are then used to estimate past TP concentrations, often referred to as diatom-inferred TP (or DI-TP). This approach is now well developed, widely used, and described in detail in numerous papers (e.g. Hall & Smol, 1992, 1996; Bennion *et al.*, 1996), as well as by Birks (this volume) and Birks *et al.* (2011).

7.5.1 Modern quantitative methods

Weighted averaging (WA) has become the most widely used method for development and application of transfer functions to reconstruct past changes in lake trophic status. Most commonly, diatom-based transfer functions have been developed to estimate epilimnetic TP concentration of surface waters (see Table 7.1) because phosphorus is the nutrient that often limits aquatic productivity (Sas, 1989) and strongly regulates communities (Schindler, 1971), although total nitrogen transfer functions also exist (Christie & Smol, 1993; Siver, 1999).

Transfer functions now exist for a number of regions around the world and the pace of their development continues to increase. The earliest transfer functions were developed during the early to mid 1990s in North America and Europe; including, western Canada (Hall & Smol, 1992; Reavie *et al.*, 1995), eastern Canada (Agbeti, 1992; Hall & Smol, 1996), midwestern USA (Fritz *et al.*, 1993), northeastern USA (Dixit & Smol, 1994), Northern Ireland (Anderson *et al.*, 1993), southern England (Bennion, 1994, 1995), Denmark (Anderson & Odgaard, 1994), European Alps (Wunsam & Schmidt, 1995), and northwestern Europe (Bennion *et al.*, 1996). Since the mid 1990s, there has been a rapid proliferation of regional diatom-based transfer functions for lake eutrophication studies, including expansion into regions of Fennoscandia, New Zealand, Australia, and China. Table 7.1 provides a summary of many of the existing diatom-based transfer functions for nutrients.

Table 7.1 Summary of some of the currently available transfer functions for reconstructing surface-water total phosphorus concentration based on diatom remains in aquatic sediments. Asterisk indicates apparent RMSE (root-mean-square error) values, calculated without bootstrapping or jackknifing.

Range ($\mu g\, l^{-1}$) (mean)	Season	Model	r^2	$r^2_{jack/boot}$	RMSEP	Variable transformation	No. of sites	Location	Other details	References
30–550 (71)	Annual	WA	n.a.	0.82	0.12	log(x)	43	S.E. China	Yangtze River floodplain	Yang *et al.* (2008)
0–675 (33)		WA-PLS2	0.67	n.a.	n.a.		73	Ireland		Taylor *et al.* (2006)
6–49	Spring	WA	n.a.	0.57	0.20	log(x)	30	Southest Ontario, Canada	Polymictic, alkaline	Werner & Smol (2005)
2–171 (14.4)	Annual	WA_{tol}	n.a.	0.50	0.20	log(x)	53	New Zealand		Reid (2005)
7–451 (64)	Annual	WA-PLS2	0.94	0.74	0.23	log(x) + 1	31	Southeast Australia	Reservoirs	Tibby (2004)
3–89 (38)	Fall	WA	n.a.	0.76	0.16	log(x)	61	South Finland		Kaupilla *et al.* (2002)
3–52 (10.3)	Summer	WA-PLS3	0.92	0.51	3.2	None	75	Quebec, Canada	Oligotrophic	Philibert & Prairie (2002)
9–1687 (96)	Annual	WA_{tol}	0.86	n.a.	0.44	ln(x)	69	Northeast Germany	Only littoral diatoms	Schönfelder *et al.* (2002)
24–1145 (164)	Annual	WA-PLS2	0.86	0.37	0.28	log(x)	29	Denmark	Shallow, eutrophic	Bradshaw *et al.* (2002)
24–1145 (164)	Annual	WA_{tol}	0.62	0.23	0.32	log(x)	29	Denmark	Only planktonic diatoms; shallow, eutrophic	Bradshaw *et al.* (2002)
7–370	Annual	WA	0.75	0.47	0.24	log(x)	43	Sweden		Bradshaw & Anderson (2001)
4–54 (14.1)	Summer	WA	0.64	0.47	10	None	64	Southeast Ontario, Canada	Alkaline	Reavie & Smol (2001)
3–83	Summer	WA	0.77	0.52	0.23	log(x)	51	Alaska, USA	Single water samples	Gregory-Eaves *et al.* (1999)
0–8740	Summer	WA	n.a.	0.55	0.79	ln(x)	238	Northeast USA		Dixit *et al.* (1999)
6–520 (73)	Spring	WA-PLS2	0.93	0.79	0.19	log(x)	72	Switzerland		Lotter *et al.* (1998)
3–24 (7.6)	Spring overturn	WA	0.62	0.41	4.2	None	54	Ontario, Canada	Oligotrophic	Hall & Smol (1996)
5–1190 (104)	Annual	WA-PLS2	0.91	0.82	0.21	log(x)	152	Northwest Europe	Small, shallow, eutrophic	Bennion *et al.* (1996)
6–42	Spring-fall	WA	0.73	0.46	0.48	None	59	West Canada	Single water samples	Reavie *et al.* (1995)
2–266 (115)	Summer	WA_{tol}	0.57	0.31	0.35	log(x)	86	Alpine (Austria, Bavaria, Italy)	Oligo-mesotrophic	Wunsam & Schmidt (1995)
25–646	Annual	WA	0.79	n.a.	0.28	log(x)	30	Southeast England	Shallow, eutrophic	Bennion (1994)
11–800		WA	0.75	n.a.	0.23	log(x)	49	Northern Ireland		Anderson *et al.* (1993)
1–51	July	WA	0.73	n.a.	0.41*	ln(x + 1)	41	Michigan, USA	Single water samples	Fritz *et al.* (1993)
5–28	Spring–fall	WA	0.86	n.a.	0.25*	ln(x + 1)	37	Western Canada	Single water samples	Hall & Smol (1992)

Recently, Hübener *et al.* (2008) implemented a new approach to reconstruct TP using diatoms along the lines suggested by Birks (1998) in the "dynamic training-set" approach. Their method attempts to improve diatom–TP reconstructions by using a "moving window" modelling approach that selects an optimal number of nearest neighbours [= a subset of samples (or, surface-sediment diatom assemblages) from a training set that possesses close modern analogs, as identified by Euclidean distance metrics obtained from detrended correspondence analysis (DCA) ordinations] for each fossil diatom sample from a large "supra-regional" training set of modern samples. As demonstrated by Hübener *et al.* (2008), this approach has some theoretical advantages and appears to outperform inferences obtained by the more traditional training-set approach, particularly for estimating TP concentrations at the low end of the range associated with pre-disturbance lake conditions. Software that can automate selection of the optimum number of nearest neighbours might allow this approach to become widely used (e.g. Juggins, 2001).

The majority of diatom–TP transfer functions are based on all taxa in sedimentary diatom assemblages, including planktonic, epiphytic, and benthic taxa. However, a few studies have attempted to improve transfer-function performance by focusing on planktonic taxa only, based on the concept that errors increase because epiphytic and benthic taxa experience nutrient concentrations that differ from those of the epilimnion. But, in practice, they tend to provide little or no improvement (Siver 1999; Bradshaw *et al.* 2002; Philibert & Prairie 2002), perhaps because of errors due to incorrect assignment of taxa to a habitat type and because many taxa may grow in more than one habitat. Additionally, weaker performance of planktonic-only transfer functions may be due to the smaller number of taxa included (Siver, 1999; Bradshaw *et al.*, 2002).

For some, an intuitive approach to track changes in P content of lakes has been to directly determine the concentration and flux rates of P in sediment profiles – but this approach is fraught with difficulties. Under favorable lake conditions, sedimentary P may serve as a useful productivity indicator (Engstrom & Wright, 1984). Unfortunately, however, it is hard to be certain when the conditions are favorable or not, and, during lake eutrophication, conditions are almost certain to shift from favorable to unfavorable (Carignan & Flett, 1981; Boyle, 2001). Changes in the duration and magnitude of hypolimnetic anoxia cause variations in the loss of sediment P. Thus, in lakes where eutrophication has stimulated changes in hypolimnetic anoxia, sediment P content can be negatively correlated (or uncorrelated) with the concentration of P in the water column (Boyle,

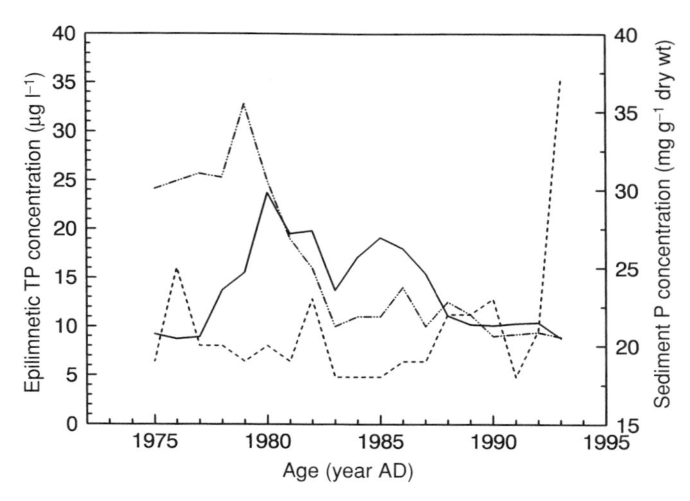

Figure 7.1. Diatom transfer function evaluation from Mondsee, Austria, showing correspondence between measured annual mean epilimnetic TP concentration (dashed & dotted line) and diatom-inferred epilimnetic TP concentration (solid line). Sediment P content (dashed line) does not reflect changes in water-column P concentrations. Sediments from Mondsee are annually laminated, permitting comparisons at an annual time-step since 1975 when water-chemistry monitoring began (modified from Bennion *et al.*, 1995 with permission of Blackwell Science Ltd.).

2001). In many lakes, reductive remobilization of P can lead to a stationary surface-sediment maximum P content, which can be falsely interpreted as indicating recent eutrophication (Carignan & Flett, 1981; Farmer, 1994; see also Figure 7.1). In eutrophic lakes, lateral fluxes of P can further complicate relationships between sedimentary profiles and water-column P concentrations (Belzile *et al.*, 1996). Because of these problems inherent in trying to estimate past lake-water TP concentrations from sediment P concentration, diatom-based transfer functions provide one of the few methods to reliably estimate past changes in lake-water TP concentrations, quantify rates of lake eutrophication, and establish goals for water-quality remediation (Anderson *et al.*, 1993; Bennion *et al.*, 1995). As a consequence, diatom transfer functions are increasingly used by scientists and lake managers to study lake eutrophication, and now represent the standard approach to reconstruct past nutrient concentrations in eutrophication studies around the world. As their use has become more widespread, there has been an increased focus on the critical ecological and statistical evaluation of diatom–TP reconstructions.

7.5.2 Assessment of diatom–TP model performance, errors, and evaluation

Use of diatom-based transfer functions developed from training sets is a powerful technique for quantitative reconstructions of lake-water chemistry and eutrophication, but it is not without problems and limitations. As with any model, diatom transfer functions will always produce a result. However, it is important to know how reliable the resulting value is. In principle, the performance of WA models is set by the quality of the training data set upon which it is based, as well as the ability of the empirical formulae to adequately model ecological distributions of the biota. As a basic rule, quantitative reconstructions are most useful when diatom-inferred TP values are quantitatively similar to measured TP (accurate), have narrow error estimates of prediction (precise), and perform well in most lakes (robust). Comprehensive reviews by Birks (1995, 1998) and Birks (this volume) provide information concerning the basic biological and statistical requirements of quantitative reconstruction procedures, as well as methods for assessing their ecological and statistical performance.

Weighted averaging is sensitive to the distribution of the environmental variables (e.g. TP) in training sets (Birks, 1995). The accuracy of the WA method depends largely on reliable estimates of WA coefficients, namely the estimated optima and tolerances, derived from modern diatom distributions in training sets. However, WA estimates of species optima will be biased when the distribution of lakes along the gradient is highly uneven, or when species are not sampled over their entire range. The bias at the ends of the gradient (e.g. at very high or low TP) is caused by truncated species responses, or so-called "edge effects," and results in inaccurate estimates of species optima that reduce accuracy and precision of diatom-inferred TP at high and low values. This has important implications for estimating reliably reference lake conditions.

Perhaps the best solution to avoid problems associated with edge effects is to sample a sufficient number of lakes along the full environmental gradient. Unfortunately, this may not always be possible in some geographic regions. For example, eutrophic lakes (i.e. with TP > 30 $\mu g\ l^{-1}$) are relatively rare in some regions (e.g. the Precambrian Shield region of Canada; Hall & Smol, 1996). Consequently, some diatom–TP training sets have a high representation of oligotrophic lakes and few or no eutrophic lakes (Hall & Smol, 1992, 1996; Reavie *et al.*, 1995). In contrast, data sets from southern England, Wales, and Denmark are dominated by eutrophic and hypereutrophic lakes, and few lakes have TP < 100 $\mu g\ l^{-1}$ (Anderson & Odgaard, 1994; Bennion, 1994). One method to extend the range of TP is to combine several regional training sets into one large data base. Researchers in Europe have done precisely this. By combining regional data sets from England, Northern Ireland, Denmark, and Sweden, they were able to extend the range of TP (5–1000 $\mu g\ l^{-1}$) and reduce the bias towards high TP concentrations (Bennion *et al.*, 1996). An added benefit of amalgamating regional data sets is that it increases the chance of having good modern analogs for fossil diatom assemblages. For example, by adding the Swedish lakes into the combined European data set, transfer functions include analogs for diatom assemblages in lakes with very low TP concentration that can be used to estimate more accurately pre-agricultural trophic status (Bennion *et al.*, 2004). The moving window method of Hübener *et al.* (2008), as mentioned above, may be helpful in this regard.

While it is valuable to assess the performance of the training set as a whole, it is even more critical to assess how well diatom transfer functions perform in the specific lake under investigation. Ultimately, the best way to assess the accuracy of TP inferences from fossil diatom assemblages is to compare directly the inferred values against available long-term records of measured lake-water TP concentration (so-called model validation or ground-truthing). Critical assessments of diatom transfer functions should include this type of validation in lakes across the entire spectrum of trophic status and human eutrophication intensity. This research is underway, but because there are few lakes for which good long-term chemical data exist, this validation approach has been not been widespread. Examples include studies by Bennion *et al.* (1995); Marchetto & Bettinetti (1995); Hall *et al.* (1997); Rippey *et al.* (1997); Lotter (1998); Bradshaw & Anderson (2001); and Marchetto *et al.* (2003). These examples indicate that in most cases diatom-inferred TP compares well to measured values, but that estimates may be sensitive to the absence of modern analogs and may tend to underestimate high TP values (Bennion *et al.*, 1995) and overestimate low TP values (Bradshaw & Anderson, 2001; Bennion & Battarbee, 2007).

The combination of long-term water-chemistry records and excellent sediment chronology allowed Bennion *et al.* (1995) to undertake a detailed evaluation of a diatom–TP transfer function. They inferred TP from diatoms in an annually laminated sediment sequence from Mondsee, Austria, and compared inferred epilimnetic TP against 20 years of annual mean TP measurements. Using this approach, they showed that diatom–TP inferences tracked measured changes in epilimnetic TP concentration (Figure 7.1). In comparison, sediment P concentrations did not reflect changes in water-column P, illustrating that redox changes and post-depositional P mobility hamper the ability to determine water-chemistry changes from sedimentary P records. A slight mismatch between the timing of the TP peak (1979 for water chemistry versus 1980 for the diatom model) was attributed to sediment dating errors due to difficulties in distinguishing some varves. The model underestimated measured TP prior to 1980, possibly as a result of poor diatom assemblage analogs in those samples. Nevertheless, the diatom transfer function closely mirrored the major trends in measured TP, namely increasing TP during a eutrophication phase in the 1970s and declining TP during the 1980s since the installation of sewage treatment (Bennion *et al.*, 1995; Figure 7.1).

In the absence of direct comparisons against measured water chemistry, several indirect methods are available to evaluate whether diatom transfer functions are likely to produce realistic estimates of past conditions. Bootstrapping can be used to estimate the magnitude of random prediction errors (RMSE) for individual fossil samples (Birks *et al.*, 1990a). Bootstrapped-derived errors can vary depending on the taxonomic composition of fossil diatom assemblages, due to differences in the abundances of taxa with stronger or weaker relationships to the environmental variable being reconstructed. Confidence is also increased when fossil diatom assemblages show a high degree of similarity to the modern training set (i.e. possess good modern analogs), and this can be quantified using a dissimilarity index (e.g. χ^2) calculated for comparisons between the taxonomic composition of all modern and fossil diatom assemblages (Birks *et al.*, 1990a). Inferences are likely to be more reliable if fossil assemblages comprise taxa that show a strong relationship to the variable of interest (i.e. show good "fit" to the environmental variable being reconstructed, *sensu* Birks *et al.*, 1990a), and this can be assessed by examining the residual fit to the first axis in canonical ordinations that are constrained to the variable being considered (see Birks *et al.*, 1990a; Hall & Smol, 1993; Simpson & Hall, 2011; Birks, this volume).

A large research effort during the past 10 years has developed and refined the numerical basis of diatom transfer functions, with the result that they now perform well and provide realistic error estimates (see also Bennion *et al.*, this volume). However, some additional factors may affect the performance of diatom–TP transfer functions. For example, TP concentrations can vary substantially during the course of a year or even a growing season. Consequently, training sets must employ a water-chemistry sampling frequency sufficient to provide an accurate estimate of the annual or seasonal mean. However, the annual range of TP concentrations is greater in more eutrophic lakes (Gibson *et al.* 1996; Bennion & Smith, 2000; Bradshaw *et al.*, 2002). Because different diatom species attain peak populations at different times of an annual cycle, they may indicate quite different TP concentrations. This feature likely introduces unavoidable noise into diatom transfer functions, with the result that prediction errors will increase under more eutrophic conditions.

Within-lake variability in diatom assemblages could introduce error into diatom inferences (Anderson, 1990a,b,c). However, the good agreement between TP inferences from six cores taken across a range of water depths (3–14 m) in a wind-stressed, shallow lake suggests that a single core located in the central deep-water basin is likely representative of the whole lake (Anderson, 1998). Variability may be higher in surface-sediment diatom assemblages than in down-core samples, indicating that surface-sediment variability may be a source of error in modern training sets and, hence, in transfer functions (Anderson, 1998; Adler & Hübener, 2007). Further research is required to assess robust errors introduced by within-lake variability in surface-sediment diatom distributions.

An obstacle to accurate inferences of past lake-water TP is the observation that diatom community composition is under multifactorial control. Total phosphorus is only one of a group of covariates (e.g. N, Si, chlorophyll *a*) that measure trophic status (Fritz *et al.*, 1993; Jones & Juggins, 1995). Potentially, diatom-inferred reconstructions can be affected by independent historical changes in covariates that are unrelated to alterations in TP (e.g. other nutrients, light, turbulence, herbivory, climate). One way to minimize misinterpreting TP trends from diatom assemblages is to analyze other bioindicators that have different susceptibilities to covariates (Hausmann *et al.*, 2002). For example, fossil pigments can be used to track changes in other algal groups (Leavitt *et al.*, 1989; Leavitt, 1993; Hall *et al.*, 1999), which may also differ in their sensitivity to shifts in N, Si, and light availability. A shift from diatoms to N-fixing cyanobacteria may supply important information concerning changes in N and Si availability that were unrelated to P. Analysis of fossil zooplankton may indicate that taxonomic shifts were also caused by changes in herbivory (Jeppesen *et al.*, 1996). Furthermore, chironomid assemblages can be used to quantify deep-water oxygen availability (e.g. Quinlan *et al.*, 1998; Little *et al.*, 2000), which may provide important information on fish habitat availability and food-web changes, as well as the potential for increased internal P supply.

7.6 Case studies

Diatoms have contributed to our knowledge of lake eutrophication in a number of different ways, and new applications continue to be developed at a steady rate. In the following section, we present some of the different types of studies that have used diatoms preserved in sediment cores to investigate lake eutrophication and recovery, pinpoint the underlying cause(s), assess regional-scale water-quality changes, and evaluate empirical water-quality models.

7.6.1 Individual lake studies

The most common paleolimnological application of diatoms has been to investigate eutrophication trends in individual lakes of interest. Such studies have successfully identified pre-impact conditions, and the timing, rate, extent, and probable causes of eutrophication and recovery (e.g. Battarbee *et al.*, 2005). Also, where eutrophication has been well documented, individual lake scenarios have been used to study the responses of diatom communities (e.g. Turkia & Lepistö, 1999). In some lakes that contain annually laminated sediments (varves), a very high degree of temporal resolution can be attained (e.g. Lotter, 1998; Brüchmann & Negendank, 2004).

7.6.1.1 Short-term (decade to century scale) eutrophication trends Intensive paleolimnological studies of Lough Augher, in Northern Ireland, provide an excellent example of the detailed information that can be obtained by studying sedimentary diatom assemblages (Anderson, 1989, 1995a; Anderson *et al.*, 1990; Anderson & Rippey, 1994). By analyzing diatoms in 12 cores from this relatively small and simple basin, Anderson and co-workers have shown in great detail how Lough Augher became eutrophic as a result of untreated sewage effluent from a local creamery that entered the lake during the period 1900–1972. Diatom-inferred TP and diatom accumulation rates increased synchronously from 1920 to 1950, in response to initial increases in nutrient loads from the creamery (Figure 7.2a). However, during the most eutrophic period, DI-TP and diatom accumulation rates became uncoupled. Diatom accumulation rates indicated that planktonic diatom production peaked during 1940–57, before the highest inferred P levels, and hence highest algal productivity, were reached (*c.* 1950–70; Anderson, 1989, 1995a). The apparent contradiction of lower diatom production at higher rates of nutrient supply could be interpreted in terms of changes in Si:P ratios as the supply of P increased. As Si:P ratios declined, increasing silica limitation would have placed diatoms at a competitive disadvantage relative to non-siliceous algae. Consistent with this hypothesis, planktonic diatom communities became dominated by lightly silicified diatoms (small *Stephanodiscus* species) which are better competitors at low Si availability (Lund, 1950; Tilman *et al.*, 1982). In contrast, littoral diatom communities were not affected to the same extent,

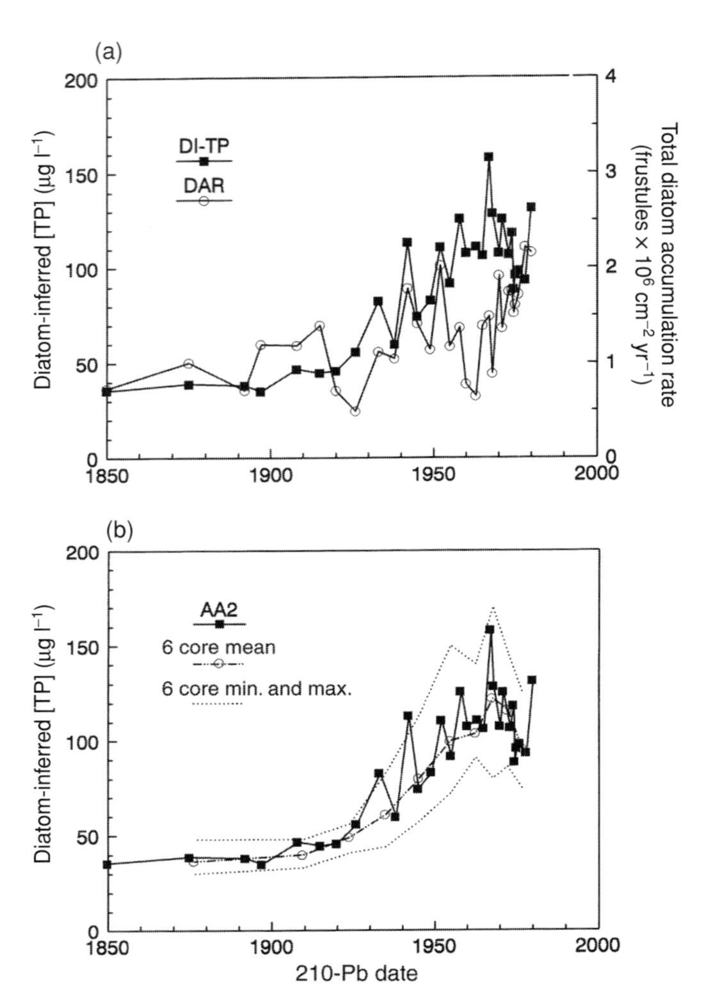

Figure 7.2 Indices of lake responses to point-source eutrophication at Lough Augher, Northern Ireland. (a) Diatom accumulation rate (DAR), and diatom-inferred epilimnetic TP concentration (DI-TP) from a central deep-water core (AA2) using weighted averaging and a surface-sediment diatom training set from Northern Ireland. (b) Comparison of DI-TP from a central deep-water core (AA2) and mean, minimum, and maximum values from six cores taken from 3–14 m water depth, to illustrate that TP inferences based on diatoms in a single deep-water core are representative of the lake basin as a whole. Chronology is derived by correlation to a dated master core. (Original: redrawn from data in Anderson, 1989; Anderson & Rippey, 1994; Anderson, 1998).

likely because they have alternative sources of nutrients (both P and Si) than the water column (e.g. sediments, macrophytes, detrital cycling). Consequently, littoral diatoms showed maximum accumulation rates during the productivity peak (1950–70).

The response of diatoms in Lough Augher to reduced nutrient loads following redirection of sewage effluent in the late 1970s showed an improvement in lake-water quality from eutrophic to mesotrophic conditions (Figure 7.2; Anderson

& Rippey, 1994). Diatom-inferred TP and diatom accumulation rates declined. Despite indications that water quality improved and evidence suggesting that P attained new equilibrium levels in Lough Augher, Anderson *et al.* (1990) showed that diatom species composition had not yet reached a new equilibrium even five years after sewage redirection. Thus, biological recovery may lag behind chemical recovery of eutrophied lakes (see also Stoermer *et al.*, 1996; Ekdahl *et al.*, 2007; Quinlan *et al.*, 2008). However, the process of biological recovery in eutrophied lakes has not yet been studied in as great detail as processes of degradation. Many lakes have now entered an oligotrophication phase following widespread reduction of nutrient loadings in recent years, especially in Europe and North America, and consequently the process of re-oligotrophication is receiving considerable attention (Bennion & Appleby, 1999; Clerk *et al*, 2004; Anderson *et al.*, 2005; Battarbee *et al.*, 2005; Augustinus *et al.*, 2006; Finsinger *et al.*, 2006; Bigler *et al.*, 2007). Paleolimnological approaches employing diatoms have much potential to determine the influence of factors such as the duration and extent of eutrophication, lake depth, flushing, and food-web structure in determining the rate, pattern, and end-point of biological recovery (e.g. Davis *et al.*, 2006).

7.6.1.2 Long-term human impacts (millennium scale) In many regions of the world with a long history of human settlement and land use (e.g. Europe, Africa, Asia, Middle East), lakes potentially have been altered by anthropogenic eutrophication over long-term, millennial timescales, and not simply since the Industrial Revolution. For example, combined diatom and pollen studies at Diss Mere, England, by Fritz (1989) and Peglar *et al.* (1989) have shown that anthropogenic activities began in the Neolithic period (5000–3500 BP) with the creation of small forest clearances. The first incidence of marked eutrophication occurred during the Bronze Age (3500–2500 BP) with marked increases in the abundance of *Synedra acus* Kützing, in response to forest clearance and the onset of cereal agriculture. Using a modern diatom–TP training set, Birks *et al.* (1995) inferred that early Bronze Age activities increased mean TP from 33 to >200 $\mu g\, l^{-1}$, altering the lake from a mesotrophic to hypereutrophic condition (Table 7.2). After this eutrophication phase, the lake never returned to pre-impact TP concentrations. Mixed pasture and crop agriculture during the Roman, Anglo-Saxon, and Medieval periods maintained Diss Mere in an eutrophic state (mean TP $\sim$ 100 $\mu g\, l^{-1}$). Major eutrophication occurred with the expansion of the town of Diss during the fifteenth and sixteenth centuries (post-Medieval) and diatom-inferred TP rose to 343 $\mu g\, l^{-1}$. Decline and eventual loss of macrophytes

Table 7.2 Mean and range of diatom-inferred lake water total phosphorus concentrations at Diss Mere, Norfolk for different historical periods during the past 7000 years. Modified from Birks *et al.* (1995), with permission.

Historical period	Age (BP)	Mean (μg TP l^{-1})	Range (μg TP l^{-1})
Modern times	0–150	495	385–675
Post-Medieval	150–500	343	297–400
Roman/Anglo-Saxon/ Medieval	500–2200	117	71–179
Iron Age/Roman	2200–2500	93	58–119
Late Bronze Age	2500–3000	90	36–114
Early Bronze Age	3000–3500	221	221
Neolithic	3500–5000	33	6–97
Mesolithic	5000–7000	20	4–35

coincided with this large increase in DI-TP (Birks *et al.*, 1995). Analysis of diatom assemblages and reconstructions of TP show the close relationship between human activities and eutrophication in this lake.

The long history of human perturbations on lakes, such as at Diss Mere and other lakes in Europe and China (e.g. Bradshaw *et al.*, 2005, 2006; Dong *et al.*, 2008), has implications for restoration and lake management. Any attempt to restore lakes with a long history of human impacts may need to consider aquatic conditions that existed thousands of years ago in order to define pre-impact or natural conditions. Such knowledge of the pre-impact state can only be obtained using the sediment record, and diatoms are likely the best indicators to use. Furthermore, restoration to the pristine condition may be extremely difficult, if not impossible, because P stored in the sediments for centuries to millennia can maintain a eutrophic state. Setting restoration targets at Dallund Sø in Denmark, a lake with a 7000 year long history of human impacts, is further complicated by climatic variations that have occurred over this time, as well as infilling of this currently 2.6 m deep lake with 11 m of sediment (Bradshaw *et al.*, 2005). Obviously, establishing appropriate reference conditions for such lakes becomes a challenge.

In the above section, we have illustrated how diatoms have been used to document eutrophication and recovery of lakes to both relatively short-term (last 100 years) and long-term (past 3500–7000 years) human impacts, including agriculture and industry. Many other studies have used diatoms to determine the effects of, for example, land clearance (e.g. Bradbury, 1975; Fritz *et al.*, 1993; Reavie *et al.*, 1995; Hall *et al.*, 1999; Davies

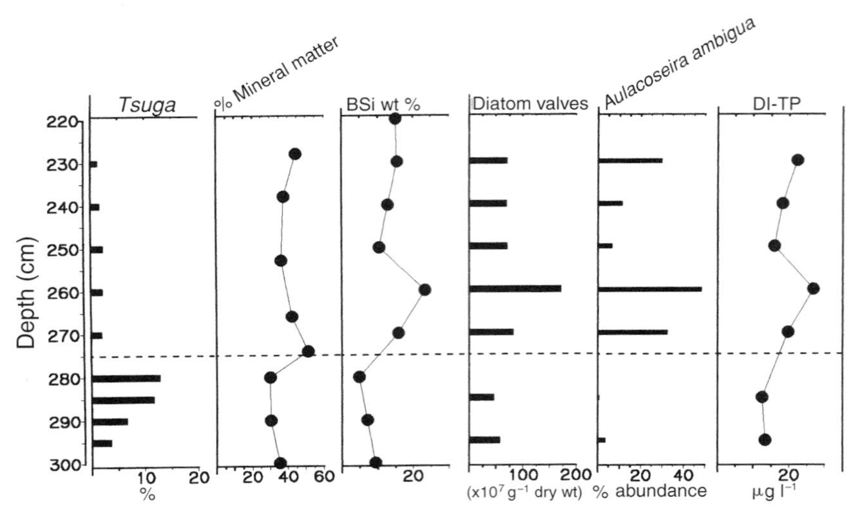

Figure 7.3 Sedimentary evidence of a natural eutrophication event in Flower Round Lake, Ontario, in response to the hemlock decline 4800 BP. The hemlock decline (*Tsuga*) increased catchment erosion (indicated by increased mineral content), leading to increased diatom production (panels with biogenic silica (BSi) and diatom valve concentration), and an increase in the percent abundance of meso- to eutrophic diatom species *Aulacoseira ambigua* (Grun. in Van Heurck) Simonsen. Diatom-inferred TP (DI-TP) increased from 14 µg l^{-1} to 30 µg l^{-1}. (Modified from Hall & Smol, 1993; with permission of Springer.)

et al., 2005; Franz *et al.*, 2006; Watchorn *et al.*, 2008), canal construction (Christie & Smol, 1996), peat drainage (e.g. Simola, 1983; Sandman *et al.*, 1990), agriculture (e.g. Håkansson & Regnéll, 1993; Anderson, 1997; Boyle *et al.*, 1999; Hausmann *et al.*, 2002; Jordan *et al.*, 2002; Miettenen *et al.*, 2005; Augustinus *et al.*, 2006) including cranberry farming in adjacent wetlands (Garrison & Fitzgerald, 2007), cage aquaculture of fish (Clerk *et al.*, 2004), industry (e.g. Ennis *et al.*, 1983; Engstrom *et al.*, 1985; Yang *et al.*, 1993), urbanization and sewage treatment (e.g. Haworth, 1984; Brugam & Vallarino, 1989; Anderson *et al.*, 1990, Brenner *et al.*, 1993; Anderson & Rippey, 1994; Bennion *et al.*, 1995; Douglas & Smol, 2000; Garcia-Rodriguez *et al.*, 2007; Michelutti *et al.*, 2007), road construction (Smol & Dickman, 1981), tourism and recreation (Garrison & Wakeman, 2000; Bigler *et al.*, 2007), long-range atmospheric N loading (Baron *et al.*, 2000), and prehistoric activities of indigenous populations (Douglas *et al.*, 2004; Ekdahl *et al.*, 2007) on lake trophic status.

7.6.1.3 Natural eutrophication events Not all eutrophication events are caused by humans. In some cases natural disturbances may accelerate nutrient supplies to lakes, though these are usually rare events, such as fire or other natural forest disturbances. For example, several studies have linked nutrient enrichment to the sudden mid-Holocene decline in hemlock trees (*Tsuga canadensis*), a former dominant component of forests in eastern North America (Boucherle *et al.*, 1986; St. Jacques *et al.*, 2000). It is widely believed that a forest pathogen caused the widespread, rapid, and synchronous death of hemlock throughout its range in North America *c.* 4800 years ago (Allison *et al.*, 1986). Increased erosional supply of nutrients, increased runoff due to lower evapotranspiration, and increased input of deciduous leaf litter are all cited as possible processes by which the hemlock decline increased aquatic productivity (Boucherle *et al.*, 1986). Hall & Smol (1993) demonstrated that diatom communities in five lakes in central Canada responded to the decline in hemlock, but that only one lake showed dramatic eutrophication (Figure 7.3). The magnitude of lake response appeared to be due to differences in catchment area and slope, which is consistent with the theory that increased nutrient supplies resulted from elevated erosion and runoff rates. For example, DI-TP doubled from 14 µg l^{-1} (oligotrophic) to 30 µg l^{-1} (meso- to eutrophic) in the lake with the largest and steepest catchment as a result of this natural forest disturbance (Figure 7.3).

7.6.1.4 Naturally productive lakes As noted earlier, estimates of pre-impact or "natural" lake trophic status can provide realistic targets for mitigation programs. Diatom-based paleolimnological techniques have identified naturally eutrophic lakes (Reavie *et al.*, 1995; Hall *et al.*, 1999; Karst & Smol, 2000; Ransanen *et al.*, 2006). Such data have important management implications, as mitigation efforts are unlikely to "restore" these naturally eutrophic lakes to an oligotrophic state (Battarbee, 1999).

Of course, situations do occur where lakes are incorrectly assumed to have been naturally eutrophic. Fortunately, paleolimnological studies, including those that employ diatoms, can play an essential role to dispel such misconceptions and stimulate appropriate initiatives for water-quality improvement. For example, paleolimnological analyses at Lake Kinneret, a warm monomictic lake in Israel, have recently corrected scientific misconceptions that were developed during the 36 year period since establishment of limnological research and monitoring in 1969 (Hambright *et al.*, 2008). Lake Kinneret has been classified as a naturally eutrophic lake based on high algal biomass recorded over the past 40 years. During this period, phytoplankton and primary production has fluctuated in highly predictable patterns. Typically, the large dinoflagellate alga *Peridinium gatunense* has accounted for about 50% of the annual primary production and usually formed distinct peaks contributing up to 95% of the phytoplankton biomass in winter and spring. Based on rather consistent algal communities for most of the period since 1969, this has been assumed to be the characteristic natural state of the lake ecosystem. Moreover, Lake Kinneret has been widely recognized as a resilient ecosystem because of this consistency during a period of substantial human disturbances and marked changes in hydrological management. Paleolimnological data, however, have shown that such preconceptions are wrong. In fact, they show that pre-industrial lake conditions were not eutrophic and that regular blooms of the large dinoflagellate are only a recent feature in this lake due to hydrological regulation of the lake for hydroelectric production and increasing water supply for urban and agricultural uses (Hambright *et al.*, 2008). Hydrological modifications, coupled with fisheries management practices, have promoted eutrophication and altered the stability of the lake ecosystem.

7.6.2 Regional assessments using diatoms

It is often instructive to assess water-quality changes on a broader, regional scale, rather than simply on a lake-by-lake basis. Regional assessments using diatoms can provide an effective tool to quantify the magnitude of water-quality changes due to human activities, determine reference conditions of lakes and to map or identify problem areas where lakes have been most severely affected.

An example of a large-scale regional monitoring project is the SNIFFER (Scotland & Northern Ireland Forum for Environmental Research) project, presented by Bennion *et al.* (2004), which developed the use of diatom paleolimnological records to assess the magnitude of post-industrial eutrophication on lakes throughout Scotland, and to define lake reference conditions and ecological status, as has recently become a requirement under the EC WFD (European Union, 2000). This project used a sediment core "top and bottom" or a "before and after" paleolimnological approach (see Smol, 2008) to quantify the magnitude of change since pre-industrial times in 26 freshwater lakes in Scotland. Diatom assemblages in sediments deposited *c.* 1850 were used to estimate pre-industrial reference conditions. Those assemblages in the uppermost core samples (0–0.5 cm) were used to determine modern conditions. The difference between DI-TP values (based on the northwestern Europe transfer function of Bennion *et al.*, 1996) for bottom and top sediment samples estimated the magnitudes of changes in TP concentration that occurred since pre-industrial times. Similarly, the difference between DCA sample scores (based on a squared chord distance coefficient) for the bottom and top sediment samples estimated the direction and magnitude of floristic change at each lake.

Using DCA, Bennion *et al.* (2004) were able to determine that large amounts of floristic compositional change, indicative of nutrient enrichment, occurred in 18 of the 26 lakes (Figure 7.4). The DCA axis 1 separates shallow, productive lakes (positioned on the left) from deep oligotrophic lakes (positioned to the right in Figure 7.4a). The second DCA axis separates lakes with diatom assemblages dominated by planktonic diatom taxa (positioned at the bottom) from lakes with assemblages dominated by non-planktonic taxa (positioned at the top; e.g. *Achnanthes*, *Cocconeis*, *Cymbella*, *Fragilaria*, and *Navicula* taxa). Several of the deep, unproductive lakes have short arrows joining reference and modern samples, indicating they have experienced relatively little compositional change. In contrast, the arrows for several of the shallow lakes are relatively long and point in a downwards direction, indicating large floristic changes since pre-industrial times that have seen non-planktonic taxa replaced by planktonic taxa. Arrows for a third group of lakes point towards the far left and identify lakes whose diatom assemblages have become dominated by small planktonic taxa associated with highly eutrophic conditions (e.g. *Aulacoseira*, *Cyclostephanos* and *Stephanodiscus* taxa). Use of diatom–TP transfer functions allowed Bennion *et al.* (2004) to identify increases in lake-water TP concentration since pre-industrial times in 19 of the 26 lakes (Figure 7.5). The TP concentration was inferred to have increased most in lakes that are currently eutrophic and hypereutrophic.

Although water management has traditionally emphasized chemical criteria, knowledge of ecological reference conditions is now a fundamental requirement of water protection

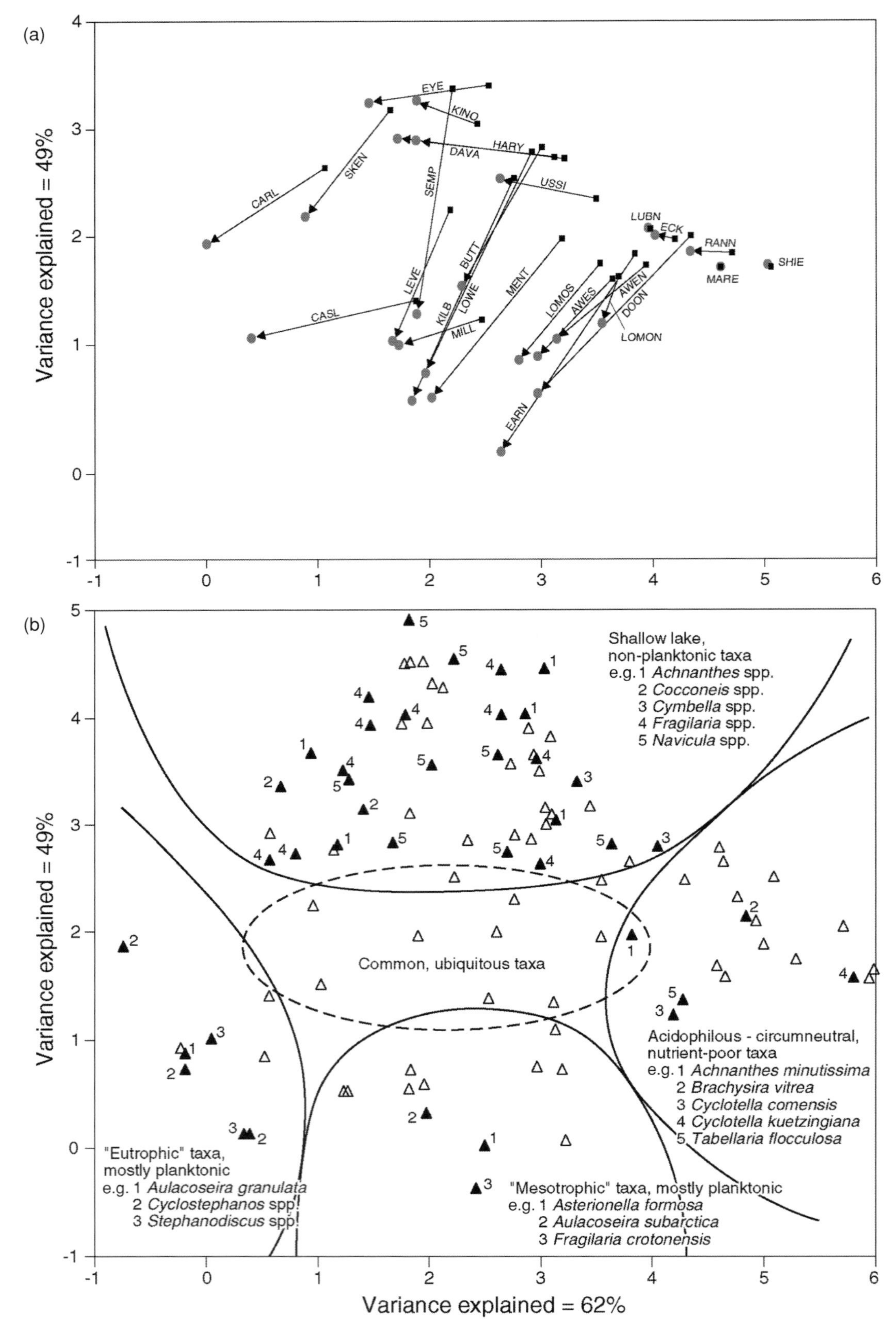

Figure 7.4 Detrended correspondence analysis biplot of (a) the site score and (b) the taxon scores for the diatom assemblages in the top (circles) and pre-1850 reference (squares) samples of the 26 lakes used to assess regional patterns of ecological and water-quality change in Scotland. Arrows connect the reference and top samples for each core in (a). The direction of the arrow indicates the direction of floristic change and its length is a measure of species compositional turnover. Taxa that commonly occur together in the assemblages lie in close proximity (from Bennion *et al*., 2004 with permission of the British Ecological Society).

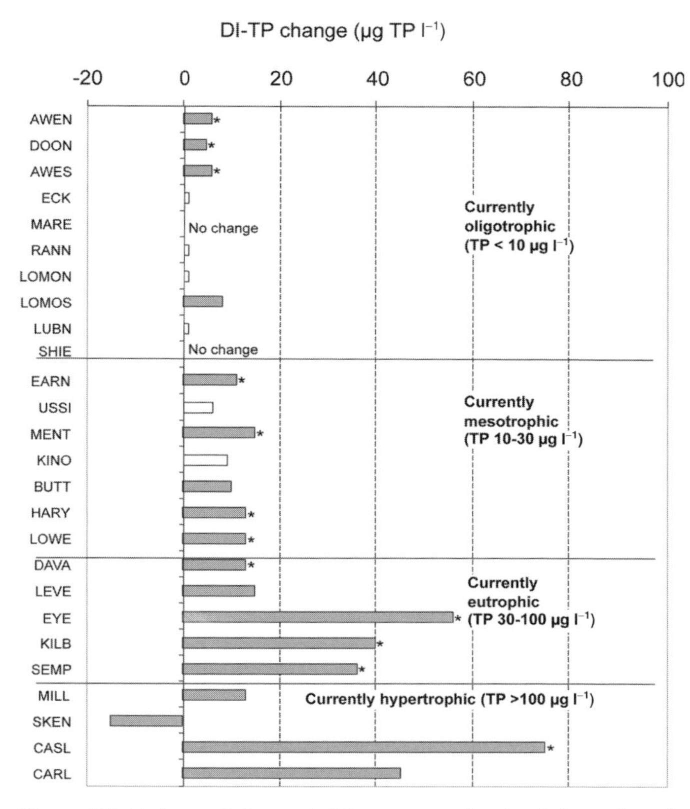

Figure 7.5 Estimated change in TP concentration (μg l⁻¹) at 26 study lakes in Scotland (present-day diatom-inferred TP minus diatom-inferred pre-1850 reference TP). The lakes are ordered according to increasing present-day measured TP. Sites where the squared chord distance dissimilarity scores are below the critical value at the 5th percentile (< 0·48) are shown as white bars or have no change in DI-TP. An asterisk indicates where the change in DI-TP is significant (i.e. change in DI-TP > RMSEP). (From Bennion *et al.*, 2004 with permission of the British Ecological Society).

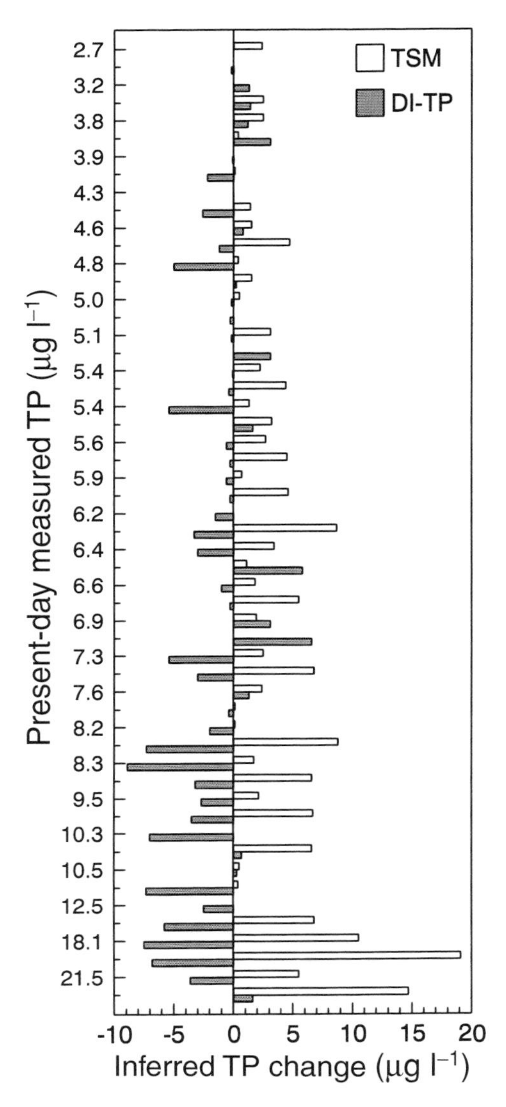

Figure 7.6 A comparison of inferred changes in post-industrial lake-water TP concentrations along a productivity gradient using two independent approaches; a diatom–TP transfer function (shaded bars) and an empirical TP model (TSM; open bars). The empirical model does not infer a decline in TP in any of the lakes, whereas the diatom model infers TP declines in many of the more productive lakes in the region (original figure).

policies in many jurisdictions (e.g. European Union, 2000). Bennion *et al.* (2004) applied two-way indicator species analysis (TWINSPAN) to the pre-1850 diatom assemblages as a way to identify the reference diatom floras of different lake types that exist in Scotland. This analysis identified four lake types that each possess distinctive diatom flora: (1) large, deep oligotrophic, slightly acid lakes; (2) large, mostly deep, oligo-mesotrophic lakes; (3) shallow mesotrophic lakes; and (4) shallow, meso- to eutrophic lakes. Lake depth and trophic status are the main factors driving the differences among reference lake types in Scotland.

The top and bottom paleolimnological approach, described above, provides an efficient method for assessing regional water-quality changes in large numbers of lakes because it requires analysis of far fewer samples than studies employing continuous temporal samples. For example, Hall & Smol (1996) used this approach to assess how Precambrian Shield lakes in a relatively remote, rural region of central Ontario have responded to a combination of shoreline development (mainly cottages and past logging) and acidic deposition since pre-industrial times. Lakes in this region are generally unproductive with present-day TP ranging from 2 to 30 μg l⁻¹. Diatoms inferred relatively small changes in lake-water TP in most lakes, despite considerable cottage development and past logging (Figure 7.6). Surprisingly, diatom-inferred TP declines in about half of the lakes with relatively high present-day TP (10–24 μg l^{-1}), indicating that presently mesotrophic lakes tended to have higher natural TP prior to cottage development. Based on these findings, Hall & Smol (1996) concluded that TP loads to lakes have been reduced by regional acidification, past logging, and recent droughts, and that the magnitude of these reductions exceeded any increases in TP loads from cottages and other lakeshore developments (see also Quinlan *et al.*, 2008). This sediment-core top and bottom approach has become quite widely used to assess regional patterns of water-quality and floristic change since pre-industrial times (e.g. Dixit & Smol, 1994 (northeastern USA); Ransanen *et al.*, 2006 (south Finland); Bradshaw *et al.*, 2006 (Denmark); Leira *et al.*, 2006 (Ireland)).

7.6.3 Eutrophication management using diatoms in combination with other approaches

Studies combining diatoms with other methods (e.g. historical records, empirical P models) provide an effective approach to assess causes of eutrophication and to guide restoration, because they can determine targets for recovery and identify critical loads for protecting against water-quality deterioration (e.g. O'Sullivan, 1992). Rippey & Anderson (1996) and Jordan *et al.* (2001), for example, have developed an innovative approach that combines information from diatom–TP transfer functions, P sedimentation rates, and lake-flushing rates to quantify past changes in TP loading to lakes. A strength of this approach is that it can be used to quantify past changes in TP loads from diffuse sources such as agriculture, which are notoriously difficult to measure using alternate methods (Jordan *et al.*, 2001). Also, it can be used to quantify how changes in external TP loads are variably directed as fluxes to the outflow, the sediments, or lake storage (Rippey & Anderson, 1996). It can also be used to identify when fertilizer applications began to exceed threshold soil P concentrations at which soluble P increases in runoff due to desorption; this can substantially improve our understanding of critical P loads to catchments and the mechanisms promoting lake eutrophication in rural catchments (Jordan *et al.*, 2001). It is a very powerful approach because few lakes in the world have P loading data for more than a decade or two (Rippey, 1995).

To reconstruct P loads, the technique applies assumptions of lake-chemistry mass-balance approaches. Total phosphorus inputs are assumed to be the sum of TP losses via the outflow and TP losses to the sediments. These are quantified by rearranging the Vollenweider (1975) steady-state P model for lakes with an additional term that accounts for periods of non-steady state changes in lake P storage (Jordan *et al.*, 2001):

$$L_i = (TP_i z\rho_i + TP_i z\sigma_i) + [(TP_i z - TP_{i-1}z)/(i - (i - 1))]$$

where L_i is the external TP loading to the lake (g m^{-2} yr^{-1}) in a time interval i, $TP_i z\rho_i$ is the TP loss through the outflow (g m^{-2} yr^{-1}), $TP_i z\sigma_i$ is the accumulation of TP in the sediments (g m^{-2} yr^{-1}), and the last term $[(TP_i z - TP_{i-1}z)/(i - (i-1))]$ accounts for non-steady state periods of changing lake TP storage between time intervals (expressed relative to the lake surface area). Past TP outflow losses and periods of changing TP storage require an estimate of lake-water TP concentration, which can be provided by diatom–TP transfer functions from analysis of sedimentary diatom assemblages. Mean depth (z, in metres) is estimated from basin morphometry. Changes in flushing rates (ρ, yr^{-1}) can be estimated from precipitation records, and the whole-basin mean sedimentary TP accumulation rate ($TP_i z\sigma_i$) is determined from dated multiple sediment cores analyzed for sedimentary TP concentration.

Rippey *et al.* (1997) used this approach to demonstrate that a period of rapid eutrophication (1973–1979) in White Lough (a small eutrophic lake in an agricultural catchment in Northern Ireland) was initiated by a reduction in surface runoff and stream flow during a drought period of warm, dry summers in the early 1970s. They identified two factors responsible for the eutrophication event. First, lake-water P concentrations increased when longer hydraulic residence time caused increased retention within the lake basin of P released from the sediments. Second, rapid eutrophication resulted when sediment P release rates increased as a consequence of elevated water-column P concentration and increased anoxia. An important implication arising from this study is that dry climatic periods, by altering the hydrological regime, may cause significant eutrophication in anthropogenically modified meso- to eutrophic lakes. Findings by Rippey *et al.* (1997) indicate that lakes with water residence times of about one year may be most susceptible. In this era of climate warming, effective ecosystem stewardship will increasingly require knowledge of such human–climate interactions.

Jordan *et al.* (2001, 2002) used the above approach to identify that diffuse TP loads started increasing in the Friary Lough catchment (a eutrophic rural lake in Northern Ireland) after 1946, at a time when the catchment was converted to intensive grassland agriculture. They showed that reconstructed TP loads during the period 1991–1995 closely matched values based on hydrochemical monitoring of the lake inflow, indicating the technique performs well. Importantly, their results identified

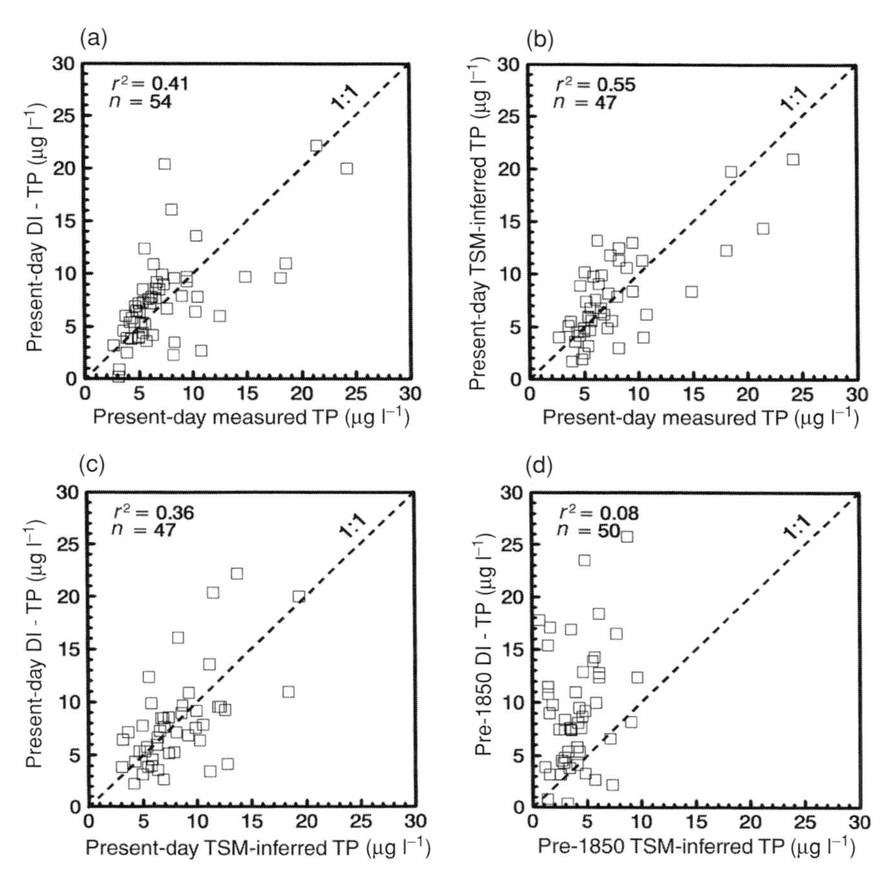

Figure 7.7 Using diatom transfer functions to assess predictive abilities of empirical models to estimate preindustrial TP. (a) Relationship between present-day diatom-inferred and measured TP in 54 lakes in south-central Ontario, (b) relationship between Trophic Status Model (TSM) and measured present-day TP in 47 of the lakes, (c) relationship between TP estimates from surface sediment diatoms and the TSM, and (d) the relationship between preindustrial estimates of TP concentrations using diatoms and the TSM (original figure).

that post-1946 eutrophication of the lake resulted from a shift to increases in P loading that originated from diffuse catchment loads in runoff (and a shift to export of soluble P). They suggested that agricultural activities elevated soil P concentrations beyond the threshold that promotes loss of soluble P via desorption (Jordan *et al.*, 2002).

7.6.4 Using diatoms to evaluate empirical eutrophication models

Empirical or mass-balance nutrient models are increasingly used by managers and scientists to quantify lake responses to changes in land use and to develop policy tools to guide decisions about shoreline development (e.g. Dillon & Rigler, 1975; Reckow & Simpson, 1980; Canfield & Bachmann 1981; Hutchinson *et al.*, 1991; Soranno *et al.*, 1996). Predictive abilities of most models, however, have never been rigorously assessed. Diatom transfer functions provide one of the only quantitative methods available to evaluate independently the ability of empirical models to estimate pre-disturbance conditions in lakes (Smol, 1995).

Diatoms were used to evaluate the ability of the Ontario Ministry of Environment & Energy's Trophic Status Model (TSM: Dillon *et al.*, 1986; Hutchinson *et al.*, 1991) to infer pre-industrial TP concentrations in lakes by comparing TSM estimates with values inferred from sedimentary diatom assemblages deposited before 1850, based on Hall & Smol's (1996) diatom–TP transfer function. Model evaluation was performed in 47 lakes in the area where the TSM was developed. The ability of diatoms ($r^2 = 0.41$) and the TSM ($r^2 = 0.55$) to estimate present-day measured TP was comparable (Figure 7.7a,b), and there was a similarly strong relationship between present-day TP inferred from the diatom model and the TSM ($r^2 = 0.36$; Figure 7.7c). However, the correspondence between diatom and TSM estimates of pre-industrial TP concentrations was extremely poor ($r^2 = 0.08$; Figure 7.7d). Diatoms consistently inferred higher background TP than the TSM (Figure 7.7d).

The diatom transfer function undoubtedly produced some errors, but the poor agreement between pre-industrial TP estimates provided reason to reconsider some key assumptions and parameters in the TSM. For example, paleolimnological studies demonstrated that acid deposition and past logging activities reduced P concentrations in many of the lakes (Hall & Smol, 1996; Figure 7.6). However, the TSM only considered human activities that increase lake water TP, and did not consider changes in regional acidification or forestry that could have caused TP to decline (Figure 7.6, 7.7). As a consequence, the TSM overestimates pre-industrial TP in Ontario lakes. Based on comparisons with diatoms, the assumption that P dynamics have remained constant during the past 150 years appears to be too simplistic to permit reliable estimates of pre-industrial conditions using the TSM. These findings have been reinforced by analyses of Reavie *et al.* (2002) and Quinlan *et al.* (2008) and suggest that diatom-inferred TP is a preferred approach to estimate reference water quality. Similarly, Bennion *et al.* (2005) provided a detailed comparison of diatom–TP transfer functions and P export coefficient models to assess past changes in 62 lakes in Great Britain. They showed that discrepancies between the two approaches were largest in shallow and productive lakes, but both methods tracked measured lake-water P concentrations in deep lakes reasonably well.

7.6.5 Using diatoms to quantify factors regulating lake eutrophication

Lake eutrophication is under complex, multifactorial control – regulated to a large extent by human activities, climate, morpho-edaphic conditions of the watershed, and food-web interactions. Consequently, eutrophication still poses major challenges to applied ecologists and managers, and mechanisms that regulate long-term water quality in lakes have proven difficult to quantify using experimental and standard monitoring practices. Lake-sediment diatom records can supply long-term information on water-quality trends, but without long-term records of possible control factors, cause and effect relationships cannot be assessed rigorously and quantitatively. For example, indirect gradient ordination, rate-of-change numerical techniques (e.g. Lotter *et al.*, 1992), and diatom transfer functions provide effective numerical approaches to identify and describe temporal patterns of community change and inferred water quality, but they do not lend much insight into the factors that cause the observed changes.

Factors regulating water quality, however, can be quantified using an approach developed by Borcard *et al.* (1992), known as variation partitioning analysis (VPA). This analysis employs canonical ordination (e.g. CCA, RDA; see Birks, this volume), and can be used to examine correlations between fossil diatom assemblages from sediment cores and historical records of possible control factors (e.g. geographic position, climate, resource use, and urbanization). As with all correlational approaches, VPA cannot prove causal mechanisms, but it can provide an efficient screening technique to identify potential management strategies or to generate testable hypotheses. The VPA analysis has been performed with two (e.g. Borcard *et al.*, 1992; Zeeb *et al.*, 1994) and three (e.g. Qinghong & Bråkenhielm, 1995; Jones & Juggins, 1995; Hall *et al.*, 1999) sets of predictor or explanatory variables.

In one example, Hall *et al.* (1999) were able to use available historical records for the period CE 1920–1994, as well as diatoms preserved in sediments of eutrophic lakes of the Qu'Appelle Valley (Prairie region of central Canada), and VPA to determine the relative importance of climatic variability (C), resource use (R), and urbanization (U) as controls of diatom communities. Prior to using VPA, Hall *et al.* (1999) identified three distinct biological assemblages since *c.* 1775 from sedimentary analyses of diatoms, pigments, and chironomids at Pasqua Lake. This is the first lake in a series of six lakes situated along the Qu'Appelle River and the first to receive nutrients in sewage from the main urban centres of Regina and Moose Jaw. Before the onset of agriculture ~1890, the lake was naturally eutrophic with abundant diatom taxa that are indicative of productive waters, as well as cyanobacterial pigments, and anoxia-tolerant chironomids. Distinct assemblages formed *c.* 1930–1960 that were characterised by an elevated abundance of eutrophic *Stephanodiscus hantzschii* Grun., elevated algal biomass (inferred as β-carotene) and a low abundance of deep-water zoobenthos. Sedimentary assemblages deposited after 1977 were variable and indicated water quality had not improved despite a three-fold reduction in P loading due to tertiary sewage treatment.

At Pasqua Lake, VPA captured 91% of the variation in fossil diatom assemblage composition using 14 significant environmental variables that were placed into three categories of possible control factors (C, R, U; Table 7.3). Resource use (cropland area, livestock biomass) and urbanization (nitrogen in sewage) were stronger determinants of diatom assemblage change than were climatic factors (temperature, evaporation, river discharge). Covariation among resource-use and urban activities (R, U, RU; Table 7.3), independent of climate, accounted for 38.2% of the total variation in the diatom assemblages since 1920. In particular, the analysis demonstrated that the long-term influence of resource use on diatoms was mediated mainly through changes in terrestrial practices involving livestock or crops (Hall *et al.*, 1999). While the effects of climatic

Table 7.3 Results of variation partitioning analysis of diatom assemblages from sediments deposited 1920–1993 in Pasqua Lake, Saskatchewan, Canada. (Modified from Hall *et al.* (1999) with permission of American Society of Limnology and Oceanography.)

Component	% of variation explained
Climate effects, independent of resource-use and urban factors (C)	10.4%
Resource-use effects, independent of urban and climatic factors (R)	24.7%
Urban effects, independent of resource-use and climatic factors (U)	9.4%
Covariation between effects of climatic and resource-use factors independent of urban (CR)	22.1%
Covariation between effects of climatic and urban factors independent of resource use (CU)	6.8%
Covariation between effects of resource-use and urban effects independent of climate (RU)	4.1%
Covariation among effects of climatic, resource-use and urban factors (CRU)	13.2%
Unexplained variation	9.3%

variables on diatom assemblages were also important, climate impacts were mediated by human activity, as demonstrated by the high proportion of variation attributable to the covariation of climatic factors with resource-use and urban activities (i.e. CU, CR, CRU = 42.1%), rather than to unique effects of climate (C = 10.4%; Table 7.3). The use of VPA and information contained in century-long sediment cores allowed Hall *et al.* (1999) to formulate specific recommendations for prairie lake managers. This included (1) that Pasqua Lake should not be managed for low productivity, because fossil analyses demonstrated that Pasqua Lake was naturally eutrophic; (2) management strategies should further investigate the role of sewage inputs, agriculture, and reservoir hydrology on water quality, because variables reflecting nutrient export from sewage from the city of Regina (Regina population, TP and TN fluxes, TN:TP), cropland area, livestock biomass, and discharge volume from the Lake Diefenbaker reservoir consistently accounted for significant amounts of variation in the fossil assemblages; and (3) nutrient abatement programs should reduce N inputs to Qu'Appelle lakes because paleolimnological data showed that inferred water quality had not substantially improved in Pasqua Lake since 1977 despite tertiary sewage treatment which reduced P loading to levels of the 1930s.

Variation partitioning analysis also has been used by Zeeb *et al.* (1994) to quantify the influence of measured water-chemistry changes on diatom assemblages in annually laminated sediments in Lake 227 (Experimental Lakes Area, Canada) during 20 years of whole-lake eutrophication experiments (1969–1989). Lotter (1998) used VPA to contrast the effects of cultural eutrophication and climatic changes on Baldeggersee (Switzerland), which possesses annually varved sediments. Similarly, Hausmann *et al.* (2002) used VPA to quantify the unique effects of human land use and climatic variability, and their covariation, on diatom assemblages deposited during the past 2600 years in Seebergsee, a lake situated near treeline in the Swiss Alps. In the latter study, late-wood density in tree-ring series was used as an independent proxy for climate and the sum of grazing-indicator pollen taxa was used as a proxy for human land-use intensity, indicating the value of including multiple paleoenvironmental proxies along with diatom records (Birks & Birks, 2006). In this way, Hausmann *et al.* (2002) were able to demonstrate that climate, independent of land use (and time), explained only 1.32% of the diatom data, while land use independent of climate (and time) explained 15.7%. They concluded that land use strongly influenced the trophic status of Seebergsee, but that land use was not strongly influenced by climatic variations.

7.6.6 Multiple indicator (or multiproxy) studies

Ecologists and ecosystem managers are becoming increasingly aware of the importance of complex food-web interactions in determining ecosystem structure and function (e.g. Polis & Winemiller, 1996). Food-web interactions undoubtedly play a role in lake responses to eutrophication (Carpenter *et al.*, 1995). One important question for applied diatom studies is: to what extent are diatoms regulated by resources versus herbivores along a lake eutrophication gradient? Studies of diatoms alone will not be able to determine the role of complex interactions among different communities and trophic levels, because diatoms are only one of many algal groups in lakes, and they comprise only one of several trophic levels. Moreover, eutrophication affects the various lake habitats (littoral, pelagic) and biotic communities (algae, invertebrates, benthos, fish) at different rates and thresholds during both degradation and recovery phases. Fortunately, many other aquatic and terrestrial biota leave fossil remains in lake sediments that can supply important additional information.

As predicted in our first edition of this chapter (Hall & Smol, 1999), paleolimnological studies of lake eutrophication increasingly are combining analysis of diatoms with other bioindicators to assess past ecological and environmental change. This so-called multiproxy approach is often very powerful to identify, for example, differential responses of epilimnetic and profundal environments (e.g. Little *et al.*, 2000). It provides important additional information (as well as redundancies of information) that allows scientists to eliminate one or more possible competing hypotheses concerning mechanisms that could account for observed changes in sedimentary diatom assemblages. (For example, Baron *et al.* (2000) combined analysis of diatoms and stable isotopes of nitrogen to determine that lakes of the Colorado Front Range were eutrophied by atmospheric N deposition; Hall *et al.* (1999) combined analyses of diatoms, pigments, and chironomids to identify that urban sewage and agricultural activities, not climate, were responsible for further degradation of water quality in lakes of the northern Great Plains, Canada).

The combined use of fossil pigments and diatoms provides one of the only ways to identify when algal communities become N-limited rather the P-limited during eutrophication, because pigments from N-fixing cyanobacteria are preserved in direct proportion with the standing crop of past populations which produced them (Leavitt, 1993). Furthermore, pigments from most groups of algae and photosynthetic bacteria preserve well in lake sediments, and can be used to reconstruct the biomass of all major algal groups, as well as total algal biomass (Leavitt *et al.*, 1989; Leavitt & Findlay, 1994). In conjunction with diatom studies, analysis of biogenic Si (Schelske *et al.*, 1986; Schelske 1999) and fossil pigments are helping to elucidate the relative roles of Si, P, and N limitation during eutrophication (e.g. Marchetto *et al.*, 2003).

Zooplankton fossils may also be used to quantify past changes in the density of planktivorous fish (Jeppesen *et al.*, 1996). By combining diatom-inferred TP, zooplankton-inferred planktivorous fish density and analysis of plant macrofossils, McGowan *et al.* (2005) were able to identify factors causing a switch between alternate stable states in Danish lakes from a clear-water, macrophyte-dominated state to a turbid, plankton-dominated state. This area of research is of considerable interest to ecologists, as well as lake and fisheries managers.

Transfer functions also exist that can quantify changes in other important limnological conditions affected by eutrophication, including deep-water oxygen availability from chironomid remains (e.g. hypolimnetic anoxia as the Anoxic Factor (Quinlan *et al.*, 1998)). As an example, Little *et al.* (2000) used quantitative multiproxy paleolimnological methods (diatom–TP, chironomid–Anoxia Factor transfer functions) to demonstrate that reduced nutrient loads and restoration of epilimnetic TP concentration to pre-disturbance values cannot be assumed to translate into similar rates of recovery of benthic communities or greater deep-water oxygen availability over the multidecadal timescales of remediation programs.

Numerous other multiproxy paleolimnological studies have enhanced our understanding of the effects of eutrophication and the processes responsible for degradation and recovery in many lakes. Some examples include; Garcia-Rodríguez *et al.*, 2002 (impacts of tourism on a lake in Uruguay); Marchetto *et al.*, 2003 (Lake Maggiore, Italy); Clerk *et al.*, 2004 (effects of cage aquaculture in Canada); Bradshaw *et al.*, 2005 (millennial-scale human impacts in Denmark); Brüchmann & Negendank, 2004 (Lake Holzmaar, Germany); Davies *et al.*, 2005 (effects of Spanish settlement in Mexico); Miettenen *et al.*, 2005 (agricultural impacts in Karelian Republic, Russia); Augustinus *et al.*, 2006 (urban lake in New Zealand); Leahy *et al.*, 2005 (oxbow lake in Australia); Finsinger *et al.*, 2006 (land use and eutrophication in Piedmont, Italy); Franz *et al.*, 2006 (Lake Iznik, northwest Turkey); Dreßler *et al.*, 2007 (epilimnetic TP, hypolimnetic anoxia and development of deepwater purple-sulfur bacteria in a lake in northern Germany); Ekdahl *et al.*, 2007 (impacts of Iroquoian and Euro-Canadian eutrophication on a lake in Canada). When it comes to eutrophication, multiproxy paleolimnological studies are now probably more common than studies employing a single proxy. This multiproxy approach is especially suited to shallow lakes, rich in fossil material (see Bennion *et al.*, this volume).

7.7 Future directions

Tremendous progress has been made in the ways we have learned to glean information concerning lake eutrophication from diatoms. Undoubtedly, the momentum of the past two decades will continue to spur many interesting and novel developments of eutrophication research, and the use of diatoms in lake management will continue to increase.

As identified by Battarbee *et al.* (2005), eutrophication studies will increasingly combine paleolimnological and observational data sets to establish the past and present status of lakes as is needed to identify reference conditions, to quantify trends and changes in ecosystem state (including responses to nutrient reduction strategies), and to determine the potential role of other factors (e.g. additional stressors, climate change) that might confound predictions of future state. Stimulated by new water-quality policies that emphasize protection of the ecological status of water bodies (e.g. EC WFD), integrated analyses of diatoms growing in lakes, on artificial substrates, and preserved in sediments will provide a powerful approach that will be increasingly used in long-term monitoring programs and decision-making procedures. Evidence of this trend can be seen in the increasing number of trained diatomists and paleolimnologists being hired by government agencies and the environmental consultancy sector. Our tools are no longer confined to research projects launched from within universities; instead, they are increasingly being utilized to solve problems of society. This situation is certainly true for North America, and many parts of Europe and Australia. Also, we anticipate the coming decades will see remarkable expansion of studies in Asia, Africa, South America, and the Middle East. This expansion will be driven by the need to assess the vulnerability of lakes to climate change in arid and semi-arid regions, to further our understanding of how changes in hydrology, temperature, and human activities may exacerbate eutrophication, and to provide innovative solutions for the looming water crises in these regions. Certainly, diatom analyses will continue to play a strong role because diatoms are amongst the most easily sampled and cost-effective organisms that can be used to seamlessly integrate water-quality and biological monitoring data over broad temporal (weeks to millennia) and spatial (lake zones (e.g. littoral, pelagic) to entire landscapes) scales, as is required for effective management of eutrophication in the face of multiple stressors. Recognition that littoral-zone diatoms can provide sensitive information about impacts of human land use around lakes, and the development of legislation that emphasizes the need to define the ecological status of the reference state will undoubtedly ensure that the use of diatoms will continue to grow for decades to come.

7.8 Summary

Diatoms respond rapidly to eutrophication and provide detailed information on ecological changes that occur during eutrophication and recovery, in both deep-water and near-shore habitats. Because diatoms are well preserved in most lake sediments, they can be used to generate long time series of data that track lake responses to eutrophication and recovery. Moreover, recent advances in quantitative methods (e.g. ordination, transfer functions) permit reliable estimates of past and recent water-chemistry changes during eutrophication from diatom assemblages. Diatoms in sediment cores can be used to establish baseline or pre-disturbance water chemistry as targets for lake rehabilitation. Natural variability can also be assessed. In this chapter, we have provided some of the many examples of how diatoms are being used as a powerful tool for eutrophication research and management. The 1990s were largely devoted to developing and refining diatoms as quantitative indicators of lake eutrophication. The numerical and experimental tools are now well developed and sophisticated. As witnessed during the past decade (2000s), the stage is set for these methods to be applied in novel ways that undoubtedly will further our ecological understanding of lake eutrophication and improve water-quality management.

Acknowledgments

Most of our lake eutrophication research has been funded by grants from the Natural Sciences and Engineering Research Council of Canada. We thank Helen Bennion, John Birks, and members of our laboratories for helpful comments on the manuscript. We would like to dedicate this chapter to the memory of our friend and colleague, John C. Kingston (1949–2004).

References

Adler, S. & Hübener, T. (2007). Spatial variability of diatom assemblages in surface lake sediments and its implications for transfer functions. *Journal of Paleolimnology*, **37**, 573–90.

Agbeti, M. D. (1992). Relationship between diatom assemblages and trophic variables: a comparison of old and new approaches. *Canadian Journal of Fisheries and Aquatic Sciences*, **49**, 1171–5.

Agbeti, M. D. & Dickman, M. (1989). Use of fossil diatom assemblages to determine historical changes in trophic status. *Canadian Journal of Fisheries and Aquatic Sciences*, **46**, 1013–21.

Alefs, J. & Müller, J. (1999). Differences in the eutrophication dynamics of Ammersee and Starnberger See (southern Germany), reflected by the diatom succession in varve-dated sediments. *Journal of Paleolimnology*, **21**, 395–407.

Allison, T. D., Moeller, R. E., & Davis, M. B. (1986). Pollen in laminated sediments provides evidence for a mid-Holocene forest pathogen outbreak. *Ecology*, **64**, 1101–5.

Anderson, N. J. (1989). A whole-basin diatom accumulation rate for a small eutrophic lake in Northern Ireland and its palaeoecological implications. *Journal of Ecology*, **77**, 926–46.

Anderson, N. J. (1990a). Variability of diatom concentrations and accumulation rates in sediments of a small lake basin. *Limnology and Oceanography*, **35**, 497–508.

Anderson, N. J. (1990b). Spatial pattern of recent sediment and diatom accumulation in a small, monomictic, eutrophic lake. *Journal of Paleolimnology*, **3**, 143–60.

Anderson, N. J. (1990c). Variability of sediment diatom assemblages in an upland, wind-stressed lake (Loch Fleet, Galloway, Scotland). *Journal of Paleolimnology*, **4**, 43–59.

Anderson, N. J. (1995a). Using the past to predict the future: lake sediments and the modelling of limnological disturbance. *Ecological Modelling*, **78**, 149–72.

Anderson, N. J. (1995b). Naturally eutrophic lakes: reality, myth or myopia? *Trends in Ecology and Evolution*, **10**, 137–8.

Anderson, N. J. (1997). Reconstructing historical phosphorus concentrations in rural lakes using diatom models. In *Phosphorus Loss to Water From Agriculture*, ed. H. Tunney, P. C. Brookes & A. E. Johnson. Wallingford, UK: CAB International.

Anderson, N. J. (1998). Variability of diatom-inferred phosphorus profiles in a small lake basin and its implications for histories of lake eutrophication. *Journal of Paleolimnology*, **20**, 47–55.

Anderson, N. J., Jeppesen, E. & Sondergaard, M. (2005). Ecological effects of reduced nutrient loading (oligotrophication) on lakes: an introduction. *Freshwater Biology*, **50**, 1589–193.

Anderson, N. J. & Odgaard, B. V. (1994). Recent palaeoecology of three shallow Danish lakes. *Hydrobiologia*, **275/276**, 411–22.

Anderson, N. J. & Rippey, B. (1994). Monitoring lake recovery from point-source eutrophication: the use of diatom-inferred epilimnetic total phosphorus and sediment chemistry. *Freshwater Biology*, **32**, 625–39.

Anderson, N. J., Rippey, B. & Gibson, C. E. (1993). A comparison of sedimentary and diatom-inferred phosphorus profiles: implications for defining pre-disturbance nutrient conditions. *Hydrobiologia*, **253**, 357–66.

Anderson, N. J., Rippey, B. & Stevenson, A. C. (1990). Change to a diatom assemblage in a eutrophic lake following point source nutrient re-direction: a paleolimnological approach. *Freshwater Biology*, **23**, 205–17.

Augustinus, P., Reid, M., Andersson, S., Deng, Y., & Horrocks, M. (2006). Biological and geochemical record of anthropogenic impacts in recent sediments from Lake Pupuke, Auckland City, New Zealand. *Journal of Paleolimnology*, **35**, 789–805.

Barbour, M. T., Swietlik, W. F., Jackson, S. K., *et al.* (2000). Measuring the attainment of biological integrity in the USA: a critical element of ecological integrity. *Hydrobiologia*, **422/423**, 453–64.

Baron, J. S., Rueth, H. M., Wolfe, A. M., *et al.* (2000). Ecosystem responses to nitrogen deposition in the Colorado Front Range. *Ecosystems*, **3**, 352–68.

Battarbee, R. W. (1978). Biostratigraphical evidence for variations in the recent patterns of sediment accumulation in Lough Neagh, Northern Ireland. *Internationale Vereinigung für theoretische und angewandte Limnologie, Verhandlungen*, **20**, 625–9.

Battarbee, R. W. (1999). The importance of palaeolimnology to lake restoration. *Hydrobiologia*, **395/396**, 149–59.

Battarbee, R. W., Anderson, N. J., Jeppesen, E., & Leavitt, P. R. (2005). Combining palaeolimnological and limnological approaches in assessing lake ecosystem response to nutrient reduction. *Freshwater Biology*, **50**, 1772–80.

Battarbee, R. W., Jones, V. J., Flower, R. J., *et al.* (2001) Diatoms. In Smol, J. P. Birks, H. J. B. & Last, W. M., (ed. *Tracking Environmental Change Using Lake Sediments, Volume 3: Terrestrial, Algal, and Siliceous Indicators*. Dordrecht: Kluwer Academic Publishers, 155–202.

Belzile, N., Pizarro, J., Filella, M., & Buffle, J. (1996). Sediment diffusive fluxes of Fe, Mn, and P in a eutrophic lake: contribution from lateral vs bottom sediments. *Aquatic Sciences*, **58**, 327–54.

Bennion, H. (1994). A diatom–phosphorus transfer-function for shallow, eutrophic ponds in southeast England. *Hydrobiologia*, **275/276**, 391–410.

Bennion, H. (1995). Surface-sediment diatom assemblages in shallow, artificial, enriched ponds, and implications for reconstructing trophic status. *Diatom Research*, **10**, 1–19.

Bennion, H. & Appleby, P. (1999). An assessment of recent environmental change in Llangorse Lake using palaeolimnology. *Aquatic Conservation: Marine and Freshwater Ecosystems*, **9**, 361–75.

Bennion, H. & Battarbee, R. (2007). The European Union Water Framework Directive: opportunities for palaeolimnology. *Journal of Paleolimnology*, **38**, 285–95.

Bennion, H., Fluin, J., & Simpson, G. K. L. (2004). Assessing eutrophication and reference conditions for Scottish freshwater lochs using subfossil diatoms. *Journal of Applied Ecolology*, **41**, 124–38.

Bennion, H., Johnes, P., Ferrier, R. Phillips, G., & Haworth, E. (2005). A comparison of diatom phosphorus transfer functions and export coefficient models as tools for reconstructing lake nutrient histories. *Freshwater Biology*, **50**, 1651–70

Bennion, H., Juggins, S., & Anderson, N. J. (1996). Predicting epilimnetic phosphorus concentrations using an improved diatom-based transfer function and its application to lake management. *Environmental Science & Technology*, **30**, 2004–7.

Bennion, H. & Smith, M. A. (2000). Variability in the water chemistry of shallow ponds in southeast England, with special reference to the seasonality of nutrients and implications for modelling trophic status. *Hydrobiologia*, **436**, 145–58.

Bennion, H., Wunsam, S., & Schmidt, R. (1995). The validation of diatom–phosphorus transfer functions: an example from Mondsee, Austria. *Freshwater Biology*, **34**, 271–83.

Bigler, C., von Gunten, L., Lotter, A. F., Hausmann, S., Blass, A., Ohlendorf, C., & Sturm, M. (2007). Quantifying human-induced eutrophication in Swiss mountain lakes since AD 1800 using diatoms. *The Holocene*, **17**, 1141–54.

Birks, H. H. & Birks, H. J. B. (2006). Multi-proxy studies in palaeolimnology. *Vegetation History and Archaeobotany*, **15**, 235–51.

Birks, H. J. B. (1995). Quantitative palaeoenvironmental reconstructions. In *Statistical Modelling of Quaternary Science Data*, Technical Guide 5, ed. D. Maddy & J. S. Brew, Cambridge: Quaternary Research Association, pp. 161–254

Birks, H. J. B. (1998). Numerical tools in palaeolimnology – progress, potentialities, and problems. *Journal of Paleolimnology* **20**, 307–32.

Birks, H. J. B., Anderson, N. J. & Fritz, S. C. (1995). Post-glacial changes in total phosphorus at Diss Mere, Norfolk inferred from fossil diatom assemblages. In *Ecology and Palaeoecology of Lake Eutrophication*, ed. S. T. Patrick & N. J. Anderson, pp. 48–9. Copenhagen, DK: Geological Survey of Denmark DGU Service Report no. 7.

Birks, H. J. B., Juggins, S., & Line, J. M. (1990a). Lake surface-water chemistry reconstructions from palaeoecological data. In *The Surface Waters Acidification Programme*, ed. B. J. Mason, Cambridge: Cambridge University Press, pp. 301–11.

Birks, H. J. B., Juggins, S., Lotter, A. & Smol, J. P. (eds). (2011). *Tracking Environmental Change Usingnts. Volume 5: Data Handling and Statistical Techniques*, Dordrecht: Springer, in press.

Birks, H. J. B., Line, J. M., Juggins, S., Stevenson, A. C., & ter Braak, C. J. F. (1990b). Diatoms and pH reconstruction. *Philosophical Transactions of the Royal Society of London*, **B 327**: 263–78.

Borcard, D., Legendre, P., & Drapeau, P. (1992). Partialling out the spatial component of ecological variation. *Ecology*, **73**, 1045–55.

Bothwell, M. L. (1988). Growth rate response of lotic periphytic diatoms to experimental phosphorus additions. *Canadian Journal of Fisheries and Aquatic Sciences*, **45**, 261–70.

Boucherle, M. M., Smol, J. P., Oliver, T. C., Brown, S. R., & McNeely, R. (1986). Limnological consequences of the decline in hemlock 4800 years ago in three Southern Ontario lakes. *Hydrobiologia*, **143**, 217–25.

Boyle, J. F. (2001). Inorganic geochemical methods in paleolimnology. In *Tracking Environmental Change Using Lake Sediments, Volume 2: Physical and Geochemical Methods*, ed. W. M. Last and J. P. Smol, Dordrecht: Springer, pp. 83–142.

Boyle, J. F., Rose, N. L., Bennion, H., Yang, H., & Appleby, P. G. (1999). Environmental impacts in the Jianghan Plain: evidence from lake sediments. *Water, Air, and Soil Pollution*, **112**, 21–40.

Bradbury, J. P. (1975). *Diatom Stratigraphy and Human Settlement*. The Geological Society of America Special Paper no. 171.

Bradshaw, E. G. & Anderson, N. J. (2001). Validation of a diatom-phosphorus calibration set for Sweden. *Freshwater Biology*, **46**, 1035–48.

Bradshaw, E. G., Anderson, N. J., Jensen, J. P., & Jeppesen, E. (2002). Phosphorus dynamics in Danish lakes and the implications for diatom ecology and palaeoecology. *Freshwater Biology*, **47**, 1963–75.

Bradshaw, E. G., Nielsen, A.B., & Anderson, N. J. (2006). Using diatoms to assess the impacts of prehistoric, pre-industrial and modern land-use on Danish lakes. *Regional Environmental Change*, **6**, 17–24.

Bradshaw, E. G., Rasmussen, P., Nielsen, H., & Anderson, N. J. (2005). Mid- to late-Holocene land-use change and lake development at Dallund Sø, Denmark: trends in lake primary production as reflected by algal and macrophyte remains. *The Holocene*, **15**, 1130–42.

Brenner, M., Whitmore, T. S., Flannery, M. S, & Binford M. W. (1993). Paleolimnological methods for defining target conditions in lake restoration: Florida case studies. *Lake and Reservoir Management*, **7**, 209–17.

Brüchmann, C. & Negendank, J. F. W. (2004). Indication of climatically induced natural eutrophication during the early Holocene period, based on annually laminated sediment from Lake Holzmaar, Germany. *Quaternary International*, **123–125**, 117–34.

Brugam, R. B. & Vallarino, J. (1989). Paleolimnological investigations of human disturbance in Western Washington lakes. *Archiv für Hydrobiologie*, **116**, 129–59.

Burkhardt, M. A. (1996). Nutrients. In *Algal Ecology: Freshwater Benthic Ecosystems*, ed. R. J. Stevenson, M. L. Bothwell & R. L. Lowe, San Diego, CA: Academic Press, pp. 184–228.

Canfield, D. E., Jr. & Bachmann, R. W. (1981). Prediction of total phosphorus concentration, chlorophyll a and Secchi depths in natural and artificial lakes. *Canadian Journal of Fisheries and Aquatic Sciences*, **38**, 414–23.

Carignan, R. & Flett, R. J. (1981). Postdepositional mobility of phosphorus in lake sediments. *Limnology & Oceanography*, **26**, 361–6.

Carpenter, S. R., Christensen, D. L., Cole, J. J., *et al.* 1995. Biological control of eutrophication in lakes. *Environmental Science & Technology*, **29**, 784–6.

Cattaneo, A. (1987). Periphyton in lakes of different trophy. *Canadian Journal of Fisheries and Aquatic Sciences*, **44**, 296–303.

Charles, D. F. & Smol, J. P. (1994). Long-term chemical changes in lakes: Quantitative inferences using biotic remains in the sediment record. In *Environmental Chemistry of Lakes and Reservoirs: Advances in Chemistry, Series 237*, ed. L. Baker, Washington, DC: American Chemical Society, pp. 3–31.

Charles, D. F., Smol, J. P., & Engstrom, D. R. (1994). Paleolimnological approaches to biomonitoring. In *Biological Monitoring of Aquatic Systems*, ed. S. Loeb & D. Spacie, Ann Arbor, MI: Lewis Press, pp. 233–93.

Cholnoky, B. J. (1968). *Die Okologie der Diatomeen in Binnengewässern*. Weinheur: J. Cramer.

Christie, C. E. & Smol, J. P. (1993). Diatom assemblages as indicators of lake trophic status in southeastern Ontario lakes. *Journal of Phycology*, **29**, 575–86.

Christie, C. E. & Smol, J. P. (1996). Limnological effects of 19th century canal construction and other disturbances on the trophic state history of Upper Rideau Lake, Ontario. *Lake and Reservoir Management*, **12**, 78–90.

Clerk, S., Selbie, D. T., & Smol, J. P. (2004). Cage aquaculture and water-quality changes in the LaCloche Channel, Lake Huron, Canada: a paleolimnological assessment. *Canadian Journal of Fisheries and Aquatic Sciences*, **61**, 1691–701.

Cottingham, K. L, Rusak, J. A., & Leavitt, P.R. (2000). Increased ecosystem variability and reduced predictability following fertilization: evidence from palaeolimnology. *Ecology Letters*, **3**, 340–8.

Davies, S. J., Metcalfe, S. E., Bernal-Brooks, F., *et al.* (2005). Lake sediments record sensitivity of two hydrologically closed upland lakes in Mexico to human impact. *Ambio*, **34**, 470–5.

Davis, R. B., Anderson, D. S., Dixit, S. S., Appleby, P.G., & Schauffler, M. (2006). Responses of two New Hampshire (USA) lakes to human impacts in recent centuries. *Journal of Paleolimnology*, **35**, 669–97.

Dillon, P. J. & Rigler, F. H. (1975). A simple method for predicting the capacity of a lake for development based on lake trophic status. *Journal of the Fisheries Research Board of Canada*, **32**, 1519–31.

Dillon, P. J., Nicholls, K. H., Scheider, W. A., Yan, N. D., & Jeffries, D. S. (1986). *Lakeshore Capacity Study, Trophic Status*. Research and Special Projects Branch, Ontario Ministry of Municipal Affairs and Housing. Toronto, ON: Queen's Printer for Ontario.

Dixit, S. S. & Smol, J. P. (1994). Diatoms as indicators in the Environmental Monitoring and Assessment Program – Surface Waters (EMAP-SW). *Environmental Monitoring and Assessment*, **31**, 275–306.

Dixit, S. S., Smol, J. P., Kingston, J. C., & Charles, D. F. (1992). Diatoms: powerful indicators of environmental change. *Environmental Science & Technology*, **26**, 23–33.

Dixit S. S., Smol, J. P., Charles, D. F., *et al.* (1999). Assessing water quality changes in the lakes of the northeastern United States using sediment diatoms. *Canadian Journal of Fisheries and Aquatic Sciences*, **56**,131–52.

Dodds, W. K., Bouska, W. W., Eitzmann, J. L., *et al.* (2009). Eutrophication of U.S. freshwaters: analysis of potential economic damages. *Environmental Science & Technology*, **43**, 12–19.

Dong, X., Bennion, H., Battarbee, R., Yang, X., Yang, H., & Liu, E. (2008). Tracking eutrophication in Taihu Lake using the diatom record: potential and problems. *Journal of Paleolimnology*, **40**, 413–29.

Douglas, M. S. V. & Smol, J. P. (2000). Eutrophication and recovery in the High Arctic: Meretta Lake (Cornwallis Island, Nunavut, Canada) revisited. *Hydrobiologia*, **431**, 193–204.

Douglas, M. S. V., Smol, J. P., Savelle, J. M., & Blais, J. M. (2004). Prehistoric Inuit whalers affected Arctic freshwater ecosystems. *Proceeding of the National Academy of Sciences*, **101**, 1613–7.

Dreßler, M., Hübener, T., Gors, S., Werner, P., & Selig, U. (2007). Multi-proxy reconstruction of trophic state, hypolimnetic anoxia and phototrophic sulphur bacteria abundance in a dimictic lake in Northern Germany over the past 80 years. *Journal of Paleolimnology*, **37**, 205–19.

Ekdahl, E. J., Teranes, J. L., Wittkop, C. A., *et al.* (2007). Diatom assemblage response to Iroquoian and Euro-Canadian eutrophication of Crawford Lake, Ontario, Canada. *Journal of Paleolimnology*, **37**, 233–46.

Engstrom, D. R. & Wright, H. E. Jr., (1984). Chemical stratigraphy of lake sediments as a record of environmental change. In *Lake Sediments and Environmental History*, ed. E. Y. Haworth & J. W. G. Lund, Minneapolis, MN: University of Minnesota Press, pp. 1–67.

Engstrom, D. R., Swain, E. B., & Kingston, J. C. (1985). A palaeolimnological record of human disturbance from Harvey's Lake, Vermont: geochemistry, pigments and diatoms. *Freshwater Biology*, **15**, 261–88.

Ennis, G. L., Northcote, T. G., & Stockner, J. G. (1983). Recent trophic changes in Kootenay Lake, British Columbia, as recorded by fossil diatoms. *Canadian Journal of Botany*, **61**, 1983–92.

European Union (2000). Directive 2000/60/EC of the European Parliament and of the Council of 23 October 2000 on establishing a framework for community action in the field of water policy. *Journal of the European Community*, **L327**, 1–72.

Fairchild, G. W. & Sherman, J. W. (1992). Linkage between epilithic algal growth and water column nutrients in softwater lakes. *Canadian Journal of Fisheries and Aquatic Sciences*, **49**, 1641–9.

Fairchild, G. W. & Sherman, J. W. (1993). Algal periphyton response to acidity and nutrients in softwater lakes: lake comparison vs. nutrient enrichment approaches. *Journal of the North American Benthological Society*, **12**, 157–67.

Fairchild, G. W., Lowe, R. L., & Richardson, W. T. (1985). Algal periphyton growth on nutrient-diffusing substrates: An *in situ* bioassay. *Ecology*, **66**, 465–72.

Farmer, J. G. (1994). Environmental change and the chemical record in Loch Lomond sediments. *Hydrobiologia*, **290**, 39–49.

Finsinger, W., Bigler, C., Krahenbuhl, U., Lotter, A. F., & Ammann, B. (2006). Human impacts and eutrophication patterns during the past ~ 200 years at Lago Grande di Avigliana (N. Italy). *Journal of Paleolimnology*, **36**, 55–67.

Foy, R. H., Lennox, S. D., & Gibson, C. E. (2003). Changing perspectives on the importance of urban phosphorus inputs as the cause of nutrient enrichment in Lough Neagh. *Science of the Total Environment*, **310**, 87–99.

Franz, S. O., Schwark, L., Bruchmann, C., *et al.* (2006). Results from a multi-disciplinary sedimentary pilot study of tectonic Lake Iznik (NW Turkey) – geochemistry and paleolimnology of the recent past. *Journal of Paleolimnology*, **35**, 715–36.

Fritz, S. C. (1989). Lake development and limnological response to prehistoric and historic land-use in Diss, Norfolk, U.K. *Journal of Ecology*, **77**, 182–202.

Fritz, S. C., Kingston, J. C., & Engstrom, D. R. (1993). Quantitative trophic reconstructions from sedimentary diatom assemblages: a cautionary tale. *Freshwater Biology*, **30**, 1–23.

Garcıa-Rodríguez, F., Mazzeo, N., Sprechmann, P., *et al.* (2002). Paleolimnological assessment of human impacts in Lake Blanca, SE Uruguay. *Journal of Paleolimnology*, **28**, 457–68.

Garcıa-Rodrıguez, F., Anderson, R., & Adams, J. B. (2007). Paleolimnological assessment of human impacts on an urban South African lake. *Journal of Paleolimnology*, **38**, 297–308.

Garrison, P. J. & Fitzgerald, S. A. (2007). The role of shoreland development and commercial cranberry farming in a lake in Wisconsin, USA. *Journal of Paleolimnology*, **33**, 169–88.

Garrison, P. J. & Wakeman, R. S. (2000). Use of paleolimnology to document the effect of lake shoreland development on water quality. *Journal of Paleolimnology*, **24**, 369–93.

Gibson, C. E., Foy, R. H., & Bailey-Watts, A. E. (1996). An analysis of the total phosphorus cycle in temperate lakes: the response to enrichment. *Freshwater Biology*, **35**, 525–32

Goldman, C. R. (1981). Lake Tahoe: two decades of change in a nitrogen deficient oligotrophic lake. *Internationale Vereinigung für theoretische und angewandte Limnologie, Verhandlungen*, **24**, 411–5.

Gregory-Eaves, R., Smol, J. P., Finney, B. P., & Edwards, M. E. (1999). Diatom-based transfer functions for inferring past climatic and environmental changes in Alaska, USA. *Arctic Antarctic and Alpine Research*, **31**, 353–65.

Håkansson, H. & Regnéll, R. (1993). Diatom succession related to land use during the last 6000 years: a study of a small eutrophic lake in southern Sweden. *Journal of Paleolimnology*, **8**, 49–69.

Hall, R. I. & Smol, J. P. (1992). A weighted-averaging regression and calibration model for inferring total phosphorus concentration from diatoms in British Columbia (Canada) lakes. *Freshwater Biology*, **27**, 417–34.

Hall, R. I. & Smol, J. P. (1993). The influence of catchment size on lake trophic status during the hemlock decline and recovery (4800 to 3500 BP) in southern Ontario lakes. *Hydrobiologia*, **269/270**, 371–90.

Hall, R. I. & Smol, J. P. (1996). Paleolimnological assessment of long-term water-quality changes in south-central Ontario lakes affected by cottage development and acidification. *Canadian Journal of Fisheries and Aquatic Sciences*, **53**, 1–17.

Hall, R. I. & Smol, J. P. (1999). Diatoms as indicators of lake eutrophication. In *The Diatoms: Applications for the Environmental and Earth Sciences*, ed. E. F. Stoermer & J. P. Smol, Cambridge, UK: Cambridge University Press, pp. 128–68.

Hall, R. I., Leavitt, P. R., Smol, J. P. & Zirnhelt, N. (1999). Comparison of diatoms, fossil pigments and historical records as measures of lake eutrophication. *Freshwater Biology*, **38**, 401–17.

Hall, R. I., Leavitt, P. R., Dixit, A. S., Quinlan, R. & Smol, J. P. (1999). Effects of agriculture, urbanization and climate on water quality in the northern Great Plains. *Limnology & Oceanography*, **44**, 739–56.

Hambright, K. D., Zohary, T., Eckert, W., *et al.* (2008). Exploitation and destabilization of a warm, freshwater ecosystem through engineered hydrological change. *Ecological Applications*, **18**, 1591–603.

Hansson, L. A. (1988). Effects of competitive interactions on the biomass development of planktonic and periphytic algae in lakes. *Limnology and Oceanography*, **33**, 121–8.

Hansson, L. A. (1992). Factors regulating periphytic algal biomass. *Limnology and Oceanography*, **37**, 322–8.

Hargesheimer, E. E. & Watson, S. B. (1996). Drinking water treatment options for taste and odour control. *Water Research*, **30**, 1423–30.

Hargrave, B. T. (1969). Epibenthic algal production and community respiration in the sediments of Marion Lake. *Journal of the Fisheries Research Board of Canada*, **26**, 2003–26.

Harper, D. (1992). *Eutrophication of Freshwaters*. London: Chapman Hall.

Hausmann, S., Lotter, A. F., van Leeuwen, J. F. N., *et al.* (2002). Interactions of climate and land use documented in the varved sediments of Seebergsee in the Swiss Alps. *The Holocene* **12**, 279–89.

Hawes, I. H. & Smith, R. (1992). Effect of localised nutrient enrichment on the shallow epilithic periphyton of oligotrophic Lake Taupo. *New Zealand Journal of Marine and Freshwater Research*, **27**, 365–72.

Haworth, E. Y. (1984). Stratigraphic changes in algal remains (diatoms and chrysophytes) in the recent sediments of Blelham Tarn, English Lake District. In *Lake Sediments and Environmental History*, ed. E. Y. Haworth & J. W. G. Lund, Leices ter: Leicester University Press, pp. 165–90.

Hayes, C. R. & Greene, L. A. (1984). The evaluation of eutrophication impact in public water supply reservoirs in East Anglia. *Water Pollution Control*, **83**, 45–51.

Hickman, M., Schweger, C. E., & Klarer, D. M. (1990). Baptiste Lake, Alberta – a late Holocene history of changes in a lake and its catchment in the southern boreal forest. *Journal of Paleolimnology*, **4**, 253–67.

Hübener, T., Dreßler, M., Schwarz, A., Langner, K. & Adler, S. (2008). Dynamic adjustment of training sets ('moving-window' reconstruction) by using transfer functions in paleolimnology – a new approach. *Journal of Paleolimnology*, **40**, 79–95.

Hutchinson, N. J., Neary, B. P., & Dillon, P. J. (1991). Validation and use of Ontario's Trophic Status Model for establishing lake development guidelines. *Lake and Reservoir Management*, **7**, 13–23.

Jeppesen, E., Madsen, E. A., & Jensen, J. P. (1996). Reconstructing the past density of planktivorous fish and trophic structure from sedimentary zooplankton fossils: a surface sediment calibration data set from shallow lakes. *Freshwater Biology*, **36**, 115–27.

Jeziorski, A., Yan, N. D., Paterson, A. M., *et al.* (2008). The widespread threat of calcium decline in fresh waters. *Science*, **322**, 1374–7.

Jones, V. J. & Juggins, S. (1995). The construction of a diatom-based chlorophyll *a* transfer function and its application at three lakes on Signy Island (maritime Antarctic) subject to differing degrees of nutrient enrichment. *Freshwater Biology*, **34**, 433–45.

Jordan, P., Rippey, B., & Anderson, N. J. (2001). Modeling diffuse phosphorus loads from land to freshwater using the sedimentary record. *Environmental Science & Technology*, **35**, 815–20.

Jordan, P., Rippey, B., & Anderson, N. J. (2002). The 20th century whole-basin trophic history of an inter-drumlin lake in an agricultural catchment. *The Science of the Total Environment*, **297**, 161–73.

Juggins, S. (2001). The European Diatom Database. User Guide. Version 1.0. October 2001. See http://craticula.ncl.ac.uk/Eddi/jsp/help.jsp.

Kann, J. & Falter, C. M. (1989). Periphyton indicators of enrichment in Lake Pend Oreille, Idaho. *Lake and Reservoir Management*, **5**, 39–48.

Karr, J. R. & Dudley, D. R. (1981). Ecological perspective on water quality goals. *Environmental Management*, **5**, 55–68.

Karst, T. L. & Smol, J. P. (2000). Paleolimnological evidence of limnetic nutrient concentration equilibrium in a shallow, macrophyte-dominated lake. *Aquatic Science*, **62**, 20–38.

Kauppila, T., Moisio, T., & Salonen, V. P. (2002). A diatom-based inference model for autumn epilimnetic total phosphorus concentration and its application to a presently eutrophic boreal lake. *Journal of Paleolimnology*, **27**, 261–73.

Kilham, S. S., Theriot, E. C., & Fritz, S. C. (1996). Linking planktonic diatoms and climate change in the large lakes of the Yellowstone ecosystem using resource theory. *Limnology and Oceanography*, **41**, 1052–62.

King, L., Clarke, G., Bennion, H., Kelly, M. & Yallop, M. (2006). Recommendations for sampling littoral diatoms in lakes for ecological status assessments. *Journal of Applied Phycology*, **18**, 15–25.

Kireta, A. R., Reavie, E. D., Danz, N. P., *et al.* (2007). Coastal geomorphic and lake variability in the Laurentian Great Lakes: implications for a diatom-based monitoring tool. *Journal of Great Lakes Research*, **33 (Special Issue 3)**, 36–153.

Kotak, B. G., Lam, A. K-Y., Prepas, E. E., Kenefick, S. L., & Hrudey, S. E. (1995). Variability of the hepatotoxin microcystin-LR in hypereutrophic drinking water lakes. *Journal of Phycology*, **31**, 248–63.

Krammer, K., and Lange-Bertalot, H. (1986–1991). *Bacillariophyceae. Süsswasserflora von Mitteleuropa*, Band 2 (1–4), vol. 1–4, Stuttgart: Gustav Fischer Verlag.

Kreis, R. G., Jr., Stoermer, E. F., & Ladewski, T. B. (1985). *Phytoplankton Species Composition, Abundance, and Distribution in Southern Lake Huron, 1980; Including a Comparative Analysis with Conditions in 1974 Prior to Nutrient Loading Reductions*. Great Lakes Research Division Special Report no. 107. Ann Arbor, MI: The University of Michigan.

Lambert, D., Cattaneo, A., & Carignan, R. (2008). Periphyton as an early indictor of perturbation in recreational lakes. *Canadian Journal of Fisheries and Aquatic Sciences*, **65**, 258–65.

Leahy, P. J., Tibby, J., Kershaw, A. P., Heijnis, H., & Kershaw, J. S. (2005). The impact of European settlement on Bolin Billabong, a Yarra River floodplain lake, Melbourne, Australia. *River Research and Applications*, **21**, 131–49.

Leavitt, P. R. (1993). A review of factors that regulate carotenoid and chlorophyll deposition and fossil pigment abundance. *Journal of Paleolimnology*, **9**, 109–27.

Leavitt, P. R., Carpenter, S. R., & Kitchell, J. F. (1989). Whole-lake experiments: the annual record of fossil pigments and zooplankton. *Limnology and Oceanography*, **34**, 700–17.

Leavitt, P. R. & Findlay, D. L. (1994). Comparison of fossil pigments with 20 years of phytoplankton data from eutrophic Lake 227, Experimental lakes Area, Ontario. *Canadian Journal of Fisheries and Aquatic Sciences*, **51**, 2286–99.

Leira, M., Jordan, P., Taylor, D., Dalton, C., Bennion, H., Rose, N., & Irvine, K. (2006). Assessing the ecological status of candidate reference lakes in Ireland using palaeolimnology. *Journal of Applied Ecology*, **43**, 816–27.

Likens, G. E. (ed.) (1989) *Long-Term Studies in Ecology*. New York, NY: Springer-Verlag.

Little, J. L., Hall, R. I., Quinlan, R., & Smol, J. P. (2000). Quantifying past trophic status and hypolimnetic anoxia in Gravenhurst Bay, Ontario: differential responses of diatoms and chironomids following nutrient diversion. *Canadian Journal of Fisheries and Aquatic Sciences*, **57**, 333–41.

Lotter, A. F. (1998). The recent eutrophication of Baldeggersee (Switzerland) as assessed by fossil diatom assemblages. *The Holocene*, **8**, 395–405.

Lotter, A. F., Ammann, B., & Sturm, M. (1992). Rates of change and chronological problems during the late-glacial period. *Climate Dynamics*, **6**, 233–9.

Lotter, A. F., Birks, H. J. B., Hofmann, W., & Marchetto, A. (1998). Modern diatom, cladocera, chironomid, and chrysophyte cyst assemblages as quantitative indicators for the reconstruction of past environmental conditions in the Alps. II. Nutrients. *Journal of Paleolimnology*, **19**, 443–63

Lowe, R. L. (1996). Periphyton patterns in lakes. In *Algal Ecology: Freshwater Benthic Ecosystems*, ed. R. J. Stevenson, M. L. Bothwell, & R. L. Lowe, San Diego, CA: Academic Press, pp. 57–77.

Lowe, R. L. & Hunter, R. D. (1988). Effect of grazing by *Physa integra* on periphyton community structure. *Journal of the North American Benthological Society*, **7**, 29–36.

Lund, J. W. G. (1950). Studies on *Asterionella* Hass. I. The origin and nature of the cells producing seasonal maxima. *Journal of Ecology*, **38**, 1–35.

Lund, J. W. G. & Reynolds, C. S. (1982). The development and operation of large limnetic enclosures in Blelham Tarn, English Lake District, and their contribution to phytoplankton ecology. *Progress in Phycological Research*, **1**, 1–65.

Marchetto, A. & Bettinetti, R. (1995). Reconstruction of the phosphorus history of two deep, subalpine Italian lakes from sedimentary diatoms, compared with long-term chemical measurements. *Memoire dell'Istituto Italiano di Idrobiologia*, **53**, 27–38.

Marchetto, A., Lamia, A., Musazzi, S., *et al.* (2003). Lake Maggiore (N. Italy) trophic history: fossil diatom, plant pigments, and chironomids, and comparison with long-term limnological data. *Quaternary International*, **113**, 97–110.

Marsden, M. W. (1989). Lake restoration by reducing external phosphorus loading: the influence of sediment phosphorus release. *Freshwater Biology*, **21**, 139–62.

Mason, C. F. (1991). *Biology of Freshwater Pollution*, 2nd edition, London: Longman Group (FE) Ltd.

McGowan, S., Leavitt, P. R., Hall, R. I., *et al.* 2005. Controls of algal abundance and community composition during alternative stable state change. *Ecology*, **86**, 2200–11.

Michelutti, N., Hermanson, M. H., Smol, J. P., Dillon, P. J., & Douglas, M. S. V. (2007). Delayed response of diatom assemblages to sewage inputs in an Arctic lake. *Aquatic Science*, **69**, 523–33.

Miettinen, J. O., Simola, H., Gronlund, E., Lahtinen J., & Niinioja, R. (2005). Limnological effects of growth and cessation of agricultural land use in Ladoga Karelia: sedimentary pollen and diatom analyses. *Journal of Paleolimnology*, **34**, 229–43.

Niederhauser, P. & Schanz, F. (1993). Effects of nutrient (N, P, C) enrichment upon the littoral diatom community of an oligotrophic high-mountain lake. *Hydrobiologia*, **269/270**, 453–62.

Niemi, G. J., Kelly, J. R., & Danz, N. P. (2007). Environmental indicators for the coastal region of the North American Great Lakes: introduction and prospectus. *Journal of Great Lakes Research*, **33 (Special Issue 3)**, 1–12.

Niemi, G. J. & McDonald, M. 2004. Application of ecological indicators. *Annual Review of Ecology and Systematics*, **35**, 89–111.

Nygaard, G. (1949). Hydrobiological studies on some Danish ponds and lakes. II: the Quotient hypothesis and some new or little known phytoplankton organisms. *Det Kongelinge Dansk Videnskabernes Selskab Biologiske Skrifter*, **7**, 1–193.

O'Sullivan, P. E. (1992). The eutrophication of shallow coastal lakes in southwest England – understanding and recommendations for restoration, based on palaeolimnology, historical records, and the modelling of changing phosphorus loads. *Hydrobiologia*, **243/244**, 421–34.

Patrick, R. & Reimer, C. (1966). *The Diatoms of the United States*, vol. 1, Monograph 3, Philadelphia, PA: Academy of Natural Sciences, pp. 1–668.

Peglar, S. M., Fritz, S. C., & Birks, H. J. B. (1989). Vegetation and land-use history in Diss, Norfolk, England. *Journal of Ecology*, **77**, 203–22.

Philibert, A. & Prairie, Y. (2002). Is the introduction of benthic species necessary for open-water chemical reconstruction in diatom-based transfer functions? *Canadian Journal of Fisheries and Aquatic Sciences*, **59**, 938–51.

Polis, G. A. & Winemiller, K. O. (ed.) (1996). *Food Webs: Integration of Patterns & Dynamics*. New York, NY: Chapman & Hall.

Pollard, P. & Huxham, M. (1998). The European Water Framework Directive: a new era in the management of aquatic ecosystem health? *Aquatic Conservation: Marine and Freshwater Ecosystems*, **8**, 773–792.

Qinghong, L. & Bråkenhielm, S. (1995). A statistical approach to decompose ecological variation. *Water, Air and Soil Pollution*, **85**, 1587–92.

Quinlan, R., Hall, R. I., Paterson, A. M., Cumming, B. F., & Smol, J. P. (2008). Long-term assessments of ecological effects of anthropogenic stressors on aquatic ecosystems from paleoecological analyses: challenges to perspectives of lake management. *Canadian Journal of Fisheries and Aquatic Sciences*, **65**, 933–44.

Quinlan, R., Smol, J. P., & Hall, R. I. (1998). Quantitative inferences of past hypolimnetic anoxia in south-central Ontario lakes using fossil chironomids (Diptera: Chironomidae). *Canadian Journal of Fisheries and Aquatic Sciences*, **54**, 587–96.

Rasanen, J., Kauppila, T., & Salonen, V.-P. (2006). Sediment-based investigation of naturally or historically eutrophic lakes – implications for lake management. *Journal of Environmental Management*, **79**, 253–65.

Reavie, E. D. (2007). A diatom-based water quality model for Great Lakes coastlines. *Journal of Great Lakes Research*, **33 (Special Issue 3)**, 86–92.

Reavie, E. D., Axler, R. P., Sgro, G. V., *et al.* (2006). Diatom-based weighted-averaging transfer functions for Great Lakes coastal water quality: relationships to watershed characteristics. *Journal of Great Lakes Research*, **32**, 321–47.

Reavie, E. D., Hall, R. I., & Smol, J. P. (1995). An expanded weighted-averaging model for inferring past total phosphorus concentrations from diatom assemblages in eutrophic British Columbia (Canada) lakes. *Journal of Paleolimnology*, **14**, 49–67.

Reavie, E. D., Kireta, A. R., Kingston, J. C., *et al.* (2008). Comparison of simple and multimetric diatom-based indices for Great Lakes coastline disturbance. *Journal of Phycology*, **44**, 787–802.

Reavie, E. D. & Smol, J. P. (2001). Diatom–environmental relationships in 64 alkaline southeastern Ontario (Canada) lakes: a diatom-based model for water quality reconstructions. *Journal of Paleolimnology*, **25**, 25–42.

Reavie, E. D., Smol, J. P., & Dillon, P. J. (2002). Inferring long-term nutrient changes in southeastern Ontario lakes: comparing paleolimnological and mass-balance models. *Hydrobiologia*, **481**, 61–74.

Reckow, K. H. & Simpson, J. T. (1980). A procedure using modelling and error analysis for the prediction of lake phosphorus concentration from land use information. *Canadian Journal of Fisheries and Aquatic Sciences*, **37**, 1439–48.

Reid, M. (2005). Diatom-based models for reconstructing past water quality and productivity in New Zealand lakes. *Journal of Paleolimnology*, **33**, 13–38.

Revsbeck, N. P. & Jørgensen, B. B. (1986). Microelectrodes: their use in microbial ecology. *Advances in Microbial Ecology*, **9**, 749–56.

Revsbeck, N. P., Jørgensen, B. B., Blackburn, T. H., & Cohen, Y. (1983). Microelectrode studies of photosynthesis and O_2, H_2S and pH profiles of a microbial mat. *Limnology and Oceanography*, **28**, 1062–74.

Reynolds, C. S. (1984). *The Ecology of Freshwater Phytoplankton*. Cambridge: Cambridge University Press.

Rippey, B. (1995). Lake phosphorus models. In *Ecology and Palaeoecology of Lake Eutrophication*, ed. S. T. Patrick & N. J. Anderson, DGU Service Report no. 7, Copenhagen: Geological Survey of Denmark, pp. 58–60.

Rippey, B. & Anderson, N. J. (1996). Reconstruction of lake phosphorus loading and dynamics using the sedimentary record. *Environmental Science & Technology*, **30**, 1786–8.

Rippey, B., Anderson, N. J. & Foy, R. H. (1997). Accuracy of diatom-inferred total phosphorus concentration, and the accelerated eutrophication of a lake due to reduced flushing and increased internal loading. *Canadian Journal of Fisheries and Aquatic Sciences*, **54**, 2637–46.

Rosenberger, E. E., Hampton, S. E., Fradkin, S. C., & Kennedy, B. P. (2008). Effects of shoreline development on the nearshore environment in large deep oligotrophic lakes. *Freshwater Biology*, **53**, 1673–91.

Round, F. E., Crawford, R. M., & Mann, D. G. (1990). *The Diatoms: Biology and Morphology of the Genera*. Cambridge: Cambridge University Press.

Sandman, O., Lichu, A., & Simola, H. (1990). Drainage ditch erosion history as recorded in the varved sediment of a lake in East Finland. *Journal of Paleolimnology*, **3**, 161–9.

Sas, H. (1989) *Lake Restoration by Reduction of Nutrient Loading: Expectations, Experiences, Extrapolations*, St. Augustin: Cademia Verlag, 497 pp.

Schelske, C. L. (1999) Diatoms as mediators of biogeochemical silica depletion in the Laurentian Great Lakes. In *The Diatoms: Applications for the Environmental and Earth Sciences*, ed. Stoermer, E. F. & Smol, J. P., Cambridge: Cambridge University Press, pp. 73–84.

Schelske, C. L., Rothman, E. D., Stoermer, E. F., & Santiago, M. A. (1974). Responses of phosphorus limited Lake Michigan phytoplankton to factorial enrichments with nitrogen and phosphorus. *Limnology and Oceanography*, **19**, 409–19.

Schelske, C. L. & Stoermer, E. F. (1971). Eutrophication, silica depletion and predicted changes in algal quality in Lake Michigan. *Science*, **173**, 423–4.

Schelske, C. L., Stoermer, E. F., Fahnenstiel, G. L., & Haibach, G. L. (1986). Phosphorus enrichment, silica utilization, and biogeochemical silica depletion in the Great Lakes. *Canadian Journal of Fisheries and Aquatic Sciences*, **43**, 407–15.

Schindler, D. W. (1971). Carbon, nitrogen and phosphorus and the eutrophication of freshwater lakes. *Journal of Phycology*, **7**, 321–9.

Schindler, D. W. (1977). Evolution of phosphorus limitation in lakes. *Science*, **195**, 260–2.

Schönfelder, I., Gelbrecht, J., Schönfelder, J., & Steinberg, C. E. W. (2002). Relationships between littoral diatoms and their chemical environment in northeastern German lakes and rivers. *Journal of Phycology*, **38**, 666–82.

Simola, H. (1983). Limnological effects of peatland drainage and fertilization as reflected in the varved sediment of a deep lake. *Hydrobiologia*, **106**, 43–57.

Simpson, G. L. & Hall, R. I. (2011). Human impacts – applications of numerical methods to evaluate surface-water acidification and eutrophication. In *Tracking Environmental Change Using Lake Sediments, Volume 5: Data Handling and Statistical Techniques*, Dordrecht: Springer (in press).

Siver, P. A. (1999). Development of paleolimnological inference models for pH, total nitrogen and specific conductivity based on planktonic diatoms. *Journal of Paleolimnology*, **21**, 45–59.

Smith, V. H., Joye, S. B., & Howarth, R. W. (2006). Eutrophication of freshwater and marine ecosystems. *Limnology and Oceanography*, **51**, 351–5.

Smol, J. P. (1995). Paleolimnological approaches to the evaluation and monitoring of ecosystem health: providing a history for environmental damage and recovery. In *Evaluating and Monitoring the Health of Large-Scale Ecosystems: NATO ASI Series, Vol.* 128, ed. D. J. Rapport, C. L. Gaudet & P. Calow, Berlin: Springer-Verlag, pp. 301–18.

Smol, J. P. (2008). *Pollution of Lakes and Rivers: A Paleoenvironmental Perspective*, 2nd Edition. Blackwell Publ. 383 pp.

Smol, J. P. & Dickman, M. D. (1981). The recent histories of three Canadian Shield lakes: A paleolimnological experiment. *Archiv für Hydrobiologie*, **93**, 83–108.

Soranno, P. A., Hubler, S. L., Carpenter, S. R., & Lathrop, R. C. (1996). Phosphorus loads to surface waters: a simple model to account for spatial pattern of land use. *Ecological Applications*, **6**, 865–78.

St. Jacques, J.-M., Douglas, M. S. V., & McAndrews, J. H. (2000). Mid-Holocene hemlock decline and diatom communities in van Nostrand Lake, Ontario, Canada. *Journal of Paleolimnology*, **23**, 385–97.

Stockner, J. G. (1971). Preliminary characterization of lakes in the Experimental Lakes Area, north-western Ontario using diatom occurrence in lake sediments. *Journal of the Fisheries Research Board of Canada*, **28**, 265–75.

Stoermer, E. F., Emmert, G., Julius, M. L. & Schelske, C. L. (1996). Paleolimnologic evidence of rapid change in Lake Erie's trophic status. *Canadian Journal of Fisheries and Aquatic Sciences*, **53**, 1451–8.

Stoermer, E. F. & Yang, J. J. (1970). *Distribution and Relative Abundance of Dominant Planktonic Diatoms in Lake Michigan*. Great Lakes Research Division Publication no. 16. Ann Arbor, MI: University of Michigan.

Taylor, D., Dalton, C., Leira, M., *et al.* (2006). Recent histories of six productive lakes in the Irish Ecoregion based on multiproxy palaeolimnological evidence. *Hydrobiologia*, **571**, 237–59.

Tibby, J. (2004). Development of a diatom-based model for inferring total phosphorus in south-eastern Australian water storages. *Journal of Paleolimnology*, **31**, 23–36.

Tilman, D. (1977). Resource competition between planktonic algae: an experimental and theoretical approach. *Ecology*, **58**, 338–48.

Tilman, D., Kilham, S. S., & Kilham, P. (1982). Phytoplankton community ecology: the role of limiting nutrients. *Annual Review of Ecology and Systematics*, **13**, 349–72.

Tilman, D., Kiesling, R., Sterner, R., Kilham, S. S., & Johnson, F. A. (1986). Green, bluegreen and diatom algae: taxonomic differences in competitive ability for phosphorus, silicon and nitrogen. *Archiv für Hydrobiolgie*, **106**, 473–85.

Turkia, J. & Lepistö, L. (1999). Size variation of planktonic *Aulacoseira* in water and sediment from Finnish lakes of varying trophic status. *Journal of Plankton Research*, **21**, 757–70.

Turner, M. A., Howell, E. T., Robinson, G. G. C., *et al.* (1994). Role of nutrients in controlling growth of epilithon in oligotrophic lakes of low alkalinity. *Canadian Journal of Fisheries and Aquatic Sciences*, **51**, 2784–93.

Vaughn, J. C. (1961). Coagulation difficulties of the south district filtration plant. *Pure Water*, **13**, 45–9.

Verschuren, D., Johnson, T. C., Kling, H. J., *et al.* (2002) The chronology of human impact on Lake Victoria, East Africa. *Proceedings of the Royal Society, London*, **B 269**, 289–94.

Vinebrooke, R. D. (1996). Abiotic and biotic regulation of periphyton in recovering acidified lakes. *Journal of the North American Benthological Society*, **15**, 318–31.

Vollenweider, R. A. (1975). Input–output models with special reference to the phosphorus loading concept. *Schweizerische Zeitschrift für Hydrologie*, **37**, 58–83.

Watchorn, M. A., Hamilton, P. B., Anderson, T. W., Roe, H. M., & Patterson, R. T. (2008). Diatoms and pollen as indicators of water quality and land-use change: a case study from the Oak Ridges Moraine, southern Ontario, Canada. *Journal of Paleolimnology*, **39**, 491–509.

Webster, K. E., Kratz, T. K., Bowser, C. J., & Magnusson, J. J. (1996). The influence of landscape position on lake chemical responses to drought in northern Wisconsin. *Limnology and Oceanography*, **41**, 977–84.

Werner, D. (ed.) (1977). *The Biology of Diatoms*. Berkeley, CA: University of California Press.

Werner, P. & Smol, J. P. (2005). Diatom–environmental relationships and nutrient transfer functions from contrasting shallow and deep limestone lakes in Ontario, Canada. *Hydrobiologia*, **533**, 145–73.

Wessels, M., Mohaup, K., Kummerlin, R., & Lenhard, A. (1999). Reconstructing past eutrophication trends from diatoms and biogenic silica in the sediment and the pelagic zone of Lake Constance, Germany. *Journal of Paleolimnology*, **21**, 171–92.

Wetzel, R. G. (1964). A comparative study of the primary productivity of higher aquatic plants, periphyton and phytoplankton in a large shallow lake. *Internationale Revue der Gesamten Hydrobiologie* **48**, 1–61.

Wetzel, R. G. (2001). *Limnology*, 3rd edition, San Diego, CA: Academic Press.

Wetzel, R. G., Rich, P. H., Miller, M. C., & Allen, H. L. (1972). Metabolism of dissolved and particulate detrital carbon in a temperate hard-water lake. *Memoire dell'Istituto Italiano di Idrobiologia Supplement*, **29**, 253–76.

Weyhenmeyer, G. A., Westöö, A.-K., & Willén, E. (2008). Increasingly ice-free winters and their effects on water quality in Sweden's largest lakes. *Hydrobiologia*, **599**, 111–18.

Whitmore, T. J. (1989). Florida diatom assemblages as indicators of trophic state and pH. *Limnology and Oceanography*, **34**, 882–95.

Whittier, T. R. & Paulsen, S. G. (1992). The surface waters component of the Environmental Monitoring and Assessment Program (EMAP): an overview. *Journal of Aquatic Ecosystem Health*, **2**, 119–26.

Wunsam, S. & Schmidt, R. (1995). A diatom–phosphorus transfer function for alpine and pre-alpine lakes. *Memoire dell'Istituto Italiano di Idrobiologia*, **53**, 85–99.

Yang, J.-R., Duthie, H. C., & Delorme, L. D. (1993). Reconstruction of the recent environmental history of Hamilton Harbour (Lake Ontario, Canada) from analysis of siliceous microfossils. *Journal of Great Lakes Research*, **19**, 55–71.

Yang, J.-R., Pick, F. R., & Hamilton, P. B. (1996). Changes in the planktonic diatom flora of a large mountain lake in response to fertilization. *Journal of Phycology*, **32**, 232–43.

Yang, X., Anderson, N. J., Dong, X., & Shen, J. I. (2008). Surface sediment diatom assemblages and epilimnetic total phosphorus in large, shallow lakes of the Yangtze floodplain: their relationships and implications for assessing long-term eutrophication. *Freshwater Biology*, **53**, 1273–90.

Zeeb, B. A., Christie, C. E., Smol, J. P., *et al.* (1994). Responses of diatom and chrysophyte assemblages in Lake 227 sediments to experimental eutrophication. *Canadian Journal of Fisheries and Aquatic Sciences*, **51**, 2300–11.

8 Diatoms as indicators of environmental change in shallow lakes

HELEN BENNION, CARL D. SAYER, JOHN TIBBY, AND HUNTER J. CARRICK

8.1 Introduction

Historically, limnological and paleolimnological research has focused on large and typically deep lakes but in the last two decades there has been a growing interest in smaller and shallower water bodies. Shallow lakes are justifiably considered as a separate lake type, distinguished physically from deeper waters by the fact that they are permanently mixed (polymictic) with a consequent lack of stratification of temperature or oxygen and with increased potential for nutrient recycling and redistribution of seston by physical water circulation patterns (Carrick *et al.*, 1994). Whilst this is a useful distinction, there is no single definition of a shallow lake (Padisák & Reynolds, 2003). Scheffer (1998), in his classic text book, acknowledged a fundamental difference in the behavior, ecological functioning, and biotic communities of shallow waters and arbitrarily selected a mean depth of less than 3 m to define shallowness. For the purposes of this chapter we have chosen to adopt this definition and thereby to focus on lakes where, under a favorable light climate, benthic algae and/or rooted submerged macrophytes may occupy the majority of the lakebed (see also Jeppesen *et al.*, 1997). Under enriched conditions, however, the mechanisms that stabilize the macrophyte communities of shallow lakes may often break down and a transition to pelagic production with phytoplankton dominance occurs (Scheffer *et al.*, 1993; Vadeboncoeur *et al.*, 2003). Importantly, because of these characteristics, shallow lakes are, for the most part, more vulnerable to a given pollutant load than large lakes.

Shallow lakes can be either artificial in origin, for example farm ponds and ornamental lakes, or natural, for example floodplain lakes and shallow depressions in glaciated and polar regions (see Douglas & Smol, this volume). They are ubiquitous features of lowland landscapes and there are millions of shallow lakes and ponds worldwide, often providing hot spots of biodiversity. Shallow lakes and their associated wetlands (see Gaiser & Rühland, this volume) may also be of great socio-economic importance (Moss, 2001). Whilst many shallow aquatic systems have been impacted over long timescales (100s or 1000s of years), sadly, they are under increased threat from eutrophication, both decreased and increased pH, salinization, elevated erosion and siltation and, of course, climate change. Hence, efforts to better manage and restore these systems have increased over the last decade, ranging from reductions in pollutant loading (e.g. Jeppesen *et al.*, 2007) to biomanipulation (e.g. Søndergaard *et al.*, 2007). Given that sound management requires a good understanding of ecosystem processes and knowledge of pre-disturbance conditions to set realistic targets for recovery, the need for information on causes, timing, rates, and nature of environmental change is greater than ever.

In this chapter, we first describe the diverse range of diatom communities found in shallow lakes. We then explore the use of modern diatom assemblages as biomonitoring tools in shallow systems including indices to assess trophic status, many of which have recently been developed to assist with implementation of key pieces of environmental legislation such as the European Union Water Framework Directive (EU WFD). The remainder of the chapter focuses on paleoecological applications where diatoms have been used as indicators of lake development, erosion, alkalinization, acidification, salinization, and climate change, and especially eutrophication. We will examine the particular problems of employing diatom records to assess environmental change in shallow lakes, including issues of mixing, resuspension, dissolution, and, particularly, the application of transfer functions (e.g. Bennion *et al.*, 2001; Sayer, 2001). Finally we will review the role of diatoms in multiproxy studies and the contribution that such studies can make to our understanding of broader ecosystem changes in shallow lakes.

The Diatoms: Applications for the Environmental and Earth Sciences, 2nd Edition, eds. John P. Smol and Eugene F. Stoermer. Published by Cambridge University Press.

8.2 Ecology and distribution of diatom communities in shallow lakes

Shallow lakes generally have greater potential for benthic algal growth than deep systems because most have an extensive littoral zone relative to the pelagic zone (Wetzel, 1983; 2001). As such, these ecosystems support a range of habitats that promote a diverse and productive diatom assemblage which, in turn, fuels unique food-web pathways (e.g. Lindeman, 1942). The contribution of benthic algae to overall lake production varies as a function of lake morphometry and trophic status where estimates of percentage contribution range from 1 to nearly 100% (Vadeboncoeur *et al.*, 2003). With increasing nutrient loading it is now well known that pelagic production tends to increase, while benthic production contracts as light starts to limit the production of bottom-dwelling algae and submerged macrophytes (Scheffer & van Nes, 2007).

8.2.1 Pelagic diatom dynamics

One of the major selective pressures that impinges on diatom populations in pelagic environments is sinking (Smetacek, 1985). Most diatom species experience relatively high settling velocities because their cell density is larger than other planktonic algal species, so that sinking represents their major loss rate under non-mixing conditions. Algae with cell densities similar to water (1 g ml^{-1}) do not experience appreciable settling velocities, while most diatom species have higher cell densities due to their siliceous cell wall (1.1 to 1.4 g ml^{-1}), which translates into two- to sixfold faster settling rates (Sommer, 1988). Thus, if diatom cells cannot maintain buoyancy, then mechanisms to reduce grazing or out-compete other individuals for nutrients are of diminished value.

The seasonal dominance by diatoms in most deep lakes is restricted to relatively discrete, seasonal blooms (spring and autumn periods) that gradually transition into dominance by other phytoplankton groups (Reynolds, 1984). In shallow lakes, the environment is unique because it is often subject to intermittent mixing events that can promote the growth of considerable planktonic diatom populations through resuspension. Mixing events are initiated by wind and local weather patterns (Padisák *et al.*, 1988; Carrick *et al.*, 1993) or disturbance imposed by internal forces such as fish movement (Roozen *et al.*, 2007). Seasonal cycles in the phytoplankton can be explained by destratification of the water column associated with wind turbulence that mixes meroplankton up into the photic zone to seed new growth (e.g. Lewis, 1978). This process has been confirmed with very elegant field studies to induce mixing in two lakes of moderate depth (mean depth 6.8 and 9.9 m, respectively) (Lund, 1971; Reynolds *et al.*, 1983). Moreover, high diversity in plankton assemblages appears to be maintained by complex multi-species interactions and the prevalence of disturbance (non-equilibrium conditions, Scheffer *et al.*, 2003). Either way, diatoms contribute significantly to the plankton of shallow lakes throughout the year.

Several diatom species that are typically thought of as planktonic diatoms can exist in lakes as meroplankton as a result of their unique adaptation, i.e. their ability to form resting cells that withstand periods of prolonged darkness (e.g. Lund, 1954; Carrick *et al.*, 2005). For instance, *Aulacoseira* (formerly *Melosira*) *italica* (Ehrenb.) Simonsen uncovered from lake sediments dating back 20 years, was capable of carbon fixation within 1–8 hours of exposure to moderate light (50 μEinst m^{-2} s^{-1}) (Sicko-Goad, 1986). Along these lines, the classic work by Nipkow (1950) and later by Lund (1954, 1971) showed that diatom resting cells reside on the bottom of lakes and are reintroduced into the water column during high-wind periods, thereby reseeding the plankton. More recent work shows that this phenomenon has a marked effect on phytoplankton biomass and taxonomic composition, and therefore can explain a significant portion of temporal variation in shallow, productive lakes (Carrick *et al.*, 1993). In Lake Okeechobee, Florida, higher water-column nutrients, turbidity, and phytoplankton biomass are associated with resuspension of sediments by wind (Maceina & Soballe, 1990), while a model for Lake Tamnaren, Sweden, indicates that resuspension causes increased turbidity leading to decreased algal production through light attenuation (Hellström, 1991). In shallow Lake George, Uganda, concentrations of phytoplankton that sink to deeper water at midday are redistributed more uniformly in the water column by diurnal destratification (Ganf, 1974). This process appears to account for substantial changes in surface-water phytoplankton biomass on some occasions.

8.2.2 Benthic diatom dynamics

Substantial populations of benthic diatoms occur in clear-water shallow lakes (Liboriussen & Jeppesen, 2003). The major selective pressures that act on benthic diatom populations are, typically, light, nutrients, grazing, and physical disturbance (Lowe, 1996). Light availability is primary, in that it sets limits on all forms of aquatic plant production, whereby the spatial distribution of production is largely dictated by concentrations of particulate and/or dissolved organic carbon (DOC) concentrations in the overlying water column (Falkowski & Raven, 2007).

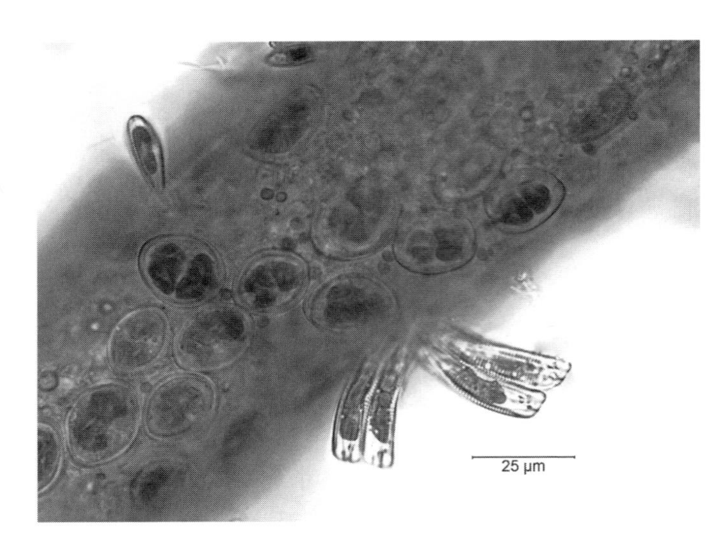

Figure 8.1 Light microscope photograph of epiphytic diatoms, *Rhoicosphenia* and *Cocconeis* spp., attached to *Cladophora glomerata* (L.) Kütz. in Pitsford Water, England, in August 2006. (Photo by Chris Carter.)

In clear-water shallow lakes, light, by definition, is in plentiful supply, and hence other internal factors may become more important regulators of diatom distribution (Scheffer & van Nes, 2007).

The abundance of benthic diatoms may be closely tied to the available substrata, and the seasonal occurrence of host plants (Burkholder, 1996) (see Figure 8.1). This may explain the unimodal seasonal distribution of both epipelic and epiphytic diatoms, with the peak numbers being realized in the early spring and summer along with vascular plant growth (Round, 1981). This seasonal pattern appears to be typical for temperate ecosystems, as the spring peak in epiphytic diatom abundance has been observed in a range of lakes (Burkholder & Wetzel, 1989; Muller, 1995). Moreover, diatom species composition varied little among three vascular plant species that grow in the littoral zone of Lake Erie (Millie & Lowe, 1983), and Burkholder *et al.* (1990) found considerable similarity between plastic and naturally occurring host plants in a temperate lake (Lawrence Lake, Michigan, USA). Research in Lake Memphremagog (Canada/USA) shows that when grazers are excluded through the use of *in situ* screens, that diatoms remain dominant throughout the year (June through August) (Cattaneo & Kalff, 1986). In the presence of grazers, diatoms gave way to dominance by less palatable taxa, namely cyanobacteria (Cattaneo, 1983).

In subtropical shallow lakes, there are relatively small seasonal changes in diatom assemblages (wet versus dry periods), for example in the periphyton of the Florida Everglades (McCormick *et al.*, 1998). High diatom biomass throughout the year appears to be coincident with relatively low temporal variation in climates typified by low annual variability in temperature (range 15–32 °C) (Beaver *et al.*, 1981). In contrast, seasonal changes in periphyton biomass are typically large in temperate systems, with amounts varying by orders of magnitude over a period of months due to periphyton recolonization that replaces seasonal die-backs (e.g. Hudon & Bourget, 1983; Burkholder & Wetzel, 1989). In general, benthic algal biomass reflects trophic status. Early work by Moss (1968) suggested predictable patterns in the development of benthic algal biomass and habitat stability; the highest standing stocks were measured in more quiescent waters (namely those occurring as benthic mats in shallow lakes). His estimates were the first to show accrual in excess of 1000 mg m^{-2}. Data from continuous-flow microcosms demonstrate predictable relationships between benthic chlorophyll *a* with additions of dissolved phosphorus (P) concentrations, producing a range of concentrations (3 to 375 mg m^{-2}; Krewer & Holm, 1982). Work by Cattaneo (1987) shows a predictable relationship between benthic chlorophyll and total phosphorus (TP) concentrations among ten lakes in eastern Quebec. The uptake of nutrients by benthic algal activity is quantitatively significant such that they are used to remove nutrients from overlying waters in order to remediate excessive runoff (Reddy *et al.*, 1999).

8.3 Diatoms as contemporary biological indicators in shallow lakes

Diatoms are often the most abundant and diverse group of algae within the phytobenthos or periphyton of aquatic systems, living on or attached to substrate or other organisms. The short generation times of algae allow them to respond more rapidly to changes in water quality than macrophytes or invertebrates (McCormick & Stevenson, 1998) and to better integrate changes in environmental conditions than infrequent water-chemistry samples. These factors, in addition to the ease of sampling littoral periphyton, render them useful organisms for monitoring purposes. Periphytic diatoms have been most widely used for monitoring the quality of rivers and streams, where the epilithic diatom community (attached to stones) is usually the preferred substrate as it is ubiquitous and relatively inert (Round, 1991; Kelly *et al.*, 1998). Numerous river studies have produced diatom indices for assessing trophic status (e.g. Prygiel & Coste, 1993; Kelly & Whitton, 1995; Ponader *et al.*, 2007). The periphyton also play an important role in small, shallow lakes where the littoral zone can be a major, if not the dominant, component of total primary productivity (Wetzel,

1996; Gaiser *et al.*, 2006). Therefore, littoral diatom communities can act as valuable biological indicators in shallow standing waters. Several studies have demonstrated that the distribution of shallow-water periphyton communities can be significantly explained by water chemistry variables such as TP, total nitrogen (TN), dissolved inorganic carbon (DIC), pH, calcium (Ca), chlorine (Cl), specific conductivity, and DOC (e.g. Hofmann, 1994; King *et al.*, 2000; Schönfelder *et al.*, 2002).

As for rivers, many periphytic diatom studies in lakes have been based on the epilithon and, given the prevalence of nutrient enrichment as the key pressure on shallow, lowland waters, these studies too have primarily resulted in the development of diatom indices, metrics, and models which allow trophic status to be assessed (e.g. van Dam *et al.*, 1994). For example, King *et al.* (2000) and DeNicola *et al.* (2004) demonstrated a good relationship between nutrient variables and epilithic algal communities in the littoral zone of English and Irish lakes, respectively, and produced weighted-averaging models for the prediction of TP. However, not all shallow lakes have littoral zones comprised of hard substrates such as rocks and stones, and in such cases alternative substrates must be found. King *et al.* (2006) recommended that, where epilithic habitats may be lacking or limited and do not represent a significant component of the lake phytobenthos, epiphytic samples (attached to plants), preferably from emergent macrophytes, should be collected. Several studies have demonstrated the value of using epiphytic diatoms to assess water quality of shallow lakes including Ács *et al.* (2005) in Hungary and Blanco *et al.* (2004) in the Mediterranean. Indeed Poulícková *et al.* (2004) considered epiphytes on reeds to be better indicators of eutrophication than epipelic assemblages associated with the muds and silts. The nature of epipelon (including motile and sessile taxa) is such that diatoms may take up nutrients from the sediment as well as the overlying water, thereby limiting their value as biological indicators (e.g. Hansson, 1989). Furthermore, the epipelic diatom community of most shallow lakes is relatively homogeneous, often dominated by small *Fragilaria (sensu lato)* taxa. These taxa respond to the favorable light conditions in the shallow water of littoral zones but are poor indicators of water quality, being able to thrive over a broad range of nutrient and other chemical conditions (e.g. Bennion *et al.*, 2001; Tibby *et al.*, 2007). For these reasons, the epipelon has been less commonly employed as a biomonitoring tool.

The use of epiphytic diatom assemblages as biological indicators is not without its own problems. Epiphytic algae are likely to take up nutrients leached from their macrophyte hosts (Allen, 1971). Denys (2004) tested the usefulness of abundance-weighted averages of the indicator values provided by van Dam *et al.* (1994) for estimating environmental conditions in a large set of small, shallow standing freshwaters in Belgium using epiphytic and surface-sediment diatom assemblages. In relation to trophic status and organic loading, epiphyte-derived scores inferred a less impaired situation than those calculated from surface-sediment assemblages in the most impacted sites. The author attributed this to a range of possible factors including the different pH, osmotic conditions, concentration of organic substances, and the availability of nutrients at the surface of physiologically active macrophytes compared with those at the sediment surface. In a subsequent study, Denys (2007) demonstrated that fewer water-column variables can be inferred reliably from epiphyton compared to littoral sediment assemblages, adding lower *in situ* species diversity and less effective spatial integration (i.e. lower recruitment of phytoplankton) in the former as possible explanations. Nevertheless, Denys (2007) was able to develop robust transfer functions for inferring pH, DIC, and TP from the epiphytic diatom assemblages.

Whilst restriction of sampling to a single substratum to avoid variation in diatom assemblages associated with different substrate types has been recommended by several authors (e.g. Hofmann, 1994; King *et al.*, 2006), other studies have worked with a combination of substrates. In a study of shallow-water bodies in the Czech Republic, Kitner & Poulícková (2003) found no significant differences in the estimation of trophic status between the epilithon, epiphyton, and epipelon and, therefore, advocated assessment based on the average of all three substrates. Schönfelder *et al.* (2002) used a range of substrates in German lowland lakes and found strong relationships between littoral diatom communities and measured environmental variables, and Gaiser *et al.* (2006) employed various substrates in shallow water-bodies in the Florida Everglades to produce periphyton-based metrics that proved to be reliable indicators of P enrichment.

Since 2000 the EU WFD has created a statutory obligation for EU Member States to monitor phytobenthos in lakes and rivers (European Union, 2000). In contrast to earlier pieces of legislation which were concerned with pollution intensity (e.g. Urban Wastewater Treatment Directive (UWWTD; 91/271/EEC) and Nitrates Directive (91/676/EEC)), the EU WFD requires an assessment of ecological status based on comparison of the observed state of a water body with that expected in the absence of anthropogenic disturbance, known as a reference condition. The ecological status must be defined as one of five classes: high, good, moderate, poor, or bad, measured by the degree of deviation from the reference condition. The existing diatom tools were not designed to assess ecological status in this way and there has, therefore, been a new wave of research to develop ecological assessment tools based on phytobenthos, focusing largely on diatoms. Kelly *et al.* (2008) have recently shown that such tools generated from diatoms alone are as powerful as those generated from diatoms and non-diatoms combined and, hence, their findings support the use of diatoms as proxies for phytobenthos in ecological status assessments. As with the earlier littoral diatom studies, much of this new work has centered on rivers (see Stevenson *et al.*, this volume), but in the last few years several countries have produced EU WFD-compliant phytobenthos tools for lakes and several more are in development. For example, Stenger-Kovacs *et al.* (2007) developed the Trophic Diatom Index for Lakes (TDIL) in Hungary using epiphytic diatom samples from 83 shallow lakes, Kelly *et al.* (2007) developed the Lake Trophic Diatom Index (LTDI) in the UK using epilithic and epiphytic diatom samples from 177 predominantly shallow lakes, and Schaumburg *et al.* (2004, 2007) used a variety of substrates to develop a diatom index for German lakes (DISeen). These tools, for the most part, involved adaptations of existing metrics into a measure of ecological status. This has required determination of reference sites, which must be subject to minimal anthropogenic impact, in order to define reference diatom communities for each lake type. For small, shallow lowland lakes this has proved difficult as almost all low-lying European catchments have experienced some degree of industrialization, urbanization, and agricultural intensification. It is in these cases that paleoecological methods have proved invaluable for assessing reference conditions.

8.4 Diatoms as paleoecological indicators in shallow lakes

In many parts of the world, shallow lakes are more abundant than deepwater systems but traditionally paleolimnological studies have focused on the latter. This can largely be attributed to the problems that shallow lakes pose for paleolimnology, primarily associated with their low retention times, sediment resuspension, and sediment–water interface mixing (e.g. Anderson & Odgaard, 1994; Denys, 2007). Bioturbation by fish, benthic invertebrates, and wind-induced resuspension events are common in shallow waters and can destroy much of the integrity of the stratigraphic record. In shallow lakes, resuspension can take place over most of the lake area (e.g. Hellström, 1991). Mixing, in particular, causes difficulty in interpretation of ^{210}Pb profiles and the application of the standard chronological models (Oldfield & Appleby, 1984). For example, in a study of three shallow Danish lakes, Anderson & Odgaard (1994) reported almost homogenous profiles of ^{137}Cs over the surface 20 cm of cores from two of the lakes, and the absence of expected peaks owing to weapons testing and Chernobyl fallout, as evidence of considerable mixing.

Crucially, frequent resuspension cycles and grazing by benthic invertebrates cause breakage of diatom frustules and diffusion of dissolved silica, making them more liable to dissolution (Jewson *et al.*, 1981; Beyens & Denys 1982; Rippey, 1983). Furthermore, many lowland, shallow lakes are relatively alkaline and warm for much of the year, and it is well established that diatom preservation declines with increasing pH and temperature as the opaline silica becomes susceptible to dissolution under such conditions (Flower, 1993; Barker *et al.*, 1994). The short retention time of shallow-water bodies is a further characteristic that may affect silicon (Si) dissolution as the recycling of silica (SiO_2) is likely to be accelerated (Marshall & Warakomski, 1980). Hence, downcore preservation can be problematic in shallow-lake sediment records. For example, in Tai Hu, a large, shallow lake (<2 m mean depth) in central China, Dong *et al.* (2008) found the preservation status of diatoms in fossil assemblages of cores to be generally poor. This lake is highly alkaline, warm, well flushed and there is extensive wind-induced resuspension of sediments, all factors known to enhance diatom dissolution. The diverse array of habitats in shallow lakes and shifts in their availability over time due to changes in macrophyte abundance poses a challenge for the reconstruction of water-chemistry variables in these systems (Sayer, 2001; Denys, 2006) but equally makes them useful for assessing changes in ecological structure and function (e.g. Sayer *et al.*, 2010a).

Whilst the range of factors described above may complicate the use of the sediment record, paleolimnological approaches have, nonetheless, been used with success in numerous shallow lakes. The choice of coring location and method of dating may need more careful consideration than in a deep, stratified lake with an obvious central basin, but core profiles from shallow-water bodies can be interpreted in a meaningful way and even a smoothed record is valuable for management purposes, as the major trends can still be detected (Anderson, 1993). Qualitative diatom records can be extracted from even the most difficult shallow-lake situations, if investigators are willing to undertake laborious techniques (e.g. Stoermer *et al.*, 1992; Donar *et al.*, 2009). Indeed, the sediment records of shallow lakes may even offer some advantages over deeper systems. For example, shallow, productive lakes typically have high sediment-accumulation rates of up to several centimeters per year that can compensate for some of the mixing problems and can potentially allow for high resolution studies. Additionally, the sediments of shallow waters typically include material from a wider variety of habitats than the basins of deep lakes and are, therefore, comprised of a richer variety of communities, which may include a diverse epilithon and epiphyton, compared with the tendency towards plankton dominance in the sediment sequences of deeper waters. Furthermore, shallow lakes occur predominantly in lowland productive catchments, many of which have been impacted over long timescales. Unfortunately, historical data are unavailable or incomplete for the vast majority of shallow lakes and, therefore, paleoecology is often the only way of obtaining information on pre-impact conditions.

The following sections provide examples of paleoecological applications where diatoms have been used as indicators of environment change in shallow lakes, including lake development, land-use related impacts (e.g. erosion, alkalinization, acidification, salinization), and particularly eutrophication, which has received most attention. We discuss the utility of diatom records for the monitoring and reconstruction of chemical and ecological conditions in these systems as well as outlining some of the potential problems.

8.4.1 Lake development

Shallow lakes have a variety of origins, including tectonic activity, glaciation, riverine activity, other geomorphic processes, and human activity (Cohen, 2003). Interestingly, despite the theoretical potential for the ecology of shallow lakes to be substantially affected by processes of lake development, there has been relatively little research that examines, specifically, the evolution of shallow lakes in the same manner as deep-lake studies such as those of Fritz (1989) and Engstrom *et al.* (2000). This may, in part, be due to the often long influence of catchment land use on shallow lakes. Nevertheless, understanding the natural development of shallow-lake ecosystems is important and those created by riverine processes provide a useful opportunity since such ecosystems are often created (e.g. through channel avulsion) with relatively little wide-scale disturbance, and in environments where terrestrial vegetation is already established (in contrast to, for example, recently deglaciated landscapes). Floodplain shallow lakes can exist for many thousands of years (Ogden *et al.*, 2001). Diatom records from many such sites, such as those in southeast Australia, display a substantial degree of stability. This is particularly true of Yarra River catchment shallow lakes, which, for a period of approximately 1000 years, were strongly dominated by the planktonic diatom *Cyclotella stelligera* (Cleve & Grunow in Cleve) Van Heurck, indicative of low nutrient conditions (Leahy *et al.*, 2005; Gell *et al.*, 2005b). Although diatom variation is slightly greater in meander cut-offs on the Goulburn River, there is still a strong long-term dominance by two diatom species, *Aulacoseira ambigua* (Grunow in Van Heurck) Simonsen and *Eunotia incisa* W. Sm. ex Greg. (Reid, 2002). This long-term stability was mirrored by stability in reconstructed pH (between 6.5 and 6.7 for 3000 years) (Tibby *et al.*, 2003) (Figure 8.2). Indeed, stability in shallow-lake diatom records where the influence of humans is minimal has been documented in many other parts of the world (e.g. Schmidt *et al.*, 1990: Douglas *et al.*, 1994).

8.4.2 Eutrophication

There have been major successes worldwide in reducing enrichment problems but eutrophication remains one of the foremost environmental issues threatening the quality of surface waters (Smith *et al.*, 2006). Whilst point sources of pollution from sewage and industrial waste have been reduced over the last few decades, nutrient input from diffuse agricultural sources has remained high, particularly in intensively cultivated regions (van der Molen & Portielje, 1999). It is in these productive, lowland catchments that shallow lakes predominate and, given that they are generally more vulnerable to a given pollutant load than large lakes, many have experienced severe eutrophication problems. Typical symptoms include high phytoplankton densities and loss of submerged macrophyte vegetation with consequent reductions in species and habitat diversity and undesirable changes to the food-web (Phillips, 1992; Jeppesen, 1998). The loss of aquatic plants from shallow lakes has received most attention from ecologists and managers owing to their importance for nature conservation and the crucial role that submerged vegetation plays in maintaining healthy ecosystem structure and function (Jeppesen *et al.*, 1997). Whilst considerable efforts have been made to combat eutrophication, recovery of shallow lakes following reductions in nutrient loading is often delayed by internal P loading or biological resistance (e.g. Jeppesen *et al.*, 2005).

Diatoms are arguably the most widely used biological group in paleolimnological studies of eutrophication (Hall & Smol, this volume) and have been employed in a variety of ways to assess trophic changes in shallow systems. Some of the earliest paleoecological diatom work on eutrophication was conducted on shallow lakes (e.g. Battarbee, 1978, 1986; Haworth, 1972; Moss, 1980; Hickman 1987). These studies focused on shifts in diatom species composition over time in lake-sediment

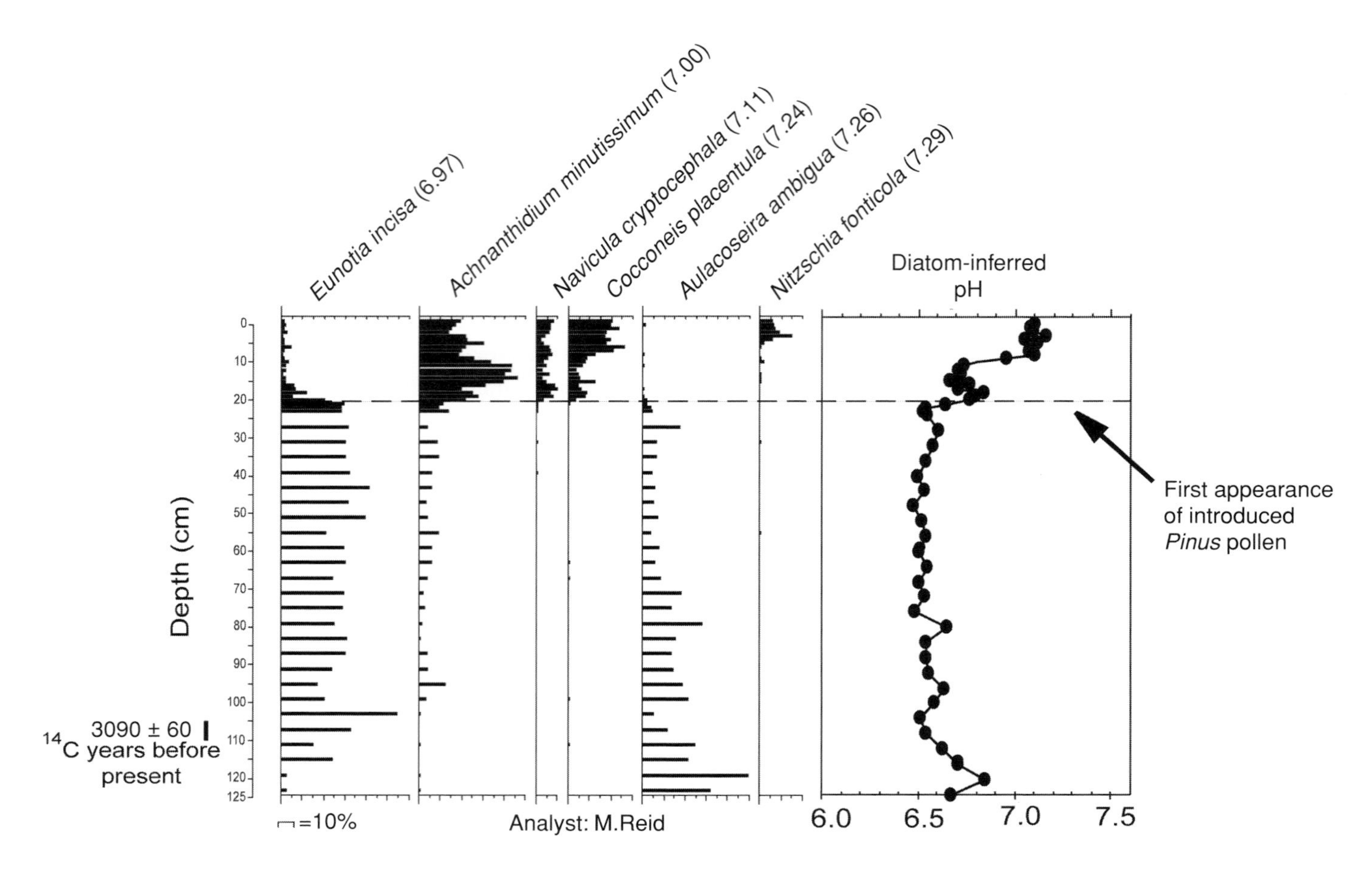

Figure 8.2 Summary diatom stratigraphy (% data) from Callemondah Billabong, southeast Australia. The dashed line represents the first appearance of introduced *Pinus* pollen in the sediment. Note that from 105 cm to this point there is approximately 3000 years of stability in the diatom flora and inferred pH. Values in parentheses are diatom pH optima. (Modified from Tibby *et al.* (2003), with permission from the American Chemical Society.)

cores. For example, Battarbee (1978) assessed the timing of cultural eutrophication in Lough Neagh, a large, shallow lake in Northern Ireland, using multiple cores. He found that, in spite of issues of diatom dissolution and variable sediment accumulation rate across the basin, all cores contained an almost identical diatom sequence with a major compositional shift from *Cyclotella* to *Stephanodiscus* taxa and a decline in benthic species in the early 1900s, recording the start of enrichment. In addition to compositional data, Battarbee (1978) inferred diatom productivity from diatom influx data which suggested an increase in production since the 1950s associated with a second phase of eutrophication. Whilst these early studies were largely qualitative in nature, they nevertheless illustrated the power of diatom records for assessing the timing and extent of nutrient enrichment.

In the 1990s, given the advances in statistical approaches for reconstructing environmental change made during the lake-acidification work (Battarbee *et al.*, this volume), the potential existed for quantifying the relationship between diatoms and nutrients in a similar way to that for pH. Several training sets were developed specifically for shallow lakes to determine the relationship between diatom assemblages and nutrient concentrations (typically TP) (e.g. Anderson *et al.*, 1993; Brenner *et al.*, 1993; Anderson & Odgaard, 1994; Bennion, 1994; Bennion *et al.*, 1996; Bradshaw *et al.*, 2002). The studies were conducted on relatively small lakes in lowland regions of Europe and the USA but more recently similar data sets have been generated for shallow lakes in Canada (Werner & Smol, 2005) and much larger (>1 km^2 surface area) shallow lakes in China (Yang *et al.*, 2008). Whilst the performance of the resulting transfer functions is generally poorer than those for pH, the models enable TP to be inferred from fossil diatom assemblages with a reasonable degree of accuracy (diatom-inferred TP (DI-TP) versus measured TP correlation coefficients of $r^2 > 0.60$). These models have been subsequently applied to fossil diatom assemblages in shallow-lake cores to track changes in trophic status (e.g. Anderson *et al.*, 1993; Anderson

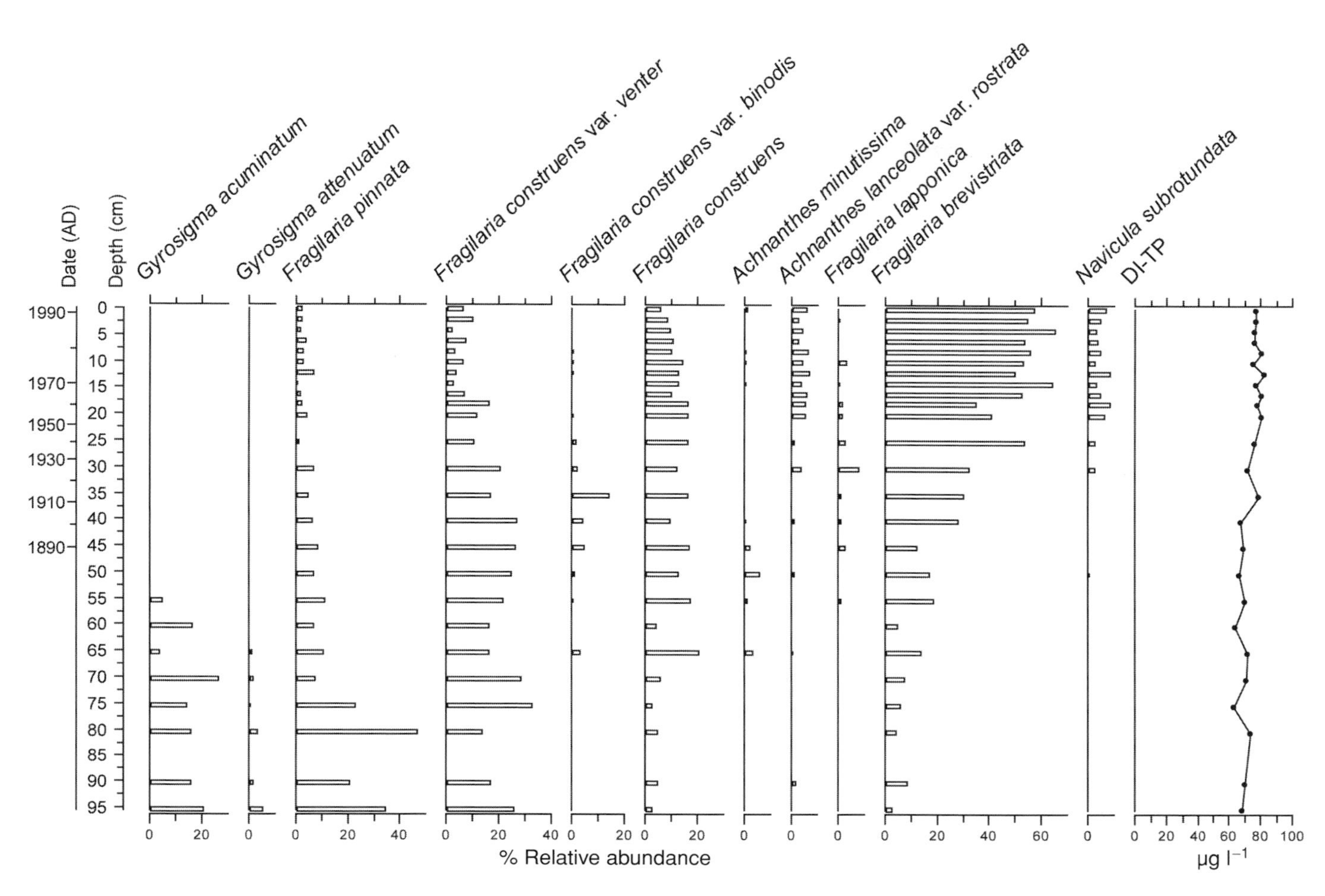

Figure 8.3 Diatom stratigraphy of Eleven Acre Lake, England, showing shifts in the common non-planktonic taxa (% relative abundance) and relatively little change in DI-TP (μg l⁻¹) (Bennion, 1993).

& Odgaard, 1994; Bennion *et al.*, 1996; Bennion & Appleby, 1999; Brenner *et al.*, 1999; Dong *et al.*, 2008). Most of these studies focus on the last few hundred years, demonstrating recent eutrophication associated with increasing populations and agricultural expansion and intensification in lake catchments. However, several studies have reconstructed nutrient history over thousands of years, most notably Bradshaw *et al.* (2005a, b) at Lake Dallund Sø, Denmark. This study provided evidence of early enrichment during the forest clearances of the Late Bronze Age (1000–500 BC) and the expansion of arable agriculture during the Iron Age period (500 BC–AD 1050), and more marked eutrophication in the Medieval period (AD 1050–1536) as a result of further intensification of agriculture and the use of the lake for hemp and flax retting as a town developed in the locality.

Unfortunately, application of diatom TP transfer functions in shallow lakes has proved more problematic than in deeper, stratifying systems. One difficulty arises because of the dominance of non-planktonic taxa (as high as 90% of the assemblage), most commonly *Staurosira*, *Pseudostaurosira* and *Staurosirella* spp. (formerly classified in the genus *Fragilaria* spp. (Williams & Round, 1987)), in the fossil assemblages of many shallow lakes (e.g. Anderson *et al.*, 1993; Bennion, 1995; Brenner *et al.*, 1999; Bennion *et al.*, 2001; Sayer, 2001; Werner & Smol, 2005). The balance of planktonic and non-planktonic taxa in shallow-lake sediments depends on habitat availability and structure (e.g. presence of plants, illuminated surface sediments, open water), which in turn are related to nutrients; hence diatom–nutrient relationships are often indirect. Furthermore, in many of the training sets, the *Fragilaria* spp. are distributed along the whole length of the TP gradient and their TP optima essentially lie in the centre of that gradient. This wide tolerance to nutrient concentrations makes them poor indicators of lake trophic status as highlighted previously. These problems were illustrated by Bennion (1993) at Eleven Acre Lake (Figure 8.3) and Bennion *et al.* (2001) in a set of shallow lakes in eastern England where the relatively small changes in inferred TP produced by the diatom transfer function did not mirror the pattern of progressive enrichment suggested by the diatom species shifts or as reported in the literature. In a paleolimnological study of another English shallow lake, Sayer (2001) showed a clear discrepancy between DI-TP estimates and evidence for eutrophication based on historic aquatic plant and pollen records.

In spite of the problems associated with benthic taxa, Bradshaw *et al.* (2002), in a comparison of a TP inference model for shallow Danish lakes based on all taxa with one based only on planktonic forms, concluded that stronger models were generated using the maximum number of species and including diatoms from all habitats. Similarly, Philibert and Prairie (2002) found that inclusion of both benthic and planktonic species in a set of Canadian lakes resulted in the transfer function with the most predictive power. Indeed, the model based on only benthic taxa out-performed one based solely on planktonic species, which the authors attributed to the richer and more diverse diatom flora of the benthos. In a training set of polymictic, shallow Canadian lakes, Werner & Smol (2005) observed significant changes in the benthic diatom assemblages with changing epilimnetic TP concentrations. They suggested that this was most likely due to the high benthic–pelagic coupling in their study lakes through low macrophyte abundances. It could be argued, however, that regardless of the strength of the benthic–pelagic coupling, or the exact mechanism by which benthic algae respond to epilimnetic chemistry, the co-occurrence of the benthic and planktonic habitats implies that there is some correspondence in their chemical preferences (e.g. Blumenshine *et al.*, 1997; Philibert & Prairie, 2002).

In several of the shallow-lake training sets, the diatom–TP models underestimate measured TP at the upper end of the gradient (e.g. Anderson *et al.*, 1993; Bennion, 1994). This has been attributed to the uniformity of diatom assemblages in hypertrophic waters (e.g. $>300\ \mu g\ TP\ l^{-1}$) due to the limited number of diatom species physiologically adapted to high P and low silica to phosphorus (Si:P) ratios (e.g. *Stephanodiscus*) (Anderson *et al.*, 1993; Anderson, 1997). Algal communities at such high nutrient concentrations are typically dominated by green algae (Chlorophyta) and blue-green algae (Cyanobacteria). Another issue with diatom training sets in shallow lakes concerns ability to capture the variability in lake-water chemistry. In contrast to the winter maxima commonly observed in deep, stratifying lakes, shallow isothermal waters usually experience highly fluctuating nutrient concentrations throughout the year, with typical summer peaks due to P release from the sediments (e.g. Jeppesen *et al.*, 1997; Scheffer, 1998). Such marked intra-annual variability tends to be greatest at the more productive sites and implies that more samples are needed to define accurately the maximum or mean (e.g. Gibson *et al.*, 1996; Bennion & Smith, 2000). Furthermore, annual mean TP concentrations may not necessarily reflect the in-lake concentrations at the time when the diatoms are growing. In many of the plankton-dominated, shallow Danish lakes, diatom biovolume is at a maximum when TP concentration is at the lowest levels of the year (Bradshaw *et al.*, 2002).

In summary, the problems encountered during transfer-function development arise largely because of the greater complexity of eutrophication impacts on diatom communities in shallow lakes than in their deeper counterparts. Consequently, in recent years there has been greater emphasis on deriving ecology-based targets in addition to chemical targets (e.g. Sayer & Roberts, 2001). It is in understanding shifts in ecological structure and function of shallow-lake ecosystems that diatoms can perhaps play one of their most important roles. Several paleoecological studies have demonstrated the marked habitat shift from benthic and epiphytic to planktonic dominance in diatom records of shallow lakes in response to eutrophication (e.g. Brenner *et al.*, 1999; Sayer *et al.*, 1999; Schelske *et al.*, 1999; Sayer *et al.*, 2010a) providing a powerful approach for tracking benthic–pelagic shifts in primary production (Figure 8.4). Further, in a study of several shallow lakes in England, Sayer & Roberts (2001) were also able to produce information on changes in lake functioning by classifying the fossil diatom assemblages according to their seasonal preferences, as established from contemporary plankton sampling. In the pre-enrichment phase, samples were dominated by planktonic species whose growth was restricted to a single late-winter–spring pulse. However, with enrichment there was a shift towards a balance between late-winter–spring- and late-summer–autumn-adapted species indicating an "opening up" of the summer season and providing evidence for a transition from submerged macrophyte to phytoplankton dominance. When this kind of information from diatoms is coupled with evidence of ecological shifts based on other biological groups preserved in the sediment record, a more complete picture of changes in ecosystem structure and function is possible, and the problems associated with the traditional single-indicator approach can be overcome (e.g. Brooks *et al.*, 2001). The value of the multiproxy approach is discussed later.

In recent years there has been growing interest in how climate change, particularly increased temperatures, may confound recovery from eutrophication. The extent to which climate change may impact shallow lakes in temperate regions is still poorly understood. Jeppesen *et al.* (2007) suggest that

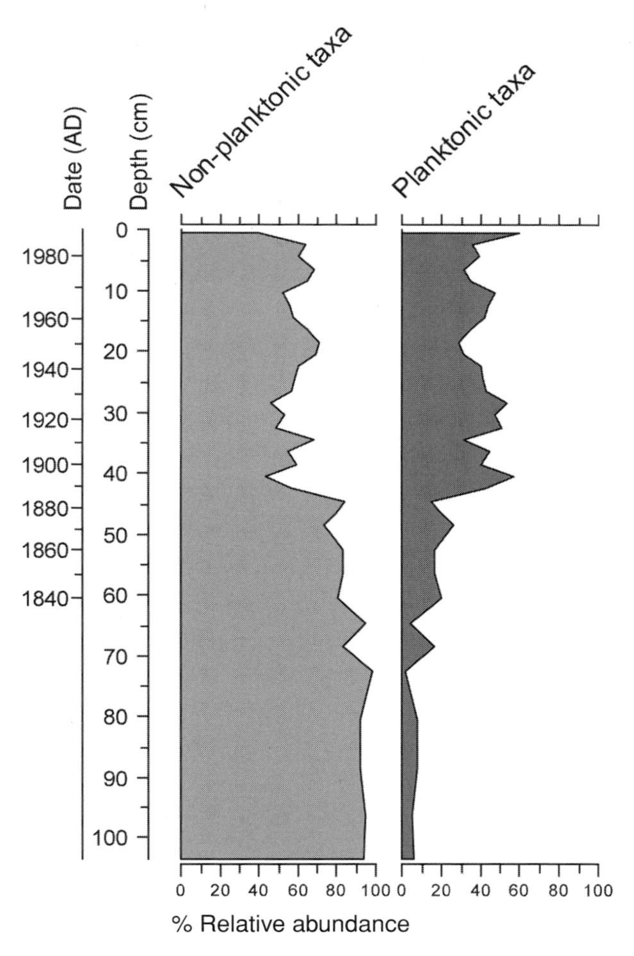

Figure 8.4 The relative abundance of non-planktonic and planktonic diatoms in a core from Groby Pool, England. (Modified from Sayer *et al.*, 1999.)

the most likely outcome, at least in Europe, is that shallow lakes will become more turbid and this is supported by studies of climate–nutrient interactions which have shown that warming generally exacerbates symptoms of eutrophication in freshwaters (Jeppesen *et al.*, in press). Changes in fossil diatom assemblages in the last decade in sediment cores from several lakes have been attributed to climate-driven processes involving dynamics of lake-water temperature and nutrients. However, these paleolimnological studies have also demonstrated the difficulty of uncoupling the effects of climate and nutrients on the diatom communities and other biota (Jeppesen *et al.*, in press).

8.4.3 Impacts of land-use change

Land-use change has a fundamental impact on shallow-lake ecology. This is particularly the case where intensive agriculture is rapidly imposed on a landscape in which it was previously absent, as in Australia (Tibby, 2003). Hence, many of our case studies are drawn from this continent. Removal of terrestrial vegetation has a range of indirect effects, of which mobilization of sediment in response to reductions in interception of rain-splash erosion is, perhaps, the most immediate. Indeed, changes in core-sediment composition may be registered before those associated with pollen as an indicator of vegetation change (Leahy *et al.*, 2005). In southeast Australia, the importance of these changes to shallow lakes on the floodplains of the Yarra (Leahy *et al.*, 2005; Gell *et al.*, 2005a; Leahy, unpublished), Goulburn (Reid, 2002) and Murray (Gell *et al.*, 2005a, b) rivers is highlighted by the observation that the most substantial changes in diatom records covering >800 years occur with the first evidence of catchment clearance. In the Yarra catchment, Leahy *et al.* (2005) demonstrate that the diatom community established in Bolin Billabong in the 1840s, following initial clearance, is essentially maintained through to the present day, despite substantial changes in management regime that include considerable upstream diversions for water supply and increased application of artificial fertilizer in the second half of the twentieth century. However, in contrast to Bradshaw *et al.* (2005a, b) (see above) dramatic increases in nutrient concentrations are not associated with vegetation clearance in the diatom stratigraphies from Yarra River floodplains (Leahy *et al.*, 2005; Gell *et al.*, 2005b).

On the Murray River floodplain, Reid *et al.* (2007) utilize diatoms along with other indicators to assess explicitly the causes of submerged plant loss in these shallow lakes (previously inferred from Cladocera remains by Ogden, 2000). The perspective provided by diatoms is useful here since there is little direct evidence from seeds or leaves for aquatic plant abundance in the sites. Reid *et al.* (2007) were able to demonstrate, using the absolute abundance of epiphytic diatoms, that plant loss in the Upper Murray River occurs soon after European settlement and well before the onset of other possible causes such as major irrigation abstraction or fish introduction.

Vegetation clearance has profound effects on water quality beyond that associated with increases in turbidity. Reid (2002) and Tibby *et al.* (2003) have demonstrated that, resulting from landscape clearance following settlement of the Goulburn River catchment (southeast Australia), there have been dramatic increases in pH in floodplain shallow lakes. A diatom-based model for inferring pH has shown that pH has increased by more than 0.5 units since European settlement (Figure 8.2). This change occurs due to increased loading of eroded base cations from the catchment, decreased delivery of acidic organic matter, a localized increase in discharge of alkaline groundwater, or a combination of these factors. Increased pH has been observed in a number of other southeast Australian shallow lakes (Gell *et al.*, 2005b). However, interestingly, a number of shallow lakes in the same region are threatened by decreases in pH as a result of the drainage of acid sulfate soils. These soils result from the exposure of organic matter deposited under anaerobic conditions created by river regulation that has until recently maintained water levels in shallow lakes at a constant height. Although quantitative pH reconstructions have not yet been published from any such shallow lakes, increases in indicator taxa are providing insights into the timing of these shifts. As an example, *Haslea spicula* (Hickie) Bukht., an indicator of both saline and acid waters, has experienced a dramatic increase in Loveday Wetland, located in the South Australian River in an irrigation district, since the 1960s (Gell *et al.*, 2007). In temperate regions, diatom records have been used to track declines in pH resulting from atmospheric deposition and subsequent recovery from acidification (Battarbee *et al.*, this volume) with notable shallow-lake examples being moorland pools in Europe (van Dam, 1996). Isolated pools in these acid-sensitive regions have been strongly acidified, with associated decreases in diatom diversity.

Increased lake salinity threatens shallow lakes in a wide variety of localities. This threat results from climate change and sea-level rise, and also from vegetation clearance and irrigation practices (both the application of excess water to naturally saline soils and the diversion of fresh irrigation water). Human-induced landscape salinization has been argued to be the most important environmental management issue in Australia, with over two million hectares affected and with more than five million hectares under threat. Salinization is predominantly caused by the removal of deep-rooted trees, which leads to an elevation of the water table, and eventual transport of naturally occurring salts to the surface. The transition from vegetation clearance to salinization of the landscape (including shallow lakes) may take more than a century. As a result, diatom-based studies that define pre-impact shallow-lake salinities, and the degree and timing of salinity increases, are invaluable. A diatom model, originally designed to document climate change (Gell, 1997), has been used to define the extent of salinization, although a specific shallow-lake model is now available (Tibby *et al.*, 2007). These reconstructions have focused on shallow lakes in the Murray–Darling Basin where, although there is substantial pre-settlement variability in some sites (e.g., Gell *et al.*, 2007), a number of lakes are now more saline than at any time in the past.

Outside the confines of the Murray–Darling Basin, where problems of salinization are particularly acute, Barry *et al.* (2005) demonstrated the combined effects of land use and climate change on the salinity of a shallow (<1.5 m) crater lake, Tower Hill, Western Victoria. From 1990 to 2000 diatom-inferred salinity increased from between 1–2 g l^{-1} to *c.* 13 g l^{-1} (Figure 8.5). This increase is associated with a rainfall deficit that commenced *c.* 1990. Interestingly, comparison of this short-term record with a long diatom record that extended to the late glacial suggests that the site has never been more saline than at present (D'Costa *et al.*, 1989). This evidence points not to the most extreme climates during this time, but to an amplification of the climate signal through vegetation clearance and the raising of the lake's outlet level contributing to the retention of more salt. By contrast, Taffs (2001) has demonstrated that, in some locations in the southeast of South Australia, increases in runoff as a result of vegetation clearance and redirection of drainage have resulted in decreases in diatom-inferred salinity, with the latter also demonstrated in the broader region by Haynes *et al.* (2007) and further afield in a number of North African lakes (Flower *et al.*, 2001). These studies highlight that in naturally saline shallow lakes, decreased salinity may represent a threat equal to that posed by increased salinity in low-salinity systems.

8.4.4 Ecological reference conditions and management targets

Paleolimnological methods for defining baselines and restoration targets are well established (Battarbee, 1999; Smol, 2008). As mentioned previously, the arrival of the EU WFD in 2000 has created a need to assess the ecological status of European water bodies based on a comparison of the present state with a pre-disturbance reference condition. Paleoecology along with expert judgement, spatial-state schemes, historical data, and modelling are stated as potential methods for deriving reference conditions (Pollard & Huxham, 1998; European Union, 2000). Of the biological elements relevant to the WFD, diatoms represent components of both phytoplankton and phytobenthos. The WFD has, therefore, provided the impetus for new paleoecological research which specifically aims to establish reference conditions and assess ecological status of lakes (see the review in Bennion & Battarbee, 2007). Analyses of diatom remains in sediment records, combined with transfer functions, have proved to be powerful techniques for defining ecological and chemical reference conditions, and assessing deviation from the reference state. For shallow, lowland lakes the sediment record often affords the only available means of determining the composition of past communities as there are few examples of unimpacted sites in the current population to act as

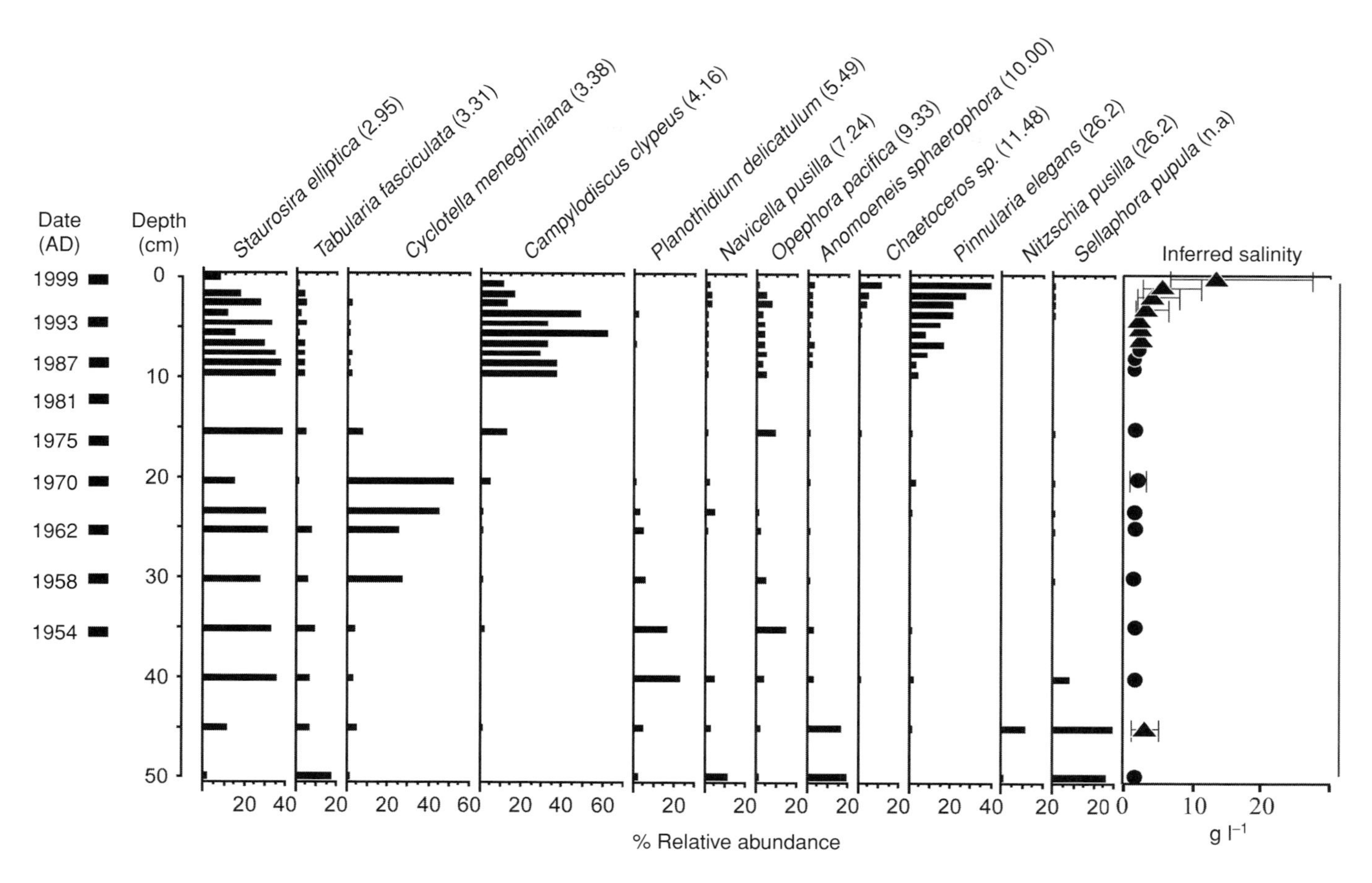

Figure 8.5 Summary diatom stratigraphy from Tower Hill Main Lake, Australia. Values in parentheses are diatom salinity optima. (Modified from Barry *et al.* (2005), with permission from E. Schweizerbart'sche Verlagsbuchhandlung OHG (Naegele u. Obermiller) Science http://www.schweizerbart.de.)

reference lakes (Bennion *et al.*, 2003). This arises because many European shallow lowland lakes have been impacted over long timescales (e.g. Bradshaw *et al.*, 2006; Leira *et al.*, 2006) and particularly during the twentieth century (e.g. Bennion *et al.*, 2004).

A number of recent diatom studies have illustrated the value of the paleolimnological record for defining chemical and ecological reference conditions for shallow lakes in the context of the WFD. For example, diatom–TP transfer functions have been applied to sediment cores to estimate TP reference conditions in shallow lakes in the UK (Bennion *et al.*, 2004; Bennion & Simpson, in press), Finland (Miettinen *et al.*, 2005; Räsänen *et al.*, 2006), Ireland (Leira *et al.*, 2006), and Denmark (Bjerring *et al.*, 2008). Most of these studies have used the so-called "top–bottom approach" whereby the core bottom is taken to represent the reference condition and this is compared with the core top which represents present-day conditions. In addition to determining the nutrient reference condition, these studies have also defined diatom assemblages characteristic of reference conditions in shallow waters. For example, Bennion *et al.* (2004) described the diatom assemblages in reference samples of a set of shallow Scottish lochs as being largely comprised of a non-planktonic community (e.g. *Fragilaria*, *Cymbella*, *Cocconeis*, *Achnanthes*, *Navicula* spp.). A simple measure of floristic change such as the squared chord distance (SCD) coefficient can then be employed to assess how far the diatom community has deviated from this reference assemblage and hence to determine which lakes are still in reference condition and which have been impacted and to what degree. An example is given in Figure 8.6 for a set of 29 high-alkalinity, shallow lakes in the UK, illustrating that only ten sites have experienced low floristic change. This and other EU WFD diatom studies have shown that the majority of shallow lakes have undergone significant shifts in their diatom communities and cannot, therefore, be described as "good status." Indeed, many of the studied sites were already nutrient rich in ~1850 AD, a date often taken to represent pre-impact conditions (Leira *et al.*, 2006; Räsänen *et al.*, 2006; Bjerring *et al.*, 2008). Paleoecological studies indicate that true baselines for some lakes may

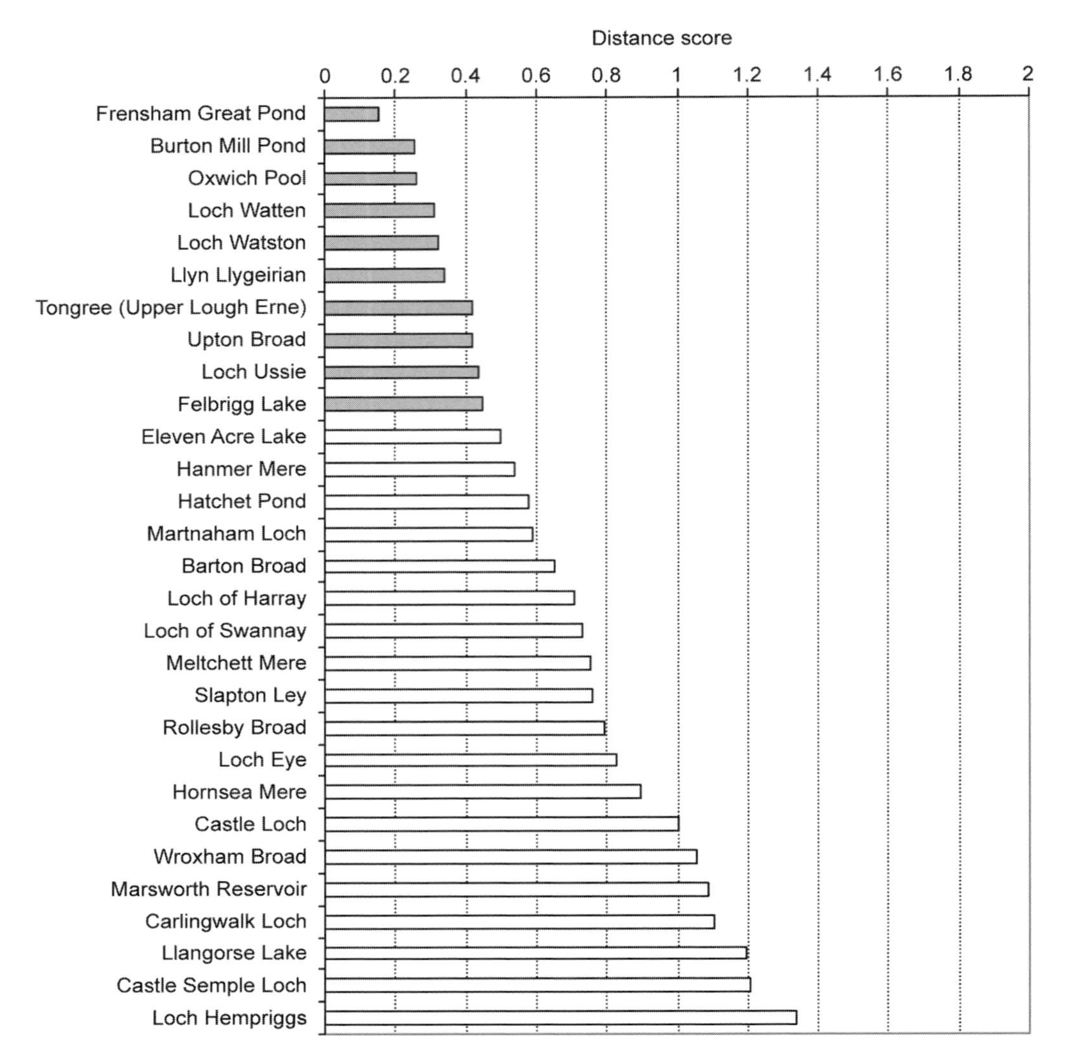

Figure 8.6 Histogram of the squared chord distance dissimilarity scores (distance score) between core bottoms and tops for a set of high-alkalinity, shallow lakes in the UK. Sites where the scores are below the critical value at the fifth percentile (<0.475) are shown as shaded bars, indicating low floristic change (Bennion *et al.*, unpublished).

date back many hundreds or thousands of years (e.g. Bradshaw *et al.*, 2006), and thereby suggest that even the most stringent management measures are unlikely to bring about a return to such conditions. With the recent implementation of programs of measures in an attempt to restore lakes to good status, there comes a great opportunity for paleoecological records to track the degree of ecological recovery and assess the extent to which changes are reversible.

Beyond Europe, paleoecological information has proved to be a powerful complement to reference sites for estimating potential ecological integrity, most notably as part of the United States Environmental Protection Agency (USEPA) Environmental Monitoring and Assessment Program for Surface Waters (EMAP-SW) (Dixit *et al.*, 1999; Hughes *et al.*, 2000). Diatom inference models were applied to core top and bottom diatom data from 257 lakes distributed in three ecoregions of north-eastern USA to reconstruct lake pH, TP, and chloride concentrations (Dixit *et al.* 1999) and hence determine deviation from the reference state. Elsewhere, diatom-based paleolimnology is providing information on the natural state of degraded Ramsar-listed wetlands (Davis & Brock, 2008) and on the possible trajectories for rehabilitation (e.g. Saunders *et al.*, 2007).

Analog matching techniques can also be used to establish ecological reference conditions from paleo-data by matching fossil assemblages present in sediments deposited in a pre-disturbance period with those assemblages found in modern surface sediments of pristine or less-impacted lakes. These methods have already been successfully applied to acidified lakes using diatoms (Flower *et al.*, 1997; Simpson *et al.*, 2005), and Sayer & Roberts (2001) have illustrated that they also offer great potential for deriving ecological targets for shallow lakes subject to eutrophication. Further work is in progress to extend the method to a larger data set of European lakes.

8.4.5 Multiproxy studies

Shallow, macrophyte-filled lakes support high biological diversity and have complex, highly interconnected food webs. Increasingly it is being realized that, aside from diatom algae, a substantial proportion of this diversity and food-web structure is preserved in sediment cores, with macrophytes (Davidson *et al.*, 2005), zooplankton (Jeppesen *et al.*, 2001), mollusca (Walker *et al.*, 1993; Ayres *et al.*, 2008), chironomids (Brodersen *et al.*, 2001; Langdon *et al.*, 2010), and fish (Davidson *et al.*, 2003; Sayer *et al.*, 2006), among an array of other invertebrate groups, all leaving readily identified remains. In so-called "multiproxy" studies a range of these groups (the proxies) are enumerated for the same site with the aim of determining whole-ecosystem responses to natural and anthropogenic drivers (Birks & Birks, 2006).

Diatoms are frequently incorporated in multiproxy studies of shallow-lake cores. Some have used diatom–TP reconstructions to explain the responses of other biological groups to nutrient enrichment (e.g., Hall *et al.*, 1997; Brodersen *et al.*, 2001; McGowan *et al.*, 2005). For example, in Danish Lake Søbygaard, McGowan *et al.* (2005) used variance partitioning to determine the relative importance of DI-TP, cladoceran-inferred zooplanktivorous fish densities and macrophyte macrofossil abundances as historical determinants of algal pigment composition before, during, and after macrophyte loss. With this approach diatoms have been utilized very much as "proxies" and DI-TP is assumed to be independent of other core data. While attractive, it runs into dangers of circularity because diatom changes (and hence DI-TP) may also be highly influenced by changes in other biological groups, particularly the availability of macrophyte habitat. Other studies have summarized changes in different proxies, including diatoms, using detrended correspondence analysis (DCA) sample scores (Birks *et al.*, 2000; Birks & Birks, 2001; Bradshaw *et al.*, 2005b), or have converted diatom and other proxy profiles into diversity indices (Sayer *et al.*, 1999). With such approaches detailed and complex data are simplified and the timing, magnitude, and coherence of biotic responses can be easily compared.

One of the most innovative multiproxy studies incorporating diatoms to date is that of Bradshaw *et al.* (2005a, b) who tracked the Holocene ecological history of Dallund Sø, a shallow lake in Denmark. In this study canonical correspondence analysis (CCA) was used to relate different "predictor" variables (pollen, macrofossils, *Pediastrum*, and sediment loss-on-ignition data) to diatom community changes (used as an indicator of internal ecosystem "response") covering the last 6000 years (Figure 8.7). Although cause and effect are impossible to discern in such relationships using this approach, it was possible to identify links between "land and lake" and thus likely shifts between internal and catchment drivers of ecosystem change. In many ways the future of diatom analysis in paleolimnology may well lie in multiproxy studies and it can be expected that such studies will greatly increase in number over the coming decade.

8.5 Future directions

8.5.1 Ecological studies

The high diversity of diatom species found in shallow lakes makes them potentially powerful indicators of past environmental conditions. However, our ability to use them in paleolimnological studies remains limited by a generally poor understanding of diatom ecology and taxonomy. Currently, we know relatively little regarding the niche dimensions and survival strategies of the majority of diatom species, particularly the periphytic taxa which often dominate in shallow lakes. More studies are required of diatom life cycles; while there are a few excellent examples (e.g. Jewson, 1992a, b – *Stephanodiscus* species; Jewson *et al.*, 2006 – *Cymbellonitzschia diluviana* Hust.), these are all too few. A further little-explored area of diatom ecology is the influence of grazing; while we know that zooplankton and invertebrate populations influence diatom densities, we have only limited insights into effects on community structure (e.g. Jones *et al.*, 2000). Additionally, more needs to be known regarding diatom habitats. While we have good ideas on which species are planktonic or attached to surfaces, for many taxa, beyond this basic level, we know relatively little. In particular, species' preferences for plant, sand, and silt substratum is imperfectly understood with the same true of relationships between diatoms and different macrophyte structures. For example, although "tight" plant–diatom relationships are probably rare for diatoms (Eminson & Moss, 1980), sometimes they do exist, with, for example, strong known associations between *Lemnicola hungarica* (Grunow) Round et P.W. Basson (synonym: *Achnanthes hungarica* (Grunow) Grunow in Cleve & Grunow) and free-floating lemnid plants (Goldsborough, 1993), and *Cocconeis pediculus* Ehrenb. and *Cladophora* sp. (Reavie & Smol, 1997). Other species–habitat associations are almost bound to occur and where they exist diatoms can be used as

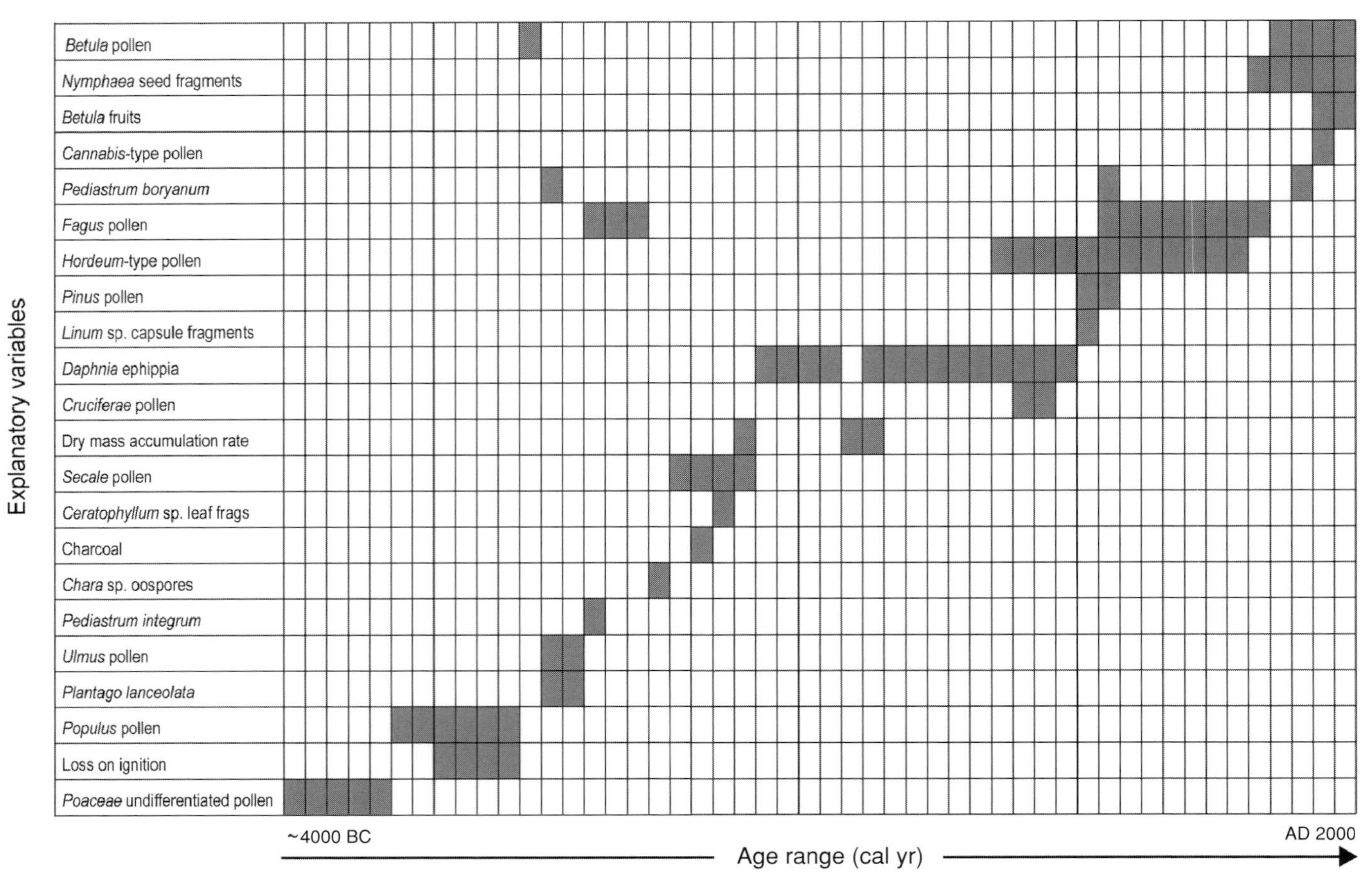

Figure 8.7 Matrix showing the results of canonical correspondence analysis of 50 subsets of samples from the Dallund Sø record. (Modified from Bradshaw *et al.*, 2005b.)

indicators of changing habitat structure. A final key area for future study is diatom seasonality. In the cool temperate zone there is evidence to suggest that some planktonic diatoms variously prefer spring, summer, or autumn periods (Kiss & Padisák, 1990; Sayer & Roberts, 2001; Sayer, 2001; Bradshaw & Anderson, 2003). Despite this, however, we know little regarding the degree to which diatoms are consistently (across multiple lakes) tied to particular seasons. Moreover, for periphytic species, relatively little is known regarding seasonal dynamics, although one or two excellent studies provide some insights (e.g. Liboriussen & Jeppesen, 2003). An enhanced knowledge of both diatom–habitat and diatom–seasonality relationships would much improve the interpretation of sediment-core sequences both for studies of eutrophication and of climate change.

A classic illustration of the need for more studies of diatom ecology to inform paleolimnological interpretation concerns the small chain-forming fragilarioid species (particularly *Fragilaria brevistriata* Grunow in Van Heurck, *F. pinnata* Ehrenb., *F. construens* (Ehrenb.) Grunow, *F. elliptica* Schum., and varieties etc.) which often dominate sediment assemblages in shallow lakes. Some hundreds of paleolimnological studies have variously associated species in this group with clear or turbid water conditions and with epiphytic, benthic, and/or facultative planktonic habitats (e.g. Karst & Smol, 2000; Bennion, 1995; Sayer, 2001; Gell *et al.*, 2005b; Anderson *et al.*, 2008). Actual studies of living populations, are, however, scant to non-existent and so comparatively little is known regarding the factors that may promote *Fragilaria* group prevalence or decline. Additionally, in sediment sequences there are often relatively large shifts between different *Fragilaria* species (see Figure 8.3), but the ecological significance of these shifts are also elusive. Clearly there is an urgent need for carefully constructed ecological studies of these small benthic *Fragilaria* species in shallow lakes.

8.5.2 Beyond diatom-transfer functions

In recent years, by far the dominant use of diatoms in paleolimnological studies of shallow lakes has been to generate nutrient reconstructions using transfer functions. Yet, it can be argued that reducing huge amounts of species composition data to a series of numbers is throwing away much valuable ecological information. An alternative way to view a diatom profile, once armed with good ecological data, would be to pose the following questions: How has the spatial organization (water column *vs.* submerged plants *vs.* sediment surface?) of diatom production changed? And: What is the seasonal patterning of diatom populations in these different habitats? All the changes we see in sedimentary diatom assemblages from shallow lakes will have been through a "seasonal-habitat filter" and this may be a more direct way to interpret them. Then, once changes in the spatial–seasonal distribution of diatom species can be visualized, the question can be posed: Why have things changed? In this latter respect, having data from other fossil groups (particularly cladoceran remains and plant macrofossils) can be extremely helpful. Such an approach moves away from the current fixation with using diatoms as proxies for something else and seeks to get into the black box of ecological processes that transfer functions tend to ignore.

Gaining an understanding of long-term changes in the location and seasonality of algal production would much improve our understanding of shallow-lake responses to nutrient enrichment and would allow for the testing of theories of long-term shifts in lake productivity (e.g. Vadeboncoeur *et al.*, 2003). For example, Sayer *et al.* (2010a) used diatoms and plant macro-remains to infer long-term changes in the seasonal patterning of algae and macrophyte production in Felbrigg Hall Lake, England. Over the last century a substantial reduction in the seasonal duration of plant dominance was inferred. In turn this was accompanied by an increase in planktonic diatom production either side of a shortening macrophyte-covered period (the development of a so-called "plant sandwich"). Despite dynamic changes inferred from diatom and other fossil data for the site, due to the overwhelming prevalence of small *Fragilaria* taxa in the core, a diatom–TP transfer function suggested minimal change in nutrient concentrations over this same period (Sayer *et al.*, 2010b). This study shows the merits of a more ecological approach to environmental reconstruction and the need to think beyond transfer functions.

8.6 Summary

Shallow lakes support a diverse range of diatom communities. The importance of littoral diatoms in shallow waters makes them valuable biological indicators, and modern assemblages have been widely used to assess changes in trophic status based on simple indices. Shallow-lake paleolimnology has grown as a discipline in the last decade and, whilst factors such as resuspension, sediment mixing, and diatom dissolution present difficulties, diatom records have been successfully employed at numerous shallow lakes to assess environmental change with respect to lake development, erosion, alkalinization, acidification, salinization, climate change, and particularly eutrophication. Since the early 1990s, the principal use of diatoms in paleolimnological studies of shallow lakes has been to generate nutrient reconstructions using transfer functions, a process which in benthic-dominated systems has proved somewhat problematic. However, in the last few years, there has been renewed interest in the role of diatoms in understanding shifts in ecological structure and particularly in their use as one component of multiproxy studies which seek to determine whole-ecosystem responses to environmental change. This has been partly fueled by the advent of new legislation such as the EU WFD which places the emphasis on ecological status of waters. In this context, diatom paleostudies are already proving valuable for assessing ecological reference conditions, and have been particularly important in shallow, lowland lakes where there are few reference sites in the current population. There is great potential to extract yet more ecological information from the diatom record to improve our understanding of shallow-lake responses to perturbations but presently this is hampered by limited knowledge of diatom ecology, particularly for the periphytic taxa. Once available, such ecological information can be used to enhance our interpretation of paleoecological records, thereby gaining greater insights into ecosystem response to a range of pressures and providing valuable information to managers of these vulnerable systems.

Acknowledgments

We would like to thank Emily Bradshaw for allowing use of her Dallund Sø data and Chris Carter for kindly providing his photograph of epiphyton. We are also grateful to Luc Denys and the editors for their valuable comments on the chapter.

References

Ács, É., Reskóné, N. M., Szabó, K., Taba, Gy., & Kiss, K. T. (2005). Application of epiphytic diatoms in water quality monitoring of Lake Velence – recommendations and assignments. *Acta Botanica Hungarica*, **47**, 211–23.

Allen, H. L. (1971). Primary productivity, chemoorganotrophy, and nutritional interactions of epiphytic algae and bacteria on macrophytes in the littoral of a lake. *Ecological Monographs*, **4**, 97–127.

Anderson, N. J. (1993). Natural versus anthropogenic change in lakes: the role of the sediment record. *Trends in Ecology and Evolution*, **8**, 356–61.

Anderson, N. J. (1997). Reconstructing historical phosphorus concentrations in rural lakes using diatom models. In *Phosphorus Loss from Soil to Water*, ed. H. Tunney, O. T. Carton, P. C. Brookes, & A. E. Johnston, Oxford, UK: CAB International, pp. 95–118.

Anderson, N. J. & Odgaard, B.V. (1994). Recent palaeolimnology of three shallow Danish lakes. *Hydrobiologia*, **275/276**, 411–22.

Anderson, N. J., Rippey, B., & Gibson, C. E. (1993). A comparison of sedimentary and diatom-inferred phosphorous profiles: implications for defining pre-disturbance nutrient conditions. *Hydrobiologia*, **253**, 357–366.

Anderson, N. J., Brodersen, K. P., Ryves, D. B., *et al.* (2008). Climate versus in-lake processes as controls on the development of community structure in a low-Arctic Lake (south-west Greenland). *Ecosystems*, **11**, 307–24.

Ayres, K., Sayer, C. D., Perrow, M., & Skeate, E. (2008). Palaeolimnology as a tool to inform shallow lake management: an example from Upton Great Broad, Norfolk, UK. *Biodiversity and Conservation*, **17**, 2153–68.

Barker, P., Fontes, J., Gasse, F., & Druart, J. (1994). Experimental dissolution of diatom silica in concentrated salt solutions and implications for paleoenvironmental reconstruction. *Limnology and Oceanography*, **39**, 99–110.

Barry, M. J., Tibby, J., Tsitsilas, A., *et al.* (2005). A long term lake salinity record and its relationship to *Daphnia* populations. *Archiv für Hydrobiologie*, **163**, 1–23.

Battarbee, R. W. (1978). Observations on the recent history of Lough Neagh and its drainage basin. *Philosophical Transactions of the Royal Society, London*, **B 281**, 303–45.

Battarbee, R. W. (1986). The eutrophication of Lough Erne inferred from the changes in the diatom assemblages of ^{210}Pb- and ^{137}Cs-dated sediment cores. *Proceedings of the Royal Irish Academy*, **B 86**, 141–68.

Battarbee, R. W. (1999). The importance of palaeolimnology to lake restoration. *Hydrobiologia*, **395/396**, 149–59.

Beaver, J. R., Crisman, T. L., & Bays, J. S. (1981). Thermal regimes of Florida lakes. *Hydrobiologia*, **83**, 267–73.

Bennion, H. (1993). A diatom–phosphorus transfer function for eutrophic ponds in south-east England. Unpublished Ph.D. thesis, University of London.

Bennion, H. (1994). A diatom–phosphorus transfer function for shallow, eutrophic ponds in southeast England. *Hydrobiologia*, **275/276**, 391–410.

Bennion, H. (1995). Surface-sediment diatom assemblages in shallow, artificial, enriched ponds, and implications for reconstructing trophic status. *Diatom Research*, **10**, 1–19.

Bennion, H. & Appleby, P. G. (1999). An assessment of recent environmental change in Llangorse Lake using palaeolimnology. *Aquatic Conservation: Marine and Freshwater Ecosystems*, **9**, 361–75.

Bennion, H. & Battarbee, R. (2007). The European Union Water Framework Directive: opportunities for palaeolimnology. *Journal of Paleolimnology*, **38**, 285–95.

Bennion, H. & Simpson G. L. (in press). The use of diatom records to establish reference conditions for UK lakes subject to eutrophication. *Journal of Paleolimnology*.

Bennion, H. & Smith, M. A. (2000). Variability in the water chemistry of ponds in south-east England, with special reference to the seasonality of nutrients and implications for modelling trophic status. *Hydrobiologia*, **436**, 145–58.

Bennion, H., Appleby, P. G., & Phillips, G. L. (2001). Reconstructing nutrient histories in the Norfolk Broads: implications for the application of diatom–phosphorus transfer functions to shallow lake management. *Journal of Paleolimnology*, **26**, 181–204.

Bennion, H., Fluin, J., & Simpson, G. L. (2004). Assessing eutrophication and reference conditions for Scottish freshwater lochs using subfossil diatoms. *Journal of Applied Ecology*, **41**, 124–38.

Bennion, H., Juggins, S., & Anderson, N. J. (1996). Predicting epilimnetic phosphorus concentrations using an improved diatom-based transfer function and its application to lake eutrophication management. *Environmental Science and Technology*, **30**, 2004–7.

Bennion, H., Simpson, G. L., Hughes, M., Phillips, G., & Fozzard, I. (2003). The role of palaeolimnology in identifying reference conditions and assessing ecological status of lakes. In *How to Assess and Monitor Ecological Quality in Freshwaters*, Tema Nord 2003:547, ed. M. Ruoppa, P. Heinonen, A. Pilke, S. Rekolainen, H. Toivo, & H. Vuoristo, Copenhagen: Nordic Council of Ministers, pp. 57–63.

Beyens, L. & Denys, L. (1982). Problems in diatom analysis of deposits: allochthonous valves and fragmentation. *Geol Mijnbouw*, **61**, 159–62.

Birks, H. H. & Birks, H. J. B. (2001). Recent ecosystem dynamics in nine North African lakes in the CASSARINA Project. *Aquatic Ecology*, **35**, 461–78.

Birks, H. H. & Birks, H. J. B. (2006). Multi-proxy studies in palaeolimnology. *Vegetation History and Archaeobotany*, **15**, 235–51.

Birks, H. H., Battarbee, R. W., & Birks, H. J. B. (2000). The development of the aquatic ecosystem of Kråkenes Lake, western Norway, during the late-glacial and early Holocene – a synthesis. *Journal of Paleolimnology*, **23**, 91–114.

Bjerring, R., Bradshaw, E. G., Amsinck, S. L., *et al.* (2008). Inferring recent changes in the ecological state of 21 Danish candidate reference lakes (EU Water Framework Directive) using palaeolimnology. *Journal of Applied Ecology*, **45**, 1566–75.

Blanco, S., Ector, L., & Bécares E. (2004). Epiphytic diatoms as water quality indicators in Spanish shallow lakes. *Vie Milieu*, **54**, 71–9.

Blumenshine, S. C., Vadeboncoeur, Y., Lodge, D. M., Cottingham, K. L., & Knight, S. E. (1997). Benthic–pelagic links: responses of benthos to water-column nutrient enrichment. *Journal of the North American Benthological Society*, **16**, 466–79.

Bradshaw, E. G. & Anderson, N. J. (2003). Environmental factors that control the abundance of *Cyclostephanos dubius* (Bacillariophyceae) in Danish lakes, from seasonal to century scale. *European Journal of Phycology*, **38**, 265–76.

Bradshaw, E. G., Anderson, N. J., Jensen, J. P., & Jeppesen, E. (2002). Phosphorus dynamics in Danish lakes and the implications for diatom ecology and palaeoecology. *Freshwater Biology*, **47**, 1963–1975.

Bradshaw, E. G., Nielsen, A. B., & Anderson, N. J. (2006). Using diatoms to assess the impacts of prehistoric, pre-industrial and modern land-use on Danish lakes. *Regional Environmental Change*, **6**, 17–24.

Bradshaw, E. G., Rasmussen, P., Nielsen, H., & Anderson N. J. (2005a). Mid- to late-Holocene land-use change and lake development at Dallund Sø, Denmark: trends in lake primary production as reflected by algal and macrophyte remains. *The Holocene*, **15**, 1130–42.

Bradshaw, E. G., Rasmussen, P., & Odgaard, B. V. (2005b). Mid- to late-Holocene change and lake development at Dallund Sø, Denmark: synthesis of multiproxy data, linking land and lake. *The Holocene*, **15**, 1152–62.

Brenner, M., Whitmore, T. J., Flannery, M. S., & Binford, M. W. (1993). Paleolimnological methods for defining target conditions in lake restoration: Florida case studies. *Lake and Reservoir Management*, **7**, 209–17.

Brenner, M., Whitmore, T. J., Lasi, M. A., Cable, J. E., & Cable, P. H. (1999). A multiproxy trophic state reconstruction for shallow Orange Lake, Florida, USA: possible influence of macrophytes on limnetic nutrient concentrations. *Journal of Paleolimnology*, **21**, 215–33.

Brodersen, K. P., Odgaard, B. V., Vestergaard, O., & Anderson N. J. (2001). Chironomid stratigraphy in the shallow and eutrophic lake Søbygaard, Denmark: chironomid–macrophyte co-occurrence. *Freshwater Biology*, **46**, 253–67.

Brooks, S. J., Bennion, H., & Birks, H. J. B. (2001). Tracing lake trophic history with a chironomid–total phosphorus inference model. *Freshwater Biology*, **46**, 513–33.

Burkholder, J. M. (1996). Interactions of benthic algae with their substrata. In *Algal Ecology*, ed. R. J. Stevenson, R. L. Lowe, & M. Bothwell. New York, NY: Academic Press, pp. 253–297.

Burkholder, J. M., & Wetzel, R. G. (1989). Microbial colonization on natural and artificial macrophytes in a phosphorus-limited, hardwater lake. *Journal of Phycology*, **25**, 55–65.

Burkholder, J. M., Wetzel, R. G., & Klomparens, K. L. (1990). Direct comparison of phosphate uptake by adnate and loosely attached microalgae within an intact biofilm matrix. *Applied Environmental Microbiology*, **56**, 2882–90.

Carrick, H. J., Aldridge, F. J., & Schelske, C. L. (1993). Wind influences phytoplankton biomass and composition in a shallow, productive lake. *Limnology and Oceanography*, **38**, 1179–92.

Carrick, H. J., Moon, J., & Gaylord B. (2005). Phytoplankton dynamics and hypoxia in Lake Erie: a hypothesis concerning benthic–pelagic coupling in the Central Basin. *Journal of Great Lakes Research*, **31**, 111–24

Carrick, H. J., Worth, D., & Marshall, M. L. (1994). The influence of water circulation on chlorophyll–turbidity relationships in Lake Okeechobee as determined by remote-sensing. *Journal of Plankton Research*, **16**, 1117–35.

Cattaneo, A. (1983). Grazing on epiphytes. *Limnology and Oceanography*, **28**, 124–32.

Cattaneo, A. (1987). Periphyton in lakes of different trophy. *Canadian Journal of Fisheries and Aquatic Sciences*, **44**, 296–303.

Cattaneo, A., & Kalff, J. (1986). The effect of grazer size manipulation on periphyton communities. *Oecologia*, **69**, 612–617.

Cohen, A. S. (2003). *Paleolimnology: the History and Evolution of Lake Systems*. New York, NY: Oxford University Press.

Davidson, T., Sayer, C., Bennion, H., *et al.* (2005). A 250 year comparison of historical, macrofossil and pollen records of aquatic plants in a shallow lake. *Freshwater Biology*, **50**, 1671–86.

Davidson, T. A., Sayer, C. D., Perrow, M. R., & Tomlinson, M. L. (2003). Representation of fish communities by scale sub-fossils in shallow lakes: implications for inferring cyprinid–percid shifts. *Journal of Paleolimnology*, **30**, 441–9.

Davis, J. & Brock, M. (2008). Detecting unacceptable change in the ecological character of Ramsar wetlands. *Ecological Management & Restoration*, **9**, 26–32.

D'Costa, D. M., Edney, P., Kershaw, A. P., & DeDeckker, P. (1989). Late Quaternary palaeoecology of Tower Hill, Victoria, Australia. *Journal of Biogeography*, **16**, 461–82.

DeNicola, D. M., de Eyto, E., Wemaere, A., & Irvine, K. (2004). Using epilithic algal communities to assess trophic status in Irish lakes. *Journal of Phycology*, **40**, 481–95.

Denys, L. (2004). Relation of abundance-weighted averages of diatom indicator values to measured environmental conditions in standing freshwaters. *Ecological Indicators*, **4**, 255–75.

Denys, L. (2006). Calibration of littoral diatoms to water chemistry in standing fresh waters (Flanders, lower Belgium): inference models for historical sediment assemblages. *Journal of Paleolimnology*, **35**, 763–87.

Denys, L. (2007). Water-chemistry transfer functions for epiphytic diatoms in standing freshwaters and a comparison with models based on littoral sediment assemblages (Flanders, Belgium). *Journal of Paleolimnology*, **38**, 97–116.

Dixit, S. S., Smol, J. P., Charles, D. F., *et al.* (1999). Assessing water quality changes in the lakes of the northeastern United States using sediment diatoms. *Canadian Journal of Fisheries and Aquatic Sciences*, **56**, 131–52.

Donar, C., Stoermer, E. F., & Brenner, M. (2009). The Holocene paleolimnology of Lake Apopka, Florida. *Nova Hedwigia, Beiheft*, **135**, 57–71.

Dong, X., Bennion, H., Battarbee, R. W., Yang, X., Yang, H., & Liu, E. (2008). Tracking eutrophication in Taihu Lake using the diatom record: potential and problems. *Journal of Paleolimnology*, **40**, 413–29.

Douglas, M. S. V., Smol, J. P., & Blake, W., Jr. (1994). Marked post-18th century environmental change in High-Arctic ecosystems. *Science*, **266**, 416–19.

Eminson, D. F. & Moss, B. (1980). The composition and ecology of periphyton communities in freshwaters. 1. The influence of host type and external environment on community composition. *British Phycological Journal*, **15**, 429–46.

Engstrom, D. R., Fritz, S. C., Almendinger, J. E., & Juggins, S. (2000). Chemical and biological trends during lake evolution in recently deglaciated terrain. *Nature*, **408**, 161–6.

European Union (2000). Directive 2000/60/EC of the European Parliament and of the Council of 23 October 2000 establishing a framework for Community action in the field of water policy. *Official Journal of the European Communities*, **L327**, 1–73.

Falkowski, P. G., & Raven, J. A. (2007). *Aquatic Photosynthesis*, 2nd edition, Princeton, NJ: Princeton Press.

Flower, R. J. (1993). Diatom preservation: experiments and observations on dissolution and breakage in modern and fossil material. *Hydrobiologia*, **269/270**, 473–84.

Flower, R. J., Dobinson, S., Ramdani, M., *et al.* (2001). Recent environmental change in North African wetland lakes: diatom and other stratigraphic evidence from nine sites in the CASSARINA project. *Aquatic Ecology*, **35**, 369–88.

Flower, R. J., Juggins, S. J., & Battarbee, R. W. (1997). Matching diatom assemblages in lake sediment cores and modern surface sediment samples: the implications for conservation and restoration with special reference to acidified systems. *Hydrobiologia*, **344**, 27–40.

Fritz, S. C. (1989). Lake development and limnological response to prehistoric and historic land-use in Diss, Norfolk, England. *Journal of Ecology*, **77**, 182–202.

Gaiser, E. E., Childers, D. L., Jones, R. D., *et al.* (2006). Periphyton responses to eutrophication in the Florida Everglades: cross-system patterns of structural and compositional change. *Limnology and Oceanography*, **51**, 617–30.

Ganf, G. G. (1974). Diurnal mixing and the vertical distribution of phytoplankton in a shallow equatorial lake (Lake George, Uganda). *Journal of Ecology*, **62**, 611–29.

Gell, P. A. (1997). The development of diatom database for inferring lake salinity, Western Victoria, Australia: towards a quantitative approach for reconstructing past climates. *Australian Journal of Botany*, **45**, 389–423.

Gell, P., Bulpin, S., Wallbrink, P., Bickford, S., & Hancock, G. (2005a). Tareena Billabong – a palaeolimnological history of an ever-changing wetland, Chowilla Floodplain, lower Murray–Darling Basin, Australia. *Marine and Freshwater Research*, **56**, 441–56.

Gell, P., Tibby, J., Fluin, J., *et al.* (2005b). Accessing limnological change and variability using fossil diatom assemblages, south-east Australia. *River Research and Applications*, **21**, 257–69.

Gell, P., Tibby, J., Little, F., Baldwin, D., & Hancock, G. (2007). The impact of regulation and salinisation on floodplain lakes: the Lower River Murray, Australia. *Hydrobiologia*, **591**, 135–46.

Gibson, C. E., Foy, R. H., & Bailey-Watts, A. E. (1996). An analysis of the total phosphorus cycle in some temperate lakes: the response to enrichment. *Freshwater Biology*, **35**, 525–32.

Goldsborough, L. G. (1993). Diatom ecology in the phyllosphere of a common duckweed (*Lemna minor* L.). *Hydrobiologia*, **269/270**, 463–71.

Hall, R. I., Leavitt, P. R., Smol, J., & Zirnhelt, N. (1997). Comparison of diatoms, fossil pigments and historical methods as measures of lake eutrophication. *Freshwater Biology*, **38**, 401–17.

Hansson, L. A. (1989). The influence of a periphytic biolayer on phosphorus exchange between substrate and water. *Archiv für Hydrobiologie*, **115**, 21–6.

Haworth, E. Y. (1972). The recent diatom history of Loch Leven, Kinross. *Freshwater Biology*, **2**, 131–41.

Haynes, D., Gell, P., Tibby, J., Hancock, G., & Goonan, P. (2007). Against the tide: the freshening of naturally saline coastal lakes, southeastern South Australia. *Hydrobiologia*, **591**, 165–83.

Hellström, T. (1991). The effect of resuspension on algal production in a shallow lake. *Hydrobiologia*, **213**, 183–90.

Hickman, M. (1987). Paleolimnology of a large shallow lake; Cooking Lake, Alberta, Canada. *Archiv für Hydrobiologie*, **111**, 121–36.

Hofmann, G. (1994). Aufwachs-Diatomeen in Seen und ihre Eignung als Indikataren der Trophie. *Bibliotheca Diatomologica*, **30**, 1–241.

Hudon, C. & Bourget, E. (1983). The effects of light on the vertical structure of epibenthic diatom communities. *Botanica Marina*, **26**, 317–30.

Hughes, R. M., Paulsen, S. G., & Stoddard, J. L. (2000). EMAP – Surface Waters: a multiassemblage, probability survey of ecological integrity in the U.S.A. *Hydrobiologia*, **422–23**, 429–43.

Jeppesen, E. (1998). *The Ecology of Shallow Lakes – Trophic Interactions in the Pelagial*. NERI Technical Report No. 247, D.Sc. dissertation, Ministry of Environment and Energy, Silkeborg, Denmark.

Jeppesen, E., Jensen, J. P., Søndergaard, M., *et al.* (1997). Top-down control in freshwater lakes: the role of nutrient state, submerged macrophytes and water depth. *Hydrobiologia*, **342/343**, 151–64.

Jeppesen, E., Leavitt, P., De Meester, L., & Jensen, J. P. (2001). Functional ecology and palaeolimnology: using cladoceran remains to reconstruct anthropogenic impact. *Trends in Ecology & Evolution*, **16**, 191–98.

Jeppesen, E., Moss, B., Bennion, H., *et al.* (in press). Interaction of climate change and eutrophication. In *Climate Change Impacts on Freshwater Ecosystems: Direct Effects and Interactions with Other Stresses*, ed. M. Kernan, B. Moss, & R. W. Battarbee, Chichester: Wiley Blackwell, ch. 6.

Jeppesen, E., Søndergaard, M., Jensen, J. P., *et al.* (2005). Lake responses to reduced nutrient loading – an analysis of contemporary long-term data from 35 case studies. *Freshwater Biology*, **50**, 1747–71.

Jeppesen, E., Søndergaard, M., Meerhoff, M., Lauridsen, T. L., & Jensen, J. P. (2007). Shallow lake restoration by nutrient loading reduction – some recent findings and challenges ahead. *Hydrobiologia*, **584**, 239–52.

Jeppesen, E., Søndergaard, M., Søndergaard, M., & Christoffersen, K. (1997). *The Structuring Role of Submerged Macrophytes in Lakes*. Ecological Studies Series 131. New York, NY: Springer-Verlag.

Jewson, D. H. (1992a). Size reduction, reproductive strategy and the life cycle of a centric diatom. *Philosophical Transactions of the Royal Society of London*, **B 336**, 191–213.

Jewson, D. H. (1992b). Life cycle of a *Stephanodiscus* sp. (Bacillariophyta). *Journal of Phycology*, **28**, 856–66.

Jewson, D. H., Lowry, S. F., & Bowen, R. (2006). Co-existence and survival of diatoms on sand grains. *European Journal of Phycology*, **41**, 131–46.

Jewson, D. H., Rippey, B. H., & Gilmore, W. K. (1981). Loss rates from sedimentation, parasitism, and grazing during the growth, nutrient limitation, and dormancy of a diatom crop. *Limnology and Oceanography*, **26**, 1045–56.

Jones, J. I., Moss, B., Eaton, J. W., & Young, J. O. (2000). Do submerged aquatic plants influence periphyton community composition for the benefit of invertebrate mutualists? *Freshwater Biology*, **43**, 591–604.

Karst, T. L. & Smol, J. P. (2000). Paleolimnological evidence of limnetic nutrient concentration equilibrium in a shallow, macrophyte-dominated lake. *Aquatic Sciences*, **62**, 20–38.

Kelly, M. G., Cazaubon, A., Coring, E., *et al.* (1998). Recommendations for the routine sampling of diatoms for water quality assessments in Europe. *Journal of Applied Phycology*, **10**, 215–24.

Kelly, M. G., Juggins, S., Bennion, H., *et al.* (2007). Use of diatoms for evaluating ecological status in UK freshwaters. Environment Agency Science Report No. SC030103, Bristol: Environment Agency.

Kelly, M. G., King, L., Jones, R. I., Barker, P. A., & Jamieson, B. J. (2008). Validation of diatoms as proxies for phytobenthos when assessing ecological status in lakes. *Hydrobiologia*, **610**, 125–9.

Kelly, M. G. & Whitton, B. A. (1995). The Trophic Diatom Index: a new index for monitoring eutrophication in rivers. *Journal of Applied Phycology* **7**, 433–44.

King, L., Barker, P., & Jones, R. I. (2000). Epilithic algal communities and their relationship to environmental variables in lakes of the English Lake District. *Freshwater Biology*, **45**, 425–42.

King, L., Clarke, G., Bennion, H., Kelly, M., & Yallop, M. (2006). Recommendations for sampling littoral diatoms in lakes for ecological status assessments: a literature review. *Journal of Applied Phycology*, **18**, 15–25.

Kiss, K. T. & Padisák, J. (1990). Species succession of Thalassiosiraceae: quantitative studies in a large shallow lake (Lake Balaton, Hungary). In *Proceedings of the 10th International Diatom Symposium*, ed. H. Simola, pp. 481–90. Königstein: Koeltz Scientific Books.

Kitner, M. & Poulícková, A. (2003). Littoral diatoms as indicators for the eutrophication of shallow lakes. *Hydrobiologia*, **506–9**, 519–24

Krewer, J. A. & Holm, H. W. (1982). The phosphorus–chlorophyll relationship in periphytic communities in a controlled ecosystem. *Hydrobiologia*, **94**, 173–6.

Langdon, P. G., Ruiz, Z., Wynne, S., Sayer, C. D., & Davidson, T. A. (2010). Ecological influences on larval chironomid communities in shallow lakes: implications for palaeolimnological interpretations. *Freshwater Biology*, **55**, 531–45.

Leahy, P. J., Tibby, J., Kershaw, A. P., Heijnis, H., & Kershaw, J. S. (2005). The impact of European settlement on Bolin Billabong, a Yarra River floodplain lake, Melbourne, Australia. *River Research and Applications*, **21**, 131–49.

Leira, M., Jordan, P., Taylor, D., *et al.* (2006). Assessing the ecological status of candidate reference lakes in Ireland using palaeolimnology. *Journal of Applied Ecology*, **43**, 816–27.

Lewis, W. M., Jr. (1978). Dynamics and succession of the phytoplankton in a tropical lake: Lake Lanao, Philippines. *Journal of Ecology*, **66**, 849–80.

Liboriussen, L. & Jeppesen, E. (2003). Temporal dynamics in epipelic, pelagic and epiphytic algal production in a clear and turbid shallow lake. *Freshwater Biology*, **48**, 418–31.

Lindeman, R. L. (1942). The trophic aspect of ecology. *Ecology*, **23**, 399–417.

Lowe, R. L. (1996). Periphyton patterns in lakes. In *Algal Ecology*, ed. R. J. Stevenson, R. L. Lowe, & M. Bothwell, New York, NY: Academic Press, pp. 57–76.

Lund, J. W. G. (1954). The seasonal cycle of the plankton diatom, *Melosira italica* Kütz. *subarctica* O. Mull. *Journal of Ecology*, **42**, 151–79.

Lund, J. W. G. (1971). An artificial alteration of the seasonal cycle of the plankton diatom *Melosira italica* subsp. *subarctica* in an English lake. *Journal of Ecology*, **59**, 521–33.

Maceina, M. J., & Soballe, D. M. (1990). Wind-related limnological variation in Lake Okeechobee, Florida. *Lake Reservoir Management*, **6**, 93–100.

Marshall, W. L. & Warakomski, J. M. (1980). Amorphous silica solubilities II. Effect of aqueous salt solutions at 25°C. *Geochimica et Cosmochimica Acta*, **44**, 915–24.

McCormick, P. V. & Stevenson, R. J. (1998). Periphyton as a tool for ecological assessment and management in the Florida Everglades. *Journal of Phycology*, **4**, 726–33.

McCormick, P. V., Shuford, R. B., Backus, J. G., & Kennedy, W. C. (1998). Spatial and temporal patterns of periphyton biomass and productivity in the Florida Everglades, Florida USA. *Hydrobiologia*, **362**, 185–208.

McGowan, S., Leavitt, P. R., Hall, R. I., *et al.* (2005). Controls of algal abundance and community composition during ecosystem state change. *Ecology*, **86**, 2200–11.

Miettinen, J. O., Kukkonen, M., & Simola, H. (2005). Hindcasting baseline values for water colour and total phosphorus concentration in lakes using sedimentary diatoms – implications for lake typology in Finland. *Boreal Environmental Research*, **10**, 31–43.

Millie, D. F. & Lowe, R. L. (1983). Studies on Lake Erie's littoral algae: host specificity and temporal periodicity of epiphytic diatoms. *Hydrobiologia*, **99**, 7–18.

Moss, B. (1968). The chlorophyll *a* content of some benthic algal communities. *Archiv für Hydrobiologie*, **65**, 51–62.

Moss, B. (1980). Further studies on the palaeolimnology and changes in the phosphorus budget of Barton Broad, Norfolk. *Freshwater Biology*, **10**, 261–79.

Moss, B. (2001). *The Broads*. London: Harper Collins.

Muller, U. (1995). Vertical zonation and production rates of epiphytic algae on *Phragmites australis*. *Freshwater Biology*, **34**, 69–80.

Nipkow, F. (1950). Rufeformen planktischer Kieselalgan im geschichteten Schlamm des Zurichsees. *Schweizerische Zeitschrift für Hydrologie*, **12**, 263–70.

Ogden, R. W. (2000). Modern and historical variation in aquatic macrophyte cover of billabongs associated with catchment development. *Regulated Rivers – Research & Management*, **16**, 487–512.

Ogden, R., Spooner, N., Reid, M., & Head, J. (2001). Sediment dates with implications for the age of conversion from palaeochannel to modern fluvial activity on the Murray River and tributaries. *Quaternary International*, **83**–85, 195–209.

Oldfield, F. & Appleby, P. G. (1984). Empirical testing of ^{210}Pb-dating models for lake sediments. In *Lake Sediments and Environmental History*, ed. J. W. G. Lund, & E. Y. Haworth, pp. 93–124. Leicester: University of Leicester Press.

Padisák, J. & Reynolds, C. S. (2003). Shallow lakes: the absolute, the relative, the functional and the pragmatic. *Hydrobiologia*, **506–9**, 1–11.

Padisák, J., Toth, L. G., & Rajczy, M. (1988). The role of storms in the summer succession of the phytoplankton community in a shallow lake (Lake Balaton, Hungary). *Journal of Plankton Research*, **10**, 249–65.

Philibert, A. & Prairie, Y. T. (2002). Is the introduction of benthic species necessary for open-water chemical reconstruction in diatom-based transfer functions? *Canadian Journal of Fisheries and Aquatic Sciences*, **59**, 938–51.

Phillips, G. L. (1992). A case study in restoration: shallow eutrophic lakes in the Norfolk Broads. In *Eutrophication of Freshwaters*, ed. D. Harper, London: Chapman & Hall, pp. 251–278.

Pollard, P. & Huxham, M. (1998). The European Water Framework Directive: a new era in the management of aquatic ecosystem health? *Aquatic Conservation: Marine and Freshwater Ecosystems*, **8**, 773–92.

Ponader, K. C., Charles, D. F., & Belton, T. J. (2007). Diatom-based TP and TN inference models and indices for monitoring nutrient enrichment of New Jersey streams. *Ecological Indicators*, **7**, 79–93.

Poulícková, A., Duchoslav, M., & Dokulil, M. (2004). Littoral diatom assemblages as bioindicators of lake trophic status: a case study from perialpine lakes in Austria. *European Journal of Phycology*, **39**, 143–52.

Prygiel, J. & Coste, M. (1993). The assessment of water quality in the Artois-Picardie water basin (France) by the use of diatom indices. *Hydrobiologia*, **269–270**, 343–349.

Räsänen, J., Kauppila, T., & Salonen, V. P. (2006). Sediment-based investigation of naturally or historically eutrophic lakes – implications for lake management. *Journal of Environmental Management*, **79**, 253–65.

Reavie, E. D. & Smol, J. P. (1997). Diatom-based model to infer past littoral habitat characteristics in the St. Lawrence River. *Journal of Great Lakes Research*, **23**, 339–48.

Reddy, K. R., O'Conner, G. A., & Schelske, C. L. (1999). *Phosphorus Biogeochemistry in Subtropical Ecosystems*. Washington DC: Lewis Publishers.

Reid, M. (2002). A diatom-based palaeoecological study of two billabongs on the Goulburn River floodplain, southeast Australia. In *Proceedings of the 15th International Diatom Symposium*, ed. J. John, Königstein: Koeltz Scientific Books, pp. 237–253.

Reid, M., Sayer, C. D., Kershaw, A. P., & Heijnis, H. (2007). Palaeolimnological evidence for submerged plant loss in a floodplain lake associated with accelerated catchment soil erosion (Murray River, Australia). *Journal of Paleolimnology*, **38**, 191–208.

Reynolds, C. S. (1984). *The Ecology of Freshwater Plankton*. Cambridge: Cambridge University Press.

Reynolds, C. S., Wiseman, S. W., Gadfrey, B. M., & Butterwick, C. (1983). Some effects of artificial mixing on the dynamics of phytoplankton populations in large limnetic enclosures. *Journal of Plankton Research*, **5**, 203–34.

Rippey, B. (1983). A laboratory study of silicon release processes from a lake sediment (Lough Neagh, Northern Ireland). *Archiv für Hydrobiologie*, **96**, 417–33.

Roozen, C. F. J. M., Lurling, M., Vlek, H., *et al.* (2007). Resuspension of algal cells by benthivorous fish boosts phytoplankton biomass and alters community structure in shallow lakes. *Freshwater Biology*, **52**, 977–87.

Round, F. E. (1981). *The Ecology of Algae*. Cambridge: Cambridge University Press.

Round, F. E. (1991). Use of diatoms for monitoring rivers. In *Use of Algae for Monitoring Rivers*, ed. B. A. Whitton, E. Rott, & G. Friedrich, Innsbruck: Universitat Innsbruck, pp. 25–32.

Saunders, K. M., McMinn, A., Roberts, D., Hodgson, D. A., & Heijnis, H. (2007). Recent human-induced salinity changes in Ramsar-listed Orielton Lagoon, south-east Tasmania, Australia: a new approach for coastal lagoon conservation and management. *Aquatic Conservation: Marine and Freshwater Ecosystems*, **17**, 51–70.

Sayer, C. D. (2001). Problems with the application of diatom–total phosphorus transfer functions: examples from a shallow English Lake. *Freshwater Biology* **46**, 743–57.

Sayer, C. D., Burgess, A., Kari, K., *et al.* (2010a). Long-term dynamics of submerged macrophytes and algae in a small and shallow, eutrophic lake: implications for the stability of macrophyte dominance. *Freshwater Biology*, **55**, 565–83.

Sayer, C. D., Davidson, T. A., Jones, J. I., & Langdon, P. G. (2010b) Combining contemporary ecology and palaeolimnology to understand shallow lake ecosystem change. *Freshwater Biology*, **55**, 487–99.

Sayer, C. D., Jackson, M. J., Hoare, D. J., *et al.* (2006). TBT causes regime shift in shallow lakes. *Environmental Science & Technology*, **40**, 5269–75.

Sayer C. D. & Roberts, N. (2001). Establishing realistic restoration targets for nutrient-enriched shallow lakes: linking diatom ecology and palaeoecology at the Attenborough Ponds, UK. *Hydrobiologia*, **448**, 117–42.

Sayer, C. D., Roberts, N., Sadler, J., David, C., & Wade, M. (1999). Biodiversity changes in a shallow lake ecosystem: a multi-proxy palaeolimnological analysis. *Journal of Biogeography*, **26**, 97–114.

Schaumburg, J., Schranz, C., Hofmann, G., *et al.* (2004). Macrophytes and phytobenthos as indicators of ecological status in German lakes – a contribution to the implementation of the Water Framework Directive. *Limnologica*, **34**, 302–14.

Schaumburg, J., Schranz, C., Stelzer, D., & Hofmann, G. (2007). Action instructions for the ecological evaluation of lakes for implementation of the EU Water Framework Directive: Makrophytes and Phytobenthos. Augsburg: Bavarian Environment Agency.

Scheffer, M. (1998). *Ecology of Shallow Lakes*. London: Chapman and Hall.

Scheffer, M., Hosper, S. H., Meijer, M.-L., Moss, B., & Jeppesen, E. (1993). Alternative equilibria in shallow lakes. *Trends in Ecology and Evolution*, **8**, 275–9.

Scheffer, M., Rinaldi, S., Huisman, J., & Weissing, F. J. (2003). Why plankton have no equilibrium: solutions to the paradox. *Hydrobiologia*, **491**, 9–18.

Scheffer, M. & van Nes, E. H. (2007). Shallow lakes theory revisited: various alternative regimes driven by climate, nutrients, depth, and lake size. *Hydrobiologia*, **584**, 455–66.

Schelske, C. L., Donar, C. M., & Stoermer, E. F. (1999). A test of paleolimnologic proxies for the planktonic/benthic ratio of microfossil diatoms in Lake Apopka. In *Proceedings of the 14th International Diatom Symposium*, ed. S. Mayama, M. Idei, & I. Koizumi. Königstein: Koeltz Scientific Books, pp. 367–82.

Schmidt, R., Mäusbacher, R. M., & Müller, J. (1990). Holocene diatom flora and stratigraphy from sediment cores of two Antarctic lakes (King George Island). *Journal of Paleolimnology*, **3**, 55–74.

Schönfelder, I., Gelbrecht, J., Schönfelder, J., & Steinberg, C. E. W. (2002). Relationships between littoral diatoms and their chemical environment in northeastern German lakes and rivers. *Journal of Phycology*, **38**, 66–82.

Sicko-Goad, L. (1986). Rejuvenation of *Melosira granulata* (Bacillariophyceae) resting cells from the anoxic sediments of Douglas Lake, Michigan. II. Electron microscopy. *Journal of Phycology*, **22**, 28–35.

Simpson, G. L., Shilland, E. M., Winterbottom, J. M., & Keay, K. (2005). Defining reference conditions for acidified waters using a modern analogue approach. *Environmental Pollution*, **137**, 119–33.

Smetacek, V. S. (1985). Role of sinking in diatom life-history cycles: ecological, evolutionary and geological significance. *Marine Biology*, **84**, 239–51.

Smith, V. H., Joye, S. B., & Howarth, R. W. (2006). Eutrophication of freshwater and marine ecosystems. *Limnology and Oceanography*, **51**, 351–5.

Smol, J. P. (2008). *Pollution of Lakes and Rivers: a Paleoenvironmental Perspective*, 2nd Edition. Oxford: Blackwell Publishing.

Sommer, U. (1988). Growth and survival strategies of planktonic diatoms. In *Growth and Reproductive Strategies of Freshwater Phytoplankton*, ed. C. D. Sandgren, Cambridge: Cambridge University Press, pp. 388–433.

Søndergaard, M., Jeppesen, E., Lauridsen, T., *et al.* (2007). Lake restoration: successes, failures and long-term effects. *Journal of Applied Ecology*, **44**, 1095–105.

Stenger-Kovacs, C., Buczko, K., Hajnal, E., & Padisák, J. (2007). Epiphytic, littoral diatoms as bioindicators of shallow lake trophic status: Trophic Diatom Index for Lakes (TDIL) developed in Hungary. *Hydrobiologia*, **589**, 141–54.

Stoermer, E. F., Andresen, N. A., & Schelske, C. L. (1992). Diatom succession in the recent sediments of Lake Okeechobee, Florida, USA. *Diatom Research*, **7**, 367–86.

Taffs, K. H. (2001). Diatoms as indicators of wetland salinity in the Upper South East of South Australia. *The Holocene*, **11**, 281–90.

Tibby, J. (2003). Explaining lake and catchment change using sediment derived and written histories: an Australian perspective. *The Science of the Total Environment*, **310**, 61–71.

Tibby, J., Gell, P., Fluin, J., & Sluiter, I. (2007). Diatom–salinity relationships in wetlands: assessing the influence of salinity variability on the development of inference models. *Hydrobiologia*, **591**, 207–218.

Tibby, J., Reid, M., Fluin, J., Hart, B. T., & Kershaw, A. P. (2003). Assessing long-term pH change in an Australian river catchment using monitoring and palaeolimnological data. *Environmental Science and Technology*, **37**, 3250–5.

Vadeboncoeur, Y., Jeppesen, E., Vander Zanden, M. J., *et al.* (2003). From Greenland to green lakes: cultural eutrophication and the loss of benthic pathways in lakes. *Limnology and Oceanography*, **48**, 140–818.

van Dam, H. (1996). Partial recovery of moorland pools from acidification: indications by chemistry and diatoms. *Netherlands Journal of Aquatic Ecology*, **30**, 203–18.

van Dam, H., Mertens, A., & Sinkeldam, J. (1994). A coded checklist and ecological indicator values of freshwater diatoms from the Netherlands. *Netherlands Journal of Aquatic Ecology*, **28**, 117–33.

van der Molen, D. T., & Portielje, R. (1999). Multi-lake studies in the Netherlands: trends in eutrophication. *Hydrobiologia*, **409**, 359–65.

Walker, M. J. C., Griffiths, H. I., Ringwood, V., & Evans, J. G. (1993). An early-Holocene pollen, mollusc and ostracod sequence from lake marl at Llangorse Lake, south Wales, UK. *The Holocene*, **3**, 138–49.

Werner, P. & Smol, J. P. (2005). Diatom–environmental relationships and nutrient transfer functions from contrasting shallow and deep limestone lakes in Ontario, Canada. *Hydrobiologia*, **533**, 145–73.

Wetzel, R. G. (1983). *Periphyton of Aquatic Ecosystems*. The Hague: B. V. Junk Publishers.

Wetzel, R. G. (1996). Benthic algae and nutrient cycling in lentic freshwater ecosystems. In *Algal Ecology: Freshwater Benthic Ecosystems*, ed. R. J. Stevenson, M. L. Bothwell, & R. L. Lowe, New York, NY: Academic Press, pp. 641–667.

Wetzel, R. G. (2001). *Limnology: Lake ànd River Ecosystems*, 3rd edition, New York, NY: Springer-Verlag.

Williams, D. M. & Round, F. E. (1987). Revision of the genus *Fragilaria*. *Diatom Research*, **2**, 267–88.

Yang, X., Anderson, N. J., Dong, X., & Shen, J. (2008). Surface sediment diatom assemblages and epilimnetic total phosphorus in large, shallow lakes of the Yangtze floodplain: their relationships and implications for assessing long-term eutrophication. *Freshwater Biology*, **53**, 1273–90.

9 Diatoms as indicators of water-level change in freshwater lakes

JULIE A. WOLIN AND JEFFERY R. STONE

9.1 Introduction

Water-level changes result from a variety of geological, biological, and/or climatic processes. Many of these changes occur over long periods; others may be rapid or result from catastrophic events. In aquatic environments, diatoms are highly sensitive indicator organisms and their microfossils, deposited in lake sediments, can be used to infer environmental changes (Smol, 2008). Unambiguous diatom signals can be reconstructed from lakes isolated from marine or brackish waters (e.g. Fritz *et al.*, this volume; Horton & Sawai, this volume). However, in freshwater systems lake-level changes are often recorded as increases in planktonic (free-floating) diatoms – although as discussed below, interpretation of this signal should be supported by additional evidence.

In the Laurentian Great Lakes region of North America (e.g. Finkelstein & Davis, 2006; Wolfe *et al.*, 2000; Wolin, 1996) and the North Sea and Baltic regions of Europe (e.g. Digerfeldt, 1998), freshwater lake levels are commonly affected by geological processes of isostatic rebound, subsidence, and outlet incision following glaciation, and lake isolation by coastal-sediment transport (e.g. Karrow & Calkin, 1985; Larsen & Schaetzl, 2001; Lewis *et al.*, 2008).

Biological processes, such as vegetation succession, can alter drainage patterns and groundwater flow, which in turn affect water levels. As vegetation develops following glaciation, surface runoff patterns change and this can moderate water levels on a seasonal or short-term basis. Natural deposition processes of internal (autochthonous) plant and animal remains, and mineral and organic inputs from the catchment (allochthonous) result in a shallower lake over time (Wetzel, 2001). These processes are accelerated by human activities such as forest clearance, farming, and nutrient inputs (e.g. Coops *et al.*, 2003; Gaillard *et al.*, 1991; Wolin & Stoermer 2005), while dams and channelization by humans or other animals can result in rapid water-level changes.

Hydrology, however, is the most important factor controlling water level. Lake levels are a balance of moisture gains and losses. Inputs include: stream inflow, watershed runoff, groundwater inflow, and lake surface precipitation. Losses occur through stream outflow, lake surface evaporation, groundwater outflow, and, in some cases, deep seepage. Most of these hydrologic responses are associated with climatic or ecological change (e.g. Dearing, 1997).

The ability to determine past water-level changes in freshwater is particularly important for reconstructing past climate variation and developing predictive models for future climate change. Lake levels usually rise during wet periods and fall during dry phases. The magnitude of the corresponding sedimentary signal is dependent on lake type and its sensitivity to climate (some classic reviews include Richardson, 1969 and Street-Perrott & Harrison, 1985). Shallow, closed-basin lakes usually record the strongest signals, while deep open-drainage and small surface-area lakes have the weakest. Unfortunately, from a paleo-reconstruction point-of-view, freshwater (open-drainage) lakes are predominant in many regions of the world and diatom-inferred reconstructions for these lakes are more problematic. Yu & Harrison (1995) and Tarasov *et al.* (1994) compiled two of the first extensive lake-level climate-record data bases for European and northern Eurasian lakes, respectively. Many of the freshwater lake studies found in these data bases use qualitative changes in diatom communities to indicate high or low lake levels.

Several modern diatom and environmental factor data bases, useful in climate-response lake-level reconstruction, have been compiled and are available electronically through the European Diatom Database (EDDI) (Juggins, 2001) and the Diatom Paleolimnology Data Cooperative, the Academy of Natural Sciences,

The Diatoms: Applications for the Environmental and Earth Sciences, 2nd Edition, eds. John P. Smol and Eugene F. Stoermer. Published by Cambridge University Press.

Philadelphia; the National Lakes Assessment diatom data base, US Environmental Protection Agency (USEPA), should be available by 2010. In many cases, human activities can alter or obscure climatic signals present in the record (e.g. Battarbee, 2000; Heinsalu *et al.*, 2008). This is particularly true in areas where human activity has existed for centuries (Voigt *et al.*, 2008). As the following examples will indicate, with proper consideration of limitations, it is possible to reconstruct water-level changes in freshwater systems. Given the number of freshwater lakes in regions where the only hydrologic signals will be found in such sedimentary archives, it is important to utilize these records to determine the effects of past climate change, geological events, or human manipulations.

9.2 Types of diatom indicators

Water-level fluctuations result in corresponding changes in available habitat, light, chemical conditions, stratification, and mixing regimes that indirectly affect freshwater diatom communities. When lake levels are low, usually more habitat is available for taxa living attached to substrates (benthos) or aquatic vegetation (epiphytes) because a greater area of the lake bottom is exposed to sunlight. Exceptions occur in steep-sided lakes or in shallow lakes where conditions may lead to greater turbidity or productivity-driven light limitation (e.g. Jeppesen *et al.*, 2000; Vadeboncoeur *et al.*, 2008), hence, it is necessary to use multiple lines of evidence to support any diatom-inferred water-level reconstructions and exclude factors that may result in comparable signals (Digerfeldt, 1986; Smol & Cumming, 2000). Such supporting data include: sediment stratigraphy, fine-grained particle analysis, stable-isotope analysis, chemical content, charcoal, chironomids, macrofossils, pollen, and ostracods. Wind exposure can affect thermal structure, and, in shallow lakes, this can result in a total loss of stratification (e.g. Bennion *et al.*, this volume). As nutrient and chemical concentrations increase, corresponding changes in pH occur (Fritz *et al.*, this volume).

9.2.1 Growth habit

Numerous paleolimnological studies have used the planktonic-to-benthic (P:B) ratio to qualitatively infer water-level changes (e.g. Björck *et al.*, 2000 in Denmark; Bergner & Trauth, 2004; Ekblom & Stabell, 2008 in Africa; Finkelstein *et al.*, 2005 in Canada; Tarasov *et al.*, 2000; Xue *et al.*, 2003 in Asia; Tapia *et al.*, 2003 in South America). Planktonic diatoms often dominate open-water environments, while benthic diatoms, epiphytes, and tychoplankton (accidental plankton) typically dominate littoral habitats. Shallow-water conditions expand benthic habitat and macrophyte growth producing a greater percentage of benthic and epiphytic forms preserved in lake sediments. Conversely, during deeper lake phases, benthic substrates are often reduced, producing a greater percentage of plankton preserved in the sediment. The difficulty in using growth-habit ratio to reconstruct past environments is that the same signals may be caused by other factors.

The ratio of lake size to lake depth can affect benthic productivity, and, by extension, the size of the benthic diatom community. Vadeboncoeur *et al.* (2008) modeled benthic algal productivity across lake-size gradients using published lake to depth ratios and primary production estimates. They found that the contribution of benthic periphyton to whole-lake productivity was highest in high-latitude and oligotrophic shallow lakes (mean depth <5 m), and that such productivity declined as nutrients increased. In moderate-size oligotrophic lake systems (mean depth = 5–100 m), benthic productivity contributed a substantially higher portion to whole-lake productivity, but decreased in importance with increasing nutrients. In agricultural regions, benthic productivity in eutrophic shallow lakes approaches zero as planktonic productivity increases and light limitation occurs.

In a shallow Estonian lake, Heinsalu *et al.* (2008) compared the sedimentary diatom record in a well-dated core to historic water-level fluctuations. They found close correlation of diatom growth-habit ratios with pre-eutrophication water levels, while yearly post-eutrophication water levels were only weakly or negatively correlated to percent planktonic diatoms (Figure 9.1). However, a strong correlation was found between post-eutrophication planktonic biomass and high spring water levels as short, mild winters facilitated earlier and prolonged planktonic diatom productivity.

Lake levels in early post-glacial lake sediments can be underestimated due to the abundance of benthic diatoms characteristic of nutrient-poor lakes and prolonged ice cover (Brugam *et al.*, 1998; Lotter *et al.*, this volume). In alpine and high Arctic lakes, changes in growth habit ratio and increases in Fragilariacea often are indicative of ice-cover duration (Smol, 1988; Lotter & Bigler, 2000; Lotter *et al.*, this volume; Douglas and Smol, this volume). An investigation into abiotic influences on the distribution of *Fragilaria* and *Staurosira* species in oligotrophic and ultra-oligotrophic Austrian alpine lakes by Schmidt *et al.* (2004) found evidence of species size variation with depth, in addition to depth correlation with benthic and planktonic taxa. Planktonic *Fragilaria gracilis* Østrup had longer valves in deeper lakes, a potential competitive response to increase buoyancy.

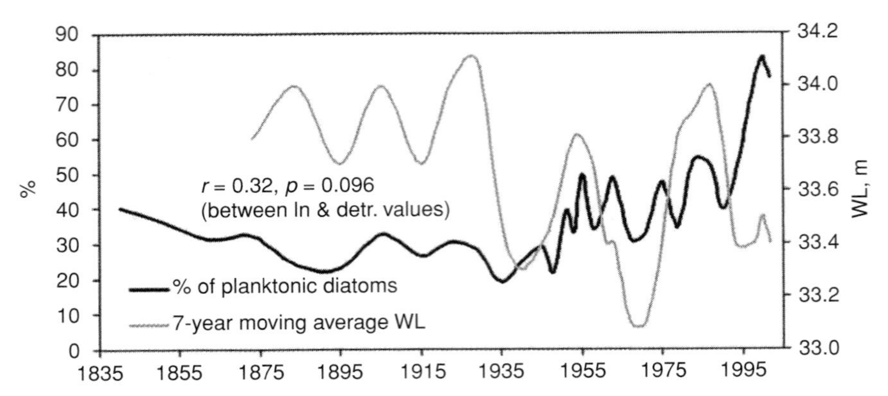

Figure 9.1 Relationship between measured water level (WL: seven-year moving average) in Lake Vörtsjärv and the percentage of planktonic forms among diatoms in the sediment-core layers over the years 1873–2002. (Reprinted with kind permission from Springer Science+Business Media: *Hydrobiology*, Water level changes in a large shallow lake as reflected by the plankton: periphyton-ratio of sedimentary diatoms. Vol. 599, 2008, page 28, Heinsalu, A., Luup, H., Alliksaar, T., Nõges, P. & Nõges, T., Figure 5a. © Springer Science+Business Media B. V. 2007.)

Further study on the correlation between species size variation and depth may prove useful for water-level reconstruction.

9.2.2 Diversity

Several investigators have noted an increase in diversity with increases in benthic diatom populations (e.g. Brugam *et al.*, 1998; Moos *et al.*, 2005). Although highly dependent upon lake morphometry, as lake levels decline, planktonic species typically are replaced by a more diverse benthic community. Wolin (1996) reported increases in benthic/epiphytic diatoms in Lower Herring Lake, Michigan, as lake levels fell; however this increase did not persist during low-water periods, but was repeated as water levels rose again. Two studies on contemporary lake-level change give insight into mechanisms behind such diversity changes.

Turner *et al.* (2005) investigated the effect of lake drawdown on algal communities, and reported macrophyte diversity and cover decline with the loss of emergent vegetation and increased percentage of barren substrate as littoral area declined. An initial decrease in diatom diversity occurred following drawdown. Little change was observed in the plankton, but benthic algae declined due to disruption of habitat and loss of colonizable area. Although epiphytes were not considered separately, they could also provide an additional lake-level signal.

In Laguna de La Caldera, an oligotrophic mountain lake in the Spanish Sierra Nevada, Sánchez-Castillo *et al.* (2008) documented modern lake-level, nutrient, and diatom-community changes following a severe 10-year drought (1985–1995). In 1996, a dramatic rise in lake level (from 1.5 m to 15 m) resulted in increased bare substrate and a pulse of nutrients as mineralization of organic matter occurred. This resulted in a drop in diatom diversity due to an initial dominance (93.3%) by a pioneering benthic species. In the following two years, benthic diversity increased as communities developed and algal nutrient utilization increased.

9.2.3 Physical and chemical environment

Other diatom signals can be present as a result of environmental conditions caused by changes in water levels. These signals often reflect physical changes related to stability of the metalimnion, susceptibility to turbulence, nutrient or thermal conditions, or changes in pH (e.g. Bradbury *et al.*, 2002). In deep-water cores from large lakes, where littoral influence is minimal, water-level fluctuations can be signalled by changes in the abundance of dominant planktonic taxa as they respond to such physical or chemical variables (e.g. Rioual *et al.*, 2007). Most studies use these signals in conjunction with some form of littoral/planktonic shifts in reconstructing lake-level change.

9.2.4 Turbulence

Diatoms associated with turbulence regimes have proved useful in lake-level reconstructions, particularly in freshwater African lakes (e.g. Stager *et al.*, 1997; Verschuren *et al.*, 2000; Bergner & Trauth, 2004; Nguetsop *et al.*, 2004; Stager *et al.*, 2005). *Aulacoseira* species are heavily silicified diatoms with high sinking rates. Their ecology requires turbulence to remain in the photic zone (e.g. Bradbury *et al.*, 2002) and they are often found in shallow eutrophic lakes in Europe (Rioual *et al.*, 2007). Increased turbulence and corresponding nutrient increases during low-water stages in a lake can favor this genus over other planktonic species.

The 15,000-year record of varved lake sediments from Elk Lake, Minnesota, is dominated by planktonic diatoms (Bradbury *et al.*, 2002). Within this community, increases in the percentage of *Aulacoseira* indicate low-water conditions during a dry prairie period between 8500 and 4000 years ago. A combination of low-water conditions and high wind exposure provide the turbulent, high-nutrient conditions favorable to this taxon.

While turbulence is beneficial to heavy diatoms with high nutrient requirements, certain diatom species benefit from stable stratification during high-water periods in clear lakes (e.g. *Cyclotella michiganiana* Skvortzow, *Fragilaria crotonensis* Kitton). Diatoms living in the deep chlorophyll maximum (DCM) are adapted to low light and capture nutrients diffusing from the hypolimnion (Bradbury *et al.*, 2002; Fahnenstiel & Scavia, 1987). Wolin (1996) used increases in taxa associated with the DCM to infer high water levels in Lower Herring Lake, Michigan.

9.2.5 Nutrient environment

Nutrient signals in diatom assemblages are commonly used to indicate human impacts within a lake basin (*see* Hall & Smol, this volume). However, nutrient concentration also increases during low-water levels (Moos *et al.*, 2009). If one can separate anthropogenic signals, these same indicators can be useful for climatic signals. In a small eutrophic lake in Northern Ireland, diatom-inferred total phosphorus increases resulted from climatic rather than anthropogenic forcings (Rippey *et al.*, 1997). A dry period in the 1970s resulted in reduced flushing and increased nutrients due to sediment phosphorus release. In a Michigan lake, Wolin & Stoermer (2005) found a nutrient-response-related lake-level signal in *Cyclotella stelligera* (Cleve et Grunow) Van Heurck during a record low-water phase in the 1930s. While in Lake 239, Experimental Lakes Area increased percent abundances in the dominant planktonic *C. stelligera* was associated with higher lake levels (Laird & Cumming 2008; 2009), this may be associated with the DCM and more stable stratification. *Aulacoseira subarctica* (O. Müll.) Haworth and *F. crotonensis* were found indicative of increased nutrient conditions in Lake 239 during large-magnitude lake-level drawdown in the mid Holocene (Moos *et al.*, 2009). Verschuren *et al.* (2000) also found *Aulacoseira* responses to lake-level induced changes in nutrient and light levels in Lake Naivasha, Kenya. During deep-lake/lower-nutrient conditions *Aulacoseira ambigua* (Grunow) Simonsen was dominant while *Aulacoseira granulata* (Ehrenb.) Simonsen dominated during shallow-lake/higher-nutrient and lower-light stages.

In shallow lakes, response to water-level disturbance is often non-linear (Coops *et al.*, 2003) and can result in clear water and turbid lake phases, independent of nutrient enrichment and top-down effects. An overall decline in diatom productivity can occur during low-water phases, particularly in shallow lakes due to light limitation in benthic communities and nutrient limitation in plankton. Nõges *et al.* (2003) found that light limitation in the plankton community affected the abundance of diatom phytoplankton during high water levels, but that light and nutrient limitations during low water resulted in a community shift from diatoms to cyanobacteria.

9.2.6 pH

Climate changes, or more precisely wetter/drier shifts, are usually inferred indirectly in paleolimnological studies from evidence for water-level changes. Complex hydrologic interactions can result in pH shifts depending on the prevailing source. In Batchawana Lake, Ontario, early work by Delorme *et al.* (1986) connected moisture changes (i.e. lake level) determined by tree pollen data with prehistoric diatom-inferred pH changes (see Battarbee *et al.*, this volume). Increases in humification and lower diatom-inferred pH prior to anthropogenic-induced causes were attributed to decreases in moisture, while high water levels resulted in increased circumneutral and alkaliphilous taxa (higher pH). Krabbenhoft & Webster (1995) provide mechanistic explanations for climate influence on pH. They found that during drought conditions, reduced groundwater inflow containing high base-cation concentrations (Ca^{2+}, Mg^{2+}) likely resulted in the acidification of Nevins Lake, Michigan. Increased percentages of acidophilic taxa can also indicate marginal wetland expansion during low-lake phases (e.g. Bunting *et al.*, 1997; Shkarpa, 2006).

In African lakes, diatom-inferred pH signals usually result in the opposite effect. Cation concentrations increase with drier climate, raising pH (e.g. Bergner & Trauth, 2004; Fritz *et al.*, this volume). In the rainforest region of Cameroon, West Africa, Nguestop *et al.* (2004) found a more complex interaction in Lake Ossa where the pH of prevailing hydrological sources determined the dominant diatom community. During the rainy season, hydrology is dominated by low-pH precipitation ($\sim$5 to 6.8) and acidiphilic (low-pH) diatom taxa prevail. During the dry season, alkalinity and pH increase in the deepest regions of the lake with increased evaporation, but low-pH groundwater and wetland discharge maintain

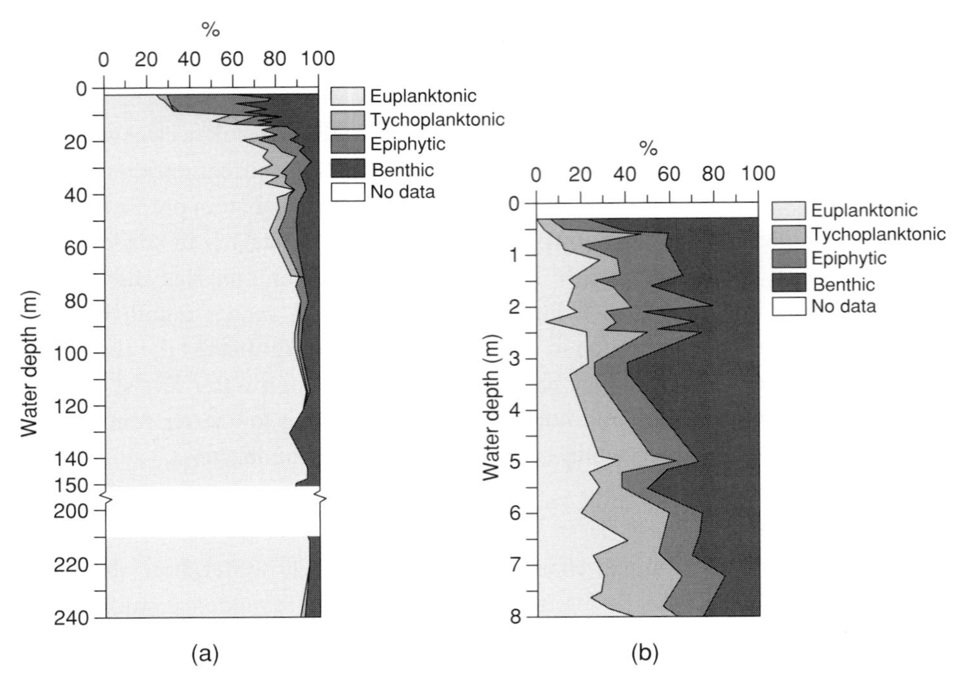

Figure 9.2 Relative abundance (%) of diatom life forms from surficial sediment samples collected along depth gradients in (a) Lake Ontario, and (b) East Lake (Yang & Duthie 1995).

low-pH conditions (<7) in littoral regions. Quantitative lake-level reconstructions were based on growth-habit ratios and diatom-inferred pH.

9.3 Quantitative reconstructions

Diatom/depth distribution studies, such as Round's (1961) pioneering work in the English Lake District, remain uncommon. The development of quantitative models for freshwater diatom-inferred lake-level change has increased since the publication of Wolin & Duthie (1999). New modeling techniques provide greater insight for interpreting changes in P:B diatom signals (Stone & Fritz, 2004) and researchers have increasingly focused on quantitative methods to successfully reconstruct lake-level changes (e.g. Laird & Cumming, 2008, Moser *et al.*, 2000; Nguestop *et al.*, 2004; Rioual *et al.*, 2007).

9.3.1 Regression

Early development of quantitative diatom-inferred depth models relied on regression to determine correlations between diatoms and depth. In most cases, taxa were subjectively assigned growth-habit categories based on the literature. In the North American Great Lakes, researchers (e.g. Stevenson & Stoermer, 1981; Kingston *et al.*, 1983) quantified benthic diatom distributions along depth transects in Lake Michigan. Yang & Duthie (1995) explored the distribution of euplanktonic, tychoplanktonic, epiphytic, and benthic life forms in transects along a 240 m depth gradient in western Lake Ontario, and an 8 m depth gradient in East Lake, an isolated embayment (Figure 9.2). Multiple regression was used to explore the relationship between measured water depth of surface samples and sample life-form composition. Strong relationships were found in both lakes, though the stronger relationship in Lake Ontario was probably due to its greater depth, distance from the shoreline of the deeper samples, and a larger number of samples. The main problem with this approach was the subjectivity involved in the assignment of diatom taxa to life-form categories. Category assignments for each taxon were based on literature references and not all references agreed on life-form designations. Barker *et al.* (1994) developed an early quantitative model in Lake Sidi Ali, Morocco, from depth-transect surface samples using pre-assigned habitat types: planktonic, tycoplanktonic, and littoral. They found a logarithmic relationship between depth (D) and the planktonic/littoral ratio (P/L); however, the literature-based growth-habit classification proved problematic in certain taxa (i.e. *Fragilaria*).

9.3.2 Ordination

Several researchers have continued the initial work of Lotter (1988) and Barker *et al.* (1994) using ordination methods (see Birks, this volume) to analyze contemporary diatom communities along surface-sediment depth transects, and identify depth-specific assemblages (e.g. Rautio *et al.*, 2000). These

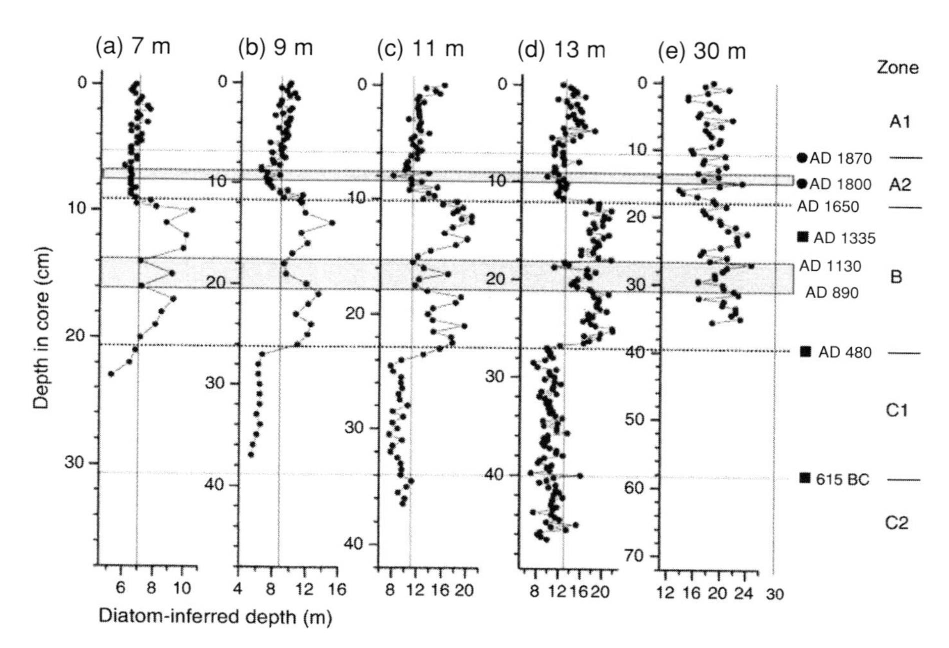

Figure 9.3 Diatom-inferred depth for each of the five gravity cores taken from depths ranging between 7 and 30 m. Vertical lines indicate present-day depth at the respective coring sites. Gray bars indicate drought intervals. Solid circles indicate ^{210}Pb chronology, solid squares indicate ^{14}C dating. Zones are based on a constrained cluster analysis of diatom assemblage data. (Reprinted from *Journal of Paleolimnology*, Vol. 42, 2009, Laird, K. R. & Cumming, B. F., Diatom-inferred lake level from near-shore cores in a drainage lake from the Experimental Lakes Area, northwestern Ontario, Canada., Fig. 5, page 72, © 2008 with permission from Elsevier.)

assemblages have then been correlated to subfossil assemblages in cores and used to infer water depth (Lotter, 1988). In Lake Ossa, Cameroon, Nguestop *et al.* (2004) reconstructed high and low lake levels using correspondence analysis (CA) of 74 modern diatom samples from local lakes to determine depth-defined assemblages.

9.3.3 Weighted-averaging analysis

The optimum abundance of a diatom along a depth or other environmental gradient such as temperature, pH, or nutrient status can be determined by weighted-averaging (WA) analysis (Birks, this volume). Yang & Duthie (1995) were the first to use WA analysis to determine optimum water depths for freshwater diatom species and develop training sets for lake-level reconstruction. Duthie *et al.* (1996) then applied this training set to reconstruct lake-level fluctuations in Hamilton Harbor, Lake Ontario. Yang & Duthie (1995) cautioned, however, that WA can underestimate high levels and overestimate low levels. This trend, reported by other researchers (e.g. Brugam *et al.*, 1998), is attributed to functional interactions among different environmental variables not accounted for by simple inferences (Birks, this volume).

Two primary approaches have been used in developing training sets for freshwater lake-level reconstruction. One is to construct training sets from multiple depths across a small number of regionally close lakes (e.g. Brugam *et al.*, 1998) or from larger sets covering a greater lake size and geographic expanse (e.g. Moser *et al.*, 2000). The other is to construct lake-specific training sets (e.g. Barker *et al.*, 1994; Moos *et al.*, 2005; Laird & Cumming, 2008, 2009).

9.3.4 Photic zone and diatom habitat availability

Multiple authors sampling surface sediments along depth transects from the near shore to deep environs have observed a rapid transition zone where benthic diatoms dominate, and above which planktonic diatom deposition increases (see Figure 9.3) (e.g. Brugam *et al.*, 1998; Rautio *et al.*, 2000; Moos *et al.*, 2005; Punning & Puusepp, 2007; Laird & Cumming, 2008, 2009). Moos *et al.* (2005) found this corresponded to the region of 1% incident-light penetration and proposed this near-shore location should be the most sensitive for constructing lake-level inferences. Subsequent studies indicate that nearshore cores can provide such sensitive records (Laird & Cumming, 2008; Laird & Cumming, 2009). Several authors have emphasized the necessity of understanding light regime and habitat availability when using benthic and planktonic ratios as proxies for lake-level change. Early work by Earle *et al.* (1988), investigating patterns in sedimentary diatoms of three morphometrically different Canadian lakes, concluded that spatial heterogeneity

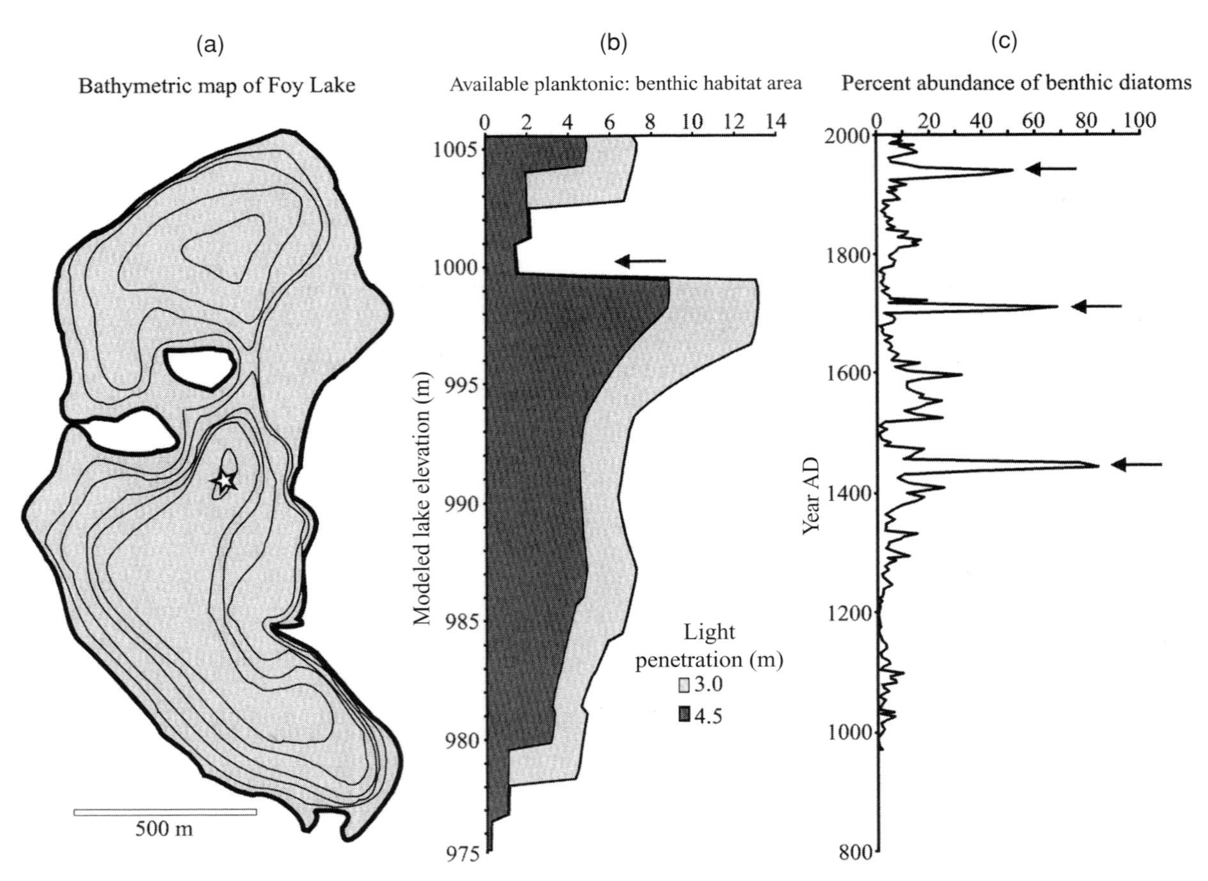

Figure 9.4 Modeling of available diatom habitat areas; modified from Stone & Fritz (2004). (a) A bathymetric map of Foy Lake, Montana (contour interval = 20 feet/6.1 m); the star indicates where core was taken. (b) Modeled changes in ratio of available diatom habitat area with depth at Foy Lake; the arrow indicates a benthic "notch," marking a rapid transition between available planktonic and benthic diatom habitat areas. (c) Percent abundance of benthic diatoms from sediment archives of Foy Lake; the arrows indicate spikes in the percentage of benthic diatoms related to specific lake elevations based on modeled changes in available diatom habitat areas.

in sample composition was a function of both water depth and basin morphometry.

The shape of a lake basin can have a profound influence on the habitat area available for diatom growth. Stone & Fritz (2004) explored this relationship by developing a three-dimensional modeling technique that calculated incremental changes in physical parameters of a lake (e.g. volume, planar surface area, lake bottom surface area) based on bathymetric and topographic maps of the basin and surrounding catchment (Figure 9.4a). Once these parameters were established for a given depth interval, the diatom habitat area available for planktonic and benthic diatom growth were calculated by making a few assumptions. The total planar surface area (water–air interface) greater than a depth of 1.5 m was used to estimate the available planktonic habitat area. Because benthic diatom habitat requires sunlight and a substrate, the total area available for benthic diatom production was estimated by calculating the surface area of the basin (sediment–water interface) exposed to sunlight, under different values of light penetration. Then, for all discrete potential lake elevations, available planktonic to benthic diatom habitat area ratios were estimated (Figure 9.4b).

Modeled diatom habitat area ratios are probably not directly comparable to ratios of planktonic and benthic diatoms from sedimentary archives; however, they can provide context for changes observed in P:B in the sedimentary record. In lake systems with complex bathymetry, as in Foy Lake (Stone & Fritz, 2004), P:B from sedimentary archives probably was a better indicator of mean lake depth than maximum lake depth. At some lake-surface elevations, a small change in lake level has the potential to produce a dramatic shift in relative ratio of available planktonic and benthic diatom habitat area. Rapid transitions appear as spikes or notches in the curve of modeled habitat availability with depth (Figure 9.4b) and represent a sharp break in basin slope, or a decoupling of mean depth from maximum depth. These transitions have the potential to produce "counter-intuitive" responses in diatom P:B in fossil sediments. By creating a sudden increase or decrease in the available habitat area for benthic diatom growth, rising lake levels could potentially increase the available benthic habitat area and lower the P:B. Conversely, a small drop in lake level has the potential to substantially decrease the available benthic habitat area, producing an increase in P:B, potentially confounding the standard interpretation of P:B in lake-level reconstructions.

Interestingly, in systems where rapid transitions occur, they can also provide a useful indicator of past lake level. Because spikes or notches in the modeled habitat area profile are tied to specific depths, when lake-level changes pass through these depths, a substantial change in P:B is likely to be observed in the sediment archives (Figure 9.4c). Well-constrained reconstruction of paleoclimate from fossil diatom assemblages in any lake system should incorporate an explicit consideration of lake-basin shape and how available planktonic and benthic habitat areas change with lake level; this is particularly true for lakes with complex bathymetry.

9.4 Applications of diatom-inferred depth changes in lakes

Ideally, the most sensitive signals of shoreline and water-level change will be found in marginal lake sediments that provide a sharp transition between planktonic and benthic assemblages (e.g. Battarbee, 2000; Digerfeldt, 1998). A multicore approach sampled along a depth gradient (e.g. Digerfeldt, 1998) should provide a stronger P:B response signal than that found in a single deepwater core. Models constructed from multiple lakes are not always as sensitive due to interactions between diatoms, habitat, and light availability and lake-specific training sets can produce a stronger, more applicable predictive model in many cases, from which robust diatom-inferred reconstructions can be made (Laird & Cumming, 2008, 2009). Unfortunately, appropriate marginal lake sediments are not always present in lakes, nor is it always feasible to take multiple transects when constructing multilake training sets. Significant diatom–depth relationships with high predictive ability have been developed from multilake training sets based on single deepwater cores (Brugam *et al.*, 1998; Moser *et al.*, 2000; Rioual *et al.*, 2007), but again, these must be interpreted in relationship to the additional factors that influence P:B ratios.

9.4.1 Lake-specific quantitative modeling

Laird & Cumming (2008) used multiple lines of evidence (fossil diatoms, scaled chrysophyte:diatom valve ratio and percent organic matter) to reconstruct past lake levels in Lake 239, Experimental Lakes Area (ELA), Ontario, Canada. Lake 239 was selected due to its predicted sensitivity to lake-level change based on size, depth, and catchment to lake area ratios (Dearing, 1997). Diatom species data from surface-sample depth transects (Moos *et al.*, 2005) and additional samples were used to develop a quantitative depth model. This model was based on the modern analogue technique (MAT) (Birks, this volume; Juggins, 2007) because the relationship between diatoms and depth was not linear. Although the model was developed from samples at depths ~2–32 m, it performed best when estimating depths less than ~10–12 m, due to the predominance of *C. stelligera* in the deeper samples. Diatom-inferred depths identified two periods of climate-driven low lake-level declines of 1–3 m during the late Holocene and ≥8 m during the arid mid Holocene. An extended period of high lake levels in the late Holocene was not adequately quantified with single-core analysis, however Laird & Cumming (2009) were later able to quantify the high lake-level stands using multiple-core analysis.

9.4.2 Multilake quantitative modeling

Rioual *et al.* (2007) provide an excellent example of freshwater diatom-inferred depth in their high-resolution quantitative climate reconstruction from fossil Eemian lake deposits in the French Massif Central (Figure 9.5). Using the extensive EDDI (Juggins, 2001), they established transfer functions for six environmental variables, including maximum depth, based on modern data from 351 European lakes. Water levels were reconstructed using the locally weighted-averaging technique (Juggins, 2001) which generates a "local" training set for each sample based on the 30 closest analogs in the data base. Diatom-inferred chemical variables (i.e. nutrients, silica, conductivity, pH) were compared with diatom-inferred maximum depth, autecology of major diatom taxa, and pollen data to identify three main climate phases during the Eemian interglacial; an early climatic optimum (~8000 years), a cooler wetter period (~7000 years) and a final warming (~2000 years). Rioual *et al.* (2007) clearly demonstrate that detailed information on long-term climate can be obtained through the application of freshwater diatom-inferred depth reconstructions in conjunction with other environmental variables.

9.5 Summary

The ability to determine water-level changes in freshwater systems is an important tool for use in reconstructing past climate and predicting future climate change, and in geologic and environmental studies. Although complex interactions affect

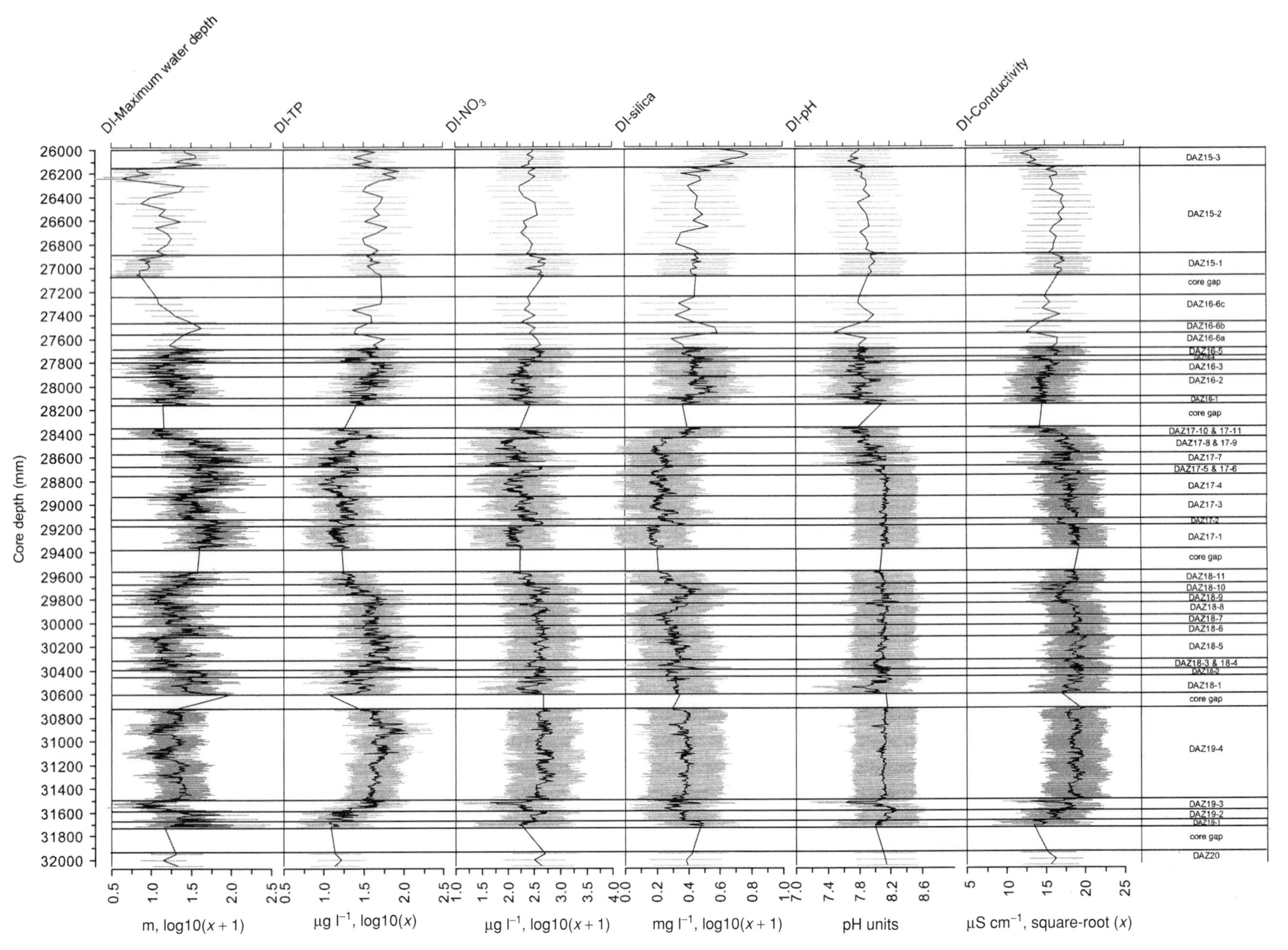

Figure 9.5 Diatom-based reconstructions for the whole sequence. The error bars correspond to specific errors of prediction calibrated under bootstrap cross-validation (Birks *et al.*, 1990). (Reprinted from *Quaternary Science Reviews*, Vol. 26, 2007, Rioual, P., Andrieu-Ponel, V., de Beaulieu, J.-L., Reille, M., Svobodova, H. & Battarbee, R. W., Diatom responses to limnological and climatic changes at Ribains maar (French Massif Central) during the Eemian and early Würm. Fig. 8, page 1568. © 2007 with permission from Elsevier.)

diatom signals in freshwater lakes, multiple responses can be utilized to quantitatively reconstruct water levels from freshwater diatom assemblages. Growth-habit changes based on depth distributions (i.e. benthic vs. planktonic) produce the most reliable evidence of water-level change. Autecological diatom response signals to changes in the physical–chemical environment can also be used. The development of quantitative diatom-inferred depth models to reconstruct water level in freshwater systems has greatly improved our understanding of past climatic and environmental change. However, it must be emphasized that depth reconstructions from freshwater diatom data should be implemented with caution. Many of the authors cited have stressed the importance of understanding the system under investigation and use of independent corroborating evidence before valid reconstructions can be made. The diatom signals observed are open to alternative interpretations. This difficulty is partly responsible for the few quantitative reconstructions found in freshwater systems. Increased efforts with more nearshore and multiple-core analysis are needed to strengthen the interpretation of the diatom data, and more diatom–depth calibration studies are needed, as well as investigations of diatom nutrient, light, and temperature interactions in relation to depth. We still lack large regional data bases for many freshwater lake regions and these should include lake depth. Diatom-inferred water-level reconstructions from multiple lakes can then be combined into regional climate reconstructions (Digerfeldt, 1998) and regional data assembled into wider-ranging continental reconstructions (e.g. Harrison *et al.*, 1996).

Acknowledgments

We would like to thank past collaborators: Jingrong Yang, John Anderson, Philip Barker, the late J. Platt Bradbury, Regine Jahn, André Lotter, Curt Stager for information, reprints, and discussions on their work for the first edition and, most importantly, to previous co-author Hamish Duthie for his invaluable insight, writing, and his significant contributions in developing quantitative methods for freshwater lake-level reconstruction. Thank you to our reviewers for their valuable comments.

References

Barker, P. A., Roberts, N., Lamb, H. F., Van Der Kaars, S., & Benfaddour, A. (1994). Interpretation of Holocene lake-level change from diatom assemblages in Lake Sidi Ali, Middle Atlas, Morocco. *Journal of Paleolimnology*, **12**, 223–4.

Battarbee, R. W. (2000). Palaeolimnological approaches to climate change, with special regard to the biological record. *Quaternary Science Reviews*, **19**, 107–24.

Bergner, A. G. N. & Trauth, M. H. (2004). Comparison of the hydrological and hydrochemical evolution of Lake Naivasha (Kenya) during three highstands between 175 and 60 kyr BP. *Palaeogeography, Palaeoclimatology, Palaeoecology*, **215**, 17–36.

Birks, H. J. B., Line, J. M., Juggins, S., Stevenson, A. C., & ter Braak, C. J. F. (1990). Diatoms and pH reconstruction. *Philosophical Transactions of the Royal Society London*, **B 327**, 263–78.

Björck, S., Noe-Nygaard, N., Wolin, J., *et al.* (2000). Eemian lake development, hydrology and climate – a multi-stratigraphic study of the Hollerup site in Denmark. *Quaternary Science Reviews*, **19**, 509–36.

Bradbury, J. P., Cumming, B., & Laird, K. (2002). A 1500-year record of climatic and environmental change in Minnesota III: measures of past primary productivity. *Journal of Paleolimnology*, **27**, 321–40.

Brugam, R. B., McKeever, K., & Kolesa, L. (1998). A diatom-inferred water depth reconstruction for an Upper Peninsula, Michigan, lake. *Journal of Paleolimnology*, **20**, 267–76.

Bunting, M. J., Duthie, H. C., Campbell, D. R., Warner, B. G., & Turner, L. J. (1997). A paleoecological record of recent environmental change at Big Creek Marsh, Long Point, Lake Erie. *Journal of Great Lakes Research*, **23**, 349–68.

Coops, H., Beklioglu, M., & Crisman, T. L. (2003). The role of water-level fluctuations in shallow lake ecosystems – workshop conclusions. *Hydrobiologia*, **506–509**, 23–7.

Dearing, J. A. (1997). Sedimentary indicators of lake-level changes in the humid temperate zone: a critical review. *Journal of Paleolimnology*, **18**, 1–14.

Delorme, L. D., Duthie, H. C., Esterby, S. R., Smith, S. M., & Harper, N. S. (1986). Prehistoric inferred pH changes in Batchawana Lake, Ontario from sedimentary diatom assemblages. *Archiv für Hydrobiologie*, **108**, 1–22.

Digerfeldt, G. (1986). Studies on past lake level fluctuations. In *Handbook of Holocene Palaeoecology and Palaeohydrology*, ed. B. E. Berglund, New York: John Wiley & Sons, pp. 127–43.

Digerfeldt, G. (1998). Reconstruction and of Holocene lake-level changes in southern Sweden: technique and results. *Paläoklimaforshung*, **25**, 87–98.

Duthie, H. C., Yang, J-R., Edwards, T. W. D., Wolfe, B. B., & Warner, B. G. (1996). Hamilton Harbour, Ontario: 8300 years of environmental change inferred from microfossil and isotopic analyses. *Journal of Paleolimnology*, **15**, 79–97.

Earle, J. C., Duthie, H. C., Glooschenko, W. A., & Hamilton, P. B. (1988). Factors affecting the spatial distribution of diatoms on the surface sediments of three Precambrian Shield lakes. *Canadian Journal of Fisheries and Aquatic Sciences*, **45**, 469–78.

Ekblom, A. & Stabell, B. (2008). Paleohydrology of Lake Nhaucati (southern Mozambique), ~400 AD to present. *Journal of Paleolimnology*, **40**, 1127–41.

Fahnenstiel, G. L. & Scavia, D. (1987). Dynamics of Lake Michigan phytoplankton: the deep chlorophyll layer. *Journal of Great Lakes Research*, **13**, 285–95.

Finkelstein, S. A. & Davis, A. M. (2006). Paleoenvironmental records of water level and climatic changes from the middle to late Holocene at a Lake Erie coastal wetland, Ontario, Canada. *Quaternary Research*, **65**, 33–43.

Finkelstein, S. A., Peros, M. C., & Davis, A. M. (2005). Late Holocene paleoenvironmental change in a Great Lakes coastal wetland: integrating pollen and diatom datasets. *Journal of Paleolimnology*, **33**, 1–12.

Gaillard, M.-J., Dearing, J. A., El-Dauoushy, F., Enell, M., & Håkansson, H. (1991). A late Holocene record of land-use history, soil erosion, lake trophy and lake-level fluctuations at Bjäresjösjön (south Sweden). *Journal of Paleolimnology*, **6**, 51–81.

Harrison, S. P., Yu, G., & Tarasov, P. E. (1996). Late Quaternary lake-level record from northern Eurasia. *Quaternary Research*, **45**, 138–59.

Heinsalu, A., Luup, H., Alliksaar, T., Nõges, P., & Nõges, T. (2008). Water level changes in a large shallow lake as reflected by the plankton: periphyton-ratio of sedimentary diatoms. *Hydrobiologia*, **599**, 23–30.

Jeppesen, E., Jensen, J. P., Sondergaard, M., Lauridsen, T., & Landkildehus, F. (2000). Trophic structure, species richness and biodiversity in Danish lakes: changes along a phosphorus gradient. *Freshwater Biology*, **45**, 201–218.

Juggins, S. (2001). The European Diatom Database. University of Newcastle, Newcastle. See http://craticula.ncl.ac.uk/Eddi/jsp/help.jsp.

Juggins, S. (2007). C2 Version 1.5 User guide. Software for ecological and palaeoecological data analysis and visualization. Newcastle upon Tyne: Newcastle University.

Karrow P. F. & Calkin P. E. (eds.) (1985). *Quaternary Evolution of the Great Lakes*. Geological Association of Canada Special Paper 30.

Kingston, J. C., Lowe, R. E., Stoermer, E. F., & Ladewski, T. B. (1983). Spatial and temporal distribution of benthic diatoms in northern Lake Michigan. *Ecology*, **64**, 1566–80.

Krabbenhoft, D. P. & Webster, K. E. (1995). Transient hydrological controls on the chemistry of a seepage lake. *Water Resources Research*, **31**, 2295–305.

Laird, K. R. & Cumming, B. F. (2008). Reconstruction of Holocene lake level from diatoms, chrysophytes and organic matter in a drainage lake from the Experimental Lakes Area (northwestern Ontario, Canada). *Quaternary Research*, **69**, 292–305.

Laird, K. R. & Cumming, B. F. (2009). Diatom-inferred lake level from near-shore cores in a drainage lake from the Experimental Lakes Area, northwestern Ontario, Canada. *Journal of Paleolimnology*, **42**, 65–80.

Larsen, C. E. & Schaetzl R. (2001). Origin and evolution of the Great Lakes. *Journal of Great Lakes Research*, **27**, 518–46.

Lewis, C. F. M, Karrow P. F., Blasco S. M., *et al.* (2008). Evolution of lakes in the Huron basin: deglaciation to present. *Aquatic Ecosystem Health & Management*, **11**, 127–36.

Lotter, A. (1988). Past water-level fluctuations at Lake Rotsee (Switzerland), evidenced by diatom analysis. In *Proceedings of Nordic Diatomist Meeting, Stockholm, 1987, Department of Quaternary Research (USDQR) Report* 12, Stockholm: University of Stockholm, pp. 47–55.

Lotter, A. F. & Bigler, C. (2000). Do diatoms in the Swiss Alps reflect the length of ice-cover? *Aquatic Sciences*, **62**, 125–41.

Moos, M. T., Laird, K. R., & Cumming, B. F. (2005). Diatom assemblages and water depth in Lake 239 (ELA, Ont.): implications for paleoclimatic studies. *Journal of Paleolimnology*, **34**, 217–27.

Moos, M. T., Laird, K. R., & Cumming, B. F. (2009). Climate-related eutrophication of a small boreal lake in northwestern Ontario: a palaeolimnological perspective. *The Holocene*, **19**, 359–67.

Moser, K. A., Korhola, A., Weckström, J., *et al.* (2000). Paleohydrology inferred from diatoms in northern latitude regions. *Journal of Paleolimnology*, **24**, 93–107.

Nõges, T., Nõges, P., & Laugaste, R. (2003). Water level as the mediator between climate change and phytoplankton composition in a large shallow temperate lake. *Hydrobiologia*, **506–509**, 257–63.

Nguetsop, V. F., Servant-Vildary, S., & Servant, M. (2004). Late Holocene climatic changes in west Africa, a high resolution diatom record from equatorial Cameroon. *Quaternary Science Review*, **23**, 591–609.

Punning, J.-M. & Puusepp, L. (2007). Diatom assemblages in sediments of Lake Juusa, southern Estonia with an assessment of their habitat. *Hydrobiologia*, **586**, 27–41.

Rautio, M., Sorvari, S., & Korhola, A. (2000). Diatom and crustacean zooplankton communities, their seasonal variability and representation in the sediments of subarctic Lake Saanajärvi. *Journal of Limnology*, **59** (suppl. 1), 81–96.

Richardson, J. L. (1969). Former lake-level fluctuations – their recognition and interpretation. *Mitteilungen der Internationalen Vereiningung für Theoretische und Angewandte Limnologie*, **17**, 78–93.

Rioual, P., Andrieu-Ponel, V., de Beaulieu, J.-L., *et al.* (2007). Diatom responses to limnological and climatic changes at Ribains maar (French Massif Central) during the Eemian and early Würm. *Quaternary Science Reviews*, **26**, 1557–609.

Rippey, B., Anderson, N. J., & Foy, R.H. (1997). Accuracy of diatom-inferred total phosphorus concentrations and the accelerated eutrophication of a lake due to reduced flushing and increased internal loading. *Canadian Journal of Fisheries and Aquatic Sciences*, **54**, 2637–46.

Round, F. E. (1961). Studies on bottom-living algae in some lakes of the English Lake District. Part VI. The effect of depth on the epipelic algal community. *Journal of Ecology*, **49**, 245–54.

Sánchez-Castillo, P. M., Linares-Cuesta, J. E., & Fernández-Moreno, D. (2008). Changes in epilithic diatom assemblages in a Mediterranean high mountain lake (Laguna de La Caldera, Sierra Nevada, Spain) after a period of drought. *Journal of Limnology*, **67**, 49–55.

Schmidt, R., Kamenik, C., Lange-Bertalot, H. & Klee, R. (2004). *Fragilaria* and *Staurosira* (Bacillariophyceae) from sediment surfaces of 40 lakes in the Austrian Alps in relation to environmental variables, and their potential for palaeoclimatology. *Journal of Limnology*, **63**, 171–89.

Shkarpa, V. (2006). Climate and human settlement effects on Silver Lake Ohio. Unpublished M.Sc. thesis, Cleveland State University, Cleveland, OH.

Smol, J. P. (1988). Paleoclimate proxy data from freshwater arctic diatoms. *Verhandlungen der Internationalen Vereinigung für Theoretische und Angewandte Limnologie*, **23**, 837–44.

Smol, J. P. (2008). *Pollution of Lakes and Rivers: A Paleoenvironmental Perspective*, 2nd edition, Hoboken: John Wiley & Sons.

Smol, J. P. & Cumming, B. F. (2000). Tracking long-term changes in climate using algal indicators in lake sediments. *Journal of Phycology*, **36**, 986–1011.

Stager, J. C., Cumming, B., & Meeker, L. (1997). A high-resolution 11,400-year diatom record from lake Victoria, East Africa. *Quaternary Research* **47**, 81–9.

Stager, J. C., Ryves, D., Cumming, B. F., Meeker, L. D., & Beer, J. (2005). Solar variability and the levels of Lake Victoria, East Africa, during the last millennium. *Journal of Paleolimnology*, **33**, 243–51.

Stevenson, R. J. & Stoermer, E. F. (1981). Quantitative differences in benthic algal communities along a depth gradient in Lake Michigan. *Journal of Phycology*, **17**, 29–36.

Stone, J. R. & Fritz, S. C. (2004). Three-dimensional modeling of lacustrine diatom habitat areas: improving paleolimnological interpretation of planktic : benthic ratios. *Limnology & Oceanography*, **49**, 1540–8.

Street-Perrott, F. A. & Harrison, S. P. (1985). Lake levels and climate reconstruction. In *Paleoclimate Analysis and Modelling*, ed. A.D. Hecht, New York, NY: John Wiley and Sons, pp. 291–340.

Tapia, P. M., Fritz, S. C., Baker, P. A., Seltzer, G. O., & Dunbar, R. B. (2003). A late quaternary diatom record of tropical climatic history from Lake Titicaca (Peru and Bolivia). *Palaeogeography, Palaeoclimatology, Palaeoecology*, **194**, 139–64.

Tarasov, P. E., Harrison, S. P., Saarse, L., *et al.* (1994). *Lake Status Records from the Former Soviet Union Mongolia: Data Base Documentation*. Paleoclimatology Publications Series Report no. 2. Boulder, CO: World Data Center-A for Paleoclimatology, NOAA Paleoclimatology Program.

Tarasov, P., Dorofeyuk, N., & Metel'tseva, E. (2000). Holocene vegetation and climate changes in Hoton-Nur basin, northwest Mongolia. BOREAS, **29**, 117–26.

Turner, M. A., Heubert, D. B., Findlay, D. L., *et al.* (2005). Divergent impacts of experimental lake-level drawdown on planktonic and benthic plant communities in a boreal forest lake. *Canadian Journal of Fisheries and Aquatic Sciences*, **62**, 991–1003.

Vadeboncoeur, Y., Peterson, G., Vander Zanden, M. J., & Kalff, J. (2008). Benthic algal production across lake size gradients: interactions among morphometry, nutrients and light. *Ecology*, **89**, 2542–52.

Verschuren, D., Laird, K. R., & Cumming, B. F. (2000). Rainfall and drought in equatorial east Africa during the past 1,100 years. *Nature*, **403**, 410–14.

Voigt, R., Grüger, E., Baier, J. & Meischner, D. (2008). Seasonal variability of Holocene climate: a palaeolimnological study on varved sediments in Lake Jues (Harz Mountains, Germany). *Journal of Paleolimnology*, **40**, 1021–52.

Wetzel, R. G. (2001). *Limnology Lake and River Ecosystems*, 3rd edition, San Diego, CA: Academic Press.

Wolfe B. B., Edwards T. W. D., & Duthie, H. C. (2000). A 6000-year record of interaction between Hamilton Harbour and Lake Ontario: quantitative assessment of recent hydrologic disturbance using ^{13}C in lake sediment cellulose. *Aquatic Ecosystem Health Management*, **3**, 47–54.

Wolin, J. A. (1996). Late Holocene lake-level fluctuations in Lower Herring Lake, Michigan, U.S.A. *Journal of Paleolimnology*, **15**, 19–45.

Wolin, J. A. & Duthie, H. C. (1999). Diatoms as indicators of water level change in freshwater lakes, In *The Diatoms: Applications for Environmental and Earth Sciences*, ed. E. F. Stoermer & J. P. Smol, Cambridge: Cambridge University Press, pp. 183–202.

Wolin, J. A. & Stoermer, E. F. (2005). Response of a Lake Michigan coastal lake to anthropogenic catchment disturbance. *Journal of Paleolimnology*, **33**, 73–94.

Xue, B., Qu, W., Wang, S., Ma, Y., & Dickman, M. D. (2003). Lake level changes documented by sediment properties and diatom of Hulum Lake, China since the late Glacial. *Hydrobiologia*, **498**, 133–41.

Yang, J-R. & Duthie, H. C. (1995). Regression and weighted averaging models relating surficial diatom assemblages to water depth in Lake Ontario. *Journal of Great Lakes Research*, **21**, 84–94.

Yu, G. & Harrison, S.P. (1995). *Lake Status Records from Europe: Data Base Documentation*. Paleoclimatology Publications Series Report no. 3. Boulder, Colorado:World Data Center-A for Paleoclimatology, NOAA Paleoclimatology Program.

10 Diatoms as indicators of hydrologic and climatic change in saline lakes

SHERI C. FRITZ, BRIAN F. CUMMING, FRANÇOISE GASSE, AND KATHLEEN R. LAIRD

10.1 Introduction

Lakes are intricately tied to the climate system in that their water level and chemistry are a manifestation of the balance between inputs (precipitation, stream inflow, surface runoff, groundwater inflow) and outputs (evaporation, stream outflow, groundwater recharge) (Mason *et al.*, 1994). Hence, changes in a lake's hydrologic budget caused by climatic change have the potential to alter lake level and lake chemistry. These changes, in turn, may affect the physiological responses and species composition of the lake's biota, including diatoms. Here we review the use of diatoms as indicators of hydrologic and climatic change, with an emphasis on environmental reconstruction in arid and semi-arid regions. First we discuss linkages among climate, hydrology, lake hydrochemistry, and diatoms that form the foundation for environmental reconstruction, and then we review selected examples of diatom-based studies.

10.1.1 Lake hydrology and hydrochemistry

Lakes vary in their hydrologic sensitivity to climatic change (Winter, 1990). In basins with a surface outlet, lake-level increase is constrained by topography, and any change in input is usually balanced by outflow. Thus, in open basins, lake level fluctuates relatively little, unless hydrologic change is sufficiently large to drop water level below the outlet level. In contrast, closed-basin lakes without surface outflow often show changes in level associated with changes in the balance between precipitation and evaporation (P – E). The magnitude of response to fluctuations in P – E depends on lake volume and the relative contribution of groundwater inflow and outflow to the hydrologic budget; lake-level change is greatest in terminal basins, which have neither surface nor groundwater outflow. In closed-basin lakes, seasonal and interannual changes in P – E alter ionic concentration (salinity) and ionic composition through dilution or concentration (Figure 10.1). Although lake-level change can affect diatom communities directly, here we consider primarily linkages between diatoms and changes in the concentration and composition of total dissolved solids (salinity). Wolin and Stone (this volume) describe in detail how diatoms track changes in lake level.

Closed-basin lakes are common in semi-arid and arid regions, where moisture balance (P – E) is negative. Although lakes in arid/semi-arid regions commonly receive far less attention than their temperate counterparts, on a global basis they contain a nearly equal volume of water (Meybeck, 1995) and occur on every continent. Furthermore, because water resources are often marginal in regions of net negative water balance, the status of these lakes is intricately tied to human activities and culture.

10.1.2 Lake salinity and ionic composition

Lake salinity is a product of the source waters, which determine the initial conditions, and subsequent modification of that water by climatic processes. Most lakes of moderate to high salinity occur in areas of negative water balance, where evaporation concentrates dissolved salts. Occasionally saline basins occur in humid and sub-humid regions where source waters flow through salt deposits. A variety of classification schemes have been used to categorize lakes based on salinity. The limit separating "freshwater" from "saline" water is commonly set at 3 g l^{-1} (Williams, 1981). Although this limit is somewhat arbitrary, it also has biological validity in that a number of organisms, including species of diatoms, have distribution limits at about 3 g l^{-1}. For example, diatom data from the North American Great Plains (Fritz *et al.*, 1993) show that some species, such as *Stephanodiscus minutulus* Kütz. Cleve & Möller, *Stephanodiscus niagarae* Ehrenb., and *Fragilaria capucina* var. *mesolepta* (Rabenh.) Grun. in Van Heurck are rare or absent above 3 g l^{-1}, whereas

The Diatoms: Applications for the Environmental and Earth Sciences, 2nd Edition, eds. John P. Smol and Eugene F. Stoermer. Published by Cambridge University Press.

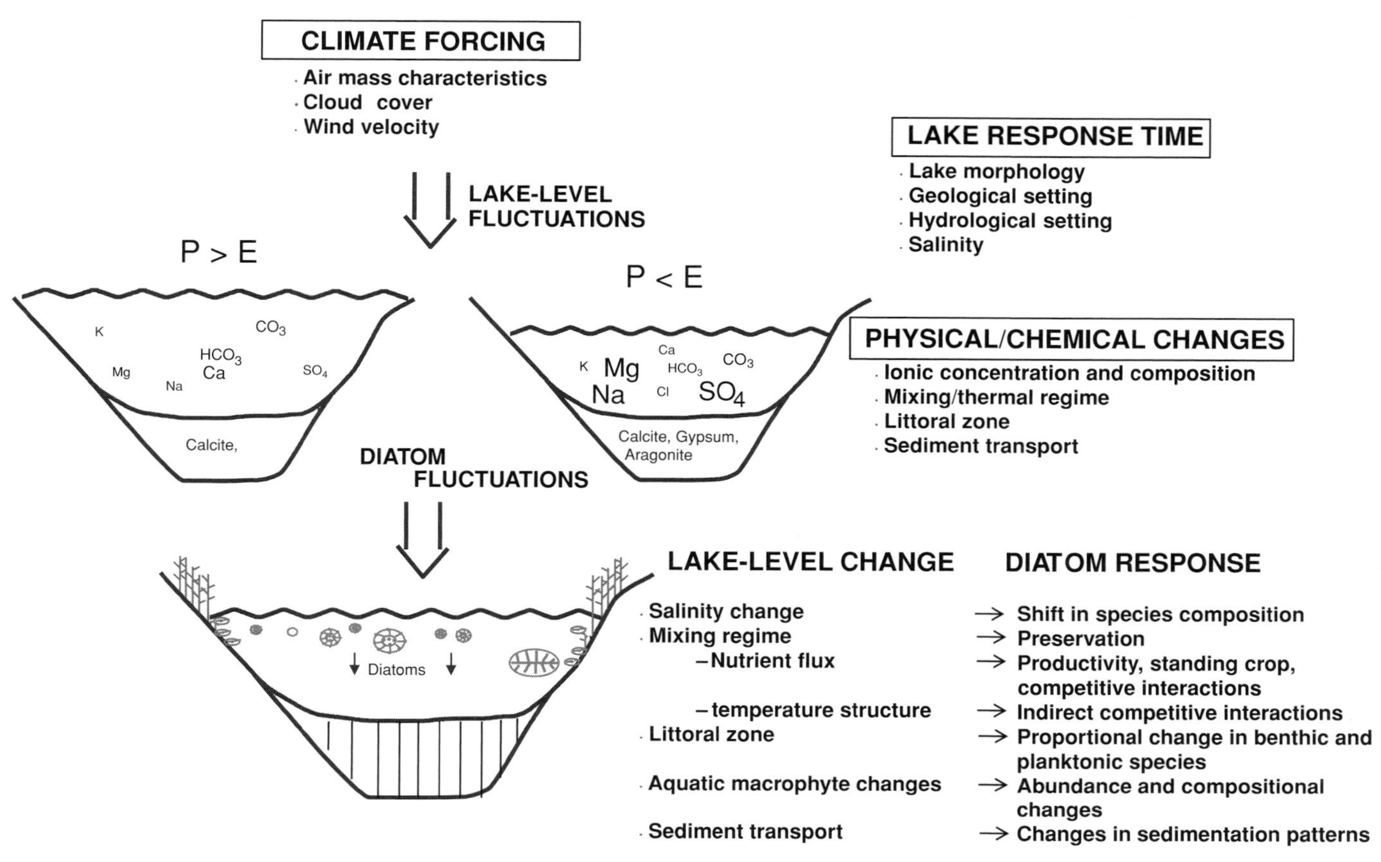

Figure 10.1 A diagram illustrating the potential influences of shifts in the balance between precipitation and evaporation on lake-water hydrochemistry and on diatom assemblages. Modified from Cumming *et al.* (1995).

other species, including *Cyclotella quillensis* Bailey, *Chaetoceros elmorei* Boyer, and *Navicula* cf. *fonticola*, only occur above this limit.

In contrast to marine waters, lakes are highly variable in ionic composition and may be dominated by carbonates, sulfates, or chlorides in combination with major cations (Ca, Mg, Na). The initial solute composition is acquired *via* chemical weathering reactions and the interaction of inflowing ground and surface waters with bedrock and surface deposits. However, significant alteration of ionic composition in closed-basin lakes can occur *via* precipitation of minerals during evaporative concentration (Eugster & Jones, 1979).

Within any geographic region, lakes may vary spatially in ionic concentration and composition as a result of variability in groundwater sources and flow (Fritz, 2008). A lake fed by a deep aquifer may differ in salinity and ionic composition relative to an adjacent basin fed by surficial flow (Bennett *et al.*, 2007; Zlotnik *et al.*, 2007). Similarly, a chain of lakes connected by groundwater flow is often progressively more saline, because of progressive concentration of groundwater recharge from up-gradient to down (Donovan, 1994; Yu *et al.*, 2002). Thus, ionic concentration and composition of closed-basin lakes are a result of the interplay between the chemistry of source waters, groundwater flow, and climate.

10.1.3 Diatoms and lake-water chemistry

Diatoms show distributional patterns based on both salinity and brine composition. The earliest scheme to classify diatoms based on salinity preference was developed by Kolbe (1927) and modified by Hustedt (Hustedt, 1953). This "halobion" system of classification, however, was developed primarily for marine water. It is not really appropriate for inland waters, where lakes are often characterized by brines other than the NaCl type and span a far greater range of salinity, from dilute freshwater to hypersaline brines, often >100 g l^{-1}. It is the series of species replacements along the salinity gradient (Figure 10.2b) that makes diatoms powerful indicators of chemical change driven by changes in hydrology and climate. Although many diatom species occur in lakes or rivers of varied ionic composition, others show clear specificity and may be characteristic of carbonate, chloride, or sulfate (e.g. Fritz *et al.*, 1993;

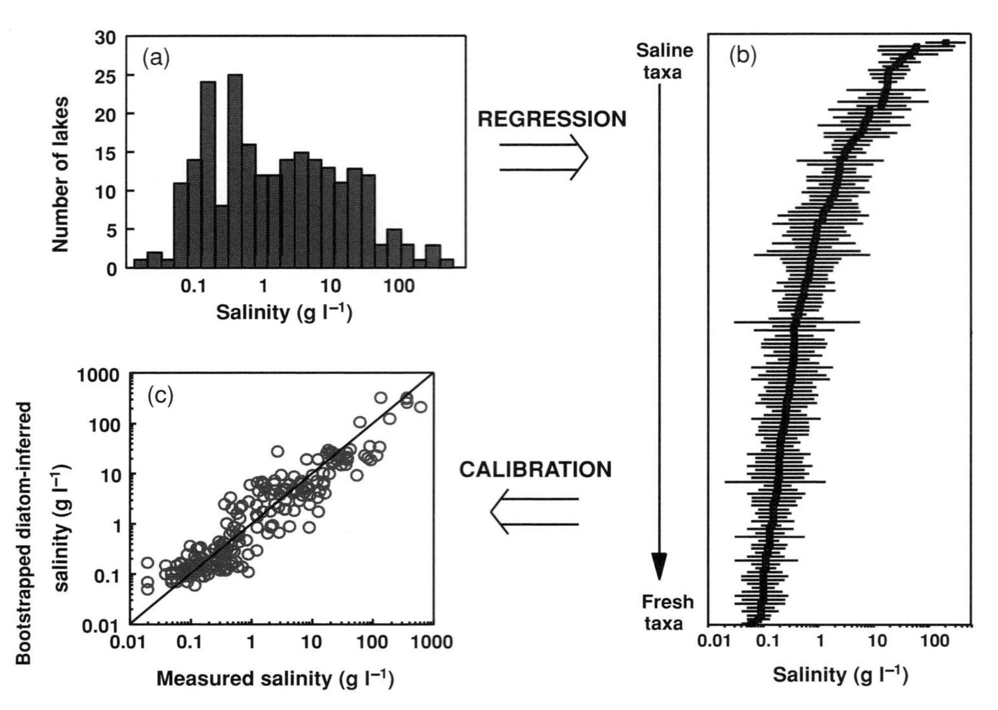

Figure 10.2 Illustration of steps involved in the development of a diatom-based salinity inference model based on the 219-lake modern British Columbia data set (Wilson *et al.*, 1996). (a) The frequency distribution of the 219 lakes along the salinity gradient. (b) The estimated salinity optima (solid squares) and tolerances (bars) of diatom taxa arranged according to salinity optima. (c) The relationship between measured and diatom-inferred salinity after bootstrapping. (Modified from Wilson *et al.* (1996).)

Wilson *et al.*, 1996; Saros & Fritz, 2002; Potapova & Charles, 2003).

The mechanisms that link diatom distribution to ionic concentration and composition have not been well studied. Clearly, increased salinity represents an osmotic stress, but it is not clear whether distributional patterns are a result of differential responses to salinity directly or to other physiochemical process correlated with salinity (Saros & Fritz, 2000a). Accumulation of proline or glycerol for osmoregulation has been investigated in very few diatom taxa (Shobert, 1974; Fisher, 1977). In the eurytopic species, *Cyclotella meneghiniana* Kütz., experimental studies suggest that high salinity affects the thickness of the silica cell wall (Tuchman *et al.*, 1984), and studies by Bhattacharyya & Volcani (1980) suggest that, at least for marine diatoms, high salinity may affect nutrient transport across the cell membrane. Physiological experiments with several common diatom taxa from North American saline lakes indicate that brine composition (sulfate versus carbonate) is associated with differences in nitrogen metabolism (Saros & Fritz, 2000b). This may occur, because high sulfate concentrations inhibit uptake of molybdate, which is required in many enzyme systems involved in nitrogen utilization. Brine composition also may affect inorganic carbon availability and, thus, short-term growth rates of diatoms (Saros & Fritz, 2002). Thus, the impact of salinity on diatom distributions may only indirectly reflect salinity and may be driven by salinity effects on nutrient uptake or some other physiological process.

10.2 Tools for environmental reconstruction

The diatom species composition of a lake is related to chemical, physical, and biological variables. Relationships between diatom taxa and their "preferred" environmental conditions can be estimated by a surface-sediment calibration set, or "training set" (Charles & Smol, 1994; Birks, 1995; Birks, this volume). The basic approach is to choose a suite of lakes, commonly referred to as the training set (Figure 10.2a), that spans the limnological gradients of interest (e.g. lake-water salinity). Given unlimited resources and time, the ideal training set of lakes would span all biologically relevant limnological gradients (e.g. pH, nutrients, salinity, ionic composition, light, etc.). Limnological variables are measured from each lake in the training set and related to diatom composition of surface-sediment samples (the uppermost 0.5 or 1 cm of sediment, representing the last few years of sediment accumulation), or to living planktonic and periphytic diatom assemblages. Ideally, multiple estimates of limnological conditions should be taken throughout the year, or sampling should be restricted to a specific period (e.g. spring turnover or late summer), to minimize noise introduced due to seasonal changes in water chemistry (Birks, this volume). Based on the distribution of taxa preserved in the surface sediments or living in training-set lakes, techniques can be used to quantify the response of each diatom taxon to a given environmental variable (Figure 10.2b). A number of different techniques can be used, ranging from simple methods, such as weighted-averaging (WA) to mathematically more complex techniques, such as Gaussian logit regression (Birks, this volume).

Subsequently, environmental conditions may be calculated from the species composition of fossil samples, based on quantitative estimates of species responses derived from the training set (Birks, this volume). This calibration step assumes that the environmental variable to be reconstructed is highly correlated with diatom distribution, that the ecological preferences of taxa can be assessed accurately based on their occurrences in the training set, that the ecological preferences of taxa have not changed over time, and that taxa present in the fossil samples are well represented in the training set. An assessment of whether downcore assemblages are represented by the contemporary training-set assemblages can be done with analog matching techniques or a number of multivariate approaches (Birks, this volume). Similarly, the predictive ability of a training set can be evaluated in several ways, including computer-intensive approaches, such as jackknifing and bootstrapping (Figure 10.2c) (Birks, this volume).

10.2.1 Diatoms and salinity gradients in modern calibration data sets

Regional calibration data sets clearly show that the distribution of diatom taxa is highly correlated with lake-water salinity (and the related variable, conductivity) and that quantitative estimates of past changes in salinity can be inferred from the species composition of sediment cores on a variety of timescales, from decades to millennia. Large calibration data sets (>150 lakes) have been developed to quantify the relationship of diatom taxa to ionic concentration and composition in North America and Africa, and smaller data sets have been developed for arid and semi-arid regions on every continent.

Regional data sets from Northwest Africa (125 samples, Khelifa, 1989; Gasse unpublished), East Africa (167 samples, Gasse, 1986), and Niger (20 samples, Gasse, 1987) were combined to create a large modern African data set. Samples include benthic, epiphytic, and epipelic diatom assemblages and are from a diverse array of aquatic environments, including lakes, swamps, bogs, and springs, which cover vast gradients in ionic concentration (40 to approximately 100,000 $\mu S\ cm^{-1}$) and exhibit virtually every possible combination of ionic composition. Environmental variables that account for significant variation in the diatom assemblage data include conductivity, pH, and cation and anion composition. Based on this data set, diatom models were developed to infer conductivity (bootstrapped coefficient of determination of 0.81), as well as pH, alkali and alkaline earth metals, and the anion ratio of carbonate + bicarbonate : sulfate + chloride (Gasse *et al.*, 1995). This African data set has now been used in inferring conductivity and other limnological variables of diatoms from sediment cores in many African lakes (e.g. Barker *et al.*, 2002; Chalié & Gasse, 2002; Legesse *et al.*, 2002; Barker & Gasse, 2003; Bergner & Trauth, 2004; Vallet-Coulomb *et al.*, 2006; Ryner *et al.*, 2007).

A 219-lake data set from British Columbia (Canada) includes lakes from the Cariboo/Chilcotin region, Kamloops region, southern Interior Plateau, and southern Rocky Mountain Trench. The lakes range from subsaline through hypersaline (Figure 10.2a) and include both bicarbonate/carbonate-dominated systems in combination with magnesium and sodium, as well as sodium sulfate dominated lakes (Cumming & Smol, 1993; Wilson *et al.*, 1994; Cumming *et al.*, 1995; Wilson *et al.*, 1996). In general, the freshwater lakes are dominated by bicarbonate/carbonate in combination with calcium although, in a few cases, they also have a high proportion of sulfate. From these data, the optima and tolerance ranges of the dominant diatom taxa (those occurring at >1% relative abundance in at least two lakes, Figure 10.2b) were estimated using a WA approach, and a salinity inference model was developed using WA calibration (Figure 10.2c, r^2 boot = 0.87).

The strong association between diatom species distributions and lake-water ionic concentration and composition has been demonstrated in a number of smaller regional data sets from around the world. An early study on 55 lakes from the northern Great Plains (NGP), which was subsequently expanded to include over 100 lakes (Saros *et al.*, 2000; Fritz, unpublished), was one the first attempts to develop a calibration to quantify modern ecological characteristics of diatoms and generate a quantitative inference model (Fritz, 1990; Fritz *et al.*, 1991; Fritz *et al.*, 1993). Lakes in the NGP data set range in salinity from <1 to 270 $g\ l^{-1}$ and are largely dominated by magnesium and sodium sulfates or sodium carbonate, although a few lakes are sodium chloride systems. Ordination techniques show that lake-water ionic concentration and composition (sulfate versus carbonate) accounted for major directions of variation in the diatom species composition in these lakes. Based on the strong relationship between diatom species composition and lake-water salinity, a WA model was developed with an apparent coefficient of determination of 0.83. Since this time, additional diatom data sets have been developed in North America, including a 57-lake calibration set from Sierra Nevada, California, that developed highly significant salinity inference models using WA-PLS (partial least squares) techniques (Bloom *et al.*, 2003).

In Central Mexico, Metcalfe (1988) investigated diatom assemblages collected from plankton, periphyton, and bottom sediments from 47 aquatic environments including lakes, springs, marshes, and rivers. The majority of these sites are dominated by carbonate/bicarbonate. Davies *et al.* (2002) documented the strong relationship between modern diatom assemblages and ionic concentration and composition from 31 lakes from central Mexico and developed a strong inference model ($r^2 > 0.9$) for conductivity from diatom assemblages. Servant and Roux (Servant-Vildary & Roux, 1990; Roux *et al.*, 1991) sampled 14 saline lakes in the Bolivian Altiplano, variously dominated by sodium carbonate, sulfate, and chloride salts, and related diatom distribution to chemical variables, including salinity and ion composition. Sylvestre *et al.* (2001) developed quantitative models for ionic composition and concentration from a larger data set from the region and applied them to the reconstruction of late-glacial climate change (Sylvestre *et al.*, 1999).

Studies in Australian lakes also have shown statistically significant relationships. In a study of 32 lakes from southeastern Australia (conductivity range: 1000 to 195,000 $\mu S\ cm^{-1}$), Gell & Gasse (1990) found that diatom species collected from the plankton and scrapings from floating and submerged objects were clearly related to lake-water conductivity. Similarly, Blinn (1995) showed that conductivity was one of the most important predictors of epipelic and epilithic diatom communities in a set of 19 lakes in Western Victoria (conductivity range: 1500 to 262,000 $\mu S\ cm^{-1}$). Tibby *et al.* (2007) investigated diatom–conductivity relationships in eight seasonally monitored wetlands and found that dramatic short-term seasonal fluctuations resulted in little change in inferred species optima, whereas regression results were more affected by variability at a subdecadal scale.

Modern studies have been conducted in a number of regions of Europe, including Spain (Reed, 1998), Russia (see Juggins, 2001), and Turkey (Kashima, 1994; 2003). Kashima (1994) quantified the ecological preferences of diatom taxa from the central Anatolian Plateau using WA techniques (apparent $r^2 = 0.9$) using 51 periphyton scrapes and water samples collected from 23 lakes, ponds, and rivers (salinity range: 8 to 100 $g\ l^{-1}$). Kashima (2003) then applied the transfer function to demonstrate higher salinities in Turkish lakes during interglacial stages and lower salinity during glacial times. Reconstructions of lake-water conductivity have also been inferred at sites in Turkey (Reed *et al.*, 1999; Roberts *et al.*, 2001) using a combined salinity transfer function that integrates lakes from multiple parts of Europe (Juggins, 2001). Reed (1998) developed a diatom-conductivity transfer function (apparent $r^2 = 0.91$) from 70 Spanish lakes (conductivity range: 150 to >338,000 $\mu S\ cm^{-1}$), which has been applied to reconstruct past lake-water conductivity from Holocene diatom assemblages (e.g. Reed *et al.*, 2001). Modern diatom–salinity relationships have also been quantified from Asia (Xiangdong *et al.*, 2003). For example, 40 freshwater to hypersaline lakes on the eastern Tibetan Plateau showed that conductivity was the most important measured environmental variable, which allowed strong ($r^2_{jack} = 0.9$) and significant diatom-based inference models to be developed using WA and WA-PLS models.

A number of studies on saline lakes have been conducted in high-latitude regions. For example, Roberts & McMinn (1996) showed that diatom assemblages were significantly correlated with lake-water salinity in a data set of 33 lakes from the Vestfold Hills in Antarctica. This data set was subsequently expanded with 14 lakes and ponds from east Antarctica to produce a diatom inference model that was used to infer the paleosalinity from a regional Holocene sequence (Roberts *et al.*, 2004). Similarly, in a study of diatom–conductivity relationships in 40 lakes in Greenland, Ryves *et al.* (2002) found that conductivity was the strongest explanatory variable for changes in diatom assemblages, despite a relatively short conductivity gradient (24 to 4072 $mS\ cm^{-1}$). The resultant WA-PLS models that were developed were similar in explanatory power to those reported above (Ryves *et al.*, 2002) and were used to document Holocene changes in lake-water salinity (Ryves *et al.*, 2002; McGowan *et al.*, 2003).

10.3 Environmental reconstructions

One of the primary uses of stratigraphic diatom studies in arid and semi-arid regions is in reconstruction of past climate and hydrology. These reconstructions can be undertaken at a variety of temporal scales, from multiple decades to millennia, dependent on the nature of the stratigraphic record. Case studies from Africa and the Americas illustrate how diatoms can record long-term climatic variation induced by changes in solar insolation, sea-surface temperature gradients, and other modes of land-surface–ocean–atmosphere interaction.

10.3.1 African records of Quaternary climate and hydrology

Climates of tropical and subtropical Africa are characterized by large interannual to millennial-scale rainfall variability, which have induced spectacular fluctuations in lake volume and salinity that are well registered in diatom records. A striking example comes from the Sahara–Sahel during the early–mid Holocene (*c.* 11.5 to 6–4 ka BP), when many lakes occupied presently dry depressions (Gasse & Roberts, 2004; Hoelzmann *et al.*, 2004). Many of them were the surface expression of aquifers that now lie several meters or tens of meters below ground surface. Although most of the lacustrine remains have been eradicated by wind deflation during the dry late Holocene, diatom records are available from a number of sites. Examples below are taken from Gasse (2002), unless otherwise indicated.

At the scale of the whole western Sahara–Sahel, the diatom flora responds to large differences in lake-water chemistry, depending on regional and local geological and hydrological factors. In the south are species presently living in dilute, sodium–calcium bicarbonate waters, such as *Cyclotella pseudostelligera* Hust., *Cyclotella glomerata* Bach., *Aulacoseira ambigua* (Grun.) O. Müller, and *Aulacoseira granulata* (Ehr.) Simonsen and varieties. During aridification phases, the development of *C. meneghiniana*, *Navicula elkab* O. Müller, and *Anomoeoneis sphaerophora* (Ehr.) Pfitzer indicates evolution toward sodium carbonate water, with the deposition of trona and natron and a slight chloride enrichment of residual brines (Figure 10.3a). Several records from the central and northern Sahara reflect lakes of the sodium–calcium/chloride–sulfate type, with abundant *Cyclotella choctawhatcheeana* Prasad, *Chaetoceros* resting spores, *Navicula pseudocrassirostris* Hust., *Nitzschia stompsii* Choln., then marine-like species, such as *Melosira moniliformis* (Grun.) Hustedt and *Mastogloia* spp., euryhaline species that tolerate hypersaline water (e.g. *Amphora tenerrima* Aleem et Hust., *Navicula salinicola* Hust.), and finally precipitation of gypsum and halite (Figure 10.3b).

Despite these regional features, all diatom records suggest a long-lasting early- to mid-Holocene humid period. This wetting reflects a northward shift of the Intertropical Convergence Zone and enhanced intensity of summer monsoon rainfall, in response to orbitally induced changes in summer insolation and complex feedbacks from the ocean and vegetation (Braconnot *et al.*, 2007). Some basins south of 20° S (Figure 10.3a) indicate a first wet pulse at *c.* 15–14 ka BP but desiccated during, or at, the end of the subsequent dry Younger Dryas chronozone (YD; 12.8–11.7 ka BP). The lakes refilled rapidly at 11.5–11.0 ka BP at the onset of the Holocene. These changes occurred in phase with the major post-glacial warming–cooling events in high northern latitudes. In the heart of the Sahara, lacustrine environments often established somewhat later, and abruptly, by 11–10 ka BP (Figure 10.3b). This lag possibly represents the time needed to fill the aquifers and to elevate the water table to the level of the depression floor. Refilling by inflowing freshwater may initially produce a saline water body, due to the weathering of salt crusts deposited during the previous arid phase. The lacustrine optimum (the deepest and most dilute

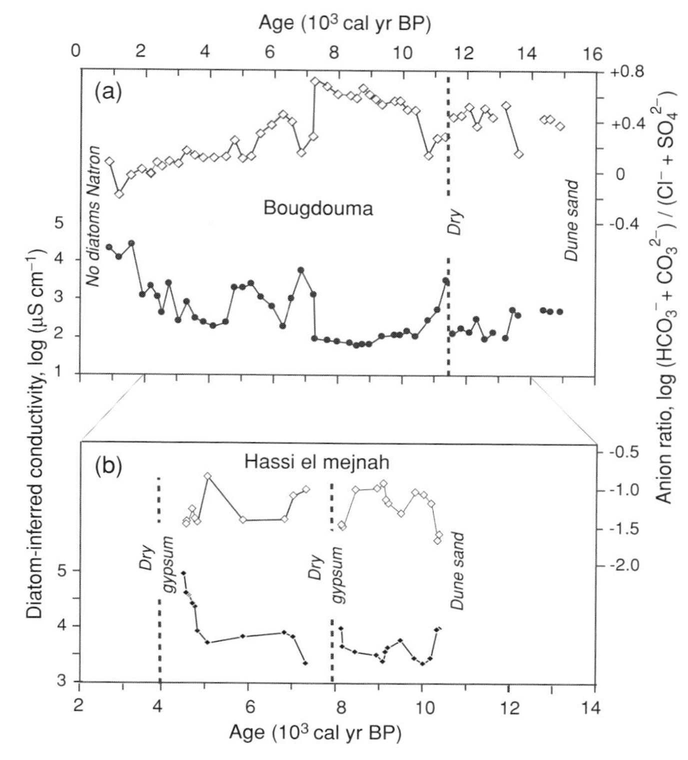

Figure 10.3 Diatom-inferred conductivity and anion ratio in two interdunal Holocene lakes from West Africa. In each core, gaps in the sequence of reconstructed values represent core sections that did not contain sufficient diatoms for quantitative water-chemistry reconstruction; the sedimentology of these intervals is indicated on the figure. (a) Bougdouma, in the semi-arid Sahel (South Niger, Manga Plateau). The circles are conductivity, and the open squares are the anion ratio. The modern site is a seasonal hyperalkaline water body, supplied by a phreatic aquifer recharged by local rainfall. (b) Hassi el Mehnah, northern Sahara (south Algeria, northern margin of the Great Western Erg). The closed squares are the conductivity data, and the open squares are the anion ratio. The paleolake of the sodium–calcium/chloride sulfate type was supplied by underflows of wadis from the Atlas mountains and local rainfall. See Gasse (2002) for further details.

conditions) was established by 9.5–8.5 ka BP. At that time, Neolithic populations lived around freshwater lakes in the heart of the Sahara, for example, in the Ténéré desert where the diatom-inferred conductivity was lower than that of modern rainfall and in ephemeral wadis from the neighboring Aïr highlands. The duration of the lacustrine optimum is site specific. Several records show two Holocene lacustrine episodes (Figure 10.3b), separated by a drastic drought around 8 ka BP. Such a drought, possibly related to the 8.2 ka event identified at high northern latitudes, has been recorded at large geographic scales in the African and Indian monsoon domains (Gasse & van Campo, 1994) and reflects a perturbation of the global climate system (Mayewski *et al.*, 2004) (see discussion below). The nature of aridification of the Sahara–Sahel following the mid Holocene is still a matter of debate in terms of whether change was abrupt (de Menocal *et al.*, 2000) or gradual (Kröpelin *et al.*, 2008). The final lake desiccation is generally extremely rapid at Saharan sites (Figure 10.3b), by 6–5 ka BP or earlier. This abruptness may reflect the threshold effect of the water table falling below the lake bottom under progressive rainfall decrease, rather than an abrupt change in precipitation. Climate model simulations suggest a similar explanation for the rapid vegetation changes (Liu *et al.*, 2007). In the hyperarid northern Chad, Lake Yoa is still supplied by groundwater flowing from the neighboring mountains; water conductivity inferred from both diatoms (Figure 10.4b) and chironomids shows a sudden increase at *c.* 4 ka BP, while terrestrial proxies indicate a progressive desertification from 5.6 to 2.7 ka BP (Kröpelin *et al.*, 2008). This was interpreted as a threshold effect of a gradual climatic change on the aquatic ecosystem. In residual Sahelian water bodies, inferred changes in conductivity suggest a stepwise P – E deterioration after 7.0–5.5 ka BP (Figures 10.3a, 10.4a) and at least one episode drier than today over the past 4 ka, e.g. at 1.5–2.0 ka BP at Kajemarum Oasis (Figure 10.4a). However, from the same sequence, maximum evaporative concentration inferred from the ostracod $\delta^{18}O$ record took place later, at *c.* 1 ka BP (Holmes *et al.*, 1997, 1998). This apparent discrepancy is not yet explained but suggests, as in the example of Lake Yoa, that caution should be taken when directly interpreting diatom-inferred salinity changes in terms of paleoclimate.

In contrast to the relatively shallow lakes of the Sahara–Sahel, East African rift lakes have undergone water-level fluctuations up to hundreds of meters during the Late Quaternary (Cohen *et al.*, 2007). In the northern tropics, the timing of post-glacial evolution resembles that of the Sahara–Sahel lakes and for the same climatic reasons. This is exemplified by the

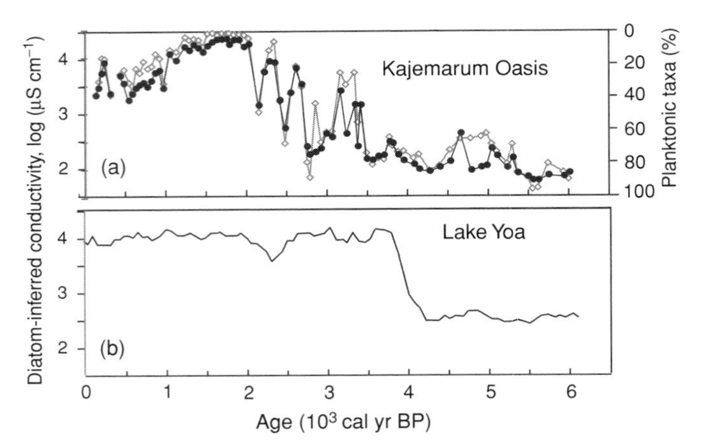

Figure 10.4 Aridification of the Sahara–Sahel since the mid Holocene. (a) The stepwise conductivity increase (antiphased with the shift in planktonic percentages) at Kajemarum Oasis (Sahel, Northern Nigeria). See Gasse (2002) and Holmes *et al.* (1997, 1998) for details concerning the chronology. (b) An abrupt increase in conductivity at Lake Yoa, northern Chad, central Sahara. (Redrawn from Kröpelin *et al.* (2008).)

diatom record from the river-fed Lake Abiyata (Figure 10.5a), a presently closed, shallow, hyperalkaline lake in the Ziway-Shala lake system (Ethiopian Rift; Chalié and Gasse, 2002). Lake-level fluctuations were reconstructed independently (Gasse and Street, 1978; Gillespie *et al.*, 1983; Figure 10.5b). Besides abundant *Aulacoseira* and *Stephanodiscus* spp., freshwater high stands show species endemic in the large East African lakes, such as *Nitzschia lancettula* O. Müller and *Surirella engleri* O. Müller. The lake-level reconstruction shows two regressive events punctuating the wet Holocene period, around 8 ka BP and 6.7 ka BP. These events also are clear in both the diatom and inferred lake-level data from Lake Abhé in the neighboring Afar Rift (Gasse, 1977, 2000; Gasse *et al.*, 1998). These events are likely blurred in the diatom record by reworking of freshly emerged sediments. As in the Sahel, the ionic concentration of sodium carbonate increases after *c.* 6.0–5.5 ka BP, and the stepwise aridification is punctuated by wet spells and extreme droughts. Saline episodes are characterized by high percentages of *C. meneghiniana*, *Thalassiosira faurii* (Gasse) Hasle, *Thalassiosira rudolfi* (Bach.) Hasle, *Rhopalodia gibberula* (Her.) O. Müller, and euryhaline–hyperhaline *Nitzschia* spp. On a glacial/interglacial timescale, diatom records from rift lakes from equatorial East Africa (Holdship, 1976; Barker *et al.*, 2002; Barker & Gasse, 2003; Bergner & Trauth, 2004) show comparable trends, despite some differences in the timing of high stands, linked to periods of maximum insolation during the rainy seasons close to the Equator.

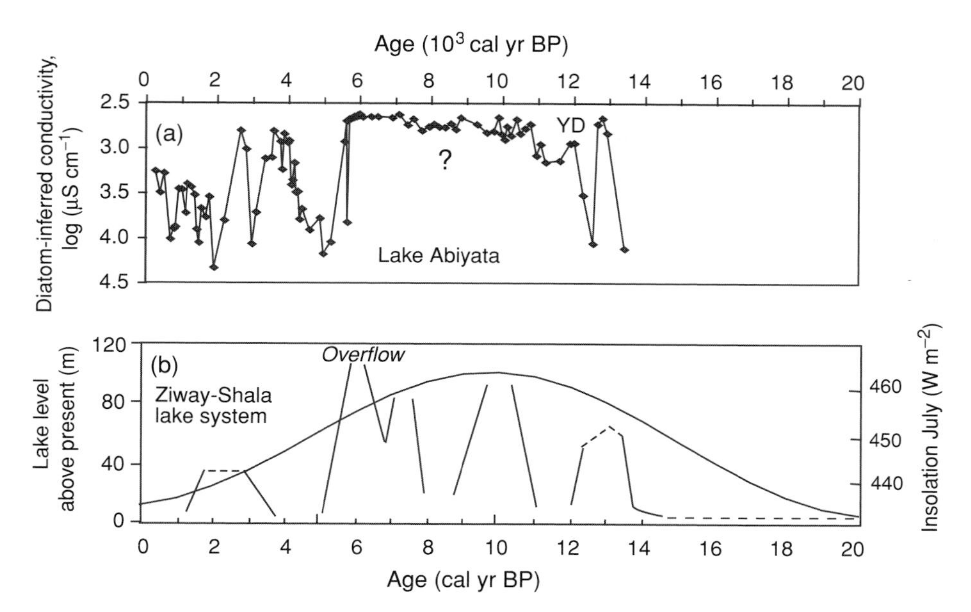

Figure 10.5 (a) Diatom-inferred conductivity of Lake Abiyata, a lake of the Ziway-Shala lake system, Ethiopian Rift (Chalié and Gasse, 2002). (b) Lake-level fluctuations in the Ziway-Shala system (in m above the level of the modern lowest lake of the system, Lake Shala) (Gasse & Street, 1978; Gillespie *et al.*, 1983) compared with changes in summer insolation at 10° N (Laskar *et al.*, 2004).

To date, few paleoclimatic studies have been applied directly to issues of water management. In a core from a small closed saline lake from southwest Madagascar (Lake Ihotry) that is supplied by both direct rainfall and a shallow aquifer, diatom data document a relatively deep, freshwater episode from *c.* 3.3 to 2.3 ka BP, followed by a desiccation trend punctuated by large variations in salinity, and finally the onset of modern conditions by 0.7–0.6 ka BP (mean conductivity = 13.3 mS cm^{-1}). Sensitivity experiments were carried out, based on a combined water and chloride balance model established from instrumental data (two years of measurements). They show that, whatever the rainfall and/or evaporation changes, the evolution of this local waterbody was primarily controlled by the regional water table through its successive connection/disconnection to the lake (Vallet-Coulomb *et al.*, 2006). It is now possible to estimate the timing of the lake's total disappearance under decreasing regional rainfall or enhanced water consumption by irrigation in the catchment.

Important progress has been achieved in African diatom research over the past decades. In particular, diatom records show the abruptness of natural changes in water availability and quality. We can now understand more completely the response of diatoms and individual lakes to natural or human-induced environmental change and analyze threshold mechanisms. Diatom records show that dramatic droughts have occurred over the past millennium, and, in the context of both global warming and increasing population pressure, many semi-arid African regions may become drier. Thus, there is an urgent need for more applied research. One approach is hydrological modeling of water and salt balances constrained by past hydrological changes inferred from diatoms. Simulation experiments could then be used to predict the ecosystem behavior under varied scenarios of future change.

10.3.2 North and South American records of Quaternary climate and hydrology

In North and South America, Quaternary hydrology and climate have been inferred from the diatom stratigraphy of closed-basin lakes that fluctuate in salinity in response to variations in precipitation and temperature. In glaciated regions of North America, lake sediments record the pattern of limnological change following retreat of the Laurentide Ice Sheet *c.* 12 ka BP. In the northern Great Plains, freshwater diatoms, such as *Cyclotella bodanica* Grun., *Cyclotella michiganiana* Skvortzow, *A. granulata*, *Stephanodiscus* spp., and benthic *Fragilaria*, dominate in the centuries following lake formation and reflect the cool, moist early-Holocene climate. Subsequent replacement of freshwater species by taxa indicative of saline waters (*C. quillensis*, *C. choctawhatcheeana*, *Chaetoceros elmorei/muelleri*) suggests hydrologic closure of lakes in response to climatic warming/drying and occurs coincident with a shift in vegetation from forest to grassland. Timing of the transition from freshwater to saline and from forest to grassland varies among sites and occurs as early as 9.5 ka BP in some places (Radle *et al.*, 1989; Laird *et al.*, 2007) but not until after *c.* 8 ka BP at other sites (Laird *et al.*, 1996a). At some sites the transition is apparently quite rapid and occurs over a period of several hundred years (Radle *et al.*, 1989; Fritz *et al.*, 1991; Laird *et al.*, 2007), whereas in other cases, changes in hydrochemistry occur gradually over one to two millennia (Laird *et al.*, 1996a). The timing and abruptness of response undoubtedly is, in part, a function of lake morphometry and the amount of change in the lake's water budget required to drop lake level below the outlet threshold (Kennedy, 1994; Stone & Fritz, 2004). In any case, the timing of hydrologic change in these parts of the northern Great Plains suggests interaction between insolation and a retreating ice sheet in regional climates during the Late-Glacial and early-Holocene periods. Further to the west, near the eastern edge of the Rocky Mountains, the timing of maximum aridity may have occurred earlier, closer to the summer insolation maximum between 11 and 9 ka BP, because of the reduced influence of the Laurentide Ice Sheet (Schweger & Hickman, 1989; Stone & Fritz, 2006).

Most diatom records suggest that the mid Holocene was very arid in the continental interior, as reflected by high diatom-inferred lake-water salinity. Some sites (Fritz *et al.*, 1991; Stone & Fritz, 2006; Laird *et al.*, 2007) show considerable century-scale variability in salinity and hence in moisture, whereas at other sites, salinity appears less variable (Laird *et al.*, 1996a). To some extent, the differences in pattern are probably a result of morphometic and hydrochemical differences rather than differences in local climate. For example, the more saline lakes are often dominated by taxa with very broad salinity tolerances, particularly *C. quillensis* and *C. elmorei/muelleri*, and, in these systems, large shifts in hydrochemistry are required to cross the salinity tolerance thresholds of the common taxa. In contrast, a number of species have thresholds near the freshwater/saline water boundary of 3 g l^{-1} and within the hyposaline range, and fluctuations across this gradient result in shifts in species composition. Clearly, an arid climate dominated the continental interior throughout much of the Holocene, and it is only within the last few millennia that freshwater conditions, indicative of increased moisture, have recurred with greater frequency (Fritz *et al.*, 1991; Laird *et al.*, 1996a; Stone & Fritz, 2006; Laird *et al.*, 2007).

Closed-basin lakes also are common in non-glaciated regions of western North America, particularly in intermontane basins west of the Rocky Mountains. These basins often contain a much longer record of climatic change, in some cases extending back as far as the late Tertiary. Pluvial lakes, such as Walker Lake, Nevada, were part of the extensive Lake Lahontan system, which covered much of the Great Basin in the late Pleistocene. Diatom assemblages dominated by *Stephanodiscus excentricus* Hust. and *Surirella nevadensis* Hanna & Grant indicate that it was large and hyposaline prior to 30 ka BP, maintained by a climate much moister than today (Bradbury *et al.*, 1989). Lake level dropped at about 25 ka BP as the climate dried to form a shallow saline lake with *Anomoeoneis costata* (Kütz.) Hust. and *Navicula subinflatoides* Hust. This shallow lake persisted until *c.* 16 ka BP and then desiccated either permanently or intermittently, as indicated by sediment texture and the absence of microfossils. Diatom abundance remained low until the mid Holocene (5 ka BP), when the Walker River rapidly refilled the lake basin. The lake sediments during the filling event show a progression from saline species (*N. subinflatoides*) to species with progressively fresher salinity optima (*C. elmorei*, *C. meneghiniana*, *S. excentricus*, *S. niagarae*, *Cyclotella ocellata* Pantocsek) as the fresh river water mixed with saline lake waters. Subsequent oscillations between freshwater and saline taxa document short-term variations in climate during the late Holocene.

Saline lakes also are present on the Colorado Plateau, formed by dissolution and collapse in a karstic terrain. Diatom-based studies of one site in north-central Arizona have been used to infer fluctuations in hydrology and climate during the Holocene (Blinn *et al.*, 1994). Attached diatoms characteristic of alkaline freshwater dominate the late glacial and early Holocene, whereas pulses of *A. sphaerophora*, a taxon characteristic of carbonate-dominated hyposaline systems, suggest periods of reduced water level and evaporative concentration during the mid Holocene. Prehistoric occupation of the lake's catchment coincided with a late-Holocene dry period, as indicated by increases in *A. sphaerophora* from *c.* 0.6 to 1.2 ka BP, and it may be that native peoples were driven to settle near the few open water sources during this interval of drought.

In the tropical Andes of South America, lacustrine diatom stratigraphy has been instrumental in understanding the nature of climatic variability at orbital, millennial, and sub-millennial scales. Changes in relative abundance among planktonic, benthic, and saline diatom taxa show that Lake Titicaca was deep and fresh during glacial periods, whereas during interglacial periods, lake level fell below the outlet threshold and salinity increased (Tapia *et al.*, 2003; Fritz *et al.*, 2007). Seismic reflection data coupled with increased abundance of the saline taxon *Chaetoceros muelleri* Lemmermann suggest that lake level dropped more than 80 m and salinity approximately doubled

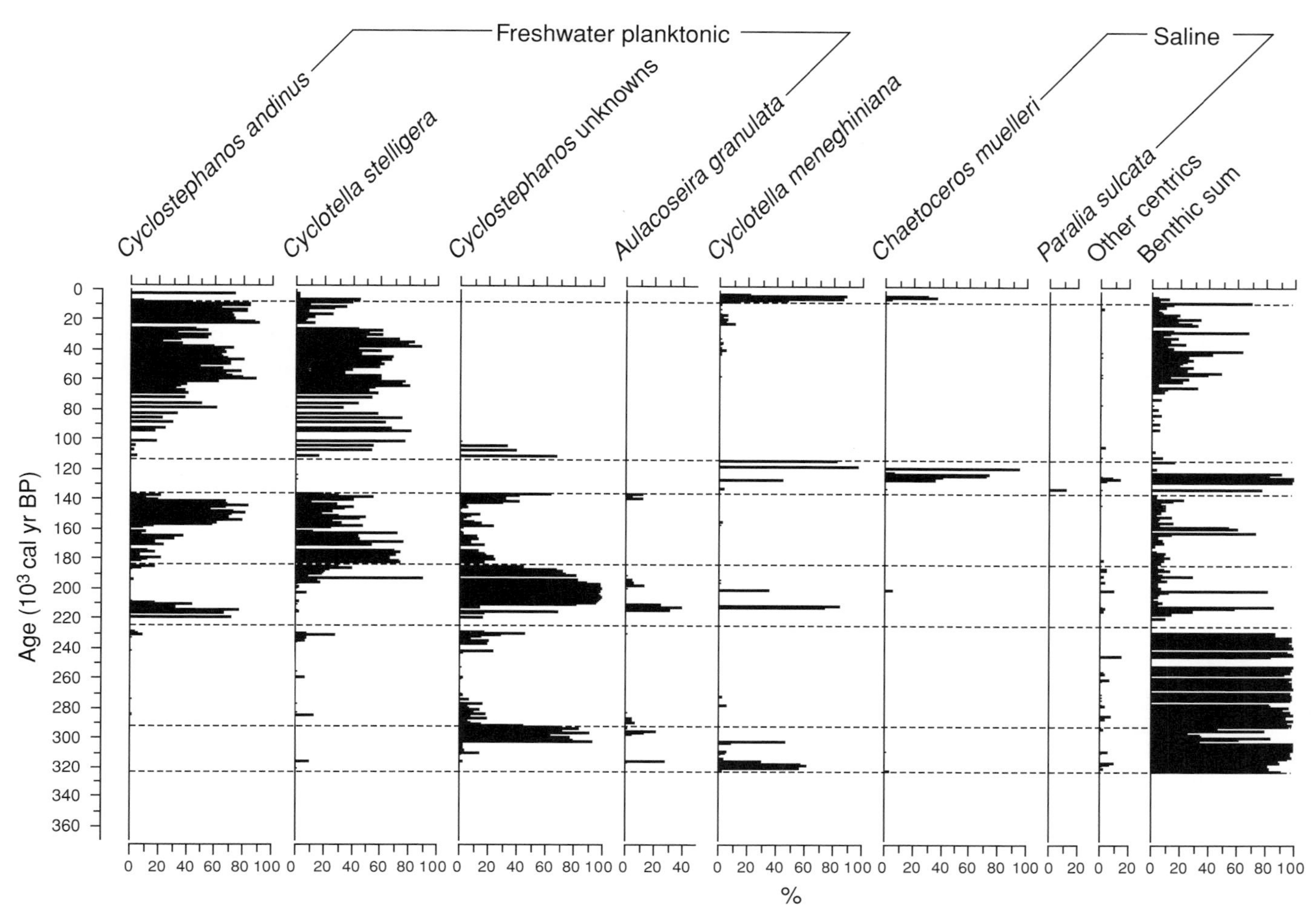

Figure 10.6 Diatom stratigraphy of a drill core spanning the last ~360,000 years from Lake Titicaca (Bolivia/Peru) showing long-term fluctuations in lake level. Periods dominated by freshwater planktonic diatoms are times when the lake was fresh and overflowing. The increase in saline taxa during the mid Holocene and MIS5e suggests major lake-level decline and elevated salinity. These hydrologic shifts are driven by changes in precipitation associated with the South American summer monsoon. See Fritz *et al.* (2007) for more details.

during the mid Holocene. Limnological change during the last interglacial (marine isotope stage 5e, MIS5e) (~125 ka BP) was even more extensive, and the high abundance of saline diatoms (Figure 10.6) coupled with geochemical evidence suggest that this was the most arid period of the last ~400 ka BP (Fritz *et al.*, 2007). These massive hydrological changes were likely driven by global and regional temperature change and by insolation forcing of the South American summer monsoon. Higher-frequency lake-level variation in Lake Titicaca (Baker *et al.*, 2005), as well as in other lakes in the Lake Titicaca drainage basin (Ekdahl *et al.*, 2008), suggests that millennial and sub-millennial hydrologic variation is forced by sea-surface temperature gradients in the tropical Atlantic and Pacific.

10.3.3 High-resolution studies of climatic variability

The spatial and temporal patterns of change in high-resolution records can help discern climatic processes that occur on decadal through millennial timescales and the sensitivity of the climatic system to various high-frequency forcing factors, such as variability of the linked atmospheric–oceanic system, solar activity, volcanic eruptions, and greenhouse gases (Rind & Overpeck, 1993; Bond *et al.*, 2001; Mayewski *et al.*, 2004). The number of diatom records of sufficiently high resolution to understand natural climatic cycles on decadal and centennial scales and of long enough duration to understand millennial-scale dynamics has increased in the last decade, but they are still relatively rare. This is probably the result of several factors, including the time-consuming nature of such analyses and the small number of systems with sufficiently high sediment accumulation rates. Even with such records, examining coherency

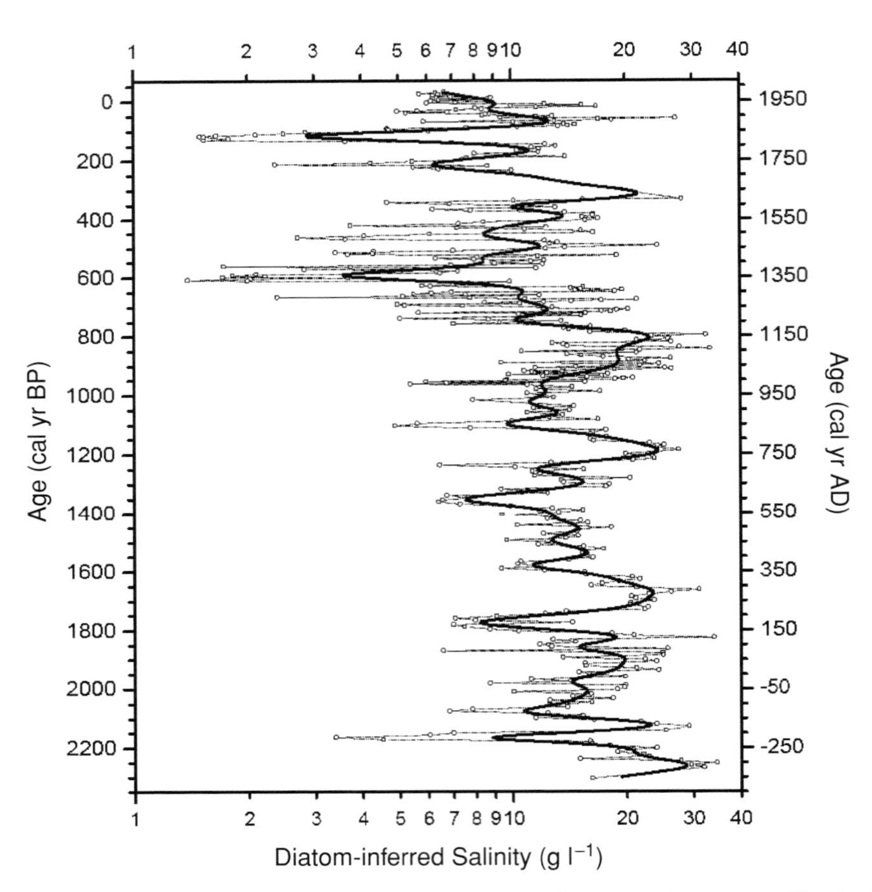

Figure 10.7 Reconstruction of diatom-inferred salinity for Moon Lake (ND) over the last ~2300 years (dotted line). The solid line is a fast Fourier transformation of interpolated five-year equal-interval diatom-inferred salinity values using a ten-point smoothing window. Chronology is based on four accelerator mass spectrometry (AMS) radiocarbon measurements calibrated to calendar years. (Modified from Laird *et al.* 1996b.)

at decadal to centennial scale is often difficult due to limited dating control and the errors inherent in radiocarbon dates (Telford *et al.*, 2004a; Telford *et al.*, 2004b; Oswald *et al.*, 2005). Yet, despite the difficulties in correlating among sites at short temporal scales, the high-resolution diatom records discussed below clearly display how dynamic the Holocene has been on decadal, centennial, and millennial scales.

The first high-resolution study of salinity changes and inferred climatic dynamics from a closed-basin lake examined changes in the northern Great Plains, USA (Laird *et al.*, 1996b; Laird *et al.*, 1998a). This sub-decadal record of diatom-inferred salinity for the past 2300 years from Moon Lake provides a long-term record of drought conditions and a means of evaluating how representative the twentieth century is of the natural variation in drought intensity and frequency (Figure 10.7). High-salinity intervals were dominated by taxa, such as *C. quillensis*, *C. elmorei/muelleri*, and *Nitzschia* cf. *fonticola*, whereas fresher intervals were dominated by taxa, such as *S. minutulus* and *Stephanodiscus parvus* Stoermer & Håkansson. The Moon Lake record indicates that recurring severe droughts of greater intensity than those during the 1930s "Dust Bowl" were the norm prior to AD 1200. These extreme and persistent droughts, for which we have no modern equivalents, were most pronounced between AD 200–370, 700–850, and 1000–1200. The latter drought period is coeval with the "Medieval Warm Period" (MWP, AD 1000–1200; Lamb, 1982). A pronounced shift in climate at AD 1200 coincided with the end of the MWP and the onset of the "Little Ice Age" (LIA, AD 1300–1850; Porter, 1986). The Moon Lake record provides support for a hydrologically complex LIA interval (Bradley & Jones, 1992), with periods of wet conditions interspersed with short episodes of drier conditions, at times comparable to the droughts of the 1930s. During the LIA, Moon Lake attained low salinity values that had not been recorded by the diatoms since the early Holocene (Laird *et al.*, 1998a). Analyses of five other lake records in the Canadian and US prairies (Fritz *et al.*, 2000; Laird *et al.*, 2003) show large-scale shifts coincident with those apparent in the Moon Lake diatom record, as do other high-resolution diatom records from the northern Rocky Mountains (Stevens *et al.*, 2006; Bracht

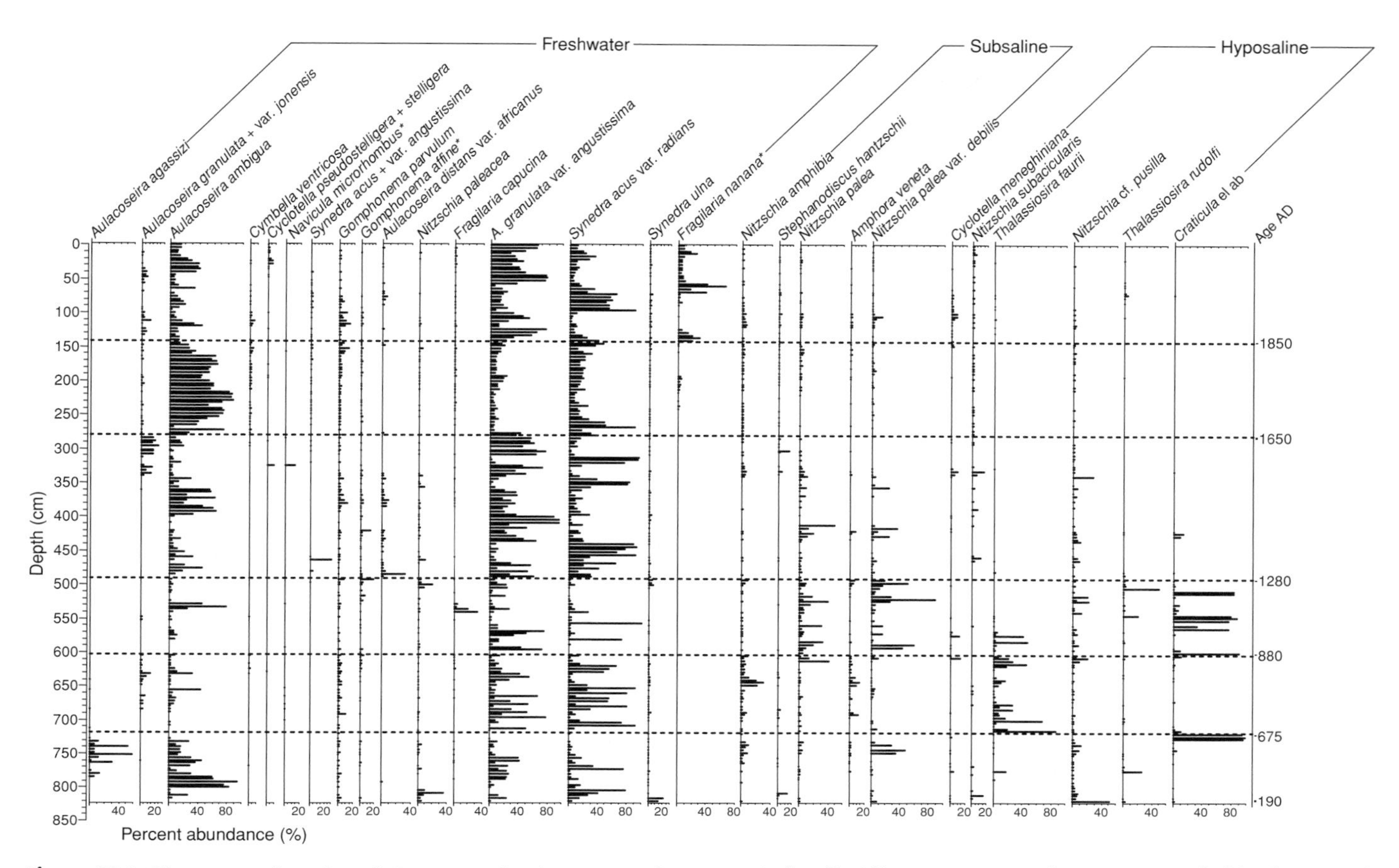

Figure 10.8 Diatom stratigraphy of Crescent Island Crater, Lake Naivasha, Kenya. Taxa are ordered according to increasing salinity optima, based on the calibration data set of Gasse *et al.* (1995).

et al., 2008), and synthesis of tree-ring data from western North America (Cook *et al.*, 2004). Thus, multiple records throughout western North America show coherent changes that are abrupt and large in magnitude, particularly at the onset of the LIA. Such changes are suggestive of a shift in mean climatic state or regime (Alley *et al.*, 2003) caused by large-scale changes in atmospheric circulation.

A high-resolution record of diatom salinity from East Africa also indicates pronounced multidecadal to centennial-scale variability over the past 1100 years (Vershuren *et al.*, 2000). A multiproxy study from Crescent Island Crater in Lake Naivasha (Kenya) suggests generally drier conditions during the MWP from ~AD 1000–1270, as indicated by pronounced salinity increases evident in increases in hyposaline taxa, such as *Craticula* (formerly *Navicula*) *elkab* and *Thalassiosira* spp., as well as several subsaline and hyposaline *Nitzschia* taxa (Figure 10.8). The diatom data also show pronounced salinity increases prior to this period. In contrast, the LIA period from ~AD 1270 to 1850 was fairly wet, although it was interrupted by several periods of aridity more severe than any recorded in the twentieth century. The freshwater periods were dominated primarily by fluctuations in *A. ambigua*, *Aulacoseira granulata* var. *angustissima* O. Müller Simonsen, and *Synedra acus* var. *radians* (Kütz.) Hust. (Figure 10.8), probably driven by changes in nutrients and light. Comparison of the Crescent Island Crater paleo-record to pre-colonial historical records from ~AD 1400 onward suggests a strong linkage between low stands of the lake and drought-induced famine, and between times of prosperity and record high-stand intervals (Verschuren *et al.*, 2000). Such data highlight the value of these high-resolution records in reconstructing changes in climate at timescales relevant to human civilization.

High-resolution records covering the past five to six millennia or longer enable examination of millennial-scale dynamics, which are often superimposed on variation at other temporal scales. One such diatom study from the interior of British Columbia indicates a strong millennial-scale pattern in inferred lake depth over the past ~5500 years (Cumming *et al.*, 2002). Abrupt shifts in species assemblages and diatom-inferred depth occurred approximately every 1200 years, similar in timing to glacial fluctuations and North Atlantic dynamics. Coherency at the millennial-scale among three high-resolution salinity records from the Canadian prairies spanning the past six millennia suggested that the abrupt shifts may be related to changes in ocean–atmosphere interactions that influence the mean position of the jet stream (Michels *et al.*, 2007). A strong millennial-scale component, similar in pacing to the ice-rafted debris flux in the North Atlantic (Bond *et al.*, 2001), was found in two decadal-scale diatom records from the South American Altiplano (Ekdahl *et al.*, 2008). Significant periodicities in selected diatom assemblages suggested the impact of Atlantic and Pacific sea-surface temperature variation on precipitation associated with the South American summer monsoon at a variety of timescales.

Foy Lake in northwestern Montana, USA, is currently fresh with an intermittent outlet, but it was likely a closed basin in the mid Holocene, as indicated by the presence of diatoms with higher salinity optima during this period (Stone & Fritz, 2006). Modeling of the changes in benthic to pelagic habitat associated with changes in lake level was used to generate a semi-quantitative reconstruction of mean lake depth from the percentage of benthic diatoms (Stone & Fritz, 2004). The subdecadal to decadal resolution of this 13,000-year record enabled the analysis of decadal, centennial, and millennial-scale dynamics by multi-taper and evolutive spectral analysis and suggests that the Holocene climatic variation was characterized by two mean states, with the warmer mid Holocene having strong periodic recurrence of severe aridity (Stone & Fritz, 2006).

Lake Victoria in East Africa is one of the few other decadal-scale diatom records from a closed-basin lake that spans the entire Holocene (Stager *et al.*, 1997; Stager *et al.*, 2003). Although Lake Victoria has an outlet that flows into the Nile River, > 90% of the water enters and leaves through the atmosphere, resulting in discernable changes in lake level and mixing regimes as climate changes. Application of correspondence analysis (CA) to diatom records from two cores located in different areas of the lake indicated that diatom assemblage changes were related to (1) precipitation:evaporation (P:E) ratios and/or lake depth, and (2) water-column stability. Comparison of the diatom records with the high-resolution ice-core records from Greenland and Antarctica suggests that large-scale high-latitude forcing at centennial scales drove regional hydrology and lake dynamics (Stager & Mayewski, 1997).

In some cases, changes in diatom assemblages in closed-basin lakes may be more strongly influenced by nutrient dynamics related to water-column stability than by salinity change itself. Direct nutrient influences may be most common in lakes that are fresh for extended periods of time and only increase significantly in salinity during long intervals of severe drought. Thus, rapid changes in diatom assemblages caused by changing phosphorus concentrations related to the timing and extent of water-column mixing have been observed in several high-resolution diatom records from forested regions adjacent to prairie or steppe, where climatic gradients are strong (Bradbury *et al.*, 2002; Bracht *et al.*, 2008; St. Jacques *et al.*, 2009). Experimental studies also suggest that nutrient influences may be important in saline lake systems, but, in such cases, the mechanisms may be through ionic influences on nutrient availability rather than direct climate impact on nutrient cycling (see discussion above).

10.4 Problems in interpretation of hydrology and climate from fossil diatom records

There are a number of problems associated with unambiguous reconstruction of climate from fossil diatom records (Gasse *et al.*, 1996). These problems relate to each of the logical steps in the reconstruction process: preservation and taphonomy of fossil assemblages, the adequacy of salinity reconstruction, and the degree of coupling of lake-water salinity and climate.

10.4.1 Taphonomy

Distorted or inaccurate environmental reconstructions can be produced by mixing and transport before burial. A mixture of different source communities in relatively large, permanent saline lakes has been thoroughly analyzed in southeastern Australia (Gell, 1995; Gasse *et al.*, 1996). The occurrence of littoral forms in plankton samples and differences in assemblage composition between phytoplankton tows and sediment traps indicate spatial heterogeneity, transport of diatoms from their source, and/or resuspension from sediment. These taphonomic processes may result in distortions and biases to the fossil assemblages (Anderson & Battarbee, 1994). Temporal and spatial homogenization of diatom assemblages may have benefits in that surface sediments include the whole hydrological cycle and the entire basin. However, these mixing processes limit the accuracy of predictive models if calibration data sets integrate assemblages from different biotopes with vastly different salinity. In addition, inputs of diatoms to the sediment represent a mixture of successional communities, whose proportion in the sediment depends on productivity. Unless seasonal fluxes are fully measured, assemblages from surficial bulk sediment cannot be regarded as reflecting "mean annual surface-water salinity" accurately.

An extreme situation is the inflow of allochthonous diatoms from rivers. Such a contamination explains the occurrence of *C. ocellata*, a taxon well known for living in freshwater but which occurs in samples of the hyperalkaline Lake Bogoria, Kenya (conductivity = 50 mS cm^{-1}) and is integrated in the modern African calibration data set (Gasse *et al.*, 1995). Although the species occurs in 28 other water bodies with lower salinities, the Bogoria samples introduce a strong bias in the calibration, because of the very high conductivity, alkalinity, and pH values of host waters.

Detection of allochthonous inputs in fossil records is sometimes possible when assemblages contain a mixture of diatom species with non-compatible ecological requirements. This is the case in the sedimentary sequence collected from the hypersaline–alkaline Lake Magadi, at the Kenya–Tanzania border (Barker, 1990; Taieb *et al.*, 1991; Gasse *et al.*, 1996). Reconstruction of the late Pleistocene sequence is problematic as it contains diatoms representative of environments from freshwater (*Fragilaria* spp., *Cymbella* spp., *Gomphonema* spp.) to hyperalkaline (*Anomoeneis sphaerophora, Craticula elkab, Thalassiosira faurii*) waters. Taphonomy may have incorporated spatially separate and yet contemporaneous diatom assemblages, as would occur if a stream carrying freshwater diatoms were to discharge into a saline lake. Several short paleochannels of unknown age in the vicinity of the core sites could have been the source of this freshwater. Proceeding under this hypothesis, the total diatom assemblage was split into its fresh and saline components to estimate the pH and conductivity of the hypothesized lake and river groups separately. The difference between these extreme ranges (2 pH units, about 15,000 μS cm^{-1}) demonstrates the large error that can be introduced into environmental reconstructions if the taphonomy of the assemblage has not been considered and only mean values have been used.

10.4.2 Dissolution

In saline waters, a severe problem in interpretation of some diatom records is dissolution and diagenesis of diatom silica, which selectively removes certain taxa from the record (Barker, 1990; Barker *et al.*, 1994; Ryves *et al.*, 2001). These phenomena occur during sedimentation and from contact between fossil diatoms and interstitial brines.

Experimental dissolution of lacustrine diatom silica in concentrated salt solutions (Barker, 1990) has been undertaken to understand the role of the solution in which the silica is immersed, as well as the shape and structure of the diatom valves themselves. Dissolution of diatom silica from sediments and live diatoms in 3 molar salt solutions (NaCl, Na_2CO_3, KNO_3, $CaCl_2$, $LiNO_3$, $MgCl_2$) and in distilled water indicates that the composition of salt solutions affects not only rates of dissolution but also the final silica concentration at which saturation and molecular diffusion are achieved. As expected, dissolution rate and level of saturation is maximal in Na_2CO_3 solutions where $pH > 9$ enhances dissociation of silicic acid. The diminishing levels of dissolution in other solutions – from NaCl, to solutions of strong bases and weak acids (KNO_3 and $LiNO_3$), to weakly acidic solutions containing divalent cations ($CaCl_2$, $MgCl_2$) – are more difficult to explain, as solubility is independent of pH within the range of pH values in these solutions (4–7.5). Whatever the cause (hydration number of cations, ionic dissociation even for pH values close to neutrality, influence of the ionic strength of the solution on the activity coefficient of silicic acid), these results suggest that the final level of dissolution in saline lakes depends on brine type.

Experimental data also show clear interspecific differences in diatom dissolution rates, as well as differential susceptibility of different components of an individual valve (Barker, 1990; Ryves *et al.*, 2001). The gross valve surface area:volume ratio can be a useful index of a particular diatom species' propensity to dissolve. In one set of experiments, for example, the assemblage resulting from 46 days of immersion in Na_2CO_3 solution was composed mainly of partially dissolved *Stephanodiscus neoastraea* Håk. & Hickel (80%), which represented only 12% of the initial community. Conversely, the predominant taxa at the starting point, *Fragilaria capucina* Desm. and *Fragilaria crotonensis* Kitton, had totally disappeared. Different components of valve structure also are differentially susceptible to dissolution: pore fields and areas of high striae density generally dissolve more rapidly than central or apical areas. Thus, species that have distinctive valve components that are resistant to dissolution, such as *Campylodiscus clypeus* W. Smith, *Mastogloia* spp., *R. gibberula*, and *Epithemia argus* (Ehr.) Kütz., can be more easily recognized in corroded samples than taxa with a more delicate or indistinct structure. The outcome of dissolution experiments can be used to interpret the integrity of fossil sequences and to derive indices that can be used to quantify the magnitude of dissolution in a fossil sample (Figure 10.9) and evaluate and improve the robustness of salinity reconstructions (Ryves *et al.*, 2001, 2006, 2009).

A diatom dissolution index was applied to modern surface samples from the northern Great Plains and Greenland to

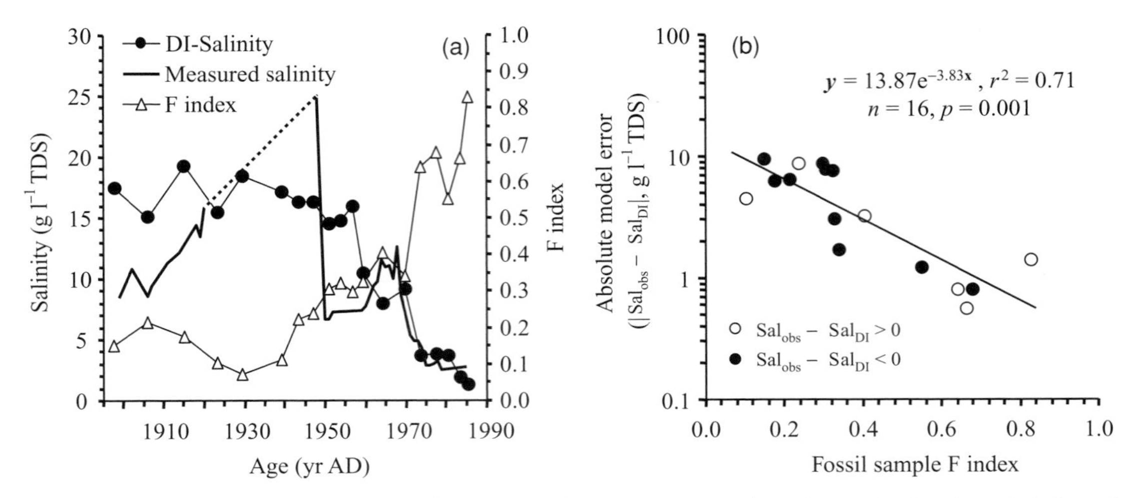

Figure 10.9 (a) Comparison of diatom-inferred (DI) salinity, measured salinity, and a diatom dissolution index (F index) in a sediment core from Devils Lake, ND. A value of 1 indicates that all valves are perfectly preserved, whereas a value of 0 indicates complete dissolution of valves. No salinity measurements were recorded between 1923 and 1948. Core dates are based on ^{210}Pb and are calculated with the constant rate of supply (c.r.s.) model. (b) Comparison of the diatom dissolution index with the deviation of the diatom-inferred salinity from the measured salinity. (From Ryves *et al.* (2006).)

evaluate the physical and chemical conditions associated with diatom dissolution (Ryves *et al.*, 2006). Multivariate statistical analysis of the data suggested that, in both regions, salinity is the variable most strongly correlated with high diatom dissolution. In lakes of the northern Great Plains, where brine type is more variable, carbonate concentration and the presence or absence of permanent stratification (meromixis) were also significant explanatory variables associated with extensive diatom dissolution.

Experimental results also suggest varying capacities of brines to induce formation of diagenetic silicates from diatom silica in divalent cation solutions (e.g. Mg^{2+} clays or zeolites), which decrease the stability of this amorphous silica phase. Neogenesis of smectites from diatom silica has been demonstrated from the Bolivian Altiplano (Badaut & Risacher, 1983). At Bougdouma (Niger), observations in scanning electron microscopy show that the original internal structure of *Aulacoseira* is replaced *in situ* by clay-like particles. This alteration is suspected to have occurred during early diagenesis, because similar changes in morphology have been observed in floating frustules of dead *C. elkab* in the neighboring alkaline pond of Guidimouni (Gasse, 1987). In several levels of the Late Quaternary record of the hyperalkaline Lake Bogoria (Gasse & Seyve, 1987), dissolution and neogenesis from *Thalassiosira rudolfii* towards zeolite mineral species (analcime) led to the total disappearance of diatoms.

10.4.3 Salinity reconstruction

The high correlation between diatom-inferred salinity and measured lake-water salinity (see above) suggests that quantitative inference techniques can be used with some confidence to estimate past salinity. However, despite the statistical robustness of these methods, ambiguities may still arise. Calibration data sets often are composed of a single surface-sediment sample from each site, which integrates diatom production over one to several years, dependent on rates of sediment accumulation. Estimates of the salinity "optimum" of individual taxa from surface sediments can be made with confidence in lakes where salinity fluctuates within a modest range or where the full range of seasonal variation in salinity is known. However, in some cases, calibration data sets are based on limited water-chemistry sampling, and in these cases a single measurement of salinity may not be representative of the water chemistry in which the diatom lived. Problems associated with the estimation of salinity optima, however, can be minimized by sampling a large number of sites (Wilson *et al.*, 1996) and by modest efforts to characterize the range of chemical variability.

A more serious limitation of the method occurs at the high end of the salinity gradient, where species diversity is low, many of the taxa have large salinity ranges, and dissolution may be higher. In these cases, the error surrounding any salinity estimate may be quite large. Devils Lake, North Dakota, represented an ideal situation for evaluating the accuracy of diatom-based salinity reconstruction (Figure 10.9), because a diatom-inferred history of salinity change from a sediment core encompassing the last 100 years could be compared with measured salinity values from the same time interval (Fritz, 1990).

Application of a dissolution index to core samples showed that dissolution was greatest during times of high salinity and that errors in paleosalinity inferences are strongly correlated with diatom dissolution (Ryves *et al.*, 2006).

An accurate inference of salinity also may be difficult, because factors in addition to salinity influence species distributions or because taxonomic differentiation in some groups is difficult. Some broadly distributed taxa, such as *C. ocellata*, occur in shallow hyposaline lakes in Spain (Reed, 1998) but are most commonly associated with very dilute freshwater lakes, such as in the Alps (Lotter *et al.*, 1998). In this case, the differences may be a product of the difficulty of taxonomic differentiation in this morphologically variable taxon. *Cyclotella quillensis*, which is restricted to saline lakes (Fritz *et al.*, 1993), can be difficult to distinguish from *C. meneghiniana* (Battarbee *et al.*, 1982), a common dominant diatom taxon in saline lakes but also in shallow eutrophic freshwater lakes (Brugam, 1983). With such species, the influences of salinity relative to nutrients may be difficult to disentangle. Other species, particularly within the Fragilariaceae, occur in a variety of lake types and may not provide clear ecological information.

The ability of a calibration data set to reconstruct past salinity depends on having adequate modern analogs for fossil assemblages. The adequacy of a calibration data set in this regard can be evaluated by multivariate statistical techniques (Figure 10.10). Poor analog situations exist where species occur in stratigraphic records in percentages greater than those in modern samples or in some cases where individual fossil species are not extant in the modern landscape. This is the case for *C. choctawhatcheeana*, which dominates some African sections of Holocene age, as well as core sequences in British Columbia (Wilson *et al.*, 1996), but has no modern analog in over several hundred modern samples from Africa (Gasse *et al.*, 1995) and only limited occurrences in British Columbia. However, this taxon occurs in modern sulfate-dominated saline lakes in the North American Great Plains (Fritz *et al.*, 1993) and in Australia (Gell & Gasse, 1990), and thus combining regional data sets from widespread geographic areas may enhance our ability to interpret fossil environments (Juggins *et al.*, 1994).

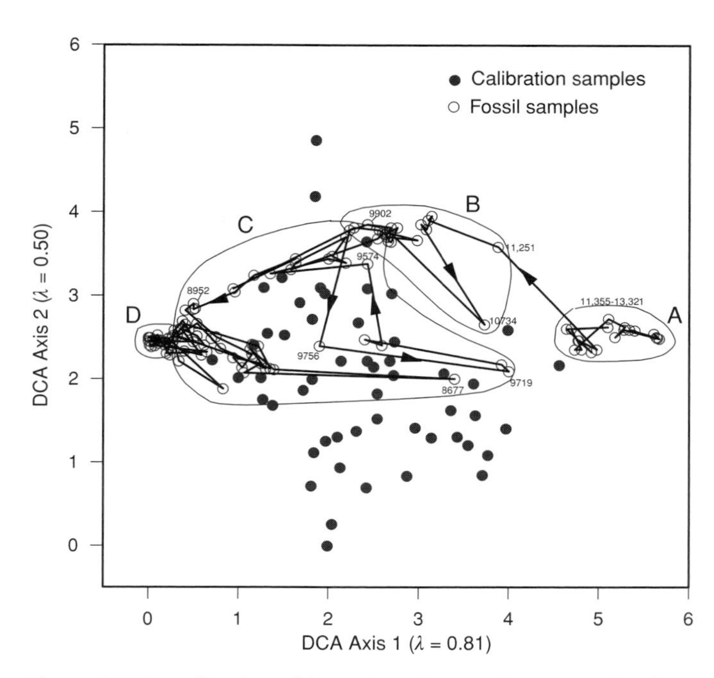

Figure 10.10 Ordination of the early-Holocene diatom samples from Moon Lake (ND) with the 55-lake calibration data set of Fritz *et al.* (1993) using detrended correspondence analysis (DCA). Circled areas with lettered designations correspond to diatom zones (A–D). See Laird *et al.* (1998b) for further details.

10.4.4 Climatic reconstruction

The problems associated with reconstruction of climate from salinity records are more difficult to address than those associated with salinity reconstruction itself and require evaluation of the nature of the hydrologic response to fluctuations in moisture. For example, salinity may not be linearly coupled to lake level or to the balance between P and E. In Devils Lake, North Dakota, USA, the lake-water salinity for any given lake stage was higher when the lake level was falling than during refilling. Presumably salts are lost to the system either by deflation or by precipitation and do not redissolve when the lake refills. Alternatively, in some lakes, dissolution of evaporites during wet intervals can lead to elevated salinity (Gasse *et al.*, 1996), contrary to what one would predict if salinity were controlled by P – E. A related situation is hypothesized for areas within the Cuenca de Mexico (Bradbury, 1989), a complex of lakes, small pools, and marshes, where expansion of saline lakes during periods of increased precipitation may have inundated marginal springs and pools with saline waters and elevated their salinity, whereas during arid periods the pools and springs were isolated and fed by fresh groundwater discharge. In addition, the chemical composition of groundwater inflow affects the geochemical response of a lake to climate. Thus, increased groundwater inflow as a result of elevated precipitation and recharge can either raise or lower lake salinity dependent on groundwater chemistry, and vice versa, during times of drought (Smith *et al.*, 1997; Telford *et al.*, 1999).

Lakes also vary greatly in their sensitivity and response times to changes in moisture balance. Devils Lake, North Dakota, for example, is connected to several upstream basins during times of high precipitation and isolated during intervals of extreme drought. Thus, its water budget and the magnitude of hydrological response to changes in moisture balance will differ during wet and dry periods. Likewise a lake that fluctuates from open to closed hydrology may exhibit large shifts in salinity that are not proportionate to climatic forcing (Radle *et al.*, 1989; Van Campo & Gasse, 1993). A meromictic system may respond differently to changes in precipitation than a holomictic system, and, as a result, periods of limnological stability or high variability do not necessarily reflect climate variability (Laird *et al.*, 1996a; McGowan *et al.*, 2003; Aebly & Fritz, 2009). Lakes with a large groundwater component to their hydrologic budget may show very small responses to changes in P – E, whereas the same climatic forcing may more strongly affect a system with limited groundwater inflow (Reed *et al.*, 1999; Bennett *et al*, 2007). Thus, one cannot simply equate the magnitude of salinity change and the magnitude of climatic forcing; some understanding of basin hydrology is required in climatic interpretation of paleosalinity records.

The coupling between climate and salinity can be assessed, to some extent, by comparison of measured climate with diatom-inferred salinity fluctuations from cores that span the period of historic record. In the northern Great Plains, for example, meteorological measurements of precipitation and temperature are available for approximately the last 100 years. Comparison of diatom-inferred salinity fluctuations with a measure of moisture availability, such as P – evapotranspiration (ET) (Laird *et al.*, 1996a) or the Bhalme and Mooley Drought Index (BMDI) (Figure 10.11), can be used to assess the relationship between moisture balance and salinity. In lakes where the correspondence is strong, one can infer climatic patterns from diatom-inferred salinity with reasonable confidence, although extrapolating the correspondence back in time assumes no major changes in morphometry and basin hydrology over time, which may not be the case. Another tool for understanding how climate and salinity are coupled is to have an independent measure of hydrological change, such as well-dated paleoshorelines that delineate large-scale changes in climate-driven lake level, which can be correlated with inferences of salinity generated from diatoms (Aebly & Fritz, 2009).

Additional problems of interpretation can occur in settings where hydrologic change may be related to non-climatic factors. In the late-Quaternary record from Walker Lake, Nevada (Bradbury *et al.*, 1989), changes in lake-water salinity were driven both by changes in P – E and by inflow or diversion of the freshwater Walker River into the lake. In other regions, hydrologic change may be caused by geomorphic or tectonic

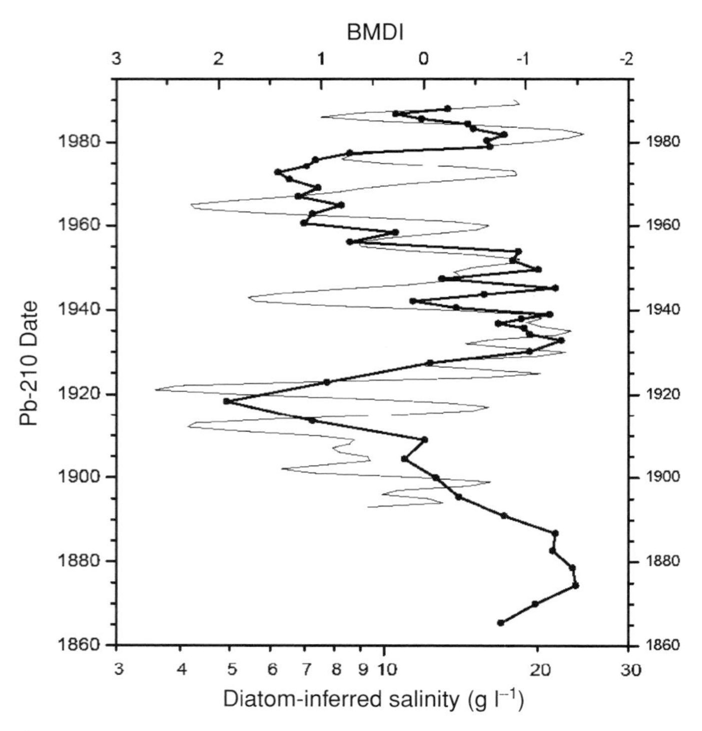

Figure 10.11 Comparison of diatom-inferred salinity (black solid circle line) of Moon Lake (ND) based on a short core covering the period of historic record with the BMDI (gray line) calculated from precipitation measurements at a climate station near the lake. Chronology is based on ^{210}Pb dates calculated with the constant rate of supply model. (From Laird *et al.* (1998a).)

processes (e.g. Fritz *et al.*, 2004), and thus the geologic setting must be carefully evaluated prior to interpretation of paleosalinity records. Finally, in many parts of the world, human activities have been extensive for millennia, and it may be difficult to distinguish climate-driven variation from the influences of land-use change.

10.5 Summary

The distribution of diatom species is clearly related to ionic concentration and composition. Presently, the mechanisms that relate diatoms and ions are not well studied, and the extent to which salinity exerts a direct versus an indirect impact on individual taxa or species assemblages is not known. None-the-less, these relationships can be exploited to reconstruct past changes in lake-water salinity and brine composition driven by hydrologic and climatic change. The short generation times and rapid response of diatoms to hydrochemical change make them a particularly useful indicator of high-frequency environmental variation.

The primary value of these studies is to establish natural patterns of climatic variability at a variety of spatial and temporal scales. At orbital and millennial timescales, diatom-based studies have been important in establishing shifts in moisture balance associated with glacial/interglacial cycles across broad geographic regions. These data can be used to evaluate the relative importance of various climatic forcings and drivers, including the extent to which the influence of insolation is modulated by local or regional variables, such as oceanic circulation, topography, land-surface characteristics, or aerosols. Efforts to discern decadal- to centennial-scale patterns of climatic variability and their causal mechanisms from high-resolution studies have intensified in the last decade. The frequency of change in these high-resolution records, as well as the spatial distribution of coherent patterns of change, may provide clues regarding the relative roles of processes, such as variability of the linked atmospheric–oceanic system, solar and lunar activity, volcanic eruptions, and greenhouse-gas concentration in causing decadal to centennial variation.

Diatom-inferred hydrochemistry data can also be used to examine responses of lake systems to independently documented climatic change. Thus, comparisons of diatom-inferred salinity with instrumental climate data may be useful not only to choose suitable sites for climatic reconstruction (Laird *et al.*, 1996a) but also as a tool to better understand the relative importance of various components of the hydrologic budget in different lake types and geographic regions (Aebly & Fritz, 2009), particularly the importance of groundwater in controlling lake hydrochemistry (Crowe, 1993).

Stratigraphic records are critical to our understanding of modern climate and the extent of human modification of the climate system. Diatom-based climatic reconstructions provide a baseline for evaluating current climatic patterns, enabling us to determine whether the modern behavior of the climate system falls outside the range of natural variability. Stratigraphic records also are tools for understanding ecosystem response under boundary conditions that fall outside the modern norm, such as may occur in the future. Despite the increased number of paleolimnological studies from arid and semi-arid regions, we still have an insufficient number of records to clearly portray spatial and temporal patterns of climate variability at decadal to millennial scales, particularly for periods prior to the late Holocene. As is true of many types of diatom-based inferences, our ability to resolve patterns of environmental change is often limited by our knowledge of the mechanisms that link diatoms, lake hydrochemistry, and climate. Thus, not only do we need more stratigraphic records to discern patterns and mechanisms of climatic change, but we also still need more modern process-based studies, particularly in the areas of lake hydrology and diatom physiology.

Acknowledgments

We thank Jane Reed for thoughtful comments on the manuscript.

References

Aebly, F. & Fritz, S. C. (2009). The paleohydrology of west Greenland for the past 8000 years. *The Holocene*, **19**, 91–104.

Alley, R. B., Marotzke, J., Nordhaus, W. D., *et al.* (2003). Abrupt climate change. *Science*, **299**, 2005–10.

Anderson, N. J. & Battarbee, R. W. (1994). Aquatic community persistence and variability: a palaeolimnological perspective. In *Aquatic Ecology: Scale, Pattern, and Process*, ed. P. S. Giller, A. G. Hildrew, & D. G. Raffaelli. Oxford: Blackwell Scientific Publications, pp. 233–59.

Badaut, D. & Risacher, F. (1983). Authigenic smectite on diatom frustules in Bolivian saline lakes. *Geochimica et Cosmochimica Acta*, **47**, 363–75.

Baker, P. A., Fritz, S. C., Garland, J., & Ekdahl, E. (2005). Holocene hydrologic variation at Lake Titicaca, Bolivia/Peru and its relationship to North Atlantic climate variation. *Journal of Quaternary Science*, **20**, 655–62.

Barker, P. A. (1990). Diatoms as palaeolimnological indicators: a reconstruction of Late Quaternary environments in two East African salt lakes. Unpublished Ph.D. Thesis, Loughborough University.

Barker, P., Fontes, J.-C., Gasse, F. & Druart, J.-C. (1994). Experimental dissolution of diatom silica in concentrated salt solutions and implications for palaeoenvironmental reconstruction. *Limnology and Oceanography*, **39**, 99–110.

Barker, P. & Gasse, F. (2003). New evidence for a reduced water balance in East Africa during the Last Glacial Maximum: implications for model-data comparisons. *Quaternary Science Reviews*, **22**, 823–37.

Barker, P., Telford, R., Gasse, F., & Thevenon, F. (2002). Late Pleistocene and Holocene palaeohydrology of Lake Rukwa, Tanzania, inferred from diatom analysis. *Palaeogeography Palaeoclimatology, Palaeoecology*, **187**, 295–305.

Battarbee, R. W., Keister, C. M., & Bradbury, J. P. (1982). The frustular morphology and taxonomic relationships of *Cyclotella quillensis* Bailey. In *7th International Diatom Symposium, Proceedings*, ed. D. G. Mann, Stuttgart: J. Cramer, pp. 173–184.

Bennett, D. M., Fritz, S. C., Holz, J., Holz, A., & Zlotnik, V. (2007). Evaluating climatic and non-climatic influences on natural and man-made lakes in Nebraska, USA. *Hydrobiologia*, **591**, 103–15.

Bergner, A. G. N. & Trauth, M. H. (2004). Comparison of hydrological and hydrochemical evolution of Lake Naivasha (Kenya) during three highstands between 175 and 60 kyr BP. *Palaeogeography, Palaeoclimatology, Palaeoecology*, **215**, 17–36.

Bhattacharyya, P. & Volcani, B. E. (1980). Sodium-dependent silicate transport in the apochlorotic marine diatom *Nitzschia alba*.

Proceedings, Academy of Natural Sciences of Philadelphia, **77**, 6386–90.

Birks, H. J. B. (1995). Quantitative palaeoenvironmental reconstructions. In *Statistical Modeling of Quaternary Science Data*, ed. D. Maddy & J. S. Brew, Cambridge: Quaternary Research Association, pp. 161–254.

Blinn, D. (1995). Diatom community structure along salinity gradients in Australian saline lakes: biogeographic comparisons with other continents. In *A Century of Diatom Research in North America: a Tribute to the Distinguished Careers of C. W. Reimer and R. Patrick*, ed. J. P. Kociolek & M. J. Sullivan, Champaign, Illinois: Koeltz Scientific Books, pp. 156–67.

Blinn, D. W., Hevly, R. H., & Davis, O. K. (1994). Continuous Holocene record of diatom stratigraphy, palaeohydrology, and anthropogenic activity in a spring-mound in southwestern United States. *Quaternary Research*, **42**, 197–205.

Bloom, A. M., Moser, K. A., Porinchu, D. F., & MacDonald, G. M. (2003). Diatom-inference models for surface-water temperature and salinity developed from a 57-lake calibration set from the Sierra Nevada, California, USA. *Journal of Paleolimnology*, **29**, 235–55.

Bond, G., Kromer, B., Beer, J., *et al.* (2001). Persistent solar influence on North Atlantic climate during the Holocene. *Science*, **294**, 2130–6.

Bracht, B. B., Stone, J. R., & Fritz, S. C. (2008). A diatom record of late Holocene climate variation in the northern range of Yellowstone National Park, USA. *Quaternary International*, **188**, 149–55.

Braconnot, P., Otto-Bliesner, B., Harrison, S. P., *et al.* (2007). Results of PMIP2 coupled simulations of the mid-Holocene and Last Glacial Maximum – Part 2: feedbacks with emphasis on the location of the ITCZ and mid and high latitudes heat budget. *Climate of the Past*, **3**, 279–96.

Bradbury, J. P. (1989). Late Quaternary lacustrine paleoenvironments in the Cuenca de Mexico. *Quaternary Science Reviews*, **8**, 75–100.

Bradbury, J. P., Forester, R. M., & Thompson, R. S. (1989). Late Quaternary paleolimnology of Walker Lake, Nevada. *Journal of Paleolimnology*, **1**, 249–67.

Bradbury, P., Cumming, B., & Laird, K. (2002). A 1500-year record of climatic and environmental change in Elk Lake, Minnesota III: measures of past primary productivity. *Journal of Paleolimnology*, **27**, 321–40.

Bradley, R. S. & Jones, P. D. (eds.) (1992). *Climate Since A.D. 1500*. New York: Routledge.

Brugam, R. (1983). The relationship between fossil diatom assemblages and limnological conditions. *Hydrobiologia*, **98**, 223–35.

Chalié, F. & Gasse, F. (2002). A 13,500 year diatom record from the tropical East African Rift Lake Abiyata (Ethiopia). *Palaeogeography, Palaeoclimatology, Palaeoecology*, **187**, 259–84.

Charles, D. F. & Smol, J. P. (1994). Long-term chemical changes in lakes: quantitative inferences from biotic remains in the sediment record. In *Environmental Chemistry of Lakes and Reservoirs*, ed. L. Baker. Washington, DC: American Chemical Society, pp. 3–31.

Cohen, A. S., Stone, J. R., Beuning, K. R. & Park, L. E. (2007). Ecological consequences of early Late Pleistocene megadroughts in tropical Africa. *Proceedings of the National Academy of Science*, **104**, 16422–7.

Cook, E. R., Woodhouse, C. A., Eakin, C. M., Meko, D. M., & Stahle, D. W. (2004). Long-term aridity changes in the western United States. *Science*, **306**, 1015–18.

Crowe, A. S. (1993). The application of a coupled water-balance-salinity model to evaluate the sensitivity of a lake dominated by groundwater to climate variability. *Journal of Hydrology*, **141**, 33–73.

Cumming, B. F., Laird, K. R., Bennett, J. R., Smol, J. P., & Salomon, A. K. (2002). Persistent millennial-scale shifts in moisture regimes in western Canada during the past six millennia. *Proceedings of the National Academy of Science*, **99**, 16117–21.

Cumming, B. F. & Smol, J. P. (1993). Development of diatom-based salinity models for paleoclimatic research from lakes in British Columbia (Canada). *Hydrobiologia*, **269/270**, 179–96.

Cumming, B. F., Wilson, S. E., Hall, R. I. & Smol, J. P. (1995). *Diatoms from British Columbia (Canada) Lakes and Their Relationship to Salinity, Nutrients, and Other Limnological Variables*. Stuttgart, Germany: Koeltz Scientific Books.

Davies, S. J., Metcalfe, S. E., Caballero, M. E., & Juggins, S. (2002). Developing diatom-based transfer functions for Central Mexican lakes. *Hydrobiologia*, **467**, 199–213.

de Menocal, P., Ortiz, J., Guilderson, T., *et al.* (2000). Abrupt onset and termination of the African humid period: rapid climate responses to gradual insolation forcing. *Quaternary Science Reviews*, **19**, 347–61.

Donovan, J. J. (1994). Measurement of reactive mass fluxes in evaporative groundwater-source lakes. In *Sedimentology and Geochemistry of Modern and Ancient Saline Lakes*, ed. R. Renaut & W. M. Last, Tulsa, OK: SEPM Society for Sedimentology Geology, pp. 33–50.

Ekdahl, E. J., Fritz, S. C., Baker, P. A., Rigsby, C. A., & Coley, K. (2008). Holocene multidecadal- to millennial-scale hydrologic variability on the South American Altiplano. *The Holocene*, **18**, 867–76.

Eugster, H. P. & Jones, B. F. (1979). Behavior of major solutes during closed-basin brine evolution. *American Journal of Science*, **279**, 609–31.

Fisher, N. S. (1977). On the differential sensitivity of estuarine and open-ocean diatoms to exotic chemical stress. *American Naturalist*, **111**, 871–95.

Fritz, S. C. (1990). Twentieth-century salinity and water-level fluctuations in Devils Lake, N. Dakota: a test of a diatom-based transfer function. *Limnology and Oceanography*, **35**, 1771–81.

Fritz, S. C. (2008). Deciphering climatic history from lake sediments. *Journal of Paleolimnology*, **39**, 5–16.

Fritz, S. C., Baker, P. A., Lowenstein, T. K., *et al.* (2004). Hydrologic variation during the last 170,000 years in the southern hemisphere tropics of South America. *Quaternary Research*, **61**, 95–104.

Fritz, S. C., Baker, P. A., Seltzer, G. O., *et al.* (2007). Quaternary glaciation and hydrologic variation in the South American tropics as reconstructed from the Lake Titicaca drilling project. *Quaternary Research*, **68**, 410–20.

Fritz, S. C., Ito, E., Yu, Z., Laird, K. R., & Engstrom, D. R. (2000). Hydrologic variation in the northern Great Plains during the last two millennia. *Quaternary Research*, **53**, 175–84.

Fritz, S. C., Juggins, S., & Battarbee, R. W. (1993). Diatom assemblages and ionic characterization of lakes of the Northern Great Plains, North America: a tool for reconstructing past salinity and climate fluctuations. *Canadian Journal of Fisheries and Aquatic Sciences*, **50**, 1844–56.

Fritz, S. C., Juggins, S., Battarbee, R. W., & Engstrom, D. R. (1991). Reconstruction of past changes in salinity and climate using a diatom-based transfer function. *Nature*, **352**, 706–8.

Gasse, F. (1977). Evolution of Lake Abhe (Ethiopia and T.F.A.I.) from 70,000 B.P. *Nature*, **265**, 42–5.

Gasse, F. (1986). *East African Diatoms: Taxonomy, Ecological Distribution*, Stuttgart: Cramer.

Gasse, F. (1987). Diatoms for reconstructing palaeoenvironments and palaeohydrology in tropical semi-arid zones: examples of some lakes in Niger since 12000 BP. *Hydrobiologia*, **154**, 127–63.

Gasse, F. (2000). Hydrological changes in the African tropics since the Last Glacial Maximum. *Quaternary Science Reviews* **19**, 189–211.

Gasse, F. (2002). Diatom-inferred salinity and carbonate oxygen isotopes in Holocene waterbodies of the western Sahara and Sahel (Africa). *Quaternary Science Reviews*, **21**, 737–67.

Gasse, F., Bergonzini, L., Chalié, F., Gibert, E., Massault, M., & Mélières, F. (1998). Paleolakes and paleoclimates around the western Indian Ocean since 25 kyrs BP. In *Hydrologie et Géochimie Isotopique, Proceedings of the International Symposium in Memory of Jean-Charles Fontes*, Paris, June 1st and 2nd 1995, ed. C. Causse & F. Gasse, Paris: ORSTOM, pp. 147–76.

Gasse, F., Gell, P., Barker, P., Fritz, S. C., & Chalie, F. (1996). Diatom-inferred salinity of palaeolakes: an indirect tracer of climate change. *Quaternary Science Reviews*, **16**, 547–63.

Gasse, F., Juggins, S., & Ben Khelifa, L. (1995). Diatom-based transfer functions for inferring hydrochemical characteristics of African palaeolakes. *Palaeogeography, Palaeoclimatology, Palaeoecology*, **117**, 31–54.

Gasse, F. & Roberts, N. (2004). Late Quaternary hydrologic changes in the arid and semi-arid belt of northern Africa. Implications for past atmospheric circulation. In *The Hadley Circulation: Present, Past and Future*, ed. H. F. Diaz & R. S. Bradley, Dordrecht: Kluwer Academic Publishers, pp. 313–45.

Gasse, F. & Seyve, C. (1987). Sondages du Lac Bogoria: Diatomees. In *Le demi-graben de Baringo-Bogoria, Rift Gregory, Kenya. 30.000 ans d'histoire hydrologique et sedimentaire*, ed. J. J. Tiercelin & A. Vincens, Bulletin du Centre de Recherche et d'Exploration-Production Elf-Aquitaine Boussens, pp. 414–37.

Gasse, F. & Street, F. A. (1978). Late Quaternary lake level fluctuations and environments of the northern Rift Valley and Afar region (Ethiopia and Djibouti). *Palaeogeography, Palaeoclimatology, Palaeoecology*, **25**, 145–50.

Gasse, F. & Van Campo, E. (1994). Abrupt post-glacial climate events in west Asia and north Africa monsoon domains. *Earth and Planetary Science Letters*, **126**, 435–65.

Gell, P. A. (1995). The development and application of a diatom calibration set for lake salinity, Western Victoria, Australia. Unpublished Ph.D. thesis, Monash University.

Gell, P. A. & Gasse, F. (1990). Relationships between salinity and diatom flora from some Australian saline lakes. In *Proceedings of the 11th International Diatom Symposium*, ed. J. P. Kociolek. San Francisco: California Academy of Sciences, pp. 631–47.

Gillespie, R., Street-Perrott, F. A., & Switsur, R. (1983). Post-glacial arid episodes in Ethiopia have implications for climate prediction. *Nature*, **306**, 680–3.

Hoelzmann, P., Gasse, F., Dupont, L. M., *et al.* (2004). Palaeoenvironmental changes in the arid and subarid belt (Sahara–Sahel–Arabian Peninsula) from 150 ka to present. In *Past Climate through Europe and Africa*, ed. R. W. Battarbee, F. Gasse & C. S. Stickley, Developments in Paleoenvironmental Research Dordrecht: Springer, pp. 219–56.

Holdship, S. A. (1976). The palaeolimnology of Lake Manyara, Tanzania: a diatom analysis of a 56 meter sediment core. Unpublished Ph.D. thesis, Duke University.

Holmes, J. A., Fothergill, P. A., Street-Perrott, F. A., & Perrott, R. A. (1998). A high-resolution Holocene ostracod record from the Sahel Zone of northeastern Nigeria. *Journal of Paleolimnology*, **20**, 369–80.

Holmes, J. A., Street-Perrott, F. A., Allen, M. J., *et al.* (1997). Holocene paleolimnology of Kajemarum Oasis, northern Nigeria: an isotopic study of ostracods, bulk carbonate and organic carbon. *Journal of the Geological Society, London* **154**, 311–19.

Hustedt, F. (1953). Die Systematik der Diatomeen in ihren Beziehungen zur Geologie und Ökologie nebst einer Revision des Halobien-systems. *Botanisk Tidskkrift*, **47**, 509–19.

Juggins, S. (2001). The European Diatom Database, User Guide, Version 1.0. (October 2001). See http://craticula.ncl.ac.uk/Eddi/jsp/help.jsp.

Juggins, S., Battarbee, R. W., Fritz, S. C., & Gasse, F. (1994). The CASPIA project: diatoms, salt lakes, and environmental change. *Journal of Paleolimnology*, **2**, 191–6.

Kashima, K. (1994). Sedimentary diatom assemblages in freshwater and saline lakes of the Anatolia Plateau, central part of Turkey: an application of reconstruction of palaeosalinity change during the Late Quaternary. In *Proceedings of the 13th International Diatom Symposium*, ed. D. Marino & M. Montresor, Bristol: Biopress Ltd., pp. 93–100.

Kashima, K. (2003). The quantitative reconstruction of salinity changes using diatom assemblages in inland saline lakes in the central part of Turkey during the Late Quaternary. *Quaternary International*, **105**, 13–19.

Kennedy, K. A. (1994). Early-Holocene geochemical evolution of saline Medicine Lake, South Dakota. *Journal of Paleolimnology*, **10**, 69–84.

Khelifa, L. B. (1989). *Diatomees continentales et paleomillieux du Sud-Tunisien aux Quaternaire superieur*. Unpublished Ph.D. thesis, Universite de Paris-Sud.

Kolbe, R. W. (1927). Zur Ökologie, Morphologie und Systematik der Brackwasser-Diatomeen. *Pflanzenforschung*, **7**, 1–146.

Kröpelin, S., Verschuren, D., Lézine, A.-M, et al. (2008). Climate-driven ecosystem succession in the Sahara: the past 6000 years. *Science*, **320**, 765–8.

Laird, K., Fritz, S. C., Grimm, E. C., & Mueller, P. G. (1996a). The paleoclimatic record of a closed-basin lake in the northern Great Plains: Moon Lake, Barnes Co., N.D. *Limnology and Oceanography*, **40**, 890–902.

Laird, K. R., Fritz, S. C., Maasch, K. A., & Cumming, B. F. (1996b). Greater drought intensity and frequency before AD 1200 in the Northern Great Plains, USA. *Nature*, **384**, 552–5.

Laird, K. R., Fritz, S. C., & Cumming, B. F. (1998a). A diatom-based reconstruction of drought intensity, duration, and frequency from Moon Lake, North Dakota: a sub-decadal record of the last 2300 years. *Journal of Paleolimnology*, **19**, 161–79.

Laird, K. R., Fritz, S. C., Grimm, E. C., & Cumming, B. F. (1998b). Early Holocene limnologic and climatic variability in the northern Great Plains. *The Holocene*, **8**, 275–85.

Laird, K. R., Cumming, B. F., Wunsam, S., et al. (2003). Lake sediments record large-scale shifts in moisture regimes across the northern prairies of North America during the past two millennia. *Proceedings of the National Academy of Science*, **100**, 2483–8.

Laird, K. R., Michels, A., Stuart, C., et al. (2007). Examination of diatom-based changes from a climatically sensitive prairie lake (Saskatchewan, Canada) at different temporal perspectives. *Quaternary Science Reviews*, **26**, 3328–43.

Lamb, H. H. (1982). *Climate, History, and the Modern World*. London: Methuen.

Laskar, J., Robutel, P., Joutel, F., et al. (2004). A long-term numerical solution for the insolation quantities of the Earth. *Astronomy and Astrophysics*, **428**, 261–85.

Legesse, D., Gasse F., Radakovitch, O., et al. (2002). Environmental changes in a tropical lake (Lake Abiyata, Ethiopia) during recent centuries. *Palaeogeography, Palaeoclimatology, Palaeoecology*, **187**, 233–58.

Liu, Z., Wang, Y., Gallimore, R., et al. (2007). Simulating the transient evolution and abrupt change of Northern Africa atmosphere–ocean–terrestrial ecosystem in the Holocene. *Quaternary Science Reviews*, **26**, 1818–37.

Lotter, A.F., Birks, H.J.B., Hofmann, W., & Marchetto, A. (1998). Modern diatom, cladocera, chironomid, and chrysophyte cyst assemblages as quantitative indicators for the reconstruction of past environmental conditions in the Alps. II. Nutrients. *Journal of Paleolimnology*, **19**, 443–63.

Mason, I. M., Guzkowska, M. A. J., & Rapley, C. G. (1994). The response of lake levels and areas to climatic change. *Climatic Change*, **27**, 161–97.

Mayewski, P. A., Rohling, E. E., Stager, J. C., et al. (2004). Holocene climate variability. *Quaternary Research*, **62**, 243–55.

McGowan, S., Ryves, D. B., & Anderson, N. J. (2003). Holocene records of effective precipitation in West Greenland. *The Holocene*, **13**, 239–50.

Metcalfe, S. E. (1988). Modern diatom assemblages in Central Mexico: the role of water chemistry and other environmental factors as indicated by TWINSPAN and DECORANA. *Freshwater Biology*, **19**, 217–33.

Meybeck, M. (1995). Global distribution of lakes. In *Physics and Chemistry of Lakes*, ed. A. Lerman, D. Imboden, & J. Gat, Berlin: Springer-Verlag, pp. 1–35.

Michels, A., Laird, K. R., Wilson, S. E., et al. (2007). Multi-decadal to millennial-scale shifts in drought conditions on the Canadian prairies over the past six millennia: implications for future drought assessment. *Global Change Biology*, **13**, 1295–307.

Oswald, W. W., Anderson, P. M., Brown, T. A., et al. (2005). Effects of sample mass and macrofossil type on radiocarbon dating of arctic and boreal lake sediments. *The Holocene*, **15**, 758–67.

Porter, S. C. (1986). Pattern and forcing of northern hemisphere glacier variations during the last millennium. *Quaternary Research*, **26**, 27–48.

Potapova, M. & Charles, D.F. (2003). Distribution of benthic diatoms in U.S. rivers in relation to conductivity and ionic composition. *Freshwater Biology*, **48**, 1311–28.

Radle, N. J., Keister, C. M., & Battarbee, R. W. (1989). Diatom, pollen, and geochemical evidence for the paleosalinity of Medicine Lake, S. Dakota, during the Late Wisconsin and early Holocene. *Journal of Paleolimnology*, **2**, 159–72.

Reed, J. M. (1998). A diatom–conductivity transfer function for Spanish salt lakes. *Journal of Paleolimnology*, **19**, 399–416.

Reed, J. M., Roberts, N., & Leng, M. J. (1999). An evaluation of the diatom response to Late Quaternary environmental change in two lakes in the Konya Basin, Turkey by comparison with stable isotope data. *Quaternary Science Reviews*, **18**, 631–47.

Reed, J. M., Stevenson, A. C., & Juggins, S. (2001). A multi-proxy record of Holocene climate change in southwest Spain: the Laguna de Medina, Cádiz. *The Holocene*, **11**, 705–17.

Rind, D. & Overpeck, J. (1993). Hypothesized causes of decade to century scale climate variability: climate model results. *Quaternary Science Reviews*, **12**, 357–74.

Roberts, D. & McMinn, A. (1996). Relationships between surface sediment diatom assemblages and water chemistry gradients of the Vestfold Hills, Antarctic. *Antarctic Science*, **8**, 331–41.

Roberts, D., McMinn, A., Cremer, H., Gore, D. B., & Melles, M. (2004). The Holocene evolution and palaeosalinity history of Beall Lake, Windmill Islands (East Antarctica) using an expanded diatom-based weighted average model. *Palaeogeography, Palaeoclimatology, Palaeoecology*, **108**, 121–40.

Roberts, N., Reed, J., Leng, M. J., et al. (2001). The tempo of Holocene climatic change in the eastern Mediterranean region: new high-resolution crater-lake sediment data from central Turkey. *The Holocene*, **11**, 721–36.

Roux, M., Servant-Vildary, S., & Servant, M. (1991). Inferred ionic composition and salinity of a Bolivian Quaternary lake, as estimated from fossil diatoms in the sediments. *Hydrobiologia*, **210**, 3–18.

Ryner, M., Gasse, F., Rumes, B., & Verschuren, D. (2007). Climatic and hydrological instability in semi-arid equatorial East Africa during the late glacial to Holocene transition. *Palaeogeography, Palaeoclimatology, Palaeoecology*, **248**, 440–58.

Ryves, D. B., Battarbee, R. W., & Fritz, S. C. (2009). The dilemma of disappearing diatoms: incorporating dissolution into paleoenvironmental modeling and reconstructions. *Quaternary Science Reviews*, **28**, 120–36.

Ryves, D. B., Battarbee, R. W., Juggins, S. J., Fritz, S. C., & Anderson, N. J. (2006). Physical and chemical predictors of diatom dissolution in freshwater and saline lake sediments in North America and West Greenland. *Limnology and Oceanography* **51**, 1342–54.

Ryves, D. B., Juggins, S., Fritz, S. C., & Battarbee, R. W. (2001). Experimental diatom dissolution and the quantification of microfossil preservation in sediments. *Palaeogeography, Palaeoclimatology, Palaeoecology*, **172**, 99–113.

Ryves, D. B., McGowan, S., & Anderson, N. J. (2002). Development and evaluation of a diatom–conductivity model from lakes in west Greenland. *Freshwater Biology*, **47**, 995–1014.

Saros, J. E. & Fritz, S. C. (2000a). Nutrients as a link between ionic concentration/composition and diatom distributions in saline lakes. *Journal of Paleolimnology*, **23**, 449–53.

Saros, J. E. & Fritz, S. C. (2000b). Changes in the growth rate of saline-lake diatoms in response to variation in salinity, brine type, and nitrogen form. *Journal of Plankton Research*, **22**, 1071–83.

Saros, J. E. & Fritz, S. C. (2002). Resource competition among saline-lake diatoms: shifts in taxon abundance in response to variations in salinity, brine type, and N:P ratios. *Freshwater Biology*, **46**, 1–9.

Saros, J. E., Fritz, S. C., & Smith, A. (2000). Shifts in mid- to late-Holocene anion composition in Elk Lake (Grant Co., MN): comparison of diatom and ostracode inferences. *Quaternary International* **67**, 37–46.

Schweger, C. E. & Hickman, M. (1989). Holocene paleohydrology of central Alberta: testing the general-circulation-model climate simulations. *Canadian Journal of Earth Sciences*, **26**, 1826–33.

Servant-Vildary, S. & Roux, M. (1990). Multivariate analysis of diatoms and water chemistry in Bolivian saline lakes. *Hydrobiologia*, **197**, 267–90.

Shobert, B. (1974). The influence of water stress on the metabolism of diatoms. I. Osmotic resistance and proline accumulation in *Cyclotella meneghiniana*. *Zeitschrift fur Pflanzenphysiologie*, **74**, 106–20.

Smith, A. J., Donovan, J. J., Ito, E., & Engstrom, D. R. (1997). Ground-water processes controlling a prairie lake's response to middle Holocene drought. *Geology* **25**, 391–4.

Stager, J. C. & Mayewski, P. A. (1997). Abrupt early to mid-Holocene climatic transition registered at the equator and the poles. *Science*, **276**, 1834–6.

Stager, J. C., Cumming, B. F., & Meeker, L. D. (1997). A high-resolution 11,400-yr diatom record from Lake Victoria, East Africa. *Quaternary Research*, **47**, 81–9.

Stager, J. C., Cumming, B. F., & Meeker, L. D. (2003). A 10,000-year high-resolution diatom record from Pilkington Bay, Lake Victoria, East Africa. *Quaternary Research*, **59**, 172–81.

Stevens, L. R., Stone, J. R., Campbell, J., & Fritz, S. C. (2006). A 2200-yr record of hydrologic variability from Foy Lake, Montana, USA, inferred from diatom and geochemical data. *Quaternary Research*, **65**, 264–74.

St. Jacques, J., Cumming, B. F., & Smol, J. P. (2009). A 900-year diatom and chrysophyte record of spring mixing and summer stratification from varved Lake Mina, west-central Minnesota, USA. *The Holocene*, **19**, 537–47.

Stone, J. R. & Fritz, S. C. (2004). Three-dimensional modeling of lacustrine diatom habitat areas: improving paleolimnological interpretation of planktonic:benthic ratios. *Limnology and Oceanography*, **49**, 1540–8.

Stone, J. R. & Fritz, S. C. (2006). Multidecadal drought and Holocene climate instability in the Rocky Mountains. *Geology*, **34**, 409–412.

Sylvestre, F., Servant, M., Servant-Vildary, S., *et al.* (1999). Lake-level chronology on the southern Bolivian Altiplano (18–23° S) during late-glacial time and the early Holocene. *Quaternary Research*, **51**, 54–66.

Sylvestre F., Servant-Vildary S., & Roux M. (2001). Diatom-based ionic concentration and salinity models from the south Bolivian Altiplano (15–23°S). *Journal of Paleolimnology*, **25**, 279–95.

Taieb, M., Barker, P., Bonnefille, R., *et al.* (1991). Histoire paleohydrologique du lac Magadi (Kenya) au Pleistocene superieur. *Comptes Rendus de l'Academie des Sciences, Paris series II*, **313**, 339–46.

Tapia, P. M., Fritz, S. C., Baker, P. A., Seltzer, G. O., & Dunbar, R. B. (2003). A late Quaternary diatom record of tropical climate history from Lake Titicaca (Peru and Bolivia). *Palaeogeography, Palaeoclimatology, Palaeoecology*, **194**, 139–64.

Telford, R. J., Heegaard, E., & Birks, H. J. B. (2004a). All age-depth models are wrong: but how badly? *Quaternary Science Reviews*, **23**, 1–5.

Telford, R. J., Heegaard, E., & Birks, H. J. B. (2004b). The intercept is a poor estimate of a calibrated radiocarbon age. *The Holocene*, **14**, 296–8.

Telford, R. J., Lamb, H. F., & Mohammed, M. U. (1999). Diatom-derived palaeoconductivity estimates for Lake Awassa, Ethiopia: evidence for pulsed inflows of saline groundwater? *Journal of Paleolimnology* **21**, 409–21.

Tibby, J., Gell, P. A., Fluin, J., & Sluiter, I. R. K. (2007). Diatom-salinity relationships in wetlands: assessing the influence of salinity variability on the development of inference models. *Hydrobiologia*, **591**, 207–18.

Tuchman, M. L., Theriot, E., & Stoermer, E. F. (1984). Effects of low level salinity concentrations on the growth of *Cyclotella meneghiniana* Kütz. (Bacillariophyta). *Archiv für Protistenkunde*, **128**, 319–26.

Vallet-Coulomb, C., Gasse, F., Robison, L., *et al.* (2006). Hydrological modeling of tropical closed Lake Ihotry; sensitivity analysis and implications for paleohydrological reconstructions over the past 4000 years. *Journal of Hydrology*, **331**, 257–71.

Van Campo, F. & Gasse, F. (1993). Pollen and diatom-inferred climatic and hydrological changes in Sumxi Co. Basin (western Tibet) since 13,000 yr B.P. *Quaternary Research*, **39**, 300–13.

Verschuren, D., Laird, K. R., & Cumming, B. F. (2000). Rainfall and drought in equatorial East Africa during the past 1,100 years. *Nature*, **403**, 410–14.

Williams, W. D. (1981). Inland salt lakes: an introduction. *Hydrobiologia*, **81**, 1–14.

Wilson, S. E., Cumming, B. F., & Smol, J. P. (1994). Diatom-based salinity relationships in 111 lakes from the Interior Plateau of British Columbia, Canada: the development of diatom-based models for paleosalinity and paleoclimatic reconstructions. *Journal of Paleolimnology*, **12**, 197–221.

Wilson, S. E., Cumming, B. F., & Smol, J. P. (1996). Assessing the reliability of salinity inference models from diatom assemblages: an examination of a 219 lake data set from Western North America. *Canadian Journal of Fisheries and Aquatic Sciences*, **53**, 1580–94.

Winter, T. C. (1990). Hydrology of lakes and wetlands. In *Surface Water Hydrology*, ed. M. G. Wolman & H. C. Riggs, Boulder, CO: Geological Society of America, pp. 159–88.

Xiangdong, Y., Kamenik, C., Schmidt, R., & Wang, S. (2003). Diatom-based conductivity and water-level inference models from eastern Tibetan (Qinghai-Xizang) Plateau lakes. *Journal of Paleolimnology*, **30**, 1–19.

Yu, Z., Ito, E., & Engstrom, D. R. (2002). Water isotopic and hydrochemical evolution of a lake chain in the northern Great Plains and its paleoclimatic implications. *Journal of Paleolimnology*, **28**, 207–17.

Zlotnik, V. A., Burbach, M., Swinehart, J., *et al.* (2007). A case study of direct push methods for aquifer characterization in dune-lake environments of the Nebraska Sand Hills. *Environmental and Engineering Geology*, **13**, 205–16.

11 Diatoms in ancient lakes

ANSON W. MACKAY, MARK B. EDLUND, AND GALINA KHURSEVICH

11.1 Introduction

We define ancient lakes as those that contain sedimentary records that span timescales since at least the last interglacial (*c.* 128 ka before present (BP)). We use this definition because by far the majority of diatom applications and reconstructions are undertaken on lakes that have formed since the end of the last glaciation (Termination 1). Ancient lakes are commonly found within grabens in active rift zones. Important examples include lakes Baikal (Russia), Biwa (Japan), Hövsgöl (Mongolia), Kivu (Democratic Republic of Congo, Rwanda), Malawi (Malawi, Mozambique, Tanzania), Ohrid (Albania and Macedonia) and Prespa (Greece, Albania and Macedonia), Tanganyika (Burundi, Congo, Tanzania), Titicaca (Bolivia, Peru), Tule (USA), and Victoria (Kenya, Tanzania, Uganda). These extant lakes contain sedimentary archives that often span at least the full Quaternary period (*c.* 2.6 million years (Ma)). Other ancient lakes with significantly long sedimentary archives include those associated with volcanic activity, e.g. Lake Albano (Italy), karst landscapes, e.g. Ioannina (Greece), and meteorite-impact craters such as El'gygytgyn (Russia), Pingualuit (Canada), Bosumtwi (Ghana), and Tswaing (formerly known as the Pretoria Salt Pan) (South Africa).

Diatom records from ancient lakes provide potentially powerful insights into mechanisms of environmental change over glacial–interglacial (G–IG) timescales, with most studies focusing on interpretation of paleoclimate records. However, records from ancient lakes can also provide useful insights into ecology and evolution of diatoms over long timescales (Khursevich *et al.* 2001). Interpretation of diatom records from ancient lakes is therefore challenging and complex, and needs to take into account short-term limnological processes superimposed over long-term gradual changes such as orbital variations and plate tectonics (Bradbury, 1999). However, tectonic activity can obfuscate climate signals through the formation of earthquake-induced seismites in soft sediments (Wagner *et al.*, 2008), increasing the need for careful interpretation. Applied diatom studies in ancient lakes need also to address contemporary processes related to modern diatoms including evolution and biodiversity, ecology, transport, and dissolution at surface sediment–water interfaces.

There is a vast range of diatom applications to earth sciences, as is borne out by the chapters in this book. The majority of these studies cover time periods since the late Pleistocene. The themes considered here represent on-going challenges in the application of diatom studies from ancient lakes. We have arranged the themes as follows:

- biodiversity and evolution of diatoms in ancient lakes
- integrity of diatom records in ancient lake sediments
- applications of ancient lake diatoms to reconstruct (i) Quaternary G–IG climate change; (ii) last interglacial climate variability; (iii) abrupt climate events; (iv) human impacts.

11.2 Biodiversity and evolution of diatoms in ancient lakes

One characteristic that draws researchers to ancient lakes is their extraordinary biodiversity and endemism at many trophic levels. The gammarids, turbellarians, and cottid fishes of Lake Baikal, and the gastropods, ostracods, and cichlid fishes of the African Rift Lakes are well-known examples of high diversity and high endemism in ancient lake biota (Brooks, 1950; Martens *et al.*, 1994, Rossiter and Kawanabe, 2000). Diatoms similarly exhibit intriguing patterns of diversity, endemism, and adaptations in ancient lakes (Figure 11.1). They deserve comment in the context of applied studies on ancient lakes because these qualities add complexity to analysis and interpretation of sediment records.

The Diatoms: Applications for the Environmental and Earth Sciences, 2nd Edition, eds. John P. Smol and Eugene F. Stoermer. Published by Cambridge University Press.

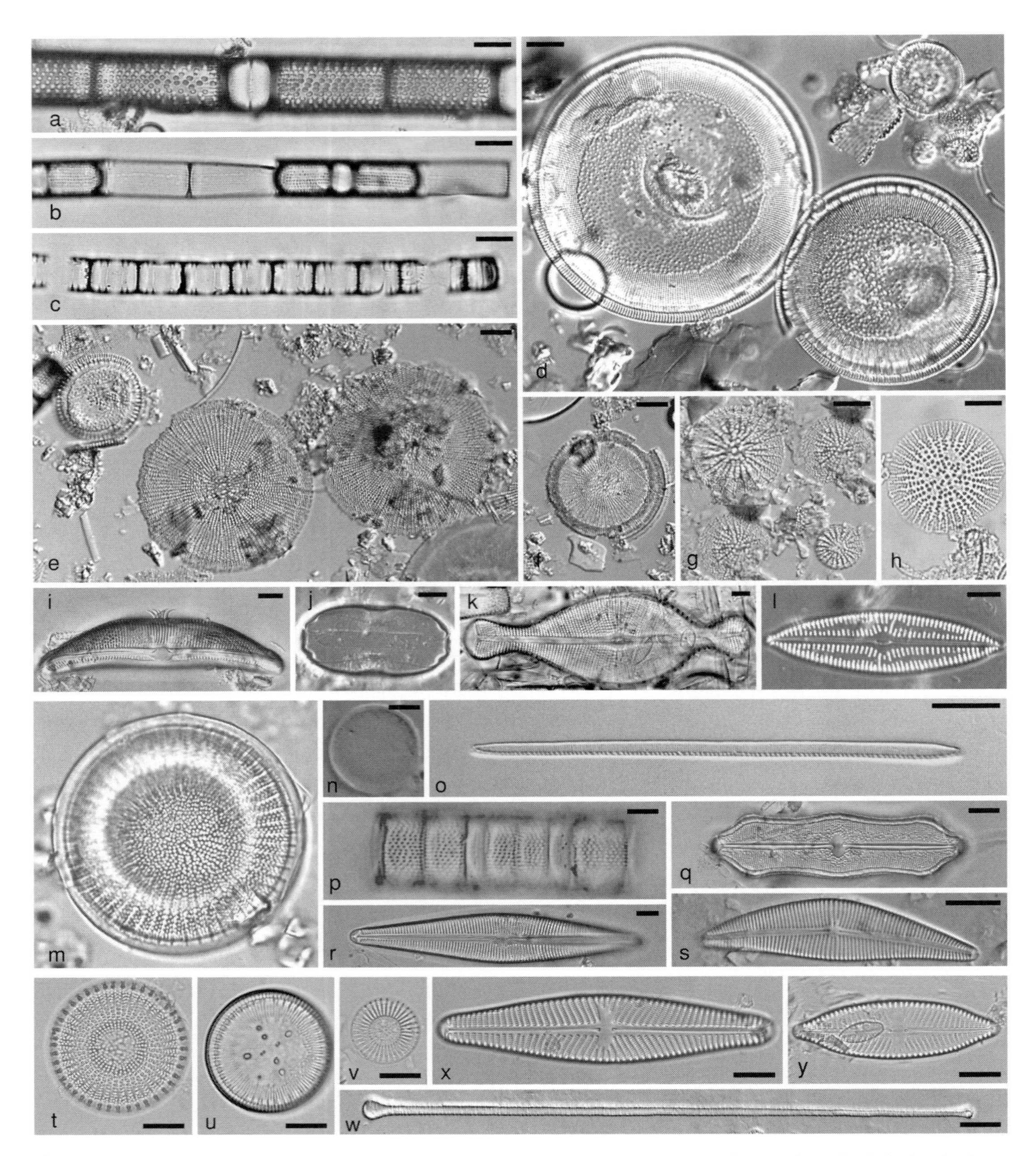

Figure 11.1 Diatoms from ancient lakes. Scale bars = 10 μm. (a)–(l) Lake Baikal. (a) *Aulacoseira baicalensis*. (b) *A. skvortzowii*. (c) *Stephanodiscus meyeri*. (d) *Cyclotella baicalensis-ornata-minuta* species flock. (e) Extinct *Stephanodiscus grandis* flora. (f) *S. grandis*. (g) Shaitan interstade *Stephanodiscus* flora. (h) *Stephanodiscus flabellatus* Khursevich and Loginova. (i)–(l) Endemic benthic diatoms from Lake Baikal. (i) *Amphora cristodentata* Skabitch. (j) *Amphora obtusa* W. Greg. (k) *Didymosphenia dentata* (Dorogostaisky) Skvortzov. (l) *Navicula lacus-baikali* var. *lanceolata* Skvortzov. (m)–(s) African Rift Lakes. (m) *Cyclostephanos damasii* (Hustedt) Stoermer & Håkansson. (n) *Thalassiosira faurei* (Gasse) Hasle. (o) *Gomphonitzschia ungeri* Grunow. (p) *Aulacoseira nyassensis* (O. Müller) Simonsen. (q) *Neidium* sp. (r) An astigmate *Gomphonema*. (s) *Afrocymbella beccarii* (Grunow) Krammer. (t)–(y) Lake Hövsgöl. (t) *Stephanodiscus mongolicus* Edlund, Soninkhishig & R. M. Williams. (u) *Cyclotella ocellata*-complex. (v) *Stephanodiscus jamsranii* Edlund, R. M. Williams & Soninkhishig. (w) *Asterionella edlundii* Stoermer & Pappas. (x) *Navicula hoevsgoelreinhardtii* Edlund & Soninkhishig. (y) *Navicula apiculatoreinhardtii* f. *biseriata* Edlund & Soninkhishig.

Several patterns of diversity stand out among ancient-lake diatoms. First is diversification and endemism characteristic of individual lake systems. For example, whereas the genera *Placoneis* and *Sellaphora* are very species rich, *Gomphonitzschia* and *Afrocymbella* (Figure 11.1) are limited in distribution to the African Rift Lakes (Hustedt, 1949), *Denticula* is diverse in the Indo-Malaysian ancient lakes (Hustedt, 1937–1939), and *Didymosphenia* and *Amphorotia* are most diverse in Lake Baikal (Skvortzow, 1937; Metzeltin & Lange-Bertalot, 1995; Kociolek *et al.*, 2000; Williams & Reid, 2006). Within each of these lakes are a high proportion of endemic taxa and groups of taxa that are likely monophyletic; these monophyletic groups have been termed species flocks (Brooks, 1950; Edlund *et al.*, 2006a).

Second, are diatom groups that are commonly diverse and endemic in multiple ancient lakes. For example, the Surirellaceae show exceptionally high diversity in Lake Ohrid (Jurilj, 1949), the Indo-Malaysian ancient lakes (Hustedt, 1937–1939; Bramberger *et al.*, 2006), Lake Baikal (Skvortzow, 1937), and the African Rift Lakes (Cocquyt *et al.*, 1993); whereas the amphoroid diatoms are very diverse in Lake Ohrid, the African Rift Lakes, Lake Biwa, Lake Hövsgöl and Lake Baikal (Figure 11.1; see Levkov, 2009). However, rarely are any species in these groups common among all the ancient lakes; each lake has its own characteristic group of species and species flocks. Edlund *et al.* (2006a, b) highlighted a special case of diversity of *Navicula reinhardtii* (Grunow) Grunow species flocks in three ancient lakes: Hövsgöl, Baikal, and Prespa (Figure 11.1). Each of these lakes, in addition to this widespread taxon *N. reinhardtii*, has an endemic flock of one to seven species that share the apomorphic character of heteromorphic terminal raphe endings characterizing *N. reinhardtii*.

The third pattern of diversity in ancient lakes is characterized by planktonic diatom communities that have very low species richness but relatively high endemism (Stoermer and Edlund, 1999; Bondarenko *et al.*, 2006). For example, Lake Baikal's plankton biomass is currently dominated by five endemic species (Figure 11.1; Popovskaya *et al.*, 2002; Edlund, 2006) and Lake Hövsgöl's plankton comprises six species including three endemics (Figure 11.1; Edlund *et al.*, 2003; Pappas and Stoermer, 2003). (Recently, Genkal & Bondarenko (2006) proposed that many of the common endemic planktonic diatoms in Lake Baikal be reclassified as relicts, as morphologically similar species can be found in mountain lakes in the Zabaikalye region to the northeast of Lake Baikal. It is clear, therefore, that the nature of extant diatom endemicity in Lake Baikal warrants further investigation.) Modern increases in planktonic richness are linked to anthropogenic impacts and introduction of widespread species (Tuji & Houki, 2001; Verschuren *et al.*, 2002; Edlund, 2006). Many hypotheses have been offered to explain low planktonic diversity in ancient lakes, including relative habitat isolation and lengthy lacustrine duration (Bradbury & Krebs, 1982), competitive exclusion or "the paradox of the plankton" (Hutchinson 1961; Stoermer & Edlund, 1999), and a model of stable diversity best predicted by resource gradient and competition theory (Edlund, 2006). Analysis of long sediment records from ancient lakes has allowed testing of competing hypotheses (Kuwae *et al.*, 2002; Khursevich *et al.*, 2005; Edlund, 2006). Although planktonic diatoms usually dominate sediment assemblages, benthic diversity can also be sampled using sediment cores. Williams *et al.* (2006) used a 600 m long core to examine the fossil diversity and duration of the genus *Tetracyclus* in Lake Baikal, describing two new species and adding to the temporal and biogeographic extent of six known taxa.

Because of these patterns of diatom diversity, analysis of sediment records in ancient lakes requires special considerations. In most post-glacial landscapes, the development of regional diatom calibration sets allows quantitative and qualitative reconstructions of lake histories based on knowledge of analogous lake systems and diatom communities (Birks, this volume). In ancient lakes, recent and fossil diatom communities typically have no modern analogs outside of the immediate lake and these conventional techniques cannot be readily applied, although Mackay *et al.* (2003) did assess the feasibility of transfer functions for Lake Baikal by exploring relationships between within-lake environmental gradients and surface-sediment diatoms. Although applicable to recent sediment records these species–environment relationships have also been used to predict the impact of future climate change on Baikal's endemic diatoms (Mackay *et al.*, 2006). Modern autecological information on individual diatom species are also critical when interpreting sedimentary records, e.g. Likhoshway *et al.* (1996); Jewson *et al.* (2008). Finally, as down-core assemblages are generally dominated by extinct and historically endemic taxa (Khursevich *et al.*, 2005), researchers are increasingly dependent on ecological interpretations based on modern species "analogs" and modern autecological characteristics to interpret fossil assemblages (e.g. Bradbury *et al.*, 1994; Mackay *et al.*, 2005).

Although recent research has shown that diatoms are capable of speciation on short timescales (<10,000 years; Theriot *et al.*, 2006), the long sediment records recovered from ancient lakes and their modern biota provide a rare opportunity to study evolutionary processes and outcomes in inland water bodies. An active descriptive research program continues to document the amazing diversity of ancient lakes (e.g. Levkov *et al.*, 2007; Edlund & Soninkhishig, 2009; Levkov, 2009), but fewer researchers have focused on phylogenetic analysis of ancient-lake diatoms. The Lake Baikal record has been most intensely studied. Initial research resolved biostratigraphic zonation of long records and identified persistent, high levels of endemism, species originations, and species extinctions. This eventually led to the use of stratigraphic records to hypothesize possible evolutionary relationships and lineages (e.g. Khursevich *et al.*, 2005). Results of these studies provided clear examples of the control of evolutionary processes by G–IG cycles (Khursevich *et al.*, 2005; Edlund, 2006), evidence of extinct and endemic genera and lineages (Khursevich & Fedenya, 2006), as well as possible morphological character evolution and character states in extinct Baikal lineages (Khursevich, 2006). But we have only begun to utilize these records. For example, in all studies on long cores a common result is that glacial cycles bring about species extinctions and interglacials result in the "recolonization" of Lake Baikal by new endemic species that are seemingly highly adapted to Baikal's unique environment (Figure 11.1). Higher-resolution core studies may help resolve the origin and evolution of these new inhabitants and lineages (*sensu* Theriot *et al.*, 2006).

One of the few studies to test formally the phylogeny of Baikalian endemics using the fossil record investigated the monophyly and phylogeny of the *Cyclotella baicalensis–minuta–ornata* species flock (Figure 11.1). These three taxa appear sequentially in the fossil record, which provides a stratigraphic record of environmental/climate induced speciation and a phylogenetic hypothesis. Julius & Stoermer (1999) used morphometric analysis to show similar ontogenetic relationships among the taxa, cladistic analysis and stratocladistics to propose and support a phylogeny, and suggested that speciation may have resulted from ontogenetic modification similar to proportional dwarfism and gigantism. Julius and Curtin (2006) further analyzed their morphometric data to show that selective forces were acting independently on each species and that the taxa were sexually differentiated based on auxospore size.

Molecular studies are also underway on ancient-lake diatoms. Edgar & Theriot (2004) included two Baikal endemics, *Aulacoseira baicalensis* (Meyer) Simonsen and *A. skvortzowii* Edlund, Stoermer & Taylor (Figure 11.1), in a phylogenetic analysis of the genus using molecular and morphological data. Both Baikal taxa were included in the *A. islandica*-clade, which was also populated by many fossil taxa. It was hypothesized that *A. islandica* was ancestral to *A. skvortzowii*, whereas *A. baicalensis* was sister to several fossil taxa. Scherbakova *et al.* (1998) analyzed 18S rRNA (a fundamental component of ribosomal RNA in eukaryotic cells) in the Baikal endemics and proposed that the split between *A. skvortzowii* and *A. baicalensis* occurred <25 Ma, well within the age of Lake Baikal. Work to determine the origin and evolutionary processes at work on ancient lake diatoms is in its infancy, but should soon begin to reveal how speciation, relict floras, climate change, and evolutionary adaptations (e.g. resting spores; Jewson *et al.*, 2008) have produced the high diversity and high endemism that characterizes ancient-lake floras.

11.3 Integrity of diatom records in ancient-lake sediments

Sedimentary diatom records in deep, ancient lakes can be impacted by dissolution processes, especially if the water column is undersaturated with respect to silica and/or if the bottom waters are oxygenated. For example, by comparing accumulation rates of diatoms through the water of Lake Baikal with diatom accumulation rates in surficial sediments, Ryves *et al.* (2003) were able to demonstrate that only *c.* 1% of diatom valves preserve in the lake's sedimentary record. This is largely because the water column of Lake Baikal is undersaturated with respect to silica, a situation common to the marine environment where diatom preservation is comparably low (e.g. Tréguer *et al.*, 1995). In Lake Baikal, very few diatoms persist during glacial periods. However, this has undoubtedly contributed to the strength of Baikal's long, paleoclimatic record because biogenic silica (a proxy of primary productivity which is derived almost entirely from diatoms) is highly sensitive to G–IG cycles (Williams *et al.*, 1997). Dissolution is also extensive during interglacial periods. For example, valves that are less heavily silicified are susceptible to a greater degree of dissolution. If dissolution is extensive, these valves can be more difficult to identify, thus increasing taxonomic uncertainty. In the Lake Baikal record this has resulted in considerable bias (Ryves *et al.*, 2003); some very finely silicified species (e.g. *Nitzschia*) are removed from the sedimentary record almost entirely, while other more heavily silicified species (e.g. *Aulacoseira*) are over-represented (Battarbee *et al.*, 2005). Attempts to compensate for this bias using correction factors reveal that over at least the last 1000 years, significant improvements can be made to diatom-inferred paleoclimatic interpretations (Mackay *et al.*, 2005).

While extensive diatom dissolution is not restricted to Lake Baikal, few experimental studies have been carried out on other ancient lakes to assess the problem. Haberyan (1990) investigated the misrepresentation of planktonic diatoms in Lake Malawi sediments by comparing fluxes in the water column with fluxes in sediment traps and surficial sediments. In Lake Malawi (as in Lake Baikal), heavily silicified species such as *Aulacoseira* were over-represented, while finely silicified species such as *Nitzschia* were under-represented. Certain species of diatom are preferentially grazed by zooplankton. As zooplankton fecal pellets are transported more quickly through the water column into surficial sediments, preferential preservation of ingested diatom species can occur. Haberyan (1985) found that *c.* 40% of diatoms being deposited into sediments from Lake Tanganyika came from copepod fecal pellets. From studying the contents of fecal pellets from a 16 ka record, it was evident that they preferentially grazed certain species within the genera *Aulacoseira*, *Fragilaria*, and *Gomphonema*, whereas elongate, finely silicified species such as *Nitzschia* cf. *spiculum* were most often found in non-pellet sedimentary material. However, whereas preferential dissolution of diatom valves in Lake Baikal causes significant variation in sedimentary diatom assemblages (Battarbee *et al.*, 2005), the bias created from preferential preservation of diatoms in copepods in Lake Tanganyika was found to be small and not significant (Haberyan 1985).

Other studies also acknowledge that dissolution can bias sediment records. Studies from Lake Titicaca, for example, note that dissolution processes (due to high lake-water pH) impact negatively on finely silicified pelagic taxa such as *Cyclostephanos tholiformis* Stoermer, Håkansson & Theriot and *Fragilaria crotonensis* Kitton (e.g. Tapia *et al.*, 2003). In Lake Victoria, there is extensive dissolution of *Nitzschia acicularis* (Kützing) W. Smith before it becomes buried into bottom sediments as suggested by a 30-fold decline in its fossil abundance. In contrast, other species in the genera *Cyclostephanos* and *Aulacoseira* decline in abundance by only 1.5–1.2 times, respectively (Verschuren *et al.*, 2002).

11.4 Applications of ancient-lake diatoms

11.4.1 Quaternary glacial–interglacial climate change

Despite many of the world's largest ancient lakes being tectonic in origin, there are few published diatom records that span the full Quaternary, in part due to technological challenges of extracting long sediment archives from deep lakes. Quaternary diatom studies are most advanced from Lake Baikal due to the Baikal Drilling Project (BDP) which began extracting long records in the early 1990s (Williams *et al.*, 2001). Other ancient lakes in tectonic basins have recently been cored under the auspices of the Intercontinental Scientific Drilling Program (http://www.icdp-online.de/front_content.php), e.g. Lake Titicaca in 2001, Lake Malawi in 2005 and Lake El'gygytgyn in 2009. At Lake Malawi for example, sedimentary records were retrieved that extend back over 1.5 million years, although long diatom records have yet to be produced. Tule Lake, on the other hand, was the first lake in western USA to provide a long, continuous diatom record of paleolimnological change from late Pliocene–Pleistocene sediments, spanning *c.* 3 Ma (Bradbury, 1991). In this section therefore, diatom applications of Quaternary climate change are inferred from few records. Here we provide considerable detail for Lake Baikal, supplemented by diatom-inferred change from other long sequences at Tule Lake and Lake Biwa where relevant.

Lake Baikal is situated within an active tectonic region of the Eurasian plate (51° 28′–55° 47′ N and 103° 43′–109° 58′ E), which first began to form over 30 million years ago. It has long been recognized as one of the world's most unusual freshwater ecosystems, especially in terms of its age (in excess of 20 million years), depth (*c.* 1640 m) and volume (23,600 km^3). Although Quaternary glaciations have had major impacts on the lake, its bottom sediments have never been glaciated, resulting in a long sedimentary record of over 7.5 km. Another important feature that sets Lake Baikal apart from other deep, ancient lakes (e.g. Tanganyika, Malawi) is that its entire water column is saturated with oxygen due to regular renewal of the deep waters every spring and autumn. This results in the oxidation of even the deepest bottom waters, allowing an extensive, and almost wholly, endemic deepwater fauna to survive.

Diatoms are important primary producers and crucial to the ecosystem functioning of Lake Baikal. Planktonic diatoms comprise the majority of spring phytoplankton biomass and they usually determine annual pelagic productivity (Popovskaya, 2000). There is significant spatial variability in diatom species across Lake Baikal, dependant on a number of factors including climate (e.g. ice and snow cover) and hydrological variables such as fluvial input and water depth (Mackay *et al.*, 2006). The most common extant planktonic diatoms in Lake Baikal are a mixture of endemic (e.g. *Aulacoseira baicalensis*, *Cyclotella minuta* (Skvortzow) Antipova, and *Stephanodiscus meyeri* Genkal & Popovskaya) and cosmopolitan species (e.g. *Nitzschia acicularis* and *Synedra acus* Kützing and its varieties).

The sediments of Lake Baikal consist of alternating layers of diatom-rich and inorganic (clayey) diatom-poor sediments, which are linked to changes in orbital parameters and northern-hemisphere insolation (Williams *et al.*, 1997; Grachev *et al.*, 1998). While the majority of these early studies focused on using biogenic silica (BSi) as a proxy for diatom production, diatom assemblages themselves have also been investigated in detail (see Mackay (2007) for a review). An exceptionally long diatom record for Lake Baikal has been constructed back to the

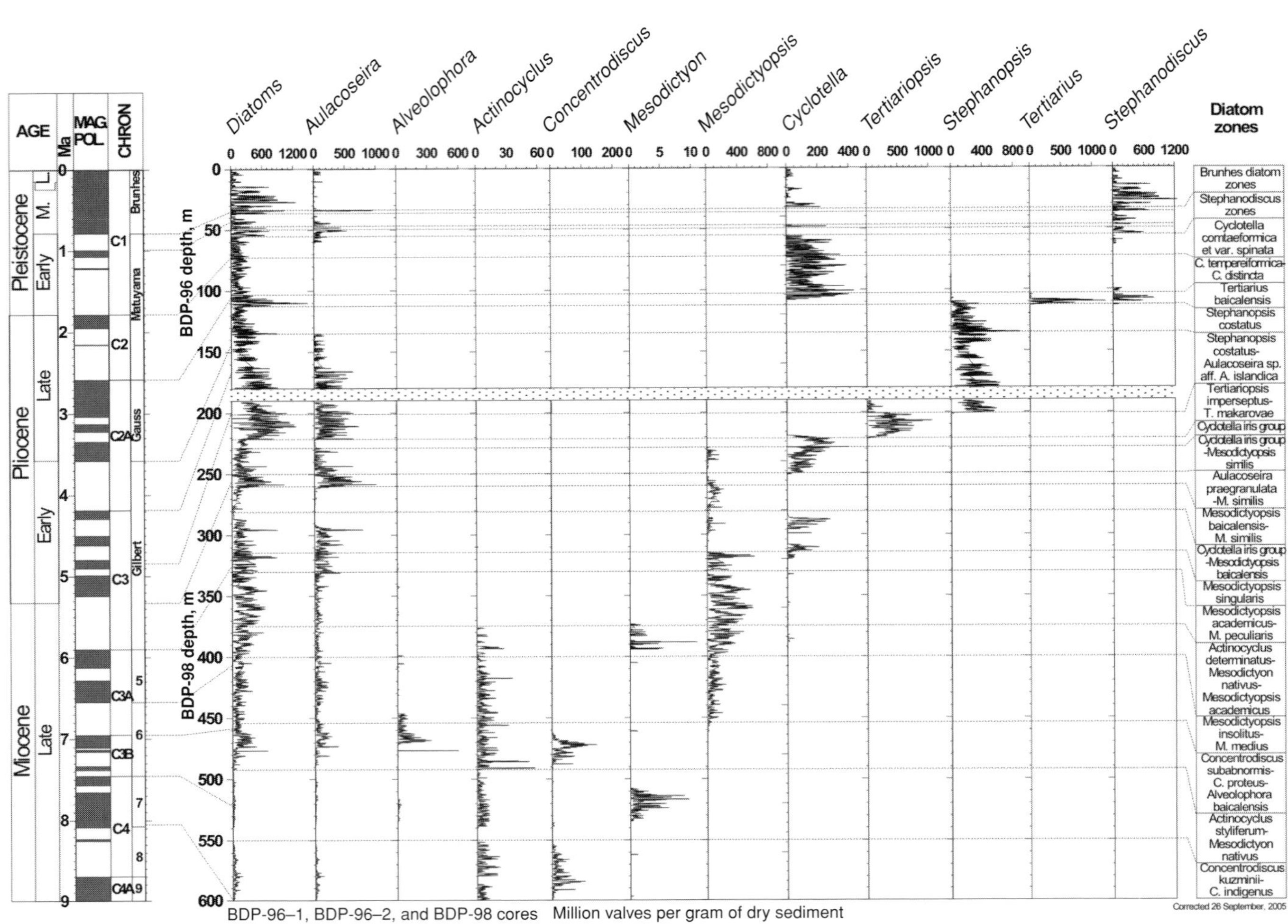

Figure 11.2 Upper Cenozoic (Upper Miocene to Holocene) diatom record from Lake Baikal formed from composite profiles analysed from BDP-96–1, BDP-96–2, and BDP-98 cores.

Miocene, *c.* 9 Ma (Khursevich & Fedenya, 2006) (Figure 11.2). This record contains a diverse diatom flora of the class Coscinodiscophyceae, which is represented by 3 subclasses, 5 orders, 6 families, 16 genera, comprising 81 species and 22 intraspecific taxa. Among them, 1 endemic family (*Thalassiobeckiaceae*), 8 genera (3 endemic for the Lake Baikal), 53 species (43 Baikalian endemics) and 15 intraspecific taxa (9 Baikalian endemics) are extinct (Khursevich, 2006).

Pliocene–Pleistocene BSi records from Lake Baikal detail paleoclimate variability going back 5 Ma. Diatom assemblages suggest ecosystem responses to a warm climate were established between 5 Ma and *c.* 3 Ma with the complete disappearance of the genus *Mesodictyopsis* at the start of the Pliocene. Between *c.* 4.7 Ma and 2 Ma there is the successive appearance and extinction of genera belonging to *Tertiariopsis* (3 species), *Stephanopsis* (1 species represented by 7 morphotypes), *Tertiarius* (1 species) (Figure 11.3) and *Thalassiobeckia* (1 species; not shown). The extinctions of *Stephanopsis* and *Tertiarius* coincide with the onset of the Northern Hemisphere Glaciation (NHG) and concomitant development of early mountain glaciation in the Baikal region at *c.* 2.75 Ma. The first substantial minima in diatom abundance and biogenic silica (as well as palynological expansion of non-arboreal steppe vegetation) and lithological data (the presence of fine clay interlayers with the ice- and iceberg-rafted detritus) reflect the first profound cold event *c.* 2.5 Ma. After this strong cooling episode, the *Cyclotella* flora became especially developed between 2.5 Ma and 1.25 Ma (Grachev *et al.*, 1998; Williams *et al.*, 2001); the abundance of *Cyclotella* taxa and biogenic silica content repeatedly alternated between high and low values (Figure 11.3). At this time, eccentricity and obliquity orbital forcings were weak, although obliquity strengthened after the Pliocene–Pleistocene boundary between 1.8 Ma and 0.8 Ma.

Some of the most significant changes in diatom assemblage composition occurred with the onset of the Mid-Pleistocene Revolution (MPR) (*c.* 0.8 Ma). Biogenic silica and diatom

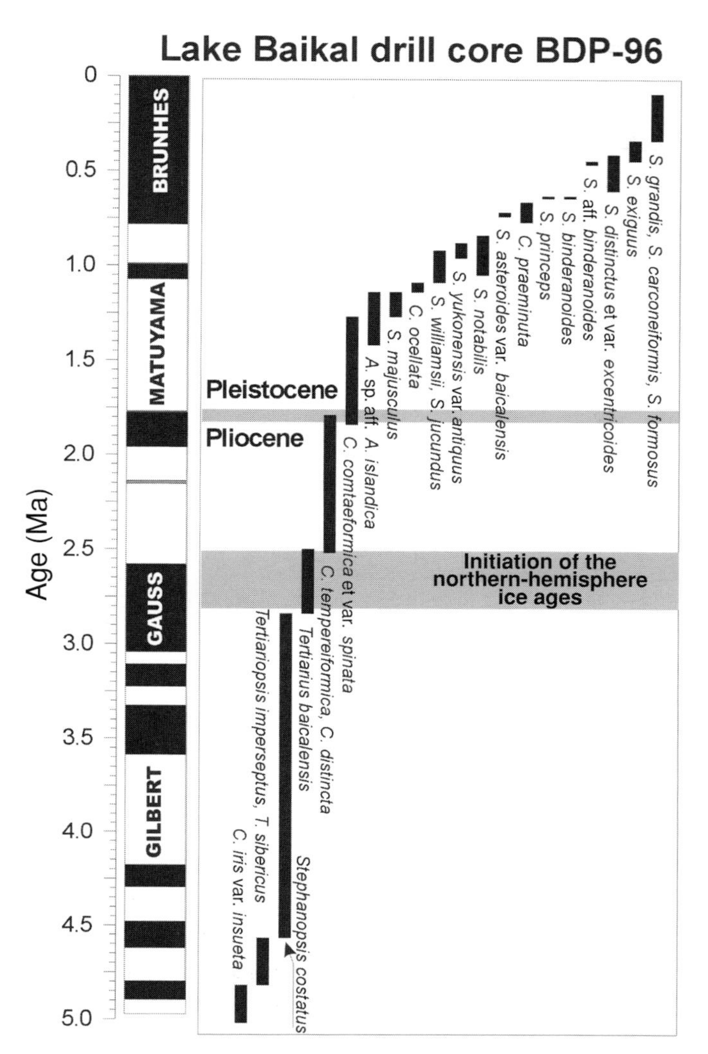

Figure 11.3 The biostratigraphic framework from Lake Baikal planktonic diatoms for the last five million years in relation to the paleomagnetic event/reversal timescale. (From Williams *et al.*, 2001.)

abundance records reveal 19 distinct intervals in the BDP-96 sedimentary sequence (Figure 11.4). These consist of alternating diatomaceous (10) and diatom-barren (9) intervals corresponding to Marine Isotope Stages 1–19 (Khursevich *et al.*, 2001) representing interglacial and glacial periods respectively.

It is worth noting here that the majority of the BDP studies relied heavily on comparisons to (i) the SPECMAP record (which details non-linear responses to global ice volume) from Ocean Drilling Program (ODP) site 677 (extracted from the Panama Basin; Shackleton *et al.*, 1990), and (ii) summer temperature reconstructions from energy-balance models (EBMs) (driven by insolation) constructed for central Asia (Short *et al.*, 1991). Although peaks in BSi were correlated to peaks in $\delta^{18}O$ in marine sediments, global-ice-volume minima occurred many thousands of years after each solar insolation maximum. Goldberg *et al.* (2000) tried, to a limited extent, to compensate for this bias by absolute dating (using U-series analysis) of two peaks of BSi corresponding to interglacials MIS 5 and MIS 7. From these dates, they estimated that global changes in ice volume lagged warming by approximately 8–9 ka. In all the BDP records, therefore, the tuning of the BSi record to $\delta^{18}O$ potentially loses important information in the form of leads and lags in the Lake Baikal record, as may occur due to its remoteness from changes in ice volume. Second-order effects of dissolution on the BSi record and associated spectral analysis were also apparent. Dissolution had the effect of truncating, or "clipping" the BSi record, especially during colder periods, including glacial inception. This was a direct result of the many "zero" values in the BSi typically recorded during glacial periods, which had an effect of shifting predicted 41 ka and 23 ka cycles into the 100 ka cycle frame (Williams *et al.*, 1997). A new astronomically tuned age model has been proposed by Prokopenko *et al.* (2006), based on correlating timing of September perihelia with peak biogenic silica concentrations. This new chronology offers some substantial improvement on the established age model, as it is no longer tied to e.g. ODP-correlated timescales. It is therefore able to provide an independent chronological framework with which to compare long Quaternary records from Lake Baikal with other continental and marine sequences (Prokopenko *et al.*, 2006).

The paleoenvironmental record also suggests the occurrence of rapid diatom speciation and extinction processes. It seems, therefore, that insolation changes and diatom evolution on relatively short, and often abrupt, timescales are closely linked (Khursevich *et al.*, 2001), although other evolutionary drivers may also be important (see above). During this period, the intensity of insolation amplitude also had a direct impact on species present in the sedimentary archive. For example, the almost continuous presence of *Stephanodiscus distinctus* Khursevich and benthic taxa between MIS 15a and MIS 11 corresponded to low amplitudes of insolation (Prokopenko *et al.*, 2002a) (Figure 11.5). Conversely, when amplitudes were larger, (e.g. as occurred between MIS 19 and MIS 15e), each precessional cycle was dominated by a different species that subsequently became extinct. During MIS 15a–MIS 11, high diatom silica levels suggest warm climatic conditions during a period of stable sedimentation rates (as inferred from low magnetic susceptibility measurements).

Since the MPR, 16 new species of *Stephanodiscus* were recorded in Lake Baikal sediments in comparison to only 4 for *Cyclotella* (Khursevich *et al.*, 2001), suggesting differential speciation rates between the genera. That is, during the past *c.* 0.8 Ma,

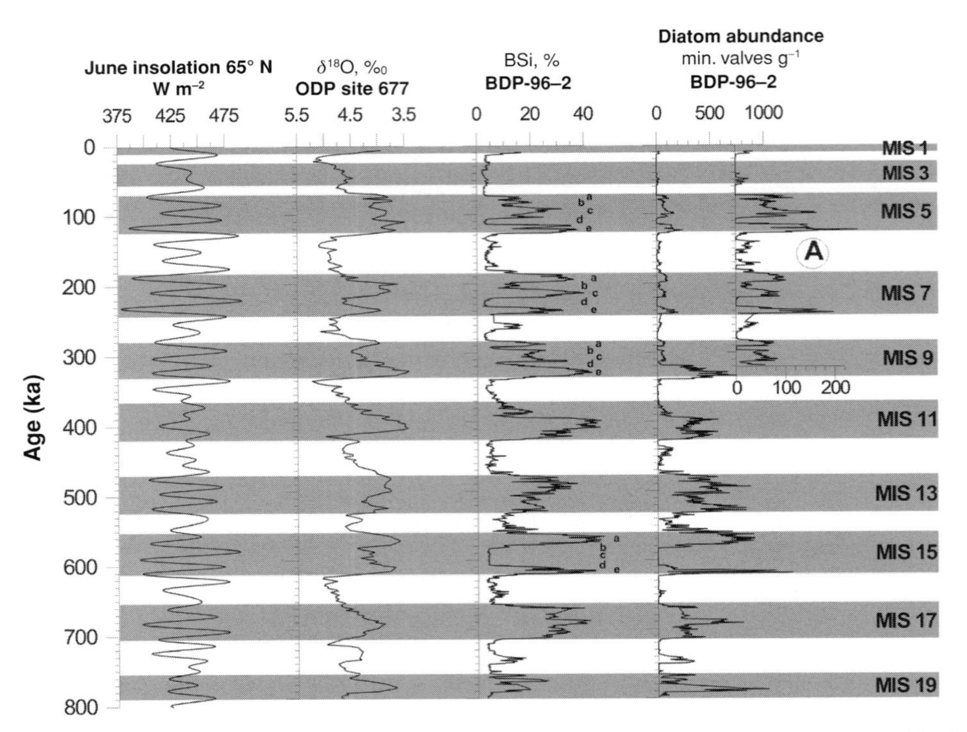

Figure 11.4 Comparison of biogenic silica (BSi) and diatom records from BDP-96–2 to insolation climate forcing during the Brunhes chron with the marine isotope stage (MIS) record of ODP-677. The timescale for the Baikal records is derived from tuning to the June insolation target. The detailed BSi record allows identification of the intervals equivalent to individual marine oxygen isotope stages and substages. The BSi and diatom profiles appear very similar for the MIS 19–9e interval. Above, diatom abundance becomes significantly lower due to the change in size of dominant diatom species. However, inset A demonstrates that the cyclic pattern of diatom abundance continues to parallel the BSi profile throughout the MIS 9c–1 interval. (From Khursevich *et al.*, 2001.)

the average speciation rate for the genus *Stephanodiscus* was *c.* 53.3 ka per new species but for the genus *Cyclotella* the rate is substantially lower, at *c.* 160 ka per new species. This difference suggests that *Stephanodiscus* species are more sensitive to paleoenvironmental changes associated with insolation, and may support the argument of Theriot *et al.* (2006) for rapid speciation of *Stephanodiscus* on millennial and even centennial timescales. However, despite the high degree of endemicity in Lake Baikal, the interglacial phytoplankton species diversity remained poor (Edlund, 2006).

One other paleolimnological record providing a comparably long diatom history of late Pliocene–Pleistocene change, is that of Tule Lake (41.95° N; 121.48° W), in northern California, USA. Tule Lake exists in a lake district formed by graben development coupled with damming by regional lava flows. The lake has been drained for agricultural purposes, but for the most part of the last 3 Ma it existed as a relatively deep and extensive lake. Like the record from Lake Baikal, the sheer age of sediments from Tule Lake allows documentation of diatom evolution through the Cenozoic, as well as providing an extensive record of paleoclimate variability (Bradbury, 1991).

The base of the core from 3 Ma to 2.9 Ma is dominated by the heavily silicified *Stephanodiscus asteroides* and the now extinct *Cyclotella elgeri* Hustedt, both of which suggest that Tule Lake was fresh, deep, and seasonally stratified (Figure 11.6). These two species are succeeded by *Aulacoseira solida* Eulenstein, which is indicative perhaps of warm winters and abundant summer precipitation (Bradbury, 1991). Between 2.65 Ma and 2.35 Ma, there is a marked and continued increase in fragilarioid taxa (i.e. *Staurosira construens* (Ehrenberg) D. M. Williams & Round, *Staurosirella pinnata* (Ehrenberg) D. M. Williams & Round, and *Pseudostaurosira brevistriata* (Grunow in Van Heurck) D. M. Williams & Round). These species are associated with extensive macrophyte vegetation habitats fringing the lake. Therefore the occurrence of these fragilarioid taxa suggests a significantly cooler, more arid climate associated with the onset of NHG. Significantly, there is a peak in fragilarioid taxa at between 2.6 Ma and 2.4 Ma. This peak suggests very low lake levels associated with cold, arid climates, and is likely to be the same cold episode identified in Lake Baikal at 2.5 Ma.

During the late-Pliocene period fragilarioid species increased between 2.0 Ma and 1.7 Ma, implying successively cooler, drier summers that correlate to the occurrence of

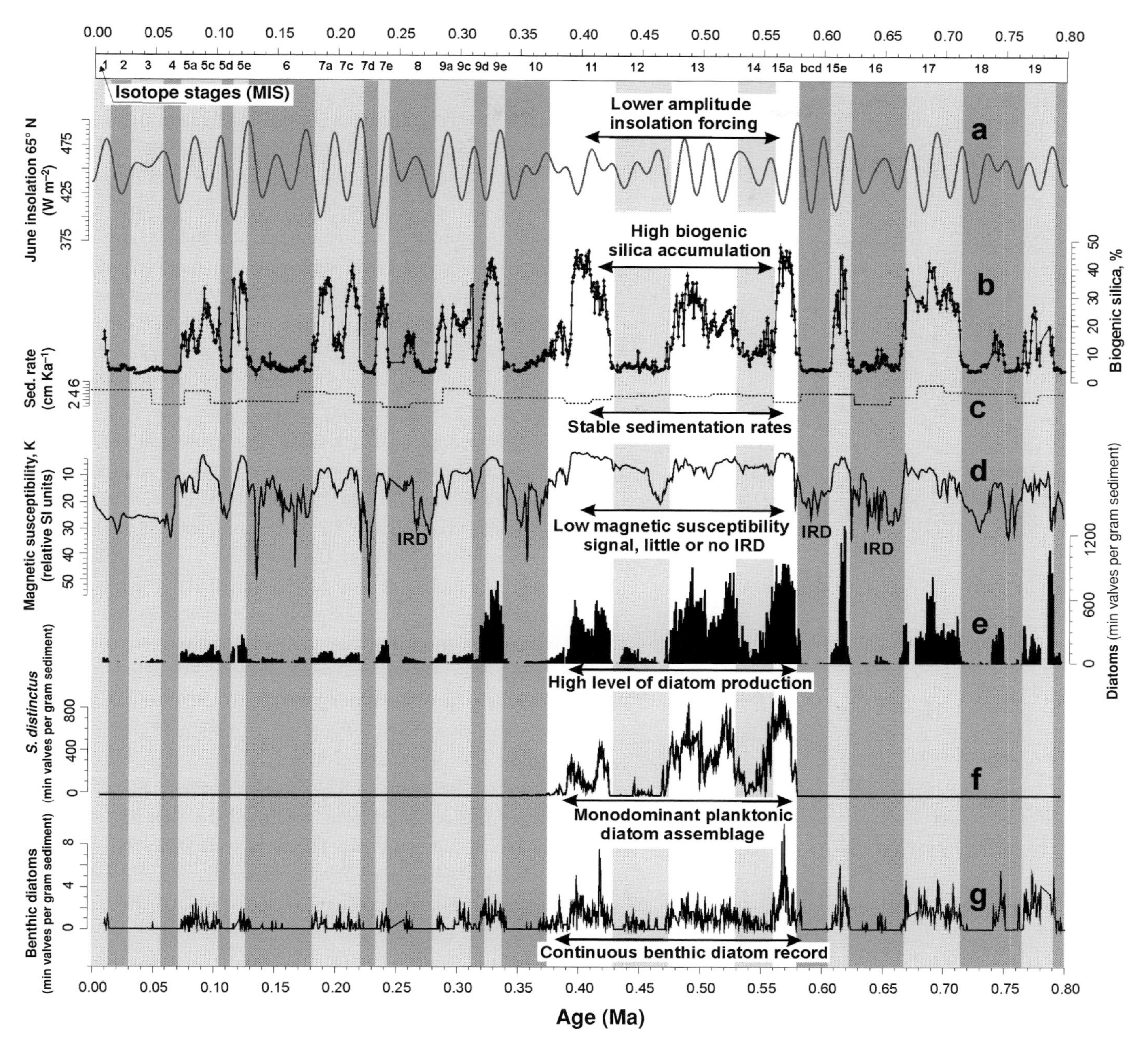

Figure 11.5 Lake Baikal paleoclimate proxy records of the Brunhes chron. All records are from drill core BDP-96–2, plotted to an age scale in millions of years before present (Ma). The June insolation curve at 65° N (a), Laskar 1.1 solution, was used as a tuning target for the age scale of the biogenic silica record (b). Darker shading indicates glacial periods in the Lake Baikal region, including even-numbered marine oxygen isotope stages and substages 5d, 7d, 9d and 15bcd. The interval from MIS 15a to MIS 11 with light shading shows high interglacial levels of BSi accumulation, indicative of warm-climate optima. Apparent sedimentation rates are relatively stable in this interval (c). The magnetic susceptibility record (d) reveals particularly low "interglacial" values during MIS 15a–11, and reflects a lack of glacial ice-rafted debris (IRD) signals during MIS 14 and 12. Supportive of the lack of extreme climatic contrasts during MIS 15a–11 in comparison with other Pleistocene glacials are high levels of diatom production in Lake Baikal (e), the monospecific interglacial planktonic diatom assemblage (f), and the mainly uninterrupted development of the littoral benthic diatom assemblage (g). (From Prokopenko *et al.*, 2002a).

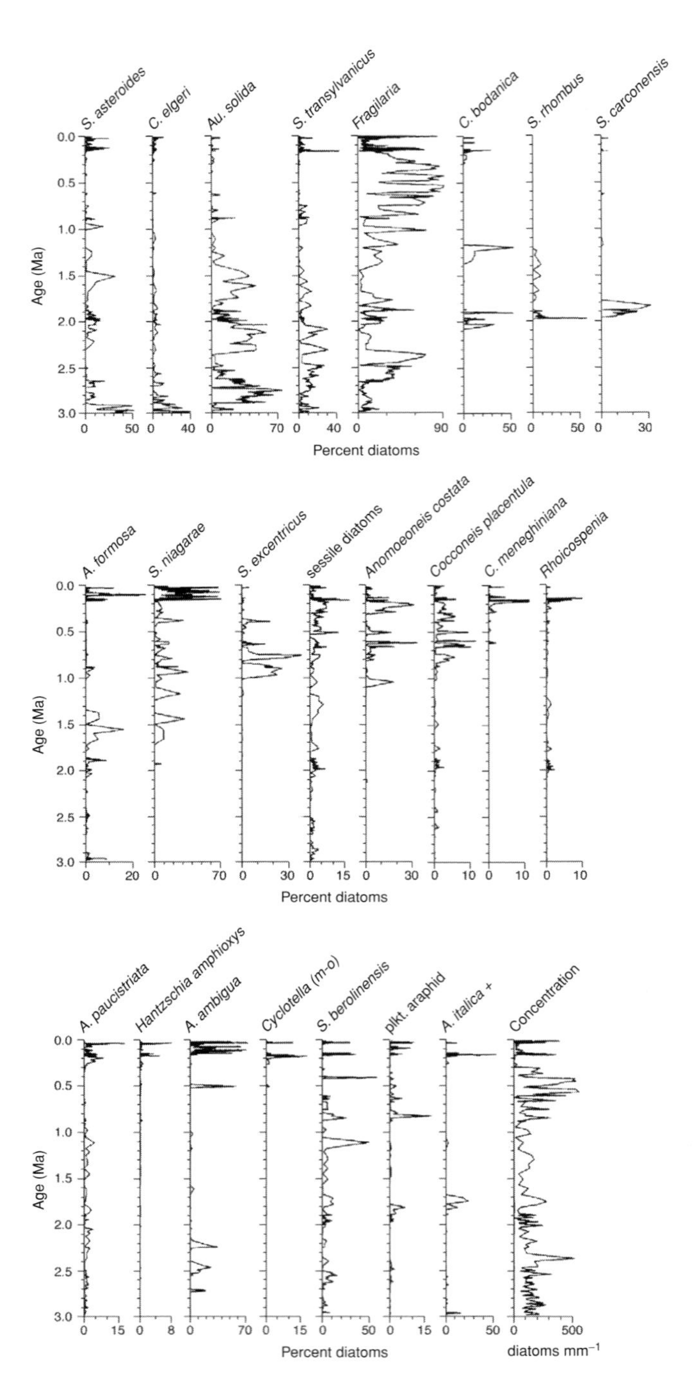

Figure 11.6 Selected diatom percentage profiles vs. time in the Tule Lake core for the last 3 Ma. *Cyctotella* (m-o) indicates mesotrophic and oligotrophic species of *Cyclotella* (*C. kuetzingiana* + *C. michiganiana* Skvortzov + *C. ocellata* Pant. + *C. comta* (Ehr.) Kützing). plkt. araphid. means planktonic araphid diatoms (*Fragilaria crotonensis* + *F. vaucheriae* (Kützing) Petersen + *Synedra acus*). The " +" following *A. italica* (Ehr.) Simonsen indicates that this taxon includes *Aulacoseira subarctica* (O. Müller) E.Y. Haworth. (With kind permission from Springer Science+Business Media: *Journal of Paleolimnology*, The late Cenozoic diatom stratigraphy and paleolimnology of Tule Lake, Siskiyou Co. California, vol. 6, 1991, 205–255, J. Platt Bradbury.)

glacial environments. This transition between the Pliocene and Pleistocene epoch is also marked by alternating *Stephanodiscus* species (*S. carconensis* Grunow and *S. rhombus* Mahood) and peaks of *Cyclotella bodanica* Eulenstein and fragilarioid taxa (Figure 11.6). Again, although the precise ecological requirements of these *Stephanodiscus* spp. are not certain, given their morphologies it is likely they represent strong seasonal nutrient fluxes associated with cooler climates and strengthened periods of circulation. Early Pleistocene climates were milder with long periods of lake circulation during summer months, as inferred from *Stephanodiscus niagarae* Ehrenberg, *S. asteroides* and *Asterionella formosa* Hassall. Few fragilarioid species are found in the record between *c.* 1.6 Ma and 1.5 Ma, suggesting the persistence of warm, wet climate. After 1.4 Ma, however, fragilarioid species become increasingly important in the sedimentary record, with major peaks coincident with even-numbered oxygen isotope stages, i.e. glacial periods. Thus, diatoms records from both Lake Baikal and Tula Lake reflect G–IG cycles during the Quaternary, although the responses are very different. In deep Lake Baikal, diatom productivity declines during glacial periods due to lower temperatures and strengthening lake-water circulation, and together with dissolution results in very poor diatom preservation (Mackay, 2007). In shallow Tule Lake, however, diatom assemblage responses are linked to changing lake levels associated with prevailing moisture balances, which decline during glacial cycles.

During the past *c.* 0.8 Ma the sedimentary record at Tule Lake is characterized by marked fragilarioid peaks, with variable associations with other diatom assemblages. For example, fragilarioid peaks exhibit approximate 100 ka cycles, i.e. may be driven by orbital eccentricity (Bradbury, 1991). Diatom evidence shows that interglacial periods are usually shorter and more variable than glacial periods, although there are exceptions to this, e.g. during MIS 7. The resolution of diatom sampling at Tule Lake during the past *c.* 0.8 Ma is much lower than at Lake Baikal. Therefore, detailed comparisons between e.g. the muted diatom response in Lake Baikal between MIS 15a and MIS 11 and Tule Lake are difficult. However, diatom concentrations in Tule Lake between 0.6 Ma and 0.4 Ma (MIS 15a–MIS 11) are at their highest for the whole record, and this is likely related to the very slow sedimentation rates recorded. Bradbury (1991) suggests that these low sedimentation rates are linked to prevailing interglacial conditions, i.e. warm temperatures and low precipitation, resulting in lower lake levels, increased fragilarioid species, and increased sediment erosion. Thus diatoms in both regions appear to be responding to prolonged mild conditions.

Since the chapter by Bradbury (1999) in the first edition of this book, there have been a number of new studies on long diatom records from other ancient lakes that extend over several G–IG cycles. For example, Bradbury highlighted that the Lake Biwa diatom record did not reflect changing G–IG cycles. However, this was probably because then the sedimentary diatom record was of rather poor resolution, and potential links with changing insolation was not an initial objective. Since then, however, detailed diatom records spanning the last 300 ka have been produced (e.g. Kuwae *et al.*, 2002, 2004). Lake Biwa is the largest lake in Japan, located on its main island (35° 20′ N; 136° 10′ E). Long sediment cores have been taken from its deep, northern, monomictic basin. Like Lake Baikal, marine isotope stages correspond well with biogenic silica (Xiao *et al.*, 1997) and diatom concentration (Kuwae *et al.*, 2002) records. Low biogenic silica and low diatom concentrations are coincident with cold, dry conditions while higher values correlate with warmer, wetter climates. Diatom concentrations are driven mainly by changes in planktonic taxa (especially *Aulacoseira* and *Stephanodiscus* spp.), with higher concentrations linked to greater productivity during warmer, wetter climates during MIS 1, 5, 7.3, and 9. Low concentrations (and therefore low productivity periods) corresponded with marine oxygen isotope events with even decimal fractions, i.e. MIS 2.2, 3.2, 4.2 back to MIS 10.4. Diatom productivity in this ancient lake is therefore clearly being influenced by changes in orbital parameters, although which specific climatic factors are responsible for driving productivity are still uncertain.

A subsequent, more highly resolved study at Lake Biwa, investigating diatom concentration and flux responses to orbital forcing over the last 140 ka, was undertaken at a centennial-scale resolution (Kuwae *et al.*, 2004). Spectral analyses of diatom fluxes at this resolution clearly highlight distinct cyclicities of 45.6 ka and 24.9 ka, related to obliquity and precession, which in this region drive upwelling intensity of the South China Sea and the associated East Asia summer monsoon (EASM). The autecological characteristics of modern species were determined with a view to interpreting the paleo-diatom record more comprehensively by comparing monthly monitoring data between the period 1978–1993 with meteorological and limnological (e.g. water temperature and phosphorus concentration) data. For example, *Aulacoseira nipponica* (Skvortzow) Tuji & Houki was found to have a positive relationship with snow cover and winter air and water temperatures, factors which drive vertical mixing of Lake Biwa's water column. Conversely, *Stephanodiscus suzukii* Tuji & Kociolek emend. Kato, Tanimura, Fukusawa, and Yasuda (and total plankton records) were more closely related to summer precipitation and P concentrations in the water, which are partly controlled by summer rainfall and concomitant in-wash of P from the catchment into the lake. Using these relationships, *Aulacoseira* flux records became a proxy for winter vertical mixing, while *Stephanodiscus* flux records were a proxy for variations in summer rainfall. Both together describe changes in the East Asian summer and winter monsoons. For example, *A. niponnica* disappeared during 140–101 ka, highlighting a decline in snow levels, relatively high temperatures, and weak vertical mixing in the lake. These are all consistent with a reduced activity in the East Asian winter monsoon.

It is evident, therefore, that when records of diatom silica and diatom concentration are undertaken at high enough resolution, impacts of orbital forcing on lacustrine ecosystems can be determined. In both Baikal and Biwa lakes, low diatom productivity is associated with cooler glacial stages, and although diatom dissolution has been implicated in significantly affecting the Lake Baikal record (see Mackay 2007 for a review), this does not seem to be the case for Lake Biwa (Kuwae *et al.*, 2002). At the shallow-water site of Lake Tule, diatom-inferred G–IG cycles are also recorded, but these are related to changes in lake level and available moisture impacting on habitat extent within the lake. Long diatom records from other ancient lakes are currently being investigated, e.g. Lakes Hövsgöl, Malawi. Almost certainly within the next five to ten years diatom responses to Quaternary G–IG cycles will be much more comprehensive, covering regions of the world where records are currently few.

11.4.2 Last-interglacial climate variability

Application of diatom analysis can greatly further our understanding of interglacial climate variability. Much work undertaken on interglacial climate change is done with a view to determining the length and stability of previous interglacials. Here we focus on the application of diatom analysis to characterize the last interglacial environments in the region of Lake Baikal. The last interglacial is more commonly known as the Eemian in northwest Europe and the Kazantsevo in southeast Siberia. Frequently the Kazantsevo/Eemian is equated with MIS 5e. However, these periods are not directly synchronous (e.g. Shackleton *et al.*, 2003), and here we only refer to the Kazantsevo interglacial *sensu stricto*. Several studies have used diatom analysis to investigate the length of the last interglacial from Lake Baikal sediments, and to determine whether the interglacial was a period characterized by stability or not. The main diatom proxies used include BSi (Karabanov *et al.*, 2000; Prokopenko *et al.*, 2002b), diatom species composition, and derived

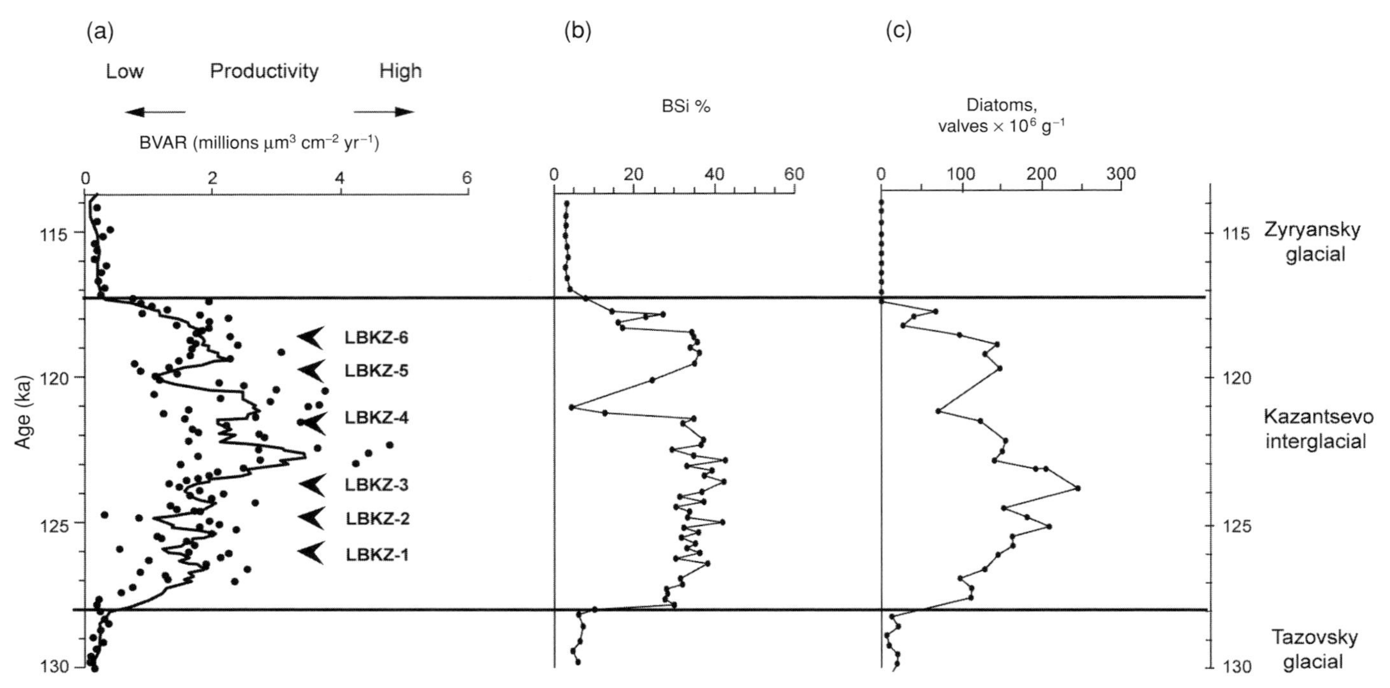

Figure 11.7 (a) Diatom biovolume accumulation rates, (b) BSi, and (c) diatom abundance records of last interglacial (Kazantsevo) productivity in Lake Baikal. Profile (a) is from the Continent Ridge core CON01–603-5 in the northern basin (Rioual & Mackay, 2005), and profiles (b) and (c) are from core BDP-96–2 extracted from the Academician Ridge (Karabanov *et al.*, 2000). The solid line in (a)represents a five-point moving average, and the arrows with corresponding codes LBKZ-1 to LBKZ-6 indicate episodes of cooler climate.

productivity from biovolume accumulation rates (BVAR) (Rioual & Mackay, 2005). In the diatom record, biovolume changes are mainly controlled by the very large extinct endemic *Stephanodiscus grandis* Khursevich & Loginova (Rioual & Mackay, 2005). Several paleoenvironmental studies have suggested that the last interglacial was, overall, a relatively stable period, e.g. the western Chinese Loess Plateau (Chen *et al.*, 2003). In Lake Baikal, diatom analysis reveals the continual presence of large *Stephanodiscus* spp. during the Kazantsevo. These large-celled species generally require high phosphorus, moderate silica levels, and low light (e.g. Bradbury *et al.*, 1994), and therefore climatic conditions must have ensured deep spring mixing. The presence of *Cyclotella* cf. *operculata* moreover suggests that summer conditions were sunny and warm enough to maintain fairly stable summer stratification of the water column (Rioual & Mackay, 2005).

However, many paleoecological studies have also reported the occurrence of an intra-Eemian cold (and possibly arid) event that lasted approximately 400 years, e.g. in European lakes (Sirocko *et al.*, 2005) and the Chinese loess (Zhisheng & Porter, 1997). Karabanov *et al.*, (2000) describe a short-lived drop in BSi values at *c.* 122 ka BP–120 ka BP, indicative of a period of productivity decline. Rioual & Mackay (2005) also report a short period of reduced productivity at *c.* 120.3 ka BP–119.5 ka BP as inferred from the BVAR record (Figure 11.7), and concluded that this event is likely to be similar to the one documented by Karabanov *et al.* (2000). However, BVAR do not drop as dramatically as BSi values and do not return to peak values found earlier during the climatic optimum at *c.* 124 ka–120.3 ka. European terrestrial records indicate that this cooler period was of low amplitude (Tzedakis *et al.*, 2003), which agrees with recent pollen-inferred precipitation reconstructions in the Lake Baikal region (Tarasov *et al.*, 2005). The decline in productivity seems to be associated with reduced insolation and increased ice volume resulting in a synchronous attenuation of monsoon circulation.

11.4.3 Abrupt climate events

Abrupt change in climate can occur within a matter of decades, and can result in non-linear responses crossing climatic thresholds. Over the last 115 ka, two major forms of abrupt changes in climate are apparent. The first are Heinrich events linked to ice-rafting pulses from the Laurentide Ice Sheet into the North Atlantic. These events were extremely cold and punctuated the paleoenvironmental record (e.g. Heinrich, 1988). Over the last *c.* 70 ka, Heinrich events have been shown to occur approximately every 7.2 ka (+/–2.4 ka), up until the start of the Holocene (Sarnthein *et al.*, 2001). Dansgaard–Oeschger

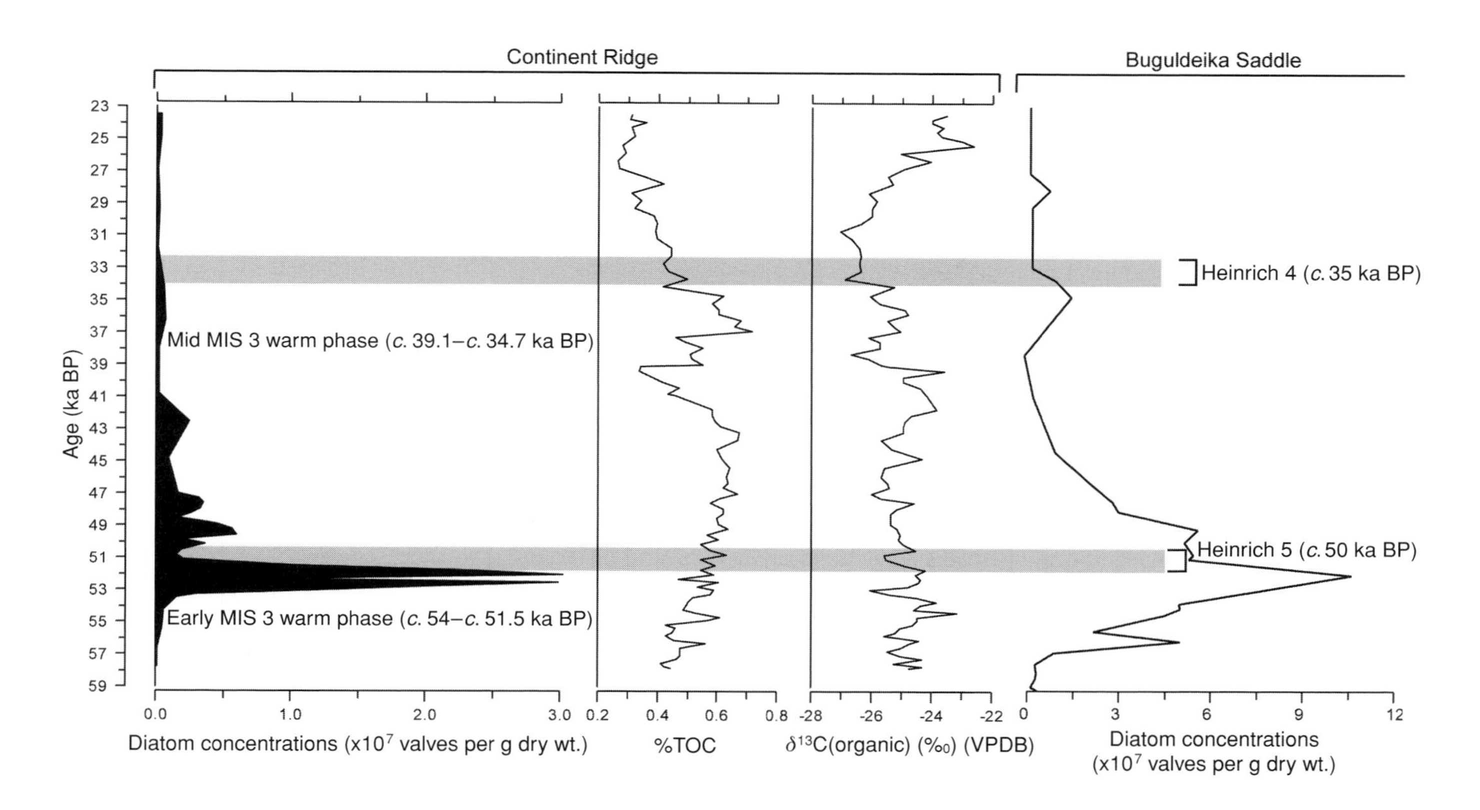

Figure 11.8 Correlation of diatom concentrations, $\delta^{13}C_{(organic)}$, and % total organic carbon (TOC) at Continent Ridge core CON01–603-5 (Swann *et al.*, 2005) with diatom concentrations at the Buguldeika Saddle (Prokopenko *et al.*, 2001) during the early (54 ka BP–51.5 ka BP) and mid (39.1 ka BP–34.7 ka BP) MIS 3 warm phases in Lake Baikal and during "Kuzmin"/Heinrich events 5 (*c.* 50 ka BP) and 4 (*c.* 35 ka BP) (shaded gray). (From Swann *et al.* (2005).)

(D–O), events on the other hand, although related to Heinrich events, are smaller in amplitude and occur more frequently, *c.* every 1.5 ka (1,470 yrs) (Dansgaard *et al.*, 1993). These D–O events also penetrate into interglacial periods, most notably the Holocene (Bond *et al.*, 1997; Campbell *et al.*, 1998) and the last interglacial (Müller *et al.*, 2005; Mackay, 2007) (Figure 11.7). It seems likely that D–O events may be triggered by the product of two solar cycles which impact on freshwater input into the North Atlantic (Braun *et al.*, 2005), although their exact cause is still uncertain.

In recent years, there has been a considerable amount of work in detecting these abrupt changes in climate in the Lake Baikal sediment archive (Prokopenko *et al.*, 2001; Swann *et al.*, 2005). Here, we focus on the evidence for abrupt climate impacts on the Lake Baikal ecosystem during interstadial MIS 3 (also called the Karginsky interstadial), which lasted *c.* 59 ka BP–29 ka BP, and was punctuated by Heinrich events H2–H5 (Prokopenko *et al.*, 2001; Swann *et al.*, 2005) (Figure 11.8). Heinrich events in the North Atlantic result in a decline in Lake Baikal diatom productivity, resulting in decreased total diatom abundances and changes to the C/N composition of bulk sediments (Prokopenko *et al.*, 2001). However, there is also considerable spatial variability within Lake Baikal with respect to responses to these cool events. In environments dominated by pelagic sedimentation, e.g. at the Continent Ridge in the north basin, North Atlantic cooling results in a drop in diatom productivity expressed as a decline in diatom concentrations (Figure 11.8). Because the influence from fluvial input was small at this site, C/N ratios did not change (Swann *et al.*, 2005). At the Continental Ridge, H5 (*c.* 51.5–50.2 ka BP) is expressed only by a fall in diatom concentration, coincident with glacier re-advancement in the catchment. The H5 event marks the very abrupt end to the early MIS 3 warm phase (*c.* 54–51.5 ka BP). In Lake Baikal this is characterized by high diatom concentrations, dominated by the endemic *Cyclotella gracilis* Nikiteeva & Likhoshway. In diatom profiles from the Buguldeika Saddle, this early MIS 3 warm phase is also dominated by *C. gracilis* (Edlund & Stoermer, 2000) although increases in fluvial sediments brought in via the Selenga River are also apparent (Prokopenko *et al.*, 2001).

Neither Tule Lake nor Lake Biwa diatom records make any link between low diatom numbers and Heinrich events over the last 115 ka. However, spectral analyses of diatom flux time series in Lake Biwa during the last 140 ka show several statistically significant periodicities, including one at 1400–1500 years, similar to D–O events and Bond cycles (Kuwae *et al.*, 2004). As D–O events are related to changes in thermohaline circulation in the North Atlantic, it is likely that past temperatures in the North Atlantic must have been linked to summer rainfall in East Asia via teleconnection processes.

Perhaps one of the best characterized and the most widespread cool periods associated with D–O events occurs during Termination 1 at the end of the last glacial. This event is commonly known as the Younger Dryas (YD). In the following section, we present diatom responses to the YD from ancient lakes Victoria and Titicaca. Lake Victoria (01° 00′ S 33° 00′ E) is one of the world's largest tropical lakes formed by tectonic activity *c.* 400 ka BP. Despite being an open lake, Lake Victoria is fed mainly by precipitation. Evaporative processes account for the majority of water loss from the lake, *c.* 90%, whereas its outlet into the Nile River accounts for only *c.* 10%. Lake levels in Lake Victoria are therefore very sensitive to changes in precipitation–evaporation. The lake is monomictic and annual overturn is driven by trade winds during the dry months of June to September. Overturn ensures nutrient regeneration, which in turn allows diatom populations to thrive in the lake, driving primary productivity. Large, shallow lakes such as Lake Victoria have a high ratio of benthic to planktonic taxa, because of an extensive littoral region where benthic diatoms can dominate. Using this knowledge, impacts of climate variability on lake-level change can be inferred.

Stager *et al.* (2002) analysed diatoms from a core taken from the Damba Channel in the north of Lake Victoria. Interpretations of past lake levels during Termination 1 were based on a combination of qualitative (i.e. species autecological characteristics) and quantitative (i.e. ordination) techniques. Correspondence analysis (CA) axis-2 diatom sample scores revealed that relative lake levels were much lower during the YD than at the start of the Holocene (Figure 11.9). During the YD, benthic taxa such as *Fragilaria* species dominate diatom assemblages, which are indicative of low lake levels. Furthermore, significant periodicities in the diatom record were evident, possibly related to D–O cycles mediating precipitation trends in tropical Africa (Stager *et al.*, 1997). In this region of tropical Africa, therefore, diatom-inferred arid periods are concurrent with weakening of Afro-Asian monsoon systems, linked to changes in thermohaline circulation associated with e.g. millennial-scale variability and melt-water pulses into the North Atlantic.

Lake Titicaca lies between Bolivia and the north of the Altiplano of Peru, the world's second-highest plateau. The lake

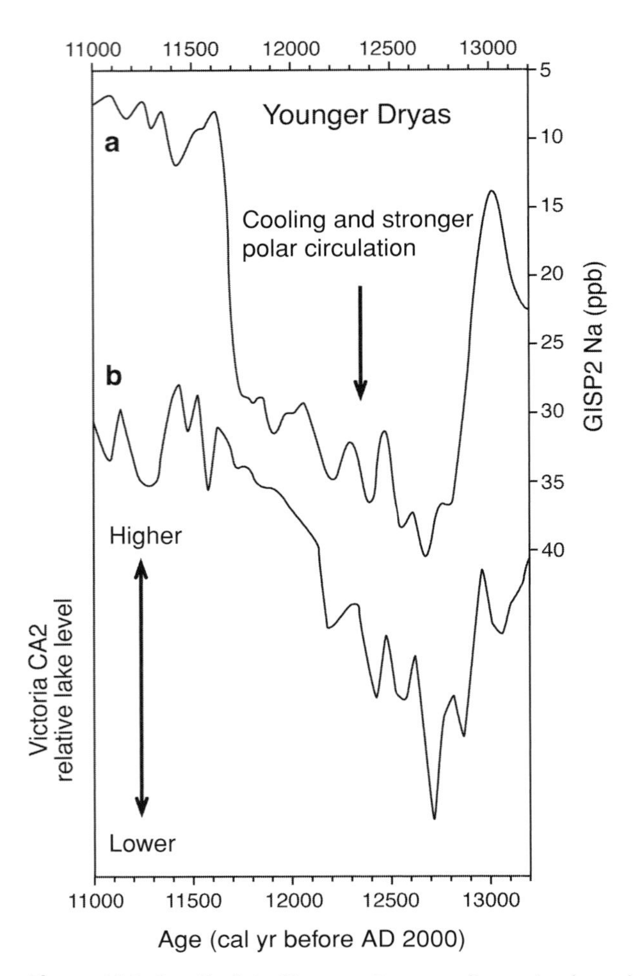

Figure 11.9 Detail of the Younger Dryas cooling episode registered in the (a) second Greenland Ice Sheet Project (GISP2) Na (ppb) record and (b) Lake Victoria's relative lake level, inferred from CA axis-2 diatom assemblage sample scores. (From Stager *et al.* (1997).)

was formed by tectonic activity *c.* 3 Ma. Lake Titicaca is the only large, deep, freshwater lake in South America and is sensitive to changes in precipitation (Tapia *et al.*, 2003). It is therefore an important site for Quaternary studies, and its sedimentary archive extends back through to at least the last glaciation (>30 ka). Detailed diatom analysis of late-Quaternary sediments from Lake Titicaca have contributed significantly to the debate on tropical precipitation trends since the last glaciation, especially within the Amazon Basin. In general terms, diatom-based reconstructions have demonstrated that during the late glacial period the climate over the Altiplano was wet, whereas a significant shift occurred to a more prevalent, drier climate during the early to middle Holocene (Baker *et al.*, 2001). Lake levels have been inferred from changing proportions of planktonic and littoral diatoms, indicative of changing extents of available habitats. For example, prevailing arid climates result in lower lake levels, leading to high proportions of littoral taxa. Conversely, prevailing wetter climates lead to higher lake levels, increased lake volume, and concomitant high proportions of planktonic taxa. On this basis, changing diatom assemblages highlight that Lake Titicaca has been sensitive to changes in the North Atlantic thermohaline circulation (THC) during both the YD and during Bond cycles through the Holocene (Baker *et al.*, 2001).

In Lake Titicaca, planktonic diatoms, especially *Cyclotella stelligera* Cleve & Grunow complex and *Cyclotella andina* E. C. Theriot, Carney & Richerson, are indicative of periods of deep-water linked to wetter climate (Tapia *et al.*, 2003). Saline-indifferent taxa (e.g. *Cyclotella meneghiniana* Kützing) and saline taxa (including *Chaetoceros muelleri* Lemmermann), on the other hand, are indicative of low lake levels and corresponding paleoclimate aridity (Tapia *et al.*, 2003). Using these groups, Tapia *et al.*, (2003) developed a paleoclimatic synthesis for the late glacial–Holocene. Between *c.* 26 ka BP–12.5 ka BP, *C. andina* and *C. stelligera* were dominant, suggesting lake levels were high. Prevailing climates were therefore wetter, especially during the period coincident with the YD event reconstructed elsewhere in the northern hemisphere. High lake-level stands and wetter climate also correlate with high lake levels in the Chilean Altiplano, highlighting the regional nature of this response. At the end of the YD, lake levels in Titicaca declined markedly, as inferred from an increase in benthic taxa persisting for *c.* 1000 years. Lower lake levels suggest a major phase of aridity causing millennial-scale drought until *c.* 10 ka BP.

11.4.4 Human impacts

Human impacts on lake ecosystems are extensive and often detrimental, and these are detailed in several chapters in this book (e.g. Battarbee *et al.*, Bennion *et al.*, Hall & Smol). However, human impacts pose considerable threats to biological diversity in ancient lakes because many of these ecosystems contain especially high densities of endemic species (often in flocks) (Rossiter & Kawanabe, 2000). Humans have introduced "exotic" species into many ancient lakes, which have had serious, negative impacts on biological diversity. For example, there has been a serious decline in endemic cichlid fish in Lake Victoria due to the introduction of the Nile perch, while in Lake Titicaca the introduction of trout has led to the decline in the endemic pupfish belonging to the genus *Orestias* (Kawanabe *et al.*, 1999). Human threats to other ancient lakes include deterioration of water quality. For example, the period of stratification in Lake Victoria has been increasing due to eutrophication (Verschuren *et al.*, 2002), with diatom analysis revealing that algal production is increasingly determined by groups indicative of more stable water conditions (e.g. *Nitzschia* spp.) and cyanobacteria. Between *c.* the mid-nineteenth century to the mid-twentieth century, the diatom community in Lake Victoria was rather stable, and was dominated by *Cyclostephanos* and *Aulacoseira* species. However, diatom production in Lake Victoria increased between the 1930s and 1980s due to the increased supply of nutrients into the lake. Initially *Cyclostephanos*, *Aulacoseira* and *Nitzschia* spp. all increased in abundance, with *Nitzschia* achieving dominance by the late 1970s/early 1980s (Verschuren *et al.*, 2002). However, with greater diatom production came greater diatom burial into the bottom sediments of Lake Victoria. This has resulted in the decline of available dissolved Si, and a concomitant decline in diatom production since the 1980s (Verschuren *et al.*, 2002).

Whilst Lake Victoria is a hydrologically open system, the Aral Sea (43° 24′–46° 56′ N and 58° 12′–61° 59′ E) is a closed, terminal lake in the semi-arid Turanian depression of central Asia. The lake has been the focus of recent attempts to disentangle anthropogenic and climate influences over the past few millennia. The Aral Sea is very sensitive to changes in hydrological inputs caused by natural climatic variability and/or anthropogenic activity. The Aral Sea is perhaps best known for the most recent regression in its lake level since the 1960s, linked to large-scale irrigation strategies implemented by the former Soviet Union and the subsequent diversion of water from the Amu Darya and Syr Darya river inflow. Responses are primarily in the form of changes in water level and chemistry, particularly conductivity and ionic composition, which impact upon the species composition of the lake's biota. Diatoms in the Aral Sea are especially sensitive to changes in lake-water chemistry and in particular conductivity, which varies inversely with depth. The first quantitative reconstruction of past diatom-inferred conductivity levels (a proxy for fluctuation lake levels) in the Aral Sea was undertaken by Austin *et al.* (2007). They identified three periods of severe lake-level regression (i) starting from AD 200–400, (ii) *c.* AD 1195–1355 and (iii) *c.* AD 1780–present day (Figure 11.10). The earliest regression is characterized by the planktonic species *Cyclotella choctawhatcheeana* Prasad, *Thalassiosira incerta* Makarova and *Thalassiosira proschkinae* Makarova. However a greater number of planktonic and tychoplanktonic diatoms (e.g. *Actinocyclus octonarius* Ehrenberg, *Chaetoceros wighamii* Brightwell, *C. choctawhatcheeana*, and *Nitzschia liebetruthii* Rabenhorst, *Nitzschia fonticola* Grunow in Van Heurck), characterize the middle regression *c.* 800 years ago. The most recent regression is dominated by *C. choctawhatcheeana* and *N. fonticola* (Figure 11.10). Climate played a pivotal role in the earlier

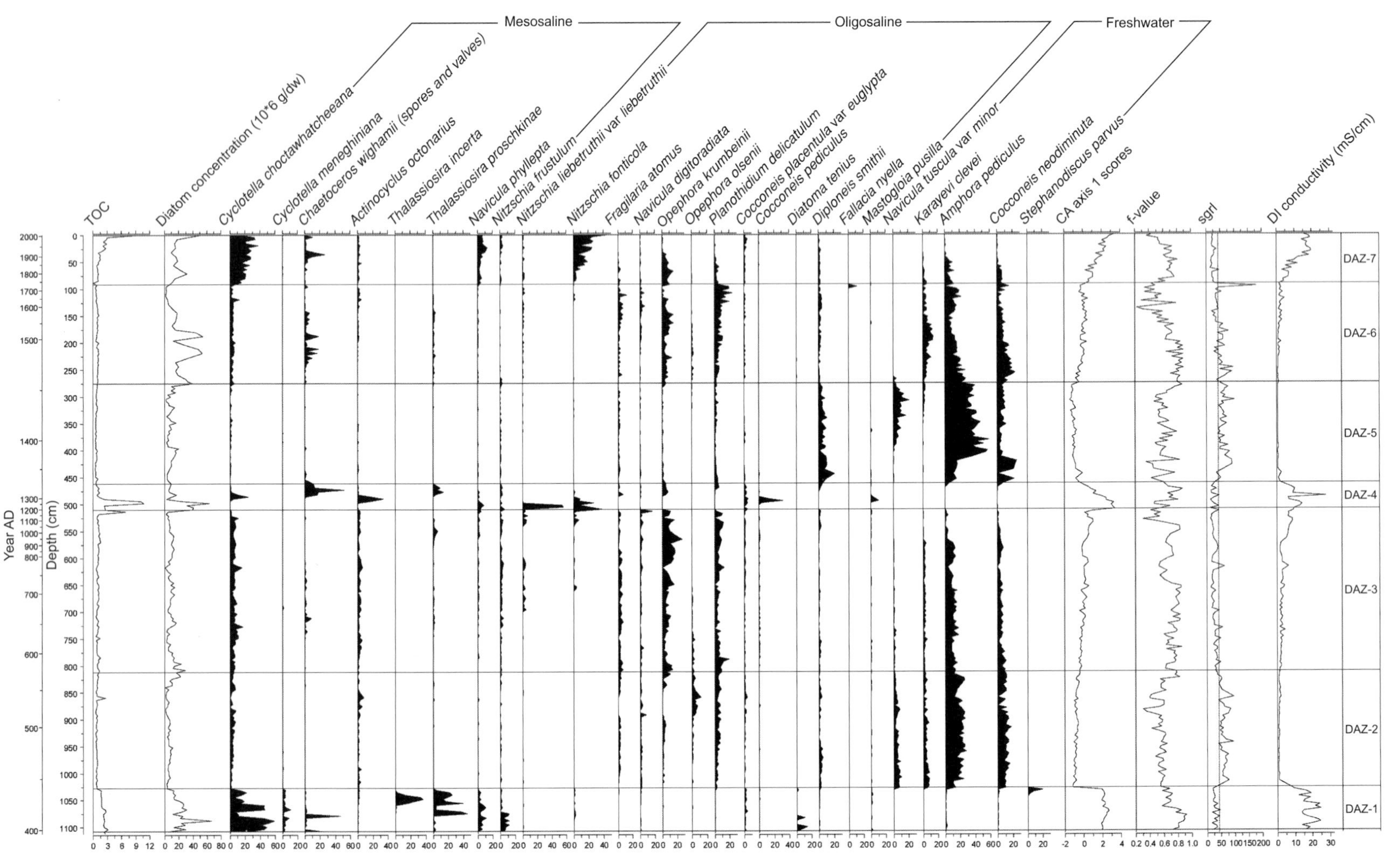

Figure 11.10 Summary of the diatom stratigraphy of Chernyshov Bay, Aral Sea. Species shown are those that appear with an abundance of >10% on one occasion or more. Also shown are the CA axis-1 diatom assemblage sample scores, the F-index of diatom dissolution, the square residual length and diatom-inferred conductivity. (From Austin *et al.* (2007).)

two regressions, as both are associated with the natural diversion of the Amu Darya linked to a decline in penetration by Mediterranean cyclones to more inland regions. However, it is likely that human impact was still significant, given that archeological evidence highlights that irrigation facilities around the lake were destroyed during each of these episodes, and that their destruction was likely to have enhanced the severity of the regressions observed.

11.5 Summary

Ancient lakes provide opportunities to investigate environmental change over long timescales, defined here as being at least since the last interglacial. Many ancient lakes, especially those associated with tectonic activity, have sedimentary records spanning the full Quaternary. Diatom studies from ancient lakes therefore provide great potential in deepening our understanding of long-term environmental change and its impact on diatom evolution. Three distinct patterns in diatom biodiversity in ancient lake systems are evident. For example, in the African Rift Lakes, the genera *Placoneis* and *Sellaphora* are very species rich. Diatom groups such as the Surirellaceae on the other hand are diverse in many ancient lake systems, although the flora in each lake is usually endemic. In deep, ancient lakes planktonic forms predominate and, although few species are apparent, high levels of endemism are common, such as in Lake Baikal. Due to all these factors, fossil diatom assemblages in ancient lakes rarely have modern analogs from which to infer past environmental conditions. Researchers therefore make use of ecological interpretations based on modern species "analogs" and modern autecological assumptions. Furthermore, there has been a history of improving bias present in sedimentary diatom records from ancient lakes through the investigation of taphonomy and transport of diatoms through the water column and into the sedimentary archive. Studies carried out in both Lake Baikal and the African Rift Lakes highlight that in general finely silicified species are under-represented and heavily silicified species are over-represented in sedimentary records.

Few long-term records of diatom-inferred change which span the full Quaternary are available in the published literature. Notable exceptions include the extensive literature published from Lake Baikal (Russia) and Tule Lake (USA). Both records reflect global cooling with the onset of NHG and a deep cool event at *c*. 2.5 Ma. Both records also exhibit changes in diatom assemblages that reflect orbital forcing of G–IG cycles. However the nature of the records are very different. The Lake Baikal record is essentially a proxy for within-lake productivity, significantly modified by diatom dissolution processes (i.e. during glacial periods, productivity declines, and relative diatom dissolution increases). The Tule Lake record is one of habitat change linked to low glacial temperatures and associated declining lake levels.

Since the original chapter on applications of ancient diatoms from continental archives by Bradbury (1999), there has been much progress on teasing out long paleoenvironmental reconstructions from ancient lakes, e.g. Lakes Baikal and Biwa. There is no doubt that the spatial distribution of long diatom records from ancient lakes will increase significantly in the next decade as recent technological advances have allowed deep drilling of other ancient lake systems in recent years. This will allow continued testing of new hypotheses on the nature of diatom evolution in continental water bodies, especially how relict floras, speciation, and climate change interact to drive the high levels of diversity and endemism seen in ancient lake systems.

Acknowledgments

AWM would like to thank various UK funding agencies (The Royal Society, Natural Environment Research Council, Leverhulme Trust) and EU Fifth Framework Programme, which have allowed his participation in research on Lake Baikal. Material covered in this chapter is based in part upon work supported by the National Science Foundation (NSF) under grants DEB-0316503 and DEB-0431529 to MBE. Any opinions, findings, and conclusions or recommendations expressed in this material are those of the authors and do not necessarily reflect the views of the NSF.

References

Austin, P., Mackay, A. W., Palushkina, O., & Leng, M. (2007). A high-resolution diatom-inferred palaeoconductivity and sea-level record of the Aral Sea for the last ca. 1600 years. *Quaternary Research*, **67**, 383–93.

Baker, P., Seltzer, G., Fritz, S., *et al.* (2001). The history of South American tropical precipitation for the past 25,000 years. *Science*, **291**, 640–3.

Battarbee, R. W., Mackay, A. W., Jewson, D., Ryves, D. B., & Sturm, M. (2005). Differential dissolution of Lake Baikal diatoms: correction factors and implications for palaeoclimatic reconstruction. *Global & Planetary Change*, **46**, 75–86.

Bond, G., Showers, W., Cheseby, M., *et al.* (1997). A pervasive millennial scale cycle in North Atlantic Holocene and glacial climates. *Science*, **278**, 1257–6.

Bondarenko, N. A., Tuji, A., & Nakanishi, M. (2006). A comparison of phytoplankton communities between the ancient lakes Biwa and Baikal. *Hydrobiologia*, **568**, 25–29.

Bradbury, J. P. (1991). The late Cenozoic diatom stratigraphy and paleolimnology of Tule Lake, Siskiyou Co. California. *Journal of Paleolimnology*, **6**, 205–55.

Bradbury, J. P. (1999). Continental diatoms as indicators of long-term environmental change. In *The Diatoms: Applications for the Environmental and Earth Sciences*, ed. E. F. Stoermer & J. P. Smol. Cambridge: Cambridge University Press, pp. 167–82.

Bradbury, J. P. & Krebs, W. N. (1982). Neogene and Quaternary lacustrine diatoms of the Western Snake River Basin, Idaho–Oregon, USA. *Acta Geologica Academiaea Scientiarum Hungaricae*, **25**, 97–122.

Bradbury, J. P., Bezrukova, Ye. V., *et al.* (1994). A synthesis of post-glacial diatom records from Lake Baikal. *Journal of Paleolimnology*, **10**, 213–52.

Bramberger, A. J., Haffner, G. D., Hamilton, P. B., Hinz F., & Hehanussa, P. E. (2006). An examination of species within the genus *Surirella* from the Malili Lakes, Sulawesi Island, Indonesia, with descriptions of 11 new taxa. *Diatom Research*, **21**, 1–56.

Braun, H., Christl, M., Rahmstorf, S., *et al.* (2005). Possible solar origin of the 1,470-year glacial climate cycle demonstrated in a coupled model. *Nature*, **438**, 208–11.

Brooks, J. L. (1950). Speciation in ancient lakes. *Quarterly Review of Biology*, **25**, 30–60, 131–76.

Campbell, I. D., Campbell, C., Apps, M., Rutter, N. W., & Bush, A. B. G. (1998). Late Holocene ~1500 yr climatic periodicities and their implications. *Geology*, **26**, 471–3.

Chen, F. H., Qiang, M. R., Feng, Z. D., Wang, H. B., & Bloemendal, J. (2003). Stable East Asian monsoon climate during the last interglacial (Eemian) indicated by paleosol S1 in the western part of the Chinese Loess Plateau. *Global and Planetary Change*, **36**, 171–9.

Cocquyt, C., Vyverman, W., & Compère, P. (1993). A check-list of the algal flora of the East African Great Lakes. *Scripta Botanica Belgica*, **8**, 1–55.

Dansgaard, W., Johnsen, S., Clausen, H. B., *et al.* (1993). Evidence for general instability of past climate from a 250-kyr ice core record. *Nature*, **364**, 218–20.

Edgar, S. M. & Theriot, E. C. (2004). Phylogeny of *Aulacoseira* (Bacillariophyta) based on molecules and morphology. *Journal of Phycology*, **40**, 772–88.

Edlund, M. B. (2006). Persistent low diatom plankton diversity within the otherwise highly diverse Lake Baikal ecosystem. *Nova Hedwigia Beiheft*, **130**, 339–56.

Edlund, M. B. & Stoermer, E. F. (2000). A 200,000-year, high-resolution record of diatom productivity and community makeup from Lake Baikal shows high correspondence to the marine oxygen-isotope record of climate change. *Limnology and Oceanography*, **45**, 948–62.

Edlund, M. B., Williams, R. M., & Soninkhishig, N. (2003). The planktonic diatom diversity of ancient Lake Hövsgöl, Mongolia. *Phycologia*, **42**, 232–60.

Edlund, M. B., Levkov, Z., Soninkhishig, N., Krstic, S., & Nakov, T. (2006a). Diatom species flocks in large ancient lakes: the *Navicula reinhardtii* complex from Lakes Hövsgöl (Mongolia) and Prespa (Macedonia). In *Proceedings of the 18th International Diatom Symposium*, ed. A. Witkowski, Bristol: Biopress Ltd., pp. 61–74.

Edlund, M. B., Soninkhishig, N., & Stoermer, E. F. (2006b). The diatom (Bacillariophyta) flora of Lake Hövsgöl National Park, Mongolia. In *The Geology, Biodiversity and Ecology of Lake Hövsgöl (Mongolia)*, ed. C. Goulden, T. Sitnikova, J. Gelhaus, & B. Boldgiv, Leiden: Backhuys Publishers, pp. 145–77.

Edlund, M. B. & Soninkhishig, N. (2009). The *Navicula reinhardtii* species flock (Bacillariophyceae) in ancient Lake Hövsgöl, Mongolia: description of four taxa. *Nova Hedwigia, Beiheft*, **135**, 239–56.

Genkal S. I. & Bondarenko N. A. (2006). Are the Lake Baikal diatoms endemic? *Hydrobiologia*, **568**, 143–53.

Goldberg, E. L., Phedorin, M. A., Grachev, M. A., *et al.* (2000). Geochemical signals of orbital forcing in the records of paleoclimates found in the sediments of Lake Baikal. *Nuclear Instruments and Methods in Physics Research A*, **448**, 384–93.

Grachev, M. A., Vorobyova, S. S., Likhoshway, Y. V., *et al.* (1998). A high-resolution diatom record of the palaeoclimates of east Siberia for the last 2.5 My from Lake Baikal. *Quaternary Science Reviews*, **17**, 1101–6.

Haberyan, K. A. (1985). The role of copepod fecal pellets in the deposition of diatoms in Lake Tanganyika. *Limnology and Oceanography*, **30**, 1010–23.

Haberyan, K. A. (1990). The misrepresentation of the planktonic diatom assemblage in traps and sediments: southern Lake Malawi, Africa. *Journal of Paleolimnology*, **3**, 35–44.

Heinrich, H. (1988). Origin and consequences of cyclic ice rafting in the northeast Atlantic Ocean during the past 130,000 years. *Quaternary Research*, **29**, 142–52.

Hustedt, F. (1937–1939). Systematische und ökologische Untersuchungen über die Diatomeen-Flora von Java, Bali und Sumatra nach dem Material der Deutschen Limnologischen Sunda-Expedition. *Archiv für Hydrobiologie, Supplement*, **15**, 131–77, 187–95, 393–506, 638–790; **16**, 1–155, 274–394.

Hustedt, F. (1949). Süsswasser-Diatomeen aus dem Albert-Nationalpark in Belgisch Kongo. *Exploration du Parc National Albert Mission H. Damas (1935–1936)*, **8**, 1–199.

Hutchinson, G. E. (1961). The paradox of the plankton. *The American Naturalist*, **95**, 137–45.

Jewson, D. H., Granin, N. G., Zhdarnov, A. A., *et al.* (2008). Resting stages and ecology of the planktonic diatom *Aulacoseira skvortzowii* in Lake Baikal. *Limnology and Oceanography*, **53**, 1125–36.

Julius, M. L. & Stoermer, E. F. (1999). Morphometric relationships in the *Cyclotella baicalensis* species complex. Abstract, Fifteenth North American Diatom Symposium, September 22–25, Pingree Park, Colorado.

Julius, M. L. & Curtin, M. (2006) Morphological patterns in three endemic *Cyclotella* species from Lake Baikal. *Geological Society of America, Abstracts with Programs*, **38**, 264.

Jurilj, A. (1949). *Nove dijatomeje–Surirellaceae–iz Ohridskog jezera i njihovo filogenetsko znacenenje*. Zagreb: Jugoskavenska Akademija Znanosti i Umjetnosti, pp. 1–94.

Karabanov, E. B., Prokopenko, A. A., Williams, D. F., & Khursevich, G. K. (2000). Evidence for mid-Eemian cooling in continental climatic record from Lake Baikal. *Journal of Paleolimnology*, **23**, 365–71.

Kawanabe, H., Coulter, G. W., & Roosevelt, A. C. (eds.) (1999). *Ancient Lakes: Their Cultural and Biological Diversity*. Ghent: Kenobi Productions, pp. 1–340.

Khursevich, G. (2006). Evolution of the extinct genera belonged to the family Stephanodiscaceae (Bacillariophyta) during the last eight million years in Lake Baikal. In *Advances in Phycological Studies: Festschrift in Honour of Prof. Dobrina Temniskova-Topalova*, ed. N. Ognjanova-Rumenova & K. Manoylov, Sofia, Bulgaria: PENSOFT Publishers and St. Kliment Ohridski University Press, pp. 73–89.

Khursevich G. K. & Fedenya S. A. (2006). Fossil centric diatoms from Upper Cenozoic sediments of Lake Baikal: change over time. In *The 19th International Diatom Symposium: Abstracts* (Listvyanka, 28 August – 2 September, 2006), ed. Ye. V. Likhoshway & R.M. Crawford,– Irkutsk: V.B. Sochava Institute of Geography SB RAS. p. 71.

Khursevich, G. K., Karabanov, E. B., Prokopenko, A. A., *et al.* (2001). Insolation regime in Siberia as a major factor controlling diatom production in Lake Baikal during the past 800,000 years. *Quaternary International*, **80–81**, 47–58.

Khursevich, G. K., Prokopenko, A. A., Fedenya, S. A., Tkachenko, L. I., & Williams, D. F. (2005). Diatom biostratigraphy of Lake Baikal during the past 1.25 Ma: new results from BDP-96–2 and BDP-99 drill cores. *Quaternary International*, **136**, 95–104.

Kociolek, J. P., Reid, G., & Flower, R. J. (2000). Valve ultrastructure of *Didymosphenia dentata* (Bacillariophyta): an endemic diatom species from Lake Baikal. *Nova Hedwigia*, **71**, 113–20.

Kuwae, M., Yoshikawa, S., & Inouchi, Y. (2002). A diatom record for the past 400 ka from Lake Biwa in Japan correlates with global paleoclimatic trends. *Palaeogeography, Palaeoclimatology, Palaeoecology*, **183**, 261–74.

Kuwae, M., Yoshikawa, S., Tsugeki, N., & Inouchi, Y. (2004). Reconstruction of a climate record for the past 140 kyr based on diatom valve flux data from Lake Biwa, Japan. *Journal of Paleolimnology*, **32**, 19–39.

Levkov, Z. (2009). *Amphora* sensu lato (*Amphora* sensu stricto & Halamphora). In *Diatoms of Europe: Diatoms of European Inland Waters and Comparable Habitats Elsewhere*, ed. H. Lange-Bertalot, Königstein: Koeltz Scientific Books, volume 5, pp. 1–916.

Levkov, Z., Krstic, S., Metzeltin, D., & Nakov, T. (2007). Diatoms from Lakes Prespa and Ohrid. *Iconographia Diatomologica*, **16**, 1–603.

Likhoshway, Ye. V., Kuzmina, A. Ye., Potyemkina, T. G., Potyemkim, V. L., & Shimaraev, M. N. (1996). The distribution of diatoms near a thermal bar in Lake Baikal. *Journal of Great Lakes Research*, **22**, 5–14.

Mackay, A. W. (2007). The paleoclimatology of Lake Baikal: a diatom synthesis and prospectus. *Earth-Science Reviews*, **82**, 181–215.

Mackay, A. W., Battarbee, R. W., Flower, R. J., *et al.* (2003). Assessing the potential for developing internal diatom-based transfer functions for Lake Baikal. *Limnology and Oceanography*, **48**, 1183–92.

Mackay, A. W., Ryves, D. B., Battarbee, R. W., *et al.* (2005). 1000 years of climate variability in central Asia: assessing the evidence using Lake Baikal diatom assemblages and the application of a diatom-inferred model of snow thickness. *Global and Planetary Change*, **46**, 281–97.

Mackay, A. W., Ryves, D. B., Morley, D. W., Jewson, D. J., & Rioual, P. (2006). Assessing the vulnerability of endemic diatom species in Lake Baikal to predicted future climate change: a multivariate approach. *Global Change Biology*, **12**, 2297–315.

Martens, K., Coulter, G., & Goddeeris, B. (eds.) (1994). *Speciation in Ancient Lakes*. Advances in Limnology, vol. 44, pp. 1–508.

Metzeltin, D. & Lange-Bertalot, H. (1995). Kritische Wertung der Taxa in *Didymosphenia* (Bacillariophyceae). *Nova Hedwigia*, **60**, 381–405.

Müller, U. C., Klotz, S., Geyh, M. A., Pross, J., & Bond, G. (2005). Cyclic climate fluctuations during the last interglacial in central Europe. *Geology*, **33**, 449–52.

Pappas, J. L. & Stoermer, E. F. (2003). Morphometric comparison of the neotype of *Asterionella formosa* Hassall (Heterokontophyta, Bacillariophyceae) with *Asterionella edlundii* from Lake Hövsgöl, Mongolia. *Diatom*, **19**, 55–65.

Popovskaya, G. I. (2000). Ecological monitoring of phytoplankton in Lake Baikal. *Aquatic Ecosystem Health and Management*, **3**, 215–25.

Popovskaya, G. I., Genkal, S. I., & Likhoshway, Y. V. (2002). *Diatoms of the Plankton of Lake Baikal: Atlas & Key*. Novosibirsk, Russia: Nauka, pp. 1–168.

Prokopenko, A. A., Hinnnov, L. A., Williams, D. F., & Kuzmin, M. I. (2006). Orbital forcing of continental climate during the Pleistocene: a complete astronomically tuned climatic record from Lake Baikal, SE Siberia. *Quaternary Science Reviews*, **25**, 3431–57.

Prokopenko, A. A., Karabanov, E. B., Williams, D. F., & Khursevich, G. K. (2002b). The stability and the abrupt ending of the last interglaciation in southeastern Siberia. *Quaternary Research*, **58**, 56–9.

Prokopenko, A. A., Williams, D. F., Karabanov, E. B., & Khursevich, G. K. (2001). Continental response to Heinrich events and Bond cycles in sedimentary record of Lake Baikal, Siberia. *Global and Planetary Change*, **28**, 217–26.

Prokopenko, A. A., Williams, D. F., Kuzmin, M. I., *et al.* (2002a). Muted climate variations in continental Siberia during the mid-Pleistocene epoch. *Nature*, **418**, 65–8.

Rioual, P. & Mackay, A. W. (2005). A diatom record of centennial resolution for the Kazantsevo interglacial stage in Lake Baikal (Siberia). *Global and Planetary Change*, **46**, 199–219.

Rossiter, A. & Kawanabe, H. (eds.) (2000). *Ancient Lakes: Biodiversity, Ecology and Evolution*, Advances in Ecological Research, vol. 31.

Ryves, D. B., Jewson, D. H., Sturm, M., *et al.* (2003). Quantitative and qualitative relationships between planktonic diatom communities and diatom assemblages in sedimenting material and surface sediments in Lake Baikal, Siberia. *Limnology and Oceanography*, **48**, 1643–61.

Sarnthein, M., Stattegger, K., Dreger, D., *et al.* (2001). Fundamental modes and abrupt changes in North Atlantic circulation and climate over the last 60 ky – concepts, reconstruction, and numerical modeling. In *The Northern North Atlantic: a Changing Environment*, ed. P. Schäfer, W. Ritzrau, M. Schlüter, J. Thiede, Berlin: Springer-Verlag, pp. 365–410.

Scherbakova, T. A., Kiril'chik, S. V., Likhoshway, Ye. V., & Grachev, M. A. (1998). Phylogenetic position of some diatom algae of the genus *Aulacoseira* from Lake Baikal following from comparison of sequences of the 18S rRNA gene fragments. *Molecular Biology*, **32**, 735–40. (In Russian, English summary).

Shackleton, N. J., Berger, A., & Peltier, W. R. (1990). An alternative astronomical calibration of the lower Pleistocene timescale based on ODP Site 677. *Transactions of the Royal Society of Edinburgh: Earth Science*, **81**, 251–61.

Shackleton, N. J., Sánchez-Goñi, M. F., Pailler, D., & Lancelot, Y. (2003). Marine isotope stage 5e and the Eemian interglacial. *Global and Planetary Change*, **36**, 151–5.

Short, D. A., Mengel, J. G., Crowley, T. J., Hyde, W. T., & North, G. R. (1991). Filtering of Milankovitch cycles by Earth's geography. *Quaternary Research*, **35**, 157–71.

Sirocko, F., Seelos, K., Schaber, K., *et al.* (2005). A late Eemian aridity pulse in central Europe during the last glacial inception. *Nature*, **436**, 833–6.

Skvortzow, B. W. (1937). Bottom diatoms from Olkhon Gate of Baikal Lake, Siberia. *Philippine Journal of Science*, **62**, 293–377.

Stager, J. C., Cumming, B., & Meeker, L. (1997). A high-resolution 11,400-yr diatom record from Lake Victoria, East Africa. *Quaternary Research*, **47**, 81–9.

Stager, J. C., Mayewski, P. A., & Meeker, L. D. (2002). Cooling cycles, Heinrich event 1, and the desiccation of Lake Victoria. *Palaeogeography, Palaeoclimatology, Palaeoecology*, **183**, 169–78.

Stoermer, E. F. & Edlund, M. B. (1999). No paradox in the plankton? – diatom communites in large lakes. In *Proceedings of the XIVth International Diatom Symposium*, ed. S. Mayama, M. Idei & I. Koizumi. Königstein: Koeltz Scientific Books, pp. 49–61.

Swann, G. E. A., Mackay, A. W., Leng, M., & Demoray, F. (2005). Climatic change in central Asia during MIS 3: a case study using biological responses from Lake Baikal sediments. *Global and Planetary Change*, **46**, 235–53.

Tapia, P. M., Fritz, S. C., Baker, P. A., Seltzer, G. O., & Dunbar, R. B. (2003). A late Quaternary diatom record of tropical climatic history from Lake Titicaca (Peru and Bolivia). *Palaeogeography, Palaeoclimatology, Palaeoecology*, **194**, 139–64.

Tarasov, P., Granoszewski, W., Bezrukova, E., *et al.* (2005). Quantitative reconstruction of the last interglacial climate based on the pollen record from Lake Baikal, Russia. *Climate Dynamics*, **25**, 625–37.

Theriot, E. C., Fritz, S. C., Whitlock, C., & Conley, D. J. (2006). Late Quaternary rapid morphological evolution of an endemic diatom in Yellowstone Lake, Wyoming. *Paleobiology*, **32**, 38–54.

Tréguer, P., Nelson, D. M., van Bennekom, A. J., *et al.* (1995). The silica balance in the world ocean: a re-estimate. *Science*, **268**, 375–9.

Tuji, A. & Houki, A. (2001). Centric diatoms in Lake Biwa. *Lake Biwa Study Monographs*, **7**, 1–90.

Tzedakis, P. C., Frogley, M. R. & Heaton, T. H. E. (2003). Last interglacial conditions in southern Europe: evidence from Ionnina, northwest Greece. *Global and Planetary Change*, **36**, 157–70.

Verschuren, D. Johnson, T. C., Kling, H. J., *et al.* (2002). History and timing of human impact on Lake Victoria, East Africa. *Proceedings of the Royal Society of London*, **B 269**, 289–94.

Wagner, B., Lotter, A. F., Nowacyk, N., *et al.* (2008). A 40,000-year record of environmental change from ancient Lake Ohrid (Albania and Macedonia). *Journal of Paleolimnology*, **41**, 407–30.

Williams, D. M., Khursevich, G. K., Fedenya, S. A., & Flower, R. J. (2006). The fossil record in Lake Baikal: comments on the diversity and duration of some benthic species, with special reference to the genus *Tetracyclus*. In *Proceedings of the 18th International Diatom Symposium*, ed. A. Witkowski, Bristol: Biopress Ltd., pp. 465–78.

Williams, D. F., Kuzmin, M. I., Prokopenko, A. A., *et al.* (2001). The Lake Baikal Drilling Project in the context of a global lake drilling initiative. *Quaternary International*, **80–81**, 3–18.

Williams, D. F., Peck, J., Karabanov, E. B., *et al.* (1997). Lake Baikal record of continental climate response to orbital insolation during the past 5 million years. *Science*, **278**, 1114–17.

Williams, D. M. & Reid, G. (2006). *Amphorotia nov. gen., a New Genus in the Family Eunotiaceae (Bacillariophyceae), Based on Eunotia clevei Grunow in Cleve et Grunow*, Diatom Monographs, vol. **6**, ed. A. Witkowski, Ruggell: A. R. G. Gantner Verlag.

Xiao, J., Inouchi, Y., Kumai, H., *et al.* (1997). Biogenic silica record in Lake Biwa of central Japan over the past 450,000 years. *Quaternary Research*, **47**, 277–83.

Zhisheng, A. & Porter, S. C. (1997). Millennial scale climatic oscillations during the last interglacial in central China. *Geology*, **25**, 603–6.

Part III

Diatoms as indicators in Arctic, Antarctic, and alpine lacustrine environments

12 Diatoms as indicators of environmental change in subarctic and alpine regions

ANDRÉ F. LOTTER, REINHARD PIENITZ, AND ROLAND SCHMIDT

12.1 Introduction

Subarctic and mountain regions are characterized by strong gradients that make their terrestrial and aquatic ecosystems very sensitive to environmental change. The terrestrial Arctic can be delimited by the northern tree line, the 10 °C July isotherm, or the southern extent of discontinuous permafrost which, in the eastern Canadian Arctic for example, currently extends to the southern end of Hudson Bay. In this chapter, we focus on the subarctic region, which, depending on local climates, roughly falls between 50° N and 70° N latitude and includes the transition from boreal forest (taiga) in the south to tundra landscapes in the north, whereas the chapter by Douglas and Smol (this volume) discusses diatom-based studies from the High Arctic. In mountain regions the same steep climatic and environmental gradients are present but over much shorter distances, with the timber line also representing the most prominent ecotone. It is characterized by the transition from closed forest to the most advanced solitary trees (i.e. timber line), to single tree islands (i.e. tree line), and eventually to open, unforested vegetation. This biological boundary can vary in width from tens of meters in mountain regions to many kilometers in the Subarctic. In northern Europe it is formed by deciduous trees such as *Betula*, *Alnus*, and *Populus*, whereas coniferous trees (e.g. *Pinus*, *Picea*, *Larix*, *Juniperus*) form the tree line in the European Alps, northern North America, and Eurasia. This ecotone is primarily related to cold temperatures with a complex set of different microclimatic factors, as well as the specific adaptation of trees, defining the forest limit (e.g. Tranquillini, 1979). In the European Alps, for instance, the timber line represents the transition between the subalpine and the alpine elevational belts (Ozenda, 1985). The lower boundary of the alpine belt, however, is difficult to locate as human impact, grazing, and climatic oscillations have lowered the natural tree limit by several hundred meters, at least during the past three to four millennia (e.g. Tinner *et al.*, 1996).

Aquatic habitats suitable for diatoms at these latitudes and altitudes are manyfold. High-latitude and high-altitude lakes and rivers are characterized by special limnological features and are usually ice free for only a short period in summer, with cold water temperatures prevailing (Battarbee *et al.*, 2002b). Lakes influenced by glacial meltwaters are turbid with high concentrations of silt and clay, whereas in the absence of glaciers, water transparency is either high or, if mires are present in the catchment, the input of humic acids will result in brown water (dystrophic) conditions. Furthermore, light availability in the water column is strongly restricted by ice and snow cover, and at high latitudes also by the polar winter. During summer, however, the light conditions change significantly, especially in high-latitude lakes: the photoperiod is long and the angle of the sun is higher.

Climatic control of ultraviolet (UV) radiation and underwater UV exposure have been identified as important factors regulating lake production and algal community composition in high-altitude and high-latitude lakes (e.g. Pienitz and Vincent, 2000). For example, UV radiation may have an inhibitory effect on periphytic littoral algae, such as *Achnanthes minutissima* Kützing, if dissolved organic matter is at low concentrations (e.g. Vinebrooke and Leavitt, 1996). There are many mechanisms linking UV exposure to climate in the aquatic environment: some are caused by climate-induced changes in the stratosphere, while many are associated with temperature-related effects on the main UV-attenuating components of natural waters, notably changes in snow and ice cover, and in chromophoric (colored) dissolved organic matter (CDOM) associated with shifts in terrestrial vegetation in the catchment of lakes. There are

The Diatoms: Applications for the Environmental and Earth Sciences, 2nd Edition, eds. John P. Smol and Eugene F. Stoermer. Published by Cambridge University Press.

also a variety of hydrological effects of climate with major implications for underwater UV such as changes in snow and rainfall (with runoff, vegetation, and albedo effects), evaporation and cloud cover, pathways of runoff, and degree of interaction between soil and water. Finally, there are more subtle effects such as changes in wind-induced mixing and climate-induced shifts in species composition towards more or less UV-tolerant species. Many of these effects are especially pronounced in high-latitude and high-altitude aquatic ecosystems where there is a precarious balance between freezing and melting throughout summer, and where small changes in CDOM concentrations can have large effects on UV transparency.

Lakes on opposite sides of the tree line exhibit striking differences in water chemistry and physical conditions (Pienitz *et al.*, 1997a, 1997b; Sommaruga and Psenner, 1997; Fallu and Pienitz, 1999; Rühland *et al.*, 2003), which are reflected by abrupt changes in diatom-community structure and composition. For example, differences in lake mixing regimes, nutrients, and lake-water transparency are most apparent at this ecotonal boundary marked by sharp changes in abiotic (e.g. changes in albedo and permafrost) and biotic variables (e.g. atmospheric deposition of macronutrients by pollen; see Lee *et al.*, 1996). Furthermore, sediment accumulation rates may change drastically across this ecotone: accumulation rates in lakes above the tree line without glacial meltwater influx are often up to an order of magnitude lower than in comparable boreal or temperate lake basins (Laing *et al.*, 2002). The dearth of alpine and northern-tree-line diatom studies is primarily related to logistical problems. Nevertheless, early diatom studies from high-altitude and high-latitude regions usually provided extended taxon lists (e.g. Krasske, 1932; Cleve-Euler, 1934; Hustedt, 1943, 1942; Foged, 1955) and often classified taxa according to their plant–geographical distribution (e.g. nordic–alpine). Freshwater diatoms, like most algae, are usually not considered sensitive indicators of temperature directly (e.g. Battarbee, 1991; Anderson, 2000), even though clear latitudinal and altitudinal patterns in their distributions are apparent (Foged, 1964; Vyverman, 1992; Pienitz and Smol, 1993; Lotter *et al.*, 1997), and classifications according to their thermal requirements (Backman and Cleve-Euler, 1922; Hustedt, 1939, 1956) have long existed. Several experimental studies (e.g. Patrick, 1971, 1977; Eppley, 1977; Hartig and Wallen, 1986; Dauta *et al.*, 1990; Montagnes and Franklin, 2001) indicate a temperature dependency of diatom growth and community composition. Under natural conditions, Stoermer and Ladewski (1976) and Kingston *et al.* (1983) found a relationship between temperature and the occurrence of certain diatoms. Kilham *et al.* (1996), however, associated the climatic distribution of diatoms with resource-related competitive interactions. The diatom–temperature relationship is a complex one, as temperature changes may have multiple, profound indirect impacts on the chemical and physical properties of the aquatic environment and available habitats, as well as on biological factors such as the competitive abilities of the biota. However, from a paleoclimatic perspective, it is not relevant if the relationship between diatoms and temperature is direct or indirect, rather how reliably the variable can be reconstructed, whether other environmental variables than climate have negligible influence (Birks, 1995), and whether the relationship between climate and indirect factors remains linear over time (Smol and Cumming, 2000). In fact, the use of diatoms as quantitative proxies of past temperatures has become common use and proved successful in several paleolimnological studies in subarctic and mountain regions of the northern hemisphere (e.g. Pienitz *et al.*, 1999; Korhola *et al.*, 2000; Hausmann and Lotter, 2001; Joynt and Wolfe, 2001; Bigler and Hall, 2002, 2003; Bigler *et al.*, 2003; Schmidt *et al.*, 2008; von Gunten *et al.*, 2008).

12.2 Diatoms as paleoenvironmental indicators at the tree line

Tree-line vegetation will be strongly impacted by future climatic warming (ACIA, 2004; IPCC, 2007). Furthermore, aquatic environments and ecosystems, water supplies, and fisheries will likewise be affected (Wrona *et al.*, 2006). Global warming will result in increased heat uptake of lakes which, in turn, will lead to earlier break-up and later onset of ice cover and, consequently, to changes in stratification and oxygen regimes (e.g. Schindler *et al.*, 1996; Livingstone, 1997; Sommaruga-Wögrath *et al.*, 1997; Catalan *et al.*, 2002a; Vincent *et al.*, 2008). Ultra-sensitive lakes, i.e. those which are colder than can be expected from air temperatures, will warm up distinctly during global warming (Thompson *et al.*, 2005a). Moreover, thawing of permafrost catchment soils will have an impact on nutrient, CDOM, and heavy-metal concentrations in such lakes (e.g. Smol, 2002; Baron *et al.*, 2005; Kamenik *et al.*, 2005).

Because the location of the northern tree line results from and also influences the mean position of major atmospheric boundaries (i.e. the Polar Front; Pielke and Vidale, 1995), it is important to understand the linkages between climate and vegetation along the northern edge of the boreal forests. The sensitivity of the arctic and alpine tree-line position to climatic factors makes this ecotone an ideal location for investigating the effects and the timing of climatic change (e.g. Pienitz and Smol,

1993; Lotter, 2005; Tinner and Ammann, 2005). Paleoecological records provide a means of reconstructing the impact of past climatic variations on tree-line vegetation, thereby allowing a better understanding of the causes and dynamics of past and future climatic change. Moreover, latitudinal changes in the position of the Polar Front through time provide important boundary conditions that may be used in testing and evaluating global circulation models.

Paleolimnological data on past climatic changes in tree-line regions are particularly important, as many of the more traditional, terrestrial-based paleoecological techniques, such as palynology, may reach their methodological limits at sites above the tree line (e.g. Gajewski *et al.*, 1995; Smol *et al.*, 1995). These restrictions do not apply to diatoms or other aquatic organisms, which are extremely abundant and ecologically diverse. Their short lifespan and fast migration rates enable them to respond quickly to environmental changes. Their potential as paleoindicators mostly relies on their good preservation and abundance in lake sediments.

Many variables affect the abundance of diatoms and species composition of diatom communities. Studies using diatom records to infer past climates have generally yielded qualitative results rather than quantitative paleoclimate estimates. Climate either directly (e.g. via changes in lake-water temperature, mixing regime, length of ice cover) or indirectly influences diatoms by controlling, for example, habitat availability and quality, catchment and aquatic vegetation, water color and transparency, or nutrient supply (e.g. Moser *et al.*, 1996).

Diatom–temperature transfer functions were applied on the fossil record preserved in the sediments of Lake Tsuolbmajävri (northern Finland, 68° 41′ N, 22° 05′ E), a shallow, well-mixed water body with a mean July surface-water temperature of *c.* 14° C (Korhola *et al.*, 2000). The diatom-inferred mean July water temperatures revealed a detailed paleoclimate record with variations within the range of 2 °C during the lake's history (Figure 12.1), including the mid-Holocene "thermal maximum" (*c.* 8–5.5 ka BP), the Medieval Warm Period, as well as the Little Ice Age (LIA) cooling event. Diatoms also suggest a water temperature increase of *c.* 1.5 °C since the termination of the LIA, which corresponds to estimates of post-LIA air-temperature increases (Overpeck *et al.*, 1997) for arctic regions in general. Because of the very close positive correlation between summer air and water temperature in Finnish Lapland it was not surprising that the obtained water-temperature curve closely tracked the air-temperature reconstruction for the same lake sediment record (Figure 12.1; see Korhola and Weckström, 2004).

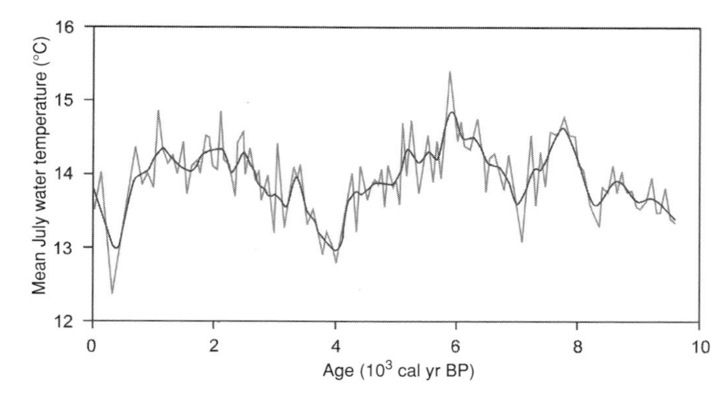

Figure 12.1 Holocene mean July water temperature reconstructions based on a fossil diatom record from Lake Tsuolbmajävri, Finnish Lapland. The bold line represents a LOESS smoother (span 0.05). (Modified after Korhola and Weckström, 2004.)

A striking and common feature of the Holocene sedimentary records of subarctic and alpine lakes is the dominance of pioneering benthic *Fragilaria* species (now grouped within the genera *Pseudostaurosira*, *Staurosira*, and *Staurosirella*) in initial diatom communities, sometimes making up more than 80% of the diatom flora (Pienitz *et al.*, 1991; Lotter *et al.*, 1997; Seppä and Weckström, 1999; Bigler *et al.*, 2002; Solovieva and Jones, 2002; Rosén *et al.*, 2004). *Fragilaria* spp. are commonly associated with high environmental instability and are known to tolerate broad environmental gradients (from brackish to freshwater) and poor light conditions (Haworth, 1976; Smol, 1988; Denys, 1990). The ability of small *Fragilaria* species to quickly reproduce and adapt to short-term environmental fluctuations makes them very competitive, which is of vital benefit especially in unstable limnological conditions such as the ones prevailing during the early Holocene.

Diatom studies in alpine, boreal, and subarctic lakes have shown relationships between the composition of diatom assemblages and water chemistry (e.g. Tynni, 1976; Koivo and Ritchie, 1978; Niederhauser and Schanz, 1993; Lotter *et al.*, 1998). These surveys also illustrate the need for more precise, quantitative diatom autecological data to refine or replace the traditional classification schemes first introduced by Kolbe (1927), Hustedt (1939, 1956), and Cholnoky (1968).

Over the past two decades, multivariate analyses of modern diatom assemblages preserved in surface lake sediments have been used to develop regional empirical models (transfer functions or surface-sediment calibration data sets; see Birks, this volume) based on the relationships between the present-day species composition and environmental conditions (e.g. Rühland and Smol, 2002; Bloom *et al.*, 2003; Schmidt *et al.*,

2004b). Application of these models to fossil diatom assemblages in long sedimentary sequences (cores) allows quantitative reconstructions of past lake conditions over centuries to millennia, such as air and lake-water temperature (e.g. Pienitz *et al.*, 1995; Wunsam *et al.*, 1995; Lotter *et al.*, 1997; Rosén *et al.*, 2000; Bigler and Hall, 2003), length of ice-cover (Thompson *et al.*, 2005b), mixing dates (Schmidt *et al.*, 2004a), pH (Weckström *et al.*, 1997; Rosén *et al.*, 2000), total phosphorus (e.g. Wunsam and Schmidt, 1995; Lotter *et al.*, 1998; Kauppila *et al.*, 2002), dissolved organic matter (DOC) and total organic matter (TOC) content (e.g. Pienitz and Smol, 1993; Fallu and Pienitz, 1999; Rosén *et al.*, 2000), and water color and optical regime (e.g. Pienitz and Vincent, 2000; Dixit *et al.*, 2001; Fallu *et al.*, 2002).

The impact of the steep climatic and vegetation gradients associated with the northern-tree-line ecotone on lake-water chemistry, in particular alkalinity, dissolved inorganic carbon (DIC), conductivity, and related parameters was also evidenced by changing diatom distributions in calibration studies. Substantial differences in lakes from the boreal forest biome to the arctic tundra were observed, with usually decreasing ionic concentrations from south to north (Pienitz *et al.*, 1997a, 1997b; Rühland and Smol, 1998; Fallu and Pienitz, 1999; Gregory-Eaves *et al.*, 2000; Fallu *et al.*, 2002; Rühland *et al.*, 2003). This aspect is of particular interest within the context of arctic warming, as accelerated permafrost melting results in major changes in hydrological processes and external inputs into lakes from rapidly eroding catchment soils.

In the following sections, we shall provide some examples of case studies that showcase the potential of applied diatom studies for addressing important questions in research on climate and global environmental-change impacts in subarctic and alpine regions.

12.3 The Subarctic

The pronounced impacts of past displacements (migration) of the northern tree line as a result of changes in Holocene climate have been documented for subarctic Canada (e.g. Pienitz *et al.*, 1999; MacDonald *et al.*, 2000) and Eurasia (e.g. Laing *et al.*, 1999). For example, following deglaciation about 8 ka BP the advance and subsequent retreat of the northern tree line in the central Canadian Subarctic in response to shifts in the mean summer position of the Arctic frontal zone was accompanied by profound limnological and hydrological changes (MacDonald *et al.*, 1993; Wolfe *et al.*, 1996). Consequently, changes in the composition of fossil diatom assemblages (in particular in the proportion of periphytic versus planktonic taxa), diatom concentrations, and diatom-inferred DOC concentrations showed three distinct and abrupt successional shifts in the history of Queen's Lake (Pienitz *et al.*, 1999).

12.3.1 Effects of landscape and climate change on underwater light exposure

To address the potential impact of long-term climate change relative to that of stratospheric ozone depletion, Pienitz and Vincent (2000) combined paleolimnological analyses with bio-optical models based on present-day conditions in subarctic lakes of northern Canada. This new paleo-optical approach allowed estimating past underwater light conditions from DOC concentrations that were inferred from fossil diatom assemblages preserved in Holocene sedimentary deposits from a lake near the northern tree line in the central Northwest Territories (Canada). Saulnier-Talbot *et al.* (2003) used a similar diatom-based paleo-optical approach to estimate past depths of UV penetration in coastal Lake Kachishayoot (northwestern Québec, Canada; 55° 20′ N, 77° 37.4′ W; 102 m above sea level, a.s.l.) after its isolation from the marine waters. Studies of optical environments in coastal systems have revealed that shifts from marine to freshwater conditions are accompanied by increased DOC, changes in UV radiation attenuation, and declines in UV penetration (Conde *et al.*, 2000). Consistent with these modern observations, Saulnier-Talbot *et al.*'s (2003) multiproxy investigation revealed abrupt increases in diatom-inferred DOC concentrations and water color that coincided with the retreat of post-glacial marine waters (induced by isostatic rebound) and the arrival of spruce trees in the landscape and the catchment of the study site (Figure 12.2). Their investigation also revealed large changes in the underwater irradiance environment over the course of the Holocene, from extremely high UV exposure following the initial formation of the lake and its isolation from the sea, to an order-of-magnitude lower exposure associated with the development of spruce forests in the catchment. Furthermore, the analysis of plant macrofossils revealed that UV penetration remained low even following forest retreat due to the development of alternative DOC sources in the catchment, such as moss-dominated wetlands (e.g. *Sphagnum*; Figure 12.2). Many freshwater ecosystems presently located in ecotonal regions of the Subarctic will likely experience similar important shifts in underwater spectral irradiance through the effects of climate change on catchment vegetation and organic matter loading. Moreover, the results of Saulnier-Talbot *et al.* (2003) show that major shifts in the physical and chemical lake conditions are likely to

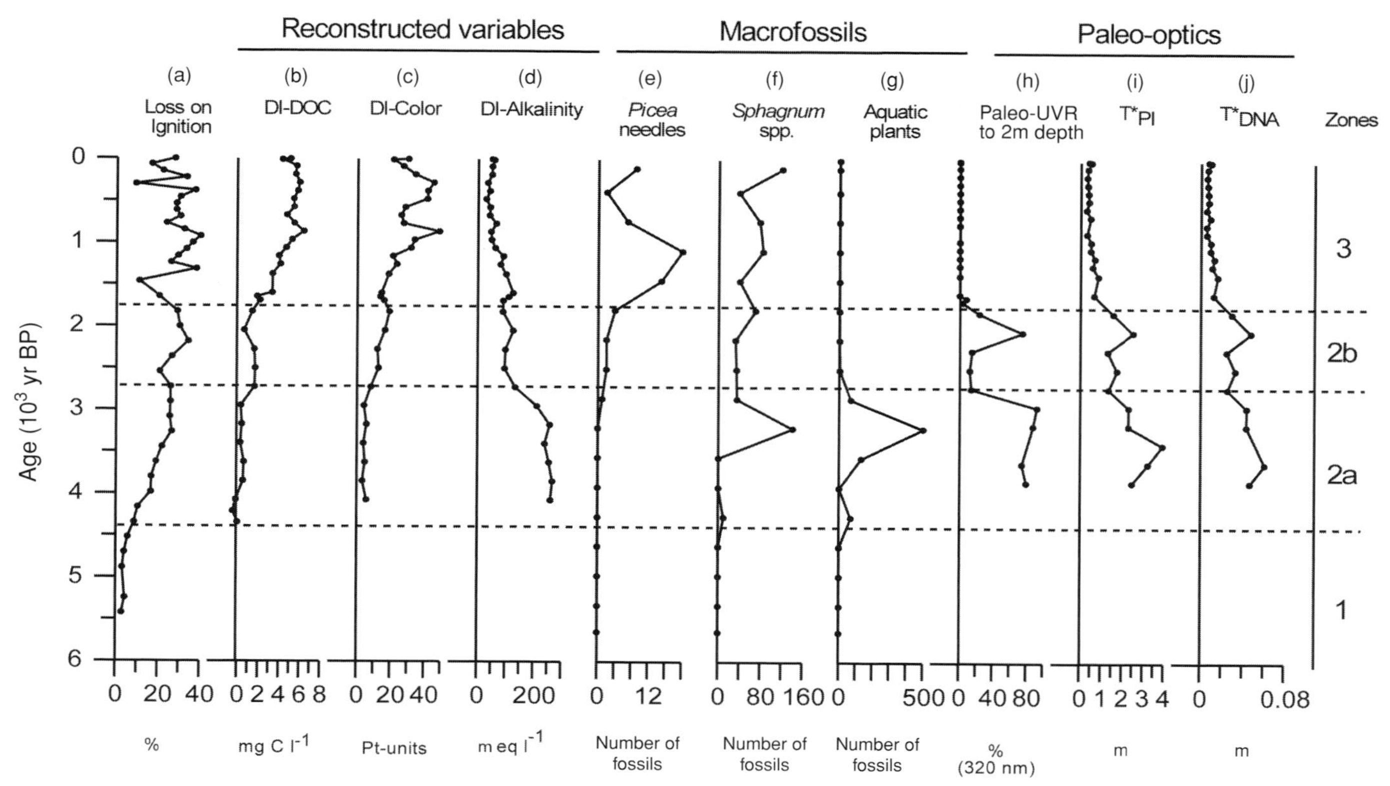

Figure 12.2 Changes in the catchment vegetation and limnological properties of Lake Kachishayoot (subarctic Québec, Canada) over the 5000 years following deglaciation. Concomitant with the gradual development of organic soils (a) and the invasion and establishment of spruce (*Picea*) (e) there was a rise in diatom-inferred DOC (b) and lake-water color (c), and a sharp decrease in underwater UV radiation (depth of paleo-UVR) (h) and biological transparency weighted for UV-photoinhibition, T^*_{PI} (i), or for UV-photo-damage of DNA, T^*_{DNA} (j). (Modified after Saulnier-Talbot *et al.*, 2003.)

be accompanied by profound changes in the composition of aquatic food webs, which may alter the overall structure and dynamics of these sensitive ecosystems.

12.3.2 Effects of landscape and climate change on underwater light exposure

Numerous studies of diatom records from subarctic lakes concurrently report increases in the relative abundances of small planktonic *Cyclotella* species during the past century that are accompanied by decreases of more heavily silicified, tychoplanktonic diatoms of the genus *Aulacoseira* and/or small, benthic *Fragilaria* (e.g. Sorvari and Korhola, 1998; Sorvari *et al.*, 2002; Rühland *et al.*, 2003; Rühland and Smol, 2005). The observed shifts in diatom community structure were speculated to be associated with the effects of intense warming in subarctic regions and were considered a "bellwether" of what could eventually occur at more southern latitudes with continuing warming. In their comparative analysis of more than 200 paleolimnological records from arctic, alpine, and temperate ecozones throughout the northern hemisphere, Rühland *et al.* (2008) provided compelling evidence that ecologically important changes are already underway in temperate lakes, similar to those observed in the rapidly warming Arctic. Specifically, they examined the impact of various climate-related changes in the physical properties (e.g. length of open-water season, duration and strength of thermal stratification, timing and strength of spring thaw and resulting flooding, light availability, nutrient cycling) on the structure and dynamics of *Cyclotella–Aulacoseira–Fragilaria* diatom assemblages. Their synthesis showed remarkably similar taxon-specific shifts since the middle of the nineteenth century in non-acidified, nutrient-poor freshwater ecosystems from throughout the northern hemisphere. Aquatic ecosystem changes occurred earlier in the highly sensitive circumpolar Arctic (by about 100 years) and in the alpine lakes (by about 50 years) than in lakes in the temperate regions. Their findings suggest that many of these lakes are currently crossing or have already crossed important climate-induced ecological thresholds. For example, several circumpolar tree-line lakes including Slipper Lake (64° 35′ 65″ N, 110° 50′ 07″ W, 460 m a.s.l.), TK20 (64° 09′ 00″ N, 107° 49″ 00″ W), TK54 (64° 38′ 68″ N, 112° 41′ 47″ W) in the

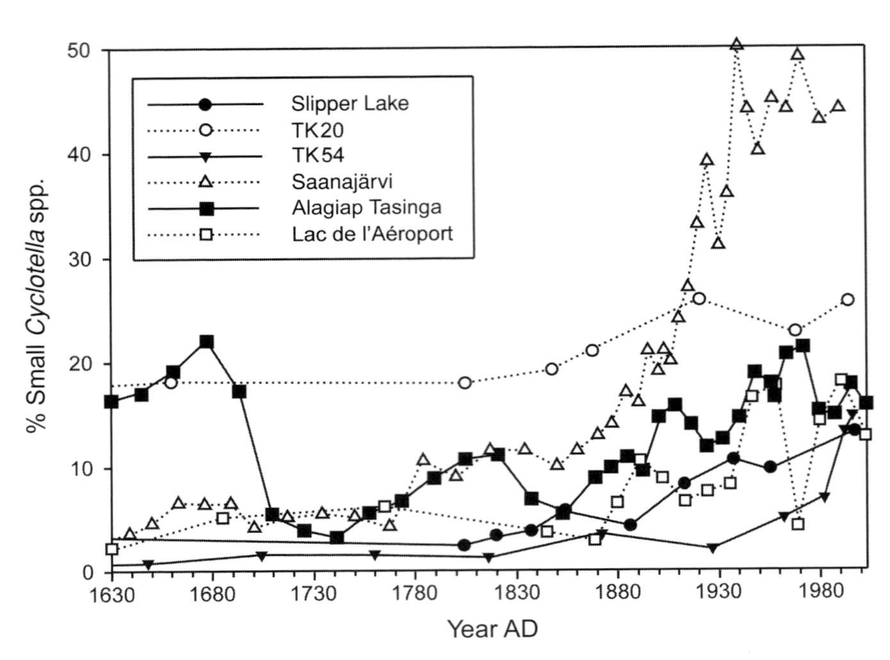

Figure 12.3 Comparison between the timing of the onset of increases in the relative abundance of small *Cyclotella* taxa (*C. stelligera*, *C. pseudostelligera*, *C. glomerata*, *C. comensis*, and *C. gordonensis*) in sedimentary records from subarctic lakes in central Canada (Slipper Lake and lakes TK20 and TK54), Finland (Lake Saanajärvi) and northern Québec (lakes Aeroport and Alagiap Tasinga). (Modified after Rühland *et al.*, 2008.)

Canadian central subarctic region, Lake Saanajärvi (69° 05′ N, 20° 52′ E) in Finnish Lapland, and Lake Birgervatnet (79° 50′ N, 11° 58′ E) in Svalbard, recorded a marked increase in the relative abundances of planktonic *Cyclotella* taxa (including *C. stelligera* (Cleve & Grunow in Cleve) Van Heurck, *C. pseudostelligera* Hustedt, *C. glomerata* Bachmann, *C. comensis* Grunow in Van Heurck, and *C. gordonensis* Kling & Hakansson) and concomitant decreases in diatoms belonging to the genera *Aulacoseira* and *Fragilaria* (Rühland *et al.* 2008). More recently, diatom profiles from Lake Aeroport and Lake Alagiap Tasinga in northern Québec, Nunavik, Canada (Saulnier-Talbot and Pienitz, unpublished data) are also showing interesting trends in small planktonic *Cyclotella* taxa. These taxon-specific shifts have yielded strikingly coherent, yet geographically asynchronous, large-scale trends of ecological change (see Figure 12.3). Such shifts are explained by competitive advantages created for less-heavily silicified (lighter) planktonic cells in a thermally stratified and hence stable water column with less mixing. The common understanding from empirical studies is that these conditions, in addition to a shortened duration of ice cover and thus a longer growing season, provide favorable habitats especially for small, fast-growing planktonic *Cyclotella* spp. (Raubitschek *et al.*, 1999; Rautio *et al.*, 2000; Pannard *et al.*, 2008). In contrast, thickly silicified (heavy) planktonic *Aulacoseira* species that require turbulence-induced resuspension into the photic zone of the water column (Kilham *et al.*, 1996) are at a disadvantage during such periods of strong stratification (Pannard *et al.*, 2008), while small benthic *Fragilaria* species also seem to be less competitive under these conditions (Lotter and Bigler, 2000).

12.4 Mountain regions: steep gradients

Studies in different mountain ranges have demonstrated a zonation of diatom assemblages along the altitudinal gradient, which also incorporates a strong gradient of water temperature (Servant-Vildary, 1982; Vyverman, 1992; Vyverman and Sabbe, 1995; Bloom *et al.*, 2003). Direct influences of temperature on the physiology of algal growth have been presented, e.g. by Raven and Geider (1988) and Montagnes and Franklin (2001), whereas indirect effects due to prolonged ice cover and changes in turbulent mixing have been discussed by Smol (1988), Smol *et al.* (1991), Anderson (2000), Laing and Smol (2000), Lotter and Bigler (2000), Battarbee *et al.* (2002a), Catalan *et al.* (2002a, 2002b), Koinig *et al.* (2002), Lotter *et al.* (2002), Schmidt *et al.* (2004b), Karst-Riddoch *et al.* (2005), and Smol and Douglas (2007).

12.4.1 Himalaya

The Himalayas are characterized by steep and high (with the earth summit, Mt. Everest 8848 m a.s.l., Nepal), mainly northwest to southeast striking, mountain ranges that surround plateaus of deserts and grasslands at altitudes above 4000 meters. Major reasons for differences in vegetation between subtropical forests and deserts are the relief and the gradual influence of the Asian monsoon. The Tibetan Plateau, for example, shows a strong gradient in precipitation. Due to the decreasing influence of the Indian summer monsoon the annual precipitation ranges from *c.* 800 mm in the south of Tibet (28° N) to less than 50 mm, with enhanced evaporation, in the north of the Quinghai province of China (40° N). Weighted averaging (WA) and WA partial least squares (WA-PLS) regression and calibration models from diatoms of 40 freshwater to hypersaline lakes (salinity: 0.1 to 91.7 g l^{-1}) on the Tibetan Plateau show a high predictive power for conductivity (cross validated $r^2 = 0.92$, RMSEP = 0.22) and water depth (cross validated $r^2 = 0.89$, RMSEP = 0.26; Yang *et al.*, 2003). Conductivity and chloride explained most of the variance in the diatom assemblages. Diatoms that are characteristic for the saline lakes (>1.9‰) included those typical for Na–SO_4–Cl lakes plus *Campylodiscus clypeus* Ehrenberg and *Surirella peisonis* Pantocsek, whereas diatom assemblages characteristic for the freshwater to slightly oligosaline lakes (<1.5‰) resembled those typical for Ca–Mg–CO_3 lakes plus *Cyclotella ocellata* Pantocsek and *Navicula radiosa* Kützing. Most of the freshwater diatoms of this data set, at least at the species levels, are commonly distributed or known from alpine and subarctic habitats of the Holarctic region, and endemic species appear to be rare (Li *et al.*, 2006, 2003). This is corroborated by several floristic and taxonomical studies from the Himalaya region (e.g. Mereschkowsky, 1906; Hustedt, 1922; Ioriya, 1995; Li *et al.*, 2003, 2004, 2007). However, the recent findings of several new diatoms require further taxonomic work (e.g. Jüttner *et al.*, 2000, 2004).

Numerous streams, differing in size, velocity, and habitats are to be found on the steep slopes of the southern mountain ranges. Their diatom assemblages along a gradient of more remote to increasingly human-impacted areas at lower altitudes are comparable with streams in other Holarctic mountain ranges (Jüttner *et al.*, 1996, 2003; Rothfritz *et al.*, 1997; Cantonati *et al.*, 2001).

Long sediment sequences from lakes that used qualitative diatom analyses to trace changes in salinity and lake levels (e.g. Gasse *et al.*, 1991, van Campo and Gasse, 1993; Fan and Gasse, 1994; Gasse, 1996; Fan *et al.*, 1996; Li *et al.*, 1999) suggest distinct changes in the intensity of the monsoonal climate since the end of the last ice age. In a diatom study of a peat profile in the Indian Himalayas, Rühland *et al.* (2006) were able to track changes in monsoon intensity and rapid melting of Himalayan glaciers at higher elevations. Recently developed diatom-based quantitative inference models (Yang *et al.*, 2003, 2004) provided a powerful tool to track changes in the monsoon climate for a region where the Asian monsoon provides summertime life, sustaining rains for hundreds of millions of people.

12.4.2 Alps

A combined data set of diatoms from surface-sediment samples of 106 Swiss lakes (Figure 12.4) spanning an altitudinal gradient from 334 to 2815 m a.s.l. and a July air-temperature gradient of 5 to 21.4 °C demonstrates the importance of climate to diatoms and other aquatic organisms (chironomids, cladocera, chrysophytes), and has been used to develop multi-proxy temperature and nutrient inference models (Lotter *et al.*, 1997, 1998; Bigler *et al.*, 2006; Hausmann and Kienast, 2006). Using several independent lines of evidence the power of reconstructing past climate change may be amplified (Lotter, 2003; Birks and Birks, 2006). Heegaard *et al.* (2006) analyzed the cumulative rate of compositional change along the altitudinal gradient in this data set and identified a region of major compositional change (an "aquatic ecotone") in diatom assemblages just below the alpine tree line (today between 1900 and 2000 m a.s.l.) at 1600 to 1700 m a.s.l. and a second one above the tree line at 2000 to 2100 m a.s.l.

As air temperatures most closely correspond to surface water temperatures during summer (Livingstone and Lotter, 1998; Livingstone *et al.*, 1999), instrumental air-temperature time series have been used as a basis for the July temperature inference models. However, above a critical altitude, a decoupling between summer air and water temperatures can be observed (Livingstone *et al.*, 2005). Canonical correspondence analysis (CCA) of the original calibration data set of Lotter *et al.* (1997) showed that catchment (14.6%; geology, land use, vegetation type), climate (13.4%; temperature, precipitation), and limnological variables (11.5%; water depth, surface area, catchment area) had the largest, statistically significant independent explanatory powers, whereas water chemistry (conductivity, pH, alkalinity, DOC, nutrients, metals) explained also a large (14.6%) but not significant part of the total variance in the diatom data. A WA-PLS model for diatoms and July air temperature provided a cross-validated r^2 of 0.80 and a RMSEP of 1.6 °C. A vast majority of the diatoms (72.5%) that occurred in 20% or more of the samples showed statistically significant relationships to July temperature, either as a unimodal or a sigmoidal response.

Below elevations of 1000 m a.s.l. planktonic diatoms are generally dominant, whereas above 1000–1500 m a.s.l. small periphytic (in some cases probably also tychoplanktonic) taxa

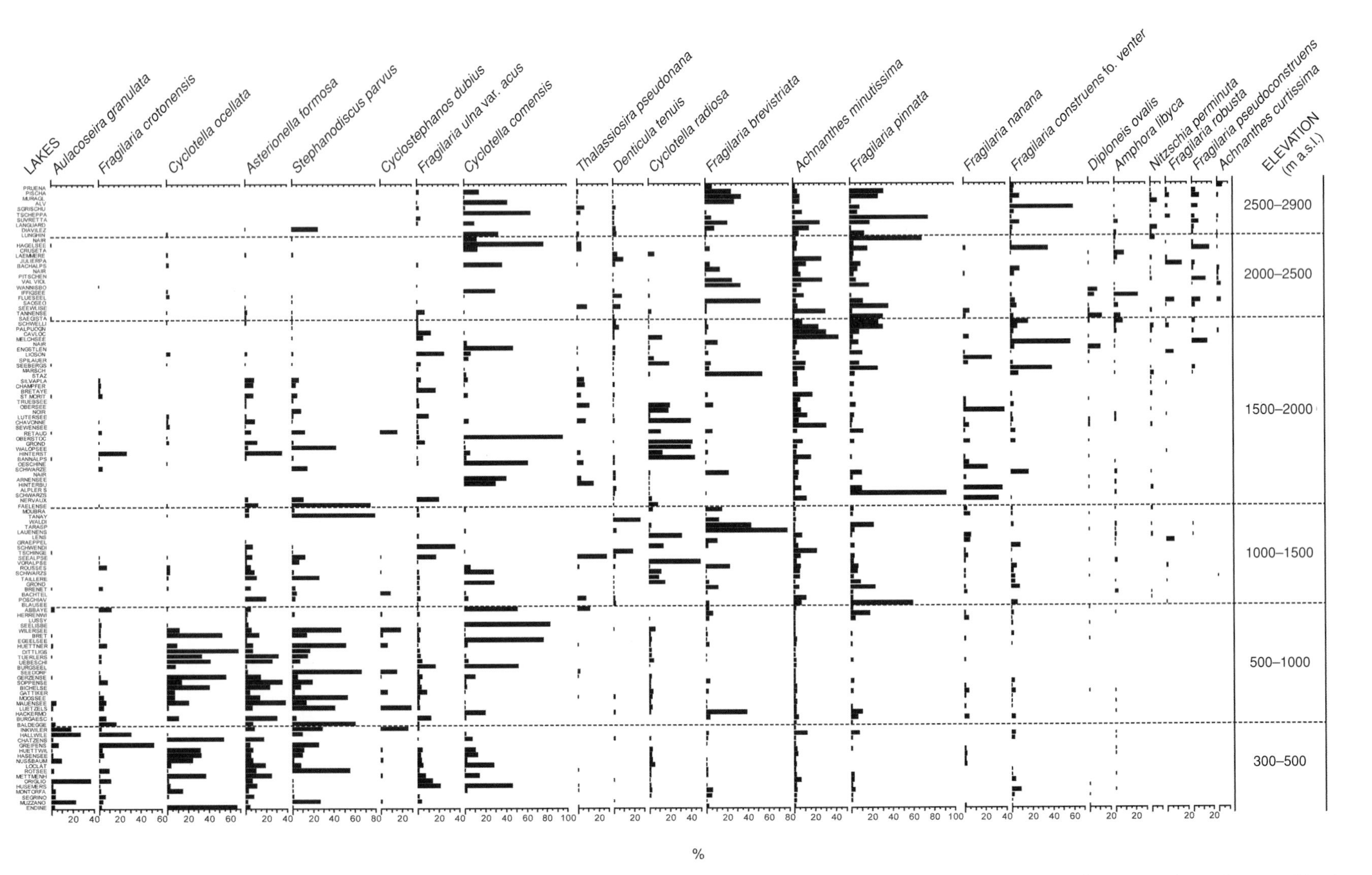

Figure 12.4 Distribution of diatoms in surficial sediments of 106 lakes along an altitudinal gradient in the Swiss Alps (only the most prominent diatoms are shown. Data combined from Lotter *et al.* (1997), Bigler *et al.* (2006), and Hausmann and Kienast (2006).

such as *Fragilaria* spp. and *Achnanthes minutissima* become more important (Figure 12.4). There are, however, exceptions, mainly involving small centric taxa (e.g. *Cyclotella comensis*, *Thalassiosira pseudonana*). Low numbers of planktonic diatoms and increasing abundances of *Fragilaria* spp. is a phenomenon often observed with increasing altitude or latitude. It may be related, on the one hand, to the fact that the growing season at these altitudes or latitudes is considerably shorter, due to prolonged snow and ice cover. As ice-melt starts at the shores, these marginal areas provide suitable habitats for the development of periphyton (Lotter and Bigler, 2000). Also, water transparency is generally high, thus favoring periphytic diatoms even in deeper lakes. On the other hand, alpine lakes are characterized by short-term fluctuations that may possibly favor *Fragilaria* spp. that are more adaptable and competitive.

Another diatom–temperature training set from the Alps, based exclusively on *Cyclotella* species, yielded an apparent r^2 of 0.62, with a cross-validated RMSEP of 1.3 °C (Wunsam *et al.*, 1995). Species temperature optima and tolerances were estimated by a WA model with *C. comensis* at the higher end (19.7 °C) and *C. styriaca* Hustedt at the lower end (11.9 °C) of the summer-temperature gradient. The high morphological variability, especially within the small *Cyclotella* taxa, calls for molecular genetic studies, which may clarify the taxonomy of the various morphotypes that have been found to date along nutrient and temperature gradients (Wunsam *et al.*, 1995; Hausmann and Lotter, 2001).

High seasonality, mainly caused by large temperature differences between summer and winter, is a major feature of freshwater ecosystems from the temperate zone (e.g. Wetzel, 2001), with strong impacts on alpine lakes (e.g. Ventura *et al.*, 2000; Catalan *et al.*, 2002a). The date of autumn mixing (in Julian days; defined as the day in autumn when the average daily temperature has declined to 4 °C) explained a statistically significant amount of variance in diatom composition from the surface sediments of 45 lakes in the Austrian Central Alps (Niedere Tauern), covering an altitudinal gradient from approximately 1500 to 2300 m a.s.l. A WA-PLS model for the date of autumn mixing yielded a cross-validated r^2 of 0.71 and a RMSEP of 0.006 $\log_{10}$ Julian days (Schmidt *et al.*, 2004a). The resulting diatom-inferred mixing dates, together with a chrysophyte cyst-based spring mixing model (Kamenik and Schmidt, 2005), enabled an estimation of Holocene ice-cover duration and mean seasonal air-temperature anomalies (Schmidt *et al.*, 2007). Of all the diatoms recorded in this training set, the small diatom *Staurosira (Fragilaria) microstriata* (Marciniak) Lange-Bertalot showed the lowest July water-temperature optimum (6.7 °C) and the highest ice-cover-duration tolerance. However, small valve size can also be the result of lake warming. For example, in the Niedere Tauern training set the minimum valve length of *Fragilaria exiguiformis* Lange-Bertalot corresponded with the mean July water-temperature optimum (9.4 °C) of this taxon as estimated from the transfer function. Elevated temperatures may have stimulated enhanced cell division (e.g. Montagnes and Franklin, 2001), or cells did not reproductive sexually (Schmidt *et al.*, 2004b).

Diatom-based calibration sets for nutrients (see also Hall and Smol, this volume), in particular for total phosphorus (TP), are also available from the Alps (Wunsam and Schmidt, 1995; Wunsam *et al.*, 1995; Lotter *et al.*, 1998; Hausmann and Kienast, 2006). Yet, the majority of the calibration sites are located below the tree line, in areas that are generally characterized by intensive agriculture and/or pasturing. With some exceptions (e.g. Hausmann *et al.*, 2002), application of these TP inference models to tree-line lakes may, however, be limited by the generally low nutrient concentrations (Marchetto *et al.*, 1995; Müller *et al.*, 1998; Schmidt *et al.*, 2004b) and the different diatom floras of the tree-line lakes.

12.4.3 Changing environments: the last glacial termination

Most of the inference models that have recently been developed using arctic/alpine diatom assemblages have yet to be applied to Holocene records. Paleoenvironmental records from the Alps illustrate how the application of these models may help to better understand environmental changes at the end of the last ice age.

Multidisciplinary paleolimnological techniques were used to decipher the environmental history of a long sediment core from the Längsee (548 m a.s.l.), a small, meromictic kettle-hole lake situated in the southeastern Alps (Carinthia, Austria, see Schmidt *et al.*, 1998). Due to its location at the southern slope of the Alps, close to the Würmian pleniglacial ice margin, the lake's catchment became deglaciated as early as 19 ka ago. Diatom-based transfer functions (Wunsam and Schmidt, 1995; Wunsam *et al.*, 1995) have been applied down-core to infer TP concentrations and summer lake surface temperatures. Recently, a mean summer epilimnetic water temperature (SEWT) inference model (cross-validated $r^2 = 0.89$, RMSEP = 1.82 °C, see Huber *et al.*, 2010) using a locally weighted WA regression and calibration technique (Juggins, 2007) was also applied to the Längsee sediment core. The model includes water-temperature measurements from three different data sets, totalling 116 lakes from the Alps and surrounding areas (Huber *et al.*, 2010). Although water temperature was not independent from pH, TP, and conductivity, SEWT

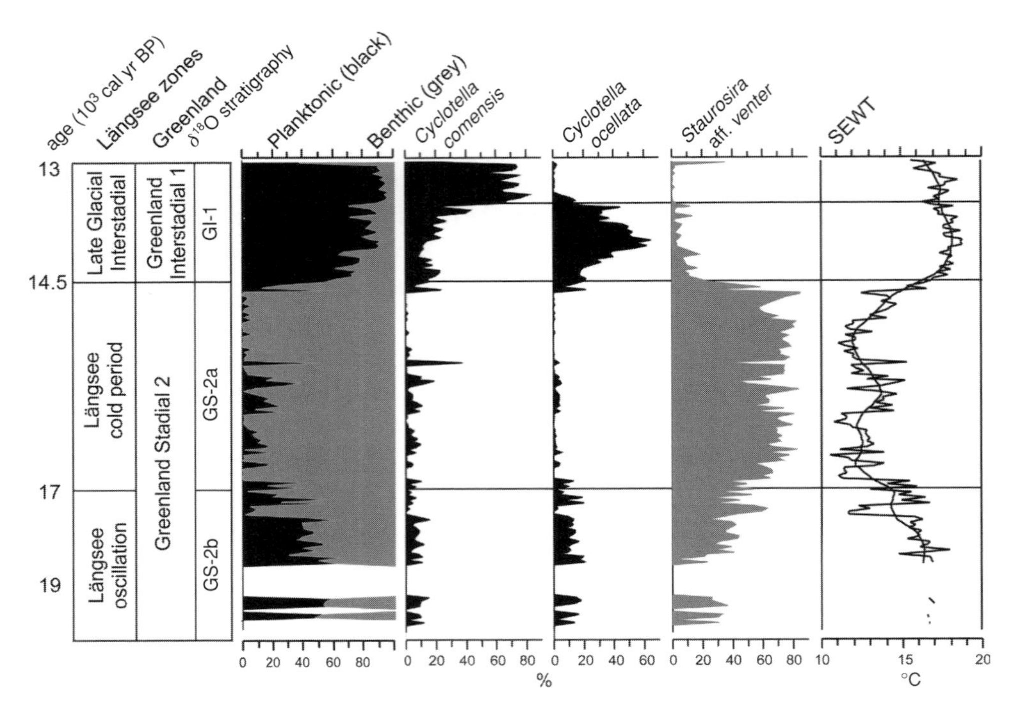

Figure 12.5 Diatom stratigraphy from a sediment core from Längsee (Austria) spanning the time between ca. 19 to 13 ka BP (modified after Huber *et al.*, 2010). The ratio between planktonic and benthic diatoms (including Fragilariaceae), the relative abundance of selected diatoms, and the diatom-inferred mean summer epilimnetic water temperatures (SEWT) are shown. Three major phases of climate change are indicated during the Last Glacial termination and correlated with the stable oxygen isotope ($\delta^{18}O$) Greenland ice-core event stratigraphy (Walker *et al.*, 1999).

captured most of the variance in the diatom data. A warm period between c. 19 and 17 ka BP (Längsee oscillation) favored the development of planktonic diatoms (*Cyclotella ocellata* Pantocsek, *Cyclotella comensis*, *Cyclotella distinguenda* var. *unipunctata* (Hustedt) Håkansson & Carter, *Cyclotella quadrijuncta* (Schröter) von Keissler, *Stephanodiscus alpinus* Hustedt). For this phase a SEWT range of 11.6 to 18 °C (mean 15.8 °C) was inferred. Consistent with the subsequent climatic cooling between c. 17 and 14.5 ka BP, SEWT dropped. This cool period represented in the Längsee sediment core likely corresponds to the northern hemispheric cooling, evidence of which is recorded by oxygen isotopes from Greenland ice cores (Greenland Stadial GS-2a), North Atlantic ice-rafting events (Heinrich 1), and a glacier re-advance in the Alps (Gschnitz). Lower temperatures leading to prolonged ice cover may explain the low percentage of planktonic diatoms. The contemporaneous increase in Fragilariaceae (mainly *Staurosira* aff. *venter*, which was more frequent in lakes with low SEWT in the modern training set), explains the decrease in inferred SEWT (mean inferred temperature = 12.9 °C). Parallel with the reforestation at about 14.5 ka BP, diatom-inferred SEWT increased. For the following Late Glacial Interstadial (= Greenland Interstadial GI-1) a mean SEWT of 17.5 °C was inferred (Figure 12.5), which is c. 4 °C colder than today. Concurrent with this temperature increase Längsee became meromictic. Strong meromixis during warm and dry climate conditions likely resulted in a nutrient gradient between the phosphorus-enriched monimolimnion and the oligo- to mesotrophic conditions in the epilimnion. This phase is characterized by the dominance of *C. comensis*, a diatom with oligo- to mesotrophic affinities. However, at times of increased nutrient concentrations, either from internal loading through enhanced mixing or from external loading, coupled with increased minerogenic matter flux from catchment erosion, *C. ocellata* appeared to have been more competitive against other planktonic diatoms (Schmidt *et al.*, 2002).

12.4.4 Indirect effects: a temperature-pH relationship in alpine lakes

High-altitude areas in the Alps are affected by precipitation with pH values between 4.8 and 5.2, and consequently acidification of crystalline bedrock sites has been reported. In these areas, soil and vegetation have less influence on biogenic acidification, because only a small fraction of the drainage area is covered by soil. High mountain lakes are therefore very sensitive to climatic and hydrological changes as well as to

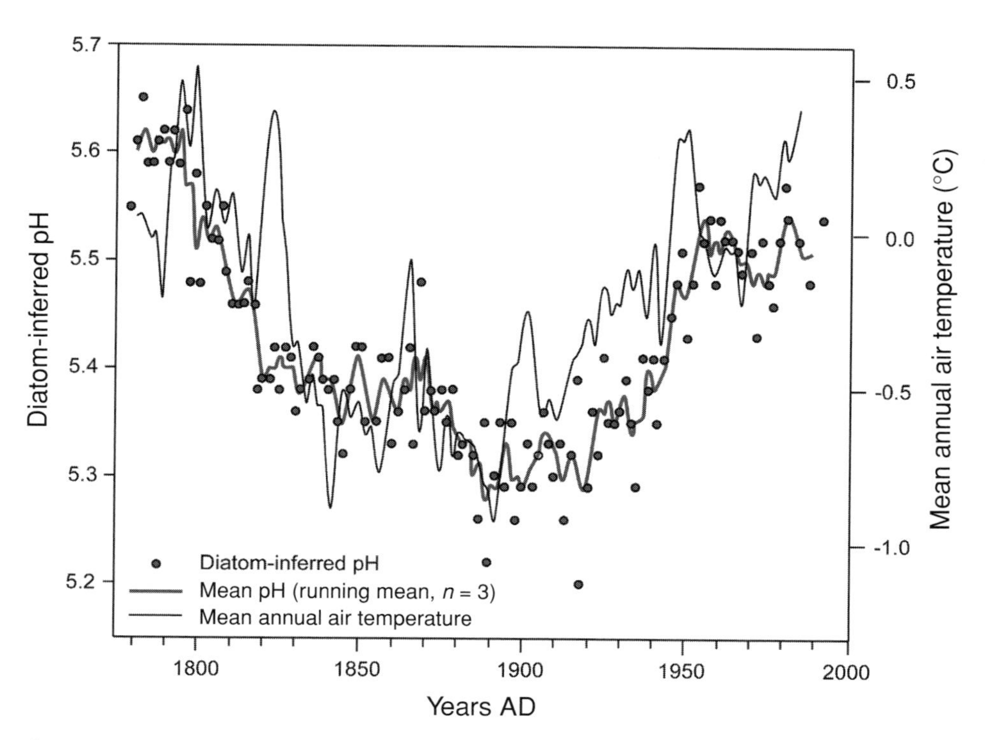

Figure 12.6 Diatom-inferred pH history of Schwarzsee ob Sölden (2796 m a.s.l., Austria) during the past two centuries, compared to the Austrian mean annual air temperatures (after Koinig *et al.*, 1997). The chronology is based on ^{210}Pb (constant rate of supply (CRS) model) and ^{137}Cs dating (see Koinig *et al.*, 1997; Sommaruga-Wögrath *et al.*, 1997).

increasing acidity of precipitation (see Battarbee *et al.*, this volume).

Paleolimnological investigations in Schwarzsee ob Sölden (2796 m a.s.l., Tyrol, Austria) showed a diatom-inferred decline from pH 6 to 5 (Arzet *et al.*, 1986). Using a WA model based on a regional alpine calibration set (Marchetto and Schmidt, 1993), the diatom-inferred pH changes in a core from Schwarzsee dating back to the eighteenth century were predominantly due to changes in the relative abundances of the dominant *Aulacoseira* species (*Aulacoseira alpigena* (Grunow) Krammer, *Aulacoseira nygaardii* Camburn & Kingston, *Aulacoseira distans*, including var. *nivalis* (W. Smith) Haworth, *Aulacoseira perglabra* (Østrup) Haworth; see Koinig *et al.*, 1997). The lowest inferred pH values of 5.2–5.3 occurred between AD 1880 and 1920. Towards 1970, the inferred pH values increased again to 5.5. A comparison of these pH reconstructions with the mean Austrian air temperature showed a strong correlation throughout the past ~200 years (Figure 12.6). These findings support the idea of a climate-driven pH control in high-alpine lakes (Psenner and Schmidt, 1992), implying that climatic cooling may cause decreases in pH, and rising temperatures increases in pH. The pH decline started at the onset of the LIA, with glacier re-advances in the Alps culminating around AD 1850. Koinig *et al.* (1997, 1998) and Sommaruga-Wögrath *et al.* (1997) suggested that pH increases during warm episodes may be a result of enhanced weathering rates, increased in-lake alkalinity production, longer water retention times, and higher amounts of dust deposition (see also Marchetto *et al.*, 1995). However, the hypothesis that recent climatic warming counteracts the influence of modern acidification is not valid for all lakes. Based on diatom assemblage changes at Lake Rassas (2682 m a.s.l., southern Tyrol, Italy), for instance, the onset of anthropogenically derived acid precipitation at the beginning of the twentieth century (Schmidt and Psenner, 1992; Psenner and Schmidt, 1992) led to a decoupling of the temperature–pH relationship, and thus to lake acidification.

12.5 Common features to subarctic and mountain lake diatom assemblages

Many of the above-mentioned calibration sets have revealed interesting similarities in diatom assemblage composition across the tree line, namely the presence of assemblages composed of large benthic and planktonic taxa in the more nutrient-enriched forest lakes, and assemblages dominated by small benthic taxa in the more dilute lakes above the tree line. This distinct trend in diatom community structure and cell size with changing latitude and altitude, in patterns unrelated to water depth, may be related to differences in the physico-thermal properties of the lakes and the length of the growing season. For example, there is evidence that climatic warming, with its likely consequences of a longer ice-free season and enhanced thermal stratification, would give a competitive advantage to planktonic forms, keeping these diatoms in suspension in the photic zone for longer periods of time (e.g. Smol, 1988; Fee *et al.*, 1992).

Differences observed between diatom cell size in lakes below and above the tree line could be related to nutrient availability, length of growing season, and diatom growth rates. The maximum specific growth rates in algae decrease with increasing cell size (Raven and Geider, 1988), and the high surface-to-volume ratio of small cells provides them with a competitive advantage under low-nutrient conditions. Small benthic *Fragilaria* spp. fit into the classic model of organisms favored by r-selection (i.e. small size, high reproductive rate, large ecological amplitude). Because of their higher turnover rates, these opportunistic, small-celled diatoms may quickly form blooms and (temporarily) out-compete larger diatom species with slower growth rates during the brief growing season in mountain and tundra lakes.

12.6 Summary

Diatoms in subarctic and mountain regions are important and sensitive indicators of environmental change. The subarctic–arctic and subalpine–alpine tree line is an important ecotone, with its past, present, and future position largely depending on climatic factors. Climatic fluctuations at this ecotone are either directly or indirectly influencing diatom communities through alterations in physical and chemical limnetic properties. Therefore, diatoms have great potential as environmental indicators and contribute to and complement ecological as well as paleoecological studies in tree-line regions (e.g. Lotter and Psenner, 2004). Reconstructing changes in past tree-line position provides essential boundary conditions for global circulation models to hindcast past or predict future climate change.

The potential for inferring past DOC/TOC levels from fossil diatom records combined with recent advances in bio-optical modeling in northern lakes led to the development of lake paleo-optics, which offers opportunities for integrating studies of the present day with historical properties of lakes (Pienitz and Vincent, 2000). This approach is of special interest for lakes in tree-line regions, since lakes situated above the tree line are more sensitive to small changes in DOC and rising UV radiation associated with stratospheric ozone depletion than lakes of the boreal forest (Vincent and Pienitz, 1996; Laurion *et al.*, 1997).

In all studies involving diatom calibration data sets, pH or pH-related variables (e.g. alkalinity, DIC, Ca) have been shown to influence significantly diatom distributions. Apart from the predominant control exerted by pH and related variables, two major gradients appear to emerge from these studies: the concentration of lake water DOC/TOC (which is related to catchment vegetation and tree line), and temperature (which is related to latitude and/or altitude). Temperature is a complex variable that is highly correlated with other characteristics of the environment, such as mixing regime and duration of icecover (e.g. Ohlendorf *et al.*, 2000; Thompson *et al.*, 2005b). The conversion of diatom-inferred mixing dates into altitude-dependent air-temperature anomalies offers opportunities to reconstruct autumn climates (Schmidt *et al.*, 2007). The distinct influence exerted by the temperature gradient on diatom distribution as shown in the studies from subarctic regions and different mountain ranges is of ecological and paleoecological significance, as it strengthens the assumption that the relationship between organisms and climate is clearest at ecotonal boundaries; consequently, ecotones are optimal areas for studying climatic change (e.g. Smol *et al.*, 1991; Heegaard *et al.*, 2006). A growing number of fossil diatom records from paleolimnological studies reveal spatially coherent trends of climate-driven, taxon-specific changes across subarctic regions of the northern hemisphere, implying that freshwater ecosystems have crossed critical ecological thresholds in recent times.

The results obtained from diatom studies in tree-line regions confirm a general relationship between the zonal distribution of diatoms and ecoclimatic or vegetational zones. Despite the extremely high degree of floristic diversity that characterizes oligotrophic boreal lakes compared with lakes in temperate regions (e.g. Lange-Bertalot and Metzeltin, 1996), a surprisingly high degree of floristic similarity can be observed among calibration data sets from different tree-line regions. Efforts are underway to develop supra-regional data sets for these regions within the Circumpolar Diatom Database (CDD; http://www.cen.ulaval.ca/paleo/index.html), which aims at expanding and improving existing diatom calibration sets by combining the existing data sets from North America, Scandinavia, and Siberia into one large data set for the whole of circumpolar, northern hemispheric regions. Nevertheless, all transfer-function approaches have to be assessed critically (see Birks, this volume) before using their results for hindcasting past environmental change at the arctic or alpine ecotone.

Acknowledgments

We would like to thank K. Rühland and É. Saulnier-Talbot for valuable comments on earlier versions of the manuscript. Work presented here has been supported by the Dutch Science Foundation (NWO-ALW grant 818.01.001), by the Natural Sciences and Engineering Research Council (NSERC) of Canada, the Fonds québécois de recherche sur la nature et les technologies (FQRNT, Québec), and the Austrian Science Fund (FWF, P18595-B17).

References

ACIA (Arctic Climate Impact Assessment) (2004). *ACIA, Impacts of a Warming Arctic: Arctic Climate Impact Assessment*. New York: Cambridge University Press.

Anderson, N. J. (2000). Diatoms, temperature and climatic change. *European Journal of Phycology*, **35**, 307–14.

Arzet, K., Steinberg, C., Psenner, R., & Schulz, N. (1986). Diatom distribution and diatom inferred pH in the sediment of four alpine lakes. *Hydrobiologia*, **143**, 247–54.

Backman, A. L. & Cleve-Euler, A. (1922). Die fossile Diatomeenflora in Österbotten. *Acta Forestalia Fennica*, **22**, 5–73.

Baron, J. S., Nydick, K. R., Rueth, H. M., Lafrancois, B. M., & Wolfe, A. P. (2005). High elevation ecosystem responses to atmospheric deposition of nitrogen in the Colorado Rocky Mountains, USA. In *Global Change and Mountain Regions. An Overview of Current Knowledge*, ed. U. M. Huber, H. K. M. Bugmann, & M. A. Reasoner, Dordrecht: Springer, pp. 429–36.

Battarbee, R. W. (1991). Palaeolimnology and climate change. In *Evaluation of Climate Proxy Data in Relation to the European Holocene*, ed. B. Frenzel, B. Gläser, Stuttgart: Fisher, pp. 149–57.

Battarbee, R. W., Grytnes, J. A., Thompson, R., *et al.* (2002a). Comparing palaeolimnological and instrumental evidence of climate change for remote mountain lakes over the last 200 years. *Journal of Paleolimnology*, **28**, 161–79.

Battarbee, R. W., Thompson, R., Catalan, J., Grytnes, J. A., & Birks, H. J. B. (2002b). Climate variability and ecosystem dynamics of remote alpine and arctic lakes: the MOLAR project. *Journal of Paleolimnology*, **28**, 1–6.

Bigler, C., Grahn, E., Larocque, I., Jeziorski, A., & Hall, R. (2003). Holocene environmental change at Lake Njulla (999 m a.s.l.), northern Sweden: a comparison with four small nearby lakes along an altitudinal gradient. *Journal of Paleolimnology*, **29**, 13–29.

Bigler, C. & Hall, R. I. (2002). Diatoms as indicators of climatic and limnological change in Swedish Lapland: a 100-lake calibration set and its validation for paleoecological reconstructions. *Journal of Paleolimnology*, **27**, 79–96.

Bigler, C. & Hall, R. I. (2003). Diatoms as quantitative indicators of July temperature: a validation attempt at century-scale with meteorological data from northern Sweden. *Palaeogeography, Palaeoclimatology, Palaeoecology*, **189**, 147–60.

Bigler, C., Heiri, O., Krskova, R., Lotter, A. F., & Sturm, M. (2006). Distribution of diatoms, chironomids and cladocera in surface sediments of thirty mountain lakes in south-eastern Switzerland. *Aquatic Sciences*, **68**, 154–71.

Bigler, C., Larocque, I., Peglar, S. M., Birks, H. J. B., & Hall, R. I. (2002). Quantitative multiproxy assessment of long-term patterns of Holocene environmental change from a small lake near Abisko, northern Sweden. *The Holocene*, **12**, 481–96.

Birks, H. H. & Birks, H. J. B. (2006). Multi-proxy studies in palaeolimnology. *Vegetation History and Archaeobotany*, **15**, 235–51.

Birks, H. J. B. (1995). Quantitative palaeoenvironmental reconstructions. In *Statistical Modelling of Quaternary Science Data*, ed. D. Maddy & J. S. Brew, Cambridge: Quaternary Research Association, pp. 161–254.

Bloom, A. M., Moser, K. A., Porinchu, D. F., & MacDonald, G. M. (2003). Diatom-inference models for surface-water temperature ation set from the Sierra Nevada, California, USA. *Journal of Paleolimnology*, **29**, 235–55.

Cantonati, M., Corradini, G., Jüttner, I., & Cox, E. J. (2001). Diatom assemblages in high mountain streams of the Alps and Himalaya. *Nova Hedwigia, Beiheft*, **123**, 37–61.

Catalan, J., Pla, S., Rieradevall, M., *et al.* (2002a). Lake Redò ecosystem response to an increasing warming in the Pyrenees during the twentieth century. *Journal of Paleolimnology*, **28**, 129–45.

Catalan, J., Vetura, M., Brancelj, A., *et al.* (2002b). Seasonal ecosystem variability in remote mountain lakes: implications for detecting climatic signals in sediment records. *Journal of Paleolimnology*, **28**, 25–46.

Cholnoky, B. J. (1968). *Die Ökologie der Diatomeen in Binnengewässern*, Lehre: J. Cramer.

Cleve-Euler, A. (1934). The diatoms of Finnish Lapland. *Societas Scientiarum Fennica Commentationes Biologicae*, **IV**, 1–154.

Conde, D., Aubriot, L., & Sommaruga, R. (2000). Changes in UV penetration associated with marine intrusions and freshwater discharge in a shallow coastal lagoon of the southern Atlantic Ocean. *Marine Ecology Progress Series*, **207**, 19–31.

Dauta, A., Devaux, J., Piquemal, F., & Boumnich, L. (1990). Growth rate of four freshwater algae in relation to light and temperature. *Hydrobiologia*, **207**, 221–6.

Denys, L. (1990). *Fragilaria* blooms in the Holocene of the western coastal plain of Belgia. In *Proceedings of the 10th International Diatom Symposium, Joensuu, Finland*, ed. H. Simola, Koeltz Scientific Books, Königstein, pp. 397–406.

Dixit, S. S., Keller, W., Dixit, A. S., & Smol, J. P. (2001). Diatom-inferred dissolved organic reconstructions provide assessments of past UV-B penetration in Canadian shield lakes. *Canadian Journal of Fisheries and Aquatic Sciences*, **58**, 543–50.

Eppley, R. W. (1977). The growth and culture of diatoms. In *The Biology of Diatoms*, ed. D. Werner, Botanical Monographs, Oxford: Blackwell, pp. 24–64.

Fallu, M.-A., Allaire, N., & Pienitz, R. (2002). Distribution of freshwater diatoms in 64 Labrador (Canada) lakes: species–environment relationships along latitudinal gradients and reconstruction models for water colour and alkalinity. *Canadian Journal of Fisheries and Aquatic Sciences*, **59**, 329–49.

Fallu, M.-A. & Pienitz, R. (1999). Diatomées lacustres de Jamésie-Hudsonie (Québec) et modele de reconstitution des concentrations de carbone organique dissous. *Écoscience* **6**, 603–20.

Fan, H. & Gasse, F. (1994). A late-Pleistocene–Holocene diatom record and palaeoenvironment of Bangong lake, west Tibet. *Acta Geographica Sinica*, **49**, 33–45.

Fan, H., Gasse, F., Huc, A., *et al.* (1996). Holocene environmental changes in Bangong Co basin (western Tibet). Part 3: biogenic remains. *Palaeogeography, Palaeoclimatology, Palaeoecology*, **120**, 65–78.

Fee, E. J., Shearer, J. A., DeBruyn, E. R., & Schindler, E. U. (1992). Effects of lake size on phytoplankton photosynthesis. *Canadian Journal of Fisheries and Aquatic Sciences*, **49**, 2445–59.

Foged, N. (1955). Diatoms from Peary Island, north Greenland. *Meddelelser om Grönland*, **128**, 1–90.

Foged, N. (1964). Freshwater diatoms from Spitsbergen. *Tromsö Museums Skrifter*, **11**.

Gajewski, K., Garneau, M., & Bourgeois, J. C. (1995). Paleoenvironments of the Canadian High Arctic derived from pollen and plant macrofossils: problems and potentials. *Quaternary Science Reviews*, **14**, 609–29.

Gasse, F., Arnold, M., Fontes, J. C., *et al.* (1991). A 13,000-year climate record from western Tibet. *Nature*, **353**, 742–5.

Gasse, F., Fontes, J. C., Van Campo, E., & Wie, K. (1996). Holocene environmental changes in Bangong Co basin (Western Tibet). Part 4: discussions and conclusions. *Palaeogeography, Palaeoclimatology, Palaeoecology*, **120**, 79–92.

Gregory-Eaves, I., Smol, J. P., Finney, B., Lean, D. R. S., & Edwards, M. E. (2000). Characteristics and variation in lakes along a north–south transect in Alaska. *Archiv für Hydrobiologie*, **147**, 193–223.

Hartig, J. H. & Wallen, D. G. (1986). The influence of light and temperature on growth and photosynthesis of *Fragilaria crotonensis* Kitton. *Journal of Freshwater Ecology*, **3**, 371–82.

Hausmann, S. & Kienast, F. (2006). A diatom-inference model for nutrients screened to reduce the influence of background variables: application to varved sediments of Greifensee and evaluation with measured data. *Palaeogegraphy, Palaeoclimatology, Palaeoecology*, **233**, 96–112.

Hausmann, S. & Lotter, A. F. (2001). Morphological variation within *Cyclotella comensis* Grunow and its importance for quantitative temperature reconstructions. *Freshwater Biology*, **46**, 1323–33.

Hausmann, S., Lotter, A. F., van Leeuwen, J. F. N., *et al.* (2002). Interactions of climate and land use documented in the varved sediments of Seebergsee in the Swiss Alps. *The Holocene*, **12**, 279–89.

Haworth, E. Y. (1976). Two late-glacial (Late Devensian) diatom assemblage profiles from northern Scotland. *New Phytologist*, **77**, 227–256.

Heegaard, E., Lotter, A. F., & Birks, H. J. B. (2006). Aquatic biota and the detection of climate change: are there consistent aquatic ecotones? *Journal of Paleolimnology*, **35**, 507–18.

Huber, K., Weckström, K., Drescher-Schneider, R., *et al.* (2010). Climate changes during the last glacial termination inferred from diatom-based temperatures and pollen in a sediment core from Längsee (Austria). *Journal of Paleolimnology*, **43**, 131–47.

Hustedt, F. (1922). Bacillariales aus Innerasien, gesammelt von Dr. Sven Hedin. *Botany*, **6**, 107–52.

Hustedt, F. (1939). Systematische und ökologische Untersuchungen über die Diatomeen-Flora von Java, Bali und Sumatra nach dem Material der Deutschen limnologischen Sunda-Expedition. *Archiv für Hydrobiologie Beiheft*, **II**, 1–155, 274–394.

Hustedt, F. (1942). Diatomeen aus der Umgebung von Abisko in Schwedisch-Lappland. *Archiv für Hydrobiologie*, **39**, 82–174.

Hustedt, F. (1943). Die Diatomeenflora einiger Hochgebirgsseen der Landschaft Davos in den Schweizer Alpen. *Int. Rev. Gesellschaft Hydrobiologie*, **43**, 124–97.

Hustedt, F. (1956). *Kieselalgen (Diatomeen)*. Stuttgart: Kosmos.

Ioriya, T. (1995). Achnanthales and Cymbellales (Bacillariophyceae) from Kathmandu Valley. In *Cryptogams of the Himalayas*, ed. M. Watanabe & H. Hagiwara, Tsukuba, Japan: National Science Museum, pp. 19–28.

IPCC (2007). *Climate Change 2007: Impacts, Adaptation and Vulnerability. Contribution of Working Group II to the Fourth Assessment Report of the Intergovernmental Panel on Climate Change*, ed. M. L. Parry, O. F. Canziani, J. P. Palutikof, P. J. van der Linden, & C.E. Hanson, Cambridge: Cambridge University Press.

Joynt, E. H. & Wolfe, A. P. (2001). Paleoenvironmental inference models from sediment diatom assemblages in Baffin Island lakes (Nunavut, Canada) and reconstruction of summer water temperature. *Canadian Journal of Fisheries and Aquatic Sciences*, **58**, 1222–43.

Juggins, S. (2007). C2 Version 1.5 User guide. Software for ecological and palaeoecological data analysis and visualisation. Newcastle University, Newcastle upon Tyne.

Jüttner, I., Cox, E. J., & Ormerod, S. J. (2000). New or poorly known diatoms from Himalayan streams. *Diatom Research*, **15**, 237–62.

Jüttner, I., Reichardt, E., & Cox, E. J. (2004). Taxonomy and ecology of some new *Gomphonema* species common in Himalayan streams. *Diatom Research*, **19**, 235–64.

Jüttner, I., Rothfritz, H., & Ormerod, S. J. (1996). Diatoms as indicators of river quality in the Nepalease Middle Hills with consideration of the effects of habitat-specific sampling. *Freshwater Biology*, **36**, 475–86.

Jüttner, I., Sharma, S., Manidahal, B., *et al.* (2003). Diatoms as indicators of stream quality in the Kathmandu Valley and Middle Hills of Nepal and India. *Freshwater Biology*, **48**, 2065–84.

Kamenik, C., Koinig, K. A., & Schmidt, R. (2005). Potential effects of pre-industrial lead pollution on algal assemblages from an alpine lake. *Verhandlungen Internationale Vereinigung für Limnologie*, **29**, 535–8.

Kamenik, C. & Schmidt, R. (2005). Chrysophyte resting stages: a tool for reconstructing winter/spring climate from Alpine lake sediments. *Boreas*, **34**, 477–89.

Karst-Riddoch, T. L., Pisaric, M. F. J., & Smol, J. P. (2005). Diatom responses to 20th century-related environmental changes in high-elevation mountain lakes of the northern Canadian Cordillera. *Journal of Paleolimnology*, **33**, 265–82.

Kauppila, T., Moisio, T., & Salonen, V.-P. (2002). A diatom-based inference model for autumn epilimnetic total phosphorus concentration and its application to a presently eutrophic boreal lake. *Journal of Paleolimnology*, **27**, 261–273.

Kilham, S. S., Theriot, E. C., & Fritz, S. C. (1996). Linking planktonic diatoms and climate in the large lakes of the Yellowstone ecosystem using resource theory. *Limnology and Oceanography*, **41**, 1052–62.

Kingston, J. C., Lowe, R. L., Stoermer, E. F., & Ladewski, T. B. (1983). Spatial and temporal distribution of benthic diatoms in northern Lake Michigan. *Ecology*, **64**, 1566–80.

Koinig, K. A., Kamenik, C., Schmidt, R., *et al.* (2002). Environmental changes in an alpine lake (Gossenköllesee, Austria) over the last two centuries – the influence of air temperature on biological parameters. *Journal of Paleolimnology*, **28**, 147–160.

Koinig, K. A., Schmidt, R., Sommaruga-Wögrath, S., Tessadri, R., & Psenner, R. (1997). Climate change as the primary cause for pH shifts in a high alpine lake. *Water, Air, and Soil Pollution*, **104**, 167–80.

Koinig, K. A., Sommaruga-Wögrath, S., Schmidt, R., Tessadri, R., & Psenner, R. (1998). Acidification processes in high alpine lakes. In *Headwaters: Water Resources and Soil Conservation*, ed. M. J. Haigh, J. Krecek, G. S. Raijwar, & M. P. Kilmartin, Rotterdam: A.A. Balkema Publishers, pp. 45–54.

Koivo, L. K. & Ritchie, J. C. (1978). Modern diatom assemblages from lake sediments in the boreal–arctic transition area. *Canadian Journal of Botany*, **56**, 1010–20.

Kolbe, R. W. (1927). *Zur Ökologie, Morphologie und Systematik der Brackwasser-Diatomeen. Die Kieselalgen des Sperenberger Salzgebiets.*, Pflanzenforschung, vol. 7, Jena: G. Fischer.

Korhola, A. & Weckström, J. (2004). Paleolimnological studies in Arctic Fennoscandia and the Kola Peninsula (Russia). In *Long-term environmental change in arctic and antarctic lakes*, ed. R. Pienitz, M. S. V. Douglas, & J. P. Smol, Developments in Paleoenvironmental Research, Dordrecht: Springer, pp. 381–418.

Korhola, A., Weckström, J., Holmström, L., & Erästö, P. (2000). A quantitative Holocene climatic record from diatoms in northern Fennoscandia. *Quaternary Research*, **54**, 284–94.

Krasske, G. (1932). Beiträge zur Kenntnis der Diatomeenflora der Alpen. *Hedwigia*, **72**, 92–134.

Laing, T. E., Pienitz, R., & Payette, S. (2002). Evaluation of limnological responses to recent environmental change and caribou activity in the Rivière Geroge region, northern Québec, Canada. *Arctic, Antarctic, and Alpine Research*, **54**, 454–64.

Laing, T. E., Rühland, K. M., & Smol, J. P. (1999). Past environmental and climatic changes related to tree-line shifts inferred from fossil diatoms from a lake near the Lena River Delta, Siberia. *The Holocene*, **9**, 547–557.

Laing, T. E. & Smol, J. P. (2000). Factors influencing diatom distributions in circumpolar treeline lakes of northern Russia. *Journal of Phycology*, **36**, 1035–48.

Lange-Bertalot, H. & Metzeltin, D. (1996). Indicators of oligotrophy. *Iconographia Diatomologica*, **2**, 1–390.

Laurion, I., Vincent, W. F., & Lean, D. R. S. (1997). Underwater ultraviolet radiation: development of spectral models for northern high latitude lakes. *Photochemistry and Photobiology*, **65**, 107–14.

Lee, E. J., Kenkel, N., & Booth, T. (1996). Atmospheric deposition of macronutrients by pollen in the boreal forest. *Ecoscience*, **3**, 304–9.

Li, S. F., Wang, F. B., & Zhang, J. (1999). Diatom-based reconstruction of Holocene environmental changes in Angren Lake, southern Tibet. *Chinese Science Bulletin*, **44**, 1123–6.

Li, Y., Gong, Z., Xie, P., & Shen, J. (2006). Distribution and morphology of two endemic gomphonemoid species, *Gomphonema kaznakowi* Mereschkowsky and *G. yangtzensis* Li nov. sp. in China. *Diatom Research*, **21**, 313–24.

Li, Y., Gong, Z., Xie, P., & Shen, J. (2007). Diatoms of eight lakes from Yunnan Province, China. *Journal of Freshwater Ecology*, **22**, 169–71.

Li, Y., Xie, P., Gong, Z., & Shi, Z. (2003). Gomphonemaceae and Cymbellaceae (Bacillariophyta) from Hengduan Mountains region (southwest China). *Nova Hedwigia*, **76**, 307–36.

Li, Y., Xie, P., Gong, Z., & Shi, Z. (2004). A Survey of the Gomphonemaceae and Cymbellaceae (Bacillariophyta) from the Jolmolungma Mountain (Everest) region of China. *Journal of Freshwater Ecology*, **19**, 189–194.

Livingstone, D. M. (1997). Break-up dates of Alpine lakes as proxy data for local and regional mean surface air temperatures. *Climatic Change*, **37**, 407–39.

Livingstone, D. M. & Lotter, A. F. (1998). The relationship between air and water temperatures in lakes of the Swiss Plateau: a case study with palaeolimnological implications. *Journal of Paleolimnology*, **19**, 181–98.

Livingstone, D. M., Lotter, A. F., & Kettle, H. (2005). Altitude-dependent differences in the primary physical response of mountain lakes to climatic forcing. *Limnology and Oceanography*, **50**, 1313–25.

Livingstone, D. M., Lotter, A. F., & Walker, I. R. (1999). The decrease in summer surface water temperature with altitude in Swiss Alpine lakes: a comparison with air temperature lapse rates. *Arctic, Antarctic, and Alpine Research*, **31**, 341–52.

Lotter, A. F. (2003). Multi-proxy climatic reconstructions. In *Global Change in the Holocene*, ed. A. W. Mackay, R. W. Battarbee, H. J. B. Birks, & F. Oldfield, London: E. Arnold, pp. 373–83.

Lotter, A. F. (2005). Palaeolimnological investigations in the Alps: the long-term develpment of mountain lakes. In *Global Change and Mountain Regions. An Overview of Current Knowledge*, ed. U. M. Huber, H. K. M. Bugmann, & M. A. Reasoner, Dordrecht: Springer, pp. 105–112.

Lotter, A. F., Appleby, P., Bindler, R., *et al.* (2002). The sediment record of the past 200 years in a Swiss high-alpine lake: Hagelseewli (2339 m a.s.l.). *Journal of Paleolimnology*, **28**, 111–27.

Lotter, A. F., & Bigler, C. (2000). Do diatoms in the Swiss Alps reflect the length of ice-cover? *Aquatic Sciences*, **62**, 125–41.

Lotter, A. F., Birks, H. J. B., Hofmann, W., & Marchetto, A. (1997). Modern diatom, cladocera, chironomid, and chrysophyte cyst assemblages as quantitative indicators for the reconstruction of past environmental conditions in the Alps. I. Climate. *Journal of Paleolimnology*, **18**, 395–420.

Lotter, A. F., Birks, H. J. B., Hofmann, W., & Marchetto, A. (1998). Modern diatom, cladocera, chironomid, and chrysophyte cyst assemblages as quantitative indicators for the reconstruction of past environmental conditions in the Alps. II. Nutrients. *Journal of Paleolimnology*, **19**, 443–63.

Lotter, A. F., & Psenner, R. (2004). Global change impacts on mountain waters: lessons from the past to help define monitoring targets for the future. In *Global Environmental and Social Monitoring*, ed. C. Lee & M. Schaaf, Paris: UNESCO, pp. 102–14.

MacDonald, G. M., Felzer, B., Finney, B. P., & Forman, S. L. (2000). Holocene lake sediment records of Arctic hydrology. *Journal of Paleolimnology*, **24**, 1–14.

MacDonald, G. M., Edwards, T. W. D., Moser, K. A., Pienitz, R., & Smol, J. P. (1993). Rapid response of treeline vegetation and lakes to past climate warming. *Nature*, **361**, 243–6.

Marchetto, A., Mosello, R., Psenner, R., *et al.* (1995). Factors affecting water chemistry of alpine lakes. *Aquatic Sciences*, **57**, 81–9.

Marchetto, A., & Schmidt, R. (1993). A regional calibration data set to infer lakewater pH from sediment diatom assemblages in alpine lakes. *Memorie dell'Istituto Italiano di Idrobiologia*, **51**, 115–25.

Mereschkowsky, C. (1906). Diatomées du Tibet. *C.R. Societé Impériale de Russe de Geographie*, **8**, 1–383.

Montagnes, D. J. S., & Franklin, D. J. (2001). Effect of temperature on diatom volume, growth rate, and carbon and nitrogen content: reconsidering some paradigms. *Limnology and Oceanography*, **46**, 2008–18.

Moser, K. A., MacDonald, G. M., & Smol, J. P. (1996). Applications of freshwater diatoms to geographical research. *Progress in Physical Geography*, **20**, 21–52.

Müller, B., Lotter, A. F., Sturm, M., & Ammann, A. (1998). The influence of catchment quality and altitude on the water and sediment composition of 68 small lakes in central Europe. *Aquatic Sciences*, **60**, 316–37.

Niederhauser, P. & Schanz, F. (1993). Effects of nutrient (N, P, C) enrichment upon the littoral diatom community of an oligotrophic high-mountain lake. *Hydrobiologia*, **269/270**, 453–62.

Ohlendorf, C., Bigler, C., Goudsmit, G. H., *et al.* (2000). Causes and effects of long periods of ice cover on a remote high Alpine lake. *Journal of Limnology*, **59**, 65–80.

Overpeck, J., Hughen, K., Hardy, D., *et al.* (1997). Arctic environmental change of the last four centuries. *Science*, **278**, 1251–6.

Ozenda, P. (1985). *La végétation de la chaîne alpine dans l'espace montagnard européen*. Paris: Masson.

Pannard, A., Bormans, M., & Lagadeuc, Y. (2008). Phytoplankton species turnover controlled by physical forcing at different time scales. *Canadian Journal of Fisheries and Aquatic Sciences*, **65**, 47–60.

Patrick, R. (1971). The effects of increasing light and temperature on the structure of diatom communities. *Limnology and Oceanography*, **16**, 405–21.

Patrick, R. (1977). Ecology of freshwater diatoms and diatom communities. In *The Biology of Diatoms*, ed. D. Werner, Botanical Monographs, Oxford: Blackwell, pp. 284–332.

Pielke, R. A. & Vidale, P. L. (1995). The boreal forest and the polar front. *Journal of Geophysical Research*, **100**, 25755–8.

Pienitz, R., Lortie, G., & Allard, M. (1991). Isolation of lacustrine basins and marine regression in the Kuujjuaq area (northern Québec), as inferred from diatom analysis. *Géographie physique et Quaternaire*, **45**, 155–74.

Pienitz, R. & Smol, J. P. (1993). Diatom assemblages and their relationship to environmental variables in lakes from the boreal-tundra ecotone near Yellowknife, Northwest Territories, Canada. *Hydrobiologia*, **269/270**, 391–404.

Pienitz, R., Smol, J. P., & Birks, H. J. B. (1995). Assessment of freshwater diatoms as quantitative indicators of past climatic change in the Yukon and Northwest Territories, Canada. *Journal of Paleolimnology*, **13**, 21–49.

Pienitz, R., Smol, J. P., & Lean, D. R. S. (1997a). Physical and chemical limnology of 24 lakes located between the Yellowknife and Contwoyto Lake, Northwest Territories (Canada). *Canadian Journal of Fisheries and Aquatic Sciences*, **54**, 347–58.

Pienitz, R., Smol, J. P., & Lean, D. R. S. (1997b). Physical and chemical limnology of 59 lakes located between the southern Yukon and the Tuktoyaktuk Peninsula, Northwest Territories (Canada). *Canadian Journal of Fisheries and Aquatic Sciences*, **54**, 330–46.

Pienitz, R., Smol, J. P., & MacDonald, G. M. (1999). Paleolimnological reconstruction of Holocene climatic trends from two boreal treeline lakes, Northwest Territories, Canada. *Arctic, Antarctic, and Alpine Research*, **31**, 82–93.

Pienitz, R. & Vincent, W. F. (2000). Effect of climate change relative to ozone depletion on UV exposure in subarctic lakes. *Nature*, **404**, 484–7.

Psenner, R. & Schmidt, R. (1992). Climate-driven pH control of remote alpine lakes and effects of acid deposition. *Nature*, **356**, 781–3.

Raubitschek, S., Lücke, A., & Schleser, G. H. (1999). Sedimentation patterns of diatoms in Lake Holzmaar, Germany – (on the transfer of climate signals to biogenic silica oxygen isotope proxies). *Journal of Paleolimnology*, **21**, 437–48.

Rautio, M., Sorvari, S., & Korhola, A. (2000). Diatom and crustacean zooplankton communities, their seasonal variability and representation in the sediemnts of subarctic Lake Saanajärvi. *Journal of Limnology*, **59**, 81–96.

Raven, J. A. & Geider, R. J. (1988). Temperature and algal growth. *New Phytologist*, **110**, 441–61.

Rosén, P., Hall, R., Korsman, T., & Renberg, I. (2000). Diatom transfer-functions for quantifying past air temperature, pH and total organic carbon concentration from lakes in northern Sweden. *Journal of Paleolimnology*, **24**, 109–23.

Rosén, P., Segerström, U., Eriksson, L., & Renberg, I. (2004). Do diatom, chironomid, and pollen records consistently infer Holocene July air temperature? A comparison using sediment cores from four alpine lakes in northern Sweden. *Arctic, Antarctic and Alpine Research*, **35**, 279–90.

Rothfritz, H., Jüttner, I., Suren, A. M., & Ormerod, S. J. (1997). Epiphytic and epilithic diatom communities along environmental gradients in the Nepalese Himalaya: implications for the assessment of biodiversity and water quality. *Archiv für Hydrobiologie*, **138**, 465–82.

Rühland, K. M., Paterson, A. M., & Smol, J. P. (2008). Hemispheric-scale patterns of climate-related shifts in planktonic diatoms from North American and European lakes. *Global Change Biology*, **14**, 2740–54.

Rühland, K., Phadtare, N. R., Pant, R. K., Sangode, S. J., & Smol, J. P. (2006). Accelerated melting of Himalayan snow and ice triggers pronounced changes in a valley peatland from northern India. *Geophysical Research Letters*, **33**, DOI: 10.1029/2006GL026704.

Rühland, K. & Smol, J. P. (1998). Limnological characteristics of 70 lakes spanning the arctic treeline from Coronation Gulf to Great Slave Lake in the central Northwest Territories, Canada. *Internationale Revue der gesamten Hydrobiologie*, **83**, 183–203.

Rühland, K. M. & Smol, J. P. (2002). Freshwater diatoms from the Canadian arctic treeline and development of paleolimnological inference models. *Journal of Phycology*, **38**, 249–64.

Rühland, K. M. & Smol, J. P. (2005). Diatom shifts as evidence for recent subarctic warming in a remote tundra lake, NWT, Canada. *Palaeogeography, Palaeoclimatology, Palaeoecology*, **226**, 1–16.

Rühland, K. M., Smol, J. P., Wang, X., & Muir, D. C. G. (2003). Limnological characteristics of 56 lakes in the central Canadian Arctic treeline region. *Journal of Limnology*, **62**, 9–27.

Saulnier-Talbot, É., Pienitz, R., & Vincent, W. F. (2003). Holocene lake succession and palaeo-optics of a subarctic lake, northern Québec, Canada. *The Holocene*, **13**, 517–26.

Schindler, D. W., Bayley, S. E., Parker, B. R., *et al.* (1996). The effects of climatic warming on the properties of boreal lakes and streams at the Experimental Lakes Area, northwestern Ontario. *Limnology and Oceanography*, **41**, 1004–17.

Schmidt, R., Kamenik, C., Kaiblinger, C., & Hetzel, M. (2004a). Tracking Holocene environmental changes in an alpine lake sediment core: application of regional diatom calibration, geochemistry, and pollen. *Journal of Paleolimnology*, **32**, 177–96.

Schmidt, R., Kamenik, C., Lange-Bertalot, H., & Klee, R. (2004b). *Fragilaria* and *Staurosira* (Bacillariophyceae) from sediment surfaces of 40 lakes in the Austrian Alps in relation to environmental variables, and their potential for palaeoclimatology. *Journal of Limnology*, **63**, 171–89.

Schmidt, R., Kamenik, C., & Roth, M. (2007). Siliceous algae-based seasonal temperature inference and indicator pollen tracking ca. 4,000 years of climate/land use dependency in the southern Austrian Alps. *Journal of Paleolimnology*, **38**, 541–54.

Schmidt, R. & Psenner, R. (1992). Climate changes and anthropogenic impacts as causes for pH fluctuations in remote high alpine lakes. *Documenta Istituto Italiano di Idrobiologia*, **32**, 31–57.

Schmidt, R., Psenner, R., Müller, J., Indinger, P., & Kamenik, C. (2002). Impact of late glacial climate variations on stratification and trophic state of the meromictic lake Längsee (Austria): validation of a conceptual model by multi proxy studies. *Journal of Limnology*, **61**, 49–60.

Schmidt, R., Roth, M., Tessadri, R., & Weckström, J. (2008). Disentangling late-Holocene climate and land use impacts on an Austrian alpine lake using seasonal temperature anomalies, ice-cover, sedimentology, and pollen tracers. *Journal of Paleolimnology*, **40**, 453–69.

Schmidt, R., Wunsam, S., Brosch, U., *et al.* (1998). Late and post-glacial history of meromictic Längsee (Austria), in respect to climate change and anthropogenic impact. *Aquatic Sciences*, **60**, 56–88.

Seppä, H. & Weckström, J. (1999). Holocene vegetational and limnological changes in the Fennoscandian tree-line area as documented by pollen and diatom records from Lake Tsuolbmajävri, Finland. *Ecoscience*, **6**, 621–35.

Servant-Vildary, S. (1982). Altitudinal zonation of mountainous diatom flora in Bolivia: application to the study of the Quaternary. *Acta Geologica Academiae Scientiarum Hungaricae*, **25**, 179–210.

Smol, J. P. (1988). Paleoclimate proxy from freshwater arctic diatoms. *Verhandlungen Internationale Vereinigung für Limnologie*, **23**, 837–44.

Smol, J. P. (2002). *Pollution of Lakes and Rivers*. London: Arnold.

Smol, J. P. & Cumming, B. F. (2000). Tracking long-term changes in climate using algal indicators in lake sediments. *Journal of Phycology*, **36**, 986–1011.

Smol, J. P., Cumming, B. F., Douglas, M. S. V., & Pienitz, R. (1995). Inferring past climate changes in Canada using paleolimnological techniques. *Geoscience Canada*, **21**, 113–18.

Smol, J. P. & Douglas, M. (2007). From controversy to consensus: making the case for recent climate change in the Arctic using lake sediments. *The Ecological Society of America*, **5**, 466–74.

Smol, J. P., Walker, I. R., & Leavitt, P. R. (1991). Paleolimnology and hindcasting climatic trends. *Verhandlungen Internationale Vereinigung für Limnologie*, **24**, 1240–6.

Solovieva, N. & Jones, V. J. (2002). A multiproxy record of Holocene environmental changes in the central Kola Peninsula, northwest Russia. *Journal of Quaternary Science*, **17**, 303–18.

Sommaruga, R. & Psenner, R. (1997). Ultraviolet radiation in a high mountain lake of the Austrian Alps: air and underwater measurements. *Photochemistry and Photobiology*, **65**, 957–63.

Sommaruga-Wögrath, S., Koinig, K. A., Schmidt, R., *et al.* (1997). Temperature effects on the acidity of remote alpine lakes. *Nature*, **387**, 64–7.

Sorvari, S. & Korhola, A. (1998). Recent diatom assemblage changes in subarctic Lake Saanajärvi, NW Finnish Lapland, and their paleoenvironmental implicatinons. *Journal of Paleolimnology*, **20**, 205–15.

Sorvari, S., Korhola, A., & Thompson, R. (2002). Lake diatom response to recent arctic warming in Finnish Lapland. *Global Change Biology*, **8**, 153–63.

Stoermer, E. F. & Ladewski, T. B. (1976). Apparent optimal temperatures for the occurrence of some common phytoplankton species in southern Lake Michigan. *University of Michigan, Great Lakes Research Division Publication*, **18**.

Thompson, R., Kamenik, C., & Schmidt, R. (2005a). Ultra-sensitive alpine lakes and climate change. *Journal of Limnology*, **64**, 139–52.

Thompson, R., Price, D., Cameron, N., *et al.* (2005b). Quantitative calibration of remote mountain-lake sediments as climatic recorders of air temperture and ice-cover duration. *Arctic, Antarctic, and Alpine Research*, **37**, 626–35.

Tinner, W. & Ammann, B. (2005). Long-term responses of mountain ecosystems to environmental changes: resilience, adjustment, and vulnerability. In *Global Change and Mountain Regions. An Overview of Current Knowledge*, ed. U. M. Huber, H. K. M. Bugmann, & M. A. Reasoner, Dordrecht: Springer, pp. 133–43.

Tinner, W., Ammann, B., & Germann, P. (1996). Treeline fluctuations recorded for 12,500 years by soil profiles, pollen, and plant macrofossils in the *Arctic and Alpine Research*, **28**, 131–47.

Tranquillini, W. (1979). *Physiological Ecology of the Alpine Timberline*. Berlin: Springer.

Tynni, R. (1976). Über Finnlands rezente und subfossile Diatomeen. *Bulletin Geological Survey of Finland*, **284**, 1–37.

van Campo, E. & Gasse, F. (1993). Pollen- and diatom-inferred climatic and hydrological changes in Sumxi Co basin (western Tibet) since 13,000 yr B.P. *Quaternary Research*, **39**, 300–13.

Ventura, M., Camarero, L., Buchaca, T., *et al.* (2000). The main features of seasonal variability in the external forcing and dynamics of a deep mountain lake (Redò, Pyrenees). *Journal of Limnology*, **59**, 97–108.

Vincent, W. F., Hobbie, J. E., & Laybourn-Parry, J. (2008). Introduction to the limnology of high latitude lake and river ecosystems. In *Polar Lakes and Rivers – Limnology of Arctic and Antarctic Aquatic Ecosystems*, ed. W. F. Vincent & J. Laybourn-Parry, Oxford: Oxford University Press, pp. 1–23.

Vincent, W. F., & Pienitz, R. (1996). Sensitivity of high-latitude freshwater ecosystems to global change: temperature and solar ultraviolet radiation. *Geoscience Canada*, 23, 231–6.

Vinebrooke, R. D. & Leavitt, P. R. (1996). Effects of ultraviolet radiation on periphyton in an alpine lake. *Limnology and Oceanography*, **41**, 1035–40.

von Gunten, L., Heiri, O., Bigler, C., *et al.* (2008). Seasonal temperatures for the past ~400 years reconstructed from diatom and chironomid assemblages in a high-altitude lake (Lej da la Tscheppa, Switzerland). *Journal of Paleolimnology*, **39**, 283–99.

Vyverman, W. (1992). Altitudinal distribution of non-cosmopolitan desmids and diatoms in Papua New Guinea. *British Phycological Journal*, **27**, 49–63.

Vyverman, W. & Sabbe, K. (1995). Diatom-temperature transfer functions based on the altitudinal zonation of diatom assemblages in Papua New Guinea: a possible tool in the reconstruction of regional palaeoclimatic changes. *Journal of Paleolimnology*, **13**, 65–77.

Walker, M. J. C., Björck, S., Lowe, J. J., *et al.* (1999). Isotopic "events" in the GRIP ice core: a stratotype for the Late Pleistocene. *Quaternary Science Reviews*, **18**, 1143–50.

Weckström, J., Korhola, A., & Blom, T. (1997). The relationship between diatoms and water temperature in thirty subarctic Fennoscandian lakes. *Arctic and Alpine Research*, **29**, 75–92.

Wetzel, R. G. (2001). *Limnology*, San Diego, CA: Academic Press.

Wolfe, B. B., Edwards, T. W. D., Aravena, R., & MacDonald, G. M. (1996). Rapid Holocene hydrologic change along boreal treeline revealed by $\delta^{13}C$ and $\delta^{18}O$ in organic lake sediments, Northwest Territories, Canada. *Journal of Paleolimnology*, **15**, 171–81.

Wrona, F. J., Prowse, T. D., Reist, J. D., *et al.* (2006). Effects of ultraviolet radiation and contaminant-related stressors on arctic freshwater ecosystems. *AMBIO*, **35**, 388–401.

Wunsam, S. & Schmidt, R. (1995). A diatom–phosphorus transfer function for alpine and pre-alpine lakes. *Memorie dell'Istituto Italiano di Idrobiologia*, **53**, 85–99.

Wunsam, S., Schmidt, R., & Klee, R. (1995). *Cyclotella*-taxa (Bacillariophyceae) in lakes of the Alpine region and their relationship to environmental variables. *Aquatic Sciences*, **57**, 360–86.

Yang, X., Kamenik, C., Schmidt, R., & Wang, S. (2003). Diatom-based conductivity and water-level inference models from eastern Tibetan (Qinghai-Xizang) plateau lakes. *Journal of Paleolimnology*, **30**, 1–19.

Yang, X., Wang, S., Kamenik, C., *et al.* (2004). Diatom assemblages and quantitative reconstruction for paleosalinity from a sediment core of Chencuo Lake, southern Tibet. *Science in China*, **47**, 522–8.

13

Freshwater diatoms as indicators of environmental change in the High Arctic

MARIANNE S. V. DOUGLAS AND
JOHN P. SMOL

13.1 Introduction

High Arctic environments continue to receive increased attention from the scientific community, policy makers, and the public at large because polar regions are considered to be especially sensitive to the effects of global climatic and other environmental changes (Rouse *et al.*, 1997; ACIA, 2004; IPCC, 2007). Polar lakes and ponds, and the biota they contain, are important sentinels of environmental changes (Pienitz *et al.*, 2004; Schindler & Smol, 2006) and have thus been the focus of many research programs (Vincent & Laybourn-Parry, 2008).

There is considerable potential for using living and fossil diatom assemblages to track environmental trends in High Arctic regions (Smol & Douglas, 1996; Douglas *et al.*, 2004a). A growing number of studies have examined the taxonomy, ecology, and paleoecology of High Arctic diatoms, as lakes and ponds are dominant features of most Arctic landscapes. Given the diversity and vastness of these regions, many exciting research opportunities exist. For example, about 18% (by area) of Canada's surface waters are situated north of 60 °N (Statistics Canada, 1987), and Sheath (1986) estimated that tundra ponds cover approximately 2% of the Earth's surface. The heightened interest in High Arctic environments, coupled with an increased accessibility of these remote regions (e.g. by helicopter), has resulted in a recent surge of interest in Arctic diatom research. Whilst some proxy techniques, such as palynology and dendroecology, have limited applicability in some High Arctic regions due to the paucity of higher plants (Gajewski *et al.*, 1995), paleolimnological approaches using diatoms have become especially important for studies of long-term global environmental change.

In this chapter, we summarize some of the ways in which diatoms have been used to track environmental changes in High Arctic freshwater ecosystems. We focus primarily on the Canadian High Arctic islands (i.e. islands north of the Canadian mainland), northern Greenland, and Spitsbergen (Figure 13.1). Lotter *et al.* (this volume) review diatom studies in lower Arctic regions. In addition, several chapters in Pienitz *et al.* (2004) reviewed a wide array of paleolimnological studies (many of which include diatom indicators) from various Arctic regions, such as from the Middle and High Canadian Arctic (Wolfe & Smith, 2004), the North American subarctic (Finney *et al.*, 2004), Greenland and the North Atlantic islands (Anderson *et al.*, 2004), northern Russia (MacDonald *et al.*, 2004), and Fennoscandia and the Kola Peninsula (Korhola & Weckström (2004). Hodgson & Smol (2008), summarize some of the paleolimnological applications used to study environmental change in polar regions. In this chapter, we first present a brief description of Arctic limnology and historical uses of diatoms in the High Arctic. Thereafter, we mainly discuss applications of diatoms to the study of past climatic change. Other applications, such as eutrophication, contaminants, and archeological studies, are mentioned; however, other chapters in this volume deal with these issues in more detail. Most of the examples cited in this chapter are from lakes and ponds, as little historical work has been completed on High Arctic diatoms from fluvial habitats (Hamilton & Edlund, 1994; Sheath *et al.*, 1996). However recent studies on diatom assemblages from flowing waters (e.g. Antoniades & Douglas, 2002; Antoniades *et al.* 2009a), and their potential uses for paleoenvironmental reconstructions (e.g. Stewart *et al.*, 2008), have suggested considerable research promise.

13.2 Limnological setting

The High Arctic is characterized by cold temperatures, extended snow and ice cover, and extremes in irradiance, with the sun below the horizon for several months during the polar night, followed by 24-hour periods of continuous light during the

The Diatoms: Applications for the Environmental and Earth Sciences, 2nd Edition, eds. John P. Smol and Eugene F. Stoermer. Published by Cambridge University Press.

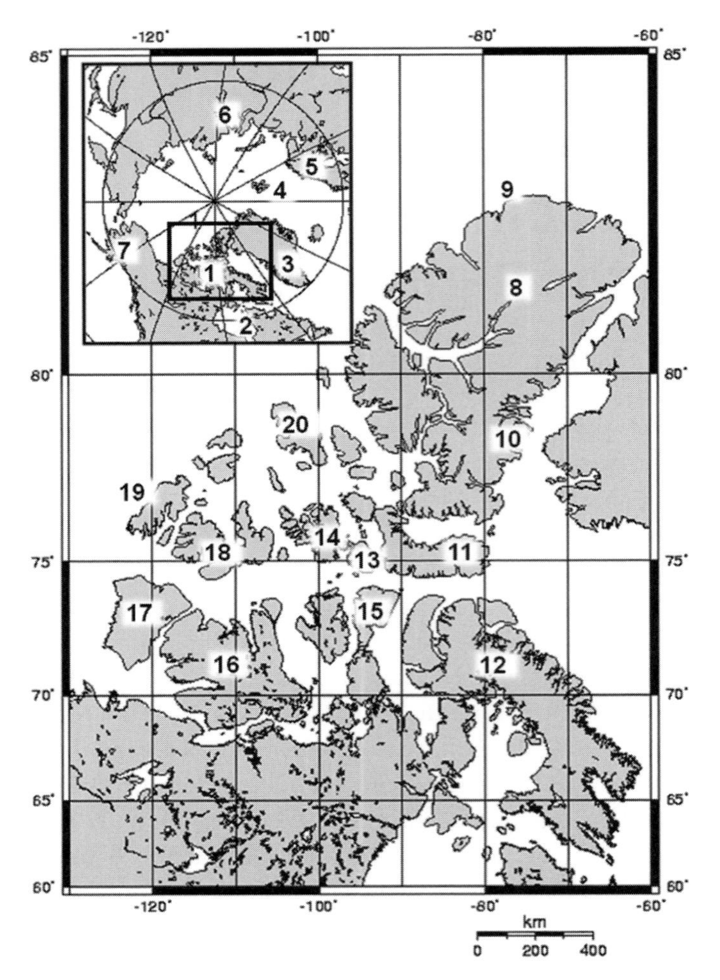

Figure 13.1 Map showing a circumpolar projection of some of the main High Arctic islands referred to in this chapter. Inset: (1) Canadian Arctic Islands (details shown on main map) (2) Belcher Islands, (3) Greenland, (4) Svalbard, (5) Fennoscandia, (6) Russia, and (7) Alaska. Main map: (8) Ellesmere Island, (9) Ward Hunt Island, (10) Cape Herschel, Ellesmere Island, (11) Devon Island, (12) Baffin Island, (13) Cornwallis Island, (14) Bathurst Island, (15) Somerset Island, (16) Victoria Island, (17) Banks Island, (18) Melville Island, (19) Prince Patrick Island, and (20) Ellef Ringnes Island.

summer months (Rouse *et al.*, 1997). Continuous permafrost results in poor drainage, such that summer snowmelt collects in a myriad of lakes and ponds throughout much of the High Arctic, outside of that covered by glaciers and perennial snow cover. Common lacustrine environments include shallow ponds inside ice-wedge polygons, thermokarst lakes, depressions in bedrock, glacial basins, and water bodies impounded by moraines (Vincent & Laybourn-Parry, 2008).

Arctic ponds and lakes are often distinguished based on their depth and ice thickness: ponds are water bodies that freeze completely to the bottom each winter, whereas lakes are sufficiently deep to maintain a layer of liquid water under their ice cover. As noted below, some deeper High Arctic lakes maintain a permanent ice pan, at least in some years, even during summer. Ponds, on the other hand, with their lower heat capacities, thaw completely and more rapidly. Moreover, because of their shallow depths (e.g. typically <2 m and often <1 m deep), their entire water column and substrates can be exploited for algal growth (Figure 13.2). Shallow ponds are very common in High Arctic regions, and have been shown to be especially sensitive bellwethers of climatic and other environmental changes (Douglas & Smol, 1994; Douglas *et al.*, 1994; Smol & Douglas 2007a, b).

While several reviews have summarized the historical development of Arctic limnology (e.g. Hobbie, 1984; Vincent & Laybourn-Parry, 2008), only a few High Arctic lakes (e.g. Char Lake; Rigler, 1978) and ponds (e.g. Douglas and Smol 1994; Smol & Douglas, 2007a, b) have been studied in considerable detail. Indeed, less limnological and phycological data exist for high latitudes such as the High Arctic and the Antarctic (see Spaulding *et al.*, this volume) than for most other ecoregions on Earth.

13.3 Historical review

Douglas *et al.* (2004a) reviewed the historical development of freshwater algal research in the High Arctic, especially as it relates to paleolimnological studies. The earliest records of High Arctic diatom study come from the nineteenth century, when ships exploring the region frequently returned with collections from a variety of habitats (Douglas *et al.*, 2004a). Many of these early collections were examined by C. G. Ehrenberg (1853) and P. T. Cleve (e.g. 1864, 1873, 1883, 1896), which included samples from the Russian Arctic (e.g. Cleve, 1898).

The earliest use of diatoms from the High Arctic to help resolve a scientific question was probably initiated by the Norwegian Arctic explorer, Fridtjof Nansen (1897). At that time, knowledge of polar regions was still very limited, and Nansen set out to determine if ice drifted over the North Pole. One piece of evidence he used to bolster the drift theory was that both marine and freshwater diatoms (identified by P. T. Cleve) collected from the Greenland pack ice were the same species that had been collected earlier from ice near the Bering Strait.

The 1913–18 Canadian Arctic Expedition produced some of the earliest surveys of diatoms from the western Canadian Arctic. Lowe (1923) examined material from small brackish ponds in both Alaska and the Northwest Territories. Because these collections were not specifically oriented towards diatoms, the list is brief and he does not provide illustrations of taxa or

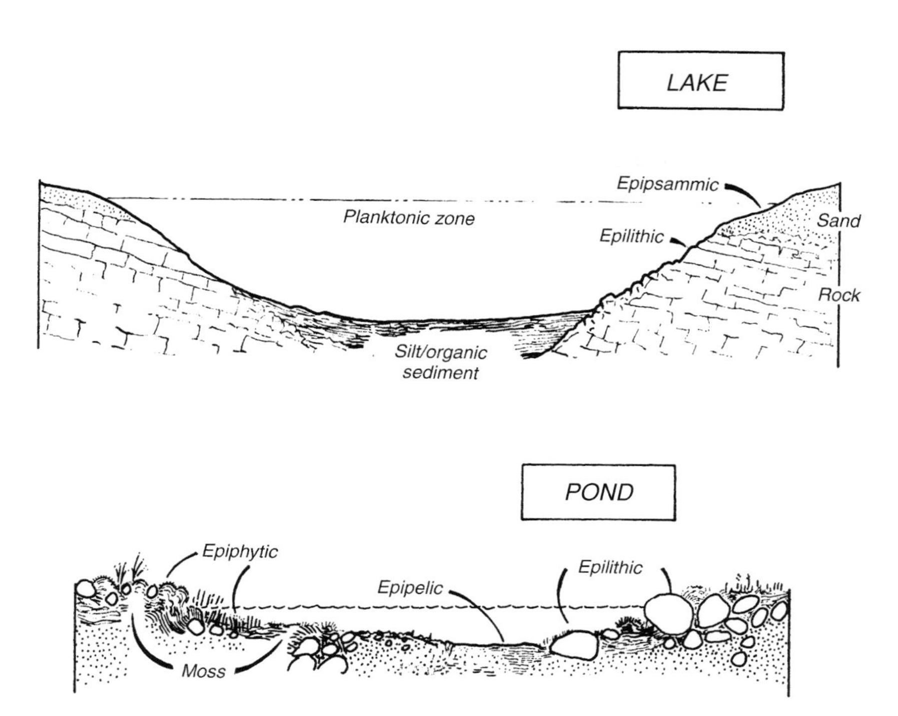

Figure 13.2 Comparison of lake and pond habitats. All periphytic habitats can occur in both pond and lake systems. Planktonic diatoms are generally only found in lake systems.

discuss ecology in any detail. A contemporary, more detailed study was completed by Petersen (1924), who recorded 52 diatom taxa (several with illustrations and taxonomic notes) from the north coast of Greenland. Other early diatom work from Greenland includes the studies by Østrup (1897a,b; 1910).

A second Canadian Arctic Expedition was undertaken in 1938–1939, with a focus on the Eastern Arctic, including several High Arctic sites. Work on the material (collected by Nicholas Polunin) was delayed by World War II, but eventually Ross (1947) published descriptions of 192 freshwater diatom taxa that Polunin collected, along with their geographic distributions. His report included 11 plates of illustrations.

One of the most important pioneer researchers working on Arctic diatoms was the Danish phycologist Niels Foged. He published extensively on diatoms from many regions (see full list in Håkansson, 1988), but much of his research focused on polar regions, such as Greenland (1953, 1955, 1958, 1972, 1973, 1977), Iceland (1974), Spitsbergen (1964), and Alaska (1981).

Foged noted that earlier investigators (reviewed in his publications) provided short floral lists, but little information about ecology. In his studies from Greenland, Foged usually considered the pH and halobion spectra of diatom communities (e.g. Foged, 1953). Where possible, he provided a more detailed analysis of water chemistry, as it related to differences among petrographical regions (Foged, 1958). He also began using paleoecological approaches, and described changes in diatom assemblages from several post-glacial deposits in Greenland (Foged, 1972, 1977, 1989). The main stratigraphic horizon of interest was the marine–lacustrine transition (see Horton & Sawai, this volume), from which he deduced emergence patterns of lakes from the sea. Foged (1982) also attempted to use diatoms as part of a forensic investigation (see Peabody & Cameron, this volume), in order to determine if the eight so-called "Greenland mummies" (who died around the year 1460, and whose mummified bodies were discovered in 1972) perished as a result of drowning, through examination of the diatoms contained in their lung tissues.

Following some early work by researchers such as Cleve (1864) and Lagerstedt (1873), Foged (1964) produced a detailed volume on Spitsbergen (Svalbard) diatoms. He noted that many species had variable morphologies and that their forms differed from those in other geographic areas. However, he considered the differences so slight that it would have been "rash" to describe each one as a separate taxon. Foged listed temperature and pH as the major factors controlling populations.

Following Foged's pioneering work, a few phycological surveys were conducted on the more accessible parts of the Arctic islands, such as southern Baffin Island (e.g. Moore, 1974a,b,c; Hickman, 1974), Svalbard (e.g. Skulberg, 1996), Bear Island (Metzeltin & Witkowski, 1996), and southeastern Greenland (e.g. Denys & Beyens, 1987). Until recently, most High Arctic diatom floras were poorly documented because of their geographic isolation. The next section reviews the recent advances in describing the limnological setting, and diatom taxonomic and autecological studies in the High Arctic.

13.4 Limnological surveys

Arctic ponds and lakes are typically oligotrophic. Nonetheless, regional differences exist, and it is important to capture the combined effects that local geology and climate have on aquatic conditions. Several recent surveys across the Canadian Arctic and elsewhere have provided descriptions of high-latitude limnology (e.g. Alert, Ellesmere Island and Mould Bay, Prince Patrick Island (Antoniades *et al.*, 2003a); Axel Heiberg Island (Michelutti *et al.*, 2002a); Banks Island (Lim *et al.*, 2005); Bathurst Island (Lim *et al.*, 2001a); Haughton Crater, Devon Island (Lim & Douglas, 2003); Isachsen, Ellef Ringnes Island (Antoniades *et al.*, 2003b); Melville Island (Keatley *et al.*, 2007a); Northern Ellesmere Island and High Arctic oases (Keatley *et al.*, 2007b); Victoria Island (Michelutti *et al.*, 2002b); Cornwallis Island (Michelutti *et al.*, 2007a); and regional comparisons (Hamilton *et al.*, 2001)). Similar studies have been conducted from Svalbard (e.g. Birks *et al.*, 2004). These baseline data are critical for our understanding of the ecology and biogeographic distribution of freshwater diatoms in these polar regions.

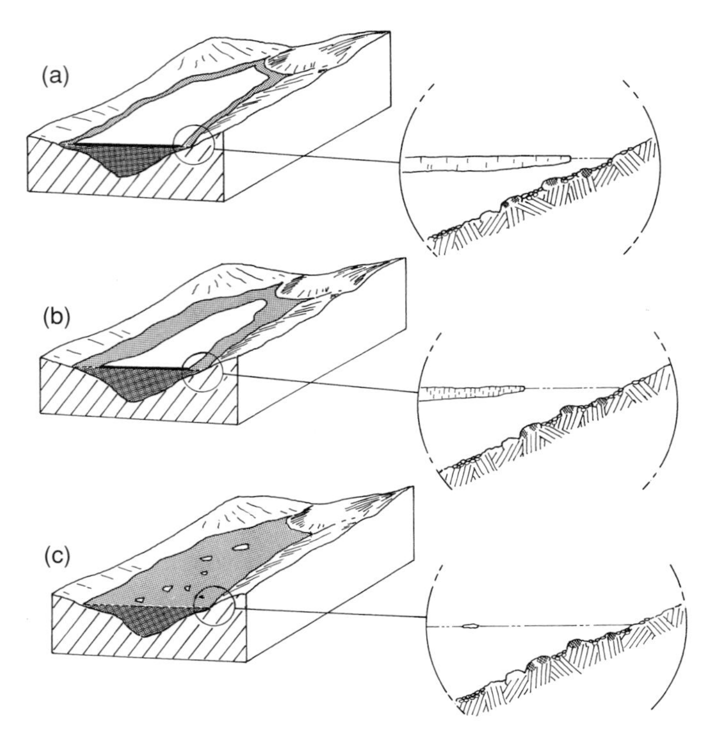

Figure 13.3 Diagrammatic representation of ice and snow conditions on a High Arctic lake during (a) cold, (b) moderate, and (c) warm summers. As temperatures increase, deeper areas of the lake are available for diatom growth. (From Smol, 1988, used with permission.)

13.5 Diatom taxonomy, autecology, and regional calibrations

The regional limnological studies described above provided the foundation to explore the autecological characteristics of freshwater diatoms from High Arctic lakes and ponds. An extensive taxonomic and ecological treatment of the freshwater diatoms from the Canadian High Arctic has recently been compiled by Antoniades *et al.* (2008) and new species have been described (Antoniades *et al.*, 2009b), as have various diatom species complexes (Paull *et al.*, 2008). High Arctic diatom assemblages are often overwhelmingly dominated by benthic taxa, reflecting the near absence of planktonic diatoms in shallow-water bodies (i.e. ponds) and in lakes with extensive ice covers (Smol, 1988). However, planktonic diatoms have been described from northeast Greenland (Cremer & Wagner, 2004; Cremer, 2006; Cremer *et al.*, 2005), and some other high-latitude regions (Smol *et al.*, 2005).

Regional surface sediment calibration sets have now been developed in several circum-Arctic regions, including Alert, Ellesmere Island and Mould Bay, Prince Patrick Island (Antoniades, 2005a), Axel Heiberg Island (Michelutti *et al.*, 2006a), Baffin Island (Joynt & Wolfe, 2001), Banks Island (Lim *et al.*, 2007), Bathurst Island (Lim *et al.*, 2001b), Cornwallis Island (Michelutti *et al.*, 2007a), Devon Island (Ng & King, 1999; Lim, 2004), Ellef Ringnes Island (Antoniades *et al.*, 2004, 2005b), Melville Island (Keatley *et al.*, 2008a), and Svalbard (Jones & Birks, 2004). In addition to linking species' distributions to physical and chemical variables, certain studies have also explored the microhabitat specificity of certain taxa (e.g. Hamilton *et al.*, 1994a; Douglas & Smol, 1995a; Lim *et al.*, 2001c; Michelutti *et al.*, 2003b; Van de Vijver *et al.*, 2003), which can provide important paleoecological information on, for example, past ice cover (see below). A few preliminary attempts at categorizing the biogeographic distribution of taxa have also been initiated (e.g. Michelutti *et al.*, 2003b; Bouchard *et al.*, 2004).

13.6 Applications and case studies

13.6.1 Climate, ice cover, and related environmental variables

The High Arctic environment imposes some important ecological constraints on diatom populations. In deep lakes, a dominant feature is extended ice and snow cover, which dramatically influences lake systems by determining available habitats (Figure 13.3) and other limnological variables (Table 13.1).

Table 13.1 Overall trends that are often recorded in High Arctic ponds and lakes with colder or warmer temperatures.

Variable	Colder	Warmer
Ice cover	↑	↓
Growing season	↓	↑
Plankton	↓	↑
Mosses	↓	↑
Diversity	↓	↑
Nutrients	↓	↑
Production	↓	↑
pH	↓	↑
Conductivity	↓	↑

Note: Most of these changes can be related to the extent of ice and snow cover and other climate-related changes. These are general trends, and exceptions do occur.

Ice cover on most Arctic lakes begins to form in early to mid September – the exact timing of which is mainly dependent on temperature, wind, and cloud cover (Hobbie, 1984). Snow cover, along with air temperature, determines the rate and duration of freezing. Ice thicknesses exceeding 5 m have been observed (Blake, 1989), although thinner ice covers are more common (Hobbie, 1984). Some High Arctic sites maintain their snow and ice cover throughout the brief summer, and only a shallow "moat" of open water becomes ice free. The depth and area of snow cover, as well as the ice type (i.e. black vs. white), affects the transmission of light, and hence the amount of light available for photosynthesis by planktonic and periphytic diatom communities (e.g. Belzile *et al.*, 2001).

Recognition of the overriding influence of ice and snow cover on High Arctic lakes led Smol (1983, 1988) to suggest that, because the extent of ice cover can influence which habitats are available for algal growth, past changes in diatom assemblages can be used to track past lake ice-cover extents, and therefore infer past climatic changes (Figure 13.3). A simple model was proposed: during colder years, ice cover is more extensive, and only a narrow moat develops in the littoral zone (Figure 13.3a). Overall diatom production tends to be less, and taxa characteristic of very shallow littoral and semi-terrestrial environments tend to be relatively more common. During warmer years (Figure 13.3b, c), ice cover is less extensive, overall production is higher, taxonomic diversity and complexity often increases, and taxa characteristic of deeper water substrates and planktonic habitats may be relatively more abundant. Nutrient inputs may also be higher from the surrounding catchment (see discussion below).

The above model was first used to try to infer past climatic changes using diatom species composition from a lake core from Baird Inlet on eastern Ellesmere Island (Smol, 1983), later from a core in glacial Tasikutaaq Lake on Baffin Island (Lemmen *et al.*, 1988), and then from northwestern Greenland (Blake *et al.*, 1992). During warmer years, ice cover is less extensive, and total algal production is often higher (Figure 13.3). Smol (1983) recorded higher diatom concentrations during proposed warmer periods in his study on Ellesmere Island, as did Blake *et al.* (1992) on Greenland. Smith (2002) linked changes in ice cover to total diatom biomass from his paleolimnological study of lakes from the Hazen Plateau (Ellesmere Island). Smith (2000) also used the diatom sedimentary record to establish evidence for lake ice rafting along glacial margins and to distinguish diamicts from tills in High Arctic lake sediments. Williams (1990a) used diatom-concentration data to interpret past climatic changes (including ice-cover changes) on Baffin Island, and extended these ideas to interpreting past sea-ice conditions from diatoms in marine sediment cores (Andrews *et al.*, 1990; Williams, 1990b; Short *et al.*, 1994; Williams *et al.*, 1995). Interestingly, Doubleday *et al.* (1995), working near the Alert military base on northern Ellesmere Island, noted the near-total disappearance of diatoms in core sections, despite generally good preservation elsewhere. They speculated that, since this lake is deep (>30 m) and continued to support an extensive ice cover even in summer, a slight cooling may have completely frozen over the lake, sealing it off, and precluding any sizeable diatom populations from growing. A similar situation was noted by Guilizzoni *et al.* (2006) on western Spitsbergen and Antoniades *et al.* (2007) in Canada's northernmost lake, on Ward Hunt Island.

The extent and duration of ice and snow cover, which are closely linked to temperature, has other implications for lake and pond environments (Table 13.1). Because many of these changes are accentuated in shallow ponds, we originally concentrated our High Arctic research on long-term monitoring work of a series of 36 ponds on Cape Herschel (78° 37′ N, 74° 42′ W), from the east–central coast of Ellesmere Island (Douglas *et al.*, 2000). We have since expanded the spatial distribution of this work to other High Arctic islands and other regions.

Although diatom diversity is often relatively low within a single Cape Herschel pond, a surprisingly diverse flora exists on the 2 × 5 km cape (Douglas & Smol, 1993, 1995a). When we wrote this chapter for the first edition of this book in 1999, we

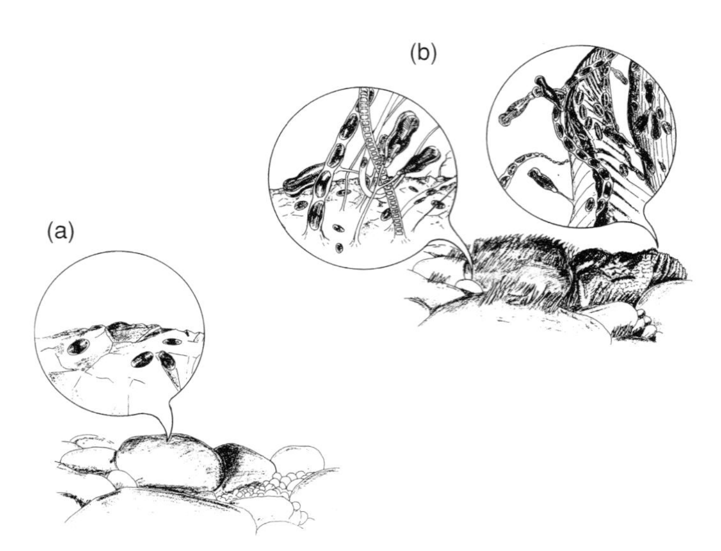

Figure 13.4 Diagrammatic representation of habitat availability and diatom growth during (a) cold and (b) warmer growing seasons. In (a) very cold environments, extended ice cover and a very short growing season result in lower production of predominantly adnate diatoms living on rocks and sediment substrates. With (b) warmer conditions, additional substrates become available (e.g. mosses) and more complex diatoms communities can develop.

noted that the Cape Herschel ponds typically began to thaw in late June or early July, and then began to re-freeze by late August (Douglas & Smol, 1994). However, over the last decade, with warmer temperatures, the period of open-water conditions has been increasing (Smol & Douglas, 2007a,b). Nonetheless, a slight cooling or warming can dramatically affect the duration of ice cover. With warmer summer temperatures, and hence a lengthened growing season, there is more opportunity for overall algal production to be higher, and for more complex and diverse aquatic communities to develop (Table 13.1) (e.g. Douglas *et al.*, 1994; Wolfe, 1994). For example, with longer ice-free seasons, new substrates, such as mosses, may become available, whilst during colder periods, only rock and sediment substrates may be present (Figure 13.4). High Arctic diatoms show some specificity to these different substrates (Hamilton *et al.*, 1994a; Douglas & Smol, 1995a; Wolfe, 1996a), and this information can be used to track past environmental shifts in paleolimnological studies (Douglas *et al.*, 1994). For instance, with extended growing seasons, secondary and tertiary growths of diatoms, including many stalked and tube-dwelling taxa, can grow attached to the original communities of adnate diatoms (Figure 13.4). The resultant overall diatom diversity is higher (Table 13.1).

With warmer temperatures, several other changes also typically occur in water chemistry (Table 13.1). For example, nutrient inflows tend to be higher with increased snowmelt, a deeper active layer develops with permafrost melting, and runoff from the catchment increases. This often results in a corresponding increase in aquatic production. Because water in shallow Arctic ponds closely tracks ambient air temperatures (Douglas & Smol, 1994), higher water temperatures also tend to enhance overall production. Pond-water pH also tends to rise, partially in response to the increased photosynthesis, but also possibly through the effects on water renewal and the relative yields of base cations and acid anions, as well as possible sulfate removal (see below for more discussion of pH and climate relationships, as well as Wolfe & Härtling, 1996, Wolfe, 1996b, and Lotter *et al.*, this volume). Specific conductivity also tends to increase with warming due to a variety of mechanisms, including evaporation (Smol & Douglas, 2007a). Melting permafrost further alters limnological characteristics (Kokelj *et al.*, 2005). As noted in many other lake regions, pH and specific conductivity exert strong influences on diatom assemblages in High Arctic ponds (Douglas & Smol, 1993).

Because many tundra ponds are so shallow (e.g. many are <0.5 m deep), increased periods of warmer climates have resulted in further lowering of water levels, and even total desiccation of some shallower sites (Smol & Douglas, 2007a). High Arctic pond sediments, such as those at Cape Herschel, appear to be sensitive archives of past environmental shifts, tracking changes over several millennia, and thus can help put these recent environmental changes into a temporal context. For example, some ponds on Cape Herschel have been accumulating sediments for over 8000 years, following their isolation from the sea due to glacioisostatic uplift (Blake, 1992). Despite their shallowness (e.g. <2 m deep), and the fact that they are frozen solid, in some cases, for 9–10 months of the year, these ponds' sedimentary profiles are not greatly disturbed by cryoturbation or other mixing processes (discussed in Douglas *et al.*, 2000).

As one case study, we present the diatom species changes from a core from Elison Lake, Cape Herschel, which is discussed in more detail in Douglas *et al.* (1994). The basal sediments of this pond have been radiocarbon dated at 3850 +/– 100 ^{14}C years (GSC-3170), and dating of the recent sediments indicated that the entire unsupported ^{210}Pb inventory occurs in the upper 3 cm of the core, reflecting the slow sedimentation rates characteristic of this region (Douglas *et al.*, 1994). The diatom flora exhibited marked species changes over the last ~150 years (Figure 13.5), characterized mainly by the striking relative increase in the moss epiphyte *Hygropetra balfouriana* (Grunow *ex* Cleve) Krammer & Lange-Bertalot (formerly

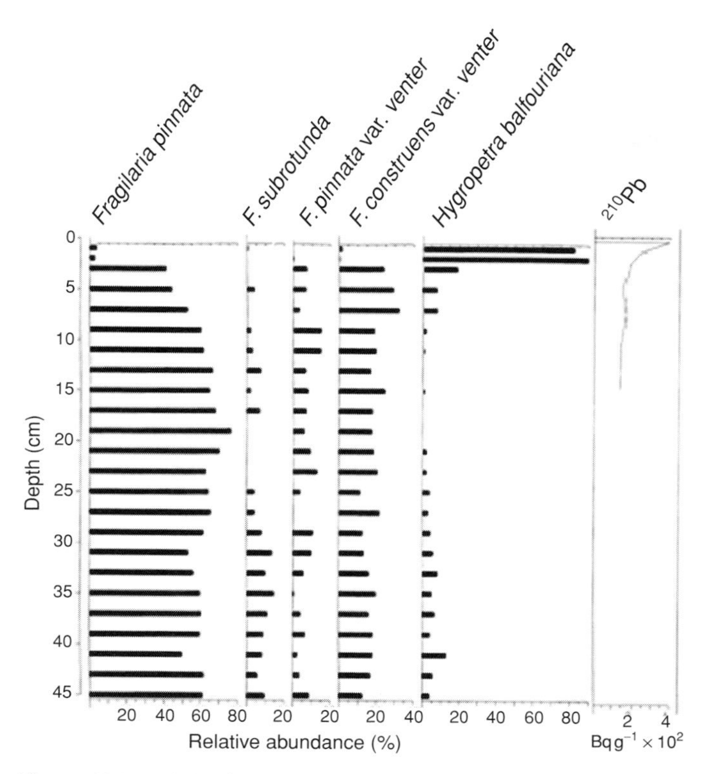

Figure 13.5 Relative frequency diagrams of the dominant diatom taxa recorded in Elison Lake, Cape Herschel, Ellesmere Island. ^{210}Pb data are to the right. (Modified from Douglas *et al.*, 1994.)

Pinnularia balfouriana) and other taxa, replacing the assemblage of small benthic *Fragilaria* species that thrived in that pond since its inception. Similarly, dramatic and synchronous stratigraphic changes occurred in all the other Cape Herschel cores that we studied (Douglas *et al.*, 1994; Douglas, 1993). These successional changes, which all occurred during the last two centuries, were unprecedented in the ponds' histories, and are interpreted to reflect species changes consistent with what would be expected with climatic warming in this region. Following the initial Douglas *et al.* (1994) study, other researchers have recorded similar changes in recent diatom assemblages in cores from other Arctic regions (e.g. Wolfe, 2000; Hamilton *et al.*, 2000; Antoniades *et al.*, 2005b, 2007; Keatley *et al.*, 2006; Lim *et al.*, 2008). For example, Perren *et al.* (2003) inferred a rapid lacustrine response to recent High Arctic warming in Sawtooth Lake (Ellesmere Island) by an increase in diatom diversity over the past ~75 years, which distinguished this period from the previous 2500 years of sediment accumulation. Working on a varved sediment core from Lower Murray Lake on northern Ellesmere Island, Besonen *et al.* (2008) recorded the sudden appearance of *Campylodiscus* diatoms in the very surface sediments, suggesting a new ecological threshold (related to decreased ice cover) had been crossed in the lake. Increased diatom diversity and/or production have also been observed to some extent on the Boothia Peninsula of mainland Arctic Canada (LeBlanc *et al.*, 2004), and from Prescott Island (Finkelstein & Gajewski, 2007) and Russell Island (Finkelstein & Gajewski, 2008) in the central Canadian Arctic Archipelago.

The large number of paleolimnological studies completed from Arctic regions, using similar methods, prompted Smol *et al.* (2005) to undertake a meta-analysis of 55 paleolimnological profiles (which included 42 diatom profiles) from 44 sites in Arctic Canada, Finland, Svalbard, and northern Russia (Figure 13.6). They concluded that Arctic regions that were expected to have warmed the most also showed the greatest degree of compositional change in diatoms and other paleoindicators, and that the ecological characteristics of the stratigraphic changes were related primarily to climate warming, via direct and/or indirect mechanisms (e.g. changes in ice cover, thermal stability). Of course a large number of other environmental factors can influence diatom species changes. However, by using carefully designed, comparative approaches, many of these confounding factors can be explored (reviewed in Smol & Douglas, 2007b). For example, by examining diatom changes in deep High Arctic lakes that still support extensive ice covers, diatom species changes should be muted if ice cover is indeed the dominant controlling factor affecting species abundances (Michelutti *et al.*, 2003b). By comparing two linked lakes that differed mainly by the extent of ice cover duration (due to shading of one of the basins by a nearby hill), Keatley *et al.* (2008b) were able to discount the influence of airborne contaminants or atmospheric nutrient inputs as a controlling factor influencing the observed species changes.

Arctic regions have always been considered to be bellwethers of environmental change, and so are expected to show the first signs of climatic change, and to the greatest degree. However, accelerated warming is now also occurring in temperate regions. Rühland *et al.* (2008) explored the diatom responses to this recent warming, coupled with the associated limnological changes by undertaking a meta-analysis of diatom records from over 200 lakes from within the northern hemisphere. They showed that similar diatom species changes were now also occurring in temperate regions, which they linked to climate-related changes such as decreased ice cover and/or increased thermal stability. As expected, the changes occurring in the temperate lakes occurred much later (by about a century in some cases) than those recorded in the Arctic (Smol *et al.*, 2005).

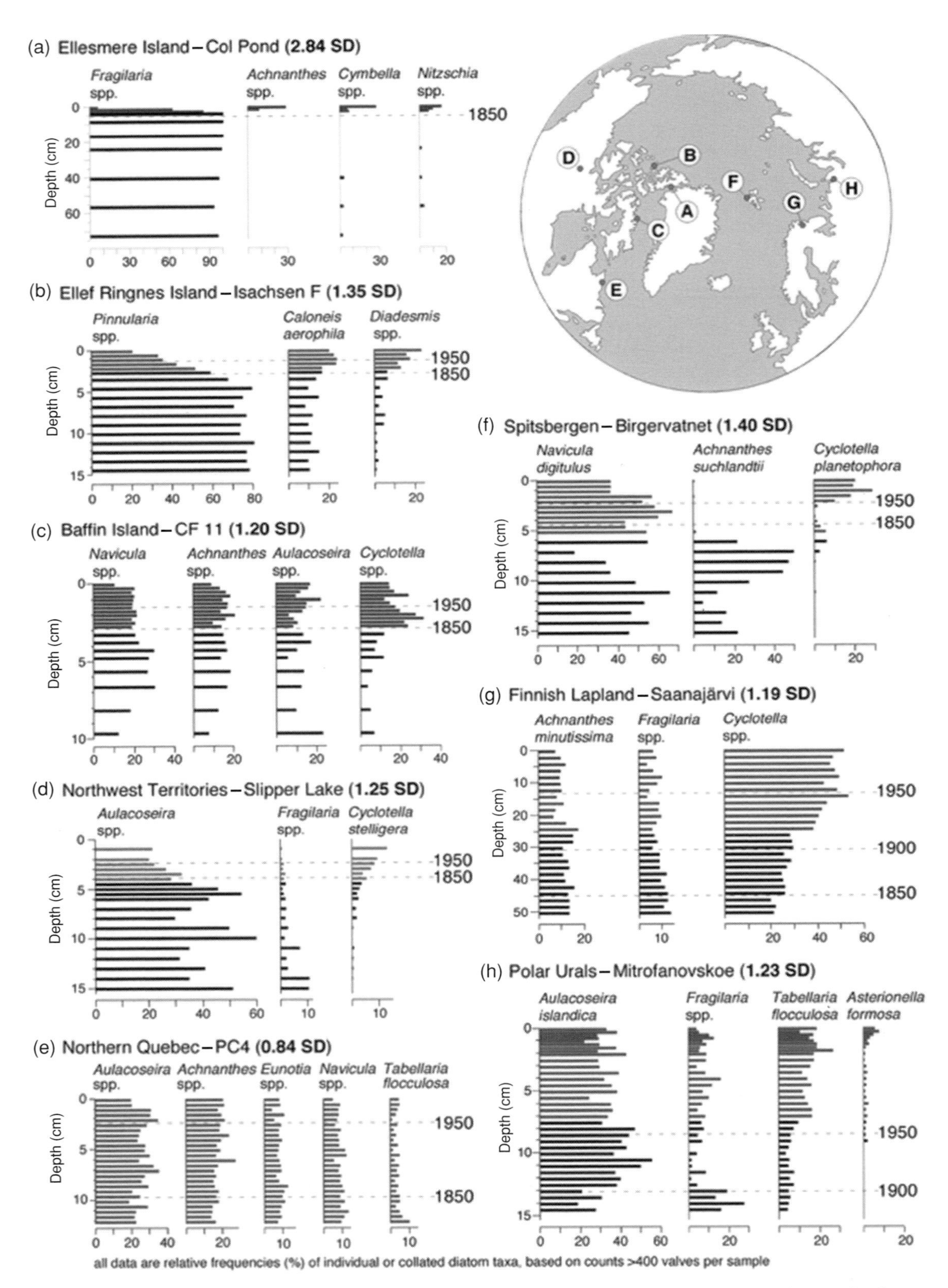

Figure 13.6 Representative diatom relative frequency diagrams of recent (*c.* last 200 years) changes in diatom assemblages from various regions of the circumpolar Arctic. The amount of assemblage change is further quantified by β-diversity measurements (SD = standard deviation units). Many sites from the High Arctic recorded the greatest changes in assemblages, such as those from the Canadian High Arctic islands (e.g. Ellesmere Island (a); Ellef Ringnes Island (b)) and the European High Arctic (e.g. Spitsbergen (f)), which were consistent with interpretations of recent warming. Less change was recorded from sites farther south, such as from Baffin Island (c) and from the Canadian mainland (e.g. Slipper Lake in the Northwest Territories (d), and from Finnish Lapland (g) and the polar Urals (h)). Meanwhile, areas such as northern Québec (e.g. lake PC4 (e) have recorded little change, as expected for that time period. (Modified from Smol *et al.*, 2005.)

13.6.2 pH–climate dynamics

The emerging relevance of all dimensions of Arctic research has led to the completion of numerous diatom-based training sets for the explicit purpose of quantitative environmental reconstructions. Although diatom–temperature transfer functions have been developed for some Arctic regions (e.g. Joynt & Wolfe, 2001), typically diatoms are responding to variables indirectly related to climate such as physical constraints on habitat availability and the length of the growing season imposed by the duration and extent of ice cover (Smol & Cumming, 2000; Smol & Douglas, 2007b). These changes, such as the length and degree of ice cover, which are modulated by climate, have cascading influences on the physical and chemical limnology of Arctic freshwaters, including changes in gas exchange, nutrient levels, and mixing regimes.

Several diatom calibration studies have identified lake-water pH as the variable accounting for the greatest amount of variation in species assemblages (see Battarbee *et al.*, this volume). The development of diatom-based pH inference models is of particular interest, as pH has been demonstrated to have a first-order relationship with climate in poorly buffered lakes from Arctic and alpine regions (Psenner & Schmidt, 1992; Koinig *et al.*, 1998; Wolfe, 2002; Michelutti *et al.*, 2006b; see also Lotter *et al.*, this volume). In these studies, shifts in lake-water pH were interpreted to be at least partly regulated by within-lake dissolved inorganic carbon (DIC) dynamics, which are intimately associated with lake ice cover, and ultimately climate. For example, cool periods have been linked with lower pH levels due to an extended ice cover that traps respired CO_2 in the water column. Conversely, reduced ice cover during warm periods allows respired CO_2 to escape more freely to the atmosphere, and thus also allows for greater photosynthetic activity which increases the drawdown of limnetic CO_2, thereby increasing pH.

Using a diatom-based pH inference model (Joynt & Wolfe, 2001), Wolfe (2002) and Michelutti *et al.* (2007c) reconstructed lake-water pH in poorly buffered lakes on Baffin Island. They demonstrated a close coupling of lake-water pH to Holocene climatic fluctuations, as inferred by numerous independent paleoclimatic proxies. These data confirm earlier work initiated in alpine regions (e.g. Psenner & Schmidt 1992; Sommaruga-Wögrath *et al.*, 1997; Koinig *et al.*, 1998), that climate, through its modulation of lake ice cover, has a first-order influence on primary production and the regulation of DIC speciation, and hence on lake-water pH. It should be noted that climate can also influence pH in well-buffered sites, for example via the increased mineralization in catchments that occurs with warming, which leads to a greater influx of base cations into the lake (e.g. Larsen *et al.* 2006). This process may explain the close agreement recorded by Antoniades *et al.* (2005b) between diatom-inferred pH and measured climate data in a well-buffered lake located near Alert, Ellesmere Island.

13.6.3 Athalassic High Arctic lakes

Athalassic or inland saline lakes are dealt with in detail by Fritz *et al.* (this volume). Although such lakes have been identified in a few Subarctic regions (e.g. Veres *et al.*, 1995; Pienitz *et al.*, 2000; McGowan *et al.*, 2003), they appear to be rare in High Arctic settings. However, recently Paul (2008) investigated the current limnology and the post-glacial diatom changes from a subsaline pond on Stygge Nunatak on Ellesmere Island. She found that diatoms tracked changes in habitat and conductivity, that were likely climate related via shifts in ratios of evaporation to precipitation (see Fritz *et al.*, this volume). In particular, recent warming was indicated by an increase in halophilic species, reflecting a net increase in evaporation.

13.6.4 Extending the diatom paleoclimatic record to pre-Holocene times

Most of the diatom-based paleolimnological studies completed to date have focused on the last few centuries of environmental change, but clearly much important information can be gained from longer temporal perspectives (e.g. Wolfe, 1996b, 2003; Briner *et al.*, 2006; Michelutti *et al.*, 2006c; Podritske & Gajewski, 2007). Unfortunately, records of pre-Holocene diatom assemblages are very sparse in the High Arctic regions, in large part due to glaciers that effectively scoured lake sediments deposited in previous interglacials. However, a site located near Clyde River on east central Baffin Island has recently been identified as containing organic sequences representing multiple interglaciations (Briner *et al.*, 2007). Radiocarbon and optically stimulated luminescence dating confirm that this site, informally named Lake CF8, contains sediment with well-preserved diatom valves (Axford *et al.*, 2009; Wilson, 2009), representing at least three interglacial periods dating to beyond 200 ka (MIS 7, MIS 5, MIS 1; Briner *et al.*, 2007).

The existence of pre-Holocene organic sequences in Lake CF8 is explained by the presence of non-erosive cold-based glacial ice that preserved landscape features, including lake sediments from prior interglacials (Davis *et al.*, 2006; Briner *et al.*, 2005). The Lake CF8 record challenges the general assumption that pre-Holocene diatom-based records are only available from sites distal to glacial margins. More importantly, the Lake CF8 record will allow for the examination of diatom

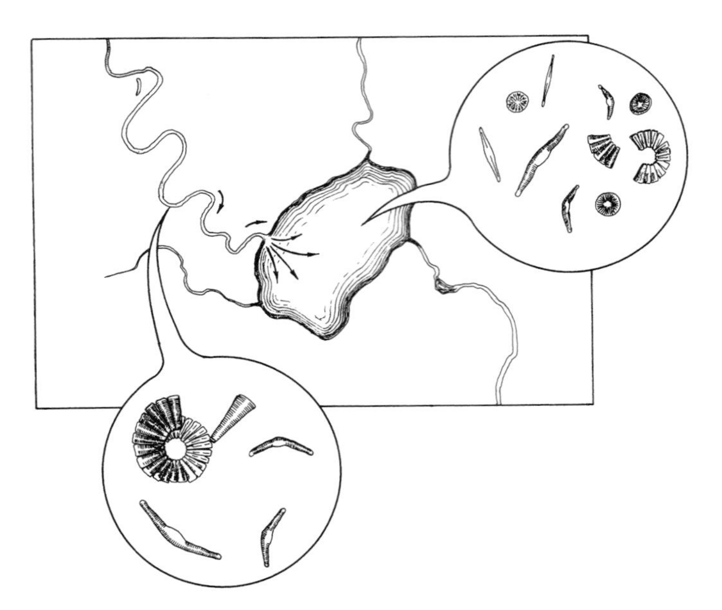

Figure 13.7 Diagrammatic representation of river diatoms being deposited in lake basins. Changes in abundances of lotic diatoms in lake sediment profiles can be used to infer past river discharge, which may be related to climate.

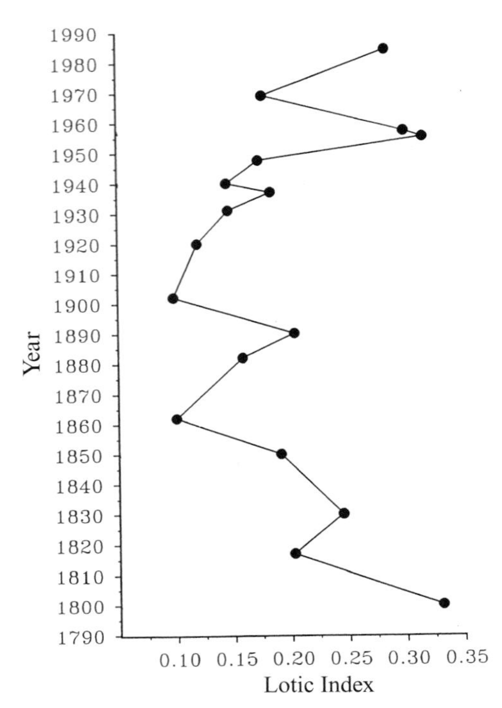

Figure 13.8 Profile of the Lotic Index in a sediment core from Lake C2, Taconite Inlet, Northern Ellesmere Island. (Modified from Ludlam *et al.*, 1996.)

assemblages in periods as warm or warmer than present, but in the absence of anthropogenic overprinting. Diatom work completed from this core (Axford *et al.*, 2009; Wilson, 2009) will help place recent limnological changes associated with the Holocene/Anthropocene boundary (e.g. Smol *et al.* 2005) within the context of the prior 200 ka of natural variability.

13.6.5 River inflows

Sedimentary diatoms may also be useful in tracking past hydrological responses to climate change (e.g. river inflows to lakes), which themselves may be climatically modulated (Figure 13.7). For example, Ludlam *et al.* (1996) noted that certain taxa (e.g. *Meridion* and *Hannaea* species) characterized river environments (i.e. lotic taxa), whilst other diatoms characterized the littoral zone (i.e. lentic taxa) of a lake on northern Ellesmere Island. They proposed a "Lotic Index," which was a ratio of the percentage of *Hannaea* plus *Meridion* valves divided by the total number of pennate diatoms. They used the Lotic Index to infer past river discharge from the diatoms preserved in a varved lake-sediment core, covering the last two centuries (Figure 13.8). They noted a period of declining runoff beginning about two centuries ago that ended in the late 1800s. The Lotic Index also showed a clear, positive relationship to sedimentation rate, as recorded in the varves (Ludlam *et al.*, 1996).

As stream and river diatom communities continue to be better characterized throughout the Canadian Arctic (e.g. Antoniades and Douglas, 2002; Antoniades *et al.* 2009a; Stewart *et al.*, 2005), the potential for using these indictors of past hydrological changes will continue to increase (e.g. Moser *et al.*, 2000; Stewart *et al.*, 2008). Moreover, by using diatoms preserved in deltaic lakes, such as those fed by the Mackenzie River (Hay *et al.*, 1997, 2000; Michelutti *et al.*, 2001) or the Peace-Athabaska River (Wolfe *et al.*, 2005), past inferences of flooding can potentially be reconstructed.

13.6.6 High-resolution diatom lake sediment records

A potential problem with paleolimnological studies of High Arctic lakes is the relatively coarse temporal resolution, because the sedimentation rates in many northern lakes and ponds are comparatively low (Wolfe & Smith, 2004). There are, however, exceptions, and high-resolution studies are possible from a limited number of sites. For example, Bradley (1996) compiled a series of ten papers documenting the environmental history (including two papers dealing with diatoms: Douglas *et al.*, 1996 and Ludlam *et al.*, 1996) from annually laminated lake sediment cores from northern Ellesmere Island. On Devon Island, Gajewski *et al.* (1997) established the annual nature of laminae in a sediment core over the last *c.* 150 years from a non-meromictic lake. Diatom concentrations increased by two orders of magnitude during the twentieth century, with major increases in the 1920s and 1950s. Varve thickness also increased at these times, coincident with increases in snowmelt percentages on the Devon Ice Cap (Koerner, 1977), both proxies for climate warming. Hughen *et al.* (1996) discussed the potential of using diatoms in laminated sediment cores from tidewater lakes on Baffin Island. Although annually laminated sediments are ideal for high-resolution studies, important data on past environmental changes can still be deciphered from non-laminated sediments, especially if careful geochronological control can be achieved using ^{210}Pb and ^{137}Cs dating.

13.6.7 Peat deposits

Diatoms preserved in peat profiles also offer many research opportunities (see Gaiser & Rühland, this volume). At present, High Arctic summers are believed to be too cold, brief, dry, and irregular to allow for extensive peat development (Janssens, 1990). Peat deposits did, however, accumulate in the past in some locations, when conditions were presumably warmer and/or wetter (Ovenden, 1988; LaFarge-England *et al.* 1991; Ellis & Rochefort, 2006). As noted by Brown *et al.* (1994), these fossil peat profiles may contain a suite of siliceous indicators, including diatom frustules, that can be used to interpret past environmental conditions. Because peat accumulation tends to be faster than sediment accumulation in many High Arctic lakes (perhaps an order of magnitude faster in some regions; Brown *et al.*, 1994), peat deposits may potentially provide relatively higher-resolution records of past environmental change. Although some work has been completed on current moss-dwelling diatoms in the Arctic (e.g. Beyens & de Bock, 1989; Hamilton *et al.*, 1994b; Douglas & Smol, 1995a,b), relatively little ecological calibration data are yet available to interpret paleo-peat diatom assemblages.

13.6.8 Snow, ice, frost, and ice cores

The Arctic is characterized by extensive frost, ice, and snow cover. Surprisingly, even these habitats include diatoms; some possibly living as part of the snow flora (Lichti-Federovich, 1980), whilst others have been used to infer various dispersal phenomena (e.g. Lichti-Federovich, 1984, 1985, 1986; Harper & McKay, this volume). Both marine and freshwater diatom frustules have been recovered from Greenland and Arctic island glacier ice cores (e.g. Gayley & Ram, 1984; Harwood, 1986a,b; Gayley *et al.*, 1989), and potentially can be used to augment environmental inferences.

Diatoms have been reported from other cryospheric habitats. For example, Koivo & Seppala (1994) observed living freshwater diatoms on an ice-wedge furrow in the Ungava Peninsula (northern Québec). Cryoecosystems, such as the meltwater lakes and streams of ice shelves and the surficial meltwater puddles and ponds on glaciers (i.e. cryoconites) are typically colonized by algae and other microbiota. However, following detailed examinations of cryoconites from a glacier on Axel Heiberg Island (Canada), Mueller *et al.* (2001) and Mueller & Pollard (2004) did not record any living diatoms (just empty frustules). Interestingly, the researchers did find living diatoms in cryoconites from Antarctica.

13.6.9 Marine/lacustrine transitions

Paleolimnological analyses of High Arctic diatom assemblages offer a powerful tool to help delineate Holocene emergence patterns from glacio-isostatic uplift (see Horton & Sawai, this volume). Hyvärinen (1969), Foged (1977), Young & King (1989), Williams (1990a), and Douglas *et al.* (1996) used diatom assemblages to document the time of lake-water freshening using sediment cores from Spitsbergen, Greenland, Devon, Baffin, and Ellesmere islands, respectively. Ng & King (1999) developed a transfer function (based on diatoms present in the surface sediments of 93 Devon Island lakes and ponds) to quantitatively infer past salinity levels.

13.6.10 Monitoring airborne contaminants and local disturbances

Smol & Douglas (1996) and Douglas *et al.* (2004a) have summarized some of the ways that High Arctic diatoms may be used to monitor human-induced environmental changes in tundra ecosystems. These include disturbances such as airborne contaminants and acidic deposition (see Battarbee *et al.*, this volume). Moreover, as Inuit populations, tourism, and resource extraction continue to increase rapidly, local disturbances (e.g. sewage inputs from communities) may potentially affect local water resources through eutrophication (see Hall & Smol, this volume). For example, Douglas & Smol (2000) completed a study of the diatoms preserved in a sediment core from Meretta Lake, Resolute Bay, Cornwallis Island (Canada). This lake has been receiving local sewage input since 1949. The diatom flora record some taxonomic shifts coincident with cultural eutrophication, but the species changes were different and quite muted compared to changes occurring with eutrophication in more temperate regions, likely reflecting the overriding effects of climate and other related variables (Douglas & Smol, 2000). With recent declines in sewage inputs, the diatom species changes noted in the lake's most recent sediments (Michelutti *et al.*, 2002c), and in archived periphytic samples (Michelutti *et al.*, 2002d), both track the decline in nutrient levels over the 1990s. Working farther south on the Belcher Islands, Michelutti *et al.* (2007b) similarly recorded a delayed diatom species response to cultural eutrophication in a sewage lake used by the community at Sanikiluaq.

Not all eutrophication in the High Arctic is recent. For example, Gregory-Eaves & Keatley (this volume) summarize how guano from Arctic seabirds can affect coastal pond ecosystems. Moreover, diatoms, paired with isotopic analyses of $\delta^{15}N$, can be used in archeological studies (see Juggins & Cameron, this volume) to track the limnological effects of past cultures on Arctic ecosystems. For example, Douglas *et al.* (2004b) demonstrated that the overwintering communities of nomadic paleo-Inuit, the Thule, affected nearby water bodies. The Thule developed the technology to hunt bowhead whales using open boats and harpoons. These whales were subsequently butchered onshore, and the oil, blubber, and whale meat were used by the Thule for lighting, fuel, and food, whilst the bones were used for structural support of their semi-subterranean houses. Paleolimnological analyses of the diatoms confirmed that eutrophication and enhanced moss growth coincided with the activities of the Thule on Somerset Island (Douglas *et al.*, 2004b). Comparable diatom-based paleolimnological work on Bathurst and Ellesmere islands showed similar eutrophication and associated limnological changes (Hadley, 2010a,b).

The High Arctic is subject to a variety of environmental stressors, and likely many new applications of diatoms will emerge in the future. For example, increased exposure to ultraviolet radiation is affecting microbial communities, and Arctic diatoms may be especially sensitive (van Donk *et al.*, 2001). Given the logistical problems of frequently sampling sites in Arctic regions, biomonitoring of lakes, ponds, and rivers using diatoms could be a very cost-effective way of assessing environmental change (Smol & Douglas, 1996).

13.7 Summary

Much of the Arctic landmass that is not under ice sheets is dotted with lakes and ponds. Organisms living in tundra lakes and ponds must often survive low temperatures (including total freezing of the water column in shallow sites), extended ice and snow covers, extremes in photoperiods including long periods of continual darkness followed by 24-hour periods of daylight, and often low nutrient availability. Despite these extreme conditions, diatoms often dominate the benthic algal communities in High Arctic environments.

High Arctic diatoms provide environmental and earth scientists with many potential applications for paleoclimatic and paleoenvironmental research and for biomonitoring, that are otherwise unavailable from traditional proxy techniques that are themselves fraught with difficulties and limitations in these extreme environments. Some of these same approaches can be applied to Antarctic studies (see Spaulding *et al.*, this volume). Because even slight climatic changes can dramatically affect the length of ice and snow cover on Arctic lakes and ponds, as well as other climate-related variables (e.g. water chemistry, habitat availability), diatoms can potentially be used to infer paleoclimatic histories in dated lake and pond sediment cores. Past river inflows into lakes, which are also influenced by climate, may be tracked by the deposition of lotic diatoms. Some of these analyses can be undertaken using high-resolution techniques in varved sediment cores. Other potential applications include interpreting past environmental changes from fossil peat profiles; dispersal and wind patterns from diatoms in snow, frost, and ice-core samples; as well as using diatoms to reconstruct past patterns of glacioisostatic uplift. Considerable potential also exists for using diatoms to monitor anthropogenic disturbances, especially considering the high logistical costs of more typical monitoring programs.

Acknowledgments

Our Arctic research is funded primarily by the Natural Sciences and Engineering Research Council of Canada and the Polar Continental Shelf Program. We thank John Glew for drawing some of the figures. Helpful comments were provided by Neal Michelutti, Kathleen Rühland, Dermot Antoniades, Derek Mueller, Alexandra Rouillard, Daniel Selbie, and members of our labs. We especially thank I. Rod Smith for his thorough review of this chapter.

References

ACIA (2004). *Impacts of a Warming Arctic. Arctic Climate Impact Assessment.* Cambridge: Cambridge University Press.

Anderson, N. J., Ryves, D. B., Grauert M., & McGowan S. (2004). Holocence paleolimnology of Greenland and the North Atlantic islands (north of 60° N). In *Long-term Environmental Change in Arctic and Antarctic Lakes*, ed. R. Pienitz, M. S. V. Douglas and J. P. Smol, Dordrecht: Springer, pp. 319–47.

Andrews, J. T., Evans, L. W., Williams, K. M., *et al.* (1990). Cryosphere/ocean interactions at the margin of the Laurentide Ice

Sheet during the Younger Dryas Chron: SE Baffin Shelf, Northwest Territories. *Paleoceanography*, **5**, 921–35.

Antoniades, D., Crawley, C., Douglas, M. S. V., *et al.* (2007). Abrupt environmental change in Canada's northernmost lake inferred from fossil diatom and pigment stratigraphy. *Geophysical Research Letters*, **34**, L18708, DOI:10.1029/2007GL030947.

Antoniades, D. & Douglas, M. S. V. (2002). Characterization of high arctic stream diatom assemblages from Cornwallis Island, Nunavut, Canada. *Canadian Journal of Botany*, **80**, 50–8.

Antoniades, D., Douglas, M. S. V., & Smol, J. P. (2003a). Comparative physical and chemical limnology of two Canadian High Arctic regions: Alert (Ellesmere Island, NU) and Mould Bay (Prince Patrick Island, NWT). *Archiv für Hydrobiologie*, **158**, 485–516.

Antoniades, D., Douglas, M. S. V., & Smol, J. P. (2003b). The physical and chemical limnology of 24 ponds and one lake from Isachsen, Ellef Ringnes Island, Canadian High Arctic. *International Review of Hydrobiology*, **88**, 519–38.

Antoniades, D., Douglas, M. S. V., & Smol, J. P. (2004). Diatom species–environment relationships and inference models from Isachsen, Ellef Ringnes Island, Canadian High Arctic. *Hydrobiologia*, **529**, 1–18.

Antoniades, D., Douglas, M. S. V., & Smol, J. P. (2005a). Benthic diatom autecology and inference model development from the Canadian High Arctic Archipelago. *Journal of Phycology*, **41**, 30–45.

Antoniades D., Douglas M. S. V., & Smol J. P. (2005b). Quantitative estimates of recent environmental changes in the Canadian High Arctic inferred from diatoms in lake and pond sediments. *Journal of Paleolimnology* **33**, 349–60.

Antoniades, D., Hamilton, P. B., Douglas, M. S. V., & Smol, J. P. (2008). *Diatoms of North America: the Freshwater Floras of Prince Patrick, Ellef Ringnes and Northern Ellesmere Islands from the Canadian Arctic Archipelago*, Königstein: Koeltz Scientific Books, .

Antoniades, D., Hamilton, P. B., Hinz, F., Douglas, M. S. V. & Smol, J. P. (2009b). Seven new species of freshwater diatoms (Bacillariophyceae) from the Canadian Arctic Archipelago. *Nova Hedwigia*, **88**, 57–80.

Antoniades, D., Smol, J. P., & Douglas, M. S. V. (2009a). Biogeographic distributions and environmental controls of stream diatoms in the Canadian Arctic Archipelago. *Botany*, **87**, 443–54.

Axford, Y., Briner, J. P., Cooke, C. A., *et al.* (2009). Recent changes in a remote Arctic lake are unique within the past 200,000 years. *Proceedings of the National Academy of Sciences of the USA*, **106**, 18443–6.

Belzile, C., Vincent, W. F., Gibson, J. A. E., & Van Hove, P. (2001). Bio-optical characteristics of the snow, ice, and water column of a perennially ice-covered lake in the High Arctic. *Canadian Journal of Fisheries and Aquatic Sciences*, **58**, 2405–18.

Besonen, M. R., Patridge, W., Bradley, R. S., *et al.* (2008). A record of climate over the last millennium based on varved lake sediments from the Canadian High Arctic. *The Holocene*, **18**, 169–80.

Beyens, L. & de Bock, P. (1989). Moss dwelling diatom assemblages from Edgeøya (Svalbard). *Polar Biology*, **9**, 423–30.

Birks, H. J. B., Monteith, D. T., Rose, N. L., Jones, V. J., & Peglar, S. M. (2004). Recent environmental change and atmospheric contamination on Svalbard as recorded in lake sediments – modern limnology, vegetation and pollen deposition. *Journal of Paleolimnology*, **31**, 411–31.

Blake, W., Jr. (1989). Inferences concerning climatic change from a deeply frozen lake on Rundfjeld, Ellesmere Island, Arctic Canada. *Journal of Paleolimnology*, **2**, 41–54.

Blake, W., Jr. (1992). Holocene emergence at Cape Herschel, east–central Ellesmere Island, Arctic Canada: implications for ice sheet configuration. *Canadian Journal of Earth Sciences*, **29**, 1958–80.

Blake, W., Jr., Boucherle, M. M., Fredskild, B., Janssens, J. A., & Smol, J. P. (1992). The geomorphological setting, glacial history and Holocene development of 'Kap Inglefield Sø', Inglefield Land, north-west Greenland. *Meddelelser om Grönland, Geosciences*, **27**, 1–42.

Bouchard, G., Gajewski, K., & Hamilton, P. B. (2004). Freshwater diatom biogeography in the Canadian Arctic Archipelago. *Journal of Biogeography*, **31**, 1955–73.

Bradley, R. S. (ed.) (1996). Taconite Inlet lakes project. *Journal of Paleolimnology*, **16**, 97–255.

Briner, J. P., Axford, Y., Forman, S. L., Miller, G. H., & Wolfe, A. P. (2007). Multiple generations of interglacial lake sediment preserved beneath the Laurentide Ice Sheet. *Geology*, **35**, 887–90.

Briner, J. P., Michelutti, N., Francis, D. R., *et al.* (2006). A multi-proxy lacustrine record of Holocene climate change on northeastern Baffin Island, Arctic Canada. *Quaternary Research*, **65**, 431–42.

Briner, J. P., Miller, G. H., Davis, P. T., & Finkel, R. (2005). Cosmogenic exposure dating in arctic glacial landscapes: implications for the glacial history of northeastern Baffin Island, Arctic Canada. *Canadian Journal of Earth Sciences*, **42**, 67–84.

Brown, K. M., Douglas, M. S. V., & Smol, J. P. (1994). Siliceous microfossils in a Holocene High Arctic peat deposit (Nordvestø Northwest Greenland). *Canadian Journal of Botany*, **72**, 208–16.

Cleve, P. T. (1864). Diatomaceer från Spetsbergen. Öfversight af Kongl. *Vetenskaps-Akademiens Forhandlinger*, **24**, 661–9.

Cleve, P. T. (1873). On diatoms from the Arctic Sea. Bihang Till Kongl. *Svenska Vetenskaps-Akademiens Forhandlinger*, **1**, 1–28.

Cleve, P. T. (1883). Diatoms, collected during the expedition of the Vega. *Ur Vega-Expedpeditionens Vettenskapliga Iakttagelser*, **3**, 457–517.

Cleve, P. T. (1896). Diatoms from Baffin Bay and Davis Strait. *Bihang Till Kongliga Svenska Vetenskaps-Akademiens Handlingar*, **22**, 1–22.

Cleve, P. T. (1898). Diatoms from Franz Josef Land. *Bihang Till Kongliga Svenska Vetenskaps-Akademiens Handlingar*, **24**, 1–26.

Cremer, H. (2006). The planktonic diatom flora of a High Arctic lake in East Greenland. *Nordic Journal of Botany*, **24**, 235–44.

Cremer, H., Bennike, O., Hakansson, L., *et al.* (2005). Hydrology and diatom phytoplankton of High Arctic lakes and ponds on Store Koldewey, northeast Greenland. *International Review of Hydrobiology*, **90**, 84–99.

Cremer, H. & Wagner, B. (2004). Planktonic diatom communities in High Arctic lakes (Store Koldewey, Northeast Greenland). *Canadian Journal of Botany*, **82**, 1744–57.

Davis, P. T., Briner, J. P., Coulthard, R. D., Finkel, R. W., & Miller, G. H. (2006). Preservation of Arctic landscapes overridden by cold-based ice sheets. *Quaternary Research*, **65**, 156–63.

Denys, L. & Beyens, L. (1987). Some diatom assemblages from the Angmagssalik region, south-east Greenland. *Nova Hedwigia*, **45**, 389–413.

Doubleday, N., Douglas, M. S. V., & Smol, J. P. (1995). Paleoenvironmental studies of black carbon deposition in the High Arctic: a case study from Northern Ellesmere Island. *The Science of the Total Environment*, **160/161**, 661–68.

Douglas, M. S. V. (1993). Diatom ecology and paleolimnology of high arctic ponds. Unpublished Ph.D. thesis, Queen's University, Kingston, Ontario.

Douglas, M. S. V., Ludlam, S., & Feeney, S. (1996). Changes in diatom assemblages in Lake C2 (Ellesmere Island, Arctic Canada): response to basin isolation from the sea and to other environmental changes. *Journal of Paleolimnology*, **16**, 217–26.

Douglas, M. S. V. & Smol, J. P. (1993). Freshwater diatoms from High Arctic ponds (Cape Herschel, Ellesmere Island, N.W.T.). *Nova Hedwigia*, **57**, 511–52.

Douglas, M. S. V. & Smol, J. P. (1994). Limnology of High Arctic ponds (Cape Herschel, Ellesmere Island, N.W.T.). *Archiv für Hydrobiologie*, **131**, 401–34.

Douglas, M. S. V. & Smol, J. P. (1995a). Periphytic diatom assemblages from High Arctic ponds. *Journal of Phycology*, **31**, 60–9.

Douglas, M. S. V. & Smol, J. P. (1995b). Paleolimnological significance of chrysophyte cysts in Arctic environments. *Journal of Paleolimnology*, **13**, 79–83.

Douglas, M. S. V. & Smol, J. P. (2000). Eutrophication and recovery in the High Arctic: Meretta Lake (Cornwallis Island, Nunavut, Canada) revisited. *Hydrobiologia*, **431**, 193–204.

Douglas, M. S. V., Smol, J. P., & Blake, W., Jr. (1994). Marked post-18th century environmental change in High Arctic ecosystems. *Science*, **266**, 416–19.

Douglas, M. S. V., Smol, J. P., & Blake, W., Jr. (2000). Paleolimnological studies of High Arctic ponds: summary of investigations at Cape Herschel, east–central Ellesmere Island. *Geological Survey of Canada Bulletin*, **529**, 257–69.

Douglas, M. S. V., Smol, J. P., Pienitz, R., & Hamilton, P. (2004a). Algal indicators of environmental change in Arctic and Antarctic lakes and ponds. In *Long-term Environmental Change in Arctic and Antarctic Lakes*, ed. R. Pienitz, M. S. V. Douglas, & J. P. Smol, Dordrecht: Springer, pp. 117–157.

Douglas, M. S. V., Smol, J. P., Savelle, J. M., & Blais, J. M. (2004b). Prehistoric Inuit whalers affected Arctic freshwater ecosystems. *Proceedings of the National Academy of Sciences of the USA*, **101**, 1613–17.

Ehrenberg, C. G. (1853). Über neue Anschaunungen des kleinsten nördlichen Polarlebens. *Deutsche Akademie der Wissenschaften zu Berlin, Monatsberichte*, **1853**, 522–9.

Ellis, C. J. & Rochefort, L. (2006). Long-term sensitivity of a High Arctic wetland to Holocene climate change. *Journal of Ecology*, **94**, 441–54.

Finkelstein, S. A. & Gajewski, K. (2007). A palaeolimnological record of diatom community dynamics and late-Holocene climatic changes from Prescott Island, Nunavut, central Canadian Arctic. *The Holocene*, **17**, 803–12.

Finkelstein, S. A. & Gajewski, K. (2008). Responses of Fragilarioid-dominated diatom assemblages in a small Arctic lake to Holcene climatic changes, Russell Island, Nunavut, Canada. *Journal of Paleolimnology*, **40**, 1079–95.

Finney, B. P., Rühland, K., Smol, J. P., & Fallu, M.-A. (2004). Paleolimnology of the North America Subarctic. In *Long-term Environmental Change in Arctic and Antarctic Lakes*, ed. R. Pienitz, M. S. V. Douglas & J. P. Smol, Dordrecht: Springer, pp. 269–318.

Foged, N. (1953). Diatoms from West Greenland. *Meddelelser om Grönland*, **147**, 1–86.

Foged, N. (1955). Diatoms from Peary Land, North Greenland. *Meddelelser om Grönland*, **194**, 1–66.

Foged, N. (1958). The diatoms in the basalt area and adjoining areas of Archean rock in West Greenland. *Meddelelser om Grönland*, **156**, 1–146.

Foged, N. (1964). Freshwater diatoms from Spitsbergen. *Tromsö Museums Skrifter*, **11**, 1–204, 22 pls.

Foged, N. (1972). The diatoms in four postglacial deposits in Greenland. *Meddelelser om Grönland*, **194**, 1–66.

Foged, N. (1973). Diatoms from Southwest Greenland. *Meddelelser om Grönland*, **194**, 1–84.

Foged, N. (1974). Freshwater diatoms in Iceland. *Bibliotheca Phycologia*, **15**, 1–118.

Foged, N. (1977). Diatoms from four postglacial deposits at Godthabsfjord, West Greenland. *Meddelelser om Grönland*, **199**, 1–64.

Foged, N. (1981). Diatoms in Alaska. *Bibliotheca Phycologica*, **53**, 1–31.

Foged, N. (1982). Diatoms in human tissues – Greenland ab. 1460 AD – Funen 1981–82 AD *Nova Hedwigia*, **36**, 345–379.

Foged, N. (1989). The subfossil diatom flora of four geographically widely separated cores in Greenland. *Meddelelser om Grönland (Bioscience)*, **30**, 1–75.

Gajewski, K., Garneau, M., & Bourgeois, J. C. (1995). Paleoenvironments of the Canadian High Arctic derived from pollen and plant macrofossils: problems and potentials. *Quaternary Science Reviews*, **14**, 609–29.

Gajewski, K., Hamilton, P. B., & McNeely, R. N. (1997). A high resolution proxy-climate record from an Arctic lake with laminated sediments on Devon Island, Nunavut, Canada. *Journal of Paleolimnology*, **17**, 215–25.

Gayley, R. I. & Ram, M. (1984). Observations of diatoms in Greenland ice. *Arctic*, **37**, 172–3.

Gayley, R. I., Ram, M., & Stoermer, E. F. (1989). Seasonal variations in diatom abundance and provenance in Greenland ice. *Journal of Glaciology*, **35**, 290–2.

Guilizzoni, P., Marchetto, A., Lami, A., *et al.* (2006). Records of environmental and climatic changes during the late Holocene from

Svalbard: palaeolimnology of Kongressvatnet. *Journal of Paleolimnology*, **36**, 325–51.

Hadley, K. R., Douglas, M. S. V., McGhee, R. H., Blais, J. M., & Smol, J. P. (2010a). Ecological influences of Thule Inuit whalers on high Arctic pond ecosystems: a comparative paleolimnological study from Bathurst Island (Nunavut, Canada). *Journal of Paleolimnology*, **44**, 85–93.

Hadley, K. R., Douglas, M. S. V., Blais, J. M., & Smol, J. P. (2010b). Nutrient enrichment in the High Arctic associated with Thule Inuit whalers: a paleolimnological investigation from Ellesmere Island (Nunavut, Canada). *Hydrobiologia*, **649**, 129–38.

Håkansson, H. (1988). Obituary: Niels Aage Johannes Foged 1906–1988. *Diatom Research*, **3**, 169–74.

Hamilton, P. B., Douglas, M. S. V., Fritz, S. C., *et al.* (1994a). A compiled freshwater diatom taxa list for the Arctic and Subarctic regions of North America. In *The Proceedings of the Fourth Arctic–Antarctic Diatom Symposium (Workshop)*, ed. P. B. Hamilton, Canadian Technical Report of Fisheries and Aquatic Sciences, **1957**, pp. 85–102.

Hamilton, P. B. & Edlund, S. A. (1994). Occurrence of *Prasiola fluviatilis* (Chlorophyta) on Ellesmere Island in the Canadian Arctic. *Journal of Phycology*, **30**, 217–21.

Hamilton, P., Gajewski, K., Atkinson, D., & Lean, D. R. S. (2001). Physical and chemical limnology of lakes from the Canadian Arctic Archipelago. *Hydrobiologia* **457**, 133–48.

Hamilton, P. B., Gajewski, K., & McNeely, R. N. (2000). Physical, chemical and biological characteristics of lakes from the Sidre Basin on the Fosheim Peninsula. *Geological Survey of Canada, Bulletin*, **529**, 234–48.

Hamilton, P. B., Poulin, M., Prevost, C., Angell, M., & Edlund, S. A. (1994b). Americanarum Diatomarum Exsiccata: Fascicle II (CANA), voucher slides representing 34 lakes, ponds and streams from Ellesmere Island, Canadian High Arctic, North America. *Diatom Research*, **9**, 303–27.

Harwood, D. M. (1986a). Do diatoms beneath the Greeenland ice sheet indicate interglacials warmer than present? *Arctic*, **39**, 304–8.

Harwood, D. M. (1986b). The search for microfossils beneath the Greenland and west Antarctic ice sheets. *Antarctic Journal of the United States*, **21**, 105–6.

Hay, M., Michelutti, N., & Smol, J. P. (2000). Ecological patterns of diatom assemblages from Mackenzie Delta lakes, Northwest Territories, Canada. *Canadian Journal of Botany*, **78**, 19–33.

Hay, M. B., Smol, J. P., Pipke, K., & Lesack, L. (1997). A diatom-based paleohydrological model for the Mackenzie Delta, Northwest Territories, Canada. *Arctic and Alpine Research*, **29**, 430–44.

Hickman, M. (1974). The epipelic diatom flora of a small lake on Baffin Island, Northwest Territories. *Archiv für Protistenk*, **116S**, 270–9.

Hobbie, J. E. (1984). Polar limnology. In *Ecosystems of the World: Lakes and Reservoir*, ed. F. B. Taub, Amsterdam: Elsevier, pp. 63–106.

Hodgson, D. A. & Smol, J. P. (2008). High-latitude paleolimnology. In *Polar Lakes and Rivers*, ed. W. Vincent & J. Laybourn-Parry, Oxford: Oxford University Press, pp. 43–64.

Hughen, K. A., Overpeck, J. T., Anderson, R. F., & Williams, K. M. (1996). The potential for paleoclimate records from varved Arctic lake sediments: Baffin Island, Eastern Canadian Arctic. *Geological Society of London, Special Publication*, **116**, 57–71

Hyvärinen, H. (1969). Trullvatnet: a Flandrian stratigraphical site near Murctlandet, Spitsbergen. *Geografiska Annaler*, **51A**, 42–5.

IPCC (2007). Summary for policymakers. In: *Climate Change 2007: The Physical Science Basis. Contribution of Working Group I to the Fourth Assessment Report of the Intergovernmental Panel on Climate Change*, ed. S. Solomon, D. Qin, M. Manning, *et al.*, Cambridge: Cambridge University Press.

Janssens, J. A. (1990). Methods in Quaternary ecology. No. 11. Bryophytes. *Geoscience Canada*, **17**, 13–24.

Jones, V. J. & Birks, H. J. B. (2004). Lake-sediment records of recent environmental change on Svalbard: results of diatom analysis. *Journal of Paleolimnology*, **31**, 445–66.

Joynt, E. H., III & Wolfe, A. P. (2001). Paleoenvironmental inference models from sediment diatom assemblages in Baffin Island lakes (Nunavut, Canada) and reconstruction of summer water temperature. *Canadian Journal of Fisheries and Aquatic Sciences*, **58**, 1222–43.

Keatley, B. E., Douglas, M. S. V., & Smol, J. P. (2006). Early-20th century environmental changes inferred using subfossil diatoms from a small pond on Melville Island, NWT, Canadian High Arctic. *Hydrobiologia*, **553**, 15–26.

Keatley, B. E., Douglas, M. S. V., & Smol, J. P. (2007a). Physical and chemical limnological characteristics of lakes and ponds across environmental gradients on Melville Island, Nunavut/NWT, High Arctic Canada. *Fundamental and Applied Limnology*, **168**, 355–76.

Keatley, B. E., Douglas, M. S. V., & Smol, J. P. (2007b). Limnological characteristics of a high arctic oasis and comparisons across northern Ellesmere Island. *Arctic*, **60**, 294–308.

Keatley, B. E., Douglas, M. S. V., & Smol, J. P. (2008a). Evaluating the influence of environmental and spatial variables on diatom species distributions from Melville Island (Canadian High Arctic). *Botany*, **86**, 76–90.

Keatley, B. E., Douglas, M. S. V. & Smol, J. P. (2008b). Prolonged ice cover dampens diatom community responses to recent climatic change in High Arctic lakes. *Arctic, Antarctic and Alpine Research*, **40**, 364–72.

Koerner, R. (1977). Devon Island ice cap: core stratigraphy and paleoclimate. *Science*, **196**, 15–18.

Koinig, K. A., Schmidt, R., Sammaruga-Wögrath, S., Tessadri, R., & Psenner, R. (1998). Climate change as the primary cause for pH shifts in a High Arctic lake. *Water Air and Soil Pollution*, **104**, 167–80.

Koivo, L. & Seppala, M. (1994). Diatoms from an ice-wedge furrow, Ungava Peninsula, Quebec, Canada. *Polar Research*, **13**, 237–41.

Kokelj, S. V., Jenkins, R. E., Burn, C. R., & Snow, N. (2005). The influence of thermokarst disturbance on the water quality of small upland lakes, Mackenzie Delta region, Northwest Territories, Canada. *Permafrost and Periglacial Processes*, **16**, 343–53.

Korhola, A. & Weckström, J. (2004). Paleolimnological studies in Arctic Fennoscandia and the Kola peninsula (Russia). In *Long-term Environmental Change in Arctic and Antarctic Lakes*, ed. R. Pienitz, M. S. V. Douglas, & J. P. Smol, Dordrecht: Springer, pp. 381–418.

Lafarge-England, C., Vitt, D. H., & England, J. (1991). Holocene soligenous fens on a High Arctic fault block, Northern Ellesmere

Island (82° N), NWT, Canada. *Arctic and Alpine Research* **23**, 80–98.

Lagerstedt, N. G. W. (1873). Sötvattens-Diatomaceer från Spetsergen och Beeren Eiland. *Bihang Till Kongliga Svenska Vetenskaps-Akademiens Handlingar*, **1**, 1–52.

Larsen, J., Jones, V. J., & Eide, W. (2006). Climatically driven pH changes in two Norwegian alpine lakes. *Journal of Paleolimnology*, **36**, 175–87.

LeBlanc, M., Gajewski, K., & Hamilton, P.B. (2004). A diatom-based Holocene palaeoenvironmental record from a mid-Arctic lake on Boothia Peninsula, Nunavut, Canada. *The Holocene*, **14**, 417–25.

Lemmen, D. S., Gilbert, R., Smol, J. P., & Hall, R. I. (1988). Holocene sedimentation in glacial Tasikutaaq Lake, Baffin Island. *Canadian Journal of Earth Sciences*, **25**, 810–23.

Lichti-Federovich, S. (1980). Diatom flora of red snow from Isbjørneø, Carey Øer, Greenland. *Nova Hedwigia*, **33**, 395–431.

Lichti-Federovich, S. (1984). Investigations of diatoms found in surface snow from the Sydkap ice cap, Ellesmere Island, Northwest Territories. *Current Research, Part A, Geological Survey of Canada*, **Paper 84–1A**, 287–301.

Lichti-Federovich, S. (1985). Diatom dispersal phenomena: diatoms in rime frost samples from Cape Herschel, central Ellesmere Island, Northwest Territories. *Current Research, Part B, Geological Survey of Canada*, **Paper 85–1B**, 391–9.

Lichti-Federovich, S. (1986). Diatom dispersal phenomena: diatoms in precipitation samples from Cape Herschel, east–central Ellesmere Island, Northwest Territories – a quantitative assessment. *Current Research, Part B, Geological Survey of Canada*, **Paper 86–1B**, 263–9.

Lim, D. S. S. (2004). Limnology and diatom palaeoecology of lakes and ponds on Banks Island, N.W.T. and Devon Island, Nunavut, Canadian Arctic. Unpublished Ph.D. thesis, University of Toronto.

Lim, D. S. S. & Douglas, M. S. V. (2003). Limnological characteristics of 22 lakes and ponds in the Haughton Crater region of Devon Island, Nunavut, Canadian High Arctic. *Arctic Antarctic and Alpine Research*, **35**, 509–19.

Lim, D. S. S., Douglas, M. S. V., & Smol, J. P. (2005). Limnology of 46 lakes and ponds on Banks Island, NWT, Canadian Arctic Archipelago. *Hydrobiologia*, **545**, 11–32.

Lim, D. S. S., Douglas, M. S. V., Smol, J. P., & Lean, D. R. S. (2001a). Physical and chemical limnological characteristics of 38 lakes and ponds on Bathurst Island, Nunavut, Canadian High Arctic, *International Review of Hydrobiology*, **86**, 1–22.

Lim, D. S. S., Kwan, C., & Douglas, M. S. V. (2001c). Periphytic diatom assemblages from Bathurst Island, Nunavut, Canadian High Arctic: an examination of community relationships and habitat preferences. *Journal of Phycology*, **37**, 379–92.

Lim, D. S. S., Smol, J. P., & Douglas, M. S. V. (2001b). Diatoms and their relationship to environmental variables from lakes and ponds on Bathurst Island, Nunavut, Canadian High Arctic. *Hydrobiologia*, **450**, 215–230.

Lim, D. S. S., Smol, J. P., & Douglas, M. S. V. (2007). Diatom assemblages and their relationships to lakewater nitrogen levels and other limnological variables from 36 lakes and ponds on Banks Island, NWT, Canadian Arctic. *Hydrobiologia*, **586**, 191–211.

Lim, D. S. S., Smol, J. P., & Douglas, M. S. V. (2008). Recent environmental changes on Banks Island (NWT, Canadian Arctic) quantified using fossil diatom assemblages. *Journal of Paleolimnology*, **40**, 385–98.

Lowe, C. W. (1923). Report of the Canadian Arctic Expedition 1913–18. Part A: freshwater algae and freshwater diatoms, Southern Party 1913–1916. Ottowa: F. A. Acland, Printer of the King's Most Excellent Majesty.

Ludlam, S. D., Feeney, S., & Douglas, M. S. V. (1996). Changes in the importance of lotic and littoral diatoms in a High Arctic lake over the last 191 years. *Journal of Paleolimnology*, **16**, 184–204.

MacDonald, G. M., Edwards, T. W. D., Gervais, B., *et al.* (2004). Paleolimnological research from northern Russian Eurasia. In *Long-term Environmental Change in Arctic and Antarctic Lakes*, ed. R. Pienitz, M. S. V. Douglas, & J. P. Smol, Dordrecht: Springer., pp. 349–80.

McGowan, S., Ryves D. B., & Anderson, N. J. (2003). Holocene records of effective precipitation in west Greenland. *The Holocene*, **13**, 239–49.

Metzeltin, D. & Witkowski, A. (1996). Diatomeen der Bären-Insel. *Iconographia Diatomologica*, **4**, 1–287.

Michelutti, N., Douglas, M. S. V., Muir, D. C. G., Wang, X. W., & Smol, J. P. (2002a). Limnological characteristics of 38 lakes and ponds on Axel Heiberg Island, High Arctic Canada. *International Review of Hydrobiology*, **87**, 385–399.

Michelutti, N., Douglas, M. S. V., Lean, D. R. S., & Smol, J. P. (2002b). Physical and chemical limnology of 34 ultra-oligotrophic lakes and ponds near Wynniatt Bay, Victoria Island, Arctic Canada. *Hydrobiologia*, **482**, 1–13.

Michelutti, N., Douglas, M. S. V., & Smol, J. P. (2002c). Tracking recent recovery from eutrophication in a High Arctic lake (Meretta Lake, Cornwallis Island, Nunavut, Canada) using fossil diatom assemblages. *Journal of Paleolimnology*, **28**, 377–81.

Michelutti, N., Douglas, M. S. V., & Smol, J. P. (2002d). Tracking recovery in a eutrophied High Arctic lake (Meretta Lake, Cornwallis Island, Canadian Arctic) using periphytic diatoms. *Verhandlungen der Internationale Vereinigung von Limnologie*, **28**, 1533–7.

Michelutti, N., Douglas, M. S. V., & Smol, J. P. (2003a). Diatom response to recent climatic change in a High Arctic lake (Char Lake, Cornwallis Island, Nunavut). *Global and Planetary Change*, **38**, 257–71.

Michelutti, N., Douglas, M. S. V., & Smol, J. P. (2007a). Evaluating diatom community composition in the absence of marked limnological gradients in the High Arctic: a surface sediment calibration set from Cornwallis Island (Nunavut, Canada). *Polar Biology*, **30**, 1459–73.

Michelutti, N., Douglas, M. S. V., Wolfe, A. P., & Smol, J. P. (2006b). Heightened sensitivity of a poorly buffered High Arctic lake to late-Holocene climatic change. *Quaternary Research*, **65**, 421–30.

Michelutti, N., Hay, M., Marsh, P., Lesack, L., & Smol, J. P. (2001). Diatom changes in lake sediments from the Mackenzie Delta, N W.T., Canada: paleohydrological applications. *Arctic, Antarctic, and Alpine Research*, **33**, 1–12.

Michelutti, N., Hermanson, M. H., Smol, J. P., Dillon, P. J., & Douglas, M. S. V. (2007b). Delayed response of diatom assemblages to sewage inputs in an Arctic lake. *Aquatic Sciences*, **69**, 523–33.

Michelutti, N., Holtham, A. J., Douglas, M. S. V., & Smol, J. P. (2003b). Periphytic diatom assemblages from ultra-oligotrophic and UV transparent lakes and ponds on Victoria Island and comparisons with other diatom surveys in the Canadian Arctic. *Journal of Phycology*, **39**, 465–80.

Michelutti, N., Smol, J. P., & Douglas, M. S. V. (2006a). Ecological characteristics of modern diatom assemblages from Axel Heiberg Island (High Arctic Canada) and their application to paleolimnological inference models. *Canadian Journal of Botany*, **84**, 1695–713.

Michelutti, N., Wolfe, A. P., Briner, J. P., & Miller, G. H. (2007c). Climatically controlled chemical and biological development in Arctic lakes. *Journal of Geophysical Research-Biogeosciences*, **112**, G03002, doi:10.1029/2006JG000396.

Moore, J. W. (1974a). Benthic algae of southern Baffin Island. I. Epipelic communities in rivers. *Journal of Phycology*, **10**, 50–7.

Moore, J. W. (1974b). Benthic algae of southern Baffin Island. II. The epipelic communities in temporary pools. *Journal of Ecology*, **62**, 809–19.

Moore, J. W. (1974c). Benthic algae of southern Baffin Island. III. Epilithic and epiphytic communities. *Journal of Phycology*, **10**, 456–62.

Moser, K.A., Korhola, A., Weckstrom,J., *et al.* (2000). Paleohydrology inferred from diatoms in northern latitude regions, *Journal of Paleolimnology*, **24**, 93–107.

Mueller, D. R. & Pollard, W. H. (2004). Gradient analysis of cryoconite ecosystems from two polar glaciers. *Polar Biology*, **27**, 66–74.

Mueller, D. R., Vincent, W. F., Pollard, W. H., & Fritsen, C. H. (2001). Glacial cryoconite ecosystems: a bipolar comparison of algal communities and habitats. *Nova Hedwigia, Beiheft*, **123**, 173–97.

Nansen, F. (1897). *Farthest North*. Westminster: Archibald Constable and Company.

Ng, S. L. & King, R. H. (1999). Development of a diatom-based specific conductivity model for the glacio-isostatic lakes of Truelove Lowland: implications for paleoconductivity and paleoenvironmental reconstructions in Devon Island lakes, N.W.T., Canada. *Journal of Paleolimnology*, **22**, 367–82.

Østrup, E. (1897a). Ferskvands-Diatomeer fra Öst Grönland. *Meddelelser om Grönland*, **15**, 251–90.

Østrup, E. (1897b). Kyst-Diatoméer fra Grönland. *Meddelelser om Grönland*, **15**, 305–62.

Østrup, E. (1910). Diatoms from North-East Greenland. *Meddelelser om Grönland*, **43**, 199–256.

Ovenden, L. (1988). Holocene proxy-climate data from the Canadian Arctic. *Geological Survey of Canada Paper*, **88–22**.

Paul, C. A. (2008). Paleolimnological assessment of Holocene climatic and environmental change in two lakes located in different regions of the Canadian Arctic Tundra. Unpublished M. Sc. thesis, Queen's University, Kingston, Ontario.

Paull, T. M., Hamilton, P. B., Gajewski, K., & LeBlanc, M. (2008). Numerical analysis of small Arctic diatoms (Bacillariophyceae) representing the *Staurosira* and *Staurosirella* species complexes. *Phycologia*, **47**, 213–24.

Perren, B. B., Bradley, R.S., & Francus, P. (2003). Rapid lacustrine response to recent High Arctic warming: a diatom record from Sawtooth Lake, Ellesmere Island, Nunavut. *Arctic, Antarctic, and Alpine Research*, **35**, 271–78.

Petersen, J. B. (1924). Fresh water algae from the north coast of Greenland collected by the late Dr. Th. Wulff. Den II Thule Exped. til Groenlands Nordkyst (1916–18). *Meddelelser om Grønland*, **64**, 307–19.

Pienitz, R., Smol, J.P., Last, W., Leavitt, P. R., & Cumming, B. F. (2000). Multiproxy Holocene palaeoclimatic record from a saline lake in the Canadian Subarctic. *The Holocene*, **10**, 673–86.

Pienitz, R., Douglas, M. S. V., & Smol, J. P. (eds.) (2004). *Long-Term Environmental Change in Arctic and Antarctic Lakes*. Dordrecht: Springer.

Podritskie, B. & Gajewski, K. (2007). Diatom community response to multiple scales of Holocene climate variability in a small lake on Victoria Island, NWT, *Quaternary Science Reviews*, **26**, 3179–96.

Psenner, R. & Schmidt, R. (1992). Climate-driven pH control of remote alpine lakes and effects of acid deposition. *Nature*, **356**, 781–3.

Rigler, F. H. (1978). Limnology in the High Arctic: a case study of Char Lake. *Verhandlungen Internationale Vereingung Limnologen*, **20**, 127–40.

Ross, R. (1947). Freshwater diatomae (Bacillariophyta). In *Botany of the Canadian Eastern Arctic. Part II: Thallophyta and Bryophyta*, ed. N. Polunin, Ottawa: National Museum of Canada, pp. 178–233.

Rouse, W., Douglas, M., Hecky, R., *et al.* (1997). Effects of climate change on fresh waters of Region 2: Arctic and Sub-Arctic North America. *Hydrologic Processes*, **11**, 873–902.

Rühland, K., Paterson, A. M., & Smol, J. P. (2008). Hemispheric-scale patterns of climate-induced shifts in planktonic diatoms from North American and European lakes. *Global Change Biology*, **14**, 2740–45.

Schindler, D. W. & Smol, J. P. (2006). Cumulative effects of climate warming and other human activities on freshwaters of Arctic and Subarctic North America. *Ambio*, **35**, 160–8.

Sheath, R. G. (1986). Seasonality of phytoplankton in northern tundra ponds. *Hydrobiologia*, **138**, 75–83.

Sheath, R. G., Morgan, V., Hambrook, J. A., & Cole, K. M. (1996). Tundra stream macroalgae of North America: composition, distribution and physiological adaptations. *Hydrobiologia*, **336**, 67–82.

Short, S., Andres, J., Williams, K., Weiner, J., & Elias, S. (1994). Late Quaternary marine and terrestrial environments, northwestern Baffin Island, Northwest Territories. *Geographie Physique et Quaternaire*, **48**, 85–95.

Skulberg, O. M. (1996). Terrestrial and limnic algae and cyanobacteria. In *A Catalogue of Svalbard Plants, Fungi, Algae and Cyanobacteria, Part 9*, ed. A. Elvebakk & P. Prestrud, Oslo: Norsk Polarinstitutt Skrifter 198, pp. 383–95.

Smith, I. R. (2000). Diamictic sediments within High Arctic lake sediment cores: evidence for lake ice rafting along the lateral glacial margin. *Sedimentology*, **47**, 1157–79.

Smith, I. R. (2002). Diatom-based Holocene paleoenvironmental records from continental sites on northeastern Ellesmere Island, High Arctic, Canada. *Journal of Paleolimnology*, **27**, 9–28.

Smol, J. P. (1983). Paleophycology of a High Arctic lake near Cape Herschel, Ellesmere Island. *Canadian Journal of Botany*, **61**, 2195–204.

Smol, J. P. (1988). Paleoclimate proxy data from freshwater Arctic diatoms. *Verhandlungen Internationale Vereingung Limnologen*, **23**, 837–44.

Smol, J. P. & Cumming, B. F. (2000). Tracking long-term changes in climate using algal indicators in lake sediments. *Journal of Phycology*, **36**, 986–1011.

Smol, J. P. & Douglas, M. S. V. (1996). Long-term environmental monitoring in Arctic lakes and ponds using diatoms and other biological indicators. *Geoscience Canada*, **23**, 225–30.

Smol, J. P. & Douglas, M. S. V. (2007a). Crossing the final ecological threshold in High Arctic ponds. *Proceedings of the National Academy of Sciences of the USA*, **104**, 12395–7.

Smol, J. P. & Douglas, M. S. V. (2007b). From controversy to consensus: making the case for recent climate using lake sediments. *Frontiers in Ecology and the Environment*, **5**, 466–74.

Smol, J. P., Wolfe, A. P., Birks, H. J. B., *et al.* (2005). Climate-driven regime shifts in the biological communities of Arctic lakes. *Proceedings of the National Academy of Sciences of the USA*, **102**, 4397–402.

Sommaruga-Wögrath, R., Koinig, K., Schmidt, R., *et al.* (1997). Temperature effects on the acidity of remote alpine lakes. *Nature*, **387**, 64–7.

Statistics Canada (1987). *Canada Year Book 1988; A Review of Economic, Social and Political Development in Canada*. Ottawa: Statistics Canada.

Stewart, K. A., Lamoureux, S. F., & Forbes, A. C. (2005). Hydrological controls on the diatom assemblage of a seasonal Arctic river: Boothia Peninsula, Nunavut, Canada. *Hydrobiología*, **544**, 259–70.

Stewart, K. A., Lamoureux, S. F., & Finney, B. P. (2008). Multiple ecological and hydrological changes recorded in varved sediments from Sanagak Lake, Nunavut, Canada. *Journal of Paleolimnology*, **40**, 217–33.

Van de Vijver, B., Van Kerckvoorde, A., & Beyens, L. (2003). Freshwater and terrestrial moss diatom assemblages of the Cambridge Bay area, Victoria Island (Nunavut, Canada). *Nova Hedwigia*, **76**, 225–43.

van Donk, E., Faafeng, B. A., de Lange, H. J., & Hessen, D. O. (2001). Differential sensitivity to natural ultraviolet radiation among phytoplankton species in Arctic lakes (Spitsbergen, Norway). *Plant Ecology*, **154**, 247–59.

Veres, A. J., Pienitz, R., & Smol, J. P. (1995). Lakewater salinity and periphytic diatom succession in three subarctic lakes (Yukon Territory, Canada). *Arctic*, **48**, 63–70.

Vincent, W. F. & Laybourn-Parry, J. (eds.) (2008). *Polar Lakes and Rivers – Limnology of Arctic and Antarctic Aquatic Ecosystems*. Oxford: Oxford University Press.

Williams, K. M. (1990a). Paleolimnology of three Jackman Sound lakes, southern Baffin Island, based on down-core diatom analyses. *Journal of Paleolimnology*, **4**, 203–17.

Williams, K. M. (1990b). Late Quaternary paleoceanography of the western Baffin Bay region: evidence from fossil diatoms. *Canadian Journal of Earth Sciences*, **27**, 1487–94.

Williams, K. M., Short, S. K., Andrews, J. T. *et al.* (1995). The Eastern Canadian Arctic at ca.6 Ka BP – a time of transition. *Géographie Physique et Quaternaire*, **49**, 13–27.

Wilson, C. R. (2009). A lacustrine sediment record of the last three interglacial periods from Clyde Foreland, Baffin Island, Nunavut: biological indicators from the past 200,000 years. Unpublished M.Sc. thesis, Queen's University, Kingston, Ontario.

Wolfe, A. P. (1994). Late Wisconsinan and Holocene diatom stratigraphy from Amarok Lake, Baffin Island, N.W.T., Canada. *Journal of Paleolimnology*, **10**, 129–39.

Wolfe, A. P. (1996a). Spatial patterns of modern diatom distribution and multiple paleolimnological records from a small non-glacial Arctic lake, Baffin Island, Northwest Territories. *Canadian Journal of Botany*, **74**, 345–59.

Wolfe, A. P. (1996b). A high resolution late-glacial and early Holocene diatom record from Baffin Island, Northwest Territories. *Canadian Journal of Earth Sciences*, **33**, 928–37.

Wolfe, A. P. (2000). A 6500 year diatom record from southwestern Fosheim Peninsula, Ellesmere Island, Canadian High Arctic. *Geological Survey of Canada, Bulletin*, **529**, 249–56.

Wolfe, A. P. (2002). Climate modulates the acidity of arctic lakes on millennial time scales. *Geology*, **30**, 215–18.

Wolfe, A. P. (2003). Diatom community responses to late-Holocene climatic variability, Baffin Island, Canada: a comparison of numerical approaches. *Holocene*, **13**, 29–37.

Wolfe, A. P. & Härtling, J. W. (1996). The late Quaternary development of three ancient tarns on southwestern Cumberland Peninsula, Baffin Island, Arctic Canada: paleolimnological evidence from diatoms and sediment chemistry. *Journal of Paleolimnology*, **15**, 1–18.

Wolfe, A. P. & Smith, I. R. (2004). Paleolimnology of the Middle and High Canadian Arctic. In *Long-term Environmental Change in Arctic and Antarctic Lakes*, ed. R. Pienitz, M. S. V. Douglas, & J. P. Smol, Dordrecht: Springer, pp. 241–268.

Wolfe, B. B., Karst-Riddoch, T. L., Vardy, S. R., *et al.* (2005). Impacts of climatic variability and river regulation on hydro-ecology of a floodplain basin, Peace-Athabasca Delta, Canada: 1700–present. *Quaternary Research*, **64**, 147–62.

Young, R. B. & King, R. H. (1989). Sediment chemistry and diatom stratigraphy of two high arctic isolation lakes, Truelove Lowland, Devon Island, N.W.T., Canada. *Journal of Paleolimnology*, **2**, 207–25.

14

Diatoms as indicators of environmental change in Antarctic and subantarctic freshwaters

SARAH A. SPAULDING, BART VAN DE VIJVER, DOMINIC A. HODGSON, DIANE M. MCKNIGHT, ELIE VERLEYEN, AND LEE STANISH

14.1 Introduction

The polar regions, both Arctic and Antarctic, show strong evidence of climate change affecting freshwater species, communities, and ecosystems, and are expected to undergo rapid and continued change in the future (IPCC, 2007). Diatoms in the freshwater and brackish habitats of inland waters of the Antarctic provide valuable records of their historic and modern environmental status. Antarctic habitats also contain a unique biodiversity of species many of which are found nowhere else on Earth. In this chapter, we review investigations using diatoms as indicators of environmental change in Antarctic and subantarctic island habitats, including lakes and ponds, streams and seepage areas, mosses and soils, cryoconite holes, brine lakes, and remarkable subsurface glacial lakes.

The Antarctic continent holds the vast majority of the Earth's freshwater, but the water is largely inaccessible because it is in the form of ice. Life is dependent upon liquid water, a substance scarce in Antarctica. Less than 0.4% of the continent is ice free, and it is within these ice-free regions that freshwater lakes and ephemeral streams form, fed by the melting of snow and glacial ice and occasional precipitation. These ice-free regions are located primarily near the Antarctic coastline (Figure 14.1). Of these regions, the "desert oases" of East Antarctica are considered to be the coldest, driest regions on Earth. In the limited parts of these oases where liquid water is available, even if present for only a few short weeks of the year, there is life (McKnight *et al.*, 1999).

Cyanobacteria (also known as blue-green algae) are the most widely distributed and abundant freshwater organisms, occurring as individuals, small colonies, or as cyanobacterial mats in lakes, ponds, and meltwater streams, and on moist soils. Higher plants occur only in lower latitudes of the Antarctic Peninsula. Diatoms are present in nearly all moist, wet to semi-dry, and some frozen habitats of the subantarctic islands and Antarctica. As in many other regions of the world, diatoms are often one of few organisms that are well preserved as subfossils.

The Diatoms: Applications for the Environmental and Earth Sciences, 2nd Edition, eds. John P. Smol and Eugene F. Stoermer. Published by Cambridge University Press.

14.1.2 Antarctic diatoms

The ability to use diatoms as environmental indicators to their greatest utility is dependent on accurate taxonomy, with the lowest prediction errors occurring in analyses using high and consistent taxonomic precision (Birks, 1994). The subantarctic and Antarctic diatom floras are remarkably regional, and it is important that future works rely on the latest primary literature from the Antarctic, rather than from other sources. We summarize the important resources in a historical and habitat context below.

Early exploration of the continent resulted in numerous collections of freshwater algae. Although diatoms were included in these surveys, they were often not of primary interest (Van Heurck, 1909; West and West, 1911). The most accessible subantarctic and maritime islands of Kerguelen and South Georgia were investigated early on (Reinsch, 1890; Carlson, 1913). Although the flora of Kerguelen Island had been the most studied of any of the Antarctic regions (Germain, 1937; Bourrelly and Manguin, 1954; Germain and Le Cohu, 1981; Le Cohu, 1981; Le Cohu and Maillard, 1983, 1986; Riaux-Gobin, 1994; Van de Vijver *et al.*, 2001), a recent intensification of work on diatom taxonomy and biogeography has occurred. There has been renewed interest in taxonomic investigations of diatoms

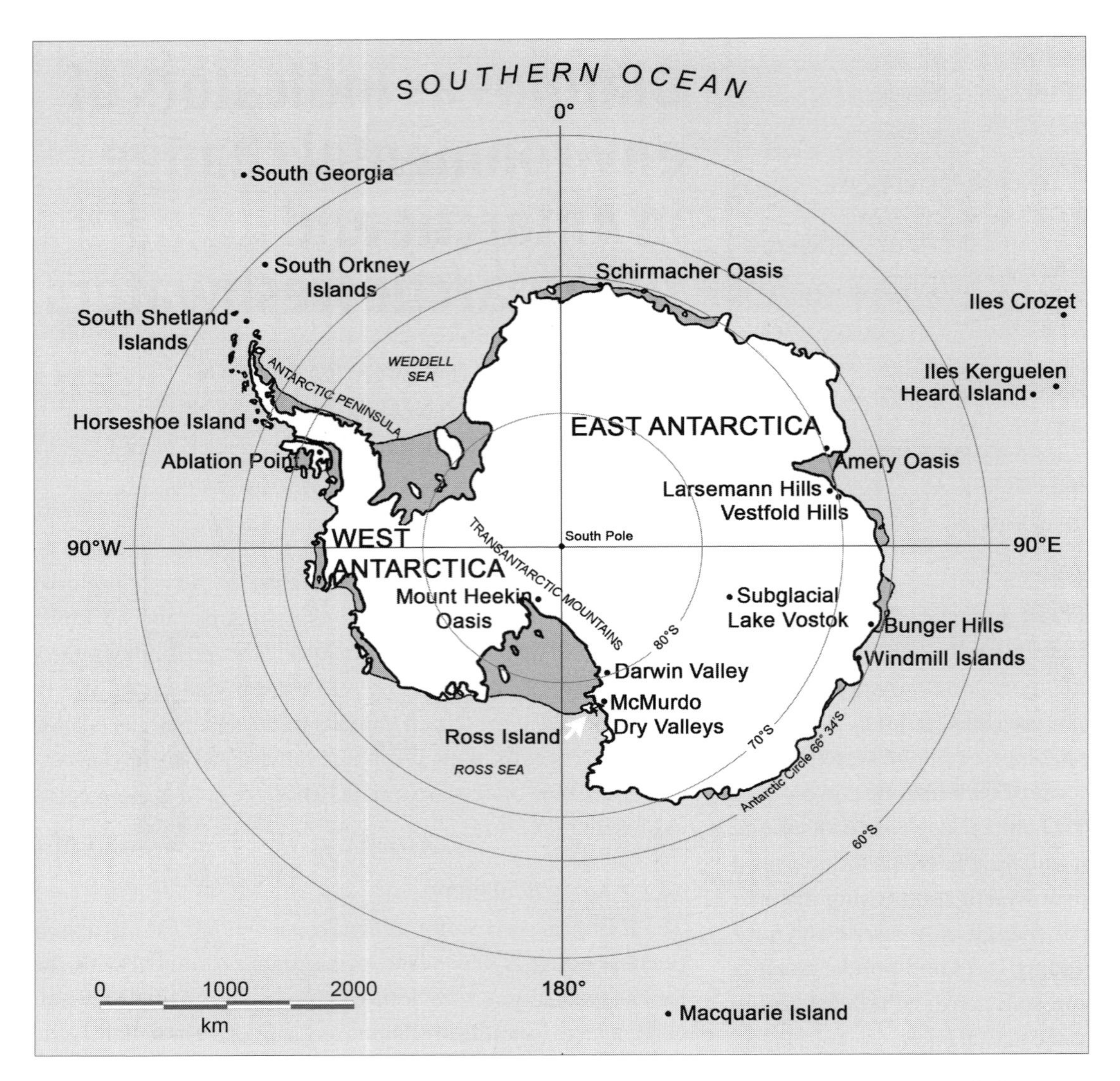

Figure 14.1 Map of the Antarctic continent showing ice-free regions of the continent and subantarctic islands.

in the subantarctic islands (Kawecka and Olech, 1993; Kopalová *et al.*, 2009; Van de Vijver and Beyens, 1999; Van de Vijver *et al.*, 2001, 2002a, 2002b, 2004), maritime Antarctic (Jones and Juggins, 1995), Antarctic Peninsula (Wasell and Håkansson, 1992; Kociolek and Jones, 1995), the McMurdo Dry Valleys (Prescott, 1979; Seaburg *et al.* 1979; Spaulding *et al.*, 1999; Esposito *et al.*, 2008), Bunger Hills (Gibson *et al.*, 2006), Windmill Islands (Roberts *et al.*, 2001), Vestfold Hills (Roberts and McMinn, 1999), Rauer Islands and Larsemann Hills (Hodgson *et al.*, 2001b; Sabbe *et al.* 2003, 2004), Amery Oasis (Cremer *et al.*, 2004) and Schirmacher Oasis (Pankow *et al.* 1991). The most recent bibliography of Antarctic works (Kellogg and Kellogg, 2002) underscores the need of improved taxonomic resolution to facilitate reconstructions of environmental change.

Species diversity of diatoms in Antarctica is low when compared to temperate regions (Jones, 1996), or even when compared to the Arctic (Douglas *et al.*, 2004). This has been confirmed using a global taxonomically consistent data set at the genus level, which showed that at comparable latitudes, lakes in the Austral region consistently have lower local and regional diatom richness than similar water bodies in the northern hemisphere (Vyverman *et al.*, 2007). Low diversity of the Antarctic flora is believed to be due, in part, to physical isolation of the continent and, thus, a limitation to dispersal. Antarctica is surrounded by circumpolar oceans, creating a barrier to colonization (Heywood, 1977). Furthermore, the number of algal species decreases with an increase in latitude (Jones, 1996), although a direct correlation between latitude and severity of the environment may not exist (Heywood, 1977). Nevertheless, areas that are conspicuously lacking in diatoms have been reported (Edward VII Peninsula (Broady,

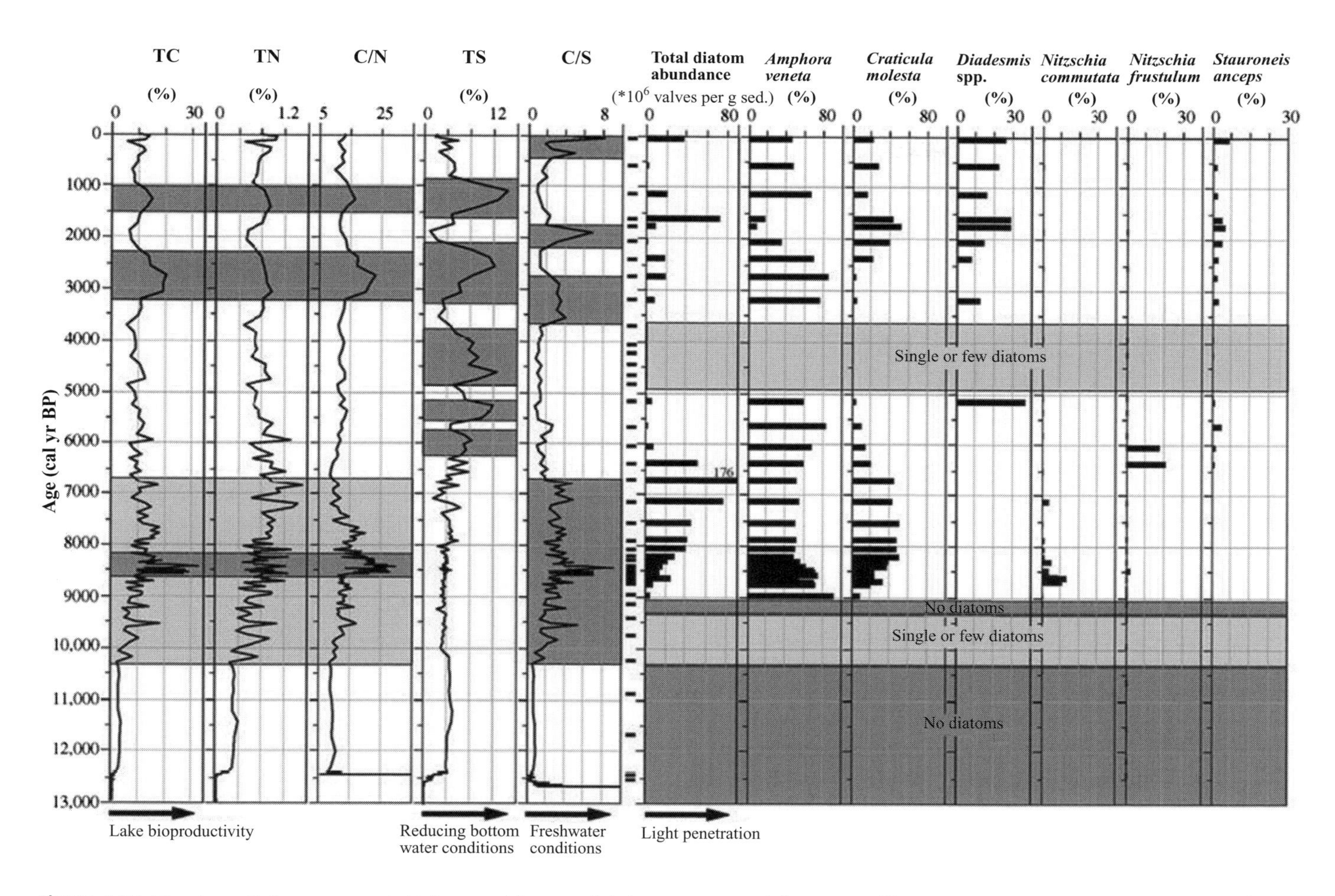

Figure 14.2 The late Pleistocene and Holocene history of Lake Terrasovoje, Amery Oasis from sediment biogeochemistry, total diatom abundance and most abundant diatom taxa (>5% of relative abundance) versus sediment ages. Sediment biogeochemistry shown includes total carbon (TC), total nitrogen (TN), carbon to nitrogen ratio (C/N), total sulfur (TS), and carbon to sulfur ratio (C/S). Shading of the biogeochemistry regions was considered of particular interest (see Wagner *et al.* (2004)). The short bars between sediment biogeochemistry and diatom abundance indicate strata that were analyzed for diatom composition. The lower gray shaded areas show the period when the lake was overridden by glacial ice (prior to *c.* 12,400 cal yr BP), followed by the onset of biogenic sedimentation with production substantially reduced by long-lived snow fields and perennial ice cover limiting light penetration, and finally the establishment of a productive diatom assemblage (10,200 cal yr BP) due to early Holocene warming and a reduction in lake ice cover. The middle gray shaded area shows a period between 4900 and 3700 cal yr BP when diatom production was again restricted by a cool period and the return of snow-covered perennial lake ice (Wagner *et al.*, 2004).

1989), Dufek Massif (Hodgson *et al.*, 2010) and Mt. Heekin Lakes (Spaulding, unpublished data)) and there are periods in the development of some lakes where diatoms are absent, or near absent, on account of unfavorable conditions such as perennial snow and ice cover (Wagner *et al.*, 2004, Figure 14.2). It will be of great interest to determine if these initial observations are supported by further investigation. Indeed, if diatoms do not occur in the highest latitudes of Antarctica, it may provide an important insight into diatom physiology and ecology.

The diatoms of the Antarctic Peninsula are more diverse than other regions, reflecting both its proximity to South America and more diverse aquatic habitats. An affiliation between the subantarctic and South American flora was recognized by several investigators (Frenguelli, 1924; Bourrelly and Manguin, 1954; Schmidt *et al.*, 1990; Jones *et al.*, 1993). For instance, many of the *Achnanthes* species found in King George Island lakes in moss habitats (Schmidt *et al.*, 1990) have also been reported in similar Chilean habitats (Krasske, 1939). Furthermore, the diatom flora of the Antarctic is quite different from that of the Arctic; the polar regions have few taxa in common (Verleyen *et al.*, 2009). While environmental conditions of the two poles may be extreme, the geographical isolation of Antarctica may limit the number of taxa to a greater extent than extreme environment (Vyverman *et al.*, 2007). The importance of this geographical isolation in structuring

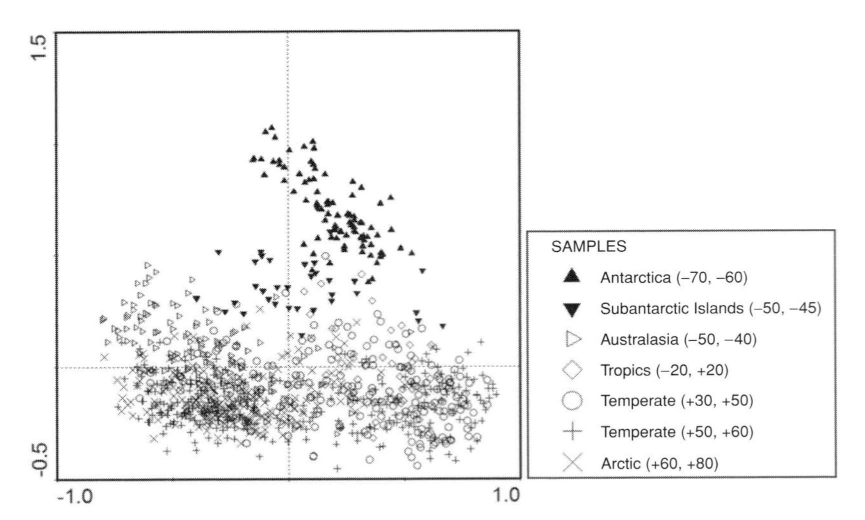

Figure 14.3 Biplot showing the first two axes in a principal component analysis of the diatom genera in 1039 lakes from both hemispheres (From Verleyen *et al.* (2009).) The species-level data were transformed to genus-level in order to ensure taxonomic consistency between globally distributed data sets. This plot shows that the diatom composition in Antarctic lakes is clearly different from the other water bodies in both hemispheres. The Australasian lakes are situated on the top left corner of the diagram and relatively separated from the water bodies in temperate regions of the northern hemisphere. Subantarctic lakes are positioned in between the Australasian and the Antarctic sites. No clear geographic grouping can be observed for the temperate lakes in the northern hemisphere. The Arctic water bodies form a distinct cluster within the other lakes of the northern hemisphere at the negative side of the first ordination axis. The tropical lakes are situated at the positive side of this axis. Numbers in parentheses in the key correspond to latitude ranges (negative for southern hemisphere).

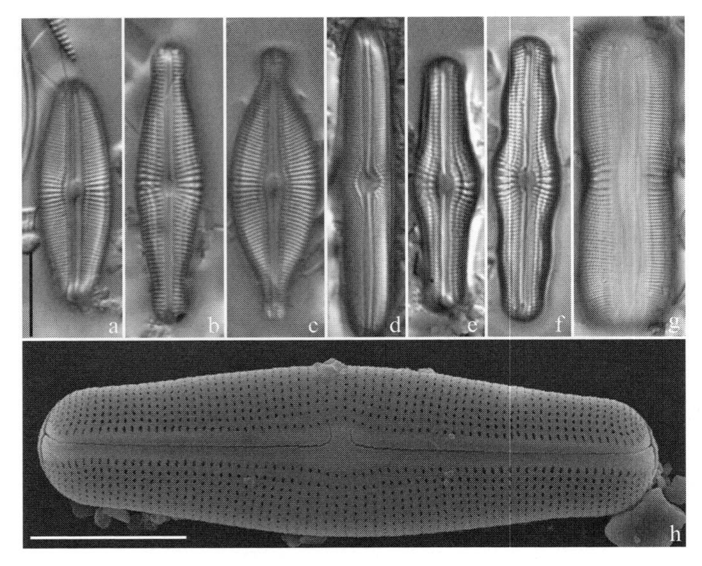

Figure 14.4 Light and scanning electron micrographs of taxa within the genus *Muelleria* (Frenguelli) Frenguelli. Scale bars equal 10 μm. (a) *M. sabbei* Van de Vijver and Spaulding from the South Shetland Islands, (b) *M. tumida* Van de Vijver and Spaulding from the South Shetland Islands, (c) *M. rostrata* Van de Vijver and Spaulding from the South Shetland Islands, (d) *M. taylorii* Van de Vijver and Coquyt from South Africa, (e) *M. cryoconicola* Stanish and Spaulding from cryoconite holes of McMurdo Dry Valleys, (f) *M. supra* Spaulding and Esposito from McMurdo Dry Valleys, (g) girdle view of *Muelleria* sp. showing deep mantles and girdle bands, and (h) *M. peraustralis* (West and West) Spaulding and Stoermer from McMurdo Dry Valleys showing unilaterally deflected proximal raphe ends and divergent distal raphe ends.

Antarctic diatom communities has been revealed by multivariate analysis of diatom genera in 1039 freshwater bodies from both hemispheres (Verleyen *et al.*, 2009), which showed that the diatom composition in Antarctic lakes is distinct from the other water bodies in both hemispheres, including Arctic regions (Figure 14.3).

Antarctic regions hold a large percentage of diatoms endemic to the continent, subregions, or even islands. The ratio of endemic to cosmopolitan taxa of Ross Island and several eastern Antarctic coastal regions was greatest between 68 and 77 degrees south (Hirano, 1965; Fukushima, 1970). Sixty three percent of the diatoms of Antarctic Peninsula lakes were found to be endemic (Schmidt *et al.*, 1990), while in the Larsemann Hills endemics account for about 40% of all freshwater and brackish taxa (Sabbe *et al.*, 2003). While the presence of species endemic to a particular geographic region implies isolating processes that lead to species evolution, the presence of endemic lineages within the Antarctic and subantarctic implies geographic stability and the presence of glacial refugia during glacial maxima in which species could survive (Spaulding *et al.*, 1999). At the same time, other work indicates diatoms are capable of rapid speciation (Theriot *et al.*, 2006). Species within the genera *Diadesmis*, *Luticola*, *Muelleria* (Figure 14.4) and

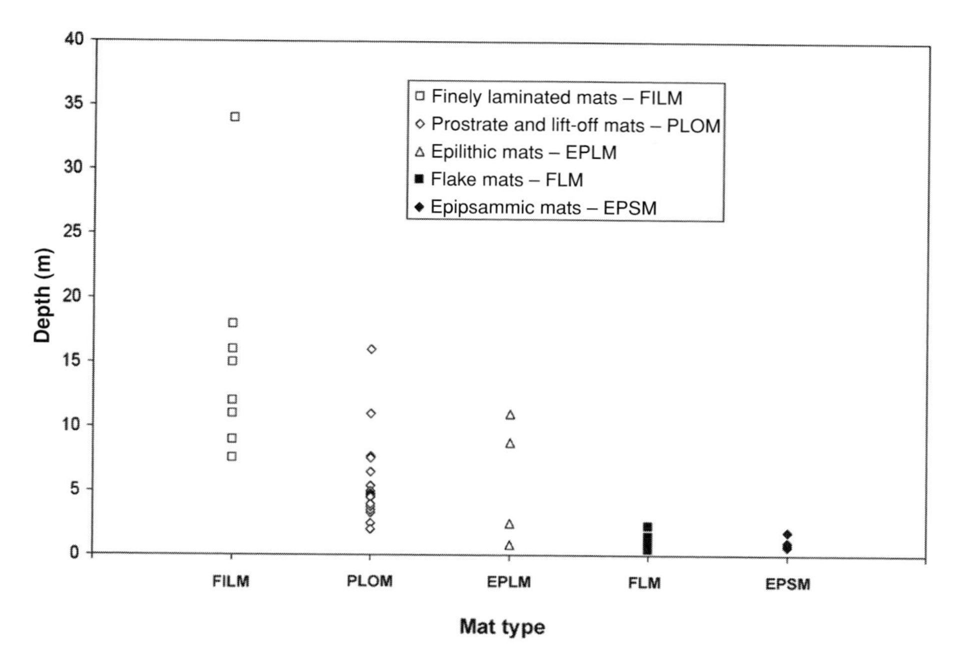

Figure 14.5 Depth zonation of microbial mat types in the lakes from the Larsemann Hills and Bølingen Islands (East Antarctica) showing the ecological threshold of 4 m of maximal lake-water depth above which well-developed microbial mats occur. Lakes shallower than 4 m are characterized by disturbed microbial mats and a different diatom flora due to the physical and chemical stress associated with yearly lake-ice formation (Sabbe *et al.* 2004).

Stauroneis are considered to have evolved within geographically restricted regions, with eleven species of *Diadesmis* from Crozet Archipelago (Van de Vijver *et al.*, 2002a) and twenty species of *Luticola*, several of which are restricted to the South Shetland Islands (Van de Vijver and Mataloni, 2008).

In addition to diatoms, the Antarctic Peninsula and East Antarctica form distinct biogeographical regions for several groups of other organisms (Andrássy, 1998; Gibson *et al.* 2008). Recent molecular data have shown that the modern Antarctic terrestrial biota have been continuously isolated on timescales of multimillion years, and pre-dates the break-up of Gondwana (Convey *et al.*, 2008). This work supports the persistence of coastal ice-free habitats as refugia through multiple Cenozoic glacial cycles. Furthermore, convergence across taxonomic groups implies ancient distributions and also presents a paradox, because speciation of diatoms in other parts of the world has been tied to evolution within ancient water bodies or landscapes (Kociolek and Spaulding, 2000; Rossiter and Kawanabe, 2000; Mackay *et al.*, this volume). Yet aerophilous habitats,which are common in the Antarctic, are habitats that are characterized by low moisture or periodic drying. Their existence is temporary and, therefore, the habitats are not geologically old. The high endemism of aerophilous diatoms in the Antarctic presents a question that is yet to be resolved.

14.2 Lakes and ponds

Extent and duration of ice cover are typically the most influential variables controlling physical, geochemical, and biological features of high-latitude lakes and ponds (Smol, 1988; Wharton *et al.*, 1993; Douglas *et al.*, 2004; Foreman *et al.*, 2004; Hodgson and Smol, 2008). Depending on latitude and elevation, the duration of ice-cover ranges from several months (e.g. northern Antarctic Peninsula lakes) to perennial (e.g. McMurdo Dry Valley lakes). Ice cover prevents wind-induced mixing and creates a stable water column, leading to perennial stratification (Spaulding *et al.*, 1997). Ice cover reduces transmission of incident light, depending on the clarity of the ice. For example, the clear ice of Lake Vanda transmitted 18% of incident light, while the opaque ice of Lake Hoare allowed less than 2% of incident light (Hawes and Schwarz, 2000). Not only does ice cover reduce the photosynthetic flux density, it alters the spectrum of transmitted light available to diatoms in benthic-mat communities (Hawes and Schwarz, 2000; Moorhead *et al.*, 2005). In addition, the physical and chemical stress associated with the yearly formation of lake ice largely determines the structure of the benthic microbial mats and associated diatom floras inhabiting the lakes (Sabbe *et al.*, 2004; Figure 14.5). Well-developed, finely laminated and undisturbed mats typically form in coastal East Antarctic lakes with a lake-water depth exceeding 4 m.

In most cases, perennial ice cover prevents the development of a planktonic diatom community and allows benthic communities to dominate in Antarctic waters (Spaulding *et al.*, 1997; Hodgson *et al.*, 2001a). Diatoms, with their silica cell walls, require water-column mixing to maintain the dense cells in suspension. In ice-covered lakes, turbulence caused by wind mixing is reduced or absent (Light *et al.*, 1981; Wharton *et al.*, 1993; Spaulding *et al.*, 1997). Although planktonic diatoms are also expected to be absent in perennially ice-covered lakes, a limited number of planktonic taxa have been reported from Antarctic freshwaters, under ice (for review see Jones, 1996). It might be expected that a planktonic flora would develop where open water is present for an extended period of time. However, even lakes that are ice free for a few months may not develop a plankton flora (Jones, 1996).

Based on investigations of diatoms in Arctic lakes, Smol (1988) proposed that the extent and duration of ice cover can be inferred from species composition and abundance of fossil diatoms (see Douglas and Smol, this volume). This model appears to be generally valid in Antarctic lakes. For example, changes in the diatom assemblages in the sediments of "Tiefersee" and "Mondsee" (unofficial names) on King George Island in the South Shetland Islands reflect differences in the extent of moat formation during the austral summer (Schmidt *et al.*, 1990). Tiefersee, located away from a glacier, develops an ice-free moat. Correspondingly, the abundance of periphytic diatoms in the sediments is greater, and also more variable than in the more persistent ice cover of Mondsee, which possesses consistently lower diatom abundance. Further, fluctuation in the extent of moat development from year to year was interpreted based on the abundance of large pennate diatoms in sediment cores. Tiefersee showed stronger indication of an ice-free moat (more variation in periphyton abundance) than Mondsee (more constant diatom abundance). On a longer timescale, the benthic diatom community and presence of sediment deposits indicated that the ice cover of Lake Hoare (McMurdo Dry Valleys) has been perennial for at least 2000 years (Spaulding *et al.*, 1997). In Lake Reid (Larsemann Hills, East Antarctica), a nearly monospecific diatom community of *Stauroforma inermis* Flower, Jones and Round occurred during the Last Glacial Maximum, likely related to perennial ice cover during this period (Hodgson *et al.*, 2005).

In addition to ice cover, diatoms respond to a range of other environmental variables. By examining subfossil remains of diatoms in lake sediments, changes in environment can be tracked over time. In some cases, reconstructions are based on absolute diatom abundance or the ecological preferences of particular diatom species. For example, absolute diatom abundance calculated from identification and enumeration of frustules combined with analyses of preserved diatom pigments (fucoxanthin, diatoxanthin, diadinoxanthin) has been used to estimate total diatom primary production. Analysis of sediments from cores of isolation lakes in the Larsemann Hills showed that transitions from marine to lacustrine habitats were accompanied by a 2- to 26-fold decrease in marine diatom production and that these declines are not entirely compensated by increased production of benthic taxa in freshwater environments during lake evolution (Verleyen *et al.*, 2004a). The ecological preferences of particular diatom species have also been used to qualitatively infer environmental changes. For example, the transitions between freshwater, brackish, and marine diatom assemblages in coastal lakes have been used to reconstruct relative sea-level changes from the isolation of lake basins at different attitudes, from which changes in ice-sheet configuration can be inferred (Wasell and Håkansson, 1992; Zwartz *et al.*, 1998; Verkulich *et al.*, 2002; Bentley *et al.*, 2005; Verleyen *et al.*, 2005; Hodgson *et al.*, 2009). Deglaciation and subsequent warm periods are associated with increased nutrient supply to lakes, which strongly influences diatom species composition. Therefore, the onset and progression of glacial melt can be inferred from the colonization of lakes by benthic diatoms and the establishment of productive diatom communities (e.g. Björck *et al.*, 1993; Björck *et al.*, 1996; Hodgson *et al.*, 2005; Wagner *et al.*, 2004). In other cases, diatoms associated with specific substrates, such as mosses, have been used to infer the presence of these substrates in sediment cores, even when moss macrofossils are not preserved. For example, the diatom assemblage preserved in sediments deposited during the last interglacial (Marine Isotope Stage (MIS) 5e, 125–115 ka BP) in the Larsemann Hills, East Antarctica, includes species that are associated with soil and moss habitats (e.g. *Diadesmis costei* Le Cohu and Van de Vijver, *Diatomella balfouriana* Grev.) and aquatic mosses (e.g. *Psammothidium manguinii* Hustedt) of the warmer maritime–Antarctic biome. One of the species (*D. costei*) is considered endemic to the subantarctic (Hodgson *et al.*, 2006b, Figure 14.6). This species assemblage is absent from the region and other ice-free oases in east Antarctica today (Verleyen *et al.* 2003) and suggests that temperatures were not only higher during the last interglacial, but present-day conditions have not warmed sufficiently for colonisation by species associated with the maritime and subantarctic biomes.

Changes in entire species assemblages may occur in response to environmental variables and can be used qualitatively to reconstruct changes through time using inference

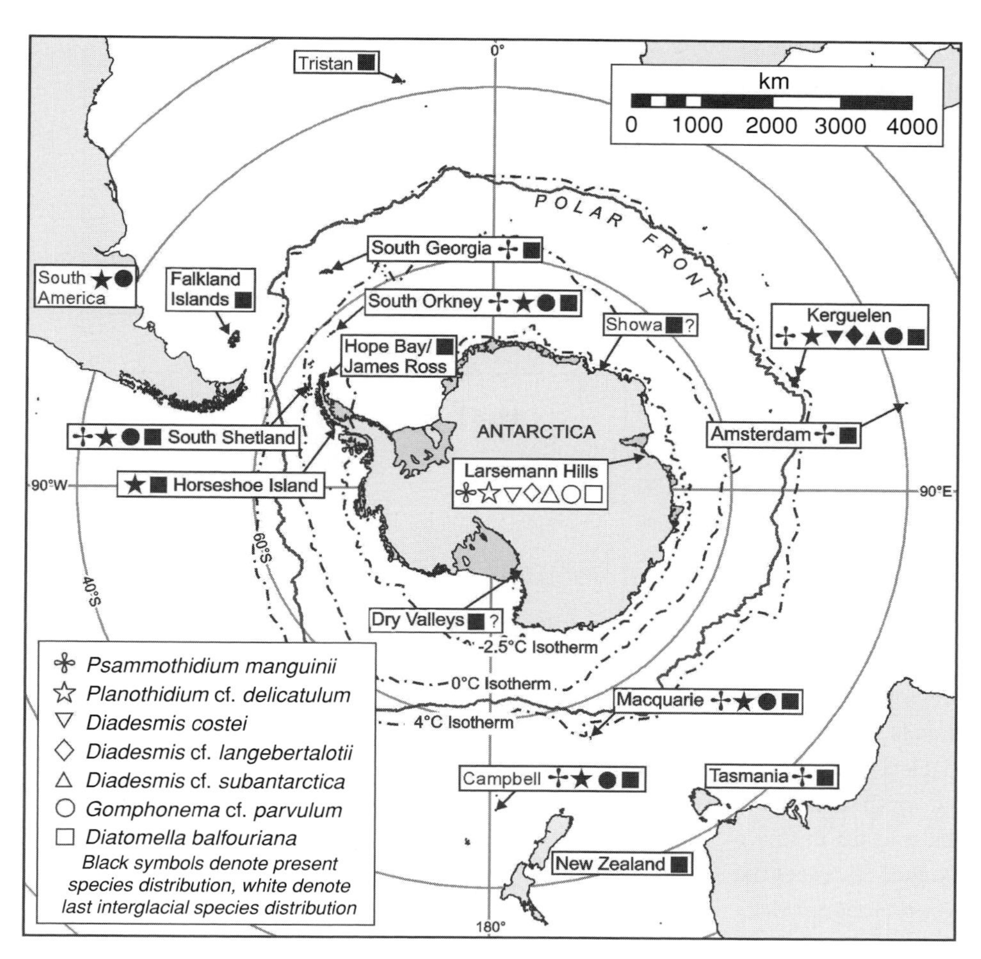

Figure 14.6 The diatom flora of Progress Lake in the Larsemann Hills, East Antarctica, during the last interglacial (white symbols) and its present day distribution in the maritime and subantarctic biomes, based on current biogeographic and morphological taxonomic evidence (black symbols). The position of the Polar Front and mean annual sea surface temperature isotherms are included to illustrate latitudinal temperature gradients (Hodgson *et al.*, 2006b).

models, or transfer functions, if the response relationships are well understood. In the Antarctic, diatom transfer functions have been developed to reconstruct changes in lake salinity (Roberts and McMinn, 1998; Verleyen *et al.*, 2003), water depth (Verleyen *et al.*, 2003), ammonium and chlorophyll *a* concentration (Jones and Juggins, 1995) and conductivity, soluble reactive phosphate, and silicate (Saunders *et al.*, 2008). Transfer functions have then been applied to reconstruct historical conditions, such as lake-water salinity (or conductivity). In closed-basin lakes, the hydrologic balance reflects input through precipitation and snow and ice melt, and loss by evaporation, sublimation and ablation. The response of particular lakes depends on the Antarctic region. In most cases, when East Antarctic lakes have become less saline, warmer periods have been inferred due to increased precipitation and snow accumulation (Roberts and McMinn, 1999; Roberts *et al.*, 2000; Roberts *et al.*, 2004; Verleyen *et al.*, 2004b, 2004c; Hodgson *et al.*, 2005). Alternatively, warmer periods have been inferred when lakes become more saline, whereby reduced snow cover has led to decreased albedo, which in turn has caused increased evaporation of the lake water and sublimation of the lake ice (Hodgson *et al.*, 2006a). In the McMurdo Dry Valleys, cold temperatures limit stream flow and lake water becomes more concentrated during colder periods (Lawrence and Hendy, 1989; Chinn, 1993; Doran *et al.*, 2008). In Lake Fryxell, carbonate concentrations were associated with halophilic diatom taxa and allowed reconstruction of paleo lake levels (Whittaker *et al.*, 2008). Peaks in stream discharge were also interpreted based on the occurrence of marine diatom fragments in lake sediments, because meltwater streams erode through ancient marine deposits, carrying marine fragments to the lake during periods of stream downcutting. Unlike diatom-inference models used in other parts of the world, Antarctic models are typically dependent on benthic diatoms alone (planktonic species are largely absent). Consequently, some of the major planktonic shifts associated with climate change and water-column warming in lower latitude environments cannot be resolved in the Antarctic using fossil diatom assemblages. Further descriptions of the methods and applications of diatoms in Antarctic paleolimnology are summarized in two recent reviews (Hodgson et al., 2004; Hodgson and Smol, 2008).

14.3 Streams

Streams and rivers are unusual habitats in the Antarctic, because they are limited in their extent and hydrologically very different from flowing waters of the rest of the planet (McKnight et al., 1999). The McMurdo Dry Valleys contain waters that flow for less than 1 km to 30 km (Onyx River of the Wright Valley). While snowfall accumulates at high elevations, it sublimates or contributes to soil moisture but does not lead to runoff to feed stream flow (Chinn, 1993). Lacking precipitation and terrestrial runoff, the ephemeral streams are fed by melting glaciers, and flow in some regions for no more than 10 weeks of the year. Furthermore, flow is dependent on solar radiation and aspect of the source glacier; flow can vary 5 to 10-fold on a daily basis (Conovitz et al. 1998). Sediments are underlain by an active layer of permafrost of 0.1 m to 0.5 m depth, which freezes and thaws annually and influences the exchange of water and solutes across the hyporheic zone (McKnight et al., 1999). Finally, because streams drain to closed-basin lakes, hydrologic processes are reflected in lake-water level and solute chemistry (Chinn, 1993; House et al., 1995; Doran et al., 2008).

Diatoms are common in McMurdo Dry Valley streams and grow in benthic mats in association with cyanobacteria, chlorophytes, nematodes, tardigrades, and rotifers (Alger et al., 1997; Esposito et al., 2008). The benthic mats are colorful oranges, blacks, greens, and browns, reflecting photosynthetic and photoprotective pigments and an accumulation of biomass at values considered high even in comparison to temperate-zone biomass (McKnight and Tate, 1997). The algal mats are perennial and survive the long periods of the year when streams do not flow, overwintering in a freeze-dried state (McKnight et al., 1999). Diatoms within these ephemeral streams tolerate not only annual periods of desiccation and freezing, but can persist for decades or centuries without liquid water. Of the 38 species of diatoms identified in McMurdo Dry Valley streams (Alger et al., 1997) 60% were considered endemic. Common taxa include *Diadesmis contenta* (Grunow) Mann, *Diadesmis perpusilla* (Grunow) Mann, *Hantzschia abundans* Lange-Bertalot,

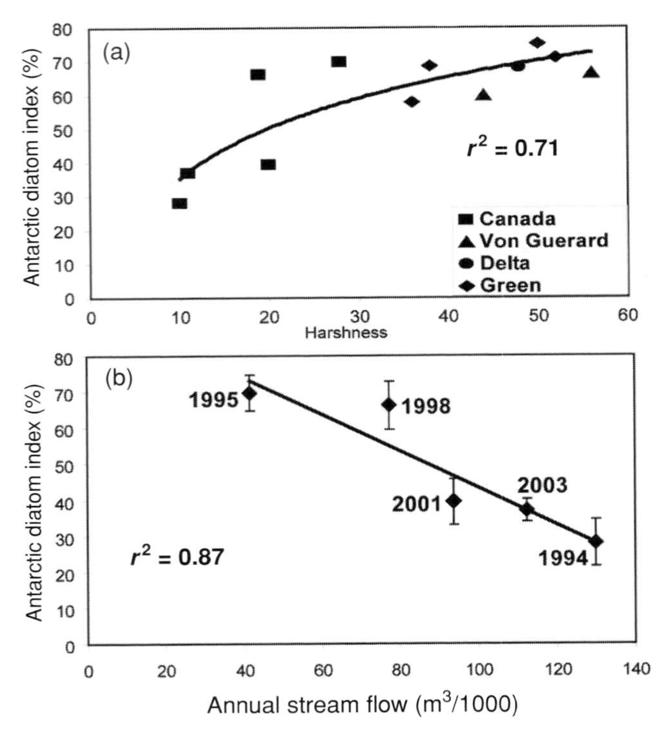

Figure 14.7 (a) Average percentage of Antarctic diatoms as a function of harshness score. Points are coded according to the legend by stream name (Canada, Von Guerard, Delta, and Green streams). The trendline and r^2 value are based on a logarithmic relationship. (b) Average percentage of Antarctic diatoms in Canada Stream as a function of stream flow. Each point represents a yearly value, as indicated. The strong negative relationship is shown with the trendline and r^2 value (Esposito et al., 2006).

Luticola austroatlantica Van de Vijver et al., *Luticola gaussii* (Heiden) Mann, *Muelleria peraustralis* (West and West) Spaulding and Stoermer, *Psammothidium chlidanos* (Hohn and Hellerman) Lange-Bertalot, and *Stauroneis latistauros* Van de Vijver and Lange-Bertalot. Diatom species composition, diversity, and endemism appear to be strongly tied to annual stream flow (Esposito et al., 2006), an observation that has direct application to quantifying climatic shifts. Under conditions of decreased solar radiation, temperature, and resulting reduction in stream flow, endemic Antarctic species become more dominant in stream assemblages (Figure 14.7). That is, cooling and the associated increase in ecological harshness of streams, favors Antarctic species over species that are more widespread. The McMurdo Dry Valleys experienced cooling trends over the 1990s (Doran et al., 2002; Torinesi et al., 2003), when diatom stream assemblages changed, a trend that may be reversed if warming occurs, as predicted (Shindell and Schmidt, 2004).

14.4 Mosses and soils

Diatoms are abundant in wet, semi-wet soils, and non-aquatic moss ecosystems, habitats that are particularly abundant in subantarctic regions. On many of the subantarctic and maritime Antarctic islands of the southern Indian Ocean, mosses are the principal vegetation, covering large areas and rich in bryophyte species (Seppelt, 2004; Ochyra *et al.*, 2008). Although moss habitats contribute a large proportion of terrestrial biomass, the diatoms living on these mosses and soils were relatively unknown until recently. Early studies of the diatom flora of Macquarie Island (Bunt, 1954) and Campbell Island (Hickman and Vitt, 1973) documented some taxa, but recent work reveals a well-developed, rich flora of more than 250 species (Van de Vijver and Beyens, 1999; Van de Vijver *et al.*, 2002b, 2004; Gremmen *et al.*, 2007). Other works have contributed to the discovery of the terrestrial diatom flora of King George Island (Van de Vijver 2008), Livingston Island (Chipev and Temniskova-Topalova, 1999, Zidarova, 2008) and Deception Island (Fermani *et al.*, 2007, Van de Vijver and Mataloni, 2008). The taxonomy of many of the subantarctic groups has recently been revised (Van de Vijver, 2008; Van de Vijver and Mataloni, 2008), demonstrating a diatom flora of regionally endemic species.

After accounting for regional differences, the composition and species richness of diatoms in mosses and soils is strongly influenced by moisture content (Chipev and Temniskova-Topalova, 1999; Van de Vijver and Beyens, 1999; Van de Vijver *et al.*, 2002b, 2004). In dry habitats, species numbers are consistently lower than in wet habitats. Furthermore, only a limited number of species and genera appear to be capable of surviving xeric conditions (Van de Vijver and Beyens, 1999) and in those taxa, valve size typically decreases with decreasing moisture content. For example, valve size of *Pinnularia borealis* Ehrenb. on the island of South Georgia was significantly lower in the driest mosses than in wetter mosses. The environmental variables that influence the species composition differ across habitat types. That is, the species composition of diatom communities in moss habitats is related to habitat type (Van de Vijver and Beyens, 1999), while the species of diatom communities in soil habitats is influenced by nutrient concentrations (phosphate and nitrate). Animal-influenced habitats, such as the upper layer of the soil in penguin rookeries, are a clear example of this influence (Van de Vijver *et al.*, 2002a; Fermani *et al.*, 2007). All of these factors provide great potential for use of moss diatoms in interpreting shifts in moisture and habitat availablility.

The Antarctic Convergence, or Antarctic Polar Front, is considered the climatic boundary between the Antarctic continent and several oceanic fronts (Broecker, 1996; Bard *et al.*, 1997; Domack and Mayewski, 1999) and the subantarctic islands are located within this turbulent zone of the Antarctic Convergence. In this geographic position the islands are influenced by circumpolar winds, the Southern Westerlies, whose fluctuations influence ocean circulation and global climate. As a result, understanding species and environment relationships has allowed paleoclimatic reconstruction in this important region. The diatom species composition of moss habitats on Kerguelen Island has been related to elevation, a proxy for temperature (Gremmen *et al.*, 2007). Regardless of the moss species composition, diatoms growing within moss habitats were related to altitude, with diatoms having colder optima also favoring higher altitudes. Accordingly, peat cores obtained from Ile de la Possession (Crozet Archipelago) led to an environmental reconstruction of the past 6200 years using both moss remains and diatoms. Results indicate a major shift at 2800 yr BP to a wetter, windier climate attributed to an intensification, or latitudinal shift, or both, of the Southern Westerlies (Van der Putten *et al.*, 2009).

14.5 Eolian transport of diatom valves

The presence of eolian marine and freshwater diatoms in ice cores of the polar plateau (Kellogg and Kellogg, 1996) can be used as tracers of air masses and circulation patterns. Atmospheric transport of freshwater diatom frustules is known to occur over long distances (Lichti-Federovich, 1984; Gayley *et al.*, 1989; Chalmers *et al.*, 1996); however, transport of valves does not imply that cells are viable. In the Antarctic, diatom valves were present in ice cores obtained from high elevations of the polar plateau, several hundred kilometers from freshwater sources (Burckle *et al.*, 1988; Kellogg and Kellogg, 1996). In these studies, the diatoms were considered eolian, because no melt pools were known from the surface of the ice cap. However, the occurrence of diatoms in melt pools on the surface of glaciers is well established and can serve as a source of living populations (this chapter). The species found on the polar plateau were 98% lacustrine taxa, including *Luticola muticopsis* (Van Heurck) Mann, *Navicula shackletoni* West and West, *Luticola cohnii* (Hilse) Mann, *N. deltaica* Kellogg (Burckle *et al.*, 1988), species known from freshwater and hypersaline lakes and deposits in Antarctica. The reconstruction of lake level in Lake Fryxell during the Holocene (Whittaker *et al.*, 2008) was aided by recognition of a period in which deposition of eolian diatoms dominated the accumulation of sediments. Both sands and diatoms were blown onto the lake surface, and migrated through the lake ice cover.

Marine forms are not likely to be transported by wind over long distances (Burckle *et al.*, 1988), and glacial ice is dominated by freshwater forms (lakes make up only 0.32% of the Antarctic terrestrial surface area) rather than marine forms. Yet, a crucial component of eolian transport is that subfossil diatoms of terrestrial deposits are most easily transported (Sancetta *et al.* 1992), as compared to live cells. While empty diatom valves of marine and freshwater species may be transported considerable distance, living forms may not be so easily entrained in the atmosphere. For example, onshore winds were found to be effective in lifting marine diatoms from the sea surface, but there was no support for transport beyond the nearshore region (Maynard, 1968). A similar study from the Ross Sea, Antarctica, found that while marine particles were transported with aerosols, fragments or whole diatom valves were not included (Saxena *et al.*, 1985). The transport of living marine diatoms for any distance has not been demonstrated, presumably because living cells have difficulty becoming airborne, as they often form long chains, have external organic coatings, and contain dense internal protoplasm (Burckle *et al.*, 1988). The combination of diatom species that are endemic to narrow regions and eolian transport has the potential to be utilized to a greater extent.

14.6 Cryoconite holes

Cryoconite holes are depressions on glacier surfaces filled with water and sediment, named for the "cold rock dust" they contain (Nordenskjöld 1875). Wind-blown sediment is transported onto glacier surfaces and collects in depressions in the ice. The lower albedo of the sediment causes the ice to melt more quickly than its surroundings, forming a hole. Over time, wind-blown sediment and organic matter accumulates and enhances the melting process. Organisms transported into cryoconite holes can increase expansion of the holes through absorption of solar energy and metabolic activity (Gerdel and Drouet, 1960; McIntyre, 1984). Cryoconite holes may be significant on glaciers, accounting for up to 6% of the surface area of the accumulation zone (Gerdel and Drouet 1960; Fountain *et al.*, 2004; Hodson *et al.*, 2007), ranging from 5–145 cm in diameter to 4–56 cm in depth. In the Antarctic, meltwater from cryoconite holes generates as much as 13% of glacial runoff (Fountain *et al.*, 2004). In addition to their importance in glacial mass balance, cryoconite holes may serve as refugia for biota and play an important role in re-seeding neighboring lakes or streams (Wharton and Vinyard, 1983; Wharton *et al.*, 1981; Porazinska *et al.*, 2004). In lieu of the extreme physical and chemical conditions within cryoconite holes, researchers consider these systems to be analogs of refugia on Earth during cold periods and other icy planets (Tranter *et al.*, 2004; Säwström *et al.*, 2002).

Although the organisms in cryoconite holes typically reflect the aquatic biota of surrounding areas, containing cyanobacteria, green algae, metazoans, and diatoms (Wharton 1981, 1985; Mueller *et al.*, 2001; Christner *et al.*, 2003; Porazinska *et al.*, 2004; Stibal *et al.*, 2006), there may be important differences. The microhabitat appears to favor species that can withstand the extreme changes in physical and chemical conditions and, in the McMurdo Dry Valleys, contain high abundances of the diatom *Muelleria cryoconicola* Stanish and Spaulding, which appears to be endemic to these microhabitats (Van de Vijver *et al.*, 2010). Although these microhabitats have not been used in terms of interpreting environmental conditions yet, the unique species within them are of potential use in detecting environmental change to glacial surfaces.

14.7 Brine lakes

While a few brine, or hypersaline, lakes are extant in the McMurdo Dry Valleys (Doran *et al.*, 2003), many closed-basin lakes (including lakes Vanda, Bonney, and Fryxell) were reduced to brine lakes in the past (Lyons *et al.*, 2006; Whittaker *et al.*, 2008). The present brine lakes include Lake Vida, which is considered to be at the extreme end of lacustrine types (Doran *et al.*, 2003). The ice cover of Lake Vida is 19 m thick and the water column is composed of a sodium chloride (NaCl) brine, seven times as saline as seawater. The brine is considered to have been isolated for at least 2800 years, based on ^{14}C dates. Because of its extreme salinity, the brine remains liquid below -10 °C. Although the surface ice contains viable microbial mats, there has been no effort to date to determine if diatoms are present. Large glacial lakes existed in the McMurdo Dry Valleys during the last glacial maximum (Stuiver *et al.*, 1981; Hall and Denton, 2000; Whittaker *et al.*, 2008) and an understanding of present-day brines may help understand sediment records of the past. While sediments of Lake Vanda contain well-preserved diatoms (Spaulding, personal observation), to date they have not been used to add to interpretation of the paleoclimate of the region.

14.8 Subglacial lakes

In the future, diatoms near, and in, subglacial lake sediments will be of great interest in interpreting past conditions. Evidence from perennially ice-covered lakes, or periods in lake-sediment records when lakes were perennially ice covered, shows that diatom photosynthesis is substantially inhibited in low-light environments. As a result, diatoms communities are typically absent during periods of thick perennial snow and ice cover (Hodgson *et al.*, 2006b; Wagner *et al.*, 2004, Figure 14.3) or consist of monospecific assemblages under permanent lake ice cover (Hodgson *et al.*, 2005). Diatoms are therefore not anticipated to be growing in subsurface lakes where there is insufficient light for photosynthesis, although heterotrophic growth could be present. Nevertheless, a wide range of other microorganisms (bacteria, yeasts, fungi, microalgae) are present within the deep glacial ice above Lake Vostok, up 2750 m in depth (Abyzov *et al.*, 1998). Scanning electron micrographs of Lake Vostok accretion ice at a depth of 3590 m show fragments of extant diatom species (Doran *et al.*, 2004; see page 491, figures 10e–f), some recognizable as marine forms. Further recovery of frustules from around Lake Vostok may reveal the types and extent of diatom habitats, and therefore characterize the conditions present during different periods in lake development.

14.9 Fossil diatoms in glacial deposits to reconstruct ice-sheet and climate history

The presence of subfossil marine diatoms has provided crucial evidence for the extent of the Antarctic ice sheets, and the age of these diatoms (determined by biostratigraphy and cosmogenic ^{10}Be concentrations in the sediment matrix) indicates time periods when the ice sheet was absent. This is particularly pertinent for the West Antarctic Ice Sheet, whose partial or complete collapse in previous interglacials may have resulted in substantial contributions to global sea-level rise (Denton and Hughes, 2002). Knowledge of how this ice-sheet configuration changes during periods of relatively well-known paleoclimatic conditions critically advises our ability to predict future behavior. Several studies have proposed West Antarctic Ice Sheet collapse at least once in the last 600 ka (Hearty *et al.*, 1999; Scherer *et al.*, 1998), with candidate periods being MIS 11, 400 ka, an unusually long interglacial when the Southern Ocean was warmer than at present, and even (for partial collapse) MIS 5e (Scherer *et al.*, 1998), when temperatures were between 2 and 5 °C higher than today and global sea levels were 4–6 m higher (Jansen *et al.*, 2007).

A long-standing controversy continues over the East Antarctic Ice Sheet and whether it was stable, or warmed and melted, raising global sea level by more than 25 m during the Pliocene. The debate concerning the paleoclimate of Antarctica during the Pliocene centers on the interpretation of diatoms in the glacially deposited Sirius Formation (Webb *et al.*, 1984; Denton *et al.*, 1991; Webb and Harwood, 1991; LeMasurier *et al.*, 1994; Wilson *et al.*, 2002). The "dynamic" hypothesis proposes that the Sirius deposits contain reworked marine diatoms that were deposited within basins in East Antarctica during a warm interval in the Pliocene, when the East Antarctic Ice Sheet retreated to about one-third of its present size (Webb *et al.*, 1984; Webb and Harwood, 1991). According to this hypothesis, grounded ice scoured the marine basins during a later cold period, incorporated marine sediments and diatoms, and deposited the reworked material in the Transantarctic Mountains (in the Sirius Formation). Alternatively, the "stable" hypothesis proponents contend that the East Antarctic Ice Sheet has remained essentially unchanged for millions of years (Denton *et al.*, 1991; LeMasurier *et al.*, 1994), and that diatoms in the Sirius deposits are the result of surface (eolian) contamination (Burckle and Potter, 1996). Burckle and Potter discovered Pliocene–Pleistocene diatoms within fractures of Devonian rocks, and concluded that the younger diatoms were transported by wind and deposited within the much older rocks. Mahood and Barron (1996) point out that controversies such as that surrounding the Sirius Formation could be resolved by closer taxonomic scrutiny of diatom taxa used as markers of biostratigraphy and environmental conditions. With such detailed examination of taxa, misunderstandings of the identity, age, and source of diatoms would be far less likely. Indeed, the most recent efforts to resolve the age and origin of glacial deposits of the Sirius Formation has revealed a complex glacial history supporting temperate climatic conditions in the Mount Feather region until the Late Miocene (Wilson *et al.*, 2002). The non-marine diatoms of the Mount Feather Diamicton are dominated by *Anomoeoneis costata* Hust. and *Stephanodiscus* sp., species whose presence implies a temperate climate. These taxa no longer grow under the current glacial conditions of the Antarctic continent. Additional work to characterize the diatom assemblages promises to further identify and constrain the sources of glacial deposits.

14.10 Summary

Diatoms of the Antarctic and subantarctic islands are part of a unique flora that are exposed to high gradients of solar radiation, short growing seasons, extreme temperature ranges, changing amounts of moisture, strong salinity gradients, and vulnerability to climate change. The species diversity of diatoms of inland habitats in Antarctica is low compared to both temperate and Arctic regions, due, in part, to physical isolation of the continent. Within the continent, diatoms of the Antarctic Peninsula are more diverse than other regions, reflecting both the proximity to South America and the more diverse aquatic habitats of the warmer peninsula region. Despite the cold conditions, Antarctic regions have also been recognized as sites that contain a large percentage of diatoms endemic to the continent, small subregions, or even islands.

Diatoms have been used to determine environmental change in the Antarctic from a variety of habitats, including lakes, streams, mosses and soils, and fossil deposits. In some sites, the history of lake ice cover has been inferred. Because perennial ice cover prevents the development of a planktonic diatom community, the composition of diatom assemblages in the sediments can be used to detect differences in the extent of melting of ice cover and the formation of a moat during the austral summer. The sediments of Lake Hoare (McMurdo Dry Valleys) revealed a benthic diatom community and the presence of sediment deposits indicated that the ice cover has been perennial for at least 2000 years. Several of the coastal lakes have been influenced by marine inflows, or were entirely marine. Lake sediment cores allow estimates of marine diatom production and shifts to increased production of benthic taxa in freshwater environments as glaciers retreated during the Holocene. Deglaciation and subsequent warm periods are associated with increased nutrient supply to lakes, which strongly influences diatom species composition. Therefore, the onset and progression of glacial melt can be inferred from the colonization of lakes by benthic diatoms and the establishment of productive diatom communities. Diatom transfer functions have been developed to reconstruct changes in lake salinity, water depth, ammonium and chlorophyll *a* concentration, conductance, soluble reactive phosphate, and silicate.

The response of particular lakes to changes in temperature depends on the Antarctic region. In most cases, when East Antarctic lakes have become less saline, warmer periods have been inferred due to increased precipitation and snow accumulation reflecting input through precipitation and snow and ice melt, and loss by evaporation and ablation. Alternatively, warmer periods have been inferred when lakes become more saline, whereby reduced snow cover has led to decreased albedo, which in turn has caused increased evaporation of the lake water and sublimation of the lake ice. Unlike diatom-inference models used in other parts of the world, Antarctic models are typically dependent on benthic diatoms alone because planktonic species are largely absent.

In streams, diatom species composition, diversity, and endemism appear to be strongly tied to annual stream flow, an observation that has direct application to quantifying climatic shifts. Under conditions of decreased solar radiation, temperature, and resulting reduction in stream flow, endemic Antarctic species become more dominant in stream assemblages. In moss and stream habitats, the diatom species composition on Kerguelen Island has been related to elevation, a proxy for temperature. Regardless of the moss species composition, diatoms growing within moss habitats were related to altitude, with diatoms having colder optima also favoring higher altitudes.

A debate concerning the paleoclimate of Antarctica during the Pliocene centers on the interpretation of diatoms in the glacially deposited Sirius Formation. The dynamic hypothesis proposes that the Sirius deposits contain reworked marine diatoms that were deposited within basins in East Antarctica during a warm interval. According to this hypothesis, grounded ice scoured the marine basins during a later cold period, incorporated marine sediments and diatoms, and deposited the reworked material in the Transantarctic Mountains. Alternatively, the stable-hypothesis proponents contend that the East Antarctic Ice Sheet has remained essentially unchanged for millions of years.

Future work may utilize the understanding of eolian transport, cryoconite holes, brine lakes, and subglacial lakes to apply diatoms as tools for understanding aspects of Earth's history. The polar regions, both Arctic and Antarctic, show strong evidence of climate change in freshwater species, communities, and ecosystems, and are expected to undergo rapid and continued change in the future. Diatoms in the freshwater and brackish habitats of inland waters of the Antarctic hold valuable records of historic and modern environmental status, as well as containing a unique biodiversity of species found nowhere else on Earth.

Acknowledgments

We thank Wim Vyverman, Koen Sabbe, and HOLANT project partners for useful discussions. Thanks also to Jill Baron for comments on earlier drafts. The Fort Collins Science Center of the US Geological Survey provided support to S. A. Spaulding. E. Verleyen is a post doctoral research fellow with the Fund for Scientific Research – Flanders (Belgium). Any use of trade, product, or firm names is for descriptive purposes only and does not imply endorsement by the US Government.

References

Abyzov, S. S., Mitskevich, I. N., & Poglazova, M. N. (1998). Microflora of the deep glacial horizons of central Antarctica. *Microbiology*, 67, 66–73.

Alger, A. S., McKnight, D. M., Spaulding, S. A., *et al.* (1997). *Ecological Processes in a Cold Desert Ecosystem: the Abundance and Species Distribution of Algal Mats in Glacial Meltwater Streams in Taylor Valley, Antarctica*. Boulder, CO: Institute of Arctic and Alpine Research Occasional Paper, no. 51.

Andrássy, I. (1998). Nematodes in the sixth continent. *Journal of Nematode Systematics and Morphology*, **1**, 107–86.

Bard, E., Rostek, F. & Sonzogni, C. (1997). Interhemispheric synchrony of the last deglaciation inferred from alkenone palaeothermometry. *Nature*, **385**, 707–10.

Bentley, M. J., Hodgson, D. A., Smith, J. A., & Cox, N. J. (2005). Relative sea level curves for the South Shetland Islands and Marguerite Bay, Antarctic Peninsula. *Quaternary Science Reviews*, **24**, 1203–16.

Birks, H. J. B. (1994). The importance of pollen and diatom taxonomic precision in quantitative palaeoenvironmental reconstructions. *Review of Palaeobotany and Palynology*, **83**, 107–17.

Björck, S., Håkansson, H., Olsson, S., Barnekow, L., & Janssens, J. (1993). Palaeoclimatic studies in South Shetland Islands, Antarctica, based on numerous stratigraphic variables in lake sediments. *Journal of Paleolimnology*, **8**, 233–72.

Björck, S., Olsson, S., Ellis-Evans, C., *et al.* (1996). Late Holocene palaeoclimatic records from lake sediments on James Ross Island, Antarctica. *Palaeogeography, Palaeoclimatology, Palaeoecology*, **121**, 195–220.

Bourrelly, P. & Manguin, E. (1954). Contribution à la flore algale d'eau douce des îles Kerguelen. *Mémoires de l'Institut Scientifique de Madagascar*, **5**, 7–58.

Broady, P. A. (1989). Survey of algae and other terrestrial biota at Edward VII Peninsula, Marie Byrd Land. *Antarctic Science*, **1**, 215–24.

Broecker, W. (1996). Paleoclimatology. *Geotimes*, **41**, 40–1.

Bunt, J. S. (1954). A comparative account of the terrestrial diatoms of Macquarie Island. *Proceedings of the Linnean Society of New South Wales*, **LXXIX**, parts 1–2, 34–57.

Burckle, L. H., Gayley, R. I., Ram, M., & Petit, J. R. (1988). Diatoms in Antarctic ice cores: some implications for the glacial history of Antarctica. *Geology*, **16**, 326–9.

Burckle, L. H. & Potter, N. Jr. (1996). Pliocene–Pleistocene diatoms in Paleozoic and Mesozoic sedimentary and igneous rocks from Antarctica: a Sirius problem solved. *Geology*, **24**, 235–8.

Carlson, G. W. (1913). Süsswasseralgen aus der Antarktis, Süd-Georgien und den Falkland Inseln. *Wissenshafliche Ergebnisse schwedische Süd-Polar-Expedition 1901–1903. Botanik*, **4**, 1–94.

Chalmers, M. O., Harper, M. A., & Marshall, W. A. (1996). *An Illustrated Catalogue of Airborne Microbiota from the Maritime Antarctic*. Cambridge: British Antarctic Survey.

Chinn, T. H. (1993). Physical hydrology of the Dry Valley lakes. In *Physical and Biogeochemical Processes in Antarctic Lakes, Antarctic Research Series Vol. 59*, ed. W. J. Green & E. I. Friedmann. Washington DC: American Geophysical Union, pp. 1–51.

Chipev, N. & Temniskova-Topalova, D. (1999). Diversity dynamics and distribution of diatom assemblages in land habitats on Livingston Island (Antarctica). In *Bulgarian Antarctic Research. Life Sciences*, ed. V. Golemansky and N. Chipev, vol. 2, pp. 32–42.

Christner, B. C., Kvitko, B. H., & Reeve, J. N. (2003). Molecular identification of bacteria and eukarya inhabiting an Antarctic cryoconite hole. *Extremophiles*, **7**, 177–83.

Conovitz, P. A., McKnight, D. M., MacDonald, L. H., Fountain, A. G. & House, H. R. (1998). Hydrologic processes influencing streamflow variation in Fryxell Basin, Antarctica. In *Ecosytem Processes in a Polar Desert: The McMurdo Dry Valleys, Antarctica, Antarctic Research Series, Vol. 72*. ed. J. C. Priscu. Washington DC: American Geophysical Union, pp. 93–108.

Convey, P., Gibson, J. A. E., Hillenbrand, C., Hodgson, D. A., Pugh, P. J. A., Smellie, J. L. & Stevens, M. I. (2008). Antarctic terrestrial life – challenging the history of the frozen continent? *Biological Reviews*, **83**, 103–17.

Cremer, H., Gore, D., Hultzsch, N., Melles, M., & Wagner, B. (2004). The diatom flora and limnology of lakes in the Amery Oasis, East Antarctica. *Polar Biology*, **27**, 513–31.

Denton, G. H. & Hughes, T. J. (2002). Reconstructing the Antarctic Ice Sheet at the last glacial maximum. *Quaternary Science Review*, **21**, 193–202.

Denton, D. H., Prentice, M. L. & Burkle, L. H. (1991). Cenozoic history of the Antarctic Ice Sheet. In *The Geology of Antarctica*, ed. R. J. Tingey. Oxford: Oxford University Press, pp. 365–433.

Domack, E. W. & Mayewski, P. A. (1999). Bi-polar ocean linkages: evidence from late-Holocene Antarctic marine and Greenland ice-core records. *The Holocene*, **9**, 247–51.

Doran, P. T., Fritsen, C. H., McKay, C. P., Priscu, J. C., & Adams, E. E. (2003). Formation and character of an ancient 19 m ice cover and underlying trapped brine in an "ice-sealed" East Antarctic lake. *Proceedings of National Academy of Sciences*, **100**, 26–31.

Doran, P. T., McKay, C. P., Fountain, A. G., *et al.* (2008). Hydrologic response to extreme warm and cold summers in the McMurdo Dry Valleys, East Antarctica. *Antarctic Science*, **20**, 499–509.

Doran, P. T., Priscu, J. C., Berry Lyons, W., *et al.* (2004). Paleolimnology of ice-covered environments. In *Long-term Environmental Change in Arctic and Antarctic Lakes*, ed. R. Pienitz, M. S. V. Douglas, & J. Smol. Dordrecht: Springer, pp. 475–508.

Doran, P. T., Priscu, J. C., Lyons, W. B., *et al.* (2002). Antarctic climate cooling and terrestrial ecosystem response. *Nature*, **415**, 517–20.

Douglas, M. S. V., Hamilton, P. B., Pienitz, R., & Smol, J. P. (2004). Algal indicators of environmental change in Arctic and Antarctic lakes and ponds. In *Long-term Environmental Change in Arctic and Antarctic Lakes*, ed. R. Pienitz, M. Douglas, & J. Smol, Dordrecht: Springer, pp. 117–58.

Esposito, R. M. M., Horn, S. L., McKnight, D. M., *et al.* (2006). Antarctic climate cooling and response of diatoms in glacial meltwater streams. *Geophysical Research Letters*, **33**, L07406.

Esposito, R. M. M., Spaulding, S. A., McKnight, D. M., *et al.* (2008). Inland diatoms from the McMurdo Dry Valleys and James Ross Island, Antarctica. *Botany*, **86**, 1378–92.

Fermani, P., Mataloni, G., & Van de Vijver, B. (2007). Soil microalgal communities on an Antarctic active volcano (Deception Island, South Shetlands). *Polar Biology*, **30**, 1381–93.

Foreman, C. M., Wolf, C. F., & Priscu, J. C. (2004). Impact of episodic warming events on the physical, chemical and biological relationships of lakes in the McMurdo Dry Valleys, Antarctica. *Aquatic Geochemistry*, **10**, 239–68.

Fountain, A. G., Tranter, M., Nylen, T. H., Lewis, K. J., & Mueller, D. R. (2004). Evolution of cryoconite holes and their contribution to meltwater runoff from glaciers in the McMurdo Dry Valleys, Antarctica. *Journal of Glaciology*, **50**, 2004, 35–45.

Frenguelli, J. (1924). Diatomeas de Tierra del Fuego. *Anales de la Sociedad Cientifica Argentina*, **97**, 87–118, 231–66.

Fukushima, H. (1970). Notes on the diatom flora of Antarctic inland waters. In *Antarctic Ecology*, ed. M. W. Holdgate, London: Academic Press, pp. 628–31.

Gayley, R. I., Ram, M., & Stoermer, E. F. (1989). Seasonal variations in diatom abundance and provenance in Greenland ice. *Journal of Glaciology*, **35**, 290–2.

Gerdel, R. W. & Drouet, F. (1960). The cryoconite of the Thule area, Greenland. *Transactions of the American Microscopic Society*, **79**, 256–72.

Germain, H. (1937). Diatomées d'une tourbe de l'Ile Kerguelen. *Bulletin de la Societé Française de Microscopie*, **6**, 11–16.

Germain, H. & LeCohu, R. (1981). Variability of some features in a few species of *Gomphonema* from France and the Kerguelen Islands (South Indian Ocean). In *Proceedings of the 6th Symposium on Recent and Fossil Diatoms*, ed. R. Ross, Königstein: Koeltz Scientific Books, pp. 167–77.

Gibson, J. A. E., Roberts, D., & Van de Vijver, B. (2006). Salinity control of the distribution of diatoms in lakes of the Bunger Hills, East Antarctica. *Polar Biology*, **29**, 694–704.

Gibson, J. A. E., Wilmotte, A., Taton, A., *et al.* (2008). Biogeographical trends in Antarctic lakes. In *Trends in Antarctic Terrestrial and Limnetic Ecosystems*, ed. D. Bergstrom, A. Huiskes, & P. Convey, Amsterdam: Springer, pp. 71–99.

Gremmen, N. J. M., Van de Vijver, B., Frenot, Y., & Lebouvier, M. (2007). Distribution of moss-inhabiting diatoms along an altitudinal gradient at sub-Antarctic Îles Kerguelen. *Antarctic Science*, **19**, 17–24.

Hall, B. L. & Denton, G. H. (2000). Radiocarbon chronology of Ross Sea drift, eastern Taylor Valley, Antarctica: evidence for a grounded ice sheet in the Ross Sea at the last glacial maximum. *Geografiska Annaler*, **82**, 305–36.

Hawes, I. & Schwarz, A. M. (2000). Absorption and utilization of irradiance by cyanobacterial mats in ice-covered Antarctic lakes with contrasting light climates. *Journal of Phycology*, **37**, 5–15.

Hearty, P. J., Kindler, P., Cheng, H., & Edwards, R. L. (1999). A +20 m middle Pleistocene sea-level highstand (Bermuda and the Bahamas) due to partial collapse of Antarctic ice. *Geology*, **27**, 375–8.

Heywood, R. B. (1977). Antarctic freshwater ecosystems: review and synthesis. In *Adaptations within Antarctic Ecosystems*, ed. G. A. Llano, Washington, DC: Smithsonian Institution, pp. 801–28.

Hickman, M. & Vitt, D. H. (1973). The aerial ephytic diatom flora of moss species from subantarctic Campbell Island. *Nova Hedwigia*, **24**, 443–58.

Hirano, M. (1965). Freshwater algae in the Antarctic regions. In *Biogeography and Ecology in Antarctica*, ed. J. van Mieghem & P. van Oye, The Hague: Junk, pp. 28–191.

Hodson, A., Anesio, A. M., Ng, F., *et al.* (2007). A glacier respires: quantifying the distribution and respiration CO_2 flux of cryoconite across an entire Arctic supraglacial ecosystem. *Journal of Geophysical Research*, **112**, G04S36. DOI:10.1029/2007JG000452.

Hodgson, D. A., Convey, P., Verleyen, E., *et al.* (2010). The limnology and biology of the Dufek Massif, Transantarctic Mountains 82° South. *Polar Science*, in press.

Hodgson, D. A., Doran, P. T., Roberts, D., & McMinn, A. (2004). Paleolimnological studies from the Antarctic and subantarctic islands. In *Long-Term Environmental Change in Arctic and Antarctic Lakes*, ed. R. Pienitz, M. S. V. Douglas, & J. P. Smol, Dordrecht: Springer, pp. 419–74.

Hodgson, D. A., Noon, P. E., Vyverman, W., *et al.* (2001a). Were the Larsemann Hills ice-free through the last glacial maximum? *Antarctic Science*, **13**, 440–54.

Hodgson, D. A., Roberts, D., McMinn, A., *et al.* (2006a). Recent rapid salinity rise in three East Antarctic lakes. *Journal of Paleolimnology*, **36**, 385–406.

Hodgson, D. A. & Smol, J. P. (2008). High latitude paleolimnology. In *Polar Lakes and Rivers – Limnology of Arctic and Antarctic Aquatic Ecosystems*, ed. W. F. Vincent & J. Laybourn-Parry, Oxford: Oxford University Press, pp. 43–64.

Hodgson, D. A., Verleyen, E., Sabbe, K., *et al.* (2005). Late Quaternary climate-driven environmental change in the Larsemann Hills, East Antarctica, multi-proxy evidence from a lake sediment core. *Quaternary Research*, **64**, 83–99.

Hodgson, D. A., Verleyen, E., Squier, A. H., *et al.* (2006b). Interglacial environments of coastal east Antarctica: comparison of MIS 1 (Holocene) and MIS 5e (last interglacial) lake-sediment records. *Quaternary Science Reviews*, **25**, 179–97.

Hodgson, D. A., Verleyen, E., Vyverman, W., *et al.* (2009). A geological constraint on relative sea level in Marine Isotope Stage 3 in the Larsemann Hills, Lambert Glacier region, East Antarctica (31 366–33 228 cal yr BP). *Quaternary Science Reviews*, **25**, 2689–96.

Hodgson, D. A., Vyverman, W., & Sabbe, K. (2001b). Limnology and biology of saline lakes in the Rauer Islands, eastern Antarctica. *Antarctic Science*, **13**, 255–70.

House, H. R., McKnight, D. M., & von Guerard, P. (1995). The influence of stream channel characteristics on stream flow and annual water budgets for lakes in Taylor Valley. *Antarctic Journal of the United States*, **30**, 284–7.

IPCC (2007). *Climate Change 2007: Impacts, Adaptation and Vulnerability. Contribution of Working Group II to the Fourth Assessment Report of the Intergovernmental Panel on Climate Change*, ed. M. L. Parry, O. F. Canziani, J. P. Palutikof, P. J. van der Linden, & C. E. Hanson, Cambridge: Cambridge University Press.

Jansen, E., Overpeck, J., Briffa, K. R., *et al.* (2007). Palaeoclimate. In *Climate Change 2007: The Physical Science Basis. Contribution of Working Group I to the Fourth Assessment Report of the Intergovernmental Panel on Climate Change*, ed. S. Solomon, D. Qin, M. Manning, *et al.* Cambridge: Cambridge University Press, pp. 433–97.

Jones, V. J. (1996). The diversity, distribution and ecology of diatoms from Antarctic inland waters. *Biodiversity and Conservation*, **5**, 1433–49.

Jones, V. J. & Juggins, S. (1995). The construction of a diatom-based chlorophyll *a* transfer function and its application at three lakes on Signy Island (maritime Antarctic) subject to differing degrees of nutrient enrichment. *Freshwater Biology*, **34**, 433–45.

Jones, V. J., Juggins, S., & Ellis-Evans, J. C. (1993). The relationship between water chemistry and surface sediment diatom assemblages in maritime Antarctic lakes. *Antarctic Science*, **5**, 339–48.

Kawecka, B. & Olech, M. (1993). Diatom communities in the Vanishing and Ornithologist Creek, King George Island, South Shetland Islands, Antarctica. *Hydrobiologica*, **269–270**, 327–33.

Kellogg, D. E. & Kellogg, T. B. (1996). Diatoms in South Pole ice: implications for eolian contamination of Sirius Group deposits. *Geology*, **24**, 115–18.

Kellogg, T. B. & Kellogg, D. E. (2002). *Non-Marine and Littoral Diatoms from Antarctic and Sub-Antarctic Locations. Distribution and Updated Taxonomy*, Diatom Monographs, vol. 1, ed. A. Witkowski, Ruggell: A. R. Ganter Verlag.

Kociolek, J. P. & Jones, V. (1995). *Gomphonema signyensis* sp. nov., a freshwater diatom from maritime Antarctica. *Diatom Research*, **10**, 269–76.

Kociolek, J. P. & Spaulding, S. A. (2000). Freshwater diatom biogeography. *Nova Hedwigia*, **71**, 223–41.

Kopalová, K., Elster, J., Nedbalová, L., & Van de Vivjer, B. (2009). Three new terrestrial diatom species from seepage areas on James Ross Island (Antarctic Peninsula Region). *Diatom Research*, **24**, 113–22.

Krasske, G. (1939). Zur Kieselalgenflora Südchiles. *Archiv für Hydrobiologie*, **35**, 349–468.

Lawrence, M. J. F. & Hendy, C. H. (1989). Carbonate deposition and Ross Sea Ice advance, Fryxell Basin, Taylor Valley, Antarctica. *New Zealand Journal of Geology and Geophysics*, **32**, 267–77.

Le Cohu, R. (1981). Les espèces endémiques de diatomées aux îles Kerguelen. *Colloque sur les Ecosyèmes Subantarctiqes*, **51**, 35–42.

Le Cohu, R. & Maillard, R. (1983). Les diatomées monoraphidées des îles Kerguelen. *Annales Limnologie*, **19**, 143–67.

Le Cohu, R. & Maillard, R. (1986). Les diatomées d'eau douce des îles Kerguelen (l'exclusion des Monoraphidées). *Annales Limnologie*, **22**, 99–118.

LeMasurier, W. E., Harwood, D. M., & Rex, D. C. (1994). Geology of Mount Murphy Volcano; an 8 m.y. history of interaction between rift volcano and the West Antarctic ice sheet. *Geological Survey of American Bulletin*, **106**, 265–80.

Lichti-Federovich, S. (1984). Investigation of diatoms found in surface snow from Sydkap Ice Cap, Ellesmere Island, Northwest Territories. *Geological Survey of Canada Paper*, **84**–1A, 287–301.

Light, J. J., Ellis-Evans, J. C. & Priddle, J. (1981). Phytoplankton ecology in an Antarctic lake. *Freshwater Biology*, **11**, 11–26.

Lyons, W. B., Laybourn-Parry, J., Welch, K. A., & Priscu, J. C. (2006). Antarctic lake systems and climate change. In *Trends in Antarctic Terrestrial and Limnetic Ecosystems: Antarctica as a Global Indicator*, ed. D. M. Bergstrom, P. Convey, & A. H. L. Huiskes, Dordrecht: Springer, pp. 273–98.

Mahood, A. D. & Barron, J. A. (1996). Comparative ultrastructure of two closely related *Thalassiosira* species: *Thalassiosira vulnifica* (Gombos) Fenner and *T. fasiculata* Harwood et Maruyuma. *Diatom Research*, **11**, 284–95.

Maynard, N. G. (1968). Significance of air-borne algae. *Zeitschrift für Allgemeine Mikrobiologie*, **8**, 225–6.

McIntyre, N. F. (1984). Cryoconite hole thermodynamics. *Canadian Journal of Earth Sciences*, **21**, 152–6.

McKnight, D. M. & Tate, C. M. (1997). Canada Stream: a glacial meltwater stream in Taylor Valley, South Victoria Land, Antarctica. *Journal of the North American Benthological Society*, **16**, 14–17.

McKnight, D. M., Niyogi, D. K., Alger, A. S., *et al.* (1999). Dry Valley streams in Antarctica: ecosystems waiting for water, *Bioscience*, **49**, 985–95.

Moorhead, D., Schmeling, J., & Hawes, I. (2005). Contributions of benthic microbial mats to net primary production in Lake Hoare, Antarctica. *Antarctic Science*, **17**, 33–45.

Mueller, D. R. & Pollard, W. H. (2004). Gradient analysis of cryoconite ecosystems from two polar glaciers. *Polar Biology*, **27**, 66–74.

Mueller, D. R., Vincent, W. F., Pollard, W. H., & Fritsen, C. H. (2001). Glacial cryoconite ecosystems: a bipolar comparison of algal communities and habitats. *Nova Hedwigia*, **123**, 173–97.

Nordenskjöld, A. E. (1875). Cryoconite found 1870, July 19–25, on the inland ice, east of Auleitsivik Fjord, Disco Bay, Greenland. *Geological Magazine, Decade 2*, **2**, 157–62.

Ochyra, R., Bednarek-Ochyra, G., & Lewis-Smith, R. (2008). *Illustrated Moss Flora of Antarctica*, Cambridge: Cambridge University Press.

Pankow, H., Haendel, D. & Richter, W. (1991). Die Algenfora der Schirmacheroase (Ostantarktika). *Nova Hedwigia, Beiheft*, **103**, 1–197.

Porazinska, D. L., Fountain, A. G., Nylen, T. H., *et al.* (2004). The biodiversity and biogeochemistry of cryoconite holes from McMurdo Dry Valley glaciers, Antarctica. *Arctic, Antarctic, and Alpine Research*, **36**, 84–91.

Prescott, G. W. (1979). A contribution to a bibliography of Antarctic and subantarctic algae. *Bibliotheca Phycologica*, **45**, 1–312.

Reinsch, P. E. (1890). Die Süsswasseralgenflora von Süd-Georgien. In *Die deutschen Expeditionen und ihre Ergebnisse, 1882–1883*, ed. G. Neumeyer, Berlin: A. Asher, pp. 329–65.

Riaux-Gobin, C. (1994). A check-list of the *Cocconeis* species (Bacillariophyceae) in Antarctic and Subantarctic areas, with special focus on Kerguelen Islands. *Cryptogamie Algologie*, **15**, 135–46.

Roberts, D. & McMinn, A. (1998). A weighted-averaging regression and calibration model for inferring lake water salinity from fossil diatom assemblages in saline lakes of the Vestfold Hills: a new tool for interpreting Holocene lake histories in Antarctica. *Journal of Paleolimnology*, **19**, 99–113.

Roberts, D. & McMinn, A. (1999). A diatom-based palaeosalinity history of Ace Lake, Vestfold Hills, Antarctica. *The Holocene*, **9**, 401–8.

Roberts, D., McMinn, A., Cremer, H., Gore, D., & Melles, M. (2004). The Holocene evolution and palaeosalinity history of Beall Lake,

Windmill Islands (East Antarctica) using an expanded diatom-based weighted averaging model. *Palaeogeography, Palaeoclimatology, Palaeoecology*, **208**, 121–40.

Roberts, D., McMinn, A., Johnston, N., *et al.* (2001). An analysis of the limnology and sedimentary diatom flora of fourteen lakes and ponds from the Windmill Islands, East Antarctica. *Antarctic Science*, **13**, 410–19.

Roberts, D., McMinn, A., & Zwartz, D. (2000). An initial palaeosalinity history of Jaw Lake, Bunger Hills based on a diatom-salinity transfer function applied to sediment cores. *Antarctic Science*, **12**(2), 172–6.

Rossiter, A. & Kawanabe, H. (2000). *Ancient Lakes: Biodiversity, Ecology and Evolution*, San Diego: Academic Press.

Sabbe, K., Verleyen, E., Hodgson, D. A., Vanhoutte, K., & Vyverman, W. (2003). Benthic diatom flora of freshwater and saline lakes in the Larsemann Hills and Rauer Islands, East Antarctica. *Antarctic Science*, **15**, 227–48.

Sabbe, K., Hodgson, D. A., Verleyen, E., *et al.* (2004). Salinity, depth and the structure and composition of microbial mats in continental Antarctic lakes. *Freshwater Biology*, **49**, 296–319.

Sancetta, C., Lyle, M., Heusser, L., Zahn, R., & Bradbury, J. P. (1992). Late-glacial to Holocene changes in winds, upwelling, and seasonal production of the Northern California current system. *Quaternary Research*, **38**, 359–70.

Saunders, K. M., Hodgson, D. A., & McMinn, A. (2008). Quantitative relationships between benthic diatom assemblages and water chemistry in Macquarie Island lakes and their potential to reconstruct past environmental changes. *Antarctic Science*, **21**, 35–49.

Säwström, C., Mumford, P., Marshall, W., Hodson, A., & Laybourn-Parry, J. (2002). The microbial communities and primary productivity of cryoconite holes in an Arctic glacier (Svalbard 79° N). *Polar Biology*, **25**, 591–6.

Saxena, V. K., Curtin, T. B., & Parungo, F. P. (1985). Aerosol formation by wave action over the Ross Sea. *Journal de Recherches Atmospheriques*, **19**, 213–24.

Scherer, R. P., Aldahan, A., Tulaczyk, S., *et al.* (1998). Pleistocene collapse of the West Antarctic Ice Sheet. *Science*, **281**, 82–5.

Schmidt, R., Mäusbacher, R., & Müller, J. (1990). Holocene diatom flora and stratigraphy from sediment cores of two Antarctic lakes (King George Islands). *Journal of Paleolimnology*, **3**, 55–74.

Seaburg, K. C., Parker, B. C., Prescott, G. W., & Whitford, L. A. (1979). The algae of southern Victorialand, Antarctica. A taxonomic and distributional study. *Bibliotheca Phycologica*, **46**, 1–170.

Seppelt, R. D. (2004). *The Moss Flora of Macquarie Island*. Kingston, Tasmania: Australian Antarctic Division.

Shindell, D. T. & Schmidt, G. A. (2004). Southern hemisphere climate response to ozone changes and greenhouse gas increases. *Geophysical Research Letters*, **31**, L18209, DOI:10.1029/2004GL020724.

Smol, J. P. (1988). Paleoclimate proxy data from freshwater Arctic diatoms. *Verhandlungen des Internationalen Verein Limnologogie*, **23**, 837–44.

Smol, J. P., Wolfe, A. P., Birks, H. J. B., *et al.* (2005). Climate-driven regime shifts in the biological communities of arctic lakes. *Proceedings of the National Academy of Sciences of the USA*, **102**, 4397–402.

Spaulding, S. A., Kociolek, J. P., & Wong, D. (1999). A taxonomic and systematic revision of the genus *Muelleria* (Bacillariophyta). *Phycologia*, **38**, 314–41.

Spaulding, S. A., McKnight, D. M., Stoermer, E. F., & Doran, P. T. (1997). Diatoms in sediments of perennially ice-covered Lake Hoare, and implications for interpreting lake history in the McMurdo Dry Valleys of Antarctica. *Journal of Paleolimnology*, **17**, 403–20.

Stibal, M., Šabacká, M., & Kaštovská, K. (2006). Microbial communities on glacier surfaces in Svalbard: impact of physical and chemical properties on abundance and structure of cyanobacteria and algae. *Microbial Ecology*, **52**, 644–54.

Stuiver, M., Denton, G. H., Hughes, T. J., & Fastook, J. L. (1981). History of the marine ice sheet in West Antarctica during the last glaciation: a working hypothesis. In *The Last Great Ice Sheets*, ed. G. H. Denton & T. J. Hughes, New York, NY: John Wiley and Sons, pp. 319–436.

Theriot, E. C., Fritz, S. C., Whitlock, C., & Conley, D. J. (2006). Late Quaternary rapid morphological evolution of an endemic diatom in Yellowstone Lake, Wyoming. *Paleobiology*, **32**, 38–54.

Torinesi, O., Fily, M., & Genthion, C. (2003). Variability and trends of the summer melt period of Antarctic ice margins since 1980 from microwave sensors. *Journal of Climatology*, **16**, 1047–60.

Tranter, M., Fountain, A. G., Fritsen, C. H., *et al.* (2004). Extreme hydrochemical conditions in natural microcosms entombed within Antarctic ice. *Hydrological Processes*, **18**, 379–87.

Van de Vijver, B. (2008). Distribution of moss-inhabiting diatoms along an altitudinal gradient at sub-Antarctic Îles Kerguelen. *Diatom Research*, **22**, 221–32.

Van de Vijver, B. & Beyens, L. (1999). Moss diatom communities from Ile de la Possession (Crozet, sub-Antarctica) and their relationship with moisture. *Polar Biology*, **22**, 232–40.

Van de Vijver, B., Beyens, L., & Lange-Bertalot, H. (2004). The genus *Stauroneis* in the Arctic and Antarctic regions. *Bibliotheca Diatomologica*, **51**, 1–317.

Van de Vijver, B., Frenot, Y. & Beyens, L. (2002a). Freshwater diatoms from Ile de la Possession (Crozet archipelago, sub-Antarctica). *Bibliotheca Diatomologica*, **46**, 1–412.

Van de Vijver, B., Gremmen, N., & Smith, V. (2008). Diatom communities from the sub-Antarctic Prince Edward Islands: diversity and distribution patterns. *Polar Biology*, **7**, 795–808.

Van de Vijver, B., Ledeganck, P., & Beyens, L. (2001). Habitat preference in freshwater diatom communities from sub-Antarctic Îles Kerguelen. *Antarctic Science*, **13**, 28–36.

Van de Vijver, B., Ledeganck, P., & Beyens, L. (2002b). Three new species of *Diadesmis* from soils of Ile de la Possession (Crozet Archipelago, sub-Antarctic). *Cryptogamie Algologie*, **23**, 333–41.

Van de Vijver, B. & Mataloni, G. (2008). New and interesting species in the genus *Luticola* D. G. Mann (Bacillariophyta) from Deception Island (South Shetland Islands). *Phycologia*, **47**, 451–67.

Van de Vijver, B., Mataloni, G., Stanish, L., & Spaulding, S. A. (2010). New and interesting species of the genus *Muelleria* (Bacillariophyta) from the Antarctic region and South Africa. *Phycologia*, **1**, 22–41.

Van Der Putten N., Hébrard, J. P., Verbruggen, C., *et al.* (2009). An integrated palaeoenvironmental investigation of a 6200 year old peat sequence from Ile de la Possession, Iles Crozet, sub-Antarctica. *Palaeogeography, Palaeoclimatology, Palaeoecology*, **270**, 179–95.

Van Heurck, H. (1909). Diatomées. In *Expédition Antarctique Belge, Résultats du Voyage du S. Y. Belgica en 1897–1899*, Antwerp: Buschmann, pp. 1–126

Verkulich, S. R., Melles, M., Pushina, Z. V. & Hubberten, H. W. (2002). Holocene Environmental changes and development of Figurnoye Lake in the southern Bunger Hills, East Antarctica. *Journal of Paleolimnology*, **28**, 253–67.

Verleyen, E., Hodgson, D. A., Leavitt, P. R., Sabbe, K. & Vyverman, W. (2004a). Quantifying habitat-specific diatom production: a critical assessment using morphological and biochemical markers in Antarctic marine and lake sediments. *Limnology and Oceanography*, **49**, 1528–39.

Verleyen, E., Hodgson, D. A., Milne, G. A., Sabbe, K., & Vyverman, W. (2005). Relative sea level history from the Lambert Glacier region (East Antarctica) and its relation to deglaciation and Holocene glacier re-advance. *Quaternary Research*, **63**, 45–52.

Verleyen, E., Hodgson, D. A., Sabbe, K., Vanhoutte, K., & Vyverman, W. (2004c). Coastal oceanographic conditions in the Prydz Bay region (East Antarctica) during the Holocene recorded in an isolation basin. *The Holocene*, **14**, 246–57.

Verleyen, E., Hodgson, D. A., Sabbe, K. & Vyverman, W. (2004b). Late Quaternary deglaciation and climate history of the Larsemann Hills (East Antarctica). *Journal of Quaternary Science*, **19**, 361–75.

Verleyen, E., Hodgson, D. A., Vyverman, W., *et al.* (2003). Modelling diatom responses to climate induced fluctuations in the moisture balance in continental Antarctic lakes. *Journal of Paleolimnology*, **30**, 195–215.

Verleyen, E., Vyverman, W., Sterken, M., *et al.* (2009). The importance of regional and local factors in shaping the taxonomic structure of diatom metacommunities, *Oikos*, **118**, 1239–49.

Vyverman, W., Verleyen, E., Sabbe, K., *et al.* (2007). Historical processes constrain patterns in global diatom diversity. *Ecology*, **88**, 1924–31.

Wagner, B., Cremer, H., Hultzsch, N., Gore, D. B. & Melles, M. (2004). Late Pleistocene and Holocene history of Lake Terrasovoje, Amery Oasis, East Antarctica, and its climatic and environmental implications. *Journal of Paleolimnology*, **32**, 321–39.

Wasell, A. & Håkansson, H. (1992). Diatom stratigraphy in a lake on Horseshoe Island, Antarctica: a marine–brackish–fresh water transition with comments on the systematics and ecology of the most common diatoms. *Diatom Research*, **7**, 157–94.

Webb, P. N. & Harwood, D. M. (1991). Late Cenozoic glacial history of the Ross Embayment, Antarctica. *Quaternary Science Reviews*, **10**, 215–23.

Webb, P. N., Harwood, D. M. McKelvey, B. C., Mercer, J. C., & Stott, L. D. (1984). Cenozoic marine sedimentation and ice-volume variation on the East Antarctic craton. *Geology*, **12**, 287–91.

West, W. & West, G. S. (1911). Freshwater algae. In *Biology, Vol. 1 Reports on the Scientific Investigations, British Antarctic Expedition 1907–9*, ed. J. Murray, London: Heinemann, pp. 264–87.

Wharton, R. A., Jr., McKay, C. P., Clow, G. D., & Anderson, D. T. (1993). Perennial ice covers and their influence on Antarctic lake ecosystems. In *Physical and Biogeochemical Processes in Antarctic Lakes, Antarctic Research Series No 59*, ed. W. J. Green & E. I. Friedmann, Washington, DC: American Geophysical Union, pp. 53–70.

Wharton, R. A., McKay, C. P., Simmons, G. M., Jr., & Parker, B. C. (1985). Cryoconite holes on glaciers. *BioScience*, **35**, 499–503.

Wharton, R. A. & Vinyard, W. C. (1983). Distribution of snow and ice algae in western North America. *Madroño*, **30**, 201–9.

Wharton, R. A., Vinyard, W. C., Parker, B. C., Simmons, G. M., & Seaburg, K. G. (1981). Algae in cryoconite holes on Canada Glacier in Southern Victorialand, Antarctica. *Phycologia*, **20**, 208–11.

Whittaker, T. E., Hall, B. L., Hendy, C. H., & Spaulding, S. A. (2008). Holocene surface-level changes and depositional environments at Lake Fryxell, Antarctica. *Holocene*, **18**, 775–86.

Wilson, G. S., Barron, J. A., Ashworth, A. C., *et al.* (2002). The Mount Feather Diamicton of the Sirius Group: an accumulation of indicators of Neogene Antarctic glacial and climatic history. *Palaeogeography, Palaeoclimatology, Palaeoecology*, **182**, 117–31.

Zidarova, R. (2008). Algae from Livingston Island (S Shetland Islands): a checklist. *Phytologia Balcanica*, **14**, 19–35.

Zwartz, D., Bird, M., Stone, J., & Lambeck, K. (1998). Holocene sea-level change and ice-sheet history in the Vestfold Hills, East Antarctica. *Earth and Planetary Science Letters*, **155**, 131–45.

Part IV

Diatoms as indicators in marine and estuarine environments

15

Diatoms and environmental change in large brackish-water ecosystems

PAULINE SNOEIJS AND
KAARINA WECKSTRÖM

15.1 Introduction

15.1.1 Classification of brackish waters

Brackish waters comprise a range of exclusive habitats that can be subdivided into three major categories: transition zones between freshwater and marine habitats, transition zones between hyperhaline water and marine habitats, and inland waters without marine water exchange. Salinities of brackish-water habitats vary from relatively stable (e.g. some large saline lakes; see Fritz *et al.*, this volume) to extremely instable in time and space (e.g. estuaries bordering tidal seas; see Trobajo & Sullivan, this volume). In the past, many efforts have been made to classify brackish waters according to salinity and the occurrence of biological species (Kolbe, 1932; Segerstråle, 1959; den Hartog, 1964). The more detailed such classifications are, the less well they appear to fit with all types of brackish waters. Based on salinity, defined as the total concentration of ionic components in g per kg water, generally accepted approximate limits are: limnetic (freshwater) <0.5 practical salinity units (psu) = parts per thousand (ppt), oligohaline 0.5–5 psu, mesohaline 5–18 psu, polyhaline 18–30 psu, euhaline 30–40 psu, hyperhaline >40 psu (known as the "Venice System": Anonymous, 1959).

15.1.2 Large brackish-water ecosystems

Earth's longest salinity gradient comprises the continental microtidal Baltic Sea (Leppäranta & Myrberg, 2009: surface area 377,000 km^2, water volume 21,000 km^3, mean depth 58 m, maximum depth 459 m) and its transition area to the North Sea. This over 2000-km-long salinity gradient stretches from 0.5 psu in the northern Bothnian Bay to fully marine (35 psu) in the Skagerrak due to a net outflow from the Baltic Sea (caused by freshwater discharge from the large drainage area) through narrow and shallow connections with the North Sea (Figure 15.1a). Other northern brackish sea areas, such as the White Sea (Russia), Hudson Bay (Canada), and the Gulf of St. Lawrence (Canada), have less stable salinity gradients because they are strongly affected by tidal fluctuations. Another large brackish sea with net outflow is the intercontinental non-tidal Black Sea, including the Sea of Azov (Kosarev & Kostianoy, 2008; surface area 423,000 km^2, water volume 555,000 km^3, mean depth 1315 m, maximum depth 2258 m). The Black Sea has a narrow and shallow connection with the Mediterranean Sea and a salinity of 14–20 psu (Figure 15.1b). The Sea of Azov has a narrow and shallow connection with the Black Sea and a salinity gradient from 0 to 13 psu (Kosarev *et al.*, 2008; Figure 15.1b). Oxygen deficiency of deep bottoms widely occurs in the Black and Baltic seas. The Caspian Sea has no connection to the ocean and is therefore technically a lake (Kosarev, 2005; surface area 390,000 km^2, water volume 78,000 km^3, mean depth 208 m, maximum depth 1025 m). At present it is the Earth's largest land-locked water body, but once it was a part of the ancient Thetis Ocean which connected the Atlantic and the Pacific oceans. Over five million years ago the Caspian Sea was closed off from the ocean by continental drift. It is still saline due to its oceanic past and salinity ranges from 0 psu at the Volga Delta to 14 psu in the open waters (Figure 15.1c). Some almost closed-off coastal areas, e.g. the very shallow Kara-Bogaz-Gol Gulf, are hyperhaline (up to salinity 350 psu) due to evaporation. This chapter contains more examples from the Baltic Sea than from the Black and Caspian seas because the number of international research publications dealing with diatoms on the latter two seas is low.

The Diatoms: Applications for the Environmental and Earth Sciences, 2nd Edition, eds. John P. Smol and Eugene F. Stoermer. Published by Cambridge University Press.

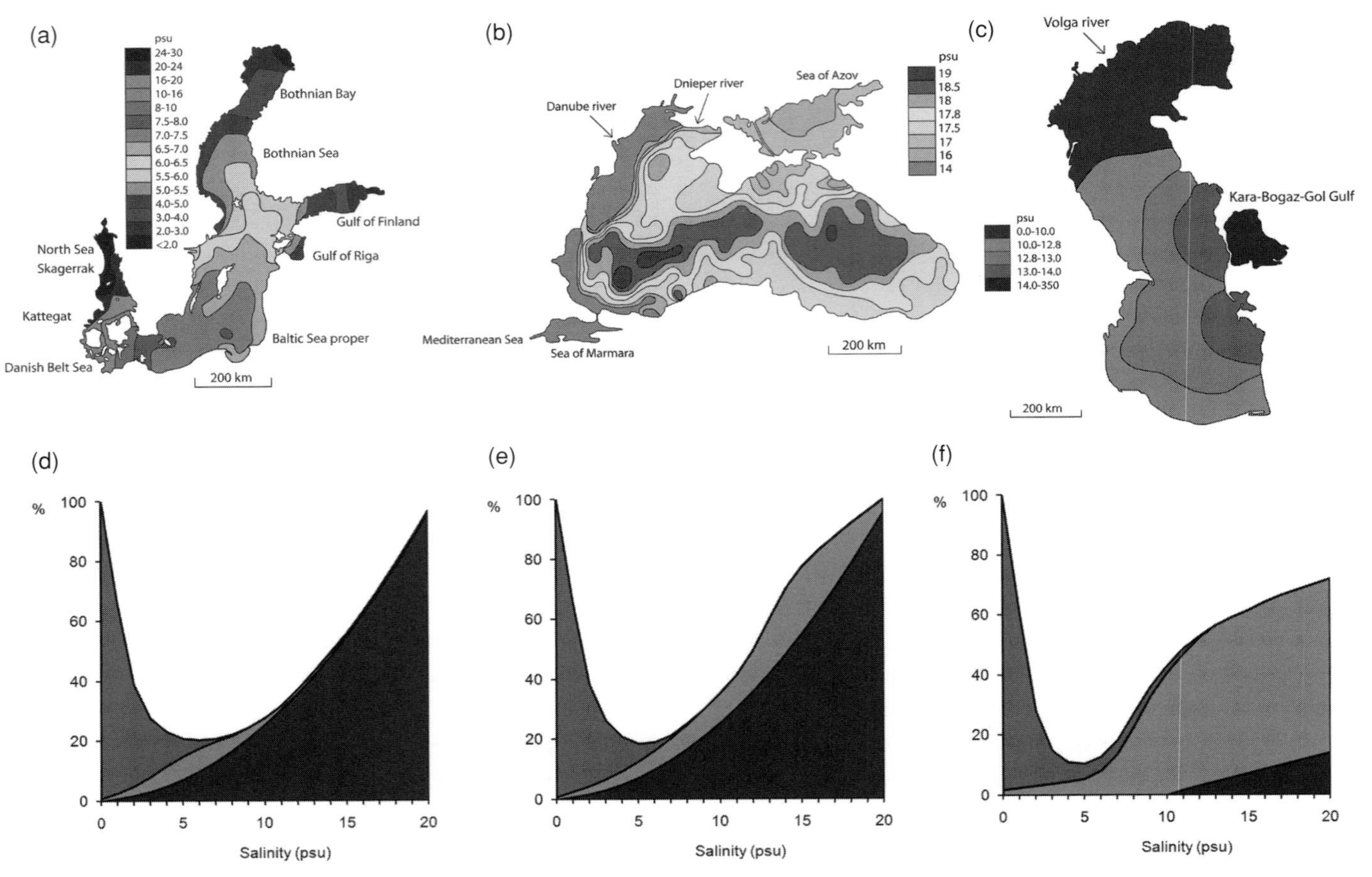

Figure 15.1 Maps showing the surface salinity gradients of the Baltic Sea (a), the Black Sea (b), and the Caspian Sea (c); and theoretical biological diversity distributions subdivided into freshwater species (light gray), brackish-water species (dark gray) and marine species (black) for the Baltic Sea (d), the Black Sea (e), and the Caspian Sea (f).

15.1.3 Large-scale environmental change

The Earth's ecosystems are being transformed under the pressure of large-scale human-induced environmental changes. Global climate change, eutrophication, discharges of hazardous substances, introductions of alien species, and overexploitation of natural resources affect the health of organisms, community composition and food-web interactions in aquatic systems with accelerating speed. Earth's largest brackish-water bodies are particularly sensitive to such changes. Their drainage areas are large and water turnover times extremely low. Harmful and hazardous substances discharged to these ecosystems circulate in the water, sediments, and biota for long times. Anthropogenic impacts have changed the Baltic, Black, and Caspian Sea ecosystems drastically during the last ~100 years (Kosarev & Yablonskaya, 1994; Falandysz *et al.*, 2000; Zonn, 2005; Zonn *et al.*, 2008), and in some cases the drastic changes have turned into regime shifts (Folke *et al.*, 2004).

15.1.4 Diatoms and environmental change

There are basically three main ways in which diatoms can be used to reflect environmental change: as constituents of live benthic and pelagic communities, as (sub)fossil diatom assemblages, and as indicators of hazardous substances. Live microphytobenthic and phytoplankton communities are species-rich and often dominated by diatoms. These communities are sensitive indicators of environmental change and small modifications of environmental factors usually result in measurable species shifts, a feature which is used in environmental monitoring programs worldwide. Such compositional ecological data should, however, always be interpreted with care and an appreciation for the confounding influences of biotic interactions and multivariate effects. Remnants of diatom communities in the (sub)fossil record of sediments (diatom assemblages) are powerful tools in the reconstruction of paleoenvironments and long-term environmental change (Smol, 2008). In the same way, living diatom assemblages on artificial substrata can be used as tools for measuring environmental change without necessarily reflecting all aspects of natural communities (Snoeijs, 1991). The composition of (sub)fossil diatom assemblages is affected by physical processes (currents, upwelling, resuspension, erosion), biological processes (bioturbation), and different degrees of dissolution of frustules of different diatom taxa in some systems. Artificial substrata are not colonized by all types of diatom life forms. Diatom assemblages thus do not reflect natural species interactions, but they do have strong indicator value for the directions and rates of change caused by changing environmental factors. Finally, diatoms can be used to reflect environmental change by measuring the chemical composition of the diatoms themselves in relation to the chemical composition of the environment. Some potentially hazardous substances, e.g. metals, readily attach to diatom frustules (Pedersén *et al.*, 1981) and accumulate in microphytobenthic communities. Information on the levels and transport of hazardous substances in the whole ecosystem can then be obtained by using previously tested mathematical models.

15.2 Brackish-water diatoms

15.2.1 What is a brackish-water diatom species?

Diatom taxa occurring in transitional zones between marine and freshwater habitats can be divided into two affinity groups: one contains taxa with primarily a marine affinity, and the other contains taxa with a freshwater affinity. Round & Sims (1981) considered the distribution of diatom genera in the light of diatom evolution, and found that more than 90% of genera are confined to either marine or freshwater habitats. Most likely this percentage is actually higher, because genera previously found in both marine and freshwater habitats are now, or will be, re-named as new taxa and so with the new groupings more genera will fit this pattern (Round *et al.*, 1990; Mann & Evans, 2007). Round & Sims (1981) considered only a few small genera to have a true brackish-water distribution (e.g. *Brebissonia* Grunow, *Dimidiata* Hajos, *Scoliopleura* Grunow, *Scoliotropis* Cleve). An important and somewhat controversial question is how to define a "brackish-water species." If this is a species exclusively occurring in brackish water, then there are none or extremely few in transitional zones between marine and freshwater. All diatoms living in these transitional zones should then be considered as marine or freshwater taxa with different degrees of euryhalinity (Carpelan, 1978a). Real brackish-water species would then only be endemics found in land-locked brackish waters, as evolutionary results of genetic bottlenecks. But if a brackish-water species is one with its abundance optimum in brackish waters, then there are many in transitional zones, all with their own typical salinity range and optimum, which however may differ in different types of brackish-water habitats.

15.2.2 Evolution

Of the large brackish-water ecosystems included in this chapter, the Caspian Sea contains the most brackish-water species and the Baltic Sea the fewest (Figure 15.1d–f). A number of brackish-water species living in the Black and Caspian Seas are retained from the Pliocene low-salinity Pontian Sea-Lake (Finenko, 2008). These species are referred to as Ponto–Caspian relicts. In the late Pontian stage (7.5–5.0 Ma, Matoshko *et al.*, 2009), the Earth's crust began to rise in the northern Caucasus, gradually isolating the Caspian Sea from the Pontian basin. From that period onwards the Caspian Sea and the Black Sea developed separately, although temporary links between them were formed from time to time. More endemic brackish-water species are believed to have evolved in the Caspian Sea (Karpinsky, 2005), while the Black Sea became influenced by marine Mediterranean species. Karayeva & Makarova (1973) list 16 Ponto–Caspian and 16 Caspian endemic brackish-water diatom taxa that occur in the present Caspian Sea. The Baltic Sea contains a mixture of freshwater and marine Atlantic species (Snoeijs *et al.*, 1993–1998). However, the flora and fauna of all three large brackish-water ecosystems include many ecotypes with special adaptations to their typical brackish environments. Adaptive trade-offs were already recognized by Juhlin-Dannfelt (1882) for Baltic diatom taxa: "the species living both in brackish and in fresh water do not seem to follow any rule as to their relative dimensions" (*op. cit.*, p. 13). When a freshwater species is transported in a freshwater stream to an estuary bordering a tidal marine environment, it will experience a strong salinity shock and die off when it meets lethal salinity levels. Such a freshwater species carried to the northern Baltic Sea by land-runoff meets a large habitat (hundreds of kilometers) with extraordinary stable low salinity. In these waters, some freshwater species can survive and grow, and may develop into Baltic ecotypes and endemic species genetically adapted to a narrower salinity range (more stenohaline). Probably, the Baltic diatom species that have adapted to a stable low salinity have a lower genetic variability than typical estuarine species which need to cope with a constantly changing environment. Snoeijs & Potapova (1998) showed that discontinuities in phenotypic characters exist between some well-known and widely distributed freshwater *Diatoma* spp. and their Baltic populations. The Baltic morphotypes proved to be constant over time in the Bothnian Sea, and they were consistently different from populations in freshwater habitats. Some of the Baltic *Diatoma* forms may be defined as ecotypes (e.g. a northern Baltic form of *Diatoma moniliformis* Kütz.), others may be species of their own which perhaps are endemic (e.g. *Diatoma constricta* Grunow and *Diatoma bottnica* Snoeijs).

15.2.3 Identification of brackish-water diatoms

The floristic diatom composition of the Baltic Sea is similar to that of estuaries elsewhere, e.g. in the Skagerrak (Kuylenstierna, 1989–1990), South Africa (Archibald, 1983), and Western Australia (John, 1983). Since Juhlin-Dannfelt published his diatom flora of the Baltic Sea in 1882, many subsequent floristic works have appeared (e.g. Cleve-Euler, 1951–1955; Mölder & Tynni, 1967–1973; Tynni, 1975–1980; Snoeijs *et al.*, 1993–1998). Similarly, floristic documentation exists on the diatoms of the Black Sea (Proschkina-Lavrenko, 1955, 1963; Petrov & Nevrova, 2007) and the Caspian Sea (Proschkina-Lavrenko & Makarova, 1968; Anonymous, 1985). However, many diatom species from all three areas, especially small-sized ones, have not yet been formally described.

15.3 Salinity

15.3.1 Different types of diatom responses to salinity

Three factors should always be taken into account when using diatoms as environmental markers for salinity. (1) In environments with fluctuating salinity, the species are selected more according to their ability to cope with changing salinity (euryhalinity) than to their salinity optima. (2) In environments with stable salinity, evolutionary processes have resulted in various degrees of endemism depending on the geological age and stability of the water body and the degree of isolation from other populations of the same species. (3) Species perform differently in different brackish waters because they are also affected by other environmental constraints, such as alkalinity, water temperature, light regime, nutrient concentrations, exposure to wave action, hazardous substances, and biotic interactions. Thus, when it is observed in an estuary that a certain species has a lower salinity limit of 15 psu, whereas the same species is living in the Baltic Sea down to 3 psu, this may be a matter of a physiological adaptation, ecological preference for sites exposed to wave action (heavily-exposed sites with salinity 3 psu are absent in the estuary), or an ongoing evolutionary process in the northern Baltic Sea (genetic differentiation into a Baltic ecotype). Such a multitude of possible responses to salinity makes the use of biological species as indicators of environmental change a more complicated matter in brackish-water habitats than in either marine or freshwater habitats. Furthermore, nothing overly general can be stated about the relationship between salinity and diatom primary production. This community parameter is determined more by environmental factors other than salinity, such as nutrient concentrations, temperature, light availability, exposure to water movement, and substratum.

15.3.2 From a halobion system to salinity calibration models

Traditionally, diatom species occurring in brackish waters have been classified into five or more groups in a so-called "halobion system," according to their salinity ranges. This system has been widely used, e.g. in the interpretation of paleoecological diatom stratigraphies where shifts in salinity were involved. The first halobion system was established by Kolbe (1927, 1932), originally based on the diatom taxa occurring in the Sperenberger Saltz Gebiet (Germany). Later revisions were made by Hustedt (1953), Simonsen (1962), and Carpelan (1978b). The development of new statistical methods over the last two decades for analyzing complicated multivariate data (e.g. ter Braak, 1986; ter Braak & Juggins, 1993; Birks, 1995) has opened the possibility for more precise analyses and calibrations of species performances along environmental gradients. This application of diatoms is widely employed for reconstructing paleoenvironments by utilizing calibration data sets containing individual salinity optima and ranges for each species. Such calibrations are better adapted to fit a specific geographic region or type of brackish water than the halobion system. Salinity transfer functions and inference models have been presented by e.g. Juggins (1992) for the Thames estuary (England), by Wilson *et al.*, (1996) for lakes in western North America, by Ryves *et al.* (2004) for Danish fjords, and by Saunders *et al.* (2007) for Australian coastal lakes and lagoons.

15.3.3 Reconstructing paleosalinity in large brackish-water systems

Contemporary patterns of diatom taxon abundances in relation to salinity can be used to reconstruct changes in salinity through previous time periods. It is necessary that the taxa in a contemporary calibration data set used for reconstructing paleoenvironments are the same biological entities as in the fossil data and that their ecological responses to the environmental variable(s) of interest have not changed significantly over the time span represented by the fossil assemblage (Birks, 1995). It may be argued that evolutionary effects in large brackish-water ecosystems interfere with this assumption. For example, some common freshwater species show unimodal responses along the Baltic Sea salinity gradient, e.g. *Achnanthidium minutissimum* (Kütz.) Czarnecki, *Ctenophora pulchella* (Ralfs) Williams & Round, and *Rhoicosphenia curvata* (Kütz.) Grunow in a calibration data set ranging from salinity 0.4 to 11.4 psu (Ulanova *et al.*, 2009). If this was an evolutionary effect typical of the northern Baltic Sea, the salinity optima calculated for these species may not exactly reflect those of the same species elsewhere or even in fossil diatoms in the Baltic Sea area. Also, salinity ranges may be narrower for contemporary brackish-water species in the whole Baltic Sea as an evolutionary consequence of the stability of the salinity gradient. On the other hand, the unique Baltic Sea salinity gradient yields individual response models with extremely high precision (*c.* 1 psu; Snoeijs, 1994) that cannot be obtained in brackish-water areas with fluctuating salinity. The marginal evolutionary changes that may have occurred in the Baltic Sea area during the Holocene (*c.* the last 10,000 years) are not likely to interfere with the practical application of diatom models for reconstructing past environments during this time period, but they are likely to have an effect over longer geological timescales of 10^5–10^6 years.

15.3.4 Community composition as a tool in salinity studies

Salinity is often the dominant environmental factor influencing diatom community composition in large brackish-water systems. It is the quantitative diatom community composition which is typical of brackish water with a given salinity, not the occurrence of one or two particular "indicator" species, that provides the most realistic assessments. Each species in the community contributes information by its typical salinity range and optimum in combination with its abundance at a particular time and site. However, Ulanova *et al.* (2009) found that small (<1,000 μm³) and large (≥1,000 μm³) diatoms respond to salinity in similar ways when the salinity gradient is sufficiently long (0.4–11.4 psu). Hence, counts of large diatom taxa alone may be sufficient for indicating salinity changes in coastal environments. If this is found to be a general feature in diatom calibration data sets covering environmental gradients of sufficient length, in the future diatom studies can be facilitated; large diatoms are easier to identify and counting frustules will be faster (Snoeijs *et al.*, 2002; Busse & Snoeijs, 2002, 2003; Ulanova *et al.*, 2009).

15.3.5 Unimodal and monotonic species responses to salinity

The different responses of individual diatom species to salinity can be described using a variety of approaches, e.g. the hierarchical set of five response models proposed by Huisman *et al.* (1993): I = a null model with no relationship with salinity, II = a monotonic response model limited by 0 and 100% taxon abundance with two parameters, III = a monotonic response model with a plateau just below 100% taxon abundance with three parameters, IV = a symmetric Gaussian unimodal response model with three parameters, V = a skewed unimodal response model with four parameters (Figure 15.2). The percentage of unimodal responses can be used to compare and validate diatom-environment calibration data sets because this percentage increases with the length of an environmental gradient. For the Baltic Sea, Ulanova *et al.* (2009) found 59% unimodal and 40% monotonic responses for a 0.4–11.4 psu salinity gradient (living diatom communities on stones) while Weckström & Juggins (2005) found 27% and 35% (and 33% no relationship with salinity) for a 0.7–6.4 psu gradient (subfossil surface-sediment diatom assemblages).

15.3.6 Diatom diversity along salinity gradients

Macroalgal and macrofaunal species richness are extremely low in the Baltic Sea, with many marine species living at the limit of their tolerance to low salinity. No minimum for species richness was recorded in the Baltic Sea for epiphytic diatoms between 3 and 14 psu (Snoeijs, 1995). Simonsen (1962) and Wendker (1990) also found that the number of species of benthic diatoms was not correlated with salinity in the Schlei estuary (western Baltic Sea, salinity 3–18 psu). This implies that Remane's brackish-water rule based on aquatic fauna (Remane, 1940, 1958) is not valid for benthic diatoms. According to this rule, a gradual decrease occurs in the number of species as conditions become increasingly brackish with minimum species richness at salinity 5 psu, which is the critical salinity for physiological stress in many animals. The absence of a minimum suggests that benthic diatoms as a group are not stressed by salinity in the Baltic Sea. There are few groups of organisms like pennate diatoms (most of them benthic species), which to the same degree inhabit both freshwater and marine environments. These groups can thus penetrate into brackish water from both habitats, which probably explains the absence of a species minimum for epiphytic diatoms in the Baltic Sea. Taxa with freshwater affinities that penetrate into the Baltic Sea southward compensate for the northward loss of taxa with marine affinities. However, species richness of pelagic diatoms (many of them centric diatom species), does show a decreasing trend with decreasing salinity from the Kattegat to the northern Bothnian Bay (Snoeijs, 1999); the number of species decreases successively along the salinity gradient by *c.* 50%. The decrease in pelagic diatoms is not as drastic as that in

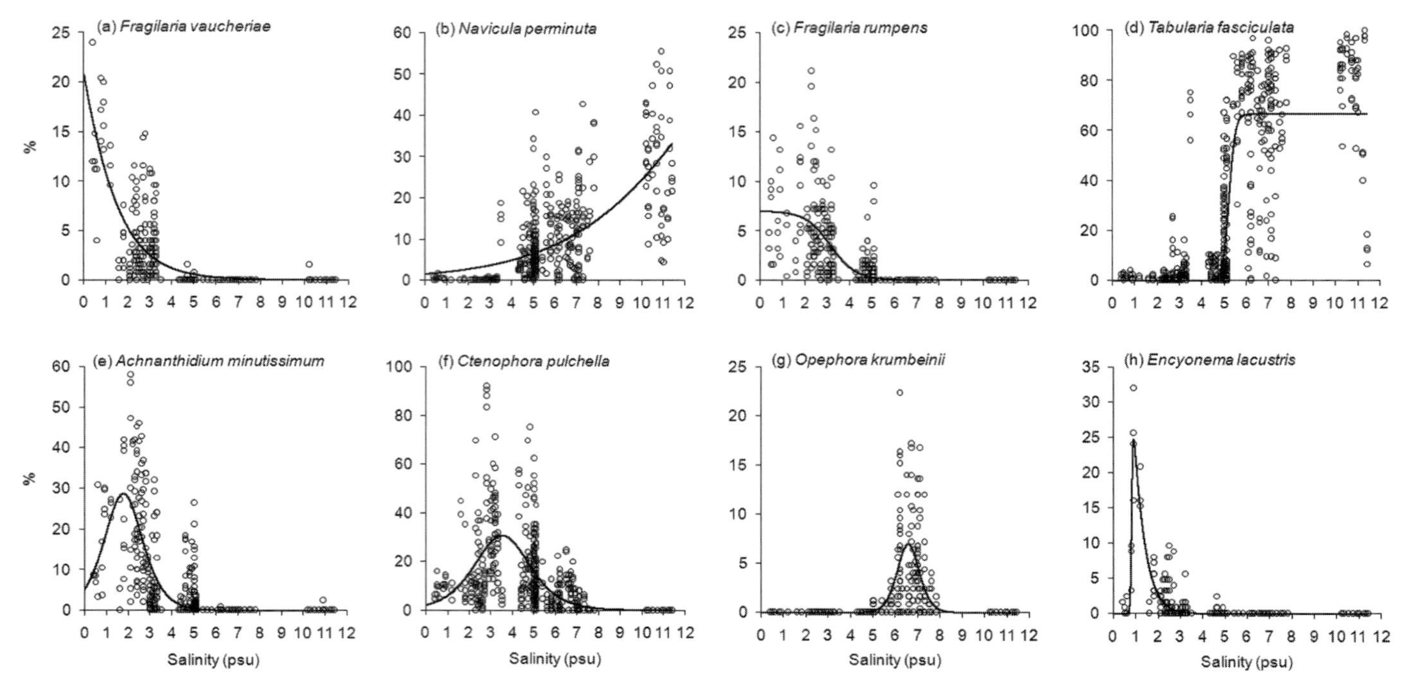

Figure 15.2 Distributions of eight diatom species along the salinity gradient of the Baltic Sea area. The dots indicate the relative abundance of the species plotted against salinity. The curves show modeled optima and tolerance ranges of the species using the taxon response models proposed by Huisman *et al.* (1993): (a, b) Monotonic response models with two parameters; (c, d) monotonic plateau response models with three parameters; (e, f, g) symmetric Gaussian unimodal response models with three parameters; and (h) a skewed unimodal response model with four parameters. (Adapted after Ulanova *et al.* (2009).)

red, brown, and green macroalgae (most of them with marine origin), which in the same salinity range decrease by *c.* 95%, 90%, and 70%, respectively (Snoeijs, 1999), or in macroscopic animals (decreasing by >95%; Bonsdorff, 2006).

15.3.7 Community composition along salinity gradients

In the Baltic Sea area from the southern Öresund to the northern Bothnian Bay (Figure 15.1a), three distributional discontinuities in epiphytic diatom community composition are found (Snoeijs, 1992, 1994, 1995). One is situated at the entrance to the Baltic Sea (salinity 10 psu), the second between the Baltic Sea proper and the Bothnian Sea (salinity 5 psu), and the third between the Bothnian Sea and the Bothnian Bay (salinity 3 psu). These discontinuities do not only fit with the thresholds between the different Baltic Sea sub-basins, but also with subdivisions of the meso- and oligohaline zones, based on other biological observations in the Baltic: α-mesohaline (salinity 10–18 psu), β-mesohaline (5–10 psu), α-oligohaline (3–5 psu), β-oligohaline (0.5–3 psu) (Anonymous, 1959). Typical epiphytic species in these different zones are: *Licmophora hyalina* (Kütz.) Grunow, *Tabularia fasciculata* (C. Agardh) Williams & Round (whole mesohaline range), *Licmophora communis* (Heiberg) Grunow, *Licmophora oedipus* (Kütz.) Grunow (α-mesohaline), *Tabularia waernii* Snoeijs (β-mesohaline), *Ctenophora pulchella* (whole oligohaline range), *Diatoma constricta* (α-oligohaline), *Synedra ulna* (Nitzsch) Ehrenb., and *Synedra acus* Kütz. (β-oligohaline). The discontinuity at salinity 5 psu is most drastic (Figure 15.3), and is characterized by a marked transition from an epiphytic flora dominated by diatom taxa with marine affinities (mostly lightly silicified species) to one dominated by taxa with freshwater affinities (mostly heavily silicified species). The transition between salinity 5 and 7 psu represents a lower limit of salinity tolerance for many marine taxa, resulting in the absence or very low abundance of these taxa below salinity 5 psu.

15.3.8 Fossil diatom assemblages reflect salinity shifts over geological time

During and after deglaciation of the Weichselian Ice Sheet in Northwest Europe, the Baltic Basin repeatedly lost and re-established its connection with the ocean. This dynamic period was driven by the global climate development affecting global sea levels, and crustal uplift of the Scandinavian Peninsula (Ignatius *et al.*, 1981; Björck, 1995; Lambeck, 1999). The post-glacial history of the Baltic Sea area can be

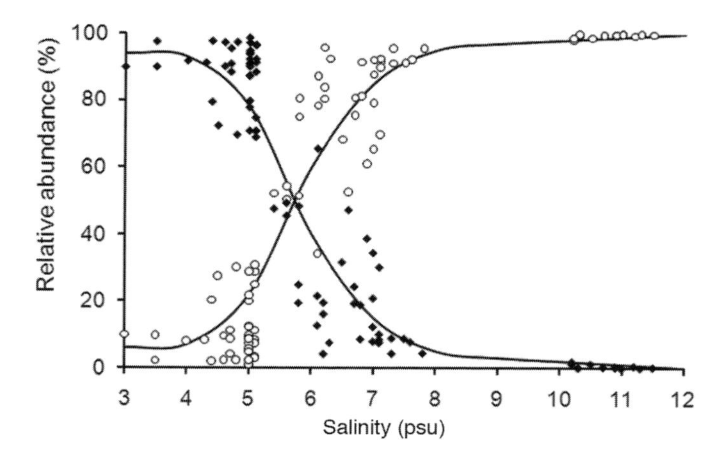

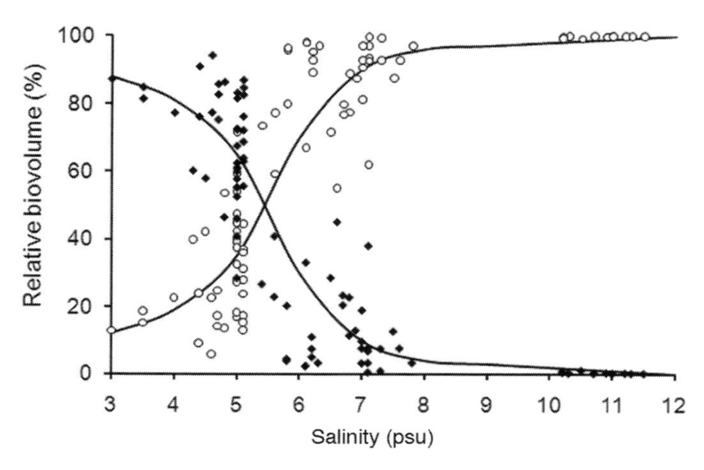

Figure 15.3 Epiphytic diatom taxa with marine affinity (open symbols) and freshwater affinity (shaded symbols) plotted against salinity ($n = 88$ sampling sites), based on relative abundance and relative biovolume. (Adapted after Snoeijs (1995).)

subdivided into four major stages: the Baltic Ice Lake (*c.* 15,000–11,600 cal yr BP; cold, fresh melting water), the Yoldia Sea (*c.* 11,600–10,700 cal yr BP; cold, mixture of marine water and fresh melting water), the Ancylus Lake (*c.* 10,700–9,800 cal yr BP; freshwater), and the Litorina Sea *sensu lato* (*c.* 9,800 cal yr BP–present; brackish, *c.* 5–25‰), see Björck (2008) and references therein. The transition from the Ancylus Lake into the Litorina Sea is complex (Björck *et al.*, 2008); the end of this transition phase has occasionally been called the Mastogloia Sea (Sundelin, 1922) based on a specific diatom assemblage, dominated by *Mastogloia* spp., found in coastal areas. However, currently, this transitional stage is referred to as the Initial Litorina Sea (Andrén *et al.*, 2000a,b) or Early Litorina Sea (Berglund *et al.*, 2005). These stages are clearly visible in diatom stratigraphies, which were first discovered more than 100 years ago (Munthe, 1892; Cleve, 1899; Alhonen, 1971; Ignatius & Tynni, 1978; Gudelis & Königsson, 1979; Miller, 1986). Several proxies have been used to infer paleosalinity in the Baltic Sea (e.g. Munthe, 1894; Punning *et al.*, 1988; Winn *et al.*, 1988). However, diatoms as a proxy appear to be the least ambiguous in recording the Baltic Holocene history (e.g. Sohlenius *et al.*, 1996). As diatom assemblages differ between the open sea and coastal waters, examples of diatom taxa from both areas typical of different stages in the development of the Baltic Sea are given in Table 15.1.

The majority of diatom studies on the different salinity stages of the Baltic Sea have been carried out in coastal basins (e.g. Alhonen, 1971; Miller, 1986; Berglund *et al.*, 2005). Transitions between the major stages are well established, as changes in diatom assemblages are generally marked (see, however, above discussion about the transition from the Ancylus Lake into the Litorina Sea). Environmental changes within these stages are less well known. Overall, it is difficult to assess if these coastal salinity reconstructions are representative for the whole Baltic Sea. The work carried out to date in the open Baltic shows similar developments to the coastal studies but with a more regional depiction (e.g. Abelman, 1985; Andrén *et al.*, 2000a,b; Witkowski, 1994). Some more recent studies have attempted to reconstruct Holocene paleosalinity using diatoms in combination with other proxy records in the open sea area of the Baltic Sea. Westman & Sohlenius (1999) found major shifts in diatom assemblages both at the major stage boundaries and within the Litorina stage. Diatoms indicated that salinity was highest in the northwestern Baltic Sea proper from *c.* 7800–7400 to 6300–5100 cal yr BP (equivalent to *c.* 7000–6500 to 5500–4500 ^{14}C yr BP; no correction for the reservoir effect was made due to lack of data), thereafter decreasing to the present day. The high salinity at the time can partly be explained by the cross-section areas of the Baltic Sea inlets being at their widest and thus increasing the inflow from the North Sea. However, the main cause appears to have been the low freshwater supply to the Baltic Basin during this period (Gustafsson & Westman, 2002). The shifts in diatoms were further correlated with changes in production and bottom anoxia indicated by organic carbon and biogenic silica concentrations as well as sedimentological evidence of anoxia (laminations). Similar results were also obtained by Andrén (1999a), who found the highest salinity phase a little later (*c.* 5600–4500 cal yr BP) in the same area (Gotland Basin). Emeis *et al.* (2003) used diatoms as a qualitative proxy record of salinity and the isotopic composition of organic matter ($^{13}C/^{12}C$ isotope ratio) as a quantitative proxy record of salinity. Their estimates generally follow the trends found by Westman & Sohlenius (1999) and Andrén (1999a), but

Table 15.1 Diatom taxa typical of different developmental stages of the Baltic Sea (From Alhonen, 1971, 1986; Ignatius & Tynni, 1978; Gudelis & Königsson, 1979; Miller & Robertsson, 1979; Robertsson, 1990; Risberg, 1991; Witkowski, 1994; Kabailiene, 1995). Affininies: F = freshwater taxon, B = brackish taxon, BM = brackish-marine taxon, MB = marine-brackish taxon (according to Snoeijs *et al.*, 1993–1998).

Approximate time period (years before present)	Developmental stage	Typical diatom species	Affinity
c. 15,000–11,600	**Baltic Ice Lake**	*Aulacoseira islandica* ssp. *helvetica* O. Müll.	F
	Cold freshwater (melting water)	*Aulacoseira alpigena* (Grunow) Krammer	F
		Stephanodiscus rotula (Kütz.) Hendey	F
		Tabellaria fenestrata (Lyngb.) Kütz.	F
c. 11,600–10,700	**Yoldia Sea**	*Aulacoseira islandica* ssp. *helvetica* O. Müll.	F
	Marine water mixed with cold freshwater (melting water)	*Actinocyclus octonarius* Ehrenb.	B
		Diploneis didyma (Ehrenb.) Ehrenb.	B
		Diploneis interrupta (Kütz.) Cleve	B
		Diploneis smithii (Bréb.) Cleve	B
		Nitzschia obtusa W. Smith	B
		Thalassiosira baltica (Grunow) Ostenf.	B
		Tryblionella navicularis (Bréb.) Ralfs	B
		Tryblionella punctata W. Smith	B
		Grammatophora oceanica Ehrenb.	BM
		Rhabdonema arcuatum (Lyngb.) Kütz.	BM
		Rhopalodia musculus (Kütz.) O. Müll.	BM
c. 10,700–9800	**Ancylus Lake**	*Aulacoseira islandica* ssp. *helvetica* O. Müll.	F
	Freshwater	*Caloneis latiuscula* (Kütz.) Cleve	F
		Campylodiscus noricus Ehrenb.	F
		Cocconeis disculus (Schumann) Cleve	F
		Cymatopleura elliptica (Bréb.) W. Smith	F
		Diploneis domblittensis (Grunow) Cleve	F
		Diploneis mauleri (Brun) Cleve	F
		Ellerbeckia arenaria (Ralfs) Crawford	F
		Encyonema prostratum (Berk.) Kütz.	F
		Epithemia hyndmannii W. Smith	F
		Eunotia clevei Grunow	F
		Gomphocymbella ancyli (Cleve) Hustedt	F
		Gyrosigma attenuatum (Kütz.) Rabenh.	F
		Navicula jentzschii Grunow	F
		Stephanodiscus astraea (Ehrenb.) Grunow	F
		Stephanodiscus neoastraea Håkansson & Hickel	F
Transition period	**Mastogloia Sea**	*Campylodiscus clypeus* Ehrenb.	B
	Brackish water (0–5%)	*Campylodiscus echeneis* Ehrenb.	B
		Ctenophora pulchella (Ralfs) Williams & Round	B
		Diploneis smithii (Bréb.) Cleve	B
		Epithemia turgida (Ehrenb.) Kütz.	B
		Epithemia turgida var. *westermannii* (Ehrenb.) Grunow	B
		Mastogloia braunii Grunow	B
		Mastogloia elliptica (C. Agardh) Cleve	B
		Mastogloia pumila (Cleve & Möller) Cleve	B
		Mastogloia smithii Thwaites	B
		Navicula peregrina (Ehrenb.) Kütz.	B
		Rhoicosphenia curvata (Kütz.) Grunow	B

Table 15.1 *(cont.)*

Approximate time period (years before present)	Developmental stage	Typical diatom species	Affinity
c. 9800–present	**Litorina Sea** Brackish water (5–25%)	*Actinocyclus octonarius* Ehrenb.	B
		Campylodiscus clypeus Ehrenb.	B
		Cocconeis scutellum Ehrenb.	B
		Coscinodiscus asteromphalus Ehrenb.	B
		Diploneis didyma (Ehrenb.) Ehrenb.	B
		Diploneis interrupta (Kütz.) Cleve	B
		Chaetoceros diadema (Ehrenb.) Gran	BM
		Petroneis latissima (Gregory) A. J. Stickle & Mann	BM
		Rhabdonema arcuatum (Lyngb.) Kütz.	BM
		Rhabdonema minutum Kütz.	BM
		Thalassionema nitzschioides (Grunow) Grunow	BM
		Ardissonea crystallina (C. Agardh) Grunow	MB
		Chaetoceros mitra (J. W. Bailey) Cleve	MB
		Hyalodiscus scoticus (Kütz.) Grunow	MB
		Pseudosolenia calcar-avis (Schultze) Sundström	MB

differ radically for the younger part of the Litorina Sea (since *c.* 3100 cal yr BP), where salinity is proposed to have increased. While the salinity increase inferred by the $^{13}C/^{12}C$ isotope ratio seems to be consistent with the diatom stratigraphy in the analyzed core, this observation is not generally supported by diatom records from this time period (see e.g. discussion in Zillén *et al.*, 2008).

So far, diatoms (and other biological proxies such as molluscs and foraminifers) have only been used as qualitative indicators of paleosalinity in the Baltic Sea. Although the major geological stages of the Baltic are well established, the salinity estimates vary markedly between studies conducted to date. Hence there is a clear need for a quantitative approach using an open-sea diatom–salinity calibration data set, which would provide a useful means for more accurate reconstructions of actual past salinity concentrations.

15.4 Climate change

15.4.1 Global climate change

Water temperature is one of the principal factors regulating biological processes, and changes in temperature have profound effects on single organisms, species, populations, communities, and ecosystems. Natural long-term climate dynamics with changing temperature regimes, and its repercussions such as drought in equatorial areas and melting of the polar ice caps, have always affected Earth's biota. However, a significant portion of recent climatic changes can be attributed to anthropogenic emissions (IPCC, 2007). Increasing concentrations of carbon dioxide and other greenhouse gases trap outgoing heat radiation in the lower atmosphere which results in a global temperature rise. High-latitude areas are considered to be especially sensitive to the changing climate. For example, annual average surface water temperature in the Baltic Sea has increased by 1.4 °C during the last ~100 years, more than in any of the world's other 63 large marine ecosystems (MacKenzie & Schiedek, 2007). The effects of increased water temperature on living diatom communities can be studied experimentally and in discharge areas for industrial cooling water, whereas deposited sedimentary diatom assemblages can be used to trace changes in past temperatures and climate. The results of such studies can be used to predict possible future changes.

15.4.2 Diatom species diversity and temperature

The response of communities and ecosystems to increased temperature will be strongly influenced by changes in diversity and community composition. Hillebrand *et al.* (2010) found that the turnover of benthic diatoms, macrophytes, and macrofauna significantly increased with increasing temperature within the thermal plume of a nuclear power plant at the Swedish Baltic Sea coast, while the number of species did not change for any of these groups. Thus, increasing temperatures in the thermal plume increased temporal beta-diversity and decreased compositional stability of communities, but observed richness did not change at any point in time.

15.4.3 Diatom life cycles and temperature

In a three-year study with monthly samplings, Potapova & Snoeijs (1997) found that the life cycle of the epiphyte *Diatoma moniliformis* was severely disrupted at an artificially heated (+10 °C) site in the northern Baltic Sea at salinity 5 psu. At an unheated reference site, the life cycle was well synchronized. Sexualization occurred in the cold season and auxosporulation was probably triggered by a combination of low water temperature (0–3 °C) and rapidly increasing day length in late winter–early spring. Size reduction during the vegetative life cycle could be divided into two parts; first cell volume decreased and surface-to-volume ratios increased and then both parameters decreased. Thus, the cells of *D. moniliformis* had high surface-to-volume ratios, enabling fast nutrient uptake during the period of optimum population growth in late spring. At the heated site, the required low temperature of 0–3 °C never occurred and auxosporulation was not synchronized.

15.4.4 Benthic primary production along temperature gradients

In cooling-water discharge areas the benthic diatoms react to higher water temperature by an extended growing season and increased primary production (Keskitalo & Heitto, 1987; Snoeijs, 1989, 1990, 1991). The effects of anomalous temperatures on diatoms are most pronounced during the cold season. Ecological factors directly affecting the enhanced growth of the diatoms are: reduced loss of cells by reduced abrading by ice; and higher production due to higher availability of light, nutrients, and substrate in the form of macroalgae. Very large (0.5 to 1 m long) epilithic diatom colonies can also be observed in spring at 0 °C, e.g. *Berkeleya rutilans* (Trentep.) Grunow in the Bothnian Sea after a winter without ice cover (Snoeijs, 1990), or *Navicula ramosissima* (C. Agardh) Cleve under ice cover in the Baltic Sea proper (Snoeijs & Kautsky, 1989). This indicates that, during the cold season, low water temperature is not a direct limiting factor for benthic diatoms as a group. Low light intensity is only limiting around mid-winter, and inorganic nutrient concentrations are high in the same period. All these factors favor benthic diatom growth during the cold season, in comparison with growth of other benthic algal groups when an ice cover is limited or absent.

15.4.5 Pelagic primary production along temperature gradients

According to mesocosm experiments with natural plankton communities from the Kiel Bight (Baltic Sea), winter and spring warming will lead to substantial changes in the spring bloom of phytoplankton (Sommer & Lengfellner, 2008). These authors found that higher winter–spring temperatures clearly decreased phytoplankton peak biomass, mean cell size, and the abundance of diatoms. This was mainly explained by the markedly increased grazing rates of the zooplankton at higher temperatures. Wasmund *et al.* (1998), on the other hand, explain monitored decreases in diatom abundance in the Baltic Sea spring bloom since 1988–89 by stratification changes: after mild winters thermal stratification sets in without effective surface layer mixing. Diatoms need mixed waters, whereas dinoflagellates grow well in a stable water column. Alheit *et al.* (2005) explained the mild winters starting in 1989 by a changing NAO (North Atlantic Oscillation) index from negative to positive, resulting in increasing air and sea-surface temperatures in the Baltic Sea area. In agreement with Wasmund *et al.* (1998), they explained the change in phytoplankton dominance from diatoms to dinoflagellates since the late 1980s by the lack of spring mixing caused by elevated temperatures.

15.4.6 Benthic community composition along temperature gradients

Species shifts in benthic diatom communities in cooling-water discharge areas are most pronounced during the cold season (Snoeijs, 1989, 1990, 1991). They are mainly regulated by temperature optima for some species and by the absence of ice cover for others. Generally, observed trends are that the diatoms enhanced by increasing temperature are large diatoms with colonial growth in chains (*Achnanthes brevipes* C. Agardh, *Melosira lineata* (Dillwyn) C. Agardh, *Melosira moniliformis* (O. F. Müll.) C. Agardh), smaller diatoms with colonial growth in mucilage tubes (*Navicula perminuta* Grunow, *Nitzschia filiformis* (W. Smith) van Heurck), or large epiphytic diatoms (*Tabularia tabulata* (C. Agardh) Snoeijs, *Licmophora* cf. *gracilis* (Ehrenb.) Grunow). The species favored most are "cold-water and low-light species" with growth optima around 10 °C in winter and early spring (e.g. *Melosira* spp.), and "warm-water species" with growth optima above 20 °C in late summer and autumn (e.g. *Nitzschia filiformis*). Some diatom species do not increase or decrease greatly in abundance with temperature, but their occurrence and abundance in time changes, e.g. the epiphytic spring maximum of *Diatoma moniliformis* may occur up to two months earlier in the year. Few of the abundant diatom species in the area do not show any response to a raised temperature regime; one of these is the common epiphyte *Rhoicosphenia curvata*, which seems to have a very wide temperature tolerance and no clear optimum.

15.4.7 Pelagic community composition along temperature gradients

In the discharge area of the Olkiluoto nuclear power plant on the Finnish west coast (Bothnian Sea), it was found that the spring maximum of the pelagic species *Chaetoceros wighamii* Brightwell occurred two weeks earlier than normal, and that the abundance of *Pseudosolenia minima* Levander was slightly increased (Keskitalo, 1987). However, the temperature rise did not markedly affect species composition or total phytoplankton quantities in the discharge area. The same was found for phytoplankton off the Forsmark nuclear power plant on the Swedish Bothnian Sea coast (Willén, 1985). On the contrary, shifts from diatoms to cyanobacteria were found in the plankton off a nuclear power plant near St. Petersburg (Russia) at the Gulf of Finland coast (Ryabova *et al.*, 1994). This is probably explained by the nutrient status of the water: basically oligotrophic at Olkiluoto and Forsmark and eutrophic at St. Petersburg.

15.4.8 Indicators of ice cover

During an average winter, the whole northern Baltic Sea is ice covered; the mean period of ice cover for the Bothnian Bay is 120 days per year, and for the Bothnian Sea 60 days per year (Leppäranta & Myrberg, 2009). The dominant diatoms occurring in connection with ice cover, some of them also normal constituents of the Baltic phytoplankton, belong to the Arctic flora. These diatoms occur in the Arctic and the Baltic Sea, and some also in the Oslofjord in southern Norway or in the Antarctic (Huttunen & Niemi, 1986; Hasle & Syvertsen, 1990; Norrman & Andersson, 1994), but they have not been reported from the Norwegian west coast. Some examples of this Arctic flora are: *Pauliella taeniata* (Grunow) Round & Basson (Syn. *Achnanthes taeniata* Grunow), *Fragilariopsis cylindrus* (Grunow) Hasle, *Melosira arctica* Dickie, *Navicula vanhoeffenii* Gran, *Navicula pelagica* Cleve, *Nitzschia frigida* Grunow, *Thalassiosira baltica* (Grunow) Ostenf., and *Thalassiosira hyperborea* (Grunow) Hasle. These diatoms occur less in the vernal bloom of the Baltic Sea when ice cover is limited (Haecky *et al.*, 1998; Höglander *et al.*, 2004), and can thus be used as markers for decreases and increases of Baltic Sea ice cover caused by climate change.

15.4.9 Fossil diatom assemblages reflect long-term changes in temperature and climate

In the Baltic Sea, fossil diatom records have been used in several studies to elucidate Holocene climate changes in the area. Their application to date for climate inferences has been virtually non-existent in the Black Sea and the Caspian Sea. Diatoms may track changes in climate directly via temperature changes or indirectly via changes in, for example, nutrient availability (Smol & Cumming, 2000). A warmer climate appears to increase nutrient availability in the Baltic Sea, probably due to increased upwelling and/or increased river discharge and weathering (e.g. Andrén *et al.*, 2000a; HELCOM, 2007). In the northwestern Baltic proper, indications of increased productivity such as the maximum absolute abundance of diatoms and high organic carbon concentrations were recorded *c.* 1700–1400 cal yr BP and 900–800 cal yr BP (Andrén *et al.*, 2000a) coinciding with the Roman and Medieval warm periods, which are believed to have been warmer than the present (*c.* 1850–1550 cal yr BP and 1050–650 cal yr BP, respectively; Lamb, 1995). High occurrences of *Pseudosolenia calcar-avis* (Schultze) Sundström and *Chaetoceros* spp. resting spores were observed during these times. The former is a common marine planktonic subtropical and tropical taxon, which also occurs seasonally in temperate regions (Hendey, 1964). It is frequently found during the Litorina stage of the Baltic, but is not found alive at present in the Baltic Sea (Snoeijs *et al.*, 1993–1998). A high abundance of *Chaetoceros* spp. resting spores, on the other hand, has been interpreted as an indication of high productivity events (Grimm & Gill, 1994).

Andrén *et al.* (1999) recorded increased abundances of the ice diatom *Pauliella taeniata* in a sediment core from the southwestern Baltic Sea (Oder Estuary) in the eighteenth and nineteenth century. During this time, *Thalassiosira baltica* also had its maximum abundance. These occurrences coincide with the pronounced cooling of northern European climate during the Little Ice Age with its coldest phase approximately confined between the sixteenth and the mid-nineteenth century. However, the first signs of the Little Ice Age date back as far as *c.* the 12th century, when storm and flood intensity increased markedly (Lamb, 1995). Around this time a clear decrease in the abundance of *Pseudosolenia calcar-avis* and increases in *T. baltica* and *Thalassiosira hyperborea* var. *lacunosa* (Berg) Hasle found in a sediment core from further north in the Gotland Basin indicate a change towards a colder climate (Andrén *et al.*, 2000a). A similar change in the assemblage composition was observed nearby in a core from the Landsort Deep (Thulin *et al.*, 1992). Jiang *et al.* (1997) found diatom assemblages during the late glacial period around 11,920–10,650 cal yr BP (10,000–10,100 ^{14}C yr BP; no correction for the reservoir effect was made due to lack of data) to be dominated by sea-ice species such as

Fragilariopsis cylindrus, *Thalassiosira* cf. *scotia* Fryx et hoban, and *Thalassiosira antarctica* Comber resting spores indicating Arctic/subarctic conditions in the Kattegat–Skagerrak area at the time.

In the Kattegat–Skagerrak area diatoms have also been used for paleoceanographic reconstructions covering the late glacial and the Holocene (Jiang *et al.*, 1997). Given the relationship between modern diatom assemblages and present hydrographic regimes (Jiang, 1996), downcore changes in diatom assemblage composition could be interpreted as changes in local hydrography and past strength and variability of different ocean currents in the area. Quantitative diatom-based inference models using weighted-averaging techniques have been used in the same area for climate reconstructions by inferring past sea-surface salinity (Jiang *et al.*, 1998). Changes in salinity were then linked to atmospheric processes and consequent changes in precipitation.

15.5 Eutrophication

15.5.1 Eutrophication of large brackish-water ecosystems

The anthropogenic nutrient load of nitrogen and phosphorus to aquatic ecosystems has increased markedly during the past century (Conley, 2000; de Jonge *et al.*, 2002). For example, in the Baltic Sea the phosphorus (P) load has increased about eight-fold, and the nitrogen (N) load about four-fold (Larsson *et al.*, 1985). This has resulted in an estimated 30–70% increase in the pelagic primary production and a 500–1000% increase in the sedimentation of organic carbon (Elmgren, 1989; Jonsson & Carman, 1994). The increased N and P loads have led to undesirable eutrophication effects in receptive areas, which are manifested as increased plankton biomass, decreased water transparency, nuisance blooms of cyanobacteria, decreased taxonomic diversity, population explosions of some species at the expense of others, declines in macroalgal density, hypoxia of bottom waters, and mass mortalities of zoobenthos and nektobenthic fish (e.g. Bakan & Büyükgüngör, 2000; Falandysz *et al.*, 2000; Yunev *et al.*, 2007). All these signs of eutrophication can be seen in the Baltic, Black, and Caspian seas, which belong to the most polluted marine areas in the world (Fonselius, 1972: Zonn, 2005; Zonn *et al.*, 2008). Published data including diatom studies for assessing the state of eutrophication of coastal and marine waters of Europe and adjacent areas are mostly available from the Baltic Sea, the Wadden Sea, and the North Sea, whereas only few data are available from other water bodies including the Black and the Caspian seas (Yunev *et al.*, 2007).

15.5.2 Long-term trends in eutrophication – paleolimnological evidence

Changes in (sub)fossil diatom assemblage composition reflect the ongoing eutrophication process. Grönlund (1993) found signs of the onset of eutrophication *c.* 200 years ago in the eastern Gotland Basin and Witkowski (1994), Andrén (1999b), and Andrén *et al.* (1999) *c.* 100–150 years ago in the southern Baltic Sea proper (Arkona Sea, Bornholm Basin, Gulf of Gdansk and Oder Estuary). Indications of accelerated eutrophication were found by Miller & Risberg (1990) and Risberg (1990), starting in the 1960s in the northwestern Baltic Sea proper, whereas Andrén *et al.* (2000a) detected clear indications of increased production and shift in the diatom assemblage in the 1950s–1960s in sediment sequences from the Gotland Basin. This is approximately in concert with the increase in extension of bottom areas with laminated sediments in the Baltic Sea proper, which started in the 1960s when large inputs of organic material caused bottom oxygen deficiency and absence of bioturbation because the fauna was killed (Fonselius, 1970; Jonsson & Jonsson, 1988; Zillén *et al.*, 2008).

Observed indications of eutrophication in the upper sediments of the open sea areas include increased concentrations of biogenic silica (derived primarily from dead siliceous microfossils such as diatom frustules, silicoflagellates, and chrysophyte cysts), increased diatom absolute abundance, changes in diatom species composition, decreases in the number of diatom taxa, and large increases in the abundance of *Chaetoceros* spp. resting spores. The mass occurrences of resting spores in the upper sediments have been interpreted as an indication of increased growth of pelagic *Chaetoceros* spp. (as vegetative cells) during high productivity events. The spores appear to be formed as a stress response when nutrients are depleted at the termination of large phytoplankton blooms (Grimm & Gill, 1994). Shifts in species composition have been observed to advantage planktonic diatoms. Diatom taxa that have increased in abundance in the upper sediments of the open Baltic Sea include *Actinocyclus octonarius* Ehrenb., *Coscinodiscus asteromphalus* Ehrenb., *Cyclotella choctawhatcheeana* Prasad, *Thalassiosira baltica*, *Thalassiosira* cf. *levanderi* van Goor, *Thalassiosira hyperborea* var. *lacunosa*, and *Thalassiosira hyperborea* var. *pelagica* (Cleve-Euler) Hasle. Not all of these taxa are necessarily indicators of eutrophication; their presence in the upper sediment layer in greater relative amounts than in the layers below may indicate higher primary production of pelagic diatoms in general, as well as increased turbidity, which decreases the photic layer and favors planktonic taxa over benthic (see e.g. Andrén *et al.*, 1999). As a result, fewer benthic diatom taxa are transported from the coastal zone and deposited in the open-sea sediments.

Eutrophication in coastal areas of the Baltic Sea is manifested similarly to the open sea areas. The abundance of planktonic diatoms has increased, which has also been observed in other coastal areas worldwide (e.g. Cooper *et al.*, this volume) and species richness of diatom assemblages has decreased (Weckström *et al.*, 2007). Increased amounts of *Chaetoceros* spp. resting spores and increased biogenic silica concentrations of the sediments have also been observed in many of the coastal areas of the Baltic Sea, e.g. in the Kattegat, Danish fjords, Gulf of Riga, and the Archipelago Sea (Ellegaard *et al.*, 2006; Olli *et al.*, 2008; Vaalgamaa & Conley, 2008; Tuovinen *et al.*, 2010).

15.5.3 Quantitative reconstructions of past nutrient concentrations

In recent years, long-term trends in eutrophication of the Baltic coastal waters have been assessed using diatom transfer functions for nitrogen (Birks, this volume), which are based on large calibration data sets of modern diatom assemblages and measured environmental data (MOLTEN, DETECT and DEFINE, 2006). These studies were prompted by the need of the European Water Framework Directive (Anonymous, 2000) to define ecological reference conditions and the present departure from the natural state of coastal ecosystems. The studies indicate that urban coastal areas in Finland have been eutrophied for centuries (depending on the population history of their cities), but show a general recovery with improved municipal waste-water treatment (Weckström *et al.*, 2004, 2007). Although nutrient concentrations at these sites have clearly decreased, the present planktonic diatom assemblages of these embayments show only little change back to the pre-disturbance diverse benthic communities. This suggests that decreased external loading can be counteracted by internal loading from a nutrient pool accumulated in the sediments during the period of high external loading. Hence, after initial improvement following reduced nutrient loads, no further improvement occurs. In coastal areas of the Baltic, which are mostly impacted by agriculture, the changes are generally only moderate compared to urban areas (e.g. Weckström, 2006) with the exception of Denmark, where intensive agriculture has had a much more pronounced effect on coastal water quality (Clarke *et al.*, 2003; Ellegaard *et al.*, 2006). This can mainly be explained by the significantly higher percentage of field cover in the Danish coastal catchment areas and a much longer cultivation history (Bradshaw, 2001). The increased use of artificial fertilizers after World War II amplified the effects of diffuse nutrient loading from agricultural areas, causing clear changes in diatom assemblages in different parts of the Baltic (e.g. Clarke *et al.*, 2003; Weckström, 2006; Olli *et al.*, 2008). The accuracy of the previously described coastal quantitative diatom-based nutrient reconstructions has been estimated by comparing their inferences against actual measured monitoring data (Weckström, 2006; Ellegaard *et al.*, 2006; Tuovinen *et al.*, 2010). The results are promising and a number of reconstructions are in good agreement with monitoring data. However, problems were also encountered, which mostly related to the dominant diatom taxa in the core responding more strongly to other environmental variables than the one being reconstructed (i.e. nitrogen) and lack of modern analogs. In an urban embayment in Helsinki, Finland, there was an unexpected decrease in reconstructed N concentrations in the early to mid 1900s, when diatom assemblage structure (high abundance of planktonic diatoms, low species richness) and known catchment land-use changes suggested the opposite (Weckström, 2006). This decrease was mainly due to a marked decline in the relative abundance of *Cyclotella atomus* Hustedt, which has the highest N optimum in the calibration data set used, and the dominance of *Thalassiosira guillardii* Hasle, which, although a eutrophic species, clearly has a lower optimum. The observed shift between these two taxa may have been caused by silica limitation of the system with pronounced eutrophication, as the latter taxon is very lightly silicified and would gain advantage in such a situation. Another example was encountered in the Archipelago Sea, where the reconstruction was clearly driven by changes in the abundance of *Pauliella taeniata* and *Thalassiosira levanderi*. *P. taeniata* is an ice-diatom, which decreased markedly in the late 1980s to early 1990s due to exceptionally warm winters and low number of ice-days in the area. This provided *T. levanderi* with a competitive advantage and, as a result, it occurred in high abundances. The nitrogen optimum of this species is, however, low in the calibration data set used and hence the reconstructed values are low when monitoring data suggest the opposite (Tuovinen *et al.*, 2010). These examples highlight the importance of comparing obtained quantitative nutrient reconstructions with community structure indices (e.g. abundance of planktonic diatoms, diversity), changes in the abundance of individual species, and the ecological implications of these changes in order to correctly interpret reconstructions.

15.5.4 Regime shifts in pelagic phytoplankton communities

Different algal species have different nutrient requirements, but often species belonging to the same taxonomic group have similar requirements. Eutrophication effects on the composition of microalgal communities are therefore often clearly reflected at the class or division levels. The most typical compositional change following eutrophication is a decrease in the relative importance of diatoms in favor of non-silicified flagellates (Smayda, 1990; Cadée & Hegeman, 1991; Yunev *et al.*, 2007). Increased eutrophication initially enhances diatom growth. However, silicate depletion occurs with increased diatom production, because large amounts of silica are removed from the photic zone with the sinking diatoms and deposited on the sediment. As dissolution of silicate from diatom shells is slow, silica is released from sediments at a low rate (Eppley, 1977; Schelske *et al.*, 2006). In addition, silicate levels in the receptive coastal areas may be decreased by diatom growth and consequent silicate retention in sediments behind river dams (e.g. Conley *et al.*, 2000). As silica is an essential nutrient for diatoms, being incorporated into their cell walls, their growth will be limited and non-siliceous phytoplankton species will increase in importance. Usually, a decrease of diatoms is regarded negative for ecosystem health because they have a high nutritive value for micro-, meio- and macrofauna organisms (Plante-Cuny & Plante, 1986), despite observations that a diatom diet can have negative effects on copepod reproduction, known as the "paradox of diatom-copepod interactions in the pelagic food web" (Ianora *et al.*, 2003). Several groups of flagellates and cyanobacteria are known to produce toxins in brackish waters (Luckas *et al.*, 2005), but only very few diatom species, e.g. some domoic-acid-producing *Pseudo-nitzschia* spp. (Hasle *et al.*, 1996; Besiktepe *et al.*, 2008; Thessen & Stoecker, 2008), are toxic (see Villac *et al.*, this volume).

In the Black Sea, eutrophication has increased pelagic primary production manyfold, and formerly dominant diatom species have been replaced by dinoflagellates, prymnesiophytes, and other non-silicious flagellates (Leppäkoski & Mihnea, 1996; Vasiliu, 1996; Humborg *et al.*, 1997; Yunev *et al.*, 2007). Similar trends have been observed in the Baltic Sea, where diatom biomass during the spring bloom has decreased in the last decades and biomass of dinoflagellates increased (Wasmund & Uhlig, 2003). In eutrophied coastal areas of the Baltic, cryptophytes and dinophytes often dominate over diatoms (Sagert *et al.*, 2008). This decreasing trend, triggered by eutrophication, is further strengthened by the warming climate in the Baltic Sea; during mild winters the sea-surface temperatures are too high for spring mixing, hence thermal stratification sets in without effective mixing of the surface layer. This provides dinoflagellates with a further competitive advantage, as diatoms need mixed waters, whereas dinoflagellates thrive in a stable water column (Wasmund *et al.*, 1998). Due to eutrophication and its effects on algal community composition, silicate has shown a decreasing trend throughout the entire Baltic Sea in recent decades (Wulff *et al.*, 1990; Rahm *et al.*, 1996; Papush & Danielsson, 2006). Wulff *et al.* (1990) assumed, in a mathematical model, that the process is reversible; if discharges of nitrogen and phosphorus are lowered by 25% in the Gulf of Bothnia and by 50% in the Baltic Sea proper, silicate concentrations will increase again due to reduced primary production levels. In contrast to the regime shift observed in pelagic ecosystems, the silicate pool in the sediments may counteract silicate limitation of the dominant benthic diatoms in very shallow, eutrophied coastal areas. As an example of this, the relative abundance of diatoms living in and on the sandy sediments did not decrease in favor of other microalgal groups in a four-week mesocosm experiment with N and P additions (Sundbäck & Snoeijs, 1991).

15.6 Non-indigenous species

Discharge of ships' ballast water, fouling on ship hulls, and coexistence with intentionally transferred species are major vectors for introductions of non-indigenous species to new sea areas (Leppäkoski *et al.*, 2002; Spaulding & Edlund, this volume). Shipping activities and intentional transfer of species have increased dramatically during the past century and a number of diatom species have been introduced to the large brackish-water ecosystems in these ways, some of which have become highly abundant.

15.6.1 The Baltic Sea

The Indo-Pacific phytoplankton species *Coscinodiscus wailesii* Gran et Angst is a nuisance diatom in European waters (Laing & Gollash, 2002). It forms dense blooms that produce copious amounts of mucilage and is inedible to most grazing zooplankton. In the Baltic Sea area it was observed for the first time in 1983 (Leppäkoski & Olenin, 2000). Other examples of diatom species introduced to the Baltic Sea are *Odontella sinensis* (Greville) Grunow (since 1903, Indo-Pacific native), *Thalassiosira punctigera* (Castracane) Hasle (since 1979, of unknown origin), and *Pleurosigma simonsenii* Hasle (since 1987, Indian Ocean native). A special case is *Pleurosira laevis* fo. *polymorpha* (Kütz.) Compère. It was probably introduced during an experimental release of eels off the Forsmark nuclear power plant; the eels were raised in aquaria in southern Europe. This large chain-forming diatom was never observed before 1989, but since 1990 it has formed up to 0.5 meter high colonies in the power plant's brackish cooling-water discharge area each year in September–November, especially in sites with slowly flowing water (Snoeijs, personal observation).

15.6.2 The Black Sea

Reported non-indigenous diatom species in the Black Sea are *Asterionella japonica* Cleve et Moller ex Gran and *Pseudosolenia calcar-avis* (both introduced from the Pacific or Atlantic Ocean) and *Thalassiosira nordenskioeldii* Cleve (introduced from the Arctic or North Atlantic Ocean) (Gomoiu *et al.*, 2002). Some new diatom species emerged in the coastal plankton of the north-western Crimea during long-term observations 1968–2002: *Asterionellopsis glacilis* (Castracane) F. E. Round, *Chaetoceros tortissimus* Gran, *Lioloma pacificus* (Cupp) Hasle, and *Pseudo-nitzschia inflatula* Hasle (Shiganova, 2008).

15.6.3 The Caspian Sea

The major diatom invader in the Caspian Sea is *Pseudosolenia calcar-avis* through an accidental introduction during acclimatization of gray mullets (Karpinsky *et al.*, 2005). *P. calcar-avis* appeared in the Caspian Sea in 1930 and made up two-thirds of the total phytoplankton biomass by 1936 (Aladin *et al.*, 2002) and largely replaced the native *Dactyliosolen fragilissimus* (Bergon) G. R. Hasle. In some seasons *P. calcar-avis* comprises 80–92% of the total phytoplankton biomass in the Middle and Southern Caspian Sea. This large-sized diatom is, unlike *D. fragilissimus*, not used as food by the Caspian zooplankton, but *P. calcar-avis* sinks to the bottom, where it is decomposed by bacteria and is included into the benthic food chain. Thus, its introduction caused a serious rearrangement of energy fluxes in the Caspian Sea ecosystem (Karpinsky *et al.*, 2005). Further non-indigenous diatom species in the Caspian Sea are *Pseudo-nitzchia seriata* (Cleve) H. Peragallo (since 1990), *Cerataulina pelagica* (Cleve) Hendey (since 2002), *Chaetoceros pendulus* Karsten, and *Tropidoneis lepidoptera* (Gregory) Cleve (both after 2002) (Karpinsky *et al.*, 2005).

15.7 Hazardous substances

15.7.1 Metals

Natural diatom communities and diatom assemblages grown on artificial substrata can be used for monitoring metals in the environment. Epilithic diatom samples taken in 1991 outside the Rönnskär industry (Bothnian Bay, Sweden), had metal levels on the diatoms corresponding to the distance from the discharge point (Figure 15.4). Diatoms seem to act as a "sponge" attracting metal ions from the water. The (radioactive) metals attach mainly to the outside of the cell wall, as shown by

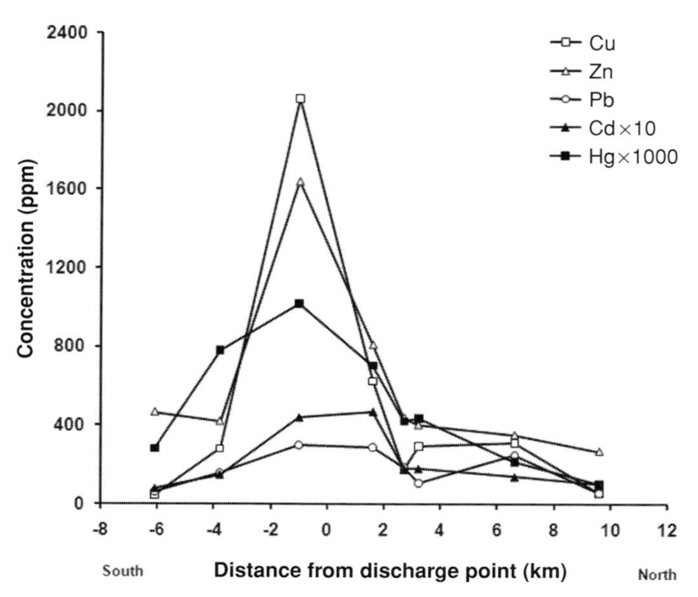

Figure 15.4 Dry-weight based concentrations of copper, zinc, lead, cadmium (shown with factor ×10), and mercury (shown with factor ×1000) in epilithic diatom samples taken outside the Rönnskär industry, which discharges heavy metals (Bothnian Bay, Sweden), in May 1991. Nine epilithic diatom samples taken in January 1992 at about the same distances from the discharge point outside the Forsmark nuclear power plant, which discharges practically no heavy metals (Bothnian Sea, Sweden), showed the following mean values: Cu 24 ± 7 ppm, Zn 170 ± 48 ppm, Pb 31 ± 14 ppm, Cd 3 ± 1 ppm, Hg 0.15 ± 0.05 ppm. (Data: P. Snoeijs (previously unpublished).)

Pedersén *et al.* (1981) and Lindahl *et al.* (1983), and only a minor portion of the metals is actually taken up into the cell. Diatoms have a thin organic casing around the silica frustule, and one of its functions may be to complex cations such as iron or aluminum in order to minimize loss of silica from the frustule by dissolution (Round *et al.*, 1990). As most radionuclides and metals are cations, it is probable that they are attached to the organic casing around the diatom frustules. Periphyton assemblages on artificial substrata dominated by diatoms can also effectively be used as pollution indicators for toxic compounds such as tri-*n*-butyl tin (TBT) in short-term toxicity tests (Molander *et al.*, 1990; Blanck & Dahl, 1996). This method, called PICT (pollution induced community tolerance), uses the selective pressure of a toxicant on sensitive species, so that the community as a whole shows an increase in tolerance for the toxicant. The increase in community tolerance is an indicator of damage to the algal community. For example, assemblages grown on artificial substrata in the field at different distances from a discharge point can in the laboratory be tested for PICT, chlorophyll *a*, photosynthesis, and species composition.

Table 15.2 Fission products from Chernobyl measured in epilithic diatom communities in Forsmark (Sweden) one week after the Chernobyl accident (May 6, 1986) in kBq kg^{-1} dry weight. Sampling sites are arranged according to water movement conditions. Adapted after Snoeijs & Notter (1993a).

	Water movement			
Radionuclide	Quiescent	Slowly flowing	Flowing	Fast flowing
^{110m}Ag *	1	1	< 1	1
^{134}Cs	14	6	3	3
^{137}Cs	25	10	6	4
^{132}Te	87	39	18	21
^{131}I	181	48	24	21
^{140}Ba	439	124	32	9
^{103}Ru	540	142	49	16
^{144}Ce	612	218	42	11
^{141}Ce	825	310	58	15
^{95}Zr	864	328	64	14
^{95}Nb	1022	339	78	18

* Discharged by both the Chernobyl accident and the Forsmark nuclear power plant.

15.7.2 Radionuclides

Littoral diatoms were shown to be good monitoring organisms for radionuclides in the Forsmark area, southern Baltic Sea (Snoeijs & Notter, 1993a). As monitoring organisms, they occupy a position between macroalgae and sediment. Diatoms have advantages over sediment, being part of the food chain and responding quickly to changes in actual concentrations in the water (Table 15.2). Compared to macrophytes and animals, diatoms have higher radionuclide concentrations, less selectivity for the type of radionuclide, and the concentrations are less dependent on physiological and seasonal cycles. Radionuclide concentrations seem to be related to the algal surface area exposed to the water. Epilithic diatoms have the highest concentrations, diatoms living in mucilage tubes slightly lower, thin filamentous macroalgae still lower, and macroalgae with broader filaments have the lowest concentrations (Snoeijs & Notter, 1993a). Since 1992, diatoms have been used as monitoring organisms for radionuclide discharges at all four Swedish nuclear power plants. Radionuclides recycle in epilithic diatom communities to a high extent. This implies that distinct discharges of the Chernobyl type can be traced in diatom samples for a long time. Concentrations of ^{137}Cs in epilithic diatom communities five years after the Chernobyl accident still reflected the original fallout pattern of the radioactive cloud that crossed the Swedish east coast (Snoeijs & Notter, 1993b; Carlson & Snoeijs, 1994). Furthermore, new generations of diatoms still continuously take up ^{134}Cs and ^{137}Cs from the environment, as these diatom communities show large yearly fluctuations in biomass, and the radiocesium isotopes only disappear from the communities according to their physical half-lives.

15.8 Summary

In this chapter we concentrate on the application of diatoms as indicators of environmental change with focus on the Earth's three largest brackish-water areas, the Baltic, Black, and Caspian seas. Ecosystems worldwide change fast under the pressure of today's large-scale human-induced environmental impacts. Global climate change, eutrophication, discharges of hazardous substances, introductions of alien species, and over-exploitation of natural resources affect the health of organisms, community composition, and food-web interactions in aquatic systems with accelerating speed. The Earth's large brackish-water bodies are particularly sensitive to such changes. Their drainage areas are extensive and water turnover times extremely low. Harmful and hazardous substances discharged into these ecosystems circulate in the water, sediments, and biota for long times. Anthropogenic impacts have changed the Baltic, Black, and Caspian Sea ecosystems drastically during the last 100 years and in some cases this has resulted in regime shifts. We consider biodiversity, productivity, composition, and other aspects of present-day living diatom communities, as well as (sub)fossil diatom assemblages, as powerful tools in the reconstruction of (paleo)environments and long-term environmental change. Periphyton assemblages on artificial substrata dominated by diatoms can effectively be used as pollution indicators for toxic compounds by (1) using the selective pressure of a toxicant on sensitive species and (2) using the capacity of diatom frustules to accumulate (radioactive) metals from the surrounding water.

We have evaluated the definition of "brackish-water species" in ecological and evolutionary perspectives and summarized the historical development of diatoms as indicators of salinity from the 1930s "halobion system" to contemporary salinity calibration models. During the past century, global climate change has increased surface-water temperatures in aquatic ecosystems worldwide, e.g. by 1.4 °C in the Baltic Sea. The effects of increased water temperature on living diatom communities can be studied experimentally and in discharge areas for industrial cooling water, whereas deposited sedimentary diatom assemblages can be used to trace changes in past temperatures and other factors related to climate. The results of such studies can be used to predict possible future climate changes. Also, the anthropogenic nutrient load of nitrogen and phosphorus to aquatic ecosystems has increased markedly during the past century. For example, in the Baltic Sea the phosphorus load has increased about eightfold, and the nitrogen load about four fold. This has resulted in an estimated 30–70% increase in the pelagic primary production, a 500–1000% increase in the sedimentation of organic carbon, and (together with other factors such as over-fishing) in a pelagic regime shift. Such long-term trends in eutrophication can be traced in the subfossil diatom record and quantitative diatom-based nutrient reconstruction models for brackish water are available. A number of diatom species originating from other areas have been introduced to the Baltic, Black, and Caspian seas, some of which have the potential to seriously alter ecosystem energy fluxes, e.g. *Pseudosolenia calcar-avis* in the Caspian Sea.

Acknowledgments

We are grateful to Elinor Andrén, John P. Smol and Eugene F. Stoermer for valuable comments on the manuscript and to Peter Rasmussen for helpful discussions. Financial support was provided by the Swedish Research Council (PS) and the Academy of Finland (KW).

References

Abelmann, A. (1985). Palökologische und ökostratigraphische Untersuchungen von Diatomeenassoziationen an holozänen Sedimenten der zentralen Ostsee. *Berichte Reports, Geologisch-Paläontologisches Institut der Universität Kiel*, **9**, 1–199.

Aladin, N., Plotnikov, I. S., & Filippov, A. A. (2002). Invaders in the Caspian Sea. In *Invasive Aquatic Species of Europe – Distribution, Impacts and Management*, ed. E. Leppäkoski, S. Olenin, & S. Gollasch, Dordrecht: Kluwer Academic Publishers, pp. 351–9.

Alheit, J., Möllmann, C., Dutz, J., *et al.* (2005). Synchronous ecological regime shifts in the central Baltic and the North Sea in the late 1980s. *ICES Journal of Marine Science*, **62**, 1205–15.

Alhonen, P. (1971). The stages of the Baltic Sea as indicated by the diatom stratigraphy. *Acta Botanica Fennica*, **92**, 3–17.

Alhonen, P. (1986). Late Weichselian and Flandrian diatom stratigraphy: methods, results and research tendencies. *Striae*, **24**, 27–33.

Andrén, E. (1999a). Holocene environmental changes recorded by diatom stratigraphy in the southern Baltic Sea. *Meddelanden från Stockholms universitets institution för geologi och geokemi*, **302**, 1–22.

Andrén, E. (1999b). Changes in the composition of the diatom flora during the last century indicate increased eutrophication of the Oder estuary, south-western Baltic Sea. *Estuarine, Coastal and Shelf Science*, **48**, 665–76.

Andrén, E., Andrén, T., & Kunzendorf, H. (2000a). Holocene history of the Baltic Sea as a background for assessing records of human impact in the sediments of the Gotland Basin. *Holocene*, **10**, 687–702.

Andrén, E., Andrén, T. & Sohlenius, G. (2000b). The Holocene history of the southwestern Baltic Sea as reflected in a sediment core from the Bornholm Basin. *Boreas*, **29**, 233–50.

Andrén, E., Shimmield, G., & Brand, T. (1999). Environmental changes of the last three centuries indicated by siliceous microfossil records from the southwestern Baltic Sea. *Holocene*, **9**, 25–38.

Anonymous (1959). Final Resolution of the Symposium on the Classification of Brackish Waters. *Archivio di Oceanografia e Limnologia (Supplement)*, **11**, 243–5.

Anonymous (1985). *Kaspiiskoye morye (Caspian Sea)*. Moscow: Nauka (in Russian).

Anonymous (2000). Directive 200/60/EC of the European Parliament and of the Council of 23 October 2000 establishing a framework for Community action in the field of water policy. *Official Journal of the European Communities*, **L 327/1**.

Archibald, R. E. M. (1983). The diatoms of the Sundays and Great Fish rivers in the eastern Cape Province of South Africa. *Bibliotheca Diatomologica*, **1**, 1–362 + 34 plates.

Bakan, G. & Büyükgüngör, H. (2000). *The Black Sea*. In *Seas at the Millenium: an Environmental Evaluation*, ed. C. R. C. Sheppard, Amsterdam: Elsevier Science Ltd, pp. 285–305.

Berglund, B. E., Sandgren, P., Barnekow, L., *et al.* (2005). Early Holocene history of the Baltic Sea, as reflected in coastal sediments in Blekinge, southeastern Sweden. *Quaternary International*, **130**, 111–39.

Besiktepe, S., Ryabushko, L., Ediger, D. Yimaz, *et al.* (2008). Domoic acid production by *Pseudo-nitzschia calliantha* Lundholm, Moestrup et Hasle (Bacillariophyta) isolated from the Black Sea. *Harmful Algae*, **7**, 438–42.

Birks, H. J. B. (1995). Quantitative palaeoenvironmental reconstructions. *Quaternary Research Association, Technical Guide*, **5**, 161–254.

Björck, S. (1995). A review of the history of the Baltic Sea 13.0–8.0 ka BP. *Quaternary International*, **27**, 19–40.

Björck, S. (2008). The late Quaternary development of the Baltic Sea basin. In *Assessment of Climate Change for the Baltic Sea Basin*, ed. the BACC author team, Berlin: Springer-Verlag, pp. 398–407.

Björck, S., Andrén, T., & Jensen, J. B. (2008). An attempt to resolve the partly conflicting data and ideas on the Ancylus–Litorina transition. *Polish Geological Institute Special Papers*, **23**, 21–6.

Blanck, H. & Dahl, B. (1996). Pollution-induced community tolerance (PICT) in marine periphyton in a gradient of tri-n-butyltin (TBT) contamination. *Aquatic Toxicology*, **35**, 57–77.

Bonsdorff, E. (2006). Zoobenthic diversity-gradients in the Baltic Sea: continuous post-glacial succession in a stressed ecosystem. *Journal of Experimental Marine Biology and Ecology*, **330**, 383–91.

Bradshaw, E. G. (2001). Linking land and lake. The response of lake nutrient regimes and diatoms to long-term land-use change in Denmark. Unpublished Ph.D. thesis, University of Copenhagen.

Busse, S. & Snoeijs, P. (2002). Gradient responses of diatom communities in the Bothnian Bay, northern Baltic Sea. *Nova Hedwigia*, **74**, 501–25.

Busse, S. & Snoeijs, P. (2003). Gradient responses of diatom communities in the Bothnian Sea (northern Baltic Sea), with emphasis on responses to water movement. *Phycologia*, **42**, 451–64.

Cadée, G. C. & Hegeman, J. (1991). Historical phytoplankton data of the Maarsdiep. *Hydrobiological Bulletin*, **24**, 111–8.

Carlson, L. & Snoeijs, P. (1994). Radiocaesium in algae from Nordic coastal waters. In *Nordic Radioecology – The Transfer of Radionuclides through Nordic Ecosystems to Man*, ed. H. Dahlgaard, Amsterdam: Elsevier Science Publishers, pp. 105–17.

Carpelan, L. H. (1978a). Evolutionary euryhalinity of diatoms in changing environments. *Nova Hedwigia*, **29**, 489–526.

Carpelan, L. H. (1978b). Revision of Kolbe's system der Halobien based on diatoms of Californian lagoons. *Oikos*, **31**, 112–22.

Clarke, A., Juggins, S. & Conley, D. (2003). A 150-year reconstruction of the history of coastal eutrophication in Roskilde Fjord, Denmark. *Marine Pollution Bulletin*, **46**, 1615–18.

Cleve, P. T. (1899). Bidrag till Kännedom om Östersjöns och Bottniska vikens postglaciala geologi. *Sveriges Geologiska Undersökningar*, **C180** (in Swedish).

Cleve-Euler, A. (1951–1955). Die Diatomeen von Schweden und Finnland I-V. *Kungliga Svenska Vetenskapsakademiens Handlingar*, **2**(1), 1–163; **3**(3), 1–153; **4**(1), 1–255; **4**(5), 1–158; **5**(4), 1–132.

Conley, D. J. (2000). Biogeochemical nutrient cycles and nutrient management strategies. *Hydrobiologia*, **410**, 87–96.

Conley, D. J., Stålnacke, P., Pitkänen, H., & Wilander, A. (2000). The transport and retention of dissolved silicate by rivers in Sweden and Finland. *Limnology and Oceanography*, **45**, 1850–53.

de Jonge, V. N., Elliott, M., & Orive, E. (2002). Causes, historical development, effects and future challenges of a common environmental problem: eutrophication. *Hydrobiologia*, **475–476**, 1–19.

den Hartog, C. (1964). Typologie des Brackwassers. *Helgoländer Wissenschaftliche Meeresuntersuchungen*, **10**, 377–90.

Ellegaard, M., Clarke, A. L., Reuss, N., *et al.* (2006). Long-term changes in plankton community structure and geochemistry in Mariager Fjord, Denmark, linked to increased nutrient loading. *Estuarine, Coastal and Shelf Science*, **68**, 567–78.

Elmgren, R. (1989). Man's impact on the ecosystem of the Baltic Sea: energy flows today and at the turn of the century. *Ambio*, **18**, 326–32.

Emeis, K.-C., Struck, U., Blanz, T., Kohly, A., & Voss, M. (2003). Salinity changes in the central Baltic Sea (NW Europe) over the last 10 000 years. *Holocene*, **13**, 411–21.

Eppley, R. W. (1977). The growth and culture of diatoms. In *The Biology of Diatoms*, ed. D. Werner, Botanical Monographs, volume 13, Oxford: Blackwell Scientific Publications, pp. 24–64.

Falandysz, J., Trzosinska, A., Szefer, P., Warzocha, J., & Draganik, B. (2000). The Baltic Sea, especially southern and eastern regions. In *Seas at the Millenium: an Environmental Evaluation*, ed. C. R. C. Sheppard, Amsterdam: Elsevier Science Ltd, pp. 99–120.

Finenko, Z. Z. (2008). Biodiversity and bioproductivity. In *The Black Sea Environment, The Handbook of Environmental Chemistry*, 5Q, ed. A. G. Kostianoy & A. N. Kosarev, Berlin: Springer-Verlag, pp. 351–74.

Folke, C., Carpenter, S., Walker, B., *et al.* (2004). Regime shifts, resilience and biodiversity in ecosystem management. *Annual Review of Ecology, Evolution and Systematics*, **35**, 557–81.

Fonselius, S. H. (1970). Stagnant sea. *Environment*, **12**, 2–11; 40–8.

Fonselius, S. H. (1972). On eutrophication and pollution in the Baltic sea. In *Marine Pollution and Sea Life*, ed. M. Ruvio, London: Fishing News (Books) Ltd, pp. 23–8.

Gomoiu, M.-T., Alexandrov, B., Shadrin, N., & Zaitsev, Y. (2002). The Black Sea – a recipient, donor and transit area for alien species. In *Invasive Aquatic Species of Europe – Distribution, Impacts and Management*, ed. E. Leppäkoski, S. Olenin, & S. Gollasch, Dordrecht: Kluwer Academic Publishers, pp. 341–50.

Grimm, K. A. & Gill, A. S. (1994). Fossil phytoplankton blooms and selfish genes: the ecological and evolutionary significance of *Chaetoceros* resting spores in laminated diatomaceous sediments. *Geological Society of America Abstract with Programs*, **26**, A170–71.

Grönlund, T. (1993). Diatoms in surface sediments of the Gotland Basin in the Baltic Sea. *Hydrobiologia*, **269–270**, 235–42.

Gudelis, V. & Königsson, L.-K. (eds.) (1979). *The Quaternary History of the Baltic*. Acta Universitatis Upsaliensis, Symposia Universitatis Upsaliensis Annuum Quingentesimum Celebrantis, vol. 1, pp. 1–279.

Gustafsson, B. G. & Westman, P. (2002). On the causes for salinity variations in the Baltic Sea during the last 8500 years. *Paleoceanography*, **17**, 1040.

Haecky, P., Jonsson, S., & Andersson, A. (1998). Influence of sea ice on the composition of the spring phytoplankton bloom in the northern Baltic Sea. *Polar Biology*, **20**, 1–8.

Hasle, G. R., Lange, C. B., & Syvertsen, E. E. (1996). A review of *Pseudo-nitzschia*, with special reference to the Skagerrak, North Atlantic, and adjacent waters. *Helgoländer Meeresuntersuchungen*, **50**, 131–75.

Hasle, G. R. & Syvertsen, E. E. (1990). Arctic diatoms in the Oslofjord and the Baltic Sea – a bio- and palaeogeographic problem? In *Proceedings of the 10th International Diatom Symposium*, ed. H. Simola, Königstein: Koeltz Scientific Books, pp. 285–300.

HELCOM (2007). Climate change in the Baltic Sea area – HELCOM Thematic Assessment in 2007. *Baltic Sea Environment Proceedings*, **111**, 1–49.

Hendey, N. I. (1964). An introductory account of the smaller algae of British coastal waters. Part V: Bacillariophyceae (diatoms). *Fishery Investigation Series*, **IV**, 1–317 + plates I–XLV.

Hillebrand, H., Snoeijs, P., & Soininen, J. (2010). Warming leads to higher species turnover in a coastal ecosystem. *Global Change Biology*, **16**, 1181–93.

Höglander, H., Larsson, U., & Hajdu, S. (2004). Vertical distribution and settling of spring phytoplankton in the offshore NW Baltic Sea proper. *Marine Ecology Progress Series*, **283**, 15–27.

Huisman, J., Olff, H., & Fresco, L. F. M. (1993). A hierarchical set of models for species response analysis. *Journal of Vegetation Science*, **4**, 37–46.

Humborg, C., Ittekkot, V., Cociasu, A., & Von Bodungen, B. (1997). Effect of the Danube River dam on Black Sea biochemistry and ecosystem structure. *Nature*, **386**, 385–8.

Hustedt, F. (1953). Die Systematik der Diatomeen in ihren Beziehungen zur Geologie und Ökologie nebst einer Revision des Halobien-Systems. *Svensk Botanisk Tidskrift*, **47**, 509–19.

Huttunen, M. & Niemi, Å. (1986). Sea-ice algae in the northern Baltic Sea. *Memoranda Societatis pro Fauna et Flora Fennica*, **62**, 58–62.

Ianora, A., Poulet, S. A., & Miralto, A. M. (2003). The effects of diatoms on copepod reproduction: a review. *Phycologia*, **42**, 351–63.

Ignatius, H. & Tynni, R. (1978). Itämeren vaiheet ja piilevätutkimus [Baltic Sea stages and diatom analysis]. *Tuurun yliopiston maaperägeologia osaston julkaisuja*, **36**, 1–26 (in Finnish, with English summary).

Ignatius, H., Axberg, S., Niemistö, L., & Winterhalter, B. (1981). Quaternary geology of the Baltic Sea. In *The Baltic Sea*, ed. A. Voipio, Amsterdam: Elsevier Science Publishers, pp. 54–105.

IPCC (2007). Climate Change 2007: Synthesis Report. Contribution of Working Groups I, II and III to the Fourth Assessment Report of the Intergovernmental Panel on Climate Change, ed. R. K. Pachauri, & A. Reisinger, Geneva: IPCC.

Jiang, H. (1996). Diatoms from the surface sediments of the Skagerrak and the Kattegat and their relationship to the spatial changes of environmental variables. *Journal of Biogeography*, **23**, 129–37.

Jiang, H., Björck, S., & Knudsen, K. L. (1997). A palaeoclimatic and palaeoceanographic record of the last 11 000 14C years from the Skagerrak–Kattegat, northeastern Atlantic margin. *Holocene*, **7**, 301–10.

Jiang, H., Björck, S., & Svensson, N.-O. (1998). Reconstruction of Holocene sea-surface salinity in the Skagerrak–Kattegat: a climatic and environmental record of Scandinavia. *Journal of Quaternary Science*, **13**, 107–14.

John, J. (1983). The diatom flora of the Swan River estuary, western Australia. *Bibliotheca Phycologica*, **64**, 1–359.

Jonsson, P. & Carman, R. (1994). Changes in deposition of organic matter and nutrients in the Baltic Sea during the twentieth century. *Marine Pollution Bulletin*, **28**, 417–26.

Jonsson, P. & Jonsson, B. (1988). Dramatic changes in Baltic sediments during the last three decades. *Ambio*, **17**, 158–60.

Juggins, S. (1992). Diatoms in the Thames estuary, England. Ecology, palaeoecology, and salinity transfer function. *Bibliotheca Diatomologica*, **25**, 1–216.

Juhlin-Dannfelt, H. (1882). On the diatoms of the Baltic Sea. *Bihang till Kungliga Svenska Vetenskapsakademiens Handlingar*, **6**, 1–52.

Kabailiene, M. (1995). The Baltic Ice Lake and Yoldia Sea stages, based on data from diatom analysis in the central, south-eastern and eastern Baltic. *Quaternary International*, **27**, 69–72.

Karayeva, N. I. & Makarova, I. V. (1973). Special features and origin of the Caspian Sea diatom flora. *Marine Biology*, **21**, 269–75.

Karpinsky, M. G. (2005). Biodiversity. In *The Caspian Sea Environment, The Handbook of Environmental Chemistry*, 5P, ed. A. G. Kostianoy & A. N. Kosarev, Berlin: Springer-Verlag, pp. 159–73.

Karpinsky, M. G., Shiganova, T. A., & Katunin, D. N. (2005). Introduced species. In *The Caspian Sea Environment, The Handbook of Environmental Chemistry, Part 5P*, ed. A. G. Kostianoy and A. N. Kosarev. Berlin: Springer-Verlag, pp. 175–90.

Keskitalo, J. (1987). Phytoplankton in the sea area off the Olkiluoto nuclear power station, west coast of Finland. *Annales Botanici Fennici*, **24**, 281–99.

Keskitalo, J. & Heitto, L. (1987). Overwintering of benthic vegetation outside the Olkiluoto nuclear power station, west coast of Finland. *Annales Botanici Fennici*, **24**, 231–43.

Kolbe, R. W. (1927). Zur Ökologie, Morphologie und Systematik der Brackwasser-Diatomeen. *Pflanzenforschung*, **7**, 1–146.

Kolbe, R. W. (1932). Grundlinien einer allgemeinen Ökologie der Diatomeen. *Ergebnisse der Biologie*, **8**, 221–348.

Kosarev, A. N. (2005). Physico-geographical conditions of the Caspian Sea. In *The Caspian Sea Environment, The Handbook of Environmental Chemistry, Part 5P*, ed. A. G. Kostianoy & A. N. Kosarev, Berlin: Springer-Verlag, pp. 5–31.

Kosarev A. N. & Kostianoy, A. G. (2008). Introduction. In *The Black Sea Environment, The Handbook of Environmental Chemistry, Part 5Q*, ed. A. G. Kostianoy and A. N. Kosarev. Berlin: Springer-Verlag, pp. 1–10.

Kosarev, A. N., Kostianoy, A. G., & Shiganova, T. A. (2008). The Sea of Azov. In *The Black Sea Environment, The Handbook of Environmental Chemistry, Part 5Q*, ed. A. G. Kostianoy & A. N. Kosarev, Berlin: Springer-Verlag, pp. 63–89.

Kosarev, A. N. & Yablonskaya, E. A. (1994). *The Caspian Sea*. The Hague: SPB Academic Publishing.

Kuylenstierna, M. (1989–1990). Benthic Algal Vegetation in the Nordre Älv Estuary (Swedish west Coast). Doctoral Dissertation, Gothenburg University, Sweden, vol. 1, text (1990), vol. 2, plates (1989).

Laing, I. & Gollasch, S. (2002). *Coscinodiscus wailesii* – a nuisance diatom in European waters. In *Invasive Aquatic Species of Europe – Distribution, Impacts and Management*, ed. E. Leppäkoski, S. Olenin & S. Gollasch, Dordrecht: Kluwer Academic Publishers, pp. 53–5.

Lamb, H. H. (1995). *Climate, History and the Modern World*, 2nd edition, London: Routledge, pp. 1–464.

Lambeck, K. (1999). Shoreline displacements in southern-central Sweden and the evolution of the Baltic Sea since the last maximum glaciation. *Journal of the Geological Society, London*, **156**, 465–86.

Larsson, U., Elmgren, R., & Wulff, F. (1985). Eutrophication and the Baltic Sea: causes and consequences. *Ambio*, **14**, 9–14.

Leppäkoski, E. & Mihnea, P. E. (1996). Enclosed seas under man-induced change: a comparison between the Baltic and Black seas. *Ambio*, **25**, 380–9.

Leppäkoski, E. & Olenin, S. (2000). Non-native species and rates of spread: lessons from the brackish Baltic Sea. *Biological Invasions*, **2**, 151–63.

Leppäkoski, E., Olenin, S. & Gollasch, S. (2002). The Baltic Sea – a field laboratory for invasion biology. In *Invasive Aquatic Species of Europe – Distribution, Impacts and Management*, ed. E. Leppäkoski, S. Olenin & S. Gollasch, Dordrecht: Kluwer Academic Publishers, pp. 253–9.

Leppäranta, M. & Myrberg, K. (2009). *Physical Oceanography of the Baltic Sea*. Berlin: Springer Praxis Books/Geophysical Sciences.

Lindahl, G., Wallström, K., Roomans, G. M., & Pedersén, M. (1983). X-ray microanalysis of planktic diatoms in *in situ* studies of metal pollution. *Botanica Marina*, **26**, 367–73.

Luckas, B., Dahlmann, J., Erler, K., *et al.* (2005). Overview of key phytoplankton toxins and their recent occurrence in the North and Baltic seas. *Environmental Toxicology*, **20**, 1–17.

MacKenzie, B. R. & Schiedek, D. (2007). Daily ocean monitoring since the 1860s shows record warming of northern European seas. *Global Change Biology*, **13**, 1335–47.

Mann, D. G. & Evans, K. M. (2007). Molecular genetics and the neglected art of diatomics. In *Unravelling the Algae – the Past, Present and Future of Algal Systematics*, ed. K. Brodie and J. Lewis, Boca Raton, FL: CRC Press, pp. 231–65.

Matoshko, A., Gozhik, P. & Semenenko, V. (2009). Late Cenozoic fluvial development within the Sea of Azov and Black Sea coastal plains. *Global and Planetary Change*, **68**, 270–87.

Miller, U. (1986). Ecology and palaeoecology of brackish water diatoms with special reference to the Baltic Basin. *In Proceedings of the 8th International Diatom Symposium*, ed. M. Ricard, Königstein: Koeltz Scientific Books, pp. 601–11.

Miller, U. & Risberg, J. (1990). Environmental changes, mainly eutrophication, as recorded by fossil siliceous micro-algae in two cores from the uppermost sediments of the north-western Baltic. *Nova Hedwigia, Beiheft*, **100**, 237–53.

Miller, U. & Robertsson, A.-M. (1979). Biostratigraphical investigations in the Anundsjö region, Ångermanland, northern Sweden. *Early Norrland*, **12**, 1–76.

Molander, S., Blanck, H., & Söderström, M. (1990). Toxicity assessment by pollution-induced tolerance (PICT), and identification of metabolites in periphyton communities after exposure to 4,5,6-trichloroguaiacol. *Aquatic Toxicity (Amsterdam)*, **18**, 115–36.

Mölder, K. & Tynni, R. (1967–1973). Über Finnlands rezente und subfossile Diatomeen I–VII. *Bulletin of the Geological Society of Finland*, **39**, 199–217; **40**, 151–70; **41**, 235–51; **42**, 129–44; **43**, 203–20; **44**, 141–49; **45**, 159–79.

MOLTEN, DETECT and DEFINE (2006). Coastal ecology and palaeoecology of the Baltic and adjacent seas. See http://craticula.ncl.ac.uk/Molten/jsp/index.jsp.

Munthe, H. (1892). Studier ofer Baltiska hafets qvartära historia. *Bihang till Kungliga Svenska Vetenskaps Akademins Handlingar*, **18 II** (in Swedish).

Munthe, H. (1894). Preliminary report on the physical geography of the Litorina Sea. *Bulletin of the Geological Institution of the University of Uppsala*, **2**, 1–38.

Norrman, B. & Andersson, A. (1994). Development of ice biota in a temperate sea area (Gulf of Bothnia). *Polar Biology*, **14**, 531–7.

Olli, K., Clarke, A., Danielsson, Å., *et al.* (2008). Diatom stratigraphy and long-term dissolved silica concentrations in the Baltic Sea. *Journal of Marine Systems*, **73**, 284–99.

Papush, L. & Danielsson, Å. (2006). Silicon in the marine environment: dissolved silica trends in the Baltic Sea. *Estuarine, Coastal and Shelf Science*, **67**, 53–66.

Pedersén, M., Roomans, G. M., Andrén, M., *et al.* (1981). X-ray microanalysis of metals in algae – a contribution to the study of environmental pollution. *Scanning Electron Microscopy*, **1981 II**, 499–509.

Petrov, A. & Nevrova, E. (2007). Database on Black Sea benthic diatoms (Bacillariophyta): its use for a comparative study of diversity pecularities under technogenic pollution impacts. In *Proceedings Ocean Biodiversity Informatics*, ed. E. Vanden Berghe, W. Appeltans, M. J. Costello, & P. Pissierssens, IOC Workshop Report, 202, Paris: UNESCO, pp. 153–65.

Plante-Cuny, M. R. & Plante, R. (1986). Benthic marine diatoms as food for benthic marine animals. In *Proceedings of the 8th International Diatom Symposium*, ed. M. Ricard, Königstein: Koeltz Scientific Books, pp. 525–37.

Potapova, M. & Snoeijs, P. (1997). The natural life cycle in wild populations of *Diatoma moniliformis* (Bacillariophyceae) and its disruption in an aberrant environment. *Journal of Phycology*, **33**, 924–37.

Proschkina-Lavrenko, A. I. (1955). *Diatomovye vodorosli planktona Chernogo morya* [Planktonic diatoms of the Black Sea], Moskva: Akademii Nauk S.S.S.R. (in Russian).

Proshkina-Lavrenko, A. I. (1963). *Diatomovye vodorosli planktona Azovskogo morya* (Planktonic Diatoms of the Sea of Azov), Moskva: Akademii Nauk S.S.S.R. (in Russian).

Proshkina-Lavrenko, A. I. & Makarova, I. V. (1968). *Vodorosli planktona Kaspiiskogo morya* (Planktonic Diatoms of the Caspian Sea), Leningrad: Nauka (in Russian).

Punning, J.-M., Martma, T., Kessel, H., & Vaikmäe R. (1988). The isotope composition of oxygen and carbon in the subfossil mollusc shells of the Baltic Sea as an indicator for paleosalinity. *Boreas*, **17**, 27–31.

Rahm, L., Conley, D., Sandén, P., Wulff, F., & Stålnacke, P. (1996). Time series analysis of nutrient inputs to the Baltic Sea and changing DSi:DIN ratios. *Marine Ecology Progress Series*, **130**, 221–8.

Remane, A. (1940). Einführung in die zoologische Ökologie der Nord- und Ostsee. *In Die Tierwelt der Nord- und Ostsee*, ed. G. Grimpe & E. Wagler, Leipzig: Akademische Verlags-gesellschaft Becker & Erler, vol. 1a, pp. 1–238.

Remane, A. (1958). Ökologie des Brackwassers. In *Die Biologie des Brackwassers*, ed. A. Remane & C. Schlieper, Stuttgart, pp. 1–216.

Risberg, J. (1990). Siliceous microfossil stratigraphy in a superficial sediment core from the northwestern part of the Baltic proper. *Ambio*, **19**, 167–72.

Risberg, J. (1991). Palaeoenvironment and sea level changes during the early Holocene on the Södertörn peninsula, Södermanland, eastern Sweden. *Stockholm University, Department of Quaternary Research, Report*, **20**, 1–27.

Robertsson, A. M. (1990). The diatom flora of the Yoldia sediments in the Närke province, south central Sweden. *Nova Hedwigia, Beiheft*, **100**, 255–62.

Round, F. E., Crawford, R. M. & Mann, D. G. (1990). *The Diatoms – Biology and Morphology of the Genera*, Cambridge: Cambridge University Press.

Round, F. E. & Sims, P. A. (1981). The distribution of diatom genera in marine and freshwater environments and some evolutionary considerations. *In Proceedings of the 6th International Diatom Symposium*, ed. R. Ross, Königstein: Koeltz Scientific Books, pp. 301–20.

Ryabova, N., Zimina, L., Zimin, V., *et al.* (1994). In *Abstracts of the International Meeting on the Urbanization and the Protection of the Biocoenosis of the Baltic Coasts, Juodrante, Lithuania, 4–8 October 1994*, ed. R. Volskis, Paris: UNESCO, Regional Office for Science and Technology for Europe, Technical Report, 22, pp. 63–9.

Ryves, D. B., Amsinck, S. L., Anderson, N. J., *et al.* (2004). Reconstructing the salinity and environment of the Limfjord and Vejlerne Nature Reserve, Denmark, using a diatom model for brackish lakes and fjords. *Canadian Journal of Fisheries and Aquatic Sciences*, **61**, 1988–2006.

Sagert, S., Rieling, T., Eggert, A., & Schubert, H. (2008). Development of a phytoplankton indicator system for the ecological assessment of brackish coastal waters (German Baltic Sea coast). *Hydrobiologia*, **611**, 91–103.

Saunders, K. M., McMinn, A., Roberts, D., Dodgson, D. A., & Heijnis, H. (2007). Recent human-induced salinity changes in Ramsar-listed Orielton Lagoon, south-east Tasmania, Australia: a new approach for coastal lagoon conservation and management. *Aquatic Conservation: Marine and Freshwater Ecosystems*, **17**, 51–70.

Schelske, C. L., Stoermer, E. F., & Kenney, W. F. (2006). Historic low-level phosphorus enrichment in the Great Lakes inferred from biogenic silica accumulation in sediments. *Limnology & Oceanography*, **51**, 728–48.

Segerstråle, S. G. (1959). Brackish water classification, a historical survey. *Archivio di Oceanografia e Limnologia (Supplement)*, **11**, 7–33.

Shiganova, T. (2008). Introduced species. In *The Black Sea Environment, The Handbook of Environmental Chemistry, Part 5Q*, ed. A. G. Kostianoy & A. N. Kosarev, Berlin: Springer-Verlag, pp. 375–406.

Simonsen, R. (1962). Untersuchungen zur Systematik und Ökologie der Bodendiatomeen der westlichen Ostsee. *Internationale Revue der gesamten Hydrobiologie, Systematische Beihefte*, **1**, 1–144.

Smayda, T. J. (1990). Novel and nuisance phytoplankton blooms in the sea: evidence for a global epidemic. In *Toxic Marine Phytoplankton – Proceedings of the 4th International Conference on Toxic Marine Plankton*, ed. W. Granéli, B. Sundström, L. Edler, & D. M. Anderson, pp. 29–40.

Smol, J. P. (2008). *Pollution of Lakes and Rivers: A Paleoenvironmental Perspective*, 2nd edition, Oxford: Blackwell Publishing.

Smol, J. P. & Cumming B. F. (2000). Tracking long-term changes in climate using algal indicators in lake sediments. *Journal of Phycology*, **36**, 986–1011.

Snoeijs, P. (1989). Ecological effects of cooling water discharge on hydrolittoral epilithic diatom communities in the northern Baltic Sea. *Diatom Research*, **4**, 373–98.

Snoeijs, P. (1990). Effects of temperature on spring bloom dynamics of epiplithic diatom communities in the Gulf of Bothnia. *Journal of Vegetation Science*, **1**, 599–608.

Snoeijs, P. (1991). Monitoring pollution effects by diatom community composition – a comparison of methods. *Archiv für Hydrobiologie*, **121**, 497–510.

Snoeijs, P. (1992). Studies in the *Tabularia fasciculata* complex. *Diatom Research*, **7**, 313–44.

Snoeijs, P. (1994). Distribution of epiphytic diatom species composition, diversity and biomass on different macroalgal hosts along seasonal and salinity gradients in the Baltic Sea. *Diatom Research*, **9**, 189–211.

Snoeijs, P. (1995). Effects of salinity on epiphytic diatom communities on *Pilayella littoralis* (Phaeophyceae) in the Baltic Sea. *Ecoscience*, **2**, 382–94.

Snoeijs, P. (1999). Marine and brackish waters. In *Swedish Plant Geography. Acta Phytogeographica Suecica*, **84**, 187–212.

Snoeijs, P. & Kautsky, U. (1989). Effects of ice-break on the structure and dynamics of a benthic diatom community in the northern Baltic Sea. *Botanica Marina*, **32**, 547–62.

Snoeijs, P. & Notter, M. (1993a). Benthic diatoms as monitoring organisms for radionuclides in a brackish-water coastal environment. *Journal of Environmental Radioactivity*, **18**, 23–52.

Snoeijs, P. & Notter, M. (1993b). Radiocaesium from Chernobyl in benthic algae along the Swedish Baltic Sea coast. *Swedish University of Agricultural Sciences Report SLU-REK*, **72**, 1–21.

Snoeijs, P. & Potapova, M. (1998) Ecotypes or endemic species? – a hypothesis on the evolution of *Diatoma* taxa (Bacillariophyta) in the northern Baltic Sea. *Nova Hedwigia* **67**: 303–48.

Snoeijs, P., Busse, S. & Potapova, M. (2002). The importance of diatom cell size in community analysis. *Journal of Phycology*, **38**, 265–72.

Snoeijs, P., Vilbaste S., Potapova, M., Kasperoviciene, J. & Balashova (1993–1998). *Intercalibration and Distribution of Diatom Species in the Baltic Sea*, Uppsala: Opulus Press, vol. 1–5.

Sohlenius, G., Sternbeck, J., Andrén, E. & Westman, P. (1996). Holocene history of the Baltic Sea as recorded in a sediment core from the Gotland Deep. *Marine Geology*, **134**, 183–201.

Sommer, U. & Lengfellner, K. (2008). Climate change and the timing, magnitude, and composition of the phytoplankton spring bloom. *Global Change Biology*, **14**, 1199–208.

Sundbäck, K. & Snoeijs, P. (1991). Effects of nutrient enrichment on microalgal community composition in a coastal shallow-water sediment system: an experimental study. *Botanica Marina*, **34**, 341–58.

Sundelin, U. (1922). Några ord angående förläggningen av L.G. i de av transgression ej drabbade delarna av det baltiska området samt angående tidpunkten för Litorinahavets inträde. *GFF*, **44**, 543–44.

ter Braak, C. J. F. (1986). Canonical correspondence analysis: a new eigenvector technique for multivariate direct gradient analysis. *Ecology*, **67**, 1167–79.

ter Braak, C. J. F. & Juggins, S. (1993). Weighted averaging partial least squares regression (WA-PLS): an improved method for reconstructing environmental variables from species assemblages. *Hydrobiologia*, **269/270**, 485–502.

Thessen, A. E. & Stoecker, D. K. (2008). Distribution, abundance and domoic acid analysis of the toxic diatom genus *Pseudo-nitzschia* from the Chesapeake Bay. *Estuaries & Coasts*, **31**, 664–72.

Thulin, B., Possnert, G. & Vuorela, I. (1992). Stratigraphy and age of two postglacial sediment cores from the Baltic Sea. *Geologiska Föreningeni Stockholm Förhandlingar*, **114**, 165–79.

Tuovinen, N., Weckström, K. & Virtasalo, J. (2010). Assessment of recent eutrophication and climate influence in the Archipelago

Sea based on the subfossil diatom record. *Journal of Paleolimnology*, **44**, 95–108.

Tynni, R. (1975–1980). Über Finnlands rezente und subfossile Diatomeen VIII–XI. *Geological Survey of Finland, Bulletin*, **274**, 1–35; **284**, 1–37; **296**, 1–55; **312**, 1–93.

Ulanova, A., Busse, S. & Snoeijs, P. (2009). Coastal diatom–environment relationships in the brackish Baltic Sea. *Journal of Phycology*, **45**, 54–68.

Vaalgamaa, S. & Conley, D. J. (2008). Detecting environmental change in estuaries: nutrient and heavy metal distributions in sediment cores in estuaries from the Gulf of Finland, Baltic Sea. *Estuarine, Coastal and Shelf Science*, **76**, 45–56.

Vasiliu, F. (1996). The Black Sea. In *Marine Benthic Vegetation – Recent Changes and the Effects of Eutrophication*, ed. W. Schramm and P. H. Nienhuis. Ecological Studies, Volume 123. Berlin: Springer-Verlag, pp. 435–47.

Wasmund, N. & Uhlig, S. (2003). Phytoplankton trends in the Baltic Sea. *ICES Journal of Marine Science*, **60**, 177–86.

Wasmund, N., Nausch, G. & Matthäus, W. (1998). Phytoplankton spring blooms in the southern Baltic Sea – spatio-temporal development and long-term trends. *Journal of Plankton Research*, **20**, 1099–117.

Weckström, K. (2006). Assessing recent eutrophication in coastal waters of the Gulf of Finland (Baltic Sea) using subfossil diatoms. *Journal of Paleolimnology*, **35**, 571–92.

Weckström, K. & Juggins, S. (2005). Coastal diatom–environment relationship from the Gulf of Finland, Baltic Sea. *Journal of Phycology*, **42**, 21–35.

Weckström, K., Juggins, S. & Korhola, A. (2004). Quantifying background nutrient concentrations in coastal waters: a case study from an urban embayment of the Baltic Sea. *Ambio*, **33**, 324–7.

Weckström, K., Korhola A. & Weckström, J. (2007). Impacts of eutrophication on diatom life forms and species richness in coastal waters of the Baltic Sea. *Ambio*, **36**, 155–60.

Wendker, S. (1990). Untersuchungen zur subfossilen und rezenten Diatomeenflora des Schlei-Ästuars (Ostsee). *Bibliotheca Diatomologica*, **20**, 1–268.

Westman, P. & Sohlenius, G. (1999). Diatom stratigraphy in five offshore sediment cores from the northwestern Baltic proper implying large scale circulation changes during the last 8500 years. *Journal of Paleolimnology*, **22**, 53–69.

Willén, T. (1985). Phytoplankton, chlorophyll *a* and primary production in the Biotest Basin, Forsmark, 1981–1982. Abstracts of the 9th BMB Symposium, Turku/Åbo, Finland, 11–15 June, Åbo Akademi, p. 123.

Wilson, S. E., Cumming, B. F. & Smol, J. P. (1996). Assessing the reliability of salinity inference models from diatom assemblages: an examination of a 219-lake data set from western North America, *Canadian Journal of Fisheries and Aquatic Science*, **53**, 1580–94.

Winn, K., Werner, F. & Erlenkeuser, H. (1988). Hydrography of the Kiel Bay, western Baltic, during the Litorina transgression. *Meyniana*, **40**, 31–46.

Witkowski, A. (1994). Recent and fossil diatom flora of the Gulf of Gdansk, southern Baltic Sea. *Bibliotheca Diatomologica*, **28**, 1–313.

Wulff, F., Stigebrandt, A. & Rahm, L. (1990). Nutrient dynamics of the Baltic Sea. *Ambio*, **19**, 126–76.

Yunev, O. A., Carstensen, J., Moncheva, S., Khaliulin, A., Ærtebjerg, G. & Nixon, S. (2007). Nutrient and phytoplankton trends on the western Black Sea shelf in response to cultural eutrophication and climate changes. *Estuarine, Coastal and Shelf Science*, **74**, 63–76.

Zillén, L., Conley, D. J., Andrén, T., Andrén, E. & Björck, S. (2008). Past occurrences of hypoxia in the Baltic Sea and the role of climate variability, environmental change and human impact. *Earth-Science Reviews*, **91**, 77–92.

Zonn, I. S. (2005). Environmental issues of the Caspian. In *The Caspian Sea Environment, The Handbook of Environmental Chemistry, Part 5P*, ed. A. G. Kostianoy and A. N. Kosarev, Berlin: Springer-Verlag, pp. 223–42.

Zonn, I. S., Fashchuk, D. Y. & Ryabinin, A. I. (2008). Environmental issues of the Black Sea. In *The Black Sea Environment, The Handbook of Environmental Chemistry, Part 5Q*, ed. A. G. Kostianoy and A. N. Kosarev, Berlin: Springer-Verlag, pp. 407–21.

16 Applied diatom studies in estuaries and shallow coastal environments

ROSA TROBAJO AND MICHAEL J. SULLIVAN

16.1 Introduction

Diatoms are an important and often dominant component of benthic microalgal assemblages in estuarine and shallow coastal environments. This chapter will be concerned mainly with motile diatom assemblages on sediments in these environments and secondarily with epiphytic diatom assemblages on submerged aquatic vegetation. Admiraal (1984) provided an excellent summary of the ecology of estuarine sediment-inhabiting diatoms and Underwood and Kromkamp (1999) an excellent review of microphytobenthos primary production in estuaries.

Many topics were covered in the above two works, including the distribution, effects of physicochemical factors, population growth, primary production, and interactions with herbivores in Admiraal's work, and a summary of the main factors affecting production and biomass of microphytobenthos (diatom-dominated) within microtidal temperate estuaries in Underwood and Kromkamp's review. The focus of the present review will be considerably narrower because only those applied studies that have focused on the autecology of particular diatom species or that have utilized structural (e.g. species diversity) and/or functional (e.g. primary production rates) attributes of benthic diatom assemblages will be considered. By "applied" we mean studies that treat benthic diatom assemblages as tools to address concerns about larger ecosystem problems such as cultural eutrophication of estuarine and shallow coastal environments. The three diatom-related research topics that will be reviewed in this chapter include eutrophication and sediment toxicity, sediment stability, and contribution of benthic diatom assemblages to global primary production and their role in food webs. These topics are important because of threats posed to estuarine and shallow coastal systems by increasing nutrient levels and reduced light transmission in the water column, both of which may significantly impact the role of algae and other primary producers in trophic dynamics and consequently affect ecosystem health. Relevant studies conducted in the Baltic Sea will not be included here as they will be part of the chapter on applied studies of diatoms in brackish waters by Snoeijs and Weckström (this volume). In discussing the various studies, the names and authorities of diatom taxa will be abbreviated according to Brummit and Powell (1992) but will be those listed by the author(s); no taxonomic revisions will be attempted.

A review of the physical properties of estuaries and their resident diatom assemblages is presented in Cooper *et al.* (this volume). Before proceeding to the actual review of specific papers that follows in this chapter, the reader may first wish to consult Round's (1979) review of estuaries from the perspective of its resident vascular plants and algae in various habitats, such as salt marshes and exposed mudflats and sandflats. An important point to be made is that the latter sedimentary environments are often described in the literature as "unvegetated" simply because vascular plants such as *Spartina alterniflora* Loisel and seagrasses are absent. The use of the adjective unvegetated is not only misleading but also inaccurate because a diverse assemblage of cyanobacteria and eukaryotic algae thrive on and within such sediments. This assemblage, often termed the microphytobenthos, was reviewed and referred to as a "secret garden" by MacIntyre *et al.* (1996) and Miller *et al.* (1996). Functional properties of benthic microalgae as an integrated whole were discussed in detail by these authors but the sedimentary diatom component is often neglected.

16.2 Eutrophication and pollution

Anthropogenic inputs of nutrients, mainly various forms of nitrogen and phosphorus, pose a serious threat to estuarine and shallow coastal waters (Valiela *et al.*, 1997; Howarth and

The Diatoms: Applications for the Environmental and Earth Sciences, 2nd Edition, eds. John P. Smol and Eugene F. Stoermer. Published by Cambridge University Press.

Marino, 2006), many of which, over the last two decades, have undergone significant eutrophication.

Round (1981) stated that the best course of action is to prevent pollution wherever possible rather than attempting to determine the amount of pollution a given system can tolerate. While prevention of pollution is a worthy goal, in reality, various forms of pollution will continue. Therefore, it is critical that the "health" of aquatic systems be constantly monitored for their protection and our own. However, although the information content of benthic diatom assemblages is high due to the large number of taxa typically present and their responses to physicochemical properties of the overlying water vary greatly amongst taxa, these organisms have been little used to infer water quality and ecosystem integrity or to predict the effects of increased nutrient levels in estuaries and shallow coastal waters.

16.2.1 Distributional studies

Before predictions of changes in benthic diatom assemblage structure in response to nutrient enrichment can be made, these responses should be well identified and studied, and preferably evaluated experimentally. Admiraal produced an excellent series of papers (1977–1984) dealing with interactions between diatom populations on mudflats of the Ems–Dollard Estuary (on the border between The Netherlands and Germany) and various environmental factors to explain their spatial and temporal distribution. With regard to nutrients, Admiraal (1977a) showed that the K_s (half saturation constant) value for orthophosphate uptake by cultures of *Navicula arenaria* Donkin was about 0.1 μM. This value was the same as that of several marine planktonic diatoms (see his Table 3) and indicated that phosphorus concentrations in the Ems–Dollard were not limiting to this diatom. Admiraal (1977a) generalized this to the resident diatom assemblages of intertidal flats of the Ems–Dollard; therefore, increases in phosphorus in this body of water could not be detected using the benthic diatom assemblages.

Admiraal (1977b) then turned his attention to inorganic nitrogen. Cultures of ten diatom species isolated from mudflats of the Ems–Dollard were exposed to varying concentrations of ammonia (NH_4^+ and NH_3), nitrate, and nitrite. The majority of species grew well at high concentrations of the last two nitrogen compounds; however, ammonia concentrations higher than 0.5 μM were strongly inhibitory to growth of the cultures, measured as chlorophyll (chl) *a*. Photosynthetic rates of cultures and field assemblages were also inhibited by this ammonia concentration and the effect was enhanced by high irradiance and high pH. Admiraal noted that tolerance for ammonia was not high amongst the species but truly sensitive species were lacking; the most tolerant diatom in the culture experiments and on the mudflats was *Navicula salinarum* Grunow. He also stated that a clear relationship between ammonia tolerance of the various species and their relative abundance on polluted and unpolluted mudflats was not obvious. To test the hypothesis that ammonium concentrations may play an important role in determining the field distribution of benthic diatom species in salt marshes (Sullivan, 1978) and in mudflats (Peletier, 1996), Underwood *et al.* (1998) manipulated ammonium concentrations in tidal mesocosms and found that certain taxa showed significant treatment-related changes in population density. Sediment cores from the Colne Estuary (UK) had their porewater increased to 450 μM NH_4^+ and, after 26 days, the population densities of *Gyrosigma fasciola* W. Sm., *Gyrosigma littorale* (W. Sm.) J. W. Griff. et Henfr., *Pleurosigma angulatum* (Queckett) W. Sm., *Navicula phyllepta* Kütz., *Cylindrotheca signata* Reimann et J. C. Lewin, *Cylindrotheca closterium* Reimann et J. C. Lewin, and *Nitzschia apiculata* (W. Greg.) Grunow were significantly reduced, while that of *Gyrosigma limosum* Sterrenburg et Underwood was unaffected, and that of *Nitzschia sigma* (Kütz.) W. Sm. was significantly higher compared with control cores. Culturing experiments also showed significant differences in growth rates of four common epipelic estuarine diatoms (*Navicula phyllepta*, *Navicula perminuta* Grunow, *Navicula salinarum*, and two clones of *C. closterium*) over a range of nitrate (10–2000 μM) and ammonium (10–4000 μM) concentrations across a salinity range of 10–25‰ (Underwood and Provot, 2000). Growth rates of *N. phyllepta* were significantly lower at ammonium concentrations >400 μM, but the other three taxa showed little evidence of ammonium toxicity, even at concentrations >1mM. On the basis of the response of each species, the following ranking of preferences for inorganic nitrogen concentrations (lowest to highest) with salinity optima in parentheses was constructed by Underwood and Provot (2000): *N. phyllepta* (10–20‰), *N. perminuta* (10–30‰), *N. salinarum* (20–35‰), and *C. closterium* (clone 1, 10–25‰, clone 2, 25–35‰).

The tolerance of benthic diatom species to free sulfide (a known toxin) was also tested by Admiraal and Peletier (1979a). *N. salinarum*, *N. arenaria*, *Nitzschia* cf. *thermalis* (Ehrenb.) Grunow, and *Gyrosigma spencerii* (W. Sm.) A. Cleve were relatively tolerant of free sulfide concentrations up to 6.8 mM for periods of 5 to 24 h. However, *Nitzschia closterium* (Ehrenb.) W. Sm., *Nitzschia* cf. *dissipata* (Kütz.) Grunow, *N. sigma* (Kütz.) W. Sm., *Surirella ovata* Kütz., and *Stauroneis constricta* (W. Sm.) A. Cleve were greatly inhibited, or killed, following 48 h of exposure to a free sulfide concentration of 0.9 mM. The high tolerance of *N. salinarum* correlated well with its dominance on polluted mudflats receiving large amounts of organic waste from potato flour and sugar mills and sewage plants bordering the Ems–Dollard Estuary. In contrast, *N. arenaria*, although relatively tolerant of free sulfide, was only dominant on well-oxygenated, relatively unpolluted, sandflats. Admiraal and Peletier (1979a) concluded that the differing abilities of these taxa to tolerate free sulfide, ammonia, low salinity, anaerobic conditions, and prolonged darkness, and the capability to grow heterotrophically (see Admiraal and Peletier 1979b for uptake of organic substrates) explain their distribution in the estuary. No one factor could explain the observed distributions but a combination of them offered a logical explanation. Admiraal and Peletier (1980) reached the same conclusions in studying a different mudflat in the Ems–Dollard Estuary exposed to large freshwater runoff and high concentrations of organic pollution from potato flour mills. Peletier (1996) revisited the same mudflat sampled by Admiraal and Peletier (1980). Since this time (1977–1980) the loading rates of organic waste from potato flour and cardboard factories have been greatly reduced. Analysis of the benthic diatom assemblages showed that the former dominants *Navicula salinarum* and *Fallacia pygmaea* (Kütz.) Stickel et D. G. Mann had disappeared from the mudflats, whereas *Navicula flanatica* Grunow and *N. phyllepta* had become the dominant taxa.

The above data indicated that species like *F. pygmaea* and *N. salinarum* are particularly tolerant of sulfide and ammonia, while *N. phyllepta* and *N. flanatica* show a lower tolerance to high ammonia and free sulfide concentration. *N. phyllepta* has been suggested as a mesotrophic species, favoring nitrate and ammonium concentrations in the range of 100–400 μM (Underwood and Barnett, 2006).

16.2.2 Studies utilizing diversity indices

Hendey (1977) studied the diversity of diatom assemblages epiphytic on *Ceramium rubrum* C. Agardh along the North Coast of Cornwall (UK), and its relationship to pollution by sewage. Hendey found that H' (Shannon's diversity index; Lloyd *et al.*, 1968) was lowest for three consecutive years when the summer tourist influx greatly stressed the sewage facilities; however, no control samples from nearby undeveloped areas of the coast were taken for comparative purposes. In a Massachusetts (USA) salt marsh, van Raalte *et al.* (1976a) applied urea and sewage sludge containing 10% nitrogen on a long-term basis to sediments populated by the cordgrass *Spartina alterniflora*. Both organic nitrogen enrichments significantly decreased H' and S (number of species in the sample or species richness) of the edaphic (i.e. within or associated with the salt-marsh sediments) diatom assemblages. The relative abundance of *Navicula salinarum* was 5–9% in control plots but increased to 20–25% in enriched plots. Sullivan (1976) studied this same microalgal assemblage beneath the same spermatophyte in a Delaware salt marsh and added inorganic nitrogen (NH_4NO_3) and inorganic phosphorus ($CaH_4(PO_4)_2$) instead of urea or sewage sludge. Phosphorus enrichment significantly decreased both H' and S, whereas nitrogen enrichment significantly decreased only S. Both nitrogen and phosphorus enrichment had a stimulatory effect on the relative abundance of *N. salinarum*; this agrees with previous work by Admiraal (1977b) and van Raalte *et al.* (1976a) for nitrogen enrichment.

Van Dam (1982) published an informative paper on the relationship between diversity, species abundance–rank data, and water-quality assessment. He pointed out the pitfalls of using diatom assemblage diversity to categorize pollution across different aquatic systems, and stressed the importance of identifying the constituent taxa and having knowledge of their autecology, particularly for those that were dominant in the assemblage. Watt (1998) reached the same conclusions from his study of the feasibility of using sediment diatom communities to monitor nutrient loading in two estuaries of contrasting trophic status in the province of KwaZulu-Natal (South Africa). He concluded that diversity indices cannot be used as a measure of the degree of nutrient enrichment and pointed out that species composition, rather than community structure, was a more suitable property of benthic diatom assemblages on which to base a water-quality monitoring program.

16.2.3 Functional studies

Other studies have utilized functional properties of the entire benthic microalgal assemblage, where diatoms were either a substantial component or completely dominant algae, to assess the effects of eutrophication. Sullivan and Daiber (1975) worked in the same spermatophyte zone as Sullivan (1976) did, and found that both inorganic nitrogen and phosphorus enrichment increased the standing crop of edaphic algae as measured by chl *a*. Van Raalte *et al.* (1976b) measured primary production rates (as ^{14}C uptake) of edaphic algae from the same plots fertilized with urea or sewage sludge described above. Primary production increased only at the highest level of nitrogen enrichment because of the shading effect of the overstory grass canopy. Light was the single most important factor limiting edaphic algal production on this Massachusetts marsh. Darley *et al.* (1981) enriched the sediments in both short and tall *Spartina alterniflora* zones with NH_4Cl and measured ^{14}C uptake by the edaphic algae (diatoms constituted 75–93% of total biomass) as well as their chl *a* levels. Both parameters increased significantly beneath the short *S. alterniflora* canopy but ammonium enrichment had little effect on edaphic algae beneath tall *S. alterniflora*. This correlated well with the more frequent inundation of the tall form by tidal waters making nitrogen limitation unlikely.

To investigate nutrient fluxes and succession, Sundbäck and colleagues (1988, 1991a,b) utilized benthic microalgal assemblages from the sediments of the Skagerrak and Kattegat (straits connecting the Baltic Sea and the North Sea). Nutrient enrichment of subtidal sediments (15 m) showed that the flux of inorganic nitrogen and phosphate between sediments and water column was mediated by benthic microalgae. Increasing irradiance resulted in increased photosynthetic rates and nutrient requirements for benthic microalgae, preventing release of phosphate and ammonium from the sediments. This indirect effect of benthic microalgae on nutrient fluxes was considered to be more important than the direct effect of their nutrient uptake. Furthermore, the diatom component of the benthic microalgal assemblage played a major role in controlling the flux of nutrients from sediments to the water column. In shallow waters (0.2 m), nutrient enrichment significantly increased the standing crops of filamentous cyanobacteria, flagellates, and diatoms, but the relative dominance of these major taxonomic groups did not change. Diatoms accounted for *c.* 50% of the benthic microalgal biomass and small *Amphora* and *Nitzschia* taxa (7–12 μm in length) were most favoured by the addition of inorganic nitrogen and phosphorus.

16.2.4 Water-quality indices

Although considerable progress has been made in utilizing diatoms to assess water quality in freshwater systems (see other chapters in this volume), the situation, as reviewed above, is quite different in estuarine and shallow coastal systems (Sullivan, 1986; Trobajo, 2007). Until the 1970s, monitoring of water quality depended mainly on the presence of certain key species (i.e. indicator species) as indicative of polluted conditions. Round (1981, pp. 552–7) provided an excellent history of this strategy, starting with the work of Kolkwitz and Marsson (1908). As pointed out by Round (1981) and other workers, the major weakness of utilizing indicator species is that those species taken to be indicative of pollution also occur in non-polluted waters (i.e. they are not restricted to polluted waters). Therefore, the presence of such indicator species does not necessarily mean that the body of water in question is polluted, as quite the contrary may be true. Lange-Bertalot (1978, 1979) developed a more practical method to assess water quality utilizing both the species identity and relative abundance of constituent diatom species in the assemblages of European rivers. His more important contribution, however, was to point out that the cornerstone of any index of water quality must be the tolerance of the individual diatom taxa to increasing levels of a particular pollutant. The affinity or fidelity for polluted waters of diatom taxa should not be used in such indices for reasons given above. What is essential to this approach is the species-specific limits of tolerance to decreasing water quality, since diatom species are usually not limited by increasing levels of water quality. Since then, and based on these concepts, more than 20 different methods have been put forward for assessing the water quality of rivers and freshwater lakes (see other chapters in this volume).

Although all the studies cited in the above sections support the hypothesis that microphytobenthos of estuarine and shallow coastal waters can be used as indicators of environmental conditions, very little work in this field has been done to date. However, the development of diatom-based water-quality indices has become more important since the establishment in 2000 of the Water Framework Directive (WFD European Union, 2000). The goal of this directive is that European water bodies (rivers, lakes, ground, transitional, and coastal waters) should be returned to "good ecological status" by 2015–27. Estuaries and shallow coastal waters are identified by the Directive (under the name of "transitional waters") as significant water bodies and, in Annex V of the WFD, it is stated that the ecological quality of transitional waters must be assessed using phytobenthos (among other biological assemblages such as phytoplankton, benthic invertebrates, and fish fauna). Therefore, the implementation of this innovative legislation offers an unique opportunity, and at the same time represents an urgent need, to develop diatom indices that allow an assessment of the ecological status of different transitional waters (i.e. different types of estuaries and shallow coastal waters) similar to that established for freshwater systems (e.g. Leira *et al.*, 2006; Kelly *et al.*, 2008a,b; other chapters in this volume). Kelly *et al.* (2008a) analyzed a data set from the English Lake District and showed not only that diatom assemblages can serve as effective proxies for the entire epilithic algal assemblage, but also that transfer functions for total phosphorus, calcium ion concentration, dissolved inorganic carbon, and conductivity developed from non-diatom algae were less effective than when developed from diatoms alone.

Because transitional waters have naturally elevated nutrient levels in comparison to some freshwater and marine systems (National Academy of Sciences, 2000; Dauvin, 2007; Elliott and Quintino, 2007), any attempt to use the diatom indices currently available for freshwater or marine ecosystems will likely result in misleading interpretations. For example, they may indicate that an estuary or coastal wetland is degraded when, in fact, it is not. In many transitional waters, which are characterized by fluctuating conditions and an intrinsic level of nutrients and organic matter, there is a naturally higher abundance of certain diatom species (such as many *Nitzschia* and *Navicula* species) that tolerate large, short-term variations in many environmental variables (e.g. salinity, nutrients, and organic matter concentrations). There is no well-defined autecological discrimination for these species when they are considered in freshwater indices. Most, in fact, are considered as brackish taxa with similar high tolerances to organic matter and dissolved nutrient concentrations.

In order to move forward in the development of diatom-based indices to assess the ecological status of estuaries and shallow coastal waters, it is crucial to overcome the paucity of information on the autecology of the most representative benthic diatoms inhabiting these systems, and to elucidate the major factors controlling diatom species composition (Underwood *et al.*, 1998; Sullivan and Currin, 2000; Trobajo, 2007).

16.2.5 Estuaries and shallow coastal waters: changing conditions, changing approaches

Estuaries and shallow coastal waters can be considered as transitional gradients between terrestrial and aquatic systems, both continental freshwater and marine. One important characteristic of these environments is that they are highly dynamic, basically because they change configuration relatively frequently due to variations in winds, tides, and river discharges. This dynamism determines one of the principal ecological characteristics of estuarine and coastal wetland systems, which is the fluctuation in, and interaction of, many of the parameters used to define them (e.g. salinity, relationship between nutrients, water turnover, primary production). They are also, in general, very open systems, actively exchanging materials and energy with adjacent systems. Because the functioning of these systems is complex, with continual changes and interactions between factors, they are difficult to study and to model conceptually. But it is precisely this difficulty in determining which factors govern fluctuating and transitional systems, such as estuaries and shallow coastal waters, that make them so challenging, since they demand a more dynamic and functional approach (Elliot and Quintino, 2007; Trobajo, 2007). The crux of any applied diatom work to assess water quality or ecological status of such systems is an understanding of their functioning and knowledge of the main factors affecting diatom species composition and distribution.

Elliott and Quintino (2007) suggested that, in order to detect and assess anthropogenic effects on these fluctuating ecosystems, a change in our approach to studying them is required, shifting to more dynamic and functional studies rather than the traditional descriptive ones. These authors pointed out that, although estuaries and other transitional waters, such as shallow coastal waters, have long been regarded as environmentally stressed areas because of the high degree of variability in their physico-chemical characteristics, their biota are well-adapted to cope with these stresses. Thus, these ecosystems characterized by naturally fluctuating conditions should be regarded not as stressful but as resilient areas. Estuarine biota have the ability to tolerate stress in estuaries without adverse effect (Elliott and Quintino (2007) call this phenomenon "environmental homeostasis"). What can be stressful for marine or freshwater-adapted biota is actually advantageous for the estuarine biota, allowing the estuarine biota to dominate estuaries. Salinity is an excellent example, since reduced and highly varying salinity represents a stress for a marine or freshwater-adapted estuarine community, but it may not be for a brackish community for which stress can result from having no salinity variation (Costanza *et al.*, 1992; Elliott and Quintino, 2007). Although salinity has often been considered as an important factor regulating diatom distribution in estuaries and coastal wetlands (e.g. Laird and Edgar, 1992; Tomàs, 1988; Watt, 1998; Bak *et al.*, 2001), some authors have pointed out that tolerance to salinity changes is a prerequisite for diatoms living in such fluctuating systems (Sullivan and Currin, 2000; Trobajo *et al.*, 2004; Underwood and Barnett, 2006). Diatom species composition and abundance is therefore determined by other factors such as those involved in the flux of system energy. In a study of benthic diatom communities of six Mediterranean shallow coastal waters, Trobajo *et al.* (2004) found the main factors to be confinement and productivity, both of which are integrative factors involving different processes (e.g. nutrient availability, temperature).

Elliot and Quintino (2007) also pointed out that the natural (i.e. non-anthropogenic) stresses present in transitional waters mimic those of anthropogenic stress and this makes it difficult to detect the latter type of stress. This difficulty is referred to as the "estuarine quality paradox." This paradox has repercussions for implementation of many environmental management systems like the ones contemplated in the WFD, the National Land and Water Resource Audit in Australia (Heap *et al.*, 2001), the Clean Water Act in the USA (USEPA, 2002), and the 1998 Water Act in South Africa (Adams *et al.*, 2002). It is of crucial importance that techniques to assess eutrophication, as well as indicators (e.g. diatom-based) developed to evaluate stress typical of freshwater or marine ecosystems, should be carefully evaluated before being applied to transitional waters. For a comprehensive updated overview of eutrophication in transitional waters see the work of Zaldívar *et al.* (2008).

16.2.6 Studies of seagrass epiphytes

Eutrophication also poses a serious threat to the world's seagrass beds (Short and Wyllie-Echeverria, 1996). Nutrient enrichment stimulates the growth of phytoplankton in the water column and microalgae epiphytic on seagrass leaves. The increased algal biomass reduces the amount of light reaching the seagrass leaves and acts as a diffusive barrier to uptake of inorganic carbon and nutrients. Thus, light effects are really secondary effects of nutrient enrichment, the net result of which may be manifested by a decline of seagrasses through reduced growth and reproductive success. Coleman and Burkholder (1994) enriched mesocosms with low-flow rates containing *Zostera marina* Linnaeus with nitrate. A significant increase in assemblage productivity was recorded, as measured by ^{14}C uptake. Cyanobacteria were mainly responsible for the increase three weeks after enrichment began and small diatoms at six weeks; the response of large diatom species was variable. The authors concluded that nitrate enrichment has a controlling influence on species composition and dominance in epiphytic algal assemblages. Coleman and Burkholder (1995) moved their experiment to field populations of *Z. marina* in Back Sound, Beaufort, North Carolina (USA) where flow rates were much higher than in their mesocosm experiments. Nitrate enrichment did not increase total epiphytic algal production but did significantly affect some dominant species. Using track light–microscope–autoradiography, they showed that the species-specific production rate of *Cocconeis placentula* Ehrenb. increased, whereas that of the crustose red alga *Sahlingia subintegra* (Rosenv.) Kornmann decreased. Again, they concluded that nitrate levels control species-specific productivity and community structure of the epiphytic algal assemblage of *Z. marina*. Meanwhile, Wear *et al.* (1999) fertilized beds of the seagrasses *Halodule wrightii* F. Asch., *Syringodium filiforme* Kütz., and *Thalassia testudinum* de Konig in Big Lagoon, Perdido Bay, Florida (USA) with a slow-release (3–4 mo), temperature-sensitive Osmocote™ fertilizer. This fertilizer contained 19% N (as ammonium nitrate), 6% P (as anhydrous phosphoric acid), and 12% K (as anhydrous potassium hydroxide) by weight. All results for epiphytic algae were independent of the seagrass species. Enrichment caused a doubling of epiphytic algal production (^{14}C uptake), and biomass (g dry wt cm^{-2}) was increased to 1.43 times that of controls. The chl *a* and fucoxanthin concentrations (measured by high performance liquid chromatography, HPLC) of the epiphytic algae tripled in response to enrichment and variation in the latter pigment explained 88% of the variation in the former pigment. Diatoms were the dominant epiphytic algal group on the three seagrasses, and accounted for virtually all of the fucoxanthin.

On the other hand, it is equally important to quantify the nutrient removal effect of the benthic vegetation (macrophytes and epiphyton). Dudley *et al.* (2001), through ^{15}N enrichments (both $^{15}NO_3^-$ and $^{15}NH_4^+$) of aquaria containing intact sediment cores with *Ruppia magacarpa* C. T. Mason collected from a mesotrophic estuary in southwestern Australia, showed that the epiphytes (dominated by diatoms) removed more nitrate and ammonium from the water column than *R. megacarpa*, despite the biomass of the former being only 25% of the latter. The authors concluded that transient increases in these two sources of anthropogenic enrichment can be reduced within 30 hours by the epiphyte/seagrass complex in estuarine habitats.

16.2.7 Sediment toxicity

Agricultural, farming, and industrial activities are responsible for the accumulation of heavy metals in estuaries and adjacent coastal wetlands. It is important to be able to detect sediment toxicity in estuarine environments and in the last few years some toxicity bioassays involving benthic diatoms have been tested and developed.

Moreno-Garrido *et al.* (2003) developed a growth inhibition test with *Cylindrotheca closterium* (Ehrenb.) J. C. Lewin et Reimann in which cells were exposed to sediment spiked with heavy metals (cadmium (Cd), copper (Cu), and lead (Pb)) and counted daily using either light or fluorescence microscopy. *C. closterium* proved to be a suitable species for monitoring toxicity effects of Cu and Pb. However, they also found this test to be suitable only for coarse sediments, since sediments with 15% silt significantly reduced the diatom population growth (after shaking the culture the silt-sized sediment particles <63 μm settled more slowly than the cells and smothered them). Later, Moreno-Garrido *et al.* (2007) studied the responses (measuring EC_{50}, i.e. the concentration of the compound able to inhibit growth in 50% of the population) of three diatoms, *C. closterium*, *Phaeodactylum tricornutum* Bohlin, and *Navicula* sp., to different percentages of sediments collected from a shallow coastal lagoon in Aveiro (northwest Portugal) that contained different levels of pollutants (tin (Sn), zinc (Zn), mercury (Hg), Cu and chromium (Cr)). The most sensitive species to the silt-size sediment effect was the non-motile one (*P. tricornutum*). *C. closterium* was slightly more sensitive to the toxic effect of the pollutants considered, although populations of the three assayed species showed comparable pattern responses when exposed to the different percentages of sampled sediment.

Adams and Stauber (2004) developed a pore-water and a whole-sediment toxicity test with *Entomoneis* cf. *punctulata* Osaka et H. Kobayasi using flow cytometry to distinguish the diatom cells (based on their chl *a* autofluorescence) from sediment particles. Since nutrient release from sediment had previously been found to stimulate freshwater algal growth (Hall *et al.*, 1996), potentially masking contaminant toxicity, the authors based the toxicity test developed for this species on inhibition of its esterase activity (acute bioassay) instead of its growth inhibition (chronic bioassay), although growth inhibition was also measured in order to compare results. The sensitivity of E. cf. *punctulata* esterase activity to a range of metals (Cu, arsenic (As), Pb, manganese (Mn), Cd, Zn) to water exposures was determined and Cu proved the most toxic metal, with 3 h and 24 h IC_{50} values of 97 and 269 μg Cu l^{-1}, respectively. Although enzyme inhibition in E. cf. *punctulata* was less sensitive to metals than growth inhibition in water-only tests, Adams and Stauber (2004) found a good correlation between the acute enzyme response (24 h exposure) and chronic growth-rate inhibition for Cu, Zn, and Cd, suggesting that this acute test may be a useful surrogate for chronic growth tests of longer duration.

Since sediment quality is critical for the health of an aquatic ecosystem, the monitoring of contaminants in sediments should be part of any integral aquatic ecosystem management plan. The above summarized works show the suitability and potentiality of selected benthic diatom species to detect estuarine and marine sediment toxicity as well as to assess sediment quality.

16.3 Sediment stability

As in terrestrial environments, stability of sediments and their resistance to erosion (caused by tidal currents, wind-induced waves, bioturbation, and other factors) has been the subject of much research in estuaries and shallow coastal environments. The results of this research have implications in different fields such as the transport, retention and magnification of particle-associated pollutants, coastal erosion, and subsidence in deltas. All are of great interest for estuarine and coastal management, especially the latter because of IPCC (Intergovernmental Panel on Climate Change) predictions of global-change scenarios which will have a major impact on estuarine and delta ecosystems (IPCC, 2007). Many sediment-inhabiting diatoms of these ecosystems produce mucilage, rich in polysaccharides and proteoglycans, known by the generic name of extracellular polymeric substances (EPS). The production of diatom EPS is a widely cited mechanism for increasing sediment stability (Underwood and Paterson, 2003). These substances not only increase the critical erosion stress required to initiate erosion of flocs from the sediment bed but also trap and bind fine sediments (Tolhurst *et al.*, 1999; Widdows *et al.*, 2000; Underwood and Paterson, 2003). The importance of diatom mucilage production was first shown experimentally by Holland *et al.* (1974), who noted that stabilization of sediment was more pronounced by cultures of diatoms that produced large quantities of mucilage.

Coles (1979) studied the relationship between intertidal sediment stability and benthic diatom populations in the Wash, a large embayment in eastern England, and showed that, even where the spermatophyte canopy was dense, the macrophytes alone were unable to provide sufficient shelter to allow mud deposition to occur under the control of physical forces alone. Trapping and binding of fine sediment particles by copious mucilage secretion of large epipelic diatom populations permitted sediment deposition and stabilization.

Grant *et al.* (1986) published some of the first scanning electron microscopy (SEM) pictures showing that mucous strands secreted by diatoms were responsible for the binding of sand grains in the field. Bacterial cells were too small to create grain-to-grain binding. That same year, Paterson (1986) employed low-temperature scanning electron microscopy (LTSEM) to produce three-dimensional images of intact benthic diatom assemblages within, and on, the surface of intertidal sediments from the Severn Estuary, UK. The LTSEM revealed an extensive extracellular matrix consisting mainly of mucopolysaccharides produced during diatom locomotion (i.e. migration). As in the work of Grant *et al.* (1986) with intertidal sands, bacterial mucilage strands had no significant effect on sediment stability.

Since these early works, numerous studies have been published, not only dealing with the contribution and mechanism of diatom EPS to sediment stability and to salt-marsh development (Vos *et al.*, 1988; Paterson, 1990; Underwood and Paterson 1993a,b; Yallop *et al.*, 1994; Underwood, 1997, 2000; Widdows *et al.*, 2004), but also elucidating the factors affecting EPS production, such as irradiance and nutrient concentrations (Staats *et al.*, 1999, Smith and Underwood, 2000; Staats *et al.* 2000a,b; Wulff *et al.*, 2000; Perkins *et al.*, 2001; Underwood and Paterson, 2003; Underwood *et al.*, 2004). There is evidence that benthic diatoms produce a number of different types of EPS which vary in structure and sugar composition. These differences are important, since they could result in different structural properties within sediment biofilms (Decho, 2000; Stal, 2003; Underwood and Paterson, 2003). For an excellent and exhaustive overview on the production, composition, and role of marine epipelic diatom EPS the reader is highly recommended to review the work of Underwood and Paterson (2003).

16.3.1 Biofilm removal

The EPS of diatom biofilms can be removed into the overlying water column during high tides, either by physical erosion of the surface sediment or by dissolution (Underwood and Smith, 1998; Orvain *et al.*, 2003). Deposit-feeding invertebrates and microbial heterotrophs may also contribute to the removal of biofilm EPS (Decho, 2000; Middelburg *et al.*, 2000; Hamels *et al.*, 2004). EPS can also be degraded by photolysis (Moran *et al.*, 2000).

The relative importance of each of the above processes is likely to vary depending on the properties of the different types of EPS. Hanlon *et al.* (2006) studied changes in concentrations of different fractions of EPS in diatom-rich biofilms of the Colne Estuary (UK) mudflats over two tidal emersion–immersion cycles. They found that losses of chl *a* and colloidal EPS (up to 50–60%) occurred mainly in the first 30 minutes after tidal cover. About half of this may be due to *in situ* bacterial degradation (extracellular hydrolysis), with "wash away" into the water column accounting for the remainder.

16.4 Contribution to primary production and food webs

16.4.1 Benthic diatom resuspension

Baillie and Welsh (1980) were the first to draw the attention of the scientific community to the resuspension of benthic diatoms and their contribution to the primary production occurring in the water column. Resuspension of mudflat sediments (33% clay, 61% silt) produced peaks of chl *a* during early flood and late ebb tide. Examination of diatoms in the water column revealed that 75% of frustules belonged to pennate diatoms and it was concluded that flooding tides were responsible for a net transport of epipelic diatoms from the mudflat to a salt marsh where feeding bivalves were located. Varela and Penas (1985) suggested that resuspended diatoms were the major food source for culture of the cockle *Cerastoderma edule* on sandflats of Ría de Arosa, northwest Spain. Shaffer and Sullivan (1988) used K-systems analysis to show that primary production in the water column of the Barataria Estuary, Louisiana (USA), was often greatly augmented by resuspension of benthic diatom cells from the sediments. K-system analysis revealed that factors composed of wave height, meteorological tides, astronomical tides, and the standing crop (as chl *a*) and primary production of benthic microalgae accounted for 97% of the variation in water-column productivity over a 30-day period. Benthic pennate diatoms represented an average of 74% of the diatom cells in water-column samples.

Jonge and his colleagues have written an excellent series of papers on resuspension in the Ems Estuary (on the border between The Netherlands and Germany). In Jonge and van Beusekom (1992), the authors used a large data set to calculate the contribution of benthic microalgae (virtually all pennate diatoms) to total chl *a* in the Ems Estuary. On average for the entire estuary, 11×10^3 kg chl *a* was present in the extensive tidal flats and 14×10^3 kg chl *a* in the water column. Of the latter amount, "real" phytoplankton were responsible for 4×10^3 kg chl *a* and resuspended benthic diatoms for 10×10^3 kg chl *a*. Immediately above the tidal flats, resuspension doubled the algal biomass available to filter feeders. The implications of this for those interested in modeling trophic dynamics in estuarine systems are obvious.

Later, Jonge and van Beusekom (1995) looked more closely at the actual resuspension event in the Ems Estuary. Mud (<55 μm fraction) and benthic diatoms were found to be suspended simultaneously from tidal flats. Resuspension of mud and chl *a* from the top 0.5 cm of sediments was a linear function of "effective wind speed," which was the wind speed averaged over the three high-water periods immediately preceding the collection of water-column samples. Tidal currents appeared to have a minor effect in this study and this factor was not considered in any of the calculations The highly significant linear regression of effective wind speed on water-column chl *a* indicated that a wind speed of 12 m s^{-1} (27 mph) would resuspend 50% of the diatom biomass into the water column, which would eventually reach the main channels of the estuary. Again, the trophic implications of resuspended diatoms are great.

16.4.2 Vertical migration

The primary production of microphytobenthos dominated by diatoms is particularly high in intertidal mudfalts and in shallow subtidal sites, representing a significant fraction of overall primary productivity in estuarine and shallow costal regions

(MacIntyre *et al.*, 1996; Underwood and Kromkamp, 1999; Miles and Sundbäck, 2000; Glud *et al.*, 2002). Investigations of diatom vertical migrations in sediments have revealed the importance of considering not only factors affecting these migratory patterns but also the species composition as a significant factor in assessing benthic diatom contribution to total primary production in estuaries and shallow coastal waters. For a comprehensive review of early works dealing with diatom vertical migration see Round (1981, pp. 179–82).

In a more recent review of vertical migration in estuarine intertidal diatom communities, Consalvey *et al.* (2004) showed endogenous rhythms of vertical migration linked to tidal and diel cycles. Saburova and Polikarpov (2003), in their study of the vertical migration and rhythms of cell division of epipelic diatoms on an intertidal sandflat in the Chernaya River Estuary (White Sea, Russia), showed that *c.* 40% of the diatoms were present in the uppermost 2 mm layer and *c.* 60% in deeper layers (maximum recorded depths were between 42 and 83 mm, depending on the type of sediment). Of the cells in deeper layers, 80 to 90% were actively dividing. Cell division was confined to these aphotic, anoxic sediments. However, experimental studies showed that these strongly reducing conditions were not necessary for diatom cell division and that nutrient properties of pore waters from the deep layers were important. The authors concluded that, during the diurnal cycle, cells photosynthesize in the photic layers accumulating organic carbon and then migrate to the deeper, aphotic layers to divide, where higher concentrations of nutrients reside (especially that of NH_4^+). Thus, migration would permit cells to consume energy and undergo division under optimal conditions in spatially disconnected zones.

Combining LTSEM and high-resolution (chlorophyll) fluorescence imaging, Underwood *et al.* (2005) experimentally studied the importance of temporal changes in the vertical distribution of microphytobenthic algae on the overall functioning of intertidal biofilms. During the early morning the diatom-rich biofilm consisted of smaller naviculoid and nitzschioid taxa or euglenophyte species. By midday, *Gyrosigma balticum* (Ehrenb.) Rabenhorst and *Pleurosigma angulatum* (Quekett) W. Sm. were dominant. Some taxa, such as *Plagiotropis vitrea* (W. Sm.) Grunow, disappeared from surface layers after midday. Species composition continued to change toward the end of the photoperiod, with *G. balticum* dominating the biofilm. Significant differences in photosystem II efficiency among species were also found. This is the first study showing differences in individual diatom species photophysiology are linked to the presence of cells at the sediment surface and patterns in primary production. It also shows that the integrated biofilm response is the consequence of the different responses to irradiances of its constituent taxa, and consequently, that species composition is an important factor in determining biofilm photosynthesis. Similarly, Forster *et al.* (2006) found significant relationships between the diversity (species richness and Shannon index) of epipelic diatom species in the Westerschelde estuary (Belgium/The Netherlands) and their ecosystem function (measured as estimated net primary production). Therefore, these two works (Underwood *et al.*, 2005; Forster *et al.*, 2006) indicate that consideration of the importance of diatom species composition and diversity can improve our understanding of estuarine biofilm functioning.

A vertical migratory pattern has also recently been found in diatoms inhabiting subtidal sediments (200 cm depth) of the Bay of Brest (on the French Atlantic coast) by Longphuirt *et al.* (2006) using pulse amplitude modulated (PAM) fluorometry. Sediment cores incubated at constant light and temperature displayed similar day/night fluorescence variation indicating that the migratory rhythm was endogenous and diel without any relation to tidal cycle. Therefore, care should be taken in selecting appropriate sampling times for accurate quantification of primary production of benthic microalgae in both intertidal and subtidal sediments.

16.4.3 Deep water primary production

Work by Cahoon and Cooke (1992), Cahoon and Laws (1993), and Cahoon *et al.* (1990, 1993, 1994) has showed a highly diverse and productive diatom assemblage within coastal shelf sediments. In fact, the chl *a* content of a square meter of shelf bottom may exceed that of the entire water column above, even at depths of 142 m (Cahoon *et al.*, 1990). Recently, McGee *et al.* (2008) found a total of 126 species of living diatoms, 90% of which were considered obligate benthic forms, at depths of 67 to 191 m on the upper continental slope of North Carolina (USA). Mid-day, photosynthetically acitve radiation values recorded at 191 m averaged 0.106 μmol photons m^{-2} s^{-1}, which was 0.028% of the surface incident radiation. These results extend the lower limit of benthic diatom production to upper slope depths and suggest that the contribution of diatoms to total global primary production is significantly greater than previously thought.

16.4.4 Stable-isotope ecology

Despite the paradigm that vascular plant primary production (primarily that of *Spartina* spp.) is the basis for salt-marsh and estuarine food webs, utilization of benthic microalgae by numerous benthic fauna that inhabit these ecosystems has been demonstrated in laboratory and field investigations. In the last decade the application of multiple stable isotope ($^{13}C/^{12}C$, $^{15}N/^{14}N$ and $^{34}S/^{32}S$) analyses have shown not only the importance of benthic microalgae (mainly diatoms) supporting secondary production, but also that, in some cases, benthic microalgae were primarily responsible for supporting salt marsh food chains. Sullivan and Currin (2000) provided a comprehensive overview of the application of natural multiple stable isotope analysis to determine the role of benthic microalgae in salt marsh food webs.

The utility of natural abundance stable isotopes is limited in resolving food-web questions when the different potential foods (i.e. macrophytes, phytoplankton, benthic microalgae) have similar natural isotope values (actually ratios) or overlapping ranges. Two approaches have been used to supplement the information supplied by natural isotope abundance: fatty acid profiles (not reviewed here as it is not experimental) and enrichment of primary producers with the heavy isotope of either carbon (^{13}C) or nitrogen (^{15}N) or both.

The first study using an enriched stable isotope tracer as a method for separating the isotope signatures of seagrass and its epiphytes for estuarine food web analysis was carried out by Winning *et al.* (1999). Using additions of $K^{15}NO_3$ in laboratory aquaria containing pieces of *Zostera capricorni* Asherson turf (including the substrate, the root system and the epiphytes attached to the leaves), the authors were able to differentially label the seagrass (25–90‰) and its epiphytes (87–713‰). These values were substantially higher than natural abundance values ($\delta^{15}N < 10$‰), highlighting the potential ability of this technique to resolve questions of the relative trophic importance of seagrass and epiphyte carbon in the field, since these two food sources often possess indistinguishable natural $\delta^{15}N$ values.

Herman *et al.* (2000) sprayed $NaH^{13}CO_3$ on intertidal sediments of the Westerschelde Estuary (the Netherlands) to study the uptake of benthic microalgae (diatom-dominated) by macrobenthos. Their results showed the central role that benthic diatoms may have as a food source as 95% of total biomass consumed by surface deposit feeders was derived from the benthic microalgae. Page and Lastra (2003) demonstrated the importance of resuspended benthic diatoms in the diets of three commercially important, intertidal bivalves (*Cerastoderma edule*, *Tapes decussates*, and *Mytilus galloprovincialis*) of the Ría de Arosa (Spain) by spraying $^{15}NH_4Cl$ on exposed sediments at low tide. Three days later they found that the stomach contents of all three bivalves had greatly enriched $\delta^{15}N$ values and there was even some ^{15}N enrichment of their muscle tissue. Based on the natural abundance stable isotope values, Galván *et al.* (2008) easily determined that the dominant macrophyte *Spartina alterniflora* was not relevant in the diets of salt marsh consumers in the Plum Island Estuary (Massachusetts, USA). However, the natural abundance values for $\delta^{13}C$ and $\delta^{15}N$ were too similar to determine the relative importance of benthic microalgal versus phytoplankton carbon to the food web. They labeled the benthic microalgae at low tide in mudflats and creek walls with $Na^{15}NO_3$ (phytoplankton therefore were not enriched in ^{15}N). The result was that both algal groups proved to be important food sources but their relative contributions varied by benthic faunal species and habitat (mudflat vs. creek wall). Middelburg *et al.* (2000) sprayed $NaH^{13}CO_3$ on intertidal sediment of the Scheldt Estuary (the Netherlands) to study the carbon transfer from benthic microalgae (diatoms dominant in assemblage) to benthic consumers (bacteria, nematodes, and macrofauna) and found that benthic microalgal carbon entered all heterotrophic components in proportion to their biomass distribution (bacteria > macrofauna > meiofauna). The transfer of label from the benthic diatoms to bacteria and nematodes was extremely rapid (1 h). These results require careful consideration of the central role of benthic diatoms when modeling carbon flow in coastal sediments.

16.5 Summary

Diatom studies of an applied nature in estuaries and shallow coastal waters have been few in number, especially when compared to the situation in freshwaters and the ocean. This may be a reflection of the intermediate position such environments occupy between purely fresh and fully marine waters and their resulting highly dynamic nature. In short, the complexity of estuaries and shallow coastal waters and the continuous change in and interaction between physicochemical factors make controlled experiments difficult to carry out. Nevertheless, some excellent applied work has been done with benthic diatom assemblages. Pollution in general, and eutrophication in particular, are major worldwide problems as human populations are often concentrated around estuaries and shallow coastal waters. The work that has been done, including what has been discussed in this chapter, has shown that diatom assemblages respond to changes in nutrient and environmental conditions and may well emerge as a sensitive biomonitoring system of water and ecosystem quality in coastal areas. However, it is of importance to develop specific diatom indices for estuaries and coastal wetlands that incorporate the dynamic nature of these habitats. The paucity of autecological information on diatom taxa from estuarine and shallow coastal systems for which no comprehensive taxonomic monograph exists hampers the formulation of diatom-based water quality or ecological status indices at the present time. Work specifically designed to determine the tolerances of benthic marine diatoms to various forms of pollution in estuaries and salt marshes should be given high priority. However, and in order to develop new biomonitoring tools, these ecophysiological experiments need to be combined with the use of modern molecular methods (with associated imaging and support systems) to standardize the identification and taxonomy of benthic diatoms, such as DNA "barcode" approaches. Moreover, it will be very important to apply mass molecular sequencing methods (e.g. 454 pyrosequencing) and to make population genetic studies of estuarine and shallow coastal diatoms, since these techniques offer new possibilities for determining and predicting the responses of benthic diatom assemblages to eutrophication and other disturbances of these systems.

Because estuaries and shallow coastal waters are complex, variable and very diverse environments, long-term microphytobenthic data sets would greatly aid efforts to use benthic diatoms as bioindicators and monitoring tools in these aquatic habitats. Applied work in the area of estuarine sediment stability has convincingly shown the paramount importance of benthic diatom assemblages and their mucilaginous secretions which, at sufficient cell densities, bind the diatoms and sediment particles into a network. This network is best developed by epipelic diatoms, whose daily upward and downward migrations leave trails of sediment-binding mucilage. It has also been shown that diatom mucilages are important precursors for vascular plant colonization in restored and newly created salt marshes. Diametrically opposed to stabilization of sediments in estuaries and shallow coastal waters is the process of resuspension, whereby sediment particles with or without diatoms and unattached diatom cells enter the water column. Applied work has revealed the magnitude and importance of this phenomenon. In some estuarine systems virtually all the so-called "phytoplankton" are, in fact, resuspended benthic diatoms from the sediments that may greatly augment the primary production of the water column and constitute an important food source for filter-feeding animals. Diatom vertical migration studies of an applied nature are also of considerable interest since the integration of migratory processes into the calculation of primary production of intertidal and subtidal sediments can greatly modify estimates. Applied studies employing stable-isotope enrichments of primary producers have significantly increased our understanding of salt-marsh and estuarine food webs and revealed the central role of benthic diatoms in these systems.

Finally, although well offshore of estuarine and shallow coastal systems, the recent discovery of obligate benthic diatoms living at depths approaching 200 meters suggests that global coastal production has been greatly underestimated and holds significant implications for oceanic productivity and biogeochemical cycling. Clearly, further studies of the depth distributions of living benthic diatoms along continental shelves and upper slopes are warranted.

Acknowledgments

The contribution of Michael J. Sullivan to this chapter was made possible through a series of funded grants provided by the Mississippi-Alabama Sea Grant Consortium, while funds provided by the Government of Catalonia (Agència Catalana de l'Aigua and Departament d'Innovació, Universitats i Empresa) supported the contribution of Rosa Trobajo.

References

Adams, J. P., Bate, G. C., Harrison, T. D., *et al.* (2002). A method to assess the freshwater inflow requirements of estuaries and applications to the Mtata Estuary, South Africa. *Estuaries*, **25**, 1382–93.

Adams, M. A. & Stauber, J. L. (2004). Development of a whole-sediment toxicity test using a benthic marine microalga. *Environmental Toxicology and Chemistry*, **23**, 1957–68.

Admiraal, W. (1977a). Influence of various concentrations of orthophosphate on the division rate of an estuarine benthic diatom, *Navicula arenaria*. *Marine Biology*, **42**, 1–8.

Admiraal, W. (1977b). Tolerance of estuarine benthic diatoms to high concentrations of ammonia, nitrite ion, nitrate ion and orthophosphate. *Marine Biology*, **43**, 307–15.

Admiraal, W. (1984). The ecology of estuarine sediment-inhabiting diatoms. *Progress in Phycological Research*, **3**, 269–322.

Admiraal, W. & Peletier, H. (1979a). Sulphide tolerance of benthic diatoms in relation to their distribution in an estuary. *British Phycological Journal*, **14**, 185–96.

Admiraal, W. & Peletier, H. (1979b). Influence of organic compounds and light limitation on the growth rate of estuarine benthic diatoms. *British Phycological Journal*, **14**, 197–206.

Admiraal, W. & Peletier, H. (1980). Distribution of diatom species on an estuarine mud flat and experimental analysis of the selective effect of stress. *Journal of Experimental Marine Biology and Ecology*, **46**, 157–75.

Baillie, P. W. & Welsh, B. L. (1980). The effect of tidal resuspension on the distribution of intertidal epipelic algae in an estuary. *Estuarine and Coastal Marine Science*, **10**, 165–80.

Bak, M., Wawrzyniak-Wydrowska, B., & Witkowski, A. (2001). Odra River discharge as a factor affecting species composition of the Szczecin Lagoon diatom flora, Poland. In *Studies on Diatoms.*

Lange-Bertalot-Festischrift, ed. R. Jahn, J. P. Kociolek, A. Witkowski, & P. Compere, Ruggell: Gantner Verlag, pp. 491–506.

Brummit, R. K. & Powell, C. E. (1992). *Authors of Plant Names*. Kew: Royal Botanic Gardens.

Cahoon, L. B., Beretich, G. R., Jr., Thomas, C. J., & McDonald, A. M. (1993). Benthic microalgal production at Stellwagen Bank, Massachusetts Bay, USA. *Marine Ecology Progress Series*, **102**, 179–85.

Cahoon, L. B. & Cooke, J. E. (1992). Benthic microalgal production in Onslow Bay, North Carolina, USA. *Marine Ecology Progress Series*, **84**, 185–96.

Cahoon, L. B. & Laws, R. A. (1993). Benthic diatoms from the North Carolina continental shelf: inner and mid shelf. *Journal of Phycology*, **29**, 257–63.

Cahoon, L. B., Laws, R. A., & Thomas, C. J. (1994). Viable diatoms and chlorophyll a in continental slope sediments off Cape Hatteras, North Carolina. *Deep-Sea Research II*, **41**, 767–82.

Cahoon, L. B., Redman, R. S., & Tronzo, C. R. (1990). Benthic microalgal biomass in sediments of Onslow Bay, North Carolina. *Estuarine, Coastal and Shelf Science*, **31**, 805–16.

Coleman, V. L. & Burkholder, J. M. (1994). Community structure and productivity of epiphytic microalgae on eelgrass (*Zostera marina* L.) under water-column nitrate enrichment. *Journal of Experimental Marine Biology and Ecology*, **179**, 29–48.

Coleman, V. L. & Burkholder, J. M. (1995). Response of microalgal epiphyte communities to nitrate enrichment in an eelgrass (*Zostera marina*) meadow. *Journal of Phycology*, **31**, 36–43.

Coles, S. M. (1979). Benthic microalgal populations on intertidal sediments and their role as precursors to salt marsh development. In *Ecological Processes in Coastal Environments*, ed. R. L. Jefferies & A. J. Davy, Oxford: Blackwell, pp. 25–42.

Consalvey, M., Paterson, D. M., & Underwood, G. J. C. (2004). The ups and downs of life in a benthic biofilm: migration of benthic diatoms. *Diatom Research*, **19**, 181–202.

Costanza, R., Norton, B. G., & Haskell, B. D. (1992). *Ecosystem Health: New Goals for Environmental Management*. Washington, DC: Island Press.

Darley, W. M., Montague, C. L., Plumley, F. G., Sage, W. W., & Psalidas, A. T. (1981). Factors limiting edaphic algal biomass and productivity in a Georgia salt marsh. *Journal of Phycology*, **17**, 122–8.

Dauvin, J.-C. (2007). Paradox of estuarine quality: benthic indicators and indices, consensus or debate for the future. *Marine Pollution Bulletin*, **55**, 271–81.

Decho, A. W. (2000). Microbial biofilms in intertidal systems: an overview. *Continental Shelf Research*, **20**, 1257–74.

Dudley, B. J., Gahnström, A. M. E., & Walker, D. I (2001). The role of benthic vegetation as a sink for elevated inputs of ammonium and nitrate in a mesotrophic estuary. *Marine Ecology Progress Series*, **219**, 99–107.

Elliott, M. & Quintino, V. (2007). The estuarine quality paradox, environmental homeostasis and the difficulty of detecting anthropogenic stress in naturally stressed areas. *Marine Pollution Bulletin*, **54**, 640–5.

Forster R. M., Creach, V., Sabbe, K., Vyverman, W., & Stal, L. J. (2006). Biodiversity-ecosystem function relationship in microphytobenthic diatoms of the Westerschelde Estuary. *Marine Ecology Progress Series*, **311**, 192–201.

Galván, K., Fleeger, J. W., & Fry, B. (2008). Stable isotope addition reveals dietary importance of phytoplankton and microphytobenthos to saltmarsh infauna. *Marine Ecology Progress Series*, **359**, 37–49.

Glud, R. N., Kühl, M., Wenzhöfer, F., & Rysgaard, S. (2002). Benthic diatoms of a high Arctic fjord (Young Sound, NE Greenland): importance for ecosystem primary production in European intertidal mudflats – a modelling approach. *Continental Shelf Research*, **20**, 1771–88.

Grant, J., Bathmann, U. V., & Mills, E. L. (1986). The interaction between benthic diatom films and sediment transport. *Estuarine, Coastal and Shelf Science*, **23**, 225–38.

Hall, N. E., Fairchild, J. F., La Point, T. W. *et al.* (1996). Problems and recommendations in using algal toxicity testing to evaluate contaminated sediments. *Journal of Great Lakes Research*, **22**, 545–56.

Hamels, I., Mussche, H., Sabbe, K., Muylaert, K., & Vyverman, W. (2004). Evidence for constant and highly specific active food selection by benthic cialiates in mixed diatoms assemblages. *Limnology and Oceanography*, **49**, 58–68.

Hanlon, A. R. M., Bellinger, B., Haynes, K., *et al.* (2006). Dynamics of extracellular polymeric substance (EPS) production and loss in an estuarine, diatom-dominated, microalgal biofilm over a tidal emersion-immersion period. *Limnology and Oceanography*, **5**, 79–93.

Heap, A., Bryce, S., Ryan, D., *et al.* (2001). Australian estuaries & coastal waterways: a geoscience perspective for improved and integrated resource management. A report to the National Land & Water Resources Audit. Theme 7: Ecosystem Health. Australian Geological Survey Organisation, Record 2001/07.

Hendey, N. I. (1977). The species diversity index of some in-shore diatom communities and its use in assessing the degree of pollution insult on parts of the north coast of Cornwall. *Nova Hedwigia, Beiheft*, **54**, 355–78.

Herman, P. M. J., Middelburg, J. J., Widdows, J., Lucas, C. H., & Heip, C. H. R. (2000). Stable isotopes as trophic tracers: combining field sampling and manipulative labelling of food resources for macrobenthos. *Marine Ecology Progress Series*, **204**, 79–92.

Holland, A. F., Zingmark, R. G., & Dean, J. M. (1974). Quantitative evidence concerning the stabilization of sediments by marine benthic diatoms. *Marine Biology*, **27**, 191–6.

Howarth, R. W. & Marino, R. (2006). Nitrogen as the limiting nutrient for eutrophication in coastal marine ecosystems: Evolving views over three decades. *Limnology and Oceanography*, **51**, 364–76.

IPCC (Intergovernmental Panel on Climate Change) (2007). *Impacts, Adaptation and Vulnerability*. IPCC Secretariat: Geneva.

Jonge, V. N. de & van Beusekom, J. E. E. (1992). Contribution of resuspended microphytobenthos to total phytoplankton in the Ems estuary and its possible role for grazers. *Netherlands Journal of Sea Research*, **30**, 91–105.

Jonge, V. N. de & van Beusekom, J. E. E. (1995). Wind- and tide-induced resuspension of sediment and microphytobenthos from tidal flats in the Ems estuary. *Limnology and Oceanography*, **40**, 766–78.

Kelly, M., Jones, R. I., Barker, P. A., & Jamieson, B. J. (2008a). Validation of diatoms as proxies for phytobenthos when assessing ecological status in lakes. *Hydrobiologia*, **610**, 125–9.

Kelly, M., Juggins, R., Guthrie, R., *et al.* (2008b). Assessment of ecological status in U.K. rivers using diatoms. *Freshwater Biology*, **53**, 403–22.

Kolkwitz, R. & Marsson, M. (1908). Ökologie der pflanzliche Saprobien. *Berichte der Deutschen Botanischen Gesellschaft*, **26**, 505–19.

Laird, K. & Edgar, R. K. (1992). Spatial distribution of diatoms in the surficial sediments of a New England salt marsh. *Diatom Research*, **7**, 267–79.

Lange-Bertalot, H. (1978). Diatomeen-Differentialarten anstelle von Leitformen: ein geeigneteres Kriterium der Gewässerbelastung. *Archiv für Hydrobiologie, Supplement* 51, *Algological Studies*, **21**, 393–427.

Lange-Bertalot, H. (1979). Pollution tolerance of diatoms as a criterion for water quality estimation. *Nova Hedwigia, Beiheft*, **64**, 285–304.

Leira, M., Jordan, P., Taylor, D., *et al.* (2006). Assessing the ecological status of candidate reference lakes in Ireland using palaeolimnology. *Journal of Applied Ecology*, **43**, 816–27.

Lloyd, M., Zar, J. H., & Karr, J. R. (1968). On the calculation of information-theoretical measures of diversity. *The American Midland Naturalist*, **79**, 257–72.

Longphuirt, S. N., Leynaert, A., Guarini, J-M., *et al.* (2006). Discovery of microphytobenthos migration in the subtidal zone. *Marine Ecology Progress Series*, **328**, 143–54.

MacIntyre, H. L., Geider, R. J., & Miller, D. C. (1996). Microphytobenthos: the ecological role of the "secret garden" of unvegetated, shallow-water marine habitats. I. Distribution, abundance and primary production. *Estuaries*, **19**, 186–201.

McGee, D., Laws, R. A., & Cahoon, L. B. (2008). Live benthic diatoms from the upper continental slope: extending the limits of marine primary production. *Marine Ecology Progress Sereies*, **356**, 103–12.

Middelburg, J. J., Barranguet, C., Boschker, H. T. S., *et al.* (2000). The fate of intertidal microphytobenthos carbon: an *in situ* ^{13}C-labelling study. *Limnology and Oceanography*, **45**, 1224–34.

Miles, A. & Sundbäck, K. (2000). Diel variation in microphytobenthic productivity in areas of different tidal amplitude. *Marine Ecology Progress Series*, **205**, 11–22.

Miller, D. C., Geider, R. J., & MacIntyre, H. L. (1996). Microphytobenthos: the ecological role of the "secret garden" of unvegetated, shallow-water marine habitats. II. Role in sediment stability and shallow-water food webs. *Estuaries*, **19**, 202–12.

Moran, M. A., Sheldon, W. M. Jr., & Zepp, R. G. (2000). Carbon loss and optical property changes during long-term photochemical and biological degradation of estuarine dissolved organic matter. *Limnology and Oceanography*, **45**, 1254–64.

Moreno-Garrido, I., Hampel, M., Lubián, L. M., & Blasco, J. (2003). Sediment toxicity tests using benthic marine microalgae *Cylindrotheca closterium* (Ehremberg) Lewin and Reimann (Bacillariophyceae). *Ecotoxicology and Environmental Safety*, **54**, 290–5.

Moreno-Garrido, I., Lubián, L. M., Jiménez, B., Soares, A. M. V. M., & Blasco, J. (2007). Estuarine sediment toxicity tests on diatoms: Sensitivity comparison for the three species. *Estuarine, Coastal and Shelf Sciences*, **71**, 278–86.

National Academy of Sciences (Committee on the Causes and Management of Coastal Eutrophication, Ocean Studies Board and Water Science and Technology Board, Commission on Geosciences, Environment, and Resources, National Research Council.) (2000). *Clean Coastal Waters: Understanding and Reducing the Effects of Nutrient Pollution*, Washington, DC: National Academy Press.

Orvain, A. M., Galois, C., Barnar, A., *et al.* (2003). Carbohydrate production in relation to microphytobenthic biofilm development: an integrated approach in tidal mesocosm. *Microbial Ecology*, **45**, 237–51.

Page, H. M. & Lastra, M. (2003). Diet of intertidal bivalves in the Ría de Arosa (NW Spain): evidence from stable C and N isotope analysis. *Marine Biology*, **143**, 519–32.

Paterson, D. M. (1986). The migratory behaviour of diatom assemblages in a laboratory tidal micro-ecosystem examined by low temperature scanning electron microscopy. *Diatom Research*, **1**, 227–39.

Paterson, D. M. (1990). The influence of epipelic diatoms on the erodibility of an artificial sediment. In *Proceedings of the 10th International Diatom Symposium*, ed. H. Simola, Königstein: Koeltz Scientific Books, pp. 345–55.

Peletier, H. (1996). Long-term changes in intertidal estuarine diatom assemblages related to reduced input of organic waste. *Marine Ecology Progress Studies*, **137**, 265–71.

Perkins, R. G., Underwood, G. J. C., Brotas, V., *et al.* (2001). Responses of microphytobenthos to light: primary production and carbohydrate allocation over an emersion period. *Marine Ecology Progress Series*, **223**, 101–12.

Round, F. E. (1979). Botanical aspects of estuaries. In *Tidal Power and Estuary Management*, ed. R. T. Severn, D. Dineley, & L. E. Hawker, Bristol: Scientechnica, pp. 195–213.

Round, F. E. (1981). *The Ecology of Algae*, Cambridge: Cambridge University Press.

Saburova, M. A. & Polikarpov, I. G. (2003). Diatom activity within soft sediments: behavioural and physiological processes. *Marine Ecology Progress Series*, **251**, 115–26.

Shaffer, G. P. & Sullivan, M. J. (1988). Water column productivity attributable to displaced benthic diatoms in well-mixed shallow estuaries. *Journal of Phycology*, **24**, 132–40.

Short, F. T. & Wyllie-Echeverria, S. (1996). Natural and human-induced disturbance of seagrasses. *Environmental Conservation*, **23**, 17–27.

Smith, D. J. & Underwood, G. J. C. (2000). The production of extracellular carbohydrate exopolymers (EPS) by estuarine benthic diatoms: the effects of growth phase and light and dark treatment. *Journal of Phycology*, **36**, 321–33.

Staats, N., de Winder, B., Stal, L. J., & Mur, L. R. (1999). Isolation and characterisation of extracellular plysaccharides from the epipelic diatoms *Cylindrotheca closterium* and *Navicula salinarum*. *European Journal of Phycology*, **34**, 161–9.

Staats, N., Stal, L. J., de Winder, B., & Mur, L. R. (2000a). Oxygenic photosynthesis as driving process in exopolysaccharide production of benthic diatoms. *Marine Ecology Progress Series*, **193**, 261–9.

Staats, N., Stal, L. J., & Mur, L. J. (2000b). Exopolysacharide production by epipelic diatom *Cylindrotheca closterium*: effects of nutrient conditions. *Journal of Experimental Marine Biology*, **249**, 3–27.

Stal, L. J. (2003). Microphytobenthos, their extracellular polymeric substances, and the morphogenesis of intertidal sediments. *Geomicrobiology Journal*, **20**, 463–78.

Sullivan, M. J. (1976). Long-term effects of manipulating light intensity and nutrient enrichment on the structure of a salt marsh diatom community. *Journal of Phycology*, **12**, 205–10.

Sullivan, M. J. (1978). Diatom community structure: taxonomic and statistical analysis of a Mississippi salt marsh. *Journal of Phycology*, **14**, 468–75.

Sullivan, M. J. (1986). Mathematical expression of diatom results: are these "pollution indices" valid and useful? In *Proceedings of the 8th International Diatom Symposium*, ed. M. Ricard, Königstein: Koeltz Scientific Books, pp. 772–6.

Sullivan, M. J. & Currin, C. A. (2000). Community structure and functional dynamics of benthic microalgae in salt marshes. In *Concepts and Controversies in Tidal Marsh Ecology*, ed. M. P. Weinstein & D. A. Kreeger, Dordrecht, Kluwer Academic Publishers, pp. 81–106.

Sullivan, M. J. & Daiber, F. C. (1975). Light, nitrogen, and phosphorus limitation of edaphic algae in a Delaware salt marsh. *Journal of Experimental Marine Biology and Ecology*, **18**, 79–88.

Sundbäck, K. & Granéli, W. (1988). Influence of microphytobenthos on the nutrient flux between sediment and water: a laboratory study. *Marine Ecology Progress Series*, **43**, 63–9.

Sundbäck, K., Enoksson, V., Granéli, W., & Pettersson, K. (1991a). Influence of sublittoral microphytobenthos on the oxygen and nutrient flux between sediment and water: a laboratory continuous-flow study. *Marine Ecology Progress Series*, **74**, 263–79.

Sundbäck, K. & Snoeijs, P. (1991b). Effects of nutrient enrichment on microalgal community composition in a shallow-water sediment system: an experimental study. *Botanica Marina*, **34**, 341–58.

Tolhurst, T., Black, K. S., Shayler, S. A., *et al.* (1999). Measuring *in situ* erosion shear stress of intertidal sediments with the cohesive strength meter (CSM). *Estuarine, Coastal and Shelf Science*, **49**, 281–94.

Tomàs, X. (1988). Diatomeas de las aguas epicontinentales saladas del litoral mediterráneo de la Península Ibérica. Ph.D. thesis, University of Barcelona, Spain.

Trobajo, R. (2007). *Ecological Analysis of Periphytic Diatoms in Mediterranean Coastal Wetlands (Empordà Wetlands, NE Spain)*, ed. A. Witkowski, Diatom Monographs, vol. **7**, Ruggel: A. R. G. Gantner Verlag.

Trobajo, R., Quintana, X. D., & Sabater, S. (2004). Factors affecting the periphytic diatom community in Mediterranean coastal wetlands (Empordà wetlands, NE Spain). *Archiv für Hydrobiologie*, **160**, 373–99.

Underwood, G. J. C. (1997). Microalgal colonisation in a saltmarsh restoration scheme. *Estuarine, Coastal and Shelf Science*, **44**, 471–81.

Underwood, G. J. C. (2000). Changes in microalgal species composition, biostabilisation potential and succession during saltmarsh restoration. In *British Salt Marshes*, ed. B. R. Sherwood, B. G. Gardiner, & T. Harris, Cardigan: Linnaean Society of London / Forrest Text, pp. 143–54.

Underwood, G. J. C. & Barnett, M. (2006). What determines species composition in microphytobenthic biofilms? In *Proceedings of the Microphytobenthos Symposium*, ed. J. Kromkamp, Amsterdam: Royal Netherlands Academy of Arts and Sciences, pp. 121–38.

Underwood, G. J. C., Boulcott, M., Raines, C. A., & Waldran, K. (2004). Environmental effects on exopolymer production by marine benthic diatoms: dynamics, changes in composition and pathways of production. *Journal of Phycology*, **40**, 293–304.

Underwood, G. J. C. & Kromkamp, J. (1999). Primary production by phytoplankton and microphytobenthos in estuaries. In *Advances in Ecological Research, Estuaries*, Vol. 29, ed. D. B. Nedwell & D. G. Raffaelli, San Diego: Academic Press, pp. 94–153.

Underwood, G. J. C. & Paterson, D. M. (1993a). Recovery of intertidal benthic diatoms after biocide treatment and associated sediment dynamics. *Journal of the Marine Biological Association of the United Kingdom*, **73**, 25–45.

Underwood, G. J. C. & Paterson, D. M. (1993b). Seasonal changes in diatom biomass, sediment stability and biogenic stabilization in the Severn Estuary. *Journal of the Marine Biological Association of the United Kingdom*, **73**, 871–87.

Underwood, G. J. C. & Paterson, D. M. (2003). The importance of extracellular carbohydrate production by marine epipelic diatoms. *Advances in Botanical Research*, 40, 183–240.

Underwood, G. J. C., Perkins, R. G., Consalvey, M., *et al.* (2005). Patterns in microphytobenthic primary productivity: species-specific variation in migratory rhythms and photosynthetic efficiency in mixed species. *Limnology and Oceanography*, **50**, 755–67.

Underwood, G. J. C., Phillips, J., & Saunders, K. (1998). Distribution of estuarine benthic diatom species along salinity and nutrient gradients. *European Journal of Phycology*, **33**, 173–83.

Underwood, G. J. C. & Provot, L. (2000). Determining the environmental preferences of four estuarine epipelic diatom taxa: growth across a range of salinity, nitrate and ammonium conditions. *European Journal of Phycology*, **35**, 173–82.

Underwood, G. C. J. & Smith, J. (1998). In situ measurement of exopolymer production by intertidal epipelic diatom-dominated biofilms in the Humber Estuary. *Geological Society of London, Special Publication*, **139**,125–34.

USEPA (United States Environmental Protection Agency) (2002). Federal Water Pollution Control Act (as amended through P.L. 107–303, November 27, 2002). See http://www.saj.usace.army.mil/Divisions/Regulatory/DOCS/wetlands/fwpca_2005.pdf.

Valiela, I., Collins, G., Kremer, J., *et al.* (1997). Nitrogen loading from coastal watersheds to receiving estuaries: new method and application. *Ecological Applications*, **7**, 358–80.

van Dam, H. (1982). On the use of measures of structure and diversity in applied diatom ecology. *Nova Hedwigia, Beiheft*, **73**, 97–115.

van Raalte, Valiela, I., & Teal, J. M. (1976a). The effect of fertilization on the species composition of salt marsh diatoms. *Water Research*, **10**, 1–4.

van Raalte, Valiela, I., & Teal, J. M. (1976b). Production of epibenthic salt marsh algae: light and nutrient limitation. *Limnology and Oceanography*, **21**, 862–72.

Varela, M. & Penas, E. (1985). Primary production of benthic microalgae in an intertidal sand flat of the Ría de Arosa, NW Spain. *Marine Ecology Progress Series*, **25**, 111–9.

Vos, P. C., de Boer, P. L., & Misdorp, R. (1988). Sediment stabilization by benthic diatoms in intertidal sandy shoals: qualitative and quantitative observations. In *Tide-Influenced Sedimentary Environments and Facies*, ed. P. L. de Boer, A. van Gelder, & S. D. Nio, Dordrecht: Reidel, pp. 511–26.

Watt, D. A. (1998). Estuaries of contrasting trophic status in KwaZulu-Natal, South Africa. *Estuarine, Coastal and Shelf Science*, **47**, 209–16.

Wear, D. J., Sullivan, M. J., Moore, A. D., & Millie, D. F. (1999). Effects of water-column enrichment on the production dynamics of three seagrass species and their epiphytic algae. *Marine Ecology Progress Series*, **179**, 201–13.

WFD European Union (2000). Directive 2000/60/EC of the European Parliament and of the Council of 23 October 2000 establishing a framework for Community action in the field of water policy. *Official Journal of the European Communities*, **L327** (43), 1–72.

Widdows, J., Blaw, A., Heip, C. H. R., *et al.* (2004). Role of physical and biological processes in sediment dynamics of a tidal flat in Werschelde Estuary, SW Netherlands. *Marine Ecology Progress Series*, **274**, 41–56.

Widdows, J., Brown, S., Brinsley, M. D., Salkeld, P. N., & Elliott, M. (2000). Temporal changes in intertidial sediment erodability: influence of biological and climate factors. *Continental Shelf Research*, **20**, 1275–89.

Winning, M. A., Connolly, R. M., Loneragan, N. R. & Bunn, S. E. (1999). ^{15}N enrichment as a method of separating the isotopic signatures of seagrass and its epiphytes for food web analysis. *Marine Ecology Progress Series*, **189**, 289–94.

Wulff, A., Wängberg, S.-Å., Sunbäch, K., Nilsson, C., & Underwood, G. J. C. (2000). Effects of UVB radiation on a marine microbenthic community growing on a sand-substratum under different nutrient conditions. *Limnology and Oceanography*, **45**, 1144–52.

Yallop, M. L., de Winder, B., Paterson, D. M., & Stal, L. J. (1994). Comparative structure, primary production and biogenic stabilization of cohesive and non-cohesive marine sediments inhabited by microphytobenthos. *Estuarine, Coastal and Shelf Science*, **39**, 565–82.

Zaldívar, J-M., Cardoso, A. C., Viaroli, P., *et al.* (2008). Eutrophication in transitional waters: an overview. *Transitional Waters Monographs*, **1**, 1–78. See http://siba2.unile.it/ese/twm.

17 Estuarine paleoenvironmental reconstructions using diatoms

SHERRI COOPER, EVELYN GAISER, AND ANNA WACHNICKA

17.1 Introduction

Paleoecology offers powerful techniques with which to study historical changes due to human influences in depositional environments, including estuaries and coastal wetlands. Diatoms are particularly useful in these endeavors, not only because they are preserved in the sediment record, but because they have a short reproductive rate and respond quickly to changes in nutrient availability and water-quality conditions. Diatoms are abundant in aquatic environments, generally cosmopolitan in distribution, and have a fairly well-studied taxonomy and ecology.

Paleoecological studies in estuarine and coastal environments have lagged behind paleolimnology, primarily because of the more dynamic nature and presumed invulnerability of coastal ecosystems. Estuaries are characterized by variability in salinity, sediment deposition, water currents and residence time, turbidity zones, and unique biogeochemistry of sediments. There is often mixture and transport of sediments after initial deposition, and differential silicification and preservation of diatom valves.

Historically, there has been a lack of appreciation for the magnitude and severity of human impacts on estuaries and other coastal ecosystems, and of how important these ecosystems are to human society. The demand for resources and the wastes generated as human populations grow will continue to cause cultural, economic, aesthetic, and environmental problems in coastal areas. Understanding the processes surrounding these problems is important for managing the continuing impacts of growing populations (see Costanza *et al.*, 1997; Clark *et al.*, 2001; Niemi *et al.*, 2004). Environmental issues relevant to estuaries include hydrological changes (e.g. channelization), eutrophication, anoxia, harmful algal blooms, industrial pollution (toxic materials), loss of habitats (e.g. wetlands and submerged aquatic vegetation areas), land-use effects on turbidity and sedimentation, over-harvesting of target species, and invasion of exotic species.

This chapter highlights the nature and significance of estuarine ecosystems, describes methods for using diatoms in estuarine paleoenvironmental reconstructions, and discusses issues relevant to these methods. Case studies exemplify methods for investigating issues specific to coastal ecosystems. Our focus is on post-glacial Holocene anthropogenic influences on water quality, trophic interactions, and habitat changes in estuaries.

17.1.1 Description of estuaries

An estuary can be defined as a semi-enclosed coastal body of water that has a free connection with the open ocean, within which seawater is diluted with fresh water derived from land drainage (Pritchard, 1967). River mouths, drowned river valleys, coastal bays or lagoons, fjords, tidal marsh systems, and sounds all fit this definition. Estuaries are transition zones between freshwater and marine habitats. They are most common in low-relief coastal regions and less extensive on uplifted coastlines. Although estuaries are abundant, many are geologically tenuous. Present day estuaries are generally less than 15,000 years old, the time when sea level began its rise after the last glaciation (Day *et al.*, 1989).

A coastal shore is held in delicate balance between forces that tend to move it either landward or seaward. The location of estuaries depends on the position of the shoreline, and is conditioned by geomorphology, coastal lithology, tidal range, river

The Diatoms: Applications for the Environmental and Earth Sciences, 2nd Edition, eds. John P. Smol and Eugene F. Stoermer. Published by Cambridge University Press.

discharge and sediment load, climate, sea-level oscillations, tectonism, and isostasy. Geomorphologic and sedimentologic changes are continuously occurring within and around estuaries. A comprehensive overview of geomorphology and sedimentology of estuaries is contained in Perillo (1995).

Human populations throughout history have used estuaries as food sources, places of navigation and settlement, and repositories for waste. It has been estimated that approximately 50 percent of the world's population live within 150 km of the coastline (Cohen *et al.*, 1997). At the end of the twentieth century, coastal urban centers were home to one billion people worldwide, with roughly half of the world's coasts threatened by development-related activities. Fragile coastal and estuarine ecosystems – such as wetlands, tidal flats, saltwater marshes, mangrove swamps, and the flora and fauna that depend on them – are especially endangered by urban land conversion and human uses (World Resources Institute *et al.*, 1996). In the United States, coastal areas are the region where the majority of new development has occurred (e.g. Culliton *et al.*, 1990).

17.1.2 Habitats and environmental history

There are four different types of estuaries based on geological origins: drowned river valleys (coastal-plain estuaries), fjords (glacially modified river valleys), bar-built estuaries, and estuaries formed by tectonic processes. There are other classification schemes that depend on level of freshwater runoff, tidal flushing, and mixing (Mann, 2000). The salinity gradient is one important physical parameter that affects the distributional patterns of plant and animal species, and this variety of habitats within an estuary contributes to high system productivity. Seagrasses, marsh plants, epiphytes, benthic microalgae, macroalgae, and phytoplankton all contribute to primary production.

Both fresh and brackish water marshes can be found along the banks of estuaries. Coastal wetlands also occupy broad expanses of some low-lying areas in the transition between upstream terrestrial environments and the sea. These wetlands are among the most threatened habitats on Earth, being "squeezed" by drainage and development to the interior and by sea-level rise along the coast. These wetland ecotones can be highly productive because they exist at the tidal confluence of upstream and coastal resources and highly diverse because of the heterogeneous mixture of habitat at this critical interface (e.g. Day and Yanez-Arancibia, 1982). They are also important in terms of flood control, water quality, and maintenance of natural chemical and biological cycles (Mitsch and Gosselink, 1986; Mann, 2000). In areas where water is shallow and relatively clear, beds of submerged grasses grow. These habitats are home to myriad organisms, and provide nursery grounds for juvenile fish. Mudflats, sand banks, and rocky shores each have their own unique flora and fauna.

The sedimentary subsystem of the estuary has unique chemistry and benthic organisms. Within sediments, respiration often exceeds the reserves and supply of oxygen, thus sediments are typically anoxic below the water–sediment interface. Detritus from the productive zones of the estuary are used, and nutrients are regenerated, recycled, and stored in the sediments by the action of bacteria, burrowing organisms, and natural chemistry (Day *et al.*, 1989).

Sedimentation patterns within estuarine systems are extremely variable in both space and time, and resuspension and mixing of sediments before final deposition is common (Brush, 1989; Sanford, 1992; Dyer, 1995). The location of the turbidity maximum, where fresh- and salt-waters meet, may change seasonally and yearly depending on river discharge (Schubel, 1968). Storms that cause high runoff from the drainage basin may flush huge amounts of sediments from rivers into estuarine areas, or scour sediments from shallow areas. Marine sediments are often transported into the mouths of estuaries with regular currents and tides (Dyer, 1995), and occasionally by storms. For these reasons, there has been doubt about the validity of using paleoecological methods within these dynamic systems (see Methods section below).

As an example, there are several different broad areas of sedimentation present from north to south in Chesapeake Bay, USA. The northern bay sediments contain proportionately more organic carbon and less sulfur than those deposited in the middle bay. Fluvial input is the dominant mechanism for introduction of particulate matter in this area (Helz *et al.*, 1985; Hennessee *et al.*, 1986). In the mid-bay region, distributions of organic carbon and sulfur are more representative of marine environments, and primary production is the dominant source of organic matter (Hennessee *et al.*, 1986). The southern part of the Chesapeake Bay is characterized by sandier sediments, a greater percentage of which are of marine origin. There are compounding factors in specific areas, such as the presence of geologically entrapped gas in sediments, that affect the bulk density and chemical properties of the sediments (Hill *et al.*, 1992). This type of information is important for site selection, and when interpreting various biogeochemical parameters.

17.2 Eutrophication and pollution in estuaries

Eutrophication of estuaries is one of the major concerns to estuarine scientists. Worldwide, estuaries receive large amounts of nutrients and toxins from wastewaters of sewage-treatment facilities and other point sources. In most cases, it is difficult to separate the relative impact of nutrient enrichment from point sources vs. non-point sources such as agricultural and atmospheric additions, and urban and rural runoff (Kennish, 1992; Puckett, 1995). Eutrophication is implicated in many coastal water-quality and ecosystem concerns, including nuisance and noxious algal blooms, hypoxic and anoxic bottom waters, fish and shellfish kills, declines in diversity of organisms, habitat losses, and shifts in trophic dynamics. These impacts of eutrophication affect economic, cultural, and human health.

Eutrophication and its consequences have been well documented in coastal waters throughout the world (Nixon, 1995; Boesch, 2002) with a national focus in the United States on Chesapeake Bay (Kemp *et al.*, 2005) and the Gulf of Mexico (Rabalais *et al.*, 2007a). The extensive research into the ecological responses of a range of estuary types to eutrophication suggests that there are substantial differences in the magnitude and trajectory of an estuary's response driven by complex, non-linear, and estuary-specific ecological interactions (Cloern, 2001). Continued population growth is likely to exacerbate eutrophication problems unless a better scientific understanding of these processes is translated into practical management strategies for watersheds and coastal waters.

17.2.1 Fisheries

For human populations, one valuable resource provided by estuaries is their fisheries. Some important estuarine-dependent species include shrimps, oysters, clams, crabs, flounders, scallops, salmon, and menhaden. The significance of estuarine-related species in commercial and recreational fisheries varies by region and estuarine development. The greatest fish yield in the United States comes from Louisiana, the state (with the exception of Alaska) with the largest area of estuaries, lagoons. and coastal marshes (Day *et al.*, 1989; Mann, 2000). Commercial fisheries are an economically recognized ecosystem service (Costanza *et al.*, 1997), relying heavily on the natural environment and natural food sources. Food sources are tied to the primary productivity of estuarine ecosystems and relationships between benthos, nekton, and plankton. Estuarine nekton has adapted to the physiologically demanding and unpredictable environments of estuaries, and often use estuaries seasonally, tied directly to food and other trophic interactions (Gross *et al.*, 1988).

Human impacts on estuarine fisheries have been extensive. The relative effect of these anthropogenic pressures again varies by region and estuarine development. Over-harvesting in particular has played a critical role in fisheries declines, but other human impacts may be just as important for future management and possible recovery (Jackson *et al.*, 2001). Paleoecology is one tool for investigating the history of trophic interactions directly related to fisheries production that are difficult to unravel using recent data, and diatom analyses can be an important component of these studies, both from a trophic standpoint as well as water quality in general. Paleoecology also allows hindcasting of models, using a variety of indicators. Clark *et al.* (2001) advocates modeling as a critical step in managing ecosystems for a sustainable future.

17.2.2 Submerged aquatic vegetation (SAV)

A common feature of estuarine and nearshore marine environments are seagrass meadows or submerged aquatic vegetation (SAV), which play a vital role in regulating physical, chemical, and biological properties across a variety of spatial and temporal scales (Bell and Westoby, 1986; Fourqurean *et al.*, 2003). An important feature of these meadows is their support of dense accumulations of epiphytic algae that contribute to gas flux, nutrient cycling, soil formation, and stabilization as well as food and habitat for consumers (Daehnick *et al.*, 1992; Frankovich and Fourqurean, 1997; Moncreiff and Sullivan, 2001). Because of their nearshore location, seagrass meadows are directly exposed to runoff of agricultural and industrial waste, altered salinities, and dredging associated with coastal development. This has led to global declines in seagrass abundance resulting in radical shifts in the function of nearshore ecosystems (Boström *et al.*, 2002). Mechanisms leading to seagrass die-off are complex, but include increased shading by phytoplankton in response to increased nutrient loading (Valiela *et al.*, 1997), sulfide toxicity resulting from increased organic matter production (Holmer *et al.*, 2003; Ruiz-Halpern *et al.*, 2008), salinity stress, and warmer water temperatures (Koch *et al.*, 2007). Algal productivity and community composition is directly affected by these physicochemical alterations as well as by reduction in substrate for epiphytic growth (Armitage *et al.*, 2005). Diatoms offer the potential for use as early indicators of impending change or for application in paleoecology to determine changes in the frequency of seagrass die-offs.

17.2.3 Geochemistry

As stated above, nutrient influx to sediments within estuaries and coastal waters has changed dramatically over past decades. Productivity, organic deposition and decomposition play significant roles. Organic matter carried to estuaries by rivers or runoff, as well as coastal and *in situ* productivity by phytoplankton, benthic algae, seagrass, epiphytes, and marshes may eventually come to rest on the sediment surface (Day *et al.*, 1989). Particle–water interactions, bacterial denitrification, phosphorus adsorption/desorption reactions, dredging, and oxidation–reduction status all modify and complicate the biogeochemistry of nutrients in estuarine environments (Jickells, 1998).

Tracers used in paleoecological studies of estuaries include fluxes of organic matter, nutrients, algal pigments, stable isotopic signatures of carbon (^{13}C) and nitrogen (^{15}N), and biogenic silica (BSi) (Zimmerman and Canuel, 2002; Verleyan *et al.*, 2004; Clarke *et al.*, 2006). If primary production has increased substantially due to nutrient loading, both ^{13}C and ^{15}N should increase in recently deposited sediments, indicating the contribution of autotrophic carbon and high rates of nitrogen recycling, respectively, in the water column (Brandenberger *et al.*, 2008b). Source-specific organic matter (such as ligneous and lignosulfonate by-products) and other soil biomarkers (such as soil nitrate) have been used to indicate alternate nitrogen enrichment, land clearance, forestry activities, soil erosion, and industrial effluents (Louchouarn *et al.*, 1999; Farella *et al.*, 2001; Houel *et al.*, 2006).

Estuarine waters and sediments are subject to changes in oxidation–reduction status depending on temperature, mixing, season, and nutrient flux. Hypoxic and anoxic conditions often occur in sediments, and may prevail in bottom waters. These conditions can be investigated using redox-sensitive trace metals (RSMs) (Brandenberger *et al.*, 2008b). Pollutants such as heavy metals may also be of special interest in sediments as markers of known anthropogenic activity and transport to estuaries.

17.3 Using diatoms in estuarine paleoecology studies

There are multiple paleoindicators that can be used in estuaries. These are generally chosen based on the history of each site, the expertise of the researchers, and the questions being addressed in each study. Available indicators include fossils of organisms found in fresh, brackish and marine waters (including diatoms, foraminifera, pollen, seeds, dinoflagellate cysts, ostracods, fish otoliths, fish scales, charcoal, and others), radioisotopes, RSMs, priority pollutants, nutrients, isotopic ratios, organic biomarkers, and BSi. This chapter concentrates on diatoms. The use of diatoms in estuarine paleoecological studies requires the application of other paleoecological methods for collection and analysis of cores, dating of sediments, and correlation with other indicators.

Diatoms are particularly useful as indicators in estuarine systems for the same reasons that they are useful in many other aquatic habitats. Each species has specific preferences for water quality; these include pH, salinity, temperature, nutrients, light availability, currents, and substrate or habitat. Each species identified contributes information about the past environment while community statistics provide additional information.

17.3.1 Diatoms in estuaries

There are a variety of diatom habitats within an estuary. Pelagic or planktonic species float with the currents and tides, epiphytic species grow attached to higher aquatic plants or other algae, epipelic species live on the surface of muddy sediments, epilithic species grow attached to rocky surfaces, and epipsammic are attached to, or moving through, sandy sediments. Intertidal marshes have their own diatom communities inhabiting the high and low marsh (Sullivan, 1982; Hemphill-Haley, 1995). In an estuary, freshwater diatoms may be brought in with river flow, and marine species may be transported into brackish areas by tidal action, while estuarine species flourish in the productive mixing zones. Sediment samples contain all of the species that have been deposited over years of sediment accumulation. However, resuspension, sediment mixing and transport exist, and very delicate or lightly silicified species may be missing in sediments. Water samples, which contain live and dead diatom cells, can be useful in assessing the magnitude of physical processes that act in estuaries, such as erosion, transport, and degree of sediment mixing (Juggins, 1992; Wachnicka, 2009). Diatom communities in coastal estuarine and brackish water environments are also discussed in several other chapters (see Snoeijs; Trabajo and Sullivan; and Horton and Sawai, this volume).

17.3.2 Valve preservation

Although diatoms are abundant and well preserved in estuarine sediments, they are not indestructible. Both chemical and mechanical destruction can occur under certain conditions and result in a biased preservation of species. Ecological interpretations may be hindered where finely silicified indicator species are missing due to chemical dissolution, whereas fractured or eroded valves may indicate intertidal exposure and abrasion (Hemphill-Haley, 1995). Excellent preservation of valves may indicate rapid burial and little post-depositional disturbance. In studies from Chesapeake Bay, Cooper (1995a, b) found that many lightly silicified and delicate planktonic species such as those of *Chaetoceros* Ehrenberg, *Leptocylindrus* Cleve, *Cerataulina* H. Peragallo, and *Rhizosolenia* (Ehrenberg) Brightwell, known to occur in Chesapeake Bay waters, were not represented in sediment samples as whole valves. However, resting spores and pieces of valves representing these genera were seen. This was true for both surface sediments and deeper core sediment samples and therefore did not affect comparisons of species abundance or diversity between samples. In a study of eutrophication effects on biota of the Adriatic Sea, Puškaric *et al.* (1990) also found that lightly silicified diatoms such as *Chaetoceros* Ehrenb. and *Skeletonema* Greville were not preserved in core samples, although they were abundant in northern Adriatic waters. These genera were not present in surface sediments from cores, but were found in sediment trap material when incorporated in fecal pellets or other organic matrices.

Another consideration in estuarine sediment samples is the difference in silicification between freshwater and marine species. Freshwater species from fluvial inputs will generally be more highly silicified and therefore better preserved than brackish-water and marine species (Conley *et al.*, 1989). Observations on the autecology of species and preservation of frustules can provide evidence of transport and reworking of sediments.

17.3.3 Silica cycling in estuaries

A certain minimum concentration of silica in solution is essential to the growth of each species of diatom. Soluble silica is added to natural waters through deforestation and sewage effluent, along with phosphorus and nitrogen (Conley *et al.*, 2008). Dissolved silica may become the limiting nutrient during spring diatom blooms with concomitant preservation of the BSi of diatom frustules in the sediments (Conley and Malone, 1992; Dortch and Whitledge, 1992; Conley *et al.*, 2009). Silica depletion may cause major changes in species composition and ultimately affect trophic interactions as well as other ecological and biogeochemical changes (Schelske, 1999). Estuarine systems in general are characterized as sinks for soluble silica during low discharge and high primary productivity (Tréguer *et al.*, 1995).

The geochemistry of silica is important to understanding preservation of diatom valves in sediments, as well as its role in diatom growth. Most of the amorphous BSi found in the sediments of lakes and estuaries is in the form of diatom frustules, although other organisms that contribute to BSi include silicoflagellates, radiolarians, and sponges. Biogenic silica appears to dissolve at the same rate as inorganic amorphous silica and is influenced by organic coating, surface area, and aluminum (Van Cappellen *et al.*, 2002; D. J. Conley, personal communication, 2009). Chemical dissolution of diatom frustules has been found to be more prevalent in extremely alkaline or saline waters, although only minor differences were found in diatom assemblages in sodium chloride (NaCl) solutions (Barker *et al.*, 1994). Biogenic silica can be a good estimate of diatom silica in sediments, but a linear relationship between BSi and diatom abundance is not necessarily expected because of the large size range and variable silica content of different diatom species within samples. Methods for measurement of BSi in sediments are described in Conley and Schelske (2001).

17.3.4 Diatoms and coastal wetlands

Coastline bathymetry and latitude are primary determinants of the dominant plant communities in coastal wetlands, with herbaceous salt marshes lining high-latitude, low-lying coastlines, and mangroves occupying coastlines of the frost-free subtropics and tropics. Both support submersed aquatic communities in shallow fringing lagoons and embayments (Kolka and Thompson, 2006). In all of these habitats, diatoms can be found in the plankton, growing epiphytically on submersed seagrass foliage or mangrove roots or living on or in the mud, sand, silt, or rocks that form the wetland or estuary floor. The composition of each of these communities reflects surrounding physicochemical conditions, which is largely influenced by high spatial and temporal variability in salinity, nutrient supply, and hydrology (Sullivan and Currin, 2000). They therefore can provide excellent indications of the condition of wetland ecosystems as they develop over time in response to long-term coastal evolution and changes imposed by modern exploitation of coastal resources.

17.4 Methods

Although sedimentation in estuaries is dynamic and variable, depositional environments can be identified. Many of these areas occur in deeper waters and channels (e.g. original river bed channels flooded as sea level has risen), marsh or mudflat depositional areas, and quiet lagoons. Existing deep water channels may be areas of good preservation and higher sedimentation rates because of anoxic sediments with less bioturbation and sediment focusing effects. Typically these areas do not require dredging and remain primarily undisturbed. Prior to coring, channel incisement and morphology of estuarine bottom sediments can be investigated using seismic reflection survey equipment (e.g. Riggs *et al.*, 1992). With careful analysis, including such tools as X-rays of cores, sediment type analysis, and observation of pollen and/or tephra horizons (Dugmore and Newton, 1992), excellent cores can be obtained from estuarine environments to study the history of climate and anthropogenic influences over the past few thousands of years. Sediment cores are obtained in estuaries following similar methods as for lakes (Glew *et al.*, 2001).

Careful dating and determination of sedimentation rates is necessary to obtain a chronology for each core with which to correlate indicators such as diatoms, especially when comparing anthropogenic influences with climatic and pre-historical changes. Typically, sedimentation rates within estuarine systems have increased with human land use of watershed areas. Radioisotope (^{210}Pb, ^{14}C, and ^{137}Cs) methods and pollen horizons are commonly used (see Brush, 1989; Appleby, 2001).

17.4.1 Diatom methods

Methods for extracting diatoms from estuarine sediments can vary, but all include steps for eliminating unwanted material from the sediment matrix similar to those used for lake sediments (see Battarbee *et al.*, 2001).

Good comprehensive references for identification of estuarine diatoms are relatively few. Hustedt's *Kieselalgen* (1927–30), his paper on Beaufort diatoms (1955), and Krammer and Lange-Bertalot's volumes in *Süßwasserflora von Mitteleuropa* (1986–1991) are useful. The recent publications on diatoms of the Baltic Sea (Snoeijs, 1993; Snoeijs and Vilbaste, 1994; Snoeijs and Potapova, 1995; Snoeijs and Kasperovičienė, 1996; and Snoeijs and Balashova, 1998) are very helpful for many brackish-water and coastal diatoms. For additional estuarine and more marine genera, Hendey (1964), Peragallo and Peragallo (1897–1908), Tomas (1997), Witkowski *et al.* (2000), and Hein *et al.* (2008) are useful, among myriad other papers and dissertations.

References used in identification of each species should be recorded. These notes may be critical for comparisons between studies, verification of species identifications, for future work, and for much needed information on the autecology of diatom species.

17.4.2 Statistical analysis of data

For studies of diatoms in sediment samples, sample sizes of 400–500 diatom valves identified have commonly been used (e.g. Laws, 1988). For statistical purposes, a minimum of 300 diatom valves identified per sample is recommended (van Dam, 1982; Sullivan and Moncreiff, 1988; Shaffer and Sullivan, 1988). These numbers can easily be obtained from estuarine sediments, and most likely will be contained in less than 1 ml of wet sediment.

Shannon's H' has been shown to be an effective indicator of diatom population variation and is a widely used diversity index in ecology (Hendey, 1976; van Dam, 1982; Washington, 1984). Shannon's H' accounts for both the total number of species as well as the evenness of species abundance. Diversity is therefore a complement to information on diatom species autecology. A steady decline in diversity of diatom communities through time, in relation to land-use activity, was seen in four cores from Chesapeake Bay (Cooper, 1995a). Patten (1962), Stockner and Benson (1967), and Hendey (1976) all found that diatom community diversity (as measured by H') decreased with eutrophication and pollution in an estuary, a lake, and the coast of Cornwall, respectively.

Cluster analysis of diatom communities may reveal differences in distribution or major trends through time in core samples. Other multivariate statistical analyses of diatom data, together with measurements of other fossils and geochemical and physical parameters, can help elucidate what factors are most closely linked to changing diatom communities. Several methods currently in use with paleoecological studies using diatoms include principal components analysis (PCA), canonical correspondence analysis (CCA), detrended correspondence analysis (DCA), and non-metric multidimensional scaling (NMDS). An overview of these techniques is contained in Birks (1995), Moser *et al.* (1996), and Birks (this volume).

Weighted-averaging (WA) and weighted-averaging partial least squares (WA-PLS) are suggested as improved methods for reconstructing environmental variables from species assemblages (ter Braak *et al.*, 1993), and have been used successfully in developing a salinity transfer function for the Thames Estuary (Juggins, 1992). The first diatom-based transfer functions for nutrients in coastal systems were developed in the Baltic Sea area (Clarke *et al.*, 2003; Weckström *et al.*, 2004), and transfer functions for Florida Bay and Biscayne Bay are also available for salinity, water total nitrogen (WTN), water total phosphorus (WTP), and water total organic carbon (WTOC) (Wachnicka, 2009).

Artificial neural networks (ANN) techniques, which have successfully been used in foraminifera-based paleoceanographic studies to infer sea-surface temperature (Malmgren and Nordlund 1997; Malmgren *et al.*, 2001; Kucera *et al.*, 2005), have lately been applied to diatom data to develop pH and TP inference models (Racca et al., 2001; Köster et al., 2004; Racca et al., 2004).

17.4.3 Calibration sets

Much progress has been made in development of quantitative approaches to infer environmental variables from diatom assemblages. One of the most widely used tools is the surface-sediment calibration set in paleolimnological studies (Smol, 2008). Surface sediment calibration sets of diatom species in estuaries have historically identified specific physical habitats or salinity and temperature responses of diatoms rather than other chemical and nutrient preferences (e.g. McIntire and Overton, 1971; Wilderman, 1986; Laws, 1988; Kuylenstierna, 1990; Juggins, 1992; Jiang, 1996). Examples of a few studies which have analyzed anthropogenic effects in estuaries include research by Wendker (1990), who used surface-sediment calibrations of physical and chemical (including some nutrient) parameters in the Schlei Estuary (Germany), and Kennett and Hargraves (1984, 1985), who investigated seasonal differences in benthic diatoms in a Rhode Island estuary (USA) due to stratification and chemical changes with hypoxic and anoxic water conditions.

The use of a diatom nutrient calibration set for coastal systems is explored in a paper by Anderson and Vos (1992), and recent calibration sets have been used for nutrient analysis of estuaries (Clarke et al., 2003; Weckström et al., 2004; Weckström and Juggins, 2005; Weckström, 2006; Ellegaard et al., 2006; Wachnicka, 2009). Continued attempts of within-system and cross-system calibration sets for estuarine systems is worthwhile, although each system will have its own unique signals due to differences in mixing, sedimentation, land use, geography, fresh water flow, residence time, etc. Specific indicator species may be useful in a cosmopolitan sense or may vary from system to system (e.g. Cooper, 1995c; Stachura-Suchoples, 2001; Cooper et al., 2004a; Weckström and Juggins, 2005). More studies on the autecology of estuarine and coastal diatoms are needed, particularly in tropical areas (Wachnicka 2009).

17.5 Diatom studies

Both ecological and paleoecological studies of diatoms and other algae have been used to assess human impacts on the environment, including eutrophication and pollution since at least the 1960s (e.g. Patten, 1962; Palmer, 1969; Patrick, 1973). In the 1990s, Cooper and Brush (1991, 1993) and Cooper (1995a) used diatoms from subsampled sediment cores collected in Chesapeake Bay, to show evidence of alterations in water quality and habitat availability related to land use in the watershed. These studies suggest that diatom assemblages have changed in response to increased eutrophication, sedimentation, turbidity, and freshwater flow to the estuary. Paleoecological analyses reshaped the Chesapeake Bay Program in its understanding of historical water-quality changes (Kemp et al. 2005). More recent studies that use paleoecological techniques employing diatoms to investigate the effects of eutrophication, pollution, and land use on estuarine systems include Cooper et al. (2004a), Clarke et al. (2006), Parsons et al. (2006), Weckström (2006), Köster et al. (2007), Rabalais et al. (2007b), Brandenberger et al. (2008b), Varekamp et al. (2010), and Wachnicka (2009). Several of these examples are included as case studies later in this chapter. For more discussion of studies using diatoms in brackish waters, please refer to Snoeijs, and Trabajo and Sullivan (this volume).

17.5.1 Diatoms and SAV

Diatoms often numerically dominate benthic microalgal communities in seagrass beds (Sullivan and Currin, 2000) and their productivity and composition respond to the same water-quality drivers associated with seagrass die-offs (Armitage et al., 2005; Frankovich et al., 2006). In healthy seagrass beds, epiphytic diatom assemblages have been shown to be two to three times more productive than phytoplankton or epipelic assemblages, suggesting an additional functional consequence of seagrass die-offs as well as a possible metric for tracking these losses in sediment assemblages (Moncreiff et al., 1992).

Several studies show evidence for substrate specificity (Pinckney and Micheli, 1998), suggesting that availability of different substrate types may also be recorded by sediment assemblages of diatoms. Wachnicka (2009) showed how epiphytic communities associated with seagrass beds can be differentiated from planktonic and benthic assemblages and that these differences persist along broad nutrient and salinity gradients. These findings are encouraging for paleoecological reconstruction of seagrass die-offs, as interpreting changes in seagrass abundance independently of biogeochemical drivers is desirable. Additionally, Cummins et al. (2004) showed how phytoplankton communities within seagrass beds differed from those in areas of bare substrate, proposing that changes in composition of planktonic assemblages could provide additional evidence for changes in seagrass abundance over time. Associations of different organic biomarkers with planktonic and epiphytic-dominated diatom communities in Florida Bay allowed Xu et al. (2006) to infer changes in habitat from the abundance of highly branched isoprenoids in sediments of Florida Bay.

Although it appears that the relative abundance and composition of epiphytic and planktonic diatoms could be used to track the disappearance and re-establishment of seagrass communities, paleoecological studies have often depended on other proxies such as seagrass-associated foraminifera, molluscs, and ostracods to make inferences about changing seagrass abundance (e.g. Brewster-Wingard and Ishman, 1999). This is an area for potential research development, as understanding natural fluctuations in seagrass abundance over time is important for providing a context for interpreting the severity of contemporary wide-spread die-offs.

17.5.2 Diatoms and coastal wetlands

Diatoms have been a particularly important tool in documenting coastal wetland response to the variety of natural and anthropogenic forces that shape coastlines, including changes in water source or quality, sea-level rise, salt-water encroachment, and storms (see also Trabajo and Sullivan, Horton and Sawai, and Gaiser and Rühland, this volume). Diatom analysis has been used in paleoecological studies to reconstruct salinity changes in estuarine and coastal systems, based on a number of classification schemes (e.g. Hustedt, 1953, 1957; Simonsen, 1962). An excellent review of the history of diatom analysis to reconstruct salinity can be found in Juggins (1992). Studies employing diatoms to reconstruct shoreline displacement and sea-level change such as occurs with climate change, tectonics (e.g. earthquakes), water-source delivery, or isostasy often include salinity reconstructions and habitat reconstructions (e.g. Sylvestre *et al.*, 2001; Hassan *et al.*, 2007).

17.5.2.1 Sea-level rise Coastal wetlands often provide excellent sediment records from which to recount rates of sea-level rise because they present a more stable accreting environment than neighboring estuaries and because fluctuations in connectivity to the sea can have a profound influence on the composition of depositing sediments and fossils. Sea levels affect wetland diatom communities by influencing salinity and inundation periods, and so adequate ecological data on the preferences of coastal wetland taxa are key to these reconstructions. In a coastal wetland in Spain, Bao *et al.* (2007) used salinity-preference data for diatoms provided in the comprehensive work by Vos and De Wolf (1993) to show how connections between the wetland and the sea varied over time with sea-level rise and gradual infilling. Saunders *et al.* (2007) explored the evolution of a coastal lagoon in Tasmania using a salinity transfer function from collections of 34 adjacent coastal wetlands. Diatom-inferred salinity trends showed that the lagoon was once an open, clear-water bay dominated by seagrasses and epiphytes that became hypersaline when a causeway was built in the 1950s that blocked the flow of freshwater and created a stagnant, phytoplankton-dominated pond.

In some cases, diatom-based interpretations of the rate of sea-level rise have been based on measurements of elevation preferences of diatoms, reflecting both tolerance to salinity and subaerial exposure. Horton *et al.* (2006) showed how polyhaline, mesohaline, and oligohaline taxa sort along an elevation gradient along the Outer Banks of North Carolina. The resultant inference model had an $R^2 = 0.81$ and a very small error of only 0.08 m, so it was applied to a core from a back-barrier marsh to reveal a 3.7 mm yr^{-1} rate of sea-level rise. In a later study, Horton *et al.* (2007) examined the zonation of diatoms along two transects through mangrove marshes in Sulawesi, Indonesia, and found a similarly strong effect of elevation on diatoms there ($R^2 = 0.84$, RMSE = 0.16 m). Similar models exist for reconstructing sea-level rise from coastal wetland diatoms of eastern Japan (Sawai *et al.*, 2004) and the UK (Gehrels *et al.*, 2001).

17.5.2.2 Salt-water encroachment In some cases, sea-level rise is accompanied by decreased freshwater delivery toward coasts, exacerbating the rate of salt-water encroachment. Nowhere is this more pronounced than in Australia, where salinization is considered the most significant modern environmental threat in the country (Tibby *et al.*, 2007). Gell *et al.* (2005) show that diatoms are key indicators of salinization and that they can be used in biomonitoring of salt in lakes and paleosalinity studies (Gasse *et al.*, 1995). Salinization of inland and coastal waters has encouraged widespread proliferation of salt-tolerant taxa (Gell *et al.*, 2007). Tibby *et al.* (2007) developed a specific conductivity transfer function for diatoms from eight wetlands along the Murray River on the Australian coast where intensive water withdrawal for horticultural irrigation causes conductivity to vary greatly. In the same estuary, Fluin *et al.* (2007) showed how drainage blockades built to retain freshwater to the interior caused diatoms to shift from a mesohaline community to one dominated by oligohaline littoral taxa, while that of the coastal lagoon shifted to tidal subsaline taxa. Working in a coastal floodplain swamp in eastern Australia, Taffs *et al.* (2008) showed how agricultural drainage caused not only increased salinities but a release of sulfates accumulated in the wetland that reduced the diatom-inferred pH to 2–3.

Clearly, Australia, partly due to its natural aridity, is suffering widespread consequences of freshwater overexploitation. However, these problems are not isolated to Australia. In North America, salt-water encroachment from reduced coast-ward freshwater flow is threatening coastal communities from New England to Florida. In the Everglades, diatoms have been used to track rates of salt-water encroachment that exceed 10 lateral meters per year (Ross *et al.*, 2001).

17.5.2.3 Storm disturbance Coastal wetland diatom communities can reflect climate in response to long-term changes in sea level but also to shorter-term climate events that impact coastlines. Zong *et al.* (2006) reconstructed the Holocene monsoon history of the Pearl River Estuary in China. Ratios of polyhaline, mesohaline, to oligohaline taxa were used to indicate the degree of freshwater delivery to the estuary system; a peak in mid-Holocene abundance of oligohalobous taxa contributed to a growing body of evidence for increased monsoonal activity during that time. Ryves *et al.* (2004) determined nutrient and salinity preferences for diatoms of brackish lakes and fjords in Jutland, Denmark, and applied them to a marsh sediment record to show increased pulses of freshwater delivery during historically recorded North Sea storms in the sixteenth and seventeenth centuries.

Storm surges associated with hurricanes can also leave a record in coastal wetland sediments. Parsons (1998) found a mixed community of freshwater, brackish, and marine diatoms in the mud layer deposited from Hurricane Andrew in a Mississippi estuary; these accumulated muds caused the lagoon to transition to a salt marsh dominated by epiphytic diatoms. In a study of mangrove swamps in French Guiana, Sylvestre *et al.* (2004) showed that, while mangrove epiphyte communities are fairly stable in time, marine planktonic taxa are deposited during high tides on top of fluid mudflats along the coast that normally contain *Gyrosigma* Hassall-dominated epipelic communities. Clearly, coastal wetland diatom communities can be used to reflect salinity variability on a variety of temporal scales.

17.5.2.4 Eutrophication of coastal wetlands Although diatom-based models for water-quality reconstruction are fairly well developed for coastal rivers and estuaries, few studies have focused on diatom response to altered water quality in coastal wetlands. In the coastal Everglades where plant communities in the estuary ecotone are being altered by changes in water management for developing Florida cities, diatoms appear to be a useful tool in tracking the degree to which hydrology or nutrients are the cause for widespread ecosystem alterations (Gaiser, 2008). Similarly, industrial development along the coasts of the Baltic Sea has caused coastal eutrophication, which can now be measured using well-defined diatom–nutrient transfer functions (Weckström and Juggins, 2005). Diatom-based studies of eutrophication appear more common in interior wetlands than coastal, and these are reviewed in detail in Gaiser and Rühland (this volume).

17.5.3 Estuarine paleoenvironmental case studies from North America

17.5.3.1 Pamlico and Neuse estuaries The Pamlico Sound is the largest sound formed behind barrier beaches along the Atlantic coast and the second largest estuarine system in the USA. Two of the major tributaries that contribute to this system are the Pamlico and Neuse river estuaries in eastern North Carolina, both drowned river valley estuaries. Pamlico Sound is connected to the Atlantic Ocean through relatively small openings in the Outer Banks. This limited access results in dampened ocean tides of less than 6 cm. Wind-driven tides often dominate. Residence time for water in the Pamlico averages about 24 days. Neuse residence time is on the order of two to six months. Both estuaries are relatively shallow. The salinity of the overlying water is variable, from < 1–20 parts per thousand (ppt) in the Pamlico and 1 to 33 ppt in the Neuse. A large phosphate mining operation on the south shore of the Pamlico River began effluent discharges to the estuary in 1964. There are no published data on phytoplankton-related studies in the Pamlico before 1966. The Cherry Point Marine Air Station is located on the south shore of the Neuse River. Generally, nutrient loading to these two estuaries is considered quite high (Quinn *et al.*, 1989), and P loadings in the Pamlico are especially high compared to other US river basins (Stanley, 1992). The diatom data from these sediment cores confirm these assessments.

Sediment cores were collected and dated using radioisotopes and pollen horizons for a study of water-quality history in the Pamlico and Neuse (Cooper *et al.*, 2004a). The sediments span a time period of over 500 years. Principal water-quality parameters investigated include eutrophication, trace-metal flux, sedimentation rates, diatom assemblage changes, and benthic vs. pelagic habitat quality. Geochemical parameters measured include total organic carbon (TOC), nitrogen (N), phosphorus (P), sulfur (S), iron (Fe), degree of pyritization of iron (DOP) and BSi, as well as trace metals. The primary organisms studied were the diatoms. Significant responses of diatom assemblages to environmental change show which historical periods and events have most shaped water-quality changes in these systems.

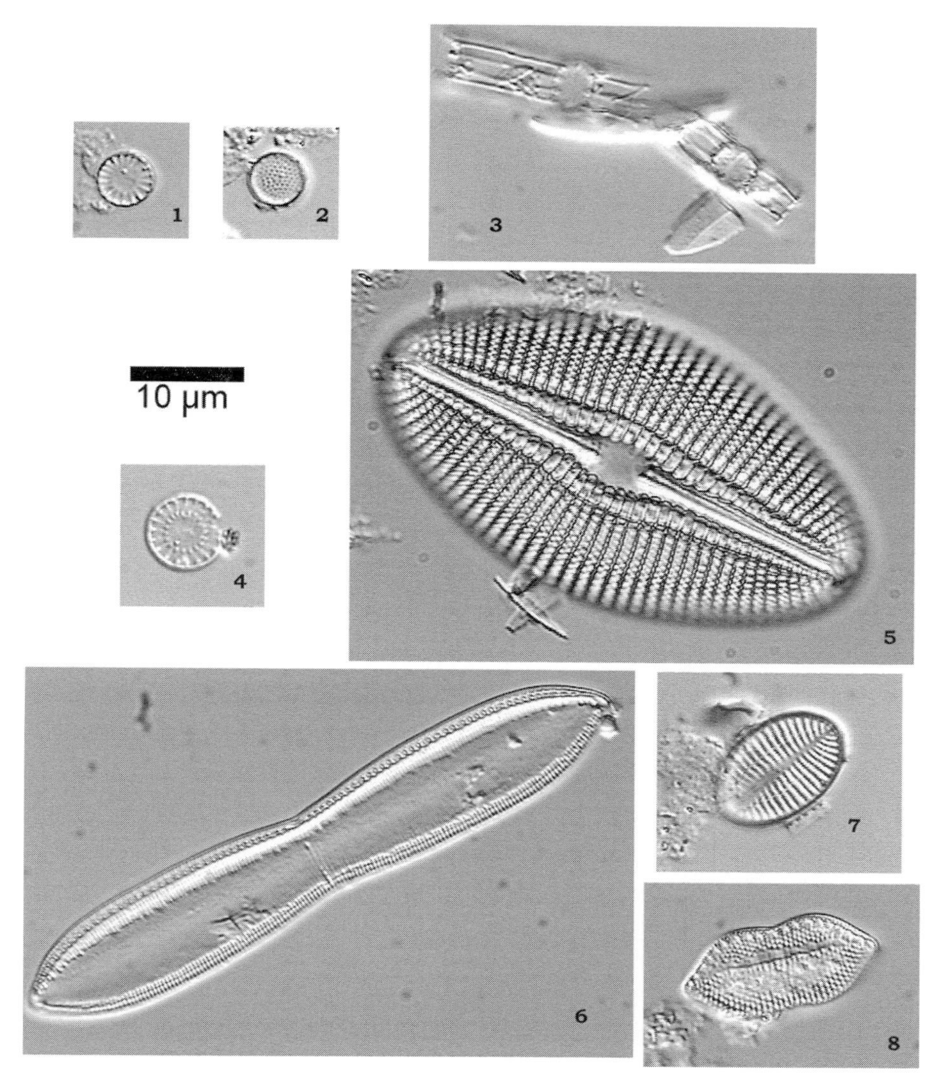

Figure 17.1 Common diatom taxa found in estuarine sediment cores collected in the Pamlico and Neuse River estuaries of North Carolina, USA. Images 1–4 are from surface sediments of the Neuse and Pamlico river cores and consist of planktonic taxa: (1) *Cyclotella atomus* Hustedt, (2) *Thalassiosira proschkinae*, (3) *Skeletonema costatum* (Greville) Cleve, and (4) *Cyclotella meneghiniana*. Images 5–8 are benthic and epiphytic taxa from deeper sediments of the Neuse and Pamlico river cores dated before European settlement of the area (pre-AD 1720): (5) *Diploneis smithii* (Brébisson) Cleve, (6) *Nitzschia marginulata* Grunow, (7) *Cocconeis peltoides* Hustedt, and (8) *Nitzschia* cf. *panduriformis delicatula* Grunow.

A chronology was developed for each core, with calculated sedimentation rates used for determining the flux of geochemical and biotic indicators. Diatoms were well preserved in the sediments, with approximately one to five million valves found in each ml of wet sediment. Over 430 taxa were identified from 49 subsamples in six cores. Diatom assemblages found in sediments deposited before the 1940s in both estuaries were composed of many pennate benthic and epiphytic species and cluster together by site. Diatom assemblages in more recent sediments have become increasingly composed of small planktonic forms, species that are often found in large blooms in higher nutrient waters (Figure 17.1 and 17.2).

Changes in diversity (Shannon's H') show a declining trend in all cores analyzed, and the opposing increase in centric:pennate (c:p) ratios indicates a shift from predominantly benthic to predominantly planktonic communities (Figure 17.2). Significant increases in diatom abundance and BSi were found through time. The increase in the c:p ratio, diatom flux, and assemblage changes are interpreted to reflect eutrophication as well as increased turbidity and sedimentation since the 1940s. Similar changes in taxonomic composition of assemblages over time were observed in both estuaries. Small planktonic diatoms such as *Thalassiosira proschkinae* Makarova showed increasing abundance through time (as high as 44.1% abundance in surface

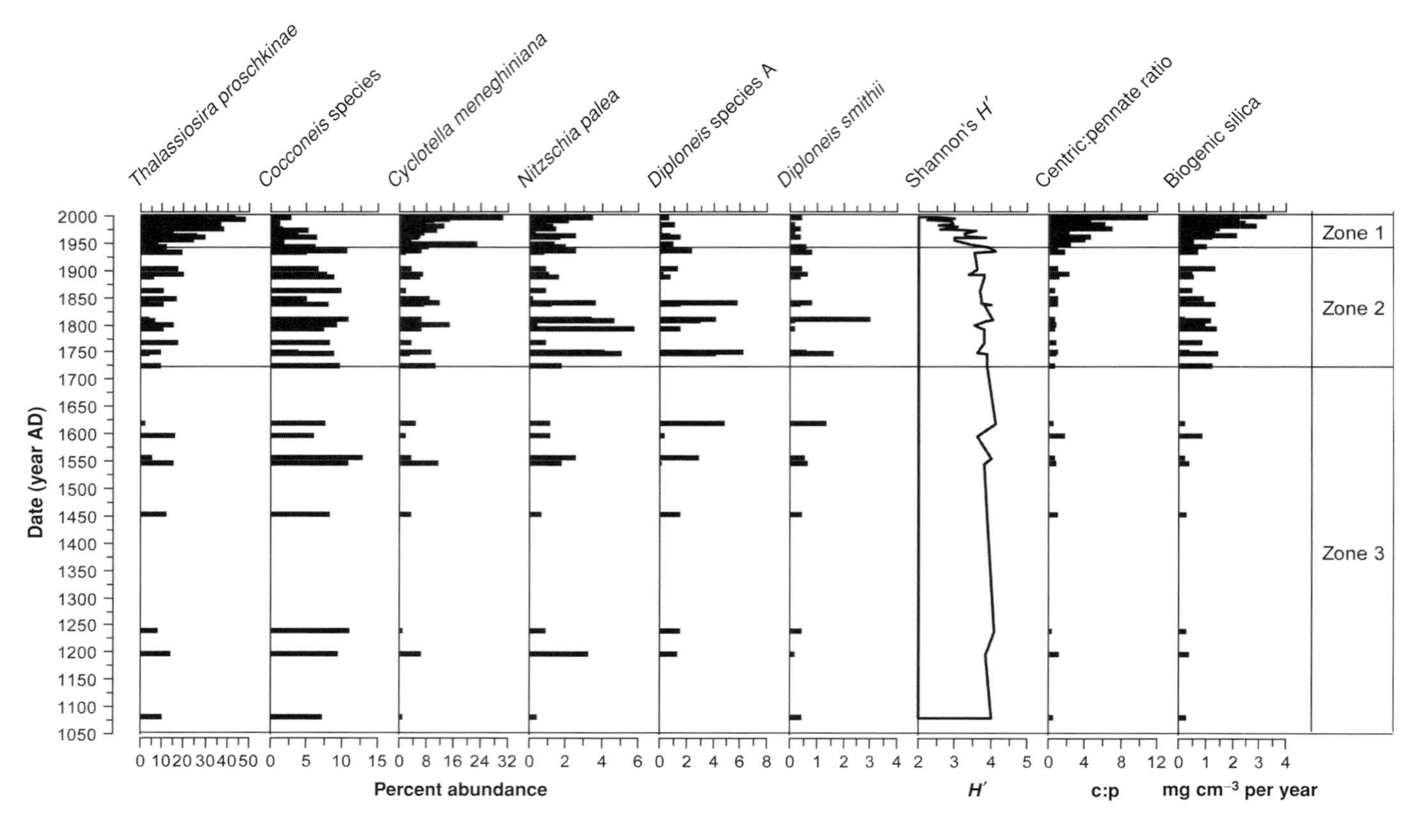

Figure 17.2 Stratigraphic diagram of selected diatom taxa as percent abundance found in all samples analyzed from six sediment cores collected in the Neuse and Pamlico river estuaries, North Carolina, USA. The diagram also shows diversity as Shannon's H', the c:p ratio for all samples and BSi in mg cm^{-3} per year. Dates were assigned to samples according to a combination of radioisotope data (^{210}Pb and ^{14}C) and pollen horizons. The zones identified include: Zone 1, recent influence of industrialization and fertilizer use (begin date 1940); Zone 2, European land clearance affecting *Ambrosia* horizon (begin date AD 1720) through to the industrial influence period; and Zone 3, pre-European influence to land and water.

sediments of Pamlico estuary), while epiphytic diatom taxa, including species in the genus *Cocconeis* Ehrenberg, declined (Figures 17.1 and 17.2). *T. proschkinae* is reported as a truly brackish water species (Muylaert and Sabbe, 1996) and has been reported in high nutrient waters from Hong Kong to the southern Baltic Sea. The reports of large blooms in Hong Kong are from eutrophic coastal lagoons with very low tidal flushing (Lee and Liu, 2002). In the Baltic Sea, *T. proschkinae* is a common taxon and thrives well in turbid waters with varying, but generally quite high, nutrient concentrations, especially high P levels (Stachura-Suchoples, 2001; Weckström and Juggins, 2005).

Cyclotella meneghiniana Kützing, a small centric diatom known to occur in lower-salinity water with high nutrient content (Weckström and Juggins, 2005) also showed a clear trend of increasing abundance with time in the Pamlico sediment cores. Recent abundances of *C. meneghiniana* in the Pamlico were as high as 31% at the most upriver site, but were never higher than 15.3% in the Neuse cores. This variation was most likely due to salinity differences at the core collection sites (Figures 17.1 and 17.2).

Principal components analysis was performed on the diatom data to summarize major trends in assemblages, as well as the influence of environmental variables. All taxa present with at least 1% abundance were included, reducing the diatom list from more than 430 to 84 taxa (Cooper *et al.*, 2004a). Results indicate that both estuaries show similar changes in diatom assemblages through time. High surface-sediment concentrations of arsenic (As), chromium (Cr), copper (Cu), nickel (Ni), and lead (Pb) found in the Neuse River may be associated with industrial and military operations. Cadmium (Cd) and silver (Ag) levels in the Pamlico most likely result from phosphate-mining effluent discharges. The Pb profiles from the core samples showed correlation with leaded-gasoline use patterns through time. The high levels of trace metals in the surface sediments of these estuaries confirm previous research on metal deposition and provide a temporal framework for these changes (Riggs *et al.*, 1989, 1991).

Historically, land use in the Albemarle–Pamlico region includes agriculture but mostly forest. Forests recently comprised approximately 47% of the total basin area, with about 36% in agricultural crops (Bricker *et al.*, 2007). The urbanized percentage of the basin is low at 4.9–9.5% (Bricker *et al.*, 2007), but growing. During the mid 1980s, animal culture surged past crop production in North Carolina (primarily in eastern North Carolina) to reverse the historic relationship of these two sectors. Hog production in particular expanded rapidly between 1991 and 1995 with an average annual growth rate of nearly 30% (Blue Ribbon Study Commission on Agricultural Waste, 1996).

The results of this paleoecological study show that many water-quality changes have occurred very recently in these North Carolina estuaries, although the overall trends are similar to those measured over a longer time frame in Chesapeake Bay. The primary differences in diatom assemblages appear to be due to differences in geomorphology (depth, salinity, and tidal flushing), high P loads, land-use history, and total human population. As research and management actions aimed at increasing water quality of the estuaries continue, knowledge of water-quality changes related to human activities in the past is essential.

17.5.3.2 Long Island Sound Long Island Sound (LIS) is a large estuary adjacent to major metropolitan areas in New York and Connecticut on the east coast of the USA, with a very developed watershed. The sound has experienced deterioration in water quality and contamination of its sediment in the past centuries due to increased discharge of effluents from city wastewater treatment plants, increased industrial release of waste products in the waters, and increased atmospheric deposition of pollutants from point and non-point sources. Specifically, an apparent increase in the duration, frequency, and severity of hypoxia and anoxia, loss of submerged aquatic vegetation, changes in the benthic fauna, and a severe lobster die-off, are matters of more recent environmental concern (Balcom and Howell, 2006; Stacey, 2007). Nutrient loadings (particularly of N) in the LIS basin have been cited as one of the most significant causes of recent deterioration of the water quality and potential ecological changes (Bricker *et al.*, 2007). In addition to these local anthropogenic impacts, there have been changes in regional and global climate, including the precipitation pattern, mean annual temperature, and seasonal temperature fluctuations (Varekamp *et al.*, 2004).

Sediment cores were collected in LIS by the US Geological Survey using a gravity corer (Buchholtz ten Brink *et al.*, 2000) and dated using radioisotopes as well as pollen horizons for ragweed (1650) and chestnut blight (1910). Paleoindicators investigated included diatoms, foraminifera, stable isotopes (^{18}O and ^{13}C), and trace-element (Mg/Ca) studies on tests of *Elphidium excavatum* Terquem, organic C and N, ^{13}C and ^{15}N, S, BSi, and heavy-metal concentrations (Thomas *et al.*, 2004a, b).

Analysis of diatom assemblage changes over the past 600 years in two cores from western LIS show shifts in diversity (declining), c:p ratios (increasing), and variable abundance, with one core showing declining abundance after 1980 (Cooper *et al.*, 2004b). The declining abundance of diatoms is matched by a decline in BSi. Most significant changes occur after AD 1800, with the lowest diatom diversity (Shannon's H′) and highest c:p ratios in the 1850s and 1970s. These dates coincide with BSi and organic C signatures indicating increased eutrophication of western LIS (Cooper *et al.*, 2004b; Varekamp *et al.*, 2010). The BSi accumulation rates are low in pre-colonial times (~0.5%), increasing to an overall high by *c.* 1970 (~3.5%), then showing a precipitous drop. Ecological studies and foraminifera results suggest that a switch from diatom dominance to dinoflagellates may have occurred in recent decades due to extreme eutrophication and silica depletion (Capriulo *et al.*, 2002; Thomas *et al.*, 2004b). Diatom analysis also shows a period of high *Pseudo-nitzschia* H. Peragallo abundance in the stratigraphic record *c.* AD 1940. These species are capable of the production of the neurotoxin domoic acid (Trainer *et al.* 2000).

Average water temperature within LIS has risen in recent decades, exacerbating the effects of stratification of the water column, hypoxia, and eutrophication. These findings have been important for understanding the history of anoxia in this region, and the historical and current environmental stressors on LIS fisheries such as the lobster. Aggressive sewage-treatment-plant N control appears to have lowered the point-source loads of N to LIS since the 1990s (Stacey, 2007).

17.5.3.3 Puget Sound and Hood Canal Hood Canal is a fjord-like sub-basin of Puget Sound, Washington State, on the west coast of the continental USA. Dissolved oxygen levels in Hood Canal have been declining, especially in the southern reaches, and nitrogen enrichment is implicated, although complex physical processes of weather, flushing, and flow are likely involved (Newton, 2007). Besides anthropogenic nutrient inputs, these areas of the Pacific northwest also receive nutrient-rich waters from coastal upwelling. Frequent fish kills and recent hypoxic conditions prompted formation of the Washington State Hood Canal Dissolved Oxygen Program. Numerous studies, including monitoring and modeling efforts, are investigating both Hood Canal and Puget Sound to develop management strategies for these coastal waters. One paleoecological study is providing a historical reconstruction of ecological indicators spanning more than 300 years in both Hood Canal and central Puget Sound (Brandenberger *et al.*, 2008b). These records are valuable additions to environmental monitoring, and can provide evidence for linkages to natural and anthropogenic factors.

The paleoecological tracers used to reconstruct records in Puget Sound and Hood Canal include diatom and foraminifera fossils, pollen, stable isotopes of C and N, biomarkers for terrestrial and marine C, and RSMs that are enriched in sediment during periods of hypoxia and anoxia. The sediments were dated using radiometric methods (Brandenberger *et al.*, 2008a, b). The diatom assemblages identified in these sediment cores include a predominance of planktonic marine and brackish water forms with some benthic and epiphytic taxa. Contrary to expected findings, diatom abundance appears to be declining in recent sediments, since around 1900. The Puget Sound region is generally not considered to be nutrient or silica limiting. The BSi concentrations measured (7.5–15.6% SiO_2 by weight) were higher than the concentrations found in the Gulf of Mexico (Rabalais *et al.*, 2007b) and the average 6–8.4% BSi found in Chesapeake Bay (Cooper, 1995a; Zimmerman and Canuel, 2000). The BSi data for the central Puget Sound basin (core PS-1) and the northern Hood Canal (core HC-5) were similar in magnitude and also illustrate a sharp decrease in BSi in the early 1900s ($\sim$9.9%) with higher concentrations of 12–13% in more recent sediments and highest levels of 13–14% in those dated pre-1900s. Other markers of productivity in the water column show similar trends, and these findings are considered to be significant as they contradict hypotheses that trends similar to Chesapeake Bay or the Gulf of Mexico should be evident (Brandenberger *et al.*, 2008b).

One conclusion from these studies is that anthropogenic nutrient inputs are not contributing to hypoxia and anoxia in Puget Sound and Hood Canal to the same extent as they have been shown to do in other coastal estuaries (e.g. Zimmerman and Canuel, 2002; Rabalais *et al.*, 2007a). Puget Sound and Hood Canal sediments do record changes in indicators of land-use change, deforestation, and urbanization, but are not all clearly linked to nutrient changes.

The record of RSMs found in these Puget Sound and Hood Canal cores indicates a century-scale oscillation in hypoxia with periods of persistent hypoxia recorded *c.* the 1700s, 1800s, and 1900s. More oxygenated conditions are recorded around the middle of each century. The cyclicity of hypoxia recorded in these cores spans four centuries with a higher frequency of hypoxic events recorded during warmer climate periods coinciding with a positive Pacific decadal oscillation. These observations suggest that Puget Sound is controlled by natural processes of ventilation (physical mixing) that may buffer or accelerate changes in oxygen levels in deep waters (Brandenberger *et al.*, 2008b). The data collected from these sediment cores cannot clearly define the increasing hypoxia trends reported by Newton (2007). More work is needed to elucidate these most recent time periods.

The diatom assemblage data may indicate more eutrophic conditions in the last several decades, as there are increasing relative abundances in species generally associated with excess anthropogenic nutrient inputs (Brandenberger *et al.*, 2008b). Examples include species in the genus *Pseudo-nitzschia* capable of producing the toxin domoic acid, and the small planktonic *Thalassiosira proschkinae*. Parsons *et al.* (2002) and Anderson *et al.* (2002) show evidence of increasing abundance of *Pseudo-nitzschia* similar to the levels found in Puget Sound with anthropogenic nutrient inputs to coastal waters of the Gulf of Mexico and elsewhere.

This study provides new information supporting the need to evaluate the status of an estuarine ecosystem with respect to hypoxia on longer timescales than available monitoring efforts. The reconstructions within the scope of this project provide a measure of the natural cyclicity within these estuaries, and the response of the paleoindicators to documented anthropogenic impacts. The results of this study may indicate fundamentally different management approaches than those being designed for coastal-plain estuaries that show hypoxia trends more correlated to anthropogenic eutrophication.

17.5.3.4 Merrymeeting Bay and Kennebec Estuary Merrymeeting Bay (MMB) and Kennebec Estuary are located in mid-coast Maine, USA. The former is a relatively shallow tidal basin formed at the head of the Kennebec Estuary by the confluence of the Kennebec, Androscoggin, Cathance, Abbagadasett, Muddy, and Eastern rivers that drain one fourth the area of Maine and parts of new Hampshire (Köster *et al.*, 2007). It exits through a narrow channel to the Kennebec Estuary, but remains primarily below 5 ppt salinity (Kistner and Pettigrew, 2001). Sediment cores were collected at two sites in MMB, and dated using radioisotopes as well as pollen horizons, spanning a chronology of about 1800 years (Köster *et al.*, 2007). Paleoindicators investigated included diatoms, pollen, total C,

N, P, ^{13}C and ^{15}N, BSi, and trace metals. Similar indicators are being used on sediment cores collected downstream in the Kennebec, but this study is not complete.

Diatom analysis identified four zones throughout the MMB cores that were characterized by differences in species composition and abundance. The results indicate changes in salinity, nutrient availability, and water clarity over time. Evidence supports the conclusion that colonial land clearance in the eighteenth century began a trend of higher sedimentation and eutrophication that intensified through the nineteenth and twentieth centuries. Indicators suggest that aquatic productivity was originally very nutrient limited in MMB, unlike Puget Sound (Brandenberger *et al.* 2008b). Diatom ratios of planktonic (including *Cyclotella meneghiniana*) to benthic species increased dramatically after 1900, along with the appearance of pollution-tolerant species such as *Navicula gregaria* Donkin, *Nitzschia palea* (Kützing) W. Smith, *Nitzschia tubicola* Grunow and *Stephanodiscus* Ehrenberg spp. (Köster *et al.*, 2007). Although benthic diatom productivity increased with primary wastewater treatment improvements following the Clean Water Act of 1972, planktonic to benthic ratios remain high and SAV has not recovered. Other geochemical parameters continue to indicate eutrophic conditions, and conform with the historical studies of Chesapeake Bay (Cooper and Brush, 1993; Zimmerman and Canuel, 2002), Long Island Sound (Thomas *et al.*, 2004b; Varekamp *et al.*, 2010), Pamlico and Neuse estuaries (Cooper *et al.*, 2004a), as well as New Bedford Harbor (Chmura *et al.*, 2004).

This mid-coast Maine estuarine system is one of the least impacted by non-point-source nutrients in New England because of the large forested watershed with little agriculture and few urban areas (Boyer *et al.*, 2002). Management could be relatively straightforward in terms of tertiary sewage treatment and point-source reduction of nutrients from paper mills and other industry, especially when compared to estuarine systems with higher non-point-source nutrient inputs (Cooper, 1995a; Cooper *et al.*, 2004a; Rabalais *et al.*, 2007b).

17.5.3.5 Florida Bay and Biscayne Bay The coastal wetlands and estuaries of South Florida have received significant paleoecological attention in attempts to (1) track rates of sea-level rise and its effects on the vast, low-lying ecosystems of the region (e.g. Ross *et al.*, 2001; Gaiser *et al.*, 2005) and (2) determine predevelopment target conditions for restoring this highly hydrologically modified system (e.g. Slate and Stevenson, 2000; Willard *et al.*, 2001; Cooper *et al.*, 2008). While freshwater wetlands in this system have depauperate diatom records, cores taken in the shallow marine estuaries of Florida Bay and Biscayne Bay contain diatom-rich sediments that offer a promising tool for reconstructing past salinity, nutrient availability, and other physical and biological parameters. However, diatom-based paleoecological applications in South Florida have been challenging due to the limited taxonomic attention given to the coastal Caribbean diatom flora and the paucity of autecological data available for benthic marine species. In lieu of diatom proxies, the majority of paleoenvironmental investigations conducted in this region have relied on foraminifera, ostracods, mollusks, stable isotopes, and organic biomarkers (Swart *et al.*, 1996; Brewster-Wingard and Ishman, 1999; Cronin *et al.*, 2001; Gaiser *et al.*, 2006; Xu *et al.*, 2006) to study salinity patterns and influx of freshwater. Salinity changes have been driven by the last century of anthropogenic modifications of the South Florida environment, and long-term climate-driven changes, such as sea-level rise.

Pyle *et al.* (1998) conducted the first diatom-based paleoecological study in South Florida, analyzing fossil assemblages preserved in a 70-cm-long sediment core collected near Pass Key in Florida Bay. They used the limited autecological information available in the literature to infer past salinity and potential productivity. Subsequent studies of two chronologically calibrated sediment cores extracted from Russell Bank and near Pass Key by Huvane and Cooper (2001) divided the most common species based on their life forms as described in the literature, and analyzed their abundance down the cores to describe past salinity and nutrient levels. In order to reinforce their findings, the authors also investigated changes in species diversity and ratio of centric to pennate valves in these cores. Both studies concluded that salinity fluctuated during the last 200 years in Florida Bay, and that there was a clear trend towards higher salinity over approximately the last 40 years.

Realizing the potential of diatom proxies to resolve temporal trends and their current spatial variability in this dynamic region, Wachnicka (2009) conducted a comprehensive calibration survey, sampling diatoms from 96 sites across Florida Bay, Biscayne Bay, adjacent coastal mangroves, and the freshwater Everglades marshes. Diatoms were collected from sediments, the water column, and seagrass leaves during two seasons, and related to a suite of physicochemical and biological measurements that characterized each site. A diverse benthic flora of over 400 species, dominated by *Mastogloia* Thwaites and *Amphora* Ehrenberg, were identified. Descriptions for over 20% of the species could not be found in the literature or collections, reinforcing the need for additional taxonomic attention in this region. In the single genus *Amphora*, 19 new species and 2 varieties were described (Wachnicka and Gaiser, 2007).

The calibration study found significant spatial and temporal differences between diatom communities occupying different habitats (freshwater, mangrove, estuarine) in both bays, and a clear separation between epiphytic, epipelic, and planktonic assemblages, especially in Biscayne Bay. For each assemblage, salinity, WTN, WTP, and WTOC were found to be correlated with assemblage composition, so optima and tolerances were determined for each taxon for each of these variables. This resulted in a series of WA and WA-PLS regression models that predicted salinity, WTN, WTP, and WTOC with $r^2 = 0.97$, 0.75, 0.75, and 0.79 in Florida Bay and $r^2 = 0.91$, 0.78, 0.76, and 0.83 in Biscayne Bay, respectively. These predictive models were used to calibrate diatom records environmentally in seven chronologically calibrated sediment cores collected in Florida Bay and Biscayne Bay. Moreover, discriminant function (DF) analysis was applied to the cores to predict habitat (freshwater, mangrove, estuarine) and life habit (plankton, epipelon, epiphyton) changes. These analyses revealed that there was a shift in diatom assemblage types in Biscayne Bay after mid 1900 (Figure 17.3), followed by changes in salinity and nutrient levels, which corresponded to the time of construction of major canals and levees in the region (Wachnicka, 2009).

In eastern Florida Bay, Wachnicka (2009) shows distinct salinity and nutrient fluctuations that correspond to wetter and drier years and hurricanes, which determine the amount of freshwater delivery to the bay from the mainland and through precipitation. Basal sediments in some of the cores (some as old as 3000 years) contained communities corresponding to contemporary freshwater Everglades marshes, which are overlaid by brackish and marine diatom assemblages, showing marine transgression related to rising sea levels. The DF analysis showed persistent fluctuation between macrophyte- and plankton-dominated states in both bays. Signals of increased magnitude of salinity and nutrient fluctuations in the last century reflect water-flow modifications that took place in South Florida in the early 1900s that influenced the quantity and quality of water entering Florida and Biscayne bays. These changes are superimposed on the long-term changes imposed by sea-level rise.

These studies provide valuable information about water quality and environmental conditions present in Florida Bay and Biscayne Bay before and after major human-introduced changes occurred on the mainland. This information will help federal and state agencies responsible for the Comprehensive

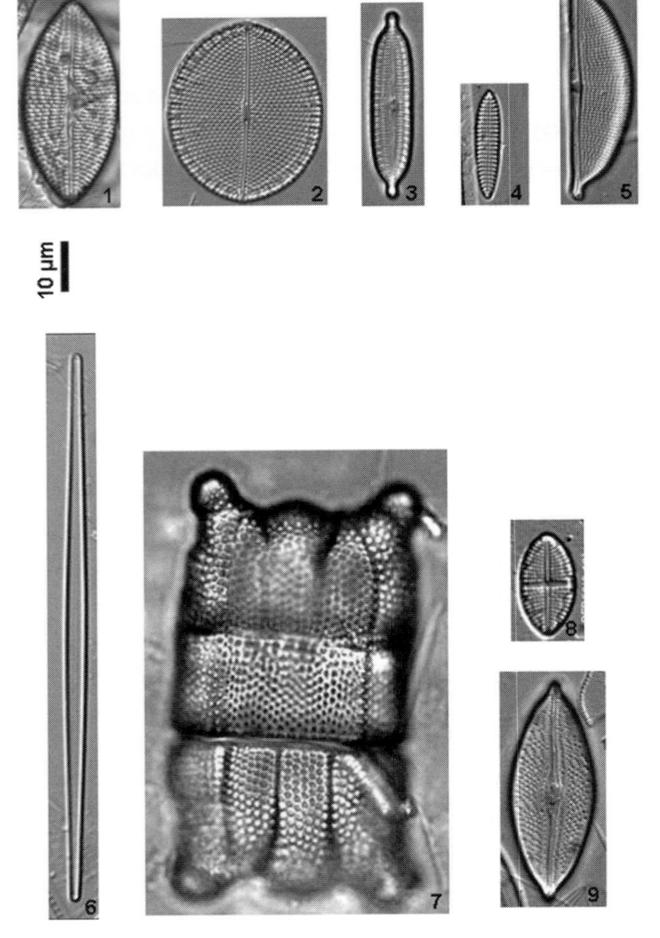

Figure 17.3 Dominant diatom taxa recorded in sediment cores collected in Biscayne Bay, Florida, USA. Images 1–5 represent taxa from sediments in the cores dated prior to major twentieth century anthropogenic changes in South Florida: (1) *Mastogloia bahamensis* Cleve, (2) *Mastogloia cribrosa* Grunow, (3) *Mastogloia corsicana* Grunow, (4) *Dimeregramma dubium* Grunow, and (5) *Amphora corpulenta* var. *capitata* Tempère & Peragallo. Images 6–9 represent taxa from younger sediments of the Biscayne Bay cores dated after major developments of water management structures in South Florida (post mid 1900s): (6) *Hyalosynedra laevigata* (Grunow) Williams & Round, (7) *Biddulphia pulchella* Gray, (8) *Mastogloia crucicula* Grunow (Cleve), and (9) *Mastogloia punctifera* Brun.

Everglades Restoration Program, which aims to increase freshwater flow into the Everglades and ultimately Florida Bay. This program also aims to increase the amount of freshwater flow into Biscayne Bay by restoring the historic overland flow through coastal wetlands, a strategy for restoring these unique ecosystems to their natural, pre-developmental states. These investigations provide important information about how communities may respond to recent sea-level changes and to the hydrologic changes expected if restorative policies do not take place.

17.6 Summary

Diatoms are particularly useful as paleoecological indicators because of their species-specific requirements for water quality and physical conditions. Each identified species adds clues to the nature of the past environment in which they lived. In addition, statistical views of diatom communities show trends in water quality and shifts in habitat availability over time. In estuarine systems, diatom fossils have been used to elucidate anthropogenic effects by determining changes in salinity, nutrient availability (e.g. eutrophication), habitat availability (e.g. benthic vs. pelagic), turbidity, sedimentation, anoxia and hypoxia, and organic pollution. These changes have in turn been related to changing land-use patterns of watersheds, and can be compared to research and monitoring results to aid management and policy. With increasing knowledge of the autecology of estuarine diatom species and their nutrient responses, the most pressing issues in estuarine ecology today – research, modeling, and management related to eutrophication and global warming – may be more pro-active and exact.

Paleoecological studies employing diatoms require the use of other paleoecological techniques (including site selection, core collection, dating of sediments, and determination of sedimentation rates), as well as expertise in diatom taxonomy. Other indicators that can be measured in sediments further strengthen these studies. Multivariate statistical techniques can be used to correlate these paleoecological indicators and allow interpretations of causes, and therefore consequences, of human impacts on estuaries.

Paleoecological techniques have been under-utilized in studying contemporary issues of water quality in estuarine systems around the world. These ecosystems are the focus of concern regarding detrimental anthropogenic influences, and decisions are being made that will affect management options and future research. Management decisions are often based on data collected through ongoing research and monitoring programs that enable only a brief glimpse in the life of an estuary. Estuarine ecosystems are extremely variable (in both time and space), and short-term studies are not always capable of determining trends that are applicable to policy decisions. A combination of research, monitoring, and paleoecological studies becomes a synergistic tool for discerning trends, causes, and consequences of watershed land use and water uses.

Acknowledgments

This chapter was enhanced by collaboration with: Bowdoin College (Rusack Funds), Bryn Athyn College, Duke University Wetland Center, the Florida Coastal Everglades Long Term Ecological Research Program (NSF #DBI-0620409 and #DEB-9910514), Maryland Sea Grant, Pacific Northwest National Marine Sciences Laboratory operated by Battelle, the NOAA Coastal Hypoxia Research Program (NOAA CHRP grant # NA05NOS4781203), the Water Resources Research Institute of North Carolina (Project # 70161), and Wesleyan University (LISRF grant #CWF 334-R). This chapter is contribution #112 of the NOAA Coastal Hypoxia Research Program and contribution #423 of the Southeast Environmental Research Center.

References

Anderson, D. M., Glibert, P. M., & Burkholder, J. M. (2002). Harmful algal blooms and eutrophication: nutrient sources, composition and consequences. *Estuaries*, **25**, 704–26.

Anderson, N. J. & Vos, P. (1992). Learning from the past: diatoms as palaeoecological indicators of changes in marine environments. *Netherlands Journal of Aquatic Ecology*, **26**, 19–30.

Appleby, P. G. (2001). Chronostratigraphic techniques in recent sediments. In *Tracking Environmental Change Using Lake Sediments, Volume 1: Basin Analysis, Coring and Chronological Techniques*, ed. W. M. Last & J. P. Smol, New York: Springer, pp. 171–203.

Armitage, A. R., Frankovich, T. A., Heck, K. L., & Fourqurean, J. W. (2005). Experimental nutrient enrichment causes complex changes in seagrass, microalgae, and macroalgae community structure in Florida Bay. *Estuaries*, **28**, 422–34.

Balcom, N. & Howell, P. (2006). Responding to a resource disaster: American lobsters in Long Island Sound 1999–2004. *Connecticut Sea Grant Publication*, **CTSG-06–02**.

Bao, R., Alonso, A., Delgado, C., & Pagés, J. L. (2007). Identification of the main driving mechanisms in the evolution of a small coastal wetland (Traba, Galicia, NW Spain) since its origin 5700 cal yr BP. *Palaeogeography, Palaeoclimatology, Palaeoecology*, **247**, 296–312.

Barker, P., Fontes, J.-C., Gasse, F., & Druart, J.-C. (1994). Experimental dissolution of diatom silica in concentrated salt solutions and implications for paleoenvironmental reconstruction. *Limnology and Oceanography*, **39**, 99–110.

Battarbee, R. W., Jones, V. J., Flower, R. J., *et al.* (2001). Diatoms. In *Tracking Environmental Change Using Lake Sediments, Volume 3: Terrestrial, Algal, and Siliceous Indicators*, ed. J. P. Smol, H. J. B. Birks & W. M. Last, Dordrecht: Kluwer Academic Publishers, pp. 155–202.

Bell, J. D. & Westoby, M. (1986). Abundance of macrofauna in dense seagrass is due to habitat preference, not predation. *Oecologia*, **68**, 205–9.

Birks, H. J. B. (1995). Quantitative paleoenvironmetal reconstructions. In *Statistical Modelling of Quaternary Science Data, Technical Guide*

No. 5, ed. D. Maddy & J. S. Brew, Cambridge: Quaternary Research Association, pp. 161–236.

Blue Ribbon Study Commission on Agricultural Waste (1996). Report to the 1995 General Assembly of North Carolina 1996 regular session, Raleigh, NC: North Carolina Legislative Research Commission.

Boesch, D. F. (2002). Challenges and opportunities for science in reducing nutrient over-enrichment of coastal ecosystems. *Estuaries*, **25**, 886–900.

Boström, C., Bonsdorff, E., Kangas, P., & Norkko, A. (2002). Long-term changes of a brackish-water eelgrass (*Zostera marina* L.) community indicate effects of coastal eutrophication. *Estuaries, Coastal and Shelf Science*, **55**, 795–804.

Boyer, E.W., Goodale, C. L., Jaworski, N. A., & Haworth, R.W. (2002). Anthropogenic nitrogen sources and relationships to riverine nitrogen export in the northeastern U.S.A. *Biogeochemistry*, **57**, 137–69.

Brandenberger, J. M., Crecelius, E. A., & Louchouarn, P. (2008a). Historical inputs and natural recovery rates of heavy metals and organic biomarkers in Puget Sound during the 20th century. *Environmental Science and Technology*, **42**, 6786–90.

Brandenberger, J. M., Crecelius, E. A., Louchouarn, P., *et al.* (2008b). Reconstructing trends in hypoxia using multiple paleoecological indicators recorded in sediment cores from Puget Sound, WA. National Oceanic and Atmospheric Administration, Pacific Northwest National Laboratory Report No. PNWD-4013.

Brewster-Wingard, G. L. & Ishman, S.E. (1999). Historical trends in salinity and substrate in central and northern Florida Bay: a paleoecological reconstruction using modern analogue data. *Estuaries*, **22**, 369–83.

Bricker, S., Longstaff, B., Dennison, W., *et al.* (2007). *Effects of Nutrient Enrichment in the Nations' Estuaries: A Decade of Change*. NOAA Coastal Ocean Program Decision Analysis Series No. 26. Silver Spring, MD: National Centers for Coastal Ocean Science.

Brush, G. S. (1989). Rates and patterns of estuarine sediment accumulation. *Limnology and Oceanography*, **34**, 1235–46.

Buchholtz ten Brink, M. R., Mecray, E. L., & Galvin, E. L. (2000). *Clostridium perfringens* in Long Island Sound sediments: an urban sedimentary record. *Journal of Coastal Research* **16**(3), 591–612.

Capriulo, G. M., Smith, G., Troy, R., *et al.* (2002). The planktonic food web structure of a temperate zone estuary, and its alteration due to eutrophication. *Hydrobiologia*, **475–476**, 263–333.

Chmura, G. L., Santos, A., Pospelova, V., *et al.* (2004). Response of three paleo-primary production proxy measures to development of an urban estuary. *Science of the Total Environment*, **320**, 225–43.

Clark, J. S., Carpenter, S. R., Barber, M., *et al.* (2001). Ecological forecasts: an emerging imperative. *Science*, **293**, 657–60.

Clarke, A., Juggins, S., & Conley, D. (2003). A 150-year reconstruction of the history of coastal eutrophication in Roskilde Fjord, Denmark. *Marine Pollution Bulletin*, **46**, 1615–29.

Clarke, A. L., Weckström, K., Conley, D. J., *et al.* (2006). Long-term trends in eutrophication and nutrients in the coastal zone. *Limnology and Oceanography*, **51**, 385–97.

Cloern, J. E. (2001). Our evolving conceptual model of the coastal eutrophication problem. *Marine Ecology Progress Series*, **210**, 223–53

Cohen, J. E., Small, C., Mellinger, A., *et al.* (1997). Estimates of coastal populations. *Science*, **278**, 1211–12.

Conley, D. J., Kilham, S. S., & Theriot, E. (1989). Differences in silica content between marine and freshwater diatoms. *Limnology and Oceanography*, **34**, 205–13.

Conley, D. J., Likens, G. E., Buso, D. C., *et al.* (2008). Deforestation causes increased dissolved silicate losses in the Hubbard Brook Experimental Forest. *Global Change Biology*, **14**, 2548–54.

Conley, D. J. & Malone, T. C. (1992). Annual cycle of dissolved silicate in Chesapeake Bay: implications for the production and fate of phytoplankton biomass. *Marine Ecology Progress Series*, **81**, 121–8.

Conley, D. J., Paerl, H. W., Howarth, R. W., *et al.* (2009). Controlling eutrophication: nitrogen and phosphorus. *Science*, **323**, 1014–15.

Conley, D. J. & Schelske, C. L. (2001). Biogenic silica. In *Tracking Environmental Change Using Lake Sediments, Volume 3: Terrestrial, Algal, and Siliceous Indicators*, ed. J. P. Smol, H. J. B. Birks, & W. M. Last, Dordrecht: Kluwer Academic Publishers, pp. 281–93.

Cooper, S. R. (1995a). Chesapeake Bay watershed historical land use: impact on water quality and diatom communities. *Ecological Applications*, **5**, 703–23.

Cooper, S. R. (1995b). Diatoms in sediment cores from the mesohaline Chesapeake Bay, U. S. A. *Diatom Research*, **10**, 39–89.

Cooper, S. R. (1995c). An abundant, small brackish water *Cyclotella* species in Chesapeake Bay, U.S.A. In *A Century of Diatom Research in North America: A Tribute to the Distinguished Careers of Charles W. Reimer and Ruth Patrick*, ed. J. P. Kociolek & M. J. Sullivan, Champaign, IL: Koeltz Scientific Books, pp. 133–40.

Cooper, S. R. & Brush, G. S. (1991). Long-term history of Chesapeake Bay anoxia. *Science*, **254**, 992–96.

Cooper, S. R. & Brush, G. S. (1993). A 2,500-year history of anoxia and eutrophication in Chesapeake Bay. *Estuaries*, **16**, 617–26.

Cooper, S. R., Goman, M., & Richardson, C. J. (2008). Historical changes in water quality and vegetation in WCA-2A determined by paleoecological analyses. In *The Everglades Experiments: Lessons for Ecosystem Restoration*, ed. C. J. Richardson, Springer Ecological Studies, vol. 201, pp. 321–50.

Cooper, S. R., McGlothlin, S. K., Madritch M., & Jones, D. L. (2004a). Water quality history of the Neuse and Pamlico estuaries of North Carolina using paleoecological methods. *Estuaries*, **27**, 619–35.

Cooper, S. R., Thomas, E., & Varekamp, J. C. (2004b). Long Island Sound: diatoms from sediment cores as part of environmental and ecological change studies. Abstracts of the Ecological Society of America Mid-Atlantic Ecology Conference, Lancaster, PA, March 27, p. 6. See http://esa.org/midatlantic/conferences/Abstracts04.pdf.

Costanza, R., d'Arge, R., de Groot, R., *et al.* (1997). The value of the world's ecosystem services and natural capital. *Nature*, **387**, 253–60.

Cronin, T. M., Holmes, C. W., Brewster-Wingard, G. L., *et al.* (2001). Historical trends in epiphytal ostracodes from Florida Bay:

implication for seagrass and macro-benthic algal variability. *Bulletin of American Paleontology*, **361**, 159–97.

Culliton, T. J., Warren, M. A., Goodspeed, T. R., *et al.* (1990). *50 Years of Population Change along the Nation's Coasts 1960–2010*. Silver Spring, MD: National Oceanic and Atmospheric Administration.

Cummins, S. P., Roberts, D. E., Ajani, P., & Underwood, A. J. (2004). Comparisons of assemblages of phytoplankton between open water and seagrass habitats in a shallow coastal lagoon. *Marine and Freshwater Research*, **55**, 447–56.

Daehnick, A. E., Sullivan M. J., & Moncreiff, C. A. (1992). Primary production of the sand microflora in seagrass beds of Mississippi Sound. *Botanica Marina*, **35**, 131–9.

Day, J. W. Jr., & Yanez-Arancibia, A. (1982). Ecology of coastal ecosystems in the Southern Gulf of Mexico: the Terminos Lagoon region. CIENC. *InterAmericana*, **22**, 11–26.

Day, J. W. Jr., Hall, C. A. S., Kemp, W. M., & Yanez-Arancibia, A. (1989). *Estuarine Ecology*. New York, NY: John Wiley & Sons.

Dortch, Q. & Whitledge, T. E. (1992). Does nitrogen or silicon limit phytoplankton production in the Mississippi River plume and nearby regions? *Continental Shelf Research*, **12**, 1293–309.

Dugmore, A. J. & Newton, A. J. (1992). Thin tephra layers in peat revealed by X-radiography. *Journal of Archaeological Science*, **19**, 163–70.

Dyer, K. R. (1995). Sediment transport processes in estuaries. In *Geomorphology and Sedimentology of Estuaries. Developments in Sedimentology 53*, ed. G. M. E. Perillo, Amsterdam: Elsevier Science Publishers, pp. 423–49.

Ellegaard, M., Clarke, A., Rauss, N., *et al.* (2006). Multi-proxy evidence of long-term changes in ecosystem structure in a Danish marine estuary, linked to increased nutrient loading. *Estuarine, Coastal and Shelf Science*, **68**, 567–78.

Farella, N., Lucotte, M., Louchouarn, P., & Roulet, M. (2001). Deforestation modifying terrestrial organic transport in the Rio Tapajos, Brazilian Amazon. *Organic Geochemistry*, **32**, 1443–58.

Fluin, J., Gell, P., Haynes, D., Tibby, J., & Hancock, G. (2007). Paleolimnological evidence for the independent evolution of neighboring terminal lakes, the Murray Darling Basin, Australia. *Hydrobiologia*, **591**, 117–34.

Fourqurean, J. W., Boyer, J. N., Durako, M. J., Hefty, L. N., & Peterson, B. J. (2003). Forecasting responses of seagrass distributions to changing water quality using monitoring data. *Ecological Applications*, **13**, 474–89.

Frankovich, J. W. & Fourqurean, J. W. (1997). Seagrass epiphyte loads along a nutrient availability gradient, Florida Bay, USA. *Marine Ecology Progress Series*, **159**, 37–50.

Frankovich, T. A., Gaiser, E. E., Zieman, J. C., & Wachnicka, A. H. (2006). Spatial and temporal distributions of epiphytic diatoms growing on *Thalassia testudinum* Banks ex König: relationships to water quality. *Hydrobiologia*, **569**, 259–71.

Gaiser, E. (2008). Periphyton as an indicator of restoration in the Everglades. *Ecological Indicators*, DOI:10.1016/j.ecolind.2008.08.004.

Gaiser, E. E., Wachnicka, A., Ruiz, P., Tobias, F. A., & Ross, M. S. (2005). Diatom indicators of ecosystem change in coastal wetlands. In *Estuarine Indicators*, ed. S. Bortone, Boca Raton, FL: CRC Press, pp. 127–44.

Gaiser, E. E., Zafiris, A., Ruiz, P. L., Tobias, F. A. C. & Ross, M. S. (2006). Tracking rates of ecotone migration due to salt-water encroachment using fossil mollusks in coastal south Florida. *Hydrobiologia*, **569**, 237–57.

Gasse, F., Juggins, S., & BenKhelifa, L. (1995). Diatom-based transfer functions for inferring past hydrochemical characteristics of African lakes. *Palaeogeography, Palaeoclimatology, Palaeoecology*, **117**, 31–54.

Gehrels, W. R., Roe, H. M., & Charman, D. J. (2001). Foraminifera, testate amoebae and diatoms as sea-level indicators in UK saltmarshes: a quantitative multiproxy approach. *Journal of Quaternary Science*, **163**, 201–20.

Gell, P., Bulpin, S., Wallbrink, P., Bickford, S., & Hancock, G. (2005). Tareena Billabong – a paleolimnological history of an everchanging wetland, Chowilla Floodplain, lower Murray–Darling Basin. *Marine and Freshwater Research*, **56**, 441–56.

Gell, P., Tibby, J., Little, F., Baldwin, D. & Hancock, G. (2007). The impact of regulation and salinisation on floodplain lakes: the lower River Murray, Australia. *Hydrobiologia*, **591**, 135–46.

Glew, J. R., Smol, J. P., & Last, W. M. (2001). Sediment core collection and extrusion. In *Tracking Environmental Change Using Lake Sediments, Volume 1: Basin Analysis, Coring and Chronological Techniques*, ed. W. M. Last & J. P. Smol, New York: Springer, pp. 73–105.

Gross, M., Coleman, R., & MacDowell, R. (1988). Aquatic productivity and the evolution of diadromous fish migration. *Science*, **239**, 1291–3.

Hassan, G. S., Espinosa, M. A., & Isla, F. I. (2007). Dead diatom assemblages in surface sediments from a low impacted estuary: the Quequén Salado River, Argentina. *Hydrobiologia*, **579**, 257–70.

Hein, M. K., Winsborough, B. M., & Sullivan, M. J. (2008). Bacillariophyta (Diatoms) of the Bahamas. *Iconographia Diatomologica*, **19**, 1–300.

Helz, G. R., Sinex, S. A., Ferri, K. L., & Nichols, M. (1985). Processes controlling Fe, Mn and Zn in sediments of northern Chesapeake Bay. *Estuarine, Coastal and Shelf Science*, **21**, 1–16.

Hemphill-Haley, E. (1995). Diatom evidence for earthquake-induced subsidence and tsunami 300 yr ago in southern coastal Washington. *GSA Bulletin*, **107**, 367–78.

Hendey, N. I. (1964). An introductory account of the smaller algae of British coastal waters. *Fishery Investigations Series IV. Part V. Bacillariophyceae (Diatoms)*. London, UK: Her Majesty's Stationery Office.

Hendey, N. I. (1976). The species diversity index of some in-shore diatom communities and its use in assessing the degree of pollution insult on parts of the north Coast of Cornwall. *Nova Hedwigia*, **54**, 355–78.

Hennessee, E. L., Blakeslee, P. J., & Hill, J. M. (1986). The distributions of organic carbon and sulfur in surficial sediments of the Maryland portion of Chesapeake Bay. *Journal of Sedimentary Petrology*, **56**, 674–83.

Hill, J. M., Halka, R. D., Conkwright, R. D., Koczot, K., & Park, J. (1992). Geologically constrained shallow gas in sediments of

Chesapeake Bay: distribution and effects on bulk sediment properties. *Continental Shelf Research*, **12**, 1219–29.

Holmer, M., Duarte, C. M., & Marba, N. (2003). Sulfur cycling and seagrass (*Posidonia oceanica*) status in carbonate sediments. *Biogeochemistry*, **66**, 223–39.

Horton, B. P., Corbett, R., Culver, S. J., Edwards, R. J., & Hillier, C. (2006). Modern saltmarsh diatom distributions of the Outer Banks, North Carolina, and the development of a transfer function for high resolution reconstructions of sea level. *Estuarine Coastal and Shelf Science*, **69**, 381–94.

Horton, B. P., Zong, Y., Hillier, C., & Engelhart, S. (2007). Diatoms from Indonesian mangroves and their suitability as sea-level indicators for tropical environments. *Marine Micropaleontology*, **63**, 155–68.

Houel, S., Louchouarn, P., Lucotte, M., Canuel, R., & Ghaleb, B. (2006). Translocation of soil organic matter following reservoir impoundment in boreal systems: implications for *in situ* productivity. *Limnology and Oceanography*, **51**(3), 1497–513.

Hustedt, F. (1927–30). Die Kieselalgen Deutschlands, Österreichs und der Schweiz (three volumes). In *Rabenhorst's Kryptogamen-Flora von Deutschland, Öterreich und der Schweiz*. Band 7. Leipzig: Akademische Verlagsgesellschaft.

Hustedt, F. (1953). Die Systematik der Diatomeen in Ihren Beziehungen zur Geologie und Ökologie nebst einer Revision des Halobien-Systems. *Svensk Botanisk Tidskrift*, **47**, 509–19.

Hustedt, F. (1955). *Marine Littoral Diatoms of Beaufort, North Carolina*. Durham, NC: Duke University Press.

Hustedt, F. (1957). Die Diatomeenflora des Fluss-Systems der Weser im Gebiet der Hansestadt Bremen. *Abhandlungen des Naturwissenschaftlichen Vereins zu Bremen*, **34**, 181–440.

Huvane, J. K. & Cooper, S. R. (2001). Diatoms as indicators of environmental change in sediment cores from Northeastern Florida Bay. *Bulletins of American Paleontology*, **361**, 145–58.

Jackson, J. B. C., Kirby, M. X., Berger, W. H., *et al.* (2001). Historical overfishing and the recent collapse of coastal ecosystems. *Science*, **293**, 629–38.

Jiang, H. (1996). Diatoms from the surface sediments of the Skagerrak and the Kattegat and their relationship to the spatial changes of environmental variables. *Journal of Biogeography*, **23**, 129–37.

Jickells, T. D. (1998). Nutrient biogeochemistry of the coastal zone. *Science*, **281**, 217–22.

Juggins, S. (1992). Diatoms in the Thames Estuary, England: ecology, paleoecology, and salinity transfer function. *Bibliotheca Diatomologica*, **25**, 1–216.

Kemp, W. M., Boynton, W. R., Adolf, J. E., *et al.* (2005). Eutrophication of Chesapeake Bay: historical trends and ecological interactions. *Marine Ecology Progress Series*, **303**, 1–29.

Kennett, D. M. & Hargraves, P. E. (1984). Subtidal benthic diatoms from a stratified estuarine basin. *Botanica Marina*, **27**, 169–83.

Kennett, D. M. & Hargraves, P. E. (1985). Benthic diatoms and sulfide fluctuations: upper basin of Pettaquamscutt River, Rhode Island. *Estuarine, Coastal and Shelf Science*, **21**, 577–86.

Kennish, M. J. (1992). *Ecology of Estuaries: Anthropogenic Effects*. Marine Science Series. Boca Raton, FL: CRC Press, Inc.

Kistner, D. A. & Pettigrew, N. R. (2001). A variable turbidity maximum in the Kennebec estuary, Maine. *Estuaries*, **24**, 680–7.

Koch, M. S., Schopmeyer, S., Kyhn-Hansen, C. & Madden, C. J. (2007). Synergistic effects of high temperature and sulfide on tropical seagrass. *Journal of Experimental Marine Biology and Ecology*, **341**, 91–101.

Kolka, R. K. & Thompson, J. A. (2006). Wetland geomorphology, soils and formative processes. In *Ecology of Freshwater and Estuarine Wetlands*, ed. D. P. Batzer & R. R. Sharitz, Los Angeles, CA: University of California Press, pp. 7–41.

Köster, D., Racca, J. M. J., & Pienitz, R. (2004). Diatom-based inference models and reconstructions revisited: methods and transformations. *Journal of Paleolimnology*, **32**, 233–45.

Köster, D., Lichter, J., Lea, P. D. & Nurse, A. (2007). Historical eutrophication in a river–estuary complex in mid-coast Maine. *Ecological Applications*, **17**, 765–78.

Krammer, K. & Lange-Bertalot, H. (1986–1991). Bacillariophyceae. In *Süßwasserflora von Mitteleuropa Band 2/1–4*, ed. H. Ettl, H. Heynig, & D. Mollenhauer, Stuttgart: Fischer.

Kucera, M., Rosell-Melé, A., Schneider, R., Waelbroeck, C., & Weinelt, M. (2005). Multiproxy approach for the reconstruction of the glacial ocean surface (MARGO). *Quaternary Science Review*, **24**, 813–19.

Kuylenstierna, M. (1990). Benthic algal vegetation in the Nordre Älv Estuary (Swedish west coast), 2 volumes. Ph.D. thesis, University of Göteborg, Göteborg, Sweden.

Laws, R. A. (1988). Diatoms (Bacillariophyceae) from surface sediments in the San Francisco Bay estuary. *Proceedings of the California Academy of Sciences*, **45**, 133–254.

Lee, A. H. M. & Liu, J. H. (2002). Marine water quality in Hong Kong in 2001. Environmental Protection Department, The Government of the Hong Kong Special Administrative Region. Report Number EPD/TR1/02.

Louchouarn, P., Lucotte, M., & Farella, N. (1999). Historical and geographical variations of sources and transport of terrigenous organic matter within a large-scale coastal environment. *Organic Geochemistry*, **30**, 675–99.

McIntire, C. D. & Overton, W. S. (1971). Distributional patterns in assemblages of attached diatoms from Yaquina Estuary, Oregon. *Ecology*, **52**, 758–77.

Malmgren, B. A., Kucera, M., Nyberg, J., & Waelbroeck, C. (2001). Comparison of statistical and artificial neural network techniques for estimating past sea surface temperatures from planktonic foraminifer census data. *Paleoceanography*, **16**, 520–30.

Malmgren, B. A. & Nordlund, U. (1997). Application of artificial neural networks to paleoceanographic data. *Palaeogeography, Palaeoclimatology, Palaeoecology*, **136**, 359–73.

Mann, K. H. (2000). *Ecology of Coastal Waters: With Implications for Management*, 2nd edition, Malden, MA: Blackwell Science, Inc.

Mitsch, W. J. & Gosselink, J. G. (1986). *Wetlands*. New York: Van Nostrand Reinhold.

Moncreiff, C. A. & Sullivan, M. J. (2001). Trophic importance of epiphytic algae in subtropical seagrass beds: evidence from multiple stable isotope analyses. *Marine Ecology Progress Series*, **215**, 93–106.

Moncreiff, C. A., Sullivan, M. J. & Daehnick, A. E. (1992). Primary production dynamics in seagrass beds of Mississippi Sound: the contributions of seagrass, epiphytic algae, sand, microflora and phytoplankton. *Marine Ecology Progress Series*, **87**, 161–71.

Moser, K. A., MacDonald, G. M., & Smol, J. P. (1996). Applications of freshwater diatoms to geographical research. *Progress in Physical Geography*, **20**, 21–52.

Muylaert, K. & Sabbe, K. (1996). The diatom genus *Thalassiosira* (Bacillariophyta) in the estuaries of the Schelde (Belgium/the Netherlands) and the Elbe (Germany). *Botanica Marina*, **39**, 103–15.

Newton, J. (2007). Hood Canal, WA: the complex factors causing low dissolved oxygen events require ongoing research, monitoring, and modeling. In *Effects of Nutrient Enrichment in the Nations' Estuaries: A Decade of Change*, ed. S. Bricker, B. Longstaff, W. Dennison, *et al.*, NOAA Coastal Ocean Program Decision Analysis Series No. 26, Silver Spring, MD: National Centers for Coastal Ocean Science, pp. 95–8.

Niemi, G., Wardrop, D., Brooks, R., *et al.* (2004). Rationale for a new generation of indicators for coastal waters. *Environmental Health Perspectives*, **112**, 979–86.

Nixon, S. W. (1995). Coastal marine eutrophication: a definition, social causes, and future concerns. *Ophelia*, **41**, 199–219.

Palmer, C. M. (1969). A composite rating of algae tolerating organic pollution. *Journal of Phycology*, **5**, 78–82.

Parsons, M. L. (1998). Salt marsh sedimentary record of the landfall of Hurricane Andrew on the Louisiana coast: diatoms and other paleoindicators. *Journal of Coastal Research*, **14**, 939–50.

Parsons, M. L., Dortch, Q., & Turner, R. E. (2002). Sedimentological evidence of an increase in *Pseudo-nitzschia* (Bacillariophyceae) abundance in response to coastal eutrophication. *Limnology and Oceanography*, **47**, 551–8.

Parsons, M. L., Dortch, Q., Turner, R. E., & Rabalais, N. R. (2006). Reconstructing the development of eutrophication in Louisiana salt marshes. *Limnology and Oceanography*, **51**, 534–44.

Patrick, R. (1973). Use of algae, especially diatoms, in the assessment of water quality. *American Society for Testing and Materials, Special Technical Publication*, **528**.

Patten, B. C. (1962). Species diversity in net plankton of Raritan Bay. *Journal of Marine Research*, **20**, 57–75.

Peragallo, H. & Peragallo, M. (1897–1908). *Diatomées marines de France et des districts maritimes voisins*. Grez-sur-Loing: Micrographie-Editeur.

Perillo, G. M. E. (1995). Geomorphology and sedimentology of estuaries. In *Geomorphology and Sedimentology of Estuaries: Developments in Sedimentology 53*, ed. G. M. E. Perillo, Amsterdam: Elsevier Science Publishers, pp. 1–16.

Pinckney, J. L. & Micheli, F. (1998). Microalgae on seagrass mimics: does epiphyte community structure differ from live seagrasses? *Journal of Experimental Marine Biology and Ecology*, **221**, 59–70.

Pritchard, D. W. (1967). Observations of circulation in coastal plain estuaries. In *Estuaries*, ed. G. H. Lauff, American Association for the Advancement of Science Publication No. 83, pp. 37–44.

Puckett, L. J. (1995). Identifying the major sources of nutrient water pollution. *Environmental Science & Technology*, **29**, 408–14.

Puškaric, S., Berger, G. W., & Jorissen, F. J. (1990). Successive appearance of subfossil phytoplankton species in Holocene sediments of the northern Adriatic and its relation to the increased eutrophication pressure. *Estuarine, Coastal and Shelf Science*, **31**, 177–87.

Pyle, L., Cooper, S. R., & Huvane, J. K. (1998). Diatom paleoecology of Pass Key core 37, Everglades National Park, Florida Bay. U. S. *Geological Survey Open-File Report* 98–522.

Quinn, H. A., Tolson, J. P., Klein, C. J., Orlando, S. P., & Alexander, C. E. (1989). *Strategic Assessment of Near Coastal Waters: Susceptibility of East Coast Estuaries to Nutrient Discharges, Albemarle/Pamlico Sound to Biscayne Bay*. Rockville, MD: National Oceanic and Atmospheric Agency.

Rabalais, N. N., Turner, R. E., Sen Gupta, B. K., *et al.* (2007a). Hypoxia in the northern Gulf of Mexico: does the science support the plan to reduce, mitigate, and control hypoxia? *Estuaries and Coasts*, **30** (5), 753–72.

Rabalais, N. N., Turner, R. E., Sen Gupta, B. K., Platon, E., & Parsons, M. L. (2007b). Sediments tell the history of eutrophication and hypoxia in the northern Gulf of Mexico. *Ecological Applications*, **17** (5), S129–S143.

Racca, J. M. J., Gregory-Eaves, I., Pienitz, R., & Prairie, Y. T. (2004). Tailoring paleolimnological transfer functions. *Canadian Journal of Fisheries and Aquatic Sciences*, **61**, 2440–54.

Racca J. M. J., Philibert, A., Racca, R., & Prairie, Y. T. (2001). A comparison between diatom-based pH inference models using artificial neural networks (ANNO) weighted averaging (WA) and weighted averaging partial least squares (WA-PLS) regressions. *Journal of Paleolimnology*, **26**, 411–22.

Riggs, S. R., Bray, J. T., Powers, E. R., *et al.* (1991). Heavy metals in organic-rich muds of the Neuse River estuarine system. Albemarle–Pamlico Estuarine Study Report No. 90–07, Raleigh, NC: Albemarle-Pamlico National Estuary Program.

Riggs, S. R., Powers, E. R., Bray, J. T., *et al.* (1989). Heavy metal pollutants in organic-rich muds of the Pamlico River estuarine system: their concentration, distribution, and effects upon benthic environments and water quality. Albemarle Pamlico Estuarine Study Report 89–06, Raleigh, NC: Albemarle-Pamlico National Estuary Program.

Riggs, S. R., York, L. L., Wehmiller, J. F., & Snyder, S. W. (1992). Depositional patterns resulting from high-frequency Quaternary sea-level fluctuations in northeastern North Carolina. *SEPM Special Publication*, **48**, 141–53.

Ross, M. S., Gaiser, E. E., Meeder, J. F., & Lewin, M. T. (2001). Multi-taxon analysis of the "white zone", a common ecotonal feature of South Florida coastal wetlands. In *The Everglades, Florida Bay, and Coral Reefs of the Florida Keys*, ed. J. Porter & K. Porter, Boca Raton, FL: CRC Press, pp. 205–38.

Ruiz-Halpern, S., Macko, S., & Fourqurean, J. W. (2008). The effects of manipulation of sedimentary iron and organic matter on sediment biogeochemistry and seagrasses in a subtropical carbonate environment. *Biogeochemistry*, **87**, 113–26.

Ryves, D. B., Clarke, A. L., Appleby, P. G., *et al.* (2004). Reconstructing the salinity and environment of the Limfjord and Vejlerne Nature Reserve, Denmark, using a diatom model for brackish lakes and fjords. *Canadian Journal of Fisheries and Aquatic Science*, **61**, 1988–2006.

Sanford, L. P. (1992). New sedimentation, resuspension, and burial. *Limnology and Oceanography*, **37**, 1164–78.

Saunders, K. M., McMinn, A., Roberts, D., Hodgson, D., & Heijnis, H. (2007). Recent human-induced salinity changes in Ramsar-listed Orielton Lagoon, south-east Tasmania, Australia: a new approach for coastal lagoon conservation and management. *Aquatic Conservation: Marine and Freshwater Ecosystems*, **17**, 51–70.

Sawai, Y., Horton, B. P., & Nagumo, T. (2004). The development of a diatom-based transfer function along the Pacific coast of eastern Hokkaido, northern Japan – an aid in paleoseismic studies of the Kuril subduction zone. *Quaternary Science Reviews*, **23**, 2467–83.

Schelske, C. L. (1999). Diatoms as mediators of biogeochemical silica depletion in the Laurentian Great Lakes. In *The Diatoms: Applications for the Environmental and Earth Sciences*, ed. E. F. Stoermer & J. P. Smol, New York, NY: Cambridge University Press, pp. 73–84.

Schubel, J. R. (1968). Turbidity maximum of the northern Chesapeake Bay. *Science*, **161**, 1013–15.

Shaffer, G. P. & Sullivan, M. J. (1988). Water column productivity attributable to displaced benthic diatoms in well-mixed shallow estuaries. *Journal of Phycology*, **24**, 132–40.

Simonsen, R. (1962). Untersuchungen zur Systematik und Ökologie der Bodendiatomeen der Westlichen Ostsee. *Internationale Revue der gesamten Hydrobiologie*, **1**, 1–144.

Slate, J. E. & Stevenson, R. J. (2000). Recent and abrupt environmental change in the Florida Everglades indicated from siliceous microfossils. *Wetlands*, **20**, 346–56.

Smol, J. P. (2008). *Pollution of Lakes and Rivers: a Paleoenvironmental Perspective*, Malden, MA: Blackwell Publishing, vol. 2.

Snoeijs, P. (1993). *Intercalibration and Distribution of Diatom Species in the Baltic Sea*, Uppsala: Opulus Press.

Snoeijs, P. & Balashova, N. (1998). *Intercalibration and Distribution of Diatom Species in the Baltic Sea*, Uppsala: Opulus Press, vol. 5.

Snoeijs, P. & Kasperovičienė, J. (1996). *Intercalibration and Distribution of Diatom Species in the Baltic Sea*, Uppsala: Opulus Press, vol. 4.

Snoeijs, P. & Potapova, M. (1995). *Intercalibration and Distribution of Diatom Species in the Baltic Sea*, Uppsala: Opulus Press, vol. 3.

Snoeijs, P. & Vilbaste, S. (1994). *Intercalibration and Distribution of Diatom Species in the Baltic Sea*, Uppsala: Opulus Press, vol. 2.

Stacey, P. E. (2007). Long Island Sound, CT & NY: point source reductions lessened hypoxia in 1990s. In *Effects of Nutrient Enrichment in the Nations' Estuaries: a Decade of Change*, NOAA Coastal Ocean Program Decision Analysis Series No. 26, ed. S. Bricker, B. Longstaff, W. Dennison, *et al.*, Silver Spring, MD: National Centers for Coastal Ocean Science, pp. 101–103.

Stachura-Suchoples, K. (2001). Bioindicative values of dominant diatom species from the Gulf of Gdansk, southern Baltic Sea, Poland, In *Lange-Bertalot-Festschrift*, ed. R. Jahn, J. P. Kociolek, A. Witkowski, & P. Compere, Gantner: Ruggell, pp. 477–90.

Stanley, D. W. (1992). Historical Trends: Water Quality and Fisheries, Albemarle-Pamlico Sounds, with Emphasis on the Pamlico River Estuary. University of North Carolina Sea Grant Publication UNC-SG-92-04.

Stockner, J. G. & Benson, W. W. (1967). The succession of diatom assemblages in the recent sediments of Lake Washington. *Limnology and Oceanography*, **12**, 513–32.

Sullivan, M. J. (1982). Distribution of edaphic diatoms in a Mississippi salt marsh: a canonical correlation analysis. *Journal of Phycology*, **18**, 130–3.

Sullivan, M. J. & Currin, C. A. (2000). Community structure and functional dynamics of benthic microalgae in salt marshes. In *Concepts and Controversies in Tidal Marsh Ecology*, ed. M. P. Weinstein & D. A. Kreeger, Dordrecht: Kluwer Academic Publishers, pp. 81–106.

Sullivan, M. J. & Moncreiff, C. A. (1988). Primary production of edaphic algal communities in a Mississippi salt marsh. *Journal of Phycology*, **24**, 49–58.

Swart, P. K., Healy, G. F., Dodge, R. E., *et al.* (1996). The stable oxygen and carbon isotopic record from a coral growing in Florida Bay: a 160-year record of climatic and anthropogenic influence. *Palaeogeography, Palaeoclimatology, Palaeoecology*, **123**, 219–37.

Sylvestre, F., Beck-Eichler, B., Duleba, W., & Debenay, J.-P. (2001). Modern benthic diatom distribution in a hypersaline coastal lagoon: the Lagoa de Araruama (R. J.), Brazil. *Hydrobiologia*, **443**, 213–31.

Sylvestre, F., Guiral, D., & Debenay, J. P. (2004). Modern diatom distribution in mangrove swamps from the Kaw Estuary (French Guiana). *Marine Geology*, **208**, 281–93.

Taffs, K. H., Farago, L. J., Heijnis, H., & Jacobsen, G. (2008). A diatom-based Holocene record of human impact from a coastal environment: Tuckean Swamp, eastern Australia. *Journal of Paleolimnology*, **39**, 71–82.

ter Braak, C. J., Juggins, S., Birks, H. J. B., & Van der Voet, H. (1993). Weighted averaging partial least squares regression (WA-PLS): definition and comparison with other methods for species environment calibration. In *Multivariate Environmental Statistics*, ed. G. R. Patil & C. R. Rao, Amsterdam: North-Holland, pp. 525–60.

Thomas, E., Abramson, I., Varekamp, J. C., & Buchhotz ten Brink, M. R. (2004a). Eutrophication of Long Island Sound as traced by benthic foraminifera. *Sixth Biennual Long Island Sound Research Conference Proceedings*, pp. 87–91.

Thomas, E., Cooper, S., Sangiorgi, F., *et al.* (2004b). The eutrophication of Western Long Island Sound. *Geological Society of America Abstracts*, **36**(5), 493.

Tibby, J., Gell, P. A., Fluin, J., & Sluiter, I. R. K. (2007). Diatom–salinity relationships in wetlands: assessing the influence of salinity variability on the development of inference models. *Hydrobiologia*, **591**, 207–18.

Tomas, C. R. (1997). *Identifying Marine Phytoplankton*, San Diego, CA: Academic Press.

Trainer, V. L., Adams, N. G., Bill, B. D., *et al.* (2000). Domoic acid production near California coastal upwelling zones, June 1998. *Limnology and Oceanography*, **45**, 1818–33.

Tréguer, P., Nelson, D. M., Van Bennekom, A. J., *et al.* (1995). The silica balance in the world ocean: a reestimate. *Science*, **268**, 375–9.

Valiela, I., McClelland, J., Hauxwell, J., *et al.* (1997). Macroalgal blooms in shallow estuaries: controls and ecophysiological and ecosystem consequences. *Limnology and Oceanography*, **42**, 1105–18.

Van Cappellen, P., Dixit, S., & van Beusekom, J. (2002). Biogenic silica dissolution in the oceans: reconciling experimental and field-based dissolution rates. *Global Biogeochemical Cycles*, **16**(4), DOI:10.1029/2001GB001431.

van Dam, H. (1982). On the use of measures of structure and diversity in applied diatom ecology. *Nova Hedwigia*, **73**, 97–115.

Varekamp, J. C., Thomas E., Altabet, M., Cooper, S. R., & Brinkhuis, H. (2010). *Environmental Change in Long Island Sound in the Recent Past: Eutrophication and Climate Change*, Hartford, CT: CT Department of Environmental Protection.

Varekamp, J. C., Thomas, E., Lugolobi, F., & Buchholtz ten Brink, M. R. (2004). The paleo-environmental history of Long Island Sound as traced by organic carbon, biogenic silica and stable isotope/trace element studies in sediment cores. *Sixth Biennual Long Island Sound Research Conference Proceedings*, Groton, CT: Connecticut Sea Grant, pp. 109–13.

Verleyen, E., Hodgson, D. A., Leavitt, P. R., Sabbe, K., & Vyverman, W. (2004). Quantifying habitat-specific diatom production: a critical assessment using morphological and biogeochemical markers in Antarctic marine and lake sediments. *Limnology and Oceanography*, **49**, 1528–39.

Vos, P. C. & De Wolf, H. (1993). Diatoms as a tool for reconstructing sedimentary environments in coastal wetlands: methodological aspects. *Hydrobiologia*, **269**, 285–96.

Wachnicka, A. (2009). Diatom-based paleoecological evidence of combined effects of anthropogenic and climatic impacts on salinity, nutrient levels and vegetation cover in Florida Bay and Biscayne Bay, USA. Unpublished Ph.D. thesis, Florida International University, Miami, FL.

Wachnicka, A. & Gaiser, E. E. (2007). Characterization of *Amphora* and *Seminavis* from South Florida. *Diatom Research*, **22**, 387–455.

Washington, H. G. (1984). Diversity, biotic, and similarity indices: a review with special relevance to aquatic ecosystems. *Water Research*, **18**, 653–94.

Weckström, K. (2006). Assessing recent eutrophication in coastal waters of the Gulf of Finland (Baltic Sea) using subfossil diatoms. *Journal of Paleolimnology*, **35**(3), 571–92.

Weckström, K. & Juggins, S. (2005). Coastal diatom–environment relationships from the Gulf of Finland, Baltic Sea. *Journal of Phycology*, **42**, 21–35.

Weckström, K., Juggins, S., & Korhola, A. (2004). Quantifying background nutrient concentrations in coastal waters: a case study from an urban embayment of the Baltic Sea. *Ambio*, **33**, 324–7.

Wendker, S. (1990). Untersuchungen zur subfossilen und rezenten Diatomeenflora des Schlei-Ästuars (Ostsee). In *Bibliotheca Diatomologica*, **20**, 1–268 (with eight plates).

Wilderman, C. C. (1986). Techniques and results of an investigation into the autecology of some major species of diatoms from the Severn River Estuary, Chesapeake Bay, Maryland, U.S.A. In *Proceedings of the 8th International Diatom Symposium*, ed. M. Ricard, Königstein: Koeltz Scientific Books, pp. 631–43.

Willard, D. A., Weimer, L. M., & Riegel, W. L. (2001). Pollen assemblages as paleoenvironmental proxies in the Florida Everglades. *Review of Palaeobotany and Palynology*, **113**, 213–35.

Witkowski, A., Lange-Bertalot, H., & Metzeltin, D. (2000). *Diatom Flora of Marine Coasts I*, ed. H. Lange-Bertalot, Königstein: Koeltz Scientific Books.

World Resources Institute, United Nations Environment Programme, United Nations Development Programme and The World Bank. (1996). *World Resources: a Guide to the Global Environment 1996–97*. New York, NY: Oxford University Press.

Xu, Y., Jaffé, R., Wachnicka A., & Gaiser, E. E. (2006). Occurrence of C25 highly branched isoprenoids (HBIs) in Florida Bay: paleoenvironmental indicators of diatom-derived organic matter inputs. *Organic Geochemistry*, **37**, 847–59.

Zimmerman, A. R. & Canuel, E. A. (2000). A geochemical record of eutrophication and anoxia in Chesapeake Bay sediments: anthropogenic influence on organic matter composition. *Marine Chemistry*, **69**,117–37.

Zimmerman, A. R. & Canuel, E. A. (2002). Sediment geochemical records of eutrophication in the mesohaline Chesapeake Bay. *Limnology and Oceanography*, **47**, 1084–93.

Zong, Y., Lloyd, J. M., Leng, M. J., Yim, W. W.-S., & Huang, G. (2006). Reconstruction of Holocene history from the Pearl River Estuary, southern China, using diatoms and carbon isotope ratios. *The Holocene*, **16**, 251–63.

18 Diatoms on coral reefs and in tropical marine lakes

CHRISTOPHER S. LOBBAN AND
RICHARD W. JORDAN

18.1 Coral reef communities as diatom habitats

Coral reefs, well known for their tremendous biodiversity and beauty (Veron, 2000; Spalding *et al.*, 2001), are the most complex ecosystems in the sea, and are often compared to rainforests, both because of the large numbers of organisms (estimates in both cases based on larger organisms and extrapolated wildly to small animals and microorganisms!) and because corals, like rainforest trees, create the structure and habitat for the wealth of other organisms. Coral reefs are formed by a highly successful yet environmentally sensitive symbiotic association between animals (cnidarians; scleractinian corals) and protists (dinoflagellate algae; zooxanthellae in the genus *Symbiodinium*). The term *coral* is generally used to denote the holobiont, i.e. both partners in the symbiosis. On a healthy coral reef, macroalgae are generally sparse and coral cover is high, but the balance can be tipped to communities dominated by fleshy algae by nutrient inputs that promote algal growth in the otherwise oligotrophic waters, or by reduction of normally high grazing pressure (Littler & Littler, 1984).

Coral reefs are vital resources for millions of humans in tropical, especially developing countries, who depend on them for fisheries (Cesar, 2000; Sadovy, 2005; Vincent, 2006), tourism income (Brander *et al.*, 2007), storm protection (UNEP-WCMC, 2006), and sometimes structural materials (Berg *et al.*, 1998; Mallik, 1999). Biodiversity of coral reefs is also recognized for its pharmacological potential (Adey, 2000). Exploitation of reefs by humans, and the fact that many reefs are in heavily populated areas or (in the case of the Great Barrier Reef – Gordon, 2007) near areas with extensive agriculture, has brought coral reefs increasingly into jeopardy (Hughes *et al.*, 2003; Pandolfi *et al.*, 2003). Overfishing of herbivorous fishes on coral reefs near coastal human populations is often exacerbated by nutrient loading from sewage and by sedimentation from poor land practices and sewage (Pastorak & Bilyard, 1985; Richmond, 1994). Coral reefs are also threatened by global temperature increases and ocean acidification resulting from greenhouse-gas emissions (Hoegh-Guldberg *et al.*, 2007).

Biodiversity in coral-reef ecosystems is patchily known: corals, fishes, the larger crustaceans, and macroalgae are relatively well known in some geographic areas, but the microbiota are virtually unknown everywhere except for foraminifera (Hallock, 1984; Paulay, 2003). Diatom studies on coral reefs are rare (e.g. Miller *et al.*, 1977; Montgomery, 1978;[1] Navarro, 1983; Navarro *et al.*, 1989, 2000; Gottschalk *et al.*, 2007; Wachnicka & Gaiser 2007; Hein *et al.*, 2008), although many taxonomists have reported diatoms based on a few samples from a given tropical locality. In this chapter, we briefly summarize the various reef and mangrove habitats, note some biogeographical considerations and their consequences when using diatoms as indicator species, and describe two particular habitats that we have studied, farmer-fish territories and meromictic marine lakes.

18.2 Reefs and mangroves as diatom habitats

Reef habitats range from sheltered shorelines of lagoons inside barrier reefs to the high-energy environments of the fore-reef slope (Figure 18.1). Some coasts have a fringing reef with a narrow, shallow lagoon as part of the reef flat, rising to intertidal on the reef crest (this is the surf line) before plunging down to great depths. Some coastlines have an offshore, barrier reef with deeper water behind it; in some cases, such as the Great Barrier Reef off the east coast of Australia, the reef is tens of

The Diatoms: Applications for the Environmental and Earth Sciences, 2nd Edition, eds. John P. Smol and Eugene F. Stoermer. Published by Cambridge University Press.

[1] CL recently digitized the 250 plates of diatom SEMs in Montgomery's (1978) dissertation and the files can be obtained from the Florida State University Library, Tallahassee.

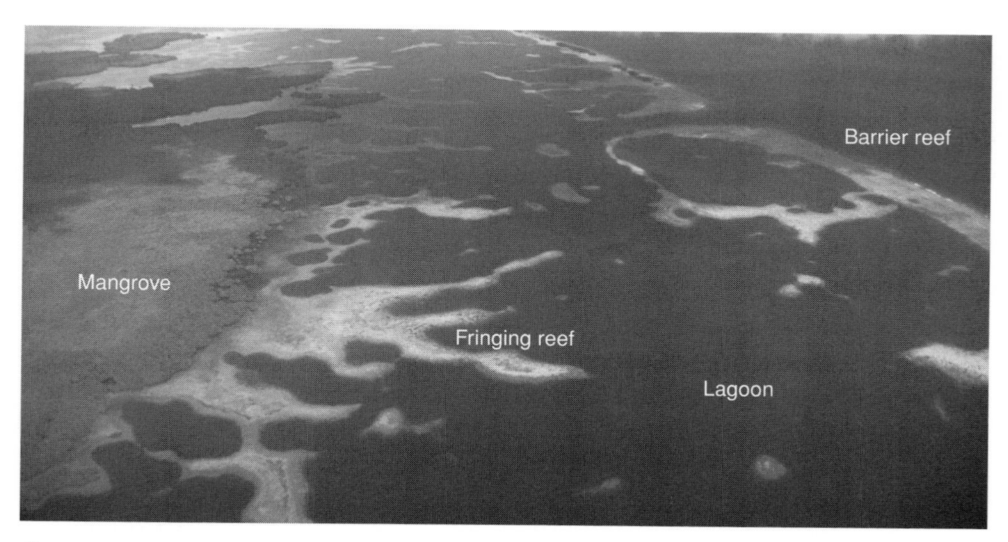

Figure 18.1 Aerial photograph of the west coast of Pohnpei, Federated States of Micronesia, showing the relationship of mangrove, fringing reef, and barrier reef. Photo by María Schefter.

kilometers offshore. Coral growth is limited by light so that living reefs extend to only some 400 m in very clear oceanic waters, much less in turbid coastal areas.

While diatoms apparently do not grow on living coral tissue, they occur on most other substrata and are among the early pioneers (Zolan, 1980), colonizing sand, sediments, newly bare coral, and other calcareous surfaces such as calcified green seaweeds. They can be very abundant as a secondary turf on filamentous seaweeds as well as non-calcified macroalgae (Montgomery, 1978; Hein *et al.*, 2008) and seagrasses (Sullivan, 1981a; Frankovich *et al.*, 2006). While fish are often seen as reducing algal cover and helping to maintain reef structure, some – farmer fish – promote the growth of filamentous algae and, as recently observed, the epiphytic communities of diatoms (Jones *et al.*, 2006; Navarro & Lobban, 2009) (discussed below).

Mangroves (mangal communities) occur in tidal waters and often fringe shores near coral reefs (Figure 18.1); they are important nurseries for coral-reef fishes as well as acting as sediment traps in estuaries (Mumby *et al.*, 2004; Mumby, 2006). The mangal community includes mangrove trees, other mangal plants such as nipa palms (*Nypa fruticans* Wurmb.), and the associated animals and microbial populations (Tomlinson, 1995; Hogarth, 2007). While mangroves are often associated with estuaries, they also occur on some atolls and around tropical marine lakes. Characteristic of many mangrove tree species is the presence of pneumorrhizae or air roots, that permit the root system to anchor in soft mud and to acquire oxygen in spite of anoxic conditions. These roots, which grow down from the trunk (e.g. *Rhizophora*) or up from the underground roots (e.g. *Sonneratia*, *Avicennia*), acquire a coating of certain species of seaweeds (e.g. the "bostrychietum" community) (Post, 1967; King, 1990; Skelton & South, 2002) and sediments that provides a special habitat for prokaryotic and eukaryotic microbes including diatoms (Navarro, 1982; Siqueiros-Beltrones *et al.*, 2005). The communities vary with position in the intertidal and with season (Siqueiros-Beltrones *et al.*, 2005). Sediment surfaces in mangrove swamps also provide a habitat for diatoms (Sylvestre *et al.*, 2004). Mangrove-associated diatoms are further discussed by Cooper *et al.* and by Horton and Sawai (this volume).

In addition to the previously described habitats, marine diatoms also occur in tropical meromictic marine lakes in the Republic of Palau. In a subsequent section, we provide an insight into the diatom communities of perhaps the most famous of these lakes, Mecherchar Jellyfish Lake. Research activities over the last decade have revealed highly diverse diatom communities on mangrove roots, seaweeds, and in the plankton, dissimilar from those in the surrounding lagoons.

18.3 Biogeography and the potential for moderate endemism

Biogeographically, coral reefs and mangals occur throughout the tropics, generally within 30 degrees north and south of the equator. The actual distributions correlate with seawater temperature minima: approximately 16–17 (18) °C for corals (Kleypas *et al.*, 1999) and 20 °C for mangroves (Hogarth, 2007). The biodiversity of coral, mangrove, and fish species is highest in the Indo-West Pacific, centered on a triangle between Indonesia and the Philippines (Hughes *et al.*, 2002; Carpenter & Springer, 2005; Hogarth, 2007), but the biogeography of

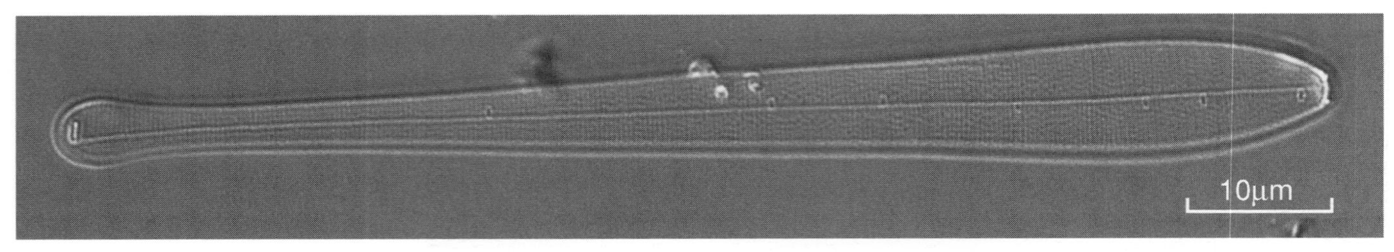

Figure 18.2 *Licmophora flabellata* from Guam showing rimoportulae and head pole spines.

seaweeds is different and much more complex (Kerswell, 2006; Tomasetti, 2007).

The distribution of tropical diatoms – or any other microbenthos – is incompletely known at present, both because of the low number of studies and also partly because of a view, still widely accepted, that there is low endemicity among microbial eukaryotes, i.e. that "everything is everywhere," although species can only grow where conditions are right (Fenchel & Finlay, 2004). The opposing view, "moderate endemicity," is that a small-to-moderate fraction of a microbial biota is endemic, i.e. that there are factors besides environmental limits that have prevented the universal distribution of some species (Foissner, 2006, 2008; Vanormelingen *et al.*, 2008). Both hypotheses depend on circumstantial evidence in a world where the absence of evidence greatly overwhelms the evidence of absence, and taking either as a working model leads to assumptions with consequences for identification. A greater problem at present, however, is that we do not yet know enough about most species, regardless of where they occur, to know when differences or similarities are significant. Accurate recognition of taxa is critical for ecological inferences, yet there is increasing evidence of cryptic speciation both within sites and within apparently global populations (Amato *et al.*, 2007; Poulíčková *et al.* 2008). Morphospecies may contain more than one genetic species, hidden by unnoticed morphological differences or unobserved differences in reproduction. Two examples will serve to show the broad outlines of the problem, which is well known to diatomists but may not be so obvious to environmental managers used to dealing with relatively large organisms.

First, *Licmophora flabellata* (Carmichael) C. Agardh is a well-known species, widespread in Europe (Hustedt, 1931; Hendey, 1964; Honeywill, 1998; Witkowski *et al.*, 2000), whereas *Licmophora pfannkucheae* Giffen (Giffen, 1970; also illustrated and described by Witkowski *et al.*, 2000) is known only from three localities in southern Africa. *L. pfannkucheae* has a series of rimoportulae along the length of the sternum, clearly drawn by Giffen (1970) and visible in the light micrograph in Witkowski *et al.* (2000: plate 18, figure 16). A similar series is also present in *L. flabellata* (Figure 18.2), but it is not always visible in the light microscope (e.g. Hustedt, 1931, figure 581; Simonsen, 1970, figures 23a vs. 24, 25; Witkowski *et al.*, 2000, plate 18, figures 2, 3) and was not mentioned in descriptions of the species until Sar & Ferrario (1990) observed them in Argentine material and checked Carmichael's type material of *Echinella flabellata* (see also Honeywill, 1998). This feature, together with the distinctive plastid type and the presence of two small spines on the head pole (again rarely visible in the light microscope), characterize *L. flabellata*. No observations with a scanning electron microscope (SEM) or of live cells are mentioned by either Giffen (1970) or Witkowski *et al.* (2000). The stria density is slightly lower in *L. pfannkucheae*. Is it truly a separate species or simply within the natural range of variation of *L. flabellata*? If a true species, is it a regional endemic or a widespread species that has been mistaken for *L. flabellata*?

The second example is the pair of species *Berkeleya scopulorum* (Bréb.) Cox and *Climaconeis scalaris* (Bréb.) Cox, formerly confounded under *Navicula scopulorum* Bréb. They are virtually indistinguishable on the basis of acid-cleaned material, even in the SEM, but clearly differentiated by plastid structure as shown by Cox (1979). In general, it is not enough to assume that specimens viewed only with a light microscope can be identified by comparison to drawings or light micrographs in monographs, yet in many situations it is not possible to observe living cells or to observe material in the SEM, or the literature lacks these observations for comparison. Detailed studies such as that by Wachnicka & Gaiser (2007) are needed for most tropical diatom genera.

18.4 Farmer fish territories as special habitats for diatoms

Farmer fish are certain damselfishes (Pomacentridae, a family of small, common demersal fish). Numerous damselfish species are both herbivorous and defend territories, but knowledge of them is based on the intense study of a few species and little work on the rest (Ceccarelli *et al.*, 2001). However, some species are known well enough to be characterized as territorial algal turf farmers versus territorial algal turf grazers

Figure 18.3 *Stegastes nigricans* farm. The algal turf "farm" can be seen on the coral (*Porites rus*) branches in the foreground compared to the background.

Figure 18.4 Detail of diatoms on a rhodophyte filament from a *Stegastes nigricans* turf in Guam. The principal diatoms seen are *Climacosphenia elongata*, *Licmophora flabellata* and *Stictocyclus stictodiscus*; scale = 500 μm.

(Lewis, 1997). The farmer fish, including *Plectroglyphidodon lacrymatus* (Quoy & Gaimard, 1825), *Stegastes nigricans* (Lacepède, 1802) (Figure 18.3), and *Hemiglyphidodon plagiometapon* (Bleeker, 1852), selectively cultivate algae on areas of coral (Hixon & Brostoff, 1983, 1996; Polunin, 1988; Myers, 1989; Ceccarelli *et al.*, 2001; see also Lobban & Harrison 1994: 115). Hixon & Brostoff (1983) called them "keystone" species in maintaining the structure of algal communities. They are common on many tropical coral reefs, occupying and defending territories that in some habitats cover more than 50% of the reef planar surface area (Polunin, 1988). Yet, because they have a low impact on algae outside their territories, their ecological role in comparison to other herbivorous fishes is still debated (Ceccarelli *et al.*, 2001). In contrast to other herbivorous fishes, they not only consume algae, they also stimulate the growth of filamentous seaweeds on the older parts of coral branches and on dead coral, thus radically altering the community composition compared to areas outside territories (Lassuy, 1980; Hixon & Brostoff, 1996). They can also increase the filamentous epiphyte load on macroalgae (Ceccarelli *et al.*, 2005). Diatoms, including large chain-forming centric diatoms, can form an extensive secondary turf on top of the seaweed turf (Figure 18.4; Navarro & Lobban, 2009).

Differences in the macroalgae maintained by different fish species have been reported, suggesting that there might also be differences in the diatom epiflora. Studies in the Gulf of California, Mexico (Montgomery, 1980a,b) showed that *Microspathodon dorsalis* (Gill, 1862) maintains and feeds on a virtual monoculture of *Polysiphonia* species, whereas *Stegastes rectifraenum* (Gill, 1862) maintains a multispecies algal mat.

There is evidently also some geographic, perhaps latitudinal, variation in the relationships between the various farmer fish and their algal communities. *S. nigricans* territories have been studied in Papua New Guinea, Guam, and Okinawa. Hata & Kato (2002) described farms in Okinawa as a "virtual monoculture" of the filamentous red alga *Polysiphonia* sp. 1 (as *Womersleyella setacea* (Hollenb.) R. E. Norris), an alga they rarely found outside the territories, and they suggested a mutual interaction. Subsequently, they presented evidence for an obligate relationship between that *Polysiphonia* species and the farmer fish (Hata & Kato, 2003, 2006). In Guam (2200 km southeast of Okinawa; 13° N compared to 26° N) there is much more diversity in the seaweeds in *S. nigricans* turfs, the *S. nigricans* territories are much larger than those described by Hata & Kato (2004), and there is an abundant coating of diatoms (Navarro & Lobban 2009; Figure 18.4). There are also geographic differences in farmer-fish species distributions within regions as well as the major differences between Indo-Pacific, eastern Pacific, and Caribbean–Atlantic.

The role of diatoms in the above-described communities is virtually unknown. Until recently, papers that described algal species composition inside and outside damselfish territories made no mention of diatoms, included diatoms only as a single category, or classified the epiphytic diatoms as detritus. (We checked the 14 papers on this topic listed by Ceccarelli *et al.* (2001) and more recent papers including Hata & Kato (2002) and Zemke-White & Beatson (2005).) In view of the dearth of knowledge of marine tropical diatom floras, it is not surprising that studies of species composition in farmer-fish turfs were restricted to seaweed taxa. However, Jones *et al.* (2006) recently documented the importance of diatoms in farmer-fish turfs and guts. The diatom species involved were not identified but Jones *et al.* (2006) found that "diatoms were the most important algal food source" for three farmer fishes in Papua New Guinea, including *P. lacrymatus* and *S. nigricans*, comprising 30–80% by volume of stomach contents. They also concluded that these damselfish facilitate, rather than deplete, diatoms in their territories. Although diatoms may not be abundant in farmer-fish turfs in all regions, their abundance in Papua New Guinea, the Great Barrier Reef, Palau, and Guam suggests a broad Indo-Pacific or circum-tropical occurrence. It is clear that knowledge of the diatom species present would permit more ecologically accurate assessment of the symbioses between farmer fish and algae. At least in a general way, we can expect that nitrogen from fish wastes stimulates both the seaweeds and the diatoms, in much the same way that anemonefish stimulate growth of endosymbiotic zooxanthellae in the sea anemones with which they associate (Roopin *et al.*, 2008; Roopin & Chadwick, 2009).

Although the authors, our colleague Edward Theriot, and our students (Jordan *et al.* 2009) only recently began to study the diatoms associated with farmer-fish turfs, we can already recognize the broad outline of the structure. It is evident even by *in situ* inspection that both the diatoms and the seaweed taxa are very heterogeneously distributed within the often-complex three-dimensional territories of the fish. Moreover, the diatom assemblages on seaweeds in the territories have a three-dimensional structure reminiscent of larger land and sea vegetation; many are attached by mucilage pads to the seaweeds. The most abundant taxa in Guam, essentially the same in Palau so far as known, are those that form extensive branched chains, including *Chrysanthemodiscus floriatus* A. Mann and *Biddulphiopsis membranacea* (Cleve) H. A. von Stosch & R. Simonsen. Straight-chain formers include moderately sized cells and chains of *Lampriscus* spp. and the huge *Stictocyclus stictodiscus* (Grunow) Ross (Figure 18.5j), whose cells can be more than 125 μm diameter and a millimeter in length, thicker than many filamentous seaweeds. Then there are the very long, straight, tapered, or spindle-shaped cells of species that do not form chains, often attached by mucilage pads or stalks, such as *Climacosphenia* spp. (Figure 18.4), *Toxarium undulatum*, *Ardissonea* spp., the *Synedra* group of genera and certain *Nitzschia* spp.; many of these exceed 500 μm in length. Several species form long zigzag chains; common taxa in this group include *Grammatophora* spp. (Figure 18.5a,b), *Cyclophora tenuis* Castracane (Figure 18.5c,d), *Neosynedra* spp., *Achnanthes brevipes* C. Ag., and *Hyalosira tropicalis* Navarro. Amongst these, and sometimes forming very extensive colonies themselves, are many *Licmophora* species, ranging from the ubiquitous *L. flabellata* (Figure 18.2) to the rarely-recorded *L. debyi* (Leuduger-Fortmorel) A. Mann (Figure 18.5h,i). Some of these attach with simple mucilage pads (e.g. *L. remulus* Grunow) while others make long branched stalks (e.g. *L. flabellata*) that in turn become epiphytized by smaller *Licmophora* spp. and very small *Amphora* spp. Occasional small colonies of several tube dwellers contribute to this layer, including *Parlibellus* spp. and *Nitzschia martiana* (C. A. Agardh) van Heurck. Below what one can think of as these "canopy" and "understory" layers are the single-celled taxa, motile or sessile, including *Podocystis spathulata* (Shadbolt) Van Heurck, *Rhopalodia* spp. (Figure 18.5f), diverse *Mastogloia* spp. [*M. fimbriata* (Brightwell) Cleve and *M. binotata* (Grun.) Cleve (Figure 18.5g) are common], *Cocconeis* spp., *Amphora* spp., and *Nitzschia* spp., along with less common taxa such as *Podosira* sp. and many very small-celled taxa. Interlaced among all of these are loosely associated colonies of species such as *Bleakeleya notata* (Grun.) Round, *Hyalosira interrupta* (Ehrenb.) Navarro (Figure 18.5e), and *Bacillaria paxillifer* (O. F. Müller) Hendey. In a detailed but not exhaustive examination of a single *S. nigricans* turf sample, we documented over 120 taxa. Several of the species listed above have been rarely recorded and little is known of their ecology. For instance, in a paper whose title included "ecology" (Round, 1978), the ecological news was that the giant diatom *Stictocyclus stictodiscus* is benthic not planktonic.

18.5 Diatoms of meromictic marine lakes

Marine lakes are enclosed bodies of ocean water, i.e. with a normal mixture of sea salts, a small subset of the thousands of saline lakes around the world (Mackenzie *et al.*, 1995). Most saline lakes are well mixed (holomictic) or turn over annually once (monomictic) or twice (dimictic), but some turn over incompletely, leaving a deep unmixed layer for long periods (meromictic – often misleadingly referred to as "permanently" stratified lakes). Hamner & Hamner (1998) counted about 60 meromictic lakes of marine origin worldwide but the great majority occur in temperate and polar latitudes. Although other tropical lakes may be found to have these characters (possibly Lake Satonda in Indonesia; Kempe & Kazmierczak, 1990), some 11 lakes in Palau are the only ones presently known to be tropical, meromictic, and marine. Here we discuss the characteristics of the Palau lakes and the diatom research that we have carried out there over the last decade.

There are about 70 marine lakes located on numerous rock islands in the Republic of Palau (Western Caroline Islands), many of which have been isolated from the surrounding lagoon

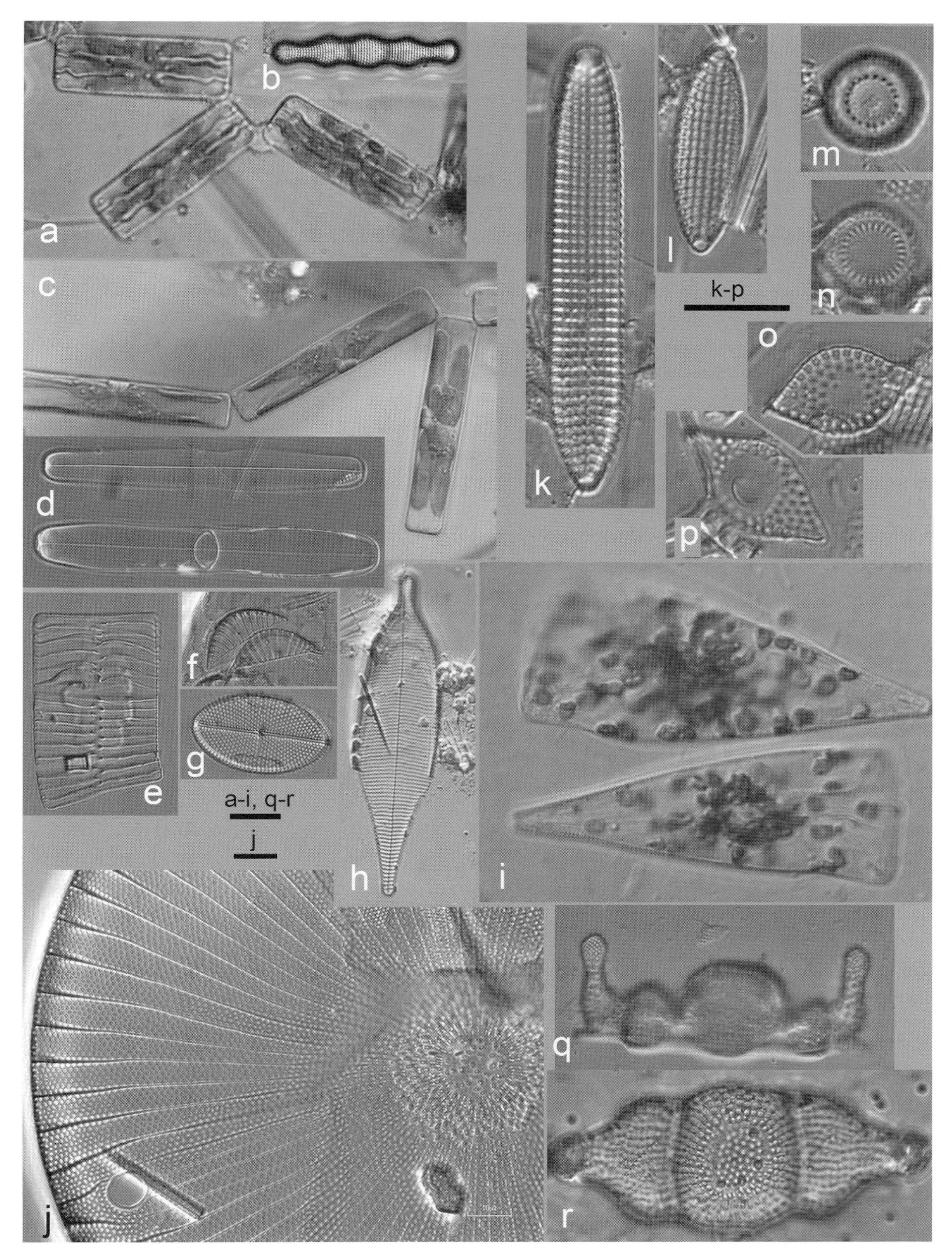

Figure 18.5 Sample diatoms in Guam farmer-fish turfs (a–j) and Palau Mecherchar Jellyfish Lake (k–q). (a, b) *Grammatophora undulata*, live cells showing the characteristic septa and cleaned valve; (c, d) *Cyclophora tenuis* live cells and cleaned valves showing characteristic cup-like thickening on one valve; (e) *Hyalosira interrupta*; (f) *Rhopalodia guettingeri*; (g) *Mastogloia binotata*. (h, i) *Licmophora debyi* valve and live cells in girdle view; (j) *Stictocyclus stictodiscus* valve showing characteristic pseudonodulus; (k,l) *Dimeregramma* spp. (p?); (m, n) *Paralia longispina*; (o, p) unidentified taxon in Cymatosiraceae; (q, r) *Biddulphia tuomeyi*. Scale bars = 10 μm.

waters throughout the Holocene (i.e. for at least 10,000 years). The islands are the result of uplifted Miocene coral reefs, and constant weathering and erosion has created a karst topography of depressions, caverns, and fissures in the limestone. After the last glacial maximum (c. 18,000 years ago), sea-level rise caused lagoon water to flood through the cracks and into the island basins. The subsequent lake formation apparently occurred in two steps, with the deeper depressions filling up first (c. 12,000 years ago), and the shallower depressions much later (c. 5000 years ago). Most of the lakes are fully marine and holomictic, but those that are meromictic are often inhabited by endemic Metazoa (Dawson, 2005; Dawson & Hamner, 2005). Yet little is known of the microbial diversity in the lakes, apart from a few papers on the resident seaweeds (Hara *et al.*, 2002), benthic foraminifera (Lipps & Langer, 1999; Kawagata *et al.*, 2005a,b), and photosynthetic bacteria (Venkateswaran *et al.*, 1993).

Hamner & Hamner (1998) identified eleven meromictic marine lakes in the Republic of Palau, of which seven are located on Mecherchar Island, including Mecherchar Jellyfish Lake. All of these lakes possess a bacterial layer in the water column overlying anoxic bottom waters. Of these lakes, we have sampled L-shaped Lake, Big Crocodile Lake, and Spooky Lake on Mecherchar Island, and Koror Jellyfish Lake on Koror Island. Here we provide a brief overview of the diatom communities of one of these meromictic marine lakes, Mecherchar Jellyfish Lake (JFL).

Jellyfish Lake is about 30 m deep, with a productive epilimnion in the top 15 m and a clear anoxic monimolimnion (a hypolimnion that doesn't mix) below, separated by a 3-m-thick layer of purple photosynthetic sulfur bacteria at the chemocline. Although lagoon waters enter the lake during high tide via three tunnels, there is no significant mixing with the lake water. The temperature and salinity of the lake's surface waters are about 31 °C and 26 psu, respectively (Hamner & Hamner, 1998). The lake is presently fringed by mangrove trees but there is evidence that mangroves were scarce or absent during drier climates, e.g. 100 years ago. Lack of water movement results in very clear water at the surface compared to many mangals. The mangrove roots are thickly coated with mussels, sponges, and green seaweeds but the diversity is low; large seaweeds are predominantly *Caulerpa fastigiata* Montagne, *Caulerpa verticillata* J. Agardh, *Avrainvillea amadelpha* (Montagne) A. Gepp & E. Gepp, and *Cladophora* spp. (Hara *et al.*, 2002), though they and the mangrove root bark directly are epiphytized by filamentous red and blue green algae. *A. amadelpha* has been found only in this lake in Palau, but both *Caulerpa* species also occur in the lagoon outside.

Previously published data on the lake's phytoplankton composition is limited to a brief comment in Hamner *et al.* (1982), "the phytoplankton is a complex mixture of chain-forming centric diatoms, of which *Chaetoceros* sp. is common, large dinoflagellates such as *Ceratium*, and myriad microflagellates." Our annual observations from 2001 to 2008 have confirmed that the dominant planktonic diatom genus is *Chaetoceros*, chiefly represented by *C. affinis* var. *affinis* Lauder and *C. affinis* var. *willei* (Grun.) Hasle. Varieties of *C. affinis* occurred in several other marine lakes, but they were never recorded in the lagoon waters outside JFL (and rarely in other lagoon waters), although other species of *Chaetoceros* (e.g. *C. peruvianus* Brightwell, *C. decipiens* P. T. Cleve) were often present in large numbers in the lagoons. There are also a number of other planktonic diatoms that appear to inhabit only these meromictic lakes, notably species of *Actinocyclus* and *Azpeitia*. Different species of these two genera are present in the lagoons, but are always uncommon. In the Republic of Palau, rainfall varies seasonally but does not strongly influence the meromictic marine lakes, with the freshwater merely sitting on top of the denser saline waters. However, these surficial waters may support freshwater or brackish species, since a number of genera (e.g. *Aulacoseira*), normally considered freshwater in origin, occur in Mecherchar Jellyfish Lake.

Like most of the meromictic marine lakes in the Republic of Palau, Mecherchar Jellyfish Lake has a muddy bottom, due to the input of leaves and debris from the mangroves and the steep forested slopes above them. The presence of the mangroves is considered to be one of the main reasons why diatom valves are well preserved in the lake sediments, due to the humic acids (e.g. tannins) accumulating in the submerged leaf litter. Although leaf litter does not always have a positive enrichment effect on benthic organisms, high concentrations of tannins may limit colonization of the surface sediments (Lee, 1999). Despite this, the benthic community of the lake is quite diverse, and is especially rich in raphid pennate genera such as *Amphora*, *Caloneis*, *Diploneis*, *Lyrella*, *Navicula*, *Nitzschia*, *Pinnularia*, *Pleurosigma*, and *Surirella*. The other meromictic lakes have similar assemblages, but the benthic communities of the lagoons appear somewhat different, and lack such genera as *Lyrella* and *Pinnularia*. A number of epiphytic diatom taxa appear to be restricted to JFL, including several species of *Triceratium* and *Dimeregramma*, and unidentified members of the Cymatosiraceae. Similarly, some diatom genera common on lagoon seaweed samples have not been found in JFL, including *Ardissonea*, *Bleakeleya*, *Licmophora*, and *Toxarium*.

Several short sediment cores obtained from Jellyfish Lake were dominated by small epiphytic diatoms that were not found in the seaweed samples. They may be associated more directly with the mangrove roots but our data are not yet complete enough to confirm their habitat. To date, there have been few investigations of mangrove root diatom communities from around the world (Gaiser *et al.*, 2005), despite the fact that subtropical coastal environments are dominated by mangrove swamps. In those that have been published (Sullivan, 1981a; Podzorski, 1985; Navarro & Torres, 1987; Nagumo & Hara, 1990; Reimer, 1996; Siqueiros-Beltrones & Castrejón, 1999; Sylvestre *et al.*, 2004), the assemblages include diatom genera such as *Amphora*, *Cocconeis*, *Cyclotella*, *Diploneis*, *Denticula*, *Hyalosynedra*, *Mastogloia*, *Navicula*, *Nitzschia*, *Rhopalodia*, and *Rhabdonema*.

Preliminary results indicate that JFL has a number of distinct habitats and diatom communities, which appear to be similar to some of the other meromictic marine lakes, especially those on Mecherchar Island, but are different from the surrounding lagoon waters and monomictic marine lakes. For example, Lake Ketau on Mecherchar Island is not meromictic and has a flora/fauna similar to the lagoon outside (Hamner & Hamner, 1998). Our observations confirm that this lake is dominated by *Pseudosolenia calcar-avis* Schulze and not *Chaetoceros* like the other Mecherchar lakes. Clearly, more data are needed on the species composition of the meromictic lake assemblages, while future paleolimnological work should include an investigation into whether diatoms can be used to study the effects of El Niño/Southern Oscillation (ENSO) events on the marine lakes (Dawson *et al.*, 2001; Martin *et al.*, 2005). According to Martin *et al.* (2005), Mecherchar Jellyfish Lake is more strongly stratified in La Niña than in El Niño conditions.

18.6 Potential uses of coral reef diatoms as indicators

How might diatoms be useful as indicators of the kinds of problems that beset coral reefs today, how do reefs respond to the changes, and have such changes occurred in the past? Since many diatom species have clearly defined and often narrow environmental requirements, they are widely used in other habitats to indicate water quality, sea level, and the paleoenvironment, as well illustrated by other chapters in this volume. There seems every reason to suppose that they can serve our understanding of reefs, but this will depend on knowing the taxonomy and ecology of at least some of the hundreds of species present either as individuals or as characteristic assemblages. Key characteristics of coral-reef environments include the relatively stable temperature, low nutrients except in situations where there is tight nutrient cycling (symbiotic associations of corals and zooxanthellae, fish, and algae), extreme small-scale heterogeneity of environment in the complex three-dimensional reef structure (especially irradiance, water motion, nutrient pulses), and extreme patchiness of communities on several scales from centimeters or less (e.g. across a farmer-fish territory) to tens or thousands of meters (across and along a reef). A particular diatom may form a dense patch only a few centimeters across that lasts for only a short time (as we have observed with several *Climaconeis* spp. at our main Guam study site), while other taxa are routinely found throughout a study site (e.g. *Biddulphiopsis* at that site). Some tropical diatoms that have been rarely recorded can be common and even abundant in particular habitats and geographic locations (e.g. *Licmophora normaniana* in the Bahamas (Hein *et al.*, 2008) or *Stictocyclus stictodiscus* in the western Pacific), while others may be rare all the time and everywhere. The trick will be to discover which of these species have occurrence patterns or environmental constraints that can be exploited to indicate changes on the reefs that we cannot already tell from other signals.

18.7 Summary

Coral reefs are structurally complex, biologically diverse ecosystems in nutrient-poor tropical seas and the reef symbiosis is finely attuned to high light and low nutrients, conditions that are increasingly deteriorating due to anthropogenic inputs. However, if modern coral-reef diatom assemblages could be clearly defined, we believe there is a strong possibility that components of past reef communities may be recognized in fossil assemblages throughout the world, when the distribution of coral reefs was much more widespread.

References

Adey, W. H. (2000). Coral reef ecosystems and human health: biodiversity counts! *Ecosystem Health*, **6**, 227–36.

Amato, A., Kooistra, W. H. C. F., Ghiron, J. H. L., *et al.* (2006). Reproductive isolation among sympatric cryptic species in marine diatoms. *Protist*, **158**, 193–207.

Berg, H., Linden, O., Ohman, M. C., & Troeng, S. (1998). Environmental economics of coral reef destruction in Sri Lanka. *Ambio*, **27**, 627.

Brander, L. M., Van Beukering, P., & Cesar, H. S. J. (2007). The recreational value of coral reefs: a meta-analysis. *Ecological Economics*, **63**, 209–18.

Carpenter, K. E. & Springer, V. G. (2005). The center of marine shore fish biodiversity: the Philippine Islands. *Environmental Biology of Fishes*, **72**, 467–80.

Ceccarelli, D. M., Jones, G. P., & McCook, L. J. (2001). Territorial damselfishes as determinants of the structure of benthic communities on coral reefs. *Oceanography and Marine Biology Annual Review*, **39**, 355–89.

Ceccarelli, D. M., Jones, G. P., & McCook, L. J. (2005). Effects of territorial damselfish on an algal-dominated coastal coral reef. *Coral Reefs*, **24**, 606–20.

Cesar, H. S. J. (ed.) (2000). *Collected Essays on the Economics of Coral Reefs*. Kalmar, Sweden: Coral Degradation in the Indian Ocean (CORDIO).

Cox, E. J. (1979). Studies on the diatom genus *Navicula* Bory. *Navicula scopulorum* Bréb. and a further comment on the genus *Berkeleya* Grev. *British Phycological Journal*, **14**, 161–74.

Dawson, M. N. (2005). Five new subspecies of *Mastigias* (Scyphozoa: Rhizostomeae: Mastigiidae) from marine lakes, Palau, Micronesia. *Journal of the Marine Biological Association UK*, **85**, 679–94.

Dawson, M. N. & Hamner, W. M. (2005). Rapid evolutionary radiation of marine zooplankton in peripheral environments. *Proceedings of the National Academy of Sciences of the United States of America*, **102**, 9235–40.

Dawson, M. N., Martin, L. E., & Penland, L. K. (2001). Jellyfish swarms, tourists, and the Christ-child. *Hydrobiologia*, **451**, 131–44.

Fenchel, T. & Finlay, B. J. (2004). The ubiquity of small species: patterns of local and global diversity. *BioScience*, **54**, 777–84.

Foissner, W. (2006). Biogeography and dispersal of micro-organisms: a review emphasizing protists. *Acta Protozoologica*, **45**, 111–36.

Foissner, W. (2008). Protist diversity and distribution: some basic considerations. *Biodiversity and Conservation*, **17**, 235–42.

Frankovich, T. A., Gaiser, E. E., Zieman, J. C., & Wachnicka, A. H. (2006). Spatial and temporal distributions of epiphytic diatoms growing on *Thalassia testudinum* Banks ex König: relationships to water quality. *Hydrobiologia*, **569**, 259–71.

Gaiser, E., Wachnicka, A., Ruiz, P., Tobias, F., & Ross, M. (2005). Diatom indicators of ecosystem change in subtropical coastal wetlands. In *Estuarine Indicators*, ed. S. A. Bortone, Boca Raton, FL: CRC Press.

Giffen, M. H. (1970). Contributions to the diatom flora of South Africa IV. The marine littoral diatoms of the estuary of the Kowie River, Port Alfred, Cape Province. *Nova Hedwigia, Beiheft*, **31**, 259–312.

Gordon, I. (2007). Linking land to ocean: feedbacks in the management of socio-ecological systems in the Great Barrier Reef catchments. *Hydrobiologia*, **591**, 29–33.

Gottschalk, S., Uthicke, S., & Heimann, K. (2007). Benthic diatom community composition in three regions of the Great Barrier Reef, Australia. *Coral Reefs*, **26**, 345–57.

Hallock, P. (1984). Distribution of selected species of living algal symbiont-bearing foraminifera on two Pacific coral reefs. *Journal of Foraminiferal Research*, **14**, 250–61.

Hamner, W. M., Gilmer, R. W., & Hamner, P. P. (1982). The physical, chemical, and biological characteristics of a stratified, saline, sulfide lake in Palau. *Limnology and Oceanography*, **27**, 896–909.

Hamner, W. M. & Hamner, P. P. (1998). Stratified marine lakes of Palau (Western Caroline Islands). *Physical Geography*, **19** (3), 175–220.

Hara, Y., Horiguchi, T., Hanzawa, N., *et al.* (2002). The phylogeny of marine microalgae from Palau's marine lakes. *Kaiyo Monthly*, **29**, 19–26 (in Japanese).

Hata, H. & Kato, M. (2002). Weeding by the herbivorous damselfish *Stegastes nigricans* in nearly monocultural algae farms. *Marine Ecology Progress Series*, **237**, 227–31.

Hata, H. & Kato, M. (2003). Demise of monocultural algal farms by exclusion of territorial damselfish. *Marine Ecology Progress Series*, **263**, 159–67.

Hata, H. & Kato, M. (2004). Monoculture and mixed-species algal farms on a coral reef are maintained through intensive and extensive management by damselfishes. *Journal of Experimental Marine Biology and Ecology*, **313**, 285–96.

Hata, H. & Kato, M. (2006). A novel obligate cultivation mutualism between damselfish and *Polysiphonia* algae. *Biology Letters*, **2**, 593–96.

Hein, M. K., Winsborough, B. M., & Sullivan, M. J. (2008). Bacillariophyta (Diatoms) of the Bahamas. *Iconographia Diatomologica*, **19**, 1–300.

Hendey, N. I. (1964). *An Introductory Account of the Smaller Algae of British Coastal Waters. Part V: Bacillariophyceae (Diatoms)*. London: Her Majesty's Stationery Office.

Hixon, M. A. & Brostoff, W. N. (1983). Damselfish as keystone species in reverse: intermediate disturbance and diversity of reef algae. *Science*, **220**, 511–13.

Hixon, M. A. & Brostoff, W. N. (1996). Succession and herbivory: effects of differential fish grazing on Hawaiian coral-reef algae. *Ecological Monographs*, **66**, 67–90.

Hoegh-Guldberg, O., Mumby, P. J., Hooten, A. J., *et al.* (2007). Coral reefs under rapid climate change and ocean acidification. *Science*, **318**, 1737–42.

Hogarth, P. J. (2007). *The Biology of Mangroves and Seagrasses*. Oxford: Oxford University Press.

Honeywill, C. (1998). A study of British *Licmophora* species and a discussion of its morphological features. *Diatom Research*, **13**, 221–71.

Hughes, T. P., Baird, A. H., Bellwood, D. R., *et al.* (2003). Climate change, human impacts, and the resilience of coral reefs. *Science*, **301**, 929–33.

Hughes, T. P., Bellwood, D. R., & Connolly, S. R. (2002). Biodiversity hotspots, centres of endemicity, and the conservation of reefs. *Ecology Letters*, **5**, 775–84.

Hustedt, F. (1931). Die Kieselalgen Deutschlands, Österreichs und der Schweiz unter Berücksichtigung der übrigen Länder Europa sowie der angrenzenden Meeresgebiete. In *Rabenhorsts Kryptogamen-Flora von Deutschland, Österreich und der Schweiz*, Leipzig: Akademische Verlagsgesellschaft. vol. 7, part 2, section 1.

Jones, G. P., Santana, L., McCook, L. J., & McCormick, M. I. (2006). Resource use and impact of three herbivorous damselfishes on coral reef communities. *Marine Ecology Progress Series*, **328**, 215–24.

Jordan, R. W., Lobban, C. S., & Theriot, E. C. (2009). Western Pacific diatoms project. ProtistCentral. See http://www.protistcentral.org/Project/get/project/id/17.

Kawagata, S., Yamasaki, M., Genka, R., & Jordan, R. W. (2005a). Shallow-water benthic foraminifers from Mecherchar Jellyfish Lake (Ongerul Tketau Uet), Palau. *Micronesica*, **37**, 215–33.

Kawagata, S., Yamasaki, M., & Jordan, R. W. (2005b). *Acarotrochus lobulatus*, a new genus and species of shallow-water foraminifer from Mecherchar Jellyfish Lake, Palau, NW Equatorial Pacific Ocean. *Journal of Foraminiferal Research*, **35**, 44–9.

Kempe, S. & Kazmierczak, J. (1990). Chemistry and stromatolites of the sea-linked Satonda Crater Lake, Indonesia: a recent model for the Precambrian sea? *Chemical Geology*, **81**, 299–310.

Kerswell, A. P. (2006). Global biodiversity patterns of benthic marine algae. *Ecology*, **87**, 2479–88.

King, R. J. (1990). Macroalgae associated with the mangrove vegetation of Papua New Guinea. *Botanica Marina*, **33**, 55–62.

Kleypas, J. A., McManus, J. W., & Menez, L. A. B. (1999). Environmental limits to coral reef development: where do we draw the line? *American Zoologist*, **39**, 146–59.

Lassuy, D. R. (1980). Effects of "farming" behavior by *Eupomacentrus lividus* and *Hemiglyphidodon plagiometapon* on algal community structure. *Bulletin of Marine Science*, **30**, 304–12.

Lee, S. Y. (1999). The effect of mangrove leaf litter enrichment on macrobenthic colonization of defaunated sandy substrates. *Estuarine, Coastal and Shelf Science*, **49**, 703–12.

Lewis, A. R. (1997). Effects of experimental coral disturbance on the structure of fish communities on large patch reefs. *Marine Ecology Progress Series*, **161**, 37–50.

Lipps, J. H. & Langer, M. R. (1999). Benthic foraminifera from the meromictic Mecherchar Jellyfish Lake, Palau (western Pacific). *Micropaleontology*, **45**, 278–84.

Littler, M. M. & Littler, D. S. (1984). Models of tropical reef biogenesis: the contribution of algae. *Progress in Phycological Research*, **3**, 323–64.

Lobban, C. S. & Harrison, P. J. (1994). *Seaweed Ecology and Physiology*. New York: Cambridge University Press.

Mackenzie, F. T., Vink, S., Wollast, R., & Chou, L. (1995). Comparative geochemistry of marine saline lakes. In *Physics and Chemistry of Lakes*, ed. A. Lerman, D. Imboden, & J. Gat, New York, NY: Springer-Verlag, 265–78.

Mallik, T. K. (1999). Calcareous sands from a coral atoll – should it be mined or not? *Marine Georesources and Geotechnology*, **17**, 27–32.

Martin, L., Dawson, M. N., Bell, L. J., & Colin, P. L. (2005). Marine lake ecosystem dynamics illustrate ENSO variation in the tropical western Pacific. *Biology Letters*, **2**, 144–7.

Miller, W. I., Montgomery, R. T., & Collier, A. W. (1977). A taxonomic survey of the diatoms associated with Florida Keys coral reefs. *Proceedings of the Third International Coral Reef Symposium*. Miami, CA: Rosenstiel School of Marine and Atmospheric Science.

Montgomery, R. T. (1978). Environmental and ecological studies of the diatom communities associated with the coral reefs of the Florida Keys, vols. I. & II. Unpublished Ph.D. thesis, Florida State University, Tallahassee. Accessible online at http://digitool.fcla.edu/R/?LOCAL_BASE=GEN01-FSU01&pds_handle=GUEST.

Montgomery, W. L. (1980a). Comparative feeding ecology of two herbivorous damselfishes (Pomacentridae: *Teleostei*) from the Gulf of California, Mexico. *Journal of Experimental Marine Biology and Ecology*, **47**, 9–24.

Montgomery, W. L. (1980b). The impact of non-selective grazing by the giant blue damselfish, *Microspathodon dorsalis*, on algal communities in the Gulf of California, Mexico. *Bulletin of Marine Science*, **30**, 290–303.

Mumby, P. J. (2006) Connectivity of reef fish between mangroves and coral reefs: algorithms for the design of marine reserves at seascape scales. *Biological Conservation*, **128**, 215–22.

Mumby, P. J., Edwards, A. J., Arias-González, J. E., *et al.* (2004). Mangroves enhance the biomass of coral reef fish communities in the Caribbean. *Nature*, **427**, 533–6.

Myers, R. (1989). *Micronesian Reef Fishes*, Barrigada, Guam: Coral Graphics.

Nagumo, T. & Hara, Y. (1990). Species composition and vertical distribution of diatoms occurring in a Japanese mangrove forest. *The Japanese Journal of Phycology* (Sôrui), **38**, 333–43.

Navarro, J. N. (1982). Marine diatoms associated with mangrove prop roots in the Indian River, Florida, U.S.A. *Bibliotheca Phycologica*, **61**, 1–151.

Navarro, J. N. (1983). A survey of the marine diatoms of Puerto Rico. VII. Suborder Raphidineae: families Auriculaceae, Epithemiaceae, Nitzschiaceae and Surirellaceae. *Botanica Marina*, **26**, 393–408.

Navarro, J. N. & Lobban, C. S. (2009). Freshwater and marine diatoms from the western Pacific islands of Yap and Guam, with notes on some diatoms in damselfish territories. *Diatom Research*, **24**, 123–57.

Navarro, J. N., Micheli, C. J., & Navarro, A. O. (2000). Benthic diatoms of Mona Island (Isla de Mona), Puerto Rico. *Acta Cientifica*, **14**, 103–43.

Navarro, J. N., Pérez, C., Arce, N., & Arroyo, B. (1989). Benthic marine diatoms of Caja de Muertos Island, Puerto Rico. *Nova Hedwigia*, **49**, 333–67.

Navarro, J. N. & Torres, R. (1987). Distribution and community structure of marine diatoms associated with mangrove prop roots in the Indian River, Florida, U.S.A. *Nova Hedwigia*, **45**, 101–12.

Pandolfi, J. M., Bradbury, R. H., Sala, E., *et al.* (2003). Global trajectories of long-term decline of coral reef ecosystems. *Science*, **301**, 955–8.

Pastorak, R. A. & Bilyard, G. R. (1985). Effects of sewage pollution on coral reef communities. *Marine Ecology Progress Series*, **21**, 175–89.

Paulay, G. (ed.) (2003). Marine biodiversity of Guam and the Marianas. *Micronesica*, **35–36**, 1–682.

Podzorski, A. C. (1985). An illustrated and annotated check list of diatoms from the Black River Waterways, St. Elizabeth, Jamaica. *Bibliotheca Diatomologica*, **7**, 1–177.

Polunin, N. V. C. (1988). Efficient uptake of algal production by a single resident herbivorous fish on the reef. *Journal of Experimental Marine Biology and Ecology*, **123**, 61–76.

Post, E. (1967). Zur Ökologie des Bostrychietum. *Hyrdrobiologia*, **29**, 263–87.

Poulíčková, A., Špačková, J., Kelly, M., Duchoslav, M., & Mann, D. G. (2008). Ecological variation within *Sellaphora* species complexes

(Bacillariophyceae): specialists or generalists? *Hydrobiologia*, **614**, 373–86.

Reimer, C. W. (1996). Diatoms from some surface waters on Great Abaco Island in the Bahamas (Little Bahama Bank). *Nova Hedwigia, Beiheft*, **112**, 343–54.

Richmond, R. (1994). Effects of coastal runoff on coral production. In *Proceedings of the Colloquium on Global Aspects of Coral Reefs: Health, Hazards and History*, ed. R. N. Ginsburg, Miami: University of Miami Press, pp. 360–4.

Roopin, M. & Chadwick, N. E. (2009). Benefits to host sea anemones from ammonia contributions of resident anemonefish. *Journal of Experimental Marine Biology and Ecology*, **370**, 27–34.

Roopin, M., Henry, R. P., & Chadwick, N. E. (2008). Nutrient transfer in a marine mutualism : patterns of ammonia excretion by anemonefish and uptake by giant sea anemones. *Marine Biology*, **154**, 547–56.

Round, F. E. (1978). *Stictocyclus stictodiscus* (Bacillariophyta): comments on its ecology, structure and classification. *Journal of Phycology*, **14**, 150–6.

Sadovy, Y. (2005). Trouble on the reef: the imperative for managing vulnerable and valuable fisheries. *Fish and Fisheries*, **6**, 167–85.

Sar, E. A. & Ferrario, M. E. (1990). *Licmophora flabellata*. Ultrastructure and taxonomy. 1. Implication. *Diatom Research*, **4**, 403–8.

Simonsen, R. (1970). Protoraphidaceae, eine neue Familie der Diatomeen. *Nova Hedwigia, Beiheft*, **31**, 377–94.

Siqueiros-Beltrones, D., López-Fuerte, F. O., & Gárate-Lizárraga, I. (2005). Structure of diatom assemblages living on prop roots of the red mangrove (*Rhizophora* mangle) from the west coast of Baja California Sur, Mexico. *Pacific Science*, **59**, 79–96.

Siqueiros-Beltrones, D. A. & Castrejón, E. S. (1999). Structure of benthic diatom assemblages from a mangrove environment in a Mexican subtropical lagoon. *Biotropica*, **31**, 48–70.

Skelton, P. A. & South, G. R. (2002). Mangrove-associated algae from Samoa, South Pacific. *Constancea*, **83.12**. See http://ucjeps.berkeley.edu/constancea/83/skelton_south/skelton_south.html; accessed August 18, 2008.

Spalding, M. D., Ravilious, C., & Green, E. P. (2001) *World Atlas of Coral Reefs*. Berkeley, CA: University of California Press.

Sullivan, M. J. (1981a). Community structure of diatoms epiphytic on mangroves and *Thalassia* in Bimini Harbour, Bahamas. In *Proceedings of the Sixth Symposium on Recent and Fossil Diatoms, Budapest*, ed. R. Ross, pp. 385–98. Königstein: Koeltz Scientific Books.

Sylvestre, F., Guiral, D., & Debenay, J. P. (2004). Modern diatom distribution in mangrove swamps from the Kaw Estuary (French Guiana). *Marine Geology*, **208**, 281–93.

Tomasetti, R. (2007). Global biogeography of marine algae. Unpublished M.Sc. thesis, University of Guam.

Tomlinson, P. B. (1995). *The Botany of Mangroves*, Cambridge: Cambridge University Press.

UNEP-WCMC. (2006) *In the Front Line: Shoreline Protection and Other Ecosystem Services from Mangroves and Coral Reefs*, Biodiversity series number 24, Cambridge: United Nations Environmental Programme – World Conservation Monitoring Centre, See http://www.unep-wcmc.org/resources/publications/UNEP_WCMC_bio_series/24.cfm, accessed August 28, 2008.

Vanormelingen, P., Verleyen, E., & Vyverman, W. (2008). The diversity and distribution of diatoms: from cosmopolitanism to narrow endemism. *Biodiversity and Conservation*, **17**, 393–405.

Venkateswaran, K., Shimada, A., Maruyama, A., *et al.* (1993). Microbial characteristics of Palau Jellyfish Lake. *Canadian Journal of Microbiology*, **39**, 506–12.

Veron, J. E. N. (2000). *Corals of the World* (three volumes), Townsville: Australian Institute of Marine Science.

Vincent, A. C. J. (2006). Live food and non-food fisheries on coral reefs, and their potential management. In *Coral Reef Conservation*, ed. I. M. Côté & J. M. Edwards, New York: Cambridge University Press, pp. 183–235.

Wachnicka, A. & Gaiser, E. E. (2007). Morphological characterization of *Amphora* and *Seminavis* (Bacillariophycea) from South Florida, U.S.A. *Diatom Research*, **22**, 387–455.

Witkowski, A., Lange-Bertalot, H., & Metzeltin, D. (2000). Diatom Flora of Marine Coasts I. In *Iconographia diatomologica*, ed. H. Lange-Bertalot, Ruggell: A.R.G. Gantner Verlag.

Zemke-White, L. W., & Beatson, E. L. (2005). Algal community composition within territories of the damselfish *Stegastes nigricans* (Pomacentridae, Labroidei) in Fiji and the Cook Islands. *The South Pacific Journal of Natural Science*, **23**, 43–7.

Zolan, W. J. (1980). Periphytic diatom assemblages on a windward fringing reef flat in Guam. Unpublished M.S. thesis, University of Guam.

19 Diatoms as indicators of former sea levels, earthquakes, tsunamis, and hurricanes

BENJAMIN P. HORTON AND YUKI SAWAI

19.1 Introduction

The significance of relative sea level during the late Quaternary is recognized by disciplines across the Earth sciences. Sea-level histories are important for calibrating and constraining geophysical models of Earth's rheology and glacio-isostatic adjustment (e.g. Peltier, 2004). Sea level is crucial to any study of coastal evolution as it serves as the ultimate baseline for continental denudation (Summerfield, 1991). For human populations, sea levels during the late Quaternary have been an important factor in sustaining coastal communities and may have profoundly influenced the very initiation of human civilization (e.g. Turney and Brown, 2007). Publication of reports from the Intergovernmental Panel on Climate Change (IPCC, 2007) re-emphasized the importance of sea level as a barometer of climate and drew attention to the potentially devastating consequences of future climate-related sea-level change (e.g. Rahmstorf, 2007). However, the IPCC also highlighted the uncertainty with which the driving mechanisms of sea-level change are understood and the disconnection between long-term geological and recent observational trends. Predictions of sea level for the twenty-first century rely on models, and the veracity of model output is based on verification against observations. Interpretation of these observations requires great care in light of the large spatial and temporal variability in relative sea-level change (Milne *et al.*, 2009).

Sea level is far from a constant, planar surface and exhibits spatial and temporal changes at a multitude of scales. To the observer, these changes are manifestations of relative sea level, a term which reflects the uncertainty in separating the often simultaneous contributions from movements of the ocean surface and land (Shennan, 2007). Relative sea-level changes record transfers of mass between oceans and continents during expansion and contraction of great ice sheets (e.g. Lambeck *et al.*, 2002), and incorporate extreme events such as storm surges (e.g. Donnelly and Woodruff, 2007) and tsunamis (e.g. Jankaew *et al.*, 2008). Further, relative sea-level changes document vertical movements in the Earth's crust over a wide range of timescales, from uplift and subsidence during great plate-boundary earthquakes (e.g. Nelson *et al.*, 2008), through century- and millennial-scale movements driven by glacio-isostatic adjustment (e.g. Milne *et al.*, 2005).

Since the last glacial maximum (*c.* 20,000 cal yr BP) approximately 50 million cubic kilometers of ice has melted from the land-based ice sheets, raising relative sea level in regions distant from the major glaciation centers (far-field sites) by *c.* 120 m (Lambeck *et al.*, 2002). In contrast, relative sea levels have dropped by many hundreds of meters in regions once covered by the major ice sheets as a consequence of the isostatic rebound of the solid Earth (e.g. Shaw *et al.*, 2002). Such changes in relative sea level are part of a complex pattern of interactions among eustatic, isostatic (glacio- and hydro-) and local factors, all of which have different response timescales. The eustatic contribution to sea-level change during deglaciation averaged 10 mm yr^{-1}; however peak rates potentially exceeded 50 mm yr^{-1} during "meltwater pulses" at 19,000 and 14,500 cal yr BP (Alley *et al.*, 2005). Empirical and glacio-isostatic modeling studies suggested a significant reduction in the eustatic contributions to relative sea-level change in the mid Holocene at *c.* 7000 cal yr BP and the Earth entered into a period of sea-level stability during which ocean volume, on average, changed only by a few meters (Milne *et al.*, 2005). Clarke *et al.* (1978) identified six types of sea-level curve (I–VI) during the mid to late Holocene, which reflect a range of relative sea-level histories recorded in coasts which have emerged, submerged, or are in transitional areas and record a combination of both uplift and subsidence (Figure 19.1). Although these curves

The Diatoms: Applications for the Environmental and Earth Sciences, 2nd Edition, eds. John P. Smol and Eugene F. Stoermer. Published by Cambridge University Press.

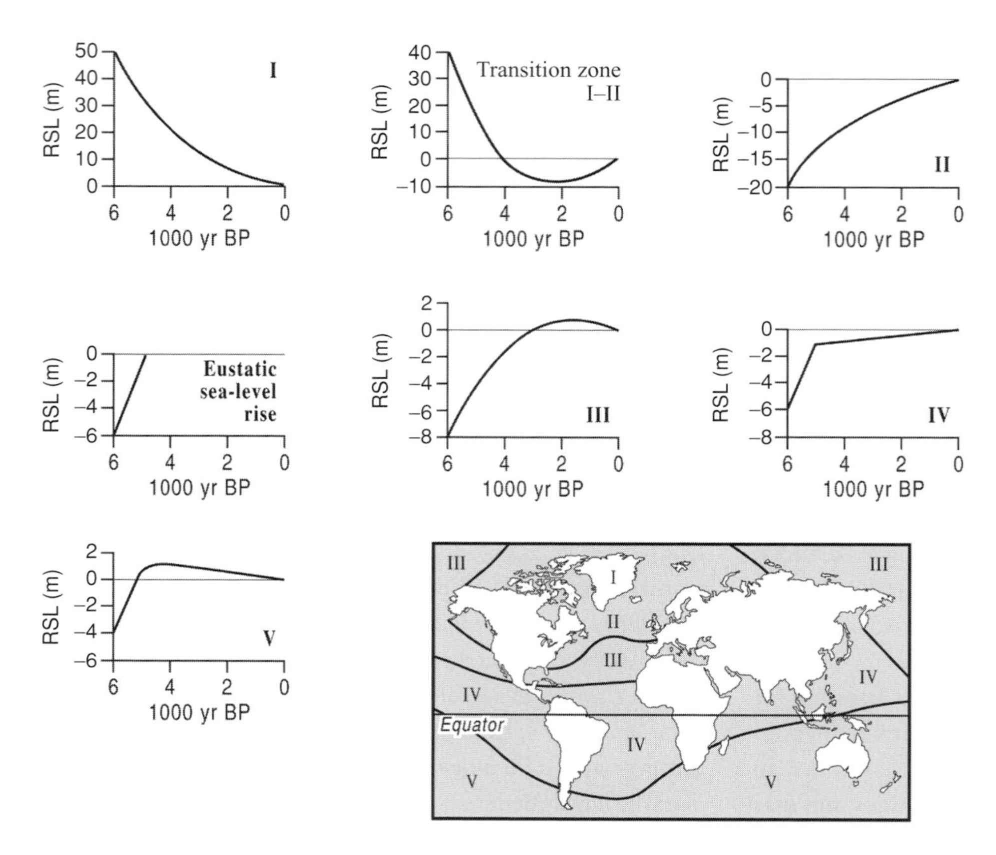

Figure 19.1 Sea-level zones and typical relative sea-level (RSL, in m) curves for each zone under the assumption that no eustatic change has occurred since 5000 yr BP. (Modified from Clark *et al.* (1978).)

provide the general impression of the rate and direction of relative sea-level change, they do not reflect the true uncertainty associated with estimates of the elevation and age of former sea levels.

In this chapter we examine the processes and patterns of late-Quaternary sea-level changes revealed by diatom analyses along the passive and active coastal margins. We provide examples of late-Quaternary relative sea-level reconstructions from salt marshes and isolation basins. We also explore the development of diatom-based transfer functions to produce high-resolution reconstructions of former sea level related to human-induced climate change and the earthquake deformation cycle. Finally, we evaluate the applicability of diatoms as proxies for tsunami and hurricane research.

19.2 Diatoms and salt-marsh environments

Many studies have sought to reconstruct variations in sea level using diatom data contained in a range of late Quaternary sedimentary deposits (e.g. Palmer and Abbott, 1986; Hemphill-Haley, 1995a,b; Zong and Tooley 1996; Dawson *et al.*, 1998; Zong, 1998; Plater *et al.*, 2000; Shennan *et al.*, 2000; Roe *et al.*, 2008). The utility of diatoms as a sea-level indicator is underpinned by the comprehensive ecological classifications of coastal diatoms, particularly for salinity (Hustedt, 1953, 1957; van der Werff and Huls, 1958–1974; Juggins, 1992; Vos and de Wolf, 1993). Polyhalobian and mesohalobian classes represent marine and brackish conditions whereas oligohalobian forms indicate freshwater environments. Changes in salinity across a salt marsh produces vertically zoned diatom assemblages with respect to the tidal frame (e.g. Nelson and Kashima, 1993; Zong and Horton, 1998; Sherrod, 1999; Patterson *et al.*, 2000, 2005; Horton *et al.*, 2006, 2007; Szkornik *et al.*, 2006; Roe *et al.*, 2008). Figure 19.2 illustrates diatoms assemblages from a salt marsh of eastern Hokkaido, northern Japan (Sawai *et al.*, 2004a). The assemblages show a clear transition from marine–brackish diatoms (e.g. *Planothidium delicatulum* and *Diploneis decipens* var. *parallela*) associated with tidal-flat settings through low marsh-vegetated environments, where a mixed diatom community are found, to a group dominated by freshwater and salt-tolerant freshwater taxa, such as *Amphora salina* in relation to high-marsh conditions (Figure 19.2a). These modern relations can be quantified and are representative of past conditions encountered in late-Quaternary intertidal sedimentary sequences. In this way, diatoms can be used to fix its former elevation relative

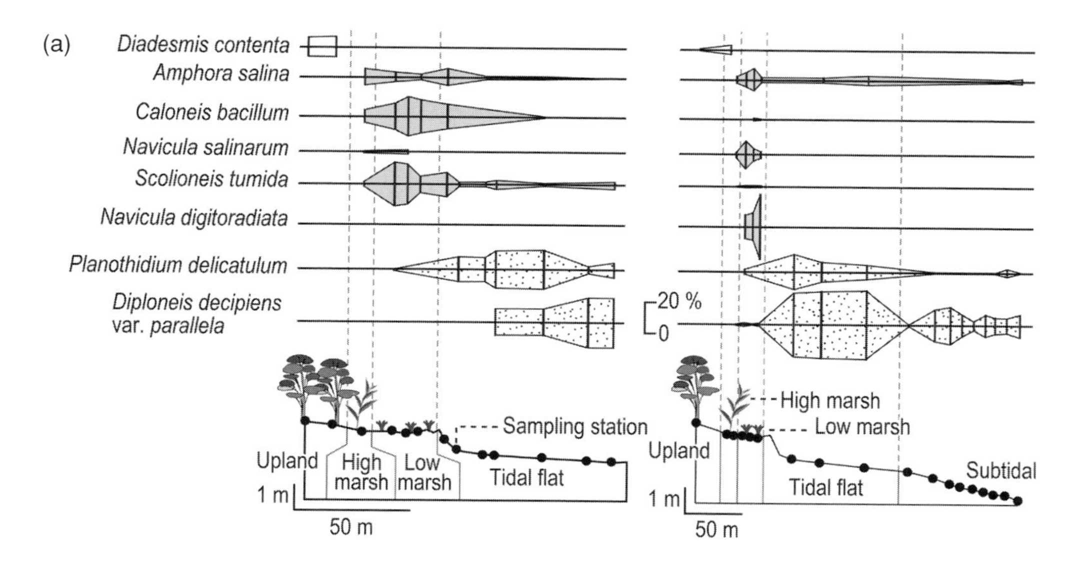

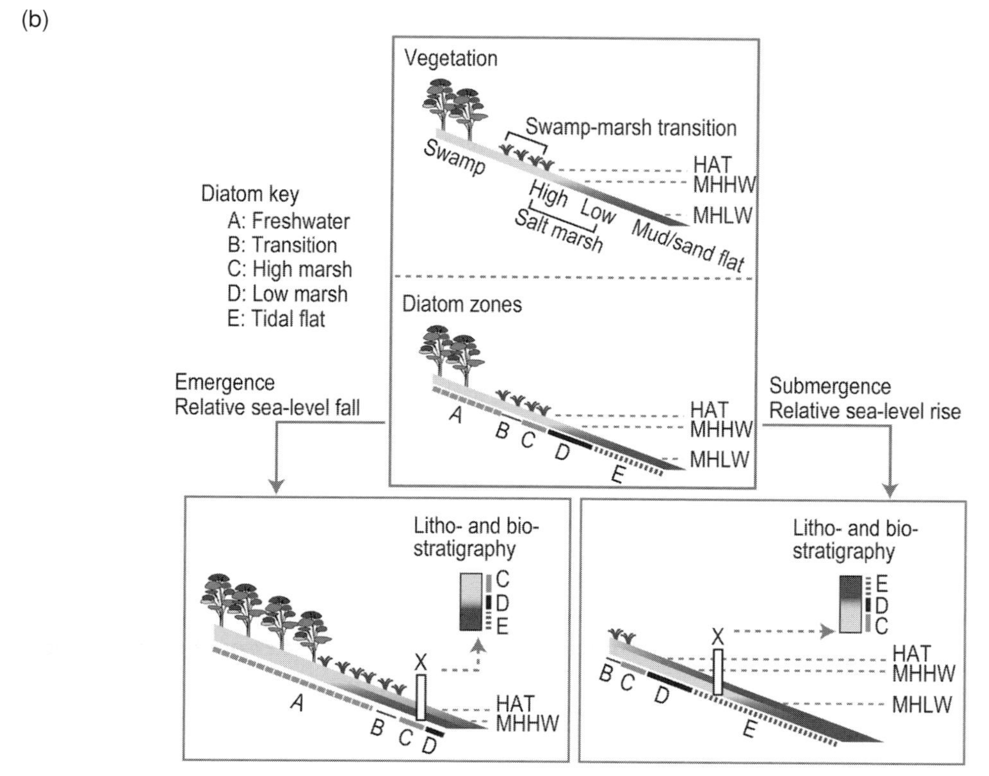

Figure 19.2 (a) Distribution of modern diatoms in a salt marsh of eastern Hokkaido, northern Japan. (Modified from Sawai *et al.* (2004a).) (b) Schematic representation of the lithostratigraphical and diatom signature of relative sea-level rise and fall in the paleoenvironment. Tidal levels: HAT = Highest Astronomical Tide; MHHW = Mean High High Water; MHLW = Mean High Low Water.

to the tidal frame at its time of formation (termed the indicative meaning; Figure 19.2b). This measure of paleomarsh-surface elevation, when combined with other lithostratigraphic data, can be used to infer the past position of local relative sea level (e.g. Hemphill-Haley, 1995b; Sherrod, 1999; Sawai *et al.*, 2004a; Kemp *et al.*, 2009a).

The relationship between diatom species and salt-marsh environments may be further illustrated through canonical correspondence analysis (CCA). This is a multivariate statistical technique that relates community composition to known environmental variations (ter Braak, 1987; ter Braak and Verdonschot, 1995). It is used to extract synthetic environmental gradients from ecological data sets. These gradients have been used to describe and illustrate succinctly the different habitat

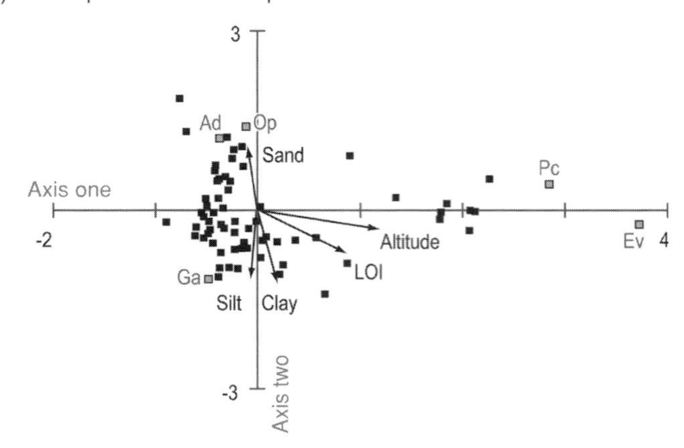

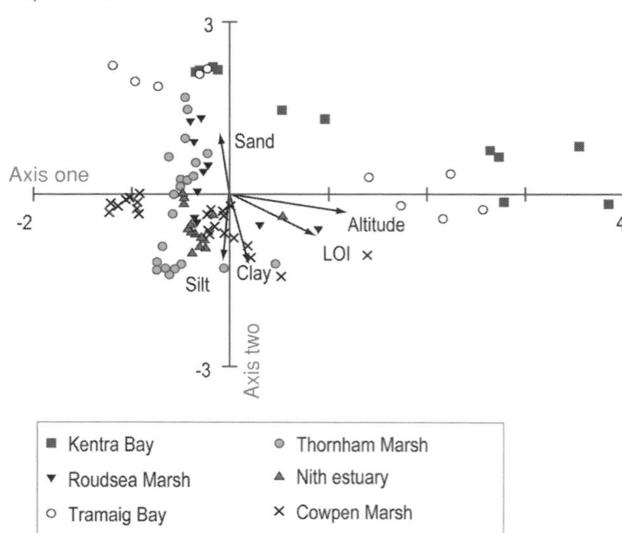

Figure 19.3 Canonical correspondence analysis biplots based on diatom abundances in six intertidal zones of southwest Scotland (Kentra Bay, Nith Estuary and Tramaig Bay) and northwest (Roudsea Marsh) and east (Cowpen and Thornham marshes) England. Species abbreviations: Ad = *Achnanthes delicatula*; Ev = *Eunotia exigua*; Ga = *Gyrosigma acuminatum*; Op = *Opephora pacifica*; Pc = *Pinnularia microstauron*. Environmental abbreviations: LOI = loss on ignition; SWLI = standardized water level index (surrogate for elevation). Only diatom taxa are included that exceeded 5% relative abundance. (Modified from Zong and Horton (1999).)

preferences (niches) of diatom taxa via an ordination diagram. Zong and Horton (1999) employed CCA to enhance understanding of the relationship between diatom assemblages and several environmental variables from six salt marshes of southwest Scotland, and northwest and east England (Figure 19.3). They identified three diatom assemblage zones. The first zone, plotted on the right of the diagram, is dominated by freshwater diatom taxa, particularly aerophilous forms (e.g. *Pinnularia microstauron*). This diatom zone coincides with the high-marsh coastal environment, where the frequency of tidal inundation is low and the duration short. Consequently, the sediments are fine grained with a high percentage of organic content. The second zone, plotted on the lower middle of the diagram, is associated mainly with a low-marsh coastal environment that is flooded by most high spring tides. Diatom assemblages in this zone are characterized by a mixed community of salt-tolerant freshwater species and brackish taxa. At sites with higher salinity, diatom assemblages can be dominated by brackish taxa. Most species in this second zone are epipelon. The third zone, on the left part of the bi plot, extends on to the tidal flat where vascular vegetation is absent and the substrate is less stable. Here, diatom assemblages are dominated by marine and brackish species (e.g. *Opephora pacifica*), most of which are either epipelic or epipsammic, depending on whether the substrates are muddy or sandy (Zong and Horton, 1999). The CCA axis one, therefore, reflects the major gradient from high marsh plotted on the right with high altitude and loss on ignition (LOI) values to tidal-flat samples plotted on the left (low altitude and LOI values). The CCA axis two reflects the grain-size gradient from sand fraction (top) through to silt and clay fractions (bottom).

Reliable sea-level reconstructions are, however, hampered by the identification of allochthonous (transported) from autochthonous taxa. In salt-marsh environments the relative abundance of allochthonous valves is generally higher than in quieter water environments such as lakes and freshwater bogs. The classical approach to address this problem is to separate allochthonous valves via life form of each species (Vos and de Wolf, 1993). For instance, planktonic diatoms can be regarded as allochthonous components in fossil diatom assemblages of coastal deposits, while benthic (e.g. epiphytic and epipsammic) are autochthonous (Simonsen, 1969; Vos and de Wolf, 1993). Fragmentation of diatom valves has often been used as an indicator of long-distance transportation (Voorrips and Jansma, 1974; Heyworth *et al.*, 1985; Vos and de Wolf, 1993). However, this approach is sometimes complicated by other processes (natural chemical dissolution, compaction, and sample treatment), which are difficult to evaluate (Vos and de Wolf, 1993). Development of a staining technique to observe living diatom specimens for modern samples under a light microscope allowed the allochthonous issue to be addressed directly. Kosugi (1985, 1986, 1987, 1989) compared living diatom populations with dead assemblages in surface sediments in central Japan. The work by Kosugi revealed a modification of diatom assemblages within sub-environments of the intertidal zone. Using this technique, Sawai (2001a) showed that *Cocconeis scutellum* could be a significant allochthonous component in intertidal deposits. *C. scutellum* lives on seagrass (*Zostera marina* and *Zostera japonica*), to which the cells are strongly attached through the raphid valve. After their death, however, non-attached rapheless valves become separated from the seagrass, whereas raphid valves remain attached. Separated rapheless valves are transported over long distances by tidal currents and should be excluded in reconstructing coastal environments (Sawai, 2001a). A further problem in reconstruction of coastal paleoenvironment using diatoms is the breakage and dissolution of diatom valves (Juggins, 1992; Ryves *et al.*, 2009). There exist several possible approaches to the general problem of poor preservation including weighting species and samples according to relative resistance to dissolution and a morphological assessment of valve preservation in samples.

19.3 Using diatoms to reconstruct sea-level change

Salt-marsh sediments provide a wealth of valuable information concerning late-Quaternary sea-level change. A variety of strategies and methodologies are employed to distil relative sea-level signals from these low-energy sediments which are underpinned by a multiproxy approach, employing a combination of litho-, bio-, and chrono-stratigraphic data (Edwards, 2007). Lithostratigraphy is usually determined from a series of sediment cores collected at a study site or in outcrops, which are described in the field and analyzed in the laboratory for grain size and organic content. Biostratigraphy (e.g. relative abundances of diatoms) is employed to establish the nature of the environment in which the sea-level indicator accumulated. Commonly, other microfossils (e.g. foraminifera or pollen) and/or geochemical are used to ensure reliability of any reconstruction (e.g. Patterson *et al.*, 2000, 2005; Gehrels *et al.*, 2001; Horton *et al.*, 2009). All reconstructions of relative sea level are dependent upon the development of an accurate and reliable chronological framework. This is commonly inferred by radiocarbon dating.

19.3.1 Sea-level index points

Critical to the reconstruction of late-Quaternary sea-level changes is the establishment of sea-level index points. A sea-level index point is a datum that can be employed to show vertical movements of sea level. The concept was developed during the International Geoscience Programme (IGCP) Projects 61 and 200, and is described in "Sea-Level Research: a Manual for the Collection and Evaluation of Data" (van de Plassche, 1986) as well as in many other papers (e.g. Tooley, 1982; Shennan, 1986; Shennan and Horton, 2002; Shennan, 2007; Engelhart *et al.*, 2009). In order to be established as an index point, a sediment sample must possess the following:

(1) **Location of a sea-level index point.** Once it is established that the sediment or morphological feature is in its location of origin, this attribute consists simply of its geographical coordinates (e.g. Engelhart *et al.*, 2009).
(2) **Age of a sea-level index point.** The majority of observations used for late Quaternary studies of sea-level changes have their ages measured by radiocarbon techniques and most modern studies use calibrated ages expressed with 95% confidence limits. The development of radiocarbon measurement using accelerator mass spectrometry (AMS) has dramatically reduced the minimum sample size for radiocarbon dating and facilitated dating of individual plant macrofossils, which, when prepared correctly, provides a means to reduce significantly contamination by younger (e.g. invasive plant roots) or older, allochthonous carbon. The zero in the BP (Before Present) scale for calibrated ages is 1950. Other widely used methods for establishing a chronology include measurements based on sediment characteristics, such as luminescence, isotopic properties, and archeological evidence. The most reliable sea-level index points should have at least one type of corroborating evidence to support the radiocarbon age and to demonstrate continuity of sedimentation or formation (e.g. Kemp *et al.*, 2009a).
(3) **Elevation of a sea-level index point.** Few sea-level index points formed exactly at paleo mean sea level (MSL). Many represent environments within the upper part of the tidal range but in total they cover the full tidal range. Thus, in order to measure relative sea-level change, it is necessary to establish the relationship of the sample to a tidal level. The relationship of a sample to a tide level, and hence sea level, is called the "indicative meaning" (van de Plassche 1986; Shennan 1986; Horton *et al.*, 2009). It comprises two parameters, namely the reference water level (e.g. mean higher high water (MHHW) and the indicative range (the vertical range over which the sample could occur).

Figure 19.4 illustrates the application of diatoms to produce four sea-level index points from Lockham on the north side of the Humber Estuary, UK (Metcalfe *et al.*, 2000). At the base of

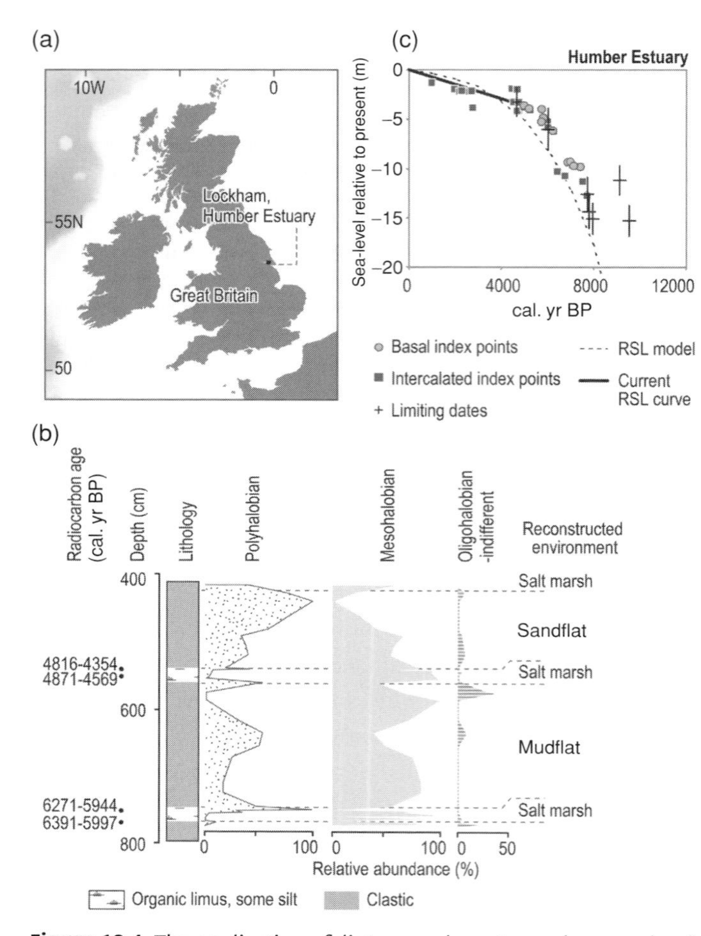

Figure 19.4 The application of diatom analyses to produce sea-level index points. (a) Location map of the UK showing the Humber Estuary, Lockham. (b) Lockham diatom diagram showing intercalated and basal peat sea-level index points. Diatoms are grouped according to their salinity classes (polyhalobous, mesohalobous, and oligohalobous; Hustedt, 1953, 1957). Calibrated radiocarbon ages and depth (cm) down-core shown on the left of the lithology column. The sediment legend is drawn according to a simplified Troels-Smith (1955) model. (Modified from Metcalfe *et al.*, 2000.) (c) Intercalated and basal sea-level index points and limiting data from the Humber Estuary. The best estimate of the late-Holocene trend is plotted as a solid line (see Shennan and Horton, 2002 for details of the methods used to define the trend). The dashed line shows the predicted relative sea level from the glacial isostatic adjustment model described by Peltier *et al.* (2002). (Modified from Shennan and Horton, 2002.)

the core, diatoms are well preserved in a peaty silt unit (771 to 755 cm) with polyhalobous and mesohalobous (e.g. *Navicula peregrina*, *Diploneis interrupta*, *Diploneis smithii*, and *Paralia sulcata*) taxa, which are indicative of a salt-marsh environment (Figure 19.4b). This basal peat produces two sea-level index points with ages of 6391–5997 and 6271–5944 cal yr BP. Up-core, the diatoms indicate increasing salinities and development of a mud flat. A regressive contact occurs at 573 cm with the return of salt-marsh diatom assemblages. Both the lower and upper parts of the salt-marsh unit have been dated to 4871–4569 and 4816–4354 cal yr BP, respectively, to produce two intercalated index points. Diatom preservation in the overlying sandy sediments is poor (450 to 426 cm) and the interpretation of this part of the sequence as a sandflat is based largely on the laminated texture. Resistant marine planktonic taxa are present, but their significance is uncertain (Metcalfe *et al.*, 2000).

The index points from Lockham are combined with other diatom-based sea-level data from the Humber Estuary to produce a Holocene sea-level curve (Figure 19.4c). The upward trend of sea level is rapid in the early Holocene, then at a much reduced rate in the mid and late Holocene, which conforms with modeled glacio-isostatic adjustment predictions (e.g. Peltier *et al.*, 2002; Shennan and Horton, 2002) for an area not far beyond the limits of the last glacial maximum ice advance. The scatter of index points includes the total influence of local-scale processes such as changes in tidal range and sediment consolidation.

19.3.2 Isolation basins

An isolation basin is a naturally enclosed rock basin that was at one time below sea level but has subsequently been raised above, and is isolated from the sea (Lloyd, 2000). Isolation basins act as sedimentary sinks and are used to reconstruct past sea-level changes in Scandinavia (e.g. Haften and Tallantire, 1978; Kjemperud, 1981; Corner *et al.*, 1999), Scotland (e.g. Lloyd, 2000; Shennan *et al.*, 2000; Lloyd and Evans, 2002), and Greenland (e.g. Long *et al.*, 1999, 2006, 2008). Figure 19.5a illustrates how the sedimentary sequence within an isolation basin records changes in environment from fully marine to fully freshwater through the process of coastal evolution. Marine–brackish diatoms are thus characteristic of pre-isolation basins, whilst brackish–freshwater and freshwater diatoms are characteristic of post-isolation basins (Kjemperud, 1981). The isolation event is marked in the stratigraphy by the contact between the marine and terrestrial deposits. Dating of the contact provides the age of isolation, whereas the height of the sill that separates the basin from the sea represents the height of the former position of sea level during isolation (Gehrels, 2007). Figure 19.5b–d summarizes the diatom, radiocarbon, and elevation data for two isolation basins from Northwest Scotland (Shennan *et al.*, 2000). The lithostratigraphy comprises a basal blue gray clay silt with a trace of sand overlain by a green gray silty organic deposit. The lithostratigraphic boundary,

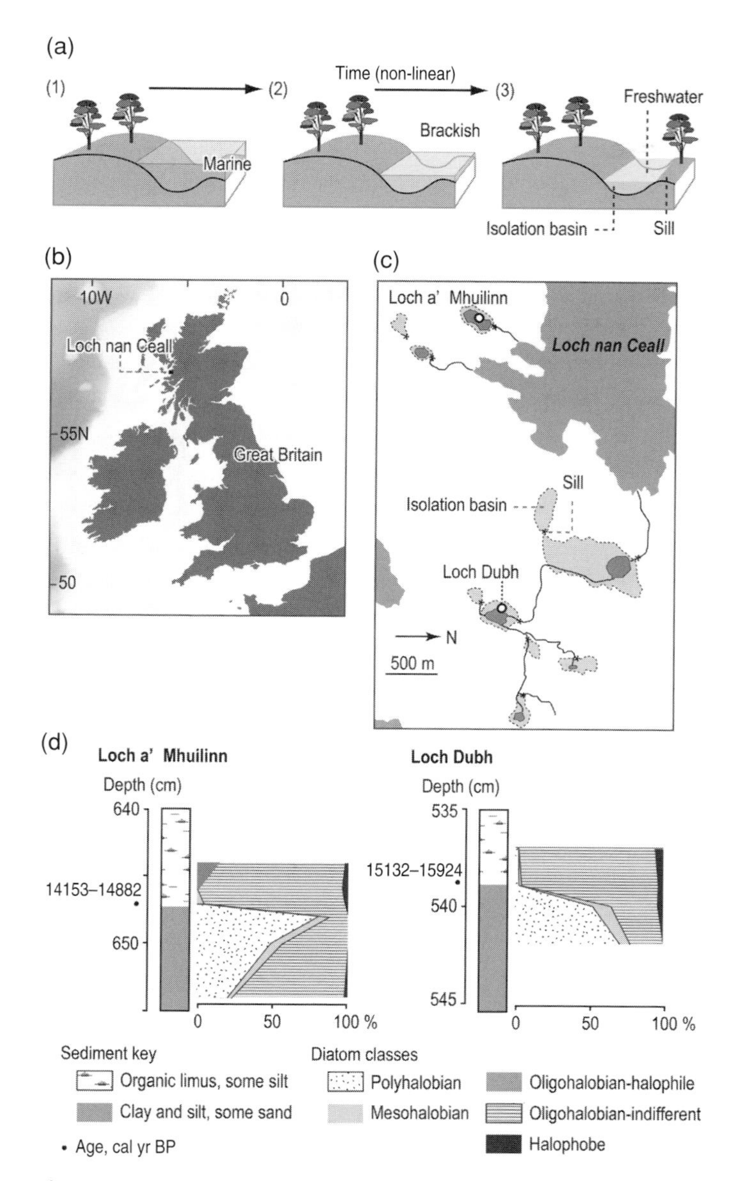

Figure 19.5 Sedimentary sequence within an isolation basin. (a) A schematic diagram showing the isolation of a tidal basin by falling relative sea level: (1) the basin is inundated at all stages of the tide and marine sediments accumulate; (2) the basin is isolated from the sea at low tides and brackish conditions are experienced; and (3) high water is now below the height of the sill and the isolation basin becomes a freshwater lake. (b) Location map of the UK showing the Loch nan Ceall study area. (c) Loch nan Ceall site map, showing the locations of isolation basins and their sill. (Modified from Shennan *et al*., 2000.) (d) Loch a'Mhuillinn and Loch Dubh diatom diagrams showing the transition from clastic sediment to organic limus to produce a sea-level index point. Diatoms are grouped according to their salinity classes (polyhalobous, mesohalobous, or oligohalobous; Hustedt, 1953, 1957). Calibrated radiocarbon ages and depth (cm) downcore shown on the left of the lithology column. The sediment legend is drawn according to a simplified Troels-Smith (1955). (Modified from Shennan *et al*., 2000.)

and therefore the dated sample and a sea-level index point, is coincident with the decline of polyhalobous diatom assemblages and replacement by oligiohalobous species (Figure 19.5d). The value of diatom analyses from isolation basins has been illustrated in the glacially eroded landscape of northwest Scotland, which has produced an unique, high-resolution 16,000 cal yr BP record of relative sea-level change from the time of local deglaciation (Shennan *et al*., 2000).

19.3.3 Transfer functions

A succession of statistical methods has been applied to salt-marsh diatom data sets for reconstructing relative sea level, with results from transfer-function analysis being one of the most objective (e.g. Zong and Horton, 1999; Gehrels *et al*., 2001; Sawai *et al*., 2004a,b; Hamilton and Shennan, 2005; Horton *et al*., 2006, 2007; Szkornik *et al*., 2006; Nelson *et al*., 2008; Kemp *et al*., 2009a). A transfer function is developed from a training set of modern data that include samples from the tidal flat to upland. The training set is analyzed to determine relations among the spatial distribution (Telford and Birks, 2009) of diatom species and the independent variable of elevation (relative to the tidal frame. These relationships are then employed to reconstruct past tide levels from subfossil assemblages contained within sedimentary sequences (Gehrels, 2007). Transfer functions have a substantial advantage over established quantitative and semi-quantitative methods of reconstructing sea-level change. The reconstructions are produced using replicable methodology and improved record comparability, and include sample-specific errors with quantified error terms (Horton and Edwards, 2006). Transfer functions employ an expanded range of salt-marsh sediments rather than relying on transgressive and regressive contacts, although the most precise reconstructions are derived from microfossils preserved in sediments from the uppermost part of the intertidal zone (Gehrels, 2007). While the transfer-function approach provides a useful addition to the range of existing techniques employed to study sea-level change, it has, like them, important limitations that preclude its use as a simple "black box" application. A firm understanding of the ecology and taphonomy of salt-marsh diatoms must remain central to the development and application of these transfer functions (Woodroffe, 2009).

Transfer functions have been used to bridge the gap between short-term instrumental records and long-term, established geological reconstructions of former sea levels. For example, Woodroffe and Long (2009) developed the first diatom-based transfer functions from salt marshes in Greenland to extend recent direct observations of ice-sheet behavior and associated glacio-sostatic responses. They demonstrate that the transfer-function method is applicable in these Arctic salt marshes and has the potential to be applied to reconstructing past sea-level change. Preliminary application of the method to a single fossil sequence suggests that from 600 to 400 cal yr BP sea level rose at 2.7 mm yr^{-1}, after which the rate of rise decreased and remained stable until the present day. Kemp *et al.* (2009a) developed and applied a diatom-based transfer function from high-marsh sediment in North Carolina, USA (Figure 19.6a). The relationship between observed and diatom-predicted elevation is very strong (Figure 19.6c), illustrating the powerful performance of the transfer function ($r = 0.95$). Indeed, these results indicate that very precise reconstructions of former sea levels are possible (potential errors <0.05 m). Kemp *et al.* (2009a) applied the transfer functions to 18 samples within a short core, which consisted of an upper 0.50 m of brown salt-marsh peat with abundant *Juncus roemerianus* remains and a lower gray sand. Diatoms are abundant throughout the core and the assemblage is dominated by polyhalobous and mesohalobous species such as *Diploneis bombus*, *Diploneis smithii*, *Opephora marina*, and *Paralia sulcata* (Figure 19.6d). The transfer function assigned each of the 18 samples a paleomarsh elevation and a sample-specific error. Relative sea level is reconstructed with a chronology based upon ^{210}Pb accumulation and a radiocarbon date. The final test of a transfer function is to validate the reconstructions against instrumental (tide-gauge) records of historical sea-level change. Since 1973, the nearest tide gauge to the study site shows a sea-level rise of 4.2 mm yr^{-1}, which is comparable to the transfer-function estimates of 4.5 mm $yr^{-1} \pm 1.5$ mm since 1963. Kemp *et al.* (2009a) were able to support a small number of other salt-marsh-based studies that infer that twentieth-century rates of sea-level rise are more rapid than the long-term (last 2000 years) rates of rise and that the timing of this acceleration may be indicative of a link with human-induced climate change (e.g. Donnelly *et al.*, 2004; Gehrels *et al.*, 2005, 2008; Kemp *et al.*, 2009b).

19.4 Diatoms and land-level change

Records of relative sea-level change archive the response of the Earth's surface to the earthquake deformation cycle, volcanic processes, and epeirogenic and isostatic tectonic processes (Nelson, 2007). For example, coasts above subduction-zone faults slowly rise or fall (recording displacement from strain accumulation above the fault) for hundreds of years between great earthquakes on the subducting plate boundary, and then are suddenly jerked downward or upward, respectively, by a similar amount during earthquakes (Figure 19.7a). Following

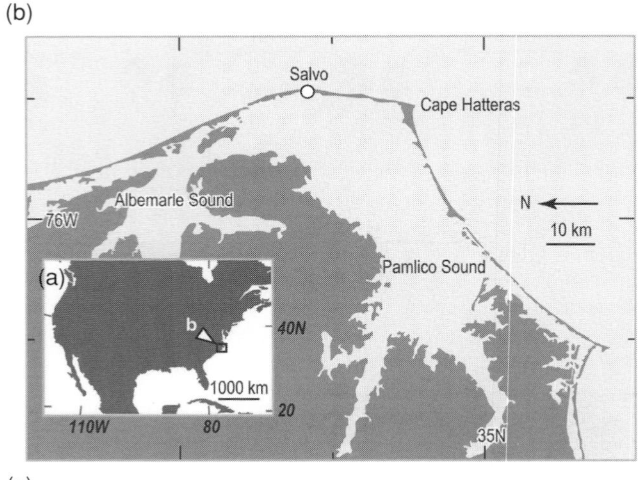

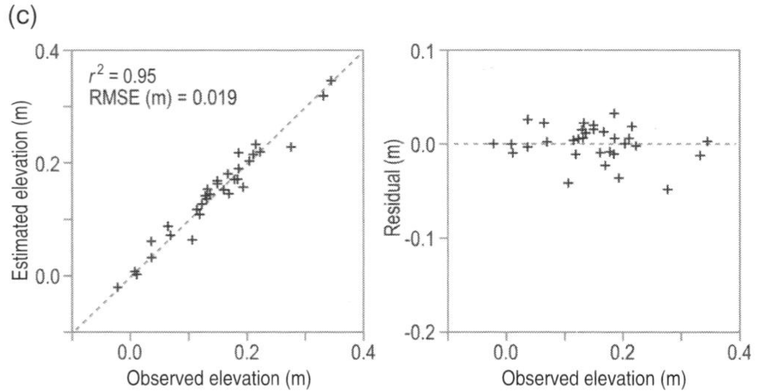

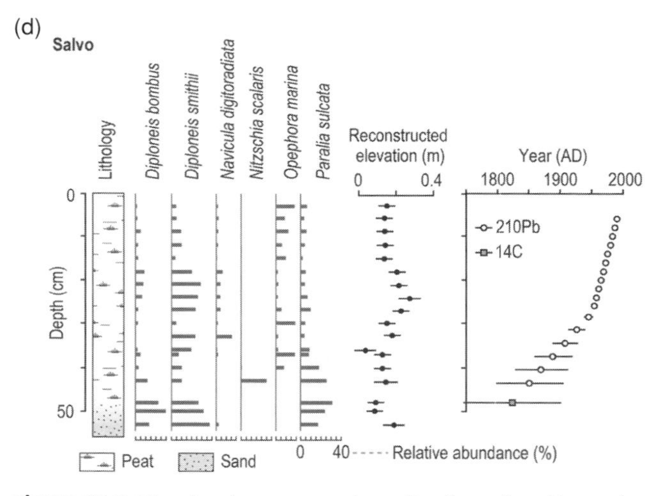

Figure 19.6 The development and application of a diatom-based transfer function to reconstruct past sea levels. (a, b) Location map of the USA showing Salvo, North Carolina, USA. (c) Performance of transfer function developed using coastal diatoms. Modern diatom data set was obtained on four transects in North Carolina (Modified from Kemp *et al.*, 2009a). (d) Salvo diatom diagram showing the transition from clastic sediment to organic limus to produce a continuous record of paleomarsh elevation. The most dominant diatom taxa are shown (i.e. relative abundance >10%). The reconstructed elevation in meters above mean sea level is based on weighted average partial least squares (WA-PLS). The sediment legend is drawn according to a simplified Troels-Smith (1955) model. (Modified from Kemp *et al.*, 2009a.)

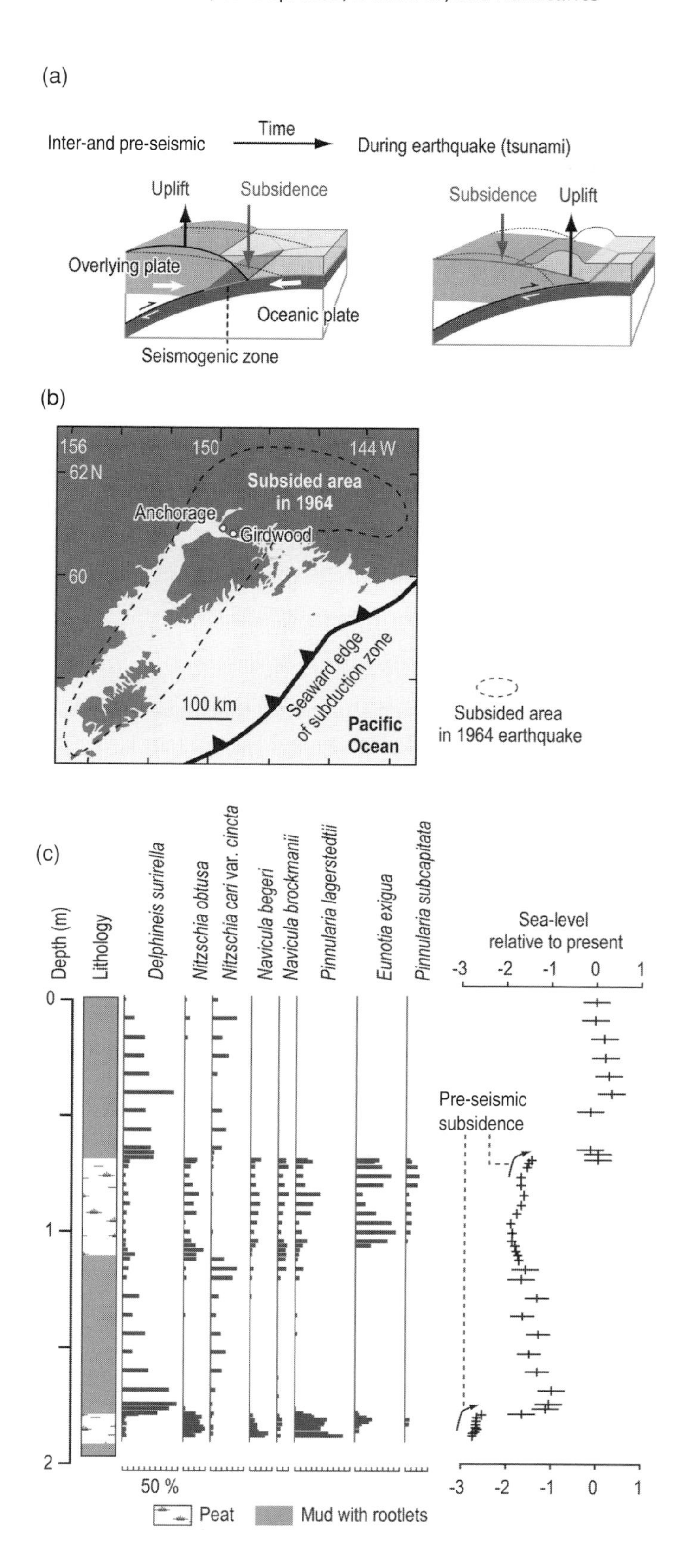

Figure 19.7 Diatom analyses during past earthquake cycles above subduction-zone faults. (a) Schematic diagram showing deformation associated with a subduction-zone thrust fault on a coastline during an earthquake cycle. (Modified from Atwater *et al.* (2005).) (b) Location map of south-central Alaska, USA, showing area subsided in the 1964 earthquake (Plafker, 1969). (c) Relative sea-level changes reconstructed from diatom species data using a WA-PLS transfer function. Two short periods of preseismic submergence immediately prior to the substantial coseismic subsidence marked by the contacts at the tops of peats are highlighted. The sediment legend is drawn according to a simplified Troels-Smith (1955) model. (Modified from Shennan and Hamilton, 2006.)

each earthquake, strain again accumulates during the interseismic period between the previous and the next earthquake on the fault (Nelson, 2007). These processes have produced geologic evidence in the salt marshes and wetlands of the Pacific northwest of the United States and Canada (e.g. Atwater *et al.*, 2005; Nelson *et al.*, 2008), Alaska (e.g. Hamilton and Shennan, 2005), Chile (e.g. Cisternas *et al.*, 2005), Japan (e.g. Sawai, 2001b), and New Zealand (e.g. Hayward *et al.*, 2004).

Early research at accessible estuarine outcrops used suddenly buried plant macrofossils within tidal wetland soils to infer sudden coseismic subsidence (e.g. Atwater, 1987), but the broad elevational range of macrofossil species limited estimates of the amount of subsidence to $>\pm 0.5$ m (Nelson *et al.*, 1996a). Diatoms superseded these analyses by providing quantitative estimates of relative sea-level change. For example, diatoms in buried soils along the Washington and Oregon coastline, USA have estimated subsidence during seven great Holocene subduction-zone earthquakes (Hemphill-Haley, 1995a; Atwater and Hemphill-Haley, 1997; Nelson *et al.*, 1996b, 1998; Shennan *et al.*, 1996, 1998). Further, Sherrod *et al.* (2000, 2001) employed diatoms to infer earthquake-induced uplift and subsidence in coastal marshes of Puget Sound, Washington, USA. Buried forest and high-marsh soils indicated abrupt changes in sea level during the late Holocene. Dramatic changes in diatom assemblages across these contacts confirmed rapid emergence of approximately 7 m (Sherrod *et al.*, 2000) and submergence of 1–3 m (Sherrod, 2001). The transfer-function technique provides a continuous record of relative sea-level change during an earthquake cycle; this is essential information for understanding the mechanics of plate-boundary ruptures in subduction zones. There have been detailed land-level reconstructions using diatoms in southern Alaska (Figure 19.7b) during parts of six earthquake cycles (Zong *et al.*, 2003; Shennan and Hamilton, 2006). Diatom-based transfer functions have produced a precise record of coastal land subsidence not only

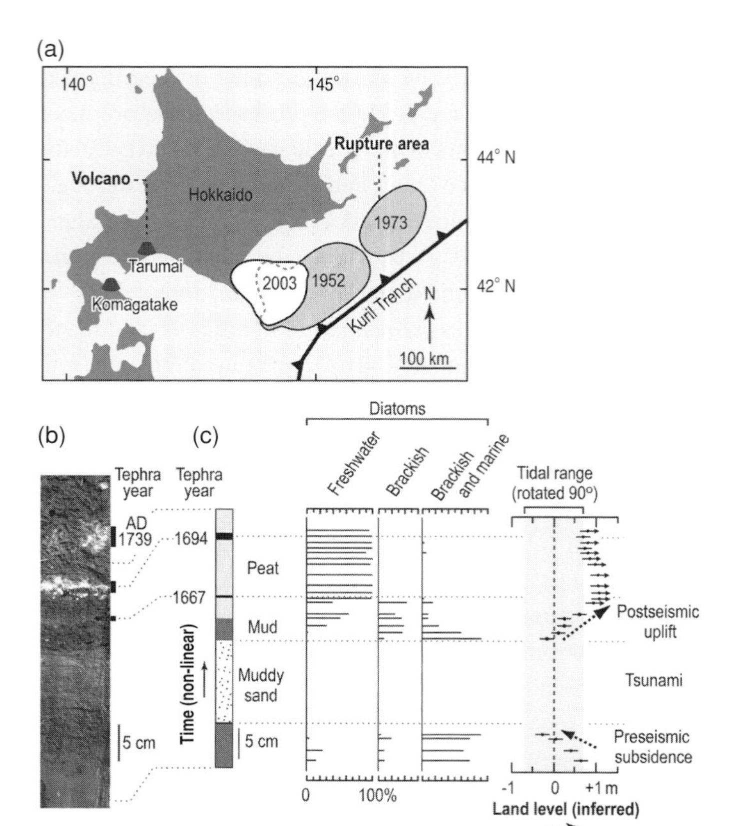

Figure 19.8 Land-level reconstructions in Hokkaido, northern Japan. (a) Location of Hokkaido and plate-tectonic setting. The solid line with triangles shows the seaward edge of the subduction zone. The location of three interplate earthquakes in the twentieth and twenty-first centuries and two volcanoes, which are the origins of the tephra layer shown in the stratigraphy, are shown. (b) Photograph of stratigraphy showing the seventeenth-century large earthquake. (c) Diatom diagram showing the schematic stratigraphy, changes in diatom assemblages, and the results of diatom-based transfer functions for land-level change. (Modified from Sawai *et al.*, 2004b.)

during a great earthquake but also in the decade or so prior to such events (Figure 19.7c). Observations and interpretations of this preseismic subsidence are complex and subtle. But if this subsidence does indeed occur, the implication is that warning signs are detectable for some time before one of these huge earthquakes occurs (Bourgeois, 2006). Relative sea-level change associated with an earthquake cycle also has been detected in stratigraphic archives on the southeast of the Kuril subduction zone (Figure 19.8a). Diatom analyses suggest tidal mudflats change into freshwater forests during the first decades after a seventeenth-century tsunami (Figure 19.8b,c). The mudflats gradually rose by a meter from decadal postseismic deformation. The tsunami and the ensuing uplift exceeded any in the region's 200 years of written history, and both resulted from a plate-boundary earthquake of unusually large size along the Kuril subduction zone (Sawai *et al.*, 2004b).

19.5 Diatoms as indicators of paleotsunamis and past storms

19.5.1 Paleotsunamis

Records of late-Quaternary tsunamis may be preserved in sedimentary deposits. These records provide a means better to understand tsunami processes by expanding the number and period of occurrences of tsunamis available for investigation. Sedimentary records tell of repeated tsunamis from past millennia from the North Sea (Bondevik *et al.* 2005), Greece (Dominey-Howes *et al.*, 2000), New Zealand (Goff *et al.* 2001), Kamchatka (Martin *et al.*, 2008), the southern Kuril Trench (Sawai, 2002; Nanayama *et al.* 2003; Iliev *et al.*, 2005), the Japan Trench (Minoura and Nakaya, 1991; Minoura *et al.*, 1994; Sawai *et al.*, 2008), Aleutian Megathrust (Shennan and Hamilton, 2006), Puget Sound (Atwater and Moore, 1992), the Cascadia subduction zone (Clague and Bobrowsky, 1994; Clague *et al.*, 2000; Hutchinson *et al.*, 1997, 2000; Witter *et al.*, 2001; Williams *et al.*, 2005; Nelson *et al.* 2008), and the region of the 1960 Chile (Cisternas *et al.* 2005) and 2004 Sumatra-Andaman (Jankaew *et al.*, 2008; Monecke *et al.*, 2008) earthquakes. Identification of tsunami deposits is often based on finding anomalous sand deposits in low-energy environments such as coastal ponds, lakes, and marshes. The anomalous deposits are diagnosed using several criteria such as regional-scale inundation, kilometre-scale landward thinning of deposit thickness, single or multiple grading, and floral (e.g. diatoms) and faunal fossils within the deposits.

In theory, marine diatoms should dominate tsunami deposits, because they are transported by rapid marine incursions. However, in practice, diatoms within tsunami deposits are generally composed of mixed assemblages, because tsunamis inundate coastal and inland areas, eroding, transporting, and redepositing brackish and freshwater taxa (Figure 19.9; e.g. Dawson *et al.*, 1996; Dawson and Smith, 2000; Bondevik *et al.*, 2005; Razzhigaeva *et al.*, 2007; Sawai *et al.*, 2009). Breakage of diatom valves within tsunami deposits has also been observed (e.g. Dawson, 2007). In the process of turbulent transportation, weakly silicified diatom species may be broken selectively. For example, in the case of the 1998 Papua New Guinea tsunami deposit, linear, sigmoid, and clavate diatoms

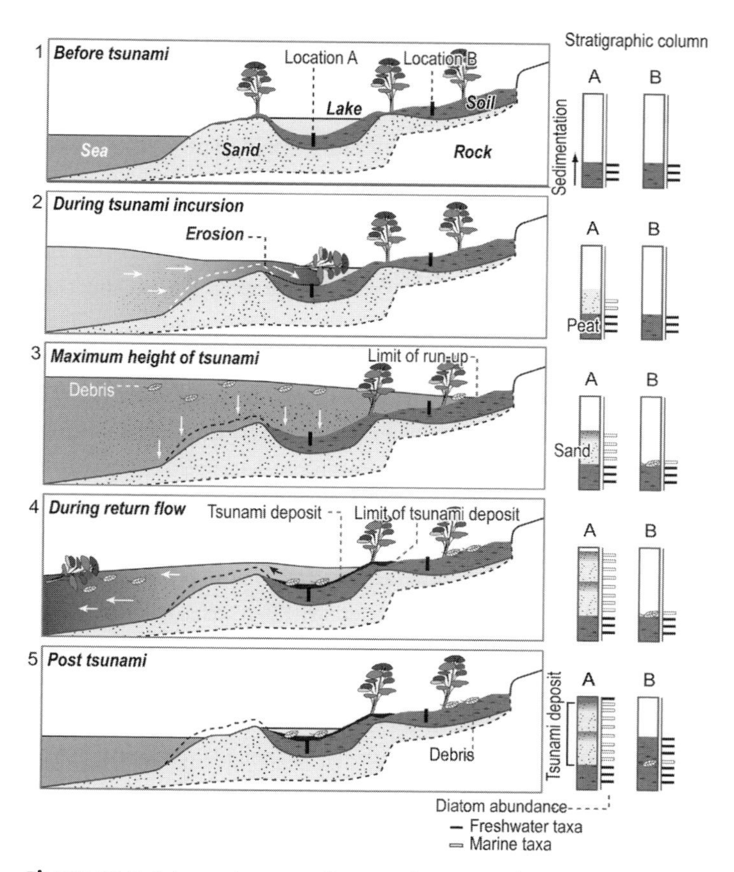

Figure 19.9 Schematic story of tsunami waves and stratigraphy at a coastal lake (A) and coastal plain (B). (1) Before the tsunami, the coastal lake stratigraphy consists of a peaty soil with freshwater diatoms. (2) During the tsunami, an incursive wave erodes coastal sand, which is transported into the coast lake. (3) In the stage of the maximum height of the tsunami, eroded fine fractions begin to settle. (4) The return flow both erodes and leaves coastal sand due to its high current velocity. (5) In the post-tsunami period, calm environments cap the stratigraphy with freshwater peaty materials. (Modified from Bondevik *et al*., 1997.)

were more readily broken due to their relatively fragile valve constructions (Dawson, 2007). In contrast, there is good preservation of diatom valves within tsunami deposits of Washington State, USA (Hemphill-Haley, 1996) and Thailand (Sawai *et al*., 2009), suggesting the quality of the preservation is related to the speed of transportation and burial.

Diatoms can also be used to estimate tsunami runup by mapping fossil diatoms transported by the tsunami. Indeed, observations suggest that the tsunami inundation area is greater than the distribution of sandy tsunami deposits. For example, in Niawaikum River, Washington State, USA, sandy deposits associated with the AD 1700 earthquake are distributed about 3 km from the river mouth, but marine diatoms, probably

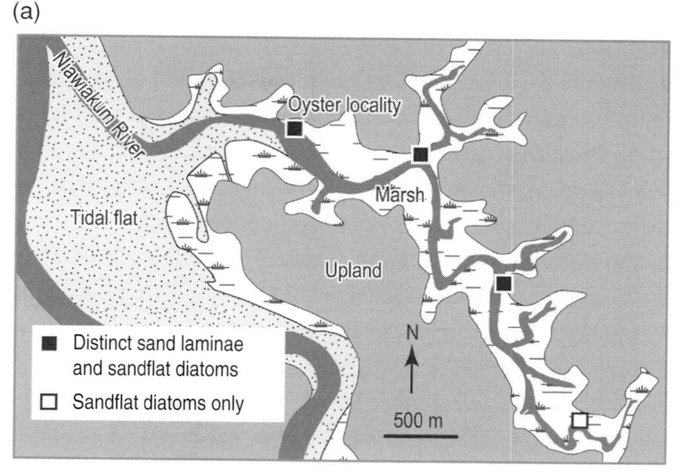

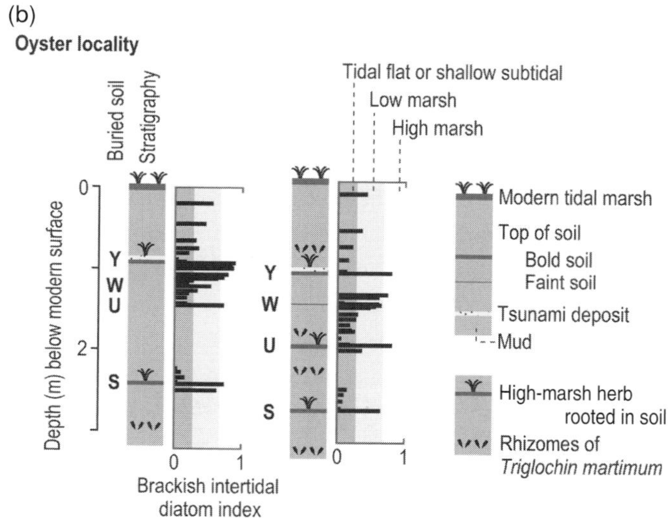

Figure 19.10 Diatom assemblages within the tsunami deposit of the AD 1700 Cascadia earthquake along the Niawiakum River. (a) Occurrence of sandflat diatoms and sand laminae (modified from Hemphill-Haley, 1996). (b) Diatom-based environmental reconstructions at Oyster locality in Niawiakum River. (Modified from Atwater and Hemphill-Haley, 1997.)

transported by the tsunami, can be traced at least 1 km farther upstream from the landward limit of the sandy deposit (Figure 19.10) (Hemphill-Haley, 1995a). This approach is very effective in freshwater environments, because marine diatoms are clearly contrasted with *in situ* freshwater diatoms (Hemphill-Haley, 1996).

19.5.2 Paleotempestology

Paleotempestology is the study of past tropical cyclone activity by means of geological proxies as well as historical documentary records. The instrumental record of storm (hurricane/typhoon) activity is too short to fully capture and determine the frequency of the occurrence of these rare but destructive events. Therefore, obtaining a record of past land-falling storms, and their extent of geological and ecological impacts, is one means to assess future risk, reveal the spatial and temporal variability of storm activity and decipher the relationship between storm frequency and intensity and global climatic change. Every storm that makes landfall has the potential to leave a signal within the sediment deposited by its storm surge. Identification of storm-surge deposits is similar to tsunamis; they are often based on anomalous sand deposits in an environment where they are unusual, such as lakes and marshes (e.g. Liu and Fearn, 1993; van de Plassche *et al.*, 2004; Donnelly and Woodruff, 2007; Horton *et al.*, 2008). Storm-surge deposits have been identified through diatom analysis where species show abrupt changes in assemblage (e.g. Bakker *et al.*, 1990; Tooley and Jelgersma, 1992) and provide an additional technique to either support existing litho-stratigraphic characteristics or, importantly, identify storm-surge sediment where lithostratigraphic characteristics are inadequate (Parsons, 1998). The species present in the storm layer reveal some information about the source of the sediment. For example, the presence of marine species would be evidence for a nearshore origin of the sediment.

Parsons (1998) investigated the diatom assemblages of a salt-marsh pond, which was inundated by the storm surges of Hurricane Andrew that made landfall in Louisiana on August 26, 1992. Analysis of pond-sediment cores identified a hurricane mud layer characterized as composite sediment, containing indicators of estuarine, brackish, and freshwater sources. The composite nature of the hurricane sediment was indicated by a higher diatom species diversity coupled with a more even species representation. Other distinguishing characteristics of the mud layer included lower marine diatom abundance, larger mean grain size, and more poorly sorted sediment. Hurricane Andrew appeared to have altered the geochemistry of the pond through the reduction of sulfide in the sediment allowing the proliferation of aquatic submerged flora (*Najas* sp.), resulting in a diatom assemblage shift towards epiphytic species. Two years after the hurricane landfall the diatom population had yet to revert to the pre-hurricane community (Parsons, 1998).

19.6 Summary

Diatoms are at the centre of new developments seeking to improve our understanding of sea-level changes. Those recorded in coastal sedimentary sequences have led to fundamental new knowledge regarding the external (such as climate change, earthquakes, and tsunamis) and internal mechanisms (including the coastal sedimentary budget) of the sea-level changes we observe and reconstruct.

Diatoms are valuable indicators of sea-level change, because modern diatom assemblages from salt-marsh environments are vertically zoned with respect to the tidal frame and they are representative of past conditions encountered in late-Quaternary sedimentary sequences. In this way, the diatoms contained within intertidal deposits can be used to fix the former elevation relative to the tidal frame at its time of formation. Diatoms preserved within low-energy sedimentary environments can produce sea-level index points from estuarine, salt-marsh, and coastal-lagoonal facies of coastlines submerging in response to glacioisostatic adjustment. Along coastlines that are uplifting, diatoms contained within isolation basins can be used to identify the contact between marine and terrestrial deposits. Diatoms are also vital to estimate vertical land-level changes in coastal sediments associated with large earthquakes at plate boundaries.

In the late 1990s, the development and refinement of statistical techniques for microfossil data revitalized studies of sea-level change on many coastlines by providing high-resolution ($\pm$0.1–0.3 m) records that can be meaningfully compared with other local, regional, and global environmental records derived from terrestrial and marine environments. Diatom-based transfer functions develop relations between the relative abundance of diatoms and environmental data, such as elevation. Calibration is then used to reconstruct past elevations from diatom assemblages in cores or outcrops.

Within high-energy deposits, abrupt changes in diatom assemblages, which may include offshore taxon, have been employed as a proxy for tsunamis and landfalling hurricanes. The intrusion of seawater into otherwise freshwater or brackish coastal lakes or marshes during a tsunami or storm surge will introduce marine diatoms and change the salinity of these coastal ecosystems. Thus, diatom analyses provide opportunities to assess hurricane and tsunami hazards more fully.

Acknowledgments

This chapter is a contribution to the International Geoscience Programme (IGCP) Project 588, "Preparing for coastal change: a detailed process-response framework for coastal change at different timescales," and to the International Union for Quaternary Research (INQUA) Commission on Coastal and Marine

Processes. This work was supported by the National Science Foundation (Award No. EAR-0717364 and EAR-0842728).

References

Alley, R. B., Clark, P. U., Huybrechts, P., & Joughin, I. (2005). Ice-sheet and sea-level changes. *Science*, **310**, 456–60.

Atwater, B. F. (1987). Evidence for great Holocene earthquakes along the outer coast of Washington State. *Science*, **236**, 942–4.

Atwater, B. F. & Hemphill-Haley, E. (1997). Recurrence intervals for great earthquakes of the past 3500 years at northeastern Willapa Bay, Washington. *US Geological Survey Professional Paper*, **1576**.

Atwater, B. F. & Moore, A. L. (1992). A tsunami 1000 years ago in Puget Sound, Washington. *Science*, **258**, 1614–17.

Atwater, B. F., Musumi-Rokkaku, S., Satake, K., *et al.* Y (eds.) (2005). *The Orphan Tsunami of 1700*, Seattle: University of Washington Press.

Bakker, C., Herman, P. M. J., & Vink, M. (1990). Changes in seasonal succession of phytoplankton induced by the storm-surge barrier in the Oosterschelde (S. W. Netherlands). *Journal of Plankton Research*, **12**, 947–72.

Bondevik, S., Mangerud, J., Dawson, S., Dawson, A. G., & Lohne, O. (2005). A record of three tsunami events in the Shetland Islands during the last 8,000 cal years. *Quaternary Science Reviews*, **24**, 1757–75.

Bondevik, S., Svendsen, J. I., & Mangerud, J. (1997). Tsunami sedimentary facies deposited by the. Storegga tsunami in shallow marine basins and coastal lakes, western Norway. *Sedimentology*, **44**, 1115–31.

Bourgeois, J. (2006). A movement in four parts? *Nature*, **440**, 430–31.

Cisternas, M., Atwater, B. F., Torrejon, F., *et al.* (2005). Predecessors of the giant 1960 Chile earthquake. *Nature*, **437**, 404–7.

Clague, J. J. & Bobrowsky, P. T. (1994). Tsunami deposits beneath tidal marshes on Vancouver Island, British Columbia. *Geological Society of America Bulletin*, **106**, 1293–303.

Clague, J. J., Bobrowsky, P. T., & Hutchinson, I. (2000). A review of geological records of large tsunamis at Vancouver Island, British Columbia, and implications for hazard. *Quaternary Science Reviews*, **19**, 849–63.

Clark, J. A., Farrell, W. E., and Peltier, W. R. (1978). Global changes in postglacial sea level a numerical calculation. *Quaternary Research*, **9**, 265–87.

Corner, G. D., Yevzerov, V. Y., Kolka, V. V., & Moller, J. J. (1999). Isolation basin stratigraphy and Holocene relative sea-level change at the Norwegian–Russian border north of Nikel, northwest Russia. *Boreas*, **28**, 146–66.

Dawson, S. (2007). Diatom biostratigraphy of tsunami deposits: examples from the 1998 Papua New Guinea tsunami. *Sedimentary Geology*, **200**, 328–35.

Dawson, S., Dawson, A. G., & Edwards, K. J. (1998). Rapid Holocene relative sea-level changes in Gruinart, Isle of Islay, Scottish Inner Hebrides. *The Holocene*, **8**, 183–95.

Dawson, S. & Smith, D. E. (2000). The sedimentology of mid-Holocene tsunami facies in northern Scotland. *Marine Geology*, **170**, 69–79.

Dawson, S., Smith, D. E., Ruffman, A., & Shi, S. (1996). The diatom biostratigraphy of tsunami sediments: examples from recent and middle Holocene events. *Physics and Chemistry of the Earth*, **21**, 87–92.

Dominey-Howes, D. T. M., Papadopoulos, G. A., & Dawson, A. G. (2000). Geological and historical investigation of the 1650 Mt. Columbo (Thera Island) eruption and tsunami, Agean Sea, Greece. *Natural Hazards*, **21**, 83–96.

Donnelly, J. P. & Woodruff, J. D. (2007). Intense hurricane activity over the past 5,000 years controlled by El Niño and the West African Monsoon. *Nature*, **447**, 465–8.

Donnelly, J. P., Cleary, P., Newby, P., & Ettinger, R. (2004). Coupling instrumental and geological records of sea-level change: evidence from southern New England of an increase in the rate of sea-level rise in the late 19th century. *Geophysical Research Letters*, **31**, 1–4.

Edwards, R. J. (2007). Sedimentary indicators of relative sea-level changes – low energy. In *Encyclopedia of Quaternary Science*, ed. S. A. Elias, Amsterdam: Elsevier, pp. 2994–3005.

Engelhart, S. E., Horton, B. P., Douglas, B. C., Peltier, W. R., & Tornqvist, T. E. (2009). Spatial variability of late Holocene and 20th century sea level rise along the US Atlantic coast. *Geology*, **37**, 1115–18.

Gehrels, W. R. (2007). Sea level studies: microfossil reconstructions. In *Encyclopedia of Quaternary Science*, ed. S. A. Elias, Elsevier, pp. 3015–23.

Gehrels, W. R., Hayward, B. W., Newnham, R. M., & Southall, K. E. (2008). A 20th century sea-level acceleration in New Zealand. *Geophysical Research Letters*, **35**, L02717.

Gehrels, W. R., Kirby, J. R., Prokoph, A., *et al.* (2005). Onset of recent rapid sea-level rise in the western Atlantic Ocean. *Quaternary Science Reviews*, **24**, 2083–100.

Gehrels, W. R., Roe, H. M., & Charman, D. J. (2001). Foraminifera, testate amoebae and diatoms as sea-level indicators in UK saltmarshes; a quantiative multiproxy approach. *Journal of Quaternary Science*, **16**, 201–20.

Goff, J. R., Chagué-Goff, C., & Nichol, S. (2001). Palaeotsunami deposits: a New Zealand perspective. *Sedimentary Geology*, **143**, 1–6.

Haften, U. & Tallantire, P. A. (1978). Palaeoecology and post-Weichselain shore-level changes on the coast of Møre, western Norway. *Boreas*, **7**, 109–22.

Hamilton, S. & Shennan, I. (2005). Late Holocene relative sea-level changes and the earthquake deformation cycle around upper Cook Inlet, Alaska. *Quaternary Science Reviews*, **24**, 1479–98.

Hayward, B. W., Cochran, U., Southall, K., *et al.* (2004). Micropalaeotological evidence for the Holocene earthquake history of the eastern Bay of Plenty, New Zealand, and a new index for determining the land elevation record. *Quaternary Science Reviews*, **23**, 1651–67.

Hemphill-Haley, E. (1995a). Diatom evidence for earthquake induced subsidence and tsunami 300 yr ago in southern Washington. *Geological Society of America Bulletin*, **107**, 367–78.

Hemphill-Haley, E. (1995b). Intertidal diatoms from Willapa Bay, Washington: applications to studies of small-scale sea-level changes. *Northwest Science*, **69**, 29–45.

Hemphill-Haley, E. (1996). Diatoms as an aid in identifying late-Holocene tsunami deposits. *The Holocene*, **6**, 439–48.

Hendey, N. I. (1964). *An introductory account of the smaller algae of British coastal waters, Bacillariophyceae (diatoms)*. Fishery Investigation Series, London: HMSO.

Heyworth, A., Kidson, C., & Wilks, P. (1985). Late-glacial and Holocene sediments at Clarach Bay, near Aberystwyth. *Journal of Ecology*, **73**, 459–80.

Horton, B. P., Corbett, R., Culver, S. J., Edwards, R. J., & Hillier, C. (2006). Modern saltmarsh diatom distributions of the Outer Banks, North Carolina, and the development of a transfer function for high resolution reconstructions of sea level. *Estuarine, Coastal, and Shelf Science*, **69**, 381–94.

Horton, B. P. & Edwards, R. J. (2006). Quantifying Holocene sea level change using intertidal foraminifera: lessons from the British Isles. *Cushman Foundation for Foraminiferal Research, Special Publication*, **40**.

Horton, B. P., Peltier, W. R., Culver, S. J., *et al.* (2009). Holocene sea-level changes along the North Carolina Coastline and their implications for glacial isostatic adjustment models. *Quaternary Science Reviews*, **28**, 172–36.

Horton, B. P., Rossi, V., & Hawkes, A. D. (2008). The sedimentary record of the 2005 hurricane season along the Gulf Coast, *Quaternary International*, **195**, 15–30.

Horton, B. P., Zong, Y., Hillier, C., & Engelhart, S. (2007). Diatoms from Indonesian mangroves and their suitability as sea-level indicators for tropical environments. *Marine Micropaleontology*, **63**, 155–68.

Hustedt, F. (1953). Diatomeen aus dem Naturschutzpark Seeon. *Archiv für Hydrobiologie*. **47**, 625–35.

Hustedt, F. (1957). Die Diatomeenflora des Fluss-systems der Weser im Gebiet der Hansestadt Bremen. *Abhandlungen vom Naturwissenschaftlichen Verein zu Bremen*, **34**, 181–440.

Hutchinson, I., Clague, J. J., & Mathewes, R. W. (1997). Reconstructing the Tsunami record on an emerging coast: a case study of Kanim Lake, Vancouver Island, British Columbia, Canada. *Journal of Coastal Research*, **13**, 545–53.

Hutchinson, I., Guilbault, J.-P., Clague, J. J., & Bobrowsky, P. T. (2000). Tsunamis and tectonic deformation at the northern Cascadia margin: a 3000-year record from Deserted Lake, Vancouver Island. *The Holocene*, **10**, 429–39.

Iliev, A. Y., Kaistrenko, V. M., Gretskaya, E. V., *et al.* (2005). Holocene tsunami traces on Kunashir Island, Kuril subduction zone. *Tsunamis, Case Studies and Recent Developments*, **23**, 171–92.

IPCC (2007). *Climate Change 2007: The Physical Science Basis. Contribution of Working Group I to the Fourth Assessment. Report of the Intergovernmental Panel on Climate Change*, ed. S. Solomon, D. Qin, M. Manning, *et al.*, Cambridge: Cambridge University Press.

Jankaew, K., Atwater, B. F., Sawai, Y., *et al.* (2008). Medieval forewarning of the 2004 Indian Ocean tsunami in Thailand. *Nature*, **455**, 1228–31.

Juggins, S. (1992). Diatoms in the Thames estuary, England: ecology, palaeoecology, and salinity transfer function. *Bibliotheca Diatomologica*, **25**, 216.

Kemp, A. C., Horton, B. P., Corbett, R., *et al.* (2009a). The relative utility of foraminifera and diatoms for reconstructing late Holocene sea-level change in North Carolina, USA. *Quaternary Research*, **71**, 9–21.

Kemp, A. C., Horton, B. P., Culver, S. J., *et al.* (2009b). The timing and magnitude of recent accelerated sea-level rise (North Carolina, USA). *Geology*, **37**, 1035–8.

Kjemperud, A. (1981). Diatom changes in sediments from basins possessing marine/lacustrine transitions in Frosta, Nord-Trøndelag, Norway. *Boreas*, **10**, 27–38.

Kosugi, M. (1985). Discrimination of living or dead cells of diatom based on the stained images – its method and significance. *Quaternary Research*, **24**, 139–47.

Kosugi, M. (1986). Transportation and sedimentation patterns on dead diatoms in a tidal area – a case study in the lower reach of the Obitsu River. *Geological Review of Japan*, **59**, 37–50.

Kosugi, M. (1987). Limiting factors on the distribution of benthic diatoms in coastal regions – salinity and substratum. *Diatom*, **3**, 21–31.

Kosugi, M. (1988). Classification of living diatom assemblages as the indicator of environments, and its application to reconstruction of paleoenvironments. *Quaternary Research*, **27**, 1–20.

Kosugi, M. (1989). Processes of formation on fossil diatom assemblages and the paleoecological analysis. *Benthos Research*, **35–36**, 17–28.

Lambeck, K., Esat, T. M., & Potter, E-K. (2002). Links between climate and sea levels for the past three million years. *Nature*, **419**, 199–206.

Liu, K. & Fearn, M. L. (1993). Lake-sediment record of late Holocene hurricane activities from coastal Alabama. *Geology*, **21**, 793–6.

Lloyd, J. M. (2000). Combined foraminiferal and thecamoebian environmental reconstruction from an isolation basin in NW Scotland: implications for sea-level studies. *Journal of Foraminiferal Research*, **30**, 294–305.

Lloyd, J. M. & Evans, J. R. (2002). Contemporary and fossil foraminifera from isolation basins in northwest Scotland. *Journal of Quaternary Science*, **17**, 431–43.

Long, A. J., Roberts, D. H., & Dawson, S. (2006). Early Holocene history of the west Greenland ice sheet and the GH-8.2 event. *Quaternary Science Reviews*, **25**, 904–22.

Long, A. J., Roberts, D. H., Simpson, M. J. R., *et al.* (2008). Late Weichselian relative sea-level changes and ice sheet history in southeast Greenland. *Earth and Planetary Science Letters*, **282**, 8–18.

Long, A. J., Roberts, D. H., & Wright, M. R. (1999). Isolation basin stratigraphy and Holocene relative sea-level change on Arveprinsen Ejland, Disko Bugt, West Greenland. *Journal of Quaternary Science*, **14**, 323–45.

Martin, M. E., Weiss, R., Bourgeois, J., *et al.* (2008). Combining constraints from tsunami modeling and sedimentology to untangle the 1969 Ozernoi and 1971 Kamchatskii tsunamis. *Geophysical Research Letters*, **35**, L01610.

Metcalfe, S. E., Ellis, S., Horton, B. P., *et al.* (2000). The Holocene evolution of the Humber estuary: reconstructing change in a dynamic environment. *Geological Society Special Publication*, **166**.

Milne, G. A., Gehrels, W. R., Hughes, C. W., & Tamisiea, M. E. (2009). Identifying the causes of sea-level change. *Nature Geoscience*, **2**, 471–8.

Milne, G. A., Long, A. J., & Bassett, S. E. (2005). Modelling Holocene relative sea-level observations from the Caribbean and South America. *Quaternary Science Reviews*, **24**, 1183–202.

Minoura, K. & Nakaya, S. (1991). Trances of tsunami preserved in inter-tidal lacustrine and marsh deposits: some examples from northeast Japan. *Journal of Geology*, **99**, 265–87.

Minoura, K., Nakaya, S., & Uchida, M. (1994). Tsunami deposits in a lacustrine sequence of the Sanriku coast, northeast Japan. *Sedimentary Geology*, **89**, 25–31.

Monecke, K., Finger, W., Klarer, D., *et al.* (2008). A 1,000-year sediment record of tsunami recurrence in northern Sumatra. *Nature*, **455**, 1232–4.

Nanayama, F., Satake, K., Furukawa, R., *et al.* (2003). Unusually large earthquakes inferred from tsunami deposits along the Kuril trench. *Nature*, **424**, 660–3.

Nelson, A. R. (2007). Tectonic locations. In *Encyclopedia of Quaternary Science*, ed. S. A. Elias, Amsterdam: Elsevier, pp. 3072–87.

Nelson, A. R., Jennings, A. E., & Kashima K. (1996b). An earthquake history derived from stratigraphic and microfossil evidence of relative sea-level change at Coos Bay, southern coastal Oregon. *Geological Society of America Bulletin*, **108**, 141–54.

Nelson, A. R. & Kashima, K. (1993). Diatom zonation in southern Oregon tidal Marshes relative to vascular plants, foraminifera and sea level. *Journal of Coastal Research*, **9**, 673–97.

Nelson, A. R., Ota, Y., Umitsu, M., Kashima, K., & Matsushima, Y. (1998). Seismic or hydrodynamic control of rapid late-Holocene sea-level rises in southern coastal Oregon, USA?, *The Holocene*, **8**, 287–99.

Nelson, A. R., Sawai, Y., Jennings, A., *et al.* (2008). Great-earthquake paleogeodesy and tsunamis of the past 2000 years at Alsea Bay, central Oregon coast, USA. *Quaternary Science Reviews*, **27**, 747–68.

Nelson, A. R., Shennan, I., & Long, A. J. (1996a). Identifying coseismic subsidence in tidal-wetland stratigraphic sequences at the Cascadia subduction zone of western North America. *Journal of Geophysical Research*, **101**, 6115–35.

Palmer, A. J. M. & Abbott, W. H. (1986). Diatoms as indicators of sea-level change. In *Sea-Level Research: a Manual for the Collection and Evaluation of Data*, ed. O. van de Plassche, Norwich: Geobooks, pp. 457–88.

Parsons, M. L. (1998). Salt marsh sedimentary record of the landfall of Hurricane Andrew on the Louisiana coast: diatoms and other paleoindicators. *Journal of Coastal Research*, **14**, 939–50.

Patterson, R. T., Dalby, A. P., Roe, H. M., *et al.* (2005). Relative utility of foraminifera, diatoms and macrophytes as high resolution indicators of paleo-sea level. *Quaternary Science Reviews*, **24**, 2002–14.

Patterson, R. T., Hutchinson, I., Guilbault, J. P., & Clague, J. J. (2000). A comparison of the vertical zonation of diatom, Foraminifera, and Macrophyte assemblages in a coastal marsh: implications for greater paleo-sea level resolution. *Micropaleontology*, **46**, 229–44.

Peltier, W. R. (2004). Global glacial isostasy and the surface of the ice-age earth: the ice-5G (VM2) model and grace. *Annual Review of Earth Planetary Science*, **32**, 111–149.

Peltier, W. R., Shennan, I., Drummond, R., & Horton, B. P. (2002). On the post-glacial isostatic adjustment of the British Isles and the shallow visco-elastic structure of the Earth. *Geophysical Journal International*, **148**, 443–75.

Plafker, G. (1969). Tectonics of the March 27, 1969 Alaska earthquake. *U.S. Geological Survey Professional Paper 543-I*, 74 pp.

Plater, A. J., Horton, B. P., Haworth, E. Y., *et al.* (2000). Holocene tidal levels and sedimentation rates using a diatom-based palaeoenvironmental reconstruction: the Tees estuary, north-eastern England. *Holocene*, **10**, 441–52.

Rahmstorf, S. (2007). A semi-empirical approach to projecting future sea-level rise. *Science*, **315**, 368–70.

Razzhigaeva, N. G., Ganzei, L. A., Grebennikova, T. A., *et al.* (2007). Tsunami deposits of the Shikotan earthquake of 1994. *Oceanography*, **47**, 579–87.

Roe, H. M., Doherty, C. T., Patterson, R. T., & Swindles, G. T. (2008). Contemporary distributions of saltmarsh diatoms in the Seymour–Belize Inlet Complex, British Columbia, Canada: implications for studies of sea-level change. *Marine Micropaleontology*, **70**, 134–50.

Ryves, D. B., Battarbee, R. W., & Fritz, S. C. (2009). The dilemma of disappearing diatoms: incorporating diatom dissolution data into palaeoenvironmental modelling and reconstruction. *Quaternary Science Reviews*, **28**, 120–36.

Sawai, Y. (2001a). Distribution of living and dead diatoms in tidal wetlands of northern Japan: relations to taphonomy. *Palaeogeography, Palaeoclimatology, Palaeoecology*, **173**, 125–41.

Sawai, Y. (2001b). Episodic emergence in the past 3000 years at the Akkeshi Estuary, Hokkaido, northern Japan. *Quaternary Research*, **56**, 231–41.

Sawai, Y. (2002). Evidence for 17th century tsunamis generated on the Kuril-Kamchatka subduction zone, Lake Tokotan, Hokkaido, Japan. *Journal of Asian Earth Science*, **20**, 903–911.

Sawai, Y., Fujii, Y., Fujiwara, O., *et al.* (2008). Marine incursions of the past 1500 years and evidence of tsunamis at Suijin-numa, a coastal lake facing the Japan Trench. *The Holocene*, **18**, 517–28.

Sawai, Y., Horton, B. P., & Nagumo, T. (2004a). Diatom-based elevation transfer function along the Pacific coast of eastern Hokkadio, northern Japan – an aid in paleo-seismic study along the coasts near Kurile subduction zone. *Quaternary Science Reviews*, **23**, 2467–83.

Sawai, Y., Jankaew, K., Martin, M. E., *et al.* (2009). Diatom assemblages in tsunami deposits associated with the 2004 Indian Ocean tsunami at Phra Thong Island, Thailand. *Marine Micropaleontology*, **73**, 70–9.

Sawai, Y., Satake, K., Kamataki, T., *et al.* (2004b). Transient uplift after a 17th-century earthquake along the Kuril subduction zone. *Science*, **306**, 1918–20.

Shaw, J., Gareau, P., & Courtney, R. C. (2002). Palaeogeography of Atlantic Canada. *Quaternary Science Reviews*, **21**, 1861–78.

Shennan, I. (1986). Flandrian sea-level changes in the Fenland, II: tendencies of sea-level movement, altitudinal changes and local and regional factors. *Journal of Quaternary Science*, **1**, 155–79.

Shennan, I. (2007). Sea level studies. In *Encyclopedia of Quaternary Science*, ed. S. A. Elias, Amsterdam: Elsevier, pp. 2967–74.

Shennan, I. & Hamilton, S. L. (2006). Co-seismic and pre-seismic subsidence associated with great earthquakes in Alaska. *Quaternary Science Reviews*, **25**, 1–8.

Shennan, I. & Horton, B. P. (2002). Relative sea-level changes and crustal movements of the UK. *Journal of Quaternary Science*, **16**, 511–26.

Shennan, I., Long, A. J., Rutherford, M. M., *et al.* (1996). Tidal marsh stratigraphy, sea-level change and large earthquakes: a 5000 year record in Washington, USA. *Quaternary Science Reviews*, **15**, 1023–59.

Shennan, I., Lambeck, K., Horton, B. P., *et al.* (2000). Late Devensian and Holocene records of relative sea-level changes in northwest Scotland and their implications for glacio-hydro-isostatic modelling. *Quaternary Science Reviews*, **19**, 1103–35.

Shennan, I., Long, A. J., Rutherford, M. M., *et al.* (1998). Tidal marsh stratigraphy, sea-level change and large earthquakes, II, submergence events during the last 3500 years at Netarts Bay, Oregon, U.S.A. *Quaternary Science Reviews*, **17**, 365–93.

Sherrod, B. L. (1999). Gradient analysis of diatom assemblages in a Puget Sound salt marsh – can such assemblages be used for quantitative paleoecological reconstructions? *Palaeogeography, Palaeoclimatology, Palaeoecology*, **149**, 213–26.

Sherrod, B. L. (2001). Evidence for earthquake-induced subsidence about 1100 yr ago in coastal marshes of southern Puget Sound, Washington. *Geological Society of America Bulletin*, **113**, 1299–311.

Sherrod, B. L., Bucknam, R. C., & Leopold, E. B. (2000). Holocene relative sea level changes along the Seattle Fault at Restoration Point, Washington. *Quaternary Research*, **54**, 384–93.

Simonsen, R. (1969). Diatoms as indicators in estuarine environments. *Velöffentliche Institut Meeresforschung Bremerhaven*, **11**, 287–91.

Sullivan, M. J. (1975). Diatom communities from a Delaware salt marsh. *Journal of Phycology*, **11**, 384–90.

Summerfield, M. A. (1991). *Global Geomorphology. An Introduction to the Study of Landfornis*, Harlow: Longman.

Szkornik, K., Gehrels, W. R., & Kirby, J. R. (2006). Using salt-marsh diatoms to reconstruct sea-level changes in Ho Bugt western Denmark. *Marine Geology*, **235**, 137–50.

Telford, R. J. & Birks, H. J. B. (2009). Evaluation of transfer functions in spatially structured environments. *Quaternary Science Reviews*, **28**, 1309–16.

ter Braak, C. J. F. (1987). The analysis of vegetation–environment relationships by canonical correspondence analysis. *Vegetatio*, **69**, 69–77.

ter Braak, C. J. F. & Verdonschot, P. F. M. (1995). Canonical correspondence analysis and related multivariate methods in aquatic ecology. *Aquatic Sciences*, **57**, 153–87.

Tooley, M. J. (1982). Sea-level changes in northern England. *Proceedings of the Geologists' Association*, **93**, 43–51.

Tooley, M. J. & Jelgersma, S. (Ed.) (1992). *Impacts of Sea-Level Rise on European Coastal Lowlands*. Blackwell: Oxford.

Troels-Smith, J. (1955). Characterisation of unconsolidated sediments. *Danmarks Geologiske Unders*, **4** (3), 1–73.

Turney, C. S. M. & Brown, H. (2007). Catastrophic early Holocene sea level rise, human migration and the Neolithic transition in Europe. *Quaternary Science Reviews*, **26**, 2036–41.

van de Plassche, O., (ed.) (1986). *Sea-Level Research: a Manual for the Collection and Evaluation of Data*. Norwich: Geobooks.

van de Plassche, O., Wright, A. J., van der Borg, K., & de Jong, A.F.M. (2004). Salt-marsh erosion associated with hurricane landfall in southern New England in the fifteenth and seventeenth centuries. *Geology*, **34**, 829–32.

van der Werff, H. & Huls, H. (1958–1974). *Diatomeeenflora van Nederland*, 8 parts, published privately, De Hoef, The Netherlands.

Voorrips, A. & Jansma, M. A. (1974). Pollen and diatom analysis of a shore section of the former Lake Wervershoof. *Geologie en Mijnbouw*, **53**, 429–35.

Vos, P. C. & de Wolf, H. (1993). Diatoms as a tool for reconstruction sedimentary environments in coastal wetlands: methodological aspects. *Hydrobiologa*, **269–270**, 285–96.

Williams, H. F. L., Hutchinson, I., & Nelson, A. R. (2005). Multiple source for late-Holocene tsunamis at Discovery Bay, Washington State, USA. *The Holocene*, **15**, 60–73.

Witter, R. C., Kelsey, H. M., & Hemphill-Haley, E. (2001). Pacific storms, El Niño and tsunamis: competing mechanisms for sand deposition in a coastal marsh, Euchre Creek, Oregon. *Journal of Coastal Research*, **17**, 563–83.

Woodroffe, S. A. (2009). Recognising subtidal foraminiferal assemblages: implications for quantitative sea-level reconstructions using a foraminifera-based transfer function. *Journal of Quaternary Science*, **24**, 215–23.

Woodroffe, S. A. & Long, A. J. (2009). Salt marshes as archives of recent relative sea level change in west Greenland. *Quaternary Science Reviews*, **28**, 1750–61.

Zong, Y. (1998). Diatom records and sedimentary responses to sea-level change during the last 8000 years in Roudsea Wood, northwest England. *The Holocene*, **8**, 219–28.

Zong, Y. & Horton, B. P. (1998). Diatom zones across intertidal flats and coastal saltmarshes in Britain. *Diatom Research*, **13**, 375–94.

Zong, Y. & Horton, B. P. (1999). Diatom-based tidal-level transfer functions as an aid in reconstructing Quaternary history of sea-level movements in Britain. *Journal of Quaternary Science*, **14**, 153–67.

Zong, Y., Shennan, I., Combellick, R. A., Hamilton, S. L., & Rutherford, M. M. (2003). Microfossil evidence for land movements associated with the AD 1964 Alaska earthquake. *The Holocene*, **13**, 7–20.

Zong, Y. & Tooley, M. J. (1996). Holocene sea-level changes and crustal movements in Morecambe Bay, northwest England. *Journal of Quaternary Science*, **11**, 43–58.

20

Marine diatoms as indicators of modern changes in oceanographic conditions

OSCAR E. ROMERO AND
LEANNE K. ARMAND

20.1 Marine diatoms as indicators of modern environment change

A substantial part of the ocean's primary productivity is provided by diatoms (Tréguer *et al.*, 1995). In general, they are the dominant primary producers in temperate and cold areas, and are very abundant in the recently upwelled waters of Eastern Boundary Currents and in diverging surface currents where nutrients are brought to the surface (Nelson *et al.*, 1995, and references therein). On an annual basis, the relative contribution of diatoms to primary productivity is highly variable: Nelson *et al.* (1995) and Tréguer *et al.* (1995) proposed upper limits of 35% in oligotrophic areas and up to 75% in coastal upwelling areas and other nutrient-rich systems. Regardless of the area, the general trend is for an increase in the relative abundance of diatoms in the phytoplankton together with primary productivity (Ragueneau *et al.*, 2000). As a general statement we may say that diatoms are the dominant primary producers in a number of oceanographic settings that offer both the required high-nutrient and turbulence conditions (e.g. coastal upwelling areas, equatorial divergences, ice-edges, river plumes; Ragueneau *et al.*, 2000). In contrast, small, non-siliceous pico- and nanoplankton are of great importance to total productivity in oligotrophic regions (Tréguer *et al.*, 1995, and references therein). Since diatoms account for more than half of the total suspended biogenic silica (BSi) and are a substantial part of the primary production of the ocean (Nelson *et al.*, 1995; Tréguer *et al.*, 1995), time series of diatom fluxes (i.e. the measured settling rate of their frustules over a known time and space) in a variety of marine habitats help improve models of the biogeochemical cycle of silicon (Ragueneau *et al.*, 2000), and contribute to development of the silicate pump model (Dugdale *et al.*, 1995; Dugdale and Wilkerson, 1998; De La Rocha *et al.*, this volume).

The Diatoms: Applications for the Environmental and Earth Sciences, 2nd Edition, eds. John P. Smol and Eugene F. Stoermer. Published by Cambridge University Press.

The fate of the BSi produced in the euphotic layer is governed by competition between recycling in and export from the upper surface waters. Globally, at least 50% of the BSi produced by diatoms in the euphotic zone dissolves in the upper 100 m (Nelson *et al.*, 1995). At a depth of 500 m only 5% of the original opal remains and at 1000 m only 1% remains. Although it has been proposed that BSi dissolution below a water depth of 1000 m is minimal (Takahashi, 1986), a reduction of 60% in the diatom export between *c.* 1500 m and 2600 m depth in the southeast Pacific Ocean has been recorded (Romero *et al.*, 2001). Finally, the portion of BSi that escapes dissolution, either in surface waters or in the water column, settles downward, ultimately reaching the sediment (see figure 1 in Tréguer *et al.*, 1995). Exceptions in high-productivity regions such as along the Antarctic continental shelf have been observed along with direct bloom fall out.

Observing, interpreting and predicting consequences of long-term biological changes can best be achieved using long-standing collections of planktonic biota from multiyear time-series surveys, sediment-trap deployments, and sediment cores. In this chapter we review the current status of diatoms as monitors for changes in modern oceanographic conditions, with particular interest focused on the results determined from sediment-trap research. The growing body of data available on export fluxes of diatoms reviewed here were collected by independent, cooperative, international research programs that deployed sediment traps from free-floating mooring platforms in almost all the world oceanic basins (Figure 20.1).

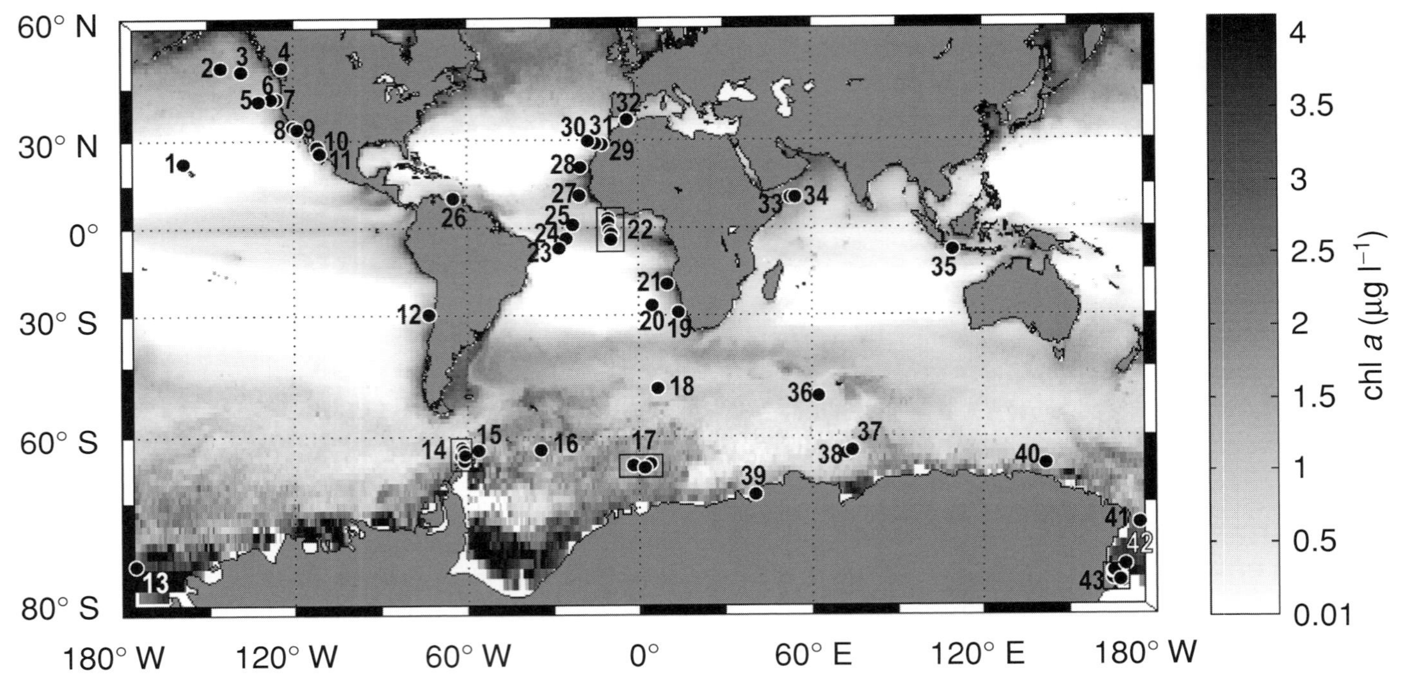

Figure 20.1 Mooring locations addressed in this chapter that detail diatom fluxes against the backdrop of mean average chl *a* from the Earth's oceans. The chl *a* climatology (1997–2007) has been derived from the MODIS AQUA sensor. Key: 1 = Aloha; 2 = PAPA; 3 = C station; 4 = Jervis Island; 5 = Gyre; 6 = Midway; 7 = Nearshore; 8 = Santa Barbara Basin (SSB); 9 = Southern California Bight (SP); 10 = Guaymas Basin; 11 = Carmen Basin; 12 = CH3–1, CH4–1, CH10–1, CH11–1; 13 = Ross Sea Site B; 14 = RACER sites R20, R39, R48, R13, R43; 15 = KG1, KG2, KG3; 16 = WS1; 17 = WS2–4; 18 = PF1; 19 = NU; 20 = CWR; 21 = WR2; 22 = EA1, GBN3, EA2, EA3, GBZ4, EA4, EA5; 23 = WA3, WA6; 24 = WA4, WA7; 25 = WA8; 26 = Cariaco Basin; 27 = Cape Verde (CV); 28 = Cape Blanc (CB); 29 = Eastern Boundary Current (EBC); 30 = Estación Europea de Series Temporales Oceánicas de Canarias (ESTOC); 31 = LaPalma; 32 = ALB-1D; 33 = MST-8; 34 = MST-9; 35 = JAM; 36 = M2; 37 = M3; 38 = PZB-1; 39 = Station A; 40 = KH94–4 site 64°S; 41 = Ross Sea Site C; 42 = Ross Sea Site A; 43 = McMurdo Sound sites B(EIT), D(HP1), E(NH1), F(NH3), I(GH), L(HP2), TI, BG, DG1, DG2, MS.

Most of these moorings collected single-year data, while for a few deployments data are available for at least two years (e.g. off northwest Africa, Cariaco Basin, Gulf of California, off southern Java, southeast Weddell Sea, Bransfield Strait). In this context, we analyze the value of diatoms as a proxy of export production from surface waters by using a wide variety of data sets, and review existing observations regarding how diatoms remove BSi and organic carbon from upper ocean layers and transport them to the ocean interior. We compare and discuss temporal flux variations from locations including coastal and open-ocean areas of the tropical, subtropical, and polar areas, covering different oceanographic settings as well as different primary productivity levels. Finally, the trap-collected diatom assemblages are compared to assemblages in the underlying surface sediments of Holocene age, and the implications of seasonal and preservational biases are discussed. Application of the knowledge gained from observations of the components and timing of seasonal cycle flux in direct context to the environmental conditions and forcings are invaluable as a method of monitoring or assessing future changes under the Earth's warming climate, or alternatively, to the deciphering of productivity events or climatologies in the past. The latter of these topics is developed further by Leventer *et al.* (this volume) and Jordan and Stickley (this volume).

20.2 Historical development

Conceptually, particle flux at a given point can be determined by simply placing a receptacle (sediment trap) in the water column to intercept the particles and microorganisms as they settle during a known time window. Direct quantification of particle flux in the ocean has only been possible in very recent decades, mainly due to development of sediment traps and *in situ* filtration systems. In particular, sediment traps allow monitoring of ocean processes and deliver continuous records corresponding to periods ranging from a day to weeks, months, and years. In spite of the high costs involved in installment and maintainance of sediment traps, these devices are most commonly used for long-time series (Buesseler *et al.*, 2007; Honjo *et al.*, 2008).

Most sediment traps that have provided flux data for this synthesis are approximations of the WHOI PARFLUX time-series sediment trap (Honjo and Doherty, 1988; Buesseler *et al.*, 2007). The trap design has evolved over the last ~20 years. However, the principle design and hydrodynamic profile of the original trap has been preserved from the model tested during the 1979 Panama Basin Intercomparison Experiment (Honjo *et al.*, 1992). The default specifications include: (1) a 0.25 m^2 or 0.5 m^2 round aperture covered by a honeycomb baffle whose individual cells have an aspect ratio of 2.5, (2) a 16° funnel slope, and (3) accommodation for an in-line mooring attachment (Honjo *et al.*, 2008).

Most of the moorings addressed in this revision are based on the Heinmiller mooring design that supports a heavily instrumented mooring cable inclusive of the sediment trap, while maintaining uniformly distributed tension throughout the line and providing optimum stability against deep-sea advection through floatation buoys (Heinmiller, 1979; Buesseler *et al.*, 2007). An acoustic release mechanism allows efficient retrieval of the time-series traps and their samples. Since Heinmiller (1979), numerous engineering improvements have steadily increased instrument stability on the mooring. More recent sediment traps are equipped with precision tilt-gauge–compass combinations, pressure loggers as a proxy for depth, or acoustic current meters to detect abnormal advection events (Buesseler *et al.*, 2007).

Time-series trap developments began in the early 1980s with completion of default design criteria for time-series sediment traps and attendant deep-sea trap moorings (Honjo and Doherty, 1988). Since then, over 400 sediment traps have been moored at *c.* 240 stations for the diverse objectives of international research groups working in all ocean regions and basins for varying periods (Honjo *et al.*, 2008). The global breadth of data sets on the seasonal and interannual dynamics of the bulk flux components provides a global picture of the operation of the biological pump system (Honjo *et al.*, 2008).

In practice, expectations of accurate interception of the rain of particles are complicated by several factors which, in some cases, threaten to invalidate the results. The three most important considerations are: (1) hydrodynamic bias in collection efficiency caused by the flow of water relative to the trap opening; (2) contamination of the sample by organisms that settle into the trap and die (swimmers), thus artificially enhancing the collection, and (3) remineralization or degradation of the particles during the interval of time between their arrival in the trap and retrieval of the sample (Asper, 1996, and references therein; Buesseler *et al.*, 2007).

20.3 Methodology

Although this chapter summarizes information collected from time-series trap studies carried out by different working groups using different types of sediment traps, a common methodology has been applied on the whole to the collection, splitting, preservation and treatment of samples for diatom studies. Before and after deployment, sampling cups are poisoned with the biocide mercuric chloride ($HgCl_2$). When deployed, the sample cups are filled with filtered seawater. Sodium chloride (NaCl) is added to increase the density (around 40‰). As soon as possible after a trap is recovered, the cup carousel is removed from the trap body, sealed in clean, opaque plastic bags, and refrigerated until formal examination. In a laboratory, the individual cup samples are carefully wet-sieved through a 1 mm screen and "swimmers" are removed by hand picking with porcelain forceps. The <1 mm fraction is then split into sample aliquots, known commonly as a "split," one of these aliquots then being made available for diatom analyses.

This aliquot split can be freeze-dried and stored until processing and analysis following standard quantitative methodologies (Fischer and Wefer, 1991). Splits of the original samples are rinsed with distilled water to remove preservatives, cleaned of organic material and calcareous components with concentrated hydrochloric acid, hydrogen peroxide and potassium permanganate, and then rinsed with distilled water (Schrader and Gersonde, 1978). Preparation of slides for qualitative and quantiative analyses involves the settling of the acid-cleaned material onto cover slips by a random settling method (Moore, 1973). The coverslips are affixed to individual glass slides with a high-refractive index mounting media. Diatom analysis is carried out on permanent slides. A standard enumeration methodology is employed, wherein a minimum of 400 frustules total are counted along random transects on the split's triplicate slides. The resulting counts yield estimates of daily fluxes of diatom valves per m^2 per d calculated according to Sancetta and Calvert (1988), as follows:

$$F = [N \times (A/a \times V) \times \text{split}]/\text{days} \times D$$

where N number of specimens (valves), in an area *a*, as a fraction of the total area of petri dish, *A*, and the dilution volume *V* in ml. This value is multiplied by the same split, representing the fraction of total material in the trap, and then divided by the number of days of deployment and the trap collection area D.

20.4 Sediment-trap diatom flux studies

In the following section we briefly canvas *c.* 50 reported diatom fluxes from sediment trap studies in most of the Earth's oceans, broken down into regions and subsequently coastal or pelagic studies. These are identified on Figure 20.1 against a backdrop of the mean annual chlorophyll observed from MODIS AQUA satellite sensor. The trap placements, duration and main findings are listed in Tables 20.1 and 20.2. Although there are more than 400 sediment trap studies (Honjo *et al.*, 2008), largely the biological component has received less attention than the more quickly processed geochemical studies that enable rapid insights to water chemistry and other gross uptake and export studies focused on the carbon or silica cycles. In presenting this review focused on the diatom composition, environmental relations and their eventual signature in the sediments, we aim to re-center attention on the goods and services that diatom analysis can bring in linking climate processes, oceanographic dynamics, biological responses, ecological requirements and chemical cycles. The summary of these detailed studies will highlight the importance of key species or genera that are our link in understanding either the changing "health" of our oceans or the capacity of the oceans to store and effectively export atmospheric CO_2 to the deeper ocean.

20.5 Low- and mid-latitude regions

20.5.1 Coastal regimes

20.5.1.1 Atlantic Ocean

Northwest Africa

Canary Islands (Sites EBC, ESTOC, and LaPalma) Three sediment traps, deployed immediately to the north of the Canary Islands, were located between a major upwelling region and the oligotrophic open-ocean. Upwelling occurs during boreal summer and early autumn (Abrantes *et al.*, 2002). Diatom flux peaked in boreal mid winter and in late boreal spring–early summer. Daily mean fluxes revealed values an order of magnitude higher at near shore areas than in the open ocean (Abrantes *et al.*, 2002; Table 20.2). The diatom assemblage was dominated by species of *Pseudo-nitzschia* H. Perag. in boreal winter, while species of *Chaetoceros* Ehrenb. and *Leptocylindrus danicus* Cleve were the most common constituents of the boreal spring and summer assemblages.

Diatom concentration in surface sediments decreased from near-shore areas into the open-ocean; this correlated well with the decrease in primary productivity in surface waters (Abrantes *et al.*, 2002). Spores of *Chaetoceros* dominated the diatom assemblage closer to the coast, while the abundance of pelagic species of *Nitzschia* Hassall increased seaward in surface sediments.

Off Cape Blanc, Mauritania (Site CB) Four-year observations of diatom fluxes at site CB showed significant variations in diatom flux and species composition on both a seasonal and interannual basis. The marked drop from 1988 to 1989 in production and export of diatoms relates to the weakened intensity of coastal upwelling, and offshore spreading of the upwelling filament, and/or reduced silicate content of waters upwelled off Cape Blanc (Figure 20.2, Romero *et al.*, 2002a). Although the composition of diatom assemblages varied seasonally off Cape Blanc, no significant variations were observed from year to year. In general, the dominance of neritic *Thalassionema nitzschioides* var. *nitzschioides* (Grunow) Van Heurck, resting spores of *Chaetoceros* spp., and *Cyclotella litoralis* Lange and Syvertsen (Figure 20.2), mirrored the continuous offshore spreading of coastal upwelling. In contrast, the occurrence of pelagic *Nitzschia bicapitata* Cleve, *Nitzschia interruptestriata* (Heiden) Simonsen, *Thalassionema nitzschioides* var. *parva* (Heiden) Moreno-Ruiz, and *Fragilariopsis doliolus* (Wall.) Medlin and Sims was linked to inshore transport of oceanic waters, generally in boreal winter (Romero *et al.*, 2002a). Freshwater diatoms and phytoliths, transported by wind from the north African continent, were found constantly in trapped assemblages (Romero *et al.*, 2003). With the exception of some fragile, pelagic diatoms, dominant species found in the water-column settled material also occurred in the underlying surface sediments.

Southern Caribbean Sea

Cariaco Basin, off Venezuela (Site Cariaco Basin) In the Cariaco Basin, trade wind-induced upwelling and western boundary surface currents drove the annual production cycle. Diatoms occurred in greatest numbers during boreal winter and early spring, while secondary maxima occurred in late boreal spring–early summer (Romero *et al.*, 2009a). Diatom flux decreased by one order of magnitude during the strong 1997–1998 El Niño/Southern Oscillation (ENSO) event. Periods of high diatom production during boreal winters were dominated by *C. litoralis* (Romero *et al.*, 2009a). A major shift in diatom composition occurred during boreal winters affected by strong ENSO events, where a mixture of pelagic and coastal planktonic species comprised the diatom assemblage. Comparison with diatom distributions from sea-floor sediments has not been undertaken.

Table 20.1 Description of mooring sites

Site	Coordinates	Trap depth (m)	Water depth (m)	Trap	Sampling period	Authors
Atlantic Ocean						
Northwest Africa						
Canary Islands						
EBC	28° 45.5′ N 13° 09.3′ W	700	996	Aquatec	Jan. 1997–Sep. 1997	Abrantes *et al.* (2002)
ESTOC	29° 11.0′ N 15° 27.0′ W	500, 750, 3000	3610	Aquatec	Dec. 1999–Sep. 1997	Abrantes *et al.* (2002)
LaPalma	29° 45.7′ N 17° 57.3′ W	900, 3700	4327	Aquatec	Jan. 1997–Sep. 1997	Abrantes *et al.* (2002)
Cape Blanc (CB)	21°08.7′ N 20°41.2′ W	3562	4108	Kiel SMT 230	Mar. 1988–Nov. 1991	Lange *et al.* (1998) Romero *et al.* (2002a)
Cape Verde (CV)	11° 29′ N 21° 01′ W	1003	4523	Kiel SMT 230	Oct. 1992–Oct. 1993	Romero *et al.* (1999)
Cariaco Basin	10° 30′ N 64° 40′ W	230, 410	1400	PARFLUX Mark 7	Nov. 1996–Apr. 1998, Jan.–Oct. 1999	Romero *et al.* (2009a)
Benguela System						
WR2	20° 03′ S 09° 09′ E	599	2169	Kiel SMT 230	Mar. 1989–Mar. 1990	Treppke *et al.* (1996a)
CWR	27° 00′ S 03° 51.3′ E	2698	2700	Technicapp PPS5	Feb. 2000–Feb. 2001	Lončarić *et al.* (2007)
NU	29° 12′ S 13° 07′ E	2516	3055	SMT 230	Jan. 1992–Jan. 1993	Romero *et al.* (2002b)
Eastern tropical Atlantic, Guinea Basin						
EA1	03° 10′ N 11° 15′ W	984	4524	Kiel SMT 230	Apr. 1991–Nov. 1991	Romero *et al.* (1999)
GBN3	01° 47′ N 11° 08′ W	853	4481	Kiel SMT 230	Mar. 1989–Mar. 1990	Treppke *et al.* (1996b)
EA2	01° 47′ N 11° 15′ W	956	4399	Kiel SMT 230	Apr. 1991 – Nov. 1991	Romero *et al.* (1999)
EA3	01° 05′ S 10° 46′ W	1097	4141	Kiel SMT 230	Apr. 1991–Nov. 1991	Romero *et al.* (1999)
GBZ4	02° 11′ S 09° 54′ W	696	3912	Kiel SMT 230	Apr. 1991–Nov. 1991	Treppke *et al.* (1996b)
EA4	02° 11′ S 10° 06′ W	1068	3906	Kiel SMT 230	Apr. 1991–Nov. 1991	Romero *et al.* (1999)
EA5	04° 20′ S 10° 16′ W	948	3490	Kiel SMT 230	Apr. 1991–Nov. 1991	Romero *et al.* (1999)
Western equatorial Atlantic						
WA8	00° 58′ N 23° 27.1′ W	718	4166	Kiel SMT 230	Aug. 1994–Mar. 1996	Romero *et al.* (1999) Romero *et al.* (2000)
WA4	03° 59.3′ S 25° 35′ W	808	5601	Kiel SMT 230	Mar. 1993–Aug. 1994	Romero *et al.* (2000)
WA7	03° 58′ S 25° 39′ W	854	5601	Kiel SMT 230	Aug. 1994–Mar. 1996	Romero *et al.* (2000)
WA3	07° 31′ S 28° 02′ W	671	5570	Kiel SMT 230	Mar. 1993–Aug. 1994	Romero *et al.* (1999) Romero *et al.* (2000)
WA6	07° 28.3′ S 28° 07.4′ W	544	5637	Kiel SMT 230	Aug. 1994–Feb. 1996	Romero *et al.* (2000)

(*cont.*)

Table 20.1 *(cont.)*

Site	Coordinates	Trap depth (m)	Water depth (m)	Trap	Sampling period	Authors
Mediterranean Sea, Alboran Sea						
ALB-1D	36° 14′ N 04° 28′ W	928	958	Technicapp PPS3	Jul. 1997–May 1998	Bárcena *et al.* (2004)
Pacific Ocean						
Subarctic						
PAPA station	50° N 145° W	1000, 3800	4200	PARFLUX Mark 5	Sep. 1982–Apr. 1986	Takahashi (1986)
C station	49.5° N 138° W	3500	3900	PARFLUX Mark 5	Mar. 1985–Apr. 1986	Takahashi *et al.* (1990)
British Columbia						
JV7	50° N 124° W	50, 200, 450	530	No information	Mar. 1985–Dec. 1987	Sancetta (1989a)
Off southern Oregon						
Nearshore	42° 08.6′ N 125.771° W	1000, 1500	2829	No information	Sep. 1987–Sep. 1990	Sancetta (1992)
Midway	42.192° N 127.578° W	1000	2830	No information	Sep. 1987–Feb. 1990	Sancetta (1992)
Gyre	41.546° N 131.998° W	1000	3664	No information	Sep. 1987–Sep. 1990	Sancetta (1992)
Santa Barbara Basin (SBB)	34° 14′ N 120° 02′ W	590	650	Mark VI	Aug. 1993–Nov. 1994	Lange *et al.* (1997, 2000)
Southern California Bight						
	33° 33′ N 118° 30′ W	500	850	Mark VI	Jan.– Jul. 1988	Sautter and Sancetta (1992)
Gulf of California						
Guaymas Basin	27° 53′ N 111° 40′ W	500	700	No information	Feb. 1991–Feb. 1992	Sancetta (1995)
Carmen Basin	26° 02′ N 110° 55′ W	500	700	No information	Jul. 1990–Aug. 1992	Sancetta (1995)
North Pacific Gyre						
ALOHA I	22° 57.3′ N 158° 06.2′ W	2800	4800	PARFLUX Mark 7–21	Jun. 1992 – Jun. 1993	Scharek *et al.* (1999a)
Aloha I	23° 07.6′ N 157°55.8′ W	4000	4800	PARFLUX Mark 7–21	Sep. 1993 – Sep. 1994	Scharek *et al.* (1999a)
Southestern Pacific, off Chile						
CH3–1	30° 01.5′ S 73° 11′ W	2333	4360	Kiel SMT230	Jul. 1993–Jan. 1994	Romero *et al.* (2001)
CH4–1	30° 00.3′ S 73° 10.3′ W	2303	4330	Kiel SMT230	Jan. 1994–Jul. 1994	Romero *et al.* (2001)
CH10–1	29° 59.9′ S 73° 16.8′ W	1492	4500	Kiel SMT230	Feb. 1997–Nov. 1997	Romero *et al.* (2001)
CH11–1	29° 58.8′ S 73° 18.1′ W	2526	4442	Kiel SMT230	Nov. 1997–Jul. 1998	Romero *et al.* (2001)
Indian Ocean						
Off Southern Java, Indonesia						
JAM	08° 17.5′ S 108° 02′ E	2200	2700	Mark VII	Nov. 2000–Jul. 2003	Romero (2009b)

Table 20.1 *(cont.)*

Site	Coordinates	Trap depth (m)	Water depth (m)	Trap	Sampling period	Authors
Somali Basin						
MST-8	10° 46.9′ N 51° 56.4′ E	1265	1533	Salzgitter/HDW "Kiel"-type trap	Jun. 1992–Feb. 1993	Koning *et al.* (2001)
MST-9	10° 41.7′ N 53° 32.7′ E	3047	4047	Salzgitter/HDW "Kiel"-type trap	Jun. 1992–Feb. 1993	Koning *et al.* (2001)
Southern Ocean and Antarctic region						
McMurdo Sound/Terra Nova Bay/Ross Sea						
B (EIT)	77° 42′ S 166° 21′ E	25, 51, 77, 103	415	Rice Uni Mk II single cup	Oct.–Dec. 1984	Leventer and Dunbar (1987) Dunbar *et al.* (1989)
D (HP1)	77° 52′ S 166° 30′ E	28, 57, 109, 161	172	Rice Uni Mk II single cup	Oct.–Dec. 1984	Leventer and Dunbar (1987) Dunbar *et al.* (1989)
E (NH1)	77° 40′ S 163° 36′ E	47, 99, 151, 203, 255	264	Rice Uni Mk II single cup	Oct.–Dec. 1984	Leventer and Dunbar (1987) Dunbar *et al.* (1989)
F (NH3)	77° 38′ S 163° 46′ E	32, 84, 186, 238	250	Rice Uni Mk II single cup	Oct.–Dec. 1984	Leventer and Dunbar (1987) Dunbar *et al.* (1989)
I (GH)	76° 56′ S 163° 13′ E	34, 127, 220, 313, 406, 499, 592, 685	715	Rice Uni Mk II single cup	Nov.– Dec. 1984	Leventer and Dunbar (1987) Dunbar *et al.* (1989)
L (HP2)	77° 51′ S 166° 37′ E	15, 37	41	Rice Uni Mk II single cup	Nov.–Dec. 1984	Leventer and Dunbar (1987) Dunbar *et al.* (1989)
TI	77° 42′ S 166° 12′ E	7, 101, 195, 288	313	Rice Uni Mk II single cup	Oct. 1986–Jan. 1987	Dunbar *et al.* (1989)
BG	77° 36′ S 166° 11′ E	25, 187, 314	345	Rice Uni Mk II single cup	Oct. 1986–Jan. 1987	Dunbar *et al.* (1989)
DG1	77° 07′ S 163° 20′ E	35, 135, 170	181	Rice Uni Mk II single cup	Oct. 1986–Jan. 1987	Dunbar *et al.* (1989)
DG2	77° 01′ S 163° 35′ E	25, 189, 331, 428	464	Rice Uni Mk II single cup	Oct. 1986–Jan. 1987	Dunbar *et al.* (1989)
MS	77° 46′ S 165° 39′ E	35, 228, 388, 480	500	Rice Uni Mk II single cup	Oct.–Dec 1986	Dunbar *et al.* (1989)
A	76° 30′S 167° 30′ E	250, 659	719	Oregon State Uni Multitracer	Feb. 1991–Feb. 1992	Leventer and Dunbar (1996) Dunbar *et al.* (1998)
B	76° 30′ S 175° W	250, 469	519	Oregon State Uni Multitracer	Feb. 1991–Feb. 1992	Leventer and Dunbar (1996) Dunbar *et al.* (1998)
C	72° 30′ S 172° 30′ E	250, 443	493	Oregon State Uni Multitracer	Jan. 1990–Feb. 1992	Leventer and Dunbar (1996) Dunbar *et al.* (1998)
Adélie Land, East Antarctic						
KH94-4 site	64° 42′ S 139° 58′ E	537, 796, 1259, 1722, 2727	2930	McLane Mark VI & VII	Dec. 1994–Jan. 1995	Suzuki *et al.* (2001)
Prydz Bay/southeast Indian Sector						
M2	52° 00′ S 61° 32′ E	1300, 4025	4600	PPS5 Technicap sed. trap	Feb. 1994–Jan. 1995	Pichon *et al.* (unpublished) Caubert (1998)
M3	63° 01′ S 70° 57′ E	1300, 3445	4000	PPS5 Technicap sed. trap	Feb. 1994–Jan. 1995	Pichon *et al.* (unpublished) Caubert (1998)
PZB-1	62° 28.63′ S 72° 58.55′ E	1400, 2400, 3400	4000	McLane	Dec. 1998–Jan. 2000	Pilskaln *et al.* (2004)

(cont.)

Table 20.1 *(cont.)*

Site	Coordinates	Trap depth (m)	Water depth (m)	Trap	Sampling period	Authors
Weddell Sea/Atlantic sector						
Northwest Weddell Sea (WS1)	62° 26.5′ S 34° 45.5′ W	863	3880	McLane PARFLUX Mark 5	Jan. 1985–Mar. 1986	Fischer *et al.* (1988) Wefer *et al.* (1990) Abelmann and Gersonde (1991) Wefer and Fischer (1991)
Southeast Weddell Sea (WS2)	64° 55.0′ S 02° 30.0′ W	4456	5053	McLane PARFLUX Mark 5	Jan.–Nov. 1987	Wefer *et al.* (1990) Abelmann and Gersonde (1991) Wefer and Fischer (1991)
Southeast Weddell Sea (WS3)	64° 53.1′ S 2° 33.7′ W	360	5053	Kiel SMT 230	Jan. 1988–Feb. 1989	Abelmann and Gersonde (1991) Wefer and Fischer (1991)
Southeast Weddell Sea (WS4)	64° 55.5′ S 2° 35.5′ W	400	5044	Kiel SMT 230	Mar. 1989–Feb. 1990	Abelmann and Gersonde (1991) Wefer and Fischer (1991)
Polar Front (PF1)	50° 09.0′ S 5° 43.8′ E	700	3779	Kiel SMT 230	Jan. 1987–Mar. 1988	Abelmann and Gersonde (1991) Wefer and Fischer (1991)
Lützow-Holm Bay (St A)	69° 00′ S 39° 35′ E	5, 25	n/a	Single trap 14–30 day redeployment	Jan. 1984–Jan, 1985	Matsuda *et al.* (1987)
Lützow–Holm Bay (St A)	69° 00′ S 39° 37′ E	20	67	Single trap	Dec. 2005–Jan. 2006	Ishikawa *et al.* (2001) Ichinomiya *et al.* (2008)
Antarctic Peninsula/Eastern Pacific sector						
Bransfield Strait 2 / King George Island 1 (KG1)	62° 15.30′ S 57° 31.70′ W	496, 1588	1952	McLane PARFLUX Mark 5	Dec. 1983–Nov. 1984	Gersonde (1986) Wefer *et al.* (1988, 1990) Wefer and Fischer (1991)
King George Island 2 (KG2)	62° 20.1′ S 57° 28.3′ W	693	1650	McLane PARFLUX Mark 6	Dec. 1984–Nov. 1985	Wefer *et al.* (1990) Abelmann and Gersonde (1991) Wefer and Fischer (1991)
King George Island 3 (KG3)	62°22.0′ S 57° 49.9′ W	687	1992	McLane PARFLUX Mark 6	Nov. 1985–May 1986	Wefer *et al.* (1990) Abelmann and Gersonde (1991) Wefer and Fischer (1991)
Bransfield Strait RACER R13	63° 25′ S 62° 23′ W	100		Moss Landing type, free-floating multi-trap	Dec. 1986–Feb. 1987	Leventer (1991)
R20	61° 55′ S 62° 00′ W	100 or 200		Moss Landing type, free-floating multi-trap	Jan.–Mar. 1987	Leventer (1991)
R39	62°30′ S 61°32′ W	100		Moss Landing type, free-floating multi-trap	Dec. 1986–Mar. 1987	Leventer (1991)
R43	64° 17′ S 61°17′ W	100 or 200		Moss Landing type, free-floating multi-trap	Dec. 1986–Mar. 1987	Leventer (1991)
R48	63° 14′ S 60° 55′ W	100, 200		Moss Landing type, free-floating multi-trap	Dec. 1986–Mar. 1987	Leventer (1991)

Table 20.2 Summary of the main observations and characterization of investigated areas (Sp = Spring, Su = Summer, W = Winter, A= Autumn)

Area	Oceanographic regime	Daily mean diatom flux (valves m^{-2} d^{-1})	Season/s of highest diatom flux	Most abundant diatom species in the traps	Most abundant diatom species in surface sediments	References
LOW-TO-MID-LATITUDE AREAS						
COASTAL REGIMES						
Atlantic Ocean						
Northwest Africa						
Canary Islands	Coastal with open-ocean influence	$1.0 \times 10^5 - 10^6$	Boreal midW, boreal Sp/early Su	*Pseudo-nitzschia* spp., *Chaetoceros* spp. resting spores, *L. danicus*	*Chaetoceros* spp. resting spores (near-shore), *Nitzschia* spp. (open-ocean)	Abrantes *et al.* (2002)
Cape Blanc	Coastal with minor open-ocean influence	6.8×10^5	Boreal S–Su	*T. nitzschioides* var. *nitzschioides*, *Chaetoceros* spp. resting spores, *C. litoralis*	*T. nitzschioides* var. *nitzschioides*, *Chaetoceros* spp. resting spores	Romero *et al.* (2002a)
South Caribbean Sea						
Cariaco Basin	Coastal with moderate open-ocean influence	7.0×10^6	Late boreal W/early Sp	*C. litoralis*, *Chaetoceros* spp. resting spores	No data	Romero *et al.* (2009b)
Benguela Upwelling System						
Walvis Ridge (WR2)	Open-ocean with strong coastal influence	4.1×10^5	Austral Sp and A	*T. nitzschioides* var. *nitzschioides*, *G. cylindrus*, *A. ostrearia*, *R. bergonii*, *Pseudo-nitzschia* spp., *N. bicapitata*	*Chaetoceros* spp. resting spores, *F. doliolus*, *Azpeitia* spp.	Treppke *et al.* (1996a) Romero *et al.* (2002b)
Central Walvis Ridge (CWR)	Open-ocean with moderate coastal influence	5.5×10^6	Austral Sp and A	*N. bicapitata*, *T. nitzschioides* var. *curvata*, *T. nitzschioides* var. *inflata*	*F. doliolus*, *A. marinus*	Lončarić *et al.* (2007)
Namibia upwelling (NU)	Open-ocean with moderate coastal influence	5.5×10^6	Austral Su and W	*F. doliolus*, *A. curvatulus*, *C. radiatus*	*F. doliolus*, *T. nitzschioides* var. *nitzschioides*, *Azpeitia* spp., *A. marinus*	Romero *et al.* (2002b)
Indian Ocean						
Off southern Java	Open-ocean with moderate coastal influence	3.5×10^6	Early Sp–late Su	*N. bicapitata*, *N. interruptestriata*, *N. sicula*, *T. nitzschioides* var. *nitzschioides*, *C. litoralis*, *Chaetoceros* spp. resting spores	No data	Romero *et al.* (2009b)

(cont.)

Table 20.2 *(cont.)*

Area	Oceanographic regime	Daily mean diatom flux (valves m^{-2} d^{-1})	Season/s of highest diatom flux	Most abundant diatom species in the traps	Most abundant diatom species in surface sediments	References
Off Somalia, NW Indian Ocean	Coastal with moderate open-ocean influence	$1.3–3.5 \times 10^{7}$	Boreal Sp	*T. nitzschioides* var. *nitzschioides*, *Chaetoceros* spp. resting spores	*T. nitzschioides* var. *nitzschioides*, *Chaetoceros* spp. resting spores	Koning *et al.* (2001)
Pacific Ocean						
British Columbia fjords	Coastal with moderate open-ocean influence	3.3×10^{7}	Boreal Sp–Su	*Minidiscus* spp., *Thalassiosira* spp.	*S. costatum*	Sancetta (1989a)
Off southern Oregon						
Nearshore and Midway	Coastal with moderate open-ocean influence	No data	Late W/early Sp–Su	*N. bicapitata*, *P. delicatissima*, *N. sicula*	*T. nitzschioides*, *Chaetoceros* spp. spores, *F.doliolus*	Sancetta (1992)
Gyre	Open-ocean with moderate coastal influence	No data	Boreal W / early Sp	*M. chilensis*, *Chaetoceros* spp. resting spores, *T. nitzschioides*	*F. doliolus*, *T. nitzschioides*	
Santa Barbara Basin	Coastal with moderate open-ocean influence	1.4×10^{6}	Boreal Sp	*Chaetoceros* spp. resting spores	*Chaetoceros* spp. resting spores	Lange *et al.* (1997, 2000)
Southern California Bight	Coastal with moderate open-ocean influence	No data	Late boreal W–late Sp	*C. radicans*, *P. delicatissima*, *Chaetoceros* spp. resting spores	*Chaetoceros* spp. resting spores	Sautter and Sancetta (1992)
Gulf of California	Coastal with moderate open-ocean influence	$1.0–2.0 \times 10^{8}$	Early boreal W, Sp	*Coscinodiscus* spp., *Rhizosolenia* spp.	*T. nitzschioides*, *Chaetoceros* spp. resting spores	Sancetta (1995)
Off northern Chile	Coastal with moderate open-ocean influence	9.9×10^{6}	Late W/early Sp	*Chaetoceros* spp. resting spores	*Chaetoceros* spp. resting spores	Romero *et al.* (2001)
PELAGIC REGIONS						
Atlantic Ocean						
Cape Verde	Open-ocean with minor coastal influence	2.2×10^{6}	Boreal A–W	*N. bicapitata*, *T. nitzschioides* var. *parva*	Freshwater diatoms, *A. curvatulus*, *P. sol*, *T. nitzschioides* var. *parva*	Romero *et al.* (1999)
North Guinea Basin	Open-ocean with moderate coastal influence	$2.2 \times 10^{6}–1.2 \times 10^{7}$	Boreal Sp–Su	*N. bicapitata*, *T. nitzschioides* var. *parva*, *N. interruptestriata*	*R. bergonii*, *Azpeitia* spp.	Treppke *et al.* (1996b) Romero *et al.* (1999)
South Guinea Basin	Open-ocean	$5.0 \times 10^{5}–4.5 \times 10^{6}$	Boreal Sp	*N. bicapitata*, *P. calcar-avis*, *T. nitzschioides* var. *parva*	*R. bergonii*, *Azpeitia* spp.	Treppke *et al.* (1996b) Romero *et al.* (1999)
Western Equatorial	Open-ocean	$4.0–9.0 \times 10^{5}$	No clear seasonal trend	*N. bicapitata*, *N. interruptestriata*, *N.* sp. cf. *sicula*, *N. capuluspalae*, *Azpeitia* spp.	*Azpeitia* spp., *R. bergonii*, freshwater diatoms	Romero *et al.* (2000)

Mediterranean Sea						
Alboran Sea	Open-ocean with coastal influence	1.2×10^6	Late Sp/early Su	*Chaetoceros* spp. resting spores	*Chaetoceros* spp. resting spores, marine planktonic, *L. danicus*	Bárcena *et al.* (2004)
Pacific Ocean						
North Pacific Gyre	Open-ocean	No data	Boreal Su	*M. woodiana*, *H. hauckii*	No data	Scharek *et al.* (1999a, b)
Subarctic northeastern Pacific						
Stations PAPA and C	Open-ocean	No data	Boreal Sp–Su	*D. seminae*	*D. seminae*, *C. marginatus*	Takahashi (1986) Takahashi *et al.* (1990)
SOUTHERN OCEAN AND ANTARCTIC REGIONS						
South Atlantic						
Lützow–Holm Bay Stn A	Coastal	2.5×10^8	Potentially encountered but only Austral Su studied	*F. kerguelensis*, *P. pseudodenticulata*, *P.* cf. *turgiduloides*, *T. australis*	Nearby sediment stations C24, C18, C10: *F. kerguelensis*, *F. curta*, *Fragilariopsis* spp., *T. gracilis*	Tanimura (1992) Ichinomiya *et al.* (2008)
Northwest Weddell Sea (WS-1)	Seasonal ice zone	7.1×10^5	Austral late Su	*F. curta*, *T. gracilis*, *F. cylindrus*, *P. lineola*	*F. kerguelensis*	Fischer *et al.* (1988) Abelmann and Gersonde (1991) Zielinski and Gersonde (1997) Gersonde and Zielinski (2000)
Southeast Weddell Sea (WS2, WS3, WS4)	Seasonal ice zone	4.9×10^6–2.7×10^7	Austral late Su or A	*F. kerguelensis*, *F. cylindrus*, *F. curta*	*F. kerguelensis*, *F. cylindrus*, *F. curta*	Abelmann and Gersonde (1991) Zielinski and Gersonde (1997)
Polar Front (PF1)	Open-ocean zone	1.5×10^7	Austral late Su	*F. kerguelensis*, *T. nitzschioides*	*F. kerguelensis*, *T. lentiginosa*	Abelmann and Gersonde (1991)
South Indian						
Indian Ocean M2	Open-ocean zone	4.8–7.1×10^7	Austral Su	*F. kerguelensis*, *P-n. lineola*, *T. nitzschioides* var. *capitulata*, *P-n. heimii*, *F. rhombica*	Not specifically determined. Modern Analog Technique testing against the Crosta *et al.* (1998) sea-floor data set undertaken	Pichon (unpublished) Caubert (1998)
Indian Ocean M3	Seasonal ice zone	6.1×10^6–1.3×10^7	Austral late Su–early A	*F. kerguelensis*, *T. gracilis*, *F. curta*, *F. cylindrus*	Not specifically determined. Modern Analog Technique testing against the Crosta *et al.* (1998) sea-floor data set undertaken.	Pichon, Tréguer, Crosta, Le Fèvre, Schmidt, Morvan and Armand (unpublished) Caubert (1998)

(cont.)

Table 20.2 *(cont.)*

Area	Oceanographic regime	Daily mean diatom flux (valves m^{-2} d^{-1})	Season/s of highest diatom flux	Most abundant diatom species in the traps	Most abundant diatom species in surface sediments	References
Prydz Bay-1	Seasonal ice zone	1.4×10^7–2.8×10^9	Austral Su	*F. cylindrus*, *P-n. lineola*, *F. kerguelensis*, *F. curta*	*F. kerguelensis*, *T. lentiginosa*, *E. antarctica*, *T. gracilis*	Pilskaln *et al.* (2004) Armand (unpublished data)
Adélie Land KH94-4	Seasonal ice zone	1.4×10^5–9.3×10^9	Only Austral Sp studied	*F. curta*, *F. kerguelensis*, other pennates	Not determined	Suzuki *et al.* (2001)
Southwest Pacific						
McMurdo Sound	Seasonal ice zone – coastal	1.0×10^5–4.0×10^7	Only Austral Sp/Su studied	*Amphiprora* sp., *Pleurosigma* sp., *N. stellata*, *F. curta*, *T. antarctica*	*Thalassiosira* spp., *F. curta*	Leventer and Dunbar (1987) Dunbar *et al.* (1989)
Ross Sea Flux Experiment	Seasonal ice zone – coastal	Site A = 5.0×10^7 Site B = 6.0–8.0×10^7 Site C = 1.0–2.5×10^7	Austral Su (mid Feb-Sites A, B)	Site A: *F. curta*, *F. cylindrus* Site B: *F. curta*, *F. cylindrus*, *F. kerguelensis*, *F. obliquecostata*, *F. ritscherii*, *T. gracilis*. Site C: *F. curta*, *F. cylindrus*, *F. ritscherii*, *F. obliquecostata*, *Chaetoceros* spp.	Vicinity of Site A: *F. curta*, *Fragilariopsis* spp., *T. antarctica*, *Chaetoceros* spp. Site B: Diverse *Fragilariopsis* spp., *T. antarctica*, *T. gracilis*. Site C: not measured.	Leventer and Dunbar (1996)
Southeast Pacific						
Bransfield Strait (KG1, KG2, KG3)	Coastal	6.0×10^6–6.2×10^8	Austral Sp–early Su	*Chaetoceros* spp. vegetative and resting spores *T. antarctica* (KG3)	*F. kerguelensis*, *T. lentiginosa*, *F. separanda*, *T. gracilis*.	Abelmann and Gersonde (1991) Zielinski and Gersonde (1997)
RACER (13)	Coastal	2.9×10^7	Only Austral Sp/Su studied	*Chaetoceros* spp. resting spores	*Chaetoceros* spp. resting spores	Leventer (1991)
RACER (20)	Seasonal ice zone	2.7–9.9×10^6	Only Austral Sp/Su studied	*F. kerguelensis*, *F. curta*, *F. separanda/rhombica*, *T. gracilis*	*F. kerguelensis*, *Thalassiosira* spp.	Leventer (1991)
RACER (39)	Coastal	3.3×10^7	Only Austral Sp/Su studied	*Chaetoceros* spp. resting spores	Not determined	Leventer (1991)
RACER (43)	Coastal	2.5–3.5×10^8	Only Austral Sp/Su studied	*Chaetoceros* spp. resting spores	*Chaetoceros* spp. resting spores	Leventer (1991)
RACER (48)	Coastal	$1.7 - 6.2 \times 10^7$	Only Austral Sp/Su studied	*Chaetoceros* spp. resting spores	*Chaetoceros* spp. resting spores	Leventer (1991)

Note: Species authorities not identified elsewhere in the chapter: *Amphora ostrearia* Brébisson; *Alveus marinus* (Grunow) Kaczmarska and G. A. Fryxell; *Amphiprora* Ehrenb. sp.; *Coscinodiscus* Ehrenb. spp.; *Coscinodiscus radiatus* Ehrenb.; *Eucampia antarctica* (Castracane) Manguin; *Fragilariopsis obliquecostata* (Van Heurck) Heiden; *Guinardia cylindrus* (Cleve) Hasle; *Minidiscus chilensis* Rivera; *Nitzschia capuluspalae* Simonsen; *Nitzschia sicula* (Castracane) Hust.; *Nitzschia stellata* Manguin; *Pleurosigma* Smith sp.; *Pseudosolenia calcar-avis* (Schulze) Sundström; *T. nitzschioides* var. *curvata* (Schrader) Moreno-Ruiz; *Thalassionema nitzschioides* var. *inflata* Kolbe.

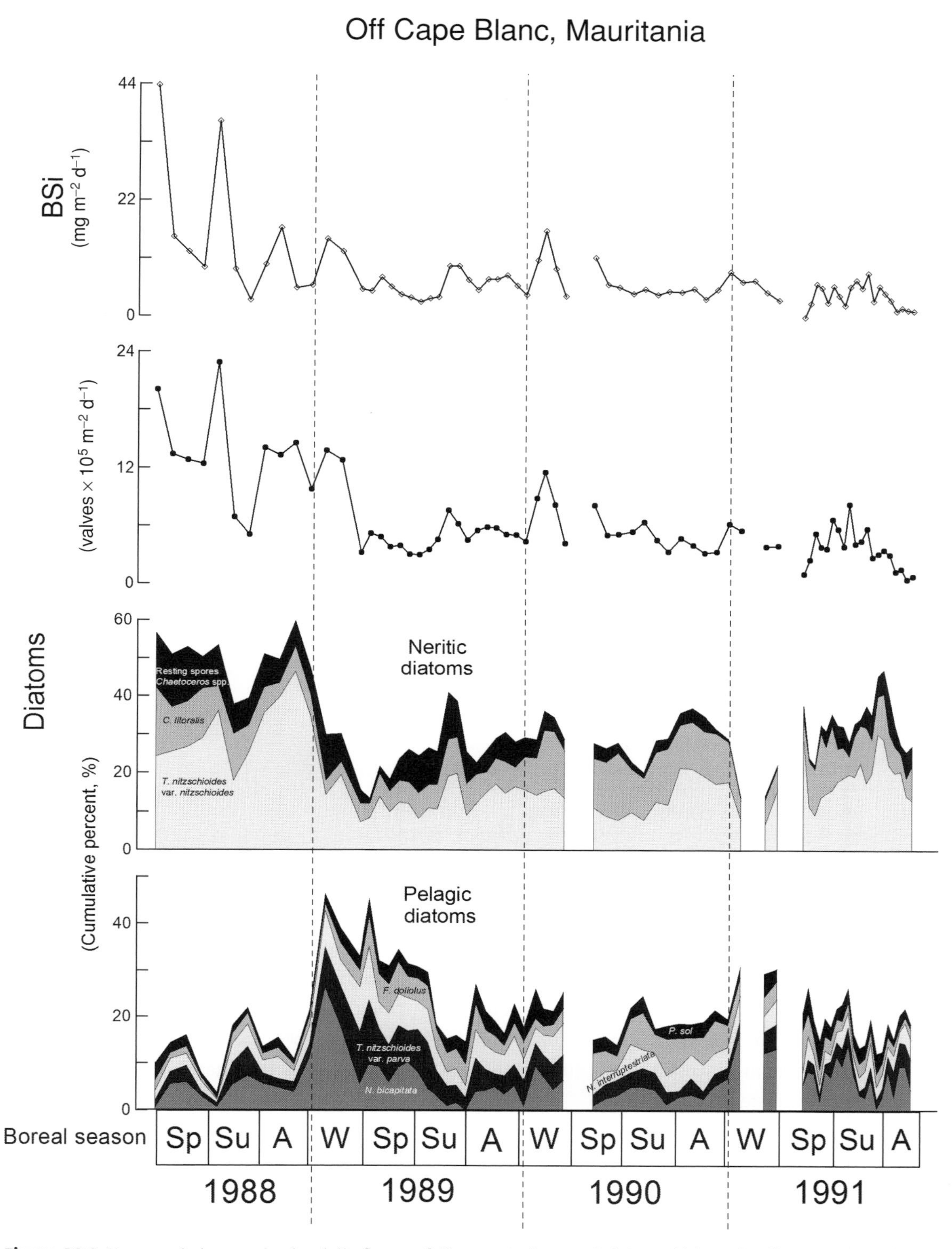

Figure 20.2 Temporal changes in the daily fluxes of diatoms (valves m^{-2} d^{-1}) and biogenic silica (mg m^{-2} d^{-1}) at the sediment trap site CB, off Mauritania, from March 1988 to November 1991. (From Romero *et al.*, 2002a; used with permission.) Cumulative percent of the most abundant neritic (*Thalassionema nitzschioides* var. *nitzschioides*, *Cyclotella litoralis*, resting spores of *Chaetoceros* spp.) and pelagic diatom species (*Nitzschia bicapitata*, *Thalassionema nitzschioides* var. *parva*, *Nitzschia interruptestriata*, *Fragilariopsis doliolus*, *Planktoniella sol* (Wallich) Schütt) at the CB site, off Mauritania, from March 1988 to November 1991. (Modified from Romero *et al.*, 2002a.)

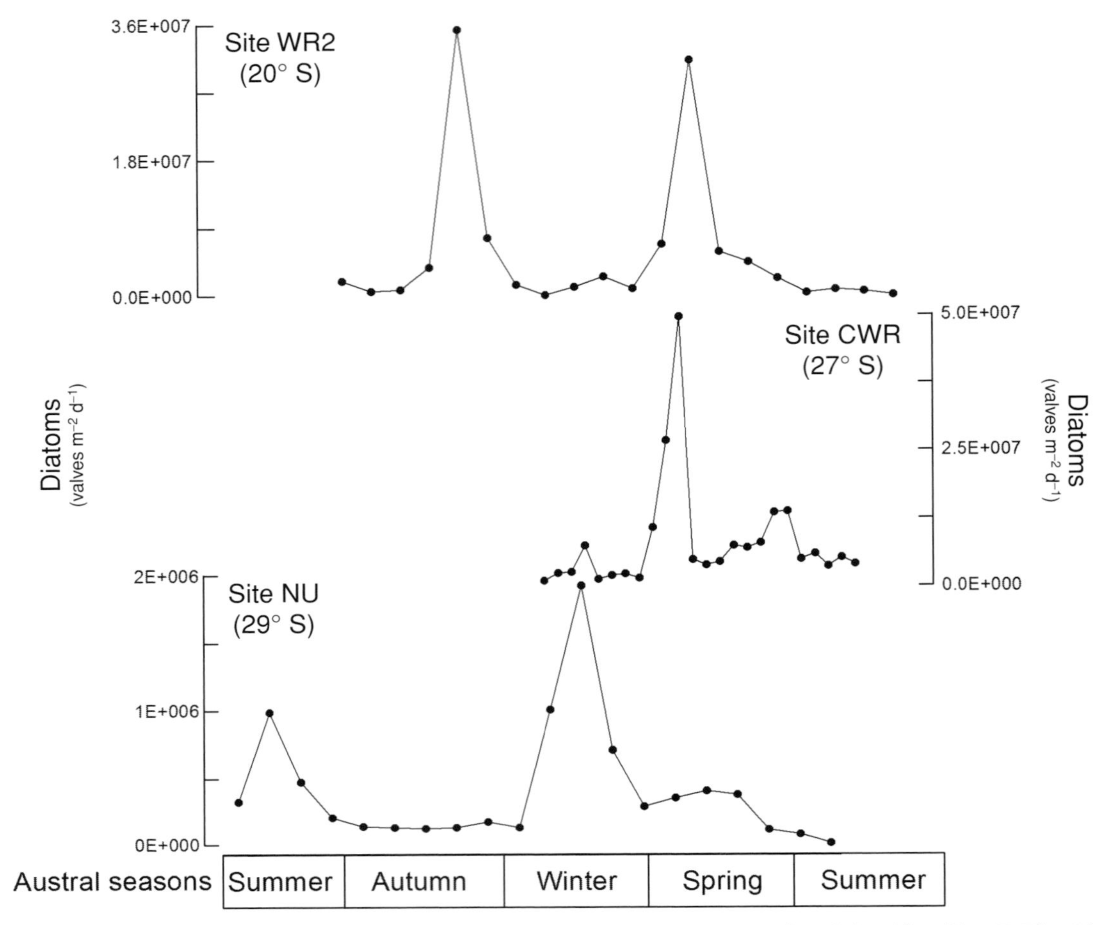

Figure 20.3 Temporal changes in the daily fluxes of diatoms (valves m^{-2} d^{-1}) at three mooring positions along the Benguela Upwelling System (Treppke *et al.*, 1996a; Romero *et al.*, 2002b; Lončarić *et al.*, 2007).

Benguela Upwelling System

Walvis Ridge (Site WR2) Diatom fluxes at site WR2 were observed as distinctly seasonal. The diatom maxima in Austral autumn and Austral spring were directly related to wind-stress variations, and were inversely related to sea-surface temperature changes (Figure 20.3; Treppke *et al.*, 1996a). *Thalassionema nitzschioides* var. *nitzschioides* and *Pseudo-nitzschia* spp. were the major contributors and followed the pulse of the main maxima, and hence were representative of the offshore input by coastal upwelling (Treppke *et al.*, 1996a). Many diatom taxa found in the WR2 trap also occurred in surface sediments, but their relative abundances differ from their mean percent values calculated in the traps. In the sediments, resting spores of *Chaetoceros* spp. were dominant (Treppke *et al.*, 1996a; Table 20.2).

Central Walvis Ridge (Site CWR) Site CWR was deployed in a semi-enclosed basin of the central Walvis Ridge in the oligotrophic South Atlantic Gyre. Austral spring samples were dominated by minute and fragile, solitary or aggregated, pennate species, mainly *N. bicapitata* (Figure 20.3; Lončarić *et al.*, 2007). The dominance of long, needle-like pennates and large centric, oligotrophic-region specialized *Ethmodiscus gazellae* (Janisch) Hust. in Austral summer suggested that changes in valve size and morphology were part of the species survival strategy in response to seasonally changing conditions. The sediment surface is relatively enriched in more heavily silicified centric diatoms; where *N. bicapitata* constituted 60% of the total diatom flux, it only comprised 20% in the underlying surface sediments (Lončarić *et al.*, 2007; Table 20.2).

Southern Benguela Upwelling System (Site NU) The pelagic NU trap was moored at the edge of the southernmost part of the central Benguela Upwelling System within the subtropical South Atlantic Gyre. The maximum export of diatoms in

Austral summer (Figure 20.3) probably responded to intensified offshore spreading of nutrient-rich waters of the Namaqua upwelling cell (~29° S), while the Austral winter maximum appeared as a consequence of *in situ* production, favored by strong mixing and low sea-surface temperatures (Romero *et al.*, 2002b). Remarkably, the mean annual diatom flux was lower at the NU site than at the CWR and WR2 sites (Table 20.2). This might be due to (1) interannual differences in surface-water productivity, or (2) the composition of the diatom assemblage at each trap site: while *N. bicapitata* constituted 60% of diatom deposition flux at CWR, it contributed only 6% at the WR2 site (Treppke *et al.*, 1996a). In place of the small *N. bicapitata*, larger taxa dominated at NU site. A good agreement exists between the biocoenotic and the thanatocoenotic communities at 29°–30°S in the Benguela Upwelling System (Romero *et al.*, 2002b; Table 20.2).

20.5.1.2 Indian Ocean

Off southern Java, Indonesia (Site JAM) A time-series sediment trap was deployed at a water depth of approximately 2200 m off southern Java in the tropical southeastern Indian Ocean between November 2000 and July 2003. The diatom flux peaked mostly during the monsoon seasons: highest diatom maxima were recorded in September 2001, February–March 2003, and June–July 2003 (Romero *et al.*, 2009b). The magnitude of diatom flux peaks also differed across years. Temporal changes in composition of the diatom community responded to the dynamics of the main hydrographic and atmospheric conditions in the southeastern Indian Ocean. During the southeast monsoon season, diatom-flux dynamics were driven by the occurrence of upwelling whereas nutrient entrainment by riverine input and vertical mixing appeared to be the mechanisms behind the diatom flux increase during the northwest monsoon season. Probably as a response to ENSO occurrence and the associated intensification of the upwelling, the contribution of warm-water, pelagic diatoms increased by the late 2002 southeast monsoon season. No sea-floor assemblage has been assessed.

Off Somalia (Sites MST-8 and MST-9) The Somali region of the northwestern Indian Ocean is strongly influenced by the southwestern monsoon from May through October. Diatom fluxes of autochthonous species were ten times higher during the southwest monsoon than during the northeast monsoon season, and five times higher on the continental slope than in the open ocean off Somalia (Koning *et al.*, 2001). A diverse, non-upwelling, assemblage was present year-round, but its relative abundance was low during the southwest monsoon. Well-silicified, upwelling species *T. nitzschioides* and *Chaetoceros* spp. resting spores comprised ~60% of the surface sediment assemblage, thereby preserving a residual upwelling signal (Koning *et al.*, 2001; Table 20.2). The burial efficiency of pre-upwelling and upwelling species, exemplified by *N. bicapitata*, was very low.

20.5.1.3 Pacific Ocean

British Columbia fjords (Site Jervis Inlet) Two elements dominated the trapped material in the fjords of British Columbia (Canada): lithics and diatoms (Sancetta, 1989a). Diatom-flux maxima coincided largely with the boreal spring bloom, while secondary maxima occurred occasionally in later boreal summer or early autumn. These observations suggest that the diatom flux was primarily controlled by the seasonal cycle of surface-water productivity (Sancetta, 1989a). The boreal spring bloom typically commenced with an increase in the flux of *Minidiscus* spp. Hasle, closely followed by several species of *Thalassiosira* (Cleve) Hasle. The surface sediment assemblage was dominated by *Skeletonema costatum* (Grev.) Cleve, also the main component of the distinct boreal autumm bloom (Sancetta, 1989b; Table 20.2).

Off southern Oregon (Sites Nearshore, Midway, and Gyre) Three years of sediment-trap data collected off southern Oregon monitored changes in a region affected by upwelling filaments (Nearshore and Midway) against an area outside this influence (Gyre). The trap sites influenced by coastal upwelling were generally of similar diatom composition and timing of their flux maxima, while the Gyre region differed in both regards (Sancetta, 1992; Table 20.2). Interannual variability in taxon composition appeared to correlate with productivity: the most consistent interannual trend was present nearshore, while the least productive (Gyre) exhibited the greatest variability. The dominance of *Chaetoceros* spores, *T. nitzschioides*, and *F. doliolus* in nearshore/midway surface sediments revealed a preserved assemblage resembling the late boreal autumn and winter trap community (Sancetta, 1992; Table 20.2). Sediments at the Gyre location were dominated by *F. doliolus* and *T. nitzschioides*.

Santa Barbara Basin, off California (Site SBB) Diatom fluxes mirrored the productivity cycle, with high values during the boreal spring–summer upwelling period and low fluxes during the boreal autumn–winter in the Santa Barbara Basin (Figure 20.4; Lange *et al.*, 2000). Diatom fluxes decreased under ENSO conditions, while the composition of diatom assemblage suffered significant changes. Warm-water species associated with ENSO showed unusually high contributions (Lange *et al.*, 2000),

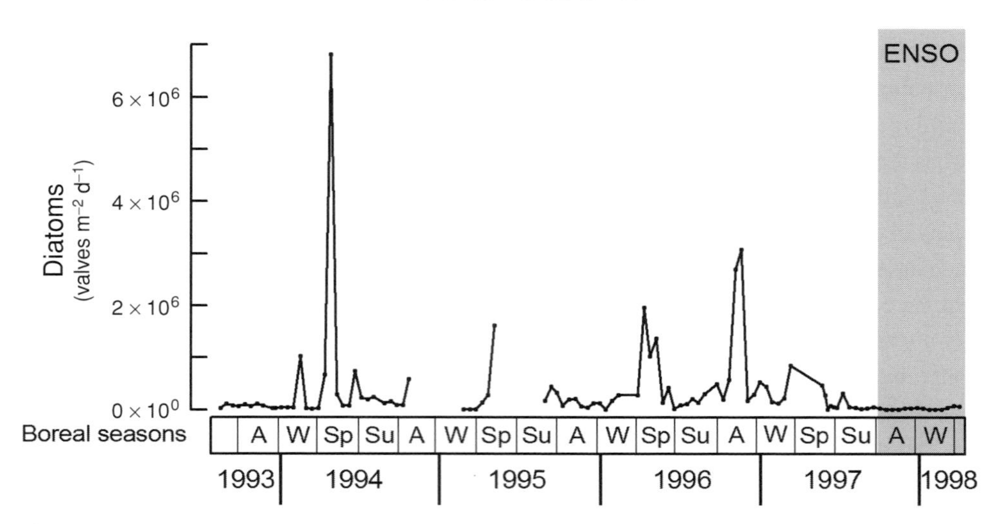

Figure 20.4 Temporal changes in the daily fluxes of diatoms (valves m^{-2} d^{-1}) at site SBB in the Santa Barbara Basin, off California, from mid boreal summer 1993 and early boreal spring 1998 (Lange *et al.*, 1997, 2000). The gray shading indicates the period under ENSO conditions.

while diatoms indicative of spring upwelling decreased by 50% (Lange *et al.*, 1997). *Chaetoceros* resting spores dominated the boreal spring upwelling peak and responded to the shoaling of the thermocline and high surface-pigment concentration. The *Chaetoceros* spore assemblage invariably dominated the sedimentary imprint (Lange *et al.*, 1997; Table 20.2).

Southern California Bight (Site SP) Close to the Santa Barbara Basin, a seven-month trap study was carried out in the southern California Bight. Maxima in diatom flux occurred from February to early March and in June (Sautter and Sancetta, 1992). During the boreal winter bloom, dominant diatoms included *Pseudo-nitzschia delicatissima* (Cleve) Heiden and *Chaetoceros radicans* Schütt, both typical of highly productive coastal environments. During the upwelling boreal spring period, *Chaetoceros* resting spores dominated. The composition of surface-sediment diatom assemblage was similar to trapped material (Sautter and Sancetta, 1992).

Gulf of California (Sites Guaymas and Carmen basins) The highest concentrations of diatoms were recorded in boreal winter and spring in the Gulf of California (Sancetta, 1995). None of the dominant taxa showed a consistent seasonal pattern. Most abundant were *Thalassionema bacillare* (Heiden) Kolbe and *T. nitzschioides*, while *N. bicapitata* had maxima in boreal winter. The diatom assemblage was consistent with warm surface water persisting into boreal winter during the ENSO 1991–1992. A higher contribution of summer–autumn diatoms during winter and early spring of 1992 than in preceding years, provided evidence of elevated surface water temperature (Sancetta, 1995).

Southeast Pacific, off northern Chile (Site CH) Off northern Chile, the diatom flux revealed a clear seasonal trend in 1993–1994: highest daily contributions of diatoms occurred in the late Austral winter of 1993 (Romero *et al.*, 2001). The second sampling period between February 1997 and July 1998 coincided with a strong ENSO event (McPhaden, 1999). The magnitude of the diatom flux was significantly decreased in 1997, yet retained the well-defined seasonal pattern observed in 1993–1994. *Chaetoceros* resting spores and *Actinocyclus curvatulus* Janisch, together with taxa typical of open-ocean conditions constituted between 70–90% of the total diatom flux (Romero *et al.*, 2001). Major components of the diatom flora off northern Chile maintained much of their regular seasonal flux cycle pattern of maxima and minima during the 1997–1998 period. The recurrence of coastal upwelling has provided a clear imprint in surface sediments between 27° and 35° S: spores of *Chaetoceros* dominated, accompanied by *T. nitzschioides* var. *nitzschioides* and *S. costatum* (Romero and Hebbeln, 2003; Table 20.2).

20.5.2 Pelagic regions

20.5.2.1 Atlantic Ocean

Cape Verde Islands (Site CV) In this area, upwelling occurs November through to February and was clearly reflected in the diatom fluxes, where maxima occured during boreal autumn and winter (Romero *et al.*, 1999). Small oceanic diatoms dominate the trapped diatom assemblage: the indicator species for oceanic equatorial upwelling, *N. bicapitata*, contributed more than 50% of the diatom winter flux peak. Although the contribution of small oceanic diatoms was much lower, the most abundant diatoms in the sediment traps were also found in the surface sediments (Romero *et al.*, 1999). Enrichment of freshwater diatoms and some pelagic species was also observed in the surface sediments.

Western equatorial Atlantic (Sites WA) Along a north–south transect in the western equatorial Atlantic, diatom fluxes exhibited, (1) an unimodal pattern at the equator characterized by maxima in January and March, coincident with the southward movement of the Intertropical Convergence Zone (ITCZ), and an enhanced input of lithogenic particles, and (2) a bimodal pattern of flux at 4° S with the maxima in February, and during the equatorial upwelling season (July–September) (Romero *et al.*, 2000). The southward flux decrease coincided with the regional occurrence of elevated chlorophyll along the north–south transect (Longhurst *et al.*, 1995). Minimal fluxes at 7° S coincided with the lowest values of organic carbon known for a wide area of the Atlantic Ocean (Fischer *et al.*, 2000). Small specimens of *N. bicapitata* dominated the diatom flux (Romero *et al.*, 2000; Table 20.2).

Dissolution of frustules, particularly those of *N. bicapitata*, in the water column and/or at the sediment–water interface removed the "productive season" signal from the annual cycle in the western equatorial Atlantic. Instead, the sediment assemblage was enriched in the more heavily silicified *Azpeitia* Perag. spp. and *Roperia tesselata* Grunow, whose higher contribution in the traps coincided with weak upwelling and the lowest chlorophyll concentration in the surface waters (Romero *et al.*, 2000).

Eastern equatorial Atlantic (Sites EA, GBN, and GBZ) Several traps in the eastern equatorial Atlantic were deployed in the Guinea Basin across a high-surface chlorophyll band within the equatorial divergence zone. Although mean values differed in magnitude between years, the pattern of increased diatom flux occurred north of the equator but not south of it (Treppke *et al.*, 1996b; Romero *et al.*, 1999; Table 20.2), as was expected from primary productivity measurements (Longhurst *et al.*, 1995). These observations may have been the result of several factors such as restricted particle sinking close to the equatorial divergence, preferential concentration and sinking of particulates further north of the equator (Yoder *et al.*, 1994), or that the majority of chlorophyll was delivered by non-siliceous pico- and nanoplankton, which were not trapped or preserved (Romero *et al.*, 1999). A clear seasonality was observed: two maxima flux peaks occurred, one in boreal spring and other in boreal summer. *Nitzschia bicapitata*, *T. nitzschioides* var. *parva* and *N. interruptestriata* were consistently present and abundant (Romero *et al.*, 1999). As described for the western equatorial Atlantic (Romero *et al.*, 2000), strong discrepancies were also observed between the sediment traps and surface sediments in the eastern basin: *Rhizosolenia bergonii* Perag., *Azpeitia* spp., and *R. tesselata* dominated the surface-sediment assemblage, yet their contributions to the overlying traps never exceeded 5% (Table 20.2).

20.5.2.2 Mediterranean Sea

Alboran Sea (Site ALB-1F) Diatom fluxes in the western Mediterranean Sea have been shown to follow mainly a bimodal pattern: the highest daily fluxes were measured in boreal spring and early summer (Bárcena *et al.*, 2004). The contribution of *Chaetoceros* spp. resting spores, representing high sea-surface productivity, exceeded 75% of the total diatom assemblage, and was elevated during May and July–October. The dominance of *Chaetoceros* spp. resting spores in the sediment traps was not reflected in the surface sediments. Instead, the marine planktonic diatoms were found to be the dominant component (Bárcena *et al.*, 2004; Table 20.2).

20.5.2.3 Pacific Ocean

North Pacific Gyre (Station ALOHA) The seasonal cycle of diatom fluxes was dominated by a boreal summer maximum at Station ALOHA in the oligotrophic North Pacific (Scharek *et al.*, 1999a). The diatom flux mainly consisted of the mixed-layer thriving *Mastogloia woodiana* Taylor and *Hemiaulus hauckii* Grunow (Scharek *et al.*, 1999a). Among the reasons for a summer proliferation of *H. hauckii* and *M. woodiana*, Scharek *et al.* (1999b) discussed the simultaneously increasing concentrations of the nitrogen (N_2)-fixing cyanobacteria *Trichodesmium* spp. Ehrenberg ex Gomont and the N_2-fixing capacity of *H. hauckii* through endosymbionts. The very high affinity of *H. hauckii* and *M. woodiana* for silicic acid, a compound relatively depleted in near surface waters of the North Pacific Gyre, is considered to favor these two species over other diatom species of the region (Scharek *et al.*, 1999b).

Subarctic northeastern Pacific Ocean (Stations PAPA and C) Between September 1982 and August 1986, Takahashi (1986, 1987, 1994) studied the flux of diatoms at Station PAPA located in the Alaskan Current Gyre. Takahashi and colleagues (1990) performed a similar study at the C Station, located between Station

PAPA and the western coast of North America between September 1985 and September 1986. In most instances, diatom fluxes at station C were lower than those at Station PAPA (Takahashi *et al.*, 1990). The two to four week delay in the flux maximum at station C compared to station PAPA was probably due to the fact that vertical mixing at station C was weak until October. At this point in time, the vertical density gradient decreased to levels less than that observed at station PAPA, which then allowed mixing to occur at the surface and subsequently the productivity flux to occur. *Denticulopsis seminae* Simonsen and Kanaya dominated the diatom assemblage in the traps (Takahashi, 1994). Approximately 2% of the diatoms collected within the sediment traps were preserved in surface sediments below Station PAPA (Takahashi, 1986). As in the traps, *D. seminae* dominated the sediment assemblage, yet a relatively large degree of dissolution of *D. seminae* in surface sediments was apparent. This was in contrast to *Coscinodiscus marginatus* Ehrenb., a more dissolution resistant taxon whose greatest production occured during boreal winters, when nutrient levels were elevated (Takahashi *et al.*, 1990; Table 20.2).

20.6 Southern Ocean and coastal Antarctic regions

20.6.1 South Atlantic

Northwest Weddell Sea (Site WS1) The WS1 site was chosen as a pelagic site capable of revealing the flux under seasonal sea-ice cover (Fischer *et al.*, 1988). Diatom flux was clearly shown to be drastically reduced by the dampening effect of Austral winter sea-ice cover (Fischer *et al.*, 1988; Abelmann and Gersonde, 1991). Species in the upper trap were dominated by sea-ice-affiliated *Fragilariopsis curta* (Van Heurck) Hust., *Fragilariopsis cylindrus* (Grunow) Krieg., and *Thalassiosira gracilis* (Karst.) Hust. (Table 20.2). Although the influence of oceanic species was low, enhancement in the deeper trap of robust species such as *Fragilariopsis kerguelensis* (O'Meara) Hust. and *Thalassiosira lentiginosa* (Janisch) G. A. Fryxell provided evidence of dissolution through the bathy- and abyssalpelagic zones (Abelmann and Gersonde, 1991). Evidence from diatom sea-floor distributions revealed the impact of dissolution with a strong representation of *F. kerguelensis* in the central Weddell Sea sediments in contrast to the emphasis of Austral summer flux sea-ice species observed at the surface (Zielinski and Gersonde, 1997; Gersonde and Zielinski, 2000).

Southeast Weddell Sea (Sites WS2, WS3, WS4) A second series of sequential annual moorings were placed in the southeast Weddell Sea (Abelmann and Gersonde, 1991). The three-year cycle determined the variability of annual flux and peak periodicity (Austral summer or autumn). *Fragilariopsis kerguelensis* remained the most dominant diatom throughout the annual cycle, although the sea-ice-related *F. curta* and *F. cylindrus* were increasingly predominant during sea-ice retreat across the mooring transect and during the major flux peak. The dominance of *F. kerguelensis* was given as 90% of the deep-trap assemblage, while the abundances of the sea-ice diatoms were ≤5% (Abelmann and Gersonde, 1991). Sea-floor diatom assemblages in the vicinity and under the mooring are described as being similar in composition to the deep trap (Abelmann and Gersonde, 1991; Zielinksi and Gersonde, 1997; Table 20.2).

Bouvet Island (Site PF1) A final German Weddell Sea mooring series focussed on the Polar Frontal zone north of Bouvet Island (Abelmann and Gersonde, 1991). The diatom assemblage was dominated by *F. kerguelensis* with a significant influx of *T. nitzschioides* in the early Austral summer. The presence of this latter species was postulated as a record of southward movement of the Polar Front across the mooring during Austral summer. Sea-floor sediments presented a typical open-ocean assemblage dominated by *F. kerguelensis* and *T. lentiginosa* (Zielinski and Gersonde, 1997; Table 20.2).

Lüdzow-Holm Bay (Site A) Station A in Lüdzow–Holm Bay marks the site of a long-term fast ice biological study region. Emphasis on the Austral summer bloom and export revealed that diatoms were released from the melting ice and then largely deposited directly to the sea floor (Matsuda *et al.*, 1987; Ishikawa *et al.*, 2001). Recently, the diatom flux was characterized identifying the early influence of centric diatoms to the flux represented by *Porosira pseudodenticulata* (Hust.) Jousé and *Thalassiosira australis* Perag., while pennate diatoms, represented by *Pseudo-nitzschia* cf. *turgiduloides* Hasle, *F. kerguelensis*, and other *Fragilariopsis* Hust. spp., dominated through the mid–late Austral summer flux (Ichinomiya *et al.*, 2008). Tanimura (1992) documented *F. kergulensis*, *F. curta*, and *T. gracilis* as dominant species in nearby sea-floor samples (Table 20.2).

20.6.2 South Indian

Indian Ocean (Sites M2 and M3) Under the French ANTARES-JGOFS program two annual moorings were placed in the Indian Ocean in contrasting environments: the permanently open ocean zone (site M2) and the seasonal ice zone in 1994/5 (site M3, winter sea-ice covered) (Figure 20.5; Pichon *et al.*, unpublished manuscript; Caubert, 1998; Armand, personal communication). The M2 assemblage consisted chiefly of *F. kerguelensis*,

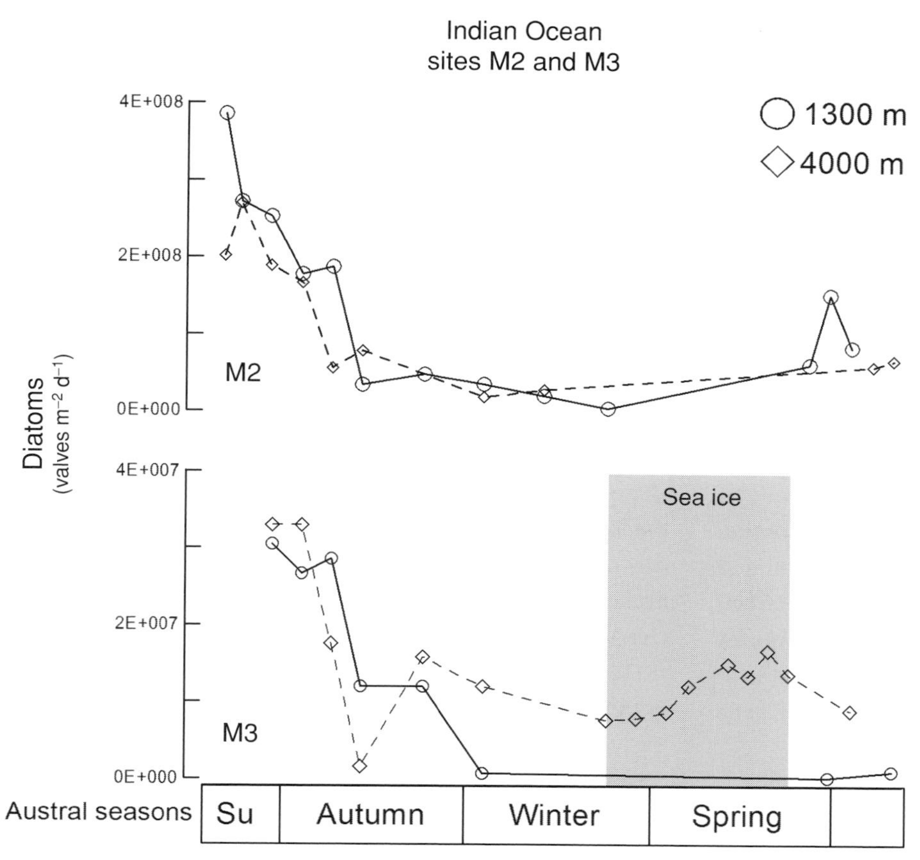

Figure 20.5 Temporal changes in the daily diatoms fluxes (valves m^{-2} d^{-1}) at sediment trap sites M2 (permanently open ocean zone) and M3 (seasonal ice zone) in the Indian Sector of the Southern Ocean between late Austral summer 1994 and early Austral summer 1995 (Pichon *et al.*, unpublished).

Pseudo-nitzschia lineola (Cleve) Hasle, *Thalassionema nitizschioides* var. *capitulata* (Castracane) Moreno-Ruiz, *Pseudo-nitzschia heimii* Manguin, and *Fragilariopsis rhombica* (O'Meara) Hust. The M3 mooring revealed clear discrepancies between the upper and lower traps with respect to their assemblages and abundances. Generally the two traps were dominated by *F. kerguelensis*, *T. gracilis*, and *F. curta*, yet the deeper trap revealed significant increases in *F. curta*, *P. lineola*, and *F. cylindrus* in Austral summer, contrary to *F. kerguelensis* abundances which decreased in summer and increased in Austral autumn. Lateral advection may have played a large role in the collection of material in the deeper trap at the winter sea-ice covered M3 site.

No direct sea-floor samples have been taken under the M2 and M3 locations. Yet the summer flux assemblages from both moorings were found to be statistically closest to surface samples from the Crosta *et al.* (1998) data set by ± 2° latitude from their current location (Pichon *et al.*, unpublished; Armand, personal communication).

Prydz Bay (Site PZB-1) The Prydz Bay mooring, in close vicinity to the French M3 mooring in the seasonal ice zone near Prydz Bay, covered an annual seasonal cycle (Table 20.1). Although three traps were placed at various depths, only the surface trap at 1400 m captured the complete diatom cycle (Pilskaln *et al.*, 2004). The greatest diatom flux occurred in mid Austral summer where fluxes of *F. cylindrus*, *P-n. lineola*, *F. kerguelensis*, and *F. curta* were the major contributors (Table 20.2). The early Austral spring flux was attributed to the first stage of sea-ice melt as observed with notable flux increases of *F. cylindrus* and *F. curta* into the trap (Armand, unpublished data).

20.6.3 Southeast Indian

Adélie Land (Site KH94-4) An Austral spring survey in the marginal sea-ice zone off Adélie Land in the East Antarctic was carried out between December 1994 to January 1995 (Suzuki *et al.*, 2001). One major bloom settling event was captured by all five traps providing a particle sinking rate of >200 m d^{-1} (Sasaki *et al.*, 1997). *Fragilariopsis curta* and *F. kerguelensis* were the dominant species (Suzuki *et al.*, 2001). An earlier biogeographic distribution of diatoms along the Adélie Land/George V Coast documents high abundances of *F. curta* captured on the sea floor nearby at 65° S (Leventer, 1992), supporting the flux observations of Sasaki *et al.* (1997).

20.6.4 South Pacific

Ross Sea

McMurdo Sound (Sites B–L, TI, BG, DG1–2, MS) Eleven sites were studied over the late-Austral autumn–summer period in McMurdo Sound in 1984 and 1986/7 (Leventer and Dunbar 1987; Dunbar *et al.*, 1989, Table 20.1). Their early work provided clear evidence of (1) the rapid recycling of thinly silicified species in the surface waters, (2) the flux of sea-ice related diatoms involving limited dissolution in the mid-water depths, enhanced dissolution in the silica-depleted nepheloid layer enhancing robust species (e.g. *F. curta*), and (3) the role of lateral advection in the deeper waters leading to the dilution of the original surface-productivity signal (introduction and increased flux of *Thalassiosira* spp.). A discrepancy was evident between the trapped Austral spring/early summer diatom flux signature and that retained in the underlying sea-floor sediments. At the sea floor the species *Thalassiosira* spp. and *F. curta* dominated the diatom assemblage by up to 90%.

Ross Sea Flux Experiment (Sites A, B, C) A complete annual time series from the Ross Sea was produced between January 1991 and February 1992 from three locations (Leventer and Dunbar, 1996; Dunbar *et al.*, 1998, Table 20.1). These traps, all affected by the presence of sea-ice cover, confirmed the major flux of diatoms occurred in Austral summer. In the southwest Ross Sea (trap A) the flux of *Fragilariopsis* species was dominant, chiefly through contributions of *F. curta* and *F. cylindrus*. The influx of *Thalassiosira antarctica* Comber and *Chaetoceros* spp. at the deeper trap was pronounced from late April, peaking in January. In the south-central Ross Sea (Site B) *F. curta's* dominance at Site A was tempered by a more diverse *Fragilariopsis* assemblage and the presence of *T. gracilis*. This diverse assemblage was considered a response to pack-ice conditions and greater wind stress. Site C in the northwestern Ross Sea, although troubled by a cup rotational fault, also revealed *F. curta* as the dominant species with additional contributions from other *Fragilariopsis* taxa and *Chaetoceros* spp. The appraoching similaries between sites A and C were attributed to congelation sea-ice cover considered conducive to the enhanced productivity of *F. curta*.

Sea-floor samples taken in the vicinity of sites A and B clearly represented the dominance of *Fragilariopsis* taxa, particularly *F. curta*. *Thalassiosira antarctica* abundances were elevated in most samples near Site A, while *T. gracilis's* presence was progressively evident in the south and central Ross Sea sediments.

20.6.5 Antarctic Peninsula

Bransfield Strait

King George Island (KG1, KG2, KG3) The study of particle flux near King George Island in the Bransfield Strait is acknowledged to be the first continuous time series in the Southern Ocean. The three successive moorings, KG1, KG2, and KG3, were deployed between December 1983 and May 1986 (Gersonde, 1986; Wefer *et al.*, 1988; 1990; Wefer and Fischer, 1991; Abelmann and Gersonde, 1991; Table 20.1). Maximum diatom fluxes were encountered in the first year during the late Austral spring/early summer and were lowest in the second year. The diatom assemblage was nonetheless similar in composition in the first two years composed chiefly by *Chaetoceros* spp. vegetative cells and resting spores (Table 20.2). In the single deep trap deployed (KG1) the abundance of *Chaetoceros* taxa was increased by up to 20%, but lateral transport was also considered a factor in this steep increase (Abelmann and Gersonde, 1991). Observations from the third year's collection (KG3) captured the major influence of *Thalassiosira antarctica* both as vegetative and resting cells during and after, respectively, the main diatom flux peak in Austral spring. The marked diatom assemblage change in the third year was hypothesized as a result of Austral-summer cold-water incursion from the western Weddell Sea into the King George Basin (Abelmann and Gersonde, 1991). The biogeographic distribution of sea-floor diatoms in this region are dominated by the open-ocean signatures of *F. kerguelensis*, *T. lentiginosa*, *Fragilariopsis separanda* Hust., and *T. gracilis* (Zielinski and Gersonde, 1997; Table 20.2).

Research on Antarctic Coastal Ecosystem Rates, RACER (Sites R13, R20, R39, R43, R48) The RACER study over Austral spring and summer of 1986/87, focused directly on identification and enumeration of the diatom flux in the southern Drake Passage (Site 20), Bransfield (sites 13, 48) and Gerlache Straits (Site 43), and on the continental shelf near Linvingston Island (Site 39) (Leventer, 1991). Five free-floating traps were used for 24–36-hour periods at monthly intervals for four months. The continental shelf and the strait-located sites were all overwhelmingly dominated by *Chaetoceros* ssp. resting spores through the Austral spring months, with abundances dropping slightly into the summer months when the influence of *Fragilariopsis* spp. and benthic species became more apparent (e.g. sites 39, 13, and 43). Diatom fluxes in these four sites were all considerably high at the start of the sample period and decreased by approximately two orders of magnitude by mid summer (Leventer, 1991). The Drake Passage site contained a distinctly open-ocean assemblage characterized principally by *F. kerguelensis*. Other species of note were *F. curta*, *F. rhombica/separanda*, and *Thalassiosira* spp.

Sediment samples were most similar to the diatom assemblages observed during the main diatom flux periods. Where deeper traps had been placed (sites 20, 48, and 43) dissolution in the epipelagic zone occurred that enhanced *Chaetoceros* spp. resting spore abundance most markedly, whereas both the *Chaetoceros* spp. resting spores and *F. kerguelensis* abundances increased only slightly with depth in the Drake Passage. These deeper flux assemblages closely resembled the sea-floor sediment assemblages (Leventer, 1991).

20.7 Synthesis

20.7.1 Marine diatom fluxes and the productivity regimes of the world ocean

In spite of some constraints, the large regional differences seen in the diatom export fluxes (Table 20.2) can be used as good representatives of present-day diverse productivity regimes in the Earth's oceans. Within Eastern Boundary Current systems (northwest Africa, Benguela Upwelling System, off California Current, the Peru–Chile Current), temporal differences are seen in the diatom production and exportation rates at timescales of days through months to years. For example, the quasi-constant input of nutrients off Cape Blanc, northwest Africa, results in an intense, almost permanent diatom export flux with low seasonality (Romero *et al.*, 2002b). On the other hand, in the Benguela Upwelling System (Treppke *et al.*, 1996b; Romero *et al.*, 2002a; Lončarić *et al.*, 2007), as in other coastal upwelling areas (Canary Island region, Abrantes *et al.*, 2002; off California, Sautter and Sancetta, 1992; Lange *et al.*, 1997, 2000; in the Gulf of California, Sancetta, 1995; off northern Chile, Romero *et al.*, 2001), marked seasonal upwelling pulses result in concomitant marked fluctuations in the export of diatoms.

In the Southern Ocean a wide variability of particle flux values are encountered. One factor is important here: sea ice. With Austral-winter sea-ice cover and the simultaneous lack of sunlight hindering the stimulation of productivity in the high latitudes, a decrease in both particle flux and subsequently diatom flux in winter evolves (Fischer *et al.*, 1988; Abelmann and Gersonde, 1991; Leventer and Dunbar, 1996; Pilskaln *et al.*, 2004, Grigorov, personal communication). The diatom flux in the seasonal sea-ice zone can vary over several orders of magnitude from lows in the marginal sea-ice zone off the Adélie Coast, Ross, and Weddell seas to the maximum diatom fluxes reported offshore the Adélie Coast and Prydz Bay (Table 20.2). The seasonal variability of diatom fluxes at any one site from the Austral-winter to main spring–summer flux is generally separated by two orders of magnitude. The stability of these seasonal cycles appears reasonably constant given the few studies that have gone beyond a single year; however the influence of upwelling along the coastal margin such as in the Bransfield Strait, latitudinal migration of the Polar Front, or the inception of polynyas in the Ross Sea can equally vary the flux and composition of the diatom surface assemblage.

20.7.2 Changes in longer-term studies

So far, a few long-term records (>3 years) of diatom fluxes in the Earth's ocean are available to assess the interannual variations of probably periodic or episodic nature. The decreasing diatom fluxes measured off Mauritania at the CB site (Figure 20.2) match the long-term flux pattern for the North Atlantic: these changes have been primarily associated to a drop in BSi (diatoms) fluxes, whereas the carbonate (coccolithophorids) fluxes remained rather constant (Deusser *et al.*, 1995). In addition to possible long-term changes in the siliceous vs. calcareous production, these long-term studies allow us to demonstrate the varying response of the diatom production to ENSO events. Diatom production and export affected by ENSOs in the subarctic Pacific Ocean decreased significantly, though the composition of the diatom flora experienced only minor changes (Takahashi, 1987). Similarly, diatom production in coastal waters off northern Chile decreased during the 1997 ENSO compared to earlier years. Spores of *Chaetoceros* dominated the diatom flora off northern Chile under both ENSO and La Niña conditions (Romero *et al.*, 2001). Yet, the qualitative composition of the diatom community in other low-latitude coastal systems has experienced significant changes during ENSO years. An increase in the relative contribution of tropical diatoms coupled with the decrease of coastal-upwelling diatoms was associated with warm-water incursions into the Santa Barbara Basin and decreased upwelling intensity due to the occurrence of ENSO (Lange *et al.*, 1997). In the Cariaco Basin, the highly diverse diatom assemblage trapped during ENSO years was composed by a mixture of pelagic and coastal planktonic diatoms, while only coastal diatoms dominated during La Niña years (Romero *et al.*, 2009a).

Mooring studies from the Southern Ocean are yet to be analyzed in comparison with physical climatic forcings, such as the Southern Annular Mode, changes with the ENSO/La Niña events, and solar cycles. Perhaps as a result of the few long-term studies actually available and largely completed prior to 2000, interpretation has remained at an understanding of the basic seasonal cycle and evolving assemblage composition with respect to basic hydrological regimes (particularly upwelling

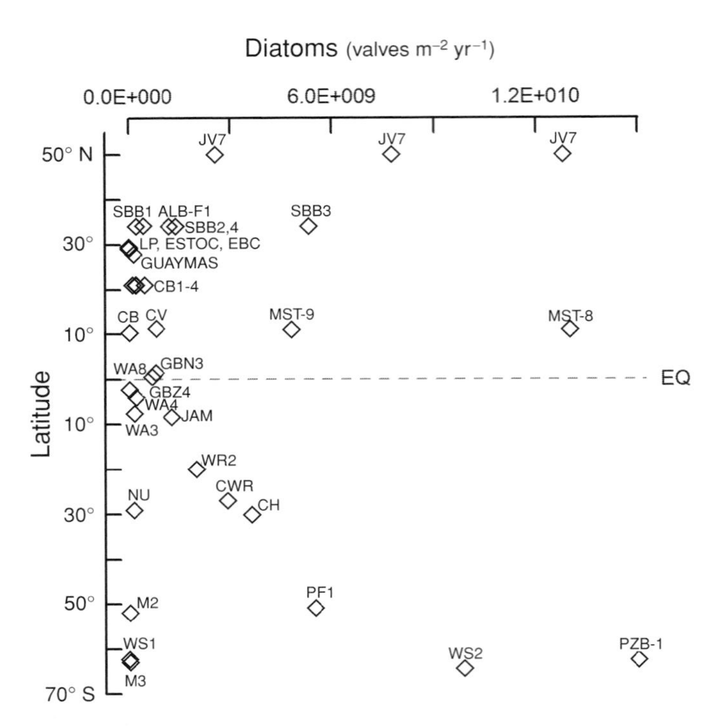

Figure 20.6 Latitudinal variations in the annual diatom fluxes measured by means of sediment traps (all data in valves m^{-2} yr^{-1}). See Table 20.1 for the detailed description of study sites and positions. Data are from reported annual diatoms fluxes or extrapolated from daily fluxes reported in Table 20.2. EQ = Equator.

events) or the sea-ice cycle. Such detailed analysis in tandem with renewed moorings in the high-latitude environment will allow for a more comprehensive understanding of the environmental forcings at play. The Palmer Long Term Ecological Research Site has shown through its mooring investigations that the annual peak of total mass flux in the Austral summer has already advanced by 40 days since 1993 (Ducklow *et al.*, 2008). Moline and Prezlin (1996) suggested the study of phytoplankton succesional patterns at Palmer Station would be a more sensitive tool for the detection of long-term trends, contrary to the current use of biomass and productivity indicies, which have large interannual variations. From these two studies alone the ability to demonstrate effectively the role of forcing on long-term collections is feasible; furthermore the seasonal flux of key diatom species is open for explotation in this manner.

20.7.3 The latitudinal pattern of marine diatom fluxes

A plot of the latitudinal distribution pattern of yearly diatom fluxes (Figure 20.6) reveals highest values ($>8.0 \times 10^9$ valves) south of the Polar Front and in the Pacific Ocean. Both the high southern latitude and the Pacific Ocean flux maxima agree well with the present-day silicic acid distribution pattern; elevations in the Pacific and the Southern oceans and decreases in the Atlantic and the Indian oceans (Conkright *et al.*, 2002). In spite of the low amount of complete yearly flux data available and the fact that fluxes have been collected in different years (Tables 20.1 and 20.2), our spatial distribution appears to reflect the conveyor-belt pathway. Therefore diatom fluxes can be suggested as strongly dependant on the silicic acid supply from the deeper waters: very low subsurface silicic acid concentrations ($\geq$250 m) can be found in the Atlantic, intermediate values in the Indian Ocean, and high values in the North Pacific and the Southern Ocean (Ragueneau *et al.*, 2000; Fischer *et al.*, 2004). Biogenic silica production and flux show a strong dependency on the supply of dissolved silicid acid ("silica pump," Dugdale *et al.*, 1995). In areas with high BSi flux such as the Bransfield Strait and in some low-latitude coastal upwelling systems with moderate to high opal fluxes (e.g. Benguela Upwelling System), a linear relationship between BSi and organic carbon fluxes has been noted (Ragueneau *et al.*, 2000; Romero *et al.*, 2002b; Fischer *et al.*, 2004).

The ballast hypothesis suggests that fluxes of organic carbon to the deep sea may be determined by the flux of minerals, and that calcium carbonate is a more efficient "carrier" of organic carbon than BSi (François *et al.*, 2002; Klaas and Archer, 2002). The validity of drawing such conclusions has been recently questioned. Laboratory experiments have yielded no evidence that calcium carbonate has a higher carrying capacity for organic matter than BSi does (De La Rocha *et al.*, 2008). It is clear that much work has to be done to improve the quantification of the processes by which organic carbon and mineral fluxes are linked.

20.7.4 Diatom species indicative of certain events

The compositon of the diatom assemblages collected with sediment traps varies according to the dominant hydrographic conditions, and may also reflect the BSi content of the subsurface waters. Fragile (*N. bicapitata*) to moderately silicified diatoms (*Azpeitia* spp., *N. interruptestriata*, *R. bergonii*, *T. nitzschioides* var. *parva*) characterize pelagic waters of low- to mid-latitude areas of the world ocean and are dominant during times of high diatom production and export (Romero *et al.*, 1999). Moderately silicified to robust diatoms (*Thalassionema nitzschioides* var. *nitzschioides*, *Chaetoceros* spp. resting spores) are typically found in areas under the direct influence of coastal upwelling (e.g. Sancetta, 1995; Romero *et al.*, 2001, 2002a; Table 20.2) or in hemipelagic areas seasonally affected by the offshore spreading of chlorophyll filaments (Abrantes *et al.*, 2002).

In the Southern Ocean, diatom flux is related to the seasonal cycle of sea-ice growth and retreat in the seasonal ice zone, whereas the hydrological condition and influx of surface meltwater in spring also enhances the flux cycle in the remainder of the Southern Ocean. In many surface moorings an evolving sequence of species has been noted. For example, the influx of centric diatoms such as *T. antarctica* and *P. pseudodenticulata* prior to the major productivity peak associated with the retreat of sea ice, otherwise dominated by *F. curta* and *F. cylindrus* or *Chaetoceros* spp., has been noted from McMurdo Sound (Leventer and Dunbar, 1987), the Bransfield Strait (Abelmann and Gersonde, 1991) and Lützow–Holm Bay (Ichinomiya *et al.*, 2008). Deeper traps, however, clearly show the influence of recycling in the upper water column with the loss of fragile species and increased dominance of robust species. In the coastal confines of the Bransfield Strait an overwhelming bias towards the resting spores of *Chaetoceros* spp. is clear (Abelmann and Gersonde, 1991; Leventer, 1991). Elsewhere along the Antarctic continental margin, and also within the marginal sea-ice edge, the dominance is balanced between *F. curta*, *F. kerguelensis*, and *T. gracilis* (Table 20.2). Northward of the sea-ice influence, the dominance of *F. kerguelensis* continues but is tempered by the influx of other diatoms such as *Thalassiothrix antarctica*, *T. lentiginosa* and *F. separanda*. The influx of northerly surface waters across the Polar Front introduced two highly silicified signature species, *T. nitzschioides* and *Azpeitia tabularis* (Grunow) G. A. Fryxell and Sims, to trapped material (Abelmann and Gersonde, 1991; Grigorov, personal communication).

20.7.5 Benefits in looking at the marine surface sediments under a trap

In several areas of the oceans, the effect of preservation removes information from the most productive season, thus leaving the sediment with no evidence of original variability. This is the case of many pelagic areas, such as the eastern and the western equatorial Atlantic, where *N. bicapitata* dominated during times of high production. However, the species' lightly silicified valves are removed by dissolution, resulting in a surface sediment assemblage enriched in increasingly dissolution-resistant diatoms (Romero *et al.*, 1999; Table 20.2).

Diatoms preserved in surface sediments underneath coastal upwelling systems record the high-productivity season of the surface waters, which are represented by moderately robust taxa. In the northwest Indian Ocean, the upwelling species *T. nitzschioides* and *Chaetoceros* resting spores have high fluxes and relatively high burial efficiencies: in surface sediments underlying the traps, these species are present in significant amounts, about 60% of the total diatom flux, and they thereby preserve the upwelling signal recorded in the sediment trap (Koning *et al.*, 2001). Similar conclusions have been drawn from trap and surface-sediment comparisons carried out along Eastern Boundary Current systems (Lange *et al.*, 1997; Romero *et al.*, 2001, 2002a; Abrantes *et al.*, 2002).

The diatom bioregionalization of sea-floor sediments in the Southern Ocean are well documented and placed in context to both sea-surface temperature and sea-ice cover (Zielinski and Gersonde, 1997; Armand *et al.*, 2005; Crosta *et al.*, 2005; Romero *et al.*, 2005). The major influence of *F. kerguelensis* in the open-ocean regions, and respectively with *F. curta* and *F. cylindrus* in the seasonal sea-ice zone provide clear, yet approximate, zonation of the regions affected by sea-ice cover. The utility of this discrimination providing a basis for the estimation of sea-ice cover in paleoclimatological reconstructions (reviewed in Armand and Leventer, 2009). Moorings with deep traps (>500 m) all revealed the influence of dissolution on more fragile species (e.g. Leventer and Dunbar, 1987; Abelmann and Gersonde, 1991; Leventer, 1991) and perhaps more importantly the influence of lateral advection on accumulation particularly in winter months (Leventer and Dunbar, 1996; Dunbar *et al.*, 1998; Pichon and colleagues, unpublished). In most cases, such deep moorings, in or northward of the sea-ice zone, showed a closer match between the deeper trap diatom assemblage and the sea-floor assemblage.

In the Antarctic coastal zone either distinct pulses (e.g. Stn A, Lutzow–Holm Bay, certain sites in McMurdo Sound), or the saturation of the sedimented record by *Chaetoceros* resting spores from the bust of summer productivity (e.g. Bransfield Strait) were the main signatures observed. The formation of laminations along the Antarctic coastline has also yet to be analyzed in terms of a mooring study. The importance of understanding the formation and diatom composition of laminations is crucial to the ground-truthing of paleoclimatic scenarios, particularly during the last deglaciation when a warmer climate than today was in effect (Armand and Leventer, 2009; Leventer *et al.*, this volume). Regardless of bias presented by the dissolution of certain diatom species in their voyage to the sea floor, the major diatom-flux signature is retained in the sediments and although a "fall dump" contribution may be valid for the open-ocean-zone accumulation of sediments (Kemp *et al.*, 2000), there is little mooring evidence to detail this contribution, either as a flux or in terms of the diatom assemblages observed to date.

20.7.6 Perspectives

Diatom-flux measurements using sediment traps are presently one of the best tools for recording the link between climate forcing at the sea surface and their records on the sea floor. Sediment-trap experiments covering all the seasons have permitted a match between diatom-flux information with the paleoceanographic timescale and information on the temporal variability of the ocean particle flux in different regions of the world's oceans.

The comprehensive elucidation of factors controlling diatom fluxes in the ocean requires that further investigations of oceanic processes are complemented with long-term data sets from the atmosphere and terrestrial environment. To resolve the associated variability in fluxes of biogeochemical elements to the oceans' interior, simultaneous long-term monitoring of the atmosphere and the surface ocean by satellite remote sensing, and the deep sea by time-series devices, such as sediment traps, will be necessary. In particular, because diatoms account for more than half of the total BSi and are a substantial part of the primary productivity of the oceans, time series of diatom fluxes in a variety of marine habitats will help to improve models of the biogeochemical cycle of silicon (Ragueneau *et al.*, 2000) as well as the role of diatoms as carrier of organic matter (De La Rocha *et al.*, 2008).

The efficiency of the marine biological pump in sequestering atmospheric carbon dioxide is linked to the production of diatoms. Shifts in the structure of diatom populations in surface waters are in turn strongly influenced by the availability of silicate. Since in some systems, export production seems to be controlled by silicon (the "silicate pump model," Dugdale and Wilkerson, 1998), a better understanding of the interaction of the silicon cycle with the cycles of other nutrients (e.g. nitrogen and phosphorus) in the ocean will be crucial to assess the efficiency of the oceanic biological pump to intaking atmospheric carbon dioxide (Ragueneau *et al.*, 2000). As an alternative to the silicate pump model, the "iron hypothesis" suggests the existence of a link between iron availability in the ocean water and growth of phytoplankton, especially in oligotrophic areas. Several ocean-based iron fertilization experiments have been undertaken to understand the enhanced bloom response phenomenon (e.g. Boyd *et al.*, 2007). Yet the underpinning role of iron to the cycle of mineral fluxes remains, with particular regard to iron source and annual variability, a continuing research effort.

20.8 Summary

Sediment-trap studies have demonstrated that diatom production and export in the oceans is markedly seasonal and episodic, and can be related to the main hydrological processes occurring in surface waters. Although the distribution of the c. 40 stations included in this review (Table 20.1) are not sufficient to draw a comprehensive and representative reconstruction of temporal and spatial variations of diatom contribution to the silicate pump, clear contrasts can be discerned in the diatom flux of different regions of the world's oceans. Our analysis shows substantial differences among oceanic basins and regions in the basic function of the diatom flux and its possible role in the silica and biological pump as diatoms remove and transport silica and carbon molecules from the epipelagic zone to the ocean's interior.

The relationhip between opal concentration in surface sediments and either total primary productivity or diatom productivity in surface waters is not simple. Dissolution is not evenly distributed over the entire diatom community, and removes the signal of the "productive season" of the annual cycle in several areas of the world's oceans (e.g. equatorial Atlantic). Moderately robust diatom frustules are more likely to escape dissolution after cell death and during sinking to the sea floor where, compared to their abundance in the water column, they become often over-represented in the underlying sediments.

The combination of sediment-trap experiments and investigations of the sediment surface and from downcore observations makes it possible to determine the seasonal and interannual variations of assemblages, and to use downcore fluctuations in them as indicators of paleoconditions for different seasons over glacial–interglacial timescales. Caution is advised for paleoproductivity studies which often interpret formation of opal-rich sediments as evidence of high diatom productivity. Combining diatoms with other proxies of export production (e.g. organic carbon, barite, benthic foraminifers, etc.) appears to be the best way to achieve a coherent reconstruction of past surface ocean history.

Acknowledgments

OER acknowledges Marco Klann, Ismene Seeberg, and Eva Kwoll for their lab work. OER's participation was supported by the German Research Foundation's projects RO3039/2–1 and 03G0156A, MARUM, University Bremen (Bremen,

Germany), and the Instituto Andaluz de Ciencias de la Tierra, Universidad de Granada (Granada, Spain). LKA's participation was supported by the CSIRO-University of Tasmania's program in Quantitative Marine Science (QMS) with additional support from the Australian Government's Cooperative Research Centres Programme through the Antarctic Climate and Ecosystems Cooperative Research Centre (ACE CRC). We thank Dr. Mathieu Mongin (CSIRO Marine and Atmospheric Research, Tasmania) for assistance with drafting Figure 20.1. We expressely acknowledge and appreciate the contributions made by international colleagues who provided us with manuscripts and/or data included in this chapter, particularly Drs. Mutsuo Ichinomiya, Amy Leventer, Ivo Grigorov, and the late Dr. Jean-Jacques Pichon. Dr. Gerhard Fischer and our editors are sincerely acknowledged for their contributions in improving and editing this work.

References

Abelmann, A. & Gersonde, R. (1991). Biosiliceous particle flux in the Southern Ocean. *Marine Chemistry*, **35**, 503–36.

Abrantes, F., Meggers, H., Nave, S., *et al.* (2002). Fluxes of micro-organisms along a productivity gradient in the Canary Islands region (29°N): implications for paleoreconstrucions. *Deep-Sea Research II*, **49**, 3599–629.

Armand, L. K., Crosta, X., Romero, O. E., & Pichon, J.-J. (2005). The biogeography of major diatom taxa in Southern Ocean sediments. 1. Ice-related species. *Palaeogeography, Palaeoclimatology, Palaeoecology*, **223**, 93–126.

Armand, L. K. & Leventer, A. (2009). Palaeo sea ice distribution and reconstruction derived from the geological record. In *Sea Ice*, 2nd edition, ed. D. N. Thomas & G. S. Dieckmann. Oxford: Wiley-Blackwell. pp. 469–529.

Asper, V. L. (1996). Particle flux in the ccean: oceanographic tools. In *Particle Flux in the Ocean*, ed. V. Ittekot, P. Schäfer, S. Honjo, & P. J. Depetris. Oxford: John Wiley and Sons. pp. 71–84.

Bárcena, M. A., Flores, J. A., Sierro, F. J., *et al.* (2004). Planktonic response to main oceanographic changes in the Alboran Sea (western Mediterranean) as documented in sediment traps and surface sediments. *Marine Micropaleontology*, **53**, 423–45.

Boyd, P. W., Jickells, T., Law, C. S., *et al.* (2007). Mesoscale iron enrichment experiments 1993–2005: synthesis and future directions. *Science*, **315**, 612–17.

Buesseler, K. O., Antia, A. N., Chen, M., *et al.* (2007). An assessment of the use of sediment traps for estimating upper ocean particle fluxes. *Journal of Marine Systems*, **65**, 345–416.

Caubert, T. (1998). Couplage et découplage des cycles de C et du Si dans le secteur indien de l'Ocean Austral. Unpublished Ph.D thesis, Université de Bretagne Occidentale.

Collier, R., Dymond, J., Honjo, S., *et al.* (2000). The vertical flux of biogenic and lithogenic material in the Ross Sea: moored sediment trap observations 1996–1998. *Deep-Sea Research II*, **47**, 3491–520.

Conkright, M. E., Garcia, H. E., O'Brien, T. D., *et al.* (2002). *World Ocean Atlas 2001, Volume 4: Nutrients*, ed. S. Levitus, NOAA Atlas NESDIS 52, Washington, DC: U.S. Government Printing Office (CD-ROM).

Crosta, X., Pichon, J.-J., & Burckle, L. H. (1998). Application of modern analog technique to marine Antarctic diatoms: reconstruction of maximum sea-ice extent at the last glacial maximum. *Paleoceanography*, **13**, 286–297.

Crosta, X., Romero, O. E., Armand, L. K., & Pichon, J.-J. (2005). The biogeography of major diatom taxa in Southern Ocean sediments. 2. Open-ocean related species. *Palaeogeography, Palaeoclimatology, Palaeoecology*, **223**, 66–92.

De La Rocha, C. L., Nowald, N., & Passow, U. (2008). Interactions between diatom aggregates, minerals, particulate organic carbon, and dissolved organic matter: further implications for the ballast hypothesis. *Global Biogeochemical Cycles*, **22**, GB4005. DOI:10.1029/2007GB003156.

Deuser, W. G., Jickells, T. D., King P., & Commeau, J. A. (1995). Decadal and annual changes in biogenic opal and carbonate fluxes to the deep Sargasso Sea. *Deep-Sea Research I*, **11–21**, 2391–3291.

Ducklow, H. W., Erikson, M., Kelly, J., *et al.* (2008). Particle export from the upper ocean over the continental shelf of the west Antarctic Peninsula: a long-term record 1992–2007. *Deep-Sea Research II*, **55**, 2118–31.

Dugdale, R. C., Wilkerson, F. P., & Minas, H. J. (1995). The role of silicate pump in driving new production. *Deep-Sea Research I*, **42**, 697–719.

Dugdale, R. C. & Wilkerson, F. P. (1998). Silicate regulation of new production in the equatorial Pacific upwelling. *Nature*, **391**, 270–3.

Dunbar, R. B., Leventer, A. R. & Mucciarone, D. A. (1998). Water column sediment fluxes in the Ross Sea, Antarctica: atmospheric and sea ice forcing. *Journal of Geophysical Research* **103**, 30,741–59.

Dunbar, R. B., Leventer, A. R., & Stockton, W. L. (1989). Biogenic sedimentation in McMurdo Sound, Antarctica. *Marine Geology*, **85**, 155–79.

Fischer, G., Fütterer, D., Gersonde, R., *et al.* (1988). Seasonal variability of particle flux in the Weddell Sea and its relation to ice cover. *Nature*, **335**, 426–8.

Fischer, G., Gersonde, R., & Wefer, G. (2002). Organic carbon, biogenic silica and diatom fluxes in the marginal winter sea-ice zone and in the Polar Front region: interannual variations and differences in composition. *Deep-Sea Research II*, **49**, 1721–45.

Fischer, G., Ratmeyer, V., & Wefer, G. (2000). Organic carbon fluxes in the Atlantic and the Southern Ocean: relationship to primary production compiled from satellite radiometer data. *Deep-Sea Research II*, **47**, 1939–59.

Fischer, G. & Wefer, G. (1991). Sampling, preparation and analysis of marine particulate matter. *Geophysical Monographs*, **63**, 391–7.

Fischer, G., Wefer, G., Romero, O.E., *et al.* (2004). Transfer of particles into the deep Atlantic and the global ocean: control of nutrient supply and ballast production. In *The South Atlantic in the Late Quaternary: Reconstruction of Material Budgets and Current Systems*, ed. G. Wefer, S. Mulitza, & V. Ratmeyer, Berlin: Springer, pp. 21–46.

François, R., Honjo, S., Krishfield, R., & Manganini, S. (2002). Factors controlling the flux of organic carbon to the bathypelagic zone of the ocean. *Global Biogeochemical Cycles*, **16**, 1087, DOI: 10.1029/2001GB001722, 2002.

Gersonde, R. (1986). Biogenic siliceous particle flux in Antarctic waters and its palaeoecological significance. *South African Journal of Science*, **82**, 500–1.

Gersonde, R. & Zielinski, U. (2000). The reconstruction of late Quaternary Antarctic sea-ice distribution – the use of diatoms as a proxy for sea-ice. *Palaeogeography,Palaeoclimatology, Palaeoecology*, **162**, 263–86.

Heinmiller, J. R. H. (1976). Mooring operations techniques of the buoy project at the Woods Hole Oceanographic Institution. WHOI Technical Reports, Woods Hole Oceanographic Institution, Woods Hole, 76–69.

Honjo, S. & Doherty, K. W. (1988). Large aperture time-series sediment traps; design objectives, construction and application. *Deep-Sea Research*, **35**, 133–49.

Honjo, S., Manganini, S. J., Krishfield, R. A., & François, R. (2008). Particulate organic carbon fluxes to the ocean interior and factors controlling the biological pump: a synthesis of global sediment trap programs since 1983. *Progress in Oceanography*, **76**, 217–85.

Honjo, S., Spencer, D. W., & Gardner, W. D. (1992). A sediment trap intercomparison experiment in the Panama Basin, 1979. *Deep-Sea Research I*, **39**, 333–58.

Ichinomiya, M., Gomi, Y., Nakamachi, M., *et al.* (2008). Temporal variations in the abundance and sinking flux of diatoms under fast ice in summer near Syowa Station, East Antarctica. *Polar Science*, **2**, 33–40.

Ishikawa, A., Wasiyama, N., Tanimura, A., & Fukuchi, M. (2001). Variation in the diatom community under fast ice near Syowa Station, Antarctica, during the austral summer of 1997/98. *Polar Biosciences*, **14**, 10–23.

Kemp, A. E. S., Pike, J., Pearce, R. B., & Lange, C. B. (2000). The "Fall dump" – a new perspective on the role of a "shade flora" in the annual cycle of diatom production and export flux. *Deep-Sea Research II*, **47**, 2129–54.

Klaas, C. & Archer, D. E. (2002). Association of sinking organic matter with various types of mineral ballast in the deep sea: implication for the rain ratio. *Global Biogeochemical Cycles*, **16**, 1116, DOI: 10.1029/2001GB001765.

Koning, E., van Iperen, J. M., van Raaphorst, W., *et al.* (2001). Selective preservation of upwelling-indicating diatoms in sediments off Somalia, NW Indian Ocean. *Deep-Sea Research I*, **48**, 2473–95.

Lange, C. B., Romero, O. E., Wefer, G., & Gabric, A. J. (1998). Offshore influence of coastal upwelling off Mauritania, NW Africa, as recorded by diatoms in sediment traps at 2195 m water depth. *Deep-Sea Research I*, **45**, 985–1013.

Lange, C. B., Weinheimer, A. L., Reid, F. M. H., Tappa, E., & Thunell, R. C. (2000). Response of siliceous microplankton from the Santa Barbara Basin to the 1997–98 El Niño Event. *California Cooperative Oceanic Fisheries Investigations, Report*, **41**, 186–93.

Lange, C. B., Weinheimer, A. L., Reid, F. H. M., & Thunell, R. C. (1997). Sedimentation patterns of diatoms, radiolarians, and silicoflagellates in Santa Barbara Basin, California. *California Cooperation on Oceanic and Fisheries Investigation, Report*, **38**, 161–70.

Leventer, A. (1991). Sediment trap diatom assemblages from the northern Antarctic Peninsula region. *Deep-Sea Research*, **38**, 1127–43.

Leventer, A. (1992). Modern distribution of diatoms in sediments from the George V Coast, Antarctica. *Marine Micropaleontology*, **19**, 315–32.

Leventer, A. & Dunbar, R. B. (1987). Diatom flux in McMurdo Sound, Antarctica. *Marine Micropaleontology*, **12**, 49–64.

Leventer, A. & Dunbar, R. B. (1996). Factors influencing the distribution of diatoms and other algae in the Ross Sea. *Journal of Geophysical Research*, **101**, 18489–500.

Lončarić, N., van Iperen, J., Kroon, D., & Brummer, G.-J. A. (2007). Seasonal export and sediment preservation of diatomaceous, foraminiferal and organic matter mass fluxes in a trophic gradient across the SE Atlantic. *Progress in Oceanography*, **73**, 27–59.

Longhurst, A. R., Sathyendranath, S., Platt, T., & Caverhill, C. (1995). An estimate of global primary production in the ocean from satellite radiometer data. *Journal of Plankton Research*, **17**, 1245–71.

Matsuda, O., Ishikawa, S., & Kawaguchi, K. (1987). Seasonal variation of downward flux of particulate organic matter under the Antarctic fast ice. *Proceedings of the National Institute of Polar Research, Polar Biology*, **1**, 23–34.

McPhaden, M. J. (1999). Genesis and evolution of the 1997–98 El Niño. *Science*, **283**, 950–4.

Moline, M.A. & Prézelin, B. B. (1996). Long-term monitoring and analyses of physical factors regulating variability in coastal Antarctic phytoplankton biomass, *in situ* productivity and taxonomic composition over subseasonal, seasonal and interannual time scales. *Marine Ecology Progress Series*, **145**, 143–60.

Moore, T. C., Jr. (1973). Method of randomly distributing grains for microscopic examination. *Journal of Sedimentary Petrology*, **43**, 904–6.

Nelson, D. M., Tréguer, P., Brzezinski, M. A., Leynaert, A., & Quéguiner, B. (1995). Production and dissolution of biogenic silica in the ocean: revised global estimates, comparison with regional data and relationship to biogenic sedimentation. *Global Biogeochemical Cycles*, **9**, 359–72.

Pilskaln, C. H., Manganini, S. J., Trull, T. W., *et al.* (2004). Geochemical particle fluxes in the Southern Indian Ocean seasonal ice zone: Prydz Bay region, east Antarctica. *Deep-Sea Research I*, **50**, 307–32.

Ragueneau, O., Tréguer, P., Leynart, A., *et al.* (2000). A review of the Si cycle in the modern ocean: recent progress and missing gaps in the application of biogenic opal as a paleoproductivity proxy. *Global and Planetary Change*, **26**, 317–65.

Romero, O. E., Armand, L. K., Crosta, X., & Pichon, J.-J. (2005). The biogeography of major diatom taxa in Southern Ocean sediments. 3. Tropical/Subtropical species. *Palaeogeography, Palaeoclimatology, Palaeoecology*, **223**, 49–65.

Romero, O. E., Boeckel, B., Donner, B., *et al.* (2002b). Seasonal productivity dynamics in the pelagic central Benguela System inferred from the flux of carbonate and silicate organisms. *Journal of Marine Systems*, **37**, 259–78.

Romero, O. E., Dupont, L., Wyputta, U., Jahns, S., & Wefer, G. (2003). Temporal variability of fluxes of eolian-transported freshwater diatoms, phytoliths, and pollen grains off Cape Blanc as reflection of land-atmosphere-ocean interactions in northwest Africa. *Journal of Geophysical Research*, **108**, DOI:10.1029/2000JC000375.

Romero, O. E., Fischer, G., Lange, C. B., & Wefer, G. (2000). Siliceous phytoplankton of the western equatorial Atlantic: sediment traps and surface sediments. *Deep-Sea Research II*, **47**, 1939–59.

Romero, O. E. & Hebbeln, D. (2003). Biogenic silica and diatom thanatocoenosis in surface sediments below the Peru-Chile Current: controlling mechanisms and relationship with productivity of surface waters. *Marine Micropaleontology*, **48**, 71–90.

Romero, O. E., Hebbeln, D., & Wefer, G. (2001). Temporal and spatial distribution in export production in the SE Pacific Ocean: evidence from siliceous plankton fluxes and surface sediment assemblages. *Deep-Sea Research I*, **48**, 2673–97.

Romero, O. E., Lange, C. B., Fischer, G., Treppke, U. F., & Wefer, G. (1999). Variability in export production documented by downward fluxes and species composition of marine planktonic diatoms: observations from the tropical and equatorial Atlantic. In *Use of Proxies in Paleoceanography, Examples from the South Atlantic*, ed. G. Fischer & G. Wefer Berlin: Springer-Verlag, pp. 365–92.

Romero, O. E., Lange, C. B., & Wefer, G. (2002a). Interannual variability (1988–1991) of siliceous phytoplankton fluxes off northwest Africa. *Journal of Plankton Research*, **24**, 1035–46.

Romero, O. E., Rixen, T., & Herunadi, B. (2009b). Effect of hydrographic and climatic forcing on the diatom production and export in the tropical southeastern Indian Ocean. *Marine Ecology Progress Series*, **384**, 69–82.

Romero, O. E., Thunell, R. C., Astor, Y., & Varela, R. (2009a). Seasonal and interannual dynamics in diatom production in the Cariaco Basin, Venezuela. *Deep-Sea Research I*, **56**, 571–81.

Sancetta, C. (1989a). Spatial and temporal trends of diatom flux in British Columbia fjords. *Journal of Plankton Research*, **11**, 503–20.

Sancetta, C. (1989b). Processes controlling the accumulation of diatoms in sediments: a model derived from British Columbian fjords. *Paleoceanography*, **4**, 235–51.

Sancetta, C. (1992). Comparison of phytoplankton in sediment trap time series and surface sediments along a productivity gradient. *Paleoceanography*, **7**, 183–94.

Sancetta, C. (1995). Diatoms in the Gulf of California: seasonal flux patterns and the sediment record for the last 15,000 years. *Paleoceanography*, **10**, 67–84.

Sancetta, C. & Calvert, S.E. (1988). The annual cycle of sedimentation in Saanich Inlet, British Columbia: implications for the interpretation of diatom fossil assemblages. *Deep-Sea Research*, **I 35** (1), 71–90.

Sasaki, H., Suzuki, H., Takayama, M., *et al.* (1997). Sporadic increase of particle sedimentation at the ice edge of the Antarctic Ocean during the austral Summer 1994–1995. *Proceedings of the National Institute of Polar Research Symposium on Polar Biology*, **10**, 50–5.

Sautter, L. R. & Sancetta, C. (1992). Seasonal associations of phytoplankton and planktic foraminifera in an upwelling region and their contribution to the seafloor. *Marine Micropaleontology*, **18**, 263–78.

Scharek, R., Latasa, M., Karl, D. M., & Bigidare, R. R. (1999b). Temporal variations in diatom abundance and downward vertical flux in the oligotrophic North Pacific Gyre. *Deep-Sea Research I*, **46**, 1051–75.

Scharek, R., Tupas, L. M., & Karl, D. M. (1999a). Diatom fluxes to the deep sea in the oligotrophic North Pacific gyre at Station ALOHA. *Marine Ecology Progress Series*, **188**, 55–67.

Schrader, H. & Gersonde, R. (1978). Diatoms and silicoflagellates. *Utrecht Micropaleontological Bulletin*, **17**, 129–76.

Suzuki, H., Sasaki, H., & Fukuchi, M. (2001). Short-term variability in the flux of rapidly sinking particles in the Antarctic Marginal Ice Zone. *Polar Biology*, **24**, 697–705.

Takahashi, K. (1986). Seasonal fluxes of pelagic diatoms in the subarctic Pacific, 1982–1983. *Deep-Sea Research*, **33**, 1225–51.

Takahashi, K. (1987). Response of subarctic Pacific diatom fluxes to the 1982–1983 El Niño disturbance. *Journal of Geophysical Research*, **93**, 14,387–92.

Takahashi, K. (1994). From modern flux to paleoflux: assessment from sinking assemblages to thanatocoenosis. In *Carbon Cycling in the Glacial Ocean: Constraints on the Ocean's Role in Global Change*, ed. R. Zahn, T. F. Pedersen, M. A. Kaminski, & L. Labeyrie, NATO ASI Series, Berlin: Springer-Verlag, vol. 117, pp. 413–24.

Takahashi, K., Billings, J. B., & Morgan, J. K. (1990). Oceanic province: assessment from the time-series diatom fluxes in the northeastern Pacific. *Limnology and Oceanography*, **35**, 154–65.

Tanimura, Y. (1992). Distribution of diatom species in the surface sediments of Lützow-Holm Bay, Antarctica. In *Centenary of Japanese Micropaleontology*, ed. Ishizaki, K. & T. Saito, Tokyo: Terra Scientific Publishing Company, pp. 399–411.

Tréguer, P., Nelson, D. M., van Bennekom, *et al.* (1995). The silica balance in the world ocean: a reestimate. *Science*, **268**, 375–9.

Treppke, U. F., Lange, C. B., Donner, B., *et al.* (1996a). Diatom and silicoflagellate fluxes at the Walvis Ridge: an environment influenced by coastal upwelling in the Benguela System. *Journal of Marine Research*, **54**, 991–1016.

Treppke, U. F., Lange, C. B., & Wefer, G. (1996b). Vertical fluxes of diatoms and silicoflagellates in the eastern equatorial Atlantic, and their contribution to the sedimentary record. *Marine Micropaleontology*, **28**, 73–96.

Yoder, J. A., Ackleson, S. G., Barber, R. T., Flament, P., & Balch, W. M. (1994). A line in the sea. *Nature*, **371**, 689–92.

Wefer, G. & Fischer, G. (1991). Annual primary production and export flux in the Southern Ocean from sediment trap data. *Marine Chemistry*, **36**, 597–613.

Wefer, G., Fischer, G., Fütterer, D., & Gersonde, R. (1988). Seasonal particle flux in the Bransfield Strait, Antarctica. *Deep-Sea Research*, **35**, 891–8.

Wefer, G., Fischer, G., Fütterer, D. K., *et al.* (1990). Particle sedimentation and productivity in Antarctic waters of the Atlantic sector. In *Geological History of the Polar Oceans: Arctic versus Antarctic*, ed. U. Bleil & J. Thiede, London: Kluwer Academic Publishers, pp. 363–79.

Zielinski, U. & Gersonde, R. (1997). Diatom distribution in Southern Ocean surface sediments (Atlantic sector): implications for paleoenvironmental reconstructions, *Palaeogeography, Palaeoclimatology, Palaeoecology*, **129**, 213–50.

21

Holocene marine diatom records of environmental change

AMY LEVENTER, XAVIER CROSTA, AND JENNIFER PIKE

21.1 Introduction

Atmospheric temperature records from central Greenland and Antarctic ice cores reveal a dramatic shift between the Late Pleistocene and the Holocene in terms of estimates and amplitudes of temperature change. Following recovery of these long, high-resolution ice cores in the early 1990s, initially it was believed that this shift into the Holocene involved a change from low mean temperatures with large, rapid oscillations on decadal to millennial timescales, to high mean temperatures with relatively little variability. More recently, other records from different regions of the world, together with our increased understanding of external climate forcings and feedbacks, have shown that this ice-core-derived picture of Holocene climate stability is not the case (Maslin *et al.*, 2001; Wanner *et al.*, 2008). Holocene climate variability appears to exhibit relatively regular patterns of change. However, these patterns of change are complex; not all changes are observed globally or synchronously (Mayewski *et al.*, 2004). And, although the oscillations in climate during the Holocene are of lower amplitudes than those of the Late Pleistocene, they are of sufficient magnitude to cause significant perturbations to our contemporary climate and to have had an impact on human civilizations.

The primary goal of this chapter is to present a detailed view of the contribution of diatom analysis from marine sedimentary records to our understanding of climatic and environmental change during the Holocene. This chapter will provide a link between other chapters in this book that deal with diatoms as indicators of recent changes in oceanographic condition (Romero and Armand, this volume) and diatoms as indicators of paleoceanographic events (Jordan and Stickley, this volume). Greater knowledge concerning ocean and climate variability during the Holocene provides insight vital to an understanding of the mechanisms and forcings involved in sub-orbital-scale climate change. In addition, given recently increased access to high-resolution marine sedimentary records, many of which are highly diatomaceous, we are able to address the role of other, predominantly higher-frequency, climate fluctuations in forcing local, regional, and global environmental change. Our ability to evaluate relatively recent climate events helps place modern-day warming into a longer-term context and provides a basis from which to evaluate the changing Anthropocene climate.

Today, most paleoenvironmental reconstruction is accomplished via multiproxy analysis, with each proxy providing a specific detail of a complicated system. In this chapter we focus on those marine records in which diatom data have played a key role in deciphering environmental history. Marine diatom records are ideal for investigating Holocene ocean and climate variability, particularly in higher latitudes where carbonate-based proxies are limited (e.g. the Antarctic margin). Diatoms can also make significant contributions to our understanding of Holocene variability at lower latitudes in sites of high diatom primary production such as regions of coastal upwelling (e.g. California margin) and semi-enclosed seas/fjords (e.g. Gulf of California; Saanich Inlet, British Columbia). Sufficient high-resolution diatom records exist to address all timescales of variability across all latitudes during the Holocene.

The use of diatom assemblage data to reconstruct variability in Holocene climate and oceanography is more straightforward than when working with older assemblages, since the data are applied over timescales when the species are extant, so that modern-day correlation tools can be applied between these modern species and their environmental preferences (see Romero and Armand, this volume). Perhaps the greatest number of diatom records are from the southern polar to subpolar setting, where well-preserved diatoms occur in high abundance throughout Holocene sediments, a consequence of

The Diatoms: Applications for the Environmental and Earth Sciences, 2nd Edition, eds. John P. Smol and Eugene F. Stoermer. Published by Cambridge University Press.

high diatom productivity and excellent preservation. Our ability to interpret those records relies on the fact that the distribution of diatom species in surface sediments of the Southern Ocean is relatively well known, based on a long history of studies. These biogeographic data bases provide the relevant background information for paleo-reconstructions. Equally strong modern data bases also have provided an avenue for Holocene reconstruction in areas of upwelling where diatoms are abundant. Finally, although in general diatoms tend to be less-well silicified and more easily dissolved in nutrient-poor Arctic waters (Machado, 1993; Kohly, 1998), similar studies of diatom species distribution in surface sediments from the northern high latitudes are abundant and serve as the basis for interpretation of the Arctic and subarctic Holocene paleo-record.

Two general approaches dominate the use of marine diatom records in Holocene paleoceanographic studies. These approaches are determined by the resolution of the sedimentary record. Historically, relatively low-resolution studies were done, through a combination of assemblage-based reconstructions and studies of single species, including both their presence versus absence and changes in their morphology. While these studies continue to provide important paleoclimatic data, investigations increasingly focus on ultra-high-resolution sedimentary records, in which annual and sub-annual data are available, often at the sub-millimeter scale of observation. Our understanding of oceanographic and geologic conditions that favor deposition and preservation of laminated and varved sediments, coupled to improved long-coring capabilities, has made acquisition of long, high-resolution sedimentary records more common.

This chapter presents a review of the current state of knowledge concerning the application of diatom-based methods that aim to reconstruct changes in Holocene climate and oceanography. The marine records discussed are divided into lower- and higher-resolution studies. Lower-resolution records are presented with a large scope, to provide an overview of the kinds of approaches that have been used to reconstruct broad-scale features of the Holocene. In contrast, higher-resolution records are presented in a more focused way, to illustrate the tremendous potential of these newly emerging studies. In particular, we focus on the application of high-resolution diatom work to evaluation and understanding of the role of solar forcing over decadal and century timescales during the Holocene.

21.2 Low-resolution studies

In the marine realm, most early studies of Holocene diatoms were based on relatively low-resolution records that were used to reconstruct broad-scale features, such as globally recognized, sub-orbital-scale changes in climate including the mid-Holocene climatic optimum, the Neoglacial, the Medieval Warm Period, and the Little Ice Age, as well as changes specific to a particular geographic region. While diatoms have been widely used as proxies for a tremendous variety of environmental factors, we focus on those most commonly addressed in evaluation of Holocene paleoclimate; these include changes in marine productivity and variability in the extent of sea ice and sea-surface temperatures. These kinds of data allow for interpretation of overall changes in the distribution of water masses and hydrography (Table 21.1). Crosta and Koç (2007) provide a comprehensive review of basic techniques and methodologies associated with diatom analyses; consequently, these topics are referred to in this chapter, but are not covered in detail here. Most significantly, selected studies are presented that demonstrate the merits of these approaches and their specific application to reconstruction of Holocene climate. Note that all ages presented in this chapter retain their initially published format.

21.2.1 Primary productivity

As the most abundant primary producer in the world's oceans, marine diatoms have been used extensively to reconstruct Holocene changes in surface-water productivity. On a global ocean scale, one of the most commonly used indicators of paleoproductivity is the distribution of *Chaetoceros* resting spores (CRS) (see discussion in Leventer *et al.*, 1996; Crosta *et al.*, 1997), an observation supported by both experimental and field data. In fact, in Holocene marine sediments associated with overlying high primary production, the relative abundance of CRS can overwhelm the abundance of other species to such a great extent that often, in order to assess the contribution of minor species to a total assemblage, a "*Chaetoceros*-free" count is completed (e.g. Williams, 1986; Koç *et al.* 1990; Leventer *et al.*, 1996).

One of the first applications of CRS as a proxy for primary productivity in the Antarctic was presented by Leventer *et al.* (1996), who noted CRS dominance in laminated sediments from the Palmer Deep, Antarctic Peninsula. These authors presented a generalized scenario to explain the high concentration of CRS in Palmer Deep sediments as a function of the occurrence of a shallow mixed layer in which massive, nutrient-depleting blooms were able to develop (Leventer *et al.*, 1996, their figure 12). Sjunneskog and Taylor (2002) expanded upon this earlier work through application of total diatom abundance counts, a value in this case controlled by the abundance of CRS, as a paleoproductivity proxy for the Holocene. Their data illustrated

Table 21.1 List of recent diatom studies relevant to Holocene reconstruction

Reference/year	Region	Method	Environmental parameter(s)
Arctic/high northern latitudes			
Sancetta, 1979	North Pacific	Factor and regression analysis	Hydrography, SST, productivity
Koç Karpuz and Schrader, 1990	Greenland, Iceland, Norwegian seas (GIN)	Transfer function	SST
Williams, 1990	Baffin Bay	Diatom species counts	Productivity
Koç Karpuz and Jansen, 1992	Norwegian Sea	Factor analysis	Hydrography
Koç *et al.*, 1993	GIN	Factor analysis	SST, Hydrography, sea ice
Schrader *et al.*, 1993	Norwegian Sea	Diatom abundance	Productivity
Williams, 1993	Southeast Greenland	Diatom species counts	Hydrography, sea ice
Koç and Jansen, 1994	GIN	Factor analysis	SST
Koç *et al.*, 1996	North Atlantic	Factor analysis	SST
Bauch and Polyakova, 2000	Laptev Sea	Diatom species counts	Sea ice, river input
Birks and Koç, 2002	Norwegian Sea	Transfer function	SST
Jiang *et al.*, 2002	North Iceland Shelf	Transfer function	SST, hydrography
Andersen *et al.*, 2004a	North Atlantic	Transfer function	SST, hydrography
Andersen *et al.*, 2004b	North Atlantic	Transfer function	SST, hydrography
Jensen *et al.*, 2004	South Greenland	Diatom species counts	Hydrography, sea ice
Knudsen *et al.*, 2004	North Iceland	Diatom species counts	Hydrography, sea ice
Polyakova and Stein, 2004	Kara Sea	Diatom species counts	Paleosalinity, sea ice
Jiang *et al.*, 2005	North Iceland Shelf	Transfer function	SST
Witak *et al.*, 2005	North Atlantic	Diatom species counts	Hydrography, sea ice
Caissie *et al.*, 2006	Bering Sea	Diatom species counts	Sea ice, productivity
Eiríksson *et al.*, 2006	North Atlantic	Transfer function	SST (from Jiang *et al.*, 2005)
Moller *et al.*, 2006	Southwest Greenland	Diatom species counts	Hydrography, sea ice
Moros *et al.*, 2006	West Greenland	Diatom species counts	Hydrography, sea ice, productivity
Ran *et al.*, 2006	North Icelandic Shelf	Factor analysis	Hydrography
Witon *et al.*, 2006	Faroe Islands	Factor analysis	Hydrography
Seidenkrantz *et al.*, 2007	Southwest Greenland	Diatom species counts	Hydrography
Justwan *et al.*, 2008	North Icelandic Shelf	Transfer function	SST
Justwan and Koç, 2008	North Atlantic	Transfer function	Sea ice
Ren *et al.*, 2009	Southwest Greenland	Diatom species counts	SST, hydrography
Antarctic			
Leventer and Dunbar, 1988	McMurdo Sound, Ross Sea	Diatom species counts	Hydrography, sea ice
Leventer *et al.*, 1993	Granite Harbor, Ross Sea	Diatom species counts	Hydrography
Leventer *et al.*, 1996	Palmer Deep, Antarctic Peninsula	Diatom species counts	Hydrography, sea ice, productivity
Shevenell *et al.*, 1996	Lallemand Fjord, Antarctic Peninsula	Centric–pennate diatom ratio	Hydrography
Rathburn *et al.*, 1997	Mac.Robertson Shelf, Prydz Bay	Diatom assemblages	
Bárcena *et al.*, 1998	Bransfield Strait, Antarctic Peninsula	Diatom species counts	Hydrography, productivity, sea ice
Cunningham *et al.*, 1999	Ross Sea	Factor analysis	Hydrography, sea ice
McMinn, 2000	Vestfold Hills, Prydz Bay	Diatom species counts	Sea ice
Cremer *et al.*, 2001	Windmill Islands, East Antarctica	Diatom species counts	Hydrography, sea ice
Hodell *et al.*, 2001	South Atlantic	Transfer function	SST, sea ice
Taylor and McMinn, 2001	Mac.Robertson Shelf, Prydz Bay	Factor analysis	Sea ice, productivity
Taylor *et al.*, 2001	Lallemand Fjord, Antarctic Peninsula	Factor analysis, diatom species counts, *Thalassiosira antarctica* types	Hydrography, sea ice, ice-shelf history

(*cont.*)

Table 21.1 *(cont.)*

Reference/year	Region	Method	Environmental parameter(s)
Bárcena *et al.*, 2002	Bransfield Strait, Antarctic Peninsula	Diatom species counts and abundance	Productivity
Leventer *et al.*, 2002	Palmer Deep, Antarctic Peninsula	Diatom species counts and *Eucampia* morphology	Hydrography, productivity, sea ice
Sjunneskog and Taylor, 2002	Palmer Deep, Antarctic Peninsula	Total diatom abundance	Hydrography, productivity
Taylor and Sjunneskog, 2002	Palmer Deep, Antarctic Peninsula	Factor analysis, diatom species counts, *T. antarctica* types	Hydrography, productivity
Taylor and McMinn, 2002	Prydz Bay	Factor analysis	Hydrography
Yoon *et al.*, 2002	Western Antarctic Peninsula	Diatom species counts	Hydrography, sea ice
Cremer *et al.*, 2003	Windmill Islands, East Ant.	Diatom species counts	Hydrography, sea ice
Bianchi and Gersonde, 2004	South Atlantic	Transfer function, diatom species counts	SST, sea ice
Nielsen *et al.*, 2004	South Atlantic	Transfer function	SST, sea ice
Domack *et al.*, 2005	Larsen Embayment	Diatom abundance	Ice-shelf history
Finocchiaro *et al.*, 2005	Ross Sea	Diatom lamination types	Hydrography
Maddison *et al.*, 2005	Palmer Deep, Antarctic Peninsula	Diatom lamination types	Hydrography
McMullen *et al.*, 2006	George V Coast, East Antarctica	Diatom species counts	Sea ice, productivity
Stickley *et al.*, 2005	Mac.Robertson Shelf, Prydz Bay	Diatom lamination types	Hydrography
Bárcena *et al.*, 2006	Bransfield Strait, Antarctic Peninsula	Diatom species counts	Hydrography, sea ice
Denis *et al.*, 2006	Adélie Land, East Ant.	Diatom lamination types	Sea ice, productivity, seasonality
Maddison *et al.*, 2006	George V Coast, East Antarctica	Diatom lamination types	Hydrography, sea ice
Bak *et al.*, 2007	Scotia Sea	Diatom species counts	Hydrography, sea ice
Heroy *et al.*, 2007	Bransfield Basin, Antarctic Peninsula	Diatom species counts, *Eucampia* morphology	Hydrography
Crosta *et al.*, 2007	Adélie Land, East Antarctica	Diatom species counts	Sea ice, productivity
Crosta *et al.*, 2008	Adélie Land, East Antarctica	Diatom species counts	SST, sea ice
West Coast North and South America			
De Vries and Schrader, 1981	Coastal Peru	Diatom species counts	Hydrography, upwelling, productivity
Sancetta, 1995	Gulf of California		
Pike and Kemp, 1996	Gulf of California	Diatom lamination types	Oceanic circulation patterns
Pike and Kemp, 1997	Gulf of California	Diatom lamination types	Oceanic circulation patterns
Chang *et al.*, 2003	British Columbia	Diatom lamination types	Productivity, ENSO
Barron *et al.*, 2004	Gulf of California	Diatom species counts	Upwelling, productivity
Barron *et al.*, 2005	Gulf of California	Diatom species counts	Hydrography, upwelling, productivity
Romero *et al.*, 2006	Coastal Chile	Diatom species and abundance	Upwelling, productivity
Hay *et al.*, 2007	British Columbia	Diatom species counts?	Hydrography, upwelling, productivity
Barron and Bukry, 2007	Gulf of California	Diatom species counts	Hydrography, productivity
Rebolledo *et al.*, 2008	Coastal Chile	Diatom abundance	River input
Eastern Pacific			
Shiga and Koizumi, 2000	Okhotsk Sea	Diatom species counts	Sea ice
Koizumi *et al.*, 2003	Okhotsk Sea	Diatom species counts	Sea ice
Shimada *et al.*, 2000	Okhotsk Sea	Diatom species counts	Sea ice

Table 21.1 *(cont.)*

Reference/year	Region	Method	Environmental parameter(s)
Shimada *et al.*, 2004	Okhotsk Sea	Diatom species counts	Hydrography, sea ice, productivity
Ryu *et al.*, 2005	Sea of Japan	Diatom species counts	Hydrography, upwelling, productivity, paleosalinity
Koizumi, 2008	Sea of Japan	Transfer function	SST
West Coast Europe and Africa			
Gil *et al.*, 2006	Portuguese margin, Skagerrak	Diatom species counts and abundance	Upwelling, productivity, paleosalinity
Lebreiro *et al.*, 2006	Iberian margin	Diatom abundance	Productivity
Bernárdez *et al.*, 2007	Northwest Iberian Peninsula	Diatom species counts	Upwelling, productivity
Bernárdez *et al.*, 2008	Northwest Iberian Peninsula	Diatom species counts	River input

millennial-scale variability, with the highest primary productivity characteristic of deglaciation (~12.8 ka) and between 9.0 and 4.4 ka, during the mid-Holocene climatic optimum. They noted that productivity over the past 500 years is the lowest recorded in the entire post-glacial period. More detailed information regarding paleoproductivity has been acquired through spectral analysis of the relative distribution of CRS versus vegetative cells in Palmer Deep sediments; these data reveal centennial-scale variability that was interpreted to represent changes in upper-ocean conditions (Leventer *et al.*, 1996), a topic discussed in greater detail later in this chapter.

Similarly, in high northern latitudes, high abundances of CRS are interpreted as indicative of elevated productivity (Birks and Koç, 2002; Caissie *et al.*, 2006) in well-stratified summer surface waters (Jensen *et al.*, 2004; Moros *et al.*, 2006). Sediments dominated by CRS also characterize laminated sediments recovered from the Santa Barbara Basin (Lange *et al.*, 1990; Grimm *et al.*, 1996, 1997) and the Gulf of California (Sancetta, 1995; Pike and Kemp, 1996, 1999; Kemp *et al.*, 2000), though extensive application of CRS as a paleoproductivity proxy in these settings has not been done. Romero *et al.* (2006), in a multiproxy study from the central Chilean margin, track the increasing contribution of an upwelling assemblage dominated by CRS, from the last glacial maximum to the early part of the Holocene.

Chaetoceros is by no means the only diatom used to trace changes in productivity during the Holocene. For example, Barron *et al.* (2004) presented a 15,000 year record from the Gulf of California, in which they utilized both diatoms and silicoflagellates to trace changes in upwelling and its seasonality during the Holocene. During the Younger Dryas, decreased abundance of *Azpetia nodulifera* (A. Schmidt) G. Fryxell & P. A. Sims (along with the silicoflagellate *Dictyocha perlaevis* Frenguelli) recorded decreased upwelling, while increased abundance of the diatom *Roperia tesselata* (Roper) Grunow ex Pelletan, a proxy for increased winter upwelling, was observed between 10.6 and 10.0 ka. More recently, higher-resolution work in the Gulf of California has used these same proxies to detect cyclical changes in coastal upwelling and associated primary productivity over the past 2000 years (Barron and Bukry, 2007).

In an early study from the Peruvian margin, De Vries and Schrader (1981) distinguished subtropical species (*Thalassiosira eccentica* (Ehrenberg) Cleve "group," *Thalassiothrix frauenfeldii* (Grunow) Grunow, *Delphineis* spp., and *Chaetoceros* resting spores) as indicative of cooler, neritic waters with upwelling, as contrasted to a suite of species characteristic of oceanic, warm tropical waters. More recent work by Abrantes *et al.* (2007) along the Peru and Chile margins, noted increased absolute abundance of diatoms associated with overlying upwelling in both the coastal upwelling areas of the South American margin and eastern equatorial Pacific. More specifically, high *Chaetoceros* was a signature of coastal upwelling, with *Paralia sulcata* (Ehrenberg) Cleve also observed in upwelling off central Chile, while *Thalassionema* characterized equatorial upwelling. Decreased upwelling, on the other hand, resulted in greater abundance of *Thalassiosira* and *A. nodulifera*.

21.2.2 Records of sea-surface temperature

The recent increase in globally averaged atmospheric temperatures and instrumental data that document recent ocean warming has spurred interest in developing records of Holocene sea-surface temperatures (SSTs) as a way to place modern events into a longer-term perspective. Many diatom-based approaches have been published, with quantitative work done in the

Arctic/subarctic, the Antarctic, and the Sea of Japan (Koizumi, 2008).

21.2.2.1 Arctic reconstructions Crosta and Koç (2007) summarized diatom-based paleotemperature work from the high northern latitudes. Birks and Koç (2002) developed two different transfer functions to reconstruct August and February SSTs from the eastern Norwegian Sea with very similar results. These records demonstrated a rapid warming after the Younger Dryas, with maximum warmth between ~9700 and 6700 cal yr BP (mid-Holocene climatic optimum); at the maximum, temperatures were about 3 to 5 °C warmer than today. From ~6700 to 3000 cal yr BP, SSTs cooled gradually, followed by a slight rise around 2750 cal yr BP. Jiang *et al.* (2002) similarly developed a transfer function to reconstruct SSTs from a sediment core recovered from the North Icelandic Shelf. Their data showed that in the mid Holocene (4600–3800 cal yr BP) SSTs were higher than those observed at the site today. Generally, cooling was observed until about 3200 cal yr BP, followed by rising SSTs until about 2400 cal yr BP, after which temperatures cooled again until about 1300 cal yr BP. An episode of warmer SSTs was observed at around 850 cal yr BP, followed by cooling temperatures. Jiang *et al.* (2002) noted that the range of SSTs is only 1–2 °C and that the patterns of changing SSTs are consistent with a qualitative view of changing climate based on diatom assemblages and their ecological preferences. Subsequently, this same transfer-function method was applied to a high-resolution core from the North Icelandic margin to reconstruct annual mean SST as part of a multiproxy analysis of the core (Eiríksson *et al.*, 2006). One larger-scale goal of this project was to utilize this suite of proxy data to evaluate surface and bottom water-temperature variability over a time period characterized by well-known climate periods, including the Roman and Medieval Warm Periods and the Little Ice Age. The diatom-based transfer function demonstrated that the total temperature range over the ~2000 year record was small, about 2 °C, with an overall cooling trend interrupted by several century-scale warm periods (Jiang *et al.*, 2002). Justwan *et al.* (2008) similarly applied a suite of diatom-based transfer functions to a core from the North Icelandic Shelf; a 25.4-meter Holocene section provided the opportunity for high-resolution study of the entire Holocene, which was completed at an average resolution of 40 years. Reconstructed August SSTs revealed warmest temperatures between 10.4 and 4.7 cal kyr BP, the Holocene climate optimum. These paleotemperature data were combined with benthic foraminiferal data to develop an understanding of Holocene changes in surface and bottom waters. Similar results were obtained on the Voring Plateau (eastern Norwegian Sea) (Anderson *et al.*, 2004a).

21.2.2.2 Antarctic reconstructions In the Antarctic, several diatom-based transfer functions have been developed for reconstructing SSTs and sea-ice presence (SIP) (reviewed in Crosta and Koç, 2007). Zielinski *et al.* (1998) developed and tested the reliability of an SST transfer function for the South Atlantic. Bianchi and Gersonde (2004) applied this transfer function to reconstruct SSTs in two cores from the South Atlantic. Their data recorded major trends in SST over the past 18 cal ka BP, beginning with deglacial warming at about 18 to 17 cal kyr BP followed by a reversal in SSTs around 13 cal kyr BP (the Antarctic Cold Reversal). An early Holocene climatic optimum was observed; in the more northerly core (50° S), temperatures for the remainder of the Holocene remained relatively stable, while SSTs at the more southerly site (53° S), remained cool until about 8 cal kyr BP.

Hodell *et al.* (2001) and Nielsen *et al.* (2004) applied the Crosta *et al.* (1998a) transfer function to develop a 12,500 year record of both summer SST and SIP in two cores from the South Atlantic. The temperature record displayed millennial- and centennial-scale variability. The Holocene was divided into three major intervals – an early Holocene climate optimum (12,500–6200 cal yr BP), a mid-Holocene climatic minimum (6500–2900 cal yr BP) and late Holocene warming (2900 cal yr BP to present), with more or less abrupt transitions between intervals. Higher-frequency variability was determined via harmonic analysis, which showed cyclicity at 1220, 1070, 400, and 150 years (Nielsen *et al.*, 2004). The authors noted that the longer cycle lengths are similar to those observed in Antarctic ice-core deuterium records (Masson *et al.*, 2000) while the shorter cycles are observed in other Antarctic marine records (for example, Leventer *et al.*, 1996; Domack *et al.*, 2001).

Qualitative SST only was provided by studies of the morphological variability of *Thalassiosira antarctica* Comber, a diatom, that exhibits variability as a function of water temperature. Laboratory culture work by Villareal and Fryxell (1983) demonstrated that at very cold temperatures (−1.5 °C), average valve diameter is smaller, areolation is finer, and valves do not exhibit marginal processes, as contrasted to the same diatoms cultured at a warmer temperature (4.0 °C). Taylor *et al.* (2001) classified these two forms as types 1 (colder) and 2 (warmer). Several studies utilized this distinction to address Holocene changes in SST (Taylor *et al.*, 2001; Taylor and Sjunneskog, 2002; Domack *et al.*, 2003; Bak *et al.*, 2007). Taylor *et al.* (2001) noted, for example, that the *T. antarctica* T2 is more abundant during the mid-Holocene climatic optimum than the subsequent Neoglacial, when *T. antarctica* T1 becomes more abundant. Domack *et al.* (2003) observed a recent increase in *T. antarctica* T2 in a core from the Antarctic Peninsula; these data document recent warming of SST in the region.

21.2.3 Distribution of sea ice

A decrease in sea-ice thickness and extent has been observed over the past several decades in the Arctic (Lemke *et al.*, 2007; Stroeve *et al.*, 2007), a consequence and potential contributor to recent global change. Though instrumental records from the Southern Ocean are not as extensive, recent changes in sea-ice extent are postulated as well (Curran *et al.*, 2003). Given the important role that sea ice plays in climate, oceanographic, and ecological systems (Armand and Leventer, 2003), increasing attention is focused on paleo-sea-ice extent. This longer-term perspective is best developed from Holocene records and often relies heavily on diatom data, for many species of diatom are associated with sea ice (see review in Leventer, 1998). Diatom-based reconstruction of sea-ice extent has a long history and is accomplished through a variety of techniques, from the relatively simple association of a single species or a group of species with sea ice, to more complex statistical approaches (Armand, 2000; Armand and Leventer, 2003, 2010).

21.2.3.1 Arctic reconstructions Reconstruction of Arctic sea ice during the Holocene, initiated by Sancetta (1979, 1981, 1982), was developed further by Williams (1986, 1990, 1993), Koç *et al.* (1993), De Sève (1999), and Justwan and Koç (2008). Most early studies relied on the use of assemblage-based diatom data to document changes in sea-ice cover; these data are qualitative and geographically discontinuous. More recent work has shifted toward the use of transfer functions to provide a quantitative assessment. Both types of work depend on the occurrence of a suite of diatoms that are affiliated with sea ice, commonly including *Thalassiosira nordenskioldii* Cleve, *Thalassiosira hyalina* (Grunow) Gran, *Thalassiosira gravida* Cleve, *Nitzschia frigida* Grunow in Cleve & Grunow, species of *Fragilariopsis* such as *Fragilariopsis cylindrus* (Grunow) Krieger in Helmcke and Krieger, *Actinocyclus curvatulus* Janisch, *Thalassiosira trifulta* G. Fryxell, and *Porosira glacialis* (Grunow) Jörgensen (Gran, 1904; Horner and Alexander, 1972; Koç Kapuz and Schrader, 1990; Williams, 1990, 1993; De Sève, 1999; Knudsen *et al.*, 2004; Jensen *et al.*, 2004; Caissie *et al.*, 2006; Moros *et al.*, 2006).

Early qualitative work by Williams (1990) linked reduced diatom abundance in cores from Baffin Bay to reduced diatom productivity at times when sea ice was more extensive than today. She linked increased sea-ice cover to the presence of low-salinity meltwater, as this would freeze more easily than surface waters of normal salinity. Koç *et al.* (1993) noted that the eastern Nordic seas were ice free by 13.4 ka (oldest part of record), with decreasing ice extent through the mid Holocene, followed by cooling and increased ice extent until today. Knudsen *et al.* (2004), in a multiproxy study from the North Iceland Shelf, use a sea-ice assemblage comprised of *F. cylindrus* and *Fragilariopsis oceanica* (Cleve) Hasle to reconstruct sea-ice conditions for the past 15,800 years. Finally, Bauch and Polyakova (2000), working in the Laptev Sea on a relatively short record (~3000 years), find that more extensive pack-ice marks the earliest part of their record.

Jensen *et al.* (2004) and Moros *et al.* (2006) both used an assemblage approach to reconstruct variability in sea-ice concentrations from the south and west of Greenland. Jensen *et al.* (2004) include *F. cylindrus*, *F. oceanica*, and *Thalassiosira nordenskioeldii* in their sea-ice assemblage; Moros *et al.* (2006) used both these species of *Fragilariopsis*. Both studies utilized their data to address how changes in sea ice and other climatic variables might have impacted Norse and paleo-eskimo societies, respectively. Jensen *et al.* (2004) identified both the Medieval Warm Period (AD 800–1250) and the Little Ice Age (AD 1580–1850), separated by a time of climatic instability, with colder and windier conditions characterized by increasing sea ice. They observed that the abandonment of Norse settlements in Greenland coincided with climatic deterioration. Moros *et al.* (2006) noted that intermittent human occupation of Greenland stretches back to 4.5 kyr BP; discontinuous settlement has been linked to changes in climate. Artefacts of the Saqqaq people, described as preferential open-water hunters, show that they lived in the area from about 4.4 to 3.4 kyr BP, a time that the diatom data indicate was warmer and characterized by less sea ice. In contrast, the Dorset people, better adapted as sea-ice hunters, inhabited the area from around 2.8 to 2.1 kyr BP. Moros *et al.* (2006) linked their initial appearance to a time of increased sea-ice cover, and their disappearance to a change in oceanographic conditions characterized by more limited sea ice and its association with reduced marine resource availability.

Justwan and Koç (2008), in a study from the North Icelandic Shelf, presented a transfer function for reconstruction of May sea-ice concentration. Their data show that sea ice was present throughout the Holocene; generally it was inversely correlated to August SST. Maximum sea-ice concentration (~50%) occurred between 11.5 and 10.4 cal kyr BP. A sharp decrease in sea-ice concentration followed (~10% sea ice) and lasted until 4.7 cal kyr BP, after which the concentration rose gradually to almost 20% at about 2.2 cal kyr BP. After a sharp drop at 2.2 cal kyr BP, sea-ice concentrations were variable.

21.2.3.2 Antarctic reconstructions As in the Arctic, a suite of Southern Ocean diatoms are identified as sea-ice associated (Burckle *et al.*, 1982; Burckle, 1984; Burckle and Mortlock, 1998; Armand, 1997; Zielinski and Gersonde, 1997; Crosta *et al.*, 1998a, 1998b; Leventer, 1998; Armand, 2000; Armand and Leventer, 2003; Armand *et al.*, 2005; Armand and Leventer, 2010), including *Actinocyclus actinochilus* (Ehrenb.) Simonsen, *Fragilariopsis curta* (van Heurck) Hustedt, *Fragilariopsis cylindrus*, *Fragilariopsis obliquecostata* (van Heurck) Heiden and Kolbe, *Fragilariopsis rhombica* (O'Meara) Hustedt, *Fragilariopsis ritscheri* (Hustedt) Hustedt, *Fragilariopsis separanda* Hustedt, *Porosira glacialis*, *Porosira pseudodenticulata* (Hustedt) Jousé, *Stellarima microtrias (Ehrenberg)* Hasle and Sims, *Thalassiosira tumida* (Janisch) Hasle, and *Thalassiosira antarctica*. These species, either singly or as groups, have been instrumental in the development of qualitative reconstructions of Holocene sea ice. Selected studies are described below; these were chosen to illustrate subtle differences in the approaches utilized. Transfer functions also have been applied to determine annual sea-ice cover and maximum summer and winter sea-ice extents in the Southern Ocean (Crosta *et al.*, 1998a, 1998b). These studies focused primarily on the late Quaternary and last glacial maximum, though the works of Hodell *et al.* (2001) and Nielsen *et al.* (2004), mentioned earlier, present Holocene sea-ice records for the South Atlantic.

Many studies that reconstruct Southern Ocean sea-ice cover are based on the abundance of one or more species of *Fragilariopsis*. For example, Crosta *et al.* (2007) identified both millennial- and centennial-scale variability in a core from Adélie Land, along the East Antarctic margin. Their interpretations were based on the anti-correlated changes in down-core abundance of the sea-ice taxa identified as the "*F. curta* group" (*F. curta* + *F. cylindrus*) versus *Fragilariopsis kerguelensis* (O'Meara) Hustedt, a species associated with summertime blooms just south of Polar Front. Based on these two groups of species, Crosta *et al.* (2007) were able to divide the Holocene into three time periods. From 8700 cal yr BP to 7700 cal yr BP, the region was cooler with more sea ice, followed by warmer conditions and less sea ice from 7700 cal yr BP to 3900 cal yr BP, the Hypsithermal (mid-Holocene climatic optimum). Finally, the youngest part of the record, from 3900 cal yr BP to 1000 cal yr BP, was characterized by increasing sea ice and colder conditions characteristic of the Neoglacial. The authors also addressed centennial-scale variability in sea ice through spectral analysis of their diatom record. These results and their coherence with the record of solar variability are discussed later in this chapter.

Leventer *et al.* (1996) and Yoon *et al.* (2002) both use the ratio of (*F. curta* + *F. cylindrus*) / *T. antarctica* as a well as overall abundance of sea-ice taxa as a qualitative measure of sea ice over the past 3700 yr BP and 15000 yr BP, respectively. Leventer *et al.* (1996) noted that high ratios in Palmer Deep sediments are associated with high concentrations of CRS, and thus attributed centennial-scale cyclical changes in paleoproductivity to associated changes in sea ice and the occurrence of a sea-ice meltwater-stabilized upper water column. Yoon *et al.* (2002) observed broader-scale changes in the nearby western Antarctic Peninsula margin, with the mid-Holocene climatic optimum between 6000 and 2500 yr BP followed by increasing sea ice during the Neoglacial. Bárcena *et al.* (1998, 2006) developed a late-Holocene record from the Bransfield Strait, using a "sea-ice taxa group" (SITG) that included *F. curta* + *F. cylindrus* + *F. vanheurckii* (Peragallo) Hustedt. In contrast to the work of Leventer *et al.* (1996), Bárcena *et al.* noted an anti-correlation between their SITG and CRS. They interpreted the Neoglacial events as periods marked by increased sea ice and decreased productivity. Similarly, on the George V Coast, McMullen *et al.* (2006) observed the transition from the end of the Holocene climatic optimum to a colder and less productive Neoglacial period with greater sea-ice cover at about 3300 yr BP. Taxa used as a proxy for sea ice include *F. curta*, *Fragilariopsis sublinearis* (Van Heurck) Heiden and *P. glacialis*; *Chaetoceros* abundance is used as a monitor for paleoproductivty.

Bianchi and Gersonde (2004) presented sea-ice proxy data from two sediment cores recovered from the South Atlantic (50° S and 53° S). These records extend back 18,000 years. The authors estimated the mean winter sea-ice extent based on abundance of *F. curta* + *F. cylindrus*, based on the work of Gersonde and Zielinski (2000) who found that in the South Atlantic, these species occurred in relative abundances of >3% if winter sea ice was present. In addition, abundances of *F. obliquecostata* >3% were used as a proxy for the permanent sea-ice edge. Bianchi and Gersonde (2004) observed the gradual retreat of winter sea ice beginning around 18 cal kyr BP, which was accompanied by increased productivity in a meltwater-stabilized water column, recorded by increased abundance of the paleoproductivity indicator, *Chaetoceros* resting spores. A climate reversal, centered around 13 cal kyr BP, was marked by cooling and increased sea ice, after which an early-Holocene climatic optimum was observed. Although sea-ice cover remained stable at the more northerly site through the Holocene, an expansion of sea ice between 8 and 7 cal kyr BP was observed at the more southerly site.

While most Southern Ocean studies utilize these commonly recognized "sea-ice" diatoms as proxies, a few have used the variable morphology of the chain-forming diatom *Eucampia antarctica* (Castracane) Manguin as a paleo-sea-ice proxy. Fryxell and Prasad (1990) suggested that, under higher light conditions (thinner and/or less sea-ice cover), more cell divisions can occur, leading to longer chains. In sediment samples, chain length can be estimated by comparing the number of intercalary versus terminal valves that occur in a sample. Only the free end of a chain, or the terminal valve, can have a pointed morphology, while valves in the interior of a chain are flat-ended. Chains of *Eucampia* growing in the lower light conditions, characteristic of waters with more sea-ice cover, are more likely to be short with a consequent higher relative proportion of terminal valves. On the basis of these observations, Kaczmarska *et al.* (1993) introduced a new paleoenvironmental proxy, the *Eucampia* index (ratio of terminal to intercalary valves), to record oscillations of winter sea ice through the Late Pleistocene on the Kerguelen Plateau. This same relationship was used by Leventer *et al.* (2002) to document early Holocene warmth and decreased sea ice in the western Antarctica Peninsula region.

21.3 Holocene marine laminated diatom ooze – ultra-high resolution records of environmental change

Holocene marine laminated diatom oozes began to gain attention as important geological archives during the 1960s, due to their potential for ultra-high resolution, seasonal-scale reconstructions of oceanic processes (see Pike and Stickley, 2007 for a review). At this time, scientists were beginning to realize that seasonally varying fluxes into, and through, the water column could, under certain circumstances, be preserved in the deep-sea sediment record. For example, in the Gulf of California, northwest Mexico, short sediment cores comprised of alternating light and dark olive-green laminations were attributed to seasonally varying diatom flux to the sea floor (Calvert, 1964; Calvert, 1966). Advances in coring technologies during the 1970s, such as the Deep Sea Drilling Project's hydraulic piston core, preceded recovery of a succession of long, high-resolution, Holocene laminated diatom oozes. During the 1980s, the study of marine varves, i.e. seasonally laminated sediments, expanded from consideration of the seasonal-scale mechanisms and processes involved in producing the varves, to a consideration of longer-timescale oceanographic and climatic processes and forcing mechanisms following the varve microcosm concept proposed by Anderson (1986). This stated that seasonal-scale processes, especially those related to insolation and solar variability, will be reflected in longer-timescale variations within the varved sequence. Indeed, it has been patterns of solar variability and subsequent ocean-climate variability during the Holocene, over decadal to centennial timescales, that have been some of the most compelling records to be derived from long sequences of laminated diatom ooze; these records are addressed at the end of this chapter. Advances in analytical methods during the 1990s brought the study of Holocene laminated diatom oozes into a new era. Techniques such as scanning electron microspcopy (SEM)-scale sediment fabric and diatom assemblage analysis (Brodie and Kemp, 1994; Grimm, 1992; Kemp, 1990), and sub-millimeter-scale sampling for quantitative diatom abundance studies (Sancetta, 1995), have been combined with the use of sediment traps (see Armand and Romero, this volume) and satellite remote sensing of the sea surface (Thunell *et al.*, 1993) to fully unlock the potential of using Holocene laminated marine diatom oozes for seasonal- to millennial-scale ocean and climate variations.

21.3.1 Formation and accumulation of diatom-rich laminated sediments

There are two main requirements for the formation of marine laminated diatom ooze (Pike and Stickley, 2007). First, there has to be a temporally varying flux of either diatoms or lithogenic grains (or both) to the sediment and, secondly, there has to be a mechanism of preserving the primary laminated sediment fabric. As proposed by Grimm *et al.* (1996) for some of the homogenous intervals intercalated with laminated sediments of the Santa Barbara Basin, California margin, if climatic seasonality is minimal, and sedimentary fluxes do not vary during the year, a primary massive fabric will result. Working out which components vary to produce sedimentary laminae is often difficult in the geological record; however, when dealing with Holocene sediments, this problem can be alleviated to a large extent by the use of sediment-trap campaigns in the modern ocean (Thunell *et al.*, 1993) (see Armand and Romero, this volume). The flux of autochthonous (and allochthonous) particles at a location can be sampled across the modern seasons and compared with the Holocene sedimentary laminations.

Laminated marine diatom oozes tend to accumulate beneath regions of high oceanic primary production, such as coastal upwelling regions along the western margin of continents. Seasonal variations in wind and nutrient regimes in these regions leads to seasonally high abundance of diatoms such as *Skeletonema costatum* (Grev.) Cleve and CRS – small, rapidly

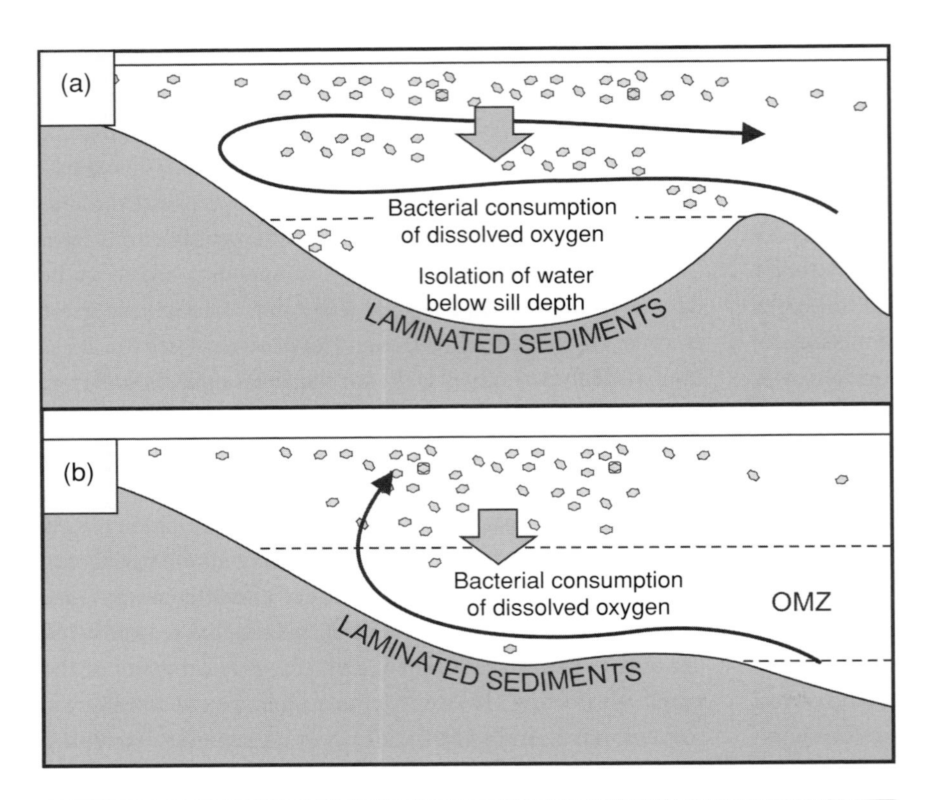

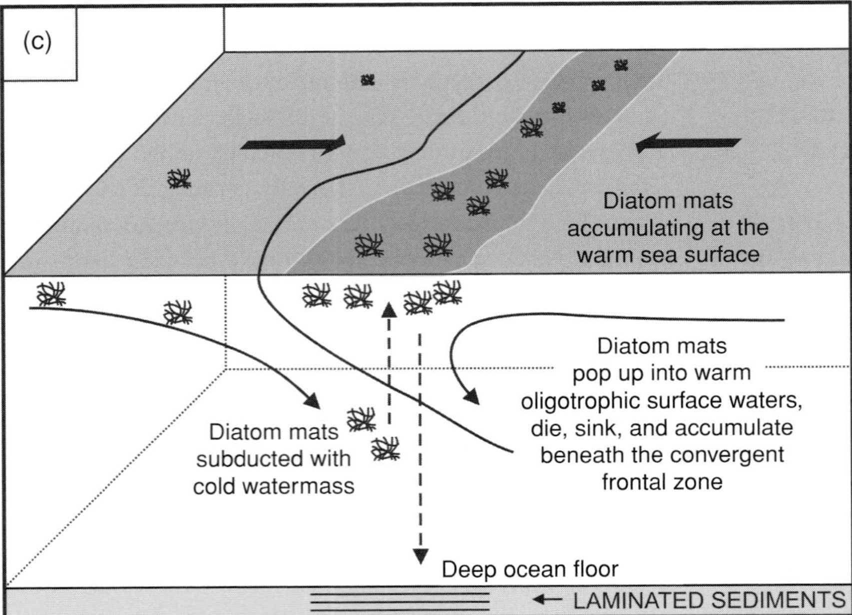

Figure 21.1 Schematic representation of different Holocene laminated diatom ooze accumulation sites. (a) A silled basin on the continental shelf; (b) An intersection of the oxygen minimum zone with the continental slope; and (c) convergent oceanic frontal zones in open ocean settings. (Reprinted from *Encyclopedia of Quaternary Science* (ed. S. Elias), Pike, J. and Stickley, C., Diatom records: Marine laminated sequences, pp. 557–567, Copyright (2007), with permission from Elsevier.)

reproducing taxa – and their subsequent flux to the sediments. Preservation of the seasonal flux signal will take place if one of two circumstances prevails (Pike and Stickley, 2007): (1) there are very low levels of benthic dissolved oxygen; or (2) impenetrable, mesh-like diatom mats comprised of *Thalassiothrix longissima* Cleve & Grunow are deposited within the laminated sequence (Figure 21.1). The first case occurs in areas beneath high primary-productivity regions. For example, in

Santa Barbara Basin, a silled continental shelf basin, water-column remineralization of the high surface-water production rapidly consumes dissolved oxygen below basin sill depth (Reimers *et al.*, 1990), thus restricting any bioturbating macrofauna (Figure 21.1a). Along open continental slopes (Figure 21.1b) and within semi-enclosed seas, the same process strengthens the oceanic oxygen minimum zone (OMZ) (Wyrtki, 1962) leading to preservation of laminations along the slopes and sides of basins (Calvert, 1964; Gardner and Hemphill-Haley, 1986). The second mechanism that promotes the preservation of primary sedimentary laminations in the deep ocean is the presence of diatom mats. In regions where large fluxes of diatom mats accumulate in the surface waters, usually oceanic frontal zones (e.g. Bodén and Backman, 1996; Grigorov *et al.*, 2002; Kemp and Baldauf, 1993), these can sink intact to the sea floor and form a barrier to bioturbating infauna even if benthic dissolved oxygen levels are sufficient to support infaunal communities (Figure 21.1c). This mechanism was much more prevalent during the Quaternary and Neogene and during the Holocene it only seems to play a minor role in preservation of laminated diatom oozes from the Gulf of California (Pike and Kemp, 1999) where *Thalassiothrix longissima*-dominated diatom mats are imported into the gulf with warm subtropical Pacific waters, especially during El Niño events.

21.3.2 Analysis of diatom-rich laminated sediments

Sampling and analysis of laminated diatom oozes has become refined over the last 20 years, which has resulted in the full exploitation of the paleoceanographic and paleoclimatic archives contained therein. Traditional methods of analyzing laminated sediments, such as X-radiography, did not provide any taxonomic or ecological information about the biological component of the varves. High-resolution discrete sampling of individual laminations (Barron *et al.*, 2005; Leventer *et al.*, 2002; Sancetta, 1995) facilitated close comparisons with sediment-trap time series. However, the real advance in the study of laminated diatom oozes came with the introduction of scanning electron microscope backscattered electron imagery (BSEI) of polished thin sections (Kemp, 1990). This method involved the sampling of intact sequences of laminated sediments, embedding them with resin, curing the embedded blocks and making lamina-perpendicular polished thin sections (for reviews see Dean *et al.*, 1999; Pearce *et al.*, 1998; Pike and Kemp, 1996). Analysis of the polished thin sections by BSEI (e.g. Maddison *et al.*, 2006) or light microscopy (LM, e.g. Denis *et al.*, 2006) permitted identification of laminae or sub-laminae tens to hundreds of micrometers thick, which further lead to the realization that marine varves commonly comprise more than a simple couplet, containing a record of many different diatom flux events per year associated with discrete seasonal processes (e.g. Dean and Kemp, 2004, showing up to 22 events per year in Saanich Inlet sediments). Combining thin-section analysis with discrete sampling has two further strengths. First, discrete samples targeted at BSEI (or LM)-identified laminae can aid in taxonomic assignments (often difficult with only two-dimensional cross-sections of frustules) and whole-assemblage characterizations (BSEI or LM often draws the eye to the bigger taxa; e.g. Maddison *et al.*, 2006). Second, BSEI can place taxa into their seasonal context, which can aid the interpretation of long time series of assemblages derived from discrete sampling (e.g. Crosta *et al.*, 2008; Pike *et al.*, 2009).

21.3.3 Sites of accumulation of Holocene diatom-rich laminated sediments

Although not abundant in the modern ocean and Holocene geological record, laminated diatom oozes are globally widespread (Table 21.2). For example, the Holocene record from the Santa Barbara Basin, California margin, has been well documented (see review in Schimmelmann and Lange, 1996), and the record of laminated diatom ooze extends back into the Late Pleistocene (Bull *et al.*, 2000). The Gulf of California and Baltic Sea represent semi-enclosed and enclosed seas, respectively, where Holocene laminated diatom oozes have accumulated (Burke *et al.*, 2002; Pike and Kemp, 1997). Gulf of California laminated sequences often show up to five laminae per year, dominated by seasonally related flux events (Pike and Kemp, 1997). Lithogenic-rich laminae are deposited during the summer, resulting from eolian deposition during convective thunderstorms in the surrounding deserts (Baumgartner *et al.*, 1991). The sequence of diatom laminae begins with the "fall dump" (Kemp *et al.*, 2000), which comprises large centric diatom species such as *Coscinodiscus asteromphalus* Ehrenb., *Coscinodiscus occulus-iridis* Ehrenb., *Coscinodiscus granii* Gough, *Rhizosolenia bergonii* H. Peragallo, and *Stephanopyxis palmeriana* (Grev.) Grunow. These are deposited as water-column stability breaks down during the onset of winter storms. This fall-dump lamina is followed by a mixed diatom assemblage lamina related to the influx of nutrients to the surface during winter water-column mixing. The onset of spring upwelling produces a bloom of *Skeletonema costatum* and *Hyalochaete Chaetoceros* spp., the CRS being preserved into the sediment record. Similar to the Holocene Santa Barbara Basin record (Lange *et al.*, 1987; Bull *et al.*, 2000), the Holocene laminated sediments of the Gulf of California also record the occurrence of El Niño events

Table 21.2 Studies of Holocene marine laminated diatom oozes

Paper	Location	Theme
	Antarctica	
Maddison *et al.*, 2005	Palmer Deep, western Antarctic Peninsula; BSEI	Seasonality during deglaciation
Leventer *et al.*, 2002	Palmer Deep, western Antarctic Peninsula; discrete sampling	Seasonality, oceanographic processes
Finocchiaro *et al.*, 2005	Cape Hallett Bay, Ross Sea; discrete sampling	Seasonality, oceanography and glaciology
Maddison *et al.*, 2006	Mertz Glacier Polynya, George V Coast; BSEI and discrete sampling	Seasonality, links with polynya dynamics
Denis *et al.*, 2006	Dumont d'Urville Trough, Adélie Land; LM	Seasonality, links with millennial-scale variations
Crosta *et al.*, 2007, 2008	Dumont d'Urville Trough, Adélie Land; discrete sampling	Sea ice, links with Holocene computer modeling results
Stickley *et al.*, 2005, 2006	Iceberg Alley, Mac.Robertson Shelf; BSEI	Seasonality, use of laminations as chronometer
Pike *et al.*, 2009	East Antarctica; BSEI, LM, and discrete sampling	Seasonality, sea-ice concentration, links with Holocene computer modeling results
Leventer *et al.*, 2006	East Antarctica; lithostratigraphy	Calving bay re-entrant mechanism for lamina genesis
Denis *et al.*, 2009a	Adélie Land; discrete sampling	Glacial oscillations, deep water activity
Denis *et al.*, 2009b	Adélie Land; discrete sampling	Productivity
	North Pacific	
Pike and Kemp, 1996, 1997, 1999	Gulf of California, northwest Mexico; BSEI	Seasonality, links to ENSO, diatom mats, and early diagenesis
Baumgartner and Christensen, 1985	Gulf of California, northwest Mexico; X-radiography and discrete sampling	Links with ENSO
Baumgartner *et al.*, 1991	Gulf of California, northwest Mexico; X-radiography and discrete sampling	Eolian mechanism for lithogenic lamina formation
Calvert, 1964, 1966; Donegan and Schrader, 1982	Gulf of California, northwest Mexico; discrete sampling	Seasonality, mechanism of varve formation
Schrader and Baumgartner, 1983	Gulf of California, northwest Mexico; discrete sampling	Decadal variations in upwelling intensity
Sancetta, 1995	Gulf of California, northwest Mexico; discrete sampling	Seasonality, comparisons with sediment trap data
Lange *et al.*, 1987; Lange and Schimmelmann, 1994; Schimmelmann and Lange, 1996	Santa Barbara Basin, California margin; X-radiography and discrete sampling	Seasonality, ENSO, varve formation mechanism
Bull and Kemp, 1995, Bull *et al.*, 2000	Santa Barbara Basin, California margin; BSEI	Seasonality, ENSO
Baumgartner *et al.*, 1994	Santa Barbara Basin, California margin; discrete sampling	Uses varves to demonstrate decadal variability in fish-scale deposition
Dean *et al.*, 2001; Dean and Kemp, 2004	Saanich Inlet, British Columbia; BSEI	Seasonality, ENSO, interannual and decadal-scale variability
Chang *et al.*, 2003, Chang and Patterson, 2005	Effingham Inlet, British Columbia; BSEI	Seasonality, links to Aleutian low pressure system
Patterson *et al.*, 2007	Seymour-Belize Inlet, British Columbia; X-radiography	Links to cooler climates on millennial timescales
Grimm, 1992	Japan sea; BSEI	Seasonality, productivity during late Quaternary
	South Pacific	
Kemp, 1990; Brodie and Kemp, 1994	Peru margin; BSEI	Seasonality during interglacials; development of BSEI technique

Table 21.2 *(cont.)*

Paper	Location	Theme
	Atlantic	
Hughen *et al.*, 1996	Cariaco Basin, Caribbean; BSEI and LM	Seasonality and longer timescale climate
Burke *et al.*, 2002	Gotland Deep, Baltic Sea	Seasonality, climate
	Red Sea	
Seeberg-Elverfeldt *et al.*, 2004	Shaban Deep, northern Red Sea; BSEI and discrete samples	Seasonality, ocean circulation, brine pools
General papers		
Kemp *et al.*, 2000	BSEI	"Fall dump" mechanism
Kemp, 2003		Abrupt climate change from annually laminated sediments
Pike and Kemp, 1996; Pearce *et al.*, 1998; Dean *et al.*, 1999; Denis *et al.*, 2006	BSEI and LM	Thin section preparation for BSEI and LM analysis
Pike and Stickley, 2007		Review of diatom records from marine laminated sequences

(Baumgartner and Christensen, 1985; Pike and Kemp, 1997). The fifth lamina that is often preserved in the annual sequence is comprised of *Thalassiothrix longissima*-dominated diatom mats, transported into the gulf with warm, subtropical Pacific Ocean surface waters during El Niño (Pike and Kemp, 1997). Frequency analysis of the occurrence of these diatom mats in the sediments revealed dominant frequencies: ~11 years, ~22–24 years, and ~50 years. These frequencies match cycles in solar activity and suggest that the pattern of El Niño in the eastern Pacific during the early Holocene was, in some way, linked to solar activity (Pike and Kemp, 1997).

High-latitude fjords are also typical sites of accumulation of Holocene laminated diatom oozes, due to restricted deep-water circulation and high surface-water primary production. Ocean Drilling Progam (ODP) drilling in Saanich Inlet, British Columbia, recovered ~60 m of laminated diatom ooze. Here, centimeter-scale annual sequences can comprise up to 22 laminations, or individual flux events (Dean and Kemp, 2004) – to date the greatest number of individual flux events identified in marine varves. In a study that compared previously published sediment-trap data with the sediment laminations, diatom ooze laminae were deposited in a typical sequence of an early-spring *Thalassiosira* spp. (including *T. pacifica* Gran & Angst) lamina, followed by a late-spring CRS lamina and a late-spring/early-summer lamina dominated by *Skeletonema costatum*, related to increased irradiance and gradual stabilization of the water column. These laminae are followed by alternating mixed CRS laminae and diatomaceous muds related to summer and autumn months. The CRS laminae are related to tidally controlled input of new nutrients and diatom seed stock into the fjord (Dean *et al.*, 2001) throughout the summer and autumn. As water-column stratification breaks down in the autumn, terrigenous-derived sediment is deposited as the final lamination of the year. Dean and Kemp (2004) used a time series of seasonal-scale observations of diatom assemblages to carry out spectral analysis and showed a relationship between the occurrence of: early-spring *Thalassiosira* spp. laminae and strong La Niña events in the Pacific; late-spring/early-summer *Skeletonema costatum* laminae and El Niño events; summer and autumn multiple CRS laminae and both negative Pacific Northwest Index regimes and the behaviour of the Aleutian low-pressure system. They also identify a decadal-scale frequency in the pattern of winter clay-rich laminae which they relate to the Pacific Decadal Oscillation, demonstrating how documentation and understanding of seasonal signals recorded in laminations can lead to the inference and understanding of longer-timescale climatic patterns in a region. More recently, similar studies have been carried out on Holocene laminated diatom ooze from Effingham Inlet (Chang and Patterson, 2005; Chang *et al.*, 2003) and similar sediments have been described from Seymour-Belize Inlet (Patterson *et al.*, 2007), demonstrating the wide potential of high-latitude fjords for intra-annual to decadal-scale ocean and climate reconstructions.

In the last decade, two further environments have revealed themselves as important sites for the accumulation of Holocene laminated diatom oozes – the continental shelf basins of the Antarctic margin and the brine pools of the Red Sea. Along the Antarctic margin, deep continental-shelf glacial troughs, in which sediments accumulated beneath the depth of iceberg scouring, have been shown to contain laminated diatom oozes, especially during the last deglaciation and intermittently during the Holocene (for example Denis *et al.*, 2006; Leventer *et al.*, 2002; Leventer *et al.*, 2006). High-resolution analysis of the seasonal-scale diatom records in these sediments has shown that annual varves are often characterized by: (1) thick laminae of CRS related to the spring sea-ice melt, development of a freshwater cap, and the associated bloom of diatoms exploiting trapped nutrients; followed by (2) a mixed diatom assemblage typical of Antarctic coastal summer production; capped by (3) a thin layer of either *Thalassiosira antarctica* resting spores or *Porosira glacialis* resting spores, deposited following the decrease in irradiance, re-freezing of sea ice, and increased salinities of autumn (for example Denis *et al.*, 2006; Maddison *et al.*, 2005; Maddison *et al.*, 2006; Stickley *et al.*, 2005). Knowledge gained about the seasonal significance of different taxa laminations in the sediments has provided new insights into ice–ocean interactions and processes (Leventer *et al.*, 2002; Leventer *et al.*, 2006), the influence of longer-timescale forcing mechanisms, i.e. solar forcing, over the Holocene (Crosta *et al.*, 2007), and the development of seasonality over the Holocene (Crosta *et al.*, 2008; Pike *et al.*, 2009).

The final environment of accumulation of Holocene laminated diatom oozes that we will consider is that of the Shaban Deep in the northern Red Sea. Here, seasonal diatom-rich laminations are preserved beneath deep sea, highly saline (260 psu) brine pools that contain <0.3 mg l^{-1} dissolved oxygen (Seeberg-Elverfeldt *et al.*, 2004). Laminated sediments are preserved during the early (10,000–11,000 yr BP) and middle Holocene (4000–6000 yr BP). Dark, diatom ooze laminae are predominantly *Rhizosolenia* spp. (*R. pungens* Cleve-Euler and *R. setigera* Brightwell), although sometimes CRS predominate, whereas light laminae are characterized by coccoliths. Such light laminae represent the flux of summer production in the northern Red Sea, whereas the *Rhizosolenia* spp.-dominated laminae are related to fall-dump processes and CRS laminae are related to winter blooms in response to high nutrient availability (Seeberg-Elverfeldt *et al.*, 2004). Seeberg-Elverfeldt *et al.* (2004) suggest that the prevalence of two diatom ooze lamina types in the sediment sequences could be used to characterize ocean stratification and mixing processes over time.

21.3.4 Advances using Holocene laminated diatom ooze studies

Seasonally laminated, or varved, marine sediments can reveal information about paleo-flux events over seasonal to centennial timescales. Laminated sediments, as analyzed using thin sections (e.g. Chang *et al.*, 2003; Stickley *et al.*, 2005; Denis *et al.*, 2006) reveal information about regional seasonal cycles and changes in diatom assemblages that can be considered like ancient sediment-trap records and compared to modern time series (e.g. Pike and Kemp, 1997; Sancetta, 1995) shedding light into the geological record. Using the varve microcosm concept developed by Anderson (1986), analysis of Holocene laminated diatom oozes has been central to the understanding of large-scale oceanographic and atmospheric processes. For example, analysis of laminated diatom oozes deposited beneath the Mertz Glacier Polynya, George V Coast, Antarctica, led to the suggestion that abundance of species such as *Corethron pennatum* (Grunow) Ostenfeld and *Rhizosolenia* spp., typical taxa associated with summer open-water conditions, could provide information about the persistence of the polynya through the Holocene and, hence, an insight into the deep-water production potential of the area through time (Maddison *et al.*, 2006). Laminated diatom oozes deposited around the East Antarctic margin during the last deglaciation also provided an insight into large-scale ice–ocean–biology interactions with the suggestion of a calving-bay re-entrant over the sites of their accumulation, trapping and providing nutrients to fuel extremely high productivity (Leventer *et al.*, 2006) and the production of CRS-dominated laminations up to 10 cm thick related to single flux seasons (Stickley *et al.*, 2005). The understanding of variations of oceanographic and climatic processes over seasonal timescales can aid interpretations of longer-timescale (intra-annual to millennial) variations forced by insolation processes. For example, accumulation of *Thalassiothrix longissima*-dominated diatom mats in Holocene sediments from the Gulf of California, brought into the gulf during strong El Niño events, suggested that circulation in the eastern Pacific related to El Niño was, in the early Holocene, also related to solar forcing (Pike and Kemp, 1997). Holocene high-latitude fjord records from the North Pacific have also revealed archives of paleo-El Niño/La Niña activity in the occurrence of *Thalassiosira* spp. and *Skeletonema costatum* laminae, respectively, as well as variability in the strength of the Aleutian low pressure system, the Pacific Decadal Oscillation, and the Quasi-biennial Oscillation; large-scale ocean–atmosphere phenomena (Dean and Kemp, 2004). The ultimate contribution of high-resolution time series of diatom assemblage changes, with sound seasonal interpretations, will be to test against computer-model hindcasts over the Holocene, as recently carried out by Crosta and colleagues (Crosta *et al.*, 2007; Crosta *et al.*, 2008; Pike *et al.*, 2009).

21.4 Application of high-resolution diatom analyses: a case study

21.4.1 Evidence for solar cycles in diatom ooze sediments

To date, only a limited number of marine cores have been investigated for their diatom content at high enough resolution and over periods long enough to allow Holocene climate reconstructions at decadal to century scales. In this section we present an example of how diatom census counts from high-resolution marine geological records have been used to assess the influence of a significant climatic forcing, solar variability. We focus on this singular contribution of high-resolution marine diatom work as representative of the kind of research that will be instrumental to continued advances in the contribution of diatom data to climate research.

Jiang *et al.* (2005) counted diatom assemblages from a core recovered from the modern position of the oceanographic Polar Front north of Iceland. Using a diatom-based transfer function they reconstructed summer SSTs and winter SSTs over the last 2000 years with a RMSEP (root-mean-square error of prediction) of ~1 °C and a temporal resolution between 6 and 30 years. They compared the SST records with high-latitude solar-insolation data. They used the solar-irradiance reconstruction for the past 1200 yr (Bard *et al.*, 2000), which is based on ^{14}C and ^{10}Be records. Their SST records showed an increasing trend of 1 °C over the last 1200 years, with abrupt variations of 0.5 °C. Summer insolation at 67° N concomitantly increased by 2 W m^{-2} with abrupt changes of about 0.5 W m^{-2}. Visual comparison of summer and winter SSTs with summer insolation at 67° N shows a good match between total solar irradiance (TSI) and climate response over the last 1200 years. In Jiang *et al.*'s study, solar forcing explains ~45% of the SST variance at timescales longer than 50 years. Jiang *et al.* (2005) reproduced the SST changes with the global ocean–atmosphere–sea-ice ECBilt-CLIO-VECODE model, in which a 2 W m^{-2} negative anomaly induced a SST drop of ~0.4 °C in summer and ~1 °C in winter, and a 2 W m^{-2} positive anomaly caused a SST warming of 0.5 °C in summer and 1 °C in winter. They conclude that "a positive and significant correlation between our SST record from the North Icelandic Shelf and inferred insolation, together with modelling experiments, supports the hypothesis that solar forcing is an important constituent of natural climate variability in the northern North Atlantic region" (Jiang *et al.*, 2005, p. 75). However, they didn't provide insight on how TSI changes were transmitted to the surface of the ocean.

Leventer *et al.* (1996) counted diatoms from the Palmer Deep Basin (western Antarctic Peninsula); counts were completed with a temporal resolution of ~40 years to ~80 years. Diatom abundances and assemblages present cyclic variations over the last ~5000 years. High diatom abundances and high values of the ratio of CRS versus *Chaetoceros* vegetative cells (CRS/Cveg) are concomitant to low values of magnetic susceptibility (MS). This was interpreted to indicate periods of very high primary production, that is, the occurrence of blooms large enough to deplete nutrient stocks. Leventer *et al.* (1996) suggested that high production occurred when stratification of the upper water column increased, a consequence of the input of low-salinity meltwater and/or the absence of storms that would deepen the mixed layer. Increases/decreases in the ratio are therefore a rough proxy of increased/decreased stratification of the upper water column. Spectral analysis of the CRS/Cveg ratio and MS data calculated cycles of ~110, ~230, and ~800 years indicating periodic variations in stratification; spectral power is greatest at the 230 years periodicity. Leventer *et al.* (1996, p. 1641) suggest that,

> The widespread geographic occurrence of similar ~200-yr cyclicity in a variety of terrestrial and oceanic proxy records, summarized below, suggests global significance of the driving mechanism. We propose a solar forcing based on the ~200-yr cycle that is a repetition of the observed "Maunder minimum" in sunspot activity (Stuiver and Braziunas, 1989). This ~200-yr cyclicity is demonstrated by secular ^{14}C fluctuations recorded in tree rings, which some researchers have attributed to solar modulation of the ^{14}C-producing radiation reaching the Earth (e.g. Suess and Linick, 1990; Stuiver and Braziunas, 1993).

They conclude (Leventer *et al.*, 1996, p. 1642)., however, that,

> At this point a proposal linking our observed productivity cycles with solar variability is speculative because we understand so little about the influence such small changes have on climate systems, especially in the southern high latitudes.

Finally, Crosta *et al.* (2007) counted diatom assemblages in a core from the Dumont d'Urville Trough (Adélie Land, East Antarctica). The core covers the 9000–1000 cal yr BP period and diatom identification was performed every 4 to 8 cm providing a 10–20 year resolution. Crosta *et al.* (2007) focused on two diatom species because of their ecological significance.

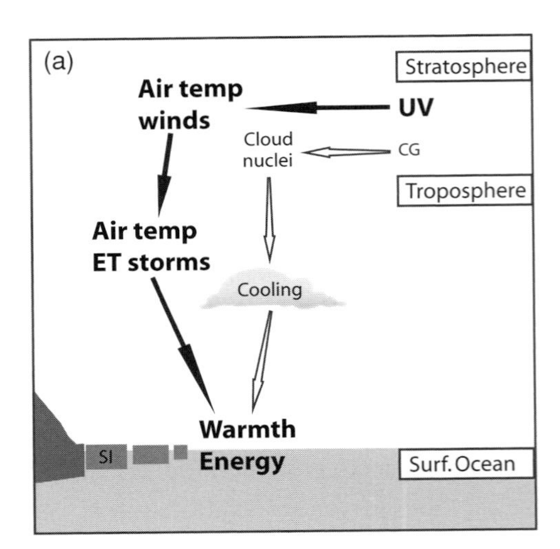

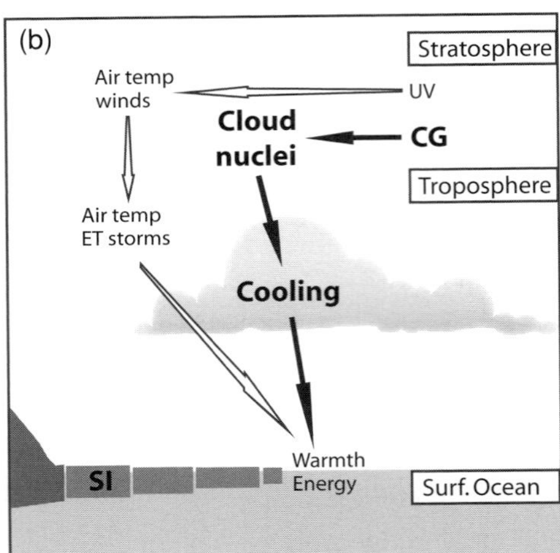

Figure 21.2 Conceptual model of processes affecting Antarctic sea-ice extent, during (a) solar maxima and (b) solar minima. Large, bold text and black arrows represent processes that are enhanced during events and that affect sea ice through positive action while small text and white arrows represent processes that are minimized during events and that affect sea ice through negative action. ET: Storms from extra-tropical origins; UV: ultraviolet rays; CG: cosmic grains; SI: sea ice. (Modified from Crosta *et al.* (2007) by permission of American Geophysical Union.)

High presence of *Fragilariopsis curta* in sediments was interpreted as indicative of heavy sea-ice cover during the winter and unconsolidated sea ice during the spring–summer growing season associated with low SSTs (Armand *et al.*, 2005). High occurrences of *Fragilariopsis kerguelensis* conversely are considered indicative of open waters during spring–summer growing season with warmer SSTs (Crosta *et al.*, 2005). Both records presented high-amplitude variations, though variations of *F. kerguelensis* and *F. curta* were greater during the mid-Holocene Hypsithermal and the Neoglacial, respectively, in agreement with the dominance of the open-ocean species during the warmer Hypsithermal and of the cryophilic species during the colder Neoglacial. Anti-phased high-amplitude variations of the two above mentioned species in core MD03–2601 demonstrate changes in sea-ice seasonality and length of the growing season off Adélie Land at decadal to century timescales.

Crosta *et al.* (2007) applied both spectral analysis and wavelet analysis to their diatom data. Spectral analysis calculated cyclicities at ~90, ~110, ~125–150, ~240, and ~580 years in both periods; cyclicities that are very close to those for solar activity. Wavelet analyses estimate the same periods but indicate non-stationary cyclicities congruent to non-stationary cyclicities in solar activity. This similarity lead Crosta *et al.* (2007, p. 9) to "suggest that a link may be likely between solar activity and the ultimate geophysical parameter influencing diatom production at high southern latitudes, namely sea ice cover and seasonal cycle," though they note that internal climate variability cannot be disregarded as it can act together with solar activity at congruent frequencies.

21.4.2 Links between solar activity and surface ocean conditions

The three high-resolution diatom records depict cyclic changes in SST and sea ice at decadal to century scales congruent with solar-activity frequencies indicating a possible casual relationship between TSI and surface ocean conditions. Crosta *et al.* (2007) investigated possible links, briefly summarized below and in Figure 21.2. A direct connection between TSI and weather and climate has been surmised for more than 100 years but generally dismissed by most scientists, who assumed that the effect of solar variations would be small. Changes in total solar irradiance are supposed to have been sufficient to account for the cooling of the Little Ice Age (Lean and Rind, 1998, 1999), but it appears that TSI alone can not explain the amplitude of climate response at the Holocene timescale (Renssen *et al.*, 2006). However, internal feedbacks amplifying the small energy changes associated with solar activity are present, including variations in the spectral irradiance in the ultraviolet (UV) and variations in the solar wind and the flux of energetic particles.

Variations in UV wavelengths, accompanying changes in solar activity, are responsible for the production and loss of ozone in the stratosphere, which may cause significant shifts in atmospheric circulation and therefore climate (van Geel *et al.*, 1999; Friis-Christensen, 2000; Reid, 2000). Solar-cycle variations in the UV band are much larger than in total irradiance. During high solar activity, UV radiation is absorbed by stratospheric ozone, which causes a stratospheric warming. The small temperature anomaly subsequently penetrates into the lower troposphere (Haig, 1999; Shindell *et al.*, 1999), which may reduce sea-ice freezing and promote sea-ice melting. Additionally, increased UV radiation strengthens and displaces poleward the subtropical jetstreams, which leads to a poleward relocation of the mid-latitude storm tracks that may inject more mid-latitude heat into polar environments, thus enhancing sea-ice melting, while stronger winds may enhance the mechanical break-up of the sea ice (Enomoto and Ohmura, 1990; Simmonds, 1996). Both processes result in a lengthening of the sea-ice-free season and phytoplankton growing season, thus promoting open-ocean diatoms (*F. kerguelensis*) at the expense of cryophilic diatoms (*F. curta*) (Figure 21.2).

Variations in solar winds and associated cosmic-ray flux also can lead to climate change through modification of the optical parameters and balance of the atmosphere. Though still under debate (Damon and Laut, 2004), it seems that global cloud cover is highly correlated with cosmic-ray flux and anti-correlated with solar activity (Svensmark and Friis-Christensen, 1997; Svensmark, 2000). During high solar activity, more turbulent solar winds scatter galactic cosmic rays and their associated energetic particles into space, thus reducing their flux to the Earth. This decreases the amount of nuclei for cloud condensation, thus reducing the cloud cover (Marsh and Svensmark, 2000). The atmosphere is then more transparent to solar radiation that can heat up the lower troposphere. The warm anomaly could reduce sea-ice formation and lengthen the sea-ice-free season promoting open-ocean diatom production.

21.5 Summary

Clearly, Holocene marine diatom records cover the whole range of timescales from seasonal to pluri-millennial, providing important insights on processes responsible for climate and oceanographic change. Most early work resolved millennial-scale changes in climatic and oceanographic signals, with the greatest contributions to Holocene records from the higher latitude and upwelling regions of the world's oceans. Development of these lower-resolution records proliferated in the 1990s, providing a strong basis for understanding broad features of Holocene climate. More recently, our improved understanding of conditions that favor the deposition and preservation of high-resolution records coupled to increased technological capabilities that permit the undisturbed recovery of these sequences, has focused our attention on higher-resolution studies that are critical toward understanding the response of the Earth's climate to stochastic or periodic changes on the timescales of decades to centuries.

Acknowledgments

Amy Leventer thanks her Antarctic colleagues and the National Science Foundation Office of Polar Programs for continued support over the years. Xavier Crosta was supported by the TARDHOL project through the national EVE-LEFE program (Institut National des Sciences de l'Univers – Centre National de la Recherche Scientifique). Jennifer Pike would like to acknowledge the Natural Environment Research Council and the UK Ocean Drilling Program for funding various aspects of the research presented.

References

Abrantes, F., Lopes, C., Mix, A., & Pisias, N. (2007) Diatoms in southeast Pacific surface sediments reflect environmental properties. *Quaternary Science Reviews*, **26**, 155–69.

Andersen, A., Koç, N., Jennings, A., & Andrews, T. (2004a). Nonuniform response of the major surface currents of the Nordic Seas to insolation forcing: implications for the Holocene climate variability. *Paleoceanography*, **19**, PA2003, DOI: 10.1029/2002PA000873.

Andersen, C., Koç, N., & Moros, M. (2004b). A highly unstable Holocene climate in the subpolar North Atlantic; evidence from diatoms. *Quaternary Science Reviews*, **23**, 2155–66.

Anderson, R. Y. (1986). The varve microcosm: propagator of cyclic bedding. *Paleoceanography*, **1**, 373–82.

Armand, L. K. (1997). The use of diatom transfer functions in estimating sea-surface temperature and sea ice in cores from the southeast Indian Ocean. Unpublished Ph.D. thesis, Australian National University, Canberra.

Armand, L. K. (2000). An ocean of ice – advances in the estimation of past sea ice in the Southern Ocean. *GSA Today*, **10**, 1–7.

Armand, L. K., Crosta, X., Romero, O., & Pichon, J.-J. (2005). The biogeography of major diatom taxa in Southern Ocean sediments. 1. Ice-related species. *Palaeogeography, Palaeoclimatology, Palaeoecology*, **223**, 93–126.

Armand, L. K. & Leventer, A. (2003). Palaeo sea ice distribution – reconstruction and palaeoclimatic significance. In *Sea Ice: Physics, Chemistry and Biology*, ed. D. Thomas & G. Dieckmann, Oxford: Blackwell Science Ltd, pp. 333–72.

Armand, L. K. & Leventer, A. (2010). Palaeo sea ice distribution and reconstruction derived from the geological record. In *Sea Ice: an Introduction to its Physics, Biology, Chemistry, and Geology*, ed. D. Thomas & G. Dieckmann, Oxford: Blackwell Science Ltd, pp. 469–530

Bak, Y.-S., Yoo, K.-C., Yoon, H.I., Lee, J.-D., & Yun, H. (2007). Diatom evidence for Holocene paleoclimatic change in the South Scotia Sea, West Antarctica. *Geosciences Journal*, **11**, 11–22.

Bárcena, M. A., Fabrés, J., Isla, E., *et al.* (2006). Holocene neoglacial events in the Bransfield Strait (Antarctica). Palaeoceanographic and palaeoclimatic significance. *Scientia Marina*, **70**, 607–19.

Bárcena, M. A., Gersonde, R., Ledsma, S., *et al.* (1998). Record of Holocene glacial oscillations in Bransfield Basin as revealed by siliceous microfossil assemblages. *Antarctic Science*, **10**, 269–85.

Bárcena, M. A., Isla, E., Plaza, A., *et al.* (2002). Bioaccumulation record and paleoclimatic significance in the western Bransfield Strait. The last 2000 years. *Deep Sea Research II*, **49**, 935–50.

Bard, E., Raisbeck, G., Yiou, F., & Jouzel, J. (2000). Solar irradiance during the last 1200 years based on cosmogenic nuclides. *Tellus*, **52B**, 985–92.

Barron, J.A. & Bukry, D. (2007). Solar forcing of Gulf of California climate during the past 2000 yr suggested by diatoms and silicoflagellates. *Marine Micropaleontology*, **62**, 115–39.

Barron, J. A., Bukry, D., & Bischoff, J. L. (2004). High resolution paleoceanography of the Guaymas Basin, Gulf of California, during the past 15 000 years. *Marine Micropaleontology*, **50**, 185–207.

Barron, J.A., Bukry, D., & Dean, W.E. (2005). Paleoceanographic history of the Guaymas Basin, Gulf of California, during the past 15 000 years, based on diatoms, silicoflagellates, and biogenic sediments. *Marine Micropaleontology*, **56**, 81–102.

Bauch, H. A. & Polyakova, Y. I. (2000). Late Holocene variations in Arctic shelf hydrology and sea ice regime: evidence from north of the Lena Delta. *International Journal of Earth Sciences*, **89**, 569–77.

Baumgartner, T. R. & Christensen, N. (1985). Coupling of the Gulf of California to large-scale interannual variability. *Journal of Marine Research*, **43**, 825–48.

Baumgartner, T. R., Ferreira-Bartrina, V., & Moreno-Hentz, P. (1991). Varve formation in the central Gulf of California: a reconsideration of the origin of the dark laminae from the 20th century varve record. In *The Gulf and Peninsular Province of the Californias*, ed. J. P. Dauphin & B. R. T. Simoneit, Tulsa, OK: American Association of Petroleum Geologists, pp. 617–35.

Baumgartner, T., Ferreira, V., Cayan, D., & Soutar, A. (1994). Interdecadal variability of sardine and anchovy populations in the California Current. *EOS, Transactions, American Geophysical Union*, **75**, 34.

Bernárdez, P., González-Álvarez, R., Francés, G., *et al.* (2007). Late Holocene history of the rainfall in the NW Iberian Peninsula – evidence from the marine record. *Journal of Marine Systems*, **72**, 366–82.

Bernárdez, P., González-Álvarez, R., Francés, G., *et al.* (2008). Palaeoproductivity changes and upwelling variability in the Galicia Mud Patch over the last 5000 years: geochemical and micropaleontologic evidence. *The Holocene*, **18**, 1207–18.

Bianchi, C. & Gersonde, R. (2004). Climate evolution at the last deglaciation: the role of the Southern Ocean. *Earth and Planetary Science Letters*, **228**, 407–24.

Birks, C. J. A. & Koç N. (2002). A high-resolution diatom record of late-Quaternary sea-surface temperatures and oceanographic conditions from the eastern Norwegian Sea. *Boreas*, **31**, 323–44.

Bodén, P. & Backman, J. (1996). A laminated sediment sequence from northern North Atlantic Ocean and its climatic record. *Geology*, **24**, 507–10.

Brodie, I. & Kemp, A. E. S. (1994). Variation in biogenic and detrital fluxes and formation of laminae in late Quaternary sediments from the Peruvian coastal upwelling zone. *Marine Geology*, **116**, 385–98.

Bull, D. & Kemp, A. E. S. (1995). Composition and origins of laminae in late Quaternary and Holocene sediments from the Santa Barbara Basin. In *Proceedings of the Ocean Drilling Program, Scientific Results*, **146 (2)**, 77–87.

Bull, D., Kemp, A. E. S., & Weedon, G.P. (2000). A 160-k.y.-old-record of El Niño southern Oscillation in marine production and coastal run-off from Santa Barbara Basin, California, USA. *Geology*, **28**, 1007–10.

Burckle, L. H. (1984). Diatom distribution and palaeoceanographic reconstruction in the Southern Ocean – present and last glacial maximum. *Marine Micropaleontology*, **9**, 241–261.

Burckle, L. H. & Mortlock, R. (1998). Sea ice extent in the Southern Ocean during the last glacial maximum: another approach to the problem. *Annals of Glaciology*, **27**, 302–4.

Burckle, L. H., Robinson, D., & Cooke, D. (1982). Reappraisal of sea ice distribution in the Atlantic and Pacific sectors of the Southern Ocean at 18,000 yr BP. *Nature*, **299**, 435–7.

Burke, I. T., Grigorov, I., & Kemp, A. E. S. (2002). Microfabric study of diatomaceous and lithogenic deposition in laminated sediments from the Gotland Deep, Baltic Sea. *Marine Geology*, **183**, 89–105.

Caissie, B. E., Brigham-Grette, J., Lawrence, K. T., & Cook, M. S. (2006). Rising temperatures, shrinking ice: the deglaciation in the Bering Sea based on diatoms, alkenones, and oxygen isotopes. 36th Annual Arctic Workshop, University of Colorado, Boulder, CO, March 16–18.

Calvert, S. E. (1964). Factors affecting distribution of laminated diatomaceous sediments in the Gulf of California. In *Marine Geology of the Gulf of California*, ed. T. H. van Andel & G. G. Shor, Tulsa, OK: American Association of Petroleum Geologists, pp. 311–30.

Calvert, S. E. (1966). Origin of diatom-rich varved sediments from the Gulf of California. *Journal of Geology*, **74**, 546–65.

Chang, A. S. & Patterson, R. T. (2005). Climate shift at 4400 years BP: evidence from high-resolution diatom stratigraphy, Effingham Inlet, British Columbia, Canada. *Palaeogeography, Palaeoclimatology, Palaeoecology*, **226**, 72–92.

Chang, A. S., Patterson, R.T., & McNeely, R. (2003). Seasonal Sediment and Diatom Record from Late Holocene Laminated Sediments, Effingham Inlet, British Columbia, Canada. *Palaios*, **18**, 477–94.

Cremer, H., Gore, D., Kirkup, H., *et al.* (2001). The late Quaternary environmental history of the Windmill Islands, east Antarctica – initial evidence from the diatom record. In *Proceedings of the 16th International Diatom Symposium, Athens 2000*, ed. A. Economou-Amilli, Athens: University of Athens, pp. 471–81.

Cremer, H., Roberts, D., McMinn, A., Gore, D., & Melles, M. (2003). The Holocene diatom flora of marine bays in the Windmill Islands, East Antarctica. *Botanica Marina*, **46**, 82–106.

Crosta, X., Debret, M., Denis, D., Courty, M. A., & Ther, O. (2007). Holocene long- and short-term climate changes off Adélie Land,

East Antarctica. *Geochemistry, Geophysics, Geosystems*, **8**, Q11009, DOI:10.1029/2007GC001718.

Crosta, X., Denis, D., & Ther, O. (2008). Sea ice seasonality during the Holocene, Adélie Land, East Antarctica. *Marine Micropaleontology*, **66**, 222–32.

Crosta, X. & Koç, N. (2007). Diatoms: from micropaleontology to isotope geochemistry. In *Proxies in Late Cenozoic Paleoceanography*, ed. C. Hillaire-Marcel & A. de Vernal, Developments in Marine Geology Series, volume 1, Amsterdam: Elsevier, pp. 327–69.

Crosta, X., Pichon, J.-J., & Burckle, L. H. (1998a). Application of modern analog technique to marine Antarctic diatoms: reconstruction of maximum sea ice extent at the last glacial maximum. *Paleoceanography*, **13**, 286–97.

Crosta, X., Pichon, J.-J., & Burckle, L.H. (1998b). Reappraisal of Antarctic seasonal sea ice at the last glacial maximum. *Geophysical Research Letters*, **25**, 2703–6.

Crosta, X., Pichon, J.-J., & Labracherie, M. (1997). Distribution of *Chaetoceros* resting spores in modern peri-Antarctic sediments. *Marine Micropaleontology*, **29**, 283–99.

Crosta, X., Romero, O., Armand, L.K., & Pichon, J.-J. (2005). The biogeography of major diatom taxa in Southern Ocean sediments. 2. Open-ocean related species. *Palaeogeography, Palaeoclimatology, Palaeoecology*, **223**, 66–92.

Cunningham, W. L., Leventer, A., Andrews, J. T., Jennings, A.E., & Litch, K. J. (1999). Late Pleistocene–Holocene marine conditions in the Ross Sea, Antarctica: evidence from the diatom record. *The Holocene*, **9**, 129–39.

Curran, M. A. J., van Ommen, T. D., Morgan, V. I., Phillips, K. L., & Palmer, A. S. (2003). Ice core evidence for Antarctic sea ice decline since the 1950s. *Science*, **302**, 1203–6.

Damon E. & Laut P. (2004). Pattern of strange errors plagues solar activity and terrestrial climate data. *Eos, Transactions, American Geophysical Union*, **39**, 370–4.

De Sève, M. A. (1999). Transfer function between surface diatom assemblages and sea-surface temperature and salinity of the Labrador Sea. *Marine Micropaleontology*, **36**, 249–67.

De Vries, T. J. & Schrader, H. (1981). Variation of upwelling/oceanic conditions during the latest Pleistocene through Holocene off the central Peruvian coast: a diatom record. *Marine Micropaleontology*, **6**, 157–67.

Dean, J. M. & Kemp, A. E. S. (2004). A 2100 year BP record of the Pacific decadal Oscillation, El Niño Southern Oscillation and Quasi-biennial Oscillation in marine production and fluvial input from Saanich Inlet, British Columbia. *Palaeogeography, Palaeoclimatology, Palaeoecology*, **213**, 207–29.

Dean, J. M., Kemp, A. E. S., Bull, D., *et al.* (1999). Taking varves to bits: scanning electron microscopy in the study of laminated sediments and varves. *Journal of Paleolimnology*, **22**, 121–36.

Dean, J. M., Kemp, A. E. S., & Pearce, R.B. (2001). Palaeo-flux records from electron microscope studies of Holocene laminated sediments, Saanich Inlet, British Columbia. *Marine Geology*, **174**, 139–58.

Denis, D., Crosta, X., Schmidt, S., *et al.* (2009a). Holocene glacier and deep water dynamics, Adélie Land region, East Antarctica. *Quaternary Science Reviews*, **28**, 1291–303.

Denis, D., Crosta, X., Schmidt, S., *et al.* (2009b). Holocene productivity changes off Adélie Land (East Antarctica). *Paleoceanography*, **24**, PA3207, DOI:10.1029/2008PA001689.

Denis, D., Crosta, X., Zaragosi, S., *et al.* (2006). Seasonal and subseasonal climate changes recorded in laminated diatom ooze sediments, Adélie Land, East Antarctica. *The Holocene*, **16**, 1137–47.

Domack, E., Duran, D., Leventer, A., *et al.* (2005). Stability of the Larsen B ice shelf on the Antarctic Peninsula during the Holocene epoch. *Nature*, **436**, 681–5.

Domack, E., Leventer, A., Dunbar, R., *et al.* (2001). Chronology of the Palmer Deep site, Antarctic Peninsula: a Holocene palaeoenvironmental reference for the circum-Antarctic. *The Holocene*, **11** (1), 1–9.

Domack, E., Leventer, A., Root, S., *et al.* (2003). Marine sediment record of natural environmental variability and recent warming in the Antarctic Peninsula. *AGU Antarctic Research Series*, **79**, 205–44.

Donegan, D. & Schrader, H. (1982). Biogenic and abiogenic components of laminated hemipelagic sediments in the central Gulf of California. *Marine Geology*, **48**, 215–37.

Eiríksson, J., Bartels-Jønsdøttir, H. B., Cage, A. G., *et al.* (2006). Variability of the North Atlantic Current during the last 2000 years based on shelf bottom water and sea surface temperatures along an open ocean/shallow marine transect in western Europe. *The Holocene*, **16**, 1017–29.

Enomoto H. & Ohmura A. (1990). The influences of atmospheric half-yearly cycle on the sea ice extent in the Antarctic. *Journal of Geophysical Research*, **95(C6)**, 9497–511.

Finocchiaro, F., Langone, L., Colizza, E., *et al.* (2005). Record of the early Holocene warming in a laminated sediment core from Cape Hallett Bay (Northern Victoria Land, Antarctica). *Global and Planetary Change*, **45**, 193–206.

Friis-Christensen, E. (2000). Solar variability and climate: a summary. *Space Science Reviews*, **94**, 411–21.

Fryxell, G. A. & Prasad, A. K. S. K. (1990). *Eucampia antarctica* var. *recta* (Mangin) stat. nov. (Biddulphiaceae, Bacillariophyceae): life stages at the Weddell Sea ice edge. *Phycologia*, **29**, 27–38.

Gardner, J. V. & Hemphill-Haley, E. (1986). Evidence for a stronger oxygen-minimum zone off central California during late Pleistocene to early Holocene. *Geology*, **14**, 691–4.

Gersonde, R. & Zielinski, U. (2000). The reconstruction of late Quaternary Antarctic sea-ice distribution – the use of diatoms as a proxy for sea-ice. *Palaeogeography, Palaeoclimatology, Palaeoecology*, **162**, 263–86.

Gil, I. M., Abrantes, F., & Hebbeln, D. (2006). The North Atlantic Oscillation forcing through the last 2000 years: spatial variability as revealed by high resolution marine diatom records from N and SW Europe. *Marine Micropaleontology*, **60**, 113–29.

Gran, H. H. (1904). Diatomaceae from the ice-floes and plankton of the Arctic Ocean. In *The Norwegian North Polar Expedition 1893–1896*, ed. F. Nansen, London: Longman, vol. 4, pp. 1–74.

Grigorov, I., Pearce, R. B., & Kemp, A. E. S. (2002). Southern Ocean laminated diatom ooze: mat deposits and potential for palaeoflux studies, ODP leg 177, Site 1093. *Deep-Sea Research II*, **49**, 3391–407.

Grimm, K. A. (1992). High-resolution imaging of laminated biosiliceous sediments and their paleoceanographic significance (Quaternary, site 798, Oki Ridge, Japan Sea). *Proceedings of the Ocean Drilling Program, Scientific Results*, **127–128**, 547–57.

Grimm, K. A., Lange, C.B., & Gill, A. S. (1996). Biological forcing of hemipelagic sedimentary laminae: evidence from ODP Site 893, Santa Barbara Basin, California. *Journal of Sedimentary Research*, **66**, 613–624.

Grimm, K. A., Lange, C. M., & Gill, A. S. (1997). Self-sedimentation of phytoplankton blooms in the geologic record. *Sedimentary Geology*, **110**, 151–61.

Haig, J. D. (1999). Modelling the impact of solar variability on climate. *Journal of Atmospheric and Solar-Terrestrial Physics*, **61**, 63–72.

Hay, M. B., Dallimore, A., Thomson, R. E., Calvert, S. E., & Pienitz, R. (2007). Siliceous microfossil record of late Holocene oceanography and climate along the west coast of Vancouver Island, British Columbia (Canada). *Quaternary Research*, **67**, 33–49.

Heroy, D. C., Sjunneskog, C., & Anderson, J. B. (2007). Holocene climate change in the Bransfield Basin, Antarctic Peninsula: evidence from sediment and diatom analysis. *Antarctic Science*, **20**, 69–87.

Hodell D. A., Kanfoush S. L., Shemesh A., *et al.* (2001). Abrupt cooling of Antarctic surface waters and sea ice expansion in the South Atlantic sector of the Southern Ocean at 5000 cal yr BP. *Quaternary Research*, **56**, 191–8.

Horner, R. A. & Alexander, V. (1972). Algal populations in Arctic sea ice: an investigation of heterotrophy. *Limnology and Oceanography*, **17**, 454–8.

Hughen, K., Overpeck, J. T., Peterson, L. C., & Anderson, R. (1996). Varve analysis and palaeoclimate from sediments of the Cariaco Basin, Venezuela. In *Palaeoclimatology and Palaeoceanography from Laminated Sediments*, ed. A. E. S. Kemp, Bath: Geological Society of London, pp. 171–83.

Jensen, K.G., Kuijpers, A., Koc, N., & Heinemeier, J. (2004). Diatom evidence of hydrographic changes and ice conditions in Igaliku Fjord. *The Holocene*, **14**, 152–64.

Jiang, H., Eiriksson, J., Schultz, M., Knudsen, K. L. & Seidenkrantz, M. S. (2005). Evidence for solar forcing of sea surface temperature on the North Icelandic Shelf during the late Holocene. *Geology*, **33**, 73–6.

Jiang, H., Seidenkrantz, M.-S., & Knudsen, K. L. (2002). Late-Holocene summer sea-surface temperatures based on a diatom record from the North Icelandic Shelf. *The Holocene*, **12**, 137–47.

Justwan, A. & Koç, N. (2008). A diatom based transfer function for reconstructing sea ice concentrations in the North Atlantic. *Marine Micropaleontology*, **66**, 264–78.

Justwan, A., Koç, N., & Jennings, A. E. (2008). Evolution of the Irminger and East Icelandic Current systems through the Holocene, revealed by diatom-based sea surface temperature reconstructions. *Quaternary Science Reviews*, **27**, 1571–82.

Kaczmarska, I., Barbrick, N. E., Ehrman, J. M., & Cant, G. P. (1993). *Eucampia* index as an indicator of the Late Pleistocene oscillations of the winter sea ice extent at the ODP Leg 119 Site 745B at the Kerguelen Plateau. *Hydrobiologia*, **269–270**, 103–12.

Kemp, A. E. S. (1990). Sedimentary fabrics and variation in lamination style in Peru continental margin upwelling sediments. *Proceedings of the Ocean Drilling Program, Scientific Results*, **112**, 43–58.

Kemp, A. E. S. (2003). Evidence for abrupt climate changes in annually laminated marine sediments. *Philosophical Transactions of the Royal Society, London*, **A 361**, 1851–70.

Kemp, A. E. S. & Baldauf, J. G. (1993). Vast Neogene laminated diatom mat deposits from the eastern equatorial Pacific Ocean. *Nature*, **362**, 141–4.

Kemp, A. E. S., Pike, J., Pearce, R. B., & Lange, C.B. (2000). The "fall dump" – a new perspective on the role of a "shade flora" in the annual cycle of diatom production and export flux. *Deep-Sea Research II*, **47**, 2129–54.

Knudsen, K. L., Jiang, H., Jansen, E., *et al.* (2004). Environmental changes off north Iceland during the deglaciation and the Holocene: foraminfera, diatoms and stable isotopes. *Marine Micropaleontology*, **50**, 273–305.

Koç, N. & Jansen, E. (1994). Response of the high latitude Northern Hemisphere to orbital climate forcing: evidence from the GIN-Seas. *Geology*, **22**, 523–526.

Koç, N., Jansen, E., & Haflidason, H. (1993). Paleoceanographic reconstructions of surface ocean conditions in the Greenland, Iceland and Norwegian seas through the last 14 ka based on diatoms. *Quaternary Science Reviews*, **12**, 115–40.

Koç, N., Jansen, E., Hald, M., & Labeyrie, L. (1996). Late glacial-Holocene sea-surface temperatures and gradients between the North Atlantic and the Norwegian Sea: implications for the Nordic heat pump. *Geological Society Special Publication*, **111**, 177–85.

Koç Karpuz, N. & Jansen, E. (1992). A high resolution diatom record of the last deglaciation from the SE Norwegian Sea: documentation of rapid climatic changes. *Paleoceanography*, **7**, 499–520.

Koç Karpuz, N. & Schrader, H. (1990). Surface sediment diatom distribution and Holocene paleotemperature variations in the Greenland, Iceland and Norwegian Sea. *Paleoceanography*, **5**, 557–80.

Kohly, A. (1998). Diatom flux and species composition in the Greenland Sea and the Norwegian Sea in 1991–1992. *Marine Geology*, **145**, 293–312.

Koizumi, I. (2008). Diatom-derived SSTs (*Td'* ratio) indicate warm seas off Japan during the middle Holocene (8.2–3.3 kyr BP). *Marine Micropaleontology*, **69**, 263–81.

Koizumi, I., Shiga, K., Irino, T., & Ikehara, M. (2003). Diatom record of the late Holocene in the Okhotsk Sea. *Marine Micropaleontology*, **49**, 139–56.

Lange, C. B., Berger, W. H., Burke, S. K., *et al.* (1987). El Niño in Santa Barbara Basin: diatom, radiolarian and foraminiferan responses to the "1983 El Niño" event. *Marine Geology*, **78**, 153–60.

Lange, C. B., Burke, S.K., & Berger, W.H. (1990). Biological production off Southern California is linked to climatic change. *Climatic Change*, **16**, 319–29.

Lange, C. B. & Schimmelmann, A. (1994). Seasonal resolution of laminated sediments in Santa Barbara Basin: its significance in paleoclimatic studies. In *Proceedings of the Tenth Annual Pacific Climate (PACLIM) Workshop, April 4–7, 1993*, ed. K. T. Redmond & V. L. Tharp. California Department of Water Resources, Interagency Ecological Studies Program, Technical Report 36, pp. 83–92.

Lean, J. & Rind, D. (1998). Climate forcing by changing solar radiation. *Journal of Climate*, **11**, 3069–94.

Lean, J. & Rind, D. (1999). Evaluating Sun–climate relationships since the Little Ice Age. *Journal of Atmospheric and Solar–Terrestrial Physics*, **61**, 1591–4.

Lebreiro, S. M., Francés, G., Abrantes, F., *et al.* (2006). Climate change and coastal hydrographic response along the Iberian margin (Tagus Prodelta and Muros Ría) during the last two millenia. *The Holocene*, **16**, 1003–15.

Lemke, P., Ren, J., Alley, R. B., *et al.* (2007). Observations: changes in snow, ice and frozen ground. In *Climate Change 2007: The Physical Science Basis. Contributions of Working Group 1 to the Fourth Assessment Report of the Intergovernmental Panel on Climate Change*, ed. S. Solomon, D. Qin, M. Manning, *et al.*, Cambridge: Cambridge University Press, pp. 337–83.

Leventer, A. (1998). The fate of sea ice diatoms and their use as paleoenvironmental indicators. In *Antarctic Sea Ice: Biological Processes*, ed. M. P. Lizotte & K. R. Arrigo, American Geophysical Union Antarctic Research Series 73, pp. 121–37.

Leventer, A., Domack, E., Barkoukis, A., McAndrews, B., & Murray, J. (2002). Laminations from the Palmer Deep: a diatom-based interpretation. *Paleoceanography*, **17**, 1–15.

Leventer, A., Domack, E., Dunbar, R., *et al.* (2006). Marine sediment record from the East Antarctic margin reveals dynamics of ice sheet recession. *GSA Today*, **16**, DOI:10.1130/GSAT01612A.1.

Leventer, A., Domack, E. W., Ishman, S. E., *et al.* (1996). Productivity cycles of 200–300 years in the Antarctic Peninsula region: understanding linkages among the sun, atmosphere, oceans, sea ice, and biota. *Geological Society of America, Bulletin*, **108**, 1626–44.

Leventer, A. & Dunbar, R. B. (1988). Recent diatom record of McMurdo Sound, Antartcica: implications for the history of sea ice extent. *Paleoceanography*, **3**, 259–74.

Leventer, A., Dunbar, R., & DeMaster, D. J. (1993). Diatom evidence for late Holocene climatic events in Granite Harbor, Antarctica. *Paleoceanography*, **8**, 373–86.

Machado, E. (1993). Production, sedimentation and dissolution of biogenic silica in the northern North Atlantic. Ph. D. thesis, Christian Albrechts University, Kiel, Germany.

Maddison, E. J., Pike, J., Leventer, A., & Domack, E. W. (2005). Deglacial seasonal and sub-seasonal diatom record from Palmer Deep, Antarctica. *Journal of Quaternary Science*, **20**, 435–46.

Maddison, E. J., Pike, J., Leventer, A., *et al.* (2006). Post-glacial seasonal diatom record of the Mertz Glacier Polynya, East Antarctica. *Marine Micropaleontology*, **60**, 66–88.

Marsh, N. D. & Svensmark, H. (2000). Low cloud properties influenced by cosmic rays. *Physical Review Letters*, **85**, 5004–7.

Maslin, M. A., Stickley, C. E., & Ettwein, V. J. (2001). Palaeoceanography: Holocene climate variability. In *Encyclopedia of Ocean Sciences*, ed. J. Steele, S. Thorpe, & K. Turekian, London: Academic Press.

Masson, V. Vimeux, F., Jouzel, J., *et al.* (2000). Holocene climate variability in Antarctica based on 11 ice-core records. *Quaternary Research*, **54**, 348–58.

Mayewski, P. A., Rohling, E. E., Stager, J. C., *et al.* (2004). Holocene climate variability. *Quaternary Research*, **62**, 243–55.

McMinn, A. (2000). Late Holocene increase in sea ice extent in fjords of the Vestfold Hills, eastern Antarctica. *Antarctic Science*, **12**, 80–8.

McMullen, K., Domack, E., Leventer, A., *et al.* (2006). Glacial morphology and sediment formation in the Mertz Trough, East Antarctica. *Palaeogeography, Palaeoclimatology, Palaeoecology*, **231**, 169–80.

Møller, H. S., Jensen, K. G., Kuijpers, A., *et al.* (2006). Late-Holocene environment and climatic changes in Ameralik Fjord, southwest Greenland: evidence from the sedimentary record. *The Holocene*, **16**, 685–95.

Moros, M., Andrews, J. T., Eberl, D. E., & Jansen, E. (2006). Holocene history of drift ice in the northern North Atlantic: evidence for different spatial and temporal modes. *Paleoceanography*, 21, DOI: 10.1029/2005PA001214.

Nielsen, S. H. H., Koç, N., & Crosta, X. (2004). Holocene climate in the Atlantic sector of the Southern Ocean: controlled by insolation or oceanic circulation? *Geology*, **32**, 317–20.

Patterson, R. T., Prokoph, A., Reinhardt, E., & Roe, H. M. (2007). Climate cyclicity in late Holocene anoxic marine sediments from the Seymour-Belize Inlet Complex, British Columbia. *Marine Geology*, **242**, 123–40.

Pearce, R. B., Kemp, A. E. S., Koizumi, I., *et al.* (1998). A lamina-scale SEM-based study of a late Quaternary diatom-ooze sapropel from the Mediterranean Ridge, ODP Site 971. *Proceedings of the Ocean Drilling Program, Scientific Results*, **160**, 333–48.

Pike, J., Crosta, X., Maddison, E. J., *et al.* (2009). Observations on the relationship between the Antarctic coastal diatoms *Thalassiosira antarctica* Comber and *Porosira glacialis* (Grunow) Jørgensen and sea ice concentrations during the late Quaternary. *Marine Micropaleontology*, **73**, 14–25.

Pike, J. & Kemp, A. E. S. (1996). Preparation and analysis techniques for studies of laminated sediments. *Geological Society of London, Special Publication*, **116**, 37–48.

Pike, J. & Kemp, A. E. S. (1997). Early Holocene decadal-scale ocean variability recorded in Gulf of California laminated sediments. *Paleoceanography*, **12**, 227–38.

Pike, J. & Kemp, A. E. S. (1999). Diatom mats from Gulf of California: Implications for silica burial and paleoenvironmental interpretation of laminated sediments. *Geology*, **27**, 311–14.

Pike, J. & Stickley, C. E. (2007). Diatom records: marine laminated sequences. In *Encyclopedia of Quaternary Science*, ed. S. Elias, Amsterdam: Elsevier, pp. 557–67.

Polyakova, Y. I. & Stein, R. (2004). Holocene paleoenvironmental implications of diatom and organic carbon records from the southeastern Kara Sea (Siberian margin). *Quaternary Research*, **62**, 256–66.

Ran, L., Jiang, H., Knudsen, K. L., Eiríksson, J., & Gu, Z. (2006). Diatom response to the Holocene climatic optimum on the North Icelandic Shelf. *Marine Micropaleontology*, **60**, 226–41.

Rathburn, A. E., Pichon, J.-J., Ayress, M. A., & DeDeckker, P. (1997). Microfossil and stable-isotope evidence for changes in late Holocene palaeoproductivity and palaeoceanographic conditions in the Prydz Bay region of Antarctica. *Palaeogeography, Palaeoclimatology, Palaeoecology*, **131**, 485–510.

Rebolledo, L., Sepúlveda, J., Lange, C. B., *et al.* (2008). Late Holocene marine productivity changes in northern Patagonia-Chile inferred from a multi-proxy analysis of Jacaf channel sediments. *Estuarine, Coastal and Shelf Science*, **80**, 314–22.

Reid, G. C. (2000). Solar variability and the Earth's climate: introduction and review. *Space Science Reviews*, **94**, 1–11.

Reimers, C. E., Lange, C. B., Tabak, M., & Bernhard, J. M. (1990). Seasonal spillover and varve formation in the Santa Barbara Basin, California. *Limnology and Oceanography*, **35**, 1577–85.

Ren, J., Jiang, H., Seidnekrantz, M.-S., & Kuijpers, A. (2009). A diatom-based reconstruction of early Holocene hydrographic and climatic change in a southwest Greenland fjord. *Marine Micropaleontology*, **70**, 166–76.

Renssen, H., Goosse H., & Muscheler, R. (2006). Coupled climate model simulation of Holocene cooling events: oceanic feedback amplifies solar forcing. *Climate of the Past*, **2**, 79–90.

Romero, O. E., Kim, J.-H., & Hebbeln, D. (2006). Paleoproductivity evolution off central Chile from the last glacial maximum to the early Holocene. *Quaternary Research*, **65**, 519–25.

Ryu, E., Yi, S., & Lee, S.-J. (2005). Late Pleistocene–Holocene paleoenvironmental changes inferred from the diatom record of the Ulleung Basin, East Sea (Sea of Japan), *Marine Micropaleontology*, **55**, 157–82.

Sancetta, C. (1979). Oceanography of the North Pacific during the last 18,000 years: evidence from fossil diatoms. *Marine Micropaleontology*, **4**, 103–23.

Sancetta, C. (1981). Oceanographic and ecologic significance of diatoms in surface sediments of the Bering and Okhotsk seas. *Deep-Sea Research*, **28**, 789–817.

Sancetta, C. (1982). Distribution of diatom species in surface sediments of the Bering and Okhotsk seas. *Marine Micropaleontology*, **28**, 221–57.

Sancetta, C. (1995). Diatoms in the Gulf of California: seasonal flux patterns and the sediment record for the past 15,000 years. *Paleoceanography*, **10**, 67–84.

Schimmelmann, A. & Lange, C. B. (1996). Tales of 1001 varves: a review of Santa Barbara Basin sediment studies. *Geological Society of London, Special Publication*, **116**, 121–41.

Schrader, H. & Baumgartner, T. (1983). Decadal variation in upwelling in the central Gulf of California. In *Coastal Upwelling, its Sediment Record. Part B: Sedimentary Records of Ancient Coastal Upwelling*, ed. J. Thiede & E. Suess, New York, NY: Plenum Press, pp. 247–76.

Schrader, H., Isrenn, K., Swanberg, N., Paetzel, M., & Sæthre, T. (1993). Early Holocene diatom pulse in the Norwegian Sea and its paleoceanographic significance. *Diatom Research*, **8**, 117–30.

Seeberg-Elverfeldt, I. A., Lange, C. B., Arz, H. W., Pätzold, J., & Pike, J. (2004). The significance of diatoms in the formation of laminated sediments of the Shaban Deep, northern Red Sea. *Marine Geology*, **209**, 279–301.

Seidenkrantz, M.-S., Aagaard-Sørensen, S., Sulsbrück, H., *et al.* (2007). Hydrography and climate of the last 4400 years in a SW Greenland fjord: implications for Labrador Sea palaeoceanography. *The Holocene*, **17**, 387–401.

Shevenell, A. E., Domack, E. W., & Kernan, G. M. (1996). Record of Holocene palaeoclimate change along the Antarctic Peninsula: evidence from glacial marine sediments, Lallemand Fjord. *Papers and Proceedings of the Royal Society of Tasmania*, **130**, 55–64.

Shiga, K. & Koizumi, I. (2000). Latest Quaternary oceanographic changes in the Okhotsk Sea based on diatom records. *Marine Micropaleontology*, **38**, 91–117.

Shimada, C., Ikehara, K., Tanimura, Y., & Hasegawa, S. (2004). Millennial-scale variability of Holocene hydrography in the southwestern Okhotsk Sea: diatom evidence. *The Holocene*, **14**, 641–50.

Shimada, C., Murayama, M., Aoki, K., *et al.* (2000). Holocene paleoceanography in the SW part of the Sea of Okhotsk: a diatom record. *Quaternary Research*, **39**, 451–60.

Shindell, D., Rind, D., Balachandran, N., Lean, J., & Lonergan, P. (1999). Solar cycle variability, ozone, and climate. *Nature*, **284**, 305–8.

Simmonds, I. (1996). Climatic role of southern hemisphere extratropical cyclones and their relationships with sea-ice. *Papers and Proceedings of the Royal Society of Tasmania*, **130**, 95–100.

Sjunneskog, C. & Taylor, F. (2002). Postglacial marine diatom record of the Palmer Deep, Antarctic Peninsula (ODP Leg 178, Site 1098). I: total diatom abundance. *Paleoceanography*, **17**, DOI: 10.1029/2000PA000563.

Stickley, C. E., Pike, J., Leventer, A., *et al.* (2005). Deglacial ocean and climate seasonality in laminated diatom sediments, Mac.Robertson Shelf, Antarctica. *Palaeogeography, Palaeoclimatology, Palaeoecology*, **227**, 290–310.

Stickley, C. E., Pike, J., & Leventer, A. (2006). Productivity events of the marine diatom *Thalassiosira tumida* (Janisch) Hasle recorded in deglacial varves from the East Antarctic margin. *Marine Micropaleontology*, **59**, 184–96.

Stroeve, J., Holland, M. M., Meier, W., Scambos, T., & Serreze. M. (2007). Arctic sea ice decline: faster than forecast. *Geophysical Research Letters*. **34**, L09501, DOI: 10.1029/2007GL029703.

Stuiver, M. & Braziunas, T. F. (1989). Atmospheric ^{14}C and century scale solar oscillations, *Nature*, **338**, 405–8.

Stuiver, M. & Braziunas, T. F. (1993). Modeling atmospheric ^{14}C influences and ^{14}C ages of marine samples to 10,000 BC. *Radiocarbon*, **35**, 137–89.

Stuiver M., Reimer P. J., & Braziunas T. F. (1998). High precision radiocarbon age calibration for terrestrial and marine samples. *Radiocarbon*, **40**, 1127–51.

Suess, H. E. & Linick, T. W. (1990). The ^{14}C record in bristlecone pine wood of the past 8000 yr based on the dendrochronology of the late C. W. Ferguson. *Royal Society of London Philosophical Transactions, ser. A*, **330**, 403–12.

Svensmark, H. (2000). Cosmic rays and Earth's climate. *Space Science Reviews*, **93**, 175–95.

Svensmark, H. & Friis-Christensen, E. (1997). Variation of cosmic ray flux and global cloud coverage: a missing link in solar–climate relationships. *Journal of Atmospheric and Solar-Terrestrial Physics*, **59**, 1225–32.

Taylor, F. & McMinn, A. (2001). Evidence from diatoms for Holocene climate fluctuation along the East Antarctic margin. *The Holocene*, **11**(4), 455–66.

Taylor, F. & McMinn, A. (2002). Late Quaternary diatom assemblages from Prydz Bay, eastern Antarctica. *Quaternary Research*, **57**, 151–61.

Taylor, F. & Sjunneskog, C. (2002). Postglacial marine diatom record of the Palmer Deep, Antarctic Peninsula (ODP Leg 178, Site 1098). 2. Diatom assemblages. *Paleoceanography*, **17**, DOI: 10.1029/2000PA000564.

Taylor, F., Whitehead, J., & Domack, E. (2001). Holocene paleoclimate change in the Antarctic Peninsula: evidence from the diatom, sedimentary and geochemical record. *Marine Micropaleontology*, **41**, 25–43.

Thunell, R., Pride, C., Tappa, E., & Muller-Karger, F. (1993). Varve formation in the Gulf of California: insights from time series sediment trap sampling and remote sensing. *Quaternary Science Reviews*, **12**, 451–64.

van Geel, B., Raspopov, O. M., Renssen, H., *et al.* (1999). The role of solar forcing upon climate change. *Quaternary Science Reviews*, **18**, 331–8.

Villareal, T. A. & Fryxell, G. A. (1983). Temperature effects on the valve structure of the bipolar diatoms *Thalassiosira antarctica* and *Porosira glacialis*. *Polar Biology*, **2**, 163–9.

Wanner, H., Beer, J., Bütikofer, J., *et al.* (2008). Mid- to late Holocene climate change: an overview. *Quaternary Science Reviews*, **27**, 1791–828.

Williams, K. M. (1986). Recent Arctic marine diatom assemblages from bottom sediments in Baffin Bay and Davis Strait. *Marine Micropaleontology*, **10**, 327–41.

Williams, K. M. (1990). Late Quaternary palaeoceanography of the western Baffin Bay region: evidence from fossil diatoms. *Canadian Journal of Earth Sciences*, **27**, 1487–94.

Williams, K. M. (1993). Ice sheet and ocean interactions, margin of the East Greenland Ice Sheet (14 ka to present): diatom evidence. *Paleoceanography*, **8**, 69–83.

Witak, M., Wachnicka, A., Kuijpers, A., *et al.* (2005). Holocene North Atlantic surface circulation and climatic variability: evidence from diatom records. *The Holocene*, **15**, 85–96.

Witon, E., Malmgren, B., Witkowski, A., & Kuijpers, A. (2006). Holocene marine diatoms from the Faeroe Islands and their paleoceanographic implications. *Palaeogeography, Palaeoclimatology, Palaeoecology*, **239**, 487–509.

Wyrtki, K. (1962). The oxygen minima in relation to ocean circulation. *Deep-Sea Research*, **9**, 11–23.

Yoon, H. I., Park, B.-K., Kim, Y., & Kim, C. Y. (2002). Glaciomarine sedimentation and its paleoceanographic implications on the Antarctic Peninsula shelf over the last 15 000 years. *Palaeogeography, Palaeoclimatology, Palaeoecology*, **185**, 235–54.

Zielinski, U. & Gersonde, R. (1997). Diatom distribution in Southern Ocean surface sediments (Atlantic sector): implications for palaeoenvironmental reconstructions. *Palaeogeography, Palaeoclimatology, Palaeoecology*, **129**, 213–50.

Zielinski, U., Gersonde, R., Sieger, R., & Futterer, D. (1998). Quaternary surface water temperature estimations: calibration of a diatom transfer function for the Southern Ocean. *Paleoceanography*, **13**, 365–83.

22 Diatoms as indicators of paleoceanographic events

RICHARD W. JORDAN AND
CATHERINE E. STICKLEY

22.1 Introduction – the importance of pre-Quaternary diatoms in paleoceanography

The marine system is vast, involving a highly diverse range of habitats; from coastal lagoons, fjords, bays, and estuaries, out over the shelf to the open ocean thousands of kilometers from any landmass. The Pacific Ocean alone covers an area of nearly 170 million km^2. In total, the world's oceans cover approximately two-thirds of the Earth's surface and since marine diatoms are the dominant marine primary producers, contributing about 40% of the total primary production in the modern oceans (Tréguer *et al.*, 1995) and over 50% of organic carbon burial in marine sediments (Falkowski *et al.*, 2004), they are key players in the marine biological carbon pump. Therefore, the study of both living and fossil marine diatoms is important for many reasons other than just their intrinsic interest – from understanding (past) marine ecological systems, through biogeochemical cycling, to links with carbon dioxide (CO_2) (e.g. Harrison, 2000), and the causes and effects of rapid climate change (e.g. Pollock, 1997).

Diatoms preserved in marine sediments are commonly used to reconstruct paleoenvironments and paleoceanographic events for the Holocene and Quaternary, but they are equally as valuable for paleoceanographic reconstructions of time periods much earlier than this – in fact for as far back as their fossil record allows, i.e. the Early Cretaceous (Gersonde & Harwood, 1990; Harwood & Gersonde, 1990). This means there are approximately 125 million years of geological time for which the diatoms can provide information, yet they are poorly studied compared to other microfossil groups. Furthermore, the use of pre-Holocene marine diatoms has typically been for biostratigraphy; hence much of the published literature is taxonomic in nature. Fossil diatoms are even central to some industrial and commercial applications (Harwood, this volume). However, unlike their Holocene and Quaternary counterparts, much less attention has been given to pre-Quaternary diatoms for paleoceanographic reconstruction.

Yet, over the past couple of decades, this situation has gradually started to improve, largely through the success of ocean drilling by the Ocean Drilling Program (ODP), Integrated Ocean Drilling Program (IODP) and a suite of marginal marine (shelf) drilling programs in Antarctica (see review by Cody *et al.*, 2008), the most recent of which is "Antarctic Geological Drilling" (ANDRILL; Naish *et al.*, 2009). Some of the most groundbreaking research on the pre-Quaternary histories of our oceans have involved, and continue to involve, Neogene (Pliocene, Miocene), Paleogene (Oligocene, Eocene, Paleocene), and even Cretaceous diatoms. In this respect, some of the more pertinent topics include production and flux estimations, diatom-based carbon-export estimations and links to past climates, paleogeographic reconstructions, evolutionary turnover-rate estimations, diversity trends, global siliceous (opal) sedimentation pattern changes and links to the silica cycle and climate change, sea-surface temperature estimates, ancient surface- and deep-water circulation patterns, and sea-ice and ice-sheet histories. In this chapter we outline what is known about diatom evolution and past sedimentation patterns as well as some of these advances in pre-Quaternary diatom research over the past ~25 years. Prior to this we summarize the principles and limitations behind the use of fossil marine diatoms including the use of modern analogs to reconstruct the past environment and preservation conditions.

22.1.1 The significance of the fossil diatom record

An advantage that diatoms (as well as other siliceous microfossil groups) have over calcareous microfossil groups and palynomorphs is that they are affected neither by carbonate

The Diatoms: Applications for the Environmental and Earth Sciences, 2nd Edition, eds. John P. Smol and Eugene F. Stoermer. Published by Cambridge University Press.

dissolution nor oxidation, rendering them important proxies for nearly every type of marine paleoenvironment. Diatoms also live in a wider range of habitats than most other microplankton – they can be planktonic or benthic (and may be attached to a variety of substrata), living in marine, brackish, or freshwater environments – making them especially useful for paleoecological reconstructions. They preserve well in areas of high productivity and are particularly important in regions where other microfossils are not so abundant, e.g. high latitudes. The absence of fossil marine diatoms from a particular sedimentary sequence may largely reflect either low abundance of the ancient living assemblage, or more likely pre- and/or post-sedimentation factors such as winnowing and/or changes to the frustules (discussed in detail in this chapter) rather than their absence from the past biocoenosis (living system) altogether.

Indeed, advances in the relatively new field of biomarker analysis (e.g. Rowland & Robson, 1990; Belt *et al.*, 2000, 2007) indicate that (particular) diatoms were present in ancient seas despite the absence of their corresponding hard-parts (valves) from the fossil assemblage (e.g. Kohnen *et al.*, 1990; Jenkyns *et al.*, 2004; Sinninghe-Damsté *et al.*, 2004; Wagner *et al.*, 2004; Shiine *et al.*, 2008). This indicates that diatoms are likely to be under-represented in the fossil record, prompting several important questions: By what degree are the marine diatoms under-represented throughout their fossil record and what are the consequences for ancient systems, e.g. the biological carbon pump? Is there a diatom signature in systems where the fossil record may suggest their absence, e.g. the Paleocene–Eocene Thermal Maximum (PETM)? How does this compare to other phytoplankton groups, e.g. calcareous nannoplankton, dinoflagellates? Do homogenous (non-laminated) fossil diatom assemblages reflect a single season (e.g. on the western Antarctic Peninsula shelf, Pike *et al.*, 2008) or represent average deposition throughout the year? These sorts of questions start to become more of an issue further back in time. Taking the example of Paleogene and Cretaceous diatom records which are patchy at best – is this patchiness simply a reflection of a relative lack of drilled material, or were diatoms relatively less abundant, and/or less well-preserved back then? The answer to this question may not be solved by merely improving the drilling technology; instead relatively new approaches, such as biomarker analysis and molecular sequencing should be embraced, along with more traditional paleontological techniques, to help address some of these issues. Given their ubiquity in nearly all aquatic systems, there may be a diatom signature in the majority of ancient marine paleoenvironments if we know how to look for them and know how to identify their sources.

Given the incompleteness of the diatom fossil record, and considering there will always be variations and/or improvements in taxonomic concepts and age models, even the way samples are prepared, studies which only use (arguably selective) information stored in data bases, e.g. the *Neptune Database* (Lazarus *et al.*, 1995; Spencer-Cervato, 1999), to promote revolutionary ideas about diatom evolution and/or diversity (e.g. Cervato & Burckle, 2003; Katz *et al.*, 2004, 2005; Rabosky & Sorhannus, 2009) should be considered with some caution (Spencer-Cervato, 1999; Cervato & Burckle, 2003), although the concept behind creating and using such data bases is novel and extremely worthwhile. Continued improvements to all data bases will greatly facilitate the way paleontologists work in the future.

As with all areas of paleontology we must work with what we have to piece together information about the past, and as far as paleoceanography is concerned, for the Early Cretaceous onwards, marine diatoms are excellent paleoceanographic indicators for the variety and quality of information they provide. As primary producers they are diverse and widespread, strongly reflecting their environment and usually the first to respond to environmental forcing, particularly changes in nutrient supply, making them excellent paleoceanographic indicators. Furthermore, they are quick and easy to extract from the sediment, and their small size means that, typically, representative sediment samples need be no more than a few grams in weight.

22.1.2 Modern marine diatoms as analogs for the past

Paleoenvironmental reconstructions using marine planktonic diatoms are by far the most common, however freshwater (terrestrial), brackish (coastal, littoral, lagoonal), and sea-ice diatoms can also be preserved in marine sediments, recording both local and regional events as well as various processes of lateral transportation versus "*in situ*" sedimentation. The job of the marine diatomist is to unravel these events and processes recorded by the fossil assemblage. It is worth summarizing what is known about modern marine diatoms since analogy with living diatoms forms the basis of the majority of paleoceanographic reconstructions. See also Romero and Armand (this volume) for more extensive information on modern marine diatoms.

In terms of frustule size, marine diatoms as a group are larger than terrestrial diatoms, but like terrestrial diatoms, can be either solitary or colonial, planktonic, or benthic.

Generally, marine diatoms living in shallow-shelf to coastal regions are known as "neritic" and include both benthic and planktonic types, whereas marine diatoms occupying surface waters in deeper oceanic regions are known as "open ocean," "offshore," or "pelagic." Unfortunately, these terms are often used somewhat loosely in the literature, depending on the (relative) paleoenvironment. This means that comparing independent studies has not always been easy. Hence, if these broad terms are going to be used for paleoenvironmental reconstruction, there is some value in careful definition on a case-by-case basis.

Holoplanktonic marine diatoms are, by definition, wholly planktonic throughout their life cycle and tend to dominate in open-ocean/offshore areas where they are dispersed passively in (near) surface currents. They rely on mixing and/or density differences to keep them suspended near the surface for photosynthesis. Some open-ocean diatoms, such as *Rhizosolenia* spp. and *Ethmodiscus rex* (Rattray) Wiseman & Hendey, can also regulate their buoyancy by migrating vertically to tap into a deep nutrient source (up to 140 m depth) and returning to the surface for photosynthesis (e.g. Villareal, 1993, 1999; Kemp *et al.*, 2000). These, and other diatoms, such as some species of *Stephanopyxis*, *Hemiaulus*, and *Coscinodiscus*, are ingeniously adapted to stratified waters (e.g. Kemp *et al.*, 2000), e.g. *Hemiaulus hauckii* Grunow in the stratified open waters of the North Pacific Subtropical Gyre is known to harbor a nitrogen-fixing, intracellular cyanobacterial symbiont (e.g. Dore *et al.*, 2008). Benthic diatoms, on the other hand, tend to be restricted to the shallow-shelf through to coastal regions within the photic zone where they either lie on the sea floor or attach to aquatic plants or rocky, sandy, or muddy substrata. Neritic diatoms found in sediments occupying depths greater than the photic zone are allochthonous, transported downslope by density flows or strong bottom currents, or dislodged/upwelled into the plankton (hence becoming what is termed as "tychoplanktonic") and carried out into the ocean by near-surface currents before settling out. Terrestrial diatoms found in marine sediments have probably been transported to the sea via rivers, but some may have been introduced by other processes, e.g. by wind (Darwin, 1846; Harper, 1999). Some benthic marine diatoms can drift around in the open ocean attached to seaweeds or driftwood, although it would be difficult to differentiate this transient transport mechanism in the fossil record from the more obvious mechanisms. Modern marine diatoms are even known to attach to the cirri of barnacles (Bigelow & Alexander, 2000; Round & Alexander, 2002), the feathers of diving seabirds (Croll & Holmes, 1982), and the skins of migratory cetaceans (e.g. Holmes, 1985; Kawamura, 1992). Cetacean diatoms, in particular, have been recorded in surface sediments off the coast of Japan, albeit in close proximity to a fish market that deals with whale meat (Nagasawa *et al.*, 1989).

All marine diatoms are sensitive to water-mass distribution and both surface and vertical oceanic circulation. Since water masses are defined by their temperature, nutrient content, and salinity, these (together with solar irradiation) are the main controls on the distribution of diatoms throughout the world's oceans and on specific characteristics of the living floral assemblages. For example, the differences between Antarctic endemic diatoms and those found exclusively in Arctic, subarctic, temperate, subtropical, and tropical regions are well known. Plus there are those that transcend boundaries, the so-called cosmopolitan diatoms or transitional floras. Others are restricted to the sea-ice zone in either hemisphere, while a few are seemingly bipolar. True sea-ice-dwelling diatoms are very rarely preserved in marine sediments but some exceptions do occur, such as in the middle-Eocene Arctic (e.g. Stickley *et al.*, 2009).

In eutrophic systems (e.g. upwelling, upper-water seasonal stratification), opportunistic diatoms tend to have a "bloom (boom) and bust" lifestyle quickly utilizing nutrients (a bloom can last up to two or three weeks only) and often forming resting spores after nutrient depletion, i.e. *Chaetoceros* spp. Resting spores have abundant stored energy in the form of photosynthetic products and tough thickened cell walls. After this initial bloom, other diatoms (or plankton groups) more tolerant of lower nutrient conditions take over the floral succession for the remainder of the growing season. Some diatoms adapted to deeper stratified waters have a different strategy for survival, growing slowly at depth in a "deep chlorophyll maximum" prior to sedimenting to the sea floor, where they often form laminated deposits (e.g. Kemp *et al.*, 2000). See Leventer *et al.* (this volume) for further details on laminated marine diatom deposits. Also, there is at least one example of a presumed symbiotic relationship involving a centric diatom from the deep chlorophyll maximum and the coccolithophorid *Reticulofenestra sessilis* (Lohmann) Jordan & Young (e.g. Gaarder & Hasle, 1962). Regardless of survival strategy, all diatoms require for growth an availability of light, silicic acid, nitrate, phosphate, and carbon; iron in particular is an important biolimiting nutrient (Coale *et al.*, 1996).

22.1.3 Application of marine diatoms to paleoceanography

Applying this knowledge of living diatoms to fossil assemblages, and with consideration of other proxies (where available), characteristics of the sediment(s), and preservation, it is possible to extract a wide range of paleoceanographic information from fossil diatoms in marine sediments. These include: past sea-surface temperatures (SSTs); paleoproductivity and flux involving both upwelling and stratified systems and a wealth of related issues regarding the availability, source and consumption of nutrients, the silica cycle and the biological carbon pump (see De La Rocha *et al.*, this volume), and relationships with atmospheric CO_2 and climate feedbacks; paleogeography; paleoecology; past current systems including deepwater and surface-water transport; water mass and oceanic frontal movements; past sea-ice patterns and ice-sheet history; and, in near-shore areas, paleosalinity and paleodepth/sea-level reconstructions. Fossil diatoms help elucidate information on short timescales such as seasonality and episodic freshwater input to the sea and/or aridity–humidity cycles, as well as long(er) timescales such as the opening/closing of ocean gateways (tectonic events), long-term warming and cooling trends, and carbonate–siliceous sediment shifts through time. All of these sorts of information are useful for feeding back into climate models (e.g. Huber *et al.*, 2004). As with most microfossil-based reconstructions, changes in relative and absolute abundance, diversity, and assemblage characteristics are used to elucidate this information. However, the opaline diatom frustule also lends itself to some important geochemical applications, e.g. δ^{30}Si which may provide information on past silicon cycling (e.g. De La Rocha *et al.*, 1998), nutrient utilization, productivity, and links with fluctuations in atmospheric CO_2 (De La Rocha, 2006); δ^{18}O which reflects the oxygen isotope composition of ocean water (and may reflect SST) (e.g. Shemesh *et al.*, 1995); the Ge/Si ratio which gives a measure of the rates of delivery of weathering products (e.g. Shemesh *et al.*, 1989); also δ^{13}C, δ^{15}N, and C/N in diatom-bound organic matter may be used to reconstruct a complexity of oceanographic processes that may also contribute to atmospheric CO_2 variations over recent geological time, such as nutrient utilization patterns (e.g. De La Rocha, 2006; Jacot Des Combes *et al.*, 2008). See Leng and Swann (this volume) for further information on the use of isotopes from diatom silica.

22.1.4 Distribution of pre-Quaternary diatoms and historical aspects

Here we briefly summarize the distribution of diatom-bearing oceanic sediments (i.e. where diatoms are preserved as opaline silica) of pre-Quaternary age and historical aspects related to their discovery and early research.

22.1.4.1 Distribution Molecular sequencing (ribosomal RNA (ssu rRNA)) suggests the origin of the diatoms is ~135 Ma (mid–Early Cretaceous) (see Sims *et al.*, 2006) and that they probably did not exist before 240 Ma (mid Triassic) (Kooistra & Medlin, 1996). The fossil record does not (yet) completely match the findings of molecular sequencing, however. We know marine diatoms probably existed in the early Jurassic (~190 Ma) (i.e. rare specimens of the simple-formed *Pyxidicula bollensis* Rothpletz and *Pyxidicula liassica* Rothpletz from the Liassic Boll Shales of Wurttemburg, Germany; Rothpletz, 1896, 1900) and in the earliest Cretaceous (~140 Ma) in non-marine deposits of Korea (Chang & Park, 2008). This relatively late "appearance" makes the diatoms the last of the three major phytoplankton groups to emerge in the Mesozoic (e.g. Katz *et al.*, 2004).

However, one might argue that the marine diatom fossil record reliably extends as far back as only 100–125 Ma (Early Cretaceous) (Gersonde & Harwood, 1990; Harwood & Gersonde, 1990; Harwood *et al.*, 2007). These earliest diatom assemblages comprise just centrics (Gersonde & Harwood, 1990), but Early Cretaceous diatom-bearing sediments are still rare (e.g. Weddell Sea, Antarctica; Queensland, Australia; Hannover, Germany; Gersonde & Harwood, 1990), and our knowledge of their existence before this is limited (see Harwood & Gersonde, 1990; Harwood *et al.*, 2007). This is partly due to the paucity of material and preservation issues and partly because little attention has been given to looking for them. Preservation may be the biggest issue since the fossil record shows that by the Early Cretaceous the diatoms were diverse and had already evolved complex valve structures (Harwood & Gersonde, 1990; Gersonde & Harwood, 1990).

In contrast, Late Cretaceous diatoms have now been found throughout the Earth, such as Arctic Canada, central Arctic, California, Siberia, Japan, Antarctica, and the Southern Ocean (Long *et al.*, 1946; Strelnikova, 1974; Hajós & Stradner, 1975; Barron, 1985a; Harwood, 1988; Takahashi *et al.*, 1999; Nikolaev *et al.*, 2001; Tapia & Harwood, 2002; Chin *et al.*, 2008; Davies *et al.*, 2009) involving both the centrics and pennates. Harwood *et al.* (2007) provide a comprehensive review of Cretaceous diatom occurrences and evolution.

Many diatom taxa survived the Cretaceous–Tertiary (K–T) boundary (see later) and their preservation, abundance, and geographic coverage improved significantly during the early Cenozoic. Most of the early literature on Paleocene diatoms focuses on the diatomaceous deposits in Russian outcrops, generally of early Paleocene age, from Sysran (central Volgaland) and the eastern slopes of the Urals, with assemblages containing a number of Late Cretaceous survivors (e.g. Jousé, 1951). The Ural Mountains represent the suture between Europe and Asia, that was once characterized by a waterway (the West Siberian Sea) overlying the West Siberian Basin, connecting the Arctic to the northern Peri-Tethys Sea via the Kara Channel in the north and the Turgay Strait to the south (Iakovleva *et al.*, 2000; 2001; Akhmetiev, 2007; Iakovleva & Heilmann-Clausen, 2007). Exactly when this waterway was open or closed has been hotly debated, since it has great implications for paleoceanographic and paleoclimate reconstructions – i.e. as one of the major routes bringing warm waters to the northern high latitudes (Akhmetiev & Beniamovski, 2009). From dinoflagellate cyst data it has been revealed that the Arctic gateway was closed until the mid to late Paleocene (Iakovleva *et al.*, 2001), but open during the early Eocene based on the distribution of diatomite units (Radionova & Khokhlova, 2000). Thus, the opening probably took place sometime during the latest Paleocene and earliest Eocene (late Thanetian and early Ypresian), although it could have been as early as mid Paleocene (Danian) (Akhmetiev & Beniamovski, 2009). Recent interpretation of late-Paleocene diatom assemblages from the Volga middle reaches suggests that the diatoms originated in a coastal setting associated with high-energy conditions and high productivity (Oreshkina & Aleksandrova, 2007). The Kara Channel appears to have closed first (early–mid Eocene), leaving the West Siberian Sea as a semi-enclosed waterbody, with the Turgay Strait eventually closing in the middle to late Eocene (e.g. late Lutetian–Bartonian, Akhmetiev, 1996, 2007; or early Lutetian according to Radionova & Khokhlova, 2000).

Paleocene diatoms have also been obtained from the North Sea, although they are often heavily pyritized (Mitlehner, 1996). The abundance of marine diatoms during the late Paleocene to early Eocene in the North Sea Basin has been attributed to restricted circulation, seasonal upwelling, eutrophication, and water-column stratification (Mitlehner, 1996).

However, after the advent of deep-sea drilling, oceanic deposits of late-Paleocene age have been encountered, albeit rarely, including the Indian Ocean (Mukhina, 1976; Fourtanier, 1991), South Atlantic (Fenner, 1991), and the Cape Basin (Gombos, 1984). The late-Paleocene assemblages are often entirely composed of heavily silicified valves, with perhaps 90% belonging to *Hemiaulus*, *Stephanopyxis*, and *Trinacria*/*Triceratium*, and sometimes lack diatoms of Late Cretaceous origin. In this respect, the assemblages are very similar to those from the Urals.

Eocene diatoms in deep-sea sediments have been obtained from high latitudes and equatorial regions (e.g. Fourtanier & Oscarson, 1994; Stickley *et al.*, 2008), as well as a number of well-known outcrops (e.g. Mors, Jutland – Fenner, 1994; or the shales in California – Kanaya, 1957). Like the Cretaceous and Paleocene assemblages, the early–middle Eocene assemblages are characterized by heavily silicified diatoms such as *Hemiaulus*, *Stephanopyxis*, and *Goniothecium* (e.g. Homann, 1991; Suto *et al.*, 2009). However, the late Eocene–early Oligocene assemblages such as those found in Oamaru, New Zealand (e.g. Edwards, 1991), exhibit a lot more thinly silicified forms like *Asterolampra* and the benthic genera *Navicula*, *Diploneis*, and *Cocconeis*. Furthermore, the resting spores of delicate taxa such as *Chaetoceros* exhibit a diversity explosion at the Eocene–Oligocene transition (Suto, 2006), while the heavily silicified Cretaceous and early Cenozoic forms of *Proboscia* lose the ridges on their probosces by the early Oligocene. Generally, Oligocene diatom deposits are relatively rare and as a result the stratigraphy for this epoch is poorly constructed. To date, notable deposits have been found in the Komandorsky Islands (Gladenkov, 1999) and in the Southern Ocean (Barron & Mahood, 1993).

By the late Miocene onwards, diatoms are often the dominant component in sediments underlying the polar seas and coastal and open ocean upwelling areas reflecting regions of high productivity, and, in terms of general distribution, not that much different from the modern distribution of diatom-rich (diatomaceous) sea-floor sediments: (1) the Southern Ocean (particularly in a belt between 45° S and 65° S), (2) northern regions such as a well-defined area in the north Pacific north of 40° N and less well-defined areas in the North Atlantic and Norwegian Sea, and (3) equatorial regions, particularly the equatorial Pacific and Indian Ocean but also the equatorial Atlantic.

22.1.4.2 Historical aspects Fossil marine diatoms were first collected from various outcrops around the planet over a hundred and fifty years ago (e.g. Ehrenberg, 1844), yet most subsequent information was taxonomic (e.g. Greville, 1861; Witt, 1886; Brun & Tempère, 1889). The use of marine diatoms in biostratigraphy, however, did not begin in a systematic way until the 1940s (e.g. Bramlette, 1946; Jousé, 1949; Kanaya, 1957), although diatomists had been working on land sections prior to that (e.g. Mann, 1921; Hanna, 1932; Reinhold, 1937; Lohman, 1938). In fact, Lohman (1931) was already discussing "marker species" and about a "*Denticula lauta* zone" in his thesis on the Upper Miocene Modelo Formation, California. In contrast, detailed studies on evolutionary trends (e.g. Gombos, 1980, 1982; Andrews & Stoelzel, 1982; Andrews, 1986) and diatom paleoceanography (e.g. Kolbe, 1954) began much later.

Results of the Deep Sea Drilling Project (DSDP; 1968–1983), the ODP (1983–2003), and the more recent IODP (2003–), have since contributed greatly to our knowledge of diatom biostratigraphy, evolution, and paleoceanography (e.g. Barron & Baldauf, 1995).

22.2 Factors affecting marine diatom preservation

One of the fundamental considerations in paleoceanography is to understand the limitations of the chosen methodology. In the case of marine diatoms as proxies for reconstructing past environments, this means understanding what processes affect the preservation of diatom valves in marine sediments. After cell death, diatoms sink below the surface waters where they may settle out and eventually reach the sea floor. However, the fossil diatom assemblage, and even the most recently sedimented sea-floor assemblage, never completely represents the (ancient) living diatom assemblage for a complexity of biological, geochemical, and physical reasons. For example, living marine diatoms are subject to grazing by meso- and micro-zooplankton and larger marine animals and, in most cases, to some degree of lateral current transport prior to sedimentation. Like all small particles, diatoms are subject to current winnowing even after deposition. Strong or prolonged current action can have adverse effects on the diatom frustule, sometimes separating the girdle bands from the valves and typically fragmenting large valves into smaller pieces. Bioturbation may have the same effect. However, dissolution of the frustule through the water column and on the sea floor/in pore waters is probably the most important contributing factor in controlling the differences between living and fossil diatom assemblages. In addition, for older sediments, all of these pre- and syn-sedimentary processes apply, but post-depositional geochemical factors (diagenesis) are an added consideration.

Diatom valves will ultimately preserve differently depending on their shape, size, degree of silicification, and sedimentary history. Here we summarize the three main factors affecting diatom preservation that should be taken into account for all marine diatom paleoenvironmental studies – dissolution, fragmentation, and diagenesis.

22.2.1 Dissolution

The frustules of living diatoms possess an organic coating that protects them from dissolution (e.g. Passow *et al.*, 2003), but after cell death the organic coating is subject to bacterial degradation leaving the siliceous frustule vulnerable to dissolution in seawater. Unlike calcium carbonate, there is no compensation depth for silica, so marine diatoms can be preserved in any depth of water (Barron, 1985b). However, in general, the ocean is undersaturated with respect to silica, so selective dissolution occurs in the water column, at the sea floor, and even within the upper few centimeters of the sea-floor sediments within pore waters if they are particularly alkaline or silica-depleted (Barron & Baldauf, 1995). Much of the water-column dissolution occurs between 0 and 1000-m water depths, but mostly within the photic zone (Brzezinski *et al.*, 1997) or surface mixed layer (Tréguer *et al.*, 1995). For example, it has been estimated that as much as ~50% of biogenic silica produced in the euphotic zone dissolves in the upper ~100 m of the water column (Nelson *et al.*, 1995). Thus, well-preserved diatomaceous deposits require rapid transport (high flux rates) through the water column and rapid burial in underlying sediments. There are two processes that can achieve high flux rates (high export production) efficiently, mass aggregation or self-sedimentation (Alldredge & Gotschalk, 1989; Passow *et al.*, 2003) and the production of zooplankton fecal pellets, which both sink at about 50–300 m d^{-1} (Schrader, 1971; Smetacek, 1985; Alldredge & Gotschalk, 1989). Both of these mechanisms play an important role in controlling carbon export from the surface oceans. How large and how buoyant a diatom is also plays a role in determining sinking efficiency. Beyond the background production, and as advocated by Kemp *et al.* (2000) and Stickley *et al.* (2005), for example, the highest diatom flux rates are relatively short-lived seasonal events occurring either as a spring bloom and flux or as a "fall dump," depending on the style of production and environment.

Selective dissolution results in the valves and frustules of lightly silicified genera (e.g. vegetative *Chaetoceros*, *Corethron*) being removed, leaving most fossil diatom assemblages composed solely of robust forms (i.e. heavily silicified resting stages) (Mikkelsen, 1977). An added complication may be when the degree of silicification is season-dependent, such as is the case for *Neodenticula seminae* (Simonsen & Kanaya) Akiba & Yanagisawa (Shimada *et al.*, 2006), which produces thicker frustules in autumn and winter than it does in spring and summer; but it is by no means certain if other diatoms respond this way. To estimate the degree of dissolution, dissolution indices or ratings based on specific diatom species or assemblages have been devised, e.g. an index based on *N. seminae* (Shimada *et al.*, 2003). The implications of this significant loss in diversity due to dissolution must be taken into account when reconstructing paleoenvironments, since as much as 95% of the living assemblage may be dissolved before reaching the sea floor (Barron, 1985b). However, it should be stressed that, under exceptional conditions, discrete intervals of lightly silicified diatoms, such as *Corethron*, can be well preserved in the sediments (Jordan *et al.*, 1991). Although perhaps more astonishing examples of excellent preservation have been the recent discoveries of Early Cretaceous marine diatoms in pieces of fossilized amber (Girard *et al.*, 2008, 2009), of the finely silicified needle-shaped sea-ice diatom *Synedropsis* in laminated middle-Eocene sediments from the central Arctic (Stickley *et al.*, 2009), and of organelles of middle-Eocene non-marine diatoms in post-eruptive lacustrine sediments (Wolfe *et al.*, 2006).

22.2.2 Fragmentation and composition of fossil diatom assemblages

Fossil diatom assemblages, unlike those of modern or some Holocene assemblages, often contain a mixture of diatom fragments (e.g. incomplete valves, isolated girdle bands, broken spines, setae, and elevations), many of which are presently unidentifiable. Most of this fragmentation will have occurred as a result of strong or prolonged current action during or after sedimentation but sometimes also as a result of sedimentary compaction or desiccation. Depending on the depositional environment, strong winnowing may also bias the assemblage towards either small and light, or large and heavy fragments. Diatoms in subglacial deposits or tills can be highly fragmented due to glacial processes (e.g. Scherer *et al.*, 2004, 2005). Nevertheless, unidentifiable fragments clearly hinder diatom-based reconstructions, since girdle bands or broken components are rarely included in diatom counts, thus reducing the amount of information available. One of the few exceptions is the case of elongate araphid diatoms such as *Thalassiothrix* and *Thalassionema*, the valves of which are invariably broken. However, systematic counts of these diatoms can be made by enumerating the apices, while ignoring the rest of the fragmented valve remains (Schrader & Gersonde, 1978). This paucity of information stems from the fact that there are few, if any, published studies dealing with isolated components. If, for instance, a detailed guide existed of *Hemiaulus* elevations based on whole valves, then a lot of "lost" information could be recovered from Paleogene sediments by observing broken specimens. Such components as girdle bands, setae, and elevations do possess species-specific differences. Isolated valves also pose another potential problem, because some diatom genera exhibit heterovalvy, either on the same frustule (e.g. *Chaetoceros*, *Corethron*, *Paralia*) or in different life-cycle stages (e.g. *Chaetoceros*, *Eucampia*, *Rhizosolenia*, *Thalassiosira*). Thus, different valves of the same fossil species could be identified and counted as separate taxa in error, thereby overestimating species diversity. In many fossil assemblages (particularly in coastal sediments), resting stages may be the dominant component (e.g. Harwood & Gersonde, 1990; Leventer *et al.*, 2006; Stickley *et al.*, 2008), since they possess more heavily silicified valves, and tend to survive dissolution to a greater degree than vegetative stages. Yet until recently, fossil resting stages such as *Chaetoceros* resting spores were lumped into a single category due to the lack of detailed studies on their valve morphology. Now, after a pioneering series of over fifteen papers by Suto (e.g. 2004, 2005), many of the resting spores found in the northern hemisphere can be identified to species level more easily.

22.2.3 Diagenesis

Diagenetic processes can have either an adverse or a favorable (protective) effect on diatom valves. Diatom valves are composed of opaline silica (amorphous, hydrous silica, or opal-A). Under heat and pressure, the opal-A of deeply buried diatom valves is converted to cristobalite-tridymite (opal-CT) and eventually to quartz (i.e. chert) (Iijima & Tada, 1981), resulting in complete destruction of delicate diatom structures. Since the transformation from opal-A to opal-CT increases with burial depth, the depth of the opal-A/opal-CT diagenetic boundary is rather critical for diatom preservation. This transformation normally occurs when the temperature in the sediments exceeds 35 °C and since the reaction rates of the mineral phases are low, the process may take tens of millions of years under modern deep-sea sedimentary environments. This means chert is normally Miocene or older in age.

For example, the Miocene–Pleistocene diatomaceous units in Bering Sea sediments are 300–725 m thick, with the top 300–400 m characterized by mildly dissolved and fragmented diatom valves. A diagenetic boundary and widespread distribution of diatom opaline silica occurs about 600 m below the sea floor, and at 600–700 m below the sea floor the silica reprecipitates as inorganic opal-A, which is then rapidly transformed into opal-CT (Hein *et al.*, 1978). Thus, longer diatom records cannot be obtained simply by drilling deeper. However, onshore sequences formed by the tectonic uplift of diatomaceous marine sediments may provide scientists with invaluable information, especially of older geological periods (e.g. Fur Formation, see Homann, 1991; Oamaru, see Edwards, 1991; Russian Urals, see Strelnikova 1974, 1975).

Another type of diagenesis and one that affects diatom valves in a manner which preserves them to a greater or lesser degree is via pyritization, with post-deposition pyrite formation occurring under anoxic or euxinic conditions, combined with the availability of iron and sulfur in sediment pore waters (Raiswell & Berner, 1985). Pyritization of diatoms can occur in sediments of any age, but are particularly common in older sediments (e.g. Paleocene–Eocene; Van Eetvelde *et al.*, 2004). Such diatoms are often preserved as internal moulds of pyrite crystals that are formed during early diagenesis, whereby the original silica is totally replaced. Alternatively, diatom frustules may be completely altered to pyrite by epigenesis, whereby all external features are preserved. In most cases diagenesis results in a loss of morphologic information, but sometimes pyritized diatoms can be identified to species level and may be useful as stratigraphic indicators (Van Eetvelde *et al.*, 2004).

It has been known for some time that diatoms can be protected in concretionary nodules (Chapman, 1906), such as in manganese nodules on the sea floor (Burns & Burns, 1978) or in calcareous nodules of the middle Miocene Kinone and lower Amatsu Formation in Boso Peninsula, central Japan (Watanabe & Takahashi, 1997).

Diatom opal can also be altered and completely destroyed beyond recognition, within just a few months, to various forms of authigenic potassium-rich and iron-rich aluminosilicates (clays) during burial, e.g. tropical deltaic systems such as the Amazon deltaic deposits (Michalopoulos *et al.*, 2000); this process may eventually lead to the formation of illites (Michalopoulos *et al.*, 2000). Such rapid conversion of diatom opal to clays indicates that not all diagenetic processes occur on geological timescales.

22.3 Past tectonic, oceanographic, and climatic events

Throughout the Cretaceous and Cenozoic, the marine biosphere has been affected by climatic and tectonic changes, both regionally and globally (e.g. Katz *et al.*, 2004; 2005). For example, the late-stage break-up of the supercontinents Pangaea and Laurasia in the late Mesozoic through early Cenozoic, and of Gondwana in the early to mid Cenozoic have played both direct and indirect roles in the evolution and diversity of all the major phytoplankton groups through impacts on sea level, destruction and/or formation of habitats, weathering, and nutrient availability (e.g. Katz *et al.*, 2005). Also the Cenozoic in particular has been a time of major climate change, evolving from a "greenhouse" world (Cretaceous–early Eocene) characterized by high atmospheric CO_2 levels, widespread global warming, and relatively high sea levels, through a "doubthouse" world (middle to late Eocene) when atmospheric CO_2 levels and deep-sea temperature started to fall and the Earth began the transit into a polar glaciated state, through a series of climatic cooling steps, to an "icehouse" world (Eocene–Oligocene transition onwards) characterized by both relatively low atmospheric CO_2 and sea levels, and major and progressive expansion of ice sheets in both hemispheres, starting first in Antarctica at ~34 Ma and followed later in the northern hemisphere at perhaps ~23 Ma (DeConto *et al.*, 2008).

Fundamental to understanding both the advantages and limitations of marine diatoms in paleoceanography is the need to comprehend, in general terms, how marine diatoms have responded, or adapted, to major tectonic, oceanographic, and climatic events in the past. For this purpose it is necessary to consider their broad evolution and paleogeographic distribution. The former, because gradual or sudden environmental change is one of the main driving forces of evolution; and the latter, because tectonic changes often result in the expansion or reduction of planktonic biogeographic zones due to reorganization of surface-water masses, plus the expansion or reduction of coastline habitats where neritic species thrive (e.g. Katz *et al.*, 2004, 2005). Figure 22.1 shows some of the climatic and tectonic events of the Cenozoic, together with the global benthic oxygen and carbon isotope curves (Zachos *et al.*, 2001), and six periods of rapid evolutionary turnover for marine diatoms. It is likely that these turnover events occurred in response to warming or cooling trends, and reorganization of water masses following the opening or closing of gateways. All of these factors have lead to progressive diversification of the diatoms.

The ~125 Ma marine diatom fossil record cannot be given justice in this chapter. There is no space, for example, to consider specific details of the evolution of marine diatoms, nor all of the paleoceanographic events for which they provide important information. We acknowledge that many of the earlier, pioneering studies by numerous diatom workers who led the way in pre-Holocene diatom research, and upon which many of the more recent works are built, are not mentioned. The purpose of this chapter is to outline some of the key subjects, providing an up-to-date reference list that the reader can refer to for more detailed information if required. We choose a number of these subjects that are inherently linked by way of providing the reader with both the necessary background of information and with a good idea of how Cretaceous and Cenozoic diatoms are helping us understand the causes and impacts of global change both on a large-scale and in more intricate detail. We

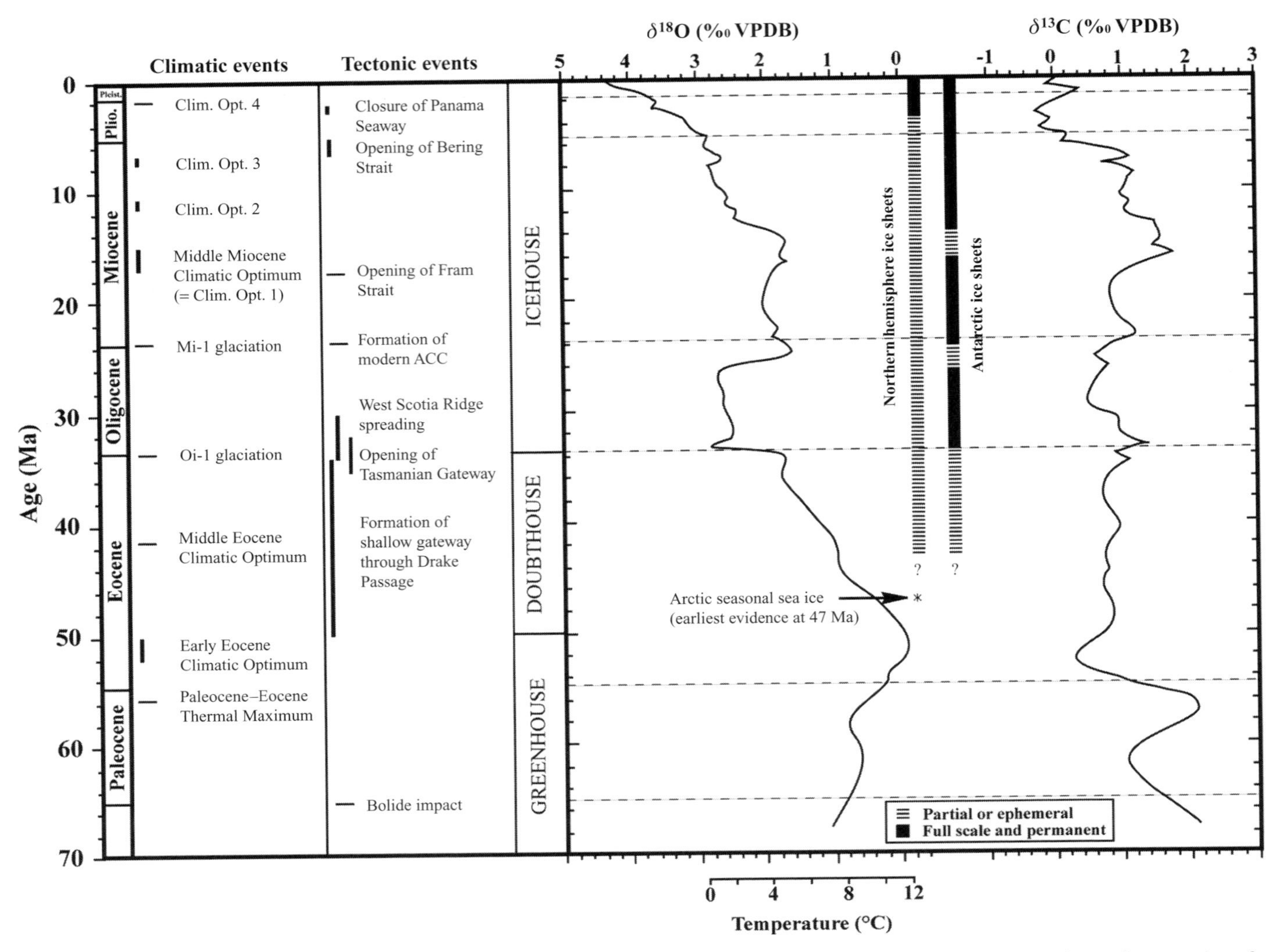

Figure 22.1 Selected climatic and tectonic events of the Cenozoic. Largely based on Zachos *et al.* (2001), but including the modification by Thomas (2008). Earliest stable Arctic sea-ice evidence from fossil sea-ice-dwelling diatoms in Stickley *et al.* (2009). Horizontal dashed lines represent some of the major diatom turnover events; bolide impact at the K–T boundary (65 Ma), Paleocene–Eocene Thermal Maximum at the Paleocene–Eocene boundary (55 Ma), Oi-1 glaciation at the Eocene–Oligocene boundary (34 Ma), Mi-1 glaciation at the Miocene–Oligocene boundary (23 Ma), West Antarctic Ice Sheet expansion at the Miocene–Pliocene boundary (5.3 Ma), and the Pliocene–Pleistocene boundary (1.77 Ma). The $\delta^{18}O$ curve can be used as a proxy for paleotemperature, while the $\delta^{13}C$ curve can be used to reconstruct past pCO_2 (partial pressure of carbon dioxide) concentration and productivity. Units are given in Vienna PeeDee Belemnite (VPDB).

provide an up-to-date overview of what has been achieved so far in more recent times and in doing so we hope to suppress some (perhaps misguided) preconceptions about the usefulness of fossil diatoms. We present: (1) an overview of evolutionary and sedimentary responses to tectonic and climatic changes for the Cenozoic, including further details on three topics: the middle Eocene as an example of paleogeography in a doubthouse world, diatom changes across the Eocene–Oligocene transition, and changes in biosiliceous (opal) sedimentation in an icehouse world; (2) changes in style of production (past stratification versus upwelling) from a greenhouse world to an icehouse world; (3) sea-ice and ice-sheet history/glacial processes; and (4) diatoms and the evolution of ocean-going mammals and birds for the Eocene–Miocene interval.

22.3.1 Evolutionary and sedimentary responses to tectonic and climatic changes: an overview

Harwood *et al.* (2007) describe four broad stages of diatom evolution through the Mesozoic and Cenozoic: Ancestral stage (origin of the diatoms, Jurassic or earlier); Archaic stage (Early Cretaceous migration into marine coastal regions dominated by benthics); Medial stage (post Cretaceous–Tertiary boundary evolutionary radiation to also now include fully planktonic diatoms during the Late Cretaceous–early Paleogene); and the Modern stage (multiple, stepwise, climate-driven evolutionary innovation and adaptive radiations through the Cenozoic from the earliest Oligocene onwards). We focus on the Cenozoic (starting at the K–T boundary), since Cenozoic marine diatom records are less piecemeal than for the Cretaceous and can therefore provide better continuity and detail of information. We refer the reader to Harwood *et al.* (2007) for a comprehensive review of Cretaceous diatom evolution.

Cretaceous–Tertiary (K–T) Boundary (~65 Ma) According to some micropaleontologists there was only a minor turnover of diatom genera at the Cretaceous–Tertiary (K–T) boundary (~65 Ma) (e.g. MacLeod *et al.*, 1997). Harwood (1988) estimates ~84% of diatom species survived the K–T boundary event at Seymour Island (Antarctica). However, as pointed out by Wornardt (1972), almost half the diatom genera present in the Maastrichtian sediments of the Moreno Shale in California did not continue into the Paleocene, while Chambers (1996), who analyzed an impressive data set from K–T sites globally, estimated only ~37% of diatom species survived through to the early Paleocene. This is quite significant and shows that diatoms were somewhat affected by the upheaval caused by several independent factors; a bolide impact, the explosive volcanism of the Deccan Traps, and the continuing tectonic changes associated with the opening of the Atlantic Ocean. At the end of the Mesozoic, some of the fifty or so genera that had characterized the Late Cretaceous assemblages seemingly became extinct, including *Gladiopsis*, *Glorioptychus*, and *Huttonia*, whilst other genera such as *Proboscia*, *Actinoptychus*, and *Triceratium* survived and are still living today. A high proportion of K–T survivors were neritic taxa capable of forming resting spores (Sims *et al.*, 2006). However, the comparative magnitude of diatom extinctions at the K–T boundary was apparently not as great as many other marine groups, perhaps attesting to their ingenious survival strategies.

Cenozoic Barron and Baldauf (1989) and Baldauf and Barron (1990) comprehensively describe a series of diatom turnover events (evolutionary appearances and disappearances) of species and genera as a response to stepwise cooling through the Cenozoic. Immediately following the end of the Cretaceous (~65 Ma), there was a ~2 ka cooling episode caused by the invasion of boreal waters into the western Tethyan Realm (Galeotti *et al.*, 2004). However, relatively warm seas still circulated around the world, including the poles, throughout the Paleocene and early Eocene (Weijers *et al.*, 2007), when CO_2 concentrations were relatively high (e.g. Zachos *et al.*, 2008), and the abundant marine life, including the marine diatoms, were characterized by low diversity (Wornardt, 1972). Paleocene and early Eocene oceanic diatom assemblages were relatively less provincial than they became in the late Cenozoic after cooling at the high latitudes brought about increased pole-to-equator thermal gradients, more intense overturning of the oceans, more efficient recycling of nutrients, and strengthening of oceanic frontal systems (Cervato & Burckle, 2003).

Diatom floras became progressively more provincial from the late middle Eocene onwards (Fenner, 1985; Barron & Baldauf, 1989, 1995; Barron, 2003), particularly during latest Miocene and Pliocene, and although no major extinctions of diatom taxa occurred during the Cenozoic, there were at least six periods of rapid evolutionary turnover: (i) around the early Eocene–middle Eocene boundary; (ii) during the earliest Oligocene; (iii) at the Oligocene–Miocene boundary; (iv) during the early middle Miocene; (v) during the latest Miocene; and (vi) during the late Pliocene (Figure 22.2; Barron & Baldauf, 1989, 1995; Barron, 2003). These turnovers coincide with rapid climatic cooling periods or major oceanographic changes in surface-water circulation due to the opening or closing of oceanic gateways (Figure 22.1; Barron & Baldauf, 1989, 1995). In Figure 22.2 the stratigraphic ranges of a number of marine diatom genera are plotted against geological time. Similar plots have been presented before (e.g. Jousé, 1978; Barron, 1993), but here the genera were selected because most of them have short ranges, are easily recognized, and are unlikely to be separated into new genera in the foreseeable future. What Figure 22.2 clearly shows is that most of the genera either begin or end their range at one of the six turnover events.

The comprehensive study of Barron (2003) clearly demonstrates increasing provincialism through the Neogene and Quaternary. He noted, for example, that in the equatorial Pacific, evolutionary turnover of species was relatively high between 18.0 and 6.0 Ma and declined after 6.0 Ma, as a response to changes in oceanic circulation. In the North Pacific, however, species turnover reached a maximum between 10.0 and 4.5 Ma, while in the Southern Ocean, evolutionary turnover of endemic diatoms was greatest between 5.0 and 1.6 Ma.

High-resolution diatom biochronologies in the methods used by Barron (2003), Cody *et al.* (2008), and Scherer *et al.* (2007a) have allowed reasonable estimation of the rates of evolutionary turnover of planktonic diatom species, as well as estimates of their diversity, in an approach which incorporates reduced preservational and recording bias. For

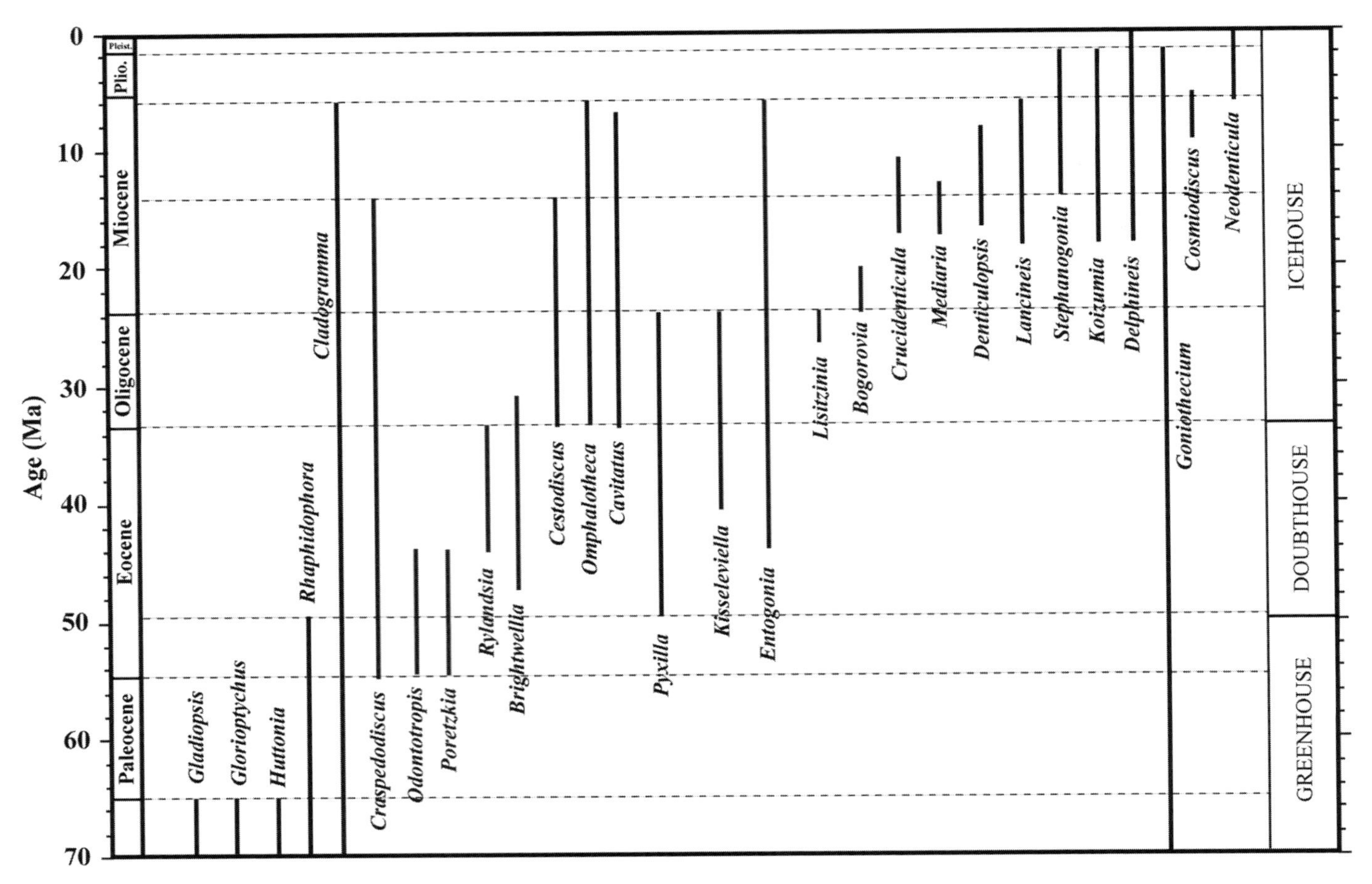

Figure 22.2 Stratigraphic ranges of selected diatom genera. See Figure 22.1 for explanation of dashed lines.

example, Barron (2003) determined that extinct planktonic diatom species have relatively short longevities of only ~2.5–3.4 Ma, which reflects rapid evolutionary response and hence their usefulness as biostratigraphic tools.

22.3.1.1 Paleogeography in a doubthouse world – the middle Eocene Diatomaceous sediments are unevenly distributed around the Earth, and are never found in stratigraphic continuity throughout the Cenozoic at any one location. This temporal and spatial patchiness is one of the main reasons why diatom paleoceanography and biostratigraphy seemingly lag behind those of other microfossil groups. During the Cenozoic, water masses were reorganized several times, and so the boundaries of the biogeographic zones, each represented by a characteristic diatom assemblage, have shifted accordingly. In addition, tectonic changes, such as the opening and closing of gateways, have led to the re-routing of surface and deep currents, and in the case of the Tethys Sea closure, to the destruction of an entire realm. Until now, diatom paleoceanographers have been reluctant to reconstruct paleogeographic maps; however, in Figure 22.3 an attempt has been made to reconcile this problem, in order to provoke discussion and stimulate future research.

Here the paleogeographic distribution of marine diatom species is presented for the middle Eocene (~49–37 Ma, Lutetian and Bartonian) as an example. At this time, the Earth started to cool (Zachos *et al.*, 2001, 2008) and the Tethyan Realm occupied most of the mid to low latitudes and there was an open connection between the Pacific, Atlantic, and Tethys. This zone was characterized by a distinct marine diatom assemblage, which included such taxa as *Craspedodiscus oblongus* (Greville) Grunow, *Triceratium inconspicuum* Greville, *Clavularia barbadensis* Greville, and *Pyxilla oligocaenica* Jousé, as well as the enigmatic siliceous microfossil *Macrora barbadensis* (Deflandre) Bukry (Figures 22.4 and 22.5). These species have been reported from locations such as Israel, Barbados, Sierra Leone Rise, and California (Greville, 1863; Kanaya, 1957; Ehrlich & Moshkovitz, 1982; Fenner, 1985). In contrast, the high latitudes represented by the Boreal and Austral realms were characterized by very different assemblages, with only rare occurrences of *P. oligocaenica* and *M. barbadensis* in the central Arctic sediments

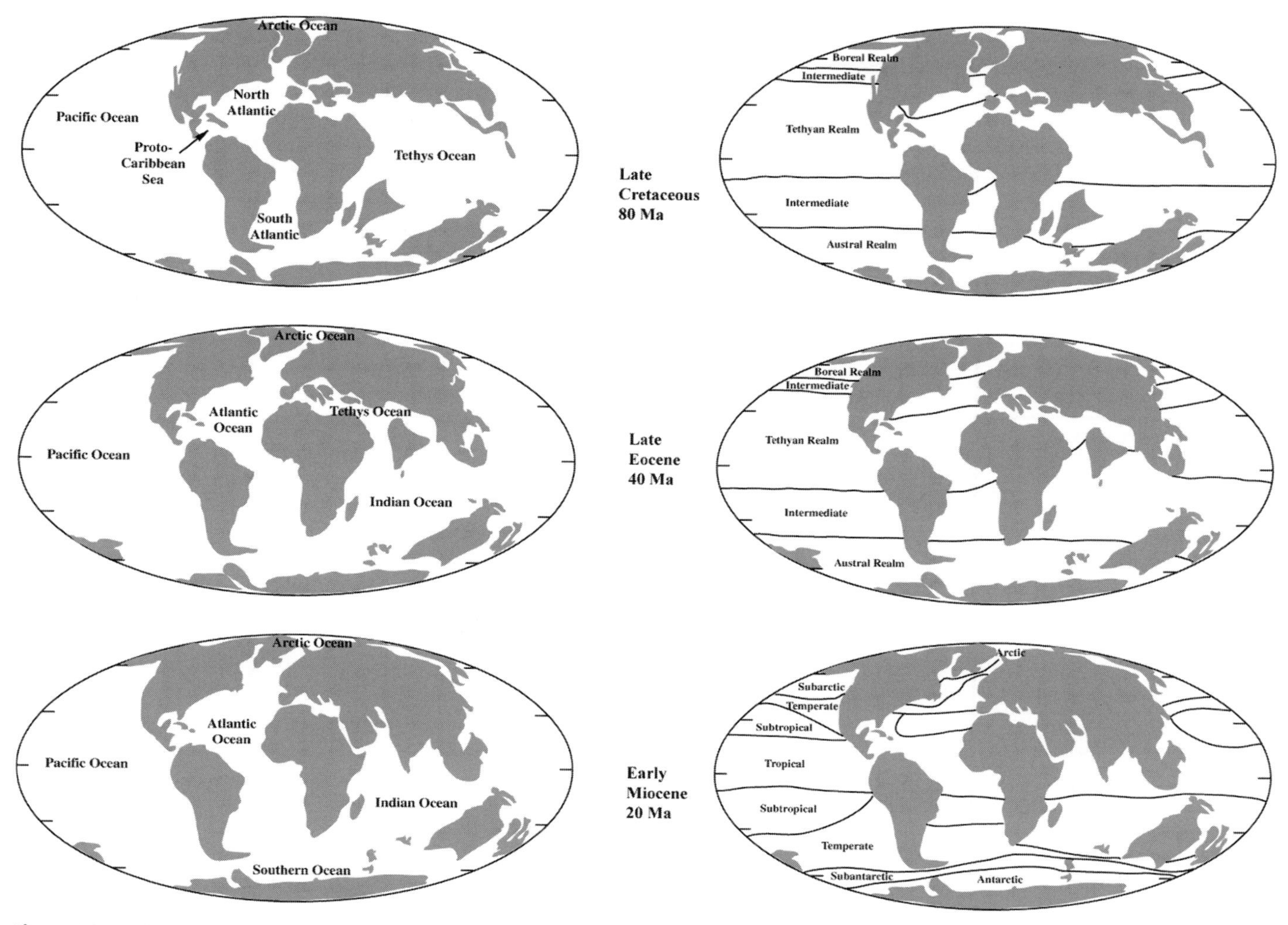

Figure 22.3 Global paleogeographic reconstructions showing the possible positions of the oceans and biogeographic zones for three time slices; Late Cretaceous (80 Ma), late Eocene (36 Ma), and early Miocene (20 Ma). Globes redrawn from images on Ron Blakey's website at http://jan.ucc.nau.edu/~rcb7/global_history.html with paleogeographic information obtained from the literature.

(Stickley *et al.*, 2008) In general, the central Arctic assemblages were endemic to the Arctic Ocean and dominated by resting stages, particularly of *Anaulus*, *Costopyxis*, *Goniothecium*, *Hemiaulus* and *Odontotropis* (Stickley *et al.*, 2008; Suto *et al.*, 2008, 2009, in press). To the north and south of the Tethyan Realm lay an Intermediate Zone, with perhaps a wider zone in the south due to fewer landmasses. Dzinoridze *et al.* (1978) stated that 38% of species comprising a Norwegian Sea assemblage could also be found in Siberia, while a comparison of central Arctic and Norwegian Sea assemblages revealed that very few were common to both zones (Backman *et al.*, 2006). In fact, a number of new taxa were recently described from the central Arctic (e.g. Suto *et al.*, 2008, 2009, in press), compliant with

Figure 22.4 Examples of middle Eocene siliceous microfossils: (1) *Craspedodiscus oblongus* (Springfield, Barbados); (2) *Triceratium inconspicuum* (Beit Guvrin, Israel); (3) *Pyxilla oligocaenica* (Chalky Mount, Barbados); and (4) *Macrora barbadensis* (Chalky Mount, Barbados).

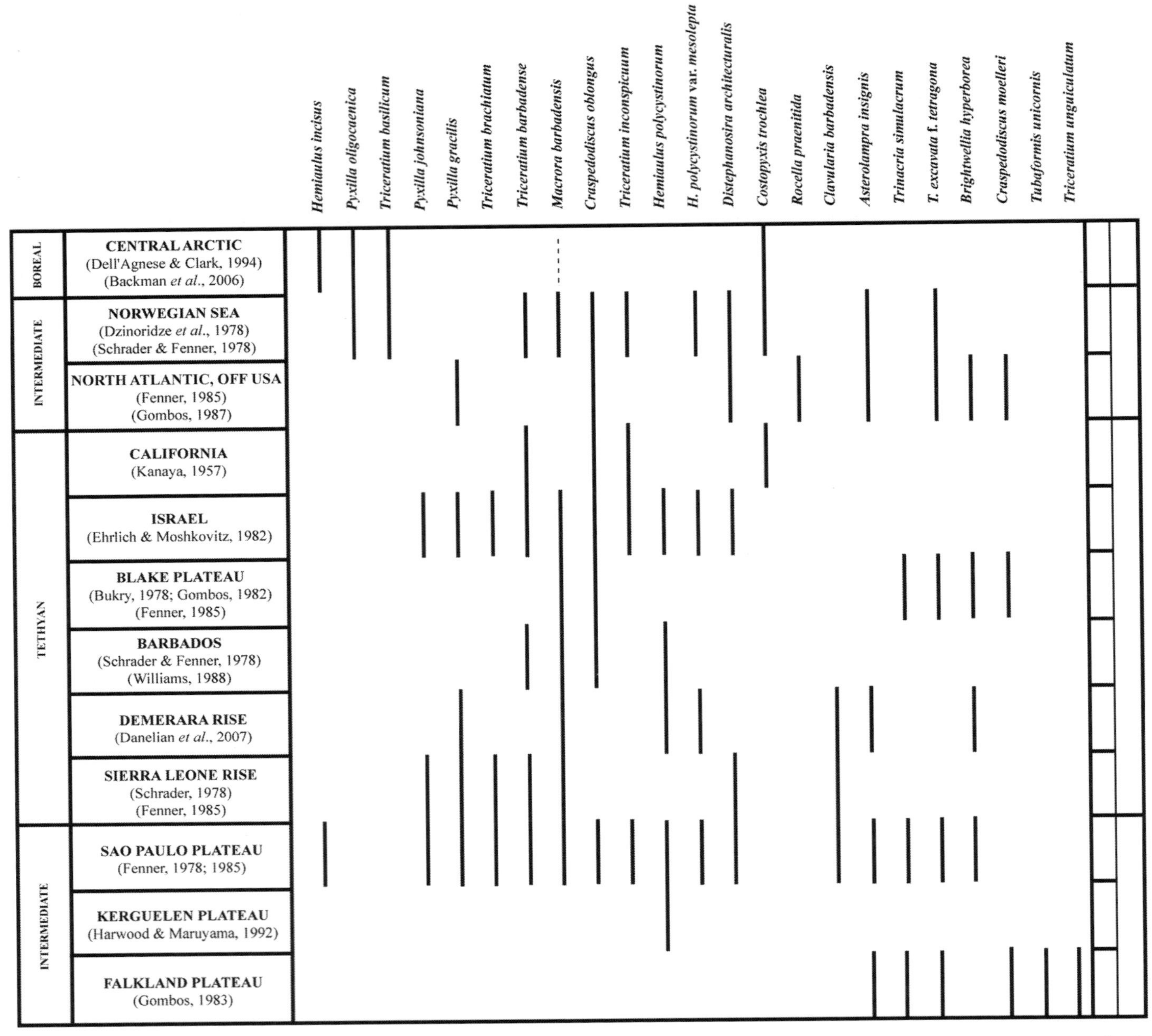

Figure 22.5 Paleogeographic distributions of middle Eocene diatoms and *M. barbadensis*, reconstructed from the literature. The rare occurrence of *M. barbadensis* in the Arctic (Stickley *et al.*, 2008) is represented by a dotted line.

tectonic reconstructions, which show that in the middle Eocene the Arctic was effectively isolated from other oceans (Backman & Moran, 2009).

Following the progressive closure of the Tethys Sea (latest Eocene and mid Miocene for the western and eastern parts, respectively; Berggren, 2002) and reorganization of surface water masses, the biogeographic zonation became more complicated with distinct polar, subpolar, temperate, and subtropical zones in each hemisphere separated by a tropical zone (Figure 22.3).

22.3.1.2 Diatom changes across the Eocene–Oligocene (E–O) transition Spencer-Cervato (1999) and Katz *et al.* (2005), using the Neptune data base report an overall increase in diatom diversity through the Cenozoic, occurring in stages, the most important of which may have occurred across the Eocene–Oligocene (E–O) transition (~34–33 Ma) – the most significant climate transition of the Cenozoic era. The peak in diatom diversity close to the (E–O) transition reported by Rabosky and

Sorhannus (2009) may partly be caused by sampling biases in their methodology. Regardless, across the (E–O) transition the global climate cooled rapidly and the Earth was plunged into its current icehouse state, characterized by rapid and major expansion of ice sheets on Antarctica (the Oi-1 event near the (E–O) boundary; Zachos *et al.*, 2001, 2008), which resulted in a rapid, major sea-level fall (e.g. Miller *et al.*, 2008), increased weathering of exposed shelf material (silicates) into the oceans (e.g. Merico *et al.*, 2008, and references therein), and more vigorous oceanic mixing, including widespread wind-driven coastal upwelling, increasing the availability of the weathered silica and other nutrients (e.g. iron) in surface waters. This resulted in enhanced and widespread diatom productivity and biogenic silica export (flux) (at the expense of calcareous plankton) leading to a global increase in carbon burial (e.g. Salamy & Zachos, 1999; Diester-Haass & Zahn, 2001). At the same time the diatoms underwent a major evolutionary turnover resulting in changes to frustule morphology, size, and degree of silicification. An increase in the concentration of silica in the oceans and reduced temperatures also meant improved potential for preservation.

The rapid radiation of C_3 grasses across the (E–O) transition has also been hypothesized as another cause, or catalyst, for the increase in diatom production and preservation (although C_3 grasses had first appeared on land at least by ~55 Ma – Jacobs *et al.*, 1999, or even in the Late Cretaceous – Prasad *et al.*, 2005) that was possibly stimulated by the evolution of mammal grazers (e.g. ungulates) and/or widespread fires associated with climate-driven aridity and increased air temperatures in the early Eocene. Therefore co-evolution of diatoms, grasses, and mammals may be more than just coincidental (e.g. Katz *et al.*, 2004). Further evidence for links between diatom and grassland radiation is in the late Miocene (6–8 Ma) when another diatom turnover phase occurred in conjunction with a further global expansion of grasses and a shift in dominance from C_3 to C_4 photosynthetic pathways (Falkowski *et al.*, 2004; Katz *et al.*, 2004). The grasses, which contain 15% dry weight silica, possess silica deposits in their cell walls (phytoliths), which are released after the cell's death. The subsequent mass transport of these phytoliths in fluvial waters led to an increase in the silica concentration of the oceans. Phytoliths may also be transported offshore by intense dust storms (Folger *et al.*, 1967). It has been estimated that phytolith dissolution releases soluble silica twice as efficiently as silicate weathering (Alexandre *et al.*, 1997). Regardless of the cause, the sudden increase in silica availability in the earliest Oligocene allowed siliceous organisms to rapidly evolve, diversify, and to switch to lighter shells

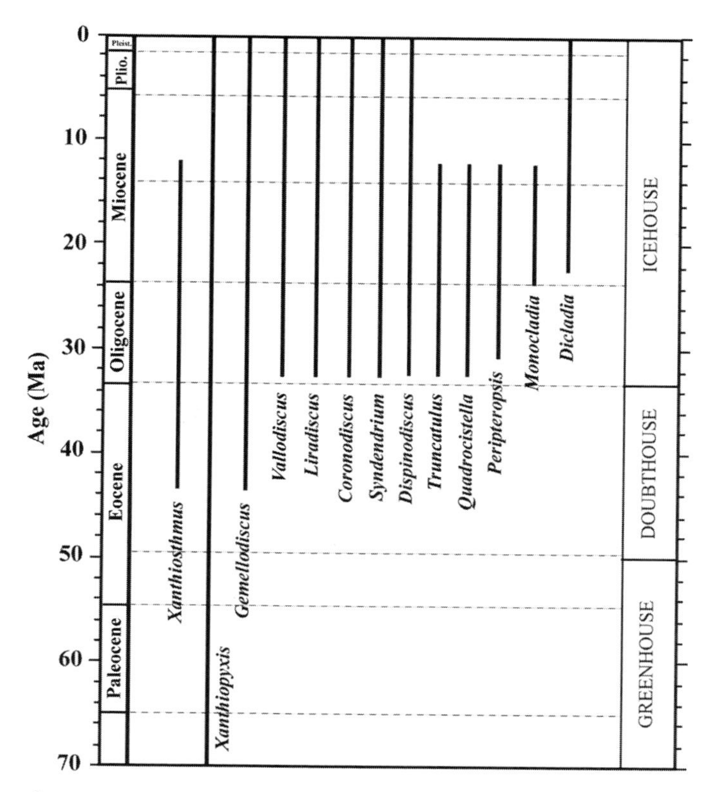

Figure 22.6 Stratigraphic ranges of resting spore morpho-genera. Based on Suto (2006), but extended to include older and younger records. See Figure 22.1 for explanation of dashed lines.

from their previously more heavily silicified forms (Falkowski *et al.*, 2004). For example, typical Cretaceous, Paleocene, and Eocene diatom assemblages are characterized by heavily silicified large genera such as *Hemiaulus*, *Pyxilla*, *Trinacria*, and *Stephanopyxis*. Diatoms became more lightly silicified after the E–O transition, coincident with the onset of the icehouse, which through more vigorous oceanic circulation drove the diatoms into uptaking silica more competitively (i.e. to use less silica in formation of their frustules). For example, diatom genera such as *Proboscia* and *Rhizosolenia* became significantly less silicified during the early Oligocene. The trend to lighter frustules continued in several steps through the icehouse as glaciation intensified and smaller and more delicate diatom genera, such as *Coscinodiscus*, *Thalassiosira*, *Chaetoceros*, *Skeletonema*, and *Thalassionema* came to characterize living oceanic assemblages today (Barron & Baldauf, 1995).

An apparently explosive diversification of the genus *Chaetoceros* across the E–O transition has been recently documented (Figure 22.6; Suto, 2006). Although diatoms are poorly preserved in the E–O transition sediments of Kamchatka, an increase in the relative abundance of particular diatom biomarkers suggests that *Rhizosolenia* spp. became dominant in the North Pacific at that time (Shiine *et al.*, 2008). Whereas it is widely accepted that diatom production and preservation increased across the E–O transition, hypotheses concerned with the driving forces behind possible concomitant diversity increases, such as the proposed coupling of terrestrial and marine systems via the silica cycle or competition between phytoplankton groups, are still debated; as is the degree of recording bias which may skew diversity-trend estimates (i.e. diatoms in one report may not be as intensely studied as in another, and methods may vary).

Regardless of diversity trends, any increase in diatom productivity would have probably caused a concomitant increase in the zooplankton, particularly copepods and krill, which some scientists believe was the trigger for cetacean evolution along the Tethys coast (Lipps & Mitchell, 1976; Gingerich *et al.*, 1983).

22.3.1.3 Changes in biosiliceous sedimentation in an icehouse world Climate change and large-scale oceanographic and tectonic reorganization have influenced global biosiliceous (opal) sedimentation patterns. Stepwise cooling through the Cenozoic has favored planktonic biogenic opal (diatoms, silicoflagellates) production by invigorating oceanic circulation and increasing the availability of nutrients through upwelling and sea-level fall associated with ice-sheet expansion in both hemispheres. However, as Cortese *et al.* (2004) state, the interpretation of opal deposition records might not always be straightforward because of the complexity of interactions of climate, tectonic, evolutionary, and oceanographic events. Barron and Baldauf (1989), Baldauf and Barron (1990), and Cortese *et al.* (2004) give detailed descriptions and causes of changes in opal deposition (all biosiliceous production including radiolarians) patterns over time.

The early Paleogene (Paleocene–early Eocene) was marked by unhindered flow of equatorial surface waters and a broad band of tropical opaline sediments (Barron & Baldauf, 1989) largely dominated by radiolarians (e.g. Moore *et al.*, 2008). Major cooling steps since the early middle Eocene (~49 Ma onwards), i.e. after the Early Eocene Climate Optimum (EECO, Zachos *et al.*, 2001; 2008) led to decoupling of both surface and deep waters, stimulating upwelling systems and increased provincialism amongst diatom floras and basin–basin fractionation. This resulted in regional partitioning in opal sedimentation (Barron & Baldauf, 1989) and switches in opal sedimentation between oceanic basins. In parallel, tectonic changes have played a major role in controlling where opaline sediments occur; the progressive deepening of the Tasmanian Gateway (Stickley *et al.*, 2004b) and Drake Passage (Barker *et al.*, 2007), for example, allowed initiation of uninterrupted circumpolar flow (the Antarctic Circumpolar Current, ACC) around Antarctica (e.g. Barker *et al.*, 2007), heralding the Southern Ocean as a new area for the deposition of diatom-rich sediments since Oligocene times. Here we summarize the most dynamic phase in global changes in diatom-rich or opal sedimentation (including all biosiliceous groups) since the evolution of the diatoms: the icehouse world, i.e. the earliest Oligocene onwards, incorporating those major climatic and tectonic factors which have influenced opal sedimentation the most.

The earliest Oligocene cooling and major ice-sheet expansion on Antarctica (~34 Ma) resulted in a relatively short-lived "silica switch" from a silica-rich equatorial and North Atlantic to a silica-rich equatorial Pacific, while at the same time opal sedimentation increased in the Southern Ocean due to ACC development (Barker *et al.*, 2007). Barron and Baldauf (1989) argue for increased basin–basin fractionation and age differences between deep water masses as the primary cause: "older", more nutrient-rich, deep Pacific waters would have favored diatom production and preservation over "younger" less nutrient-rich deep Atlantic waters. The reverse apparently occurred in the late Oligocene with opal sedimentation returning to the low-latitude North Atlantic and becoming diatom-poor in the equatorial Pacific. These switches exemplify the main hydrographic process controlling opal versus carbonate production and preservation in the world's oceans: estuarine circulation favoring opal (e.g. the modern Pacific) or anti-estuarine circulation favoring carbonate (e.g. the modern Atlantic) (Cortese *et al.*, 2004).

In their comprehensive study, Cortese *et al.* (2004) detail a series of further silica switches over the past 15 Ma (see also Barron and Baldauf, 1989 and Baldauf and Barron, 1990 for discussions). They show that over this time period the main opal "sink" has shifted: from the North Atlantic, to the Pacific, to the equatorial Pacific, to the eastern equatorial Pacific, to eastern boundary current upwelling systems of California, Namibia, Peru, and finally to the Southern Ocean. Cortese *et al.* (2004) relate these changes to a series of events including: the middle Miocene Cenozoic global cooling trend (starting ~15 Ma) and concomitant intensification of glaciation in Antarctica; the late Miocene–early Pliocene biogenic-bloom event (4.5–7 Ma); the North Pacific opal breakdown (2.7 Ma), probably related to intensification of the Northern Hemisphere Glaciation (NHG; 2.5 Ma); the progressive shoaling and eventual closure of the Panama Seaway (13–2.7 Ma); the Mid-Pleistocene Revolution (MPR; 0.9 Ma), when glacial–interglacial cycles changed from 40 ka to 100 ka periodicity; nutrient availability; evolution of diatoms and C_4 plants; and changes in continental weathering rates.

22.3.2 Changes in style of production – past stratification versus upwelling systems

As diatom production is largely a response to nutrient availability (particularly silicic acid, nitrates, and phosphates), an abundance, or continuous supply (or high pulse frequency; Katz *et al.*, 2004), of nutrients more often than not means high diatom productivity and corresponding high diatom deposition (flux). This has enormous implications not only for the silica cycle (as well as interlinked biogeochemical cycles, e.g. the nitrogen and aluminum cycles, for example) but also for the biological carbon pump (and carbon cycle) involving links with atmospheric CO_2 levels and associated climate feedbacks. To what degree, then, are diatoms implicated in the cause, as well as effect, of climate change? An understanding of their involvement in the biological pump in past oceanic systems needs to be a primary area of research, not just for the Holocene.

Diatom-rich sediments provide clues to the nature of past diatom productivity since they reflect a combination of high diatom production in the overlying surface waters and high flux rates. Traditionally, diatom-rich sediments have been used to infer the presence of upwelling where nutrient-rich intermediate and deep waters are brought to the surface for utilization, or other environments where surface nutrients are continuously renewed (e.g. proximity to river mouths). However, there is another style of diatom production, involving stratification, with some species being specially adapted to stratification. This manifests itself in the fossil record in a wholly different way to that for upwelling systems. Arguably, this has played an equally important role in controlling past diatom production and therefore the biological carbon pump and links to atmospheric CO_2 levels. In a broad sense, the stratification-style of diatom production prevailed during the Cretaceous and early Paleogene (i.e. in a warm greenhouse world; Kemp *et al.*, 1999), which was progressively superseded in dominance by upwelling-style production from the late Paleogene (late middle Eocene) onwards (i.e. in a cooling doubthouse and cool icehouse world), particularly following major ice-sheet expansion on Antarctica at the Oi-1 event. Concomitant with reorganization of the oceans during a developing icehouse world was the progressive development and strengthening of global oceanic frontal zones. These fronts became implicated in some unusual but significant diatom deposits involving giant diatom species. It is now understood that these deposits may well represent some of the most important ancient sinks for silica and carbon outside of the Southern Ocean and coastal upwelling regions in the icehouse world. Here we outline some advances in understanding past stratification and upwelling systems using diatoms, and how these systems have affected diatom production and flux since the Cretaceous. We also touch on the subject of giant diatom deposits at oceanic frontal zones.

22.3.2.1 Stratification Ocean waters can become stratified due to a variety of environmental factors: seasonal sea-ice melt (e.g. coastal Antarctica); seasonal thermocline and nutricline development in semi-enclosed and marginal marine basins enhanced by river input, creating bottom anoxia and preservation of undisturbed sedimentary laminations (e.g. the modern Black Sea, the Cretaceous and Paleogene Arctic, the Pliocene and Pleistocene Mediterranean Sea); and open-ocean settings such as the oligotrophic North Pacific Subtropical Gyre characterized by a nutrient-replete upper layer, a deep nutricline, and persistent deep chlorophyll maximum (e.g. Dore *et al.*, 2008). Diatom-rich sediments that have been deposited as a direct response to stratification are typically laminated (particularly under anoxic bottom-water conditions), and therefore of extreme value in terms of the quality and resolution of information they provide. For example, it is possible to extract seasonal to sub-seasonal information from some diatom-rich sediments as far back as the Cretaceous. See also Leventer *et al.* (this book) for Holocene examples of stratification-adapted diatoms and resultant laminated deposits.

Diatoms respond to stratification either rapidly or more slowly depending on the environment and their survival strategies. For example, stratification brought about by seasonal sea-ice melt in spring and/or summer allows blooming diatoms, such as species of *Chaetoceros*, to respond rapidly to the release of sea-ice-bound nutrients into the melt zone (e.g. deglacial sediments of the East Antarctic margin, Stickley *et al.*, 2005 and references therein). The melt zone itself is density stratified as a consequence of freshwater release from the melting ice, which acts as a "lid" sitting above normal-salinity seawater. Released nutrients are "trapped" in the upper, fresher layers allowing opportunistic species to quickly flourish, and a means by which to trace past sea-ice patterns using sea ice and sea-ice-associated diatoms (Armand & Leventer, 2010). This method is not widely used for pre-Quaternary sediments because relating extinct species to sea ice is less obvious in these older sediments, particularly for the Paleogene and Cretaceous, although compelling evidence now exists for the presence of ancient sea ice from the late-Paleogene diatom fossil record, e.g. the middle-Eocene (~47 Ma) central Arctic (Stickley *et al.*, 2009) and early-Oligocene (~31 Ma) Ross Sea, Antarctica (Olney *et al.*, 2009).

Middle-Eocene (Stickley *et al.*, 2009) and Late Cretaceous (Davies *et al.*, 2009) central-Arctic laminated sediments represent the only examples of sea-ice-driven stratification currently known from a doubthouse and greenhouse world, respectively. Both examples provide evidence for central-Arctic sea ice but only the former (middle Eocene) involved seasonal sea-ice-dwelling diatoms (*Synedropsis* spp., the oldest known sea-ice diatoms; Stickley *et al.*, 2009), while the latter (Late Cretaceous) represents intermittent winter sea ice (on sedimentary evidence only) that didn't involve sea-ice-dwelling diatoms, rather "*Chaetoceros*-type" resting spores, which may indicate the post sea-ice-melt spring bloom (Davies *et al.*, 2009).

However, the seasonal sea-ice record of Stickley *et al.* (2009) and Davies *et al.* (2009) are exceptional examples, and seasonal temperature-driven, rather than sea-ice-driven, stratification was much more widespread prior to the icehouse world of the early Oligocene onwards. Kemp *et al.* (2000, 2006) summarize the three ways specialist diatoms have adapted to this style of stratification: rapid or slow production in low-light conditions, i.e. a "shade-flora" (Kemp *et al.*, 2000); buoyancy regulation; and association with nitrogen-fixing, intracellular, cyanobacterial symbionts. For example, during sea-ice-free intervals of the Late Cretaceous central Arctic, Davies *et al.* (2009) report that "production occurred within a stratified water column, involving specially adapted species in blooms resembling those of the modern North Pacific Subtropical Gyre (e.g. Dore *et al.*, 2008), or those indicated for the Mediterranean sapropels (Kemp *et al.*, 1999)." Davies *et al.* (2009) hypothesize that the lamination-forming diatoms *Hemiaulus* and *Rhizosolenia* in Late Cretaceous sediments on the Alpha Ridge were shade-adapted species growing slowly in summer within a deep chlorophyll maximum before being deposited during a fall dump (Kemp *et al.*, 1999, 2000) after stratification broke down. Davies *et al.* (2009) suggest these ancient diatoms may have survived to contribute significant export in the Late Cretaceous central Arctic Ocean by harboring nitrogen-fixing symbionts analogous to living, bloom-forming *Hemiaulus hauckii* and *Rhizosolenia* spp., which harbor the symbiont *Richelia intracellularis* J. Schmidt (e.g. Dore *et al.*, 2008 and references therein).

Significant production involving *Hemiaulus* spp. is also implicated in temperature- and salinity-driven stratification prior to the initiation of seasonal sea ice in the middle-Eocene central Arctic (Stickley *et al.*, 2008; 2009). Likewise, mat-forming *Hemiaulus hauckii* and *Pseudosolenia calcar-avis* (Schulze) Sundström (also a slow-growing, mat-forming shade-adapted diatom able to exploit a deep-nutrient supply), are implicated in formation of the Pliocene and Quaternary organic carbon-rich Mediterranean sapropels (Kemp *et al.*, 1999), formed under anoxic conditions, underlying stratified surface waters. Kemp *et al.* (1999) suggested that almost all organic matter in the sapropel was diatom-derived, and proposed that such deeper-dwelling diatoms and this stratification style of diatom production was probably common in Paleogene and Cretaceous seas.

Giant fossil and living diatoms, which are shade-adapted and able to exploit a deeper nutrient source, thereby contributing substantial biomass to a deep chlorophyll maximum (or lower euphotic zone) during the longer growing season (e.g. summer) in a stratified water column, and which are characteristic of a fall-dump assemblage, may be equal with modern high-productivity regions (e.g. lightly silicified rapid-response spring bloom) in terms of organic-carbon and opal fluxes (Kemp *et al.*, 1999, 2000). These findings dispel the traditional view that upwelling is solely responsible for the high diatom production and export associated with diatom-rich marine sediments (Kemp *et al.*, 2006). As Davies *et al.* (2009) argue, the recent trend of rising CO_2 levels and warming has led to increased stratification, so a stratification style of production may become more prevalent in today's oceans.

22.3.2.2 Giant diatoms and ocean fronts Production and flux of giant marine diatoms such as *Thalassiothrix* spp., *Ethmodiscus rex*, and *Rhizosolenia* spp. are also implicated with (ancient) oceanic frontal zones (Kemp *et al.*, 2006) as well as the stratified systems described above. Such giant diatoms are selectively segregated and prone to concentrate at oceanic fronts, later undergoing episodic mass flux to the sea floor. These diatoms are apparently not abundant in waters moving towards such frontal systems but end up concentrating there in vast abundance (Kemp *et al.*, 2006). This process is happening today (e.g. *Rhizosolenia* spp. forming "a line in the sea" in the equatorial Pacific; Yoder *et al.*, 1994) and there is ample evidence to suggest that this has happened several times in the past throughout the world's oceans (see Kemp *et al.*, 2000 for review). Some examples are the Neogene mats of *Thalassiothrix longissima* Cleve & Grunow from the eastern equatorial Pacific (Kemp & Baldauf, 1993; Kemp *et al.*, 1995; Alhama *et al.*, 1997) and North Atlantic (Shimada *et al.*, 2008) and Quaternary mats of *Thalassiothrix antarctica* Schimper ex Karsten beneath the Polar Front in the Southern Ocean (Grigorov *et al.*, 2002). These deposits represent an enormous silica and carbon sink, with the highest pelagic sedimentation rates ever estimated from an open-ocean setting. The *T. longissima* mats of the Neogene equatorial Pacific are so extensive they can be traced on seismic sections for distances of up to 5000 km (Kemp *et al.*, 2006). Mats of *Thalassiothrix* spp. have also been documented in other parts of the world associated with an oceanic frontal concentration mechanism, e.g. late-Quaternary *T. longissima* associated with the subarctic convergence (Boden & Backman, 1996) and *T. antarctica* in late-Pliocene deposits associated with the Benguela Upwelling System (Lange *et al.*, 1999), a late-Miocene to early-Pliocene diatom ooze in the subarctic North Pacific dominated by *Thalassiothrix* spp. and *Thalassionema* spp. (Dickens & Barron, 1997), while similar frontal-associated oozes of the giant diatom *Ethmodiscus* are known from the late-Pleistocene equatorial Atlantic (e.g. Gardner & Burckle, 1975; Abrantes, 2001) and the early-Pliocene California margin (Pike, 2000).

Taking the example of the equatorial Pacific, it is clear that diatoms have made a major contribution to marine productivity, and therefore carbon export, in this region for millions of years. Extraordinary lamination-forming mats of the elongate needle-shaped marine diatom *T. longissima* lasted from 15–4.4 Ma, concomitant with major cooling and ocean reorganization (Kemp & Baldauf, 1993). The mats reflect repeated episodes of diatom production and flux "on a scale that is undocumented in the modern ocean" (Kemp & Baldauf, 1993). Mat formation is a reflection of physical surface processes associated with oceanic frontal boundaries which are happening today (Yoder *et al.*, 1994) – buoyant *T. longissima* rose through the water column to form a concentrated surface layer on the warm side of the front and following death, associated with higher temperatures and exhausted nutrients, subsequently sank to the sea floor, overwhelming the benthos (Kemp & Baldauf, 1993; Kemp *et al.*, 1995). Episodes of mat formation may record La Niña events when strong east–west frontal zones developed (Kemp *et al.*, 1995), giving unique insight into equatorial Pacific paleoceanography and biogeochemical cycling for much of the Neogene.

22.3.2.3 Upwelling Sediments underlying coastal and equatorial upwelling systems (i.e. ocean divergence) represent some of the major areas of diatom deposition, with sedimentation rates often exceeding 10 cm ka^{-1} (e.g. Juillet-Leclerc & Schrader, 1987; Abrantes, 1991; Kemp & Baldauf, 1993). The most productive upwelling cells are usually located on the western sides of continents (e.g. coast of California, Peru, and southwestern Africa), and are associated with areas of high atmospheric pressure, trade winds, deserts, nutrient-rich cool surface waters, high primary productivity, commercial fishing grounds, and thick guano deposits (Dupont *et al.*, 2005). These upwelling areas are also strongly linked to the northward flow of Antarctic Bottom Water and upwelling of Antarctic Intermediate Water (Shannon & Hunter, 1988; Romero *et al.*, 2003), so the onset and intensification of upwelling is closely related to the onset (~34 Ma) and intensification (see Zachos *et al.*, 2008; DeConto *et al.*, 2008) of Antarctic glaciation (Leckie & Webb, 1983; Cortese *et al.*, 2004). The onset perhaps (~23 Ma; DeConto *et al.*, 2008) of major northern-hemisphere glaciation and its intensification (2.5 Ma) (Cortese *et al.*, 2004) would also have had a dramatic effect on upwelling response. Modern and ancient upwelling regions act as huge sinks for silica and carbon (and therefore atmospheric CO_2) through the biological carbon pump in which diatoms are key players. However, they can be a source of CO_2; intensity of upwelling (including rate and intensity of production and nutrient recycling) and overall upwelling environment (coastal or oceanic) as well as bottom-water conditions (oxygen-rich or oxygen-depleted) determines whether net CO_2 is sequestered or outgassed (e.g. see Pollock, 1997).

In the late middle Eocene during the post-EECO long-term (middle to late Eocene) cooling trend towards the Oi-1 event (Zachos *et al.*, 2001; 2008) wind-driven coastal-upwelling-derived diatomaceous sediments started to appear, for example in New Zealand, California, Peru, and Kamchatka (e.g. Barron & Baldauf, 1989) probably as a result of progressive increase in thermal gradients between high and low latitudes. However, "modern" upwelling systems (along the western side of continents) probably became fully established only by the early Miocene (Summerhayes *et al.*, 1992) or middle to late Miocene (Pollock, 1997), coinciding with intensification of Antarctic glaciation. For instance, the modern Benguela upwelling system was probably initiated about 10 Ma (e.g. Siesser, 1980; Krammer *et al.*, 2005), while rapid desiccation in Namibia at 2.2 Ma has been associated with increased upwelling and decreasing SSTs along the coast (Dupont *et al.*, 2005). In all of these cases the depth of the thermocline and wind intensity has determined the likelihood of upwelling; a shallower thermocline favors the chances of upwelling. Changes in nutrient supply, upwelling rate, response and intensity of diatom production, and the (rate of) export and sequestration of organic carbon and its impact on atmospheric CO_2 levels are interlinked. These are key considerations for the effect and possibly impact (e.g. Pollock, 1997) of upwelling-driven diatom production on Cenozoic global climate change. Most diatom studies of coastal upwelling regions have focused on living or late-Quaternary assemblages (e.g. Schuette & Schrader, 1979, 1981; Abrantes, 1991; Lange *et al.*, 1998; Abrantes *et al.*, 2007). These studies have shown that assemblages indicative of periods of upwelling conditions are generally dominated by centric (notably chain-forming species such as *Chaetoceros* and *Thalassiosira*) or araphid (such as *Thalassionema* and *Thalassiothrix*) diatoms (e.g. Lange *et al.*, 1998; Abrantes *et al.*, 2007). Schrader and Sorknes (1990) suggested three Quaternary diatom assemblages indicative of past upwelling strength off Peru; a well-preserved delicate diatom assemblage composed of *Delphineis*, *Skeletonema* and *Ditylum* indicating the strongest upwelling, *Cyclotella* and *Actinoptychus* equating to less strong upwelling, and *Thalassionema* and *Chaetoceros* representing weaker upwelling. Today, *Delphineis karstenii* (Boden) Fryxell lives in offshore areas of the upwelling cell, while *Skeletonema costatum* (Greville) Cleve and *Ditylum brightwellii* (T. West) Grunow are common in Peruvian coastal waters (Abrantes *et al.*, 2007). As mentioned above, diatom production in surface waters of upwelling regions is very high, and is ultimately responsible for the thick deposits of siliceous ooze underlying these areas. In fact, many of the world's diatomite deposits were formed underneath active coastal upwelling centers (see Harwood, this volume). In addition, all of the world's major guano industries are alongside coastal upwelling areas (e.g. California, Peru, Chile, Namibia; Cushing, 1971), sometimes with large numbers of marine diatoms present in the guano deposits, which have been studied by diatomists for over a century (e.g. Greville, 1859; Mills, 1881).

22.3.3 Ice-sheet and sea-ice history

Our current icehouse world is characterized by the presence of large ice sheets and extensive sea ice in both polar regions, which modify the climate system via a number of important feedback processes such as the ice–albedo effect, effects on the thermocline circulation patterns, precipitation, and therefore weathering. Sea ice, for example, directly affects ocean–atmospheric exchange, while changes in ice-sheet volume affects sea level. The impact of the cryosphere is therefore global, so it is important to understand the evolution, development, and fluctuations in both sea ice and ice sheets (including glacial processes) through time for both hemispheres. Since diatoms are ubiquitous in the polar regions, they are one of the key microfossil groups that can help achieve these goals.

Polar sea-ice records and ice-sheet histories are ultimately interlinked. Since drilling in the Arctic has only recently become a possibility (Moran *et al.*, 2006) there is just one Cenozoic drillsite in the Arctic (IODP Expedition 302; Backman & Moran, 2009) that utilizes diatoms to determine ancient sea-ice history (Stickley *et al.*, 2009). A lack of Arctic long-core drillsites is likely to be the situation for several more years to come while discussions on the future, and aims, of ocean drilling beyond 2013 are still being resolved. This means there is unlikely to be any further insight into Arctic Cenozoic ice-sheet/sea-ice histories from new sites for some time beyond the information which is summarized by Backman and Moran (2009) and that by Stickley *et al.* (2009), and what is known from earlier ODP and DSDP drilling in the Norwegian and Greenland seas (Eldrett *et al.*, 2007, 2009).

Although still relatively poor compared to other oceanic regions of the world, the recovery of pre-Quaternary sediments from the Southern Ocean and the Antarctic margin is somewhat better than it is for the Arctic. For over 40 years, the DSDP and its successor, the ODP, have drilled several key locations in the Southern Ocean and circum-Antarctic margin starting with DSDP Leg 28 (Margin; Ross Sea) in the early 1970s through a series of offshore and nearshore DSDP and ODP expeditions (notably DSDP legs 28, 29, and 71 and ODP legs 113, 114, 119, 120, and 177) culminating in ODP legs 188 (Margin; Prydz Bay) and 189 (Southern Ocean; Tasmanian Gateway) in early 2000. Drilling directly on the Antarctic shelf is the best way to elucidate ice-sheet history through the Cenozoic, and in parallel with the circum-Antarctic DSDP and ODP programs, drilling continued through the 1970s, 1980s, and 1990s with a series of highly successful multinational Antarctic shelf drilling programs focusing mainly on the Ross Sea region (see Olney *et al.*, 2007 for a short review). Two of the more recent of these are the Cape Roberts Project (Barrett & Ricci, 2000) which recovered biostratigraphically useful diatomaceous sediments back to late Eocene–early Oligocene (e.g. Harwood & Bohaty, 2001; Olney *et al.*, 2005, 2007, 2009; Scherer *et al.*, 2001) from the Ross Sea in the late 1990s, and the SHALDRIL (Shallow Drilling along the Antarctic Continental Margin) project (2005 and 2006, East Antarctic Peninsula) which also recovered diatomaceous sediment (albeit stratigraphically discontinuous) back to late Eocene–early Oligocene (Anderson *et al.*, 2006). The most recent Antarctic continental-shelf drilling program is the on-going ANDRILL (Antarctic Geological Drilling) project (e.g. Naish *et al.*, 2009) in the Ross Sea region. The ANDRILL project began in 2006; this and IODP Expedition 318 to the Wilkes Land Margin, Antarctica (Expedition 318 Scientists, 2010) are the most recent ventures to elucidate Cenozoic sea-ice and ice-sheet histories of Antarctica. Cenozoic diatoms have been and continue to be of utmost importance in all of these Antarctic margin drilling programs not only for critical biostratigraphy but also to help elucidate paleoenvironments during Antarctic cryosphere evolution in an icehouse world. While a clearer picture of Neogene events is emerging, there still remains relatively little recovery of diatomaceous material older than early Oligocene in the Antarctic region. Apart from the Late Cretaceous and K–T diatomaceous deposits already noted in this chapter (Harwood, 1988; Harwood & Gersonde, 1990; Gersonde & Harwood, 1990), middle-Eocene erratics from southern McMurdo Sound (Harwood & Bohaty, 2000) and middle-Eocene sediments from ODP Leg 189 (Stickley *et al.*, 2004a) are some of the more noteworthy pre-icehouse diatom-rich deposits in the region.

22.3.3.1 Sea ice Diatoms proliferate in sea-ice and marginal sea-ice environments making them fundamental to sea-ice-related ecosystems. It has recently been discovered that sea-ice diatoms are an important source of dimethyl sulfide (DMS) (e.g. Levasseur *et al.*, 1994) which is implicated in the formation of clouds (e.g. Levasseur *et al.*, 1994) and therefore important climate feedbacks. While the influence of ancient sea-ice diatoms to past atmospheric conditions via the contribution of DMS has yet to be documented, diatom fossils themselves are widely and routinely employed as paleo-sea-ice proxies in marine core sediments (Armand & Leventer, 2010). For example, extant diatom taxa associated with both sea ice and the modern sea-ice edge are commonly used to trace sea-ice extent (summer and winter), duration, and cover for short intervals of the relatively recent past (Holocene and Quaternary) in both polar regions (Armand & Leventer, 2010) (see also Leventer *et al.*, this volume).

Evaluating pre-Quaternary sea-ice patterns is more difficult due to the uncertainty in relating extinct taxa directly to ancient sea ice, although it would be possible to infer a relationship by analogy with living taxa, including preservational characteristics, and by association with sea-ice rafted debris (Stickley *et al.*, 2009). In the Arctic, pre-Quaternary sea-ice diatom records are exceptionally rare, speculatively a reflection of a lack of drillsites across the region rather than the absence of fossil sea-ice diatoms here. There is only one report of pre-Quaternary sea-ice diatoms from the central Arctic (Stickley *et al.*, 2009); pre-Quaternary Antarctic sea-ice diatom records are somewhat better known than in the Arctic but these are still relatively rare to date.

Forty-seven-million-year-old (middle-Eocene) *Synedropsis* spp. preserved in central Arctic core sediments drilled by IODP Expedition 302, are the earliest known fossil record of sea-ice diatoms (Stickley *et al.*, 2009), pre-dating existing sea-ice-related diatom records by ~16 Ma (in Antarctica) (Stickley *et al.*, 2009). Moreover, these diatoms comprising up to ~60% of the diatom assemblage are delicate, sea-ice-dwelling species uniquely adapted to surviving the cold, dark, polar winter. The difference is that most Holocene and Quaternary sea-ice reconstructions using diatoms involve robust species which are associated with the sea-ice-edge and which bloom within the melt zone. It is very rare to find sea-ice-dwelling diatoms preserved in the fossil record, particularly in such abundance. The middle-Eocene Arctic record is therefore unique.

Fossil *Synedropsis* in Antarctic coastal sediments have been tentatively used to infer the presence of past sea ice, although they are never abundantly preserved there. One of the oldest examples is from a sedimentary erratic in the McMurdo Sound (Ross Sea region) of late-Miocene age (~8.5–6.5 Ma) containing a sea-ice diatom assemblage indicative of modern sea ice and the sea-ice margin, and dominated by *Denticulopsis delicata* Yanagisawa and Akiba and *Fragilariopsis vanheurckii* (M. Peragallo) Hustedt but which also includes *Synedropsis* spp. These diatoms indirectly indicate a glacial event (cooling) in the interior East Antarctic Ice Sheet during the late Miocene because Southern Ocean Antarctic sea ice-growth reflects conditions in the interior ice sheet (Harwood & Bohaty, 2007; DeConto *et al.*, 2007).

Synedropsis cheethamii Olney and related *Creania lacyae* Olney (Olney *et al.*, 2009), from lower-Oligocene sediments (~31 Ma) recovered by the Cape Roberts Project in the Victoria Land Basin (Ross Sea), may be the earliest known sea-ice diatoms in Antarctica (Olney *et al.*, 2009; Stickley *et al.*, 2009). This record concurs with the global climate model of DeConto *et al.* (2007), which predicts sea ice initiated rapidly in Antarctica as a response to grounded ice-sheet growth, i.e. continental ice (~34 Ma) preceded sea-ice formation in Antarctica.

A quantitative approach in elucidating pre-Quaternary sea-ice patterns was taken by Whitehead *et al.* (2005) who reconstructed sea-ice concentration at Ocean Drilling sites 1165 and 1166 (Prydz Bay and Kerguelen Plateau, Antarctica) for the Pliocene using the "*Eucampia* index" (Kaczmarska *et al.*, 1993) – the relative proportion of intercalary (mid-colony) to terminal valves of the diatom *Eucampia antarctica* (Castracane) Mangin. In winter, Antarctic sea ice reduces normal planktonic production, resulting in shorter *E. antarctica* colonies and therefore proportionally less intercalary valves, while relatively longer colonies are produced during times when sea-ice concentration is reduced, increasing the *Eucampia* index. The index was first applied by Kaczmarska *et al.* (1993) to late-Pleistocene sediments from the Kerguelen Plateau (ODP Site 745B). Whitehead *et al.* (2005) indicate that the *Eucampia* index remained relatively high through most of the Pliocene and therefore that winter sea-ice concentration was lower during the Pliocene at these drillsites than at present. Although the *Eucampia* index has not been ground-truthed for modern sediment (Armand & Leventer, 2010), these results concur with other studies around the Antarctic margin which suggest sea ice fell to minimal extent during much of the Pliocene, based on the absence of sea-ice floras in Pliocene sections. This may indicate warming in the interior and ice-sheet collapse during this phase (indeed, orbtially driven cyclic collapse and growth of the Pliocene West Antarctic Ice Sheet has recently been demonstrated; Naish *et al.*, 2009). Increases in sea-ice diatom floras in late-Pliocene and Pleistocene intervals indicate that sea-ice concentrations recovered again during these times (Harwood & Bohaty, 2007).

22.3.3.2 Ice-sheet history/glacial processes The long history of Antarctic-shelf drilling programs mostly in the Ross Sea (also the northwest Weddell Sea) region have recovered diatomaceous deposits (albeit often stratigraphically discontinuous and/or truncated) which have helped elucidate cryospheric evolution and development through the current icehouse world, including variations in both the East and West Antarctic ice sheets (e.g. Harwood *et al.*, 1989; Scherer, 1991; Scherer *et al.*, 1998, 2008). The Antarctic-shelf drilling program, ANDRILL, recovered over 1280 m of sediment core (the AND-1B core) with overall 98% recovery during its first drilling season (2006–2007) representing the longest and most complete geological record from the Antarctic continental margin to date (Naish *et al.*, 2007).

The AND-1B drilled through the thick McMurdo Ice Shelf and contains an unprecedented record of Antarctic continental-shelf sediments archiving climate change and cryospheric fluctuation through a critical phase in Earth history. Upper Miocene to Pleistocene sediments were obtained comprising alternating diatomites and glacial diamictites with episodic volcanics but only the upper ~600 m of the AND-1B core are diatom-rich (Pliocene–Pleistocene). The diamictites represent glacial advance (glacials), while the diatomites, most of which are nearly pure biogenic silica (but low diversity and often fragmented, consistent with subglacial or sedimentary compaction), record high diatom productivity in open-marine environments during climatic phases that were generally warmer than present (interglacials), with minimal summer sea ice, and when the shelves were largely ice free (Scherer *et al.*, 2007b; Naish *et al.*, 2009). The AND-1B diatomites are between 1 m and nearly 100 m in thickness and often laminated (Scherer *et al.*, 2007b). Diatom biostratigraphy, radiometric ages, and magnetostratigraphy constrain magnetic polarity stratigraphy in the AND-1B core and analysis has shown that the early-Pliocene (~3–5 Ma) interval records obliquity-paced (40 ka cycles) oscillations in the West Antarctic Ice Sheet (Naish *et al.*, 2009). Moreover, a much needed diatom biostratigraphic zonation for the Antarctic continental shelf is underway (Scherer *et al.*, 2007b), although this task will not be easy as the marine stratigraphic record on the Antarctic shelf is incomplete and correlation with established offshore Southern Ocean zonation schemes will be difficult.

As well as for age-control, the diatoms in the AND-1B core are proving useful for determining specific environments such as sea ice versus ice-free phases and SSTs (Scherer *et al.*, 2007b). Another use is by careful analysis of the abundance and style of fragmentation which may reveal glacial processes. Although the AND-1B core diamictites are largely barren of diatoms and their fragments (Scherer *et al.*, 2007b), it is known from other localities and from experimentation that valve characteristics remain identifiable even after degradation by grounded ice and subglacial transportation, i.e. sub-ice streams (Scherer *et al.*, 2004; Sjunneskog & Scherer, 2005), even if shearing is intense (Scherer, 1991; Scherer *et al.*, 2004). The relative proportion of centric to pennate diatom valves may distinguish glacially sheared sediments from undisturbed hemipelagic sediments, and from sediments fragmented by normal stress compaction on the Antarctic continental shelf (Scherer *et al.*, 2004). Unequal changes in absolute and relative diatom abundances occur with compaction and shearing, depending on the shape and size of the diatom, and it was found that the ratio of unbroken centric to pennate diatoms provides a reliable gauge of past sub-ice-stream shearing (proportionally more centrics = more shearing; Scherer *et al.*, 2004). Further, by analyzing the relative proportion and type of reworked diatoms to *in situ* diatoms, in combination with whole to fragmented valves, Sjunneskog and Scherer (2005) provide a method by which to classify and distinguish between late-Quaternary tills deposited beneath ice streams from those emplaced by slow-moving ice. This information is important in ice-sheet reconstruction and should be applicable to older sediments.

22.3.4 Diatoms and the evolution of ocean-going mammals and birds

As mentioned above, marine diatom productivity increased across the E–O transition, due to the initiation of coastal upwelling and the increased supply of silica to the oceans via C_3

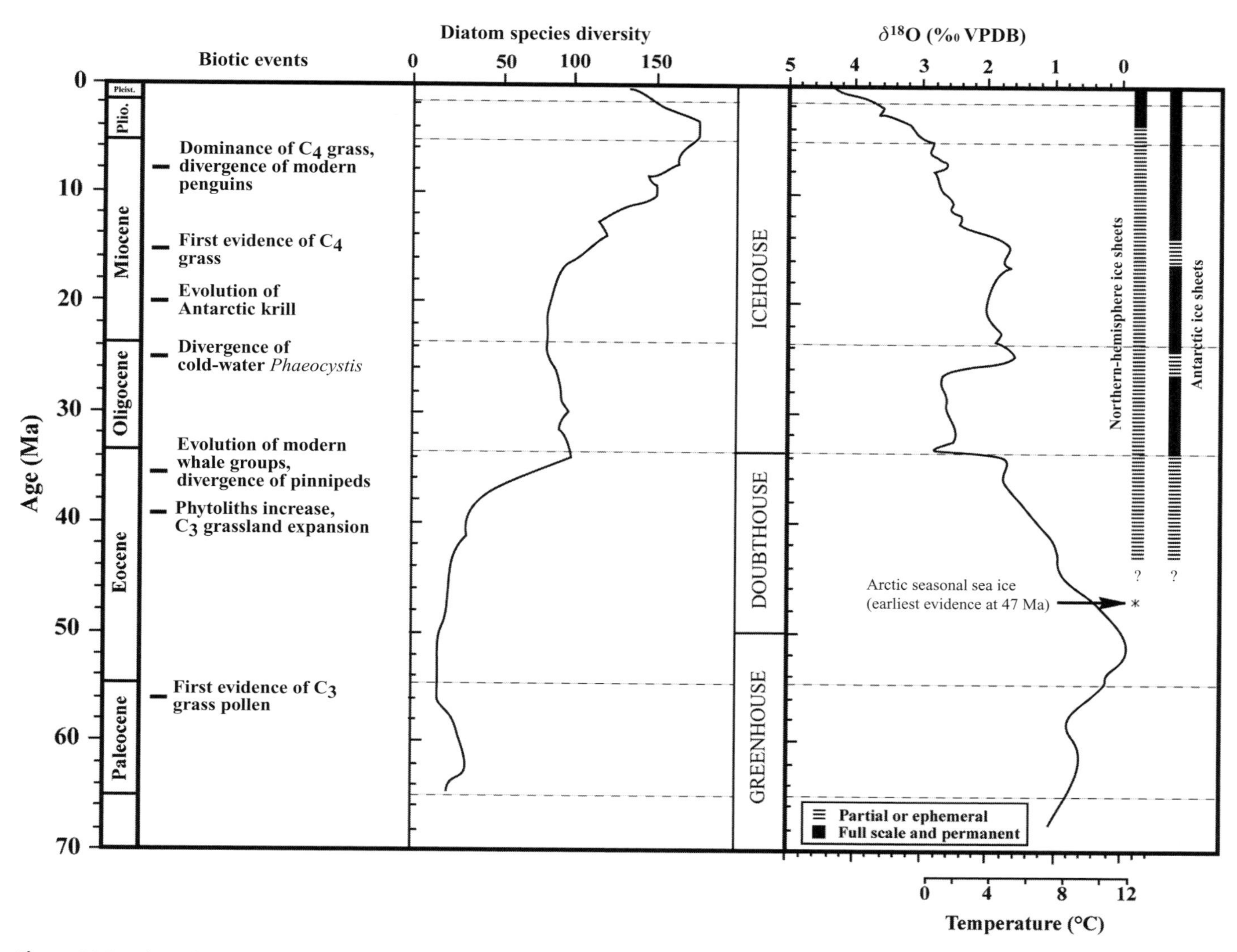

Figure 22.7 Selected biotic events plotted against diatom species diversity (the latter based on Falkowski *et al.* (2004)) and the oxygen isotope record (see Figure 22.1 for explanation).

grasses and enhanced preservation. The prolonged availability of diatoms as a food source allowed herbivorous (e.g. krill) and carnivorous (e.g. cetaceans, pinnipeds, and penguins) animals on different trophic levels to increase their numbers and rapidly diversify (Figure 22.7). For instance, Antarctic krill species (*Euphausia superba* Dana and *Euphausia crystallorophias* Holt & Tattersall), which directly feed on diatoms, diverged from sub-antarctic species (*Euphausia vallentini* Stebbing) following the formation of the ACC, according to molecular clock calculations (Patarnello *et al.*, 1996; Zane & Patarnello, 2000). The authors suggested a date of ~20 Ma, which would roughly equate with a deeper flow of the ACC through the Drake Passage, although ACC initiation probably occurred much earlier (Pfuhl & McCave, 2005; Barker *et al.*, 2007; Lyle *et al.*, 2007). However, deposition of the "modern" diatomaceous ooze belt around Antarctica intensified around 2.5 Ma (Diekmann *et al.*, 2003), following establishment of the Polar Front as an opal deposition centre (Cortese *et al.*, 2004).

Primitive whales evolved in the early Eocene around 53 Ma but did not become fully aquatic until about 40 Ma, when they developed their modern body plan (Thewissen & Williams, 2002). Their origin in the Eocene probably coincided with increased marine productivity in shallow waters associated with closure of the Tethys Sea along the shores of India and Pakistan (Gingerich *et al.*, 1983). Deepening of the Tasmanian Gateway through the latest Eocene and earliest Oligocene (Stickley *et al.*, 2004b), and later, or concomitant, opening of the Drake Passage allowed the Mysticeti and Odontoceti to diversify in the late Oligocene (Fordyce, 1980; Berger, 2007). Intensification of upwelling and associated diatom production may have been a causal factor in the subsequent early radiation of whales, while diversification of cetaceans and initiation of the pinniped lineage in the earliest Miocene were likely caused by a further intensification of upwelling along western coastlines around the planet. Furthermore, the occurrence of well-preserved fossil whale bones in diatomaceous deposits of the Miocene–Pliocene Pisco Formation has been attributed to the high diatom accumulation rate associated with the Peruvian coastal upwelling system (Brand *et al.*, 2004).

Although penguins first appeared over 60 Ma, they diversified several times in response to cooling trends during the early Eocene, the Oligocene, and the late Miocene (inferred from Clarke *et al.*, 2007). On the other hand, extant lineages are thought to have radiated about 8 Ma. Since initiation of the ACC appears to have occurred in the late Oligocene, there is a strong possibility that the timing of increases in krill, penguin, and whale diversity are closely linked. See Gregory-Eaves and Keatley (this volume) for further information on tracking past fish, seabirds, and other wildlife populations using diatoms.

22.4 Summary

The marine diatom fossil record is one of dynamic change from which a wealth of paleoceanographic and climatic information can be extracted. Climate change and oceanographic and tectonic reorganization have influenced global opal sedimentation patterns and diatom evolution since their first appearance in the fossil record. The development of continental ice sheets, ice shelves, and sea ice has affected the intensity and flow direction of global circulation, which affects oceanic portioning of nutrients essential to diatom production and preservation.

However, pre-Holocene fossil diatoms have been traditionally overlooked compared to their Holocene or even Quaternary counterparts or, in particular, calcareous microfossil groups; yet in specialized settings diatoms as old as Cretaceous can provide quality information to rival modern sediment-trap data, and on timescales from as short as sub-seasonal. Their importance in opal-accumulation studies and biostratigraphy is still highly regarded, but the realization that pre-Holocene diatoms can provide the detail of information previously thought possible only from Holocene sediments have been starting to be appreciated in more recent years.

Acknowledgments

The authors would like to thank Andrey Gladenkov and the two editors for their constructive criticisms and suggestions. We are also grateful to many of our colleagues for discussion on various topics, especially those who provided copies of their submitted chapters.

References

Abrantes, F. F. (1991). Variability of upwelling off NW Africa during the latest Quaternary: diatom evidence. *Paleoceanography*, **6**, 431–60.

Abrantes, F. (2001). Assessing the *Ethmodiscus* ooze problem: new perspective from a study of an eastern equatorial Atlantic core. *Deep-Sea Research I*, **48**, 125–35.

Abrantes, F., Lopes, C., Mix, A., & Pisias, N. (2007). Diatoms in southeast Pacific surface sediments reflect environmental properties. *Quaternary Science Reviews*, **26** (1–2), 155–69.

Akhmetiev, A. M. (1996). Ecological crises of the Paleogene and Neogene in extratropical Eurasia and their putative causes. *Paleontological Journal*, **30**, 738–48.

Akhmetiev, M. A. (2007). Paleocene and Eocene floras of Russia and adjacent regions: climatic conditions of their development. *Paleontological Journal*, **41**, 1032–9.

Akhmetiev, M. A. & Beniamovski, A. N. (2009). Paleogene floral assemblages around epicontinental seas and straits in northern central Eurasia: proxies for climatic and paleogeographic evolution. *Geologica Acta*, **7** (1–2), 297–309.

Alexandre, A., Meunier, J.-D., Colin, F., & Koud, J. M. (1997). Plant impact on the biogeochemical cycle of silicon and related weathering processes. *Geochimica et Cosmochemica Acta*, **61**, 677–82.

Alhama, F., Lopez-Sanchez, J. F., Gonzalez-Fernandez, C. F., Dickens, G. R., & Barron, J. A. (1997). A rapidly deposited pennate diatom ooze in upper Miocene–lower Pliocene sediment beneath the North Pacific Polar Front. *Marine Micropaleontology*, **31**, 177–82.

Alldredge, A. L. & Gotschalk, C. C. (1989). Direct observations of the mass flocculation of diatom blooms: characteristics, settling velocity and formation of diatom aggregates. *Deep-Sea Research*, **26**, 159–71.

Anderson, J. B., Wellner, J. S., Bohaty, S., Manley, P. L., & Wise, S. W., Jr. (2006). Antarctic Shallow Drilling Project provides key core samples. *EOS, Transactions, American Geophysical Union*, **87** (39), DOI:10.1029/2006EO390003.

Andrews, G. W. (1986). Evolutionary trends in the marine diatom genus *Delphineis* G.W. Andrews. In *Proceedings of the 9th International Diatom Symposium*, ed. F.E. Round, Bristol & Königstein: Biopress Ltd. & Koeltz Scientific Books, pp. 197–206.

Andrews, G. W. & Stoelzel, V. A. (1982). Morphology and evolutionary significance of *Perissonoë*, a new marine diatom genus. In *Proceedings of the 7th International Diatom Symposium*, ed. D.G. Mann, Königstein: Koeltz Scientific Books, pp. 225–40.

Armand, L. K. & Leventer, A. (2010). Palaeo sea ice distribution and reconstruction derived from the geological record. In: *Sea Ice An Introduction to its: Physics, Chemistry and Biology, 2nd Edition*, ed. D. N. Thomas & G. S. Dieckmann, Oxford: Wiley-Blackwell, pp. 469–530.

Backman, J. & Moran, K. (2009). Expanding the Cenozoic paleoceanographic record in the central Arctic Ocean: IODP Expedition

302: synthesis. *Central European Journal of Geosciences*, **1**, 157–75, DOI:10.2478/v10085-009-0015-6.

Backman, J., Moran, K., McInroy, D. B., Mayer, L. A., & the Expedition 302 Scientists (2006). *Proceedings of the Integrated Ocean Drilling Program*, 302: Edinburgh (Integrated Ocean Drilling Program Management International, Inc.). doi:10.2204/iodp.proc.302.2006.

Baldauf, J. G. & Barron, J. A. (1990). Evolution of biosiliceous sedimentation patterns – Eocene through Quaternary: paleoceanographic response to polar cooling. In *Geological History of the Polar Oceans: Arctic versus Antarctic*, ed. U. Bleil, J. & Thiede, J., Dordrecht: Kluwer Academic Publishers, pp. 575–607.

Barker, P. F., Filippelli, G. M., Florindo, F., Martin, E. E., & Scher, H. D. (2007). Onset and role of the Antarctic Circumpolar Current. *Deep-Sea Research II*, **54**, 2388–98.

Barrett, P. J. & Ricci, C. A. (eds.) (2000). Studies from the Cape Roberts Project, Ross Sea, Antarctica, Scientific Results of CRP-2/2A, Parts I and II. *Terra Antartica*, **7**, 211–665.

Barron, J. A. (1985a). Diatom biostratigraphy of the CESAR 6 core, Alpha Ridge. *Geological Survey of Canada, Paper* **84–22**, Report 10, 137–48.

Barron, J. A. (1985b). Miocene to Holocene planktic diatoms. In *Plankton Stratigraphy, Volume 2*, ed. H. M. Bolli, J. B. Saunders, & K. Perch-Nielsen, Cambridge: Cambridge University Press, pp. 763–809.

Barron, J. A. (1993). Diatoms. In *Fossil Prokaryotes and Protists*, ed. J. H. Lipps, Boston: Blackwell Scientific Publishers, pp. 155–67.

Barron, J. A. (2003). Appearance and extinction of planktonic diatoms during the past 18 m.y. in the Pacific and Southern oceans. *Diatom Research*, **18**, 203–24.

Barron, J. A. & Baldauf, J. G. (1989). Tertiary cooling steps and paleoproductivity as reflected by diatoms and biosiliceous sediments. In *Productivity of the Oceans: Present and Past*, ed. W. H. Berger, V. S. Smetacek, & G. Wefer, New York, NY: John Wiley, pp. 341–54.

Barron, J. A. & Baldauf, J. G. (1995). Cenozoic marine diatom biostratigraphy and applications to paleoclimatology and paleoceanography. In *Siliceous Microfossils*, ed. C. D. Blome, P. M. Whalen, & K. M. Reed, The Paleontological Society Short Course 8, pp. 107–18.

Barron, J. A. & Mahood, A. D. (1993). Exceptionally well-preserved early Oligocene diatoms from glacial sediments of Prydz Bay, East Antarctica. *Micropaleontology*, **39**, 29–45.

Belt, S. T., Allard, W. G., Massé, G., Robert, J.-M., & Rowland, S. J. (2000). Highly branched isoprenoids (HBIs): identification of the most common and abundant sedimentary isomers. *Geochimica et Cosmochimica Acta*, **64**, 3839–51.

Belt, S. T., Massé, G., Rowland, S. J., *et al.* (2007). A novel chemical fossil of palaeo sea ice: IP25. *Organic Geochemistry*, **38**, 16–27.

Berger, W. H. (2007). Cenozoic cooling, Antarctic nutrient pump, and the evolution of whales. *Deep-Sea Research II*, **54**, 2399–421.

Berggren, W. A. (2002). Paleogene of the eastern Alps. *PALAIOS*, **17**, 631–2.

Bigelow, P. R. & Alexander, C. G. (2000). Diatoms on the cirri of tropical barnacles. *Journal of the Marine Biological Association of the United Kingdom*, **80**, 737–8.

Boden, P. & Backman, J. (1996). A laminated sediment sequence from the northern north Atlantic Ocean and its climatic record. *Geology*, **24**, 507–10.

Bramlette, M. N. (1946). The Monterey Formation of California and the origin of its siliceous rocks. *United States Geological Survey Professional Paper*, **212**.

Brand, L. R., Esperante, R., Chadwick, A. V., Poma Porras, O., & Alomía, M. (2004). Fossil whale preservation implies high diatom accumulation rate in the Miocene–Pliocene Pisco Formation of Peru. *Geology*, **32**, 165–8.

Brun, J. & Tempère, J. (1889). Diatomées fossiles du Japon. Espèces Marines et Nouvelles des Calcaires Argileux de Sendai et de Yedo. *Mémoires de la Société de Physique et d'Histoire Naturelle de Genève*, **30** (9): 1–75.

Brzezinski, M., Phillip, D., Chavez, P., Frederich, G., & Dugdale, R. (1997). Silica production in the Monterey, California, upwelling system. *Limnology and Oceanography*, **42**, 1694–705.

Burns, V. M., & Burns, R. G. (1978). Authigenic todorokite and phillipsite inside deep-sea manganese nodules. *American Mineralogist*, **63**, 827–31.

Cervato, C. & Burckle, L. (2003). Pattern of first and last appearance in diatoms: oceanic circulation and the position of the Polar fronts during the Cenozoic. *Paleoceanography*, **18**, 1055, DOI:10.1029/2002PA000805.

Chambers, P. M. (1996). Late Cretaceous and Palaeocene marine diatom floras. Unpublished Ph. D thesis, University College London.

Chang, K.-H. & Park, S.-O. (2008). Early Cretaceous tectonism and diatoms in Korea. *Acta Geologica Sinica*, **82** (6), 5–61.

Chapman, F. (1906). On concretionary nodules with plant-remains found in the Old Bed of the Yarra at S. Melbourne; and their resemblance to the calcareous nodules known as 'coal-balls'. *Geological Magazine*, 5th Series, **3**, 553–6.

Chin, K., Bloch, J., Sweet, A., *et al.* (2008). Life in a temperate polar sea: a unique taphonomic window on the structure of a Late Cretaceous Arctic marine ecosystem. *Proceedings of the Royal Society B-Biological Sciences*, **275**, 2675–85.

Clarke, J. A., Ksepka, D. T., Stucchi, M., *et al.* (2007). Paleogene equatorial penguins challenge the proposed relationship between biogeography, diversity, and Cenozoic climate change. *Proceedings of the National Academy of Sciences of the USA*, **104**, 11545–50.

Coale, K. H., Johnson, K. S., Fitzwater, S. E., *et al.* (1996). A massive phytoplankton bloom induced by an ecosystem-scale iron fertilization experiment in the equatorial Pacific Ocean. *Nature*, **383**, 495–501.

Cody, R. D., Levy, R., & Harwood, D. M. (2008). Thinking outside the zone: high-resolution quantitative diatom biochronology for the Antarctic Neogene. *Palaeogeography, Palaeoclimatology, Palaeoecology*, **260** (1–2), 92–121.

Cortese, G., Gersonde, R., Hillenbrand, C.-D., & Kuhn, G. (2004). Opal sedimentation shifts in the world ocean over the last 15 Myr. *Earth and Planetary Science Letters*, **224**, 509–27.

Croll, D. A. & Holmes, R. W. (1982). A note on the occurrence of diatoms on the feathers of diving seabirds. *The Auk*, **99**, 765–6.

Cushing, D. H. (1971). Upwelling and the production of fish. In *Advances in Marine Biology*, ed. F.S. Russell & M. Yonge, London: Academic Press Inc., vol. 9, pp. 255–334.

Darwin, C. (1846). An account of the fine dust which falls on vessels in the Atlantic Ocean. *Quarterly Journal of the Geological Society (London)*, **2**, 26–30.

Davies A., Kemp, A. E. S., & Pike, J. (2009). Late Cretaceous seasonal ocean variability from the Arctic. *Nature*, **460**, 254–9.

DeConto, R., Pollard, D., & Harwood, D. (2007). Sea ice feedback and Cenozoic evolution of Antarctic climate and ice sheets. *Paleoceanography*, **22**, PA3214, DOI:10.1029/2006PA001350.

DeConto, R. M., Pollard, D., Wilson, P. A., *et al.* (2008). Thresholds for Cenozoic bipolar glaciation. *Nature*, **455**, 652–6.

De La Rocha, C. L. (2006). Opal-based proxies of paleoenvironmental conditions. *Global Biogeochemical Cycles*, **20**, GB4S09, DOI:10.1029/2005GB002664.

De La Rocha, C. L., Brzezinski, M. A., DeNiro, M. J., & Shemesh, A. (1998). Silicon-isotope composition of diatoms as an indicator of past oceanic change. *Nature*, **395**, 680–3.

Dickens, G. & Barron, J. A. (1997). A rapidly deposited pennate diatom ooze in upper Miocene-lower Pliocene sediment beneath the North Pacific Polar Front. *Marine Micropaleontology*, **31**, 177–82.

Diekmann, B., Fälker, M., & Kuhn, G. (2003). Environmental history of the south-eastern South Atlantic since the middle Miocene: evidence from the sedimentological records of ODP sites 1088 and 1092. *Sedimentology*, **50**, 511–29.

Diester-Haass L. & Zahn, R. (2001). Paleoproductivity increase at the Eocene-Oligocene climatic transition: ODP/DSDP sites 763 and 592. *Palaeogeography, Palaeoclimatology, Palaeoecology*, **172**, 153–70.

Dore, J. E., Letelier, R. M., Church, M. J., Lukas, R., & Karl, D. M. (2008). Summer phytoplankton blooms in the oligotrophic North Pacific Subtropical Gyre: historical perspective and recent observations. *Progress in Oceanography*, **76**, 2–38.

Dupont, L. M., Donner, B., Vidal, L., Pérez, E. M., & Wefer, G. (2005). Linking desert evolution and coastal upwelling: Pliocene climate change in Namibia. *Geology*, **33**, 461–4.

Dzinoridze, R. N., Jousé, A. P., Koroleva-Golikova, G. S., *et al.* (1978). Diatom and radiolarian Cenozoic stratigraphy, Norwegian Basin; DSDP Leg 38. *Initial Reports of the Deep Sea Drilling Project*, 38–39–40–41 Supplement, 289–427.

Edwards, A. R. (1991). The Oamaru Diatomite. *New Zealand Geological Survey Paleontological Bulletin*, **64**, 1–260.

Ehrenberg, C. H. (1844). Untersuchungen über die kleinsten Lebensformen im Quellenlande des Euphrats und Araxes, so wie über eine an neuen Formen sehr reiche, marine Tripelbildung von den Bermuda-Inseln. *Bericht über die zur Bekanntmachung geeigneten Verhandlungen der der Königlichen Preussischen Akademie*. Berlin: Akademie der Wissenschaften, pp. 253–75.

Ehrlich, A. & Moshkovitz, S. (1982). On the occurrence of Eocene marine diatoms in Israel. *Acta Geologica Academiae Scientiarum Hungaricae*, **25**, 23–37.

Eldrett, J. S., Greenwood, D. R., Harding, I. C., & Huber, M. (2009). Increased seasonality through the Eocene to Oligocene transition in northern high latitudes. *Nature*, **459**, 969–73.

Eldrett, J. S., Harding, I. C., Wilson, P. A., Butler, E., & Roberts, A. P. (2007). Continental ice in Greenland during the Eocene and Oligocene. *Nature*, **466**, 176–9.

Expedition 318 Scientists (2010). Wilkes Land glacial history: Cenozoic East Antarctic Ice Sheet evolution from Wilkes Land margin sediments. *IODP Preliminary Report*, 318. DOI:10.2204/iodp.pr.318.2010.

Falkowski, P. G., Katz, M. E., Knoll, A. H., *et al.* (2004). The evolution of modern eukaryotic phytoplankton. *Science*, **305**, 354–60.

Fenner, J. (1985). Late Cretaceous to Oligocene planktic diatoms. In *Plankton Stratigraphy*, vol. 2, ed. H. M. Bolli, J. B. Saunders, & K. Perch-Nielsen, Cambridge: Cambridge University Press, pp. 713–62.

Fenner, J. (1991). Taxonomy, stratigraphy, and paleoceanographic implications of Paleocene diatoms. *Proceedings of the Ocean Drilling Program, Scientific Results*, **114**, 123–54.

Fenner, J. (1994). Diatoms of the Fur Formation, their taxonomy and biostratigraphic interpretation. Results from the Harre borehole, Denmark. *Aarhus Geoscience*, **1**, 99–163.

Folger, D. W., Burckle, L. H., & Heezen, B. C. (1967). Opal phytoliths in a North Atlantic dust fall. *Science*, **155** (3767), 1243–4.

Fordyce, R. E. (1980). Whale evolution and Oligocene Southern Ocean environments. *Palaeogeography, Palaeoclimatology, Palaeoecology*, **31**, 319–36.

Fourtanier, E. (1991). Paleocene and Eocene diatom biostratigraphy of eastern Indian Ocean Site 752, ODP Leg 121. *Proceedings of the Ocean Drilling Program, Scientific Results*, **121**, 171–87.

Fourtanier, E. & Oscarson, R. (1994). Ultrastructure of some interesting and stratigraphically significant diatom taxa from the upper Paleocene to lower Eocene sediments of ODP Site 752, eastern Indian Ocean. *Proceedings of the 11th International Diatom Symposium*, ed. J. P. Kociolek, Memoirs of the California Academy of Sciences, no. 17, 399–410.

Gaarder, K. R., & Hasle, G. R. (1962). On the assumed symbiosis between diatoms and coccolithophorids in Brenneckella. *Nytt Magasin for Botanikk*, **9**, 145–9.

Galeotti, S., Brinkhuis, H., & Huber, M. (2004). Record of post-K-T boundary millennial-scale cooling from the western Tethys: a smoking gun for the impact-winter hypothesis? *Geology*, **32**, 529–32.

Gardner, J. V. & Burckle, L. H. (1975). Upper Pleistocene *Ethmodiscus rex* oozes from the eastern equatorial Atlantic. *Micropaleontology*, **21** (2), 236–42.

Gersonde, R. & Harwood, D. M. (1990). Lower Cretaceous diatoms from ODP Leg 113 Site 693 (Weddell Sea). Part 1. Vegetative cells. *Proceedings of the Ocean Drilling Program, Scientific Results*, **113**, 403–25.

Gingerich, P. D., Wells, N. A., Russell, D. E., & Shah, S. M. I. (1983). Origin of whales in epicontinental remnant seas: new evidence from the early Eocene of Pakistan. *Science*, **220**, 403–6.

Girard, V., Saint Martin, S., Saint Martin, J.-P., *et al.* (2009). Exceptional preservation of marine diatoms in upper Albian amber. *Geology*, **37**, 83–6.

Girard, V., Schmidt, A. R., Saint Martin, S., *et al.* (2008). Evidence for marine microfossils from amber. *Proceedings of the National Academy of Sciences of the USA*, **105** (45), 17426–9.

Gladenkov, A. (1999). A new lower Oligocene zone for the North Pacific diatom scale. In *Proceedings of the Fourteenth International Diatom Symposium*, ed. S. Mayama, M. Idei, & I. Koizumi, Königstein: Koeltz Scientific Books, pp. 581–90.

Gombos, A. M. (1980). The early history of the diatom family Asterolampraceae. *Bacillaria*, **3**, 227–72.

Gombos, A. M. (1982). Early and middle Eocene diatom evolutionary events. *Bacillaria*, **5**, 225–42.

Gombos, A. M. (1983). Middle Eocene diatoms from the South Atlantic. *Initial Reports of the Deep Sea Drilling Project*, **71**, 565–81.

Gombos, A. M. (1984). Late Paleocene diatoms in the Cape Basin. *Initial Reports of the Deep Sea Drilling Project*, **73**, 495–511.

Gombos, A. M. (1987). Middle Eocene diatoms from the North Atlantic, Deep Sea Drilling Project Site 605. *Initial Reports of the Deep Sea Drilling Project*, **93**, 793–9.

Greville, R. K. (1859). Descriptions of Diatomaceae observed in Californian Guano. *Quarterly Journal of Microscopical Science*, **7**, 155–66.

Greville, R. K. (1861). Descriptions of new and rare diatoms. Series I. *Transactions of the Microscopical Society of London, New Series*, **9**, 39–45.

Greville, R. K. (1863). Descriptions of new and rare diatoms. Series IX. *Transactions of the Microscopical Society of London, New Series*, **11**, 63–76.

Grigorov, I., Pearce, R. B., & Kemp, A. E. S. (2002). Southern Ocean laminated diatom ooze: mat deposits and potential for palaeoflux studies, ODP Leg 177, Site 1093. *Deep-Sea Research II: Topical Studies in Oceanography*, **49**, 3391–407.

Hajós, M. & Stradner, H. (1975). Late Cretaceous Archaeomonadaceae, Diatomaceae, and Silicoflagellatae from the South Pacific Ocean, Deep Sea Drilling Project, Leg 29, Site 275. *Initial Reports of the Deep Sea Drilling Project*, **29**, 913–1009.

Hanna, G. D. (1932). The diatoms of Sharktooth Hill, Kern County, California. *Proceedings of the California Academy of Sciences, Ser. 4*, **20**, 161–263.

Harper, M. A. (1999). Diatoms as markers of atmospheric transport. In *The Diatoms: Applications for the Environmental and Earth Sciences*, ed. E. F. Stoermer & J. P. Smol, Cambridge: Cambridge University Press, pp. 429–35.

Harrison, K. G. (2000). Role of increased marine silica input on paleo-pCO_2. *Paleoceanography*, **15** (3), 292–8.

Harwood, D. M. (1988). Upper Cretaceous and lower Paleocene diatom and silicoflagellate biostratigraphy of Seymour Island, eastern Antarctic Peninsula. *Geological Society of America, Memoirs*, **169**, 55–129.

Harwood, D. M. & Bohaty, S. M. (2000). Marine diatom assemblages from Eocene and younger erratics, McMurdo Sound, Antarctica. *Antarctic Research Series* **76**, 73–98.

Harwood, D. M. & Bohaty, S. M. (2001). Early Oligocene siliceous microfossil biostratigraphy of Cape Roberts Project Core CRP-3, Victoria Land Basin, Antarctica. *Terra Antartica*, **8**, 315–38.

Harwood, D. & Bohaty, S. M. (2007). Late Miocene sea-ice diatoms indicate a cold polar East Antarctic ice sheet event. *Geophysical Research Abstracts*, **9**, 08078.

Harwood, D. M. & Gersonde, R. (1990). Lower Cretaceous diatoms from ODP Leg 113 Site 693 (Weddell Sea) part 2, resting spores, chrysophycean cysts, and endoskeletal dinoflagellates, and notes on the origin of diatoms. *Proceedings of the Ocean Drilling Program, Scientific Results*, **113**, 403–25.

Harwood, D. M., Nikolaev, V. A., & Winter, D. M. (2007). Cretaceous records of diatom evolution, radiation and expansion. In *Pond Scum to Carbon Sink: Geological and Environmental Applications of the Diatoms*, ed. S. Starratt Paleontological Society Short Course 13, Knoxville, TN: Paleontological Society, pp. 33–59.

Harwood, D. M., Scherer, R. P., & Webb, P.-N. (1989). Multiple Miocene marine productivity events in West Antarctica as recorded in upper Miocene sediments beneath the Ross Ice Shelf (Site J-9). *Marine Micropaleontology*, **15**, 91–115.

Hein, J. R., Scholl, D. W., Barron, J. A., Jones, M. G., & Miller, J. (1978). Diagenesis of late Cenozoic diatomaceous deposits and formation of the bottom simulating reflector in the southern Bering Sea. *Sedimentology*, **25**, 155–81.

Holmes, R. W. (1985). The morphology of diatoms epizoic on cetaceans and their transfer from *Cocconeis* to two genera, *Bennettella* and *Epipellis*. *British Phycological Journal*, **20**, 43–57.

Homann, M. (1991). Die Diatomeen der Fur-Formation (Alttertiär, Limfjord/Dänemark). *Geologisches Jahrbuch Reihe A*, **123**, 3–285.

Huber, M., Brinkhuis, H., Stickley, C. E., *et al.* (2004). Eocene circulation of the Southern Ocean: was Antarctica kept warm by subtropical waters? *Paleoceanography*, PA4016, DOI:10.1029/2004PA001014.

Iakovleva, A. I., Brinkhuis, H., & Cavagnetto, C. (2001). Late Paleocene–early Eocene dinoflagellate cysts from the Turgay Strait, Kazakhstan: correlations across ancient seaways. *Palaeogeography, Palaeoclimatology, Palaeoecology*, **172**, 243–68.

Iakovleva, A. I. & Heilmann-Clausen, C. (2007). *Wilsonidium pechoricum* new species – a new dinoflagellate species with unusual asymmetry from the Paleocene/Eocene Transition. *Journal of Paleontology*, **81**, 1020–30.

Iakovleva, A. I., Oreshkina, T. V., Alekseev, A. S., & Rousseau, D.-D. (2000). A new Paleogene micropaleontological and paleogeographical data in the Petchora Depression, northeastern European Russia. *Comptes Rendus de l'Académie des Sciences – Series IIA – Earth and Planetary Science*, **330**, 485–91.

Iijima, A. & Tada, R. (1981). Silica diagenesis of Neogene diatomaceous and volcaniclastic sediments in northern Japan. *Sedimentology*, **28**, 185–200.

Jacobs, B. F., Kingston, J. D., & Jacobs, L. L. (1999). The origin of grass-dominated ecosystems. *Annals of the Missouri Botanical Garden*, **86**, 590–643.

Jacot Des Combes, H., Esper, O., De La Rocha, C. L., *et al.* (2008). Diatom $\delta^{13}C$, $\delta^{15}N$, and C/N since the last glacial maximum in the Southern Ocean: potential impact of species composition, *Paleoceanography*, **23**, PA4209, DOI:10.1029/2008PA001589.

Jenkyns, H. C., Forster, A., Schouten, A., & Sinninghe Damsté, J. S. (2004). High temperatures in the Late Cretaceous Arctic Ocean. *Nature*, **432**, 888–92.

Jordan, R. W., Priddle, J., Pudsey, C. J., Barker, P. F., & Whitehouse, M. J. (1991). Unusual diatom layers in Upper Pleistocene sediments from the northern Weddell Sea. *Deep-Sea Research*, **38**, 829–43.

Jousé, A. P. (1949). Algae diatomaceae aetatis superne cretaceae ex arenis argillaceis-systematis. *Botanicheskie Materialy Otdela Sporovykh Rastenii, Botanicheskii Institut, Akademia Nauk, SSSR*, **6**, 65–78 (in Russian).

Jousé, A. P. (1951). Diatoms of Paleocene age in the northern Urals. *Botanicheskie Materialy Otdela Sporovykh Rastenii, Botanicheskii Institut, Akademia Nauk, SSSR*, **7**, 24–42 (in Russian).

Jousé, A. P. (1978). Diatom biostratigraphy on the generic level. *Micropaleontology*, **24**, 316–26.

Juillet-Leclerc, A. & Schrader, H. (1987). Variations of upwelling intensity recorded in varved sediment from the Gulf of California during the past 3,000 years. *Nature*, **329**, 146–9.

Kaczmarska, I., Barbrick, N. E., Ehrman, J. M., & Cant, G. P. (1993). *Eucampia* index as an indicator of the late Pleistocene oscillations of the winter sea ice extent at the Leg 119 Site 745B at the Kerguelen Plateau. *Hydrobiologia*, **269–270**, 103–12.

Kanaya, T. (1957). Eocene diatom assemblages from the Kellogg and "Sidney" shales, Mt. Diablo Area, California. *The Science Reports of the Tohoku University, Sendai, Japan, 2nd Series (Geology)*, **28**, 27–124.

Katz, M. E., Finkel, Z., Grzebyk, D., Knoll, A. H., & Falkowski, P. G. (2004). Evolutionary trajectories and biogeochemical impacts of marine eukaryotic phytoplankton. *Annual Reviews of Ecology, Evolution, and Systematics*, **35**, 523–56.

Katz, M. E., Wright, J. D., Miller, K. G., *et al.* (2005). Biological overprint of the geological carbon cycle. *Marine Geology*, **217** (Special Issue), 323–38.

Kawamura, A. (1992). Notes on the pattern of diatom fouling in three southern rorqual species. *Bulletin of the Faculty of Bioresources, Mie University*, **8**, 19–26.

Kemp, A. E. S. & Baldauf, J. G. (1993). Vast Neogene laminated diatom mat deposits from the eastern equatorial Pacific Ocean. *Nature*, **362**, 141–4.

Kemp, A. E. S., Baldauf, J. G., & Pearce, R. B. (1995). Origins and paleoceanographic significance of laminated diatom ooze from the eastern equatorial Pacific Ocean (Leg 138). *Proceedings of the Ocean Drilling Program, Scientific Results*, **138**, 641–5.

Kemp, A. E. S., Pearce, R. B., Grigorov, I., *et al.* (2006). The production of giant marine diatoms and their export at oceanic frontal zones: implications for Si and C flux in stratified oceans. *Global Biogeochemical Cycles*, **20**, GB4S04-[13pp], DOI:10.1029/2006GB002698.

Kemp, A. E. S., Pearce, R. B., Koizumi, I., Pike, J., & Rance, S. J. (1999). The role of mat forming diatoms in formation of the Mediterranean sapropels. *Nature*, **398**, 57–61.

Kemp, A. E. S., Pike, J., Pearce, R. B., & Lange, C. B. (2000). The "fall dump" – a new perspective on the role of a "shade flora" in the annual cycle of diatom production and export flux. *Deep-Sea Research II*, **47**, 2129–54.

Kohnen, M. E. L., Sinninghe Damsté, J. S., Kock-van Dalen, A. C., *et al.* (1990). Origin and diagenetic transformation of C25 and C30 highly branched isoprenoid sulfur compounds: further evidence for the formation of organically bound sulfur during early diagenesis. *Geochimica et Cosmochimica Acta*, **54**, 3053–63.

Kooistra, W. H. C. F. & Medlin, L. K. (1996). Evolution of the diatoms (Bacillariophyta) IV. A reconstruction of their age from small subunit rRNA coding regions and fossil record. *Molecular Phylogenetics and Evolution*, **6** (3), 391–407.

Kolbe, R. W. (1954). Diatoms from equatorial Pacific cores. *Reports of the Swedish Deep-Sea Expedition 1947–1948*, **VI**, 1–49.

Krammer, R., Baumann, K.-H., & Henrich, R. (2005). Middle to late Miocene fluctuations in the incipient Benguela Upwelling System revealed by calcareous nannofossil assemblages (ODP Site 1085A). *Palaeogeography, Palaeoclimatology, Palaeoecology*, **230**, 319–34.

Lange, C. B., Berger, W. H., Lin, H. L., & Wefer, G. (1999). The early Matuyama Diatom Maximum off SW Africa, Benguela Current System (ODP Leg 175). *Marine Geology*, **161**, 93–114.

Lange, C. B., Romero, O. E., Wefer, G., & Gabric, A. J. (1998). Offshore influence of coastal upwelling off Mauritania, NW Africa, as recorded by diatoms in sediment traps at 2195 m water depth. *Deep-Sea Research I*, **45** (6), 985–1013.

Lazarus, D., Spencer-Cervato, C., Pika-Biolzi, M., *et al.* (1995). Revised chronology of Neogene DSDP holes from the World Ocean. Ocean Drilling Program Technical Note 24, College Station, TX: Ocean Drilling Program.

Leckie, R. M. & Webb, P.-N. (1983). Late Oligocene–early Miocene glacial record of the Ross Sea, Antarctica: evidence from DSDP Site 270. *Geology*, **11**, 578–82.

Leventer, A., Domack, E., Dunbar, R., *et al.* (2006). Marine sediment record from the East Antarctic margin reveals dynamics of ice sheet recession. *GSA Today*, **16** (12), DOI:10.1130/GSAT01612A.1

Levasseur, M., Gosselin, M., & Michaud, S. (1994). A new source of dimethylsulfide (DMS) for the Arctic atmosphere: ice diatoms. *Marine Biology*, **121**, 381–7.

Lipps, J. H. & Mitchell, E. (1976). Trophic model for the adaptive radiations and extinctions of pelagic marine mammals. *Paleobiology*, **2**, 147–55.

Lohman, K. E. (1931). Diatoms from the Modelo Formation (Upper Miocene) near Girard, Los Angeles County, California. Unpublished Masters thesis, California Institute of Technology, Pasadena, CA.

Lohman, K. E. (1938). Pliocene diatoms from the Kettleman Hills, California. *United States Geological Survey, Professional Paper* **196-B**, 55–87.

Long, J. A., Fuge, D. P., & Smith, J. (1946). Diatoms of the Moreno Shale. *Journal of Paleontology*, **20** (2), 89–118.

Lyle, M., Gibbs, S., Moore, T. C., & Rea, D. K. (2007). Late Oligocene initiation of the Antarctic Circumpolar Current: evidence from the South Pacific. *Geology*, **35**, 691–4.

MacLeod, N., Rawson, P. F., Forey, P. L., *et al.* (1997). The Cretaceous–Tertiary biotic transition. *Journal of the Geological Society*, **154**, 265–92.

Mann, A. (1921). The diatoms of the Lompoc Beds. In *The Fish Fauna of the California Tertiary*, ed. D.S. Jordan, Stanford University Publications, University Series, Biological Sciences, vol. 1, 293–8.

Merico, A., Tyrrell, T., & Wilson, P. A. (2008). Eocene/Oligocene ocean de-acidification linked to Antarctic glaciation by sea level fall. *Nature*, **452**, 979–82.

Michalopoulos, P., Aller, R. C., & Reeder, R. J. (2000). Conversion of diatoms to clays during early diagenesis in tropical, continental shelf muds. *Geology*, **28** (12), 1095–8.

Mikkelsen, N. (1977). Silica dissolution and overgrowth of fossil diatoms. *Micropaleontology*, **23**, 223–6.

Miller, K. G., Browning, J. V., Aubry, M.-P., *et al.* (2008). Eocene–Oligocene global climate and sea-level changes: St. Stephens Quarry, Alabama. *Geological Society of America Bulletin*, **120**, 34–53.

Mills, L. G. (1881). Diatoms from Guano. *Journal of the Royal Microscopical Society*, **1**, 865 + plate XI.

Mitlehner, A. G. (1996). Palaeoenvironments in the North Sea Basin around the Paleocene–Eocene boundary: evidence from diatoms and other siliceous microfossils. *Geological Society, London, Special Publications*, **101**, 255–73.

Moore, T. C., Jr., Jarrard, R. D., Olivarez Lyle, A., & Lyle, M. (2008). Eocene biogenic silica accumulation rates at the Pacific equatorial divergence zone. *Paleoceanography*, **23**, PA2202, DOI:10.1029/2007PA001514.

Moran, K., Backman, J., Brinkhuis, H., *et al.* (2006). The Cenozoic palaeoenvironment of the Arctic Ocean. *Nature*, **441**, 601–5.

Mukhina, V. V. (1976). Species composition of the Late Paleocene diatoms and silicoflagellates in the Indian Ocean. *Micropaleontology*, **22** (2), 151–8.

Nagasawa, S., Holmes, R. W., & Nemoto, T. (1989). Occurrence of Cetacean diatoms in the sediments of Otsuchi Bay, Iwate, Japan. *Proceedings of the Japan Academy, Series B*, **65** (4), 80–3.

Naish, T. R., Powell, R. D., & Levy, R. H. (eds.) (2007). Studies from the ANDRILL, McMurdo Ice Shelf Project, Antarctica – Initial Science Report on AND-1B. *Terra Antartica*, **14**, 109–328.

Naish, T., Powell, R., Levy, R., *et al.* (2009). Obliquity-paced Pliocene West Antarctic Ice Sheet oscillations. *Nature*, **458**, 322–8.

Nelson, D. M., Tréguer, P., Brzezinski, M. A., Leynaert, A., & Quéguiner, B. (1995). Production and dissolution of biogenic silica in the ocean: revised global estimates, comparison with regional data and relationship to biogenic sedimentation. *Global Biogeochemical Cycles*, **9**, 359–72.

Nikolaev, V. A., Kociolek, J. P., Fourtanier, E., Barron, J. A., & Harwood, D. M. (2001). Late Cretaceous diatoms (Bacillariophyceae) from the Marca Shale Member of the Moreno Formation, California. *Occasional Papers of the California Academy of Sciences*, **152**, 1–119.

Olney, M., Bohaty, S. M., Harwood, D. M., & Scherer, R. P. (2009). *Creania lacyae* gen. et sp. nov. and *Synedropsis cheethamii* sp. nov.: fossil indicators of Antarctic sea ice? *Diatom Research*, **24**, 357–75.

Olney, M. P., Scherer, R. P., Bohaty, S. M., & Harwood, D. M. (2005). Eocene–Oligocene paleoecology and the diatom genus *Kisseleviella* Sheshukova-Poretskaya from the Victoria Land Basin, Antarctica. *Marine Micropaleontology*, **58**, 56–72.

Olney, M. P., Scherer, R. P., Harwood, D. M., & Bohaty, S. M. (2007). Oligocene-early Miocene Antarctic nearshore diatom biostratigraphy. *Deep-Sea Research II*, **54**, 2325–49.

Oreshkina, T. & Aleksandrova, G. (2007). Terminal Paleocene of the Volga middle reaches: biostratigraphy and paleosettings. *Stratigraphy and Geological Correlation*, **15** (2), 206–30.

Passow, U., Engel, A., & Ploug, H. (2003). The role of aggregation for the dissolution of diatom frustules. *FEMS Microbiology Ecology*, **46**, 247–55.

Patarnello, T., Bargelloni, L., Varotto, V., & Battaglia, B. (1996). Krill evolution and the Antarctic ocean currents: evidence of vicariant speciation as inferred by molecular data. *Marine Biology*, **126**, 603–8.

Pike, J. (2000). Data report: backscattered electron imagery analysis of early Pliocene laminated *Ethmodiscus* ooze, ODP Site 1010, Leg 167. *Proceedings of the Ocean Drilling Program, Scientific Results*, **167**, 207–12.

Pike, J., Allen, C. S., Leventer, A., Stickley, C. E., & Pudsey, C. J. (2008). Comparison of contemporary and fossil diatom assemblages from the western Antarctic Peninsula shelf. *Marine Micropaleontology*, **67**, 274–87.

Pfuhl, H. A., & McCave, I. N. (2005). Evidence for late Oligocene establishment of the Antarctic Circumpolar Current. *Earth and Planetary Science Letters*, **235**, 715–28.

Pollock, D. E. (1997). The role of diatoms, dissolved silicate and Antarctic glaciation in glacial/interglacial climatic change: a hypothesis. *Global and Planetary Change*, **14**, 113–25.

Prasad, V., Strömberg, C. A. E., Alimohammadian, H., & Sahni, A. (2005). Dinosaur coprolites and the early evolution of grasses and grazers. *Science*, **310**, 1177–80.

Rabosky, D. L. & Sorhannus, U. (2009). Diversity dynamics of marine planktonic diatoms across the Cenozoic. *Nature*, **457**, 183–6.

Radionova, E. P. & Khokhlova, I. E. (2000). Was the North Atlantic connected with the Tethys via the Arctic in the early Eocene? Evidence from siliceous plankton. *Geologiska Foreningens i Stockholm Forhnndlingar (GFF)*, **122**, 133–4.

Raiswell, R. & Berner, R. A. (1985). Pyrite formation in euxinic and semi-euxinic sediments. *American Journal of Science*, **285**, 710–24.

Reinhold, T. (1937). Fossil diatoms of the Neogene of Java and their zonal distribution. *Geolische Serie*, **12**, 43–133 + 21 plates.

Romero, O., Mollenhauer, G., Schneider, R. R., & Wefer, G. (2003). Oscillations of the siliceous imprint in the central Benguela Upwelling System from MIS 3 through to the early Holocene: the influence of the Southern Ocean. *Journal of Quaternary Science*, **18**, 733–43.

Rothpletz, A. (1896). Über die Flysch-Fucoiden und einige andere fossile Algen, sowie über liasische, Diatomeen führende Hornschwämme. *Zeitschrift der Deutschen Geologischen Gesellschaft, Berlin*, **48**, 854–914.

Rothpletz, A. (1900). Über einen neuen jurassichen Hornschwämm und die darin eingeschlossenen. *Diatomeen. Zeitschrift der Deutschen Geologischen Gesellschaft, Berlin*, **52**, 154–60.

Round, F. E., & Alexander, C. G. (2002). *Licmosoma* – a new diatom genus growing on barnacle cirri. *Diatom Research*, **17**, 319–26.

Rowland, S. J., & Robson, J. N. (1990). The widespread occurrence of highly branched acyclic C20, C25 and C30 hydrocarbons, recent sediments biota – a review. *Marine Environmental Research*, **30**, 191–216.

Salamy, K. A., & Zachos, J. C. (1999). Late Eocene–early Oligocene climate change on Southern Ocean fertility: inferences from sediment accumulation and stable isotope data. *Palaeogeography, Palaeoclimatology, Palaeoecology*, **145**, 79–93.

Scherer, R. P. (1991). Quaternary and Tertiary microfossils from beneath Ice Stream B: Evidence for a dynamic West Antarctic Ice Sheet history. *Global and Planetary Change*, **4**, 395–412.

Scherer, R. P., Aldahan, A., Tulaczyk, S., *et al.* (1998). Pleistocene collapse of the West Antarctic Ice Sheet. *Science*, **281**, 82–5.

Scherer, R. P., Bohaty, S. M., Dunbar, R. B., *et al.* (2008). Antarctic records of precession-paced insolation-driven warming during early Pleistocene Marine Isotope Stage 31. *Geophysical Research Letters*, **35**, L03505, DOI:10.1029/2007GL032254.

Scherer, R. P., Bohaty, S. M., & Harwood, D. M. (2001). Oligocene and Lower Miocene siliceous microfossil biostratigraphy of Cape Roberts Project Core CRP-2/2A, Victoria Land Basin, Antarctica. *Terra Antartica*, **7**, 417–42.

Scherer, R. P., Gladenkov, A. Yu., & Barron, J. A. (2007a). Methods and applications of Cenozoic marine diatom biostratigraphy. In *Pond Scum to Carbon Sink: Geological and Environmental Applications of the Diatoms*, ed. S. W. Starratt, Paleontological Society Short Course 13, Knoxville, TN: The Paleontological Society, pp. 61–83.

Scherer, R., Hannah, M., Maffioli, P., *et al.* (2007b). Paleontologic characterisation and analysis of the AND-1B Core, ANDRILL McMurdo Ice Shelf Project, Antarctica. *Terra Antartica*, **14**, 223–54.

Scherer, R. P., Sjunneskog, C. M., Iverson, N., & Hooyer, T. (2004). Assessing subglacial processes from diatom fragmentation patterns. *Geology*, **32**, 557–60.

Scherer, R. P., Sjunneskog, C. M., Iverson, N., & Hooyer, T. (2005). Frustules to fragments, diatoms to dust: how degradation of microfossil micro and nanostructures can teach us how ice sheets work. *Journal of Nanoscience and Nanotechnology*, **5**, 96–9.

Schrader, H.-J. (1971). Fecal pellets: role in sedimentation of pelagic diatoms. *Science*, **174**, 55–7.

Schrader, H. J. & Gersonde, R. (1978). Diatoms and silicoflagellates. *Utrecht Micropaleontological Bulletin*, **17**, 129–76.

Schrader, H. & Sorknes, R. (1990). Spatial and temporal variation of Peruvian coastal upwelling during the latest Quaternary. *Proceedings of the Ocean Drilling Program, Scientific Results*, **112**, 391–406.

Schuette, G. & Schrader, H. (1979). Diatom taphocoenoses in the coastal upwelling area off western South America. *Nova Hedwigia*, **64**, 359–78.

Schuette, G. & Schrader, H. (1981). Diatom taphocoenoses in the coastal upwelling area off south West Africa. *Marine Micropaleontology*, **6**, 131–55.

Shannon, L. V. & Hunter, D. (1988). Notes on Antarctic Intermediate Water around southern Africa. *South African Journal of Marine Science*, **6**, 107–17.

Shemesh, A., Burckle, L. H., & Hays, J. D. (1995). Late Pleistocene oxygen isotope records of biogenic silica from the Atlantic sector of the Southern Ocean, *Paleoceanography* **10**, 179–96.

Shemesh, A., Mortlock, R. A., & Froelich, P. N. (1989). Late Cenozoic Ge/Si record of marine biogenic opal: implications for variations of riverine fluxes to the ocean, *Paleoceanography*, **4**, 221–34.

Shiine, H., Suzuki, N., Motoyama, I., *et al.* (2008). Diatom biomarkers during the Eocene/Oligocene transition in the Il'pinskii Peninsula, Kamchatka, Russia. *Palaeogeography, Palaeoclimatology, Palaeoecology*, **264**, 1–10.

Shimada, C., Hasegawa, S., Tanimura, Y., & Burckle, L. H. (2003). A new index to quantify diatom dissolution levels based on a ratio of *Neodenticula seminae* frustule components. *Micropaleontology*, **49**, 267–76.

Shimada, C., Sato, T., Toyoshima, S., Yamasaki, M., & Tanimura, Y. (2008). Paleoecological significance of laminated diatomaceous oozes during the middle-to-late Pleistocene, North Atlantic Ocean (IODP Site U1304). *Marine Micropaleontology*, **69**, 139–50.

Shimada, C., Tanaka, Y., & Tanimura, Y. (2006). Seasonal variation in skeletal silicification of *Neodenticula seminae*, a marine planktonic diatom: sediment trap experiments in the NW Pacific Ocean (1997–2001). *Marine Micropaleontology*, **60**, 130–44.

Siesser, W. G. (1980). Late Miocene origin of the Benguela upwelling system off northern Namibia. *Science*, **208**, 283–5.

Sims, P. A., Mann, D. G., & Medlin, L. K. (2006). Evolution of the diatoms: insights from fossil, biological and molecular data. *Phycologia*, **45**, 361–402.

Sinninghe Damsté, J. S., Muyzer, G., Abbas, B., *et al.* (2004). The rise of the rhizosolenid diatoms. *Science*, **304**, 584–7.

Sjunneskog, C. S. & Scherer, R. P. (2005). Mixed diatom assemblages in Ross Sea (Antarctica) glacigenic facies. *Palaeogeography, Palaeoclimatology, Palaeoecology*, **218** (3–4), 287–300.

Smetacek, V. S. (1985). Role of sinking in diatom life-history cycles: ecological, evolutionary and geological significance. *Marine Biology*, **84**, 239–51.

Spencer-Cervato, C. (1999). The Cenozoic deep-sea microfossil record: explorations of the DSDP/ODP sample set using the Neptune database (online). *Palaeontologia Electronica*, **2** (2), article 4.

Stickley, C. E., Brinkhuis, H., McGonigal, K. L., *et al.* (2004a). Late Cretaceous–Quaternary biomagnetostratigraphy of ODP Sites 1168, 1170, 1171, and 1172, Tasmanian Gateway. *Proceedings of the Ocean Drilling Program, Scientific Results*, **189**, 1–57. See http://www-odp.tamu.edu/publications/189/SR/VOLUME/CHAPTERS/111.PDF.

Stickley, C. E., Brinkhuis, H., Schellenberg, S. A., *et al.* (2004b). Timing and nature of the deepening of the Tasmanian Gateway. *Paleoceanography*, **19**, PA4027, DOI: 10.1029/2004PA001022.

Stickley, C. E., Koç, N., Brumsack, H.-J., Jordan, R. W., & Suto, I. (2008). A siliceous microfossil view of middle Eocene Arctic paleoenvironments: a window of biosilica production and preservation. *Paleoceanography*, **23**, PA1S14, DOI:10.1029/2007PA001485.

Stickley, C. E., Pike, J., Leventer, A., *et al.* (2005). Deglacial ocean and climate seasonality in laminated diatom sediments, Mac-Robertson Shelf, Antarctica. *Palaeogeography, Palaeoclimatology, Palaeoecology*, **227**, 290–310.

Stickley, C. E., St. John, K., Koç, N., *et al.* (2009). Evidence for middle Eocene Arctic sea ice from diatoms and ice-rafted debris. *Nature*, **460**, 376–9.

Strelnikova, N. I. (1974). *Diatoms of the Late Cretaceous (Western Siberia)*. Moscow: Nauka (in Russian).

Strelnikova, N. I. (1975). Diatoms of the Cretaceous Period. *Nova Hedwigia, Beiheft*, **53**, 311–21.

Summerhayes, C. P., Prell, W. L. & Emeis, K. C. (eds.) (1992). Upwelling systems: evolution since the early Miocene. *Geological Society of London, Special Publication*, **64**, 1–519.

Suto, I. (2004). Fossil marine diatom resting spore morpho-genus *Gemellodiscus* gen. nov. in the North Pacific and Norwegian Sea. *Paleontological Research*, **8**, 255–82.

Suto, I. (2005). Taxonomy and biostratigraphy of the fossil marine diatom resting genera *Dicladia* Ehrenberg, *Monocladia* Suto and *Syndendrium* Ehrenberg in the North Pacific and Norwegian Sea. *Diatom Research*, **20**, 351–74.

Suto, I. (2006). The explosive diversification of the diatom genus *Chaetoceros* across the Eocene/Oligocene and Oligocene/Miocene boundaries in the Norwegian Sea. *Marine Micropaleontology*, **58**, 259–69.

Suto, I., Jordan, R. W., & Watanabe, M. (2008). Taxonomy of the fossil marine diatom resting spore genus *Goniothecium* Ehrenberg and its allied species. *Diatom Research*, **23**, 445–69.

Suto, I., Jordan, R. W., & Watanabe, M. (2009). Taxonomy of middle Eocene diatom resting spores and their allied taxa from IODP sites in the central Arctic Ocean (the Lomonosov Ridge). *Micropaleontology*, **55**, 259–312.

Suto, I., Watanabe, M., & Jordan, R. W. (in press). Taxonomy of the fossil marine diatom resting spore genus *Odontotropis* Grunow. *Diatom Research*.

Takahashi, O., Kimura, M., Ishii, A., & Mayama, S. (1999). Upper Cretaceous diatoms from central Japan. In *Proceedings of the Fourteenth International Diatom Symposium, Tokyo*, ed. S. Mayama, M. Idei, & I. Koizumi, Königstein: Koeltz Scientific Books, pp. 146–55.

Tapia, P. M. & Harwood, D. M. (2002). Upper Cretaceous diatom biostratigraphy of the Arctic Archipelago and northern continental margin, Canada. *Micropaleontology*, **48**, 303–42.

Thewissen, J. G. M., & Williams, E. M. (2002). The early radiations of Cetacea (Mammalia): evolutionary pattern and developmental correlations. *Annual Review of Ecology and Systematics*, **33**, 73–90.

Thomas, E. (2008). Descent into the Icehouse. *Geology*, **36**, 191–2.

Tréguer, P., Nelson, D. M., Van Bennekom, A. J., *et al.* (1995). The silica balance in the world ocean: a re-estimate. *Science*, **268**, 375–9.

Van Eetvelde, Y., Dupuis, C., & Cornet, C. (2004). Pyritized diatoms: a good fossil marker in the upper Paleocene–lower Eocene sediments from the Belgian and Dieppe-Hampshire basins. *Netherlands Journal of Geosciences*, **83**, 173–8.

Villareal, T. A., Altabet, M. A., & Culver-Rymsza, K. (1993). Nitrogen transport by vertically migrating diatom mats in the North Pacific Ocean. *Nature*, **363**, 709–12.

Villareal, T. A., Joseph, L., Brzezinski, M. A., *et al.* (1999). Biological and chemical characteristics of the giant diatom *Ethmodiscus* (Bacillariophyceae) in the central North Pacific Gyre. *Journal of Phycology*, **35**, 896–902.

Wagner, T., Damsté, J. S. S., Hofmann, P., & Beckmann, B. (2004). Euxinia and primary production in Late Cretaceous eastern equatorial Atlantic surface waters fostered orbitally driven formation of marine black shales. *Paleoceanography*, **19**, DOI:10.1029/2003PA000898.

Watanabe, M. & Takahashi, M. (1997). Diatom biostratigraphy of the middle Miocene Kinone and lower Amatsu Formation in the Boso Peninsula, central Japan. *Journal of the Japanese Association for Petroleum Technology*, **62**, 215–25 (in Japanese with English abstract).

Weijers, J. W. H., Schouten, S., Sluijs, A., Brinkhuis, H., & Sinninghe Damsté, J. S. (2007). High latitude subtropical continental temperatures during the Palaeocene–Eocene Thermal Maximum. *Earth and Planetary Science Letters*, **261**, 230–8.

Whitehead, J. M., Wotherspoon, S., & Bohaty, S. M. (2005). Minimal Antarctic sea ice during the Pliocene. *Geology*, **33**, 137–40.

Witt, O. N. (1886). Über den Polierschiefer von Archangelsk, Kurojedowo im Gouv. *Simbirsk. Verhandlungen, Russischskaiserliche, Mineralogische Gesellschaft zu St. Petersburg, Series II*, **22**, 137–77.

Wolfe, A. P., Edlund, M. B., Sweet, A. R., & Creighton, S. D. (2006). A first account of organelle preservation in Eocene nonmarine diatoms: observations and paleobiological implications. *PALAIOS*, **21**, 298–304.

Wornardt, W. W., Jr. (1972). Stratigraphic distribution of diatom genera in marine sediments in western North America. *Palaeogeography, Palaeoclimatology, Palaeoecology*, **12**, 49–74.

Yoder, J. A., Ackleson, S., Barber, R. T., Flamant, P., & Balch, W. A. (1994). A line in the sea. *Nature*, **371**, 689–92.

Zachos, J. C., Dickens, G. R., & Zeebe, R. E. (2008). An early Cenozoic perspective on greenhouse warming and carbon-cycle dynamics. *Nature*, **451**, 279–83.

Zachos, J., Pagani, M., Sloan, L., Thomas, E., & Billups, K. (2001). Trends, rhythms, and aberrations in global climate 65 Ma to present. *Science*, **292**, 686–93.

Zane, L. & Patarnello, T. (2000). Krill: a possible model for investigating the effects of ocean currents on the genetic structure of a pelagic invertebrate. *Canadian Journal of Fisheries and Aquatic Science*, **57** (S3), 16–23.

23 Reconsidering the meaning of biogenic silica accumulation rates in the glacial Southern Ocean

CHRISTINA L. DE LA ROCHA, OLIVIER RAGUENEAU, AND AUDE LEYNAERT

23.1 Introduction

Accumulation rates of biogenic silica (BSi) have long been used to infer past levels of primary production (e.g. Charles *et al.*, 1991; Mortlock *et al.*, 1991; Kumar *et al.*, 1995; Schelske 1999; Frank *et al.*, 2000; Anderson *et al.*, 2002; Chase *et al.*, 2003). In some respects, BSi measurements are more useful than organic carbon accumulation rates because 3% of BSi production accumulates in marine sediments (Tréguer *et al.*, 1995), an order of magnitude more than organic carbon (Hedges & Keil, 1995). Biogenic silica is also predominantly produced by diatoms and, because diatoms play an essential role in the export of particulate organic carbon (POC) from surface waters (Goldman, 1993; Buesseler, 1998; Ragueneau *et al.*, 2006; Buesseler *et al.*, 2007), BSi accumulation should contain a signal associated with the strength of the biological pump.

In the mid 1990s, it was recognized, however, that BSi accumulation rates are not a straightforward proxy for primary production (Ragueneau *et al.*, 2000). Although there is a strong link between BSi production and BSi accumulation, variability and shifts in the ratios of BSi to POC during particle production, export, and preservation complicate the extrapolation of primary production from BSi accumulation (Pondaven *et al.*, 2000; Ragueneau *et al.*, 2000; DeMaster 2002; Nelson *et al.*, 2002; Ragueneau *et al.*, 2002; Moriceau *et al.*, 2007). Such decoupling may explain the stark difference in the estimates of primary production in the glacial Southern Ocean gained from BSi accumulation rates (Anderson *et al.*, 2002) versus that of the diatom and radiolarian species composition of the sediments (Abelmann *et al.*, 2006).

The Diatoms: Applications for the Environmental and Earth Sciences, 2nd Edition, eds. John P. Smol and Eugene F. Stoermer. Published by Cambridge University Press.

That there is information concerning production expressed in BSi accumulation rates is clear, but extracting that information requires a more sophisticated approach than has been taken previously. Recent reviews explore the mechanisms controlling spatial and temporal variations in BSi dissolution and preservation and decoupling between silicon (Si) and carbon (C) (Van Cappellen *et al.*; Ragueneau *et al.*, 2002, 2006; Sarmiento & Gruber, 2006). In this chapter, we reexamine the debate regarding the strength of the biological pump in the Southern Ocean over glacial–interglacial cycles, using what has been learned about BSi/C production ratios. What emerges is a picture of paleoproduction in the Southern Ocean that may reconcile the records of BSi accumulation rates and species composition of the sediments.

23.2 A brief overview of our current understanding of the biological pump in the glacial Southern Ocean

23.2.1 The Southern Ocean

Each year, a fraction of marine net primary production is exported to the deep ocean prior to remineralization back to carbon dioxide (CO_2) and nutrients. This exported POC is responsible for higher concentrations of dissolved inorganic carbon in deeper waters than in surface waters, and results in lower concentrations of atmospheric CO_2. Changes in the strength of this biological pumping of POC to depth contribute significantly to the waxing and waning of atmospheric CO_2 over glacial–interglacial cycles (Kohfeld *et al.*, 2005; Brovkin *et al.*, 2007).

The above is especially true of the biological pump in the Southern Ocean (Anderson *et al.*, 2002; Brovkin *et al.*,

2007). Here, mixing brings nutrient- and CO_2-rich deep waters into contact with the atmosphere, allowing for considerable ventilation of CO_2. Primary production supported by the upwelled nutrients, however, may convert CO_2 into POC, diminishing the net flux of CO_2 out of the ocean. When rates of POC production and export exceed the rate of CO_2 upwelling, the Southern Ocean serves as a sink of CO_2 rather than a source.

Currently, phytoplankton growth in the Southern Ocean is held in check by light and iron (Fe) limitation, low temperatures, and zooplankton grazing, leaving much nitrogen (N) and phosphorus (P) unutilized in surface waters each year. Fertilization of the modern Southern Ocean with Fe stimulates primary production (de Baar *et al.*, 2005) and might also increase the flux of POC to depth. Increased inputs of Fe in the past may have similarly stimulated production and contributed to the lower glacial concentrations of atmospheric CO_2 (Brovkin *et al.*, 2007). If primary or export production of the glacial Southern Ocean could be reconstructed and shown to be greater than in the modern day, this hypothesis could be confirmed.

23.2.2 The lack of a direct view of the efficiency of the biological pump

The direct way to estimate the past strength of the biological pump with respect to its ability to sequester carbon in the deep ocean is to determine the extent to which upwelled nutrients were consumed in support of primary production. In principle, this can be accomplished by any of several proxies that record shifts in stable-isotope or trace-element ratios (e.g. $\delta^{15}N$, $\delta^{13}C$, $\delta^{30}Si$, Cd/Ca) that occur as nutrients in surface waters are used up (Singer & Shemesh, 1995; François *et al.*, 1997; Rosenthal *et al.*, 1997; De La Rocha *et al.*, 1998; Elderfield & Rickaby, 2000). In actuality, these proxies have not yielded a clear picture of nutrient cycling in the glacial Southern Ocean.

Silicon and N isotopes paint conflicting pictures of nutrient drawdown (De La Rocha *et al.*, 1998; Brzezinski *et al.*, 2002; De La Rocha, 2006). Carbon isotopes in organic matter vary with the diatom species preserved in the sediments (Jacot Des Combes *et al.*, 2008). The high last glacial maximum (LGM) values of $\delta^{15}N$ are associated with peaks in the abundance of resting spores/winter stages of *Eucampia antarctica* (Castracane) Mangin and may not reflect a true extreme in nitrate drawdown (Jacot Des Combes *et al.*, 2008); and problems with diagenesis, sample cleaning, and measurement methodology mean that most published records of bulk sediment and diatom $\delta^{15}N$ may not be sound (Robinson *et al.*, 2004; De La Rocha 2006). Calcification (and incorporation of trace elements like cadmium) in foraminifera is also likely to occur in subsurface waters,

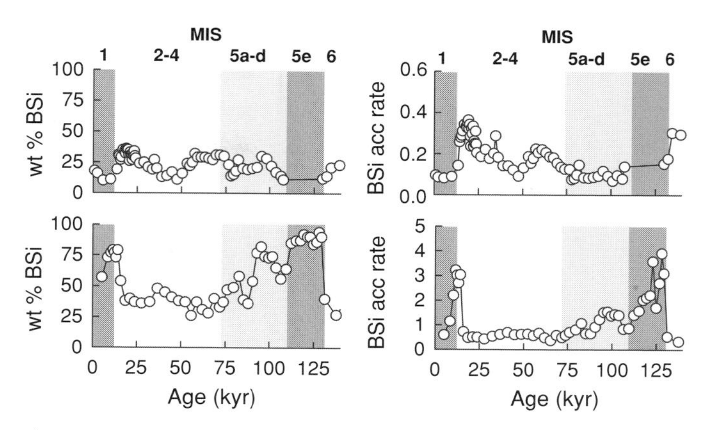

Figure 23.1 Weight percent BSi and Th-corrected BSi accumulation rates (in g cm^{-2} kyr^{-1}) from two sediment cores from the Atlantic sector of the Southern Ocean. The upper panel shows core PS2082–2 from the subantarctic zone north of the APF (43° 13′ S, 11° 44′ E, 4610 mbsl). The bottom panel shows core PS1768–8 from the Antarctic zone south of the APF (53° 36′ S, 4° 28′ E, 3270 mbsl). Data are from Frank *et al.* (2000). The numbers in bold at the top indicate marine isotope stages.

preventing them from recording nutrient conditions in the euphotic zone (Anderson *et al.*, 2002). Even if we ignored these issues, we would still not have a clear picture of the strength of the biological pump in the Southern Ocean from these proxies. Only a handful of records of each type have been reconstructed, precluding our ability to make any definitive conclusions (De La Rocha, 2006).

23.2.3 BSi as a proxy for production

The amount of BSi in Southern Ocean sediments varies dramatically over glacial–interglacial cycles (Figure 23.1) (Charles *et al.*, 1991; Mortlock *et al.*, 1991; Kumar *et al.*, 1995; François *et al.*, 1997; Frank *et al.*, 2000; Anderson *et al.*, 2002; Chase *et al.*, 2003; Hillenbrand & Cortese, 2006). Sediments south of the Antarctic Polar Front (APF) contain, on average, more BSi (in terms of weight percent) during interglacials than during glacials (Figure 23.1). Sediments to the north of the APF show the opposite pattern.

There is a temptation to interpret the weight percent BSi records in terms of production. Sedimentation rates are not constant and Southern Ocean sediments are subjected to much winnowing and focusing and weight percent BSi in the sediments must be converted to accumulation rates and corrected for focusing, via thorium (Th) isotopes, to yield a clearer picture of the delivery of material to the sea bed (Kumar *et al.*, 1995; Frank *et al.*, 2000; Anderson *et al.*, 2002). It is these Th-corrected accumulation rates of BSi (Figures 23.1 and 23.2) that are now interpreted in terms of primary production and the operation of

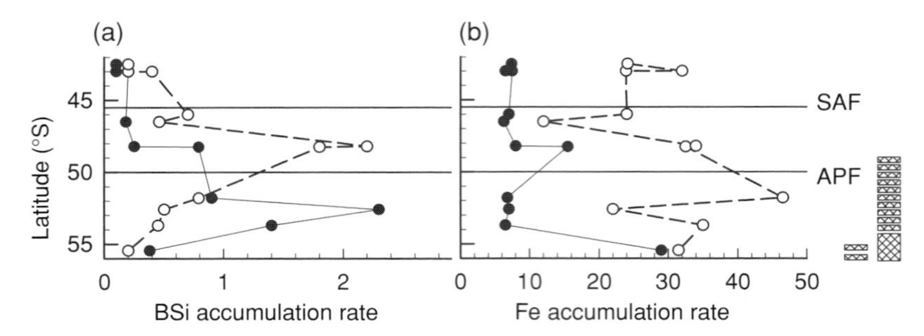

Figure 23.2 Average Th-corrected accumulation rates (in g cm^{-2} kyr^{-1}) of BSi (a) and Fe (b) in the eastern Atlantic sector for the Holocene (solid symbols and line) and LGM (white symbols and dashed line). The positions of the APF and Subantarctic Front (SAF) and the northward extent of sea ice during the Holocene and LGM are indicated to the right of the plots. Panels are modified from Frank *et al.* (2000).

the biological pump. The general view that has emerged from them is that primary production in the glacial Antarctic Ocean was lower than during the Holocene interglacial, while primary production to the north of the APF was higher, resulting in no great glacial–interglacial difference in primary production (Kumar *et al.*, 1995; François *et al.*, 1997; Frank *et al.*, 2000; Anderson *et al.*, 2002).

23.2.4 Export efficiency and accumulation rates

The idea that primary production could be extrapolated from accumulation rates stems from the view that fluxes of biogenic materials like POC and BSi through the water column can be described as a simple exponential decline with depth from primary (or export) production (Suess 1980; Martin *et al.*, 1987) (Figure 23.3). This simplification ignores the dynamic processes controlling export.

Particles in the water column occur in two distinct categories with respect to sinking and decay (Lutz *et al.*, 2002; Moriceau *et al.*, 2007). Particles of the first type decompose more quickly than they sink, do not escape the surface ocean, and are responsible for the rapid decline in flux with depth in the upper few hundred meters of the water column (Figure 23.3). Particles of the second type sink much faster than they decompose and reach the deep sea and sediments prior to decomposition.

Pulses of diatomaceous phytodetritus arriving at the sea bed (Beaulieu 2002) and the strong positive relationships between flux and seasonality (Antia *et al.*, 2001; Moriceau *et al.*, 2007) indicate that there are conditions under which a greater percentage of production winds up in large, rapidly sinking particles and is exported to depth. Diatoms, with their relatively large size, possession of the heavy, silicified cell wall,

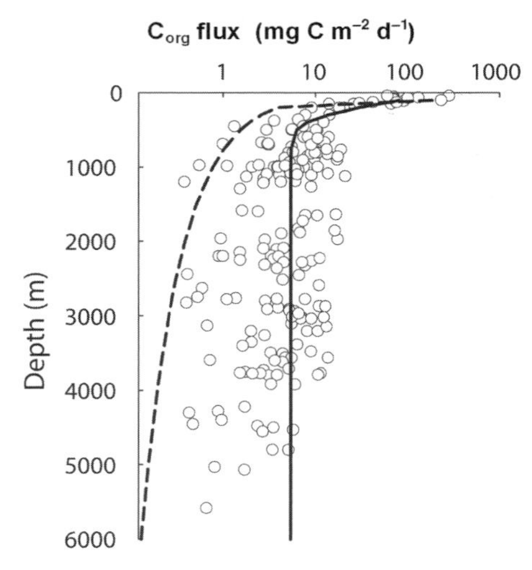

Figure 23.3 Mean annual POC flux as recorded in sediment traps. The dashed line shows the traditional "Martin curve": $C_{flux}(z) = C_{exp}(z/z_0)^{-0.858}$, where POC flux ($C_{flux}$) through any given depth (*z*) is predicted from the amount of POC exported (C_{exp}) from the base of the euphotic zone (z_0) (Martin *et al.*, 1987). The solid, better-fitting curve is based on two populations of particles with distinctly different ratios of settling velocity to degradation rate: $C_{flux}(z) = C_{fluxo}[(1-p)e^{-az} + pe^{-bz}]$, where C_{fluxo} is the initial POC flux, *p* and (1 – *p*) represented the fraction of POC in each of the two categories, and *a* and *b* are the different rates of degradation (Lutz *et al.*, 2002). POC flux data are from Lutz *et al.* (2002).

domination of the spring bloom in upwelling and coastal regions (and the Southern Ocean), and tendency to form large, rapidly sinking aggregates (Alldredge & Silver, 1988), increase the fraction of primary production exported from a region and thus play a key role in the biological pump (Goldman 1993; Buesseler 1998; Ragueneau *et al.*, 2006; Buesseler *et al.*, 2007).

Given the importance of diatoms to POC export, an increase in the amount of BSi in the sediments must, on some level, reflect an increase in the fraction of primary production exported from the surface ocean. Teasing this information out of BSi accumulation rates requires a quantitative understanding of the ecological processes controlling the fraction of production exported. Despite the work that has gone into understanding the formation, sinking, and decomposition of large particles in the ocean (including the impacts of zooplankton and bacteria), we are far from being able to model changes in export efficiency, especially between glacial and interglacial times in the Southern Ocean. For now, we must take BSi accumulation rates as a direct reflection of BSi production in surface waters (Pondaven *et al.*, 2000; Nelson *et al.*, 2002).

23.2.5 A micropaleontological view of paleoproduction

Recently, the view, based on BSi accumulation rates, that there was no great glacial increase in primary production in the Southern Ocean (Charles *et al.*, 1991; Mortlock *et al.*, 1991; François *et al.*, 1997; Frank *et al.*, 2000; Anderson *et al.*, 2002) has been challenged on the basis of micropaleontological data (Abelmann *et al.*, 2006).

Several different diatom species dominate the sedimentary mass of BSi in the Southern Ocean. There is *Fragilariopsis kerguelensis* (O'Meara, 1877) Hustedt, which is often overwhelmingly the most abundant diatom in modern sediments of open water regions south of the APF. *F. kerguelensis* is also heavily silicified, having nutrient replete Si/C ratios around 0.5 mol mol^{-1} (Hoffman *et al.*, 2007) versus the 0.1 mol mol^{-1} of average diatoms (Brzezinski, 1985). Heavily silicified resting spores of *Chaetoceros* spp. make up another significant portion of the sediments (the vegetative cells of *Chaetoceros* being too lightly silicified to be preserved). *Chaetoceros* resting spores are most common in sediments underlying neritic environments in the Southern Ocean where meltwater inputs in spring and summer stabilize sunlit surface waters, promoting *Chaetoceros* blooms (Crosta *et al.*, 1997). Formation of the resting spores is triggered by exhaustion of nitrate, by the onset of light limitation, or by a breakdown in stratification.

During glacials, the seasonal ice zone (SIZ) extended considerably further north of its interglacial position, diminishing the extent of the permanently open-ocean zone (POOZ) since the APF does not migrate. In sediments of the expanded glacial SIZ, *Chaetoceros* resting spores displaced *F. kerguelensis* as the most abundant diatom (Abelmann *et al.*, 2006). It is thus argued that lesser accumulation of BSi south of the APF during glacials is the result, not of low levels of primary or export production, but of the lightly silicified *Chaetoceros* doing the bulk of the primary production (Abelmann *et al.*, 2006). Greater occurrence of tests of the radiolarian, *Cycladophora davisiana* (Ehrenberg) 1862, in sediments of glacial age is cited as further evidence for high primary production during glacials, as the abundance of this mid-depth-dwelling radiolarian is partly tied, at least in the Sea of Okhotsk, to high levels of export production (Abelmann & Nimmergut, 2005).

The micropaleontological view is thus that POC production and export were higher in the glacial Southern Ocean due to an ecological shift from the high-Si, low-C species, *F. kerguelensis*, to low-Si, high-C species of *Chaetoceros*, the signs of which are the somewhat dissolution-resistant (Crosta *et al.*, 1997) resting spores accumulating in sediments of the expanded glacial SIZ (Abelmann *et al.*, 2006).

23.2.6 Potential impact of variability in the BSi/C ratio of production

Iron availability offers another explanation for the low glacial accumulation rates of BSi south of the APF. Diatoms growing under Fe-limiting conditions (such as in the modern Southern Ocean) have higher BSi/cell (and/or BSi/volume) than diatoms growing under Fe-replete conditions (Takeda, 1998; De La Rocha *et al.*, 2000; Leynaert *et al.*, 2004; Hoffmann *et al.*, 2007). At the same time, cellular contents of C tend to decrease under Fe limitation and Fe-limited diatoms generally have higher BSi/C ratios than cells growing with sufficient nutrients. If Fe limitation had been alleviated in the glacial Southern Ocean by the increased deposition of dust (e.g. Figure 23.2), diatom BSi/C ratios could have decreased and led to low BSi accumulation rates.

The above examples highlight a critical point with respect to using BSi accumulation rates as a primary production proxy: BSi/C production ratios are quite variable. They differ between species (e.g. the anomalously high BSi/C ratios of *F. kerguelensis*) and within a single species under different environmental conditions (e.g. limitation by Fe or other nutrients). They also vary regionally, largely with the ratio of the contributions of siliceous to non-siliceous phytoplankton to primary production (Ragueneau *et al.*, 2002).

To quantify this with respect to the Southern Ocean, the BSi/C of Fe-limited *F. kerguelensis* ($\sim$0.8 mol mol^{-1}) is almost double that of nutrient-sufficient *F. kerguelensis* ($\sim$0.5 mol mol^{-1}) (Hoffmann *et al.*, 2007) and both of these values are higher than the BSi/C of vegetative cells of a common Antarctic *Chaetoceros* species, *C. dichaeta* (Ehrenberg) 1844 (Fe-limited BSi/C $\approx$ 0.3 mol mol^{-1}; nutrient-sufficient BSi/C $\approx$ 0.1 mol mol^{-1}) (Hoffmann *et al.*, 2007). Ignoring contributions from non-siliceous phytoplankton, if an Fe-limited ecosystem of 80% *F. kerguelensis* and 20% *C. dichaeta* became an Fe-replete ecosystem of 40% *F. kerguelensis* and 60% *C. dichaeta* (roughly following the scenario of Abelmann *et al.* (2006) for south of the APF), it would go from producing 1.25 mole of POC per mole of BSi to producing 3.3 moles of POC per mole of BSi. Such a shift could accommodate the low rates of BSi accumulation in the glacial Southern Ocean, south of the APF, without any diminishment of primary production.

Even in the modern-day Southern Ocean, however, there are times and places where Fe is available. Thus, the modern annual average BSi/C production ratio, even in the *F. kerguelensis*-dominated POOZ and APF regions, is only around 0.3 mol mol^{-1} (Ragueneau *et al.*, 2002) instead of 0.8 mol mol^{-1} (although the contribution of non-siliceous phytoplankton to primary production is a also a factor in the lower than predicted ratio). This dampens the total possible range in annually integrated BSi/C production ratios due to Fe availability.

Annual BSi/C production ratios also likely differed between glacials and interglacials due to change in the contribution of non-siliceous phytoplankton to primary production. The decrease of this ratio in the modern Southern Ocean from 0.3 mol mol^{-1} in the POOZ, down to 0.1 mol mol^{-1} north of the APF is related to the extreme drop in availability of dissolved silicon (DSi) across the APF (Ragueneau *et al.*, 2002) and sets limits for what the glacial conditions could have been. If Fe fertilization diminished DSi utilization south of the APR during glacials (as suggested by the Si isotopic composition of sedimentary diatoms (De La Rocha *et al.*, 1998; Brzezinski *et al.*, 2002; De La Rocha, 2006)), and if this resulted in the export of DSi to the subantarctic (Brzezinski *et al.*, 2002), then BSi/C production ratios would have evened out somewhat between the POOZ and the subantarctic zone.

Such a scenario would mean that due to Fe fertilization and the resulting redistribution of DSi (and BSi production), the Southern Ocean south of the APF would have shifted from producing 3.3 moles of POC per mole of BSi in modern times to producing something like 5 moles of POC per mole of BSi during glacials. North of the APF, the difference would be between 10 moles of POC per mole of BSi currently to roughly half that during the glacial. Thus it could be reasonably suggested that during glacials BSi accumulation rates south of the APF could be one third lower than interglacials without necessarily reflecting any change in the primary or export production of POC. Likewise, BSi accumulation rates could have been twice as high to the north of the APF during glacials than interglacials without being related to a change in primary production.

23.2.7 Shifts in the valve size frequency distribution of *F. kerguelensis*

A novel piece of the Southern Ocean puzzle has come to light recently with the observation that the average size of *F. kerguelensis* frustules in the Southern Ocean peaks geographically at the APF and temporally at the LGM (Cortese & Gersonde, 2007). Each mitotic division of this diatom produces one daughter cell with an epitheca the same size as that of the parent cell and one daughter cell with an epitheca (and thus an entire frustule) that is slightly smaller. Accordingly, the average valve size in the population decreases with the increasing number of mitotic cell divisions from the initial parent cell. The maximum valve size can only be restored by sexual reproduction and auxospore formation. Thus, a population with a high average valve size is one whose cells undergo sexual reproduction after fewer mitotic cell divisions than a population with a low average valve size.

The addition of Fe to Fe-limited Southern Ocean waters has been observed to increase the rate of auxospore formation in *F. kerguelensis* (Assmy *et al.*, 2006). Control over the frequency of sexual versus asexual reproduction may tailor cell size to environmental conditions. Fe-limited cells are smaller than their nutrient-sufficient counterparts in keeping with the favorability of higher surface-to-volume ratio when nutrient concentrations are low (Leynaert *et al.*, 2004).

The data of Cortese and Gersonde (2007) may indicate that nutrient conditions are more favorable for growth at the modern-day APF than to the north and south of it (an observation that meshes well with the peak in chlorophyll concentrations at the APF). Their data would also indicate that nutrient conditions were also correspondingly better for diatom growth in the glacial Southern Ocean (something that fits with the idea of Fe fertilization of glacial Southern Ocean waters and the Fe accumulation rates at these sites (Figure 23.2)).

23.3 Synthesis

When the BSi accumulation rates, the sedimentary distributions of *F. kerguelensis* and *Chaetoceros* resting spores, the variability in BSi/C production ratios, and the valve size frequency distributions of *F. kerguelensis* are taken together, what is the picture of the glacial Southern Ocean that emerges?

The more northward occurrence of seasonal sea ice during the glacials would have meant the expansion of the SIZ zone north into what is currently the southern extent of the POOZ. The melt-back of ice in the spring could have favored the blooming of *Chaetoceros* species over that of *F. kerguelensis* by stratifying the surface waters and providing Fe and seed stocks of *Chaetoceros*. The net result of this would have been BSi/C production ratios as much as one third lower than during an interglacial (Figure 23.4).

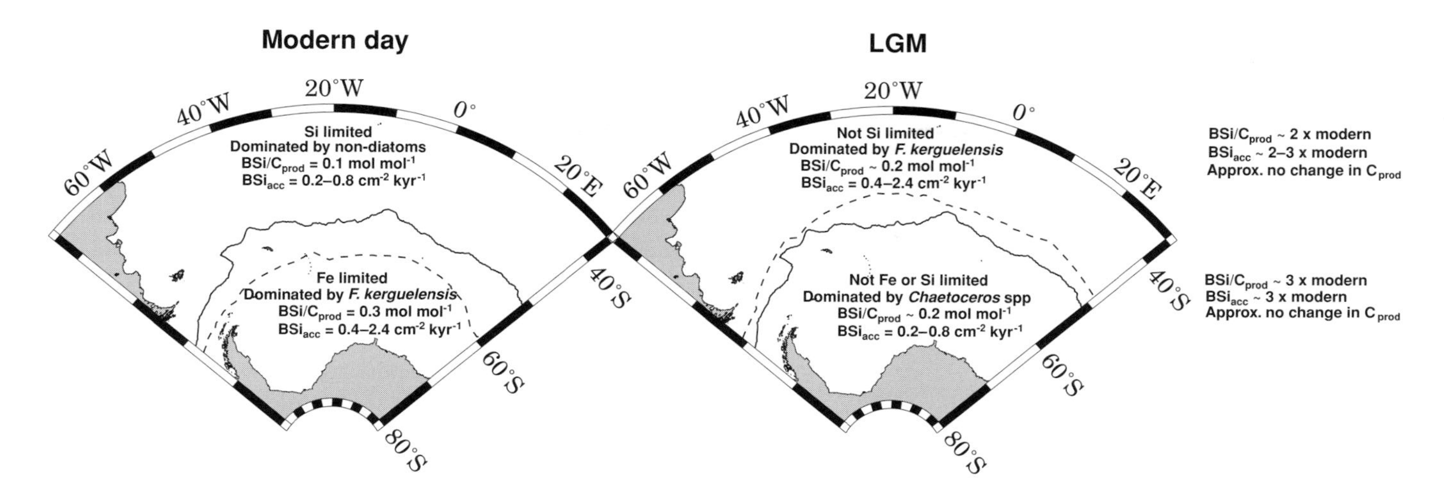

Figure 23.4 Summary of known and extrapolated conditions to the north and to the south of the APF for the modern day and LGM in the Atlantic sector of the Southern Ocean. The APF is shown as the solid black line while the maximum extent of winter sea ice is shown by the dashed line (as reported by Gersonde *et al.*, 2005).

The BSi accumulation rates at the LGM in the area that was the glacial SIZ do appear to be about 30% lower than during the Holocene (Figure 23.2), although there are not enough Th-corrected estimates of BSi accumulation to say this with certainty or precision. Despite this, it seems safe to conclude that the lower glacial rate of BSi accumulation south of the APF could entirely be due to a shift in annual BSi/C production ratios. Thus, in keeping with the suggestion of Abelmann *et al.* (2006) and Nelson *et al.* (2002), the accumulation-rate pattern (Figure 23.2) does not need to be interpreted as lower primary production here during glacials. On the other hand, it seems unlikely that an increase in primary production could be accommodated within this drop in BSi accumulation rates. A smaller difference in BSi accumulation rates between LGM and Holocene would be necessary for this to occur.

The diminished consumption of DSi south of the APF during glacials resulted in export of DSi north of the APF, relieving Si limitation there. This, and the encroachment of the SIZ into the POOZ, moved the favorable growth conditions for the heavily silicified *F. kerguelensis* to the north. This can be seen in the northward migration of both the *F. kerguelensis*-dominated sediments (Abelmann *et al.*, 2006) and the peak in average valve size of *F. kerguelensis* (Cortese & Gersonde, 2007) during glacials. Migrating northwards with the *F. kerguelensis*-dominated ecosystem was also the belt of high opal sediments that currently characterize the APF.

The net result of the above movements (and of the higher glacial availability of Fe) would have been BSi/C production ratios north of the APF being twice as high during the glacials as during modern times. It is hard to say from the limited amount of Th-corrected BSi accumulation rate data if glacial accumulation of BSi in this region was more than two times greater than during the Holocene (Figure 23.4). As with the case for south of the APF, observed differences in BSi accumulation rates can be almost completely explained by changes in BSi/C production rates.

In this region, north of the APF, no significant difference in glacial and interglacial primary production appear likely. Since N and P are in reasonable supply in this region in the modern ocean, the modern lack of DSi hinders diatom growth, not overall primary production. Because there is no reason to expect that N and P were more available north of the APF during glacials, increased availability of DSi would have stimulated BSi production without triggering an increase in primary production.

From this mixed geochemical and micropaleontological proxy approach to the BSi accumulation-rate records, we conclude that primary production in the glacial Southern Ocean, south of the APF, was not necessarily lower than it has been during the Holocene interglacial despite the significantly lower glacial rates of BSi accumulation (Figure 23.4). North of the APF, higher glacial rates of BSi accumulation likewise do not indicate higher levels of primary production. Rather, they suggest an increase in the fractional contribution of diatoms (and in particular, of the heavily silicified *F. kerguelensis*) to primary production.

What this increase in the contribution of *F. kerguelensis* means for POC export and the strength of the biological pump north of the APF during glacials is difficult to say. The high BSi/C ratios of *F. kerguelensis* could argue against an increase in export production. The shift towards diatom-domination of this ecosystem, however, could suggest an increase in the export efficiency of this region. Such reinterpretation of the BSi accumulation-rate records, keeping in mind probable variations in BSi/C production ratios, reopens the possibility that the biological pump was not diminished in the glacial Southern Ocean relative to the modern day.

23.4 Summary

Thorium-corrected rates of BSi in Southern Ocean sediments are a significant and important means for reconstructing primary production in the Southern Ocean. Recent reconstructions based instead on abundances of the radiolarian, *Cycladophora davisiana*, however, contradict the view from the BSi accumulation rates that the Southern Ocean, south of the APF, was less productive in glacial times than in the modern day (Abelmann *et al.*, 2006). It is well known that the BSi/C ratio of diatoms can vary by several-fold between species and, within some species, with nutrient availability (e.g. under Fe-limiting versus Fe-replete conditions) (Brzezinski, 1985; Hoffmann *et al.*, 2007). Taking the BSi/C variations into account is vital to the correct interpretation of BSi accumulation rate records. As shown here, for example, the lower glacial accumulation rates of BSi south of the APF in the Atlantic sector of the Southern Ocean can be entirely accounted for by shifts in diatom species composition and Fe availability and do not need to reflect lower glacial levels of primary production. Likewise, the higher glacial accumulation rates of BSi north of the APF may likely also not reflect higher glacial levels primary production.

Acknowledgments

We thank Amy Leventer for a thoughtful review, the editors for the opportunity to write this paper, and Christian Schlosser for producing Figure 23.4.

References

Abelmann, A., Gersonde, R., Cortese, G., Kuhn, G., & Smetacek, V. (2006). Extensive phytoplankton blooms in the Atlantic sector of the glacial Southern Ocean. *Paleoceanography*, **21**, PA1013, DOI:10.1029/2005PA001199.

Abelmann, A. & Nimmergut, A. (2005). Radiolarians in the Sea of Okhotsk and their ecological implication for paleoenvironmental reconstructions. *Deep-Sea Research II*, **52**, 2302–31.

Alldredge, A. L. & Silver, M. W. (1988). The characteristics, dynamics, and significance of marine snow. *Progress in Oceanography*, **20**, 41–82.

Anderson, R. F., Chase, Z., Fleisher, M. Q., & Sachs, J. (2002). The Southern Ocean's biological pump during the last glacial maximum. *Deep-Sea Research II*, **49**, 1909–38.

Antia, A. N., Koeve, W., Fischer, G., *et al.* (2001). Basin-wide particulate carbon flux in the Atlantic Ocean: regional export patterns and potential for atmospheric CO_2 sequestration. *Global Biogeochemical Cycles*, **15**, 845–62.

Assmy, P., Henjes, J., Smetacek, V., & Montresor, M. (2006). Auxospore formation in the silica-sinking, oceanic diatom, *Fragilariopsis kerguelensis* (Bacillariophyceae). *Journal of Phycology*, **42**, 1002–6.

Beaulieu, S. E. (2002). Accumulation and fate of phytodetritus on the sea floor. *Oceanography and Marine Biology*, **40**, 171–232.

Brovkin, V., Ganopolski, A., Archer, D., & Rahmstorf, S. (2007). Lowering of glacial atmospheric CO_2 in response to change in ocean circulation and marine biogeochemistry. *Paloceanography*, **22**, PA4202, DOI: 10.1029/2006PA001380.

Brzezinski, M. A. (1985). The Si:C:N ratio of marine diatoms: interspecific variability and the effect of some environmental variables. *Journal of Phycology*, **21**, 347–57.

Brzezinski, M. A., Pride, C. J., Franck, V. M., *et al.* (2002). A switch from $Si(OH)_4$ to NO_3^- depletion in the glacial Southern Ocean. *Geophysical Research Letters*, **29**, 1564, DOI:10.1029/2001GL014349.

Buesseler, K. O. (1998). The decoupling of production and particulate export in the surface ocean. *Global Biogeochemical Cycles*, **12**, 297–310.

Buesseler, K. O., Lamborg, C. H., Boyd, P. W., *et al.* (2007). Revisiting carbon flux through the ocean's twilight zone. *Science*, **316**, 567–70.

Charles, C. D., Froelich, P. N., Zibello, M. A., Mortlock, R. A., & Morley, J. J. (1991). Biogenic opal in Southern Ocean sediments over the last 450,000 years: implications for surface water chemistry and circulation. *Paleoceanography*, **6**, 697–728.

Chase, Z., Anderson, R. F., Fleisher, M. Q., & Kubik, P. W. (2003). Accumulation of biogenic silica and lithogenic material in the Pacific sector of the Southern Ocean during the past 40,000 years. *Deep-Sea Research II*, **50**, 799–832.

Cortese, G. & Gersonde, R. (2007). Morphometric variability in the diatom *Fragilariopsis kerguelensis*: implications for Southern Ocean paleoceanography. *Earth and Planetary Science Letters*, **257**, 526–44.

Crosta, X., Pichon, J.-J., & Labracherie, M. (1997). Distribution of *Chaetoceros* resting spores in modern peri-Antarctic sediments. *Marine Micropaleontology*, **29**, 283–99.

de Baar, H., Boyd, P. W., Coale, K. W., *et al.* (2005). Synthesis of iron fertilization experiments: from the Iron Age in the Age of Enlightenment. *Journal of Geophysical Research*, **110**, C09S16, DOI: 10.1029/2004JC002601.

De La Rocha, C. L. (2006). Opal-based isotopic proxies of paleoenvironmental conditions. *Global Biogeochemical Cycles*, **20**, GB4S09, DOI:10.1029/2005GB002664.

De La Rocha, C. L., Brzezinski, M. A., DeNiro, M. J., & Shemesh, A. (1998). Silicon-isotope composition of diatoms as an indicator of past oceanic change. *Nature*, **395**, 680–3.

De La Rocha, C. L., Hutchins, D. A., Brzezinski, M. A., & Zhang, Y. (2000). Effects of iron and zinc deficiency on elemental composition and silica production by diatoms. *Marine Ecology Progress Series*, **195**, 71–9.

DeMaster, D. J. (2002). The accumulation and cycling of biogenic silica in the Southern Ocean: revisiting the marine silica budget. *Deep-Sea Research II*, **49**, 3155–67.

Elderfield, H. & Rickaby, R. E. M. (2000). Oceanic Cd/P ratio and nutrient utilization in the glacial Southern Ocean. *Nature*, **405**, 305–10.

François, R., Altabet, M. A., Yu, E.-F., *et al.* (1997). Contribution of Southern Ocean surface-water stratification to low atmospheric CO_2 concentrations during the last glacial period. *Nature*, **389**, 929–35.

Frank, M., Gersonde, R., Rutgers van der Loeff, M., *et al.* (2000). Similar glacial and interglacial export bioproductivity in the Atlantic sector of the Southern Ocean: multiproxy evidence and implications for glacial atmospheric CO_2. *Paleoceanography*, **15**, 642–58.

Gersonde, R., Crosta, X., Abelmann, A., & Armand, L. (2005) Sea-surface temperature and sea ice distribution of the Southern Ocean at the EPILOG last glacial maximum – a circum-Antarctic view based on siliceous microfossil records. *Quaternary Science Reviews*, **24**, 869–96.

Goldman, J. C. (1993). Potential role of large oceanic diatoms in new primary production. *Deep-Sea Research I*, **40**, 159–68.

Hedges, J. I. & Keil, R. G. (1995). Sedimentary organic matter preservation: an assessment and speculative synthesis. *Marine Chemistry*, **49**, 81–115.

Hoffmann, L. J., Peeken, I., & Lochte, K. (2007). Effects of iron the the elemental stoichiometry during EIFEX and in the diatoms *Fragilariopsis kerguelensis* and *Chateoceros dichaeta*. *Biogeosciences*, **4**, 569–79.

Hillenbrand, C.-D. & Cortese, G. (2006). Polar stratification: a critical view from the Southern Ocean. *Palaeogeography, Palaeoclimatology, Palaeoecology*, **242**, 240–52.

Jacot Des Combes, H., Esper, O., De La Rocha, C. L., *et al.* (2008). Diatom $\delta^{13}C$, $\delta^{15}N$, and C/N since the last glacial maximum in the Southern Ocean: potential impact of species composition. *Paleoceanography*, **23**, PA4209, DOI:10.1029/2008PA001589.

Kohfeld, K. E., Le Quere, C., Harrison, S. P., & Anderson, R. F. (2005). Role of marine biology in glacial-interglacial CO_2 cycles. *Science*, **308**, 74–8.

Kumar, N., Anderson, R. F., Mortlock, R. A., *et al.* (1995). Increased biological productivity and export production in the glacial Southern Ocean. *Nature*, **378**, 675–80.

Leynaert, A., Bucciarelli, E., Claquin, P., *et al.* (2004). Effect of iron deficiency on diatom cell size and silicic acid uptake kinetics. *Limnology and Oceanography*, **49**, 1134–43.

Lutz, M., Dunbar, R. & Caldeira, K. (2002). Regional variability in the vertical flux of particulate organic carbon in the ocean interior. *Global Biogeochemical Cycles*, **16**, 1037, DOI: 10.1029/2000GB001383.

Martin, J. H., Knauer, G. A., Karl, D. M., & Broenkow, W. W. (1987). VERTEX: carbon cycling in the northeast Pacific. *Deep-Sea Research*, **34**, 267–85.

Moriceau, B., Gallinari, M., Soetaert, K., & Ragueneau, O. (2007). Importance of particle formation to reconstructed water column biogenic silica fluxes. *Global Biogeochemical Cycles*, **21**, GB3012, DOI:10.1029/2006GB002814.

Mortlock, R. E., Charles, C. D., Froelich, P. N., *et al.* (1991). Evidence for lower productivity in the Antarctic Ocean during the last glaciation. *Nature*, **351**, 220–3.

Nelson, D. M., Anderson, R. F., Barber, R. T., *et al.* (2002). Vertical budgets for organic carbon and biogenic silica in the Pacific sector of the Southern Ocean, 1996–1998. *Deep-Sea Research II*, **49**, 1645–74.

Pondaven, P., Ragueneau, O., Tréguer, P., *et al.* (2000). Resolving the 'opal paradox' in the Southern Ocean. *Nature*, **405**, 168–72.

Ragueneau, O., Dittert, N., Pondaven, P., Tréguer, P., & Corrin, L. (2002). Si/C decoupling in the world ocean: is the Southern Ocean different? *Deep-Sea Research II*, **49**, 3127–54.

Ragueneau, O., Schultes, S., Bidle, K., Claquin, P., & Moriceau, B. (2006). Si and C interactions in the world ocean: importance of ecological processes and implications for the role of diatoms in the biological pump. *Global Biogeochemical Cycles*, **20**, GB4S02, DOI:10.1029/2006GB002688.

Ragueneau, O., Tréguer, P., Leynaert, A., *et al.* (2000). A review of the Si cycle in the modern ocean: recent progress and missing gaps in the application of biogenic opal as a paleoproductivity proxy. *Global and Planetary Change*, **26**, 317–65.

Robinson, R. S., Brunelle, B. G., & Sigman, D. M. (2004). Revisiting nutrient utilization in the glacial Antarctic: evidence from a new method for diatom-bound N isotope analysis. *Paleoceanography*, **19**, PA3001, DOI:10.1029/2003PA000996.

Rosenthal, Y., Boyle, E. A., & Labeyrie, L. (1997). Last glacial paleochemistry and deepwater circulation in the Southern Ocean: evidence from foraminiferal cadmium. *Paleoceanography*, **12**, 787–96.

Sarmiento, J. L. & Gruber, N. (2006). *Ocean Biogeochemical Cycles*. Princeton: Princeton University Press.

Schelske, C. L. (1999). Diatoms as mediators of biogeochemical silica depletion in the Laurentian Great Lakes. In *The Diatoms: Applications for the Environmental and Earth Sciences*, ed. E. F. Stoermer & J. P. Smol, Cambridge: Cambridge University Press, pp. 73–84.

Suess, E. (1980). Particulate organic carbon flux in the oceans – surface productivity and oxygen utilization. *Nature*, **288**, 260–3.

Singer, A. J. & Shemesh, A. (1995). Climatically linked carbon isotope variation during the past 430,000 years in Southern Ocean sediments. *Paleoceanography*, **10**, 171–7.

Takeda, S. (1998). Influence of iron availability on nutrient consumption ratio of diatoms in oceanic waters. *Nature*, **393**, 774–7.

Tréguer, P., Nelson, D. M., Van Bennekom, A. J., *et al.* (1995). The silica balance in the world ocean: a reestimate. *Science*, **268**, 375–9.

Van Cappellen, P., Dixit, S. & Van Beusekom, J. (2002). Biogenic silica dissolution in the oceans: reconciling experimental and field-based dissolution rates. *Global Biogeochemical Cycles*, **16**, 1075, DOI:10.1029/2001GB001431.

Part V

Other applications

24 Diatoms of aerial habitats

JEFFREY R. JOHANSEN

24.1 Introduction

Although studied less than aquatic diatoms, aerial diatoms are discussed in an extensive literature. Most publications on the topic consist merely of floristic lists. Thus, our understanding of aerial diatom ecology is meager. Given the brevity of the current chapter, it is not possible to list all of the pertinent literature. This paper will summarize aerial diatom studies based on floristic literature and my own work.

The most important pioneer worker on aerial diatoms was probably Johannes Boye Petersen. Unlike many early soil phycologists, he treated diatoms with both detail and taxonomic accuracy. Petersen (1915, 1928, 1935) examined numerous aerial samples from Denmark, Iceland, and east Greenland. In all, he found 196 diatom taxa from soils, wet rocks, wet tree bark, and mosses, many of which were new to science at that time.

Other important early floristic works are those of Beger (1927, 1928), Krasske (1932, 1936, 1948), Hustedt (1942, 1949), Lund (1945), and Bock (1963). More recent studies report diatom floras associated with limestone caves, sandstone cliff faces, wet rocks, mosses, and soils. Added to these studies are numerous papers on aerial algae, which discuss diatoms to some extent. Reviews on terrestrial algae have generally slighted the diatoms, although none have ignored them (Novichkova-Ivanova, 1980; Metting, 1981; Starks *et al.*, 1981; Hoffmann, 1989; Johansen, 1993).

Petersen (1935) defined a number of categories for aerial algae based on their habitat type. Euaerial algae inhabit raised, prominent objects that receive moisture solely from the atmosphere. Terrestrial algae are those growing on the soil. Pseudoaerial algae are those living on rocks moistened by a fairly steady source of water, such as waterfall spray, springs, or seeps. Petersen divided terrestrial algae further into three categories based on their appearance in both euaerial habitats and soils (aeroterrestrial), on soils that periodically dry out (euterrestrial), or on soils that are perpetually moist (hydroterrestrial). Finally, he differentiated surface-soil algae (epiterranean) from those occurring below the surface (subterranean).

Ettl and Gärtner (1995) adopt a slightly different terminology, and for diatoms the classification system of Petersen (1935) is probably superior. However, the addition of substrate-specific adjectives which they suggest is useful, so that any of the above categories can be combined with epiphytic (on bark, leaves), epibryophytic (on mosses), epixylic (on wood), or epilithic (on rock). Subaerial, a term that has been used to describe soil, mosses, trees, and wet rocks (Stoermer 1962; Johansen *et al.* 1981; Camburn, 1982), is less descriptive and should be avoided if possible.

24.2 Systematics

Numerous diatom species, varieties, and forms have been described from aerial habitats. Petersen (1915, 1928, 1935) alone described 39 new taxa. Bock (1963), Carter (1971), Hustedt (1942, 1949), Krasske (1932, 1936, 1948), Lund (1945), Rushforth *et al.* (1984), and VanLandingham (1966, 1967) likewise published many new taxa. In the first edition of this chapter (Johansen, 1999), I reported the existence of about 340 diatom taxa from soils, 400 diatom taxa from rock substrates, and 130 taxa from mosses. There has been substantial progress in the taxonomy of aerial algae in the last ten years, and this has been primarily due to three factors. (1) Many diatom genera have been recognized to be polyphyletic (containing species that do not all belong to the same evolutionary clade), and have consequently been split into many new (or resurrected) genera. This has resulted in increased interest and study in these taxa, many of which can be found in aerial habitats (Table 24.1).

The Diatoms: Applications for the Environmental and Earth Sciences, 2nd Edition, eds. John P. Smol and Eugene F. Stoermer. Published by Cambridge University Press.

Table 24.1 Genera with significant representation (many or most species) in aerial and terrestrial environments. As many of these have been split out from large polyphyletic taxa in recent years, the old genus names are given to assist the reader in finding these taxa in older floristic accounts. A representative aerial species in each genus is given so that one with familiarity with the old names will be able to reference the genera listed. Many primarily aquatic genera have slight representation in aerial and terrestrial environments, but are not listed here.

Genus with representative species	Old genus name
Achnanthes coarctata (Bréb.) Grunow	
Adlafia bryophila (J. B. Peters.) Moser *et al.*	*Navicula*
Caloneis bacillum (Grun.) Cl.	
Chamaepinnularia soehrensis (Krasske) Lange-Bert.	*Navicula*
Cymbopleura rupicola (Grun.) Krammer	*Cymbella*
Decussata placentula (Ehrenb.) Lange-Bert. et Mezelt.	*Navicula*
Delicata delicatula (Kütz.) Krammer	*Cymbella*
Diadesmis contenta (Grun. ex V. H.) D. G. Mann	*Navicula*
Diatomella hustedtii Manguin	
Encyonema minutum (Hilse ex Rabh.) D. G. Mann	*Cymbella*
Encyonopsis microcephala (Grun.) Krammer	*Cymbella*
Eunotia muscicola Krasske	
Frankophila maillardi (Le Cohu) Lange-Bert.	*Fragilaria*
Frustulia crassinervia (Bréb.) Lange-Bert. et Krammer	
Hantzschia amphioxys (Ehrenb.) Grun.	
Kobayasiella subtillisima (Cl.) Lange-Bert.	*Navicula*
Labellicula subantarctica Van de Vijver et Lange-Bert.	
Lacunicula sardiniense Lange-Bert.	
Luticola mutica (Kütz.) D. G. Mann	*Navicula*
Mayamaea atomus (Kütz.) Lange-Bert.	*Navicula*
Microcostatus krasskei (Hust.) Johansen et Sray	*Navicula*
Muelleria terrestris (J. B. Peters.) Spaulding	*Navicula*
Navigiolum sardiniense Lange-Bert.	
Nupela chilensis (Krasske) Lange-Bert.	*Anomoeoneis*
Orthoseira roeseana (Rabh.) O'Meara	*Melosira*
Pinnularia borealis Ehrenb.	
Stauroneis obtusa Lagersgtedt	
Tetracyclus rupestris (Braun) Grun.	

(2) We have entered a period in which the old widely held notion that most algal species are cosmopolitan in similar habitats has been rejected. Concurrent with this change, species concepts of diatoms have changed such that species are now much more narrowly defined. These changes in the collective mindset of diatomists have led to the fragmentation of broadly defined species into many new species. (3) There has been intense interest and study in some aerial diatoms, particularly those in pseudoaerial and hydroterrestrial habitats. Diatoms of polar areas in particular have experienced increased scrutiny, and many of them are aerial taxa.

There are several recently published floras that demonstrate the wealth of biodiversity to be discovered in aerial and terrestrial habitats. Van de Vijver *et al.* (2002a) described 36 new species from the subantarctic island Ile de la Possession, many of which were terrestrial. Eleven new species were described in *Diadesmis* alone (Van de Vijver *et al.*, 2002a, 2002c; Le Cohu & Van de Vijver, 2002), with new taxa also in *Kobayasiella*, *Geissleria*, *Hantzschia*, and *Orthoseira* (Van de Vijver *et al.*, 2002a). Subsequently, Van de Vijver *et al.* (2004) published a monograph on polar *Stauroneis* species. Forty new species were recognized, nine from aerial and terrestrial habitats. Twenty-one new species were from small shallow pools. The island of Sardinia has many ephemeral rock pools, and the diatoms of this arid landscape often have heavily silicified valves with occluded striae (Lange-Bertalot *et al.* 2003), an obvious defense against desiccation (Lowe *et al.*, 2007). Lange-Bertalot *et al.* (2003) found 77 new diatom taxa in Sardinia, most of which were in genera with aerophilic species. Eighteen new aerophilic taxa were found in the Andes Mountains (Lange-Bertalot & Rumrich, 2000). New aerial diatom taxa continue to be described in smaller publications as well (Spaulding *et al.*, 1999; Van de Vijver, 2002, Van de Vijver *et al.*, 2002d; Van de Vijver & Le Cohu, 2003; Van de Vijver *et al.*, 2005; Van de Vijver & Kopalová, 2008; Van de Vijver & Mataloni, 2008; Van de Vijver *et al.*, 2008; Furey *et al.*, 2009; Lowe *et al.*, 2009; Veselá and Johansen, 2009).

Springs are frequently small, well oxygenated, and apparently very similar to pseudoaerial habitats such as waterfall splash zones. Work in small springs and headwater streams (often with their source in a spring) has yielded new species as well (Cantonati & Lange-Bertalot, 2006). Werum and Lange-Bertalot (2004) described ten new species of *Diadesmis*, ten new species of *Stauroneis*, and six new species of *Eunotia*. In a single wet wall habitat in Austria, Reichardt (2004) found 67 different diatom taxa, eight of which were new to science. Space does not permit listing all of the new taxa that have been described recently from aerial and pseudoaerial habitats, but the take-home message should be that no one should be surprised to find new species in any isolated spring, headwater stream, or waterfall splash zone that they take the time to examine with care. Aerial habitats have a wealth of diatom biodiversity waiting to be discovered.

24.3 Ecology

24.3.1 Factors influencing distribution

The indicator status of most aerial diatom taxa is poorly known. An aerial habitat is harsh and limiting in a number of ways. This makes it difficult to determine which environmental factors are most significant in determining species distribution. Moisture availability has frequently been thought to be the most important limiting factor (Camburn 1982, Van de Vijver *et al.* 2002b), although it is probably better to consider exposure to long periods of desiccation as more crucial. The question concerning moisture is whether or not the site is wet year round (truly hydroterrestrial or pseudoaerial), or wet to damp most of the year with brief periods of dryness (giving time for temporary establishment of hydrophilous species), or exposed to long periods of total dryness (hydrophilous species excluded). Several authors have categorized aerophiles according to their desiccation resistance (Krasske, 1932, 1936; Ito & Horiuchi 1991), although none of their work was done in arid environments, where long periods of drought are even more limiting. Many of the taxa designated by these authors as xeritic do not occur in desert soils. Several authors have commented on size diminution within species in response to dry conditions (Beger, 1927; Bock, 1963), and this was demonstrated statistically in *Pinnularia borealis* by Van de Vijver and Beyens (1997).

Associated with the effects of moisture are extremes of temperature (Lowe *et al.*, 2007). Aerial habitats experience considerably higher diel fluctuations of temperature than aquatic habitats, and this factor could be limiting for some species. Furthermore, soils experience greater extremes on a yearly cycle; temperatures in some desert soils exceed 50 °C in the summer and fall below freezing during winter. Elevated temperatures coincide with periods of dryness, and the severity of desiccation is even greater in climates with a hot, dry period. Despite the intuitive conjecture that temperature is an important limiting factor for diatoms, specific effects of temperature on aerial diatoms have not been widely tested. The most interesting attempt to relate temperature to distribution of aerial diatoms was conducted by Gremmen *et al.* (2007). They examined an altitudinal gradient in subantarctic Kerguelen Island and found that altitude explained 20% of the community variance in epibryophytic hydroterrestrial diatom communities. They assumed that the primary causal factor of community change along their altitudinal gradient was temperature. The community had a reduction in species diversity at high altitude, indicating that cold temperatures were actually limiting in this region.

It is likely that pH is nearly as critical in defining species distributions of aerial diatoms as exposure to high temperatures and desiccation. In my laboratory we have found that pH differences of 0.5 pH units can drastically change species composition in sandstone seeps from several sites in Ohio and Tennessee. Limestone caves and temperate sandstone seeps have similar moisture regimes, temperature extremes, and low nutrient availability, but markedly different floras. The sandstone seeps are generally acidic (pH 3.7–6.0), while the water dripping through a limestone cave is neutral to slightly alkaline (pH 6.5–8.0). *Diadesmis gallica* Wm. Smith and *Diadesmis laevissima* (Cl.) D. G. Mann are especially confined to caves, while *Diadesmis contenta*, *Eunotia exigua* (Bréb.) Rabenh., and *Microcostatus krasskei* occur primarily in seeps. Some of these taxa, particularly *Diadesmis* species, appear to be adapted to low light (Kawecka & Olech, 1993).

Some taxa appear to be less sensitive to pH. *Hantzschia amphioxys* Ehrenb. is the most conspicuous of these, with a published pH range of occurrence between 5.6–8.5. This, however, is more the exception than the rule. Furthermore, even taxa with wide pH ranges show distinct optima. For example, *H. amphioxys* is rare in acidic soils, but reaches high densities in neutral to slightly alkaline soils.

Substrate is also a critical limiting factor. Even given similar pH and temperature regimes, euterrestrial taxa are rarely abundant on pseudoaerial lithic substrates. For example, the cosmopolitan euterrestrial species *H. amphioxys*, *Luticola cohnii* (Hilse) D. G. Mann, *Luticola mutica* (Kütz.) D. G. Mann, and *Pinnularia borealis* Ehrenb. are absent or rare in all strictly aerial and pseudoaerial lithic substrates (Camburn, 1982; Rushforth *et al.*, 1984; VanLandingham, 1964, among others). However, if mosses are present on lithic surfaces, then all of these taxa are usually present to at least some extent (Reichardt, 1985; Van de Vijver & Beyens, 1997). Bryophytic diatom floras have been studied more extensively than either euterrestrial or lithophytic floras. These communities are often an agglomeration of euterrestrial and lithophytic species, along with a few predominantly bryophytic species. Algae inhabiting wooden substrates have not been studied sufficiently to determine if distinctive epixylic associations exist (Petersen, 1928).

Most aerial diatom species are indicative of low nutrient availability. Of the 122 diatoms listed as exclusively aerial or occurring mostly in wet, moist, or temporarily dry places in Europe, van Dam *et al.* (1994) considered only four of the taxa to be facultatively or obligately nitrogen-heterotrophic taxa, while over half of those characterized with regard to nitrogen uptake were classified as tolerating only very small concentrations of organic nitrogen. Likewise, half of the taxa scored for trophic state were oligotraphentic, with relatively few eutraphentic or hypereutraphentic species. This is not surprising, since most terrestrial, aerial, and pseudoaerial habitats have little exposure to the anthropogenic nitrogenous pollution that most surface waters in the populated world receive. In the extensive recent floristic literature dealing with new species of aerial and terrestrial taxa, the trend of finding oligotraphentic taxa has reinforced the review of van Dam *et al.* (1994).

Elevated conductivity is a feature of many aerial habitats where evaporation is high, and halophilous species commonly have been collected from both soils and pseudoaerial habitats in desert environments. Indeed, *Luticola cohnii*, *Luticola mutica*, and *Luticola nivalis* (Ehrenb.) D. G. Mann are all species which occur at 500–1000 mg $Cl^- l^{-1}$ (Van Dam *et al.*, 1994), while *D. contenta*, *Diadesmis perpusilla* (Grun.) D. G. Mann, *Hantzschia amphioxys*, *Luticola saxophila* (Bock) D. G. Mann, *Mayamaea atomus* (Kütz.) Lange-Bert., *Mayamaea excelsa* (Krasske) Lange-Bert., *Pinnularia borealis*, *Pinnularia obscura* Krasske, and *Tryblionella debilis* Arnott, some of the most commonly occurring aerial diatoms, are all considered typical of slightly brackish waters (100–500 mg $Cl^- l^{-1}$; van Dam *et al.*, 1994). Sea cliffs moistened by sea spray also have very high conductivities and can have distinctive taxa.

Although the very term aerophilous means "air-loving", it has not been determined whether aerial taxa indeed require high oxygen (O_2) concentrations. It seems likely that such a requirement may exist for many of these species. A requirement for continuously high O_2 levels may be one characteristic that excludes some of these diatoms from aquatic environments, which may lack the consistently saturated oxygen levels that likely prevail in thin films of water in terrestrial, euaerial, and pseudoaerial habitats.

In summary, a number of environmental variables likely act in concert to restrict aerial diatom distribution. Rather than considering these species to prefer aerial habitats, it may be more accurate to consider them as most competitive under the multiple stresses that characterize these habitats, especially desiccation, temperature extremes, high conductivity, and low nutrient availability. Varying substrates and pH, while not generally considered stressors, act in conjunction with the other environmental variables to determine which species can exist in a given site.

24.3.2 Biogeography

The importance of environmental factors in determining biogeography of aerial taxa is perhaps best illustrated by examining the diatom taxa present in habitats of varying severity. In the most extreme habitats studied (e.g. soils of the hot deserts of the American Southwest), we have found as few as two diatom taxa, *H. amphioxys* fo. *capitata* O. Müller and *L. mutica*. In hot deserts receiving more precipitation, such as Baja California and San Nicolas Island (USA), where coastal moisture is available, we find the above species together with *Achnanthes coarctata* (Bréb.) Grun., *H. abundans* Lange-Bert., *H. amphioxys*, *L. cohnii*, *L. nivalis* (Ehrenb.) D. G. Mann, *Muelleria gibberula* (Cl.) Spaulding et Stoermer, *Pinnularia borealis*, and *P. borealis* var. *rectangularis* Carlson (Flechtner *et al.*, 1998, 2008). In higher-elevation hot deserts receiving regular rain during the mild winter months we find even more species, including *Cyclotella meneghiniana* Kütz., *Encyonema minutum*, *Diatoma vulgare* Bory, *Epithemia zebra* (Ehrenb.) Kütz., *Fragilaria vaucheriae* (Kütz.) Peters. *Mayamaea excels*, and *Navicula lanceolata* (Ag.) Kütz. (Anderson & Rushforth, 1976). Finally, in high-elevation deserts, which receive both snow and winter rain, we see the addition of numerous taxa, including *Caloneis aerophila*, *Denticula elegans* fo. *valida* Pedicino, *Diadesmis contenta* var. *parallela* (Peters.) Hamilton, *Epithemia adnata* var. *minor* (Perag. et Hérib.) Patr., *Luticola nivalis*, and *Luticola paramutica* (Bock) D. G. Mann (Johansen *et al.*, 1981, 1984). When desert soils are compared to those of other climates, it becomes apparent that many species common in temperate and/or polar soils are absent or only rarely encountered, such as all *Eunotia* species, most *Diadesmis* species, and most of the small naviculoid genera other than *Luticola*, such as the *Mayamaea atomus* and *Eolimna minima* groups. Temperate soils clearly have more acidobiontic and acidophilic species, and much of the difference in floras between semi-arid shrub-steppe and temperate forest is likely due to the large variances in pH, conductivity, and nutrient availability rather than to differences in temperature extremes and moisture availability. We see large generic shifts between habitat types with only a few cosmopolitan taxa able to survive the full spectrum of environmental variability.

Interestingly, aerial diatom taxa are commonly found in the aquatic habitats in polar regions. In a study of 36 High Arctic ponds on Ellesmere Island (Nunavut, Canada), over two thirds of the common benthic diatom taxa collected from rocks, mosses, and sediments were known aerophiles (Douglas & Smol, 1995). These taxa were common in the very shallow ponds of Ellesmere Island. This suggests that many aerial diatom taxa may not owe their distribution pattern solely or even primarily to their ability to withstand desiccation. High Arctic ponds are oligotrophic, unpolluted, circumneutral, cold, and highly oxygenated. The occurrence of aerial taxa in such habitats indicates that some taxa may occur in aerial habitats simply because they require highly oxygenated habitats and tolerate very low nutrient levels. Additionally, some taxa may respond to substrate type (such as moss). Aerial genera and species also are found in the isolated aquatic habitats of Antarctica, particularly *Diadesmis*, *Muelleria*, and *Stauroneis* (Spaulding *et al.*, 1999, Cremer *et al.*, 2004). However, the incidence of aerial taxa in Antarctic lakes may be due to difficulty of dispersal of aquatic diatoms from other parts of the world to the isolated continent of Antarctica. Deep lakes in the High Arctic typically are dominated by species in lentic genera common in continental temperate biomes (Cremer & Wagner, 2003).

Given the high number of new taxa of aerial diatoms in all biomes, it appears that many aerial taxa may have limited distributions, i.e. they are endemic taxa. Many of the Antarctic and subantarctic taxa appear to be such regional endemics (see Spaulding *et al.*, this volume), although they are often distributed in more than one site within a region (Van de Vijver *et al.*, 2008). Kilroy *et al.* (2007) challenge the cosmopolitan model of diatom distribution (Finlay & Clarke, 1999; Finlay, 2002; Fenchel & Finlay, 2004) and argue that species ranges can be constrained by poor dispersal. Stable unproductive habitats were particularly hypothesized as likely places for endemism to develop. They did not find high regional endemism in seeps and springs in New Zealand, but did observe endemism in bogs and tarns (stable and unproductive habitats). Vyverman *et al.* (2007) provide further evidence that global diatom distributions in lentic habitats seem more constrained by historical factors than environmental factors, and challenge Bass-Becking's (1934) ubiquity hypothesis as well. Lowe *et al.* (2007) suggest that wet-wall habitats have limited vectors for dispersal, are often free from anthropogenic disturbance, and are unproductive (oligotrophic); consequently they may harbor endemic taxa. At this point in time it is difficult to know whether the increased incidence of new species from aerial habitats is because endemic taxa exist in these sites, or the aerial habitats have just been too understudied for all the biodiversity present to have been discovered. I suspect it is a combination of both.

24.4 Applications

24.4.1 Aerial diatoms as indicators

Aerial diatom species may have higher resistance to ultra-violet radiation (UVR) than aquatic species. They live in thin films of water that would be ineffective at reducing levels of UVR. Aerial algae in arid regions are exposed both to very high illuminance and UVR levels due to the dearth of shading by vascular plants. Experimental evidence indicates that stalked diatoms (*Gomphonema* and *Cymbella* spp.) have a competitive advantage in high UVR habitats (Bothwell *et al.*, 1993). These genera are also common in the highly illuminated sandstone seep walls of the arid west (Johansen *et al.*, 1983a, 1983b), even though they are quite rare on sandstone seeps in shaded temperate forests. These findings suggest a connection between UVR resistance and some aerial diatom species.

The prevalence of aerial diatoms in polar waters may be partly in response to elevated UVR resistance. If so, their relative abundance might increase with increasing levels of UVR due to destruction of ozone in the stratosphere. The apparent correlation between some aerial diatoms and elevated UVR in both desert and polar regions is intriguing. Subsequent manipulative experiments may demonstrate that aerial diatoms are useful indicators of varying levels of UVR exposure.

Aerial diatoms also may have value as clean water indicators. In a world where it becomes increasingly difficult to find unpolluted, oligotrophic lentic and lotic waters, aerial habitats may provide additional diatom taxa that can be used as indicators of clean water conditions. The only serious problem we have had in using aerial habitats for this purpose has been the difficulty in characterizing the levels of nutrients and pollutants in the thin film of water in these habitats. Soil chemistry can be determined, but is not equivalent to water chemistry. Lithic pseudoaerial habitats are especially problematic because often the liquid is impossible to separate from the algal film on the rock (Lowe *et al.*, 2007), and the chemistry of the algal film or rock itself is even more misleading than soil chemistry. Springs, many of which contain predominately aerial oligotrophic taxa, can be tested with respect to chemistry, and may provide a means for characterizing aerial taxa with respect to nutrient availability (Werum & Lange-Bertalot, 2004).

Given the extensive recent studies of aerial diatoms in polar regions, it is very possible that these diatoms may have significance in paleoclimate studies. When polar lakes and ponds become shallow or even dry, then these diatoms will clearly dominate. Some Arctic ponds are now disappearing, ostensibly due to higher evaporation associated with climate change (e.g. Smol & Douglas, 2007). Many diatoms are known to be associated with bryophytes as well. Van de Vijver *et al.* (2002b) were able to create a transfer function for soil moisture, and Gremmen *et al.* (2007) created a transfer function for altitude as a proxy for temperature. As we continue to understand the ecology of these aerial, pseudoaerial, terrestrial, and hydroterrestrial habitats, their paleoecological significance will be easier to interpret.

24.4.2 Aerial habitats as study systems

Euterrestrial diatom communities, particularly in arid environments, have a low species richness. They are similar over fairly extensive regions, with overlap even between floras from different ecoregions. Terrestrial habitats do not have dramatic periodic changes, such as spates or spring turnover, and so the temporal component of change is depressed. Although seasonal fluctuations in the densities of terrestrial algae are evident (Bristol-Roach, 1927; Petersen, 1935; Johansen *et al.*, 1993), the composition of the community shows little seasonal change (Lund, 1945; Johansen & Rushforth, 1985), or at least much less change than typically observed in aquatic habitats (Johansen *et al.*, 1993). Thus, in a number of ways, terrestrial diatom communities are simpler than the diatom communities of aquatic habitats.

Because of the simplicity of these aerial communities, aerial diatom floras may serve as better experimental systems upon which to conduct manipulative experiments. Treatments applied to terrestrial habitats are in many ways easier to do, easier to confine to a given site, and less prone to natural disturbances than treatments applied to lakes and rivers. The low species diversity of aerial habitats is desirable because community changes are clearer and easier to comprehend than in aquatic habitats where stochastic factors are more prevalent. Manipulative experiments could be conducted to test the specific effects of increasing nutrients, additional moisture, UVR reduction, contamination with pesticides and herbicides, and changing pH. Such experiments might increase our ability to delimit indicators and to determine what environmental factors are most important in controlling species distribution. Experimental work of this sort has begun (Davey & Rothery, 1992; Furey *et al.*, 2007), but much more needs to be done. Most distributional work has been correlative, and it appears that pH, moisture, salinity, and phosphate may be important (Van de Vijver & Beyens, 1998; van Kerckvoorde *et al.* 2000; Van de Vijver *et al.*, 2002b).

Given the limited extent of our understanding of aerial algae, and the potentially unique research opportunities they present, further study of euaerial, terrestrial, and pseudoaerial diatoms is warranted.

24.5 Summary

Aerial diatom communities are defined as those communities living exposed to the air outside of lentic and lotic environments. However, the diversity of habitats is fairly large, including soils of all climates, rocks, trees, bromeliads, and buildings receiving only atmospheric water, as well as rock faces and soils perpetually exposed to water from springs, waterfall spray, and sea spray. Such communities have received a great deal of attention in recent years, as they have proven to harbor many small, interesting diatom species, many of which have been described as new to science. Aerial diatoms have been shown to be especially important in High Arctic, Antarctic, and subantarctic regions, and have been especially studied in those regions. Aerial diatoms are often found in very oligotrophic conditions, and indicator status with regard to saprobity will likely be an area of future research. Transfer functions for altitude (i.e. temperature) and moisture have already been determined for some taxa. The simplicity of structure and reduced temporal variation in many aerial diatom communities makes them very suitable as study systems, and the study of the ecology of these communities is likely to increase over the next decade.

References

Anderson, D. C. & Rushforth, S. R. (1976). The cryptogamic flora of desert soil crusts in southern Utah. *Nova Hedwigia*, **28**, 691–729.

Bass-Becking, L. G. M. (1934). Geobiologie of Inleiding tot de Milieukunde. The Hague: Van Stock and Zoon.

Beger, H. (1927). Beiträge zur Ökologie und Soziologie der luftlebigen (atmophytischen) Kieselalgen. *Berichte der Deutschen Botanischen Gesellschaft*, **45**, 385–407.

Beger, H. (1928). Atmosphytische Moosdiatomeen in den Alpen. *Vierteljahrsschrift der Naturforschenden Gesellschaft in Zürich*, **73** (Beib. 15), 382–404.

Bock, W. (1963). Diatomeen extrem trockener Standorte. *Nova Hedwigia*, **5**, 199–254 (+ 3 plates).

Bothwell, M. L., Sherbot, D., Roberge, A. C., & Daley, R. J. (1993). Influence of natural ultraviolet radiation on lotic periphytic diatom community growth, biomass accrual, and species composition: short-term versus long-term effects. *Journal of Phycology*, **29**, 24–35.

Bristol-Roach, B. M. (1927). On the algae of some normal English soils. *Journal of Agricultural Science*, **17**, 563–88.

Camburn, K. E. (1982). Subaerial diatom communities in eastern Kentucky. *Transactions of the American Microscopical Society*, **101**, 375–87.

Cantonati, M. & Lange-Bertalot, H. (2006). *Achnanthidium dolomiticum* sp. nov. (Bacillariophyta) from oligotrophic mountain springs and lakes fed by dolomite aquifers. *Journal of Phycology*, **42**, 1184–8.

Carter, J. (1971). Diatoms from the Devil's Hole Cave, Fife, Scotland. *Nova Hedwigia*, **21**, 657–81.

Cremer, H., Gore, D., Hultzsch, N., Melles, M., & Wagner, B. (2004). The diatom flora and limnology of lakes in the Amery Oasis, East Antarctica. *Polar Biology*, **27**, 513–31.

Cremer, H. & Wagner, B. (2003). The diatom flora of the ultra-oliotrophic Lake El'gygytgyn, Chukotka. *Polar Biology*, **26**, 105–14.

Davey, M. C. & Rothery, P. (1992). Factors causing the limitation of growth of terrestrial algae in maritime Antarctica during late summer. *Polar Biology*, **12**, 595–601.

Douglas, M. S. V. & Smol, J. P. (1995). Periphytic diatom assemblages from High Arctic ponds. *Journal of Phycology*, **31**, 60–9.

Ettl, H. & Gärtner, G. (1995). *Syllabus der Boden-, Luft- und Flechtenalgen*. Stuttgart: Gustav Fischer Verlag.

Fenchel, T. & Finlay, B. (2004). The ubiquity of small species: patterns of local and global diversity. *BioScience*, **54**, 777–84.

Finlay, B. J. (2002). Global dispersal of free-living microbial eukaryote species. *Science*, **296**, 1061–1063.

Finlay, B. J. & Clarke, K. J. (1999). Ubiquitous dispersal of microbial species. *Nature*, **400**, 828.

Flechtner, V. R., Johansen, J. R. & Clark, W. H. (1998). Algal composition of microbiotic crusts from the central desert of Baja California, Mexico. *Great Basin Naturalist*, **58**, 295–311.

Flechtner, V. R., Johansen, J. R., & Belnap, J. (2008). The biological soil crusts from the San Nicolas Island: enigmatic algae from a geographically isolated ecosystem. *Western North American Naturalist*, **68**, 405–36.

Furey, P. C., Lowe, R. L. & Johansen, J. R. (2007). Wet wall algal community response to in-field nutrient manipulation in the Great Smoky Mountains National Park, U.S.A. *Algological Studies*, **125**, 17–43.

Furey, P. C., Lowe, R. L., & Johansen, J. R. (2009). Morphological deformities in *Eunotia* taxa from high elevation springs and streams in the Great Smoky Mountains National Park, with a description of *Eunotia macroglossa* sp. nov. *Diatom Research*, **24**, 273–90.

Gremmen, N. J. M., Van de Vijver, B., Frenot, Y., & Lebouvier, M. (2007). Distribution of moss-inhabiting diatoms along an altitudinal gradient at sub-Antarctic Îles Kerguelen. *Antarctic Science*, **19**, 17–24.

Hoffmann, L. (1989). Algae of terrestrial habitats. *The Botanical Review*, **55**, 77–105.

Hustedt, F. (1942). Aërofile Diatomeen in der nordwestdeutschen Flora. *Berichte der Deutschen Botanischen Gesellschaft*, **40**, 55–73.

Hustedt, F. (1949). Diatomeen von der Sinai-Halbinsel und aus dem Libanon-Gebiet. *Hydrobiolgia*, **2**, 24–55.

Ito, Y. & Horiuchi, S. (1991). Distribution of living terrestrial diatoms and its application to the paleoenvironmental analyses. *Diatom*, **6**, 23–44.

Johansen, J. R. (1993). Cryptogamic crusts of semiarid and arid lands of North America. *The Journal of Phycology*, **29**, 140–7.

Johansen, J. R. (1999). Diatoms of aerial habitats. In *The diatoms: Applications for the Environmental and Earth Sciences*, ed. E. F. Stoermer & J. P. Smol, Cambridge: Cambridge University Press, pp. 264–73.

Johansen, J. R., Ashley, J., & Rayburn, W. R. (1993). Effects of rangefire on soil algal crusts in semiarid shrub-steppe of the lower Columbia Basin and their subsequent recovery. *Great Basin Naturalist*, **53**, 73–88.

Johansen, J. R. & Rushforth, S. R. (1985). Cryptogamic soil crusts: seasonal variation in algal populations in the Tintic Mountains, Juab County, Utah. *Great Basin Naturalist*, **45**, 14–21.

Johansen, J. R., Rushforth, S. R., & Brotherson, J. D. (1981). Subaerial algae of Navajo National Monument, Arizona. *Great Basin Naturalist*, **41**, 433–9.

Johansen, J. R., Rushforth, S. R. & Brotherson, J. D. (1983a). The algal flora of Navajo National Monument, Arizona, U.S.A. *Nova Hedwigia*, **38**, 501–53.

Johansen, J. R., Rushforth, S. R., Obendorfer, R., Fungladda, N., & Grimes, J. (1983b). The algal flora of selected wet walls in Zion National Park, Utah, USA. *Nova Hedwigia*, **38**, 765–808.

Johansen, J. R., St. Clair, L. L., Webb, B. L., & Nebeker, G. T. (1984). Recovery patterns of cryptogamic soil crusts in desert rangelands following fire disturbance. *The Bryologist*, **87**, 238–43.

Kawecka, B. & Olech, M. (1993). Diatom communities in the Vanishing and Ornithologist Creek, King George Island, South Shetland, Antarctica. *Hydrobiologia*, **269**, 327–33.

Kilroy, C., Biggs, B. J. F., & Vyverman, W. (2007). Rules for macroorganisms applied to micoorganisms: patterns of endemism in benthic freshwater diatoms. *Oikos*, **116**, 550–64.

Krasske, G. (1932). Beiträge zur Kenntnis der Diatomeenflora der Alpen. *Hedwigia*, **72**, 92–134 (+ 2 plates).

Krasske, G. (1936). Die Diatomeenflora der Moosrasen des Wilhelmshöher Parkes. *Festschrift des Vereins für Naturkunde zu Kassel zum hundertjährigen Bestehen*, 151–64 (+ 3 tables).

Krasske, G. (1948). Diatomeen tropischer Moosrasen. *Svensk Botanisk Tidskrift*, **42**, 404–41.

Lange-Bertalot, H., Cavacini, P., Tagliaventi, N., & Alfinito, S. (2003). Diatoms of Sardinia. *Iconographia Diatomologica*, **12**, 1–438.

Lange-Bertalot, H. & Rumrich, M. (2000). Diatomeen der Anden von Venezuela bis Patagonien/Feuerland. *Iconographia Diatomologica*, **9**, 1–649.

Le Cohu, R. & Van de Vijver, B. (2002). Le genre *Diadesmis* (Bacillariophyta) dans les archipels de Crozet et de Kerguelen avec la description de cinq espèces nouvelles. *Annales de Limnologie*, **38**, 119–32.

Lowe, R. L., Furey, P. C., Ress, J. A., & Johansen, J. R. (2007). Diatom biodiversity and distribution on wetwalls in Great Smoky Mountains National Park. *Southeastern Naturalist, Special Issue*, **1**, 135–52.

Lowe, R. L., Sherwood, A. R., & Ress, J. R. (2009). Freshwater species of *Achnanthes* Bory (Bacillariophyta) from Hawaii. *Diatom Research*, **24**, 327–40.

Lund, J. W. G. (1945). Observations on soil algae. I. The ecology, size and taxonomy of British soil diatoms. *The New Phytologist*, **44**, 56–110.

Metting, B. (1981). The systematics and ecology of soil algae. *The Botanical Review*, **47**, 195–312.

Novichkova-Ivanova, L. N. (1980). *Soil Algae of the Sahara-Gobi Desert Region*. Leningrad: Nauka (in Russian).

Petersen, J. B. (1915). Studier over danske aërofile Alger. *Det Kongelige Danske Videnskabernes Selskabs Skrifter, Naturvidenskabelig og Mathematisk*, **12**(7), 271–380.

Petersen, J. B. (1928). The aërial algae of Iceland. *The Botany of Iceland*, **2**(8), 325–447.

Petersen, J. B. (1935). Studies on the biology and taxonomy of soil algae. *Danske Botanisk Arkiv*, **8**(9), 1–180.

Reichardt, E. (1985). Diatomeen an feuchten Felsen des Südlichen Frankenjuras. *Berichte Bayerische Botanische Gesellschaft*, **56**, 167–87.

Reichardt, E. (2004). Eine bemerkenswerte diatomeenassoziation in einem Quellhabitat im Grazer Bergland, Österreich. *Iconographia Diatomologica*, **13**, 419–79.

Rushforth, S. R., Kaczmarska, I., & Johansen, J. R. (1984). The subaerial diatom flora of Thurston Lava Tube, Hawaii. *Bacillaria*, **7**, 135–57.

Smol, J. P. & Douglas, M. S. V. (2007). Crossing the final ecological threshold in High Arctic ponds. *Proceedings of the National Academy of Sciences of the USA*, **104**, 12395–7.

Spaulding, S. A., Kociolek, J. P., & Wong, D. (1999). A taxonomic and systematic revision of the genus *Muelleria* (Bacillariophyta). *Phycologia*, **38**, 314–41.

Starks, T. L., Shubert, L. E., & Trainor, F. R. (1981). Ecology of soil algae: a review. *Phycologia*, **20**, 65–80.

Stoermer, E. F. (1962). Notes on Iowa diatoms. II. Species distribution in a subaerial habitat. *Iowa Academy of Science, Proceedings*, **69**, 87–91.

van Dam, H., Mertens, A., & Sinkeldam, J. (1994). A coded checklist and ecological indicator values of freshwater diatoms from the Netherlands. *Netherlands Journal of Aquatic Ecology*, **28**, 117–33.

Van de Vijver, B. (2002). *Frustulia cirisiae* sp. nov., a new aerophilous diatom from Ile de la Possession (Crozet Archipeligo, Subantarctica). *Diatom Research*, **17**, 415–21.

Van de Vijver, B. & Beyens, L. (1997). The epiphytic diatom flora of mosses from Strømness Bay area, South Georgia. *Polar Biology*, **17**, 492–501.

Van de Vijver, B. & Beyens, L. (1998). A preliminary study on the soil diatom assemblages from Ile de la Possession (Crozet, Subantarctica). *European Journal of Soil Biology*, **34**, 133–41.

Van de Vijver, B., Beyens, L., & Lange-Bertalot, H. (2004). The genus *Stauroneis* in the Arctic and (Sub-) Antarctic regions. *Bibliotheca Diatomologica*, **51**, 1–317.

Van de Vijver, B., Frenot, Y., & Beyens, L. (2002a). Freshwater diatoms from Ile de la Possession (Crozet Archipeligo, Subantarctica). *Bibliotheca Diatomologica*, **46**, 1–412.

Van de Vijver, B., Frenot, Y., Beyens, L., & Lange-Bertalot, H. (2005). *Labellicula*, a new diatom genus (Bacillariophyta) from Île de la Possession (Crozet Archipelago, Subantarctica). *Cryptogamie Algologie*, **26**, 125–33.

Van de Vijver, B., Gremmen, N., & Smith, V. (2008). Diatom communities from the sub-Antarctic Prince Edward Islands: diversity and distribution patterns. *Polar Biology*, **31**, 795–808.

Van de Vijver, B. & Kopalová. (2008). *Orthoseira gremmenii* sp. nov., a new aerophilic diatom from Gough Island (southern Atlantic Ocean). *Cryptogamie Algologie*, **29**, 105–18.

Van de Vijver, B. & Le Cohu, R. (2003). Two new species of the genus *Geissleria* Lange-Bertalot and Metzeltin (Bacillariophyceae) from the Kerguelen and Crozet archipeligos (TAAF, Subantarctica). *Nova Hedwigia*, **77**, 341–9.

Van de Vijver, B., Ledeganck, P., & Beyens, L. (2002b). Soil diatom communities from Ile de Possession (Crozet, sub-Antarctica). *Polar Biology*, **25**, 721–9.

Van de Vijver, B., Ledeganck, P., & Beyens, L. (2002c). Three new *Diadesmis* taxa on Ile de Possession (Crozet Archipeligo, Subantarctica). *Cryptogamie Algologie*, **23**, 333–41.

Van de Vijver, B., Ledeganck, P., & Lebouvier, M. (2002d). *Luticola beyensii* sp. nov., a new aerophilous diatom from Ile Saint Paul (Indian Ocean, TAAF). *Diatom Research*, **17**, 235–41.

Van de Vijver, B. & Mataloni, G. (2008). New and interesting species in the genus *Luticola* D. G. Mann (Bacillariophyta) from Deception Island (South Shetland Islands). *Phycologia*, **47**, 451–67.

Van Kerckvoorde, A., Trappeniers, K., Nijs, I., & Beyens, L. (2000). Terrestrial soil diatom assemblages from different vegetation types in Zackenberg (Northeast Greenland). *Polar Biology*, **23**, 392–400.

VanLandingham, S. L. (1964). Diatoms from Mammoth Cave, Kentucky. *International Journal of Speleology*, **1**, 517–39.

VanLandingham, S. L. (1966). Three new species of *Cymbella* from Mammoth Cave, Kentucky. *International Journal of Speleology*, **2**, 133–6.

VanLandingham, S. L. (1967). A new species of *Gomphonema* (Bacillariophyta) from Mammoth Cave, Kentucky. *International Journal of Speleology*, **2**, 405–6.

Veselá, J. & Johansen, J. R. (2009). The diatom flora of headwater streams in the Elbsandsteingebirge Region of the Czech Republic. *Diatom Research*, **24**, 443–77.

Vyverman, W., Verleyen, E., Sabbe, K., *et al.* (2007). Historical processes constrain patterns in global diatom diversity. *Ecology*, **88**, 1924–31.

Werum, M. & Lange-Bertalot, H. (2004). Diatoms in springs from central Europe and elsewhere under the influence of hydrogeology and anthropogenic impacts. *Iconographia Diatomologica*, **13**, 1–417.

25 Diatoms as indicators of environmental change in wetlands and peatlands

EVELYN GAISER AND KATHLEEN RÜHLAND

25.1 Introduction

Wetlands comprise about 6% of the Earth's surface, but their ecological importance may be disproportionately higher (Batzer and Sharitz, 2006). Existing at the interface between terrestrial and aquatic landscapes, wetlands can support more species and greater productivity than adjacent communities because they are at the confluence of species pools and resources (Gopal *et al.*, 2000; Wetzel, 2006). They are, therefore, important contributors to global biodiversity and their highly active biological communities modify nutrient and gas concentrations and soil-forming processes at a variety of scales. Organic wetlands (peatlands) store an estimated 450 gigatonnes of carbon (Gt C), equivalent to ~20% of carbon in the terrestrial biosphere (Gorham, 1991; Maltby and Immirzi, 1993; Roulet, 2000) and almost equivalent to the entire global atmospheric carbon pool (Charman, 2002). The economic value of services that all wetland types provide to humans are reported to be higher than other ecosystems (Costanza *et al.*, 1997) because they can be harvested for food, regulate availability and quality of fresh water, and protect neighboring landscapes from flooding.

Despite their importance, wetlands are being lost at an alarming rate. Almost half of the wetlands in the United States were drained or filled by 1970 (Tiner, 1984) and globally they are amongst the most threatened ecosystems on the planet. Threats come in the form of land loss and habitat degradation resulting from drainage for agricultural purposes, conversion for urban expansion, and flooding to create reservoirs for water storage or power generation. Climate change also alters productivity, species distributions, and biogeochemistry in ways that are predicted to be far greater than direct anthropogenic impacts (Gorham, 1991, 1995; Roulet, 2000). This is largely due to the fact that the greatest proportion of the world's peatlands (and carbon stores) occurs at high latitudes in boreal and subarctic regions, where the greatest temperature increases are expected over the next decades (Tarnocai, 2006). Climate impacts on northern wetlands are expected to be great, partly because of the complex feedbacks between climate and carbon flux. For example, droughts in northern peatlands that drop the water table below the peat surface can lead to substantial increases in carbon dioxide (CO_2) emissions while flooding of boreal wetlands can switch the system from a net sink to a major source of both CO_2 and methane (CH_4) (Roulet, 2000). The rapid pace of wetland loss and the inherent complexities in predicting the ecological and societal consequences of those losses have been met with a surge in scientific research as well as regulatory legislation aimed at limiting losses (Sommerville and Pruitt, 2006). In particular, both the scientific and regulatory communities have focused considerable attention on developing tools for assessing change in wetland ecosystems over the temporal and spatial scales relevant to management. However, considering the ecological consequences of human and climate-induced changes in these wetland systems to human services as well as biotic functions, much work remains to be done, particularly in northern peatland regions of the globe.

25.1.1 Importance of diatoms in wetlands

Microbial communities dominate global diversity estimates and regulate functional properties in many ecosystems (Little *et al.*, 2008). Wetlands are no exception, where microbes typically exist in three-dimensional communities living on surfaces and suspended in the shallow water column. Minimal light attenuation in shallow water and increased access to benthic nutrient supplies encourage proliferation of periphytic communities (Wetzel, 2006). Productivity estimates for benthic algal communities are highly variable, ranging anywhere from 0.2 to 5 g C m^{-2} d^{-1} (Vymazal, 1995; Ewe *et al.*, 2006), often

The Diatoms: Applications for the Environmental and Earth Sciences, 2nd Edition, eds. John P. Smol and Eugene F. Stoermer. Published by Cambridge University Press.

Table 25.1 Common terms used for various wetland types discussed in this chapter (adapted from Mitsch and Gosselink 2000; Charman, 2002; Turetsky *et al.*, 2004).

Billabong	Australian term for a wetland that is periodically flooded by a stream or river.
Bog	A highly organic (often >95% organic matter), ombrotrophic peatland that receives minerals and nutrients solely from the atmosphere; water table is ~40–60 cm below the surface, water is extremely mineral-poor and highly acidic. Dominated by *Sphagnum* mosses.
Carolina Bay	Small, isolated freshwater wetlands that dot the US southeastern Atlantic Coastal Plain.
Fen	A minerotrophic peatland that is affected by mineral seepage from the water table.
Rich Fen	Fen that is high in base cations, has pH > 5.0, has a high number of indicator species, and contains brown mosses, grasses, and sedges.
Poor Fen	Fen that is less connected to the water table, is low in base cations, has pH < 5.0, has a lower number of indicator species, and where *Sphagnum* mosses dominate.
Littoral wetlands	Marshes that form along the shallow margins of lakes and ponds
Marsh	A frequently inundated wetland characterized by herbaceous, hydrophilic vegetation.
Oxbow	Riverine wetlands; abandoned river channels that develop into a marsh or swamp.
Peatland	Organic wetland where >30–40 cm of accumulated peat (>65% organic matter) has formed. Often referred to as mires outside of North America.
Prairie pothole	Shallow marshlike pond, formed in the previously glaciated regions of mid-western North America.
Swamp	Wetlands dominated by trees or shrubs.

higher than their phytoplanktonic counterparts. This high productivity causes benthic algal communities to be ecosystem engineers: they can regulate nutrient and gas concentrations, control soil formation, limit light to plant communities, and form the basis for complex food webs (Stevenson *et al.*, 1996). Structurally, benthic periphyton communities are comprised of a mix of cyanobacteria and algae as well as fungi and bacteria, and the algal component can be numerically dominated by diatoms (Stevenson *et al.*, 1996).

The shallowly inundated surfaces of ponds, marshes, swamps, bogs, and fens offer an ideal habitat for benthic diatoms (Goldsborough and Robinson, 1996). High diversity of benthic communities results from the mix of surfaces available for attachment (i.e. rocks, plant stems, soils) that each offer a different physicochemical setting at the microspatial scales relevant to the diatom cell (Burkholder, 1996). A temporal axis of variability is also imposed by fluctuating moisture levels that increase species turnover with time. Inherent in this argument of heterogeneity-driven causes for high diatom diversity is the basic tenet of diatom ecology that each taxon has unique microhabitat preferences that dictate its distribution in space and time. As a result, the high diversity measured within and among benthic diatom communities confers an advantage in inferring habitat quality from diatom assemblage composition, as accuracy of environmental predictions tends to increase with the number of species contributing to the inference (Racca *et al.*, 2001).

25.1.2 Wetland types covered in this chapter

Wetlands are defined as any land that is saturated with water long enough to promote aquatic and wetland processes as denoted by poorly drained or waterlogged soils, and specialized biota that are adapted to these wet environments (NWWG, 1988; Charman, 2002). This chapter focuses explicitly on freshwater wetlands, while an account of diatom studies in estuarine and coastal wetlands is provided by Cooper *et al.* (this volume). Methods for collecting diatoms and interpreting distribution patterns along hydrologic and water-quality gradients are first discussed for freshwater wetlands in general. We then describe diatom studies in peat-forming wetlands, which are distinguished from other wetland types by having thick accumulations of partially decomposed plant remains (peat) and are highly organic in nature (Table 25.1). Non-peat-forming wetlands include shallow open-water ponds, vegetated marshes, or forested swamps, while peatlands include fens and bogs (Table 25.1). In this chapter, peatlands are separated from other freshwater wetland types because of inherent differences in hydromorphology that affect processes governing diatom community structure. The developmental history of wetlands often follows successional stages (called hydroseres) through time, frequently accompanied by substantial changes in vegetational composition and water chemistry. Because peatlands naturally change from one state to another, it is important to understand their history in order to interpret the effects imposed by human activity in the watershed.

25.1.3 Importance of diatoms in modern and paleoecological wetland assessment

The high diversity and habitat affinity observed for diatoms in wetlands have prompted relatively frequent application in contemporary bioassessment as well as paleoenvironmental reconstruction. Developing sound ecologically based assessments of change in wetlands is now more important than ever, given the impact of accelerating rates of land-use degradation and climate change on wetland ecosystems (Batzer and Sharitz, 2006). Conducted over long time periods through biomonitoring or paleoecology, diatom-based assessments of change can not only resolve the rates of these changes but infer their causes and functional consequences. These inferences can then guide ecosystem management and establish realistic restorative targets if trajectories of undesirable change are reversible.

Ecological studies have improved our understanding of the importance of diatoms in wetlands; these are examined in this chapter with an aim toward understanding the natural and anthropogenic drivers of patterns of diatom distribution among wetlands and changes over time. The chapter provides methodological guidance to collecting diatoms from wetlands, determining their environmental preferences, and applying this information to assess ecosystem status or recount history from diatoms in sediment records. Hydrology and water quality are explored in detail to show how diatom communities can track natural and anthropogenic changes in these two drivers over time. A section is provided on peatlands where diatoms have been a critical tool in understanding the course of natural succession, as well as understanding climate-related changes in hydrology, moisture, and fire frequency, particularly in boreal and Arctic regions. Important discoveries resulting from diatom-based applications are highlighted, and the chapter concludes with some recommendations of areas in need of greater scientific attention.

25.2 Employing diatoms in modern and paleoenvironmental assessments in wetlands

25.2.1 Measuring and employing environmental preferences of wetland diatoms

Because diatom species tend to have well-defined ecological niches, their communities can provide a good indication of habitat structure and water quality (Stevenson *et al.*, 1996). This is useful in habitats where water quality varies at frequencies that are laborious or costly to evaluate and in paleoenvironments where prior environmental conditions are not known. Knowing the ecological preferences of wetland diatoms is essential to accurate employment in environmental or paleoecological assessment. While taxonomic and autecological guides to wetland diatoms exist for some regions (e.g. Minnesota peatlands (Kingston, 1982), Carolina Bays (Gaiser *et al.*, 1998), Florida Everglades (Slate and Stevenson, 2007)), general ecological data for benthic diatoms can also be found in Lowe (1974), Lange-Bertalot (1979), van Dam *et al.* (1994) among others (see Stevenson and Bahls, 1999). In regions where diatoms are poorly described or for studies requiring precise preference data, surveys can be conducted to collect taxa and environmental data from a broad range of habitat types. To determine water-quality preferences for diatoms, a common practice in deep-water environments is to collect dead diatoms from surface sediments from the deepest depositional point, since gravitational forces should cause this sample to contain a temporally and spatially integrated reflection of lake-wide conditions (Smol, 2008). In wetlands, topographic complexity prevents a single point of sediment focusing, and slight changes in elevation can confer unique modern and depositional conditions to the diatom community. Weilhoefer and Pan (2006) examined the effect of heterogeneity on the ability to characterize diatom relationships to water quality in an Oregon wetland, and found a high degree of compositional difference (15–40%) from sampling point to sampling point. This suggests that sampling protocols in wetlands must account for microhabitat diversity in order to provide a comprehensive species account as well as appropriate characterization of habitat affinities and water-quality preferences.

Stevenson (1998) reviews numerical approaches used to develop metrics of water-quality conditions from attributes of benthic diatom communities. These attributes include the number of species or genera present in the sample, species' dominance patterns, and the abundance of species falling into predefined categories according to their sensitivity to water-quality parameters. Another approach is to quantify species' sensitivities across water-quality gradients (to derive what is sometimes called a "training set" of diatom optima and tolerances) that can be used to develop a "transfer function" or calibration model for estimating water-quality conditions from diatom assemblages in environments where water quality cannot be directly measured, particularly paleoenvironments (reviewed in detail by Birks, this volume).

25.2.2 Obtaining sediment records from wetlands

Successful reconstruction of paleoenvironments from diatoms in wetlands depends on obtaining a representative record of sufficient resolution that contains well-preserved diatoms. Choosing coring locations presents the same design challenges as sampling the modern community due to the topographic complexity inherent in wetlands. Inferring from multiple rather than single sediment cores ensures that inferences reflect the experience of the whole wetland rather than a local peculiarity (Anderson, 1990). In wetlands with significant directional elevational gradients, cores are often taken along transects to reconstruct parameters that would be differentially influenced by depth or exposure (Gaiser *et al.*, 2006a). In fluvial wetlands, erosional areas should be avoided, although it is often impossible to be sure that a currently accreting site was not formerly erosional. Storm events can further distort records by physically disturbing existing sediments or by depositing new sediments at a faster rate, complicating chronological calibration (Dyer, 1995). In temporary wetlands, periodic exposure of sediments to the atmosphere promotes sediment oxidation which can reduce resolution and complicate stratigraphy (Gaiser *et al.*, 2001). Another challenge in wetland paleoecology is bioturbation, as the shallow sediments are home to rooted grasses and trees, a rich and active burrowing invertebrate fauna, and even large animals that disturb sediments when they visit the wetlands to drink, wallow, or dig for food. For organic wetlands, sediment records retrieved from a given peatland type present different advantages, depending on the aim of the study. For example, ombrotrophic raised bogs (Table 25.1) are valuable receptacles of information about water-level changes and hence climatic changes as they receive water and minerals almost exclusively from atmospheric inputs (e.g. Charman, 2002). Cores taken from various minerotrophic peatlands (fens) (Table 25.1), on the other hand, can be studied for changes in nutrient status or changes in groundwater supply (Aaby, 1986) as these vary widely in this peatland class (Charman, 2002). However, if a more complete understanding of the peatland landscape is desired, such as the rate of transgressions and lateral site expansion, a transect of cores across various peatland types (fens, bogs, or hollows) will be desirable (Aaby, 1986; Bauer *et al.*, 2003; Myers-Smith *et al.*, 2008).

Once a coring site is established, methods for extracting sediments can follow standard paleoecological protocols (see Last and Smol, 2001). One advantage to coring in wetlands is the absence of a deep water column to penetrate. In wetlands with shallow sediments, pits can be dug into bedrock or basal sediment, revealing the stratigraphic profile that can then be systematically sectioned. One challenge to effective coring in wetlands is sediment inconsistency. Plant roots can interfere with core penetration while extraction can be complicated by highly consolidated mineral sediment. The penetrating edge of the coring tube is often cut to a serrated edge or fixed with blades so that roots can be severed as the core is twisted into the ground. A post-hole driver for penetrating sediments and a tripod to which a winch can be attached to aid extraction are valuable tools for wetland coring, especially in highly consolidated clay or silt-dominated sediments. In unconsolidated, waterlogged sediments, a vibration coring device can be used that relies on high-frequency vibration to mobilize a thin layer of soil along the inner and outer tube wall to reduce friction and allow the core tube to penetrate the substrate more easily (Thompson *et al.*, 1991).

25.2.3 Sediment accumulation in wetlands and chronological calibration

Chronological calibration of sediments is typically achieved by radiometric dating to calculate sediment accumulation rates. Accumulation rates in wetlands can vary substantially over space and time, requiring high-frequency sampling along the length of the core for radiometric analysis. Due to periodic exposure and oxidation, many wetlands contain shallow sediments, limiting the chronological resolution of the record to decadal or century timescales. However, some wetlands can contain deep peat profiles that can offer a higher resolution diatom record of environmental change than is possible from many lake-sediment records (e.g. Brown *et al.*, 1994). This is particularly appealing in areas where lake-sediment accumulation is typically low (e.g. Arctic and alpine regions). As the accumulation of organic matter (peat) is composed of partially decayed vegetation, the type of vegetation, together with the oxygen content, largely determines the rate of peat accumulation. For example, the brown moss *Drepanocladus*, commonly found in rich fen peats, is generally faster growing than *Sphagnum* moss species that typify poor fens and ombrotrophic bogs (Jasinski *et al.*, 1998). The anaerobic nature of wetlands substantially reduces the rate of organic matter decomposition along a gradient from rich fens to ombrotrophic bogs (Table 25.1), with lower oxygen levels and higher redox potentials in the latter (de Mars and Wassen, 1999; Charman, 2002). Since production minus decay equals accumulation (Charman, 2002), these peatland characteristics that exist in areas of active peat accumulation can result in large deposits of partially decayed organic matter over relatively short time periods. A low rate of decomposition, rather than high net primary productivity, is what appears to be the principal control over peat accumulation (Clymo, 1984; Vitt, 1990; Turetsky *et al.*, 2004). The end result can provide high-resolution paleoecological archives of peatland environmental history and environmental change (e.g. Brown *et al.*, 1994; Rühland *et al.*, 2000).

The highly organic nature of peat deposits (often exceeding 95% organic matter on a dry-mass basis in ombrotrophic bogs), together with the high cation-exchange capacity of *Sphagnum* moss species, provides many advantages as well as challenges for dating peat sequences (Turetsky *et al.*, 2004). For example, *Sphagnum* regulates hydrogen ion concentration through cation exchange (Rigg, 1940). As a result, negatively charged *Sphagnum* surfaces within ombrotrophic bogs and poor fens can strongly adhere to positively charged ions that are deposited into the peat, thereby immobilizing these cations within the peat column (Clymo *et al.*, 1990; Turetsky *et al.*, 2004). This characteristic of ombrotrophic peats can be advantageous for dating methods using atmospherically deposited materials that have a high potential for cation exchange and strong binding affinities to peat constituents, as this limits vertical post-depositional mobility within the peat column (Turetsky *et al.*, 2004). For example, lead-210 (^{210}Pb) was found to be quite immobile in peat and is thought to hold the most promise for dating recent peat sequences (Turetsky *et al.*, 2004). In contrast, the highly organic nature of peat deposits was found to render cesium-137 (^{137}Cs; a chronological marker of nuclear fallout) in peat highly mobile and therefore not recommended as a dating tool. For older sequences (as well as in younger deposits via wiggle-matching methods – see Turetsky *et al.*, 2004) carbon-14 (^{14}C) dating is the most commonly used dating tool for peat deposits. However, these assumptions may be violated for minerotrophic peat deposits, highlighting the need for site-specific consideration in dating-method choices (Turetsky *et al.*, 2004). Although peat is often regarded as excellent material for radiocarbon dating, some deposits of fen peat can potentially contaminate lower layers of peat with young carbon carried by the deep penetration of roots from vegetation, particularly *Equisetum* (Korhola, 1992a).

25.2.4 Extracting diatoms from wetland sediments

Diatoms can be extracted from wetland sediments using standard oxidation approaches that rid the sample of minerals and organic material that can obscure microscopic identification (see Battarbee, 1986). For peat samples, the removal of the organic matrix with strong acids (e.g. sulfuric acid (H_2SO_4) and nitric acid (HNO_3)) may not be as easily achieved as in lake sediments. Depending on the location, the type of peat, and other factors, the highly organic nature of the material may require more acid and/or a longer acid-digestion period to achieve a clean sample. Alternatively, microwave-digestion techniques require relatively small quantities of chemicals (e.g. ~10 ml of HNO_3 compared to ~15 ml of H_2SO_4 and HNO_3) under pressure and have been very successful at extracting diatoms from highly organic material such as peat (Parr *et al.*, 2004). In non-peat-forming wetlands, mineral matter can be a significant obstacle in preparing adequate slide mounts. In calcareous wetlands like the Florida Everglades, more than 80% of a sample can be calcium carbonate, requiring an acidification step before oxidation (Gaiser *et al.*, 2004a). The sample is first placed in a 10% hydrochloric acid (HCl) solution (weak enough to prevent diatom dissolution) and then the pH restored to neutral before subsequent oxidation. When sediments are dominated by siliceous mineral particles, however, there is no easy chemical solution for dissolving clay, silt, or sand without also dissolving the diatoms. A variety of settling techniques have been used to separate particles on the basis of their different specific density (Krukowski, 1988). For example, repeated washing in a soap solution (sodium polyphosphate is a common choice) that suspends small particles but allows diatoms to sink can be an effective solution to clays (Gaiser *et al.*, 2001). Conversely, a sample containing sand particles can be shaken and quickly decanted, leaving the heavy and rapidly settling sands to be discarded. The biggest challenge is with samples containing abundant silt particles, which are within the same size and density range as diatoms. The most effective solution has been to elevate the specific gravity of a sample to equal that of most diatom cells. This can be done by the use of heavy liquids such as sodium polytungstate (a good non-toxic and recoverable choice). The solution is brought to a specific gravity of 2.2 and centrifuged to separate into three layers – an upper layer containing the diatoms, a mid layer containing pure sodium polytungstate (which can be recovered and reused), and a bottom layer containing heavier silts (Gaiser *et al.*, 2004b). With all of these methods, it is important to examine any solutions before discarding to verify that diatoms are not being lost in the process.

A strategic advantage to examining diatoms preserved in wetland deposits (as in lake sediment) is that, from one diatom mount, multiple siliceous microfossils such as chrysophyte stomatocysts, testate amoebae plates, and opal phytoliths, can be readily included in the environmental assessment, thereby providing additional environmental insights (e.g. Douglas and Smol, 1987; Brown *et al.*, 1994; Gaiser *et al.*, 2004b; Rühland *et al.*, 2000). For example, diatoms together with a suite of siliceous indicators (expressed as a percentage of the sum of all siliceous microfossils) were used to examine peatland developmental history and fire/climate history in a *c.* 7200-year Siberian peat deposit (Rühland *et al.* 2000). Fluctuations in chrysopycean cysts and testate amoebae plates, together with changes in diatom assemblages, tracked shifts in the moisture regime (hydrology) throughout the peat profile whereas increases in phytolith abundances were associated with fire events corroborated by micro- and macroscopic charcoal and *Epilobium* pollen.

25.3 Hydrologic drivers of diatom assemblage structure in wetlands

One prominent feature of all types of wetlands is hydrological variability in space and time. Because of its influence on community composition, soil formation, and water quality, hydrology is the main criterion used to identify and categorize wetlands (Sommerville and Pruitt, 2006). Most wetlands exhibit discrete vegetation zones as water depth increases from the edge to the interior, or community patches that reflect an underlying topographic mosaic (Batzer and Sharitz, 2006). Hydrology can vary seasonally as a function of rainfall and runoff, and can vary over longer timescales in response to climate change, natural biologically mediated modifications (ontogeny, succession, animal activities), or due to intentional manipulations by humans (Jackson, 2006). The focus of many paleohydrological studies in wetlands is to discern how hydrology has varied over long timescales (driven by ontogenetic development or climate change) and the degree to which modern communities continue to reflect their natural hydrologic history versus contemporary changes imposed by human activity. Peatlands may start out as (or evolve into) a shallow open-water environment where water depth is important; later (or earlier) stages of peatland succession are largely driven by changes in the depth to the water table and changes in groundwater supply which play a crucial role in regulating the ecology and evolution of this ecosystem from fen to bog (or vice-versa). Due to these ontogenetic shifts in water table connectivity, the effects of hydrology on peatland diatoms can be mediated differently from other wetlands and therefore will be covered in more detail later in this chapter.

Because hydrology is the central driving variable in wetlands and prone to change over time, many diatom-based wetland paleoecological investigations include a paleohydrological component. The approaches used to infer hydrologic change from diatom assemblages depend on the magnitude and type of change experienced by the community. High-magnitude changes in water depth experienced in deeper, isolated depressions or littoral shorelines of lakes and seas as they infill, subside or fluctuate in response to climate may be reflected in changing relative abundances of shallow benthic to deeper-water planktonic diatoms. These communities may be responding directly to changing water depth or indirectly to resultant alterations in the availability of preferred substratum. Wetlands experiencing periodic drying of all or part of the basin may develop communities whose degree of desiccation resistance is reflected in their abundance. In closed-basin wetlands, changes in water depth can be associated with fluctuations in solute concentrations that indirectly drive a change in diatom composition. On the other extreme, fluvial wetlands that are adjacent to waterways may experience periodic flooding that influences diatom composition both by changing the physicochemical environment of the recipient wetland and by direct delivery of riverine diatoms. We will discuss each of these avenues of diatom-based hydrology assessment separately, and provide examples of successful paleohydrologic applications.

25.3.1 Diatom response to water depth variability in wetlands

Past changes in water levels have been inferred successfully from diatom records in lakes with bathymetries that allow littoral expansion or contraction during periods of low or high water that is then reflected in the proportion of planktonic to benthic taxa deposited in sediments (see Wolin and Stone, this volume). Even in shallow marshes, planktonic taxa can become abundant during wet phases and indicate hydrologic maxima in paleoecological records (Brugam, 1980; Earle and Duthie, 1986). However, since the ratio of planktonic to benthic forms can be influenced by changes in nutrient availability (Battarbee, 1986), water clarity (Barker *et al.*, 1994), ice cover (Smol, 1988), sedimentation (Håkanson, 1977) and wind (Dean *et al.*, 1984), there have been attempts to quantify diatom depth preferences directly, within the context of naturally occurring variability. Moisture preferences of diatoms can be obtained from autecological data accompanying taxonomic information in well-cataloged collections (i.e. van Dam *et al.*, 1994) or from systematic studies along depth gradients within lakes (Yang and Duthie, 1995; Brugam *et al.*, 1998). Bunting *et al.* (1997) categorized diatom taxa from a Lake Erie marsh according to the four moisture categories reported by van Dam *et al.* (1994), and was able to show that the steady course of succession and infilling typical of these coastal marshes was punctuated by cyclical fluctuations in water depth. Duthie *et al.* (1996) employed the quantitative water-depth optima and tolerances provided for Great Lakes diatoms by Yang and Duthie (1995) to interpret an 8300-year history of hydrology within the context of trophic and thermal changes also occurring over that timeframe.

In wetlands, shallow topographic relief reduces the length of the depth gradient, which can sometimes compromise the accuracy of estimated optima and tolerances, yet Yang and Duthie (1995) showed well-defined optima even at the low end of the gradient (1–5 m) in the littoral zone of Lake Ontario, Canada, suggesting selective assortment of diatom taxa among benthic habitats by depth. Finkelstein and Davis (2005 a,b) used depth optima from Yang and Duthie (1995) to reconstruct 1–4-m differences in inundation depths in a Lake Erie coastal wetland, corresponding to several climate-controlled transgressions over the past 2300 years. Brugam *et al.* (1998) also were able to reconstruct 0–12-m changes in water depth in a shallow Michigan lake from a weighted-averaging regression model with a 2 m resolution, enabling fairly precise predictions of change even during long wetland phases in the lake. A similar approach was taken by Punning and Puusepp (2007) in Lake Juusa, a shallow pond in southern Estonia, where depth optima were measured for 120 taxa along a 0–6-m depth gradient. They found that planktonic taxa begin to dominate at depths exceeding 3.5 m, and were able to interpret meaningfully a *c.* 5000-year record of 2–3-m water-level fluctuations.

Water-level fluctuations have not only been interpreted directly from measured depth preferences of diatom taxa but also from their association with particular substrata that orient along depth gradients. Campbell *et al.* (1997) were able to detail the post-glacial development of a kettle-hole peatland in southern Ontario from diatom and pollen records that showed transitions from taxa characteristic of a fen to those of an acidic wetland, and eventually a switch to *Sphagnum*-associated taxa characteristic of a fully developed floating bog. In wetlands across Oregon, Weilhoefer and Pan (2007) found that water depth, open-water zone width, and emergent zone width determined the distribution of diatoms among habitats (periphytic, planktonic, or tychoplanktonic) and morphological guilds (adnate, stalked). They concluded that diatoms provided a more comprehensive assessment of wetland condition than hydrogeomorphic wetland classification because they respond to smaller-scale heterogeneity not captured in larger-scale indexes. Gaiser (2009) also showed that diatoms living in periphyton mats in the Florida Everglades are sensitive to the fine-scale water-depth variation inherent in this wetland system; communities in shallow-sediment-associated periphyton mats differed from those living in deeper-water floating mats. Interpreting water-depth variability in the context of indirect influences of water depth on the diatom habitat is therefore important in modern and paleoecological assessments of hydrologic condition.

25.3.2 Diatom response to aerial exposure in temporary wetlands

Periodic declines in water depth in shallow basins can cause large areas or perhaps the whole wetland to dry, which will either reduce aquatic algal production or encourage a desiccation-resistant assemblage. Shallow billabongs and riverine wetlands (oxbows) (Table 25.1) in arid regions of Australia are among the most hydrologically variable systems in the world, and diatoms are a key tool for interpreting the drivers and extent of hydrologic changes experienced in aquatic ecosystems there (Tibby *et al.*, 2007). In a wetland complex along the Murray River, Australia, Gell *et al.* (2007) found a preponderance of aerophilic taxa, including *Diadesmis contenta* (Grun.) Mann, *Hantzschia amphioxys* Ehrb., and *Pinnularia borealis* Ehrb., interspersed with epiphytic taxa, indicating a highly variable wetland during pre-European times that changed into a planktonic community after impoundment. They used these data to propose a wetting and drying management regime to the river system to restore its natural hydrology, as permanent inundation was found to be a minor feature of pre-contact hydrology in the floodplain wetlands. In a 1200-year sediment record from Bolin Billabong near Melbourne, Leahy *et al.* (2005) found abrupt transitions from planktonic to aerophilic taxa. The peak in aerophilic diatom abundance coincided with an increase in sedimentation rate and particle size, suggesting that the source was erosion from the floodplain that resulted in a shallower wetland, rather than drying of the basin itself. In North America, diatom-based reconstructions of inundation along lake shorelines have helped resolve Holocene climate fluctuations. In a Lake Erie marsh, Finkelstein and Davis (2005a) could distinguish among low-, mid- and high-marsh conditions by the relative abundance of tychoplanktonic, benthic, epiphytic, and aerophilic diatoms. Similarly, Tinner *et al.* (2008) found excursions of aerophilic taxa at the expense of planktonic ones *c.* AD 1490–1580 in Grizzly Lake in southern Alaska, when changes in the moisture balance associated with the Little Ice Age caused pronounced lake-level fluctuations and rapid reorganization of the aquatic ecosystem.

While the presence of aerophilic taxa in sediment records can denote droughts or erosional activity in a watershed, Gaiser *et al.* (1998) showed that diatoms can also sort along hydroperiod gradients due to species-specific responses to periodic drying. They collected diatoms from Carolina Bays (Table 25.1), the small, isolated perched wetlands, that dot the southeastern Atlantic Coastal Plain of Georgia and South Carolina, and related abundances of 121 taxa to wetland hydroperiod, which ranged from <180 days to permanently flooded

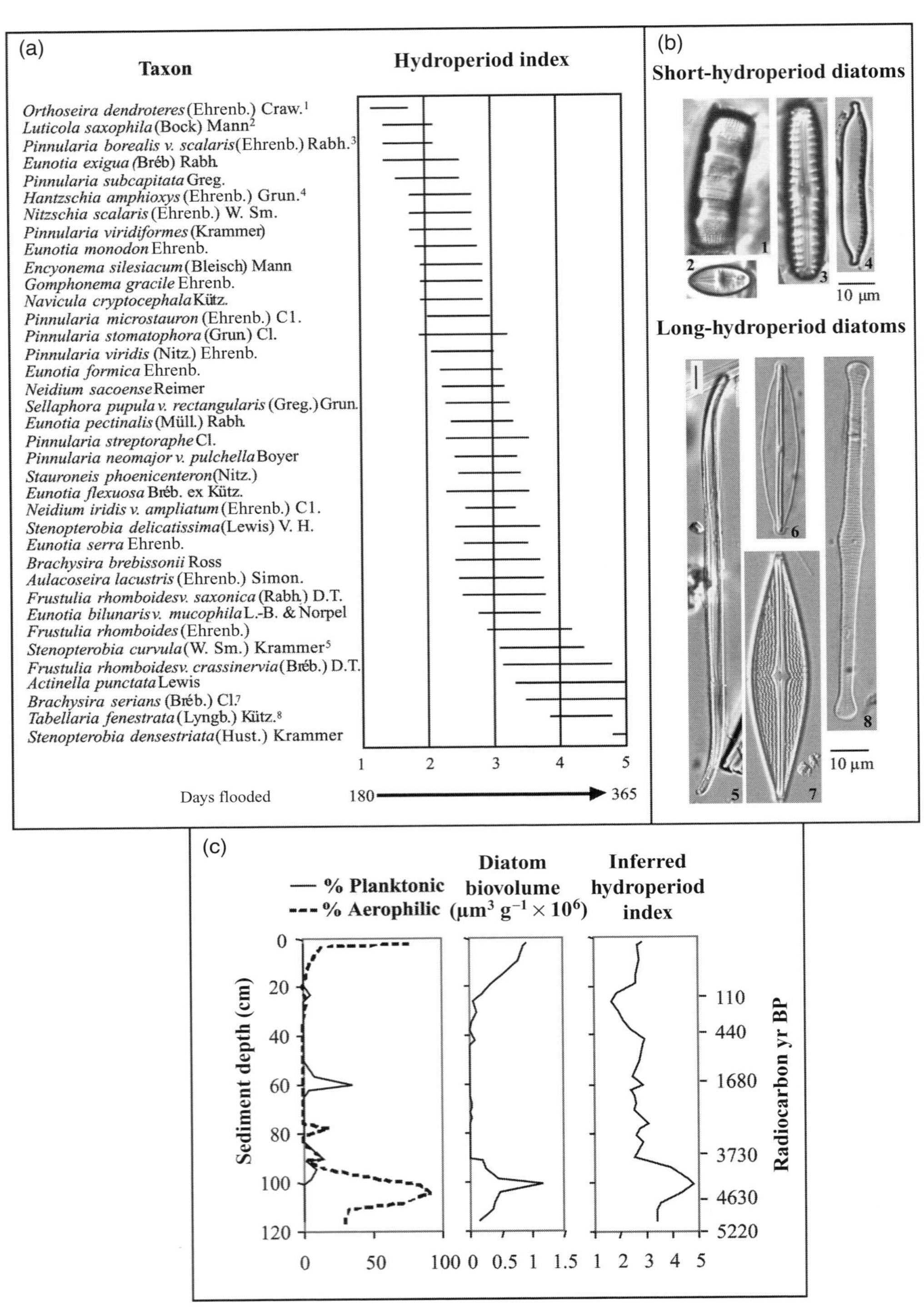

Figure 25.1 (a) Common diatom taxa from South Carolina wetlands sorted by hydroperiod tolerance; (b) Photos of short- and long-hydroperiod indicator taxa from the same wetlands, including the aerophilic taxa *Luticola saxophila* (Bock) Mann (2), *Pinnularia borealis* var. *scalaris* Ehrenb. (Rabh.) (3), and *Hantzschia amphioxys* Ehrb. (4); (c) Paleohydrology of Peat Bay wetland, South Carolina, revealing an open-water lake environment in the oldest and youngest sediments that bracket a prolonged period of episodic drying, indicated by the relative abundance of planktonic and aerophilic taxa, diatom biovolume, and diatom-inferred hydroperiod. (Redrawn from Gaiser *et al.* (1998, 2001).)

(Figure 25.1a). In the driest wetlands, aerophilic taxa including *Pinnularia borealis* var. *scalaris* Ehrenb. (Rabh.), *Hantzschia amphioxys* Ehrenb., and *Luticola saxophila* (Bock) Mann were dominant (Figure 25.1b). Importantly, there is some overlap in the list of short-hydroperiod indicators with aerophilic taxa reported in the Canadian, Australian, and Alaskan studies reported above (Finkelstein and Davis, 2005a; Leahy *et al.*, 2005; Tinner *et al.*, 2008) as well as in many peatland studies from different parts of the world (described later in this chapter). This indicates the possibility of ubiquitous application of hydroperiod optima reported in the South Carolina study. In addition, diatoms appear to exhibit a fairly narrow tolerance around well-defined hydroperiod optima so that when there is significant compositional overlap between the calibration and sediment assemblages, the hydroperiod can be inferred within ~50 days ($R^2 = 0.61$; Gaiser *et al.* 1998). This model was used to calibrate hydrology from a ~4000-year sediment record from an oxbow pond that clearly showed a transition from lacustrine to wetland conditions following a mid-Holocene hydrologic threshold of wetland formation in this region (Figure 25.1c). An increase in aerophilic taxa and generally poorly preserved diatoms bracketed the period of the Little Ice Age, indicated by an increase in inputs of the same desiccation-resistant taxa from wetlands surrounding Grizzly Lake, Alaska (Tinner *et al.* 2008) on the other side of the continent.

25.3.3 Diatom responses to changing solute concentrations in wetlands

Hydrologic changes can also impact diatoms by influencing solute concentrations that profoundly affect cell physiology and thus assemblage composition. In closed-basin lakes and wetlands, evaporative concentration of salt during droughts shifts diatom communities toward salt-tolerant taxa. Fritz *et al.* (1991) developed diatom-based salinity transfer functions for closed-basin lakes and wetlands in the North American Great Plains; an overview of these functions and their application in hindcasting and predicting drought periodicity in closed basins is provided by Fritz *et al.* (this volume). Gasse *et al.* (1995) developed similar models for African lakes and later applied them to interpret effects of climate change on lake persistence (Gasse *et al.*, 1997). However, evaporative concentration is not the only hydrologic driver of solute change. In some arid regions, salinization of surface water is also resulting from an overexploitation of freshwater (through drainage and extraction) and/or increased landward saltwater intrusion in coastal regions, resulting from rising sea levels. In Australia, surface-water salinization is considered the most severe environmental and societal threat, and many efforts are underway to measure the biological impacts of salinization in remaining aquatic ecosystems (Gordon *et al.*, 2003). Gell (1997) demonstrated the primary influence of salinity on diatom communities in lakes of western Australia, and since then, similar relationships have been reported for diatoms in lakes and wetlands throughout the continent (Taffs, 2001; Gell *et al.*, 2002, 2005; Fluin *et al.* 2007; Haynes *et al.* 2007; Tibby *et al.* 2007). All authors found that although conductivity varies seasonally, diatom salinity optima and tolerances can be used to reliably infer longer-term salinity fluctuations. Gell *et al.* (2007) applied these transfer functions to records from two wetlands, finding an increase in salt-tolerant taxa in the shallowest site as irrigation withdrawals of freshwater increased during post-European times. Coastal wetlands are particularly vulnerable to increasing salinities as sea level rises (Taffs *et al.*, 2008), and diatom-based interpretations of these coastal wetland vulnerabilities are reviewed by Cooper *et al.* (this volume).

25.3.4 Diatom responses to changes in water source and flood pulses

Floodplain wetlands (sometimes termed oxbows; Table 25.1) can occasionally connect to the main river channel during periods of high flow. These events can be reflected in algal communities not only through direct delivery of cells from the adjacent river to the wetland but as a result of population responses to changing water quality resulting from the influx of dissolved and particulate materials from the neighboring river (Engle and Melack, 1993). Scouring that occurs during flood events can disturb paleoecological records and obfuscate their interpretation; however, the occurrence of non-scouring floods may be interpretable from diatoms and related biotic and lithostratigraphic proxies. Weilhoefer *et al.* (2008) examined the diatom assemblages of wetlands exposed to high and low flooding intensities of long and short durations. They found that short/high magnitude floods delivered planktonic diatoms from the river to the wetland, causing river and wetland assemblages to converge in composition. Diatoms settling to the sediments reflected the magnitude of the flood, suggesting that sediment assemblages may contain records not only of flood frequency but also severity. Benthic diatoms inhabiting streams differ compositionally from those of shallow lentic environments, enabling the reconstruction of stream meanders and oxbow formation along stream channels. These differences allowed for changes in flooding regimes to be reconstructed from a ~160-year diatom record from a groundwater outflow in southern Ontario, Canada (Reavie *et al.*, 2001). In the mid

1800s, the diatom record showed a transition from spring-fed stream to pond taxa that persisted into the next century until a hurricane delivered a substantial sediment load to the basin and caused a shallow wetland to form. Owen *et al.* (2004) classified wetlands in Kenya by their water source (hot spring, cool spring, swamps, floodplain marshes, littoral wetlands) and found distinct diatom assemblages in each. Differences were attributed to the contrasting chemistries of source water and diatoms sorted along salinity, silica, pH, nutrient, and thermal gradients. Species preferences were defined and used to determine environmental histories at a neighboring wetland and swamp, both of which were found to have experienced significant effects of irrigation withdrawals. Although the response of diatom assemblages to changes in freshwater availability has been less frequently studied than their response to water quality, these studies demonstrate their utility at tracking a parameter that has been and will continue to be affected by climate change and local exploitation in ways that threaten their persistence.

25.4 Water-quality drivers of diatom assemblage composition in wetlands

The pool of diatom species inhabiting a wetland is compositionally constrained by geographic limitations to dispersal, underlying geology and hydrology, development history, and characteristics of the surrounding landscape. As a result, wetland diatom consortia can be described on a regional basis with measured predictability; i.e. diatoms inhabiting Kentucky wetlands (Pan and Stevenson, 1996) are different from those in the Everglades swamps (Slate and Stevenson. 2007), Kenyan papyrus marshes (Owen *et al.*, 2004), and Australian billabongs (Gell *et al.*, 1999). However, while these landscape or geological descriptors confine the species pool for a given region or wetland type, success of any species in a particular wetland is ultimately controlled by the environmental conditions currently represented in the habitat. These conditions will vary from wetland to wetland according to smaller-scale natural variability and also the level of human-induced disturbance in the landscape. The result is local variability in primary determinants of benthic algal community structure, such as nutrient concentrations, specific conductance, pH, water color, and many other variables (see Stevenson *et al.*, 1996). Because land-use changes associated with agriculture and urban development are the key factors controlling wetland-habitat loss and degradation, much effort has been allocated toward developing reliable tools for assessing wetland integrity. Algal communities have become a valuable tool in wetland assessment worldwide (Stevenson *et al.*, 1996). Each species has a definable niche described by multiple environmental axes, and since many species occur in each community, a robust concept of environmental condition can be described by them. Population sizes change over weeks to months, causing communities to reflect an integrated water-quality condition on timescales appropriate for guiding adaptive environmental management (Stevenson, 1998). As such, diatoms have been widely incorporated into wetland assessment protocols and the result is an improved ability to detect anthropogenically induced change, measure its severity, and gauge its reversibility (Gaiser, 2009). In the following subsection, we review approaches to detecting nutrient enrichment from diatoms in wetlands in the context of both risk assessment and paleoecological reconstruction, and, in the subsequent subsection on peatlands, we discuss how water-quality changes can be reflected in diatom assemblages to reveal the ontogeny and development of the contemporary physicochemical setting of organic wetlands.

25.4.1 Diatom responses to changing nutrient availability

Increased nutrient loading to wetlands in developing catchments is a global environmental problem (Zedler, 2006). Diatom assemblages are well known to sort predictably along nutrient gradients, driven directly by response to changing availability of a limiting nutrient or to ecological conditions resulting from enrichment, such as altered redox conditions, soil quality, or plant communities. Many taxonomic catalogs, monographs, and collections present nomenclatural data together with ecological affinities. For example, van Dam *et al.* (1994) provide preference classes for diatoms for water-quality parameters such as nitrogen metabolism, salinity, pH, dissolved oxygen; while Bahls (1993) categorizes diatoms into classes by their tolerance or sensitivity to pollution. For a long time, these autecological data have been used to derive indices of water quality useful in habitat assessment (see a general review in Hall and Smol, this volume). Della-Bella *et al.* (2007) show how multiple indices can be used to provide an accurate assessment of pollution in wetlands. Weighted-averaging approaches have also been employed to estimate diatom habitat preferences, based on the assumption that species will be most abundant where conditions are most favorable (describing their optima) and less abundant in less favorable conditions (with a range defining their tolerances). Stevenson (1998) advocates a multidimensional approach of creating an Autecological Index of Environmental Stressors (AIES) for wetlands from the distribution of diatoms along a gradient of impairment that can then be used to determine the level of change or stability necessary to either restore an impaired wetland or maintain an un-impacted state.

Wetlands in the state of Florida have received much attention because the abundant isolated depressions on the peninsula, and the Everglades to the south are severely threatened by rapid agricultural and urban development. Lane and Brown (2006) collected diatoms from 70 isolated marshes across the state and found that species distribution was related to a metric of land-use disturbance in the watershed. Epiphytic diatoms were more reflective of impairment than planktonic or benthic taxa, so the abundance of pollution-sensitive taxa within the epiphytic community was used to develop a metric of wetland condition (Lane and Brown, 2007). In a related study, Reiss and Brown (2007) assigned diatom, macrophyte, and invertebrate taxa to pollution tolerance classes that together provide a comprehensive index that ranks each wetland with an impairment score relative to reference wetlands. These metrics are now employed in risk assessment of Florida wetlands by the US Environmental Protection Agency (Reiss and Brown, 2007).

Diatom-based water-quality assessments in the Everglades of south Florida have been similarly motivated. McCormick and Stevenson (1998) review experimental and descriptive research that highlights the value of periphyton-based indicators in assessing nutrient conditions in the Everglades. In one of these studies, Cooper *et al.* (1999) collected diatoms along a steep phosphorus enrichment gradient in an Everglades marsh and found strong community responses to soil total phosphorus (TP), carbon, nitrogen, calcium, and biogenic silica concentrations. They provided the weighted-averaging optima and tolerances for 205 taxa that were then validated experimentally by Pan *et al.* (2000). Working in the un-enriched interior of the system, Gaiser *et al.* (2005) experimentally exposed periphyton communities to phosphorus and then compared TP optima and tolerances to those derived from natural marsh transects. They found that TP estimates for taxa vary depending on the length of the exposure gradient (a design artifact) but also according to the duration of exposure (Figure 25.2a, b). Although the taxa indicative of lower TP (i.e. *Mastogloia smithii* Thwaites, *Encyonema evergladianum* Krammer, and *Fragilaria synegrotesca* L.-Bert.) and higher TP (i.e. *Nitzschia amphibia* Grun., *Rhopalodia gibba* Ehrenb., and *Gomphonema parvulum* (Kütz.) Kütz.) conditions were the same as reported in earlier studies (Cooper *et al.*, 1999; Pan *et al.*, 2000; McCormick and O'Dell, 1996: Gaiser *et al.*, 2004c), the TP optima were lower in areas with a longer exposure history (Gaiser *et al.*, 2006b; Figure 25.2c). Through a landscape-scale, continuous monitoring program of over 150 sites distributed throughout the Everglades, Gaiser (2009) built a spatially explicit diatom data base from which baseline variability in diatom composition can be assessed and used to detect change relative to water-management activities occurring across the Everglades landscape.

Elsewhere in the USA, state and local wetland assessment programs have developed similar diatom-based evaluation tools. Pan and Stevenson (1996) collected planktonic and epiphytic diatoms from 32 Kentucky wetlands and found that assemblages sorted predictably along conductivity and TP gradients, reflective of the degree of acid-mine drainage and development in the watershed, respectively. Like Lane and Brown (2007), they also found epiphytic algae to be better at predicting water quality than planktonic communities, and advocate their use in wetland assessment. Mayer and Galatowitsch (2001) found diversity and productivity to be the best measures of impairment in prairie-pothole wetlands (Table 25.1), showing that the ratio of diversity to productivity decreases with the length of time a wetland has been under restorative practices. Lougheed *et al.* (2007) collected diatoms from depression wetlands in Michigan and evaluated (1) the compositional similarity to an undeveloped wetland, (2) the relationship among wetlands in multidimensional space, (3) the trophic index, calculated from van Dam *et al.* (1994), (4) the relative abundance of erect to stalked diatoms, and (5) a combination of all of these; they found a non-linear relationship of each index to an independent index of wetland disturbance calculated from land-use and water-quality data. They used regression tree analysis to identify the drivers and thresholds of change, revealing three levels of severity of degradation that can be used to triage wetlands for restoration. Chipps *et al.* (2006) used a similar approach in assessing water quality in Missouri river floodplain wetlands by comparing a variety of diatom metrics among "low-" and "high-" quality wetlands. The most sensitive metrics were then included in a wetland-condition index that could be used to rank wetlands on a calibrated scale.

Excessive cultural eutrophication experienced along the shorelines of the Great Lakes has engendered considerable interest in developing accurate bioindicators for the lakes and surrounding wetlands. Over 2000 diatoms were identified from a Great Lakes wetland survey by Reavie *et al.* (2006) and used to produce weighted-averaging regression models that explain 75% of the variability in water TP and that can predict within 0.22 log (μg TP l^{-1}). In another landscape-scale survey, Brazner *et al.* (2007) sampled diatoms in 276 wetlands across the Great Lakes region over a three-year period. Instead of using the weighted-averaging regression approach, they

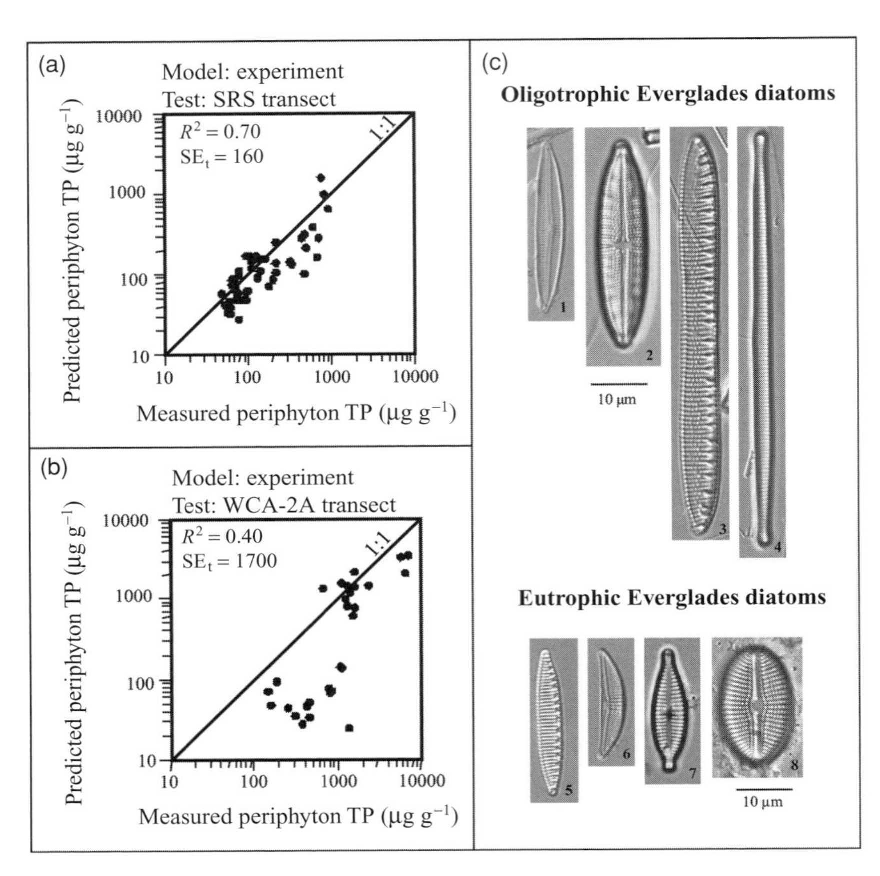

Figure 25.2 Marshes of the Florida Everglades are threatened by phosphorus runoff from surrounding agriculture. To establish a protective standard for the marsh, phosphorus was added experimentally to the oligotrophic Everglades. Total phosphorus (TP) optima and tolerances were determined for periphytic diatoms and used to infer concentrations from diatoms along marsh transects in (a) Shark River Slough (SRS), a marsh relatively unimpacted by P, and (b) Water Conservation Area 2A, a marsh with a long history of P exposure. The model performed well in the unimpacted marsh but poorly in the impacted marsh, suggesting that exposure history matters to the diatom community response. (c) Photos of native Everglades diatoms indicative of low TP concentrations that are replaced by cosmopolitan eutrophic taxa in high TP settings: (1) *Encyonema evergladianum* Krammer, (2) *Mastogloia smithii* Thwaites, (3) *Nitzschia serpentiraphe* L.-Bert., (4) *Fragilaria synegrotesca* L.-Bert., (5) *Nitzschia amphibia* Grun., (6) *Amphora veneta* Kütz., (7) *Gomphonema parvulum* Kütz., and (8) *Diploneis parma* Cleve. (Summarized from Gaiser *et al.* (2006b).)

measured a variety of metrics including diversity, percent motile and planktonic taxa, and percent abundance of five key diatom indicator species and related them, along with similar indexes derived from plant, fish, and invertebrate data, to a range of watershed disturbance and land-use characteristics at a range of scales. They determined that the percent of Stephanodiscoid diatoms was the best metric of impairment, as it was positively correlated with the abundance of row-crop agriculture in the watershed. The two approaches present a contrast of how weighted-averaging regression can provide an accurate measure of a single variable of specific interest while a multimetric approach can identify the primary variables driving distribution patterns. Further detail on the advantages, disadvantages, and specific methods of employing these and related approaches to assessment are provided by Birks (this volume).

Diatom-based nutrient metrics have been more frequently applied in contemporary than retrospective wetland assessments, and both less often than in neighboring lakes. This may be partly due to the paucity of reliable diatom records in the sediments of wetlands, relative to lakes (described in more detail above). However, where diatom preservation is adequate, nutrient-preference information has been helpful in qualitatively and quantitatively reconstructing eutrophication events or phases of higher or lower productivity. From soil cores in the Everglades, Slate and Stevenson (2000) provided evidence for post-impoundment phosphorus enrichment using nutrient-status data derived from prior studies, showing a complete displacement of native Everglades taxa by more cosmopolitan eutrophic taxa at the most enriched sites. Cooper *et al.* (2008) also did a study of soil cores in both phosphorus-enriched and un-enriched areas of the Everglades. Native taxa appeared to be replaced not only in the enriched areas but by different taxa in the un-enriched areas (a replacement of "native" taxa at both sites by different communities). Both studies also report evidence of changing pH and hydrology through time in both phosphorus-enriched and un-enriched areas that might be responsible for this pattern. Duthie *et al.* (1996) employed TP and water-level transfer functions of Yang and Duthie (1995) to show how an expansive wetland on the edge of Lake Ontario, Canada connected to the lake as water levels rose during the mid Holocene, leading to the establishment of mesotrophic Hamilton Harbour, which later eutrophied as the margin of the lake industrialized. O'Sullivan *et al.* (1991) showed how evidence of eutrophication can mimic the impact of rising water level on the plankton:benthic ratio, when taxa in a coastal English lake shifted from primarily benthic to phytoplanktonic taxa as the catchment changed from native grassland to agriculture in the 1950s, a finding verified by other biological proxies in the sediments. The opposite transition from phytoplanktonic to periphytic species was found by Leahy *et al.* (2005), as European occupation increased sedimentation rates in Bolin Billabong in Australia. They verified that the shift was due to both decreased water depths as well as eutrophication by documenting a transition from a community dominated by oligotrophic *Cyclotella stelligera* (Cl. and Grun.) Van Heurck to a periphyton community dominated by taxa characteristic of eutrophic floodplain wetlands in the region (including *Cyclotella meneghiniana* Kütz., *Gomphonema parvulum*, *Nitzschia palea* Kütz. (Sm.), and *Nitzschia frustulum* (Kütz.) Grun.).

While scientists have successfully employed diatoms in assessing wetland water quality, examples of effective transfer of this science to the regulatory community are infrequent, particularly relative to their demonstrated value as bioindicators of habitat degradation. This is particularly true for the geographically vast peatland areas that are experiencing rapid water-quality changes culminating from both local and global sources (see below).

25.5 Diatom studies in organic wetlands (peatlands)

Peatlands are a specialized type of wetland (often referred to as organic wetlands) whereby the rate of plant decomposition is much lower than plant production. The thick accumulation of highly organic, partially decayed plant detritus beneath a living plant layer is what distinguishes peatlands from mineral wetlands which lack any substantial accumulation of organic remains (Charman, 2002). Under the right set of conditions, peatland development can be initiated through either terrestrialization (the gradual infilling of an open water body with macrophytes and mosses towards a more terrestrial ecosystem) or paludification (swamping or outward spread of wet, peat-forming environments over adjacent terrestrial areas). However, peatland development often incorporates both of these pathways (Figure 25.3). The classification of peatland systems is based on a nutrient/mineral gradient from a relatively nutrient-rich, mineral-rich and wet environment (e.g., fens) to an extremely nutrient-poor, mineral-poor and dry environment (e.g., bogs) (Charman, 2002). The degree to which peatlands are connected to the water table governs these chemical and moisture gradients and is ultimately related to the assemblage of species present. A decrease in species richness and diversity follows the moisture and chemical gradient from fen to bog both for hydrophylic vegetation (e.g. grasses, sedges, mosses) (Gorham, 1956; Vitt *et al.*, 1975) and diatoms (Kingston, 1982; Poulíčková *et al.*, 2004).

Peatlands undergo natural patterns of ecological change with the replacement over time of one species or a community of species through competition. This ecological succession does not necessarily follow a unidirectional path and does not necessarily end in a climax state. Similar to the fen-to-bog gradient used in peatland classification, peatland developmental processes also typically follow this open-water-to-bog gradient (hydroseral succession) but there is considerable variation along this trajectory (Charman, 2002; Kuhry and Turunen, 2006). For example, changes in climate are understandably important to local hydrological processes and can often delay or reverse the direction of peatland development. Therefore, it is not surprising that climate-related changes in the depth of the water table play a crucial role in regulating the ecology and development of peatland systems.

25.5.1 Diatoms in peat deposits: paleoecological applications

Peatlands contain a wealth of paleoecological information that has been used to gain insights into climatic, vegetational, and anthropogenic impacts on the environment for centuries (Aaby 1986). Plant macrofossils, including the moss taxa that largely comprise the peat profile, and pollen assemblages have been, and are still, considered the quintessential indicators for tracking the successional history of peatlands (hydroseres) as well as environmental changes (e.g. von Post, 1916; Godwin, 1940; Wein *et al.*, 1987; Gignac *et al.*, 1991; Jasinski *et al.*, 1998). Testate amoebae are also standard indicators as they are often the dominant microfaunal component of *Sphagnum* peatlands (Warner, 1987) and have been successfully used to track changes in

(a) Open water/early fen

Open water
1. *Fragilaria construens* (Ehrenb.) Grun.
2. *Fragilaria pinnata* Ehrenb.
3. *Fragilaria parasitica* (W. Smith) Grun.
4. *Navicula seminulum* Grun.

Early Fen
5. *Pinnularia interrupta* W. Smith

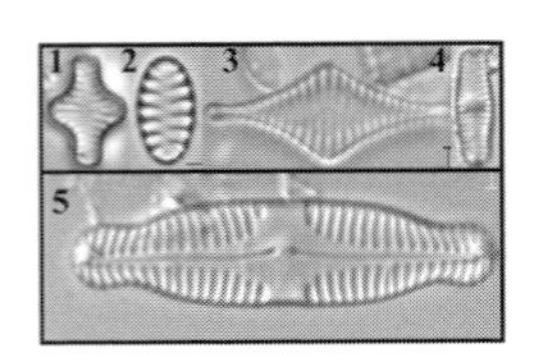

(b) Rich fen

1. *Tabellaria flocculosa* (Roth) Kütz.
2. *Eunotia bilunaris* (Ehrenb.) Mills
3. *Eunotia flexuosa* (Bréb.) Kütz.
4. *Eunotia arculus* (Grun.) L.-Bert.
5. *Eunotia glacialis* Meist.
6. *Pinnularia brevicostata* Cleve

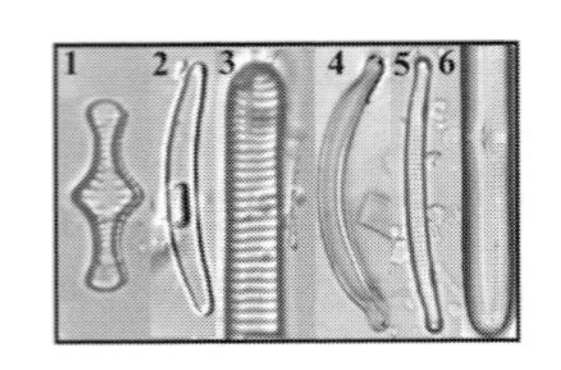

(c) Poor fen

1. *Eunotia argus* Ehrenb.
2. *Eunotia incisa* Greg.
3. *Gomphonema lagerheimii* A. Cleve
4. *Gomphonema angustatum* var. *undulatum* (Greg.) Grun.
5. *Gomphonema truncatum* Ehrenb.
6. *Nitzschia perminuta* (Grun.) M. Perag.

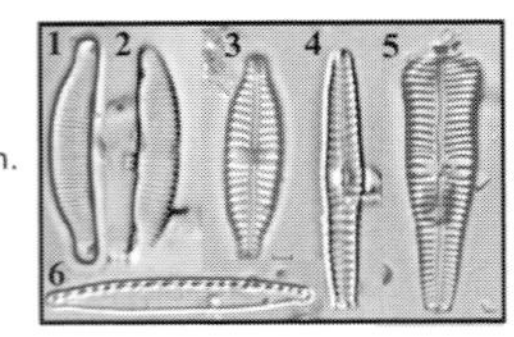

(d) Sphagnum/raised bog

1. *Pinnularia borealis* Ehrenb.
2. *Cymbella ventricosa* Kütz. var. *groenlandica* Foged
3. *Hantzschia amphioxys* Ehrenb.
4. *Eunotia exigua* (Bréb.) Rabh.
5. *Navicula ignota* var. *palustris* (Hust.) Lund
6. *Eunotia paludosa* Grun.

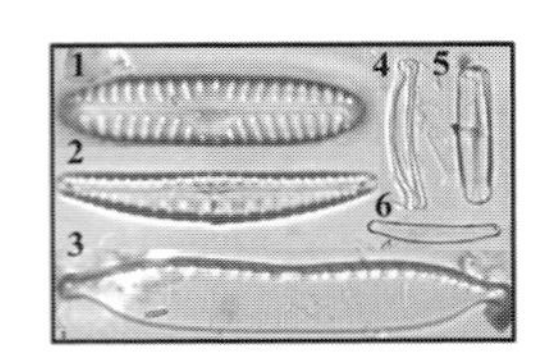

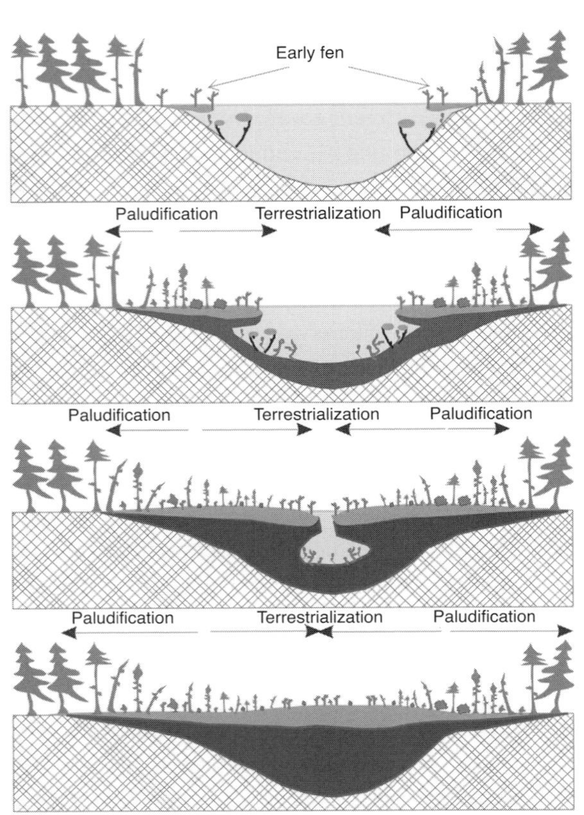

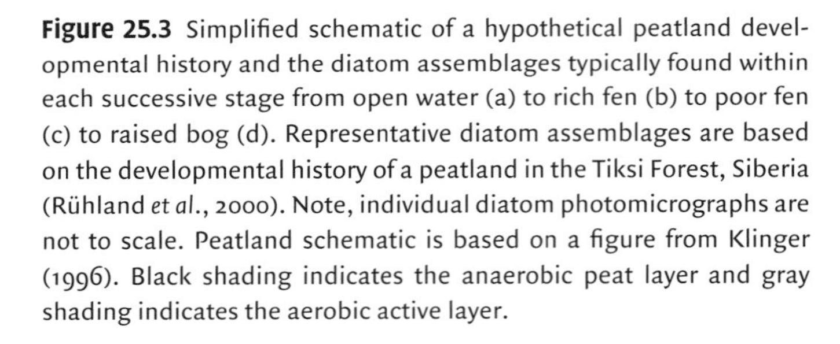

Figure 25.3 Simplified schematic of a hypothetical peatland developmental history and the diatom assemblages typically found within each successive stage from open water (a) to rich fen (b) to poor fen (c) to raised bog (d). Representative diatom assemblages are based on the developmental history of a peatland in the Tiksi Forest, Siberia (Rühland *et al.*, 2000). Note, individual diatom photomicrographs are not to scale. Peatland schematic is based on a figure from Klinger (1996). Black shading indicates the anaerobic peat layer and gray shading indicates the aerobic active layer.

moisture regimes, which are intimately linked to changes in climate (e.g. Blundell and Barber, 2005). As highlighted by Pienitz (2001), comprehensive studies of diatom flora from peatland systems remain scarce. Nevertheless, peatland diatoms have been the focus of various taxonomic studies for several decades including early work by Hustedt (1927–1966), Krasske (1932, 1936), and Foged (1951), as well as the focus of early paleoecological studies by Sears and Couch (1932) and Patrick (1954). The advantages of using multiple proxy indicators have been underscored in numerous diatom-based peatland paleoecological studies (e.g. Korhola, 1990; Rühland *et al.*, 2000; Myers-Smith *et al.*, 2008).

From a diatom's perspective, it would be fair to say that many peat environments, particularly mineral-poor bogs with minimal connectivity to the water table, would be considered an extreme environment. These are often ecosystems that are truly transitional between aquatic and terrestrial environments and can experience extremely dry periods, placing great stress on an aquatic organism. Ombrotrophic bogs can have extremely dilute water chemistry that is highly acidic, with silica in extremely low supply. For example, the surface pools of ombrotrophic peatlands (bogs) were found to have silica concentrations that were up to ten times lower than concentrations in mineral-rich (fen) peatlands (Bendell-Young, 2003). Diatoms have been found to be highly susceptible to post-burial dissolution in exceptionally organic-rich environments (Bennett *et al.*, 1991); they are often in very low abundance or damaged in lower peat layers due to mechanical breakage and silica dissolution (Kingston, 1982). The dissolution and solubility of silicate minerals was found to be greatly enhanced under anoxic, highly organic, circumneutral pH environments, conditions common to deeper layers of some peat deposits (Bennett *et al.*, 1991). Nevertheless, diatoms can tolerate surprisingly extreme conditions, including very acidic (pH as low as ~3.0), ultraoligotrophic, aerophilic, mineral-poor bogs where silica is in scarce supply. Although there is currently limited understanding of whether the functioning of diatom structures (micro- and nanoscales) in certain species are optimized to enable their survival in these extreme environments, it seems likely that diatoms able to survive and reproduce in these harsh conditions can do so through the management and minimal consumption of the extremely scarce silica supply (Sterrenburg *et al.*, 2007).

As silica is essential to diatoms, an insufficient supply in some peat environments can potentially result in preservation problems, teratological specimens (abnormal in growth or structure) (Sterrenburg *et al.*, 2007), or in the worst case scenario, the almost complete absence of a viable diatom community (particularly in the surface layers of inactive, non-aggrading peatland systems). From a paleoecological viewpoint, this is a problem, and may partially account for the paucity of studies using diatom assemblage changes in peat ecosystems. This is underscored when you consider that the ideal peatland type to study climate-related changes is characterized by one of the most physically and chemically challenging diatom habitats, ombrotrophic (raised) bogs (Aaby, 1986).

Diatoms are highly sensitive to changes in alkalinity, pH, water levels, and nutrients, characteristics which can be effectively used in paleoecological studies to track the developmental history of a peatland (Figure 25.3) and, in turn, climatic change (e.g. Kienel *et al.*, 1999; Brugam and Swain, 2000). Aside from using the peat deposit as a record of the ecological and hydrological development of the peatland itself, it is also a receptacle for information about the surrounding environment (Charman, 2002) that can be used for a wide variety of paleoenvironmental studies. For example, changes in diatom community composition within peat profiles can be used as indicators of paleotsunami events (Hemphill-Haley, 1996; Sawai *et al.*, 2004), sea-level changes and transgressions (Hamilton and Shennan, 2005), fire history (Rühland *et al.*, 2000; Myers-Smith *et al.*, 2008), as well as paleomonsoons and glacial retreat (Rühland *et al.*, 2006). As in most paleoenvironmental assessments, these interpretations can be underpinned through the use of multiple indicators (e.g. plant macrofossils, pollen, testate amoebae, phytoliths, charcoal), and by comparing results to nearby lake-sediment records (e.g. Kingston, 1984).

25.5.2 Diatom taxonomic distribution in peatlands

Perhaps one of the greatest challenges for diatom-based paleoecological studies from peat deposits is deciphering the ecological significance of a given diatom assemblage and/or species and its association with (or independence from) changes in the moss communities. Given their specialized habitats as well as their physical and chemical nature, it is no surprise that peatlands contain acidophilous, oligotrophic, and aerophilous diatom taxa. However, a better understanding of the composition and characteristics of diatom assemblages across mineral, pH, moisture, and bryophyte gradients (i.e. fens to bogs), can greatly improve paleoenvironmental interpretations and peatland habitat assessment. As mentioned earlier for mineral wetlands (i.e. non-peat forming wetlands), van Dam *et al.* (1994) provides a useful coded checklist of diatom ecological preferences for both freshwater and wetland habitats. The compilation of almost 950 diatom taxa included in this checklist is very useful for wetland (including peatland) studies as it provides indicator values important to wetland environments, particularly moisture conditions and pH. Not surprisingly, this index is often cited in diatom-focused peatland studies. Indeed, studying diatom communities from various moss environments and from the surface samples of various peatland types (rich fens to ombrotrophic bogs), provides valuable information on the diatom community structure and their associations to these specialized environments from various regions of the world, including temperate environments (e.g. van Dam *et al.*, 1981; Kingston, 1982; Foster *et al.*, 1983; Cochran-Stafira and Andersen, 1984; Walker and Paterson, 1986; Poulíčková *et al.*, 2004; Kilroy *et al.*, 2006) and Arctic and Antarctic environments (Hickman and Vitt, 1973; Beyens, 1989; Van de Vijver and Beyens, 1997, 1999; Van de Vijver *et al.*, 2001). However, much work is still required to better understand the associations between diatom assemblage composition and the chemical, physical, and vegetation gradients of the peatland types, particularly those in climate-sensitive northern latitudes. For example, diatom-based calibration studies across peatland types can refine our understanding of the autecology of peat-dwelling diatoms thereby increasing their potential as bioindicators of habitat and water-quality changes. These data are also important for generating robust and ecologically sound inference models that can be used to aid in the reconstruction of past peatland conditions. At present, comprehensive diatom-based calibration studies across peatland classes are scarce and clearly needed, particularly in remote and environmentally sensitive northern regions.

Kingston (1982) found characteristic diatom assemblages that were dependent upon the macro-vegetation type, trophic status, and the degree of connectivity to the water table in surface samples from northern Minnesota peatlands. Similar to Kingston (1982), Poulíčková *et al.* (2004) found that diatom abundance, species richness, and diversity decreased with a decrease in mineral content and moisture availability across a transect of fen sites in the Czech Republic and Slovak

Republic. Both Kingston (1982) and Poulíčková *et al.* (2004) found that particular groups of diatoms as well as particular species of diatoms were found to have associations with bryophytes in each peatland type. In Minnesota, ombrotrophic bog samples (hummocks) were found to be dominated by acidophilous and aerophilous taxa, particularly *Eunotia* (Bréb.) Rabh., prompting Kingston (1982) to raise an interesting question as to whether this taxon flourishes here because it can withstand the rigors of periodic drying (i.e. aerophilous) or because it has a symbiotic relationship with *Sphagnum* (i.e. bryophilous). Beyens (1985) suggested that the stems of certain *Sphagnum* species are able to retain moisture through capillarity, thereby providing a competitive microhabitat for some diatom species. Additionally, many diatoms found in peat environments may be able to avoid desiccation of their cell walls during dry episodes through mucilage secretion (Round, 1981; Pienitz, 2001). Several aerophilic diatom species have been reported from a range of wetland types from all parts of the world. In particular, there is considerable overlap in reports of the diatom taxa *Hantzschia amphioxys*, *Pinnularia borealis*, *Diadesmis contenta* (Grun.) Mann, *Luticola saxophila*, *Eunotia paludosa* Grun., *E. exigua*, *Eunotia praerupta* Ehrenb., *Cymbella ventricosa* Kütz. var. *groenlandica* Foged, and *Caloneis aerophila* Bock in poor fens, ombrotrophic bogs, and in the driest of mineral wetlands. Perhaps some diatom species are truly "extremophiles" (Sterrenburg *et al.*, 2007) where they can exploit microhabitat niches enabling them to survive in these harsh, dry, nutrient-deprived environments. Although many insights have been made on the characteristics, life strategies, and associations to their microhabitats, much remains to be understood about the autecology and specialization of peatland diatoms.

Several paleoecological studies have determined that, although diatom assemblages tracked hydroseral succession through time, these diatom changes are not necessarily directly dependent upon changes in the bryophyte communities within the peat environment (Brown *et al.*, 1994; Rühland *et al.*, 2000; Myers-Smith *et al.*, 2008). Clear shifts in diatom assemblages often occur within a peat profile consisting almost exclusively of one moss species (Brown *et al.*, 1994), or greatly precede shifts in the moss communities (Rühland *et al.*, 2000; Myers-Smith *et al.*, 2008). This in itself is an important point as it suggests that diatoms may be responding primarily to initial chemical cues in their environment, and secondarily to changes in the moss habitats (Rühland *et al.*, 2000; Poulíčková *et al.*, 2004). For example, in a peat profile from Siberia, the succession from rich fen to poor fen was discernable in the plant macrofossil assemblage by a change in dominant moss type from *Drepanocladus* to *Sphagnum* (Rühland *et al.*, 2000). However, a shift in the diatom assemblage occurred before *Sphagnum* moss species became dominant and may have been due to a decrease in pH in the fen environment at the initial stages of *Sphagnum* development. Diatoms can respond to relatively small changes in pH that may not have been substantial enough to initiate a response in *Drepanocladus*, as this moss can tolerate a relatively wide pH range (~5.2 to 7.4) (Gignac *et al.*, 1991). The shift from rich fen to poor fen could only occur within a specific pH range that is suitable for *Sphagnum* species to establish and flourish, pH between 3.3 to 5.5 (Gignac *et al.*, 1991), and below the levels that *Drepanocladus* can tolerate. This example demonstrates that diatom analysis, together with other peat proxies, can fine-tune peatland environmental histories.

25.5.3 Diatoms and environmental change in northern peatlands (boreal/Arctic)

Peatlands are widely distributed globally, although the vast majority are located in northern latitudes above 45° N and generally below 60° N (Charman, 2002). Here, they are an important landscape feature, particularly in subarctic and boreal regions of Canada, Siberia, Finland, and Alaska. The cool climate, generally flat topography, coupled with the presence of sporadic and discontinuous permafrost, ensures a waterlogged environment and low rates of plant decomposition that set the stage for peatland development at these latitudes. Indeed, the distribution of discontinuous permafrost is closely tied to the distribution of peatlands. Understandably, these highly specialized ecosystems are particularly sensitive to climate change because of their reliance on critical linkages between temperature, precipitation, and permafrost.

It is well established that northern latitudes are highly sensitive to climatic changes (e.g. Serreze *et al.*, 2000; ACIA, 2004). That much of the world's peatlands occur in these northern environments underscores the importance of peatland research in boreal and Arctic ecosystems. For example, in Canada, 90% of wetlands (areal extent $>1.4 \times 10^6$ km^2: NWWG, 1988) are peat-forming (Tarnocai, 2006; Roulet, 2000) and cover approximately 12% of the country's land area, almost all of which (97%) lie in boreal and subarctic regions (Tarnocai, 2006). Approximately 60% of these peatlands lie in boreal and subarctic regions that are expected to be severely affected by changes in precipitation and temperature (Lavoie *et al.*, 2005), particularly the Hudson Bay Lowlands, the Mackenzie River valley region and the northern parts of Alberta and Manitoba (Kettles and Tarnocai, 1999; Tarnocai, 2006).

In general, peatlands sequester CO_2 but they can be large sources of CH_4, with northern wetlands emitting between 30 to 50 teragrams (Tg) of CH_4 per year globally (Fung *et al.*, 1991; Bartlett and Harriss, 1993). Not only will the impact of climate change on these northern peatlands be great but changes in these peatlands will also greatly affect climate through complex feedback mechanisms. For example, increased temperatures are expected to result in permafrost degradation, changes in hydrology, and increases in wildfire frequency and extent, all of which will play an important role in determining peatland type and carbon fluxes (switches between net sinks and net sources) (Gorham, 1991; ACIA, 2004; Tarnacai, 2006). In addition to climate changes, anthropogenic landscape alterations must also be taken into account (Lavoie *et al.*, 2005). For example, one of the most extensive reservoir areas worldwide is in Canada's boreal and subarctic regions. Here, the flooding of wetlands for energy development was estimated to account for a net change in greenhouse-gas emissions of approximately 1×10^{12} g C yr^{-1}, or about 5% of Canada's current anthropogenic emissions (Roulet, 2000). Therefore, it is important to consider the complexity of changes in carbon storage and emissions fluxes for research in northern peatlands (and in underlying permafrost).

Appreciable new sources of CO_2 may be released into the atmosphere upon thawing of the upper layers and degradation through exposure of peat to aerobic conditions (Smith *et al.*, 2004). Alternatively, in regions of continuous permafrost such as in the High Arctic, continued warming could lead to the shifting of the discontinuous permafrost zone northward, resulting in the initiation of new peatland sites in the High Arctic (Turunen and Tolonen, 1996; ACIA, 2004; Tarnocai, 2006). These newly developed peatlands could then become new sinks for CO_2 and new sources of CH_4. This is somewhat analogous to the findings of MacDonald *et al.* (2006) where the rapid development of present-day circumarctic peatlands were initiated between the late-glacial and the early-Holocene periods (~12,000 to 8000 years BP) in response to abrupt warming and increased summer insolation. MacDonald *et al.* (2006) suggest that peatland development during this period would have substantially contributed to fluctuations of atmospheric CH_4 and CO_2. They further proposed that a decline in atmospheric CH_4 after ~8000 years BP is consistent with the hydroseral succession of many of these peatland systems, from minerotrophic fens to ombrotrophic bogs, as the latter is typically a much weaker source of atmospheric CH_4. Future climate scenarios anticipate that the degradation of peatlands and losses of carbon to the atmosphere will be substantial and could lead to very strong positive feedback mechanisms that could further increase warming (Oechel *et al.*, 1993) and likely outweigh the potential carbon storage of these newly developed High Arctic peatlands. Clearly, permafrost plays a key role in maintaining northern peatlands and their carbon sources and sinks and is instrumental in moderating global climate.

Although diatom-focused studies in peatlands are relatively few compared to lake sediment studies, a high proportion of the diatom-based peatland studies that do exist, particularly with a paleoecological focus, are from northern latitudes above 45° N (e.g. Brugam, 1980; Lortie, 1983; Kingston, 1984; Korhola, 1990; Korhola, 1992b; Brown *et al.*, 1994; Korhola, 1995; Kienel *et al.*, 1999; Brugam and Swain, 2000; Rühland *et al.*, 2000; Hamilton and Shennan, 2005; Myers-Smith *et al.*, 2008). One of the earliest examples of high-latitude diatom-peat studies was by Brown *et al.* (1994) who examined diatoms preserved in a rare peat deposit from the High Arctic in northwestern Greenland – a region where peat currently does not accumulate as it is too cold and dry and the region is underlain by continuous permafrost. However, for a period of at least 2000 years in the early to mid Holocene, the climate was likely warmer and wetter than it is today, and peat development occurred throughout High Arctic regions (Ovenden, 1988). Brown *et al.* (1994) found that siliceous microfossils were well preserved throughout the peat sequence, although taxon richness was typically quite low, with only 19 diatom taxa observed, dominated by *Pinnularia*, *Tabellaria*, and *Eunotia* species. Brown *et al.* (1994) suggest that a shift from more aerophilous taxa, including *Pinnularia borealis* and *Eunotia curvata* (Kütz.) Lagerst. to more aquatic taxa such as *Pinnularia intermedia* (Lagerst.) Cl. and *Fragilaria construens* var. *venter* (Ehrenb.) Grun for a brief period following *c.* 5000 years BP, signaled a change to a wetter environment with perhaps an increase in winter snow accumulation. In the *c.* 2000 years that peat was able to develop in this High Arctic site, 260 cm of organic matter accumulated, resulting in an extremely high-resolution paleoenvironmental record. This high temporal resolution is an appealing feature in the Arctic as sediment accumulation rates in lakes and ponds are typically very low.

In Alaska, permafrost degradation and fire disturbances play key roles in altering ecosystem structure and, ultimately, the carbon flux of northern wetlands (Camill *et al.*, 2001). Myers-Smith *et al.* (2008) examined changes in diatom assemblage composition in a ~600-year peat profile from the Tanana River floodplain in interior Alaska. The peat profile had three distinct successional stages from sylvic (tree-derived) peat at the base of the core, to sedge-dominated peat (fen) from *c.* AD 1900 to c. AD 1970, and then to a *Sphagnum*-dominated peat (bog) from AD 1970 to the present. Changes in diatom assemblage composition tracked these peat successional changes with more aerophilous diatom taxa, including including *Hantzschia amphioxys* var. *major* Grun. present in the initial sylvic peatland stage, changing to more circumneutral, epiphytic *Pinnularia* species in the rich-fen stage, and then to acidophilous and acidobiontic taxa, including *Eunotia rhomboidea* Hust., *Eunotia nymanniana* Grun., and *Eunotia glacialis* Meist., in the *Sphagnum* bog stage. Similar to the findings of Brown *et al.* (1994) and Rühland *et al.* (2000), diatom shifts in this Arctic peatland appeared to have responded independently to changes in bryophyte assemblages with diatom changes pre-dating the shift from fen to bog. This shift coincided with reduced growth of surrounding black spruce trees that the authors suggest was indicative of a stepwise ecosystem-level response to substantial increases in air temperatures in interior Alaska, with no change observed in precipitation. Warmer and drier summers, together with drier peatland conditions, are thought to have increased the potential for fire disturbance in the region. Indeed, in 2001, an extensive fire burned near the study site and triggered a collapse of the permafrost resulting in the lateral expansion of the peatland. A substantial change in diatom assemblage composition was coincident with the fire-induced permafrost collapse that likely accelerated the succession from fen to bog. These diatom and peatland developmental changes were likely the result of dry conditions combined with consecutive warm growing periods since the 1970s. Myers-Smith *et al.* (2008) suggest that with continued warming and increased fire frequency, permafrost collapse and peatland expansion could potentially increase carbon storage in this region. However, the authors caution that continued dry conditions could lead to drainage of the wetland and promote aerobic degradation of peatlands, thereby offsetting any of the newly developed carbon-storage gains.

25.6 Future directions

While this chapter has provided an overview of many successful diatom-based wetlands assessments, the use of diatoms as proxies of environmental change in wetlands has lagged behind other types of ecosystems. This may be partly due to a broader lag in development of wetland science and policy for wetland assessment (Batzer and Sharitz, 2006), the complicated logistics of sampling diatoms from the world's wetlands that are focused toward the poles and equator, or a misconception that diatoms always preserve poorly in wetland sediments. As a result, there are noticeable geographic, taxonomic, and conceptual gaps in wetland diatom studies that are in significant need of further attention, and some of these are outlined below.

There is a pronounced lack of wetland assessment protocols in tropical, subtropical, and polar regions. The well-developed diatom floras of Europe and the USA can support environmental assessment and paleoecological research, but although some regulatory agencies have employed diatoms to help guide restorative practices, this tool is still underutilized relative to its potential. Existing success stories, however, can be used to guide similar protocols in developing countries. For example, the taxonomy and ecology of diatoms of the Florida Everglades is fairly well described, aiding paleoecological and assessment studies aimed at understanding how water flow and quality are changing relative to water management in the growing South Florida metropolis. Many Everglades diatom taxa were once thought to be endemic to South Florida, but recent collections in similar karstic wetlands in Belize, Mexico, and Jamaica have revealed some of the same algal taxa (Rejmánková, 2001; E. Gaiser, unpublished data). As a result, the very strong relationships developed in the Everglades between diatoms and water quality and quantity could be applied in these regions to document and possibly be used to prevent ecosystem degradation resulting from rapid development. Besides the work of Metzeltin and Lange-Bertalot (1998), very little is known about the diatom flora of other tropical wetlands, including the vast wetlands of the Mekong Delta, the Bolivian Pantanal, the Okavango Delta, and large expanses of Amazonia and Pacific islands. These regions have complicated natural and human histories, in addition to growing unsustainable exploitation of freshwater resources, so they offer a last frontier of discovery in diatom taxonomy and application in water resource management.

While diatoms are an underexploited tool for tracking ecosystem responses to development and climate change in the tropics, there is a similar urgency for polar regions. Modern peatland gradients are quickly changing as permafrost melts and hydrology changes, stressing the need for studies to estimate baseline conditions. Although there are a growing number of diatom-based calibration studies for Arctic and subarctic lakes and ponds, to date, no such comprehensive diatom-based studies have been undertaken in circumpolar peatland regions. The scarcity of these diatom studies may in part be due to problems with preservation in some peat samples, particularly where the surface samples were collected from a site that is not at present actively accumulating peat (i.e. dry). There remains much to learn about the ecology of northern peatland diatoms. Although not without its own set of challenges, particularly diatom preservation, future northern peatland studies should include identifying and enumerating modern diatoms preserved in surface peat samples to estimate diatom habitat preferences along a wide spectrum of environmental and climatic gradients including permafrost boundaries and peatland classes. From this information, transfer functions can be developed that relate diatoms to environmental variables of interest that can then be used as a metric for determining the impact of climate and environmental changes on future peatlands. Similar to what has been successfully accomplished in other wetland regions, it would be of great interest to determine whether diatom-based transfer functions can be applied to boreal and subarctic peat profiles to quantitatively infer environmental change over long time periods. Given that diatom-based calibration studies exist for boreal and subarctic lakes, tremendous opportunities exist to combine peat and lake data sets that could potentially provide independent lines of evidence from two highly sensitive northern ecosystems. Undoubtedly, diatom-based peatland studies have a lot to contribute to our understanding of climatic and environmental changes in high-latitude regions, and should be better exploited.

25.7 Summary

Diatoms serve as both engineers and barometers in wetlands and peatlands. They engineer shallow-water habitats through their productivity that controls water concentrations of gases and nutrients, supplies fixed carbon to the food web, forms habitat for other organisms, and contributes to sediment accretion and stabilization. They are barometers of the state of the environment through rapid population responses to fluctuations in the primary drivers of wetland structure and function, including water quantity and quality. These responses are reflected in changes in assemblage structure that can be measured over time and space and used to recount ecosystem evolution (i.e. hydroseral succession) as well as changes imposed by human activities. Because paleoecological studies offer the opportunity to put the latter (human impacts) in the context of the former (longer-term, natural changes), reading history from diatom records in wetland and peatland profiles is a powerful tool to guide ecosystem management on local to global scales. Given the alarming rates at which the world's wetlands are being lost and the challenges involved in understanding the ecological and societal repercussions of these losses, it would seem prudent that much effort should be put into developing tools for assessing the magnitude and the global implications of changes in our wetland ecosystems. This is particularly true for the expansive wetlands in rapidly developing tropical countries and in high-latitude wetlands subject to rapid climate change, where diatom-based assessment and paleoecological studies are still in their infancy but hold much promise.

Acknowledgments

This chapter was enhanced by collaboration with the Florida Coastal Everglades Long Term Ecological Research Program (NSF #DBI-0620409 and #DEB-9910514) and the Paleoecological Environmental Assessment and Research Laboratory (PEARL) and is contribution #424 of the Southeast Environmental Research Center at Florida International University. We thank Sherri Cooper and the editors of this volume for helpful comments on earlier drafts of the chapter.

References

Aaby, B. (1986). Palaeoecological studies of mires. In *Handbook of Holocene Palaeoecology and Palaeohydrology*, ed. B. E. Berglund, Chichester: John Wiley and Sons, pp. 145–64.

ACIA (Arctic Climate Impact Assessment) (2004). *Impacts of a Warming Arctic: Arctic Climate Impact Assessment*, Cambridge: Cambridge University Press.

Anderson, N. J. (1990). Variability of diatom concentrations and accumulation rates in sediments of a small lake basin. *Limnology and Oceanography*, **35**, 497–508.

Bahls, L. L. (1993). *Periphyton Bioassessment Methods for Montana Streams.* Helena, MT: Montana State Water Quality Bureau.

Barker, P. A., Roberts, N., Lamb, H. F., & Van Der Kaars, S. (1994). Interpretation of Holocene lake-level change from diatom assemblages in Lake Sidi Ali, Middle Atlas, Morocco. *Journal of Paleolimnology*, **12**, 223–8.

Bartlett, K. B. & Harriss, R. C. (1993). Review and assessment of methane emissions from wetlands. *Chemosphere*, **26**, 261–320.

Battarbee, R. W. (1986). Diatom analysis. In *Handbook of Holocene Palaeoecology and Palaeohydrology*, ed. B. E. Berglund, Toronto: John Wiley and Sons, pp. 527–70.

Batzer, D. P. & Sharitz, R. S. (2006). *Ecology of Freshwater and Estuarine Wetlands.* Los Angeles: University of California Press.

Bauer, I. E., Gignac, D., & Vitt, D. H. (2003). Development of a peatland complex in boreal western Canada: lateral site expansion and local variability in vegetation succession and long-term peat accumulation. *Canadian Journal of Botany*, **81**, 833–47.

Bendell-Young, L. (2003). Peatland interstitial water chemistry in relation to that of surface pools along a peatland mineral gradient. *Water, Air, and Soil Pollution*, **143**, 363–75.

Bennett, P. C., Siegel, D. I., Hill, B. M., & Glaser, P. H. (1991). Fate of silicate minerals in a peat bog. *Geology*, **19**, 328–31.

Beyens, L. (1985). On the subboreal climate of the Belgian Campine as deduced from diatom and testate amoebae analyses. *Review of Palaeobotany and Palynology*, **46**, 9–31.

Beyens, L. (1989). Moss dwelling diatom assemblages from Edgeøya (Svalbard). *Polar Biology*, **9**, 423–30.

Blundell, A. & Barber, K. (2005). A 2800-year palaeoclimatic record from Tore Hill Moss, Strathspey, Scotland: the need for a multi-proxy approach to peat-based climate reconstructions. *Quaternary Science Reviews*, **24**, 1261–77.

Brazner, J. C., Danz, M. P., Trebitz, A. S., *et al.* (2007). Responsiveness of Great Lakes wetland indicators to human disturbances at multiple spatial scales: a multi-assemblage assessment. *Journal of Great Lakes Research*, **33**, 42–66.

Brown, K. M., Douglas, M. S. V., & Smol, J. P. (1994). Siliceous microfossils in a Holocene, High Arctic peat deposit (Nordvestø, northern Greenland). *Canadian Journal of Botany*, **72**, 208–16.

Brugam, R. B. (1980). Postglacial diatom stratigraphy of Kirchner Marsh, Minnesota. *Quaternary Research*, **13**, 133–46.

Brugam, R. B., McKeever, K., & Kolesa, L. (1998). A diatom-inferred water depth reconstruction for an upper peninsula, Michigan, lake. *Journal of Paleolimnology*, **20**, 267–76.

Brugam, R. B. & Swain, P. (2000). Diatom indicators of peatland development at Pogonia Bog Pond, Minnesota, USA. *The Holocene*, **10**, 453–64.

Bunting, M. J., Duthie, H. C., Campbell, D. R., Warner, B. G., & Turner, L. J. (1997). A paleoecological record of recent environmental change at Big Creek Marsh, Long Point, Lake Erie. *Journal of Great Lakes Research*, **23**, 349–68.

Burkholder, J. M. (1996). Interactions of benthic algae with their substrata. In *Algal Ecology: Freshwater Benthic Ecosystems*, ed. R. J. Stevenson, M. L. Bothwell, & R. L. Lowe. San Diego, CA: Academic Press, pp. 253–97

Camill, P., Lynch, J. A., Clark, J. S., Adams, J. B., & Jordan, B. (2001). Changes in biomass, aboveground net primary production, and peat accumulation following permafrost thaw in the boreal peatlands of Manitoba, Canada. *Ecosystems*, **4**, 461–78.

Campbell, D. R., Duthie, H. C., & Warner, B. G. (1997). Post-glacial development of a kettle-hole peatland in southern Ontario. *Ecoscience*, **4**, 404–18.

Charman, D. (2002). *Peatlands and Environmental Change*, Chichester: John Wiley and Sons.

Chipps, S. R., Hubbard, D. E., Werlin, K. B., *et al.* (2006). Association between wetland disturbance and biological attributes in floodplain wetlands. *Wetlands*, **26**, 497–508.

Clymo, R. S. (1984). The limits to bog growth. *Philosophical Transactions of the Royal Society of London*, **B 303**, 605–54.

Clymo, R. S., Oldfield, F., Appleby, P. G., *et al.* (1990). A record of atmospheric deposition in a rain-dependent peatland. *Philosophical Transactions of the Royal Society of London*, **B 327**, 331–8.

Cochran-Stafira, D. L. & Andersen, R. A. (1984). Diatom flora of a kettle-hole bog in relation to hydrarch succession zones. *Hydrobiologia*, **109**, 265–73.

Cooper, S. R., Goman, M., & Richardson, C. J. (2008). Historical changes in water quality and vegetation in WCA-2A as determined by paleoecological analyses. In *The Everglades Experiment: Lessons for Ecosystem Restoration*, ed. C. J. Richardson, New York: Springer, ch. 12, pp. 321–50.

Cooper, S. R., Huvane, J., Vaithiyanathan, P., & Richardson, C. J. (1999). Calibration of diatoms along a nutrient gradient in Florida Everglades Water Conservation Area-2A, USA. *Journal of Paleolimnology*, **22**, 413–37.

Costanza, R., d'Arge, R., deGroot, R., *et al.* (1997). The value of the world's ecosystem services and natural capital. *Nature*, **387**, 253–60.

Dean, W. E., Bradbury, J. P., Andersen, R. Y., & Barnosky, C. W. (1984). The variability of Holocene climate change: evidence from varved sediments. *Science*, **226**, 1191–4.

Della-Bella, V., Puccinelli, C., Marcheggiani, S., & Mancini, L. (2007). Benthic diatom communities and their relationship to water chemistry in wetlands of central Italy. *Annals of Limnologie*, **43**, 89–99.

de Mars, H. & Wassen, M. J. (1999). Redox potentials in relation to water levels in different mire types in the Netherlands and Poland. *Plant Ecology*, **140**, 41–51.

Douglas, M. S. V. & Smol, J. P. (1987). Siliceous protozoan plates in lake sediments. *Hydrobiologia*, **154**, 13–23.

Duthie, H. C., Yang, J. R., Edwards, T. W. D., Wolfe, B. B., & Warner B.G. (1996). Hamilton Harbour, Ontario: 8300 years of limnological and environmental change inferred from microfossil and isotopic analyses. *Journal of Paleolimnology*, **15**, 79–97.

Dyer, K. R. (1995). Sediment transport processes in estuaries. In *Geomorphology and Sedimentology of Estuaries. Developments in Sedimentology 53*, ed. G. M. E. Perillo, Amsterdam: Elsevier Science Publishers, pp. 423–49.

Earle, J. C. & Duthie, H. C. (1986). A multivariate statistical approach for interpreting marshland diatom succession. In *Proceedings of the 8th Diatom Symposium*, ed. M. Ricard, Königstein: Koeltz Scientific Books, pp. 441–58.

Engle, D. L. & Melack, J. M. (1993). Consequences of riverine flooding for seston and the periphyton of floating meadows in an Amazon floodplain lake. *Limnology and Oceanography*, **38**, 1500–20.

Ewe, S. M. L., Gaiser, E. E., Childers, D. L., *et al.* (2006). Spatial and temporal patterns of aboveground net primary productivity (ANPP) in the Florida Coastal Everglades LTER (2001–2004). *Hydrobiologia*, **569**, 459–74.

Finkelstein, S. A. & Davis, A. M. (2005a). Modern pollen rain and diatom assemblages in a Lake Erie coastal marsh. *Wetlands*, **25**, 551–63.

Finkelstein, S. A. & Davis, A. M. (2005b). Paleoenvironmental records of water level and climatic changes from the middle to late Holocene at a Lake Erie coastal wetland, Ontario, Canada. *Quaternary Research*, **65**, 33–43.

Fluin, J., Gell, P., Haynes, D., Tibby, J., & Hancock, G. (2007). Paleolimnological evidence for the independent evolution of neighbouring terminal lakes, the Murray Darling Basin, Australia. *Hydrobiologia*, **591**, 117–34.

Foged, N. (1951). The diatom flora of some Danish springs. *Natura Jutlandica*, 4–5, 1–84.

Foster, D. R., King, G. A., Glaser, P. H., & Wright, H. E. (1983). Origin of string patterns in northern peatlands. *Nature*, **306**, 256–7.

Fritz, S. C., Juggins, S., Battarbee, R. W., & Engstrom, D. R. (1991). Reconstruction of past changes in salinity and climate using a diatom-based transfer function. *Nature*, **352**, 706–8.

Fung, I. Y., Lerner, J. J., Matthews E., *et al.* (1991). Three-dimensional model synthesis of global methane cycle. *Journal of Geophysical Research*, **96**, 13033–65.

Gaiser, E. (2009). Periphyton as an indicator of restoration in the Everglades. *Ecological Indicators*. DOI:10.1016/j.ecolind.2008.08.004.

Gaiser, E. E., Brooks, M. J., Kenney, W., Schelske, C. L., & Taylor, B. E. (2004b). Interpreting the hydrologic history of a temporary pond using siliceous microfossils. *Journal of Paleolimnology*, **31**, 63–76.

Gaiser, E. E., Philippi, T. E., & Taylor, B. E. (1998). Distribution of diatoms among intermittent ponds on the Atlantic Coastal Plain: development of a model to predict drought periodicity from surface sediment assemblages. *Journal of Paleolimnology*, **20**, 71–90.

Gaiser, E. E., Richards, J. H., Trexler, J. C., Jones, R. D., & Childers, D. L. (2006b). Periphyton responses to eutrophication in the Florida Everglades: cross-system patterns of structural and compositional change. *Limnology and Oceanography*, **51**, 617–30.

Gaiser, E. E., Scinto, L. J., Richards, J. H., *et al.* (2004c). Phosphorus in periphyton mats provides best metric for detecting low-level P enrichment in an oligotrophic wetland. *Water Research*, **38**, 507–16.

Gaiser, E. E., Taylor, B. E., & Brooks, M. J. (2001). Establishment of wetlands on the southeastern Atlantic Coastal Plain: paleolimnological evidence of a mid-Holocene hydrologic threshold from a South Carolina pond. *Journal of Paleolimnology*, **26**, 373–91.

Gaiser, E. E., Trexler, J. C., Richards, J. H., *et al.* (2005). Cascading ecological effects of low-level phosphorus enrichment in the Florida Everglades. *Journal of Environmental Quality*, **34**, 717–23.

Gaiser, E. E., Wachnicka, A., Ruiz, P., Tobias, F. A., & Ross, M. S. (2004a). Diatom indicators of ecosystem change in coastal wetlands. In *Estuarine Indicators*, ed. S. Bortone, Boca Raton, FL: CRC Press, pp. 127–44.

Gaiser, E. E., Zafiris, A., Ruiz, P. L., Tobias, F. A. C., & Ross, M. S. (2006a). Tracking rates of ecotone migration due to salt-water encroachment using fossil mollusks in coastal south Florida. *Hydrobiologia*, **569**, 237–57.

Gasse, F., Barker, P. Gell, P. A., Fritz, S. C., & Chalie, F. (1997). Diatom-inferred salinity in palaeolakes: an indirect tracer of climate change. *Quaternary Science Reviews*, **16**, 547–63.

Gasse, F., Juggins, S., & BenKhelifa, L. (1995). Diatom-based transfer functions for inferring past hydrochemical characteristics of African lakes. *Palaeogeography, Palaeoclimatology, Palaeoecology*, **117**, 31–54.

Gell, P. A. (1997). The development of a diatom database for inferring lake salinity, western Victoria, Australia: towards a quantitative approach for reconstructing past climates. *Australian Journal of Botany*, **45**, 389–423.

Gell, P. J., Bulpin, S., Wallbrink, P., Bickford, S., & Hancock, G. (2005). Tareena Billabong – a paleolimnological history of an ever changing wetland, Chowilla Floodplain, lower Murray–Darling Basin. *Marine and Freshwater Research*, **56**, 441–56.

Gell, P. A., Sluiter, I. R., & Fluin, J. (2002). Seasonal and inter-annual variations in diatom assemblages in Murray River-connected wetlands in northwest Victoria, Australia. *Marine and Freshwater Research*, **53**, 981–92.

Gell, P. A., Sonneman, M., Reid, M., Illman M., & Sincock, A. (1999). An illustrated key to common diatom genera from Southern Australia. The Murray–Darling Freshwater Research Centre Identification Guide No. 26.

Gell, P., Tibby, J., Little, F., Baldwin D., & Hancock, G. (2007). The impact of regulation and salinisation on floodplain lakes: the lower River Murray, Australia. *Hydrobiologia*, **591**, 135–46.

Gignac, L. D., Vitt, D. H., Zoltai, S. C., & Bayley, S. E. (1991). Bryophyte response surfaces along climatic, chemical, and physical gradients in peatlands of western Canada. *Nova Hedwigia*, **53**, 27–71.

Godwin, H. (1940). Pollen analysis and forest history of England and Wales. *New Phytologist*, **39**, 370–400.

Goldsborough, L. G. & Robinson, G. G. C. (1996). Pattern in wetlands. In *Algal Ecology: Freshwater Benthic Ecosystems*, ed. R. J. Stevenson, M. L. Bothwell, & R. L. Lowe. San Diego, CA: Academic Press, ch. 4, pp. 77–117.

Gopal, B., Junk J. W., & Davis, J. A. (2000). *Biodiversity in Wetlands: Assessment, Function and Conservation*. Leiden: Backhuys Publishers.

Gordon, L., Dunlop, M., & Foran, B. (2003). Land cover change and water vapour flows: learning from Australia. *Philosophical Transactions of the Royal Society of London*, **B 358**, 1973–84.

Gorham, E. (1956). The ionic composition of some bog and fen waters in the English Lake District. *Journal of Ecology*, **44**, 142–52.

Gorham E. (1991). Northern peatlands: role in the carbon budget and probable responses to global warming. *Ecological Applications*, **1**, 182–95.

Gorham, E. (1995). The biogeochemistry of northern peatlands and its possible responses to global warming. In *Biotic Feedbacks in the Global Climate System. Will the Warming Feed the Warming?* ed. G. M. Woodwell & F. T. Mackenzie, New York: Oxford University Press, pp. 169–87.

Håkanson, L. (1977). The influence of wind, fetch and water depth on the distribution of sediments in Lake Vånern, Sweden. *Canadian Journal of Earth Sciences*, **14**, 397–412.

Hamilton, S. & Shennan, I. (2005). Late Holocene relative sea-level changes and the earthquake deformation cycle around upper Cook Inlet, Alaska. *Quaternary Science Reviews*, **24**, 1470–98.

Haynes, D., Gell, P., Tibby, J., Hancock, G., & Goonan, P. (2007). Against the tide: the freshening of naturally saline coastal lakes, southeastern South Australia. *Hydrobiologia*, **591**, 165–83.

Hemphill-Haley, E. (1996). Diatoms as an aid in identifying late-Holocene tsunami deposits. *The Holocene*, **6**, 439–48.

Hickman, M. & Vitt, D. H. (1973). The aerial epiphytic diatom flora of moss species from subantarctic Campbell Island. *Nova Hedwigia*, **24**, 443–58.

Hustedt, F. (1927–1966). Die Kieselalgen Deutschlands, Österreichs und der Schweiz unter Berücksichtigung der übrigen Länder Europas sowie der angrenzenden Meeresgebiete. In *Dr. L. Rabenhorst's Kryptogamen-Flora von Deutschland, Österreich und der Schweiz.*

Leipzig Akademische Verlagsgesellschaft, Part 1 (1927–1930); Part 2 (1931–1959); Part 3 (1961–1966).

Jackson, C. R. (2006). Wetland hydrology. In *Ecology of Freshwater and Estuarine Wetlands*, ed. D. P. Batzer & R. S. Sharitz, Los Angeles, CA: University of California Press, pp. 43–81.

Jasinski, J. P. P., Warner, B. G., Andreev, A. A., *et al.* (1998). Holocene environmental history of a peatland in the Lena River valley, Siberia. *Canadian Journal of Earth Sciences*, **35**, 637–48.

Kettles, I. M. & Tarnocai, C. (1999). Development of a model for estimating the sensitivity of Canadian peatlands to climate warming. *Geographie physique et Quaternaire* **53**, 323–38.

Kienel, U., Sigert, C., & Hahne, J. (1999). Late Quaternary palaeoenvironmental reconstructions from a permafrost sequence (North Siberian Lowland, SE Taymyr Peninsula) – a multidisciplinary case study. *Boreas*, **28**, 181–93.

Kilroy, C., Biggs, B. J. F., Vyverman, W., & Broady, P. A. (2006). Benthic diatom communities in subalpine pools in New Zealand: relationships to environmental variables. *Hydrobiologia*, **561**, 95–110.

Kingston, J. C. (1982). Association and distribution of common diatoms in surface samples from northern Minnesota peatlands. *Nova Hedwigia*, **73**, 333–46.

Kingston, J. C. (1984). Palaeolimnology of a lake and adjacent fen in southeastern Labrador: evidence from diatom assemblages. In *Proceedings of the 7th International Diatom Symposium*, ed. D. G. Mann, Königstein: Koeltz Scientific Books, pp. 443–53.

Klinger, L. F. (1996). The myth of the classic hydrosere model of bog succession. *Arctic and Alpine Research*, **28**, 1–9.

Korhola, A. (1990). Paleolimnology and hydroseral development of the Kotasuo bog, southern Finland, with special reference to the Cladocera. *Annales Academie Scientiarum Fennicae A III* 155, **40**.

Korhola, A. (1992a). Mire induction, ecosystem dynamics and lateral extension on raised bogs in the southern coastal area of Finland. *Fennia*, **170**, 25–94.

Korhola, A. (1992b). The Early Holocene hydrosere in a small acid hill-top basin studied using crustacean sedimentary remains. *Journal of Paleolimnology*, **7**, 1–22.

Korhola, A. (1995). The Litorina transgressions in the Helsinki region, southern Finland: new evidence from coastal mire deposits. *Boreas*, **24**, 173–83.

Krasske, G. (1932). Beiträge zur Kenntnis der Diatomoceenflora der Alpen. *Hedwigia*, **72**, 92–134.

Krasske, G. (1936). Die Diatomeenflora der Moosrasen des Wilhelmshöher Parkes.In *Festschrift des Vereins für Naturkunde zu Kassel zum hundertjährigen Bestehen*, pp. 151–64.

Krukowski, S. T. (1988). Sodium metatungstate: a new heavy-mineral separation medium for the extraction of conodonts from insoluble residues. *Journal of Paleontology*, **62**, 314–16.

Kuhry, P. & Turunen, J. (2006). The postglacial development of boreal and subarctic peatlands. In *Boreal Peatland Ecosystems*, ed. R. K. Weider & D. H. Vitt, Berlin: Springer, pp. 25–46.

Lane, C. R. & Brown, M. T. (2006). Energy-based land use predictors of proximal factors and benthic diatom composition in Florida freshwater marshes. *Environmental Monitoring and Assessment*, **117**, 433–50.

Lane, C. R. & Brown, M. T. (2007). Diatoms as indicators of isolated herbaceous wetland condition in Florida, USA. *Ecological Indicators*, **7**, 521–40.

Lange-Bertalot, H. (1979). Pollution tolerance of diatoms as a criterion for water quality estimation. *Nova Hedwigia*, **64**, 285–304.

Last, W.M. & Smol, J. P. (2001). *Tracking Environmental Change Using Lake Sediments, Volume 1: Basin Analysis, Coring, and Chronological Techniques*. Dordrecht: Kluwer Academic Publishers.

Lavoie, M., Paré, D., & Bergeron, Y. (2005). Impact of global change and forest management on carbon sequestration in northern forested peatlands. *Environmental Reviews*, **13**, 199–240.

Leahy, P. J., Tibby, J., Kershaw, A. P., Heijnis, H., & Kershaw, J. S. (2005). The impact of European settlement on Bolin Billabong, a Yarra River floodplain lake, Melbourne, Australia. *River Research and Applications*, **21**, 131–49.

Little, A. E. F., Robinson, C. J., Peterson, S. B., Raffa, K. F., & Handelsman, J. (2008). Rules of engagement: interspecies interactions that regulate microbial communities. *Annual Review of Microbiology*, **62**, 375–401.

Lortie, G. (1983). Les diatomées fossils de deux tourbières ombrotrophies du Bas-Saint-Laurent, Québec. *Géographie physique et Quaternaire*, **37**, 159–77.

Lougheed, V. L., Parker, C. A., & Stevenson, R. J. (2007). Using non-linear responses of multiple taxonomic groups to establish criteria indicative of wetland biological condition. *Wetlands*, **27**, 96–109.

Lowe, R. L. (1974). *Environmental Requirements and Pollution Tolerance of Freshwater Diatoms*. Cincinnati, OH: US Environmental Protection Agency, EPA-670/4–74–005.

MacDonald, G. M., Beilman, D. W., Kremenetski, K. V., *et al.* (2006). Rapid early development of circumarctic peatlands and atmospheric CH_4 and CO_2 variations. *Science*, **314**, 285–8.

Maltby, E. & Immirzi, P. (1993). Carbon dynamics in peatlands and other wetland soils: regional and global perspectives. *Chemosphere*, **27**, 999–1023.

Mayer, P. M. & Galatowitsch, S. M. (2001). Assessing ecosystem integrity of restored prairie wetlands from species production–diversity relationships. *Hydrobiologia*, **443**, 177–85.

McCormick, P. V. & O'Dell, M. B. (1996). Quantifying periphyton responses to phosphorus in the Florida Everglades: a synoptic-experimental approach. *Journal of the North American Benthological Society*, **15**, 450–68.

McCormick, P. V. & Stevenson, R. J. (1998). Periphyton as a tool for ecological assessment and management in the Florida Everglades. *Journal of Phycology*, **4**, 726–33.

Metzeltin, D. & Lange-Bertalot, H. (1998). Tropical diatoms of South America. I. *Iconographia Diatomologica*, **5**, 1–695.

Mitsch, W. J. & Gosselink, J. G. (2000). *Wetlands*, 3rd edition. New York, NY: John Wiley and Sons.

Myers-Smith, I. H., Harden, J. W., Wilmking, M., Fuller, C. C., McGuire, A.D. & Chapin III, F. S. (2008). Wetland succession in a permafrost collapse: interactions between fire and thermokarst. *Biogeosciences*, **5**, 1273–86.

NWWG (National Wetlands Working Group). (1988). *Wetlands of Canada*. Ecological Land Classification Series No. 24, Ottawa: Sustainable Development Branch, Environment Canada, and Montréal: Polyscience Publications, Inc.

Oechel, W. C., Hastings, S. J., Vourlitis, G., Jenkins, M., Riechers, G., & Grulke, N. (1993). Recent change of Arctic tundra ecosystems from a net carbon dioxide sink to source. *Nature*, **361**, 520–3.

O'Sullivan, P. E., Heathwaite, A. L., Appleby, P. G., *et al.* (1991). Paleolimnology of Slapton Ley, Devon, UK. *Hydrobiologia*, **214**, 115–24.

Ovenden, L. (1988). Holocene proxy-climate data from the Canadian Arctic. *Geological Survey of Canada Paper*, 88–22.

Owen, R. B., Renaut, R. W., Hover, V. C., Ashley, G. M., & Muasya, A. M. (2004). Swamps, springs and diatoms: wetlands of the semi-arid Bogoria–Baringo Rift, Kenya. *Hydrobiologia*, **581**, 59–78.

Pan, Y. D. & Stevenson, R. J. (1996). Gradient analysis of diatom assemblages in western Kentucky wetlands. *Journal of Phycology*, **32**, 222–32.

Pan, Y., Stevenson, R. J., Vaithyanathan, P., Slate J., & Richardson, C. J. (2000). Changes in algal assemblages along observed and experimental phosphorus gradients in a subtropical wetland, USA. *Freshwater Biology*, **44**, 339–53.

Parr, J. F., Taffs, K. H., & Lane, C. M. (2004). A microwave digestion technique for the extraction of fossil diatoms from coastal lake and swamp sediments. *Journal of Paleolimnology*, **31**, 383–90.

Patrick, R. (1954). The diatom flora of Bethany Bog. *Journal of Protozoology*, **1**, 34–7.

Pienitz, R. (2001). Analyse des microrestes végétaux: diatomées. In *Écologie des tourbières du Québec-Labrador*, ed. S. Payette & L. Rochefort. Québec: Les Presses de l'Université Laval, pp. 311–26.

Poulíčková, A., Hájková, P., Křenková, P., & Hájek, M. (2004). Distribution of diatoms and bryophytes on linear transects through spring fens. *Nova Hedwigia*, **78**, 411–24.

Punning, J. M. & Puusepp, L. (2007). Diatom assemblages in sediments of Lake Juusa, southern Estonia with an assessment of their habitat. *Hydrobiologia*, **586**, 27–41.

Racca, M. J., Philibert, A., Racca R., & Prairie Y. T. (2001). A comparison between diatom-based pH inference models using artificial neural networks (ANN), weighted averaging (WA) and weighted averaging partial least squares (WA-PLS) regressions. *Journal of Paleolimnology*, **26**, 411–22.

Reavie, E. D., Douglas, S. V., & Williams, N. E. (2001). Paleoecology of a groundwater outflow using siliceous microfossils. *Ecoscience*, **8**, 239–46.

Reavie, E. D., Axler, R. P., Sgro, G. V., *et al.* (2006). Diatom-based weighted-averaging transfer functions for Great Lakes coastal water quality: relationships to watershed characteristics. *Journal of Great Lakes Research*, **32**, 321–47.

Reiss, K. C. & Brown, M. T. (2007). Evaluation of Florida palustrine wetlands: application of USEPA levels 1, 2 and 3 assessment methods. *EcoHealth*, **4**, 206–18.

Rejmánková, E. (2001). Effect of experimental phosphorus enrichment on oligotrophic tropical marshes in Belize, central America. *Plant and Soil*, **236**, 33–53.

Rigg, G. B. (1940). The development of *Sphagnum* bogs in North America. *Botanical Review*, **6**, 666–93.

Roulet, N. T. (2000). Peatlands, carbon storage, greenhouse gases, and the Kyoto protocol: prospects and significance for Canada. *Wetlands*, **20**, 605–15.

Round, F. E. (1981). *The Ecology of Algae*. Cambridge: Cambridge University Press.

Rühland, K., Phadtare, N. R., Pant, R. K., Sangode, S.J., & Smol, J. P. (2006). Accelerated melting of Himalayan snow and ice triggers pronounced changes in a valley peatland from northern India. *Geophysical Research Letters*, **33**, L15709, DOI:10.1029/2006GL026704.

Rühland, K., Smol, J. P., Jasinski, J. P. P., & Warner, B. (2000). Response of diatoms and other siliceous indicators to the developmental history of a peatland in the Tiksi Forest, Siberia, Russia. *Arctic, Antarctic and Alpine Research*, **32**, 167–78.

Sawai, Y., Satake, K., Kamataki, T., *et al.* (2004). Transient uplift after a 17th-century earthquake along the Kuril subduction zone. *Science*, **306**, 1918–20.

Sears, P. B. & Couch, G. C. (1932). Microfossils in an Arkansas peat and their significance. *The Ohio Journal of Science*, **32**, 63–68.

Serreze, M. C., Walsh, J. E., Chapin, F. S., *et al.* (2000). Observational evidence of recent change in the northern high-latitude environment. *Climatic Change*, **46**, 159–207.

Slate, J. E. & Stevenson, R. J. (2000). Recent and abrupt environmental change in the Florida Everglades indicated from siliceous microfossils. *Wetlands*, **20**, 346–56.

Slate, J. E. & Stevenson, R. J. (2007). The diatom flora of phosphorus-enriched and unenriched sites in an Everglades marsh. *Diatom Research*, **22**, 355–86.

Smith L. C., MacDonald, G. M., Velichko, A. A., *et al.* (2004). Siberian peatlands a net carbon sink and global methane source since the early Holocene. *Science*, **303**, 353–6.

Smol, J. P. (1988). Paleoclimate proxy data from freshwater Arctic diatoms. *Verhundlungen der Internationalen Vereinigung von Limnologen*, **23**, 837–44.

Smol, J. P. (2008). *Pollution of Lakes and Rivers: a Paleoenvironmental Perspective*, 2nd edition, New York: Oxford University Press.

Sommerville, D. E. & Pruitt, B. A. (2006). United States wetland regulation and policy. In *Ecology of Freshwater and Estuarine Wetlands*, ed. Batzer, D. P. & R. S. Sharitz, Los Angeles, CA: University of California Press, pp. 313–47.

Sterrenburg, F. A. S., Gordon, R., Tiffany, M.-A. & Nagy, S. S. (2007). Diatoms: living in a constructal environment. In *Algae and Cyanobacteria in Extreme Environments*, ed. J. Seckbach. Dordrecht: Springer, pp. 141–72.

Stevenson, R. J. (1998). Diatom indicators of stream and wetland stressors in a risk management framework. *Environmental Monitoring and Assessment*, **51**, 107–18.

Stevenson, R. J. & Bahls, L. (1999). Periphyton protocols. In *Rapid Bioassessment Protocols for Use in Streams and Wadeable Rivers:*

Periphyton, Benthic Macroinvertebrates and Fish, 2nd edition, ed. M. Barbour, J. Gerritsen, B. Snyder, & J. Stribling, EPA 841-B-99-002, Washington, DC: United States Environmental Protection Agency, Office of Water, ch. 6.

Stevenson, R., Bothwell, M., & Lowe, R. (1996). *Algal Ecology: Freshwater Benthic Ecosystems*. Academic Press: San Francisco.

Taffs, K. H. (2001). Diatoms as indicators of wetland salinity in the upper south east of South Australia. *The Holocene*, **11**, 281–90.

Taffs, K. H., Farago, L. J., Heijnis, H., & Jacobsen, G. (2008). A diatom-based Holocene record of human impact from a coastal environment: Tuckean Swamp, eastern Australia. *Journal of Paleolimnology*, **39**, 71–82.

Tarnocai, C. (2006). The effect of climate change on carbon in Canadian peatlands. *Global and Planetary Change*, **53**, 222–32.

Thompson, T. A., Miller, C. S., Doss, P. K., Thompson, L. D. P., & Baedke, S. J. (1991). Land-based vibracoring and vibracore analysis; tips, tricks and traps. Bloomington, IN: Indiana Geological Survey, Occasional Paper 58.

Tibby, J., Gell, P. A., Fluin, J., & Sluiter, I. R. K. (2007). Diatom-salinity relationships in wetlands: assessing the influence of salinity variability on the development of inference models. *Hydrobiologia*, **591**, 207–18.

Tiner, R. W. (1984). *Wetlands of the United States: Current Status and Recent Trends*. Washington, DC: United States Department of Interior, Fish and Wildlife Service.

Tinner, W., Bigler, C., Gedye, S., *et al.* (2008). A 700-year paleoecological record of boreal ecosystem responses to climatic variation from Alaska. *Ecology*, **89**, 729–43.

Turunen, J. & Tolonen, K. (1996). Rate of carbon accumulation in boreal peatlands and climate change. In *Global Peat Resources*, ed. E. Lappalainen, Jyska: International Peat Society, pp. 21–28.

Turetsky, M. R., Manning, S. W., & Wieder, R. K. (2004). Dating recent peat deposits. *Wetlands*, **24**, 324–56.

van Dam, H., Suurmond, G., & ter Braak, C. J. F. (1981). Impact of acidification on diatoms and chemistry of Dutch moorland pools. *Hydrobiologia*, **83**, 425–59.

van Dam, H., Mertens, A., & Sinkeldam, J. (1994). A coded checklist and ecological indicator values of freshwater diatoms from the Netherlands. *Netherlands Journal of Aquatic Ecology*, **28**, 117–33.

Van de Vijver, B. & Beyens, L. (1997). The epiphytic diatom flora of mosses from Strømness Bay area, South Georgia. *Polar Biology*, **17**, 492–501.

Van de Vijver, B., & Beyens, L. (1999). Moss diatom communities from Ile de la Possession (Crozet, Subantarctica) and their relationship with moisture. *Polar Biology*, **22**, 219–31.

Van de Vijver, B., Ledeganck, P., & Beyens, L. (2001). Habitat preference in freshwater diatom communities from sub-Antarctic Îles Kerguelen. *Antarctic Science*, **13**, 28–36.

Vitt, D. H. (1990). Growth and production dynamics of boreal mosses over climatic, chemical and tropographic gradients. *Botanical Journal of the Linnean Society*, **104**, 35–59.

Vitt, D. H., Achuff, P., & Andrus, R.E. (1975). The vegetation and chemical properties of patterned fens in the Swan Hills, north central Alberta. *Canadian Journal of Botany*, **53**, 2776–95.

von Post, L. (1916). Om Skogstradspollen i sydsvendka torfmosselagerfoljder (foredragsreferat). *Geologiska Föreningen i Stockholm Förhandlingar*, **38**, 384–94.

Vymazal, J. (1995). *Algae and Elemental Cycling in Wetlands*. Boca Raton, FL: Lewis Publishers.

Walker, I. & Paterson, C. G. (1986). Associations of diatoms in the surficial sediments of lakes and peat pools in Atlantic Canada. *Hydrobiologia*, **134**, 265–72.

Warner, B. G. (1987). Abundance and diversity of testate amoebae (Rhizopoda, Testacea) in *Sphagnum* peatlands in southwestern Ontario, Canada. *Archiv für Protistenkunde*, **133**, 173–89.

Weilhoefer, C. L. & Pan, Y. (2006). Diatom-based bioassessment in wetlands: how many samples do we need to characterize the diatom assemblage in a wetland adequately? *Wetlands*, **26**, 793–802.

Weilhoefer, C. L. & Pan, Y. (2007). Relationships between diatoms and environmental variables in wetlands in the Willamette Valley, Oregon, USA. *Wetlands*, **27**, 668–82.

Weilhoefer, C. L., Pan, Y., & Eppard, S. (2008). The effects of river floodwaters on floodplain wetland water quality and diatom assemblages. *Wetlands*, **28**, 473–86.

Wein, R. W., Burzynski, M. P., Sreenivasa, B. A., & Tolonen, K. (1987). Bog profile evidence of fire and vegetation dynamics since 3000 years BP in the Acadian Forest. *Canadian Journal of Botany*, **65**, 1180–6.

Wetzel, R. G. (2006). Wetland ecosystem processes. In *Ecology of Freshwater and Estuarine Wetlands*, ed. D. P. Batzer & R. S. Sharitz, Los Angeles, CA: University of California Press, pp. 285–312.

Yang, J. R. & Duthie, H. C. (1995). Regression and weighted averaging models relating surficial sedimentary diatom assemblages to water depth in Lake Ontario. *Journal of Great Lakes Research*, **21**, 84–94.

Zedler, J. B. (2006). Wetland restoration. In *Ecology of Freshwater and Estuarine Wetlands*, ed. D. P. Batzer & R. S. Sharitz, Los Angeles, CA: University of California Press, pp. 348–406.

26

Tracking fish, seabirds, and wildlife population dynamics with diatoms and other limnological indicators

IRENE GREGORY-EAVES AND
BRONWYN E. KEATLEY

26.1 Introduction

The application of diatoms in paleoenvironmental studies has largely focused on tracking past changes in water chemistry (e.g. nutrients, salinity, pH) and habitat features (e.g. lake ice or macrophytes). When used in conjunction with other paleolimnological proxies, however, diatoms can be used to infer past changes in vertebrate populations or harvests such as fish, birds, and whales. This research is particularly insightful because the fossil record of these vertebrates is fragmented and sparsely distributed. Time series of inferred animal population dynamics also provide the much-needed long-term data required to develop sustainable management plans for these often ecologically sensitive and sometimes commercially harvested taxa (e.g. Selbie *et al.*, 2007).

Many studies that are included in this review are focused on population dynamics of migratory animals. A common thread across these studies is that large densities of migratory animals can introduce substantial nutrient loads to lakes. If the animal population is, at any time, contributing the largest source of nutrients to a study lake, then fluctuations in nutrients can be correlated to the animal's population size. Given that diatom community composition is strongly influenced by nutrient status (see Hall and Smol, this volume), the diatoms are then indirect indicators of animal population dynamics. A second field of study included in this review is focused on changes in non-anadromous fish populations. There have been numerous studies showing that fish kills, fish introductions, or human manipulations of fish community structure can influence primary producers and/or water quality.

Overall, research into interactions between wildlife and diatoms was first stimulated by limnological and contemporary phycological studies that demonstrated a correlative or experimental link. Thus, we first review these contemporary studies. We then highlight approaches used in paleolimnological studies and feature a number of case studies where these approaches have enabled researchers to develop meaningful conclusions regarding vertebrate population dynamics. We have focused our discussion on lake environments (as opposed to lotic waters) because paleolimnological research is generally limited to these ecosystems.

26.2 Contemporary studies documenting the effect of vertebrates on inland waters

One of the earliest papers that made a link between vertebrates and water quality was a study of a sockeye salmon (*Oncorhynchus nerka* Walbaum) nursery lake (Juday *et al.*, 1932). Sockeye salmon nursery lakes are impressive sites because every year large numbers of brilliantly colored mature salmon migrate from the ocean (where they put on most of their biomass) to their natal freshwater nursery lake systems to spawn, and die shortly thereafter in freshwater (Figure 26.1). When the sockeye salmon spawners are present in high enough densities, they can profoundly impact the trophic state of freshwaters. Through more recent detailed study of these systems, we now know that a 2 kg adult salmon spawner delivers ~60 g of nitrogen (N) and ~8 g of phosphorus (P) to their recipient environment, although some of this nutrient load will be exported by juvenile salmon (Moore and Schindler, 2004). Mass-balance studies

The Diatoms: Applications for the Environmental and Earth Sciences, 2nd Edition, eds. John P. Smol and Eugene F. Stoermer. Published by Cambridge University Press.

Figure 26.1 Sockeye salmon carcasses decomposing on the banks of an inflowing river to Karluk Lake, Alaska (photograph by Irene Gregory-Eaves, August 2002).

have been developed for numerous lakes and have shown that in some lakes more than 50% of the annual loading of P and N can come from sockeye salmon spawners (Krohkin, 1975; Koenings *et al.*, 1987; Naiman *et al.*, 2002).

Over the years, early natural history ideas have been tested through development of experimental studies as well as quantitative surveys over natural spatial and/or temporal gradients varying in fish or wildlife densities. Since the middle of the twentieth century, there has also been heightened interest in how changes in the trophic structure of lakes (i.e. mostly through the introduction or removal of fish) can influence algal dynamics and nutrient cycling. Natural fish kills, retrospective studies of lakes stocked with fish, and experimental manipulations of predator densities have all brought important insights into the role of how community structure can influence the basal trophic levels. Below is a review of contemporary studies, which is then followed by a discussion of the application of paleolimnological techniques to track past vertebrate population dynamics.

26.2.1 Pacific salmon

Much work on the limnology of salmon nursery lakes has been stimulated by the idea that the nutrients transferred by adult Pacific salmon from marine environments to their freshwater spawning grounds are a critical feedback that sustains the next generation of salmon. Early research had shown that larger juvenile salmon migrating to the ocean had a greater marine survival (Barnaby, 1944). Thus, Nelson and Edmundson (1955) hypothesized that lake fertilization treatments might help produce larger juveniles headed to sea and, in turn, this treatment would help rebuild salmon populations that were in decline. Between 1950 and 1953, Nelson and Edmundson (1955) applied fertilizer to Bare Lake (Alaska) and, as they predicted, these nutrient additions stimulated primary production. Since that time, whole-lake fertilization experiments have been widely applied across lakes in Alaska and British Columbia ($n > 24$) and have consistently augmented algal production (most often measured as chlorophyll *a* (chl *a*); Hyatt *et al.*, 2004). Although new research has brought into question the importance of salmon-derived nutrients in sustaining future generations of salmon (Schindler *et al.*, 2005; Uchiyama *et al.*, 2008), it is clear from the experimental as well as paleolimnological research that, where salmon-derived nutrients are a substantial proportion of the nutrient budget (e.g. Hyatt *et al.*, 2004; Schindler *et al.*, 2005; reviewed in Gregory-Eaves *et al.*, 2009) and not flushed from the system too quickly (Holtham *et al.*, 2004; Selbie *et al.*, 2009), salmon-derived nutrients are a significant predictor of algal dynamics.

Most contemporary studies that have examined the response of primary producer communities to changes in salmon-derived nutrient loading have quantified this response as changes in chl *a* concentration. Many coastal salmon nursery lakes of North America are, however, dominated by diatoms at least at some time during the growing season (Stockner and Shortreed, 1979; Kyle, 1994). Lake Kuril, a Russian salmon lake whose recent average salmon escapement densities are among the highest in the world (>20,000 spawners/km^2), also has a phytoplankton assemblage dominated by diatoms (Milovskaya *et al.*, 1998). Thus, it is clear that diatoms are an important algal group in many sockeye salmon nursery lakes.

One particular diatom genus that has been the focus of much interest in sockeye salmon nursery lakes is *Rhizosolenia* Ehrenberg, a group that is characterized by a cylindrical morphology with long spines (i.e. up to 200 μm in total length) and considered to be an inedible form. In reviewing the results from the numerous whole-lake fertilization treatments, Hyatt *et al.* (2004) found that *Rhizosolenia* can become dominant when fertilizer with a high N:P ratio is applied. When the N:P ratio of the fertilizer is reduced, the *Rhizosolenia* bloom disappears. A more recent study by McQueen *et al.* (2007) provided evidence to suggest that development of large *Rhizosolenia* blooms can reduce the efficiency with which nutrients are transferred to higher trophic levels because *Rhizosolenia* are not grazed upon by zooplankton. Therefore, if a lake enrichment program produces a *Rhizosolenia* bloom, the manipulation has the potential to be more detrimental than not treating the lake at all.

26.2.2 Birds

Birds, like salmon, can transport significant quantities of allochthonous nutrients between ecosystems and, as such, these mobile species play important roles in aquatic food-web functioning (e.g., Polis *et al.*, 1997, 2004). Seabirds, for example, feeding in the marine environment and returning to land to breed, can transport nutrients from marine to terrestrial and freshwater ecosystems. Birds can both directly and indirectly influence nutrient cycles; birds release nutrients into recipient environments in the form of guano, feathers, eggshells, and carcasses. These nutrients may then be subject to biogeochemical transformations (e.g. the volatilization of ammonia from degrading guano; e.g. Polis *et al.*, 1997; Wainright *et al.*, 1998).

In temperate and subarctic regions, ducks, geese, and gulls have been shown to deposit substantial amounts of nutrients into freshwater lakes (Manny *et al.*, 1994; Kitchell *et al.*, 1999; Payne and Moore, 2006). In one of the first empirical studies of nutrient loading by birds, Manny *et al.* (1994) calculated that migrating geese and ducks contributed 69% of the carbon (C), 27% of the N, and 90% of the P derived annually from allochthonous sources to a lake in Michigan, United States. Similarly, geese have been found to contribute up to 40% of the total nitrogen (TN) and up to 75% of the total phosphorus (TP) nutrient loading rates to small wetland ponds in New Mexico (Kitchell *et al.*, 1999).

In Arctic regions, seabirds have also been identified as important nutrient vectors among marine, terrestrial, and freshwater ecosystems (Wainright *et al.*, 1998; Evenset *et al.*, 2004, 2007; Blais *et al.*, 2005, 2007; Mallory *et al.*, 2006; Keatley *et al.*, 2009). For example, Keatley *et al.* (2009) found that bird-influenced sites were characterized by higher concentrations of TP, TN, dissolved organic carbon, and chl *a* relative to regional reference sites (Figure 26.2). Similarly, in the Antarctic, seabird-derived nutrients have also been linked to impacts on freshwaters, soils, plants, and secondary consumers (Vincent and Vincent, 1982; Lindebloom, 1984; Mizutani and Wada, 1988; Ryan and Watkins, 1989; Jones and Juggins, 1995; Erskine *et al.*, 1998; Izaguirre *et al.*, 1998; Mataloni and Tell, 2002). Despite variation in the type of birds and in the size and location of water bodies across these polar studies, it is clear that input of bird-derived nutrients resulted in elevated surface-water nutrient concentrations in what are otherwise nutrient-poor lakes and ponds.

A key focus for many studies has been whether the bird-derived nutrients become incorporated into aquatic food webs via primary producers. For example, several contemporary studies have assessed the impact of bird-derived nutrients on freshwater quality and aquatic biota (Vincent and Vincent, 1982; Hansson and Håkansson, 1992; Samuels and Mason, 1997; Izaguirre *et al.*, 1998; Wainright *et al.*, 1998; Kitchell *et al.*, 1999; Harding *et al.*, 2004; Mallory *et al.*, 2006; Lim *et al.*, 2005; Payne and Moore, 2006; Van Geest *et al.*, 2007). One of the earliest studies suggesting a link between nutrients and birds was from Samsel and Parker (1972) who found that the difference in trophic status between two lakes in Antarctica could be related to the presence of fauna both within the lake (including arctic terns, petrels, and elephant seal) and in the catchment (an Adelie penguin rookery with ~6000 penguins). Similarly, Vincent and Vincent (1982) showed that nutrient (N, P) concentrations in an Antarctic lake receiving runoff from a penguin rookery ranged between 3 and 180 times higher than a nearby lake not influenced by the penguin rookery. They further showed that seston from the penguin-influenced lake showed no response to enrichment experiments compared to their reference lake, suggesting that algae in the affected site were limited by factors other than nutrients (Vincent and Vincent, 1982). Broadly speaking, regions as diverse as Antarctica, New Mexico, and the Canadian High Arctic show largely similar patterns: lakes receiving large bird-derived subsidies are elevated in nutrients (N and P) and primary production (measured as chl *a* concentration or phytoplankton biomass) relative to lakes without bird-derived nutrient subsidies.

To date, there have been only a few studies that have documented the interactions between birds (e.g. measured as proximity to a colony or inferred from measurements of nutrients) and diatom biomass or composition. In one of the few studies from Antarctica, Kawecka and Olech (1993) documented that diatom biomass was highest in their stream sampling stations located close to a nearby penguin colony and that *Navicula atomus* (Kützing) Grunow typified the diatom assemblages of these enriched waters. Hansson and Håkansson (1992) examined diatoms across a lake-productivity gradient that was anecdotally linked to bird influence in Antarctica; they found that

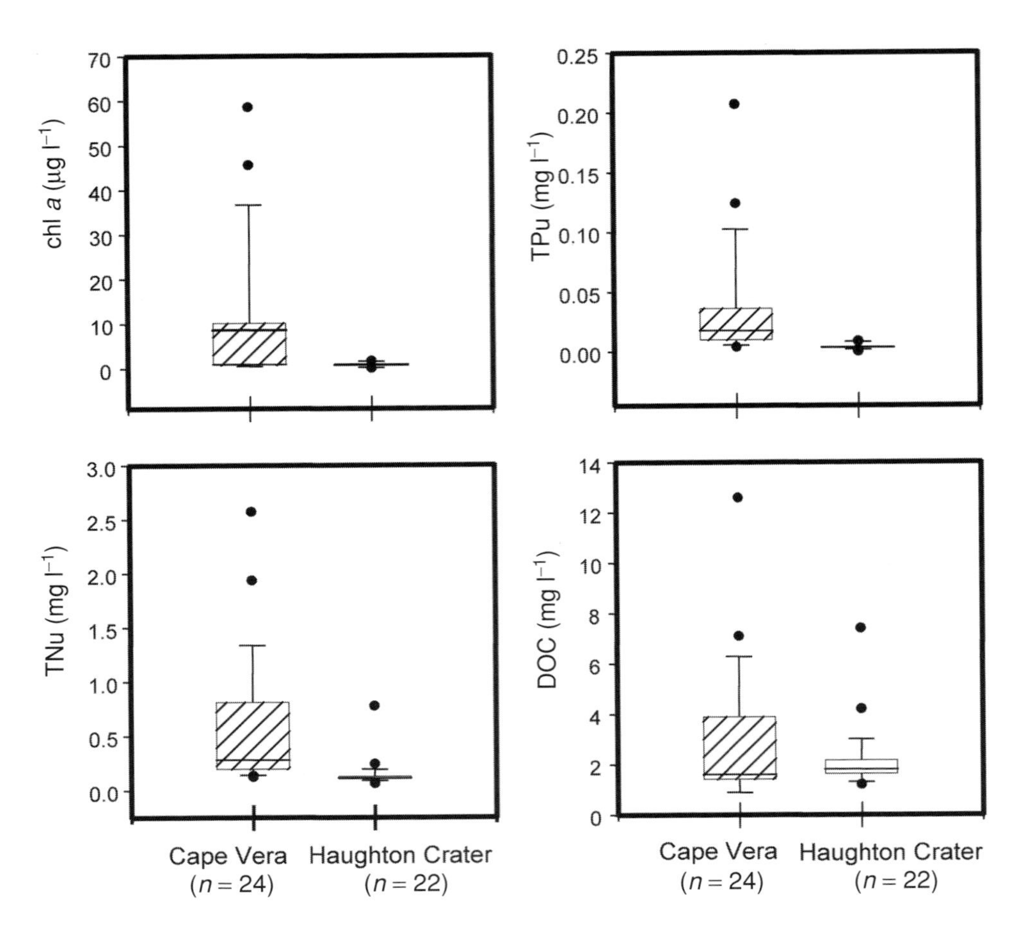

Figure 26.2 A box and whiskers comparison plot of chlorophyll *a* (chl *a*) concentrations and several other water chemistry variables from ponds around Cape Vera (Devon Island, Canadian High Arctic) and ponds sampled at a reference location on Devon Island (i.e. around the Haughton Crater). (Figure from Keatley *et al.*, 2009 (Fig. 3), reprinted with kind permission from Springer Science+Business Media). The mean value of each group is represented by a horizontal bar, the 25th and 75th percentiles are delimited by the box and the 10th and 90th percentiles are marked by the ends of the brackets. Outlying points are presented as dots. Abbreviations are as follows: TPu: total phosphorus unfiltered; TNu: total nitrogen unfiltered; DOC: dissolved organic carbon.

Caloneis cf. *bacillum*, *Navicula capitata* Ehrenberg, *Navicula salinarum* Grunow, and *Staurosira construens* var. *bidonis* (Ehrenberg) Grunow increased in abundance with increasing N and P and decreasing light. Based on a surface-sediment study from maritime lakes in Antarctica, Jones *et al.* (1993) showed that diatom distributions were related to nutrients (and salinity); high nitrogen inputs were initially thought to have originated from a nearby penguin rookery but subsequent work has concluded that most of these nutrients more likely originated from a fur seal colony (Jones and Juggins, 1995; see more details on this study below).

At the other pole, Keatley *et al.* (2009) assessed the distribution of diatoms in relation to measured environmental variables from 24 ponds that spanned a gradient of seabird influence in the Arctic. This study was set at Cape Vera, Devon Island, Canadian High Arctic, where a breeding colony of over 20,000 northern fulmars (*Fulmarus glacialis* L.) is located in coastal cliffs above a series of freshwater ponds. Due to differences in bird density, proximity to the cliffs, and drainage patterns, the 24 ponds receive varying amounts of seabird-derived nutrients (Blais *et al.*, 2005; Keatley *et al.*, 2009; Figure 26.3). In these ponds, a small but significant portion of the variance in surface-sediment diatom communities could be explained by sedimentary δ^{15}N, a geochemical tracer used to detect the strength of seabird-derived nutrient inputs (Keatley *et al.* 2009; see description of δ^{15}N below). Using a redundancy analysis (RDA), Keatley *et al.* (2009) showed that although 13.5% of the variance in diatoms could be explained by δ^{15}N, the most dominant species (e.g. *Nitzschia frustulum* (Kützing) Grunow) in these sites showed little relationship to δ^{15}N. Instead, the relationship was driven by species characteristic of the low-δ^{15}N sites (e.g. *Cymbella botellus* (Lagerstedt) Schmidt, *C. designata* Krammer, *Cymbopleura angustata* var. *spitzbergensis* Krammer, and

Figure 26.3 A view of the bird-influenced ponds taken from cliffs at Cape Vera, Devon Island (photograph courtesy of Mark Mallory, July 2005). Note the high turbidity of ponds closest to the cliffs.

Psammothidium marginulatum (Grunow) Bukhtiyarova and Round). Although $\delta^{15}N$ explained the largest percentage of variation in diatom species distributions in the Cape Vera data set, the high dominance of relatively few species precluded the generation of a robust transfer function.

26.2.3 Other migratory fauna

Evidence that other types of wildlife directly influence present-day water quality and biota in freshwaters comes largely from various types of seal colonies in Antarctica. On the Byers Peninsula, Antarctica, proximity to an elephant seal colony was associated with elevated N and P concentrations as well as algal production (measured as chl *a*) compared with other nearby oligotrophic lakes (Toro *et al.*, 2007). Heywood Lake, which is located on Signy Island, Antarctica, becomes heavily populated with fur seals (*Arctocephalus gazella Eater*); on any given summer day, hundreds of fur seals can be found in the catchment and swimming in the lake (Butler, 1999). Nutrients (measured primarily as dissolved reactive phosphorus) from the seal excrement and molted skin have been linked to phytoplankton blooms (measured as chl *a*) during the growing season (Butler, 1999). Although some diatoms were recorded at this site, most were of benthic or terrestrial origin, suggesting that other algal groups dominated the phytoplankton (Butler, 1999).

26.2.4 Non-anadromous fish

The literature is full of debate over whether non-anadromous fish have a widespread influence on algal communities (reviewed in Lampert and Sommer, 2007). In a highly cited review of contemporary studies, McQueen *et al.* (1986) provided some context for this debate and suggested that oligotrophic lakes are more sensitive to fish dynamics than eutrophic ones. Here we highlight some of the classical contemporary studies where non-anadromous fish have been shown to have a clear influence on algal communities, and below, in the paleolimnological section (Section 26.3.4) we have conducted an exhaustive review of sedimentary records that have addressed this issue.

Among other researchers, Vanni and colleagues have been very active in conducting studies to detect the presence and define the mechanisms through which non-anadromous fish can influence algal communities. One of their earlier studies that received widespread attention was based on work on Lake Mendota (Wisconsin), where they observed food-web and related algal changes in the late 1980s, when warm late-summer temperatures and hypolimnetic oxygen depletion killed most of the planktivores (i.e. primarily cisco, *Coregonus artedii* Lesueur, and yellow perch, *Perca flavescens* Mitchill; Vanni *et al.*, 1990). This fishkill had a profound effect on the zooplankton community, releasing them from intense predation pressure. Specifically, a large *Daphnia* species (*D. pulicaria* Forbes), that was capable of grazing at a much greater rate than the pre-existing community, dominated the zooplankton in the summer following the fish kill and thus greatly reduced the phytoplankton biomass. In this particular incident, no difference was detected between the lake-water nutrient concentrations during the growing season before and after the fish kill, thus the changes in phytoplankton biomass were driven by a food-web change. In a later experiment, Vanni and Layne (1997) evaluated the different mechanisms through which changes in planktivorous fish could influence algae, as the literature suggested that large fish population changes could also directly influence nutrient concentrations and ratios. From this work, Vanni and Layne (1997) concluded that total algal biomass as well as specific taxa increased in densities in the presence of increased planktivore densities. Given their control for other possible factors, this experiment allowed the authors to conclude that fish were affecting algae directly by turning over a pool of unavailable nutrients into a form that could be readily assimilated by algae at a faster rate relative to control treatments. *Asterionella formosa* Hassall was among the phytoplankton taxa that increased in biomass in treatments where only planktivore and zooplankton nutrient recycling was present. Additional work on trophic cascades and algal communities has been conducted in the littoral zones of lakes (e.g. McCollum *et al.*, 1998) and in streams (e.g. Rosemond *et al.*, 1994), but there are fewer of these studies compared to the wealth of investigations that have been conducted in the open water of lakes and ponds as well as in mesocosm experiments.

26.3 Paleolimnological studies

Paleolimnological studies using diatoms to track past vertebrate population dynamics have employed a number of different indicators including: (1) diatom species assemblage data to track past changes in nutrients (see Hall and Smol, this volume for more details) and (2) biogenic silica and/or diatom-specific pigments (i.e. diatoxanthin, fucoxanthin, and diadinoxanthin) to track past changes in diatom production trends (Conley and Schelske, 2001; Leavitt and Hodgson, 2001). All of these techniques, however, are indirectly related to past changes in wildlife population and assume that wildlife population dynamics are the main factor driving nutrient dynamics. To lend support to this assumption, most investigators have applied multiple proxy indicators and/or conducted a comparative analysis between the diatom trends and measured animal-density data (i.e. either a series of ponds along a gradient of wildlife densities or a historical time series of wildlife population dynamics).

Stable nitrogen isotopes have been widely applied in salmon and seabird studies and are measured as the ratio of ^{15}N:^{14}N present in a sample relative to the same ratio in atmospheric N (i.e. $\delta^{15}N$; Talbot, 2001). The value of $\delta^{15}N$ is a useful indicator of nitrogen inputs from salmon and seabirds because these sources have a $\delta^{15}N$ signature that differs from atmospheric and terrestrial sources of nitrogen (Kline *et al.*, 1993; Kitchell *et al.*, 1999; Blais *et al.*, 2005). For example, salmon are enriched in ^{15}N because they feed at a relatively high trophic level and the heavier isotope becomes increasingly enriched at higher trophic levels in food webs (e.g. Minagawa and Wada, 1984). Salmon also put on most of their biomass by feeding in the marine environment, an environment that tends to be enriched in ^{15}N because the main source of marine nitrogen is upwelled nitrate, which has been enriched in ^{15}N via denitrification (a process that occurs in oxygen-depleted waters; reviewed in Michener and Schell, 1994). The stable-isotope signature across different species of birds is variable and depends on diet. For example, Kitchell *et al.* (1999) found that the $\delta^{15}N$ signature of guano from snow geese that feed in corn and alfalfa fields was 2.5‰ whereas the $\delta^{15}N$ signature of guano near a northern fulmar colony (an Arctic seabird) is 20.4‰ (Blais *et al.*, 2005). Given that seabirds feed at a much higher trophic level in the marine environment, these results are not surprising.

Subfossil cladocerans have also been widely applied in studies focused on tracking anadromous or non-anadromous fish population dynamics (specific examples provided below) and provide a strong compliment to diatom analyses. In sockeye-salmon nursery-lake systems, there is a combined influence of bottom-up and top-down forces that could act on planktonic zooplankton. The adult salmon spawners can bring in a substantial load of nutrients to stimulate bottom-up production whereas the progeny of these spawners will spend several years in freshwater feeding on zooplankton (primarily *Daphnia* Müller and *Bosmina* Baird) and insect larvae and thus exert a top-down influence. With non-anadromous fish changes, most of the focus on zooplankton has been on top-down mechanisms, likely because the change in predation pressure is a direct mechanism linking the two groups.

26.3.1 Past Pacific salmon population dynamics

The longest historical time series of Pacific salmon population dynamics date back to the mid to late 1800s and are based on commercial harvests. It is clear from analyses of such harvest time-series data that Pacific salmon have undergone pronounced population changes over the past century (e.g. Beamish and Bouillon, 1993; Mantua *et al.*, 1997). Despite a concerted effort to identify the causes of these dynamics, this area of salmon research and management is still hotly debated (e.g. Mann and Plummer, 2000; Collie *et al.*, 2000). Over-fishing, habitat degradation, the release of, and competition with, hatchery fish, and changes in climate-driven ocean conditions are all factors that have been advanced. Given that most of these factors are due to human activities, which have had their greatest impact over the past century, it is very difficult to untangle the natural dynamics of salmon from those produced in response to anthropogenic stress (reviewed in Gregory-Eaves *et al.*, 2009). Applying paleolimnological techniques to this area of research, however, has been very successful, as this approach has allowed scientists to infer past changes in salmon population dynamics before and during the industrial revolution (e.g. Finney *et al.*, 2000, 2002; Gregory-Eaves *et al.*, 2003, 2004; Schindler *et al.*, 2005, 2006; Selbie *et al.*, 2007).

Diatom studies have figured prominently in the development of paleolimnological inferences of past salmon population dynamics. The importance of diatoms in this area is not because they indicate a direct connection between diatoms and salmon, but rather because they act as a second and independent proxy upon which interpretations are based. The information provided by secondary indicators in paleolimnological salmon reconstructions are critical because, although sedimentary $\delta^{15}N$ signatures provide a more direct link to salmon nutrient loading (Finney *et al.*, 2000; Selbie *et al.*, 2009), sedimentary $\delta^{15}N$ can be influenced by numerous non-salmon-related factors. Such factors include chemical transformation (e.g. denitrification and ammonia volatilization) as well as the dilution of salmon-derived nutrients through the input of terrestrially derived nitrogen or large amounts of water (i.e. fast flushing rates). Diatom species assemblages and production estimates (as well as other paleolimnological indicators such as carbon to nitrogen (C:N) ratios) can provide additional information to evaluate whether these non-salmon processes are likely to be significant. For example, diatom species assemblages are very sensitive to changes in pH (see Battarbee *et al.*, this volume), a variable that is known to alter rates of ammonia volatilization and modify $\delta^{15}N$ signatures. Specifically, large relative abundances of taxa known to be associated with alkaline conditions, but not indicative of high nutrients (e.g. *Denticula tenuis* Kützing and *Diploneis elliptica* (Kützing) Cleve; Chen *et al.* 2008) would tend to support the ammonia volatilization hypothesis rather than an inference of increased salmon abundances.

The first application where diatoms were used to track past changes in sockeye-salmon population dynamics was a multiproxy study where historical data were used to evaluate the strength of all indicators (Finney *et al.*, 2000). The primary study lake of this paper, Karluk Lake (Alaska), is a remote sockeye-salmon nursery lake whose historical escapement record (i.e. a record of the number of salmon spawners returning to the lake annually) has shown pronounced variability since the 1920s. In the early 1920s, sockeye-salmon escapement to the lake was over a million fish per year, but annual escapement then diminished to a few hundred thousand fish by the 1970s. More recently, there has been both a modest increase in sockeye salmon and a coincident lake fertilization program (with the goal of boosting lake production to produce a greater abundance of larger smolt, as discussed above). Sedimentary $\delta^{15}N$, shifts in the diatom assemblages (summarized by the first principal component axis (PCA)), and the flux of cladoceran (i.e. *Bosmina*) subfossils were all strongly correlated with the historical escapement data over the interval prior to lake fertilization. Ecologically, the changes in the diatom species assemblages were also consistent with an interpretation of changes in nutrient concentrations. The sedimentary diatom assemblage correspondingly showed marked community changes, where a known eutrophic diatom complex *Stephanodiscus minutulus/parvus* (e.g. Gregory-Eaves *et al.*, 1999) dominated when salmon spawners were very abundant, and then shifted to a diverse and largely benthic-dominated community when the number of salmon spawners were lower (Figure 26.4). There was also a notable shift in the most recent sediments to a strong increase in *Cyclotella pseudostelligera* Hustedt; a taxon that has previously been shown to be responsive to lake fertilization (reviewed in Gregory-Eaves *et al.*, 2003). By applying a TP transfer function to the fossil assemblage Gregory-Eaves *et al.* (2003) demonstrated that the inferred nutrient concentrations in Karluk Lake varied twofold over the historical record. In a neighboring lake, Frazer Lake, where sockeye salmon were unable to colonize the system naturally because of a steep waterfall at its outlet, the diatom assemblage was relatively stable and consistently contained an oligotrophic community. When sockeye salmon were introduced into Frazer Lake by the Alaska Department of Fish and Game in the 1950s, however, there was a marked increase in more nutrient-rich taxa (e.g. *Aulacoseira subarctica* (O. Müller) Haworth and *Asterionella formosa*) as well as an increase in the sedimentary $\delta^{15}N$; Gregory-Eaves *et al.*, 2003).

Following the initial Finney *et al.* (2000) study, there have been numerous records produced from other salmon nursery-lake sites along the west coast of North America, from Alaska to Idaho, and the strength of the salmon signal has been variable across these records (Gregory-Eaves *et al.*, 2004; Holtham *et al.*, 2004; Schindler *et al.*, 2005; Schindler *et al.*, 2006; Brock *et al.*, 2007; Selbie *et al.*, 2007; Hobbs and Wolfe, 2007, 2008). Schindler *et al.* (2005), for example, developed a quantitative method for inferring past salmon population dynamics. Their mixing model (similar to what has been developed for food-web studies) requires recent escapement data, data on the different $\delta^{15}N$ end-members (salmon and non-salmon sources of organic matter, the latter of which can be estimated by measuring sedimentary $\delta^{15}N$ from a control lake), and sedimentary $\delta^{15}N$ measures from the focal salmon lake. Applying the mixing model to the $\delta^{15}N$ record from Lake Nerka (Alaska), Schindler *et al.* (2005) recorded a threefold change in salmon spawner density over the past ~300 years. They also conducted fossil pigment analyses and showed that variation in salmon spawners was a significant

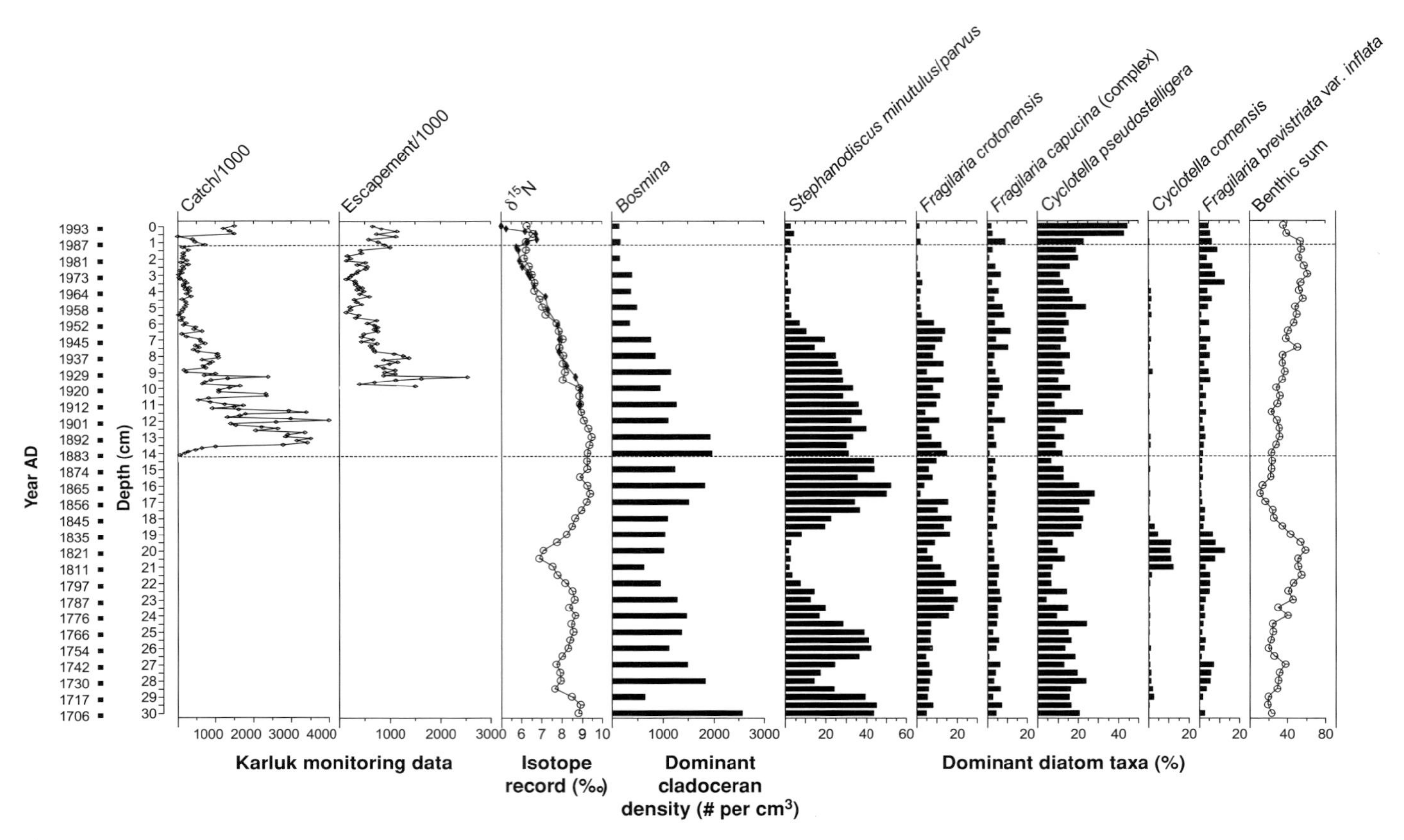

Figure 26.4 A 300-yr, historical, and multiproxy record from Karluk Lake, Alaska (from Finney *et al.*, 2000, reprinted with permission from AAAS). The commercial harvest of sockeye salmon from the estuary of the Karluk River dates back to 1882. This historical record, together with the escapement record (i.e. the number of sockeye salmon returning to spawn), provides an estimate of the system's total production in the recent past. The paleolimnological record, which will only record the nutrients that are delivered to the lake, shows a strong correspondence with the measured escapement data. Equally striking are the deviations in all indicators at AD ~1720 and ~1812. Given that no commercial harvest of sockeye salmon was occurring prior to AD 1882, these shifts provide evidence that the fish population dynamics were responding to natural events. In a larger regional study, Finney *et al.* (2000) detected shifts at these same times in several other regional salmon paleolimnological records, and in a temperature record inferred from tree-rings, and thus concluded that climatic events are important factors shaping regional sockeye-salmon population dynamics.

predictor of total algal production (measured as sedimentary beta-carotene concentrations) and diatom production (measured as sedimentary diatoxanthin concentrations). In contrast, Hobbs and Wolfe (2007, 2008), who studied several sites within the Fraser River drainage of British Columbia, found that their diatom assemblages and δ^{15}N signatures were unresponsive to salmon nutrients, despite the fact that they had explicitly selected sites where historical landslides had blocked salmon migration to their study lakes for some time. In two of their study lakes (McKinley and Fraser lakes), recent marked shifts in the diatom communities to more nutrient-rich indicators were apparent, but these did not correspond to large directional increases in salmon. Rather, the apparent diatom changes were likely driven by other factors such as logging, agriculture, and other landscape disturbances. Overall, Hobbs and Wolfe (2008) recognized that their lakes had a few features that can make it difficult to detect a strong salmon signal in lake sediments. Most notably these features are: (1) large watershed-to-surface-area ratios, which could increase the possibility of diluting a salmon signal by having an overwhelming terrestrial signal; and (2) fast flushing rates, a factor that is known to prevent the development of a strong response by phytoplankton to nutrients and for any increased production to be sedimented to the lake bottom (Holtham *et al.*, 2004; Stockner *et al.*, 2005). It is clear from the studies conducted to date that appropriate site selection is of central importance in distinguishing a salmon signal from the sediment record.

26.3.2 Past bird dynamics

Compared to salmon, data on bird densities are more rare; a reality that may be due to the fact that most birds are not

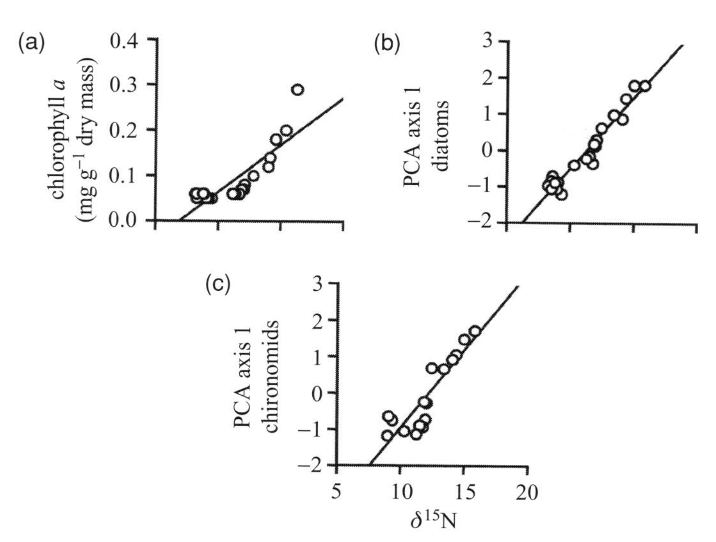

Figure 26.5 Plots showing the strong relationships between δ^{15}N (‰) and a suite of biological indicators analyzed in the sediment record from a Cape Vera (Devon Island, Canadian High Arctic) bird-influenced pond (CV9; modified from Michelutti *et al.*, 2009, Fig. 3, used with permission). The PCA (principal components analysis) axis 1 is the main direction of variation extracted from community compositional data.

commercially harvested. Given these data limitations, most paleolimnological research implemented to track past bird dynamics have been qualitative data interpretations. These interpretations are, nonetheless, very important because we know relatively little about long-term bird dynamics.

All of the published diatom-based paleolimnological research focused on tracking past seabird population dynamics has been conducted at Cape Vera in the Canadian High Arctic (Keatley, 2007; Michelutti *et al.*, 2009). As part of a larger multiproxy study, Michelutti *et al.* (2009) showed that diatoms were strongly related to δ^{15}N (a result that was also detected with numerous other proxies) in a pond strongly influenced by seabirds (Figure 26.5) and in a pond near an abandoned seabird colony. These records suggest that diatoms can reliably track seabird-derived nutrients. Further analyses by Keatley (2007) of eight sediment records from ponds at Cape Vera showed that diatoms had the greatest potential to track seabird-derived nutrients when changes in δ^{15}N within a pond sediment record were large; not surprisingly, when there was little change in δ^{15}N, there appeared to be little change in diatom species assemblages, regardless of the species comprising the assemblage. Given the sensitivity of the Arctic to climate change and the possibility that such changes will profoundly affect seabirds (Gaston *et al.*, 2005), these paleolimnological records are invaluable for understanding the response of wildlife to global change.

26.3.3 Past dynamics of large mammals: seals, whales and ungulates

There have been numerous attempts to track the population dynamics or harvesting activity (and associated nutrient leaching) of large mammals. As discussed in the previous section on contemporary studies linking wildlife to diatoms and more broadly water quality, seals are the only large mammal for which such investigations have been conducted. The paleolimnological work on seals is associated with a surface-sediment study, where Jones and Juggins (1995) applied a diatom-based transfer function to three lakes with varying degrees (low, medium, and high) of fur-seal influence. Comparisons between the diatom-inferred chl *a* data with measured water-chemistry data showed that the transfer functions faithfully reconstructed changes in primary production in both the low and medium seal-influenced lakes (Jones and Juggins, 1995). Diatom species assemblages, however, did not appear to respond to known increases in fur-seal abundance in the lake with the greatest amount of seal-derived nutrients (Heywood Lake). The lack of a diatom response in Heywood Lake clearly highlights the need for multiple indicators when using diatoms to track past changes in wildlife or even more broadly changes in whole-lake primary production. With Heywood Lake, in particular, Jones and Juggins (1995) note that light limitation, due to high turbidity from catchment erosion and/or shading from blooms of other phytoplankton groups, could have limited the ability of benthic algae (in particular the diatoms) to respond to nutrient enrichment in this highly eutrophic environment.

Whale carcasses represent another source of marine-derived nutrients that can be incorporated into freshwater ecosystems. Nutrients derived from whales and other forms of wildlife are particularly accentuated in Arctic regions where most water bodies are oligotrophic (Vincent and Hobbie, 2000; Schindler and Smol, 2006). The paleolimnological work that has been conducted to date has focused on sites where prehistoric Inuit peoples harvested whales. Beginning about 1000 years ago, Thule Inuit hunted bowhead whales for sustenance and shelter. The first study of this type was conducted by Douglas *et al.* (2004), who examined a sedimentary record to evaluate whether the Thule, who harvested, consumed, and discarded whale carcasses at their camps, had a significant impact on a nearby pond. From analyses of diatom assemblages and sedimentary δ^{15}N, together with anthropological data, Douglas *et al.* (2004) found that their indicators recorded clear limnological changes

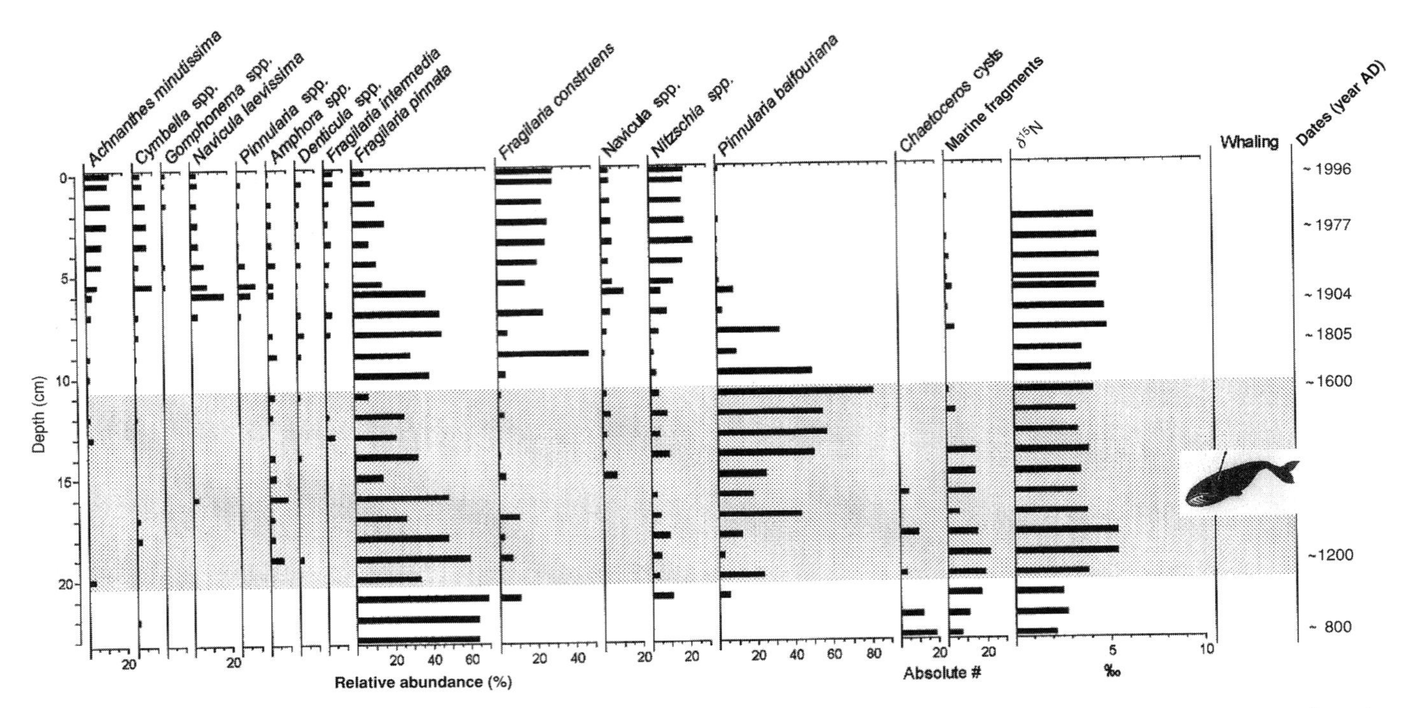

Figure 26.6 The multiproxy paleolimnological profile (including the dominant diatom taxa, the sedimentary $\delta^{15}N$, and the period of whaling (based on archeological work in the catchment; shown as a shaded band). The figure is modified from Douglas *et al.* (2004; copyright (2004) National Academy of Sciences, USA, used with permission). Marked deviations in both the sedimentary $\delta^{15}N$ and dominant diatom taxa are evident during the period of Thule occupation.

with the onset of whaling activities (Figure 26.6). Since this initial study, diatoms have been used, in conjunction with other proxies, to track Thule whaling activities from pond sediment cores from both Ellesmere Island and Bathurst Island in the Canadian High Arctic (Hadley, 2010, 2010a). These studies provide examples of how diatom-based proxies, in conjunction with other paleoindicators, can be used to track anthropogenic activities based on wildlife.

Paleolimnological methods have not been successful in tracking populations of ungulates, despite their potential to alter lake catchments. It is clear from terrestrial studies that, in some parts of North America and northern Europe, grazing by herds of ungulates (such as caribou, elk, bison, deer) can have profound effects on the vegetation (e.g. Payette *et al.* 2004), and could be large a source of erosion and/or nutrients.

To date, three studies have examined the role of different ungulates on water quality using diatom-based paleolimnological techniques. The first study was conducted by Engstrom *et al.* (1991) in Yellowstone National Park (USA), where there was some concern that increases in the number of grazing elk, bison, antelope, and deer were impairing ecosystem function; however, the lack of long-term records made it impossible to define the true impact of such animals on the ecosystem. The work by Engstrom *et al.* (1991) suggested that nearby lakes were not being affected by ecosystem alterations associated with the grazing activity of such ungulates. Based on a paleolimnological study that included analyses of diatoms and sedimentary characteristics, Engstrom *et al.* (1991) concluded that there was no change in water quality or erosion indicators. In Northern Finland, Sorvari and Korhola (1998) conducted a paleolimnological study of a lake where reindeer numbers were known to have increased recently in the vicinity. Based on their diatom analysis, however, Sorvari and Korhola (1998) suggested that the change in reindeer abundance did not influence water quality in their study lake. Similarly, in a diatom-based paleolimnological study of caribou population changes in northern Quebec, Laing *et al.* (2002) found no diatom community changes despite known increases in the herd of caribou. The lack of response in each of these studies suggest that ungulate activity (such as grazing and trampling) does not appear to impact water quality to the degree where it would be obvious in the diatom record.

26.3.4 Past dynamics of non-anadromous fish

Based on our review of the literature, we found that the strongest responses to intentional or natural changes in non-anadromous populations were evident in alpine lakes that were previously fishless (Table 26.1). This finding is consistent with the broader conclusion that introductions of non-native salmonids into lakes from western North America are one of the key stressors on remote, alpine lakes (Schindler, 2000). As such, understanding the trophic interactions that have come about as a result of these manipulations is of significant concern in these ecosystems.

Through our literature review (Table 26.1), we identified ten alpine lakes with paleolimnological records. Nine of these systems had a clear algal response to the introduction of non-native salmonids (and in the tenth study the expected pattern was noted, but not deemed significant by the original authors). Many of these sites come from a study by Drake and Naiman (2000) who examined the response of diatoms and invertebrates to fish introductions (and in some lakes, removals as well) in Mount Rainier National Park (Washington). Drake and Naiman (2000) used a dissimilarity index as their metric to evaluate whether the diatom assemblages showed a significant response to fish introductions, and most of their lakes showed a response based on this criterion. The diatom taxa that did show a change around the time of fish introductions, however, differed across lakes, suggesting that diatom responses are tempered by morphometric conditions as well as the density of fish introduced.

In contrast to the alpine-lake set, our review of fish dynamic studies from lower elevations showed that the responses of the algal community were much more variable (Table 26.1). Although it might be tempting to conclude that this is simply an affirmation of the McQueen *et al.* (1986) model, where variation in the lower elevations sites is a function of lake trophic state, we found that the wide range of responses documented in the lower elevation sites cannot easily be explained by such trophic-state differences. For example, eutrophic lakes in this data set showed both significant responses to fish manipulations (e.g. Chatwin Lake (Alberta), where average summer TP concentrations are 47 $\mu g\ l^{-1}$) and undiscernable responses (e.g. Lake Harriet (Minnesota), with average TP concentrations of 40 $\mu g\ l^{-1}$). Additional factors that might explain why temperate lakes are more variable are: (1) some of the lakes also experienced a preceding and marked change in allochthonous nutrient loading, a factor that could have overshadowed any more modest nutrient changes brought about by fish dynamics; (2) unlike the alpine lakes, many of the lower-elevation lakes had some prior fish community and thus the addition of a different type of fish or partial removal of fish would not have as profound an effect. As has been reported in the paleolimnological literature aimed at tracking past salmon population dynamics, the ability to detect a fish signal likely depends on the background nutrient loading, the lake's flushing rate, and the quantity of fish being added or removed. Overall, it seems difficult to reliably reconstruct changes in non-anadromous fish dynamics in low-elevation lakes based on sedimentary diatom analysis alone.

26.4 Future directions

Although there have been numerous exciting studies published to date, the spatial and temporal coverage of such paleolimnological work is relatively limited. For example, to our knowledge, no research of this kind has been conducted in tropical environments; and yet large populations of migratory vertebrates exist (e.g. the wildebeest and zebra migrations in the Serengeti). Even with the organisms that have been studied, increasing the spatial and temporal resolution would allow investigators to pursue numerous important questions, such as: Do fish and wildlife show a regional response to large-scale climatic patterns? How have fish and wildlife populations responded to known climatic events such as the Younger Dryas, the Mid-Holocene Warm Period and more recently, the Medieval Warm Period and the Little Ice Age? There is also research currently underway to test whether general patterns between biodiversity metrics and nutrients (e.g. Chase and Leibold, 2002) hold in systems where vertebrates are the main vectors of nutrient loading. Continued research in this area is likely to advance both our knowledge of basic ecological science and our understanding of vertebrate population dynamics, the latter of which has clear management implications.

26.5 Summary

Through this review, we have identified both the promises and the pitfalls of applying diatoms together with other indicators to track past changes in vertebrate population dynamics. There is no doubt that diatoms, when combined with analyses of other indicators such as $\delta^{15}N$ and cladocerans, can provide reliable data on the long-term dynamics of certain vertebrate populations. Given that time series are often missing or of insufficient length to answer key questions in environmental management, paleolimnological records are of great importance.

Not unlike other fields of environmental science, however, there are challenges and limitations to this technique. Firstly, contemporary and paleolimnological studies need to be conducted in parallel, so that investigators develop a strong understanding of ecosystem processes and the response of paleolimnological indicators; knowledge that is needed to draw robust inferences from sedimentary records. As highlighted

Table 26.1 Response of algal communities to fish dynamics in alpine and temperate lakes

Study	Site name	Fish introduced/ manipulation	Algal metric measured in sediment	Response to fish change	Comments
Alpine lakes					
Leavitt *et al.*, 1994	Snowflake Lake (AB)	Rainbow, Cutthroat & Brook trout	Algal (incl. diatom) pigments	X	Significant increase in all algae (beta-carotene) and diatoms (diatoxanthin)
Leavitt *et al.*, 1994	Pipit Lake (AB)	Rainbow trout	Algal (incl. diatom) pigments	X	Significant increase in all algae (beta-carotene) and diatoms (diatoxanthin)
Miskimmin *et al.*, 1995	Annette Lake (AB)	Rainbow & Brook trout, rotenone treated lake to remove native and non-native fish	Algal (incl. diatom) pigments	X	Diatom and chlorophytes replaced by chrysophytes, cryptophytes and cyanobacteria after toxaphene treatment and fish introductions
Drake and Naiman, 2000	Clover Lake (WA)	Rainbow, Cutthroat &/or Brook trout	Diatom community composition	X	Significant shift in diatom assemblage as shown by dissimilarity analyses
Drake and Naiman, 2000	Eunice Lake (WA)	Rainbow, Cutthroat &/or Brook trout	Diatom community composition	X	Significant shift in diatom assemblage as shown by dissimilarity analyses
Drake and Naiman, 2000	Tipsoo Lake (WA)	Rainbow, Cutthroat &/or Brook trout	Diatom community composition	X	Significant shift in diatom assemblage as shown by dissimilarity analyses
Drake and Naiman, 2000	Owyhigh Lake (WA)	Rainbow, Cutthroat &/or Brook trout	Diatom community composition	X	Significant shift in diatom assemblage as shown by dissimilarity analyses
Drake and Naiman, 2000	Harry Lake (WA)	Rainbow, Cutthroat &/or Brook trout	Diatom community composition	~	Shift in diatom assemblage, but change not significant
Schindler *et al.*, 2001	Bighorn Lake (AB)	Rainbow & Brook trout	Algal pigments	X	Increase in beta-carotene coincident with trout introductions
Eilers *et al.*, 2007	Diamond Lake (OR)	Rainbow trout and Tui chub	Diatom community composition	X	Diatoms shift to nutrient rich indicators following trout introduction. Tui chub may be responsible for cyanobacteria increases
Lower elevation lakes					
Brugam and Speziale 1983	Lake Harriet (MN)	Northern pike introduction		o	Previous eutrophication event large and thus fish-induced change on primary producers may have been minor in comparison
Leavitt *et al.*, 1989	Peter Lake (WI)	several fish manipulations	Algal (incl. diatom) pigments	~	Confounding influence of lake liming, but changes in pigments transient. A general pattern of piscivore presence leading to high abundances of large *Daphnia* and deep-blooming algae and bacteria

Leavitt *et al.*, 1989	Paul Lake (WI)	several fish manipulations	Algal (incl. diatom) pigments	~	Changes in pigments transient, general pattern of piscivore presence leading to high abundances of large *Daphnia* and deep-blooming algae and bacteria
Rasanen *et al.*, 1992	Lake Pyhajarvi	introduction of planktivores	Diatom community composition	o	no conclusive evidence presented that diatoms responded to fish introduction and the cladocerans showed only very minor changes
Anderson and Odgaard, 1994	Vaeng So (Denmark)	planktivorous fish biomass reduced by 50%	Diatom community composition	o	Authors suggest core might be mixed because ^{137}Cs peak not clear
Liukkonen *et al.*, 1997	Lake Vesijarvi – Laitialselka basin (Finland)	removal of coarse fish (esp. roach)	Diatom community composition	o	No obvious shift in assemblage expressed as % biomass
Liukkonen *et al.*, 1997	Lake Vesijarvi – Enonselka basin (Finland)	removal of coarse fish (esp. roach)	Diatom community composition	X	Qualitative shift in assemblage expressed as % biomass, but some of this may be explained by the diversion of waste waters
Miskimmin *et al.*, 1995	Chatwin Lake (AB)	rotenone to remove native fish and then rainbow trout introduced	Algal (incl. diatom) pigments	X	Presence or absence of trout explained 35% of variation in pigment data (significant), but this is substantially less than what was recorded in a companion high elevation lakes
St Jacques *et al.*, 2005	Lake Opeongo (ON)	stocking of cisco and lake herring	Diatom community composition	X	Significant increase in planktonic diatoms concurrent with cisco stocking
McGowan *et al.*, 2005	Lake Søbygaard (Denmark)	inferred changes in planktivorous fish density based on fossil cladoceran assemblage	Algal (incl. diatom) pigments	X	Fish and the interaction between fish + other environmental variables explained a significant fraction of the variation in the algal assemblages (as inferred from sedimentary pigment analyses)
McGowan *et al.*, 2005	Lake Lading (Denmark)	inferred changes in planktivorous fish density based on fossil cladoceran assemblage	Algal (incl. diatom) pigments	X	Fish and the interaction between fish + other environmental variables explained a significant fraction of the variation in the algal assemblages (as inferred from sedimentary pigment analyses)

X = strong response to trophic cascade
~ = weak or transient response to trophic cascade
o = no response to trophic cascade

above, experimental and/or modern spatial surveys representing a gradient of vertebrate nutrient inputs allows scientists to develop predictions which they can evaluate with the sediment record (or even test them if historical monitoring data on vertebrate dynamics exist). Careful site selection and the application of multiproxy approaches in paleolimnological studies are also of paramount importance. The confounding influence of other sources of nutrient loading, for example, can be evaluated with historical land-use/land-cover and census data where available. Multiple paleolimnological indicators are also critical because diatoms are primarily indicators of nutrients and thus represent an indirect connection to vertebrate dynamics. Stable isotopes and cladoceran analyses, however, make strong complements to the diatom analyses and thus can greatly bolster environmental and ecological interpretation. Sedimentary pigment analyses may also be incredibly insightful, particularly in sites where there is reason to believe that diatoms are only a minor fraction of the algal community (often this occurs in hypereutrophic sites; Watson *et al.*, 1997). Despite the challenges, this is an exciting area of diatom research and one that will likely see growth in the future.

Acknowledgments

This work was funded by Gordon and Betty Moore Foundation and the Natural Science and Engineering Research Council of Canada. Bronwyn Keatley was also funded by a McGill University Tomlinson post-doctoral fellowship. We would also like to acknowledge colleagues and collaborators with whom we have had stimulating discussions: Jules Blais, Guangjie Chen, Brian Cumming, Marianne Douglas, Bruce Finney, Peter Leavitt, Neal Michelutti, Kathleen Rühland, Daniel Schindler, Daniel Selbie, and John Smol.

References

Anderson, N. J. & Odgaard, B. V. (1994). Recent paleolimnology of three shallow Danish lakes. *Hydrobiologia*, **276**, 411–22.

Barnaby, J. T. (1944). Fluctuations in abundance of red salmon, *Oncorhynchus nerka* (Walbaum), of the Karluk River, Alaska. *Fishery Bulletin of the Fish and Wildlife Service*, **39**, 237–95.

Beamish, R. J. & Bouillon, D. R. (1993). Pacific salmon production trends in relation to climate. *Canadian Journal of Fisheries and Aquatic Sciences*, **50**, 1002–16.

Blais, J. M., Kimpe, L. E., McMahon, D., *et al.* (2005). Arctic seabirds transport marine-derived contaminants. *Science*, **309**, 445.

Blais, J. M., Macdonald, R. W., Mackay, D., *et al.* (2007). Biologically mediated transfer of contaminants to aquatic systems. *Environmental Science and Technology*, **41**, 1075–84.

Brock, C. S., Leavitt, P. R., Schindler, D. E., & Quay, P. D. (2007). Variable effects of marine-derived nutrients on algal production in salmon nursery lakes of Alaska during the past 300 years. *Limnology and Oceanography*, **52**, 1588–98.

Brugam, R. B. & Speziale, B. J. (1983). Human disturbance and the paleolimnological record of change in the zooplankton community of Lake Harriet, Minnesota. *Ecology*, **64**, 578–91.

Butler, H. G. (1999). Seasonal dynamics of the planktonic microbial community in a maritime Antarctic lake undergoing eutrophication. *Journal of Plankton Research*, **21**, 2393–419.

Chase, J. M. & Leibold, M. A. (2002). Spatial scale dictates the productivity-biodiversity relationship. *Nature*, **416**, 427–30.

Chen, G., Dalton, C., Leira, M., & Taylor D. (2008) Diatom-based total phosphorus (TP) and pH transfer functions for the Irish Ecoregion. *Journal of Paleolimnology*, **40**, 143–63.

Collie, J., Saila, S., Walters, C., & Carpenter, S. (2000). Of salmon and dams. *Science*, **290**, 933–34.

Conley, D. J. & Schelske, C. L. (2001). Biogenic silica. In *Tracking Environmental Change Using Lake Sediments, Volume 3: Terrestrial, Algal, and Siliceous Indicators*, ed. J. P. Smol, H. J. B. Birks, & W. M. Last, Dordrecht: Kluwer Academic Publishers, pp. 281–94.

Douglas, M. S. V., Smol, J. P., Savelle, J. M., & Blais, J. M. (2004). Prehistoric Inuit whalers affected Arctic freshwater ecosystems. *Proceedings of the National Academy of Sciences of the USA*, **101**, 1613–17.

Drake, D. C. & Naiman, R. J. (2000). An evaluation of restoration efforts in fishless lakes stocked with exotic trout. *Conservation Biology*, **14**, 1807–20.

Eilers, J. M., Loomis, D., Amand, A. S., *et al.* (2007). Biological effects of repeated fish introductions in a formerly fishless lake: Diamond Lake, Oregon, USA. *Fundamental and Applied Limnology*, **169**, 265–77.

Engstrom, D. R., Whitlock, C., Fritz, S. C., & Wright, H. E. Jr. (1991). Recent environmental changes inferred from the sediments of small lakes in Yellowstone's northern range. *Journal of Paleolimnology*, **5**, 139–74.

Erskine, P. D., Bergstrom, D. M., Schmidt, S., *et al.* (1998). Subantarctic Macquarie Island – a model ecosystem for studying animal-derived nitrogen sources using ^{15}N natural abundance. *Oecologia*, **117**, 187–93.

Evenset, A., Carroll, J., Christensen, G. N., *et al.* (2007). Seabird guano is an efficient conveyer of persistent organic pollutants (POPs) to Arctic lake ecosystems. *Environmental Science and Technology*, **41**, 1173–9.

Evenset, A., Christensen, G. N., Skotvold, T., *et al.* (2004). A comparison of organic contaminants in two High Arctic lake ecosystems, Bjørnøya (Bear Island), Norway. *The Science of the Total Environment*, **318**, 125–41.

Finney, B. P., Gregory-Eaves, I., Douglas, M. S. V., & Smol, J. P. (2002). Fisheries productivity in the northeast Pacific over the past 2,200 years. *Nature*, **416**, 729–33.

Finney, B. P., Gregory-Eaves, I., Sweetman, J., Douglas, M. S. V., & Smol, J. P. (2000). Impacts of climatic change and fishing on Pacific salmon abundance over the past three hundred years. *Science*, **290**, 795–9.

Gaston, A. J., Gilchrist, H. G., & Mallory, M. L. (2005). Variation in ice conditions has strong effects on the breeding of marine birds at Prince Leopold Island, Nunavut. *Ecography*, **28**, 331–44.

Gregory-Eaves, I., Finney, B. P., Douglas, M. S. V., & Smol, J. P. (2004). Inferring sockeye salmon (*Oncorhynchus nerka*) population dynamics and water-quality changes in a stained nursery lake over the past ~500 years. *Canadian Journal of Fisheries and Aquatic Sciences*, **61**, 1235–46.

Gregory-Eaves, I., Selbie, D., Sweetman, J. N., Finney, B.P., & Smol, J. P. (2009). Tracking sockeye salmon population dynamics from lake sediment cores: a review and synthesis. *American Fisheries Society Symposium*, **69**, 379–93

Gregory-Eaves, I., Smol, J. P., Douglas, M. S. V., & Finney, B. P. (2003). Diatoms and sockeye salmon (*Oncorhynchus nerka*) population dynamics: reconstructions of salmon-derived nutrients in two lakes from Kodiak Island, Alaska. *Journal of Paleolimnology*, **30**, 35–53.

Gregory-Eaves, I., Smol, J. P., Finney, B. P., & Edwards, M. E. (1999). Diatom-based transfer functions for inferring past changes in climatic and environmental changes in Alaska, U.S.A. *Arctic, Antarctic and Alpine Research*, **31**, 353–65.

Hadley, K. R., Douglas, M. S. V., Blais, J. M., & Smol, J. P. (2010). High Arctic nutrient enrichment associated with Thule Inuit whalers: A paleolimnological investigation from Ellesmere Island (Nunvaut, Canada). *Hydrobiologia*, **649**, 129–38.

Hadley, K. R., Douglas, M. S. V., McGhee, R. H., Blais, J. M., & Smol, J. P. (2010a). Ecological influences of Thule Inuit whalers on High Arctic pond ecosystems: a comparative paleolimnological study from Bathurst Island (Nunavut, Canada). *Journal of Paleolimnology*, **44**, 85–93.

Hansson, L. & Håkansson, H. (1992). Diatom community response along a productivity gradient of shallow Antarctic lakes. *Polar Biology*, **12**, 463–8.

Harding, J. S., Hawke, D. J., Holdaway, R. N., & Winterbourn, M. J. (2004). Incorporation of marine-derived nutrients from petrel breeding colonies into stream food webs. *Freshwater Biology*, **49**, 576–86.

Hobbs, W. O. & Wolfe, A. P. (2007). Caveats on the use of paleolimnology to infer Pacific salmon returns. *Limnology and Oceanography*, **52**, 2053–61.

Hobbs, W. O. & Wolfe, A. P. (2008). Recent paleolimnology of three lakes in the Fraser River Basin (BC, Canada): no response to the collapse of sockeye salmon stocks following the Hells Gate landslides. *Journal of Paleolimnology*, **40**, 295–308.

Holtham, A. J., Gregory-Eaves, I., Pellatt, M., *et al.* (2004). The influence of flushing rates, terrestrial input and low salmon escapement densities on paleolimnological reconstructions of sockeye salmon (*Oncorhynchus nerka*) nutrient dynamics in Alaska and British Columbia. *Journal of Paleolimnology*, **32**, 255–71.

Hyatt, K. D., McQueen, D. J., Shortreed, K. S., & Rankin, D. P. (2004). Sockeye salmon (*Oncorhynchus nerka*) nursery lake fertilization: review and summary of results. *Environmental Reviews*, **12**, 133–62.

Izaguirre, I., Vinocur, A., Mataloni, G., & Pose, M. (1998). Phytoplankton communities in relation to trophic status in lakes from Hope Bay (Antarctic Peninsula). *Hydrobiologia*, **370**, 73–87.

Jones, V. J., Juggins, S., & Ellis-Evans, J. C. (1993). The relationship between water chemistry and surface sediment diatom assemblages in maritime Antarctic lakes. *Antarctic Science*, **5**, 339–48.

Jones, V. J. & Juggins, S. (1995). The construction of a diatom-based chlorophyll a transfer function and its application at three lakes on Signy Island (maritime Antarctic) subject to differing degrees of nutrient enrichment. *Freshwater Biology*, **34**, 433–45.

Juday, C., Rich, W. H., & Mann, A. (1932). Limnological studies of Karluk Lake, Alaska, 1926–1930. *Bulletin of the Bureau of Fisheries*, **47**, 407–34.

Kawecka, B. & Olech, M. (1993). Diatom communities in the Vanishing and Ornithologist Creek, King George Island, South Shetlands, Antarctica. *Hydrobiologia*, **269–270**, 327–33.

Keatley, B. E. (2007). Limnological and paleolimnological investigations of environmental change in three distinct ecosystem types, Canadian High Arctic. Unpublished Ph.D. thesis, Queen's University, Kingston, Ontario at Kingston.

Keatley, B. E., Douglas, M. S. V., Blais, J. M., Mallory, M., & Smol, J. P. (2009). Impacts of seabird-derived nutrients on water quality and diatom assemblages from Cape Vera, Devon Island, Canadian High Arctic. *Hydrobiologia*, **621**, 191–205.

Kitchell, J. F., Schindler, D. E., Herwig, B. R., *et al.* (1999). Nutrient cycling at the landscape scale: the role of diel foraging migrations by geese at the Bosque del Apache National Wildlife Refuge, New Mexico. *Limnology and Oceanography*, **44**, 828–36.

Kline, T. C., Goering, J. J., Mathiesen, O. A., *et al.* (1993). Recycling of elements transported upstream by runs of Pacific salmon. II. δ^{15}N and δ^{13}C evidence in the Kvichak River watershed, Bristol Bay, Southwestern Alaska. *Canadian Journal of Fisheries and Aquatic Sciences*, **50**, 2350–65.

Koenings, J. P. & Burkett, R. D. (1987). An aquatic rubic's cube: restoration of the Karluk Lake sockeye salmon (*Oncorhynchus nerka*). In *Sockeye Salmon (Oncorhyncus nerka) Population Biology and Future Management, Canadian Special Publication in Fisheries and Aquatic Sciences*, ed. H. D. Smith, L. Margolis, & C. C. Wood, Ottawa, ON: National Research Council of Canada, vol. 96, pp. 419–34.

Krohkin, E. M. (1975). Transport of nutrients by salmon migrating from the sea into lakes. In *Coupling of Land and Water Systems*, ed. A. D. Hasler, New York, NY: Springer-Verlag, pp. 153–6.

Kyle, G. B. (1994). Trophic level responses and sockeye salmon production trends to nutrient treatment of three coastal Alaskan lakes. *Alaska Fishery Research Bulletin*, **1**, 153–67.

Laing, T. E., Pienitz, R., & Payette, S. (2002). Evaluation of limnological responses to recent environmental change in caribou activity in the Rivière George region northern Québec, Canada. *Arctic Antarctic and Alpine Research*, **34**, 454–64.

Lampert, W. & Sommer, U. (2007). *Limnoecology: The Ecology of Lakes and Streams*, 2nd edition, Oxford: Oxford University Press.

Leavitt, E. R., Carpenter, S. R., & Kitchell, J. E. (1989). Whole-lake experiments: The annual record of fossil pigments and zooplankton. *Limnology and Oceanography*, **34**, 700–17.

Leavitt, P. R. & Hodgson, D. A. (2001). Sedimentary pigments. In *Tracking Environmental Change Using Lake Sediments, Volume 3: Terrestrial, Algal, and Siliceous Indicators*, ed. J. P. Smol, H. J. B. Birks, & W. M. Last, Dordrecht: Kluwer Academic Publishers, pp. 295–326.

Leavitt, P. R., Schindler, D. E., Paul, A. J., Hardie, A. K., & Schindler, D. W. (1994). Fossil pigment records of phytoplankton in trout-stocked alpine lakes. *Canadian Journal of Fisheries and Aquatic Sciences*, **51**, 2411–23.

Lim, D. S. S., Douglas, M. S. V., & Smol, J. P. (2005). Limnology of 46 lakes and ponds on Banks Island, NWT, Canadian Arctic Archipelago. *Hydrobiologia*, **545**, 11–32.

Lindebloom, H. J. (1984). The nitrogen pathway in a penguin rookery. *Ecology*, **65**, 269–77.

Liukkonen, M., Kairesalo, T., & Haworth, E. Y. (1997). Changes in the diatom community, including the *appearance of Actinocyclus normanii* fo. *subsalsa*, during the biomanipulation of Lake Vesijarvi, Finland. *European Journal of Phycology*, **32**, 353–61.

Mallory, M. L., Fontaine, A. J., Smith, P. A., Wiebe Robertson, M. O., & Gilchrist, H. G. (2006). Water chemistry of ponds on Southampton Island, Nunavut, Canada: effects of habitat and ornithogenic inputs. *Archiv für Hydrobiologie*, **166**, 411–32.

Mann, C. C. & Plummer, M. L. (2000). Can science rescue salmon? *Science*, **289**, 716–19.

Manny, B. A., Johnson, W. C., & Wetzel, R. G. (1994). Nutrient additions by waterfowl to lakes and reservoirs – predicting their effects on productivity and water quality. *Hydrobiologia*, **280**, 121–32.

Mantua, N. J., Hare, S. R., Zhang, Y., Wallace, J. M., & Francis, R. C. (1997). A Pacific interdecadal climate oscillation with impacts on salmon production. *Bulletin of the American Meteorological Society*, **78**, 1069–79.

Mataloni, G. & Tell, G. (2002). Microalgal communities from ornithogenic soils at Civera Point, Antarctic Peninsula. *Polar Biology*, **25**, 488–91.

McCollum, E. W., Crowder, L. B., & McCollum, S. A. (1998). Complex interactions of fish, snails, and littoral zone periphyton. *Ecology*, **79**, 1980–94.

McGowan, S., Leavitt, P. R., Hall, R. I., *et al.* (2005). Controls of algal abundance and community composition during ecosystem state change. *Ecology*, **86**, 2200–11.

McQueen, D. J., Post, J. R., & Mills, E. L. (1986). Trophic relationships in fresh-water pelagic ecosystems. *Canadian Journal of Fisheries and Aquatic Sciences*, **43**, 1571–81.

McQueen, D. J., Hyatt, K. D., Rankin, D. P., & Ramcharan, C. J. (2007). Changes in algal species composition affect juvenile sockeye salmon production at Woss Lake, British Columbia: a lake fertilization and food web analysis. *North American Journal of Fisheries Management*, **27**, 369–86.

Michelutti, N., Keatley, B. E., Brimble, S., *et al.* (2009). Seabird-driven shifts in Arctic pond ecosystems. *Proceedings of the Royal Society*, **B 276**, 591–6.

Michener, R. H. & Schell, D. M. (1994). Stable isotope ratios as tracers in marine aquatic food webs. In *Stable Isotopes in Ecology and Environmental Sciences*, ed. K. Latjha & R. H. Michener, Oxford: Blackwell Scientific Publications, pp. 138–157.

Milovskaya, L. V., Selifonov, M. M., & Sinyakov, S. A. (1998). Ecological functioning of Lake Kuril relative to sockeye salmon production. *North Pacific Anadromous Fisheries Commission Bulletin* **1**, 434–42.

Minagawa, M. & Wada, E. (1984). Stepwise enrichment of ^{15}N along food chains: further evidence and the relation between δ^{15}N and animal age. *Geochimica et Cosmochimica Acta*, **48**, 1135–40.

Miskimmin, B. M., Leavitt, P. R., & Schindler, D. W. (1995). Fossil record of cladoceran and algal responses to fishery management-practices. *Freshwater Biology*, **34**, 177–90.

Mizutani, H. & Wada, E. (1988). Nitrogen and carbon isotope ratios in seabird rookeries and their ecological implications. *Ecology*, **69**, 340–9.

Moore, J. W. & Schindler, D. E. (2004). Nutrient export from freshwater ecosystems by anadromous sockeye salmon (*Oncorhynchus nerka*). *Canadian Journal of Fisheries and Aquatic Sciences*, **61**, 1582–9.

Naiman R. J., Bilby, R. E., Schindler, D. E., & Helfield, J. M. (2002). Pacific salmon, nutrients, and the dynamics of freshwater and riparian ecosystems. *Ecosystems*, **5**, 399–417.

Nelson, P. R. & Edmundson, W. T. (1955). Limnological effects of fertilizing Bare Lake, Alaska. *Fisheries Bulletin of the Fish Wildlife Service*, **56**, 413–36

Payette, S., Boudreau, S., Morneau, C., & Pitre, N. (2004). Long-term interactions between migratory caribou, wildfires and Nunavik hunters inferred from tree rings. *Ambio*, **33**, 482–6

Payne, L. X. & Moore, J. W. (2006). Mobile scavengers create hotspots of freshwater productivity. *Oikos*, **115**, 69–80.

Polis, G. A., Anderson, W. B., & Holt, R. D. (1997). Toward an integration of landscape and food web ecology: the dynamics of spatially subsidized food webs. *Annual Reviews of Ecological Systems*, **28**, 289–316.

Polis, G. A., Power, M. E., & Huxel, G. R. (eds). (2004). *Food Webs at the Landscape Level*. Chicago, IL: University of Chicago Press.

Rasanen, M., Salonen, V. P., Salo, J., Walls, M., & Sarvala, J. (1992). Recent history of sedimentation and biotic communities in Lake Pyhajarvi, SW Finland. *Journal of Paleolimnology*, **7**, 107–26.

Rosemond, A. D. (1994). Multiple factors limiting seasonal variation in periphyton in a forest stream. *Journal of the North American Benthological Society*, **13**, 333–44.

Ryan, P. G. & Watkins, B. P. (1989). The influence of physical factors and ornithogenic products on plant and arthropod abundance at an inland nunatak group in Antarctica. *Polar Biology*, **10**, 151–60.

Samsel, G. L. & Parker, B. C. (1972). Limnological investigations in the area of Anvers Island, Antarctica. *Hydrobiologia*, **40**, 505–11.

Samuels, A. J. & Mason, C. F. (1997). Ecology of eutrophic waterbodies in a coastal grazing marsh. *Hydrobiologia*, **346**, 203–14.

Schindler, D. E., Knapp, R. A., & Leavitt, P. R. (2001). Alteration of nutrient cycles and algal production resulting from fish introductions into mountain lakes. *Ecosystems*, **4**, 308–21.

Schindler, D. E., Leavitt, P. R., Brock, C. S., Johnson, S. P., & Quay, P. D. (2005). Marine-derived nutrients, commercial fisheries, and production of salmon and lake algae in Alaska. *Ecology*, **86**, 3225–31.

Schindler, D. E., Leavitt, P. R., Johnson, S. P., & Brock, C. S. (2006). A 500-year context for the recent surge in sockeye salmon (*Oncorhynchus nerka*) abundance in the Alagnak River, Alaska. *Canadian Journal of Fisheries and Aquatic Sciences*, **63**, 1439–44.

Schindler, D. W. (2000). Aquatic problems caused by human activities in Banff National Park, Alberta, Canada. *Ambio*, **29**, 401–407.

Schindler, D. W. & Smol, J. P. (2006). Cumulative effects of climate warming and other human activities on freshwaters of Arctic and subarctic North America. *Ambio*, **35**, 160–8.

Selbie, D. T., Finney, B. P., Barto, D., *et al.* (2009). Ecological, landscape, and climatic regulation of sediment geochemistry in North American sockeye salmon nursery lakes: insights for paleoecological salmon investigations. *Limnology and Oceanography*, **54**, 1733–45.

Selbie, D. T., Lewis, B., Smol, J. P., & Finney, B. P. (2007). Long-term population dynamics of endangered Snake River sockeye salmon: evidence of past influences on stock decline and impediments to recovery. *Transactions of the American Fisheries Society*, **136**, 800–21.

Sorvari, S. & Korhola, A. (1998). Recent diatom assemblage changes in subarctic Lake Sannajärvi, NW Finnish Lapland, and their paleoenvironmental implications. *Journal of Paleolimnology*, **20**, 205–15.

St Jacques, J. M., Douglas, M. S. V., Price, N., Drakulic, N., & Gubala, C. P. (2005). The effect of fish introductions on the diatom and cladoceran communities of Lake Opeongo, Ontario, Canada. *Hydrobiologia*, **549**, 99–113.

Stockner J., Langston, A., Sebastian, D., & Wilson, G. (2005). The limnology of Williston Reservoir: British Columbia's largest lacustrine ecosystem. *Water Quality Research Journal of Canada*, **40**, 28–50.

Stockner, J. G. & Shortreed, K. R. S. (1979). Limnological studies of 13 sockeye salmon (*Oncorhynchus nerka*) nursery lakes in British Columbia. *Fisheries and Marine Service Technical Report* **865**.

Talbot, M. R. (2001). Nitrogen isotopes in palaeolimnology. In *Tracking Environmental Change Using Lake Sediments, Volume 2: Physical and Geochemical Methods*, ed. W. M. Last & J. P. Smol, Dordrecht: Kluwer Academic Publishers, pp. 401–40.

Toro, M., Camacho, A., Rochera, C., *et al.* (2007). Limnological characteristics of the freshwater ecosystems of Byers Peninsula, Livingston Island, in maritime Antarctica. *Polar Biology*, **30**, 635–49.

Uchiyama, T., Finney, B. P., & Adkison, M. D. (2008). Effects of marine-derived nutrients on population dynamics of sockeye salmon (Oncorhynchus nerka). *Canadian Journal of Fisheries and Aquatic Sciences*, **65**, 1635–48.

Van Geest, G. J., Hessen, D. O., Spierenburg, P., *et al.* (2007). Goose-mediated nutrient enrichment and planktonic grazer control in Arctic freshwater ponds. *Oecologia*, **153**, 653–62.

Vanni, M. J. & Layne, C. D. (1997). Nutrient recycling and herbivory as mechanisms in the "top-down" effect of fish on algae in lakes. *Ecology*, **78**, 21–40.

Vanni, M. J., Luecke, C., Kitchell, J. F., *et al.* (1990). Effects on lower trophic levels of massive fish mortality. *Nature*, **344**, 333–5.

Vincent, W. F. & Hobbie, J. E. (2000). Ecology of Arctic lakes and rivers. In *The Arctic: Environment, People, Policy*, ed. M. Nuttall & T. V. Callaghan, Amsterdam: Harwood Academic Publishers, pp. 197–232.

Vincent, W. F. & Vincent, C. L. (1982). Nutritional state of the plankton in Antarctic coastal lakes and the inshore Ross Sea. *Polar Biology*, **1**, 159–65.

Wainright, S. C., Haney, J. C., Kerr, C., Golovkin, A. N., & Flint, M. V. (1998). Utilization of nitrogen derived from seabird guano by terrestrial and marine plants at St. Paul, Pribilof Islands, Bering Sea, Alaska. *Marine Biology*, **131**, 63–71.

Watson, S. B., McCauley, E., & Downing, J. A. (1997). Patterns in phytoplankton taxonomic composition across temperate lakes of differing nutrient status. *Limnology and Oceanography*, **42**, 487–95.

27

Diatoms and archeology

STEVE JUGGINS AND NIGEL G. CAMERON

27.1 Introduction

One of the goals of modern archeology is to understand how past communities interacted spatially, economically, socially, and culturally with their biophysical environment (Butzer, 1982). To this end archeologists have developed strong links with zoologists, botanists, and geologists to provide information on the environment of past societies and to help understand the complex relationships between culture and environment. This chapter reviews the role of diatom analysis in such studies, and illustrates how the technique can be applied at a range of spatial and temporal scales to place archeological material in its broader site, landscape, and cultural contexts. In particular, we examine applications to establish the provenance of archeological artefacts, the analysis of archeological sediments and processes of site formation, the reconstruction of local site environments, and the identification of regional environmental processes affecting site location and the function of site networks. We have chosen a small number of examples that best illustrate these applications; other case studies directly motivated by archeological problems may be found in reviews by Battarbee (1988), Mannion (1987), Miller & Florin (1989), and Cameron (2007), while diatom-based studies of past changes in sea level, climate, land use, and water quality that are also relevant to archeological investigation are reviewed elsewhere in this volume (e.g. Cooper *et al.*, this volume; Horton & Sawai, this volume; Fritz *et al.*, this volume; Hall & Smol, this volume).

27.2 Analysis of archeological artifacts and building materials

The direct application of diatom analysis to archeological artifacts is best represented in the field of pottery sourcing and typology. Diatoms often survive the low temperature firing process used in the manufacture of prehistoric pottery; diatom analysis therefore offers a novel and potentially powerful method for identifying clay sources and to establish the provenance of finished pottery (Jansma, 1977; Alhonen *et al.*, 1980; Jansma, 1981, 1982, 1984, 1990). Unfortunately the method is not without problems. Diatom concentrations in pottery can be very low and valves are often poorly preserved (Håkansson & Hulthén, 1986, 1988). In addition, diatoms may be derived from tempering material or other sources, rather than the clay itself (Alhonen & Matiskainen, 1980; Gibson, 1986). Perhaps the main limitation is that a considerable knowledge of the distribution and diatom content of possible source clays is required to link a pot sherd unambiguously to a clay deposit. However, when this information exists, diatom analysis of pottery sherds can aid archeological interpretation in questions of typology, technology, and transport concerning, for example, classification of pottery types (Alhonen *et al.*, 1980), preferential choice of pottery fabric (Alhonen & Matiskainen, 1980; Jansma, 1981; Matiskainen & Alhonen, 1984; Gibson, 1986), autochthonous deposition of archeological artifacts (VanLandingham, 2006), movement and trade of raw materials or manufactured goods (Jansma, 1977, 1990; Alhonen & Väkeväinen, 1981), and evidence for contact between communities and their range of movement (Alhonen & Väkeväinen, 1981).

In the absence of local sources of marine clay, pottery from inland sites containing marine diatoms can provide convincing evidence for transport and trade. The converse is also true, and analysis of pottery sherds, combined with knowledge of regional stratigraphy, can be used to demonstrate that clay or manufactured pottery was imported to island or coastal sites from the mainland or hinterland sources. For example, pottery from a number of sub-Neolithic sites on the Åland Islands in the Baltic Sea were found to contain freshwater diatoms typical of the Ancylus Lake, a freshwater lake that occupied the Baltic basin prior to the Litorina Transgression (Alhonen

The Diatoms: Applications for the Environmental and Earth Sciences, 2nd Edition, eds. John P. Smol and Eugene F. Stoermer. Published by Cambridge University Press.

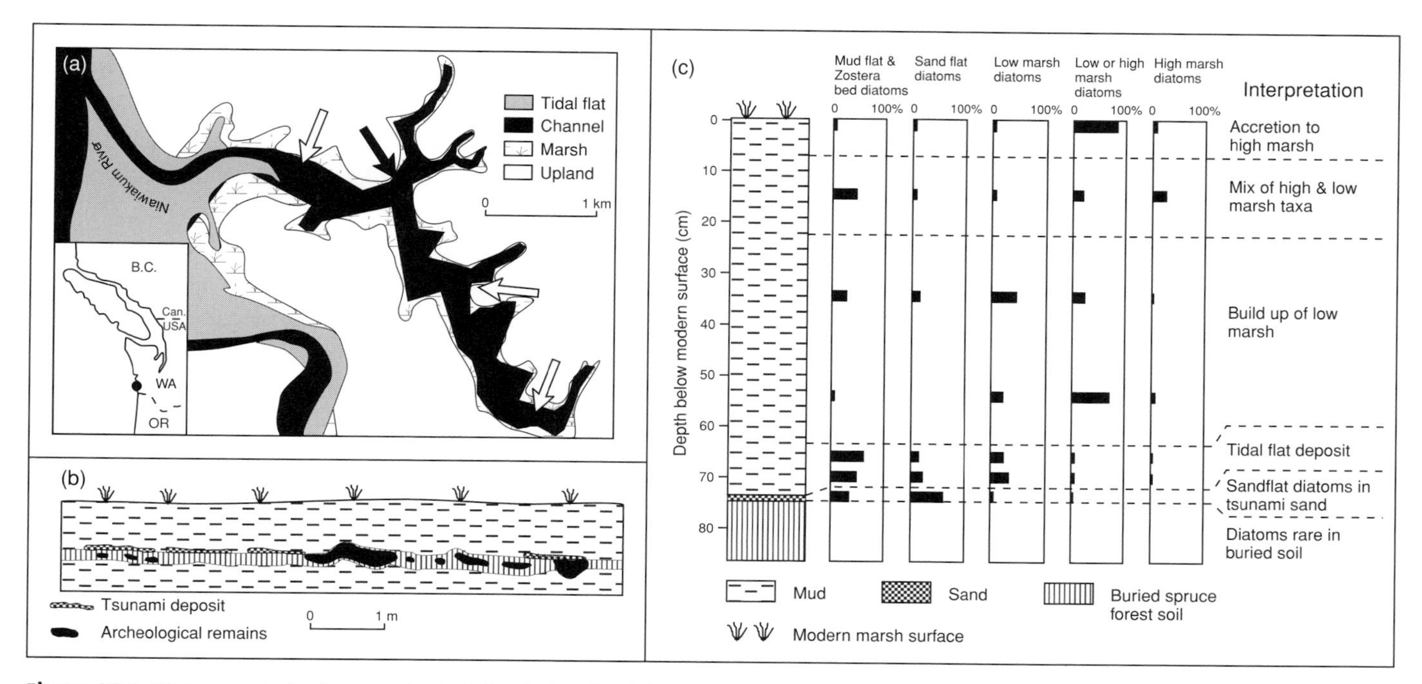

Figure 27.1 Diatom analysis of archaeological sites in the Niawiakum River Valley, southern Washington coast, USA. (a) Site location showing the Niawiakum archeological site (solid arrow) and other sampled sedimentary sequences (hollow arrows); (b) cross-section through Niawiakum archeological site; and (c) diatom stratigraphy in vicinity of the Niawiakum archeological site. (Redrawn from Cole *et al.* (1996) and Hemphill-Haley (1995).)

& Väkeväinen, 1981; Matiskainen & Alhonen, 1984). Since Ancylus clays are absent from the Åland Islands, diatom data provides clear evidence that either the finished pots or raw clays were imported from the mainland. In addition, the presence of vessels of similar style and diatom content at sites around Turku in southern Finland suggests a possible source area on the Finnish mainland.

In addition to analysis of archeological artifacts diatom analysis may also be used to identify the source of ancient building materials. For example, Flower (2006) examined the diatom content of mud bricks used in construction of the ancient Egyptian temple at Dimai in the Nile valley, dated to approximately 5000 years BP. Comparison with local sediments revealed that the bricks had almost certainly been made of a now terrestrial diatomite which had accumulated in a nearby large shallow lake.

27.3 Analysis of archeological sediments

Archeologists derive most of their basic information from excavation, and knowledge of the origin, stratigraphy, and depositional environment of archeological sediments is fundamental to all subsequent analyses. Primary or *in situ* water-lain or terrestrial sediments generally possess a diatom flora characteristic of the conditions prevailing at the time of deposition, and should be readily distinguished from that of secondary, or reworked deposits. Diatom analysis can therefore be used to identify sediment provenance and on-site depositional environments (Miller *et al.*, 1979).

When direct deposition of marine sediments or erosion from older deposits can be discounted, the presence of marine taxa may provide supporting evidence for various economic activities, such as saltworking (Juggins, 1992a), collection and use of shellfish (Denys, 1992), or algal mats (Foged, 1985). At high percentages, marine and brackish diatoms provide unambiguous evidence of flood deposits on coastal sites; fluctuations in the proportions of allochthonous marine taxa can be used to infer flooding intensity, which in turn may be related to periods of site abandonment (e.g. Groenman-van Waateringe & Jansma, 1968; Voorrips & Jansma, 1974; Jansma, 1981). This is well illustrated by the use of diatom analysis to identify coastal inundation at a series of prehistoric sites along the southern Washington (USA) coast (Cole *et al.*, 1996). Occupation horizons dating from AD 1000 to 1700 were found preserved in a buried spruce-forest soil and overlain by a thin sand horizon and up to 1 m of fine mud (Figure 27.1). The presence of marine and brackish sandflat taxa in the sand layer was taken to indicate deposition by a tsunami (Hemphill-Haley, 1996), while diatoms in the overlying sediments suggest accumulation in intertidal mudflat and salt-marsh environments following submergence of the site (Hemphill-Haley, 1995). The evidence of rapid, earthquake-induced burial and tsunami inundation approximately 300 years ago suggests that many other sites lie buried in the present intertidal zone, and helps explain the current paucity of archeological remains in an area which supported a large native population in early historic times.

Diatom analysis of lake-sediment records provides a powerful tool for examining the impact of early human societies on the environment, especially on water quality (e.g. Fritz, 1989; Whitmore *et al.*, 1996; Renberg *et al.*, 2001; Douglas *et al.*, 2004). Analysis of on-site archeological sediments can also reveal the impact of site occupants on the local environment, in particular, pollution of local water bodies adjacent to, or within, archeological sites (e.g. de Wolf & Cleveringa, 1996; Beneš *et al.*, 2002; Ekdahl *et al.*, 2004; Keevill, 2004; Ekdahl *et al.*, 2007).

Diatom analysis of archeological sediments can also help interpret the function of archeological structures. For example, Godbold & Turner (1993) used the presences of marine diatoms associated with structures in the Severn Estuary, UK, to infer their use as fish traps, and the presence of freshwater taxa allowed Neely *et al.* (1995) to identify a pit-like structure in the Tehuacan Valley of Puebla, Mexico, as a freshwater well.

27.4 Site-based paleoenvironmental reconstructions

Knowledge of the local environment of a site is essential if its socio-economic function is to be fully understood. Not surprisingly, diatom analysis has frequently been used to reconstruct paleoenvironments in the vicinity of coastal (e.g. Körber-Grohne, 1967; Groenman-van Waateringe & Jansma, 1969; Jansma, 1981; Miller & Robertsson, 1981; Jansma, 1982; König, 1983; Håkansson, 1988; Jansma, 1990; Nunez & Paabo, 1990) and inland archeological sites (Jessen, 1936; Wüthrich, 1971; Caran *et al.*, 1996; Robbins *et al.*, 1996; Castledine *et al.*, 1988; Neely *et al.*, 1990; Straub, 1990).

Estuaries in particular are often rich in archeological remains and offer potential for close collaboration between diatomists and archeologists. Estuarine wetlands have abundant natural resources and their navigable waters have provided routes for merchants and invaders, encouraging the development of trading and defensive settlements. Indeed, many of the world's cities have developed on estuaries because of their importance as ports. The historic and prehistoric development of these settlements is thus intimately linked to changing estuarine environments, and diatom analysis can provide a powerful tool for studying the interface between the sea, shipping and trading routes, and the economic and military function of shore and urban waterfronts. To this end, diatom analysis has been used to reconstruct the sedimentary environment of shipwrecks, or other isolated structures and artefacts (e.g. Foged, 1973; Marsden *et al.*, 1989; Miller, 1995; Cameron, 1997; Clark, 2004; Nayling & McGrail, 2004), to reconstruct local sea-level changes in areas adjacent to estuarine sites (e.g. Miller & Robertsson, 1981; Battarbee *et al.*, 1985; Wilkinson *et al.*, 1988; Wilkinson & Murphy, 1995), or to reconstruct changes in river water quality and tidal influence in urban areas (e.g. Foged, 1978; Boyd, 1981; Miller, 1982; Ayers & Murphy, 1983; Milne *et al.*, 1983; Jones & Watson, 1987; Juggins, 1988; Juggins, 1992b; Demiddele & Ervynck, 1993).

A particularly good example of collaboration between diatomists and archeologists comes from a study of the River Thames in central London (Milne *et al.*, 1983; Milne, 1985; Juggins, 1992b). The Thames is now tidal to Teddington Weir, 30 km above City of London, but the existence of Roman remains 4 m below the present high-tide level in the outer estuary led archeologists to assume that the river was tideless in central London during the early Roman period (first century AD; Willcox, 1975). Within the City the pre-Roman foreshore now lies buried some 100 m north of the present riverbank, and a horizontal stratigraphy of wooden and stone quays and revetments spanning the last ~2000 years is preserved beneath the modern waterfront (Figure 27.2). Initial diatom analysis of foreshore sediments associated with part of the first century AD revetment at the Pudding Lane site revealed an assemblage containing *Cyclotella striata* (Kütz.) Grunow in Cleve & Grunow, a brackish water planktonic species common in the estuary today. This taxon, together with a number of marine forms, demonstrates that the river was tidal in central London during the early Roman period, contrary to common belief (Milne *et al.*, 1983).

Subsequent work has examined additional foreshore samples from the first to twelfth centuries AD and refined the paleoenvironmental reconstructions using a salinity transfer function (Juggins, 1992b). Results show an initial rapid rise in mean half-tide salinity between the first and second centuries AD followed by a more gradual rise to the ninth to twelfth centuries (Figure 27.2). Comparison of these paleosalinity estimates with evidence for changes in the tidal range suggest that the increase in salinity between the first and twelfth centuries was primarily the result of relative sea-level rise. This would have been accompanied by a substantial upstream migration of the tidal head, particularly during the early Roman period. Progressive canalization of the upper and middle estuary after the twelfth century led to a marked increase in the tidal range of the river in central London, but it appears this had little impact

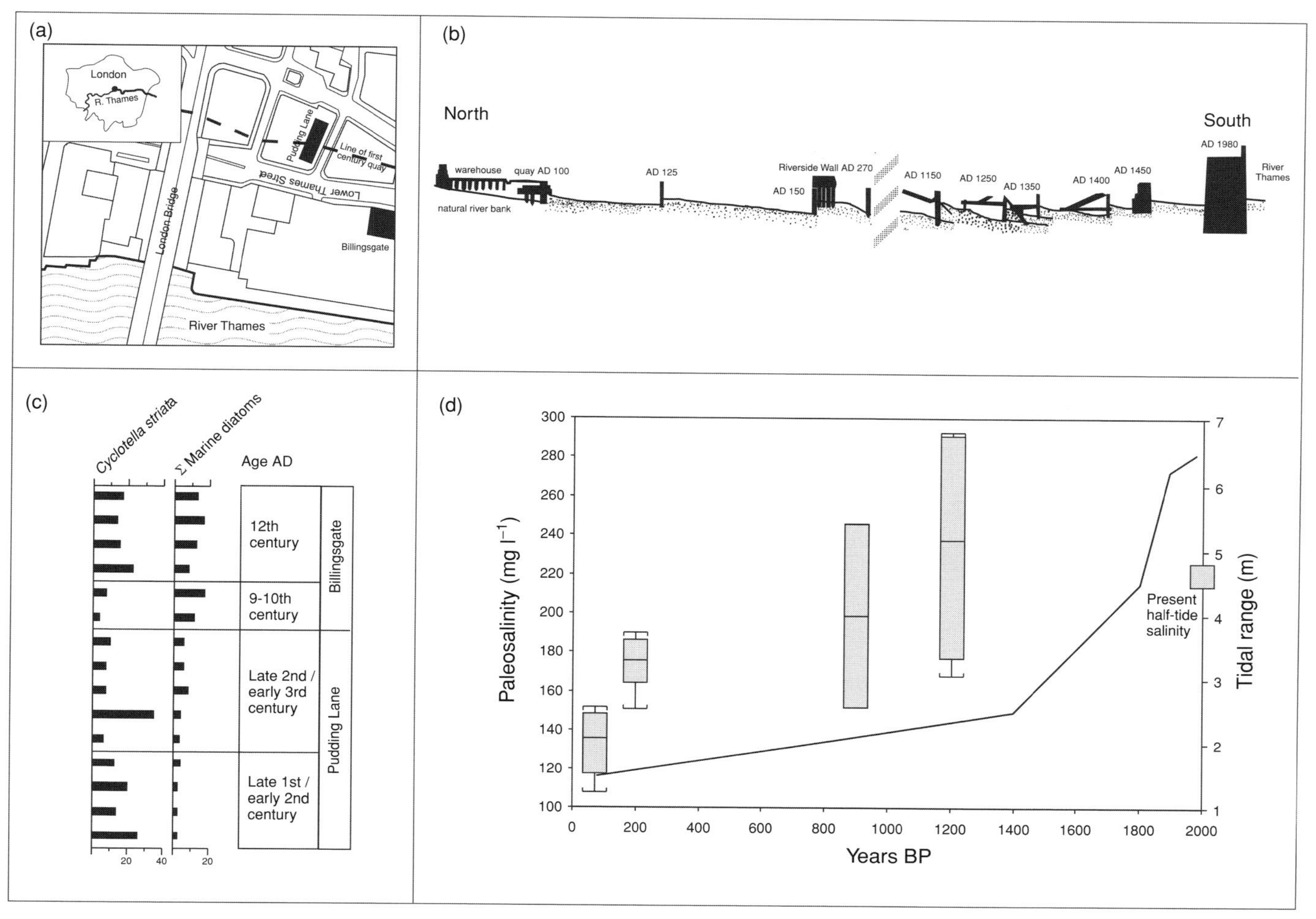

Figure 27.2 Diatom analysis of River Thames (UK) foreshore sediments. (a) Location of Pudding Lane and Billingsgate archeological sites in central London, showing line of first century quay; (b) composite cross-section showing horizontal stratigraphy of archeological structures from pre-Roman river bank to the modern river; (c) summary diatom stratigraphy showing changes in the relative proportions of *Cyclotella striata* and total marine taxa from first to twelfth centuries; and (d) diatom-based paleosalinity estimates (boxplots) and tidal range (solid line) for the River Thames in central London for the last 2000 years. (Redrawn from Milne (1985) and Juggins (1992b).)

on the salinity regime. Any tendency for further upstream penetration of brackish water was apparently balanced by increased ability of the river to impede landward flow as a result of the reduction in the tidal prism. Although more sites are needed to refine the model, these reconstructions, coupled with information on the changing river topography, provide archeologists with a preliminary framework for understanding the effects of the changing tidal regime on shipping and waterfront construction.

27.5 Regional paleoenvironmental reconstructions

In addition to reconstructing the local environment and resource space around a site, a major goal of archeology is to place sites and site networks in their landscape context, and to examine the climatic, environmental, and cultural factors that influenced regional patterns of site location and functional organization. Where there are appropriate sedimentary sequences in coastal or inland aquatic environments, diatom analysis offers the potential for the integration of on-site archeological evidence with off-site paleoenvironmental reconstructions. A clear example comes from the Baltic coast where the interaction of isostatic land uplift and eustatic sea-level rise has given rise to a complex pattern of transgressions and regressions and an altitudinal zonation of ancient shorelines and associated archeological remains. In the Stockholm region, diatom analysis of archeological sites from Mesolithic to Medieval age reveals a complex pattern of abandonment and re-occupation according to oscillations of the Baltic strandline (Florin, 1948; Miller & Robertsson, 1981).

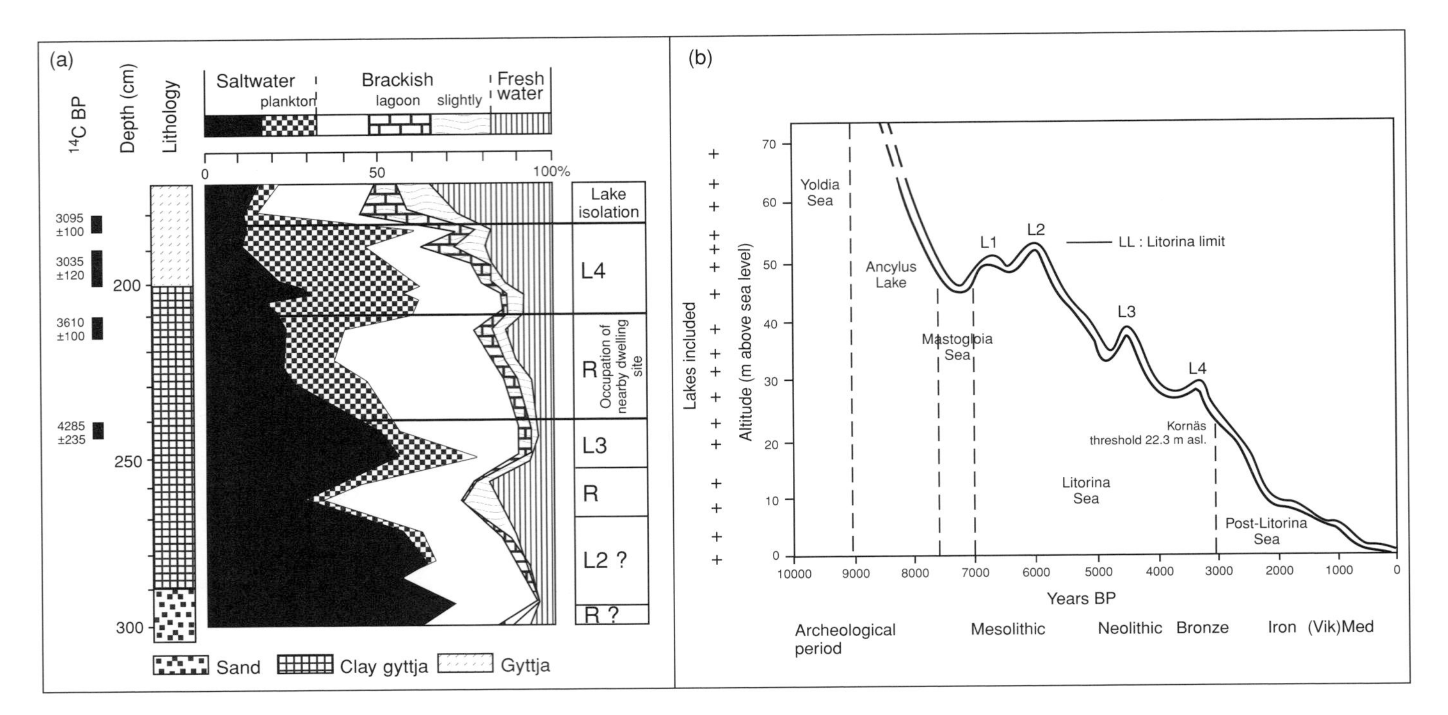

Figure 27.3 Diatom analysis and land uplift in the Baltic. (a) Summary diatom stratigraphy for Körnas bog, 30 km southwest of Stockholm (Sweden), showing inferred Litorina transgressions and lake isolation; and (b) shore displacement curve for the Stockholm region derived from the Körnas bog and other diatom stratigraphies. The curve can be used to predict the altitudes of coastal dwelling sites of different ages. (Redrawn from Miller & Florin (1989) and Miller & Robertsson (1981).)

Unfortunately none of the archeological sites contain a continuous stratigraphic record: consequently additional diatom and radiocarbon (^{14}C) analyses were carried out at a series of nearby lake basins with threshold altitudes that bracketed the archeological sites. Results from the lake sediments were used to identify both timing of the final isolation of the lake from the Baltic and the maximum altitude of the various transgressive phases, allowing the archeological data to be placed in a regional framework of relative sea-level change and shore displacement (Figure 27.3). In addition, the shoreline curve offers a tool for site prospecting by directing archeologists to the correct altitudinal contour of shorelines of different ages.

Coastal wetlands adjacent to archeological sites have been utilized in many ways by ancient societies. In many areas they were simply exploited for their natural resources but at some sites there is evidence for large-scale land reclamation and drainage. An excellent example of the integration of onsite archeological excavation of Romano-British dwellings and field systems with off site paleoenvironmental studies comes from the Severn Estuary, UK (Rippon, 2000). Here the presence of freshwater and aerophilous diatoms in coastal ditch sediments, combined with other biological indicators, revealed an agrarian landscape utilized for grazing and cereal cultivation, drained and reclaimed from the estuary and protected from tidal incursion by a coastal embankment.

Diatom analysis from inland freshwater or saline lakes can also provide valuable information for the archeologist on regional paleohydrology and paleoclimate. However, if they are to provide more than a simple deterministic environmental "backdrop" to archeological interpretation, such studies must consider the complex mosaic of environments within the landscape, and the possibility of asynchronous change in the quality of aquatic habitats in a regional site network. Such integration of diatom and other paleoecological analyses into a regional archeological investigation comes from the North American Southern High Plains. The plateau of the plains is dissected by a series of entrenched ephemeral drainage channels, or draws, that contain a rich archeological record spanning the last 12,000 years, interbedded with alluvial, lacustrine, and eolian sediments (Holliday, 1985). Diatom analysis was originally used to reconstruct local depositional environments at individual sites, including the famous Clovis Paleo-Indian site on Blackwater Draw near Portales, New Mexico (Lohman, 1935; Patrick, 1938). More recently, Winsborough (1995) has carried out diatom analysis at a network of sites as part of a wider study into regional paleoenvironments. Results show

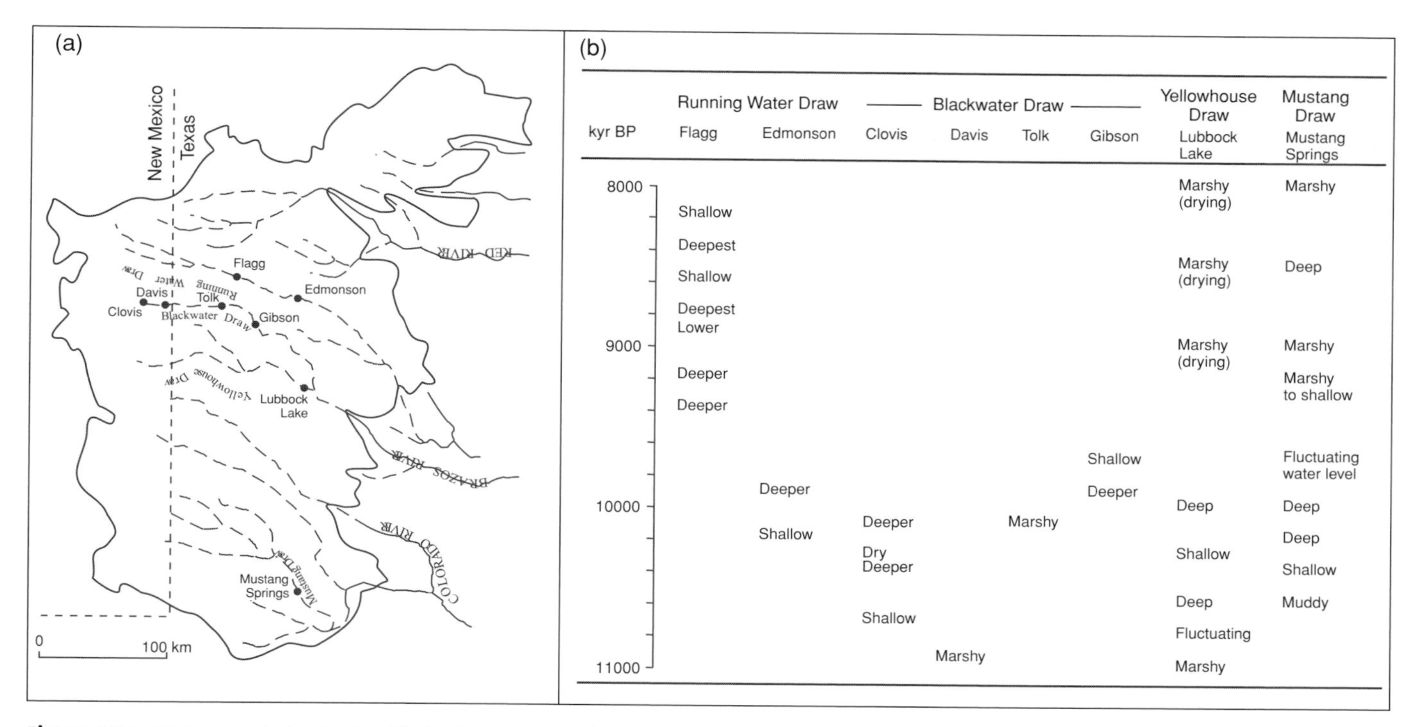

Figure 27.4 Diatom analysis of valley fills in the Southern High Plains, North America. (a) Map of Southern High Plains, showing major stream channels and site locations; and (b) summary of late Pleistocene and early Holocene paleoenvironments inferred from diatom analysis of lake and marsh sediments. (Redrawn from Winsborough (1995).)

that each of the four channels examined underwent a similar evolution; from a predominantly alluvial environment during the late Pleistocene, to a complex lacustrine and marsh environment in the early Holocene, and finally to an environment dominated by eolian sedimentation in the middle Holocene. These changes were broadly synchronous between sites, indicating regional climatic control, although there were some distinct time-transgressive relationships highlighting the importance of regionally variable and/or local environmental changes (Figure 27.4).

27.6 Conclusions

Despite the substantial potential of diatom analysis in archeological enquiry, it is still an under-used method, and one that is poorly understood by many archeologists. As a consequence, diatom analysis has often been used post hoc to address poorly framed questions from inadequately sampled material, and the results relegated to specialist appendices in archeological reports. However, there are grounds for optimism: the examples above illustrate a trend in which diatomists are becoming more closely involved in archeological research and project design, leading to integrated excavation and sampling strategies that maximizes the return of information. Such collaboration benefits both diatomists and archeologists and is essential if we are to develop new methodologies that address the particular problems of preservation, taphonomy, and interpretation posed by archeological material.

27.7 Summary

This chapter reviews the role of diatom analysis in archeology, and shows how the technique has been applied at a range of spatial and temporal scales to place archeological material in its broader site, landscape, and cultural context. The majority of published studies have used diatom analysis to provide information on local and regional paleoenvironments, and to identify environmental factors that have influenced patterns of site location and organization. However, we also note a trend towards closer collaboration between diatomists and archeologists, leading to new applications of the technique to problems of establishing the provenance of artefacts, identifying processes of site formation and abandonment, understanding the function of archeological structures, and landscape and resource utilization.

Acknowledgments

We are grateful to the many colleagues who sent us additional references and papers during the preparation of this chapter and to John Smol, Gene Stoermer, and Antoinette Mannion for helpful comments on an earlier draft of the manuscript.

References

Alhonen, P., Kokkonen, J., Matiskainen, H., & Vuorinen, A. (1980). Applications of AAS and diatom analysis and stylistic studies of Finnish Subneolithic pottery. *Bulletin of the Geological Society of Finland*, **52**, 193–206.

Alhonen, P. & Matiskainen, H. (1980). Diatom analysis from prehistoric pottery sherds – an archaeological evaluation. Proceedings of the Nordic Meeting of Diatomologists, Lammi Biological Station, Finland, May 6–7, pp. 45–62.

Alhonen, P. & Väkeväinen, L. (1981). Diatom-analytical studies of early comb ceramic vessels in Åland. *Eripainos Suomen Museo*, 67–77.

Ayers, B. & Murphy, P. (1983). A waterfront excavation at Whitefriars Street car park, Norwich, 1979. *East Anglian Archaeology Report*, **17**, 1–60.

Battarbee, R. W. (1988). The use of diatoms analysis in archaeology: a review. *Journal of Archaeological Science*, **15**, 621–44.

Battarbee, R. W., Scaife, R. G., & Phethean, S. J. (1985). Palaeoecological evidence for sea-level change in the Bann estuary in the early mesolithic period. In *Excavations at Mount Sandel 1973–78*, ed. P. C. Woodman, Belfast: HMSO, pp. 111–20.

Beneš, J., Kaštovský, J., Kočárova´, R., *et al.* (2002). Archaeobotany of the Old Prague Town defence system, Czech Republic: archaeology, macro-remains, pollen, and diatoms. *Vegetation, History and Archaeobotony*, **11**, 107–19.

Boyd, P. D. A. (1981). The micropalaeontology and palaeoecology of medieval estuarine sediments from the Fleet and Thames in London. In *Microfossils from Recent and Fossil Shelf Seas*, eds. J. W. Neale & M. D. Brasier, Chichester: Ellis Harwood, pp. 274–92.

Butzer, K. W. (1982). *Archaeology as Human Ecology*. Cambridge: Cambridge University Press.

Cameron, N. G. (1997). The diatom evidence. In *Excavations at Caldicot, Gwent: Bronze Age Palaeochannels in the Lower Nedern Valley*, ed. N. Nayling & A. Caseldine, Council for British Archaeology Research, Report 108, York: Council for British Archaeology, pp. 117–28.

Cameron, N. G. (2007). Use in archaeology. In *Encyclopedia of Quaternary Science*, ed. S. A. Elias, Elsevier: Amsterdam, pp. 523–9.

Caran, C., Neely, J., Winsborough, B., Sorensen, F., & Valastro, S., Jr. (1996). A late Paleo-Indian/Early-Archaic water well in Mexico. Possible oldest water-management feature in New World. *Geoarchaeology*, **11**, 1–36.

Castledine, C., Juggins, S., & Straker, V. (1988). Preliminary palaeoenvironmental analysis of floodplain deposits from a section near the River Exe in Exeter, Devon. In *The Exploitation of Wetlands*, ed. P. Murphy & C. French, Oxford: BAR, pp. 145–159.

Clark, P. (ed.) (2004). *The Dover Bronze Age Boat*, London: English Heritage Publications, p. 340.

Cole, S. C., Atwater, B. F., McCutcheon, P. T., Stein, J. K., & Hemphill-Haley, E. (1996). Earthquake-induced burial of archeological sites along the southern Washington coast about A. D. 1700. *Geoarchaeology*, **11**, 165–77.

Demiddele, H. & Ervynck, A. (1993). Diatomeeën als ecologische indicatoren in de Vlaamse archeologie: Romeins en middeleeuws Oudenburg (prov. West-Vlaanderen). *Archeologie in Vlaanderen, III*, 217–31.

Denys, L. (1992). On the significance of marine diatoms in freshwater deposits at archaeological sites. *Diatom Research*, **7**, 195–7.

De Wolf, H. & Cleveringa, P. (1996). The impact of the beer industry on medieval water quality. In *Proceedings of the 14th Diatom Symposium*, eds. I. Mayama & J. Koizumi, Königstein: Koeltz Scientific Books pp. 511–22.

Douglas, M. S. V., Smol, J. P., Savelle, J. M., & Blais, J. M. (2004). Prehistoric Inuit whalers affected Arctic freshwater ecosystems. *Proceedings of the National Academy of Sciences of the USA*, **101**, 1613–17.

Ekdahl, E. J., Teranes, J. L., Guilderson, T., *et al.* (2004). Prehistorical record of cultural eutrophication from Crawford Lake, Canada. *Geology*, **32**, 745–8.

Ekdahl, E. J., Terranes, J. L., Wittkop, C. A., *et al.* (2007). Diatom assemblage response to Iroquoian and Euro-Canadian eutrophication of Crawford Lake, Ontario, Canada. *Journal of Paleolimnology*, **37**, 233–46.

Florin, S. (1948). Kustförskjutningen och bebyggelseutvecklingen I östra Mellansverige under senkvartär tid. II. De baltiska strandbildningarna och stenåldersboplatsen vid Dammstugan nära Katrineholm. *Geologiska Föreningens Föingar*, **70**, 17–196.

Flower, R. J. (2006). Diatoms in ancient building materials: application of diatom analysis to Egyptian mud bricks. *Nova Hedwigia, Beiheft*, **130**, 245–63.

Foged, N. (1973). The diatoms in a wreck from the late Middle Age. *University of Lund, Department of Quaternary Geology, Report*, **3**, 39–45.

Foged, N. (1978). *The Archaeology of Svendborg. No. 1. Diatom Analysis*. Odense: Odense University Press.

Foged, N. (1985). Diatoms in a tomb from the early Bronze Age. *Nova Hedwigia*, **41**, 471–82.

Fritz, S. C. (1989). Lake development and limnological response to prehistoric and historic land-use in Diss, Norfolk, U.K. *Journal of Ecology*, **77**, 182–202.

Gibson, A. M. (1986). Diatom analysis of clays and late Neolithic pottery from the Milfield basin, Northumberland. *Proceedings of the Prehistoric Society*, **52**, 89–103.

Godbold, S. & Turner, R. C. (1993). *Second Severn Crossing Archaeological Response: Phase 1: the Intertidal Zone in Wales*. Brentwood: Westbury Press.

Groenman-van Waateringe, W. & Jansma, M. J. (1968). Diatom analysis. Appendix III to W.A. van Es, Paddepol, excavations of frustated terps, 200 B.C. – 250 A.D. *Palaeohistoria*, **14**, 286.

Groenman-van Waateringe, W. & Jansma, M. J. (1969). Diatom and pollen analysis of the Vlaardingen Creek. A revised interpretation. *Helinium*, **9**, 105–17.

Håkansson, H. (1988). Diatom analysis at Skateholm – Järavallen, southern Sweden. *Acta Regiae Societatis Humaniorum Litterarum Lundensis*, **LXXIX**, 39–45.

Håkansson, H. & Hulthén, B. (1986). On the dissolution of pottery for diatom studies. *Norwegian Archaeological Review*, **19**, 34–7.

Håkansson, H. & Hulthén, B. (1988). Identification of diatoms in Neolithic pottery. *Diatom Research*, **3**, 39–45.

Hemphill-Haley, E. (1995). Diatom evidence for earthquake-induced subsidence and tsunami 300 yr ago in southern coastal Washington. *Geological Society of America Bulletin*, **107**, 367–78.

Hemphill-Haley, E. (1996). Diatoms as an aid in identifying late-Holocene tsunami deposits. *The Holocene*, **6**, 439–48.

Holliday, V. T. (1985). New data on the stratigraphy and pedology of the Clovis and Plainview sites, Southern High Plains. *Quaternary Research*, **23**, 388–402.

Jansma, M. J. (1977). Diatom analysis of pottery. *Cingula*, **4**, 77–85.

Jansma, M. J. (1981). Diatoms from coastal sites in the Netherlands. In *Environmental Aspects of Coasts and Islands*, ed. D. Brothwell & G. Dimbleby, Oxford: BAR, pp. 145–62.

Jansma, M. J. (1982). Diatom analysis from some prehistoric sites in the coastal area of the Netherlands. *Acta Geologica Academiae Scientiarum Hungaricae*, **25**, 229–36.

Jansma, M. J. (1984). Diatom analysis of prehistoric pottery. In *Proceedings of the Seventh International Diatom Symposium*, ed. D. G. Mann, Königstein: Koeltz Scientific Books, pp. 529–36.

Jansma, M. (1990). Diatoms from a Neolithic excavation on the former Island of Schokland, Ijselmeerpolders, the Netherlands. *Diatom Research*, **5**, 301–9.

Jessen, K. (1936). Palaeobotanical report on the Stone Age site at Newferry, County Londonderry. *Proceedings of the Royal Irish Academy*, **43C**, 31–7.

Jones, J. & Watson, N. (1987). The early medieval waterfront at Redcliffe, Bristol: a study of environment and economy. In *Studies in Palaeoeconomy and Environment in South-west England*, ed. N. Balaam, B. Levitan, & V. Straker, Oxford: BAR, pp. 135–62.

Juggins, S. (1988). Diatom analysis of foreshore sediments exposed during the Queen Street Excavations, 1985. In *The Origins of the Newcastle Quayside. Excavations at Queen Street and Dog Bank*, ed. C. O'Brien, L. Bown, S. Dixon, & R. Nicholson, Newcastle upon Tyne: The Society of Antiquaries of Newcastle upon Tyne, Monograph Series No. 3, pp. 150–51.

Juggins, S. (1992a). Diatom analysis. In *Iron Age and Roman salt production and the medieval town of Droitwich: excavations at the Old Bowling Green and Friar Street*, ed. S. Woodwiss, York: Council for British Archaeology Research Report No. 81, microfiche 2:F4–5.

Juggins, S. (1992b). Diatoms in the Thames Estuary, England: ecology, palaeoecology, and salinity transfer function. *Bibliotheca Diatomologica*, **25**.

Keevill, G. (2004). Historic Royal Palaces Monograph No. 1: the Tower of London moat. In *Archaeological Excavations 1995–9*, Oxford: Oxford Archaeology with Historic Royal Palaces, p. 315.

König, D. (1983). Diatomeen des fruneolithischen Fundplatzes Siggeneben-Süd. In *Siggeneben-Süd. Ein Fundplatz der frühen Trichterbecherkultur an der holsteinischen Ostseeküste*, ed. J. Meurers-Balke, Neumünster: Karl Wachholtz, pp. 124–40.

Körber-Grohne, U. (1967). *Geobotanische Untersuchungen auf der Feddersen Wierde*. Wiesbaden: Franz Steiner Verlag.

Lohman, K. (1935). Diatoms from Quaternary lake beds near Clovis, New Mexico. *Journal of Paleontology*, **9**, 455–9.

Mannion, A. M. (1987). Fossil diatoms and their significance in archaeological research. *Oxford Journal of Archaeology*, **6**, 131–47.

Marsden, P., Branch, N., Evans, J., *et al.* (1989). A late Saxon logboat from Clapton, London borough of Hackney. *International Journal of Nautical Archaeology and Underwater Exploration*, **18**, 89–111.

Matiskainen, H. & Alhonen, P. (1984). Diatoms as indicators of provenance in Finnish Sub-Neolithic pottery. *Journal of Archaeological Science*, **11**, 147–57.

Miller, U. (1982). Shore displacement and coastal dwelling in the Stockholm region during the past 5000 years. *Annales Academiae Scientarium Fennicae, A.III*, **134**, 185–211.

Miller, U. (1995). Diatoms and submarine archaeology (siliceous microfossil analysis as a key to the environment of ship wrecks, harbour basins and sailing routes). In *Scientific Methods in Underwater Archaeology*, ed. I. Vuorela, Rixensart: Council for Europe, pp. 53–8.

Miller, U. & Florin, M.-B. (1989). Diatom analysis. Introduction to methods and applications. In *Geology and Palaeoecology for Archaeologists: Palinuro I*, ed. T. Hackens and U. Miller, Ravello: European University Centre for Cultural Heritage, pp. 133–57.

Miller, U., Modig, S., & Robertsson, A. M. (1979). The Yttersel dwelling site: method investigations. *Early Norrland*, **12**, 77–92.

Miller, U. & Robertsson, A. (1981). Current biostratigraphical studies connected with archaeological excavations in the Stockholm region. *Striae*, **14**, 167–73.

Milne, G., Battarbee, R. W., Straker, V., & Yule, B. (1983). The River Thames in London in the mid 1st century AD. *Transactions of the London and Middlesex Archaeological Society*, **34**, 19–30.

Milne, G. (1985). *The Roman Port of London*, Batsford: London.

Nayling, N. & McGrail, S. (2004). *The Barland's Farm Romano-Celtic Boat*. Council for British Archaeology Research Report 115, York: Council for British Archaeology.

Neely, J., Caran, S., & Winsborough, B. (1990). Irrigated Agriculture at Hierve el Agua, Oaxaca, Mexico. In *Debating Oaxaca Archaeology*, ed. J. Marcus, Ann Arbor, MI: Museum of Anthropology, University of Michigan.

Neely, J. A., Caran, S. C., Winsborough, B. M., Sorensen, F. R., & Valastro, Jr. S. (1995). An early Holocene hand-dug water well in the Tehuacan Valley of Puebla, Mexico. *Current Research in the Pleistocene*, **12**, 38–40.

Nunez, M. & Paabo, K. (1990). Diatom analysis. *Norwegian Archaeological Review*, **23**, 128–30.

Patrick, R. (1938). The occurrence of flints and extinct animals in pluvial deposits near Clovis, New Mexico. Part V. Diatom evidence from the Mammoth Pit. *Proceedings of the Academy of Natural Sciences of Philadelphia*, **40**, 15–24.

Renberg, I., Bindler, R., Bradshaw, E., Emteryd, O., & McGowan, S. (2001). Evidence of early eutrophication and heavy metal pollution of Lake Mälaren, central Sweden. *Ambio*, **30**, 496–502.

Rippon, S. (2000). The Romano-British exploitation of coastal wetlands: survey and excavation on the North Somerset Levels, 1993–7. *Britannia*, **XXXI**, 69–204.

Robbins, L. H., Murphy, M. L., Stevens, N. J., *et al.* (1996). Paleoenvironment and archaeology of Drotsky's Cave: western Kalahari Desert, Botswana. *Journal of Archaeological Science*, **23**, 7–22.

Straub, F. (1990). *Hauterive-Champreveyres, 4. Diatomees et reconstitution des environments prehistorique*. Archeologie Neuchatoise, 10, Saint-Blaise: Editions du Ruau.

VanLandingham, A. L. (2006). Diatom evidence for autochthonous artifact deposition in the Valsequillo region, Puebla, Mexico during the Sangamonian (*sensu lato* = 80,000 to ca. 220,000 yr BP and Illinoian (220,000 to 430,000 yr BP)). *Journal of Paleolimnology*, **36**, 101–16.

Voorrips, A. & Jansma, M. J. (1974). Pollen and diatom analysis of a shore section of the former Lake Wervershoof. *Geologie en Mijnbouw*, **53**, 429–35.

Whitmore, T. J., Brenner, M., Curtis, J. H., Dahlin, B. H., & Leyden, B. W. (1996). Holocene climatic and human influences on lakes of the Yucatan Peninsula, Mexico: an interdisciplinary, palaeolimnological approach. *The Holocene*, **6**, 273–87.

Wilkinson, T. J. & Murphy, P. L. (1995). *The Archaeology of the Essex Coast, Volume I: The Hullbridge Survey*. East Anglian Archaeology Report No. 71, Chelmsford: Essex County Council.

Wilkinson, T. J., Murphy, P., Juggins, S., & Manson, K. (1988). Wetland development and human activity in Essex estuaries during the Holocene transgression. In *The Exploitation of Wetlands*, ed. P. Murphy & C. French, Oxford: BAR, pp. 213–38.

Willcox, G. H. (1975). Problems and possible conclusions related to the history and archaeology of the Thames in the London region. *Transactions of the London and Middlesex Archaeological Society*, **26**, 185–92.

Winsborough, B. (1995). Diatoms. *Geological Society of America, Memoir*, **186**, 67–83.

Wüthrich, M. (1971). Les diatomées de la station néolithique d'Auvernier (Lac de Neuchâtel). *Schweizerische Zeitschrift für Hydrologie*, **33**, 533–552.

28

Diatoms in oil and gas exploration

WILLIAM N. KREBS, ANDREY YU. GLADENKOV, AND GARETH D. JONES

28.1 Introduction

Diatoms constitute an important rock-forming microfossil group. They evolved rapidly during the Cenozoic, and unlike other microfossils, diatoms are found in marine, brackish, and lacustrine sediments, and are thus useful for age dating and correlating sediments that accumulated in a variety of environments. Their fossil assemblages are reflective of the environments in which they lived, and their recovery in wells and outcrops can provide information on paleochemistry and paleobathymetry. They are most useful to the oil and gas industry in age dating and correlating rocks that lack or have poor recovery of calcareous microfossils, such as "cold-water" Tertiary marine rocks, and sediments that accumulated in lacustrine basins and in brackish-water settings.

Age dating and correlation of sedimentary rocks are important for hydrocarbon exploration in order to understand the geologic history of a basin as it relates to formation of source rocks, reservoirs, structures, and seals. The timing of expulsion of oil and gas is a critical factor in determining the prospect potential of a basin for hydrocarbon, and necessitates an understanding of the geochronology of events. The ability to correlate subsurface rocks is a prerequisite to mapping the distribution of reservoir facies and the interpretation of sequence stratigraphy.

Environmental data pertaining to paleobathymetry and paleochemistry provide important information on the geologic history of a basin and predictability of reservoir distribution. For example, shales that are barren of calcareous microfossils may have accumulated in deep water, beneath the carbonate compensation depth, or in brackish or freshwater. Incorrect assessments of the paleoenvironment will drastically affect the interpretation of a basin's evolution and the evaluation of its hydrocarbon potential. Inasmuch as fossil diatoms occur in all of these settings, they can provide critical evidence for the correct paleoenvironmental interpretation.

28.2 Applications

Cenozoic marine diatom biozonations and graphic correlation (Shaw, 1964; Miller, 1977; Edwards, 1989; Carney & Pierce, 1995; Aurisano *et al.*, 1995) are used within the oil and gas industry to correlate marine rocks. Marine diatom biozonations exist for high latitudes in the southern and northern hemispheres (Harwood & Maruyama, 1992; Barron & Gladenkov, 1995; Gladenkov & Barron, 1995; Koç & Scherer, 1996; Scherer & Koç, 1996; Gladenkov, 1998) as well as for temperate and equatorial latitudes (Barron, 1985a, b, 2005; Baldauf & Iwai, 1995; Barron *et al.*, 2004). Figure 28.1 is the graphic correlation plot of the Tanka-3 well in the Gulf of Suez, Egypt. Note that the line of correlation expresses the relationship between a thick section of sedimentary rock and geologic time. Biostratigrahic events, such as the first and last occurrences of diatoms and other microfossils, provide the points that allow the line of correlation to be drawn. A flat line segment represents a period of geologic time that is not preserved in the rock record. It may represent a fault plane, an unconformity caused by erosion, or even a thin rock interval comprising considerable geologic time, i.e. a condensed section. Unconformities and condensed sections naturally divide sedimentary rocks into sequences that can be mapped in the subsurface to model oil and gas plays. This approach is known as sequence stratigraphy. As Figure 28.1 reveals, diatoms are most useful in constraining the line of correlation in the younger section because they evolved rapidly during the late Tertiary and because they are most often recovered in an unaltered state at well depths less than about 5000 feet (1500+ m). However, in regions having low geothermal gradients, such as offshore Mauritania, unaltered diatoms may be recovered at greater depths. Wells drilled there

The Diatoms: Applications for the Environmental and Earth Sciences, 2nd Edition, eds. John P. Smol and Eugene F. Stoermer. Published by Cambridge University Press.

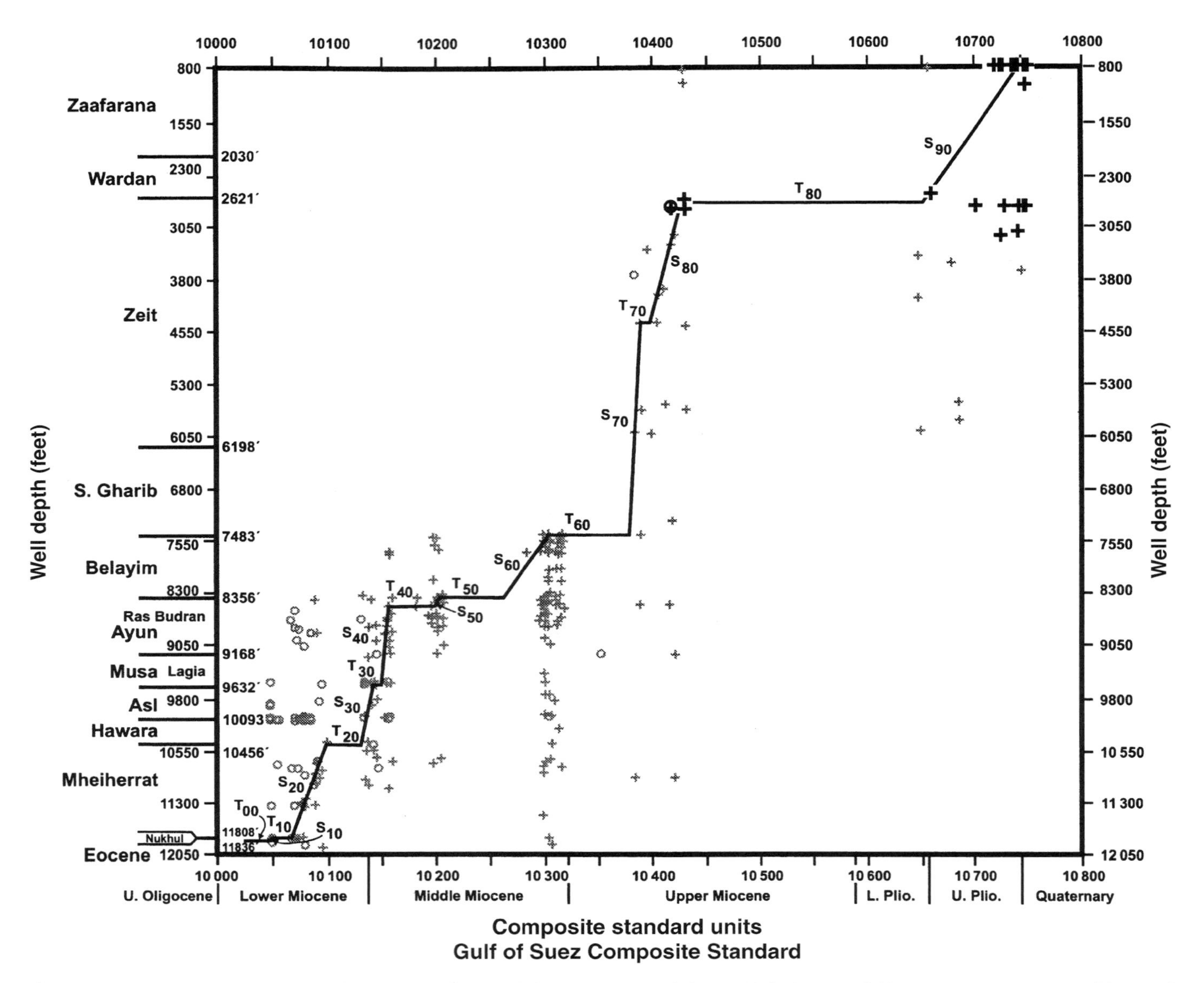

Figure 28.1 Graphic correlation plot of the Tanka-3 well, Gulf of Suez, Egypt. The vertical axis is well depth and lithostratigraphy. Geologic time in composite standard units and epochs is represented by the horizontal axis. The symbols "o" and "+" represent, respectively, the oldest and youngest occurrences of microfossil species in a rock section, or biostratigraphy data. The bold symbols are diatom data, and the light symbols represent calcareous nanofossil, planktonic, and benthic foraminifera, pollen, spore, and dinoflagellate cyst data. The diatoms are restricted to the shallow portion of the well and provide biostratigraphic control. Note that there are nine biostratigraphic sequences (S_{10} to S_{90}) separated by nine terraces (T_{00} to T_{80}). The sequences represent periods of continuous deposition, and the terraces are hiatuses in geologic time. (Modified from Krebs *et al.* (1997).)

reveal that radiolarians and diatoms are concentrated in stratigraphic intervals just atop unconformities. These intervals, rich in siliceous microfossils, formed as pelagic deposits when the influx of terrigenous sediments at the well site was minimal.

Fossil marine diatoms are particularly useful for hydrocarbon exploration in basins characterized by thick Neogene diatomaceous rock sections, such as those along coastal California, Japan, in the Bering Sea off Alaska, and Sakhalin Island, Russia. In these "cold-water" sediments, calcareous microfossils are rare or have long geologic ranges, and are therefore not useful for biostratigraphy. Diatoms, however, are often abundant in these sediments, and are especially well preserved in marine basins that were "starved" of clastic sediments. These rocks (diatomites) are common along the circum-North Pacific coast (Ingle, 1981), and one such deposit, the Monterey

Formation of California, is an important oil source and reservoir. The Hondo Field, offshore of California, for example, has a fractured chert reservoir that was derived from the diagenesis of diatomaceous sediments of the Monterey Formation, and the coeval Belridge Diatomite, Kern County, California is an oil shale (Schwartz, 1987). Diagenesis of diatomaceous sediments may have a significant impact on oil exploration by creating fractured siliceous reservoirs, and the diatomaceous sediments may themselves be important sources for petroleum (Mertz, 1984). Destruction of diatom frustules by diagenesis can be observed through the microscope (see Figure 28.2) and noted by changes in rock properties and seismic expression (Murata and Larson, 1975; Hein *et al.*, 1978; Compton, 1991; Grechin *et al.*, 1981; Iijima and Tada, 1981; Isaacs, 1981, 1982, 1983; Pisciotto, 1981; 1982, 1983; Tada and Iijima, 1983).

Sakhalin Island, in the far east of Russia, is another hydrocarbon-bearing region where fossil diatoms are used for biostratigraphy. In that region, diatomaceous Tertiary sequences contain prominent oil and gas fields that were first discovered during the 1920s. Biostratigraphic correlations were initially based on assemblages of molluscs and benthic foraminifera, but by the late 1960s fossil diatoms began to be used (Sheshukova-Poretskaya, 1967). Comparative analysis of diatom assemblages from sediments of different ages revealed their potential for correlating Cenozoic deposits in remote sections of Sakhalin, and by the mid 1970s fossil diatoms were widely used for biostratigraphy. There are now more than twenty diatom biozones for the Oligocene to the Quaternary of Sakhalin (Figure 28.3) that are used for stratigraphic correlation. The three best hydrocarbon source rocks in Sakhalin, the Oligocene Pilengskaya Formation and the Middle Miocene Pil'skaya and Kurasiiskaya formations (Bazhenova, 2002), are highly siliceous and similar to the Monterey Formation of California. Oil and gas are produced from Upper Oligocene to Upper Miocene rocks.

Although diatom frustules are infrequently found in wells beneath 1500+ m in their unaltered state, deeper recovery may occur when they are preserved in calcareous concretions that formed during early burial (Bramlette, 1946; Lagle, 1984; Bloome and Albert, 1985). Concretions form "closed systems" that protect the frustules from dissolution during diagenesis. In addition, pyritized diatoms may be obtained from greater well depths when recovered from well cutting residues that have been processed for foraminifera. In the North Sea, for example, pyritized frustules occur at a well depth of 3700 m and are used for biostratigraphy (M. Charnock, personal communication, 1990). Because of their abundance in cold-water

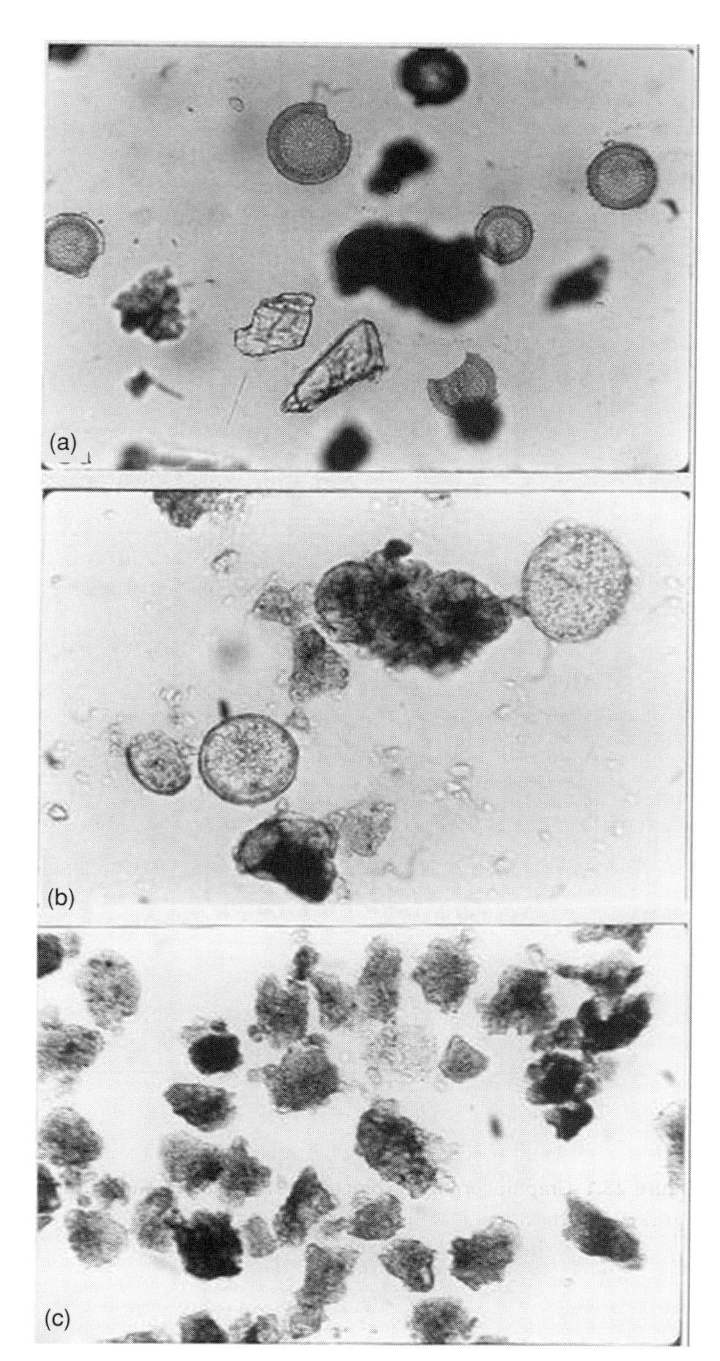

Figure 28.2 Diagenesis and dissolution of Pliocene Glenns Ferry Formation lacustrine diatoms in the Ore-Ida-1 geothermal well, Ontario, Oregon, in the western Snake River basin. (a) A well-preserved diatom assemblage from 600 feet (183 m) in the well; (b) partial frustule dissolution at 1600 ft (488 m, and (c) frustule dissolution is complete by 1800 ft (549 m).

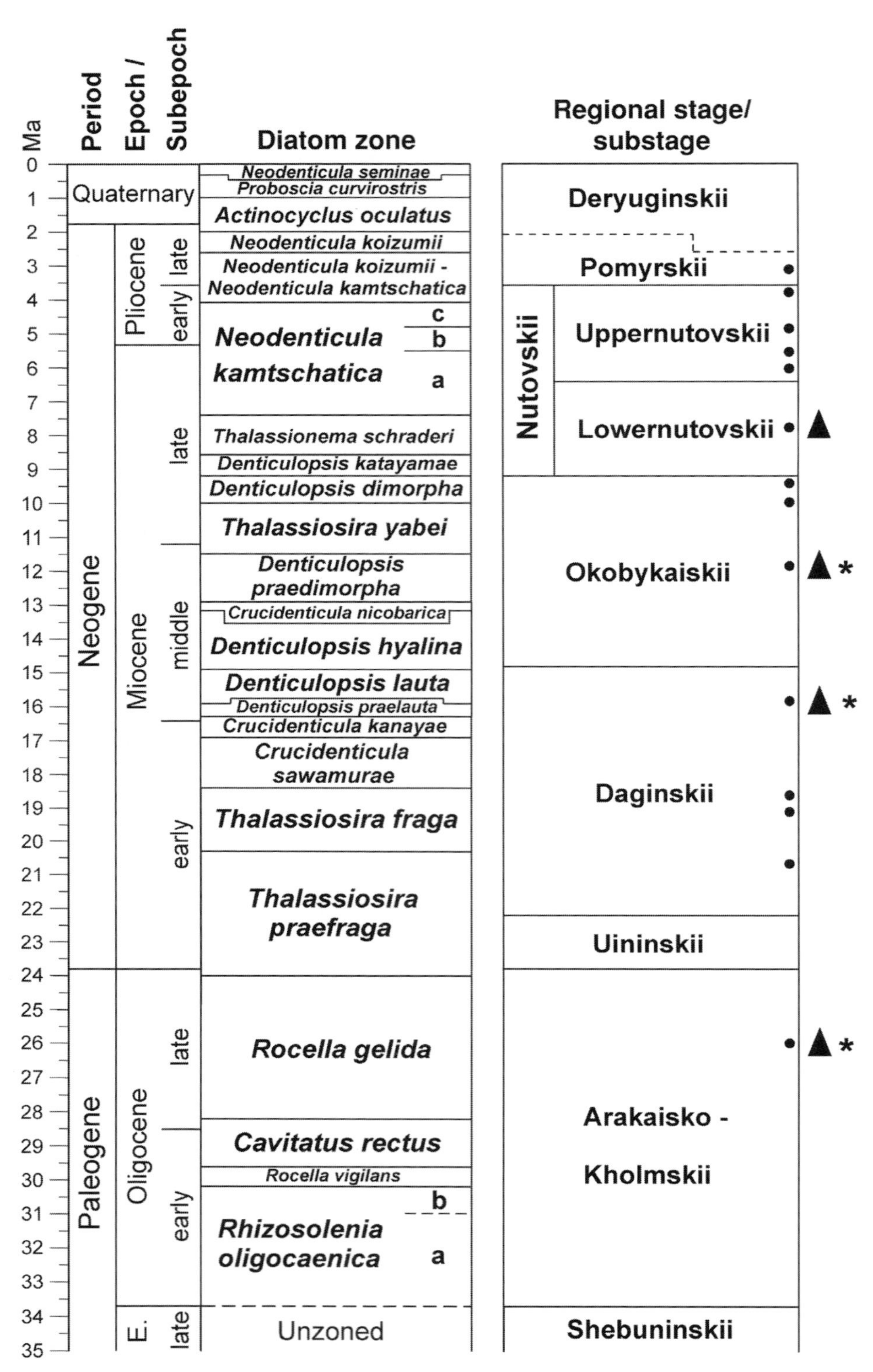

Figure 28.3 Correlation of Oligocene through Quaternary Sakhalin regional stages with the North Pacific diatom biozones and the geologic timescale of Berggren *et al.* (1995) and using data from Gladenkov (2001) and Gladenkov *et al.* (2002). Triangles indicate major hydrocarbon reservoirs, and dots signify diatom intervals. Major siliceous hydrocarbon source rocks are indicated by asterisks.

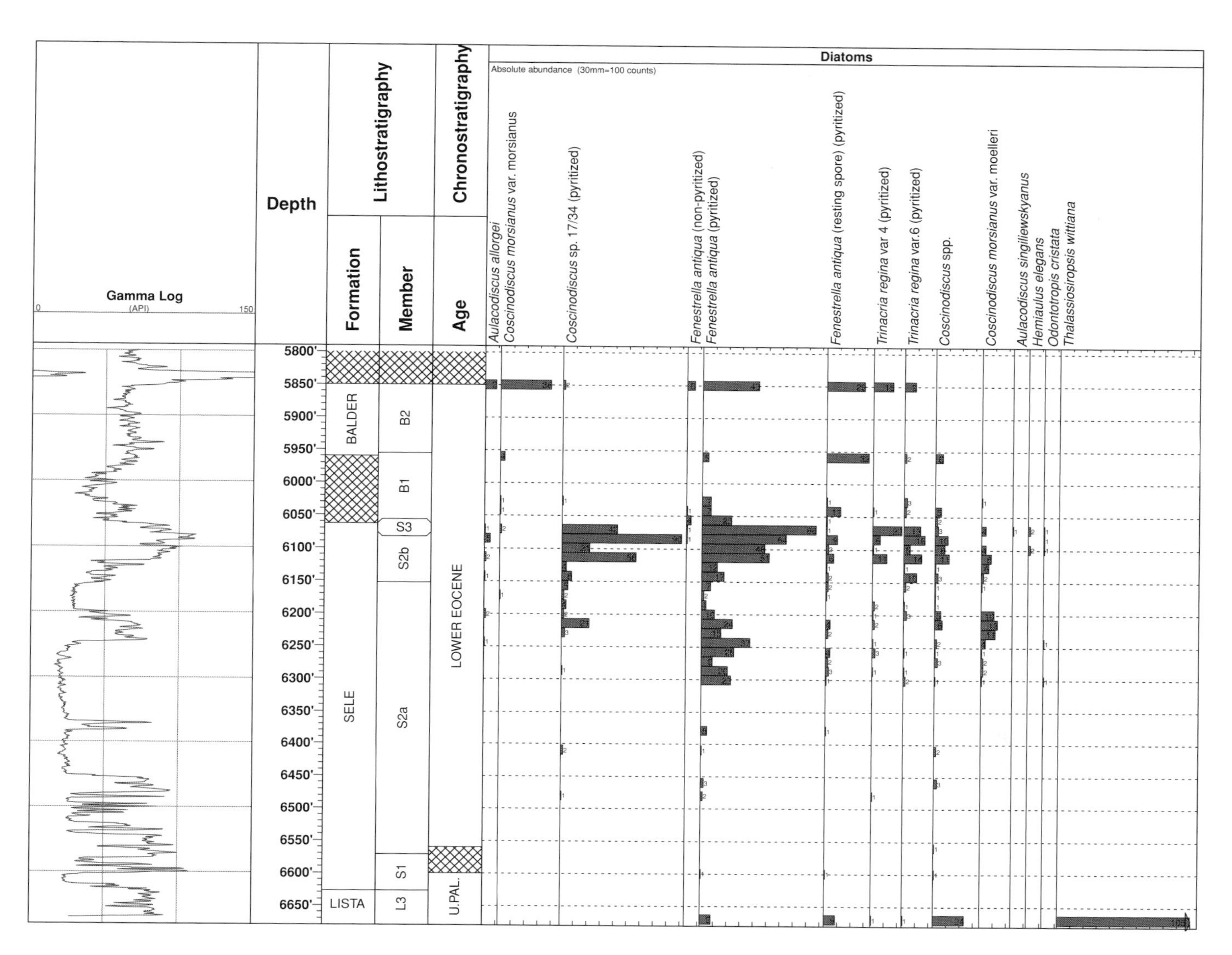

Figure 28.4 Diatom range chart (actual abundances) from the 9/28a-18 appraisal well from the Crawford Field in the central North Sea. Note influxes of *Fenestrella antiqua* and *Coscinodiscus morsianus* var. *morsianus* (Sims) Mitlehner at the top of the Balder Formation, *Fenestrella antiqua* and *Coscinodiscus morsianus* var. *moelleri* Mitlehner at the top of the Sele Formation, and *Thalassiosiropsis wittiana* (Pantocsek) Hasle at the base of the Sele Formation (samples analyzed were ditch cuttings samples and as a consequence the influx is recorded in the underlying Lista Formation). Note also their correlation with high gamma peaks representing more argillaceous intervals. (Published by kind permission of Fairfield Energy Ltd., Stratic Energy and Valiant Petroleum.)

sediments, such recovery often occurs in well intervals that are devoid of calcareous microfossils, so the presence of pyritized diatoms can be especially important in correlating deep subsurface sections that are otherwise barren.

Diatoms in the Tertiary of the North Sea are present in low numbers but are nevertheless useful to identify the Oligocene–Miocene boundary and to correlate within the Upper Paleocene to Lower Eocene Sele and Balder formations. Their greater diversity and abundance in the Upper Oligocene–Lower Miocene has been attributed to north–south wind-driven upwelling (Thyberg *et al.*, 1999), whereas their greater numbers in the Upper Paleocene–Lower Eocene formations has been interpreted as "indicative of nutrient enhancement of surface waters due to volcanic activity" (Jones *et al.*, 2003) or resulting from "restricted circulation processes including seasonal, monsoon-driven upwelling leading to sporadic basin eutrophication and water-column stratification" (Mitlehner, 1996).

Diatoms from the Upper Oligocene–Lower Miocene and Upper Paleocene–Lower Eocene often have short stratigraphic ranges that are useful for correlations, and assemblages from the Sele and Balder formations (Figure 28.4) are recognized throughout the North Sea, from west of Shetland to the onshore, e.g. Belgium (Van Eetvalde *et al.*, 2004). Because of their depth of burial and the geothermal gradient, fossil

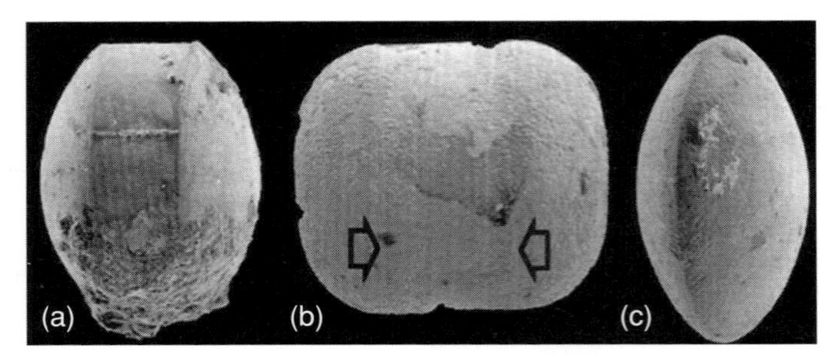

Figure 28.5 Pyritized steinkerns of three life stages of *Fenestrella antiqua* and some of their open nomenclature equivalents. (a) Vegetative cell – also known as *Coscinodiscus* sp.1. BP well 15/28a-3, 6420 feet, ?earliest Eocene, diameter 140 μm. (b) Initial cell girdle view – also known as *Coscinodiscus* "barreliformis", *Coscinodiscus* sp.1A, arrows mark valve-girdle junction. BP well 15/28a-3, 6360 feet, ?earliest Eocene. (c) Resting spore – also known as *Coscinodiscus* sp.7 or *Coscinodiscus* sp.13. BP well FC22, "Core 27", ?latest Paleocene. Original scanning electron microscope photographs by Alex Mitlehner, re-published with the permission of the Geological Society Publishing House.

diatoms from the North Sea are usually recovered as pyritized infill moulds (steinkerns) and not as frustules composed of the original amorphous silica. If not pyritized, they are sometimes preserved by calcification or re-silicified, but in any case, no internal and only limited external features are retained. Poor preservation and lack of specialists in diatom taxonomy working in the North Sea have resulted in a range of informal taxa that have been used for subsurface correlations (Bettenstaedt *et al.*, 1962; Jacqué & Thouvenin, 1975; King, 1983). For example, various stages of the life cycle of *Fenestrella antiqua* (Grun.) Swatman have been attributed to the genus *Coscinodiscus* Ehrenberg (see Figure 28.5). Mitlehner (1994) was able to identify species of normally preserved and partially pyritized diatoms from the North Sea and to correct taxonomic attributions for some taxa previously in open nomenclature (Mitlehner, 1996). Bidgood *et al.*, (1999) formally described *Fenestrella antiqua* and provided cartoon drawings of a further 46 species or varieties, but many North Sea micropaleontologists continue to use open nomenclature for consistency and because these forms are widely known by oil-industry geologists. Despite differences in nomenclature, fossil diatom assemblages have had a significant impact on our general understanding of North Sea stratigraphy (King, 1983; Mudge & Copestake, 1992a, 1992b; Jones *et al.*, 2003; Ahmali *et al.*, 2003) and several North Sea fields such as Forth/Harding, Gryphon, and Sedgwick/West Brae. Unfortunately, except for King (1983) who integrated some diatom ranges with other microfossil groups, no formal diatom ranges or biozonations have been published for the North Sea.

Figure 28.6 is an example of using diatoms for paleoenvironmental interpretations in oil and gas exploration. An outcrop section located at Wadi Abu Gaada in the Sinai, Egypt has about 170 m of massive non-calcareous mudstone that had previously been interpreted as representing a period of "deep-water" deposition beneath the carbonate compensation depth (I. Gaafar, personal communication, 1993). Calcareous marine microfossils are either absent or extremely sparse in this mudstone, but reappear at the top of the outcrop section. Recovery of marine and non-marine diatoms in the subject interval, however, revealed that deposition occurred in a brackish water/lacustrine setting (Wescott *et al.*, 2000). This revision significantly affected interpretation of the synrift history of the Gulf of Suez, the origin of its older evaporites, and interpretation of its sequence stratigraphy. The evaporites, for example, clearly formed in shallow water, not as deep-water brines, and reinterpretation of this outcrop interval as a shallow-water deposit had implications for reservoir sand distribution in the subsurface.

Fossil lacustrine diatoms can be recovered in wells that penetrate Neogene lake deposits and occasionally as allochthonous occurrences in Tertiary marine rocks. In the United States, lacustrine diatoms have proven useful for biostratigraphy in the Great Basin (Krebs *et al.*, 1987), particularly in the Snake River Basin of Idaho (Bradbury & Krebs, 1982) and in the Carson Sink, Nevada. Both basins have thick sequences of Neogene lacustrine diatomaceous rock and formations that can be characterized by their fossil diatom content. The Snake River Basin, for example, contains five formations with diatomaceous facies that range in age from middle Miocene to Pleistocene (Bradbury & Krebs, 1995). Each formation contains key fossil diatoms that can be used to correlate subsurface and outcrop sections within the Snake River Basin (Figure 28.7). In addition, fossil lacustrine diatoms provide a means of making intraformational correlations. Diatom correlations within a single lake system, such as existed during late-Miocene Chalk Hills time in the Snake River Basin, are possible because fluctuations in a lake's physical-chemical parameters are isochronous events that are reflected in its diatom assemblages.

Diatom correlations between lacustrine basins require identification and documentation of key shared marker species that have regional time significance. For example, Figure 28.8 represents the geologic ranges of non-marine *Actinocyclus* Ehrenberg species in the western United States (Krebs & Bradbury, 1995). Numerous occurrences of these species in Neogene lake deposits in the western United States have been calibrated to geologic time by means of

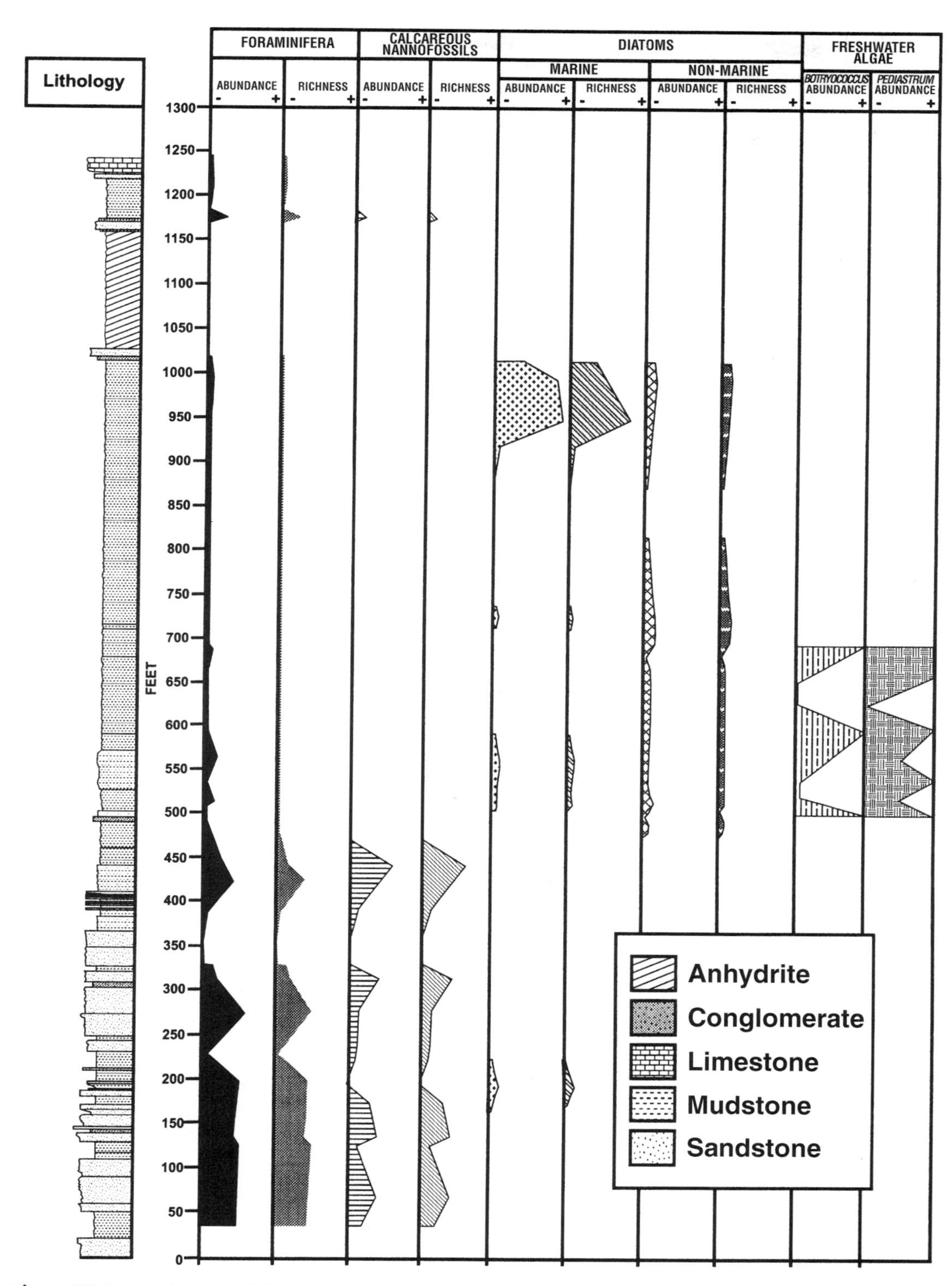

Figure 28.6 Abundance and diversity plots (biologs) of diatoms, calcareous nannofossils, foraminifera, and freshwater algae (*Botryococcus* Kützing and *Pediastrum* Meyen) from an outcrop section at Wadi Abu Gaada, Sinai, Egypt. Note the near disappearance of marine calcareous marine microfossils coincides with the appearance of non-marine diatoms. The diatomaceous mudstones had previously been interpreted as deepwater marine deposits. (Modified from Wescott *et al.* (2000).)

Age	Formation	Characteristic diatoms
Pleistocene	Bruneau Fm.	*Stephanodiscus niagarae* *Cyclotella bodanica*
Pliocene	Glenns Ferry Fm.	*Stephanodiscus carconensis* *Cyclotella pygmaea / C. hannaites* *Cyclotella elgeri*
Late Miocene	Chalk Hills Fm.	*Mesodictyon* spp. *Cyclotella* cf. *C. elgeri*
Middle Miocene	Poison Creek Fm.	*Actinocyclus venenosus* *Actinocyclus cedrus*
Middle Miocene	Sucker Creek Fm.	*Actinocyclus krasskei*

Figure 28.7 Diatoms that characterize the lacustrine facies of five Neogene–Quaternary continental formations of the western Snake River Basin, Idaho and Oregon (Bradbury & Krebs, 1995).

associated volcanic rocks that have been radiometrically dated. In some cases, other fossil groups, such as mammals and palynomorphs, provide a means of assigning relative geologic ages to these deposits and their diatom assemblages. With sufficient data, the regional geologic ranges of key fossil lacustrine diatoms can be established and used for correlations within that area. Intercontinental correlations of Neogene lacustrine rocks based upon fossil diatoms are problematic because of the provincial nature of their geologic ranges. Nevertheless, similar global patterns of lacustrine diatom succession during the Neogene have been documented (Figure 28.9) and can be useful for determining the relative geologic ages of lake deposits (Fourtanier, 1987; Fourtanier et al., 1993; Bradbury & Krebs, 1995).

28.3 Summary

The uses of diatoms in oil and gas exploration are varied. They are used as biostratigraphic markers for age dating and correlating marine and lacustrine rocks and as paleoenvironmental indicators. Diatoms are recovered from Tertiary marine and

Figure 28.8 Geologic ranges of fossil non-marine *Actinocyclus* Ehrenberg species in the western United States. The numbers refer to sites for which there are absolute ages, and the solid circles on the range line represent dated sites containing the species listed (Krebs & Bradbury, 1995).

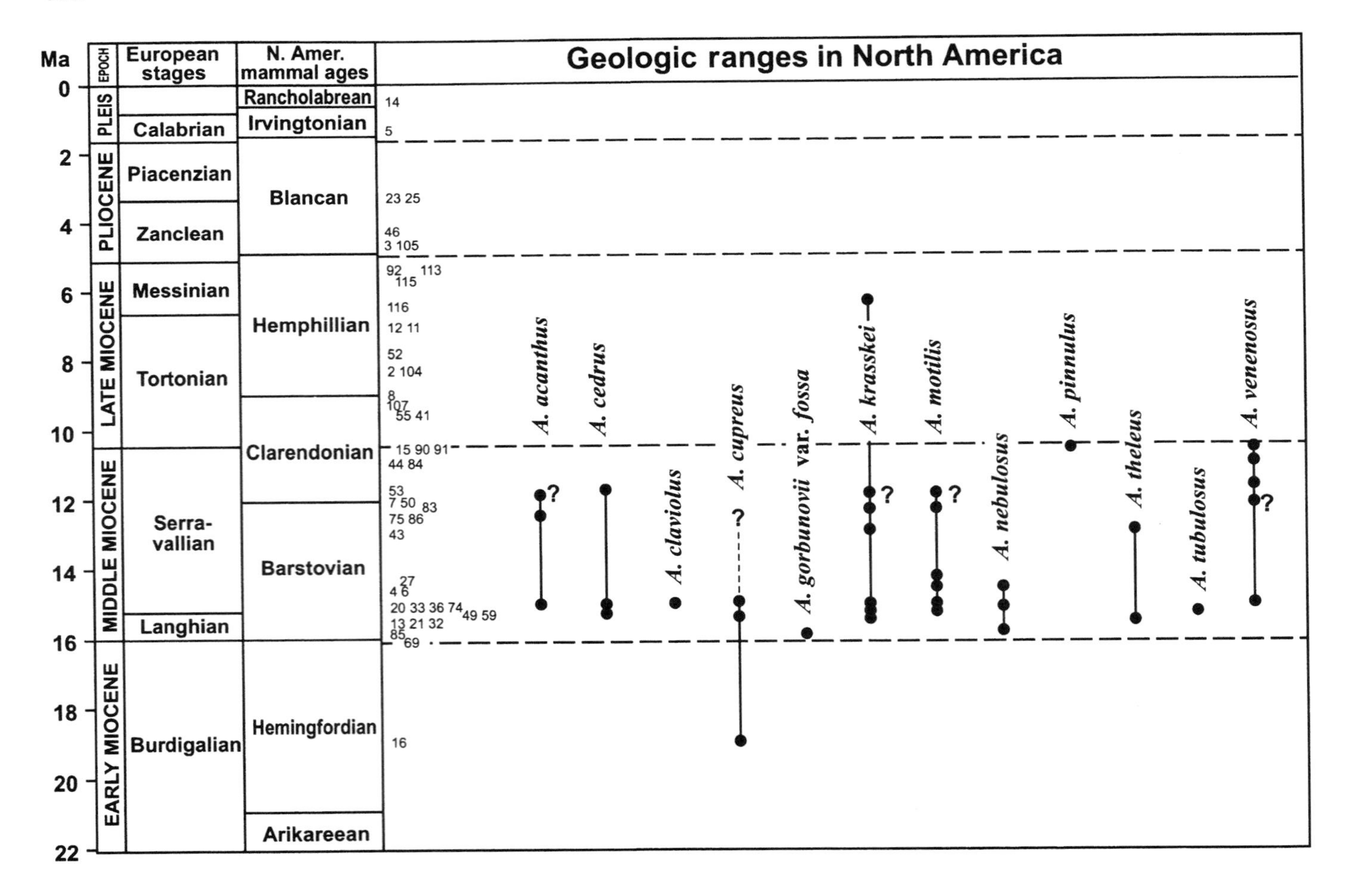

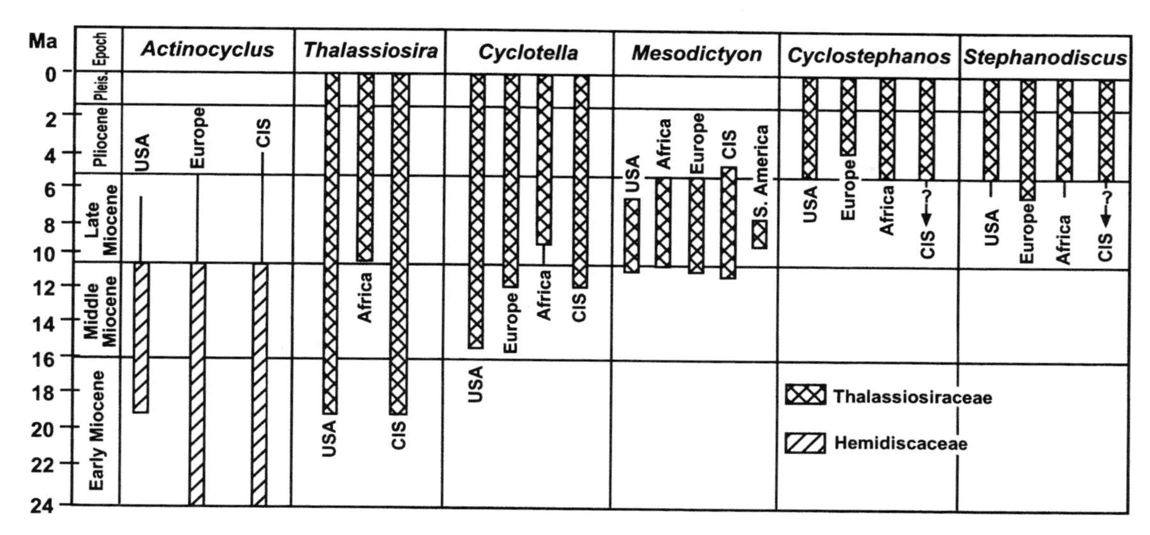

Figure 28.9 Worldwide geologic ranges of selected non-marine diatom genera. Note the turnovers during the late middle Miocene and latest Miocene. The abbreviation CIS signifies the Commonwealth of Independent States of the former Soviet Union (Bradbury & Krebs, 1995).

non-marine sediments in cores, well cuttings, and outcrops, and are particularly useful in cold-water sediments that lack calcareous microfossils and in the Upper-Cenozoic brackish water and lacustrine rocks. They are seldom recovered in their unaltered state (opaline silica) at well depths greater than 1500 m because of the burial diagenesis of silica. They may occur at greater depths in regions with low geothermal gradients, when preserved unaltered in calcareous concretions, or when pyritized, calcified, or re-silicified. Their destruction by burial diagenesis can be observed in the microscope, and it coincides with changes in rock properties and seismic definition that have important implications for the distribution of fractured siliceous reservoirs.

Acknowledgments

We thank Michael Bidgood and Alex Mitlehner for helpful discussions on North Sea diatoms, the late J. Platt Bradbury for sharing his knowledge of non-marine diatoms, and John A. Barron for his review of this manuscript.

References

Ahmali, Z., Sawyer, M., Kenyon-Roberts, S., *et al.* (2003). Palaeocene. In *The Millenium Atlas: Petroleum Geology of the Central and Northern North Sea*, ed. D. Evans, C. Graham, A. Armour, & P. Bathurst, London: Geological Society, pp. 235–59.

Aurisano, R. W., Gamber, J. H., Lane, H. R., Loomis, E. C., & Stein, J. A. (1995). Worldwide and local composite standards: optimizing biostratigraphic data. *SEPM Special Publication*, **53**, 117–30.

Baldauf, J. G. & Iwai, M. (1995). Neogene diatom biostratigraphy for the eastern Equatorial Pacific Ocean, Leg 138. *Proceedings of the Ocean Drilling Program, Scientific Results*, **138**, 105–28.

Barron, J. A. (1985a). Late Eocene to Holocene diatom biostratigraphy of the equatorial Pacific Ocean, Deep Sea Drilling Project Leg 85. In *Initial Reports of the Deep Sea Drilling Project*, **85**, 413–56.

Barron, J. A. (1985b). Miocene to Holocene planktic diatom stratigraphy. In *Plankton Stratigraphy*, ed. H. M. Bolli, J. B. Saunders, & K. Perch-Nielsen, Cambridge: Cambridge University Press, pp. 641–91.

Barron, J. A. (2005). Diatom biochronology for the early Miocene of the equatorial Pacific. *Stratigraphy*, **2** (4), 281–309.

Barron, J. A., Fourtanier, E., & Bohaty, S. M. (2004). Oligocene and earliest Miocene diatom biostratigraphy of ODP Leg 199 Site 1220, equatorial Pacific. *Proceedings of the Ocean Drilling Program, Scientific Results*, **199**. See http://www-odp.tamu/edu/publications/199_SR/2004/204.htm.

Barron, J. A. & Gladenkov, A. Y. (1995). Early Miocene to Pleistocene diatom stratigraphy of Leg 145. *Proceedings of the Ocean Drilling Program, Scientific Results*, **145**, 3–20.

Bazhenova, O. K. (2002). Oil and gas source rock potential and the presence of oil and gas in Sakhalin. In *The Cenozoic Geology and the Oil and Gas Presence in Sakhalin*, ed., Y. B. Gladenkov, Moscow: GEOS publishers, pp. 137–94 (in Russian).

Berggren, W. A., Kent, D. V., Swisher III, C. C., & Aubry, M.-P. (1995). A revised Cenozoic geochronology and chronostratigraphy. *SEPM Special Publication*, **54**, 129–212.

Bettenstaedt, F., Fahrion, H., Hiltermann, H., & Wick, W. (1962). Tertiär Norddeutschlands. In *Arbeitskreis deutscher Mikropaläontologen: Leitfossilien der Mikropaläontologie*, W. Simon & H. Bartenstein, Berlin: Gebruder Bontrager.

Bidgood, M. D., Mitlehner, A. G., Jones, G. D., & Jutson, D. J. (1999). Towards a stable and agreed nomenclature for North Sea Tertiary diatom floras – the '*Coscinodiscus*' problem. *Geological Society of London, Special Publication*, **152**, 139–53.

Bloome, C. D. & Albert, N. R. (1985). Carbonate concretions: An ideal sedimentary host for microfossils. *Geology*, **13**, 212–15.

Bradbury, J. P. & Krebs, W. N. (1982). Neogene and Quaternary lacustrine diatoms of the western Snake River Basin, Idaho-Oregon, U.S.A. *Acta Geologicamiae Scientiarum Hungaricae*, **25**, 97–122.

Bradbury, J. P. & Krebs, W. N. (1995). Fossil continental diatoms: paleolimnology, evolution, and biochronology. In *Siliceous Microfossils* ed. C. D. Blome, P. M. Whalen, & K. M. Reed, Paleontological Society Short Course 8, Knoxville, TN: The Paleontological Society, pp. 119–38.

Bramlette, M. N. (1946). The Monterey Formation of California and the origins of its siliceous rocks. *US Geological Professional Paper*, **212**.

Carney, J. L. & Pierce, R. W. (1995). Graphic correlation and composite standard databases as tools for the exploration biostratigrapher. *SEPM Special Publication*, **53**, 23–44.

Compton, J. S. (1991). Porosity reduction and burial history of siliceous rocks from the Monterey and Sisquoc formations, Point Pedernales area, California. *Geological Society of America Bulletin*, **103**, 625–36.

Edwards, L. E. (1989). Supplemented graphic correlation: a powerful tool for paleontologists and nonpaleontologists. *Palaios*, **4** (2), 127–43.

Fourtanier, E. (1987). Diatomées néogènes d'Afrique; approche biostratigraphique en milieux marin (sud-ouest africain) et continental. Ph.D thesis, Ecole Normale Supérieure de Fonteney Saint-Cloud, Fontenay Saint-Cloud, France.

Fourtanier, E., Bellier, O., Bonhomme, M. G., & Robles, I. (1993). Miocene non-marine diatoms from the western Cordilleran basins of northern Peru. *Diatom Research*, **8**, 13–30.

Gladenkov, A. Y. (1998). Oligocene and lower Miocene diatom zonation in the North Pacific. *Stratigraphy and Geological Correlation*, **6** (2), 150–63.

Gladenkov, A. Y. & Barron, J. A. (1995). Oligocene and Early Miocene Diatom Biostratigraphy of Hole 884B. *Proceedings of the Ocean Drilling Program, Scientific Results*, **145**, 21–41.

Gladenkov, Y. B. (2001). The Cenozoic of Sakhalin: modern stratigraphic schemes and correlation of geological events. *Stratigraphy and Geological Correlation*, **9** (2), 177–90.

Gladenkov, Y. B., Bazhenova, O. K., Grechin, V. I., Margulis, L. S., & Sal'nikov, B. A. (2002). *The Cenozoic Geology and Oil and Gas Presence in Sakhalin*. Moscow: GEOS publishers (in Russian).

Grechin, V. I., Pisciotto, K. A., Mahoney, J. J., & Gordeeva, S. N. (1981). Neogene siliceous sediments and rocks off southern California, Deep Sea Drilling Project Leg 63. *Initial Reports of the Deep Sea Drilling Project*, **63**, 579–93.

Harwood, D. M. & Maruyama, T. (1992). Middle Eocene to Pleistocene diatom biostratigraphy of Southern Ocean sediments from the Kerguelen Plateau, Leg 120. *Proceedings of the Ocean Drilling Program, Scientific Results*, **120**, 683–733

Hein, J. R., Scholl, D. W., Barron, J. A., Jones, A. G., & Miller, J. (1978). Diagenesis of late Cenozoic diatomaceous deposits and formation of the bottom simulating reflector in the southern Bering Sea. *Sedimentology*, **25**, 155–81.

Iijima, A. & Tada, R. (1981). Silica diagenesis of Neogene diatomaceous and volcaniclastic sediments in northern Japan. *Sedimentology*, **28**, 185–200.

Ingle, J. C. (1981). Origin of Neogene diatomites around the North Pacific Rim. *Pacific Section, SEPM Special Publication*, **15**, 159–79.

Isaacs, C. M. (1981). Porosity reduction during diagenesis of the Monterey Formation, Santa Barbara coastal area, California. *Pacific Section, SEPM Special Publication*, **15**, 257–71.

Isaacs, C. M. (1982). Influence of rock composition on kinetics of silica phase changes in the Monterey Formation. Santa Barbara area, California. *Geology*, **10**, 304–8.

Isaacs, C. M. (1983). Compositional variation and sequence in the Monterey Formation, Santa Barbara coastal area, California. *Pacific Section, SEPM Special Publication*, **28**, 117–32.

Jacqué, M. & Thouvenin, J. (1975). Lower Tertiary tuffs and volcanic activity in the North Sea. In *Petroleum and the Continental Shelf of Northwest Europe, Volume 1: Geology*, ed. A. W. Woodland, Barking: Elsevier, pp. 455–65.

Jones, E., Jones, R. W., Ebdon, C., *et al.* (2003). In *The Millenium Atlas: Petroleum Geology of the Central and Northern North Sea*, ed. D. Evans, C. Graham, A. Armour, & P. Bathurst, London: Geological Society, pp. 261–77.

King, C. (1983). *Cainozoic Micropalaeontological Biostratigraphy of the North Sea*. Institute of Geological Sciences Report 82/7. London: HMSO.

Koç, N. & Scherer, R. P. (1996). Neogene diatom biostratigraphy of the Iceland Sea Site 907. *Proceedings of the Ocean Drilling Program, Scientific Results*, **151**, 61–74.

Krebs, W. N. & Bradbury, J. P. (1995). Geologic ranges of lacustrine *Actinocyclus* species, western United States. *US Geological Survey Professional Paper*, **1543-B**, 53–73.

Krebs, W. N., Bradbury, J. P. & Theriot, E. (1987). Neogene and Quaternary lacustrine diatom biochronology, western USA. *Palaios*, **2**, 505–13.

Krebs, W. N., Wescott, W. A., Nummendal, D., Gaafar, I., & Karamat, S. (1997). Graphic correlation and sequence stratigraphy of Neogene rocks in the Gulf of Suez. *Bulletin, Société Géologique de France*, **168**(1), 63–71.

Lagle, C. W. (1984). Recovery of siliceous microfossils by the disaggregation of dolomite. *Pacific Section, SEPM Special Publication*, **41**, 185–94.

Mertz K. A., Jr. (1984). Origin of hemipelagic source rocks during the early and middle Miocene, Salinas Basin, California. *American Association of Petroleum Geologists, Bulletin*, **73** (4), 510–24.

Miller, F. X. (1977). The graphic correlation method in biostratigraphy. In *Concepts and Methods in Biostratigraphy*, ed. E. G. Kauffman & J. E. Hazel, Stroudsburg, PA: Dowden, Hutchison, and Ross, pp. 165–86.

Mitlehner, A. G. (1994). The occurrence and preservation of diatoms in the Paleogene of the North Sea Basin. Unpublished Ph.D. thesis, University College, London.

Mitlehner, A. G. (1996). Palaeoenvironments in the North Sea Basin around the Paleocene–Eocene boundary: evidence from diatoms and other siliceous microfossils. *Geological Society of London, Special Publication*, **101**, 255–73.

Mudge, D. C. & Copestake, P. (1992a). Revised lower Paleogene lithostratigraphy for the outer Moray Firth, North Sea. *Marine and Petroleum Geology*, **9**, 53–69.

Mudge, D. C. & Copestake, P. (1992b). Lower Paleogene stratigraphy of the Northern North Sea. *Marine and Petroleum Geology*, **9**, 287–301.

Murata, K. J. & Larson, R. R. (1975). Diagenesis of Miocene siliceous shale, Temblor Range, California. *US Geological Survey Journal of Research*, **3**, 553–66.

Pisciotto, K. A. (1981). Diagenetic trends in the siliceous facies of the Monterey Shale in the Santa Maria region, California. *Sedimentology*, **28**, 547–71.

Scherer, R. P. & Koç, N. (1996). Late Paleogene diatom biostratigraphy and paleoenvironments of the northern Norwegian-Greenland Sea. *Proceedings of the Ocean Drilling Program, Scientific Results*, **151**, 75–99.

Schwartz, D. E. (1987). Lithology, petrophysics, and hydrocarbons in cyclic Belridge Diatomite, south Belridge oil field, Kern Co., California. Fourth International Congress on Pacific Neogene Stratigraphy, Berkeley, CA, July 29–31, abstract volume, 102.

Shaw, A. B. (1964). *Time in Stratigraphy*. New York, NY: McGraw-Hill.

Sheshukova-Poretskaya, V. S. (1967). *Neogene Marine Diatoms from Sakhalin and Kamchatka*, Leningrad: Leningrad State University Press (in Russian).

Tada, R. & Iijima, A. (1983). Petrology and diagenetic changes of Neogene siliceous rocks in northern Japan. *Journal of Sedimentary Petrology*, **53**, 911–30.

Thyberg, B. I., Stabell, B., Faleide, J. I., & Bjørlykke, K. (1999). Upper Oligocene diatomaceous deposits of the northern North Sea – silica diagenesis and paleogeographic implications. *Norsk Geologisk Tidsskrift*, **79**, Issue 1, December, 3–18.

Van Eetvelde, Y., Dupuis, C., & Cornet, C. (2004). Pyritized diatoms: a good fossil marker in the Upper Paleocene–Lower Eocene sediments from the Belgian and Dieppe–Hampshire basins. *Geologie en Mijnbouw*, **83** (3), 173–8.

Wescott, W. A., Krebs, W. N., Bentham, P. A., & Pocknall, D. T. (2000). Miocene brackish water and lacustrine deposition in the Suez Rift, Sinai, Egypt. *Palaios*, **15**, 65–72.

29 Forensic science and diatoms

ANTHONY J. PEABODY AND
NIGEL G. CAMERON

29.1 Introduction

The legal process has used scientific procedures for many years in its various deliberations. Some of these, for instance DNA profiling of body fluids, are now essential and routine practice. The use of diatoms in forensic science is naturally much smaller, but in certain types of investigation, diatom taxonomy and ecology play a significant role. A diatomist may be able to provide investigations with evidence, which will enable the court to reach its verdict, and may be used by either the prosecution or the defence. Below we summarize some of the major applications of diatoms to forensic science.

29.2 Drowning

The most frequent application of diatoms in forensic science is in diagnosis of death by drowning. Drowning is a very common accidental cause of death, and thousands die each year in this fashion. The majority of these individuals die in circumstances that are not contentious, where there are witnesses, or strong indications of suicide, such as a note. Where circumstances surrounding an individual's death are less clear, it is often important to be as certain as possible of how death occurred.

Where a body is fresh, the pathologist may have little difficulty in reaching a verdict of drowning. However, the histopathological signs of drowning are often transient and overlaid by the grosser effects of decomposition. Additionally, in cases where an individual has been severely injured before being immersed in water, it is obviously important to determine whether death is due to these injuries or because of drowning. It is in cases such as these, where the outcome of the investigation into drowning may add years to a person's prison sentence, that forensic scientists may be asked to assist the pathologist in reaching a decision.

Timperman (1972) describes mechanisms of the drowning process and Peabody (1980) provides an overview of the problems associated with diatoms and drowning. Some authors have referred to "dry" drowning, which is more accurately described as death by a variety of causes whilst immersed in water, for instance vagal inhibition, cardiac arrest, or laryngeal spasm. These causes of death are not considered "true" drowning, because death is not the result of water entering the lungs. It is in cases of drowning where water enters the lungs that forensic scientists can make a contribution.

When an individual drowns, water enters the lungs in increasing amounts and enters the blood stream through ruptures in the peripheral alveoli. Any particulate material in the water is also carried in, and whilst the heart continues to pump, this material is carried through the body. There is probably some deposition of this particulate material in the capillary beds of major organs. In most bodies of natural water, this particulate material will include diatoms, and the numbers present and their types will be a reflection of the ecology of that particular area.

Much of the pioneer work in this area was carried out in wartime Hungary, and was reported first by Incze (1942), and later by his co-workers. This early important work is fully described in Peabody (1980).

In practice, most drowning cases considered are from a freshwater location, although from time to time saltwater drownings are submitted. In most instances, the question to be answered is "did this person drown?", rather than "where did this person drown?" In answering the first question, it is usually straightforward to answer the second, in the sense that any diatoms recovered from the body can be compared with diatoms from the drowning scene. An example of this is given by Ludes *et al.* (1999)

The Diatoms: Applications for the Environmental and Earth Sciences, 2nd Edition, eds. John P. Smol and Eugene F. Stoermer. Published by Cambridge University Press.

Case. The body of an unidentified woman was found at the mouth of an estuary. Diatoms recovered from the body indicated that she had drowned in freshwater rather than in the sea or in brackish water. When her identity was established, her home was found to be some 60 km upstream from where she was found. Following enquiries it was suggested that she had drowned near her home.

It is also common for drowning to occur in a domestic environment, particularly in the bath, or where the body of a newborn infant is found in the water of a lavatory pan following delivery. Domestic water supplies, for example in the UK, generally contain very few diatoms and consequently the forensic scientist is unable to assist in these cases. However, in regions where diatoms are present in the domestic water supply in significant numbers, it has been reported that diatom analysis can assist in the diagnosis of drowning (Pollanen, 1998).

The essential problem in sample preparation is to recover a small number of diatoms from a large organic mass. As few as five or six diatoms might be involved, and, in addition, there is danger of contamination during the extraction procedure. To compound this, many human organs normally contain diatoms, which entered the body during life via a number of possible pathways (e.g. Yen Yen & Jayaprakash, 2007). The brain, liver, and lung of 13 non-drowned subjects contained between 0 and 19 diatoms in 100 g samples. Most were to be found in the lung. On examination of kidney from the same subjects, much larger numbers of diatoms were recovered, up to 150 diatoms per 100 g of organ. Common diatoms were fragments of *Coscinodiscus* Ehrenberg, *Hantzschia amphioxys* (Ehrenberg) Grunow, and *Aulacoseira granulata* (Ehrenberg) Simonsen. The presence of these diatoms has misled many investigators unfamiliar with diatoms, and has prompted them to suggest that this line of enquiry is unproductive. The successful investigator must be able to recognize these "naturally occurring" diatoms for what they are, and to record those diatoms that may be present as a result of drowning.

Few diatomists have been concerned with the problem of diagnosing death by drowning, and some of those who have, for instance Geissler and Gerloff (1966) and Foged (1983), are of the opinion that the diatom test is invalid. Foged (1983) found quite large numbers of diatoms in his examination of organs from non-drowned subjects. It is impossible to tell whether this might be due to contamination, but at the very least should encourage care and thought in the use of the diatom test.

It is evident, therefore, that only an experienced diatomist working in conjunction with the pathologist should make this type of investigation. The diatomist's role is simply to assist the pathologist in making a diagnosis; diatomists should therefore be careful not to make thier own diagnosis but provide evidence to aid the pathologist – the latter has the ultimate job of diagnosing the cause of death.

Examination results are usually expressed in a form convenient to a pathologist, bearing in mind that one's own opinions must be free standing and capable of being defended;

> *. . . have examined the organs from John Doe and I have found diatoms in them that could have originated from River Running at Anytown. It is my opinion therefore, that drowning should be considered as a contributory cause of death.*
>
> or:
>
> *. . . have examined the organs from John Doe and I have found no diatoms in them which could have come from River Running at Anytown. I am therefore unable to indicate whether or not John Doe has drowned.*

Whilst it is possible to draw a conclusion from the presence of diatoms, the absence of diatoms cannot be taken to indicate that drowning *has not* occurred. Drowning could have occurred, but either no diatoms were infiltrated, or their numbers were so low that they were not detected.

29.2.1 Sample preparation

Many methods have been proposed for extraction of diatoms from organs (Peabody, 1980), and most have their drawbacks, including those used today. A method which minimizes contamination risks has been developed using 100 g samples of liver, lung, and brain.

The lung is included as it can be used as an indication of how many diatoms have actually reached the circulatory system. The absence of diatoms from the lung, or their presence in only low numbers, suggests that they will not be present elsewhere in the body. A large number of diatoms in the lung will indicate that water probably has reached the circulatory system, and that drowning may be confirmed by examination of the liver and brain. Examination of the lung exudate will provide a preliminary indication of how many diatoms are present. The type of diatoms present is also important in that it gives an early indication as to the site of drowning.

Often the liver is examined because it is a well-vasculated major body organ. The brain is examined because it is remote from the lungs and preserved to some extent from damage and the exterior by the cranium.

29.2.2 Extraction method

It is essential that all laboratory ware is new, unused, and cleaned before use. Likewise, all reagents must be passed through a membrane filter before use, including concentrated acid.

1. Weigh out roughly 100-g samples of liver, lung, and brain into evaporating dishes, taking account of any possible health hazards in dealing with body organs, which have an unknown history. Stringent safety precautions are advised at all stages of this extraction.
2. Transfer the dishes to a muffle furnace and char at 250 °C for 3–5 hours.
3. Ash these samples overnight at 550 °C. Allow to cool. Operating below this temperature results in incomplete ashing, whilst higher temperatures result in some destruction of any diatoms present
4. The resultant white ash can be dissolved in concentrated hydrochloric acid (HCl), followed by washing in distilled water.
5. If any organic material is evident at this point, through incomplete ashing, a standard diatom preparation using concentrated sulfuric acid (H_2SO_4) and sodium nitrate ($NaNO_3$) will suffice.
6. Mount the washed residue in a high refractive index mountant, and examine microscopically.

Whilst not perfect, the above method ensures that contamination risks are minimized, and it copes with large amounts of organic material. Parts of the process require constant attention, in particular dissolving the ash in concentrated HCl. Once samples are in the furnace, toxicological and pathological health risks are minimized, but the resultant smoke can be offensive and a possible health hazard unless ducted well away.

An enzymatic digestion method has been suggested by Ludes and Coste (1996) based on a toxicological method originally used for the extraction of drug metabolites. This type of enzymatic approach to diatom preparation has also been compared with other digestion methods (Ming *et al.*, 2007).

29.3 Diatoms on wet clothing

When individuals run or walk through a garden pond or a stream, or indeed any body of water, particulate material in the water (including diatoms) can remain on clothing afterwards. Diatoms can be recovered from the clothing and compared with diatoms from the water under consideration. It is not unknown for other fairly resistant algal material such as *Oedogonium* Hirn and *Phacus* (Ehrenberg) Dujardin to be recovered from clothing after immersion in water containing these organisms. Diatoms, however, are especially resistant and are not usually visible to the naked eye, and are therefore less likely to be removed by an alleged offender.

Case. A man gained access to a house through the garden. He stepped into a garden pond which made his jeans wet to the knee. On arrest shortly after the offence, he maintained that his jeans were wet because he had been in the sea. Large numbers of freshwater diatoms of the same types as those in the garden pond were found on his jeans. There were no marine diatoms. He received a nine-year prison sentence.

Diatoms can be removed simply by rinsing either the entire or, more specifically, the affected area in water. It is useful to examine any dried material on the garment for the presence of any other algal material. Allow the residue to settle and examine microscopically.

Case. A man denied being the driver of a car that was found submerged in a farm pond. He denied any connection with the pond. Another man was found dead in the car, having failed to escape. Despite his denials, diatoms from the farm pond were found on the clothing of the putative driver, and he received a five-year prison sentence.

The absence of diatoms from clothing, which is thought to have been through water containing diatoms, could be significant in indicating that the wearer has not in fact been in the water. The significance may depend on the type of water under consideration, the numbers of diatoms in the water, and the type of garment worn. The absence of diatoms from oilskins briefly immersed in the sea may not be significant, whereas failure to find diatoms on trousers after alleged immersion in a farm drainage ditch is noteworthy.

During the Cold War, diatom analysis played a controversial role in the official investigation into the loss at sea of a British fishing vessel, the *FV Gaul*. The *Gaul* and all 36 crewmembers were lost without trace in January 1974 in the Arctic, north of Norway. In May 1974, a lifebuoy from the *Gaul* was recovered from the sea and diatom analysis was carried out on material taken from the lifebuoy. The diatom species present on the lifebuoy indicated an absence of deep-water marine diatoms and the presence of freshwater diatoms. At the time this led to "considerable and prolonged controversy" about the fate of the *FV Gaul* with the suggestion that the lifebuoy had been planted to disguise the fact that the loss of the ship was not accidental. However, the wreck of the *Gaul* was located in 1997 in the area where the ship and her crew were presumed to have been lost in heavy seas (Steel, 2004).

There is an increasing use of ecological techniques in forensic investigations and in some of these cases diatom analysis can provide useful supporting evidence. However, although they may often be present on exhibits, diatoms are less frequently used than, for example, pollen or mineralogical analyses (e.g. Morgan *et al.*, 2006). Where the diatom assemblages associated with exhibits and control samples are in adequate numbers and of sufficient diversity quantitative techniques can be employed.

For instance, a freshwater lake in New England, USA was the scene of a serious assault. Analysis of diatoms in mud taken from the lake and from the suspect's and the victim's shoes linked the persons suspected of committing the assault to the lake (Siver *et al.*, 1994). Here diatom abundance and population ratios for a group of *Eunotia* spp. (Ehrenberg) showed marked similarities between the samples. Siver *et al.* also recorded the presence of many chrysophyte scales derived from a single species (*Mallomonas caudata* (Ivanov) Willi & Krieger) in the three groups of samples.

29.4 Safe ballasts

In the past, manufacturers of safes have used a variety of materials as insulation, fireproofing, or protection. Some old safes have diatomaceous earth (see Harwood, this volume) packed between the outer casing and the inner compartment, but since this material provides no extra security apart from being fire resistant, manufacturers have stopped using it. However, old safes are likely to be attacked, and should investigators recognize the safe for what it is, valuable evidence can be adduced (Peabody, 1971).

When such a safe is broken open, the diatomaceous earth spills out as a fine white powder, which will settle on clothing, skin, and other surfaces. It is often invisible to the naked eye, but a small sample from a suspect surface can reveal hundreds or thousands of diatom frustules. Since these ballasts are often taken from long-since defunct diatomite workings, the chances of an individual having this material on clothing by chance are extremely small.

Case. A safe was stolen from an office and taken to a house where it was opened. It contained diatomaceous safe ballast, which spilled out. The safe was then taken in a car and dumped. Three suspects had diatomaceous material on their clothing matching the safe. The same was found in the house of one of the suspects, and in the car of another. All three received prison sentences.

Cases involving diatomaceous safe ballasts provide strong evidence of association relatively easily and quickly. However, diatomaceous safe ballasts are now uncommon, and cases involving them are rarely submitted.

29.5 Other applications

Diatomaceous earth is used in a variety of materials that are often examined by forensic scientists, such as paint, polishes, and matches. Current supplies of diatomaceous earth are restricted to a few mining sites, and consequently the same material can be used in a large number of items. The examination of car and metal polish, however, can often provide good evidence of association (Peabody, 1977). The diatoms in small flakes of white paint can sometimes provide additional evidence, but the titanium dioxide (TiO_2) in paint must be dissolved first so that the diatoms can be seen. The TiO_2 can be dissolved by boiling in concentrated H_2SO_4 saturated with Na_2SO_4. Great care should be taken with this procedure.

Diatoms may be used not only to link an exhibit with a known scene of crime but also in exploring possible but unknown sites of contact for an item such as a shoe or vehicle contaminated with sediment. In a quality-assurance exercise carried out for the Forensic Science Service (FSS) in the UK a series of "blind" samples were prepared, by immersing fleece jackets briefly, and analyzed for diatoms. These samples were taken from locations unknown to the diatomist at lowland, mountain, and coastal sites in Central England and North Wales. After percentage diatom counting and analysis, qualitative predictions about the general environment and water quality from which the diatom assemblages were derived were then made. These were then used successfully to predict the general types of site from which the samples were taken, for example upland acid lake, lowland nutrient-rich moving water, slightly brackish estuary. Although the success of this demonstration is of no surprise to diatom analysts it was nevertheless of value in convincing a lay audience of the value of diatoms in forensic ecology. In this context the large diatom-environmental data bases that have been constructed for ecological and paleoecological projects may be employed to increase the accuracy of this type of forensic biogeography.

Similar in principle to the use of forensic entomology in estimating the time of death when a significant period has elapsed before the discovery of a body, experimental work on forensic pathology (Casamatta & Verb, 2000; Zimmerman & Wallace, 2008) indicates that the species diversity and sequence of diatom species and other algal taxa colonizing submerged animal carcasses has some potential for estimating the period of submersion of a body or other materials.

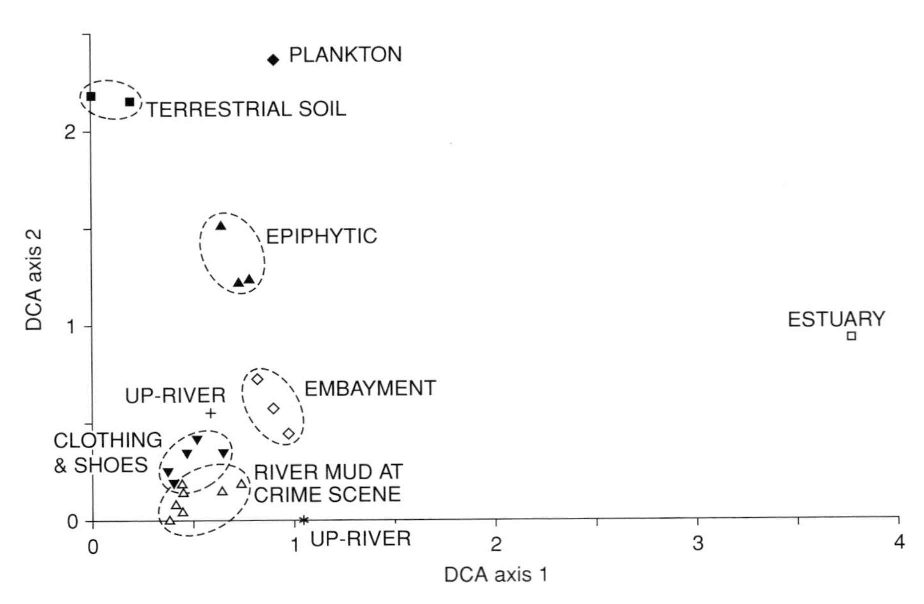

Figure 29.1 Scatterplot of diatom assemblages on the first two axes of variation from a detrended correspondence analysis (DCA). The distance between points shows the degree of similarity of the diatom assemblages. This illustrates that diatom assemblages taken from the clothing and shoes of a suspect are most similar to river mud where a murder victim was found. Diatom assemblages from other locations along the same river, the riverbank, and from vegetation and plankton samples provide poorer analogs. These diatom data suggest that the clothing and footwear was in direct contact with submerged mud in the river. (From Cameron (2004) with permission from the Geological Society of London.)

29.6 Giving evidence

A forensic scientist must be able to communicate often complicated, scientific issues to laymen who may not have any scientific background, and for whom evidence of this type is difficult to understand. This is particularly true of diatom evidence, which will almost certainly be unfamiliar to most of the individuals to whom it is presented, and who also need to understand it.

During a murder investigation, percentage diatom analysis was used to compare the diatom composition of mud on a suspect's discarded clothing and footwear with the diatom flora along a stretch of the River Avon, UK, in which the body of the victim was found (Cameron, 2004). Here ordination was employed as a means of simplifying and presenting these data graphically to the jury in order to show the best analogs from amongst a group of test samples (Figure 29.1). Multivariate analysis of diatom data was also employed by the defence team in this case to challenge the prosecution diatom evidence (Horton *et al.*, 2006). The overall concurrence of prosecution and defence diatom analytical results provided supporting evidence to assist in securing the conviction of the suspect.

29.7 Summary

Diatoms are used for a number of purposes in forensic science. The best known of these applications is in assisting the pathologist in the diagnosis of death by drowning. The importance of identifying the diatom assemblages associated with human organs and recognizing "naturally occurring" species is emphasized. Diatoms have also been used to associate or disassociate items such as clothing and motor vehicles with scenes of crime. The forensic analysis of traces of materials manufactured from diatomite is exemplified by the former use of diatomaceous earth in the lining of safes, and also paint, polishes, and matches. The communication of diatom evidence to laymen in court and the use of multivariate statistical methods to analyze diatom ecological data is discussed.

Acknowledgments

The authors wish to thank the reviewer Stefan Uitdehaag, and the editors John Smol and Eugene Stoermer for their assistance.

References

Cameron, N. G. (2004). The use of diatom analysis in forensic geoscience. *Geological Society of London, Special Publication* **232**, 277–80.

Casamatta, D. A. & Verb, R. G. (2000). Algal colonization of submerged carcasses in a mid-order woodland stream. *Journal of Forensic Sciences*, **45**, 1280–5.

Foged, N. (1983). Diatoms and drowning – once more. *Forensic Science International*, **21**, 153–9.

Geissler, U. & Gerloff, J. (1966). Das Vorkommen von Diatomeen in menschlichen Organen und in der Luft. *Nova Hedwigia*, **10** (3/4), 565–77.

Horton, B. P., Boreham, S., & Hillier, C. (2006). The development and application of a diatom-based quantitative reconstruction technique in forensic science. *Journal of Forensic Science*, **51**, 643–50.

Incze, G. (1942). Fremdkorper in Blutkreislauf Ertrunkener. *Zentralblatt für allgemeine Pathologie und pathologische Anatomie*, **79**, 176.

Ludes, B. & Coste, M. (1996). *Diatomées et médecine légale*. Paris: Tec & Doc Lavoisier.

Ludes, B., Coste, M., North, N., *et al.* (1999). Diatom analysis in victim's tissues as an indicator of the site of drowning. *International Journal of Legal Medicine*, **112**, 163–6.

Ming, M., Meng, X., & Wang, E. (2007). Evaluation of four digestive methods for extracting diatoms. *Forensic Science International*, **170**(1), 29–34.

Morgan, R. M., Wiltshire, P., Parker, A. G., & Bull, P. A. (2006). The role of forensic geoscience in wildlife crime detection. *Forensic Science International*, **162**, 152–62.

Peabody, A. J. (1971). A case of safebreaking involving diatomaceous safe-ballast. *Journal of the Forensic Science Society*, **11**, 227.

Peabody, A. J. (1977). Diatoms in forensic Science. *Journal of the Forensic Science Society*, **17**, 81–7.

Peabody, A. J. (1980). Diatoms and drowning – a review. *Medicine, Science and the Law*, **20** (4), 254–61.

Pollanen, M. S. (1998). *Forensic Diatomology and Drowning*. Amsterdam: Elsevier.

Siver, P. A., Lord, W. D., & McCarthy, D.J. (1994). Forensic limnology: the use of freshwater algal community ecology to link suspects to an aquatic crime scene in southern New England. *Journal of Forensic Sciences*, **39**, 847–53.

Steel, The Honourable Mr Justice David (2004). *Report of the re-opened formal investigation into the loss of the FV Gaul*. London: Her Majesty's Stationery Office.

Timperman, J. (1972). The diagnosis of drowning – a review. *Forensic Science*, **1**, 397–409.

Yen Yen, L. & Jayaprakash P. T. (2007). Prevalence of diatom frustules in non-vegetarian foodstuffs and its implications in interpreting identification of diatom frustules in drowning cases. *Forensic Science International*, **170** (1), 1–7.

Zimmerman, K. A. & Wallace J. R. (2008). The potential to determine a postmortem submersion interval based on algal/diatom diversity on decomposing mammalian carcasses in brackish ponds in Delaware. *Journal of Forensic Sciences*, **53** (4), 935–41.

30 Toxic marine diatoms

MARIA CÉLIA VILLAC, GREGORY J. DOUCETTE, AND IRENA KACZMARSKA

30.1 Introduction

Diatoms are major contributors to total primary production as well as many important biogeochemical processes in aquatic environments (Falkowski *et al.*, 1998; Smetacek, 1999). Nonetheless, a small number of species (<30) have been recognized as harmful to fisheries, wildlife, or people, through production of either a toxin or various exudates, or via mechanical damage due to cell morphology and/or high biomass accumulation. Fryxell & Villac (1999), and more recently Fryxell & Hasle (2003), have identified several of these harmful taxa and outlined their often devastating impacts on other organisms, ecosystems, and economies. Examples include: oily surface films associated with bird mortalities (*Coscinodiscus centralis* Ehrenb., *Coscinodiscus concinnus* W. Smith); surface accumulations on beach surf-zones affecting tourism/recreation (*Asterionellopsis glacialis* [Castracane] Round, *Anaulus australis* Drebes & D. Schulz; in Villac & Noronha, 2008); mucilage production causing a condition known as "mare sporco" (or dirty sea) (*Ceratoneis closterium* Ehrenb., *Pseudo-nitzschia pseudodelicatissima* [Hasle] Hasle) as well as clogging of bivalve gills (*Thalassiosira mala* Takano) and fishing nets (*Guinardia striata* [Stolterfoth] Hasle, *Coscinodiscus wailesii* Gran & Angst); high biomass accumulations resulting in shading and depletion of oxygen/nutrients (*C. wailesii*) as well as clogging gills of benthic shellfish and bony fish (*Cerataulina pelagica* [Cleve] Hendey); and spines/setae inflicting physical damage to fish gills leading to major financial losses for aquaculture operations (*Chaetoceros convolutus* Castracane and *Chaetoceros concavicornis* Mangin). In the case of toxin production, members of the genus *Pseudo-nitzschia* H. Peragallo (Figure 30.1) capable of synthesizing the neurotoxin domoic acid (DA) have received and continue to attract considerable attention in the literature, due likely to their profound effects on human and animal health. Here we build on the foundation of the previous version of this chapter (Fryxell & Villac, 1999) with the aim of highlighting recent advances towards a better understanding of these toxin-producing organisms.

The Diatoms: Applications for the Environmental and Earth Sciences, 2nd Edition, eds. John P. Smol and Eugene F. Stoermer. Published by Cambridge University Press.

There are over 30 described species of *Pseudo-nitzschia* of which 12 are known to produce DA (Table 30.1). Since 1999, there has been a great deal of work on DA producers and their non-toxigenic relatives within this genus. For example, the website on "Domoic Acid and *Pseudo-nitzschia* References," organized by the Department of Fisheries and Oceans (Canada), lists 1189 references, of which *c.* 60% have been published between 1998 and 2010. Details about their taxonomy, morphology, and ecophysiology are treated elsewhere (but see discussions and respective references below) and will be touched upon as we describe findings for certain species. Field and laboratory work have provided a better understanding of life cycles, biogeography, species detection, bloom dynamics, and trophic transfer of DA in marine food webs. All of these aspects can influence issues relevant to fisheries, mariculture, and tourism, as well as ecosystem and public health.

30.2 Life cycles

Diatoms spend most of their lifespan in a diploid, vegetative state with most individuals propagating themselves mitotically. Therefore, in a typical population of any diatom species, there will be many more individuals than there are genotypes, which is unlike complex animals and plants. Members of the genus *Pseudo-nitzschia* conform to the cycle of vegetative phase and sexuality (Figure 30.2). Depending on the rate of growth, degree of frustule silicification (thickness), and size of parental sex cells, the vegetative phase may last years and result in production of millions of genetically identical (clonal) cells. Due to the nature of diatom valve genesis (only smaller sibling valves are produced inside the existing parental frustule), two products

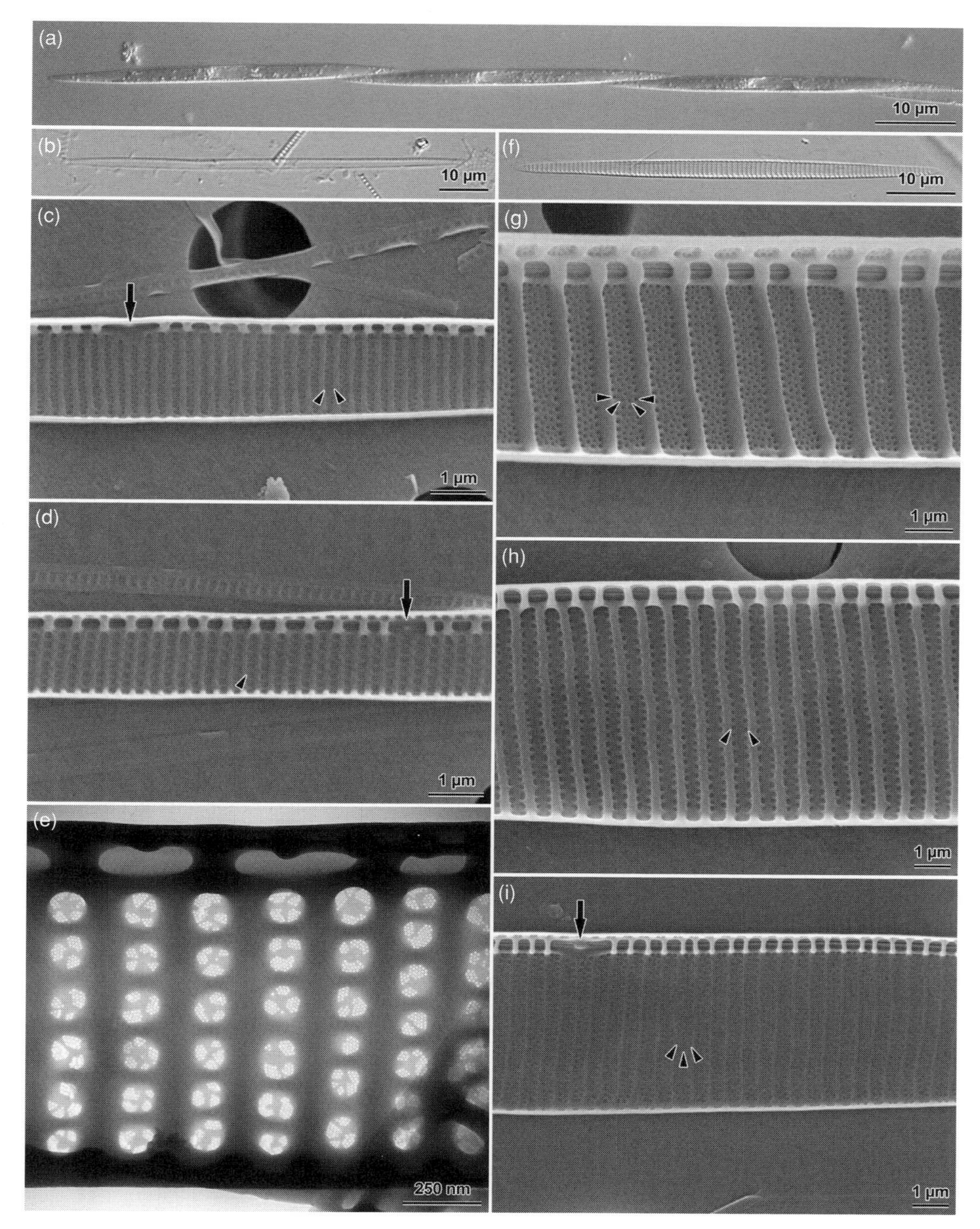

Figure 30.1 Examples of live cells and cleaned valves of some toxic *Pseudo-nitzschia* species demonstrating level of detail available with varying preparation and imaging techniques, that is, light microscopy (LM), scanning electron microscopy (SEM) and transmission electron microscopy (TEM). (a) Live *Pseudo-nitzschia* sp. cells growing in diagnostic stepped-chain colony (LM). Note two chloroplasts per cell. (b) *P. delicatissima* LM "counting category" valve, width ≤ 3 µm (Hasle, 1965) in cleaned preparation. Note striae detail is unresolvable. (c) Internal valve view of *P. delicatissima*-complex valve. Note central interspace (arrow) and two rows of small pores (arrowheads) in the striae (SEM). (d) Internal valve view of *P. pseudodelicatissima*-complex valve (SEM). Note central interspace (arrow) and a row of large, square pores (arrowhead) in the striae. (e) Pore structure of *P. pseudodelicatissima*-complex valve, typically only resolvable in TEM. Note variability of pore structure in a single stria, underscoring difficulty in species identification. (f) *P. seriata* "counting category" valve, width >3 µm (Hasle, 1965) in cleaned preparation (LM). Striae, fibulae, and presence/absence of central interspace are normally resolvable in LM preparation. (g) Internal view of *P. multiseries* valve (SEM) with multiple rows of pores in one stria (arrowheads). (h) Internal view of *P. australis* valve (SEM) with two rows of widely separated pores (arrowheads) in one stria. (i) Internal view of *P. fraudulenta* valve (SEM) with central interspace (arrow) and two (sometimes three) rows of large pores in one stria (arrowheads).

Table 30.1 List of *Pseudo-nitzschia* formally described to date, showing that some species, but not all, have been tested for domoic acid (DA) production. Those that have tested positive for DA, at least once, are noted with a "+" sign (references in Bates & Trainer, 2006). Those that have tested negative are noted by a "–" sign with a respective reference according to the following numbers in superscript: [1]Villac *et al.* (1993); [2]Quijano-Scheggia *et al.* (2009b); [3]Lundholm *et al.* (2002); [4]Lundholm *et al.* (2003); [5]Lundholm *et al.* (2006); [6]Amato & Montresor (2008); [7]Hasle & Lundholm (2005); [8]Villac & Fryxell (1998); [9]Churro *et al.* (2009); [10]Fryxell *et al.* (1991).

Pseudo-nitzschia species	DA	*Pseudo-nitzschia* species	DA
P. americana (Hasle 1964) G. A. Fryxell in Hasle 1993	– (1)	*P. micropora* Priisholm, Moestrup & Lundholm 2002	
P. arenysensis Quijano-Scheggia, Garcés & Lundholm in Quijano-Scheggia *et al.* 2009b	– (2)	*P. multiseries* (Hasle 1965) Hasle 1995	+
P. australis Freng. 1939	+	*P. multistriata* (Takano 1993) Takano 1995	+
P. brasiliana Lundholm, Hasle & G. A. Fryxell in Lundholm *et al.* 2002	– (3)	*P. obtusa* (Hasle 1965) Hasle & Lundholm 2005	– (7)
P. caciantha Lundholm, Moestrup & Hasle in Lundholm *et al.* 2003	– (4)	*P. prolongatoides* (Hasle 1965) Hasle 1993	
P. calliantha Lundholm, Moestrup & Hasle in Lundholm *et al.* 2003	+	*P. pseudodelicatissima* (Hasle 1965) Hasle 1993 *emend.* Lundholm, Hasle & Moestrup in Lundholm *et al.* 2003	+
P. cuspidata (Hasle 1965) Hasle 1993 *emend.* Lundholm, Hasle & Moestrup in Lundholm *et al.* 2003	+	*P. pungens* (Grunow ex Cleve 1882) Hasle 1993	+
P. decipiens Lundholm & Moestrup in Lundholm *et al.* 2006	– (5)	*P. pungens* var. *cingulata* Villac in Villac & Fryxell 1998	– (8)
P. delicatissima (Cleve 1897) Heiden in Heiden & Kolbe 1928	+	*P. pungens* var. *aveirensis* Lundholm, Churro, Carreira & Calado in Churro *et al.* 2009	– (9)
P. dolorosa Lundhlom & Moestrup in Lundholm *et al.* 2006	– (5)	*P. pungiformis* (Hasle 1971) Hasle 1993	
P. fraudulenta (Cleve 1897) Hasle 1993	+	*P. roundii* Hernández-Becerril 2006	
P. galaxiae Lundholm & Moestrup 2002	+	*P. seriata* (Cleve 1883) Peragallo 1897–1908	+
P. granii (Hasle 1964) Hasle 1993		*P. sinica* Y.Z.Qi & JuWang 1994	
P. granii var. *curvata* (Hasle 1964) Hasle 1993		*P. subcurvata* (Hasle 1964) G.A.Fryxell 1993	– (10)
P. heimii Manguin 1957		*P. subfraudulenta* (Hasle 1965) Hasle 1993	
P. inflatula (Hasle 1965) Hasle 1993		*P. subpacifica* (Hasle 1965) Hasle 1993	
P. linea Lundholm, Hasle & G.A.Fryxell in Lundholm *et al.* 2002		*P. turgidula* (Hust. 1958) Hasle 1993	+
P. lineola (Cleve 1897) Hasle 1993		*P. turgiduloides* (Hasle 1965) Hasle 1993	
P. manni Amato & Montresor, 2008	– (6)		

of mitosis will differ in cell size: one will replicate the parent size whereas the other will be slightly smaller. This peculiarity leads to an ever-growing number of post-sexual cell series of declining size in a typical population. The most common means of reintroducing large cell sizes into local populations is through sexual reproduction. This simultaneously introduces new combinations of genes (possibly including those associated with toxigenicity) in a cohort of sexual progeny that begins a new mitotic phase.

Factors triggering induction of sexuality are incompletely known. In culture, these factors involve a specific size range when cells are readily sexualized, possibly specific stages in a cell cycle (Armbrust *et al.*, 1990), environmental factors such as light:dark cycles (Hiltz *et al.*, 2000), irradiance and temperature (Kaczmarska *et al.*, 2008), certain bacteria (Thompson *et al.*, 2000), and likely additional factors affecting other diatom species (Edlund & Stoermer, 1997) not yet evaluated for *Pseudo-nitzschia*. Recent reports of massive sexual episodes in natural waters during *Pseudo-nitzschia* spp. blooms add new dimensions to laboratory studies, expanding our understanding of spatial, temporal, and environmental contexts of sexuality in diatom populations (Sarno *et al.*, 2010), including possible implications for toxin transfer (Holtermann *et al.*, 2010).

Most *Pseudo-nitzschia* species examined thus far (including all toxigenic ones) are heterothallic, but homothally is documented or suggested for at least three so far non-toxigenic taxa: *P. subcurvata* (Hasle) G. A. Fryxell (Fryxell *et al.*, 1991), *P. brasiliana* Lundholm, Hasle & G. A. Fryxell (Quijano-Scheggia *et al.*, 2009a), and *P. americana* (Hasle) G. A. Fryxell (I. Kaczmarska, unpublished data).

Sexual differentiation in heterothallic diatoms is associated with behavioral dimorphism, as both "female" gametes remain

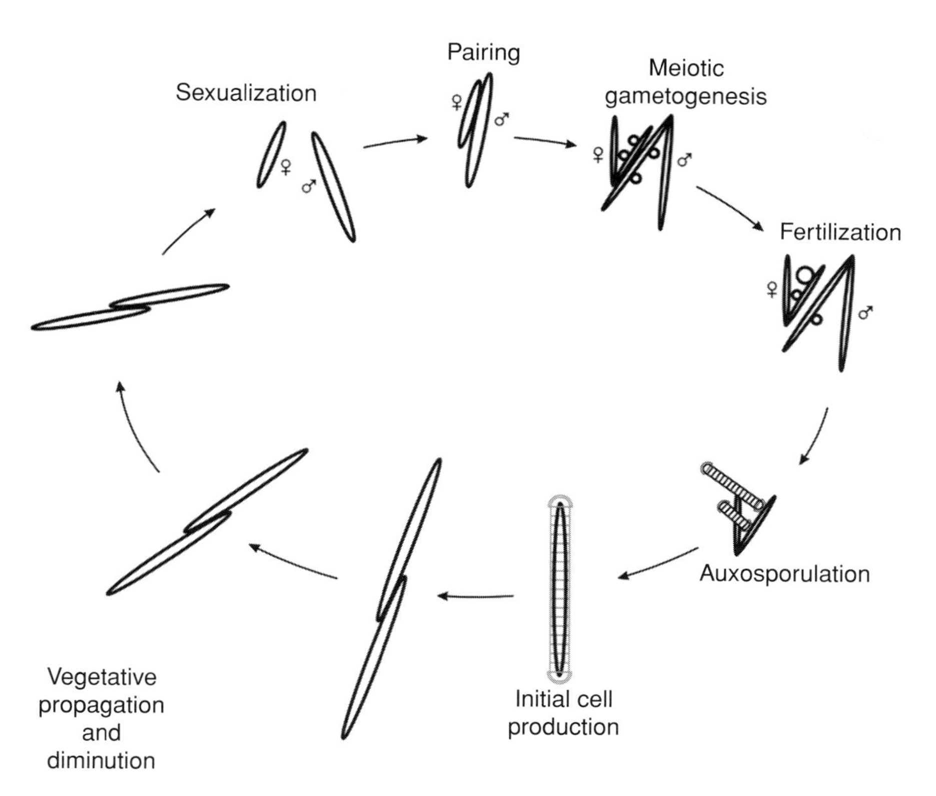

Figure 30.2 Life cycle of heterothallic *Pseudo-nitzschia* sp. Note that the sex ("male" or "female") is not related to size in any of the species thus far examined.

attached to the gametangial valve whereas "male" gametes separate from their gametangia to fuse with the "female" gamete (Davidovich & Bates, 1998). Sexual reproduction visibly begins when two individual cells pair and align with each other, and each becomes a gametangium, undergoing meiosis involving one acytokinetic division resulting in two gametes per gametangium. Following gametic fusion (but not necessarily syngamy), a specialized, post-sexual cell, unique among algae, called the auxospore develops and expands in size (typically two to four times the gametangial length), from initially spherical to cylindrical in shape. Primary (spherical) walls of auxospores rupture equatorially but may remain at the growing tips of auxospores as caps. Auxospore elongation is facilitated by deposition of rings (transverse perizonia) at each of its tips. The auxospore is a nursery for a new, genetically recombined and large cell (initial cell) that will begin the next phase of vegetative propagation of the new genotype. In heterothallic *Pseudo-nitzschia* species, two auxospores result from a pair of two sexual parents (Figure 30.2). Auxospores remain more or less perpendicularly attached to the "maternal" valve. The newly formed cells exit the auxospore walls to begin their own phase of mitotic propagation. In some species, large post-sexual cells produced more domoic acid per cell (Thompson *et al.*, 2000).

Heterothally has been documented in culture for the following: *P. delicatissima* (Cleve) Heiden in Heiden & Kolbe *sensu stricto* (Kaczmarska *et al.*, 2008); *P. arenysensis* Quijano-Scheggia, Garces & Lundholm (Quijano-Scheggia *et al.*, 2009b, as = *P. delicatissima* clade *del1 sensu* Amato *et al.* [2007]); *P. dolorosa* Lundhlom & Moestrup, *P. calliantha* Lundholm, Moestrup & Hasle, *P. caciantha* Lundholm, Moestrup & Hasle, and *P. pseudodelicatissima sensu* Lundholm *et al.* (2006) (Amato *et al.*, 2007); *P. mannii* Amato & Montresor (Amato & Montresor, 2008); *P. multiseries* (Hasle) Hasle (Davidovich & Bates, 1998); *P. pungens* (Grunow ex Cleve) Hasle (Casteleyn *et al.*, 2008); and *P. multistriata* (Takano) Takano (D'Alelio *et al.*, 2010a). Breeding experiments cited above not only elucidated sexual processes and reproductive strategies of these diatoms, but also clearly demonstrated cryptic and semi-cryptic diversity among *P. pseudodelicatissima* and *P. delicatissima* species complexes (refer to Figure 30.1c, d). Correct recognition of species has an obvious bearing on advances in our understanding of the ecology and intermittent toxigenicity of members of this genus, as well as development of species-specific (or genotype-specific) molecular tags.

Morphological similarities between members of species-complexes have long suggested that species are closely related, a notion supported by molecular markers (Amato *et al.*, 2007;

Lundholm *et al.*, 2003, 2006). Most recently, Kaczmarska *et al.* (2008) reported on sexual interactions between male clones of *P. delicatissima sensu stricto* from Atlantic Canada (subclade 2a) and five Italian strains from the Gulf of Naples (clade "*P. del2*" in Amato *et al.*, 2007, equivalent to clade 2c in Kaczmarska *et al.*, 2008). Auxosporulation was consistently induced in three Italian strains when mated with Canadian male isolates and when mated pairwise among all Italian strains. These matings produced normal but infrequent post-sexual cells. The authors concluded that the Canadian *P. delicatissima sensu stricto* and the Neapolitan females of "*P. del2*" still show some capacity for sexual interaction. Although more research is needed to determine the genetic nature of Neapolitan clones and their progeny, comparison of Internal Transcribed Spacer ribosomal DNA (ITS rDNA) sequences suggests that Neapolitan clones could be natural hybrids between *P. delicatissima sensu stricto* (Kaczmarska *et al.*, 2008) and *P. delicatissima*-like strains collectively referred to as "*P. del1*" (Amato *et al.*, 2007), clade B (Lundholm *et al.*, 2006) or clade 1 (Kaczmarska *et al.*, 2008). If this speculation is correct, then semi- and cryptic diversity among "delicatissima-type" species of *Pseudo-nitzschia* (some known to be toxic) may be constantly increasing.

30.3 Biogeography

The genus *Pseudo-nitzschia* is widely distributed in the marine realm, from polar regions to warm waters, and is an important component of coastal floras, though also well represented in oceanic environments. At the species level, *Pseudo-nitzschia* may show different distribution patterns as initially demonstrated by Hasle (1972, 2002 and references therein) and further discussed by Lundholm & Moestrup (2006). Globally consistent identification and understanding distributional patterns are key to monitoring and forecasting potentially toxic blooms and their adverse effects.

The biogeography of potentially toxic diatoms, and hence their detection, faces a challenging taxonomic issue: the diagnostic characters that differentiate *Pseudo-nitzschia* from *Nitzschia* Hassal, especially the stepped colonies in the former (debated in Kaczmarska & Fryxell, 1994; Kaczmarska *et al.*, 2008) and the presence of a conopeum on the generitype of the latter, suggest the need for taxonomic re-appraisal. Not only do some described *Pseudo-nitzschia* occur as single cells in nature (*P. americana*, *P. galaxiae* Lundholm & Moestrup, and *P. granii* [Hasle] Hasle) and at times in culture (certain clones of *P. delicatissima*), but the conopeum is absent in some *Nitzschia* with morphological affinities with *Pseudo-nitzschia* (*N. johnmartinii* G. A. Fryxell & Kaczmarska and *N. concordia* G. A. Fryxell & Kaczmarska). This has become a more critical problem since *Nitzschia navis-varingica* Lundholm & Moestrup was reported as a DA producer in ponds used for culturing marine shrimp in Vietnam (Lundholm & Moestrup, 2000; Kotaki *et al.*, 2004).

A major development in delineating species boundaries has been generated recently by molecular tools. Semi-cryptic diversity within the genus *Pseudo-nitzschia* was revealed, resulting in the description of ten species and emended descriptions of existing taxa (Lundholm & Moestrup, 2002; Priisholm *et al.*, 2002; Lundholm *et al.*, 2003, Hasle & Lundholm, 2005; Hernández-Becerril & Diaz-Almeyda, 2006; Lundholm *et al.*, 2006; Amato & Montresor, 2008; Quijano-Scheggia *et al.*, 2009b; Table 30.1). There is always some subjectivity in deciding the degree of disparity sufficient to discriminate between species, and this is true for *Pseudo-nitzschia*, either in terms of morphology (e.g. Kaczmarska *et al.*, 2005, p. 16) or ITS rDNA sequence divergence (e.g. Casteleyn *et al.*, 2008, p. 253). Species boundaries have been drawn from multiple lines of evidence (morphology, biology, and sequence data) for a number of *Pseudo-nitzschia* species. The genus has become a good model system for studying diatom speciation (Alverson, 2008). A natural system of classification could be of great practical benefit for its predictive value in a genus that has several toxic members.

Determination of whether any *Pseudo-nitzschia* species is a true cosmopolite is hindered by our still limited understanding of semi-cryptic diversity. An illustrative example may be *P. pungens*, the first *Pseudo-nitzschia* to be formally described (Table 30.1), which had long been considered as cosmopolitan with records in all latitudes, except for the Antarctic Ocean (Hasle, 1972, 2002). More recently, it was suggested as the only example known so far of a protist with a possible global gene pool (Medlin, 2007). Casteleyn *et al.* (2008), however, did find intraspecific variability within *P. pungens* and distinguished between three ITS types: one ITS variant was widespread within temperate zones of both hemispheres and corresponds to the nominal variety; another ITS variant was restricted to a number of strains from the NE Pacific and corresponds to *P. pungens* var. *cingulata* Villac; and a third, and most distantly related ITS type, comprised strains from widely separated areas of warmer waters (Vietnam, China, and Mexico). This third ITS type was later described as *P. pungens* var. *aveirensis* Lundholm, Churro, Carreira & Calado with material from the Portuguese coast (Churro *et al.*, 2009). The first two ITS types are sexually compatible in the laboratory and hybrids have been observed in nature, although hybrid fitness still remains to be determined (Casteleyn *et al.*, 2009).

30.4 Species detection and bloom ecology

The ecology of *Pseudo-nitzschia* spp. has been examined at various taxonomic levels, depending on the objective of the study and/or analytical constraints. Most commonly, and with application of light microscopy (LM), a broad taxonomic designation is used such as *Pseudo-nitzschia* spp. (Figure 30.1a) and differentiation between the "*delicatissima*"-complex, including species with valves ≤ 3 μm wide (Figure 30.1b), and the "*seriata*"-complex with species with valves >3 μm wide (Figure 30.1f). Few studies attempt species-level identification, a refinement requiring electron microscopy (EM) of species-complexes enumerated using LM. Fewer still are the studies identifying semi-cryptic species using EM and molecular markers. The value of long-term field series becomes apparent, even if with a reduced spatial coverage, such as done in the Campania region, Mediterranean Sea (Zingone *et al.*, 2006). Careful LM was carried out with field samples over a 20-year period, in which routine counts considered various morphotypes within *Pseudo-nitzschia* chains; later EM studies of the same material established seasonal trends of *P. galaxiae* (Cerino *et al.*, 2005). This is a practical and informative strategy that can be employed for routine phytoplankton monitoring that was also used to understand *Pseudo-nitzschia* dynamics in the Quoddy region, eastern Canada (Kaczmarska *et al.*, 2007). In this case, multivariate analysis demonstrated that the same general space–time variations were determined whether three morphotypes or seven corresponding species were considered in association with environmental variables, strengthening the coupling between form and function in nature. Ultimately, however, of critical interest for mariculture are the distinct types of toxin-producing microalgal species or even specific toxigenic genotypes.

A considerable effort has been made over the past decade to develop species-specific oligonucleotide probes for correct, globally consistent, and rapid identification of target species (e.g. Miller & Scholin, 1998). This is an optimal management strategy as it affords early warning of potentially toxic blooms. However, unanticipated interspecific diversity in sequences among toxigenic diatoms precluded the use of certain probes in regions geographically distinct from the source of the original target strains (e.g. Parsons *et al.*, 1999; Turell *et al.*, 2008). This underscores the need for in-depth investigations of both local and global diversity of the genus *Pseudo-nitzschia*. A powerful approach to addressing the challenge posed by interspecific diversity is to construct clone libraries for a given area using genus-specific Large Subunit (LSU) rDNA primers for polymerase-chain-reaction–based amplification of environmental samples (e.g. McDonald *et al.*, 2007), which can provide a qualitative assessment of genotypic diversity within a sample and also guide development of probes for detection of semi-cryptic species. Rapid advances are also being reported in the application of DNA probe and toxin detection technologies by remote, subsurface platforms capable of autonomously providing near real-time data on *Pseudo-nitzschia* spp. and DA concentrations (Scholin *et al.*, 2009).

The environmental window within which *Pseudo-nitzschia* proliferates and even dominates coastal phytoplankton communities is quite broad. *Pseudo-nitzschia* spp. are frequently present year round, with a more conspicuous abundance in the spring–summer of temperate regions (e.g. Orsini *et al.*, 2004; Schnetzer *et al.*, 2007) and a marked response to moderate and strong upwelling (e.g. Walz *et al.*, 1994; Palma *et al.*, 2010) as well as to nutrient input from river discharge (e.g. Dortch *et al.*, 1997; Trainer *et al.*, 2007). In contrast, high cell concentrations, even at bloom levels, have occurred in autumn or in stratified water columns when light and/or nutrient availability are likely growth-limiting (e.g. Smith *et al.*, 1990; Buck *et al.*, 1992). The concurrent presence of two to several *Pseudo-nitzschia* species can be attributed to niche partitioning that leads to co-existence and succession within the genus (Villac, 1996, pp. 145–7). Despite its apparently simple overall morphology, the *Pseudo-nitzschia* apical axis varies widely between species (from 12 μm in a short *P. brasiliana* to 174 μm in a long *P. pungens*), the ultrastructure is also very diverse (see Figure 30.1) and eco-physiological requirements are species-specific, as shown from laboratory studies (summarized in Bates, 1998; Bates *et al.*, 1998; Bates & Trainer, 2006). Shifts at the intraspecific level were found in the Gulf of Naples, Italy, where three *P. delicatissima* and *P. delicatissima*-like diatom ITS types were recovered together in pre-bloom conditions, but only one of the variants contributed substantially to the spring bloom (Orsini *et al.*, 2004). As our understanding of the composition and abundance at the species (and semi-cryptic species) level advances, distributional patterns will likely narrow down the environmental windows of occurrence and thus improve our predictive capabilities of bloom dynamics. Recognizing that the waxing and waning of populations of a given species may be regulated by life-cycle traits could also play a major role in understanding the patterns observed in nature (D'Alelio *et al.* 2010b).

A phytoplankton bloom often denotes high cell abundances with low species diversity in a parcel of water (Smayda, 1997 and references therein), either as accumulations visible at the surface or as thin subsurface layers of high cell concentration (e.g. Rines *et al.*, 2002). Andersen (1996, p. 1) defines harmful algal blooms as "events where the concentration of one or several harmful algae reach levels which can cause harm to other organisms in the sea . . . or cause accumulation of algal toxins in marine organisms which eventually harm other organisms who will eat the toxic species . . . ". Nonetheless, when toxic diatoms are ingested by planktivorous fish vectors, such as anchovies, relative abundance of the target diet-species over the feeding area is a more relevant predictor of potential harmful effects than its absolute abundance (e.g. Buck *et al.*, 1992).

Another challenge that has been addressed is the distinction of local phytoplankton succession from bloom initiation due to advection of waters that carry populations undergoing succession themselves. Increasing *Pseudo-nitzschia* abundance in some enclosed and semi-enclosed coastal systems, often associated with nutrient enrichment due to land runoff, has been thought of as derived from local seed stocks (e.g. Bates *et al.*, 1998, p. 284; Villac *et al.*, 2004). In other cases, coastal blooms result from advection onshore of blooms generated and retained in offshore cyclonic eddies (Trainer *et al.*, 2002). Meteorological events may trigger the interaction of coastal currents, river plumes, and bathymetry so that local populations in upwelling-favorable conditions are partially or totally replaced when downwelling fronts transport to the coastal site other species from offshore communities (e.g. Adams *et al.*, 2006). When considered over longer timescales (interdecadal), marked changes in abundances have been related to cultural eutrophication, as revealed by increasing *Pseudo-nitzschia* abundance in sediment cores from the Mississippi River (USA) basin (Parsons *et al.*, 2002). The sudden shift, in 2004, from a typically diatom-dominated community to a dinoflagellate-dominated community in surface waters of Monterey Bay, CA, USA, was tentatively associated with recurrent El Niño-Southern Oscillation (ENSO) events; if this becomes a long-term established trend, the alternation from diatom toxins to dinoflagellate toxins will have clear implications for monitoring strategies in the area (Jester *et al.*, 2009). These examples reinforce the case for the whole community approach to bloom dynamics studies and points out the importance of using sampling designs at scales appropriate to resolve fine structures such as subsurface phytoplankton thin layers (McManus *et al.*, 2008).

Improving our understanding of harmful algal blooms requires resolution at different spatial scales. Physical mechanisms of *Pseudo-nitzschia* retention, for example, vary from mesoscale eddies on the west coast of North America (Trainer *et al.*, 2002), to more localized salt wedge zones of estuarine areas (e.g, Dortch *et al.*, 1997), to layers of several centimeters that are often associated with pycnoclines and comprise high concentrations of actively growing diatom cells (e.g. Rines *et al.*, 2002). Although many factors can regulate bloom initiation, the onset of these events requires the introduction of and/or pre-existing presence of the causative species. The former may happen via "seeding" processes, whereas the latter simply implies the presence of viable, but non-proliferating cells, possibly as part of a hidden or cryptic flora. Bloom development likely represents an intersection/interaction of these phenomena that can occur over a range of spatio-temporal scales (see above). For example, the sinking of *Pseudo-nitzschia* populations out of the photic zone may be the starting point of a process that would ultimately allow for seeding of a bloom population following resuspension via mixing events (Fryxell *et al.*, 1997). As suggested by Smetacek (1985), although some cell death certainly occurs as phytoplankton sinks below the euphotic zone, rapid sinking does not equate to mass mortality, especially if these cells can be maintained in sufficient numbers once reaching higher nutrient levels at the pycnocline. Indeed, vertical flux of *Pseudo-nitzschia* cells (and particulate DA) has been revealed by sediment traps at various depths below the range where blooms develop (Buck *et al.*, 1992; Dortch *et al.*, 1997; Schnetzer *et al.*, 2007; Sekula-Wood *et al.*, 2009). Other cryptic microhabitats where viable or actively growing chains of *Pseudo-nitzschia* have been found are within colonies of *Chaetoceros socialis* Lauder (Rines *et al.*, 2002) and on the surface of *Phaeocystis globosa* Scherff. colonies (Sazhin *et al.*, 2007), both bloom-forming species themselves. The most inconspicuous habitat for diatom retention to date is a report of apparently viable *Pseudo-nitzschia* cells in zooplankton fecal pellets (Buck & Chavez, 1994).

30.5 Toxin trophic transfer

There have been several recent reviews of the trophic transfer and resulting impacts of algal toxins in general (e.g. Landsberg, 2002; Landsberg *et al.*, 2006; Doucette *et al.*, 2006) and DA specifically (Bates & Trainer, 2006, Bejarano *et al.*, 2008). These reports highlight the complexities and pervasiveness of DA entry into and movement through marine food webs in regions subject to toxigenic *Pseudo-nitzschia* blooms. Moreover, discoveries of new vectors of toxin transfer and studies of potential effects on species exposed frequently to DA continue to be reported and are outlined below.

Consideration of DA trophic transfer and impacts on wildlife (including commercially important fisheries) as well as humans must take into account certain intrinsic properties of DA production by *Pseudo-nitzschia* spp. These include a highly variable cellular toxicity (i.e. DA cell quota) and the partitioning of DA between particulate and dissolved phases. The former is well documented in natural populations and can range over almost three orders of magnitude as reported, for example, by Trainer *et al.* (2000; 0.1–78 pg DA per cell) off the California coast. Indeed, intermittent DA production has been reported for certain potentially toxigenic *Pseudo-nitzschia* spp. (see Bates & Trainer, 2006), as has a possible role of bacteria in enhancing toxicity (Bates *et al.*, 1995). The report of a significant positive relationship (although not confirmed as causal) between new, larger *P. multiseries* and *P. pungens* cells produced by auxosporulation and DA concentrations in razor clams along the Washington coast, USA (Holtermann *et al.*, 2010), revealed for the first time a potential influence of population-level life-cycle dynamics on toxin trophic transfer. Both laboratory and field studies have documented wide fluctuations in intracellular vs. extracellular toxin production (Bates & Trainer, 2006; and references therein), likely as a function of cell physiological status. Taken together, the variation with growth/bloom phase in DA levels contained in cells and/or dissolved in seawater will ultimately influence the efficiency of toxin entry into food webs and the effects of toxin exposure (e.g. Doucette *et al.*, 2002).

Many organisms, in addition to various species of filter-feeding bivalves, have been identified as potential vectors for transfer of DA from its algal producers to higher trophic levels (e. g. birds, marine mammals, humans). In the pelagic zone where *Pseudo-nitzschia* blooms occur these include (but are not limited to) copepods (Leandro *et al.*, 2010b), krill, and squid (Bargu *et al.*, 2008 and references therein), as well as planktivorous fish (e.g. anchovies, sardines; Scholin *et al.*, 2000; Lefebvre *et al.*, 2002). During blooms and especially as these events begin to terminate, the diatom cells can sink rapidly from the water column, generally as aggregates or flocs, and are delivered to the sea floor along with DA in the case of toxigenic species (Sekula-Wood *et al.*, 2009). Recent work revealing the widespread presence of DA in eight benthic species representing multiple feeding modes (i.e. filter feeders, predators, scavengers, and deposit feeders), frequently at elevated concentrations (>500 ppm; regulatory limit for human consumers is 20 ppm), demonstrates the importance of pelagic–benthic coupling and the potentially diverse pathways for toxin movement through benthic food webs (Kvitek *et al.*, 2008). Many of these benthic taxa are prey for various marine mammal species, some of which are listed as endangered (e.g. California sea otter). Moreover, several pelagic and benthic vector organisms, including shellfish, are harvested either commercially and/or recreationally for human consumption and thus represent a potential risk for DA exposure.

While the routes by which species occupying upper trophic levels of marine food webs are exposed to DA can comprise multiple intermediates, there are many examples involving only a single vector. Perhaps the best documented of these is the transfer of DA from *Pseudo-nitzschia australis* to the northern anchovy to the California sea lion, often resulting in mass mortalities of this pinniped species (e.g. Scholin *et al.*, 2000). Exposure of baleen whales such as the endangered blue, humpback, and North Atlantic right whales to DA via krill, fish (Lefebvre *et al.*, 2002), and calanoid copepod vectors (Leandro *et al.*, 2010a), respectively, has also been documented. In addition, the latter authors noted that, based on an effective mesh size of ~335 μm for the right whale baleen, direct ingestion of *Pseudo-nitzschia* chains, especially in areas of high cell concentrations, would be possible and provide for the direct, highly efficient transfer of toxin.

In terms of DA effects on wildlife, there is a growing body of literature not only documenting acute toxicity in a range of species but also suggesting the potential for long-term impacts at the population level (reviewed by Bejarano *et al.*, 2008). In particular, a recent study by Ramsdell & Zabka (2008) argues that *in utero* exposure of California sea lions predisposes these animals to neurological disease as they mature into adults. Furthermore, synchronization of this period of enhanced susceptibility to the developmental effects of DA with seasonal *Pseudo-nitzschia* blooms in this region only serves to increase the risk of toxin exposure and its consequences both to individuals and the population. For North Atlantic right whales, a six-year-long survey revealed the annual exposure not only to DA but also to the paralytic shellfish poisoning toxins (i.e. saxitoxins), with both algal toxins co-occurring in ~30% of these animals (G. Doucette & R. Rolland, unpublished data) thereby suggesting the possibility of synergistic effects. Efforts are now being directed at the development of risk-assessment models (e.g. Bejarano *et al.*, 2007) with the aim of projecting DA exposure trends and, taking into account data from *in vivo* studies, likely effects on the health and reproduction of marine animal populations.

Toxin transfer to humans has not been reported since the first DA outbreak in Prince Edward Island, eastern Canada, in 1987. During this event, *P. multiseries* (then identified as *P. pungens* f. *multiseries*) and its associated toxin were concentrated by the blue mussel *Mytilus edulis* L., resulting in 107 cases of human DA poisoning and three deaths (reviewed in Todd, 1993). The importance on a global scale of continuing to support existing monitoring programs cannot be understated. It is also essential to establish new monitoring efforts and to adopt emerging detection technologies, where appropriate, for continued protection of public health as well as to mitigate additional impacts of these and other harmful algae (e.g. Anderson *et al.*, 2001). Moreover, the importance of identifying segments of the human population that may exhibit enhanced susceptibility to toxin exposure (especially at sub-regulatory levels) based on their physiological traits, behavioral factors, socioeconomic status, and cultural practices, is now being recognized, as is the need to evaluate the associated risk (Bauer, 2006).

30.6 Summary

There are 30+ species of *Pseudo-nitzschia* of which 12 are known to produce the neurotoxin DA. Most *Pseudo-nitzschia* species examined thus far (including all toxigenic ones) are heterothallic. Species boundaries for a number of *Pseudo-nitzschia* are now based on morphology, breeding experiments and sequence data, revealing extensive semi-cryptic diversity within this genus. Field ecology of *Pseudo-nitzschia* spp. has been most often examined using "counting categories" of species-complexes. Thus, assessment at the species level required for accurate modeling and prediction efforts should incorporate the use of EM and molecular tools – a challenging goal from the perspective of practical application. Optimal management strategies will rely increasingly on advances in remotely deployed detection technologies capable of providing integrated, near real-time data on *Pseudo-nitzschia* spp. and DA concentrations. The study of blooms caused by *Pseudo-nitzschia* spp. clearly requires resolution at spatial scales that varies from mesoscale eddies, to more localized estuarine salt wedge zones or subsurface layers associated with pycnoclines, to cryptic microhabitats (colonies of *Chaetoceros socialis* or *Phaeocystis globosa*, zooplankton fecal pellets). Moreover, vertical flux of *Pseudo-nitzschia* cells (and particulate DA) to depths below the range where blooms develop, coupled with reports of new vectors of toxin transfer (copepods, krill, squid, various planktivorous fish, benthic species representing multiple feeding modes) demonstrate the pervasiveness of DA entry into and movement through marine food webs. Researchers are only now beginning to understand the potential for long-term impacts of DA exposure to certain wildlife populations (e.g. *in utero* exposure of California sea lions). Finally, although toxin transfer to humans has not been reported since the first DA outbreak in 1987, further work is needed to identify populations with enhanced susceptibility to DA exposure, especially at sub-regulatory levels.

Acknowledgments

We would like to thank the editors John Smol and Eugene Stoermer for the opportunity to contribute to this book, as well as Greta Fryxell, first author of the previous edition of this chapter, for her continuing generous support. We are grateful to Jim Ehrman for the SEM images and expertise in composing the figures.

Disclaimer: The National Ocean Service (NOS) does not approve, recommend, or endorse any proprietary product or material mentioned in this publication. No reference shall be made to NOS, or to this publication furnished by NOS in any advertising or sales promotion which would indicate or imply that NOS approves, recommends, or endorses any proprietary product or proprietary material mentioned herein or which has as its purpose any intent to cause directly or indirectly the advertised product to be used or purchased because of NOS publication.

References

Adams, N. G., MacFadyen, A., Hickey, B. M., & Trainer, V. L. (2006). The nearshore advection of a toxigenic *Pseudo-nitzschia* bloom and subsequent domoic acid contamination of intertidal bivalves. *African Journal of Marine Science*, **28**, 271–6.

Alverson, A. J. (2008). Molecular systematics and the diatom species. *Protist*, **159**, 339–53.

Amato, A., Kooistra, W. H. C. F., Ghiron, J. H. L., *et al.* (2007). Reproductive isolation among sympatric cryptic species in marine diatoms. *Protist*, **158**, 193–207.

Amato, A. & Montresor, M. (2008). Morphology, phylogeny, and sexual cycle of *Pseudo-nitzschia mannii* sp. nov. (Bacillariophyceae): a pseudo-cryptic species within the *P. pseudodelicatissima* complex. *Phycologia*, **47**, 487–97.

Andersen, P. (1996). *Design and Implementation of Some Harmful Algal Monitoring Systems*. Intergovernmental Oceanographic Commission Technical Series, 44. Paris: UNESCO.

Anderson, D. M., Andersen, P., Bricelj, V. M., Cullen, J. J., & Rensel, J. E. (2001). *Monitoring and Management Strategies for Harmful Algal Blooms in Coastal Waters*. APEC #201-MR-01.1, Asia Pacific Economic Program, Singapore, and Intergovernmental Oceanographic Commission Technical Series, 59. Paris: UNESCO.

Armbrust, E. V., Chisholm, S. W., & Olson, R. J. (1990). Role of light and the cell cycle on the induction of spermatogenesis in a centric diatom. *Journal of Phycology*, **26**, 470–8.

Bargu, S., Powell, C. L., Wang, Z., Doucette, G. J., & Silver, M. W. (2008). Note on the occurrence of *Pseudo-nitzschia australis* and domoic acid in squid from Monterey Bay, California (USA). *Harmful Algae*, **7**, 45–51.

Bates, S. S. (1998). Ecophysiology and metabolism of ASP toxin production. In *Physiological Ecology of Harmful Algal Blooms*, ed. D. M. Anderson, A. D. Cembella, & G. M. Hallegraeff, Heidelberg: Springer-Verlag, pp. 405–26.

Bates, S. S., Douglas, D. J., Doucette, G. J., & Léger, C. (1995). Enhancement of domoic acid production by reintroducing bacteria to axenic cultures of the diatom *Pseudo-nitzschia multiseries*. *Natural Toxins*, **3**, 428–35.

Bates, S. S., Garrison, D. L., & Horner, R. A. (1998). Bloom dynamics and physiology of domoic acid-producing *Pseudo-nitzschia* species. In *Physiological Ecology of Harmful Algal Blooms*, ed. D. M. Anderson, A. D. Cembella, & G. M. Hallegraeff, Heidelberg: Springer-Verlag, pp. 267–92.

Bates, S. S. & Trainer, V. L. (2006). The ecology of harmful diatoms. In *Ecology of Harmful Algae*, ed. E. Granéli & J. Turner, Heidelberg: Springer-Verlag, pp. 81–93.

Bauer, M. (2006). *Harmful Algal Research and Response: A Human Dimensions Strategy*. Woods Hole, MA: National Office for Marine Biotoxins and Harmful Algal Blooms, Woods Hole Oceanographic Institution.

Bejarano, A. C., Van Dolah, F. M., Gulland, F. M., Rowles, T. K., & Schwacke, L. H. (2008). Production and toxicity of the marine biotoxin domoic acid and its effects on wildlife: a review. *Human and Ecological Risk Assessment*, **14**, 544–67.

Bejarano, A. C., Van Dolah, F. M., Gulland, F. M., & Schwacke, L. (2007). Exposure assessment of the biotoxin domoic acid in California sea lions: application of a probabilistic bioenergetic model. *Marine Ecology Progress Series*, **345**, 293–304.

Buck, K. R. & Chavez, F. P. (1994). Diatom aggregates from the open ocean. *Journal of Plankton Research*, **16**, 1449–57.

Buck, K. R., Uttal-Cooke, L., Pilskaln, C. H., *et al.* (1992). Autecology of the diatom *Pseudonitzschia australis* Frenguelli, a domoic acid producer from Monterey Bay, California. *Marine Ecology Progress Series*, **84**, 293–302.

Casteleyn, G., Adams, N. G., Vanormelingen, P., *et al.* (2009). Natural hybrids in the marine diatom *Pseudo-nitzschia pungens* (Bacillariophyceae): genetic and morphological evidence. *Protist*, **160**, 343–54.

Casteleyn, G., Chepurnov, V. A., Leliaert, F., *et al.* (2008). *Pseudo-nitzschia pungens* (Bacillariophyceae): a cosmopolitan diatom species? *Harmful Algae*, **7**, 241–57.

Cerino, F., Orsini, L., Sarno, D., *et al.* (2005). The alternation of different morphotypes in the seasonal cycle of the toxic diatom *Pseudo-nitzschia galaxiae*. *Harmful Algae*, **4**, 33–48.

Churro, C. I., Carreira, C. C., Rodrigues, F. J., *et al.* (2009). Diversity and abundance of potentially toxic *Pseudo-nitzschia* Peragallo in Aveiro coastal lagoon, Portugal and description of a new variety, *P. pungens* var. *aveirensis* var. nov. *Diatom Research*, **24**, 35–62.

D'Alelio, D., Amato, A., Luedeking, A., & Montresor, M. (2010a). Sexual and vegetative phases in the planktonic diatom *Pseudo-nitzschia multistriata*. *Harmful Algae*, **8**, 225–32.

D'Alelio, D., d'Alcalà, M. R., Dubroca, L. *et al.* (2010b). The time for sex: a biennial life cycle in a marine planktonic diatom. *Limnology and Oceanography*, **55**, 106–14.

Davidovich, N. A. & Bates, S. S. (1998). Sexual reproduction in the pennate diatoms *Pseudo-nitzschia multiseries* and *P. pseudodelicatissima* (Bacillariophyceae). *Journal of Phycology*, **34**, 126–37.

Dortch, Q., Robichaux, R., Pool, S., *et al.* (1997). Abundance and vertical flux of *Pseudo-nitzschia* in the northern Gulf of Mexico. *Marine Ecology Progress Series*, **146**, 249–64.

Doucette, G. J., Maneiro, I., Riveiro, I., & Svensen, C. (2006). Phycotoxin pathways in aquatic food webs: transfer, accumulation and degradation. In *Ecology of Harmful Algae*, ed. E. Granéli & J. Turner, Heidelberg: Springer-Verlag, pp. 283–95.

Doucette, G. J., Scholin, C. A., Ryan, J. P., *et al.* (2002). Possible influence of *Pseudo-nitzschia australis* population and toxin dynamics on food web impacts in Monterey Bay, CA, USA. 10th International Conference on Harmful Algae, St. Petersburg, FL, October 21–25, p. 76 (abstract).

Edlund, M. B. & Stoermer, E. F. (1997). Ecological, evolutionary and systematic significance of diatom life histories. *Journal of Phycology*, **33**, 897–918.

Falkowski, P. G., Barber, R. T., & Smetacek, V. (1998). Biogeochemical controls and feedbacks on ocean primary production. *Science*, **281**, 200–6.

Fryxell, G. A., Garza, S. A., & Roelke, D. L. (1991). Auxospore formation in an Antarctic clone of *Nitzschia subcurvata* Hasle. *Diatom Research*, **6**, 235–45.

Fryxell, G. A. & Hasle, G. R. (2003). Taxonomy of harmful diatoms. *Monographs on Oceanographic Methodology*, **11**, 465–510.

Fryxell, G. A. & Villac, M. C. (1999). Toxic and harmful marine diatoms. In *The Diatoms: Applications for the Environmental and Earth Sciences*, ed. E. F. Stoermer & J. P. Smol, Cambridge: Cambridge University Press, pp. 419–28.

Fryxell, G. A., Villac, M. C., & Shapiro, L. P. (1997). The occurrence of the toxic diatom genus *Pseudo-nitzschia* (Bacillariophyceae) on the west coast of the U.S.A., 1920–1996: a review. *Phycologia*, **36**, 419–37.

Hasle, G. R. (1965). *Nitzschia* and *Fragilariopsis* species studied in the light and electron microscopes. II. The group *Pseudonitzschia*. *Skrifter utgitt av Det Norske Videnskaps-Akademi i Oslo. I. Mat.-Naturv. Klasse.*, **18**, 1–45.

Hasle, G. R. (1972). The distribution of *Nitzschia seriata* Cleve and allied species. *Nova Hedwigia*, **39**, 171–90.

Hasle, G. R. (2002). Are most of the domoic acid producing species of the diatom genus *Pseudo-nitzschia* cosmopolites? *Harmful Algae*, **1**, 137–146.

Hasle, G. R. & Lundholm, N. (2005). *Pseudo-nitzschia seriata* fo. *obtusa* (Bacillariophyceae) raised in rank based on morphological, phylogenetic and distributional data. *Phycologia*, **44**, 608–19.

Hernández-Becerril, D. U. & Diaz-Almeyda, E. M. (2006). The *Nitzschia bicapitata* group, new records of the genus *Nitzschia*, and further studies on species of *Pseudo-nitzschia* (Bacillariophyta) from Mexican Pacific coasts. *Nova Hedwigia*, **130**, 293–306.

Hiltz, M. F., Bates, S. S., & Kaczmarska, I. (2000). Effect of light:dark cycles and cell apical length on the sexual reproduction of *Pseudo-nitzschia multiseries* (Bacillariophyceae) in culture. *Phycologia*, **39**, 59–66.

Holtermann, K. E., Bates, S. S., Trainer, V. L., Odell, A., & Armbrust, E. V. (2010). Mass sexual reproduction in the toxigenic diatoms

Pseudo-nitzschia australis and *P. pungens* (Bacillariophyceae) on the Washington coast, USA. *Journal of Phycology*, **46**, 41–52.

Jester, R., Lefebvre, K., Langlois, G., *et al.* (2009). A shift in the dominant toxin-producing algal species in central California alters phycotoxins in food webs. *Harmful Algae*, **8**, 291–8.

Kaczmarska, I. & Fryxell, G. A. (1994). The genus *Nitzschia*: three new species from the equatorial Pacific Ocean. *Diatom Research*, **9**, 87–98.

Kaczmarska, I., LeGresley, M. M., Martin, J. L., & Ehrman, J. (2005). Diversity of the diatom genus *Pseudo-nitzschia* Peragallo in the Quoddy region of the Bay of Fundy, Canada. *Harmful Algae*, **4**, 1–19.

Kaczmarska, I., Martin, J. L., Ehrman, J. M., & LeGresley, M. M. (2007). *Pseudo-nitzschia* species population dynamics in the Quoddy region, Bay of Fundy. *Harmful Algae*, **6**, 861–74.

Kaczmarska, I., Reid, C., Martin, J. L., & Moniz, M. B. J. (2008). Morphological, biological and molecular characteristics of *Pseudo-nitzschia delicatissima* from the Canadian Maritimes. *Botany*, **86**, 763–72.

Kotaki, Y., Lundholm, N., Onodera, H., *et al.* (2004). Wide distribution of *Nitzschia navis-varingica*, a new domoic acid-producing benthic diatom found in Vietnam. *Fisheries Science*, **70**, 28–32.

Kvitek, R. G., Goldberg, J. D., Smith, G. J., Doucette, G. J., & Silver, M. W. (2008). Domoic acid contamination within eight representative species from the benthic food web of Monterey Bay, California, USA. *Marine Ecology Progress Series*, **367**, 35–47.

Landsberg, J. (2002). The effects of harmful algal blooms on aquatic organisms. *Reviews in Fisheries Science*, **10**, 113–390.

Landsberg, J., Van Dolah, F., & Doucette, G. J. (2006). Marine and estuarine harmful algal blooms: impacts on human and animal health. In *Oceans and Health: Pathogens in the Marine Environment*, ed. S. Belkin & R. R. Colwell, New York: Springer, pp. 165–215.

Leandro, L. F., Rolland, R. M., Roth, P. B., *et al.* (2010a). Exposure of the North Atlantic right whale *Eubalaena glacialis* to the marine algal biotoxin, domoic acid. *Marine Ecology Progress Series*, **398**, 287–303.

Leandro, L. F., Teegarden, G. J., Roth, P. B., & Doucette, G. J. (2010b). The copepod *Calanus finmarchicus*: a potential vector for trophic transfer of the marine algal biotoxin, domoic acid. *Journal of Experimental Marine Biology and Ecology*, **382**, 88–95.

Lefebvre, K. A., Bargu, S., Kieckhefer, T., & Silver, M. W. (2002). From sanddabs to blue whales: the pervasiveness of domoic acid. *Toxicon*, **40**, 971–7.

Lundholm, N., Hasle, G. R., Fryxell, G. A., & Hargraves, P. E. (2002). Morphology, phylogeny and taxonomy of species within the *Pseudo-nitzschia americana* complex (Bacillariophyceae) with descriptions of two new species, *Pseudo-nitzschia brasiliana* and *Pseudo-nitzschia linea*. *Phycologia*, **41**, 480–97.

Lundholm, N. & Moestrup, Ø. (2000). Morphology of the marine diatom *Nitzschia navis-varingica* sp. nov., another producer of the neurotoxin domoic acid. *Journal of Phycology*, **36**, 1162–74.

Lundholm, N. & Moestrup, Ø. (2002). The marine diatom *Pseudo-nitzschia galaxiae* sp. nov. (Bacillariophyceae): morphology and phylogenetic relationships. *Phycologia*, **41**, 594–605.

Lundholm, N. & Moestrup, Ø. (2006). The biogeography of harmful algae. In *Ecology of Harmful Algae*, ed. E. Granéli & J. Turner, Heidelberg: Springer-Verlag, pp. 23–35.

Lundholm, N., Moestrup, Ø., Hasle, G. R., & Hoef-Emden, K. (2003). A study of the *P. pseudodelicatissima/cuspidata* complex (Bacillariophyceae): what is *P. pseudodelicatissima*? *Journal of Phycology*, **39**, 797–813.

Lundholm, N., Moestrup, Ø., Kotaki, Y., *et al.* (2006). Inter- and intraspecific variation of the *Pseudo-nitzschia delicatissima* complex (Bacillariophyceae) illustrated by rRNA probes, morphological data and phylogenetic analyses. *Journal of Phycology*, **42**, 464–81.

McDonald, S. M., Sarno, D., & Zingone, A. (2007). Identifying *Pseudo-nitzschia* species in natural samples using genus-specific PCR primers and clone libraries. *Harmful Algae*, **6**, 849–60.

McManus, M. A., Kudela, R. M., Silver, M. W., *et al.* (2008). Cryptic blooms: are thin layers the missing connection? *Estuaries and Coasts*, **31**, 396–401.

Medlin, L. K. (2007). If everything is everywhere, do they share a common gene pool? *Gene*, **406**, 180–3.

Miller, P. E. & Scholin, C. A. (1998). Identification and enumeration of cultured and wild *Pseudo-nitzschia* (Bacillariophyceae) using species-specific LSU rRNA-targeted fluorescent probes and filter-based whole cell hybridization. *Journal of Phycology*, **34**, 371–382.

Orsini, L., Procaccini, G., Sarno, D., & Montresor, M. (2004). Multiple rDNA ITS-types within the diatom *Pseudo-nitzschia delicatissima* (Bacillariophyceae) and their relative abundances across a spring bloom in the Gulf of Naples. *Marine Ecology Progress Series*, **271**, 87–98.

Palma, S., Mouriño, H., Silva, A., Barão, M. I., & Moita, M. T. (2010). Can *Pseudo-nitzschia* blooms be modeled by coastal upwelling in Lisbon Bay? *Harmful Algae*, **9**, 294–303.

Parsons, M. L., Dortch, Q., & Turner, R. E. (2002). Sedimentological evidence of an increase in *Pseudo-nitzschia* (Bacillariophyceae) abundance in response to coastal eutrophication. *Limnology and Oceanography*, **47**, 551–8.

Parsons, M. L., Scholin, C. A., Miller, P. E., *et al.* (1999). *Pseudo-nitzschia* species (Bacillariophyceae) in Louisiana coastal waters: molecular probe field trials, genetic variability, and domoic acid analyses. *Journal of Phycology*, **35**, 1368–78.

Priisholm, K., Moestrup, Ø., & Lundholm, N. (2002). Taxonomic notes on the marine diatom genus *Pseudo-nitzschia* in the Andaman Sea, near the island of Phuket, Thailand, with a description of *Pseudo-nitzschia micropora* sp. nov. *Diatom Research*, **17**, 153–75.

Quijano-Scheggia, S., Garcés, E., Andree, K., Fortuño, J. M., & Camp, J. (2009a). Homothallic auxosporulation in *Pseudo-nitzschia brasiliana* (Bacillariophyta). *Journal of Phycology*, **45**, 100–7.

Quijano-Scheggia, S., Garcés, E., Lundholm, N., *et al.* (2009b). Morphology, physiology, molecular phylogeny and sexual compatibility of the cryptic *Pseudo-nitzschia delicatissima* complex (Bacillariophyta), including the description of *P. arenysensis* sp. nov. *Phycologia*, **48**, 492–509.

Ramsdell, J. S. & Zabka, T. S. (2008). *In utero* domoic acid toxicity: a fetal basis to adult disease in the California sea lion (*Zalophus californianus*). *Marine Drugs*, **6**, 262–90.

Rines, J. E. B., Donaghay, P. L., Dekshenieks, M. M., Sullivan, J. M., & Twardowski, M. S. (2002). Thin layers and camouflage: hidden *Pseudo-nitzschia* spp. (Bacillariophyceae) populations in a fjord in the San Juan Islands, Washington, USA. *Marine Ecology Progress Series*, **225**, 123–37.

Sarno, D., Zingone, A., & Montresor, M. (2010). A massive and simultaneous sex event of two *Pseudo-nitzschia* species. *Deep-Sea Research II*, **57**, 248–55.

Sazhin, A., Artigas, L., Nejstgaard, J., & Frischer, M. (2007). The colonization of two *Phaeocystis* species (Prymnesiophyceae) by pennate diatoms and other protists: a significant contribution to colony biomass. *Biogeochemistry*, **83**, 137–45.

Schnetzer, A., Miller, P. E., Schaffner, R. A., *et al.* (2007). Blooms of *Pseudo-nitzschia* and domoic acid in the San Pedro Channel and Los Angeles harbor areas of the Southern California Bight, 2003–2004. *Harmful Algae*, **6**, 372–87.

Scholin, C., Doucette, G., Jensen, S., *et al.* (2009). Remote detection of marine microbes, small invertebrates, harmful algae and biotoxins using the Environmental Sample Processor (ESP). *Oceanography*, **22**, 158–67.

Scholin, C. A., Gulland, F., Doucette, G. J., *et al.* (2000). Mortality of sea lions along the central California coast linked to a toxic diatom bloom. *Nature*, **403**, 80–4.

Sekula-Wood, E., Schnetzer, A., Benitez-Nelson, C. R., *et al.* (2009). Rapid downward transport of the neurotoxin domoic acid in coastal waters. *Nature Geoscience*, **2**, 272–5.

Smayda, T. (1997). What is a bloom? A commentary. *Limnology and Oceanography*, **42**, 1132–6.

Smetacek, V. S. (1985). Role of sinking in diatom life-history cycles: ecological, evolutionary and geological significance. *Marine Biology*, **84**, 239–51.

Smetacek, V. S. (1999). Diatoms and the ocean carbon cycle. *Protist*, **150**, 25–32.

Smith, J. C., Cormier, R., Worms, J., *et al.* (1990). Toxic blooms of the domoic acid containing diatom *Nitzschia pungens* in the Cardigan River, Prince Edward Island. In *Toxic Marine Phytoplankton*, ed. E. Granéli, B. Sundström, L. Edler, & D. M. Anderson, New York: Elsevier, pp. 227–32.

Thompson, S., Bates, S. S., Kaczmarska, I., & Léger, C. (2000). Bacteria and sexual reproduction of the diatom *Pseudo-nitzschia multiseries* (Hasle) Hasle. Symposium on Harmful Marine Algae in the US, Woods Hole Oceanographic Institution, Woods Hole, MA, December 5–9, p. 184 (abstract).

Todd, E. C. D. (1993). Amnesic shellfish poisoning – a review. *Journal of Food Protection*, **56**, 69–83.

Trainer, V. L., Adams, N. G., Bill, B. D., *et al.* (2000). Domoic acid production near California coastal upwelling zones, June 1998. *Limnology and Oceanography*, **45**, 1818–33.

Trainer, V. L., Cochlan, W. P., Erickson, A., *et al.* (2007). Recent domoic acid closures of shellfish harvest areas in Washington State inland waterways. *Harmful Algae*, **6**, 449–59.

Trainer, V. L., Hickey, B. M., & Horner, R. A. (2002). Biological and physical dynamics of domoic acid production off the Washington coast. *Limnology and Oceanography*, **47**, 1438–46.

Turrell, E., Bresnan, E., Collins, C., *et al.* (2008). Detection of *Pseudo-nitzschia* (Bacillariophyceae) species and amnesic shellfish toxins in Scottish coastal waters using oligonucleotide probes and the Jellet Rapid TestTM. *Harmful Algae*, **7**, 443–58.

Villac, M. C. (1996). Synecology of *Pseudo-nitzschia* H. Peragallo from Monterey Bay, California, USA. Unpublished Ph.D. thesis, Texas A&M University.

Villac, M. C. & Fryxell, G. A. (1998). *Pseudo-nitzschia pungens* var. *cingulata* var. nov. (Bacillariophyceae) based on field and culture observations. *Phycologia*, **37**, 269–74.

Villac, M. C., Matos, M. G., Santos, V. S., Rodrigues, A. W., & Viana, S. C. (2004). Composition and distribution of *Pseudo-nitzschia* from Guanabara Bay, Brazil: the role of salinity, based on field and culture observations. In *Harmful Algae 2002*, ed. K. A. Steidinger, J. H. Landsberg, C. R. Tomas, & G. A. Vargo, Florida Fish and Wildlife Conservation Commission, Florida Institute of Oceanography, and Intergovernmental Oceanographic Commission of UNESCO, Paris, pp. 56–58.

Villac, M. C. & Noronha, V. A. P. C. (2008). The surf-zone phytoplankton of the State of São Paulo, Brazil. I. Trends in space–time distribution with emphasis on *Asterionellopsis glacialis* and *Anaulus australis* (Bacillariophyceae). *Nova Hedwigia*, **133**, 115–29.

Villac, M. C., Roelke, D. L., Chavez, F. P., Cifuentes, L. A., & Fryxell, G. A. (1993). *Pseudonitzschia australis* and related species from the west coast of the U.S.A.: occurrence and domoic acid production. *Journal of Shellfish Research*, **12**, 457–65.

Walz, P. M., Garrison, D. L., Graham, W. M., *et al.* (1994). Domoic acid-producing diatom blooms in the Monterey Bay, California: 1991–1993. *Natural Toxins*, **2**, 271–9.

Zingone, A., Siano, R., D'Alelio, D., & Sarno, D. (2006). Potentially toxic and harmful microalgae from coastal waters of the Campania region (Tyrrhenian Sea, Mediterranean Sea). *Harmful Algae*, **5**, 321–37.

31 Diatoms as markers of atmospheric transport

MARGARET A. HARPER AND
ROBERT M. MCKAY

31.1 Introduction

"The dustiest place on earth" is the Bodélé depression in Chad which contains an extensive diatomite formed by the paleolake MegaChad (Giles, 2005). It is the source of an estimated 61,000 cubic kilometres of diatoms transported by wind from the area during the past thousand years (Bristow *et al.*, 2008). Satellite imaging of dust plumes reveals it to be the largest source of global airborne dust today (Goudie, 2008). Most natural eolian dust comes from drainage depressions in deserts (Middleton and Goudie, 2001). Many of these once held large water bodies such as MegaChad, West Africa; Owens Lake, North America; Lake Eyre, Australia; and the Aral Sea, Eurasia (Warren *et al.*, 2007).

Scientists trap eolian material for various objectives: atmospheric scientists interested in particles and aerosols, aerobiologists in disease spores and allergens, biologists in dispersal, and forensic workers interested in airborne contaminants (diatoms: Geissler and Gerloff, 1966). Collection techniques involve either passive or active traps; the first exposes funnels or adhesive surfaces and the second pumps air through filters (Lacey and West, 2006). Diatom remains are rare in most traps, and, when present, are not always recognized.

Sometimes living diatoms are aerially dispersed. They rarely grow on exposed agar plates or culture media (Kristiansen, 1996; Broady, 1996); few airborne diatoms are intact (Elster *et al.*, 2007); and motile diatoms normally avoid drying surfaces (Marshall and Chalmers, 1997). However, *Stauroneis anceps* Ehrenb., slowly dried with soil particles (>0.1 mm), survived 16 months (Hostetter and Hoshaw, 1970). Finlay *et al.* (2002) argued that all diatom species that can grow in an ecological niche are present in that niche, but most are present only in very low numbers. Most diatomists support the case for more variable distribution with some endemism, mostly below the genus level and in more isolated regions (see Vanormerlingen *et al.*, 2008).

The Diatoms: Applications for the Environmental and Earth Sciences, 2nd Edition, eds. John P. Smol and Eugene F. Stoermer. Published by Cambridge University Press.

31.2 Paleoclimatology and transport of African diatoms

Darwin (1846) first recognized that dust storms in the Atlantic came from arid parts of Africa. Most of this dust came from the Bodélé depression, the site of the former Lake MegaChad (Figure 31.1), which was larger than any other known past or present lake (Leblanc *et al.*, 2006). It was also deep and long-lived as shown by its thick diatomites (Washington *et al.*, 2006). These consist largely of *Aulacoseira granulata* (Ehrenb.) Simonsen and *Aulacoseira italica* (Ehrenb.) Simonsen with some small *Fragilaria* species and *Stephanodiscus* species (Gasse, 2002). Intense winds funneled by mountains in the region lift large diatomite pellets (about 20 mm diameter) over 0.5 m and some smaller particles about 3 km above the ground forming dust plumes (Warren *et al.*, 2007). Dust plumes containing diatoms from the Bodélé depression reach the equatorial Atlantic (Nave *et al.*, 2001). Marine sedimentation of this dust leads to valves of freshwater *Aulacoseira* being more abundant than marine diatoms at some levels in ocean cores, indicating eolian deposition during arid periods (Pokras and Mix, 1985).

In other parts of the Atlantic Ocean most freshwater valves in cores come from rivers. For example, the Zaire (Congo) River carries *Aulacoseira* valves 1000 km into the Gulf of Guinea (Gasse *et al.*, 1989). Comparison of floras indicates that *Cyclotella striata* (Kütz.) Grunow found in the Zaire fan is entirely fluvial in origin, but fragments of *Stephanodiscus* and *Aulacoseira* come from dust plumes (Pokras, 1991). *Stephanodiscus* is common in the older Plio-Pleistocene deposits of paleolake MegaChad (Gasse *et al.*, 1989). Dust plumes deposit a smaller range of taxa than rivers in the ocean (Gasse *et al.*, 1989), because airborne valves are exposed to more fragmentation processes during

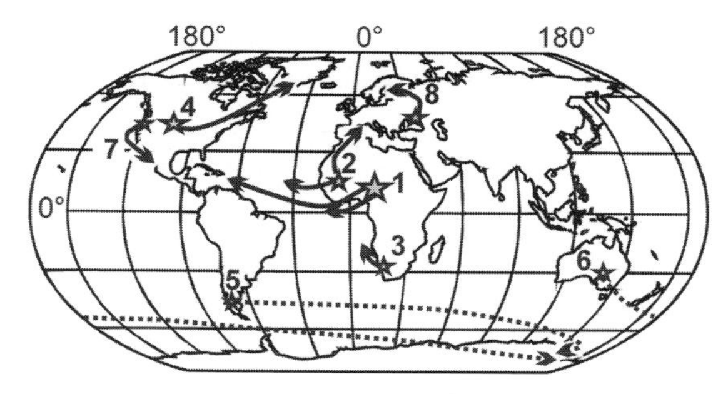

Figure 31.1 World map showing main directions of atmospheric transport of diatoms. Stars indicate likely sources, solid lines certain transport, dotted lines likely transport. Sources: (1) MegaChad diatomite, Bodélé depression; (2) Tauodeni Basin; (3) Etosha Basin; (4) American "dust-bowl"; (5) Patagonia, (6) Eyre Basin; (7) Cascade diatomite; (8) Tethys diatomite.

desiccation and saltation (Warren *et al.*, 2007). Less is known about a southern plume from the Etosha Basin (Figure 31.1), which deposits *Aulacoseira* off Angola (Pokras and Mix, 1985).

Diatoms from wind and river flow will tend to differ in seasonality. Romero *et al.* (1999) showed that the flux of airborne diatoms into sediment traps in the ocean off equatorial West Africa followed annual winter prevalence of dust storms. However, these fluxes did not correspond with mapped wind trajectories (Romero *et al.*, 2003). Satellite images often show dust plumes changing direction as they appear to dissipate, and become less visible as they drop below about 1500 m (McGowan and Clark, 2008). Changes in quantities of Saharan dust in ocean cores have also been documented by using concentrations of quartz particles over 7 μm diameter (Sarnthein *et al.*, 1981), trace-element ratios (e.g. neodynium (Nd) and strontium (Sr) isotope ratios, Grousset and Biscaye, 2005), and magnetic susceptibility (Larrasoaña *et al.*, 2003). Forms of iron oxide in dust can indicate its Saharan source (Balsam *et al.*, 2007), helium-4 has been used to trace Asian dust (Patterson *et al.*, 1999), and aluminium to trace South American dust on the Antarctic Peninsula (McConnell *et al.* 2007). All these eolian proxies need to be collected carefully to avoid contamination during sampling and processing because they can easily produce ambiguous data. So it is best to use multiple proxies and to compare their results.

Pokras and Mix (1985) found peaks of diatom abundance in Atlantic cores from 3° N to 1° S and related these to summer insolation at 30° N. Variation in summer insolation is controlled by 19- and 23-thousand-year precessional changes in the Earth's orbit (Milankovitch, 1920). DeMenocal *et al.* (1993) confirmed these periodicities over the last million years for a core at 1° S. Deflation by wind erosion reduces diatom sources which are then restored during humid periods (Pokras and Mix, 1985). The plumes from the Bodélé Basin form in winter and are centered around latitude 5° N. Lighter summer plumes come from the Taoudeni Basin (Figure 31.1) in the western Sahara and are centered around 20° N (Goudie, 2008). Summer plumes contain more clay and heavy rare earths (Moreno *et al.*, 2006) and fewer diatoms (Pokras and Mix, 1985). Bodélé dust is also deposited on the continent, for instance in equatorial Cameroon in Lake Ossa. Sediment cores from this lake contain some diatoms that appear to be reworked from paleolake MegaChad (Nguetsop *et al.*, 2004). Reworked diatoms indicated arid periods when other diatoms showed the lake became more alkaline, and Nguetsop *et al.* (2004) considered the pH of the lake water to have changed as a result of less precipitation.

Fine Bodélé dust and other North African dusts are spread over Europe, and have been recorded from America and the Arctic, and even may reach Antarctica (Goudie, 2008). The diatom flora of Saharan dust from southern France consisted of only 11% *Aulacoseira granulata*; other diatoms were aerophils (*Hantzschia amphioxys* (Ehrenb.) Grunow, *Luticola* aff. *mutica* (Kütz.) D. G. Mann) which grow on moist soil (Seyve and Fourtanier, 1985). These authors noted that this dust was coated with red clay; an indication it came from an arid place (Bullard and White, 2005). Håkansson and Nihlén (1990) found *A. granulata* was more abundant in dust carried by southerly winds to Crete, Italy, and Greece than in France. They considered that the presence of *Cyclostephanos* cf. *damasii* (Hust.) Stoermer and Håkansson and an undescribed *Cyclotella* species indicated the dust was from North Africa. However, Iordanis *et al.* (2008) suggest that *Aulacosiera* in dust from northern Greece came from fuel emissions as diatomites are associated with local lignites. Unusual westerly winds occurred in the Oligocene as *Aulacosiera* was blown out of Africa into the Indian Ocean (Fenner and Mikhelsen, 1990).

African dusts are the main source of minerals for soils in the Amazon Basin (Koren *et al.*, 2006). Some African diatoms are blown to America; for instance *A. granulata* was trapped on a Barbados sea cliff in Saharan dust (Delany *et al.*, 1967). The same species was caught at 3000 m above Florida, in air from the Atlantic (Maynard, 1968). Some valves reach snow 500 km north of Alaska (Darby *et al.*, 1974).

31.3 Deposition of wind-blown diatoms in Greenland

The Greenland Ice Core GISP2 contains a dust layer dating from 1933, which was considered by Donarummo *et al.* (2003) to come from the American "Dust Bowl" (Figure 31.1) rather than Asia. Diatoms in GISP2 included *Luticola cohnii* (Hilse) D. G. Mann and *Pinnularia borealis* Ehrenb. which were identified earlier in source dust (Page and Chapman, 1934). Other Greenland ice-cap cores contained *A. granulata* and *Stephanodiscus niagarae* Ehrenb. deposited during the spring time (Gayley *et al.*, 1989), which could also originate from North America, as spring anticyclones direct westerly winds to Greenland. However, changes in concentrations of diatom remains downcore match those of 0.38–1.31 μm microparticles (Gayley *et al.*, 1989), some of which come from Asia (Goudie, 2008).

31.4 Antarctic deposition of eolian diatoms and Pliocene warming

The atmospheric transport of diatoms in Antarctica has led to debate over the use of diatoms for providing robust age control of subaerially exposed sediments. This debate has centered on the use of a sparse Pliocene-aged marine diatom assemblage to date glacially deposited material in the Transantarctic Mountains (Denton *et al.*, 1993). The diatoms were first interpreted as an original component of the sediment, and considered to be transported by ice from Pliocene marine sediments in the East Antarctic interior. This implied there were ice-free marine basins in East Antarctica in the Pliocene (Webb *et al.*, 1984). However, a post-Miocene age conflicts with other geological evidence of relative stability of the East Antarctic Ice Sheet in the last 13 million years, and therefore the origin of these diatoms is widely disputed (e.g. Denton *et al.*, 1993). As a result of this debate, numerous studies (discussed below) have been undertaken in the Antarctic to test the viability of atmospheric transport and incorporation of the diatom remains into older consolidated sediments such as the Sirius Group.

At present, most non-marine diatoms in continental Antarctica grow in ephemeral moist patches, meltwater streams, or meltwater moats (margins of ice-covered lakes). Planktonic diatoms sink under ice cover and therefore are very rare in lakes with permanent ice (see Spaulding *et al.*, this volume). Live diatoms are absent in many parts of inland Antarctica and rare in others, though a few live *Luticola muticopsis* (Van Heurck) D. G. Mann were found in cyanobacterial mats as far south as 86° 30′S in the La Gorce Mountains (Broady and Weinstein, 1998).

Diatoms of undisputed eolian origin have been documented in ice cores from the South Pole and Dome C, and, although marine diatoms are present, these assemblages are dominated by freshwater taxa (Burckle *et al.*, 1988; Kellogg and Kellogg, 1996). Planktonic *Cyclotella stelligera* (Cleve) Van Heurck was the most common species, with 184 per litre in one sample (Kellogg and Kellogg, 1996). Each concentration peak could come from a single source area and be transported by the same wind event, but snow can blow around for years before compacting. Maximum concentrations in the South Pole ice cores were lower than Gayley *et al.* (1989) found in Greenland. This agrees with Thompson and Moseley-Thompson's (1981) microparticle data, where they find glacial and post-glacial concentrations are greater for Greenland than Antarctica.

Marine diatoms in ice cores from the South Pole provided the first evidence that the few marine diatoms found in the Sirius Group may be wind-blown (Kellogg and Kellogg, 1996). Another line of evidence supporting their eolian origin is that they occur mainly in the surface layer of the Sirius Group (Stroeven *et al.*, 1996) and there are micro-cracks allowing a few diatoms to be reworked to greater depths (Barrett *et al.*, 1997). Plio-Pleistocene species also occur in cracks in ancient rocks (Burckle and Potter, 1996) and in non-Sirius Group surface material (Barrett *et al.*, 1997). Fragments of *Aulacoseira* are relatively common (all diatoms are exceedingly sparse; less than 10 per gram) in consolidated till in the South Victoria Land sector of the Transantarctic Mountains (Barrett *et al.*, 1997). In the Sirius Group, the commonest remains are fragments of *Aulacoseira* species which are generally <10 μm in size and could come with dust transported (Figure 31.1) from Patagonia (Grousset *et al.*, 1992), Australia (Marino *et al.*, 2008), or even equatorial Africa. *Aulacoseira* fragments are typical of MegaChad diatomite (Warren *et al.*, 2007) and fine dust from it is carried over 5 km up in the atmosphere to the upper troposphere (Liu *et al.*, 2005) where it can then be transported to Antarctica by cyclonic systems (Barrett *et al.*, 1997; Robock *et al.*, 2007).

McKay *et al.* (2008) collected 108 diatoms in a wind trap placed at the end of an exposed mountain range directly at the very edge of the East Antarctic Ice Sheet (Allan Hills; latitude 76.4° S) for a period of five weeks. Fragments of *Aulacoseira* dominated the wind-trap assemblage, but many well-preserved, larger (>30 μm) non-marine and marine taxa were also noted. The wind-trap assemblage was compared to a clay veneer on the surface of a vertical cliff-face exposure of the Sirius Group. This veneer was formed by recent melting of wind-blown snow over the outcrop. The veneer also contained relatively large (>30 μm) diatoms in a well-preserved assemblage dominated by freshwater taxa. Modern meltwater processes, similar to those that formed this clay veneer, could carry diatoms into cracks in the active layers of older sediments (McKay *et al.*, 2008). Non-marine assemblages in the wind trap and clay veneer are of more local origin, as the diatoms are too large for most tropospheric transport. They could be derived from flakes of dried cyanobacterial mats which originally grew in wet places in the nearby Transantarctic Mountains.

The dominance of non-marine diatoms in studies of wind-blown diatoms in Antarctica differs markedly from the marine-dominated assemblages in the Sirius Group. This has been used as evidence that the marine assemblages in the Sirius are not wind-blown, as the assemblages are distinctly different (Harwood and Webb, 1998). One difficulty in comparing the data sets is that the use of filters, which is necessary to recover the sparse marine diatoms from the Sirius Group sediments, and results in the loss of most of the finer components of the eolian diatom signal (e.g. the dominant *Aulacoseira*). More importantly, marine diatoms could have reached the Sirius Group sediment any time since their Miocene deposition, and changes occur over time in the sources of eolian diatoms available and their assemblages. Thick diatomites were deposited in the Ross Sea during the Pliocene, when the surrounding sea and West Antarctica were substantially ice free (Naish *et al.*, 2009). These diatomites are dominated by *Thalassiothrix* and contain no benthic species (Scherer *et al.*, 2007), and so resemble the marine assemblage in the Sirius Group (Wilson *et al.*, 2002). Subareal exposure of these diatomites at the coastal margin of the Transantarctic Mountains during Pliocene sea-level lowstands could well have resulted in large volumes of marine diatoms being transported by lower atmospheric processes into nearby Transantarctic Mountain localities at this time (McKay *et al.*, 2008). These marine diatomites are now covered by the Ross Sea and the Ross Ice Shelf (Naish *et al.*, 2009), thus restricting the source of marine diatoms available to be transported by the atmosphere. This results in non-marine taxa of local ephemeral meltwater deposits being predominant in modern atmospherically transported diatom assemblages.

The problems in interpreting sparse diatom assemblages from glacial deposits in the Antarctic highlight the need for caution in using diatoms to provide chronostratigraphic age control of these sediments. These problems have made the interpretations Webb *et al.* (1984) drew from their data on the Sirius Group hard to test. Other studies (below) of glacial deposits in the Antarctic also show the need to critically assess the role of eolian-sourced diatoms.

Scherer (1991) considered that the West Antarctic Ice Sheet had collapsed at least once within the past 750 thousand years, because he found Pleistocene marine diatoms in sediments under a glacial ice-stream. Burckle *et al.* (1997) argued that Scherer's diatoms were wind-blown and concentrated under the ice by basal melting. To test if this was the case, Scherer *et al.* (1998) compared the sub-ice sediment (glacial till) with the eolian diatom component in basal ice directly overlying the till. They noted that while diatoms did occur in this basal ice (6–34 diatoms per litre), the ice was dominated by non-marine diatoms (esp. *Luticola* spp., a common genus in Antarctica), compared to the marine-dominated assemblages in the till directly beneath. They considered that recent non-marine diatoms would also be concentrated by basal melting as suggested by Burckle *et al.* (1997) and supplied to the sediment below today. The larger marine diatoms in the sediment were not selected by filtering (Scherer *et al.*, 1998). Therefore, the lack of non-marine diatoms in the sub-ice sediment indicates the marine diatoms in it do not come from the ice and are not of eolian origin.

Sparse Neogene diatom assemblages have also been recovered from glacigenic deposits of the Pagodroma Group in the Lambert Glacier region, and assessed by Whitehead *et al.* (2004) for potential contamination by eolian processes. In these sediments, the majority of diatoms are observed in mudstone beds that are intercalated with diamictite units. These mudstones have a higher abundance of diatoms, contain *in situ* molluscs, and have high (up to 12%) biogenic silicia content, pointing to a fjordal depositional environment and therefore a probable *in situ* marine diatom assemblage. However, the presence of a subordinate freshwater diatom assemblage in these deposits suggests an eolian component in the diatom assemblages of these sediments (Whitehead *et al.*, 2004).

31.5 Maritime polar and sub-polar eolian diatoms, and seasonality

Marine diatoms are rarely airborne except near beaches (Geissler and Gerloff, 1966; Delany *et al.*, 1967; Lee and Eggleston, 1989) or on stormy seas (Pokras and Mix, 1985). Maritime polar regions differ. *Fragilariopsis cylindrus* (Grunow) Krieg. was often dominant in traps on Signy Island (subantarctic) and was abundant in rime frost on Ellesmere Island (High Arctic; Lichti-Federovich, 1985). These diatoms came from dense growths on sea ice. Subarctic snow 50 km inland contained 17 marine taxa and 93 non-marine taxa (Lichti-Federovich, 1984, 1985). Traps on Signy Island, 1 km inland, collected 36 marine taxa (total 43 taxa, Chalmers *et al.*, 1996). Nearly all diatom remains were small (maximum area 300 μm^2) with a few larger ones being elongated or flimsy. The assemblage differed from those of continental Antarctica as we saw no remains of *Aulacoseira* (Chalmers *et al.*, 1996), which is common in eolian assemblages in Antarctica. Interestingly, traps recorded seasonal peaks in *Fragilariopsis* in autumn. However, these traps lacked baffles to stop the wind blowing particles out and in winter they were blocked with snow.

Regarding seasonality in non-polar regions, Tormo *et al.* (2001) noted a maximum in wind-blown *Cyclotella* in June in Spain, but Sharma *et al.* (2006) trapped only a few diatoms at all seasons in India. Dust storms are usually seasonal and may form varves as in two maar lakes in northeast China where the varves record 1400 years of variation in wind strength (Chu *et al.*, 2009).

31.6 Other cases of long-distance transport of diatomites

Dust is blown out of all continents except Antarctica (Goudie, 2008). From Asia, diatoms (*Paralia siberica* (A. W. F. Schmidt) R. M. Crawford and P.A. Sims and *Stephanopyxis* sp.) deposited by the Tethys Sea over 20 million years ago north and east of the Black Sea are blown as far as Finland (Tynni, 1970). From North America, diatoms (*Cyclotella elgeri* Hust. and *Stephanodiscus carconensis* (Eulenst.) Grunow) in the Cascade Mountain diatomites of Oregon are blown out to the Gulf of California (Sancetta *et al.*, 1992). In South America, eolian quartz and probably some diatom remains from Chile are blown into the Pacific Ocean sediments at 5° N (Abrantes *et al.*, 2007). Australian dust storms sometimes contain diatoms (Brittlebank *et al.*, 1897), but when dust was collected from New Zealand glaciers (McGowan *et al.*, 2005), M. A. Harper (unpublished data) found diatoms only in local dust and not in red dust chemically identified as Australian. There are likely to be unrecognized instances of diatoms being blown significant distances from their source. Exotic taxa are often ignored, but with checks for contamination, diatomists could recognize more instances of aerial transport. Some instances could give us information on past wind strength and direction.

31.7 Summary

Some of the evidence for past climate change comes from eolian diatoms. Satellite images have revealed that the Bodélé depression in Africa is the main source of global dust plumes. The plumes are formed from freshwater diatomites laid down by the paleolake MegaChad. In the equatorial Atlantic layers of freshwater diatoms in ocean cores indicate arid periods in the past million years. Most marine diatoms in the Sirius Group could have been blown into it from Pliocene diatomites now buried under the Ross Ice Shelf and its sediments. There are other instances of atmospheric transport of diatoms which could leave climate records in sediment or ice cores.

Acknowledgments

Thanks are due to David Walton, William Marshall, and Matthew Chalmers for the opportunity to analyse British Antarctic Survey trap data; Hamish McGowan for dust from New Zealand glaciers, and to Clement Prevost of the Terrain Sciences Division, Natural Resources of Canada, Peter Barrett, John Carter, John Harper, and others in the School of Geography, Environment and Earth Sciences, Victoria University of Wellington for general advice.

References

Abrantes, F., Lopes, C., Mix, A., & Pisias, N. (2007). Diatoms in southeast Pacific surface sediments reflect environmental properties. *Quaternary Science Reviews*, **26**, 155–69.

Balsam, W., Arimoto, R., Ji, J., Shen, Z., & Chen, J. (2007). Aeolian dust in sediment: a re-examination of methods for identification and dispersal assessed by diffuse reflectance spectrophotometry. *International Journal of Environment and Health*, **1**, 374–402.

Barrett, P. J., Bleakley, N. L., Dickinson, W. W., Hannah, M. J., & Harper, M. A. (1997). Distrbution of siliceous microfossils on Mount Feather, Antarctica, and the age of the Sirius Group. In *The Antarctic Region, Geological Evolution and Processes, Proceedings of the 7th Symposium on Antarctic Geological Sciences, Siena, Sept 10–15, 1995*, ed. C. A. Ricci, Siena: Terra Antartica Publication, pp. 763–70.

Bristow, C. S., Drake, N., & Armitage, S. (2008). Deflation in the dustiest place on Earth: the Bodélé Depression, Chad. *Geomorphology*, **105**, 50–8.

Brittlebank, C., Barnard, F. G. A., Strickland, C., & Shepherd, J. (1897). Red rain: diatoms in Australian dust fallen Dec. 27, 1896. *The Victorian Naturalist*, **13**, 125–6.

Broady, P. A. (1996). Diversity, distribution and dispersal of Antarctic terrestrial algae. *Biodiversity and Conservation*, **5**, 1307–35.

Broady, P. A. & Weinstein, R. (1998). Algae, lichens and fungi in La Gorce Mountains, Antarctica. *Antarctic Science*, **10**, 376–85.

Bullard, J. E. & White K. (2005). Dust production and release of iron oxides resulting from the aeolian abrasion of natural dune sands. *Earth Surface Processes and Landforms*, **30**, 95–106.

Burckle, L. H., Gayley, R. I., Ram, M., & Petit, J.-R. (1988). Diatoms in Antarctic ice cores: some implications for the glacial history of Antarctica. *Geology*, **16**, 326–9.

Burckle, L. H., Kellogg, D. E., Kellogg, T. B., & Fastook, J. L. (1997). A mechanism for the emplacement and concentration of diatoms in glaciogenic deposits. *Boreas*, **26**, 55–60.

Burckle, L. H. & Potter, J. R. (1996). Pliocene-Pleistocene diatoms in Paleozoic and Mesozoic sedimentary and igneous rocks from Antarctica: a Sirius problem solved. *Geology*, **24**, 235–8.

Chalmers, M. O., Harper, M. A., & Marshall, W. A. (1996). *An Illustrated Catalogue of Airborne Microbiota from the Maritime Antarctic*. Cambridge: British Antarctic Survey.

Chu, G., Sun, Q., Zhaoyan, G., *et al.* (2009). Dust records from varved lacustrine sediments of two neighbouring lakes in northeastern

China over the last 1400 years. *Quaternary International*, **194**, 108–18.

Darby, D. A., Burckle, L. H., & Clark, D. L. (1974). Airborne dust on Arctic pack ice, its composition and fallout rate. *Earth and Planetary Science Letters*, **24**, 166–72.

Darwin, C. E. (1846). An account of the fine dust which often falls on vessels in the Atlantic Ocean. *Quarterly Journal of the Geological Society, London*, **2**, 26–30.

Delany, A. C., Delany, A. C., Parkin, *et al.* (1967). Airborne dust collected at Barbados. *Geochimica et Cosmochimica Acta*, **31**, 885–909.

DeMenocal, P. B., Ruddiman, W. F., & Pokras, E. M. (1993). Influences of high- and low-latitude processes on Aftrican terrestrial climate: Pleistocene eolian records from equatorial Atlantic Ocean Drilling Program Site 633. *Paleoceanography*, **8**, 209–42.

Denton, G. H., Sugden, D. E., Marchant, D. R., Hall, B. L., & Wilch, T. I. (1993). East Antarctic Ice Sheet sensitivity to Pliocene climatic change from a Dry Valleys perspective. *Geografiska Annaler, Series A, Physical Geography*, **75**, 155–204.

Donarummo, J., Jr., Ram, M., & Stoermer, E. F. (2003). Possible deposit of soil dust from the 1930's U.S. Dust Bowl identified in Greenland ice. *Geophysical Research Letters*, **30** (6), 1269.

Elster, J., Delmas, R. J., Petit, J.-R., & Rehakova, K. (2007). Composition of microbial communities in aerosol, snow and ice samples from remote glaciated areas (Antarctica, Alps, Andes). *Biogeosciences Discussions*, **4**, 1779–813.

Fenner, J. & Mikhelsen, N. (1990). Eocene–Oligocene diatoms in the western Indian Ocean: taxonomy, stratigraphy and paleoecology. *Proceedings of the Ocean Drilling Programme*, **115**, 433–63.

Finlay, B. J., Monaghan, E. B., & Maberly, S. C. (2002). Hypothesis: the rate and scale of dispersal of freshwater diatom species is a function of their global abundance. *Protist*, **153**, 261–73.

Gasse, F. (2002). Diatom-inferred salinity and carbonate oxygen isotopes in the Holocene waterbodies of the western Sahara and Sahel. *Quaternary Science Reviews*, **21**, 737–67.

Gasse, F., Stabell, B., Fourtanier, E., & van Iperen, Y. (1989). Freshwater diatom influx in intertropical Atlantic: relationships with continental records from Africa. *Quaternary Research*, **32**, 229–43.

Gayley, R. I., Ram, M., & Stoermer E. F. (1989). Seasonal variations in diatom abundance and provenance in Greenland ice. *Journal of Glaciology*, **35**, 290–2.

Geissler, U. & Gerloff, J. (1966). Das Vorkommen von Diatomeen in menschlichen Organen und in der Luft. *Nova Hedwigia, Zeitschrift für Kryptogamenkunde*, **10**, 565–77.

Giles, J. (2005). Climate science: the dustiest place on Earth. *Nature*, **434**, 816–19.

Goudie, A. S. (2008). The history and nature of wind erosion in deserts. *Annual Review of Earth and Planetary Sciences*, **36**, 97–119.

Grousset, F. E. & Biscaye, P. E. (2005). Tracing dust sources and transport patterns using Sr, Nd and Pb isotopes. *Chemical Geology*, **222**, 149–67.

Grousset, F. E., Biscaye, P. E., Revel, M., *et al.* (1992). Antarctic (Dome C) ice-core dust at 18 k.y. B.P.: isotopic constraints on origins. *Earth and Planetary Science Letters*, **111**, 175–82.

Håkansson, H. & Nihlén, T. (1990). Diatoms of eolian deposits in the Mediterranean. *Archiv für Protistenkunde*, **138**, 313–22.

Harwood, D. M. & Webb, P. N. (1998). Glacial transport of diatoms in the Antarctic Sirius Group: Pliocene refrigerator. *GSA Today*, **8** (4), 1, 4–8.

Hostetter, H. P. & Hoshaw, R. W. (1970). Environmental factors affecting resistance to desiccation in the diatom *Stauroneis anceps*. *American Journal of Botany*, **57**, 512–8.

Iordanidis, A., Buckman, J., Triantafyllou, A. G., & Asvesta, A. (2008). ESEM-EDX characterisation of airborne particles from an industrialised area of northern Greece. *Environmental Geochemical Health*, **30**, 391–405.

Kellogg, D. E. & Kellogg, T. B. (1996). Diatoms in South Pole ice: implications for eolian contamination of Sirius Group deposits. *Geology*, **24**, 115–18.

Koren, I., Kaufman, Y. J., Washington, R., *et al.* (2006). The Bodélé depression: a single spot in the Sahara that provides most of the mineral dust to the Amazon forest. *Environmental Research Letters*, **1**, 014005.

Kristiansen, J. (1996). Dispersal of freshwater algae – a review. *Hydrobiologia*, **336**, 151–7.

Lacey, M. E. & West, J. S. (2006). *The Air Spora: a Manual for Catching and Identifying Airborne Biological Particles*. Dordrecht: Springer.

Larrasoaña, J. C., Roberts, A. P., Rohling, E. J. Winklhofer, M., & Wehausen, R. (2003). Three million years of monsoon variability over the northern Sahara. *Climate Dynamics*, **21**, 689–98.

Leblanc, M. J., Leduc, C., Stagnitti, F., *et al.* (2006). Evidence for Megalake Chad, north-central Africa, during the late Quaternary from satellite data. *Palaeogeography, Palaeoclimatology, Palaeoecology*, **230**, 230–42.

Lee, T. F. & Eggleston, P. M. (1989). Airborne algae and cyanobacteria. *Grana*, **28**, 63–6.

Lichti-Federovich, S. (1984). Investigation of diatoms found in surface snow from the Sydkap ice cap, Ellesmere Island, Northwest Territories. *Current Research, Geological Survey of Canada*, **84–01A**, 287–301.

Lichti-Federovich, S. (1985). Diatom dispersal phenomena: diatoms in rime frost samples from Cape Herschel, central Ellesmere Island, Northwest Territories. *Current Research, Geological Survey of Canada*, **85–01B**, 391–9.

Liu, X., Penner, J. E. & Herzog, M. (2005). Global modeling of aerosol dynamics: model description, evaluation, and interactions between sulfate and nonsulfate aerosols. *Journal of Geophysical Research*, **110**, DOI: D18206.1-D18206.37.

Marino, F., Castellano, E., Ceccato, D., *et al.* (2008). Defining the geochemical composition of the EPICA Dome C ice core dust during the last glacial–interglacial cycle. *Geochemistry Geophysics Geosystems*, **9**, Q10018, 1–11.

Marshall, W. A. & Chalmers, M. O. (1997). Airborne dispersal of Antarctic terrestrial algae and cyanobacteria. *Ecography*, **20**, 585–94.

Maynard, N. G. (1968). Significance of airborne algae. *Zeitschrift für Allgemeine Mikrobiologie*, **8**, 225–6.

McConnell, J. R., Aristarain, A. J., Banta, R., Edwards, P. R., & Simões, J. C. (2007). 20th-Century doubling in dust archived in an Antarctic Peninsula ice core parallels climate change and desertification in South America. *Proceedings of the National Academy of Sciences of the USA*, **104**, 5743–8.

McGowan, H. A. & Clark, A. (2008). Identification of dust transport pathways from Lake Eyre, Australia using Hysplit. *Atmospheric Environment*, **42**, 6915–25.

McGowan, H., Kamber, B., McTainsh, G.H., & Marx, S.K. (2005). High resolution provenancing of long travelled dust deposited on the Southern Alps, New Zealand. *Geomorphology*, **69**, 208–21.

McKay, R. M., Barrett, P. J., Harper, M. A., & Hannah, M. J. (2008). Atmospheric transport and concentration of diatoms in surficial and glacial sediments in the Allan Hills, Transantarctic Mountains. *Palaeogeography, Palaeoclimatology, Palaeoecology*, **260**, 168–83.

Middleton, N. J. & Goudie, A. S. (2001). Saharan dust: sources and trajectories. *Transactions of the Institute of British Geographers*, **26** (2), 165–181.

Milankovitch, M. (1920). *Theorie Mathematique des Phenomenes Thermiques produits par la Radiation Solaire*. Paris: Gauthier-Villars.

Moreno, T., Querol, X., Castillo, S., *et al.* (2006). Geochemical variations in aeolian mineral particles from the Sahara–Sahel Dust Corridor. *Chemosphere*, **65**, 261–70.

Naish, T., Powell, R., Levy, R., *et al.* (2009). Obliquity-paced Pliocene West Antarctic ice sheet oscillations. *Nature*, **458**, 322–8.

Nave, S., Freitas, P., & Abranthes, F. (2001). Coastal upwelling in the Canary Island region: spatial variability reflected by the surface sediment diatom record. *Marine Micropaleontology*, **42**, 1–23.

Nguetsop, V.F., Servant-Vilary, S., & Servant, M. (2004). Late Holocene climatic changes in West Africa, a high resolution diatom record from equatorial Cameroon. *Quaternary Science Reviews*, **23**, 591–609.

Page, L. R. & Chapman, R. W. (1934). The dust fall of December 15–16, 1933. *American Journal of Science, 5th Series*, **28**, 288–97.

Patterson, D. B., Farley, K. A., & Norman, M. D. (1999). ^{4}He as a tracer of continental dust: a 1.9 million year record of aeolian flux to the west equatorial Pacific Ocean. *Geochimica et Cosmochimica Acta*, **63**, 615–25.

Pokras, E. M. (1991). Source areas and transport mechanisms for freshwater and brackish-water diatoms deposited in pelagic sediments of the equatorial Atlantic. *Quaternary Research*, **35**, 144–56.

Pokras, E. M. & Mix, A. C. (1985). Eolian evidence for spatial variability of late Quaternary climates in tropical Africa. *Quaternary Research*, **24**, 137–49.

Robock, A., Oman, L. & Stenchikov, G.L. (2007). Nuclear winter revisited with a modern climate model and current nuclear arsenals: still catastrophic consequences. *Journal of Geophysical Research*, **112**, D13107, 1–14.

Romero, O. E., Dupont, L., Wyputta, U., Jahns, S., & Wefer, G. (2003). Temporal variability in fluxes of eolian-transported freshwater diatoms, phytoliths and pollen grains off Cape Blanc as reflection of land–atmosphere–ocean interactions in northwest Africa. *Journal of Geophysical Research*, **108** (C5), 3153–65.

Romero, O. E., Lange, C. B., Swap, R., & Wefer, G. (1999). Eolian-transported freshwater diatoms and phytoliths across the equatorial Atlantic record: temporal changes in Saharan dust transport patterns. *Journal of Geophysical Research*, **104** (C2), 3211–22.

Sancetta, C., Lyle, M., Heusser, L., Zahn, R., & Bradbury, J. P. (1992). Late-glacial to Holocene changes in winds, upwelling and seasonal production of the northern Californian current system. *Quaternary Research*, **38**, 359–70.

Sarnthein, M., Tetzlaff, G., Koopmann, B., Wolter, K., & Pflaumann, U. (1981). Glacial and interglacial wind regimes over the eastern subtropical Atlantic and North-West Africa. *Nature*, **293**, 193–6.

Scherer, R. P. (1991). Quaternary and Tertiary microfossils from beneath the Ice Stream B: evidence for a dynamic West Antarctic ice sheet history. *Palaeogeography, Palaeoclimatology, Palaeoecology, Global and Planetary Change Section*, **90**, 395–412.

Scherer, R. P., Aldahan, A., Tulaczyk, S., *et al.* (1998). Pleistocene collapse of the West Antarctic Ice Sheet. *Science*, **281** (5373), 82–5.

Scherer, R., Hannah, M., Maffioli, P., *et al.* (2007). Palaeontologic characterisation and analysis of the AND-1B Core, ANDRILL McMurdo Ice Shelf Project, Antarctica. *Terra Antartica*, **14**, 223–54.

Seyve, C. & Fourtanier, E. (1985). Contenu microfloristique d'un sédiment éolien actuel. *Bulletin – Centres pour Recherche et Exploration-Production Elf-Aquitaine*, **9**, 137–54.

Sharma, N.K., Singh, S., & Rai, A. K. (2006). Diversity and seasonal variation of visible algal particles in the atmosphere of a subtropical city in India. *Environmental Research*, **102**, 252–9.

Stroeven, A. P., Prentice, M. L., & Klemen J. (1996). On marine microfossil transport and pathways in Antarctica during the late Neogene: evidence from the Sirius Group at Mount Fleming. *Geology*, **24**, 727–30.

Thompson, L. G. & Moseley-Thompson, E. (1981). Microparticle concentration variations linked with climatic change: evidence from polar ice cores. *Science*, **212**, 812–4.

Tormo, R., Recio, D., Silva, I., & Muñoz, A. F. (2001). A quantitative investigation of airborne algae and lichen soredia obtained from pollen traps in south-west Spain. *European Journal of Phycology*, **36**, 385–90.

Tynni, R. (1970). Piilevät v. 1969 sataneessa punertavassa lumessa (Diatoms from dust-stained snowfall in 1969; Finnish, with English summary). *Geologi*, **22**, 79–81.

Vanormelingen, P., Verleyen, E., & Vyverman, W. (2008). The diversity and distribution of diatoms: from cosmopolitanism to narrow endemism. *Biodiversity and Conservation*, **17**, 393–405.

Warren, A., Chappell, A., Todd M. C., *et al.* (2007). Dust-raising in the dustiest place on earth. *Geomorphology*, **92**, 25–37.

Washington, R., Todd, M. C., Lizcano, G., *et al.* (2006). Links between topography, wind, deflation, lakes and dust: the case of the Bodélé depression, Chad. *Geophysical Research Letters*, **33**, L09401, 1–4.

Webb, P. N., Harwood, D. M., McKelvey, B. C., Mercer, J. H., & Stott, L. D. (1984). Cenozoic marine sedimentation and ice-volume variation on the East Antarctic craton. *Geology*, **12**, 287–91.

Whitehead, J. M., Harwood, D. M., McKelvey, B. C., Hambrey, M. J., & McMinn, A. (2004). Diatom biostratigraphy of the Cenozoic glaciomarine Pagodroma Group, northern Prince Charles Mountains, East Antarctica. *Australian Journal of Earth Sciences*, **51**, 521–47.

Wilson, G. S., Barron, J. A., Ashworth, A. C., *et al.* (2002). The Mount Feather Diamicton of the Sirius Group: an accumulation of indicators of Neogene Antarctic glacial and climate history. *Palaeogeography, Palaeoclimatology, Palaeoecology*, **182**, 117–31.

32 Diatoms as non-native species

SARAH A. SPAULDING, CATHY KILROY, AND MARK B. EDLUND

32.1 Introduction

The degree to which diatoms move across the Earth by natural processes is debatable (Finlay *et al.*, 2002; Vyverman *et al.*, 2007), but the inadvertent spread of diatoms in a globalized human society is apparent. In this chapter, we examine documentation of diatom introductions and their implications for aquatic ecosystems. For many organisms, especially larger ones, the ecologic, economic, and social impact of species introductions, or invasions, is relatively well known (Pimental *et al.*, 2000). On the other hand, recognition of the microscopic trespasses of diatom species and their impact on ecosystems in new geographic areas is generally far less noticed.

A species is considered to be "non-native" if it is located in a region outside of its native geographic range. Non-native species are also referred to as introduced, non-indigenous, exotic, alien, or invasive. While some non-native species cause little harm, others cause severe ecosystem damage. The use of terminology, particularly the adoption of military words to describe species geographic distributions, elicits emotional reactions that influence scientific and popular responses (Larson *et al.*, 2005). We recognize that much of the current literature employs these military metaphors, but we seek to promote an ecological perspective.

Even among well-known organisms, such as the common reed (*Phragmites australis* (Cav.) Trin. ex Steud.) growing near Lake Superior, the distinction between native and non-native status may be unclear (Willis and Birks, 2006). The biogeographic distribution of diatoms, and microscopic organisms in general, is currently under debate. One school of thought advocates that microorganisms do not exhibit biogeographic differences (Baas-Becking, 1934; Finlay *et al.*, 2002). Under this hypothesis, vast local populations of microorganisms drive global dispersal and panmixis. As a result, species invasion and extinction are rendered impossible at local and global scales. Within this context, native and non-native species do not exist. An alternate school of thought advocates that microorganisms do, in fact, exhibit biogeography. An increasing body of empirical data supports the biogeography of microorganisms (Martiny *et al.*, 2006) and of diatoms, at least in some regions (Soininen *et al.*, 2003; Telford *et al.*, 2006; Vyverman *et al.* 2007; Vanormelingen *et al.*, 2008). For example, diatom communities in Finland streams showed a strong spatial component that could not be attributed to environmental parameters other than location (Soininen *et al.*, 2003). Likewise, Telford *et al.* (2006) demonstrated that the dispersal of diatoms was limited and timescales of isolation allowed for regional genetic differences to develop and endemic taxa to evolve. In other work, historical factors explained significantly more observed geographic patterns in richness (measured at the genus level) than environmental parameters (Vyverman *et al.*, 2007). In fact, diatom species introductions may help illuminate the biogeographic discussion. The number of taxa with widespread distributions has been suggested to be related to anthropogenic change (Kociolek and Spaulding, 2000), but at the same time, the hypotheses of "everything is everywhere" is challenged by examples of anthropogenic introductions (Vanormelingen *et al.*, 2008).

Determination of the non-native status of diatoms, as a group whose basic biology, taxonomy, biogeography, and genomes are not well known, is possible through the application of long-term floristic studies and paleolimnological reconstruction. The degree to which diatoms are introduced is likely underestimated because the idea of non-native diatoms has been proposed relatively recently and includes efforts to recognize endangered, or red-list diatoms (Lange-Bertalot and Steindorf, 1996; Denys, 2000). The introduction of non-native species, extirpation of native species, and habitat alterations

The Diatoms: Applications for the Environmental and Earth Sciences, 2nd Edition, eds. John P. Smol and Eugene F. Stoermer. Published by Cambridge University Press.

that facilitate these two processes are considered biological homogenization (Rahel, 2002). Even when the number of total species increases by species introductions, the enhancement often includes taxa that are already widespread, tolerant of degraded habitats, and considered a nuisance by humans (Pimental *et al.*, 2000). We consider a few case studies of diatom introductions and their ecological implications.

32.2 Case studies of diatom introductions

32.2.1 Laurentian Great Lakes of North America

The Laurentian Great Lakes comprise five major post-glacial lake basins – Lakes Superior, Huron, Michigan, Erie, and Ontario – and contain one-fifth of the world's fresh surface waters (Herdendorf, 1982). Before European colonization of the region, great water depth, oligotrophy, unique physical limnology, geographical barriers, and the geographic extent of the Great Lakes contributed to produce food webs that included endemic fishes (Smith and Todd, 1984) and endemic diatom species such as *Stephanodiscus superiorensis* E. C. Ther. & Stoermer, *Hannaea superiorensis* Bixby & Edlund, and *Cyclotella americana* Fricke (Theriot and Stoermer, 1984; Bixby *et al.*, 2005; Jude *et al.*, 2006). However, the combined effects of overfishing, degradation of tributaries, urbanization, nutrient pollution, and finally the introduction of non-indigenous species at every trophic level irrevocably changed the structure and functioning of biological communities. Through fishing regulations, nutrient reductions, and tributary restoration, strides have been made to improve water quality and fish communities; however, the introduction and establishment of non-native species remains the most serious threat to ecological integrity of the Great Lakes (Jude *et al.*, 2006).

Over 180 non-indigenous species have been purposefully or accidently introduced to the Great Lakes (Mills *et al.*, 1993; Ricciardi, 2006). Early invaders, including the parasitic sea lamprey (*Petromyzon marinus* L.) and the planktivorous alewife (*Alosa pseudoharengus* Wilson), entered the lakes when shipping canals bypassed natural barriers (Jude and Leach, 1999). Others, such as the salmonids, were intentionally introduced to fill ecological voids caused by earlier invaders and revive a sport fishery (Jude and Leach, 1999). More recently, species introduced through ballast water from the Ponto-Caspian and Baltic regions have caused upheaval in the Great Lakes ecosystem (Stoermer *et al.*, 1996; Vanderploeg *et al.*, 2002; Getchell and Bowser, 2006). A number of high-profile invasive species introductions underscore that introduction rate is not declining, even with ballast exchange regulations (Ricciardi, 2006). In fact, species have gained entrance to the Great Lakes at all trophic levels, from viruses and bacteria to higher plants and algae.

Mills *et al.* (1993) listed 24 algae, including 16 diatoms, that were considered non-native to the Great Lakes. Subsequent introductions have added to that list (e.g. Edlund *et al.*, 2000; Lougheed and Stevenson, 2004), which includes freshwater and brackish/marine representatives that are generally found in the plankton. Several of the introduced species had limited temporal establishment or were restricted to waters with elevated conductivity (e.g. *Thalassiosira visurgis*, *Thalassiosira weissflogii* (Grunow) G. A. Fryxell & Hasle, *Pleurosira laevis* (Ehrenb.) Compère, *Terpsinoe musica* Ehrenb.). Other arrivals include widespread freshwater species that responded favorably to increased nutrients (e.g. *Cyclotella pseudostelligera* Hust., *Stephanodiscus subtilis* (Goor) A. Cleve, *Stephanodiscus binderanus* (Kütz.) Krieg.; reviewed in Edlund *et al.*, 2000).

Although monitoring programs were established in the Great Lakes, detecting the introduction of diatoms has been hindered by discontinuous funding, a pelagic focus (many introductions are first recorded nearshore in bays and harbors; Hasle, 1978), limited seasonal sampling, and taxonomic naïvety. Paleolimnological analyses have provided a retrospective on species introductions, the tempo of geographic spread, and clarified the native diatom flora (Stoermer *et al.*, 1993). Most well-dated paleorecords show that non-native diatoms were introduced and well established long before they were reported by routine survey. For example, *Thalassiosira baltica* Grunow is recorded in sediments dated AD 1988 in Lake Ontario, but this taxon was not routinely recorded in the plankton until the late 1990s (Figure 32.1). *Actinocyclus normanii* f. *subsalsa* (Juhl.-Dannf.) Hust. is present in Lake Ontario sediments by AD 1932, spread to Lake Erie by AD 1939 (Figure 32.1), and is present in Lake Huron sediments by the 1940s (Wolin *et al.*, 1988). This diatom was first reported in 1951 as "*Coscinodiscus subtilis* var." or *C. subtilis* var. *radiatus* Hohn (1952) from Lake Ontario and inland New York waterways (Hohn, 1951), but was not reported from lakes Erie and Michigan until 1960 and 1964, respectively (Wujek, 1967; Hohn, 1969; Stoermer and Yang, 1969; as *C. radiatus* Ehrenb. or *C. subsalsa* Juhl.-Dannf.). Similarly, *S. binderanus* can be found in Lake Erie sediments dated AD 1945 and Lake Ontario sediments dated 1952 A.D. (Figure 32.1), but was not reported in surveys until 1955 (Brunel, 1956).

While most non-native diatoms have had little discernable effect on the Great Lakes ecosystem, others induce dramatic impacts. *S. binderanus* blooms caused taste and odor issues in drinking water and clogged water-treatment-plant filters in the lower lakes in the 1950s–1970s (Hohn, 1969). Additional

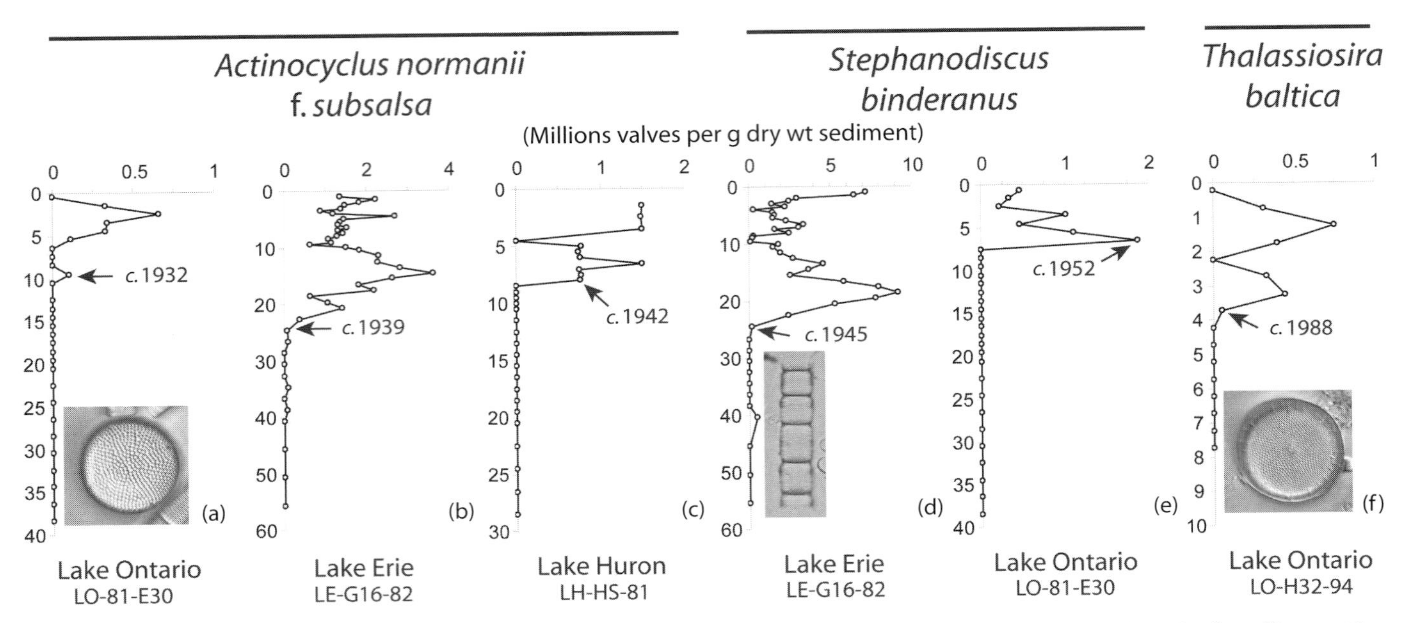

Figure 32.1 Distribution and abundance (millions of valves per g dry sediment) of select non-indigenous diatoms (inset light micrographs) and their dates of introduction (arrows) in Laurentian Great Lakes sediment cores. Depth scales (cm) vary among cores due to differences in sedimentation rates among lakes. Cores were collected in 1981 (Lakes Ontario and Huron), 1982 (Lake Erie), and 1994 (Lake Ontario; LO-H32–94). (a–c) *Actinocyclus normanii* f. *subsalsa* in Lakes Ontario, Erie, and Huron. (d, e) *Stephanodiscus binderanus* in Lakes Erie and Ontario cores. (f) *Thalassiosira baltica* in Lake Ontario sediment. Graphs modified and redrawn from Stoermer *et al.* (1985, 1987), Wolin *et al.* (1988), and Edlund *et al.* (2000).

filtration equipment was installed and monitoring programs were instituted to address problems caused by *S. binderanus* (Vaughn, 1961). Other diatoms had ecosystem-level impacts, specifically *A. normanii* f. *subsalsa*, a heavily silicified species that forms blooms in the summer. Frustule formation depleted dissolved silica from the water column, which subsequently allowed cyanobacteria to out-compete other species and bloom in bays and nearshore waters (Theriot and Stoermer, 1985). A recent invader in the Great Lakes, *T. baltica* (Edlund *et al.*, 2000) may result in a similar disruption of species succession or a decrease in biodiversity. In its native habitat, the Baltic Sea, *T. baltica* blooms control silica dynamics by removing dissolved silica from the water column and then making it inaccessible within the sediments (Olli *et al.*, 2008). Some evidence suggests that introduced species have led to extirpation of native diatoms, because introductions are coincident with such extirpations. In a sediment core from Lake Ontario, introductions of *S. binderanus* and *A. normanii* f. *subsalsa* were concomitant with local extinctions of *Stephanodiscus transilvanicus* Pant., *Cyclotella comta* (Ehrenb.) Kütz., *Cyclotella michiganiana* Skvortsov, *Cyclotella ocellata* Pant., and *Cyclotella stelligera* (Cleve & Grunow) Van Heurck (Stoermer *et al.*, 1985).

32.2.2 Volga River

Long-term floristic studies on the Volga River in Russia (Kiselev, 1948; Slynko *et al.*, 2002; Korneva, 2007) further document diatom introductions associated with anthropogenic and environmental impacts. The Volga River extends for 3530 km from its headwaters north of Moscow to its outlet in the Caspian Sea. Like the Laurentian Great Lakes, several invasive species (including bivalves, crayfish, mysids, and amphipods) were introduced by canal and ship traffic during the eighteenth and nineteenth centuries. The Volga River acted as an invasion corridor, which is a system of waterways connecting previously geographically isolated river and sea basins. As a result, aquatic species expanded beyond their historical ranges by active or passive dispersal (Slynko *et al.*, 2002). Furthermore, the Volga River has been heavily impacted by the construction of dams, which converted a free-flowing river system to a series of reservoirs. These hydrologic changes facilitated the introduction and dispersal of diatoms that favor the stable water column of impoundments. For example, in the 1950s, planktonic species (*Stephanodiscus binderanus*, *Stephanodiscus hantzschii* Grunow, and *Stephanodiscus minutulus* (Kütz.) Cleve and Möller) increased in abundance from the northern extent of the Volga River and spread in a southward direction (Korneva, 2007) until these introduced species came to dominate the spring and summer algal community of the river. In fact, the abundance of *S. binderanus* eventually became several orders of magnitude greater in reservoirs than in the unregulated sections of the river. Increase in flow of the Volga River has been attributed to climate change and resulted in rising levels of the Caspian Sea (Slynko *et al.*, 2002). With decreases in salinity of the northern basin of the Caspian Sea, brackish species spread northward (Korneva, 2007). *Skeletonema subsalsum* (Cleve-Euler) Bethge was a common species of the Caspian and Azov seas and advanced northward to the reservoirs on the river in the early 1950s. Likewise, *Thalassiosira incerta* Makarova, *Thalassiosira pseudonana* Hasle and Heimdal emend. Hasle, *Thalassiosira guillardii* Hasle, and *T. weissflogii* (Grunow) Fryxell and Hasle spread to the north. The latest advance from the Caspian Sea, in the 1980s, was of *A. normanii* (Gregory ex Greville) Hust. Overall, the total number of diatom species has increased since 1948 (Kiselev, 1948) with the new species tolerant of high electrolyte waters, high organic nutrients, and warmer temperatures. The construction of dams, loss of differences in regional water chemistry, alteration of flow, warming by thermal pollution from power plants, and eutrophication has resulted in a greatly changed river system and a more homogeneous physical environment and diatom community in the Volga River.

32.2.3 French rivers

Long-term studies of diatoms in French rivers have provided documentation of species introductions, and records of species that proliferated rapidly in France, yet were absent from other European rivers (Coste and Ector, 2000). For example, *Gomphoneis minuta* Kociolek and Stoermer, was originally described from North America, but reported in 1991 for the first time in Europe in the Ardèchec River. As in many cases, no clear vector of transport was identified. This diatom subsequently spread throughout southern France to the Loire River. *Eolimna comperei* Ector *et al.*, *Gomphoneis eriense* (Grunow) Skv. & Meyer and *Encyonema triangulum* (Ehrenb.) Kütz. are also considered to be a species introduced from outside Europe. Like the blooms in the Great Lakes and Volga River, many of the non-native diatoms in French rivers reach their greatest abundance during the summer. These include the tropical species *Hydrosera triquetra* Wall., which blooms in estuaries, and *Diadesmis confervacea* Kütz, which thrives in the warm waters of power-plant discharges.

32.2.4 Introductions to New Zealand

32.2.4.1 Asterionella formosa Analysis of sediment cores of 14 lakes in New Zealand provides evidence that *Asterionella formosa* Hass. was introduced with European settlement around 1880 (Harper, 1994). The species is frequently considered a cosmopolitan planktonic species, yet the New Zealand lake sediments showed no trace of *A. formosa* prior to European settlement. This species is now common in New Zealand, found in 45% of lakes with plankton records. The transportation of salmon eggs into New Zealand is considered the most likely vector for the introduction of *A. formosa* (Harper, 1994). Like many non-natives, *A. formosa* is a species that thrives with eutrophication. For example, in high-elevation lakes in Colorado, *A. formosa* was present in pre-settlement sediments in trace amounts and increased in abundance with elevated nitrate (NO_3^-) concentrations (Wolfe *et al.*, 2003). In these lakes, *A. formosa* was environmentally stimulated and maximum abundances occurred concomitantly with elevated nutrients, rather than via species introduction.

32.2.4.2 *Didymosphenia geminata* In many ways, the example of *Didymosphenia geminata* (Lyng.) M. Schmidt provides a different pattern for non-native species than other diatoms. Since the late 1980s, the large, stalked species seems to have undergone a transformation from a relatively rare species found in boreal and mountain regions of the northern hemisphere to a common bloom-forming organism with invasive tendencies. Although there are historical reports of *D. geminata* periodically forming thick benthic mats in its native range (e.g. Skvortzow, 1935; Heuff and Horkan, 1984; Lindstrøm and Skulberg, 2008), persistent blooms in some regions appear to be a new phenomenon.

Individual cells form small nodular colonies attached by mucopolysaccharide stalks to suitable substrates (Figure 32.2). As cell division proceeds, the colonies expand into continuous mats, which may eventually cover kilometers of streams and rivers and reach a thickness of more than 10 cm (Figure 32.2). The mats are composed primarily of stalks and resemble dirty cotton wool or housing insulation. The appearance of wet *D. geminata* mats has earned the species the moniker, "didymo." A remarkable characteristic of *D. geminata* is its ability to form blooms in low-nutrient waters (Kirkwood *et al.*, 2007). Typically, attached algal biomass in such streams is low, but stream food-web structure can be profoundly changed with blooms of *D. geminata* (Kilroy *et al.*, 2009). The species has potential economic impacts such as blocking water intakes and interfering with angling and other recreational activities (e.g. Mundie and Crabtree, 1997; Spaulding and Elwell, 2007).

Didymosphenia geminata is known to have been widespread in the northern hemisphere for at least one hundred years (Blanco and Ector, 2009, Pite *et al.*, 2009). Unusual blooms were reported in the late 1980s in rivers on Vancouver Island, Canada (Sherbot and Bothwell, 1993). Similarly, in the 1990s,

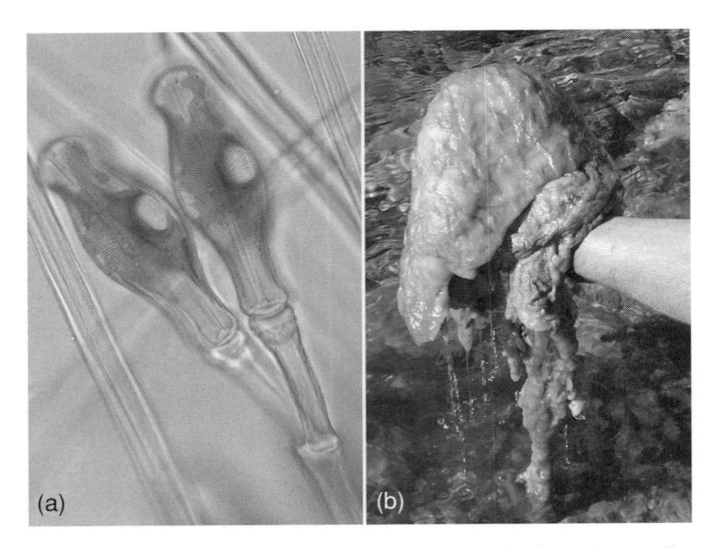

Figure 32.2 (a) A light micrograph of live *Didymosphenia geminata* cells, showing stalk attachment and branching; (b) typical appearance of a well-developed colony, with extensive cover by *D. geminata* of the river bed in the background.

D. geminata blooms were reported across rivers in Iceland (Jonssen *et al.*, 2000), central Europe (Kawecka and Sanecki, 2003), and the United States (Kumar *et al.*, 2009). While of concern to local managers, the blooms were not regarded as invasions, because *D. geminata* was assumed native to these regions. However, the sudden appearance of large blooms of *D. geminata* in rivers of New Zealand in 2004 was a watershed event: it confirmed that this species is an invasive organism and sparked international attention.

The discovery of *D. geminata* in the lower Waiau River, Southland, New Zealand, represented the first finding of the species in a location where it was almost certainly not present previously. A single record of *D. geminata* in North Island, New Zealand, in the 1920s (Cassie, 1984) was discounted as unverifiable and most likely a misidentification (Kilroy, 2004). In addition, failure to record such a distinctive species in over 10 years of sampling in the river where it was first discovered, or from any other New Zealand river, indicated that *D. geminata* was a new arrival. The fact that *D. geminata* was present in bloom proportions raised immediate concerns about its potential effect on the trout-fishing industry and tourism. In late 2004, *D. geminata* was declared an "unwanted organism" in New Zealand and a coordinated response was initiated (Vieglais, 2008). As an aquatic microorganism with unknown impacts, *D. geminata* presented a difficult case. As discussed earlier in this chapter, introductions of freshwater diatoms are rarely recognized, although the diatom flora of a remote region such as New Zealand seems likely to contain other recent arrivals. This species warranted special attention, however, because of its visible effects on streams and rivers. The response therefore included both basic and management-related research, e.g. potential distribution (Kilroy et al., 2008); assessments of social and economic consequences (e.g. Campbell, 2008); ecological effects; methodology to optimize detection (e.g. Cary *et al.*, 2008); survivability of cells, which could indicate potential transport mechanisms (Kilroy, 2008); and possible control methods.

Despite containment efforts, within a year *D. geminata* formed blooms in three other catchments, up to 400 km from the Waiau River. Subsequent surveys tracked the spread over three years, with a gradual dispersal from infected centers and occasional "jumps" to new areas (Figure 32.3). The surveys probably underestimated the presence of *D. geminata*, but were accurate with respect to the number of catchments affected. At the time of writing, some catchments are still free of *D. geminata*, including the North Island, as confirmed using sensitive genomic detection techniques (Cary *et al.*, 2008).

While there is no direct evidence for the mode of spread, the occurrence of initial blooms at sites popular with recreationalists (e.g. anglers, kayakers), frequently visited for monitoring purposes, or downstream of river crossings, strongly suggests inadvertent spread via human activities. Laboratory tests indicated that cells remained viable for weeks if maintained in wet, cool conditions (Kilroy, 2008). Thus, it is feasible that *D. geminata* could spread from catchment to catchment, and indeed from country to country on, for example, damp sports equipment. An important outcome of the *D. geminata* invasion was development by the New Zealand Government of a public education campaign, supported by recommended methods to properly clean gear.

An initial assessment of the ecological susceptibility of New Zealand's rivers to *D. geminata* blooms found the South Island to be at greater risk than the North Island (Kilroy *et al.*, 2008). However, the assessment was thought to greatly underestimate the risk to the North Island because it was based on sparse knowledge of *D. geminata*'s ecological tolerances. In particular, the assessment assumed that *D. geminata* is a cool-water species, an assumption subsequently challenged by its appearance in warm areas of the southern United States (Kumar *et al.*, 2009). Nevertheless, there are no examples to date of blooms of *D. geminata* in locations where the mean air temperature in the coldest month exceeds 5 °C (Kilroy, 2008). Other conditions considered suitable for *D. geminata* blooms included stable substrate, stable flows (in particular lake-fed or dammed rivers), and clear waters (i.e., non-humic and non-glacial) (Kilroy *et al.*,

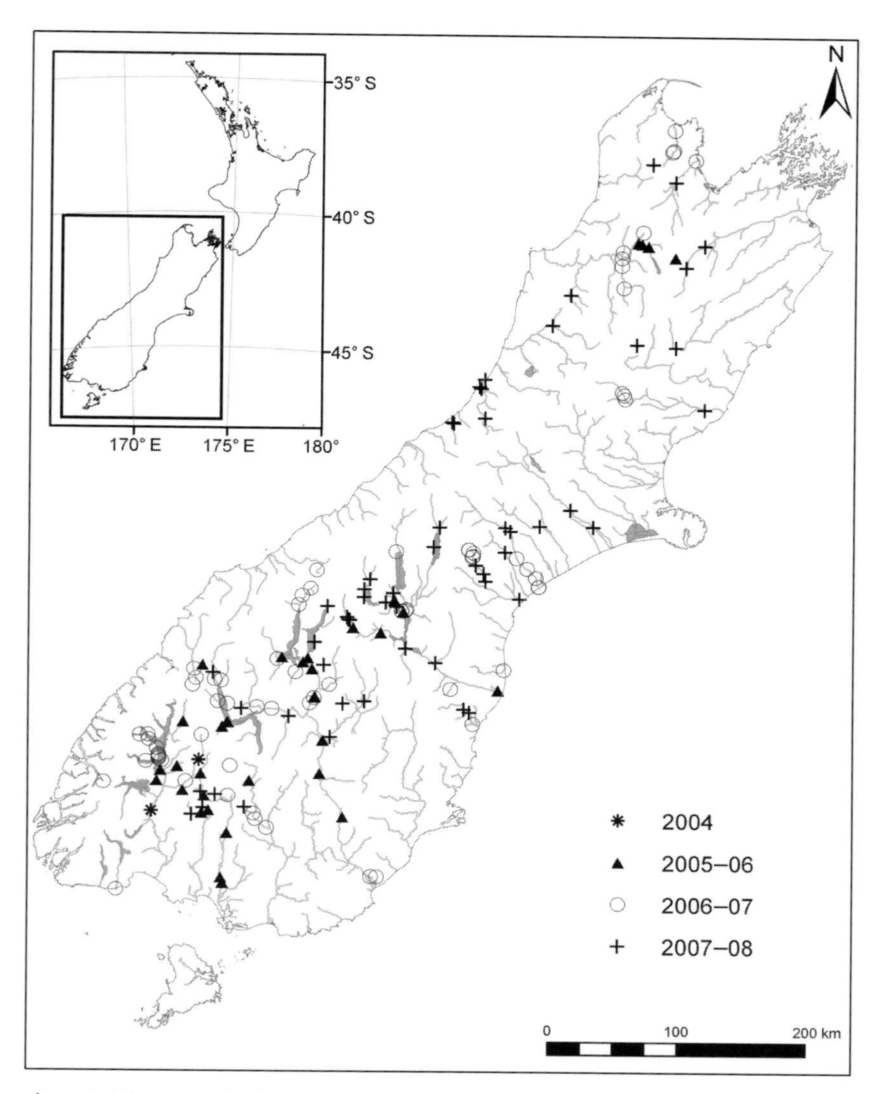

Figure 32.3 Map of *Didymosphenia geminata* distribution in South Island, New Zealand, showing the spread from 2004 to 2008. Stars show the location of the Waiau–Mararoa catchment, which is assumed to be the point of entry of the invasion in October 2004. In September 2005, *D. geminata* was detected in distant catchments and has continued to spread. Triangles: detection September 2005–August 2006; circles, September 2006–August 2007; crosses, September 2007–November 2008.

2008). Flow stability has also been shown to be important in other studies (Dufford *et al.*, 1987; Kirkwood *et al.*, 2007; Kumar *et al.*, 2009). Subsequent modeling in New Zealand has confirmed that *D. geminata* biomass is strongly related to recent hydrologic history (Kilroy, 2008).

Although the precise factors are unresolved, the distribution of *D. geminata* at the scale of river reaches in New Zealand indicates that water chemistry is an important factor determining the species' ability to become established and form blooms. Blooms have been linked to high nitrogen-to-phosphorus ratios (N:P) and high organic-to-inorganic P ratios (Ellwood and Whitton, 2007), consistent with observations of blooms in conditions of very low inorganic P (Kilroy, personal observation).

Understanding the effects of *D. geminata* on river ecosystems in New Zealand remains challenging because the organism has been established for such a short time. However, shifts in community composition, and spatial homogenization of benthic macroinvertebrate communities have been associated with *D. geminata* blooms (Kilroy *et al.*, 2009). Similar effects have been recorded in the benthic algal communities of one river (Figure 32.4). In this case, the algal community composition was determined by listing and identifying taxa in homogenized samples. Each sample was scanned for a standardized time at magnifications up to 400×. Diatom species identifications were verified with reference to permanent voucher slides of

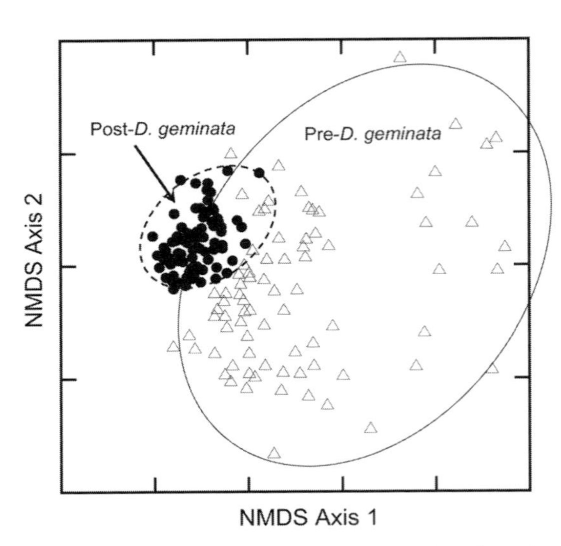

Figure 32.4 Non-metric multidimensional scaling plot (PRIMER v. 6) showing a two-dimensional (2-D) representation of Bray–Curtis similarities among algae samples from the lower Waiau River, Southland, New Zealand, collected annually in 2001, 2002, and 2004, before invasion of the river system by *Didymosphenia geminata* (open triangles) and in 2005, 2006, and 2007, after the invasion (closed circles) (2D stress = 0.23; 3D stress on the same data = 0.16). Each symbol represents the community in a single sample. Kilroy, unpublished data.

diatoms from the same sites, viewed at 1000×. The relative abundance of each taxon was assessed on an eight-point scale, using the method described in Biggs and Smith (2002). *D. geminata* was the dominant alga in all post-invasion samples and was omitted from the analysis to show its effects on other algae. In Figure 32.4, homogenization of communities is indicated by the much smaller ellipse enclosing post-invasion samples. Mean Bray–Curtis similarity among all pre-invasion samples was 27.4, compared to 49.3 for post-invasion samples (t-test, $P < 0.0001$). A general change in community composition is indicated by the separation of samples from the pre- and post-invasion periods (Figure 32.4). The community differences resulted from greatly increased densities of small diatoms (particularly *Achnanthidium minutissimum* (Kütz.) Czarn., which tends to be associated epiphytically with *D. geminata* blooms, and *Encyonema minutum* (Hilse in Rabenhorst) D. G. Mann) and to declines in species that were common prior to the invasion (e.g. *Cymbella kappii* Cholnoky).

Since the arrival of *D. geminata* in New Zealand, there have been reports of unprecedented blooms in the northern hemisphere, including eastern Canada (Simard and Simoneau, 2008), eastern United States (Kumar *et al.* 2009), Italy (Beltrami *et al.* 2008), and Spain (González *et al.*, 2008). It has been suggested that a genetic variant of *D. geminata* may be responsible (Kirkwood *et al.*, 2007), in which case the northern hemisphere may contain a true invader, which is non-indigenous (e.g. Falk-Petersen *et al.*, 2006). However, it remains to be determined if genetic alteration caused the species to abruptly become more competitive in environments where it was not previously able to become established itself or out-compete indigenous species. Bothwell *et al.* (2009) proposed that the worldwide spread of *D. geminata* is linked to the globalization of the recreational river fishing industry. This implies that the success of the species may also be due to the environmental suitability for *D. geminata* of rivers that also support fishable populations of salmonids, but does not explain the novel appearance of blooms in streams known to have supported low populations of *D. geminata* previously.

32.3 Summary

In general, the success of introduced diatoms appears to conform to patterns observed for other aquatic and terrestrial species. Attributes of successful invasive species include wide distribution, broad environmental tolerances, short generation times, rapid growth, a capability for rapid dispersal, and ability to thrive under conditions resulting from anthropogenic activity (Ricciardi and Rasmussen, 1998). The ability to form resting stages (Sicko-Goad *et al.*, 1989) or for vegetative cells to remain viable (Kilroy, 2008) outside of aquatic habitats and to survive transport are also relevant factors. Euryhaline diatoms rank high among successful Great Lakes, Volga River, and some of the European river colonizers; such diatoms are physiologically adapted to survive ballast exchange and tolerate increases in dissolved salts (Hasle and Evensen, 1975; Hasle, 1978; Korneva, 2007). *Didymosphenia geminata*, however, presents the unusual case of a species that blooms in low-nutrient waters. In New Zealand, homogenization of diatom communities in the presence of *D. geminata* has resulted in a reduction of overall aquatic biodiversity and replacement of native species by non-natives (Rahel, 2002). Another phenomenon common to many species invasions is a decline in the abundance and vigor of the invader over time (Simberloff and Gibbons, 2004). This has been observed for *D. geminata* on Vancouver Island, particularly in rivers not controlled by dams (Bothwell *et al.*, 2009) and to some extent in Iceland (Jónssen *et al.*, 2008).

Anthropogenic change not only alters environments in such a way as to favor invasive species, humans may directly deliver non-native species to the newly created habitat. Aquatic invasive species are favored in reservoirs, sites that are not only altered in physical and chemical parameters, but are more accessible to humans and potential introductions than natural lakes (Johnson *et al.*, 2008). Reservoirs of the Volga River are habitats, in particular, where invasive diatoms thrive. Downstream of reservoirs, the generation of hydroelectric power and the resulting high fluctuations in discharge was found to favor immigration of both large and weakly attached species (Peterson, 1986). The outlets of reservoirs also favor *D. geminata* (Dufford *et al.*, 1987; Kirkwood *et al.*, 2007). Dams have resulted in homogenization of flow regimes for entire regions across the United States (Poff *et al.*, 2007). There is reason to expect that the river regulation that resulted in homogenization of freshwater fish faunas and establishment of non-native and otherwise poorly adapted species also applies to diatoms.

The few studies documented here are likely not representative of the degree to which diatoms have been introduced into new habitats. Even the idea that diatoms have distinct regional distributions continues to be challenged (Kociolek and Spaulding, 2000; Finlay *et al.*, 2002; Vyverman *et al.*, 2007). Paleoecology provides a valuable historical perspective to help interpret diatom distributions and to put anthropogenic change into context.

References

Baas-Becking, L. G. M. (1934). *Geobiologie of Inleiding tot de Milieukunde*. The Hague: Van Stockum and Zoon.

Beltrami, M. E., Cappelletti, C., Ciutti, F., Hoffmann, L., & Ector, L. (2008). The diatom *Didymosphenia geminata*: distribution and mass occurrence in the Province of Trento, (Northern Italy). *Verhandlungen der Internationalen Vereinigung für theoretische und angewandte Limnologie*, **30**, 593–7.

Biggs, B. J. F. & Smith, R. A. (2002). Taxonomic richness of stream benthic algae: effects of flood disturbance and nutrients. *Limnology and Oceanography*, **47**, 1175–86.

Bixby, R. J., Edlund, M. B., & Stoermer, E. F. (2005). *Hannaea superiorensis* sp. nov., an endemic diatom from the Laurentian Great Lakes. *Diatom Research*, **20**, 227–40.

Blanco, S. & Ector, L. (2009). World distribution, ecology and nuisance effects of the freshwater invasive diatom *Didymosphenia geminata* (Lyngbye) M. Schmidt: a literature review. *Nova Hedwigia*, **88**, 347–422.

Bothwell, M. L., Lynch, D. R., Wright, H., & Deniseger, J. (2009). On the boots of fishermen. The history of didymo blooms on Vancouver Island, British Columbia. *Fisheries*, **34**, 382–8

Brunel, J. (1956). Addition du *Stephanodiscus binderanus* à la flore diatomique de l'Amérique du Nord. *Le Naturaliste Canadien (Québec)*, **83**, 91–5.

Campbell, M. L. (2008). Organism impact assessment: risk analysis for post-incursion management. *ICES Journal of Marine Science*, **65**, 795–804.

Cary, S. C., Hicks, B. J., Gemmill, C. E. G., Rueckert, A., & Coyne, K. J. (2008). A sensitive genetic-based detection and enumeration method for *Didymosphenia geminata*. *Canadian Technical Report of Fisheries and Aquatic Sciences*, **2795**, 6–9.

Cassie, V. (1984). Checklist of freshwater diatoms of New Zealand. *Bibliotheca Diatomologica*, **4**, 1–129.

Coste, M. & Ector, L. (2000). Diatomées invasives exotiques ou rare en France: principales observations effectuées au cours des dernières décennies. *Systematics and Geography of Plants*, **70**, 373–400.

Denys, L. (2000). Historical distribution of "Red List diatoms" (Bacillariophyceae) in Flanders (Belgium). *Systematics and Geography of Plants*, **70**, 409–20.

Dufford, R. G., Zimmerman, H. J., Cline, L. D., & Ward, J. V. (1987). Responses of epilithic algae to regulation of Rocky Mountain streams. In *Regulated Streams: Advances in Ecology*, ed. J. F. Craig & J. B. Kemper, New York, NY: Plenum Press, pp. 383–90.

Edlund, M. B., Taylor, C. M., Schelske, C. L., & Stoermer, E. F. (2000). *Thalassiosira baltica* (Grunow) Ostenfeld (Bacillariophyta), a new exotic species in the Great Lakes. *Canadian Journal of Fisheries and Aquatic Sciences*, **57**, 610–15.

Ellwood, N. T. W. & Whitton, B. A. (2007). Importance of organic phosphate hydrolyzed in stalks of the lotic diatom *Didymosphenia geminata* and the possible impact of atmospheric and climatic changes. *Hydrobiologia*, **592**, 121–32.

Falk-Petersen, J. Bøhn, T., & Sandlund, O. T. (2006). On the numerous concepts in invasion biology. *Biological Invasions*, **8**, 1409–24.

Finlay, B. J., Monaghan, E. B., & Maberly, S. C. (2002). Hypothesis: the rate and scale of dispersal of freshwater diatom species is a function of their global abundance. *Protist*, **153**, 261–73.

Getchell, R. G. & Bowser, P. R. (2006). Ecology of type E botulism within dreissenid mussel beds. *Aquatic Invaders*, **17**(2), 1–8.

González, E. M., Hernández, R., & Martínez, J. (2008). El 'alga chapapote': una nueva amenaza para nuestros ríos. *Guardabosques*, **42**, 29–32.

Harper, M. A. (1994). Did Europeans introduce *Asterionella formosa* Hassall to New Zealand? *Memoirs of the Californian Academy of Sciences*, **17**, 479–86.

Hasle, G. R. (1978). Some freshwater and brackish water species of the diatom genus *Thalassiosira* Cleve. *Phycologia*, **17**, 263–292.

Hasle, G. R. & Evensen, D. L. (1975). Brackish-water and fresh-water species of the diatom genus *Skeletonema*. Grev. I. *Skeletonema subsalsum* (A. Cleve) Bethge. *Phycologia*, **14**, 283–97.

Herdendorf, C. E. (1982). Large lakes of the world. *Journal of Great Lakes Research*, **8**, 378–412.

Heuff, H. & Horkan, K. (1984). Caragh. In *Ecology of European Rivers*, ed. B. A. Whitton, Blackwell Scientific Publications, pp. 364–84.

Hohn, M. H. (1951). A study of the distribution of diatoms (Bacillarieae) in Western New York State. *Cornell University Agricultural Experiment Station, Memoir* **308**.

Hohn, M. H. (1952) Contributions to the diatoms of western New York State. *Transactions of the American Microscopical Society*, **71**, 270–1.

Hohn, M. H. (1969). Qualitative and quantitative analyses of plankton diatoms, Bass Island area, Lake Erie, 1938–1965. *Bulletin of the Ohio Biological Survey*, **3**, 1–211.

Johnson, P. T. J., Olden, J. D., & Vander Zanden, M. J. (2008). Dam invaders: impoundments facilitate biological invasions into freshwaters. *Frontiers in Ecology and Environment*, **6**, 357–63.

Jonssen, G. S., Jonssen, I. R., Bjornsson, M., & Einarsson, S. M. (2000). Using regionalisation in mapping the distribution of the diatom species *Didymosphenia geminata* (Lyngbye) M. Schmidt in Icelandic rivers. *Verhandlunge, Internationale Vereinigung für theoretische und angewandte Limnologie*, **27**, 340–3.

Jónssen, I. R., Jónssen, G. S., Olafsson, J. S., Einarsson, S. M., & Antonsson, T. (2008). Occurrence and colonization pattern of *Didymosphenia geminata* in Icelandic streams. *Technical Report of Fisheries and Aquatic Sciences*, **2795**, 41–4.

Jude, D. J. & Leach, J. (1999). Fish management in the Great Lakes (revised). In *Fisheries Management in North America*, ed. C. Kohler & W. Hubert, Bethesda, MD: American Fisheries Society, pp. 623–64.

Jude, D. J., Janssen, J., & Stoermer, E. F. (2006). The uncoupling of trophic food webs by invasive species in Lake Michigan. In *The State of Lake Michigan*, ed. M. Munawar & T. Edsall, Ecovision World Monograph Series, Amsterdam: S. P. B. Academic Publishing, pp. 311–48.

Kawecka, B. & Sanecki, J. (2003). *Didymosphenia geminata* in running waters of southern Poland – symptoms of change in water quality? *Hydrobiologia*, **495**, 193–201.

Kilroy, C. (2004). A new alien diatom *Didymosphenia geminata* (Lyngbye) Schmidt: its biology, distribution, effects and potential risks for New Zealand fresh waters. NIWA Client Report CHC2004–128.

Kilroy, C. (2008). *Didymosphenia geminata* in New Zealand: distribution, dispersion and ecology of a non-indigenous invasive species. *Canadian Technical Report of Fisheries and Aquatic Sciences*, **2795**, 15–20.

Kilroy, C., Larned, S., & Biggs, B. J. F. (2009). The non-indigenous diatom *Didymosphenia geminata* alters benthic communities in New Zealand rivers. *Freshwater Biology*, **54**, 1990–2002.

Kilroy, C., Snelder, T. H., Floerl, O. Vieglais, C. C., & Dey, K. L. (2008). A rapid technique for assessing the suitability of areas for invasive species applied to New Zealand's rivers. *Diversity and Distributions*, **14**, 262–72.

Kirkwood, A. E., Shea, T., Jackson, L. J., & McCauley, E. (2007). *Didymosphenia geminata* in two Alberta headwater rivers: an emerging invasive species that challenges conventional views on algal bloom development. *Canadian Journal of Fisheries and Aquatic Science*, **64**, 1703–9.

Kiselev, I. A. (1948). To question about quantitative and qualitative composition of the Volga River reservoir phytoplankton. *State Research Institute of Zoology*, **8**, 567–84.

Kociolek, J. P. & Spaulding, S. A. (2000). Freshwater diatom biogeography. *Nova Hedwigia*, **71**, 223–41.

Korneva, L. G. (2007). Recent invasion of planktonic diatom algae in the Volga River Basin and their causes. *Biology of Inland Waters*, **1**, 28–36.

Kumar, S., Spaulding, S. A., Stohlgren, T., *et al.* (2009). Modelling the bioclimatic profile of the diatom *Didymosphenia geminata*. *Frontiers in Ecology and the Environment*, **7**, 415–20.

Lange-Bertalot, H. & Steindorf, A. (1996). Rote Liste der limnischen Kieselalgen (Bacillariophyceae) Deutschlands. *Schriftenreihe Vegetationsk*, **28**, 633–77.

Larson, B. M. H., Nerlich, B., & Wallis, P. (2005). Metaphors and biorisks: the war on infectious diseases and invasive species. *Science Communication*, **26**, 243–68.

Lindstrøm, E. A. & Skulberg, O. M. 2008. *Didymosphenia geminata* – a native diatom species of Norwegian rivers coexisting with the Atlantic Salmon. *Canadian Technical Report of Fisheries and Aquatic Sciences*, **2795**, pp. 35–40.

Lougheed, V. L. & Stevenson, R. J. (2004). Exotic marine algae reaches bloom proportions in a coastal lake of Lake Michigan. *Journal of Great Lakes Research*, **30**, 538–44.

Martiny, J. B. H., Bohannan, B. J. M., Brown, J. H., *et al.* (2006). Microbial biogeography: putting microorganisms on the map. *Nature Microbial Reviews*, **4**, 102–12.

Mills, E. L., Leach, J. H., Carlton, J. T., & Secor, C. L. (1993). Exotic species in the Great Lakes and anthropogenic introductions. *Journal of Great Lakes Research*, **19**, 1–54.

Mundie, J. H. & Crabtree, D. G. (1997). Effects on sediments and biota of cleaning a salmonid spawning channel. *Fisheries Management and Ecology*, **4**, 111–26.

Olli, K., Clarke, A., Danielsson, Å., *et al.* (2008). Diatom stratigraphy and long-term dissolved silica concentrations in the Baltic Sea. *Journal of Marine Systems*, **73**, 284–99.

Peterson, C. G. (1986). Effects of discharge reduction on diatom colonization below a large hydroelectric dam. *Journal of the North American Benthological Society*, **5**, 278–89.

Pimental, D., Lach, L., Zuniga, R., & Morrison, D. (2000). Environmental and economic costs of nonindigenous species in the United States. *BioScience*, **50**, 53–65.

Pite, D. P., Lane, K. A., Hermann, A. K., Spaulding, S. A., & Finney, B. P. (2009). Historical abundance and morphology of *Didymosphenia* species in Naknek Lake, Alaska. *Acta Botanica Croatica*, **68**, 183–97.

Poff, N. L., Olden, J. D., Merritt, D. M., & Pepin, D. M. (2007). Homogenization of regional river dynamics by dams and global diversity implications. *Proceedings of the National Academy of Sciences of the USA*, **104**, 5732–7.

Rahel, F. (2002). Homogenization of freshwater faunas. *Annual Review of Ecology and Systematics*, **33**, 291–315.

Ricciardi, A. (2006). Patterns of invasion in the Laurentian Great Lakes in relation to changes in vector activity. *Diversity and Distributions*, **12**, 425–33.

Ricciardi, A. & Rasmussen, J. B. (1998). Predicting the identity and impact of future biological invaders: a priority for aquatic resource management. *Canadian Journal of Fisheries and Aquatic Sciences*, **55**, 1759–65.

Sherbot, D. M. J. & Bothwell, M. L. (1993). *Didymosphenia geminata* (Gomphonemaceae). A review of the ecology of *D. geminata* and the physiochemical characteristics of endemic catchments on

Vancouver Island. Saskatoon, Saskatchewan: National Hydrology Research Institute, Environment Canada, NHRI Contribution No. 93005

Sicko-Goad, L., Stoermer, E. F., & Kociolek, J. P. (1989). Diatom resting cell rejuvenation and formation: Time course, species records and distribution. *Journal of Plankton Research*, **11**, 375–89.

Simard, I. & Simoneau, M. (2008). Didymo dans les rivières du Québec: état de situation. *Canadian Technical Report of Fisheries and Aquatic Sciences*, **2795**,15–20.

Simberloff, D. & Gibbons, L. (2004). Now you see them, now you don't! – population crashes of established introduced species. *Biological Invasions*, **6**, 161–72.

Skvortzow, B. W. (1935). Diatomées récoltées par le Père I. Licent au cours de ses voyages dans le Nord de la Chine au bas Tibet, en Mongolie et en Mandjourie. *Publications du Musée Hoangho Paiho de Tien Tsin. Tienstsin*, **36**, 1–43.

Slynko, Y. V., Korneva, L. G., Rivier, I. K., *et al.* (2002). The Caspian–Volga-Baltic invasion corridor. In *Invasive Aquatic Species of Europe: Distribution, Impacts and Management*, ed. E. Leppäkovski, S. Gollasch, & S. Olenin, Dordrecht: Kluwer Academic Publishers, pp. 399–411.

Smith, G. R. & Todd, T. N. (1984). Evolution of species flocks of fishes in north temperate lakes. In *Evolution of Fish Species Flocks*, ed. A. A. Echelle & I. Kornfield, Orono: University of Maine at Orono Press, pp. 45–68.

Soininen, J., Paavola, R., & Muotka, T. (2003). Benthic diatom communities in boreal streams: community structure in relation to environmental and spatial gradients. *Ecography*, **27**, 330–42.

Spaulding, S. A. & Elwell, L. (2007). Increase in nuisance blooms and geographic expansion of the freshwater diatom *Didymosphenia geminata*: recommendations for response. *USGS Open File Report* **2007–1425**.

Stoermer, E. F., Emmert, G., Julius, M. L., & Schelske, C. L. (1996). Paleolimnologic evidence of rapid recent change in Lake Erie's trophic status. *Canadian Journal of Fisheries and Aquatic Sciences*, **53**, 1451–8.

Stoermer, E. F., Kociolek, J. P., Schelske, C. L., & Conley, D. J. (1987). Quantitative analysis of siliceous microfossils in the sediments of Lake Erie's central basin. *Diatom Research*, **2**, 113–34.

Stoermer, E. F., Wolin, J. A., & Schelske, C. L. (1993). Paleolimnological comparison of the Laurentian Great Lake based on diatoms. *Limnology and Oceanography*, **38**, 1311–16.

Stoermer, E. F., Wolin, J. A., Schelske, C. L., & Conley, D. J. (1985). An assessment of ecological changes during the recent history of Lake Ontario based on siliceous algal microfossils preserved in sediments. *Journal of Phycology*, **21**, 257–76.

Stoermer, E. F. & Yang, J. J. (1969). *Plankton Diatom Assemblages in Lake Michigan*. Great Lakes Research Division Special Report No. 47, Ann Arbor, MI: University of Michigan.

Telford, R. J., Vandvik, V., & Birks, H. J. B. (2006). Dispersal limitations for microbial morphospecies. *Science*, **312**, 1015.

Theriot, E. C. & Stoermer, E. F. (1984). Principal component analysis of *Stephanodiscus*: observations on two new species from the *Stephanodiscus niagarae* complex. *Bacillaria*, **7**, 37–58.

Theriot, E. C. & Stoermer, E. F. (1985). Phytoplankton distribution in Saginaw Bay. *Journal of Great Lakes Research*, **11**, 132–42.

Vanderploeg, H. A., Nalepa, T. F., Jude, D. J., *et al.* (2002). Dispersal and emerging ecological impacts of Ponto-Caspian species in the Laurentian Great Lakes. *Canadian Journal of Fisheries and Aquatic Sciences*, **59**, 1209–28.

Vanormelingen, P., Verleyen, E., & Vyverman, W. (2008). The diversity and distribution of diatoms: from cosmopolitanism to narrow endemism. *Biodiversity and Conservation*, **17**, 393–405.

Vaughn, J. C. (1961). Coagulation difficulties of the South District Filtration Plant. *Pure Water*, **13**, 45–9.

Vieglais, C. M. C. (2008). Management approaches to didymo: the New Zealand experience. *Canadian Technical Report of Fisheries and Aquatic Sciences*, **2795**, 54–5.

Vyverman, W., Verleyen, E., Sabbe, K., *et al.* (2007). Historical processes constrain patterns in global diatom diversity. *Ecology*, **88**, 1924–31.

Willis, K. J. & Birks, H. J. B. (2006). What is natural? The need for a long-term perspective in biodiversity conservation. *Science*, **314**, 1261–5.

Wolfe, A. P., Van Gorp, A. C., & Baron, J. S. (2003). Recent ecological and biogeochemical changes in alpine lakes of Rocky Mountain National Park (Colorado, USA): a response to anthropogenic nitrogen deposition. *Geobiology*, **1**, 153–68.

Wolin, J. A., Stoermer, E. F., Schelske, C. L., & Conley, D. J. (1988). Siliceous microfossil succession in recent Lake Huron sediments. *Archiv für Hydrobiologie*, **114**, 175–98.

Wujek, D. E. (1967). Some plankton diatoms from the Detroit River and the western end of Lake Erie adjacent to the Detroit River. *The Ohio Journal of Science*, **61**(1), 32–5.

33 Diatomite

DAVID M. HARWOOD

33.1 Introduction

Diatomite, a soft, porous, fine-grained, lightweight, siliceous sedimentary rock, is produced by the accumulation and compaction of diatom (Class Bacillariophyceae) remains. The cell wall of living diatoms is impregnated with silica (amorphous hydrous, or opaline ($SiO_2.nH_2O$)), which preserves ornate and highly porous structures (Armbrust, 2009). Upon death of the organism, these siliceous elements become sedimentary particles in a diatomite deposit. The intricate structure of diatom frustules, and packing of the myriad diatom shapes into rock-forming sedimentary layers, gives diatomite deposits properties that are useful in many industrial and commercial applications. Most diatoms are 10 μm to 100 μm in size, although larger species reach more than 1 mm (Tappan, 1980) and smaller ones are <2 μm. This small size results in large concentrations of diatoms; a cubic inch of diatomite may contain 40 to 70 million diatoms (Crespin, 1946). Although the specific gravity (density) of the SiO_2 that comprises the diatom particles is nearly twice that of water, perforations and open structures in the frustule renders diatomite a considerably lower effective density (between 0.12 g cm^{-3} and 0.25 g cm^{-3}) with high porosity (from 75 to 85 percent). This sedimentary rock is able to absorb and hold up to 3.5 times its own weight in liquid (Cleveland, 1966). These properties bring industrial utility and commercial value to rock-forming accumulations of diatom remains.

Diatomite deposits of varying quality are known from freshwater and marine sedimentary environments (Moyle & Dolley, 2003). Other names for diatomite and diatomaceous earth include tripoli, kieselguhr, and infusorial earth. Diatom-rich sediments have been accumulating in marine deposits since the Late Cretaceous (~80 million years ago) (Harwood *et al.*, 2007), and in large lakes since at least the Eocene (~50 million years ago). The purity of a diatomite deposit decreases with greater amount of both clastic particles (silt and clay) and organic matter, which limit the utility of a diatomite in industrial applications. Some high-grade commercial diatomite contains up to 90% opaline SiO_2, with minor occurrence of calcium carbonate, volcanic glass (also SiO_2), and terrigenous particles (Cummins, 1960; Cressman, 1962). Diatom-bearing rocks with a higher terrigenous component (e.g. diatomaceous mudrocks or siliceous shale) are commonly interbedded with diatomite. The association of many oil fields with diatomite or diatom-bearing shale indicates that the high lipid (oil) content in diatoms is a likely source for petroleum. In natural systems of sedimentary diagenesis, primary diatom remains (comprised of amorphous opaline silica) will be transformed sequentially into other mineral phases of silica, first as disordered cristobalite/tridymite (opal-CT) and eventually into quartz (opal Q) or chert. This transformation results from progressive time, increased burial depth, burial temperature gradients, and interaction with formation fluids, which will reduce the desired physical attributes of a diatomite.

The Diatoms: Applications for the Environmental and Earth Sciences, 2nd Edition, eds. John P. Smol and Eugene F. Stoermer. Published by Cambridge University Press.

33.2 Origin of diatomite

Diatomite accumulates in areas where the rate of deposition of diatom frustules greatly exceeds that of other sediment components (Berger, 1970; Heath, 1974; Barron, 1987). In oceans, this requirement is met in areas where cold, deep, nutrient-rich water is brought to the surface by wind or currents resulting in high diatom production, and hence sedimentation of diatom frustules is also high. Areas of biogenic silica accumulation (mostly from diatoms and radiolarians) occur today in the equatorial Pacific Ocean, the circum-Antarctic, the North Pacific, and in areas where surface currents draw water away from the margin of continents (western Africa, Peru, western North America, etc.). Analysis of post-Cretaceous

Figure 33.1 Outcrop of a freshwater diatomite deposit from the Nightingale Quarry (western USA). This deposit contains an abundance of the genus *Aulacoseira*. (Photograph by E. Stoermer.)

accumulation rates of diatomaceous sediment indicates considerable variation in the timing and distribution of siliceous deposits in the world's oceans (Leinen, 1979; Barron, 1987; Baldauf and Barron, 1990), with periods of maximum opal accumulation during the Late Cretaceous, middle Eocene, and late Miocene. Heath (1974) estimated that up to 90% of biogenic silica deposited in the ocean may be laid down in estuaries and nearshore basins, but diatomite does not accumulate there, because diatoms are diluted by terrigenous sediment. Similarly, non-marine diatomite is commonly associated with lakes where clastic sediment input from the surrounding land is low. The survival of non-marine diatomite requires protection from erosion and from diagenesis resulting from contact with corrosive alkaline or silica-deficient pore waters (Conger, 1942; Lohman, 1960; Nelson, *et al.*, 1995). Potential for excellent preservation of diatom and other siliceous microfossils is present in siliceous-rich sediments, in settings of rapid burial, and in association with volcanic sedimentary components (Taliaferro, 1933).

Diatomite beds may be laminated or massive, and occur in tabular or lenticular beds that range from several centimeters to tens of meters in thickness (Figure 33.1). The total thickness of an interbedded diatomite-bearing sequence may be in excess of 300 m (Taylor, 1981), although this sequence may include chert and clay-rich intervals of little industrial value. Laminated diatomites may reflect a general absence of bioturbation in anoxic basins, seasonal variation in sediment input and biogenic deposition, and/or rapid deposition of diatom mats (Kemp and Baldauf, 1993). Sequences of thin laminated diatomites provide high resolution records for study of paleoenvironmental change (see Leventer *et al.*, this volume). Due to dissolution in the water column, it is estimated that only 1% to 10% of the diatom frustules produced at the surface waters survive settling into the sediment (Lisitzin, 1972; Calvert, 1974).

Preservation of opaline silica in diatom frustules depends on: (1) local conditions of alkalinity (pH <9 is favorable); (2) the presence of dissolved silica in pore water; (3) an association with volcanic ash (Taliaferro, 1933); and (4) inter-stratification with lithologies that may limit permeability of corrosive pore waters. With progressive time, increasing heat (above $\sim 50\,^{\circ}C$) and depth of burial (depending on thermal gradient), the amorphous, opaline silica ($SiO_2.nH_2O$, commonly called opal-A) of diatom frustules is progressively transformed to anhydrous silica or porcellanite (SiO_2 – a disordered form of crystobalite and tridymite, opal-CT), and finally to quartz (SiO_2) as chert. This transformation results in a significant decrease in porosity and increase in density and hardness (Bramlette, 1946; Issacs, 1981; Issacs and Rullkotter, 2001).

33.3 History and use of diatomite

The unique physical and chemical properties of diatomite were recognized long before diatoms were identified as the components of these rocks. The "floating bricks" of antiquity and vases made of diatomite are known from the Greeks (Ehrenberg, 1842). The Greeks were also the first to use diatomite as a mild abrasive. Reconstruction of the dome of Hagia Sofia Church in Constantinople used diatomite as a building material as early as AD 432. In 1836, C. Fischer, reported his microscopic observations of diatoms in kieselghur from a peat deposit, to Ehrenberg, who was impressed with this discovery and spread the news throughout Europe (Ehrenberg, 1836), prompting others to also search for and study diatom-rich siliceous sediments around the world. An important freshwater diatomite deposit of Quaternary age at Luneburger Heide district near Hannover (Germany) was discovered soon after Ehrenberg's report. It was the leading deposit in the world, until full-scale excavation of the marine deposit at Lompoc (California) began after World War I. Bailey (1839) reported the first deposit of diatomite in the USA and predicted the occurrence of other diatomites in many environments. Many new deposits were discovered in the middle and late nineteenth century, including that at Lompoc, California, in 1888 (Dolley & Moyle, 2003). The interest in diatomite, and the search for additional applications, led to the production of numerous descriptive studies of diatomite, diatom deposits, and industrial uses (Wahl, 1876; Card & Dun, 1897; Moss, 1898; Bigot, 1920; Eardley-Wilmont, 1928; Calvert, 1929; Hendey, 1930; Conger, 1942; Kawashimma & Shiraki, 1946).

Characteristics that make diatomite most suitable for industrial uses are low density, high porosity, low thermal conductivity, high melting point (1400–1750 °C depending on impurities), solubility only in strong alkaline solutions and hydrofluoric acid, and being chemically inert (Durham, 1973). This combination of properties has led to 300–500 commercial applications (Schroeder, 1970; Durham, 1973). Principal uses today include filtration, insulation, fine abrasion, absorption, building materials, mineral fillers, pesticides, catalysts, carriers, coatings, food additives, and as an anti-caking agent. Diverse industries make use of diatomite, including food, beverage, pharmaceutical, chemical, agricultural, paint, plastics, paper, construction, dry cleaning, recreation, and sewage treatment (Cummins, 1975).

The first commercial application, and perhaps the most notable early use of diatomite, was in the production of dynamite, as discovered in 1867 by Alfred Nobel. Nitroglycerin was discovered previously by A. Sobrero in 1847, but it was a dangerous, unstable material. Nobel's innovation was recognition that nitroglycerine could be absorbed within the solid, porous framework of diatomite, rendering it safe for transportation and handling. Dynamite made from diatomite is no longer used, except in special applications that require a softer "bang." Nobel's application of diatomite to tame the explosive power of nitroglycerine increased our ability to exploit the available natural resources that fueled the industrial revolution. Expansion of railroads and construction of canals for shipping, construction of tunnels, dams and highways, and the extraction of the coal and raw materials would not have occurred at the same pace without the convenient explosive power of diatomite-encased nitroglycerine. The lasting legacy of diatomite's role in human expansion during the late nineteenth century is the honoring of the human spirit and discovery in arts, science, literature, and peace with the Nobel prizes.

Filtration is the main current application for diatomite (~60% is used for this purpose), indeed 95% of commercial breweries use diatomite filtration (Moores, 2008). Industries that produce fruit juice, sugar, and edible-oil, as well as many public water systems for drinking and swimming, use diatomite to remove particulate materials and microbial contaminants. Diatom filtration can effectively remove particles as small as 0.1 μm size. Other commercial uses include the addition as a filler in paper, paint, cosmetics, and ceramic industries, as a thermal insulator in high temperature ovens, and as an absorbent to clean up toxic spills. It has value as a non-toxic mechanical insecticide, capable of perforating the insect exoskeleton; the insects eventually die

Figure 33.2 Mining operations of the Lompoc diatomite, an upper-Miocene marine diatomite. Excavation and transport vehicles are more than six meters high. (Photograph by E. Stoermer.)

from dehydration. Applications in the pharmaceutical industry include the filtration of human blood plasma and separation of DNA. In addition to preserving a geological history of diatom evolution (the basis for biostratigraphic correlations using diatoms, e.g. Scherer *et al.*, 2007; Cody *et al.*, 2008), and paleoenvironmental changes based on diatom distribution and time-series abundance changes (e.g. Baldauf and Barron, 1990; Gersonde, 1990) diatomites often provide an excellent host lithology for the preservation of macrofossil and vertebrate remains (e.g. diatomites in North America have preserved fossil fish, snakes, birds, bats, sirenians, and macrophytes).

33.4 Production of diatomite

World production of diatomite was estimated to average annually just below two million metric tonnes for the past five years. The United States is the largest producer of diatomite followed by China, Denmark, Japan, France, and Mexico (United States Geological Survey (USGS) Mineral Commodity Summaries, http://minerals.usgs.gov/minerals/pubs/mcs/2009/mcs2009.pdf). The upper-Miocene diatomite deposit at Lompoc, Santa Barbara County, California (Figure 33.2), is the largest producing deposit in the world, representing more than 60% of the US total production (Dolly, 2008). World resources of diatomite provide an adequate supply for the foreseeable future.

Most diatomite mining operations are surface excavations from open pits, with the selection of individual strata of different grades for specific treatment and application. The raw materials are dried, crushed, and sorted. Suspended fine particles can be passed through air separators to remove impurities and coarse materials. Diatomite is often calcined in kilns to limit the insoluble impurities, remove organic material, and oxidize iron. Increase in particle size, if desired, can be accomplished through a flux-calcined process, with the addition of a flux, commonly soda ash, before being calcined. Control of dust in the mining process is adequate due to the high moisture content of the raw materials, which may contain as much as 40 to 60% moisture, by weight. Silica dust production during processing is controlled by enclosing these areas. Production expenses break out in cost for mining (10%), processing (60 to 70%), and packing and shipping (20 to 30%). Price fluctuations in diatomite track the cost of energy consumed in these three processes (Crangle, 2008). The price of commercial diatomite varies widely depending on grade, from less than $US 9 per ton for use by the cement industry to $US 1000 per ton for specialty markets, including cosmetics, art supplies, and DNA extraction. The 2009 USGS Minerals Commodoties Index indicates an average unit value for filter-grade diatomite at $US 373 per ton.

33.5 Summary

Diatomite has a history of use by humans, spanning nearly two millennia. The ability of this porous, yet relatively strong, material to safely contain and protect nitroglycerin affected the rate of advancement of our civilization through the use of dynamite in the construction of our transportation systems and mining. The distinctive properties of high absorption, low density, and high porosity have led to the use of diatomite in diverse applications, with the greatest application today in filtration processes and as a light-weight filler. One benefit of diatom-bearing deposits is their potential as source rocks for the generation of petroleum, and in some areas as excellent reservoir rocks (e.g. the Belridge Diatomite Member of the Monterey Shale in the San Joaquin Basin is a prolific oil-producing stratigraphic unit).

Diatomite deposits with the greatest utility in industrial applications result from: (1) environmental conditions that were conducive to the growth of diatoms; (2) a depositional setting that reduced or prevented the input of volcanic, terrigenous, and biogenic carbonate; and (3) a subsequent history of shallow burial and limited diagenesis or erosion. These conditions are known from marine and non-marine settings around the globe. Most economic deposits formed during the Miocene to Quaternary.

Acknowledgments

Photographs were generously provided by Gene Stoermer, who, along with Scott Starratt and Bill Last, are thanked for their helpful comments. The paper also benefited from the wealth of information (1) collected in the collection of essays, notes, bibliography, and history of many aspects of diatomite and diatoms available in *Terra Diatomacea* from Johns-Manville (Cummins, 1975); and (2) the informative Mineral Commodity Summaries distributed by the USGS.

References

Armbrust, E. V. (2009). The life of diatoms in the world's oceans. *Nature*, **459**, 185–92.

Bailey, J. W. (1839). On fossil infusoria discovered in peat-earth at West Point, N.Y. *American Journal of Science, Series I*, **35**, 118–24.

Baldauf, J. G. & Barron, J. A. (1990). Evolution of biosiliceous sedimentation patterns – Eocene through Quaternary: paleoceanographic response to polar cooling. In *Geological History of the Polar Oceans: Arctic versus Antarctic*, ed. U. Bleil & J. Thiede, Dordrecht: Kluwer Academic Publishers, pp. 575–607.

Barron, J. A. (1987). Diatomite: environmental and geological factors affecting its distribution. In *Siliceous Sedimentary Rock-Hosted Ores and Petroleum*, ed. J. R. Hein, New York: Van Nostrand Reinhold Co., pp. 164–78.

Berger, W. H. (1970). Biogenous deep-sea sediments – fractionation by deep-sea circulation. *Geological Society of America Bulletin*, **81**, 1385–401.

Bigot, A. (1920). Industrie des silices de infusoires et de diatomees. *Revue Ingenieur et Index Technique*, **26**, 302–25.

Bramlette, M. N. (1946). The Monterey Formation of California and the origin of its siliceous rocks, *US Geological Survey Professional Paper*, **212**, 1–57.

Calvert, R. (1929). Diatomaceous Earth. *American Chemical Society Monograph*, **52**.

Calvert, S. E. (1974). Deposition and diagenesis of silica in marine sediments. *Special Publications of the International Association of Sedimentology*, **1**, 273–99.

Card, G. W. & Dun, W. S. (1897). The diatomaceous earth deposits of New South Wales. *Geological Records of New South Wales*, **5**, 128–48.

Cleveland, G. B. (1966). Diatomite. *California Division of Mines and Geology Bulletin*, **191**, 151–8.

Cody, R. D., Levy, R. H., Harwood, D. M., & Sadler, P. (2008). Thinking outside the zone: high-resolution quantitative diatom biochronology for Antarctic Neogene strata. *Palaeogeography, Palaeoclimatology, Palaeoecology*, **260**, 92–121.

Conger, P. S. (1942). Accumulation of diatomaceous deposits. *Journal of Sedimentary Petrology*, **12**, 55–66.

Crangle, J. D., Jr. (2008). Diatomite. *Mining Engineering*, Society for Mining, Metallurgy and Exploration, **60**, 27–8.

Crespin, I. (1946). Diatomite. Mineral Resources of Australia, Summary Report No. 12, Canberra, Commonwealth Government Printer, pp. 1–31.

Cressman, E. R. (1962). Non-detrital siliceous sediments. *US Geological Survey Professional Paper*, **440-T**, T1–T23.

Cummins, A. B. (1960). Diatomite. In *Industrial Minerals and Rocks*, 3rd edition, New York, NY: American Institute of Mining, Metallurgical, and Petroleum Engineers, pp. 303–19.

Cummins, A. B. (1975), *Terra Diatomacea*, Denver, CO: Johns-Manville.

Dolley, T. P. (2008). *Diatomite*. US Geological Survey Mineral Commodity Summaries, pp. 58–59. See http://minerals.usgs.gov/minerals/pubs/commodity/diatomite/mcs-2008-diato.pdf.

Dolley, T. P. & Moyle, P. R. (2003). History and overview of the U.S. diatomite mining industry, with emphasis on the Western United States. *US Geological Survey Bulletin*, **2209**, E1–E8.

Durham, D. L. (1973). Diatomite. *US Geological Survey Professional Paper* **820**, 191–5.

Eardley-Wilmont, W. L. (1928). Diatomite: its occurrence, preparation and uses. Department of Mines, Canada, Publication 691, Ottawa Mines Branch, pp. 1–14.

Ehrenberg, C. G. (1836). Uber Fossile Infusions-thiere. *Berichte der Königlichen Preussischen Akademie der Wissenschaften zu Berlin*, **1836**, 50–4.

Ehrenberg, C. G. (1842). Uber die wie Kork auf Wasser schwimmenden Mauersteine der Alten Griechen und Roemer. *Berichte der Königlichen Preussischen Akademie der Wissenschaften zu Berlin*, **1842**, Bd. 8, 132–7.

Gersonde, R. (1990). The paleontological significance of fossil diatoms from the high-latitude oceans. In *Polar Marine Diatoms*, ed. L. K Medlin & J. Priddle, Cambridge: British Antarctic Survey, pp. 57–63.

Harwood, D. M., Nikolaev, V. A., & Winter, D. M. (2007). Cretaceous records of diatom evolution, radiation, and expansion. In *From Pond Scum to Carbon Sink: Geological and Environmental Applications of the Diatoms*, ed. S. Starratt, Short Courses in Paleontology 13, Knoxville, TN: Paleontological Society, pp. 33–59.

Heath, G. R. (1974). Dissolved silica and deep-sea sediments. *SEPM Special Publication*, **20**, 77–93.

Hendey, N. I. (1930). Diatomite: its analysis and use in Pharmacy. *Quarterly Journal of Pharmacy*, **3**, 390–407.

Issacs, C. M. (1981). Porosity reduction during diagenesis of the Monterey Formation, Santa Barbara Coastal area, California. In *The Monterey Formation and Related Siliceous Rocks of California*, ed. R. E. Garrison, & R. G. Douglas, Los Angeles, CA: Pacific Section, Society of Economic Paleontoloogists and Mineralogists, pp. 257–72.

Issacs, C. M. & Rullkotter J. (eds.) (2001). *The Monterey Formation. From Rocks to Molecules*, New York, NY: Columbia University Press.

Kawashimma, C. & Shiraki, Y. (1941 to 1946). Fundamental studies of Japanese diatomaceous earths and their industrial applications. *Journal of the Japanese Ceramic Association*. Series of 18 parts from Part I (1941) in volume **49**, 14–25 to Part XVIII (1946) in volume **54**, 52–6.

Kemp, A. E. S. & Baldauf, J. G. (1993). Vast Neogene laminated mat deposits from the eastern equatorial Pacific Ocean. *Nature*, **362**, 141–3.

Leinen, M. A. (1979). Biogenic silica accumulation in the central equatorial Pacific and its implications for Cenozoic paleoceanography. *Geological Society of America Bulletin*, **90**, 1310–76.

Lisitzin, A. (1972). Sedimentation in the world ocean. *SEPM Special Publication*, **17**.

Lohman, K. E. (1960). The ubiquitous diatom – a brief survey of the present state of knowledge. *American Journal of Science*, **258-A**, 180–91.

Moores, S. (2008). Blood, sweat and beers, filtration minerals reviewed. *Industrial Minerals*, **484**, 34–40.

Moss, J. (1898). Kieselghur and other infusorial earths. *Proceedings of the British Pharmaceutical Conference for 1898*, 337–45.

Moyle, P. R. & Dolley, T. P. (2003). With or without salt – a comparison of marine and continental-lacustrine diatomite deposits. *US Geological Survey Bulletin*, **2209**, D1–D11.

Nelson, D. M., Treguer, P., Brzezinski, M. A., Leynaert, A., & Queguiner, B. (1995). Production and dissolution of biogenic silica in the ocean: revised global estimates, comparison with regional data and relationship to biogenic sedimentation. *Global Biogeochemical cycles*, **9**, 359–72.

Scherer, R. P., Gladenkov, A. Yu., & Barron, J. A. (2007). Methods and Applications of Cenozoic Marine Diatom Biostratigraphy. In *From Pond Scum to Carbon Sink: Geological and Environmental Applications of the Diatoms*, ed. S. Starratt, Paleontological Society Short Course 13, Knoxville, TN: The Paleontology Society, 61–83.

Schroeder, H. J. (1970). Diatomite in mineral facts and figures. *US Bureau of Mines Bulletin*, **650**, 967–75.

Taliaferro, N. L. (1933). The relation of volcanism to diatomaceous and associated siliceous sediments. *California University Publications, Geological Sciences*, **23**, 1–56

Tappan, H. (1980). *The Paleobiology of Plant Protists*, San Francisco, CA: W.H. Freeman.

Taylor, G. C. (1981). California's diatomite industry. *California Geology*, **34** (9), 183–92.

Wahl, W. H. (1876). Infusorial earth and its uses. *Quarterly Journal Science, London, New Series*, **6**, 336–51.

34 Stable isotopes from diatom silica

MELANIE J. LENG AND GEORGE E. A. SWANN

34.1 Introduction

Diatom silica is a form of biogenic opal ($SiO_2.nH_2O$, Figure 34.1) containing oxygen and silicon isotopes, as well as carbon and nitrogen isotopes from the intrinsic organic matter in the frustule, that can be used in lacustrine and marine paleoenvironmental studies. Since diatoms bloom following a seasonal pattern defined partly by the variability of climate, nutrient supply, mixing regimes, and in high latitudes the period of ice cover, the isotope signature acquired by diatoms will be skewed toward their major growing season specific to the lake or oceanic region under consideration. The isotope ratios (e.g. $^{18}O/^{16}O$, $^{30}Si/^{28}Si$, $^{13}C/^{12}C$, $^{15}N/^{14}N$) of diatom silica are expressed on the delta-scale (δ) in terms of per mille (‰):

$$\delta = [(R_{sample}/R_{reference}) - 1] \times 1000$$

where R is the particular isotope ratio (e.g. $^{18}O/^{16}O$, $^{30}Si/^{28}Si$), and "reference" means the appropriate universally accepted reference material. The "δ" for each element takes its name from the heavy isotope, thus $\delta^{18}O$, $\delta^{30}Si$, $\delta^{13}C$, $\delta^{15}N$. For diatom oxygen the reference is VSMOW (Vienna Standard Mean Ocean Water) calibrated through the National Bureau of Standards (NBS) quartz NBS28, for silicon it is referenced and measured alongside NBS28, for carbon the reference is VPDB (Vienna PeeDee Belemnite) calibrated against NBS19 and NBS22, and for nitrogen it is atmospheric nitrogen, commonly shortened to AIR. There are no universally accepted standard materials to analyze alongside diatoms although most laboratories use their own standard diatomites as well as NBS quartz and low-percent-C and -N organic materials.

The Diatoms: Applications for the Environmental and Earth Sciences, 2nd Edition, eds. John P. Smol and Eugene F. Stoermer. Published by Cambridge University Press.

34.2 Environmental signals from diatom isotopes

34.2.1 Oxygen isotopes

The oxygen-isotope composition of diatoms ($\delta^{18}O_{diatom}$) is primarily a function of changes in ambient water temperature and the isotopic composition of water ($\delta^{18}O_{water}$) surrounding the frustule. The temperature dependence of oxygen-isotope fractionation between diatom silica and water has not been rigorously derived, although the relationship has been estimated from analyses of diatoms from marine and freshwater sediments, coupled with estimates of the temperatures and isotope compositions of co-existing waters during silica formation (e.g. Labeyrie, 1974). The data from these calibration studies are limited, and are mainly based on bulk samples, not individual diatom species (although Brandriss *et al.* (1998) provide a notable exception). Historical estimates of the average temperature dependence based on these samples ranges from −0.3 to −0.5‰ °C^{-1} (Juillet-Leclerc and Labeyrie, 1987; Shemesh *et al.*, 1992). More recently, however, controlled experiments on lacustrine diatoms (Brandriss *et al.*, 1998; Moschen *et al.*, 2006) have shown that the true diatom-temperature coefficient is likely to be closer to −0.2‰ °C^{-1}.

Since diatoms are physically difficult to separate into single species samples due to their small size (2–200 μm), bulk samples comprised of different taxa are normally analyzed for diatom isotopes. For $\delta^{18}O_{diatom}$, a number of culture (Binz, 1987; Brandriss *et al.*, 1998; Schmidt *et al.*, 2001), sediment-trap (Moschen *et al.*, 2005), and down-core studies (Sancetta *et al.*, 1985; Juillet-Leclerc and Labeyrie, 1987; Shemesh *et al.*, 1995; Schiff *et al.*, 2009) in marine and lacustrine systems have failed to find evidence of any isotope/disequilibrium vital effect either within or between individual diatom taxa. Whilst Brandriss *et al.* (1998) documented a 0.6‰ offset between two taxa and Shemesh *et al.* (1995) found a 0.2‰ offset between

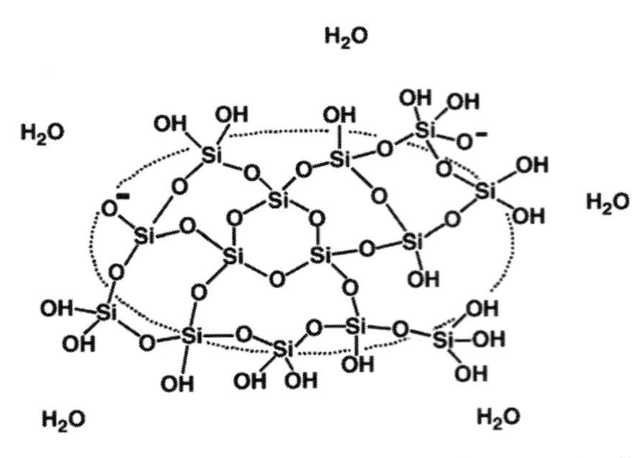

Figure 34.1 Amorphous silica (here used as an analog for diatom silica). A schematic illustration of the structure of amorphous hydrated silica showing the inner tetrahedrally bonded silica and the outer (readily exchangeable) hydrous layer. (Reprinted from Leng and Barker (2006), with permission from Elsevier.)

different size fractions of diatoms, these offsets are within the range of reproducibility routinely achieved when analyzing $\delta^{18}O_{diatom}$ (Section 34.4.1). This is in contrast to biogenic carbonates in which $\delta^{18}O$ measurements, offset from those predicted purely by thermodynamics, can occur in response to variations in kinetic or metabolic processes, e.g. changes in growth rates, nutrient availability, or rates of calcification/silicification (Duplessy *et al.*, 1970; Wefer and Berger, 1991; Spero and Lea, 1993, 1996; Spero *et al.*, 1997; Bemis *et al.*, 1998). Recent results from the northwest Pacific Ocean on sediment-core material, however, has documented isotope offsets of up to 3.5‰ between two size fractions of diatoms (Swann *et al.*, 2007, 2008). Whilst the mechanisms behind this remains unresolved, these offsets may be related to suggestions that changes in growth rates may influence oxygen-isotope fractionation in diatoms (Schmidt *et al.*, 2001).

The diatom frustule is comprised of two layers: a tetrahedrally bonded silica (–Si–O–Si) layer and an outer hydrous (–Si–OH) layer. Whereas the –Si–O–Si layer contains oxygen incorporated into the frustule during silicification, the oxygen in the –Si–OH layer is able to freely exchange with any water the diatom comes into contact with and must be removed prior to isotope analysis (see Section 34.4.1). A key assumption for $\delta^{18}O_{diatom}$ is that no oxygen isotope exchange occurs between the –Si–O–Si layer and the –Si–OH layer during or after sedimentation (Julliet, 1980a, b). Schmidt *et al.* (1997), who analyzed the oxygen isotope composition of live diatom frustules collected from the oceans, however, found no regular correlation between temperature and oxygen-isotope fractionation. Marine and lacustrine diatoms have also been shown to have significantly lower, 2‰ to 10‰, $\delta^{18}O_{diatom}$ fractionation factors that those obtained with diatoms cultured in a laboratory (Schmidt *et al.*, 1997, 2001; Brandriss *et al.*, 1998; Moschen *et al.*, 2006). These results led to several different suggestions: either that partial dissolution of the diatom frustule occurred during sedimentation, altering $\delta^{18}O_{diatom}$ (Brandriss *et al.*, 1998) or that temperature-dependent oxygen-isotope fractionation was established during early diagenesis rather than during diatom growth, where ^{16}O from the –Si–OH layer was released forming isotopically enriched –Si–O–Si (Schmidt *et al.*, 1997, 2001; Moschen *et al.*, 2006) (Figure 34.2). At present, the extent to which silica maturation affects the use of $\delta^{18}O_{diatom}$ in paleoenvironmental reconstructions remains unknown. However, as long as these isotope exchanges are systematic and predictable within and between taxa, values of $\delta^{18}O_{diatom}$ should, at the very least, provide important qualitative information for use in paleo-environmental reconstructions (see Leng and Barker, 2006; Swann and Leng, 2009).

In lacustrine environments the interpretation of $\delta^{18}O_{diatom}$ in terms of paleotemperature also requires an understanding of processes that have opposing effects on the composition of diatom silica. As well as the fractionation between diatom silica and water, described above, values of $\delta^{18}O_{diatom}$ are governed by the relationship between the isotopic composition of precipitation entering the lake (δp) and atmospheric temperature, the so-called Dansgaard relationship $\delta p/dT$ (Dansgaard, 1964). In many lake records the response of $\delta^{18}O_{diatom}$ to changes in water temperature will effectively be "damped" by the opposing effect of variations in the isotope composition of precipitation caused by the Dansgaard relationship. For diatoms, the measured isotope composition will, therefore, covary with temperature – with an increase of ~0.1 to 0.4‰ °C^{-1} (cf. Leng and Marshall, 2004). This assumes that $\delta p/dT$ always changes according to the Dansgaard relationship. Whilst it is important to establish the $\delta p/dT$ relationship at each site, approximately +0.6‰ °C^{-1} at intermediate and high latitudes and globally between +0.2 and +0.7‰ °C^{-1} (IAEA-WMO GNIP, 2006), in practice deriving any quantitative paleotemperature/δp relationship with any certainty is complicated. For example, a Dansgaard relationship is not common in coastal or monsoonal regions or in regions where rainfall comes from two or more air masses. It might therefore be more realistic to use $\delta^{18}O_{diatom}$ in areas where there are likely to have been significant changes in the isotope composition of the lake water due to changes in the precipitation/evaporation

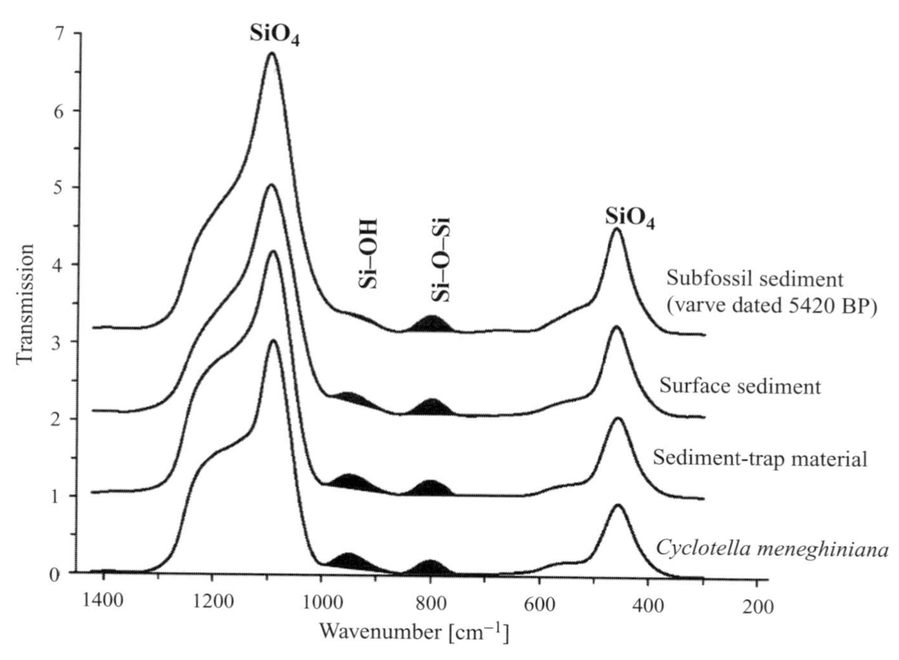

Figure 34.2 Infrared absorption spectra of diatoms from a laboratory culture and from Lake Holzmaar, Germany, showing the progressive loss of –Si–OH groups and the creation of –Si–O–Si bonds as the frustule undergoes silica maturation during sedimentation/burial. The graph of *Cyclotella meneghiniana* shows the absorption spectra of a laboratory culture which is similar in transmission to that of sediment-trap material. (Reprinted from Moschen *et al.* (2006), with permission from Elsevier.)

balance or in the source/amount of precipitation (see below), as the change in $\delta^{18}O_{diatom}$ due to these factors is normally greater than that due to temperature alone.

In closed (or terminal) lakes, for example, the effect of evaporation on $\delta^{18}O_{water}$ will usually be far greater than changes due to temperature or δp. Accordingly, $\delta^{18}O_{water}$ depends on the balance between the isotope composition of inputs (including the source and amount of precipitation, surface runoff, and groundwater inflow) and outputs (evaporation and groundwater loss). Unless there is significant groundwater seepage most closed lakes will lose water primarily through evaporation, the rate of which is controlled by wind speed, temperature, and humidity. The phase change of evaporation results in light isotopes of oxygen (^{16}O) being preferentially evaporated from water bodies, leaving water that is relatively enriched in the heavier isotope (^{18}O). In extreme circumstances, evaporation in the lake catchment area or from the surface of a lake can lead to significantly elevated $\delta^{18}O_{water}$ values (Leng *et al.*, 2005). The degree to which evaporation will increase $\delta^{18}O_{water}$ depends on the residence time of the lake (lake volume/throughput rate). Changes to a lake's residence time, caused by changes in basin hydrology or varying groundwater fluxes, will also influence the magnitude of enrichment, as will changes in the nature of catchment vegetation and soils. These factors have been considered important in the interpretation of a $\delta^{18}O_{diatom}$ record from Lake Tilo in East Africa which is interpreted as changes in aridity (Lamb *et al.*, 2005). These authors showed that the constant supply of solutes from the hydrothermal springs enabled diatoms to grow throughout the year, from which they were able to derive a Holocene lake level which they linked to the amount of evaporation.

Certain $\delta^{18}O_{diatom}$ records have been shown to be sensitive to other aspects of climate, such as the amount or source of precipitation (i.e. by recording the isotope composition of precipitation). For example, abrupt shifts of up to 18‰ in $\delta^{18}O_{diatom}$ have been found in a 14,000-year-long record from two alpine lakes on Mount Kenya (Barker *et al.*, 2001), which can not be entirely temperature related given the current knowledge of the diatom-temperature fractionation. Instead the variations have been interpreted as enhanced precipitation related to changes in Indian Ocean sea-surface temperatures through the Holocene, as measured by alkenone-based sea-surface temperature estimates.

There have been several studies of $\delta^{18}O_{diatom}$ from lakes in Northern Europe (see Leng and Barker, 2006). These lakes have similarities, in that many are open, through-flow systems with minimal evaporation. Changes in the stratigraphic record of

$\delta^{18}O_{diatom}$ are interpreted as changes in the summer isotope composition of the lake water. In the study of a pro-glacial lake covering the last 5000 years from northern Sweden, a $\delta^{18}O_{diatom}$ record combined with a sedimentary proxy for glacier fluctuations reflects changes in the isotope composition of summer precipitation (Rosqvist *et al.*, 2004).

In the marine environment, $\delta^{18}O_{diatom}$ records can be used to investigate temperature and $\delta^{18}O_{water}$ variations in a manner which is similar to that used with $\delta^{18}O$ records from foraminifera ($\delta^{18}O_{foram}$) (see Swann and Leng, 2009). This is most prominent in high-latitude regions, such as the Southern Ocean, where diatoms are abundant in the sedimentary record due, in part, to the high nutrient concentrations within the water column (see Jordan and Stickley, this volume, for further information). At these locations, changes in $\delta^{18}O_{diatom}$ are often dominated by variations in meltwater influx from the polar ice caps. For example, studies in the Southern Ocean have revealed periodic decreases of 2–3‰ during glacial periods over the last *c.* 430,000 years related to meltwater events from Antarctica, releasing water containing an inherently lower values of $\delta^{18}O$ than seawater (Shemesh *et al.*, 1994; 1995; 2002).

At locations where both diatoms and foraminifera co-occur in the sedimentary record, $\delta^{18}O$ records from the two organisms can extend the wealth of paleoenvironmental information that can be reconstructed. For example, Shemesh *et al.* (1992) combined $\delta^{18}O_{diatom}$ with $\delta^{18}O_{foram}$ data, assuming the incorporation of oxygen into the foraminifera calcite shell and diatom frustule occurred in the same season and water depth, to calculate changes in Southern Ocean surface paleo-$\delta^{18}O_{water}$ and paleotemperature. From this, a *c.* 2.0 °C sea-surface-temperature increase was observed in the shift from glacial to Holocene conditions with a concordant 1.2 ± 0.2‰ decrease in $\delta^{18}O_{water}$. In another example, combined $\delta^{18}O_{foram}$ and $\delta^{18}O_{diatom}$ analyses were made on sediments from the northwest Pacific Ocean over the onset of major northern-hemisphere glaciation, *c.* 2.75–2.73 Ma (Haug *et al.*, 2005; Swann *et al.*, 2006). The foraminifera and diatoms over this interval are thought to have lived and precipitated their shells during different seasons, thus providing inter-seasonal information. The $\delta^{18}O$ record from spring calcifying planktonic foraminifera, for example, indicate a significant, 7.5 °C, cooling of the surface waters at the onset of the northern hemisphere glaciation, while an autumnal blooming $\delta^{18}O_{diatom}$ record reveals a 4.6‰ decrease equating to a significant autumn/early winter freshening and warming of the surface waters.

34.2.2 Silicon isotopes

The availability and concentration of silicic acid (predominantly in the form of H_4SiO_4) within the water column is a critical factor in diatom cell division and growth. During biomineralization, diatoms take up silicic acid with the lighter ^{28}Si preferentially incorporated into the frustule over the heavier ^{29}Si and ^{30}Si. Accordingly, measurements of $\delta^{30}Si_{diatom}$ ($^{30}Si/^{28}Si$) and $\delta^{29}Si_{diatom}$ ($^{29}Si/^{28}Si$) (there is a mass-dependent fractionation between the two ratios of $\delta^{30}Si = 1.96 \times \delta^{29}Si$; measurement of both ratios is a good analytical indicator of sample purity but generally only $\delta^{30}Si_{diatom}$ is reported) provide information on the availability and rate of silicic acid utilization within the photic zone, which can in turn be related to the global silicon cycle. Importantly, work has indicated that the diatom silicon-isotope enrichment factor of −1.1‰ to −1.9‰ is independent of temperature, the partial pressure of CO_2 (pCO_2), or other inter-species effects (Spadaro 1983; De La Rocha *et al.*, 1997, 2000; Milligan *et al.*, 2004; Varela *et al.*, 2004).

With diatoms representing *c.* 40% of all marine primary productivity (Nelson *et al.*, 1995), the information provided by marine records of $\delta^{30}Si_{diatom}$ on diatom productivity and export production aids attempts to understand the role of the biological pump in transferring carbon into the deep ocean and controlling atmospheric concentrations of CO_2 (Brzezinski *et al.*, 2002; Matsumoto *et al.*, 2002; Sigman and Haug, 2003; Dugdale *et al.*, 2004). Caution is required, however, in interpreting $\delta^{30}Si_{diatom}$ when changes in nutrient concentrations and other environmental parameters may alter diatom silicon uptake, such as may occur during iron fertilization/nutrient limitation (Hutchins and Bruland, 1998; Takeda, 1998); and when individual water masses containing different isotope values of dissolved silicic acid ($\delta^{30}Si_{DSi}$) become mixed, e.g. highly fractionated surface water and less-fractionated deep water (Reynolds *et al.*, 2006a).

In order to fully interpret changes in $\delta^{30}Si_{diatom}$ an understanding is required of the global silicon cycle, its spatial and temporal variability as well as its impact on the isotopic controls of $\delta^{30}Si_{DSi}$. For example, changes in the marine $\delta^{30}Si_{DSi}$ are a global function of both oceanic and terrestrial silicon inputs and outputs, biogenic cycling of silicic acid, and the interaction of biogenic cycling with the global thermohaline circulation. Similarly, lacustrine $\delta^{30}Si_{DSi}$ will be a function of catchment geology, river input, weathering, water residence time, as well as the nature and timing of seasonal diatom blooms. To this end, over the past decade, research has increasingly focused on examining the contemporary global silicon-isotope cycle from both a terrestrial (De La Rocha *et al.*, 2000), riverine (Georg *et al.*, 2006a), groundwater (Georg *et al.*, 2009), marine (Varela *et al.*, 2004), and model perspective (Wischmeyer *et al.*, 2003; De La Rocha and Bickle, 2005). Within the marine system, for example, it has been shown that values of $\delta^{30}Si_{DSi}$ range from +0.5‰ to +3.2‰ (De La Rocha *et al.*, 2000; Varela *et al.*, 2004; Cardinal *et al.*, 2005; Reynolds *et al.*, 2006a). Whilst regional offsets in $\delta^{30}Si_{DSi}$ of up to 0.4‰ have been observed between Atlantic and Pacific deep waters, sediment records of $\delta^{30}Si_{diatom}$ suggest that these offsets are constant over glacial–interglacial cycles (De La Rocha *et al.*, 1998, 2002).

To date, few studies have investigated $\delta^{30}Si_{diatom}$ in lakes (Alleman *et al.*, 2005; Street-Perrott *et al.*, 2008). In contrast, a number of studies have published results from marine-sediment cores to reconstruct changes in nutrient utilization. Research to date includes evidence of a 0.7‰ lowering of $\delta^{30}Si_{diatom}$ in the Southern Ocean during the last glacial relative to the Holocene which coincides with reductions in diatom productivity and bulk organic $\delta^{13}C$ (De La Rocha *et al.*, 1998). Subsequent work has revealed similar $\delta^{30}Si_{diatom}$ trends in the Southern Ocean over the last three glacial–interglacial cycles (Brzezinski *et al.*, 2002). Evidence of a strong anti-correlation between $\delta^{30}Si_{diatom}$ and bulk-sediment $\delta^{15}N$, as well as nitrogen diatom isotope records (see Section 34.2.3.2), over this interval has raised the possibility of an iron enrichment in the Southern Ocean during the last glacial, stimulating increased diatom uptake of nitrate (NO_3^-) over silicic acid. This decreased usage of silicon during glacials would have resulted in the development of a concentrated pool of silicic acid across the Southern Ocean. Under the "silicic acid leakage hypothesis" any northward migration of this pool would have enabled diatoms to dominate at lower latitudes than today, potentially increasing the net drawdown of CO_2 into the ocean and contributing towards the lower concentrations of atmospheric pCO_2 that existed during glacials (Brzezinski *et al.*, 2002; Matsumoto *et al.*, 2002). However, whilst evidence exists for higher values of $\delta^{30}Si_{diatom}$ in the subantarctic sector of the Southern Ocean during the last glacial, supporting the silicic acid leakage hypothesis, no similar change is apparent in cores from the subtropics (Crosta *et al.*, 2005, 2007; Beucher *et al.*, 2007). This makes it unclear to what extent any silicic acid from the Southern Ocean escaped into the subtropics and raises questions over the role of the biological pump in the region in controlling long-term variations in atmospheric CO_2.

34.2.3 Organic carbon and nitrogen isotopes

During photosynthesis, organic matter, in the form of long-chained polyamides and polycationic peptides (Kröger *et al.*, 1999, 2000, 2002), is deposited within the diatom frustule which can be analyzed for both carbon ($\delta^{13}C_{diatom}$) and nitrogen ($\delta^{15}N_{diatom}$) isotopes. Provided that all external organic matter surrounding the frustule can be removed, measurements of $\delta^{13}C_{diatom}$ and $\delta^{15}N_{diatom}$ can be utilized to reconstruct changes in nutrient utilization, productivity, pCO_2, and the processes which control these changes, e.g. nutrient delivery/utilization, water-column stratification, and sea-ice coverage. Although suitable for use with lacustrine diatoms (e.g. King *et al.*, 2006), most records of $\delta^{13}C_{diatom}$ and $\delta^{15}N_{diatom}$ have to date been constrained to marine systems.

34.2.3.1 Carbon isotopes During photosynthesis dissolved inorganic carbon (DIC) from the water column, in the form of both dissolved aqueous CO_2 ($CO_{2(aq)}$) and HCO_3^-, are incorporated into the intrinsic organic matter of the diatom frustule (see Laws *et al.*, 1997; Burkhardt *et al.*, 2001; Morel *et al.*, 2002; Roberts *et al.*, 2007). The composition of $\delta^{13}C_{diatom}$ is primarily controlled by the balance between the supply and biological demand for DIC as well as the concentration of $CO_{2(aq)}$ within the photic zone. Diatoms, similar to other organisms, preferentially uptake ^{12}C over ^{13}C. As photosynthetic carbon demand from diatoms and other organisms increases, a subsequent utilisation of $^{12}C_{DIC}$ occurs leading to higher values of $\delta^{13}C_{DIC}$ and $\delta^{13}C_{diatom}$ (Laws *et al.*, 1995, 1997; Bidigare *et al.*, 1999; Rosenthal *et al.*, 2000). The extent to which diatoms preferentially use ^{12}C over ^{13}C, however, is a function of $CO_{2(aq)}$ concentration, with an increase in $CO_{2(aq)}$ leading to a decrease in $\delta^{13}C_{diatom}$ and vice-versa (Freeman and Hayes, 1992; Laws *et al.*, 1995, 1997; Rau *et al.*, 1997; Popp *et al.*, 1999). In interpreting records of $\delta^{13}C_{diatom}$, further consideration is required as to the isotopic composition of the $\delta^{13}C_{DIC}$ substrate used during photosynthesis which may be altered by changes in ocean circulation, upwelling/stratification, the influx of terrestrial/riverine material, and the dissolution of diatoms and other organisms in the photic zone, releasing ^{12}C-enriched carbon into the water column.

A number of other factors, including cell shape and size, metabolic pathways, water temperature, and other inter-species vital effects, may also lead to spatial and temporal changes in $\delta^{13}C_{diatom}$ (Laws *et al.*, 1995; Popp *et al.*, 1998; Burkhardt *et al.*, 1999, 2001; Riebesell *et al.*, 2000; Jacot des

Combes *et al.*, 2008). However, provided the influence of these factors on $\delta^{13}C_{diatom}$ can be accounted for, changes in $\delta^{13}C_{diatom}$ can be employed to reconstruct changes in paleoproductivity and $CO_{2(aq)}$, and so changes in pCO_2. For example, assuming the atmosphere and surface water are in equilibrium, Crosta and Shemesh (2002) calculated that 3‰ and 1‰ of the measured 4‰ difference in $\delta^{13}C_{diatom}$ between the last glacial and modern day are attributable to changes in productivity and pCO_2 respectively. Similarly, results in Crosta *et al.* (2005) and Schneider-Mor *et al.* (2005, 2008) detail changes in $\delta^{13}C_{diatom}$ in the Southern Ocean over the last 640 ka also related to changes in productivity, nutrient availability, and pCO_2 concentrations.

34.2.3.2 Nitrogen isotopes As with carbon, during biomineralization, the lighter ^{14}N isotope is preferentially assimilate into the diatom organic matter, relative to ^{15}N, from dissolved NO_3^- within the water column. The exact magnitude of nitrogen-isotopic discrimination by diatoms has been shown to vary significantly, from *c.* 3‰ to *c.* 15‰, according to changes in growth rates and light availability as well as cell size and shape (Karsh *et al.*, 2003; Needoba *et al.*, 2003; Needoba and Harrison, 2004), whilst tentative evidence of a possible inter-species isotope vital effect has also been identified in fossilized samples from the Southern Ocean (Jacot des Combes *et al.*, 2008). However, with increased uptake of NO_3^- resulting in higher values of $\delta^{15}N_{diatom}$, measurements of $\delta^{15}N_{diatom}$ can primarily be interpreted to reflect the magnitude of nitrate utilization in the photic zone (Shemesh *et al.*, 1993). This must again be balanced by processes that may alter the relative supply or demand of NO_3^- in the photic zone, for example changes in water-column stratification and deep-water upwelling (e.g. Robinson *et al.*, 2005; Brunelle *et al.*, 2007; Robinson and Sigman, 2008). In particular, it has been demonstrated that significant variations in diatom cellular nitrogen demand can be caused by changes in iron availability within the photic zone (Hutchins and Bruland, 1998; Takeda, 1998).

Interest in the use of $\delta^{15}N_{diatom}$ has been focused around its application as an alternative to bulk-sediment measurements of $\delta^{15}N$ to reconstruct changes in the marine nitrogen cycle. Whilst bulk $\delta^{15}N$ measurements have provided valuable environmental information (e.g. Altabet and Francois, 1994), questions have been raised as to the extent to which sedimentary records have been altered by diagenesis (Sigman *et al.*, 1999). In addition, the multitude of processes and individual organisms which may contribute towards bulk $\delta^{15}N$ can prevent both precise and accurate isotope interpretations. In contrast $\delta^{15}N_{diatom}$, as with $\delta^{13}C_{diatom}$, is believed to be protected in the diatom frustule from post-depositional changes whilst also providing an organism-specific and therefore more precise insight of changes in the nitrogen cycle (Shemesh *et al.*, 1993).

Existing records of $\delta^{15}N_{diatom}$ have primarily been generated on sediment-core material from the Southern Ocean (e.g. Crosta and Shemesh, 2002; Crosta *et al.*, 2005; Schneider-Mor *et al.*, 2005, 2008) where changes in $\delta^{15}N_{diatom}$ have been shown to be anti-correlated with those in $\delta^{30}Si_{diatom}$ (Beucher *et al.*, 2007). The increased usage of nitrate documented in these studies during glacial periods has been used to provide further support for the silicic acid leakage hypothesis (see Section 34.2.2). Elsewhere, measurements of $\delta^{15}N_{diatom}$ in the Bering Sea on samples dating back to *c.* 120 ka BP have been found to be 3‰ higher during the last glacial (Brunelle *et al.*, 2007). These changes have been linked to a stronger water-column stratification during the last glacial which reduced the overall supply and availability of nitrate to the photic zone and triggered an increase in $\delta^{15}N_{diatom}$ as the isotopically lighter ^{14}N is progressively removed by continuing biological productivity.

34.3 Concentration and purification

A pre-requisite for all diatom isotope analyses is to ensure samples are free from all sources of non-diatom contamination as most contaminants contain oxygen, silicon, and/or organic matter. Natural diatomites (>80% diatom silica) are the easiest material to work with. Although relatively rare, they do occur in some lakes and parts of the World's oceans (North Pacific and Southern Ocean) where the influence of coastlines or fluvial inputs and concentrations of mineral flux are low. Separating pure diatoms even from diatomites can be challenging, especially when diatoms are intermixed with silt, clay, tephra, carbonates, and organic matter (Figure 34.3). In particular, sediment grains can become "trapped" (Figure 34.3a) within the diatom frustule (which protects them from removal) and clays can become coated with submicron-scale fragmented/broken diatom (Figure 34.3c) material (adhering by electrostatic charge) (Brewer *et al.*, 2008). To remove these and other contaminants from diatom samples, a number of chemical and physical clean-up methodologies have been used (e.g. Juillet-Leclerc, 1986; Shemesh *et al.*, 1988; Shemesh *et al.*, 1995; Singer and Shemesh, 1995; Morley *et al.*, 2004; Rings *et al.*, 2004; Swann *et al.*, 2006; Tyler *et al.*, 2007; Crespin *et al.*, 2008). The success of the purification process is controlled by factors such as: (a) the ability of chemical reagents to fully oxidize certain components (e.g. organics and carbonates),

Figure 34.3 Scanning electron microscopy images of (a) diatom frustules and micron-scale grains contained within and around diatoms (Lake Baikal, Russia, sediments), (b) tephra within a diatom sample, which are usually difficult to remove because of electrostatic charge (Lake Tilo, Ethiopia, sediments), (c) clays and accumulations of submicron-scale material within a diatom sample (Lake Baikal sediments), (d) a pure diatom sample from ODP Site 882 in the northwest Pacific Ocean, showing good preservation and the absence of submicron fragments.

(b) the diatoms and contaminants having different grain sizes and being present in the sample as discrete grains, and (c) that there is sufficient density contrast between diatoms and the contaminants to enable density separation. In addition, a prerequisite is that the isotopic composition of the diatom is not altered by the cleaning process. Carbonates and organic materials are generally removed by chemical methods (HCl, H_2O_2, HNO_3), clays by sieving at 5 or 10 μm, and silt–sand-sized grains by differential settling techniques using specific gravity and other physical property differences between diatom silica and the contaminant. Shemesh *et al.* (1995 and subsequent papers) described a similar methodology but added a heavy-liquid settling stage to separate diatoms from remaining clastic grains. This heavy-liquid stage has been widely adopted and usually involves mixing and centrifuging samples with sodium polytungstate (SPT) at a specific gravity of *c.* 2.2 g ml^{-1}. Because diatoms have specific gravities of *c.* 2.1 g ml^{-1} while typical silicate contaminants all tend to be denser than this (quartz, feldspars, micas, clay minerals), the use of SPT allows the density separation of the two components.

An alternative approach to heavy-liquid separation for cleaning diatom samples is gravitational split-flow thin fractionation (SPLITT) (Giddings, 1985; Schleser *et al.*, 2001; Rings *et al.*, 2004). This technique works by introducing a sample into water under laminar flow which separates the sample into two components according to their density and hydrodynamic properties. Whilst SPLITT is repetitious, requiring each sample to be repeatedly passed through the device to ensure purification, the procedure results in high throughput, small losses, and sometimes the ability to isolate specific taxa (where they have different size, density, and/or shape) (Leng and Barker 2006). The use of SPLITT also avoids the need to introduce products other than water into the sample. For example, Morley *et al.* (2004) show that the heavy liquid can contaminate diatom oxygen-isotope ratios if not fully removed prior to analysis by sieving at a fine ($\leq$5 μm) size fraction.

A number of additional chemical cleaning stages may be employed as a final purification stage, such as the addition of potassium permanganate, nitric acid (HNO_3), hydrogen fluoride (HF), or alkaline solutions to remove remnants of external organic material and to etch the surficial silica layer in an attempt to disassociate any clays adhering to the diatoms (e.g. van Bennekom and van der Gaast, 1976; Juillet-Leclerc, 1986). Specifically for $\delta^{30}Si_{diatom}$, in some laboratories, samples may be dissolved before being reprecipitated using HF and triethylamine molybdate to further purify the sample as well as to remove any organic matter prior to analysis (De La Rocha *et al.*, 1996).

Given the small amount of organic matter within the diatom frustule, a series of additional steps, involving the use of oxidizing chemicals, are undertaken to ensure that all external organic matter is removed when analyzing $\delta^{13}C_{diatom}$ or $\delta^{15}N_{diatom}$. The choice of cleaning technique is not known to alter measurements of $\delta^{13}C_{diatom}$. However, values of $\delta^{15}N_{diatom}$ have been shown to vary markedly according to the method employed. For example, the work conducted by Shemesh *et al.* (1993) and Singer and Shemesh (1995) which led to the development of $\delta^{15}N_{diatom}$ as a paleoenvironmental proxy used a concentrated $HNO_3/HClO_4$ (perchloric acid) mixture. Sigman *et al.* (1999) demonstrated that this results in erroneous $\delta^{15}N_{diatom}$ values, which they ascribe to nitrogen from HNO_3 becoming incorporated into the frustule. Whilst the validity of this has been questioned by researchers continuing to use $HNO_3/HClO_4$ mixtures (e.g. Crosta and Shemesh, 2002; Schneider-Mor *et al.*, 2005), the two to three times higher diatom %N values in these studies compared to non-HNO_3 studies suggests that there are still unresolved issues associated with the use of HNO_3. As an alternative, samples can be cleaned using concentrated $HClO_4$ or H_2O_2 combined with an $HClO_4$ oxidation and dithionite–citric acid reduction cleaning stage (Robinson *et al.*, 2004, 2005; Brunelle *et al.*, 2007).

It is apparent that no single method of diatom purification is suitable for all sediment samples. Instead, the clean-up approach should be adapted to fit each type of sample with the optimal procedure usually found by a combination of trial and error, using the various techniques described above, combined with petrological examination of every sample at each stage. Traditionally, the level of contamination within a sample is assessed visually using point-counting techniques (Morley *et al.*, 2004). However, Brewer *et al.* (2008) have suggested that more reliable estimates of contamination can be obtained by analyzing the trace-element geochemistry of purified samples, for example using scanning electron microscopy (SEM) plus energy dispersive X-ray spectroscopy (EDS), X-ray fluorescence (XRF) or inductively coupled plasma atomic emission spectroscopy (ICP-AES). By measuring the concentrations of compounds such as aluminium oxide (Al_2O_3) and calcium oxide (CaO) and comparing these to (published) diatom and clay mineral elemental concentrations, the relative proportion and type of residual contamination can be established for the sample being analyed (Lamb *et al.*, 2007; Brewer *et al.*, 2008). Regardless of the method used to check sample purity, samples containing more than a few percent contamination should be re-cleaned. Where further cleaning does not improve the sample purity, mass-balance corrections can be applied to correct for the effects of non-diatom contaminants following isotope analysis. For example Morley *et al.* (2005), analyzing changes in $\delta^{18}O_{diatom}$ in Lake Baikal, showed that even after purification, diatom concentrations in "cleaned" Lake Baikal sediments ranged from 33% to 99% with $\delta^{18}O$ values between +14.4‰ and +34.3‰. By assuming that the bulk $\delta^{18}O$ value of the cleaned samples was a linear mixture of oxygen from silt and diatoms, samples were mass-balance corrected to obtain a modeled silt-free value of $\delta^{18}O_{diatom}$ using point-counting estimates of non-diatom contamination and a average $\delta^{18}O$ value of silt of +12.3‰ (Brewer *et al.*, 2008). To date, this modeling approach has not been applied to analyses of $\delta^{30}Si_{diatom}$, $\delta^{13}C_{diatom}$, or $\delta^{15}N_{diatom}$ data. Whilst mass-balance corrections contain several limitations, such as making assumptions over the amount of silt/contamination to diatom oxygen in each sample, it is one of the few non-destructive ways of correcting for the influence of contaminants so long as a small amount of the cleaned diatom sample is left after isotope analysis to measure the isotopic composition of the contaminants.

34.4 Analytical methods

34.4.1 Oxygen isotopes

Due to the presence of contaminant oxygen, the outer –Si–OH, hydrous layer of the diatom frustule needs to be removed or accounted for prior to isotope analysis, for example by using oxidizing reagents and/or high temperatures (see Section 34.2.1). Extraction of the –Si–OH layer requires the removal or exchange of 7–40% of all oxygen within a diatom (Knauth, 1973; Labeyrie, 1979; Labeyrie and Juillet, 1982; Leng *et al.*, 2001; Leng and Sloane, 2008; Swann *et al.*, 2008). Early attempts to remove the oxygen in the –Si–OH layer involved dehydrating samples under vacuum (Mopper and Garlick, 1971; Labeyrie, 1974, 1979). Although the analytical reproducibility and accuracy of the vacuum dehydration method improved with time, the $\delta^{18}O_{diatom}$ signal was thought to be contaminated in part due to partial isotopic exchange during sample dehydration (Labeyrie, 1979; Labeyrie and Juillet, 1980, 1982). At present, three reliable techniques have been established which fully account for the –Si–OH layer and permit paleoenvironmental reconstructions from $\delta^{18}O_{diatom}$: controlled isotope exchange (CIE) followed by fluorination (Labeyrie and Juillet, 1982; Juillet-Leclerc and Labeyrie, 1987), stepwise fluorination (SWF) (Haimson and Knauth 1983; Matheney and Knauth 1989), and inductive High-Temperature carbon Reduction (iHTR) (Lücke *et al.*, 2005). The amount of diatom material required for each technique varies, the critical factor being whether the mass spectrometry is online or offline and, in theory, the relative size of the diatom –Si–OH layer. Typically, between 1.5 mg and 6.5 mg of diatoms are required for a single analysis.

Under CIE, oxygen in the –Si–OH layer of the diatom is exchanged with water containing a known $\delta^{18}O$ ratio (Labeyrie and Juillet, 1982; Juillet-Leclerc and Labeyrie, 1987; Crespin *et al.*, 2008). After vacuum heating at 1000 °C to remove the oxygen in the –Si–OH layer, samples are reacted with a fluorine regent to dissociate the tetrahedrally bonded oxygen within the frustules before analysis using standard gas-source isotope ratio mass spectrometry (IRMS) techniques. Mass-balance corrections are then applied to correct for any of the labeled water not removed under vacuum.

The SWF techniques involve the use of a fluorine reagent and heat (using a furnace or laser) to extract the oxygen from the different layers of the diatom frustule in separate stages, thereby avoiding contamination between the oxygen in the –Si–OH and –Si–O–Si layers (Haimson and Knauth, 1983; Thorliefson, 1984; Matheney and Knauth, 1989). Using adaptations of the fluorination procedures established by Taylor and Epstein (1962) and Epstein and Taylor (1971), the –Si–OH layer is initially stripped away leaving behind the inner –Si–O–Si layer. A second fluorination stage is then used to liberate oxygen from the –Si–O–Si layer (see Leng and Sloane (2008)).

Whilst both CIE and SWF have produced comparable results, using either conventional furnace or laser heating, it has been suggested that measurements from CIE have to be calibrated against SWF due to concerns over incomplete oxygen exchange between the –Si–OH layer and the labeled water under CIE (Schmidt *et al.*, 1997). However, a significant problem of both the CIE and SWF techniques is the requirement for fluorine-based oxidizing reagents, which have specific health and safety requirements (Leng and Sloane, 2008). The iHTR method for analyzing $\delta^{18}O_{diatom}$ eliminates the need for a fluorine-based reagent (Lücke *et al.*, 2005). In iHTR, diatoms are mixed with graphite and heated under vacuum to volatilize any sample contaminants and remove the –Si–OH layer. Further heating of the sample to 1550 °C results in oxygen from the –Si–O–Si bonds being converted to CO for either continuous or offline mass spectrometry.

34.4.2 Silicon isotopes

Similar to $\delta^{18}O_{diatom}$, recent years have been marked by the development and refinement of a range of techniques to analyze $\delta^{30}Si_{diatom}$. Methods for the analysis of $\delta^{30}Si$ include fluorine-based conversion of purified silica into SiF_4 and IRMS (De Freitas *et al.*, 1991; De La Rocha *et al.*, 1996; Ding *et al.*, 1996; Leng and Sloane, 2008), multicollector inductively coupled plasma mass spectrometry (MC-ICP-MS) (De La Rocha, 2002; Cardinal *et al.*, 2003; Reynolds *et al.*, 2006b), acid decomposition and IRMS (Brzezinksi *et al.*, 2006), or ion microprobe analysis (e.g. Basile-Doelsch *et al.* (2005)). Analytical reproducibility for the most commonly used methods, fluorination and MC-ICP-MS, are similar at *c.* 0.1‰ (1σ) (Reynolds *et al.*, 2007).

Fluorination methods are based around techniques used for early $\delta^{30}Si$ measurements of mineral and lunar samples (Epstein and Taylor, 1970a, b). This involves the conversion of silicate into SiF_4 using a fluorine reagent and is similar to the SWF method used for analyzing $\delta^{18}O_{diatom}$. However, since $\delta^{30}Si_{diatom}$ is homogeneous across the diatom frustule, no pre-fluorination stage is required unless oxygen for $\delta^{18}O_{diatom}$ is being extracted from the same sample. A recent advancement is the development of an SWF technique which allows for the $\delta^{18}O_{diatom}$ and $\delta^{30}Si_{diatom}$ signal to be collected simultaneously from the same sample (Leng and Sloane, 2008). Following liberation and collection of the oxygen, the silicon can be collected as a by-product of the fluorination reaction as SiF_4. This is an important step given the large (milligrams) amounts of material normally required for $\delta^{18}O_{diatom}$ and $\delta^{30}Si_{diatom}$ analysis using fluorination techniques and opens the possibility of combining information on surface-water oceanographic and climate conditions ($\delta^{18}O_{diatom}$) with information on photic-zone nutrient utilization ($\delta^{30}Si_{diatom}$).

The MC-ICP-MS technique, operating in either wet or dry plasma mode, offers the option to avoid using fluorine reagents when analyzing $\delta^{30}Si_{diatom}$ (e.g. De La Rocha, 2002; Cardinal *et al.*, 2003). Through the use of standard–sample–standard bracketing, reliable $\delta^{30}Si$ measurements can be obtained with analytical errors under conventional MC-ICP-MS that are either equivalent to or lower than those achieved with IRMS (Cardinal *et al.*, 2003). Further advances in recent years have been made by using an alkaline fusion rather than HF dissolution stage to increase the sensitivity and reduce silicon fractionation during sample introduction (Georg *et al.*, 2006b; van den Boorn *et al.*, 2006). The use of MC-ICP-MS, however, is often limited by the presence of atomic interferences over the Si mass range by, amongst others, CO^+, NO^+, $N_2{}^+$ and N_2H^+. In particular, many mass spectrometers operating at normal resolution experience high-intensity interferences over mass 30, including amongst others $^{14}N^{16}O$, requiring that isotope data be reported as $\delta^{29}Si$ rather than $\delta^{30}Si$ (Cardinal *et al.*, 2003). Dual analysis of samples through both MC-ICP-MS and IRMS, however, have demonstrated a linear relationship between $\delta^{29}Si$ and $\delta^{30}Si$ (De La Rocha, 2002) allowing isotope measurements from different laboratories to be compared. The increasing availability of high-resolution MC-ICP-MS, however, eliminates the effects of interference and allows direct measurement of $\delta^{30}Si$ (Georg *et al.*, 2006b; Reynolds *et al.*, 2006a, b). Whilst yet to be widely used in diatom studies, the introduction of laser ablation systems (LA-MC-ICP-MS) is also likely to yield promising results in the near future (e.g. Chmeleff *et al.*, 2008).

In addition to IRMS and MC-ICP-MS, a number of other methods hold significant potential in analysing $\delta^{30}Si_{diatom}$. One is the development of an acid decomposition method for analysing $\delta^{30}Si$ on a Finnegan Kiel device in which samples are dissolved in HF before being converted to SiF_4 and analyzed by IRMS (Brzezinski *et al.*, 2006). Second is secondary ion mass spectrometry (SIMS). Whilst analyses to date have been restricted to quartz and cosmic spherules rather than diatom samples (e.g. Alexander *et al.*, 2002; Basile-Doelsch *et al.*, 2005; Robert and Chaussidon, 2006), the ability to analyze small grains (>50 μm) opens the possibility that it may become viable to analyze individual frustules. However, in order to achieve this, a reduction in the analytical errors associated with SIMS is required, which are currently higher than those of other techniques (Basile-Doelsch *et al.*, 2005).

34.4.3 Carbon/Nitrogen

Both $\delta^{13}C_{diatom}$ and $\delta^{15}N_{diatom}$ can be analyzed simultaneously on the same sample using an elemental analyzer (which also provides %C and %N data) attached to a mass spectrometer. Continuous-flow or cryogenic trapping methods are used where CO_2 and N_2O are passed from the analyzer to the IRMS. The exact details of the analytical protocol varies between individual laboratories (e.g. Shemesh *et al.*, 1993; Crosta and Shemesh, 2002). The coupled elemental analyser–IRMS technique has greatly simplified organic isotope measurements, has dramatically increased sample analysis throughput, while also allowing significant reduction in sample size over conventional off-line preparations. Recently a new persulfate-denitrifier technique has been developed specifically for analyzing $\delta^{15}N_{diatom}$ in which organic N in the frustule is liberated and converted to N_2O for $\delta^{15}N$ analysis (Robinson *et al.*, 2004). Importantly, the persulfate–denitrifier method increases sensitivity and so requires much smaller sample sizes, *c.* 5 mg compared to 80–100 mg of diatom for combustion methods. Surprisingly, N yields are lower and $\delta^{15}N_{diatom}$ values, typically, are 1–3‰ higher under the persulfate–denitrifier method. At present, the reason behind these discrepancies remains unresolved. One explanation is the presence of a contaminant gaseous N pool trapped within diatoms, which is analyzed during the combustion process but removed prior to isotope analysis with the persulfate–denitrifier method. However, it is also possible that chemicals in the persulfate–denitrifier method fail to react completely with all the organic nitrogen within the frustule (De La Rocha, 2006) whilst other "uncontrolled processes" during this method may result in the formation of other nitrogen compounds, limiting the accuracy of individual persulfate–denitrifier measurements (Crosta *et al.*, 2005). As such, further research is required into the analytical techniques used for $\delta^{15}N_{diatom}$.

34.6 Summary

Diatom silica is a form of biogenic opal with the structure mainly formed of tetrahedrally bonded –Si–O–Si molecules surrounded by a hydrous layer of loosely bonded –Si–OH species which can rapidly exchange oxygen with surrounding water. In addition, the frustule contains small amounts of carbon and nitrogen which are commonly thought to occur as intrinsic and submicron inclusions of organic matter. Here we show that valuable palaeoenvironmental data can be gained from the oxygen ($\delta^{18}O_{diatom}$), silicon ($\delta^{30}Si_{diatom}$), carbon ($\delta^{13}C_{diatom}$), and nitrogen ($\delta^{15}N_{diatom}$) isotope composition of diatom silica, especially since diatoms are abundant in many lakes and found over vast areas of our oceans. They are particularly important in areas where other hosts (e.g. carbonates) are absent, such as in high-latitude marine and lacustrine regions.

Analysis of the isotope composition of diatom silica requires samples that are almost pure diatomite since analytical techniques will also liberate any contaminants remaining in the sample. Recent studies of $\delta^{18}O_{diatom}$ have highlighted this, showing that even a small proportion of contaminant can have a significant influence on the isotope values. While there is a generally accepted protocol for cleaning samples involving chemistry, sieving, and settling techniques, and more recently laminar-flow separation, all sediments require their own specific procedure and every sample must be scrutinized by microscopy/SEM to check for the level of contamination prior to analysis. Where sediments cannot be purified sufficiently, a semi-quantitative or geochemical assessment of the main contaminants can facilitate mass-balance techniques.

Isotope measurements can be obtained by various fluorination and IRMS methods for O and Si. Alternative methods are being increasingly used for silicon isotopes, namely MC-ICP-MS and SIMS. Analysis of the carbon and nitrogen isotope composition from diatom-included organic matter mainly requires combustion within an elemental analyzer and IRMS and is relatively straightforward apart from dealing with the large sample sizes (*c.* 30–100 mg). Recent advances, however, have resulted in the development of novel analytical techniques for $\delta^{15}N_{diatom}$ which allow for significantly smaller sample (c. 5 mg).

Acknowledgments

Thanks are owed to Elsevier for permission to reprint Figure 34.1, and to Robert Moschen and Elsevier for permission to reprint Figure 34.2. Funding for GEAS was provided by a NERC postdoctoral fellowship award (NE/F012969/1).

References

Alexander, C. M. O'D., Taylor, S., Delaney, J. S., Ma, P., & Herzog, G. F. (2002). Mass-dependent fractionation of Mg, Si, and Fe isotopes in five stony cosmic spherules. *Geochimica et Cosmochimica Acta*, **66**, 173–83.

Alleman, L. Y., Cardinal, D., Cocquyt, C., *et al.* (2005). Silicon isotope fractionation in Lake Tanganyika and its main tributaries. *Journal of Great Lakes Research*, **31**, 509–19.

Altabet M. A. & Francois R. (1994). Sedimentary nitrogen isotopic ratio as a recorder for surface ocean nitrate utilization. *Global Biogeochemical Cycles*, **8**, 103–16.

Barker, P. A., Street-Perrot, F. A., Leng, M. J., *et al.* (2001). A 14,000-year oxygen isotope record from diatom silica in two alpine lakes on Mt. Kenya. *Science*, **292**, 2307–10.

Basile-Doelsch, I., Meunier, J. D., & Parron, C. (2005). Another continental pool in the terrestrial silicon cycle. *Nature*, **433**, 399–402.

Bemis, B. E., Spero, H., Bijma, J., & Lea, D. W. (1998). Reevaluation of the oxygen isotopic composition of planktonic foraminifera: experimental results and revised paleotemperature equations. *Paleoceanography*, **13**, 150–60.

Beucher, C. P., Brzezinski, M. A., & Crosta, X. (2007). Silicic acid dynamics in the glacial sub-Antarctic: implications for the silicic acid leakage hypothesis. *Global Biogeochemical Cycles*, **21**, GB3015, DOI:10.1029/2006GB002746.

Bidigare, R. R., Hanson, K. L., Buesseler, K. O., *et al.* (1999). Iron stimiulated change in ^{13}C fractionation and export by equatorial Pacific phytoplankton: towards a paleogrowth rate proxy. *Paleoceanography*, **14**, 589–95.

Binz, P. (1987). Oxygen-isotope analysis on recent and fossil diatoms from Lake Walen and Lake Zurich (Switzerland) and its application on paleoclimatic studies. Unpublished Ph.D. thesis, Swiss Federal Institute of Technology, Zurich.

Brandriss, M. E., O'Neil, J. R., Edlund, M. B., & Stoermer, E.F. (1998). Oxygen isotope fractionation between diatomaceous silica and water. *Geochimica et Cosmochimica Acta*, **62**, 1119–25.

Brewer, T. S., Leng, M. J., Mackay, A. W., *et al.* (2008). Unravelling contamination signals in biogenic silica oxygen isotope composition: the role of major and trace element geochemistry. *Journal of Quaternary Science*, **23**, 321–30.

Brunelle, B. G., Sigman, D. M., Cook, M. S., *et al.* (2007). Evidence from diatom-bound nitrogen isotopes for subarctic Pacific stratification during the last ice age and a link to North Pacific denitrification changes. *Paleoceanography*, **22**, DOI:10.1029/2005PA001205.

Brzezinski, M. A., Jones, J. L., Beucher, C. P., & Demarest, M. S. (2006). Automated determination of silicon isotope natural abundance by the acid decomposition of cesium hexafluosilicate. *Analytical Chemistry*, **78**, 6109–14.

Brzezinski, M. A., Pride, C. J., Franck, V. M., *et al.* (2002). A switch from $Si(OH)_4$ to NO_3^- depletion in the glacial Southern Ocean. *Geophysical Research Letters*, **29**, 1564, DOI: 10.1029/2001GL014349.

Burkhardt, S., Amoroso, G., Riebesell, U. & Sültemeyer, D. (2001). CO_2 and HCO_3^- uptake in marine diatoms acclimated to different CO_2 concentrations. *Limnology and Oceanography*, **46**, 1378–91.

Burkhardt, S., Riebesell, U., & Zondervan, I. (1999). Effects of growth rate, CO_2 concentration, and cell size on the stable carbon isotope fractionation in marine phytoplankton. *Geochimica et Cosmochimica Acta*, **63**, 3729–41.

Cardinal, D., Alleman, L. Y., de Jong, J., Ziegler, K., & André, L. (2003). Isotopic composition of silicon measured by multicollector plasma source mass spectrometry in dry plasma mode. *Journal of Analytical Atomic Spectrometry*, **18**, 213–18.

Cardinal, D., Alleman, L. Y., Dehairs, F., *et al.* (2005). Relevance of silicon isotopes to Si-nutrient utilization and Si-source assessment in Antarctic waters. *Global Biogeochemical Cycles*, **18**, GB2007, DOI: 10.1029/2004GB002364.

Chmeleff, J., Horn, I., Steinhoefel, G., & von Blanckenburg, F. (2008). *In situ* determination of precise stable Si isotope ratios by UV-femtosecond laser ablation high-resolution multi-collector ICP-MS. *Chemical Geology*, **29**, 155–66.

Crespin, J., Alexandre, A., Sylvestre, F., *et al.* (2008). IR laser extraction technique applied to oxygen isotope analysis of small biogenic silica samples. *Analytical Chemistry*, **80**, 2372–8.

Crosta, X., Beucher, C., Pahnke, K., & Brzezinski, M. A. (2007). Silicic acid leakage from the Southern Ocean: opposing effects of nutrient uptake and oceanic circulation. *Geophysical Research Letters*, **34**, L13601, DOI:10.1029/2006GL029083.

Crosta, X. & Shemesh, A. (2002). Reconciling down core anticorrelation of diatom carbon and nitrogen isotopic ratios from the Southern Ocean. *Paleoceanography*, **17**, 1010, DOI: 10.1029/2000PA000565.

Crosta, X., Shemesh, A., Etourneau, J., *et al.* (2005). Nutrient cycling in the Indian sector of the Southern Ocean over the last 50,000 years. *Global and Biogeochemical Cycles*, **19**, GB3007, DOI:10.1029/2004GB002344.

Dansgaard, W. (1964). Stable isotopes in precipitation. *Tellus*, **16**, 436–468.

De Freitas, A. S. W., McCulloch, A. W., & McInnes, A. G. (1991). Recovery of silica from aqueous silicate solutions via trialkyl or tetraalkylammonium silicomolybdate, *Canadian Journal of Chemistry*, **69**, 611–14.

De La Rocha, C. L. (2002). Measurement of silicon stable isotope natural abundances via multicollector inductively coupled plasma mass spectrometry (MC-ICP-MS). *Geochemistry, Geophysics, Geosystems*, **3**, DOI:10.1029/2002GC000310.

De La Rocha, C. L. (2006). Opal-based isotopic proxies of paleoenvironmental conditions. *Global Biogeochemical Cycles*, **20**, GB4S09. DOI:10.1029/2005GB002664.

De La Rocha, C. L. & Bickle, M. J. (2005). Sensitivity of silicon isotopes to whole-ocean changes in the silica cycle. *Marine Geology*, **217**, 267–82.

De La Rocha C. L., Brzezinski M. A., & DeNiro M. J. (1996). Purification, recovery and laser-driven fluorination of silicon from dissolved and particulate silica for the measurements of natural stable isotopes abundances. *Analytical Chemistry*, **68**, 3746–50.

De La Rocha C. L., Brzezinski M. A. & DeNiro M. J. (1997). Fractionation of silicon isotopes by marine diatoms during biogenic silica formation. *Geochimica et Cosmochimica Acta*, **61**, 5051–6.

De La Rocha C. L., Brzezinski M. A. & DeNiro M. J. (2000). A first look at the distribution of the stable isotopes of silicon in natural waters. *Geochimica et Cosmochimica Acta*, **64**, 2467–77.

De La Rocha C. L., Brzezinski M. A., DeNiro M. J., & Shemesh A. (1998). Silicon-isotope composition of diatoms as an indicator of past oceanic change. *Nature*, **395**, 680–3.

Ding T., Jiang S., Wan D., *et al.* (1996). *Silicon Isotope Geochemistry*. Beijing: Geological Publishing House.

Dugdale, R. C., Lyle, M., Wilkerson, F. P., *et al.* (2004). Influence of equatorial diatom processes on Si deposition and atmospheric CO_2 cycles at glacial/interglacial timescales. *Paleoceanography*, **19**, PA3011: DOI:10.1029/2003PA000929.

Duplessy, J. C., Lalou, C., & Vinot, A. C. (1970). Differential isotopic fractionation in benthic foraminifera and paleotemperatures revised. *Science*, **213**, 1247–50.

Epstein, S. & Taylor, H. P. (1970a). The concentration and isotopic composition of hydrogen, carbon and silicon in Apollo 11 lunar rocks and minerals. *Proceedings of the Apollo 11 Lunar Science Conference*, **2**, 1085–96.

Epstein, S. & Taylor, H. P. (1970b). Stable isotopes, rare gases, solar wind and spallation products. *Science*, **167**, 533–5.

Epstein, S. & Taylor, H. P. (1971). O18/O16, Si30/Si28, D/H and C13/C12 ratios in lunar samples. *Proceedings of the Second Lunar Conference*, **2**, 1421–41.

Freeman, K. H. & Hayes, J. M. (1992). Fractionation of carbon isotopes by phytoplankton and estimates of ancient CO_2 levels. *Global Biogeochemical Cycles*, **6**, 185–98.

Georg, R. B., Reynolds, B. C., Frank, M., & Halliday, A. N. (2006a). Mechanisms controlling the silicon isotopic compositions of river waters. *Earth and Planetary Science Letters*, **249**, 290–306.

Georg, R. B., Reynolds, B. C., Frank, M., & Halliday, A. N. (2006b). New sample preparation techniques for the determination of Si isotopic compositions using MC-ICPMS. *Chemical Geology*, **235**, 95–104.

Georg, R. B., Zhu, C., Reynolds, B. C., & Halliday, A. N. (2009). Stable silicon isotopes of groundwater, feldspars, and clay coatings in the Navajo Sandstone aquifer, Black Mesa, Arizona, USA. *Geochimica et Cosmochimica Acta*, **73**, 2229–41.

Giddings, J. C. (1985). A system based on split-flow lateral transport thin (SPLITT) separation cells for rapid and continuous particle fractionation. *Separation Science and Technology*, **20**, 749–68.

Haimson, M. & Knauth, L. P. (1983). Stepwise fluorination – a useful approach for the isotopic analysis of hydrous minerals. *Geochimica et Cosmochimica Acta*, **47**, 1589–95.

Haug, G. H., Ganopolski, A., Sigman, D. M., *et al.* (2005). North Pacific seasonality and the glaciation of North America 2.7 million years ago. *Nature*, **433**, 821–5.

Hutchins, D. A. & Bruland, K. W. (1998). Iron-limited diatom growth and Si:N uptake ratios in a coastal upwelling zone. *Nature*, **393**, 561–4.

IAEA/WMO (2006). Global network of isotopes in precipitation. The GNIP database. See http://www.iaea.org/water.

Jacot des Combes, H., Esper, O., De La Rocha, C. L., *et al.* (2008). Diatom $\delta^{13}C$, $\delta^{15}N$, and C/N since the last glacial maximum in the Southern Ocean: potential impact of species composition. *Paleoceanography*, **23**, PA4209, DOI:10.1029/2008PA001589.

Juillet, A. (1980a). Structure de la silice biogenique: nouvelles donnes apportees par l'analyse isotopique de l'oxygene. *C. R. Academy of Science, Paris*, **290D**, 1237–9.

Juillet, A. (1980b). Analyse isotopique de la silice des diatomees lacustres et marines: fractionnement des isotopes de l'oxygene en fonction de la temperature. Thèse de troisième cycle, Université Paris-Sud XI.

Juillet-Leclerc, A. (1986). Cleaning process for diatomaceous samples. In *Proceedings of the 8th Diatom Symposium*, ed. M. Ricard, Königstein: Koeltz Scientific Books.

Juillet-Leclerc, A. & Labeyrie, L. (1987). Temperature dependence of the oxygen isotopic fractionation between diatom silica and water. *Earth and Planetary Science Letters*, **84**, 69–74.

Karsh, K. L., Trull, T. W., Lourey, M. J., & Sigman, D. M. (2003). Relationship of nitrogen isotope fractionation to phytoplankton size and iron availability during the Southern Ocean Iron RElease Experiment (SOIREE). *Limnology and Oceanography*, **48**, 1058–68.

King, L., Barker, P. A., & Grey, J. (2006). Organic inclusions in lacustrine diatom frustules as a host for carbon and nitrogen isotopes. *Verhandlung der Internationalen Verein Limnologie*, **29**, 1608–10.

Knauth, L. P. (1973). Oxygen and hydrogen isotope ratios in cherts and related rocks. Unpublished Ph.D. thesis, California Institute of Technology.

Kröger, N., Deutzmann, R., & Sumper, M. (1999). Polycationic peptides from diatom biosilica that direct silica nanosphere formation. *Science*, **286**, 1129–32.

Kröger, N., Deutzmann, R., Bergsdorf, C., & Sumper, M. (2000). Species-specific polyamines from diatoms control silica morphology. *Proceedings of the National Academy of Sciences of the USA*, **97**, 14,133–8.

Kröger, N., Lorenz, S., Brunner, E., & Sumper, M. (2002). Self-assembly of highly phosphorylated silaffins and their function in biosilica morphogenesis. *Science*, **298**, 584–6.

Labeyrie, L. D. (1974). New approach to surface seawater paleotemperatures using (18)O/(16)O ratios in silica of diatom frustules. *Nature*, **248**, 40–2.

Labeyrie, L. D. (1979). La composition isotopique de l'oxygene de la silice des valves de diatomees. Mise au point d'une nouvelle methode de palaeo-climatologie. Dissertation, Universitie de Paris XI.

Labeyrie, L. D. & Juillet, A. (1980). Isotopic exchange of the biogenic silica oxygen. *Comptes Rendus Hebdomadaires des Seances de L'Academie des Sciences Serie D*, **290**, 1185–8.

Labeyrie, L. D. & Juillet, A. (1982). Oxygen isotopic exchangeability of diatom valve silica; interpretation and consequences for palaeoclimatic studies. *Geochimica et Cosmochimica Acta*, **46**, 967–75.

Lamb, A. L., Brewer, T. S., Leng, M. J., Sloane, H. J., & Lamb, H. F. (2007). A geochemical method for removing the effect of tephra on lake diatom oxygen isotope records. *Journal of Paleolimnology*, **37**, 499–516.

Lamb, A. L., Leng, M. J., Sloane, H. J., & Telford, R. J. (2005). A comparison of the palaeoclimatic signals from diatom oxygen isotope ratios and carbonate oxygen isotope ratios from a low latitude crater lake. *Palaeogeography, Palaeoclimatology, Palaeoecology*, **223**, 290–302.

Laws, E. A., Bidigare, R. R., & Popp, B. N. (1997). Effect of growth rate and CO_2 concentration on carbon isotopic fractionation by the marine diatom *Phaeodactylum tricornutum*. *Limnology and Oceanography*, **42**, 1552–60.

Laws, E. A., Popp, B. N., Bidigare, R. R., Kennicutt, M. C., & Macko, S. A. (1995). Dependence of phytoplankton carbon isotopic composition on growth rate and $[CO_2]_{aq}$: theoretical considerations and experimental results. *Geochimica et Cosmochimica Acta*, **59**, 1131–8.

Leng, M. J. & Marshall J. D. (2004). Palaeoclimate interpretation of stable isotope data from lake sediment archives. *Quaternary Science Reviews*, **23**, 811–31.

Leng, M. J. & Barker, P. A. (2006). A review of the oxygen isotope composition of lacustrine diatom silica for palaeoclimate reconstruction. *Earth Science Reviews*, **75**, 5–27.

Leng, M. J., Barker, P. A., Greenwood, P., Roberts N., & Reed J. (2001). Oxygen isotope analysis of diatom silica and authigenic calcite from Lake Pinarbasi, Turkey. *Journal of Paleolimnology*, **25**, 343–9.

Leng, M. J., Metcalfe, S. E., & Davies, S. J. (2005). Investigating late Holocene climate variability in central Mexico using carbon isotope ratios in organic matter and oxygen isotope ratios from diatom silica within lacustrine sediments. *Journal of Paleolimnology*, **34**, 413–31.

Leng, M. J. & Sloane, H. J. (2008). Combined oxygen and silicon isotope analysis of biogenic silica. *Journal of Quaternary Science*, **23**, 313–19.

Lücke, A., Moschen, R., & Schleser, G.H. (2005). High temperature carbon reduction of silica: a novel approach for oxygen isotope analysis of biogenic opal. *Geochimica et Cosmochimica Acta*, **69**, 1423–33.

Matheney, R. K. & Knauth, L.P. (1989). Oxygen-isotope fractionation between marine biogenic silica and seawater. *Geochimica et Cosmochimica Acta*, **53**, 3207–14.

Matsumoto, K., Sarmiento, J. L., & Brzezinski, M. A. (2002). Silicic acid leakage from the Southern Ocean: a possible explanation for glacial atmospheric pCO_2. *Global Biogeochemical Cycles*, **16**, 1031. DOI: 10.1029/2001GB001442.

Milligan, A. J., Varela, D. E., Brzezinski, M. A., & Morel, F. M. M. (2004). Dynamics of silicon metabolism and silicon isotopic discrimination in a marine diatom as a function of pCO_2. *Limnology and Oceanography*, **49**, 322–9.

Mopper, K. & Garlick, G. D. (1971). Oxygen isotope fractionation between biogenic silica and ocean water. *Geochimica et Cosmochimica Acta*, **35**, 1185–7.

Morel, F. M. M., Cox, E. H., Kraepiel, A. M. L., *et al.* (2002) Acquisition of inorganic carbon by the marine diatom *Thalassiosira weissflogii*. *Functional Plant Biology*, **29**, 301–8.

Morley, D. W., Leng, M. J., Mackay, A. W., *et al.* (2004). Cleaning of lake sediment samples for diatom oxygen isotope analysis. *Journal of Paleolimnology*, **31**, 391–401.

Morley, D. W., Leng, M. J., Mackay, A. W., & Sloane, H. J. (2005). Late glacial and Holocene environmental change in the Lake Baikal region documented by oxygen isotopes from diatom silica. *Global and Planetary Change*, **46**, 221–33.

Moschen, R., Lücke, A., Parplies, J., Radtke, U., & Schleser, G. H. (2006). Transfer and early diagenesis of biogenic silica oxygen isotope signals during settling and sedimentation of diatoms in a temperate freshwater lake (Lake Holzmaar, Germany). *Geochimica et Cosmochimica Acta*, **70**, 4367–79.

Moschen, R., Lücke, A., & Schleser, G. (2005). Sensitivity of biogenic silica oxygen isotopes to changes in surface water temperature and palaeoclimatology. *Geophysical Research Letters*, **32**, L07708, DOI:10.1029/2004GL022167.

Needoba, J. A. & Harrison, P. J. (2004). Influence of low light and a light:dark cycle on NO_3^- uptake, intracellular NO_3^-, and nitrogen isotope fractionation by marine phytoplankton. *Journal of Phycology*, **40**, 505–16.

Needoba, J. A., Waser, N. A., Harrison, P. J., & Calvert, S. E. (2003). Nitrogen isotope fractionation in 12 species of marine phytoplankton during growth on nitrate. *Marine Ecology Progress Series*, **255**, 81–91.

Nelson, D. M., Tréguer, P., Brzezinski, M. A., Leynaert, A., & Quéguiner, B. (1995). Production and dissolution of biogenic silica in the ocean: revised global estimates, comparison with regional data and relationship to biogenic sedimentation. *Global Biogeochemical Cycles*, **9**, 359–72.

Popp, B. N., Laws E. A., Bidigare R. R., *et al.* (1998). Effect of phytoplankton cell geometry on carbon isotopic fractionation. *Geochimica et Cosmochimica Acta*, **62**, 69–77.

Popp, B. N., Trull, T., Kenig, F., *et al.* (1999). Controls on the carbon isotopic composition of Southern Ocean phytoplankton. *Global Biogeochemical Cycles*, **13**, 827–43.

Rau, G. H., Riebesell, U., & Wolf-Gladrow, D. (1997) CO_2aq-dependent photosynthetic ^{13}C fractionation in the ocean: a model versus measurements. *Global Biogeochemical Cycles*, **11**, 267–78.

Reynolds, B. C., Aggarwal, J., Andre, L., *et al.* (2007). An inter-laboratory calibration of Si isotope reference materials. *Journal of Analytical Atomic Spectrometry*, **22**, 561–8.

Reynolds, B. C., Frank, M., & Halliday, A. N. (2006a). Silicon isotope fractionation during nutrient utilization in the North Pacific. *Earth and Planetary Science Letters*, **244**, 431–43.

Reynolds, B. C., Georg, R. B., Oberli, F., Wiechert, U., & Halliday, A. N. (2006b). Re-assessment of silicon isotope reference materials using high-resolution multi-collector ICP-MS. *Journal of Analytical Atomic Spectrometry*, **21**, 266–9.

Riebesell, U., Burkhardt, S., Dauelsberg, A., & Kroon, B. (2000). Carbon isotope fractionation by a marine diatom: dependence on the growth-rate-limiting resource. *Marine Ecology Progress Series*, **193**, 295–303.

Rings, A., Lücke, A., & Schleser, G.H. (2004). A new method for the quantitative separation of diatom frustules from lake sediments. *Limnology and Oceanography Methods*, **2**, 25–34.

Robert, F. & Chaussidon, M. (2006). A palaeotemperature curve for the Precambrian oceans based in silicon isotopes in cherts. *Nature*, **443**, 969–72.

Roberts, K., Granum, E., Leegood R. C., & Raven, J. A. (2007). Carbon acquisition by diatoms. *Photosynthesis Research*, **93**, 79–88.

Robinson, R. S., Brunelle, B. G., & Sigman, D. M. (2004). Revisiting nutrient utilisation in the glacial Antarctic: evidence from a new method for diatom-bound N isotopic analysis. *Paleoceanography*, **19**, PA3001, DOI: 10.1029/2003PA000996.

Robinson, R. S. & Sigman, D. M. (2008). Nitrogen isotopic evidence for a poleward decrease in surface nitrate within the ice age Antarctic. *Quaternary Science Reviews*, **27**, 1076–90.

Robinson, R. S., Sigman, D. M., DiFiore, P. J., *et al.* (2005). Diatom-bound $^{15}N/^{14}N$: new support for enhanced nutrient consumption in the ice age subantarctic *Paleoceanography*, **20**, PA3003, DOI:10.1029/2004PA001114.

Rosenthal, Y., Dahan, M., & Shemesh, A. (2000). Southern Ocean contributions to glacial-interglacial changes of atmospheric pCO_2: an assessment of carbon isotope record in diatoms. *Paleoceanography*, **15**, 65–75.

Rosqvist, G., Jonsson, C., Yam, R., Karlén, W., & Shemesh, A. (2004). Diatom oxygen isotopes in pro-glacial lake sediments from northern Sweden: a 5000 year record of atmospheric circulation. *Quaternary Science Reviews*, **23**, 851–9.

Sancetta, C., Heusser L., Labeyrie L., Sathy Naidu A., & Robinson S. W. (1985). Wisconsin–Holocene paleoenvironment of the Bering Sea: evidence from diatoms, pollen, oxygen isotopes and clay minerals. *Marine Geology*, **62**, 55–68.

Schiff, C., Kaufman, D. S., Wolfe, A. P., Dodd, J., & Sharp, Z. (2009). Late Holocene storm-trajectory changes inferred from the oxygen isotope composition of lake diatoms, south Alaska. *Journal of Paleolimnology*, **41**, 189–208.

Schleser, G. H., Lücke, A., Moschen, R., & Rings, A. (2001). Separation of diatoms from sediment and oxygen isotope extraction from their siliceous valves: a new approach. *Terra Nostra*, **2001/3**, 187–91.

Schmidt, M., Botz, R., Rickert, D., Bohrmann, G., Hall, S. R., & Mann, S. (2001). Oxygen isotope of marine diatoms and relations to opal-A maturation. *Geochimica et Cosmochimica Acta*, **65**, 201–11.

Schmidt, M., Botz, R., Stoffers, P., Anders, T., & Bohrmann, G. (1997). Oxygen isotopes in marine diatoms: a comparative study of analytical techniques and new results on the isotopic composition of recent marine diatoms. *Geochimica et Cosmochimica Acta*, **61**, 2275–80.

Schneider-Mor, A., Yam, R., Bianchi, C., *et al.* (2005). Diatom stable isotopes, sea ice presence and sea surface temperature records of the past 640 ka in the Atlantic sector of the Southern Ocean. *Geophysical Research Letters*, **32**, L10704, DOI:10.1029/2005GL022543.

Schneider-Mor, A., Yam, R., Bianchi, C., *et al.* (2008). Nutrient regime at the siliceous belt of the Atlantic sector of the Southern Ocean during the past 660 ka. *Paleoceanography*, **23**, PA3217, DOI:10.1029/2007PA001466.

Shemesh, A., Burckle, L. H., & Hays, J. D. (1994). Meltwater input to the Southern Ocean during the last glacial maximum. *Science*, **266**, 1542–4.

Shemesh, A., Burckle, L. H., & Hays, J. D. (1995). Late Pleistocene oxygen isotope records of biogenic silica from the Atlantic sector of the Southern Ocean. *Paleoceanography*, **10**, 179–96.

Shemesh, A., Charles, C. D., & Fairbanks R. G. (1992). Oxygen isotopes in biogenic silica: global changes in ocean temperature and isotopic composition. *Science*, **256**, 1434–6.

Shemesh, A., Hodell, D., Crosta, C., Kanfoush, S., Charles, C. & Guilderson, T. (2002). Sequence of events during the last deglaciation in Southern Ocean sediments and Antarctic ice cores. *Paleoceanography*, **17**, 1056, DOI: 10.1029/2000PA000599.

Shemesh, A., Macko, S.A., Charles, C.D., & Rau, G.H. (1993). Isotopic evidence from reduced productivity in the glacial Southern Ocean. *Science*, **262**, 407–10.

Shemesh, A., Mortlock, R. A., Smith R. J., & Froelich, P. N. (1988). Determination of Ge/Si in marine siliceous microfossils: separation, cleaning and dissolution of diatoms and radiolaria. *Marine Chemistry*, **25**, 305–23

Sigman, D. M., Altabet, M. A., Francois, R., McCorkle, D. C., & Gaillard, J. F. (1999). The isotopic composition of diatom-bound nitrogen in the Southern Ocean sediments. *Paleoceanography*, **14**, 118–34.

Sigman, D. M. & Haug, G. H. (2003). The biological pump of the past. In *Treatise on Geochemistry, Volume 6*, ed. H. Elderfield, Amsterdam: Elsevier.

Singer, A.J. & Shemesh, A. (1995). Climatically linked carbon isotope variation during the past 430,000 years in Southern Ocean sediments. *Paleoceanography*, **10**, 171–7.

Spadaro, P. A. (1983). Silicon isotope fractionation by the marine diatom *Phaeodactylum tricornutum*. Unpublished M.Sc. thesis, University of Chicago.

Spero, H. J., Bijma, J., Lea, D. W., & Bemis, B. (1997). Effect of seawater carbonate chemistry on planktonic foraminiferal carbon and oxygen isotope values. *Nature*, **390**, 497–500.

Spero, H. J. & Lea, D. W. (1993). Intraspecific stable isotope variability in the planktonic foraminifera *Globigerinoides sacculifer*: results from laboratory experiments. *Marine Micropaleontolgy*, **22**, 221–34.

Spero, H. J. & Lea, D. W. (1996). Experimental determination of stable isotope variability in *Globigerina bulloides*: implications for paleoceanographic reconstructions. *Marine Micropaleontology*, **28**, 231–46.

Street-Perrott, F. A., Barker, P. A., Leng, M. J., *et al.* (2008). Towards an understanding of late Quaternary variations in the continental biogeochemical cycle of silicon: multi-isotope and sediment-flux data for Lake Rutundu, Mt Kenya, East Africa, since 38 ka BP. *Journal of Quaternary Science*, **23**, 375–87.

Swann, G. E. A. & Leng, M. J. (2009). A review of diatom $\delta^{18}O$ in palaeoceanography. *Quaternary Science Reviews*, **28**, 384–98.

Swann, G. E. A., Leng, M. J., Sloane, H. J., & Maslin, M. A. (2008). Isotope offsets in marine diatom $\delta^{18}O$ over the last 200 ka. *Journal of Quaternary Science*, **23**, 389–400.

Swann, G. E. A., Leng, M. J., Sloane, H. J., Maslin, M. A., & Onodera, J. (2007). Diatom oxygen isotopes: evidence of a species effect in the sediment record. *Geochemistry, Geophysics, Geosystems*, **8**, Q06012, DOI:10.1029/2006GC001535.

Swann, G. E. A., Maslin, M. A., Leng, M. J., Sloane, H. J., & Haug, G. H. (2006). Diatom $\delta^{18}O$ evidence for the development of the modern halocline system in the subarctic northwest Pacific at the onset of major northern hemisphere glaciation. *Paleoceanography*, **21**, PA1009, DOI: 10.1029/2005PA001147.

Takeda, S. (1998). Influence of iron availability on nutrient consumption ratio of diatoms in oceanic waters. *Nature*, **393**, 774–7.

Taylor, H. P. & Epstein, S. (1962). Relationships between $^{18}O/^{16}O$ ratios in coexisting minerals of igneous and metamorphic rocks, part I, Principles and experimental results. *Bulletin of the Geological Society of America*, **73**, 461–80.

Thorliefson, J.T. (1984). A modified stepwise fluorination procedure for the oxygen isotopic analysis of hydrous silica. Unpublished M.Sc. thesis, Arizona State University.

Tyler, J. J., Leng, M. J., & Sloane, H. J. (2007). The effects of organic removal treatment on the integrity of $\delta^{18}O$ measurements from biogenic silica. *Journal of Paleolimnology*, **37**, 491–7.

van Bennekom, A. J. & van der Gaast, S. J. (1976). Possible clay structures in frustules of living diatoms. *Geochimica et Cosmochimica Acta*, **40**, 1–6.

van den Boorn, S. H. J. M, Vroon, P. Z., van Belle, C. C., *et al.* (2006). Determination of silicon isotope ratios in silicate materials by high-resolution MC-ICP-MS using a sodium hydroxide sample digestion method. *Journal of Analytical Atomic Spectrometry*, **21**, 734–42.

Varela, D. E., Pride, C. J., & Brzezinski, M. A. (2004). Biological fractionation of silicon isotopes in Southern Ocean surface waters. *Global Biogeochemical Cycles*, **18**, GB1047, DOI: 10.1029/2003GB002140.

Wefer, G. & Berger, W. H. (1991). Isotope paleontology: growth and composition of extant calcareous species. *Marine Geology*, **100**, 207–48.

Wischmeyer, A. G., De La Rocha, C. L., Maier-Reimer, E., & Wolf-Gladrow, D. A. (2003). Control mechanisms for the oceanic distribution of silicon isotopes. *Global Biogeochemical Cycles*, **17**, 1083, DOI:10.1029/2002GB002022.

35 Diatoms and nanotechnology: early history and imagined future as seen through patents

RICHARD GORDON

35.1 History of diatom nanotechnology

> As to *form*, the Diatoms present an infinite variety of size and outline. Mathematical curves of the most exquisite perfection, combinations which the designer would grasp with eagerness on the planning of his models, surfaces adorned with the most unlimited profusion of style and ornamentation, are everywhere presented (Bailey, 1867).

> The origin of individual organization is one of those stubborn problems to which each generation of biologists has addressed itself anew (Hall, 1969).

Diatom nanotechnology started in 1863, when Max Schultze (Figure 35.1) noted that structures manufactured from silica vapor precipitating "in the form of minute spherules or lenticular particles" resembled diatoms (Schultze, 1863a, b) (Figure 35.2). He further showed "that neither in the artificial siliceous pellicles nor in the diatom valves are the peculiar forms due to a crystalline structure," i.e. both consisted of amorphous silica. Schultze's controlled assembly at microscopic levels pre-dated the "father" of nanotechnology (Feynman, 1960), by almost 100 years. Schultze also did pioneering work on diatom motility, including discovery of the raphe (Schultze, 1858a, b, 1865; Goodale, 1885), but like most other diatomists until recently (Gordon *et al.*, 2009), made his living in other fields, such as cytology, anatomy, histology, microscopy and vision, and is noted for clarifying the cell theory (Nordenskiöld, 1928; Hall, 1969; Werner *et al.*, 1987; Nyhart, 1995; Brewer, 2006).

The Diatoms: Applications for the Environmental and Earth Sciences, 2nd Edition, eds. John P. Smol and Eugene F. Stoermer. Published by Cambridge University Press.

Schultze's discovery of what may be the basic mechanism of diatom morphogenesis, amorphous precipitation, was prescient in another fundamental way: there was no accepted explanation for pattern formation during precipitation for another century (Mullins and Sekerka, 1963). In retrospect, Schultze had the basic idea that precipitation occurred "chiefly on the elevations," i.e. positive feedback (Milsum, 1968): protuberances that are a little closer to the source of the precipitating material grow faster, enhancing their size relative to their neighbors, because, when there is no convection, there is a concentration gradient that is low near the precipitate and higher further away (Gordon and Drum, 1994).

When computer simulations of precipitation were started (Gordon *et al.*, 1980, Witten and Sander, 1981) to see the nonlinear consequences beyond the linear mathematical analysis (Mullins and Sekerka, 1963), the patterns formed were called "diffusion limited aggregates," and the concept is now called DLA (diffusion limited aggregation), a field with over 2000 publications so far, including a few books (Stanley and Ostrosky, 1986; Vicsek, 1992; Kaandorp, 1994; Kassner, 1996; Flake, 1998; Meakin, 1998). A DLA simulation and a "matching" centric diatom are shown in Figure 35.3.

Schultze's diatom morphogenesis work was forgotten after 1893 (Slack, 1870, 1871, 1874; Morehouse, 1876; Mills and Deby, 1893), until I rediscovered it (Gordon and Drum, 1994) when the late Charlie Reimer (Gaul *et al.*, 1993; Kociolek and Sullivan, 1995; Potapova, 2008) kindly allowed me to browse through his card catalog of diatom literature overnight at the Philadelphia Academy of Natural Sciences.

In 1988, I was invited by a former postdoctoral student, Rangaraj M. Rangayyan, to give my first ever talk at an engineering conference, and, almost at a loss as to what might interest

Figure 35.1 Max Schultze (1825–1874), father of diatom nanotechnology, from (Brewer, 2006), reproduced by courtesy of Heinz Schott of the Medizinhistorische Institut Bonn.

engineers, came up with the idea that diatoms could be used for "microfabrication" (Gordon and Aguda, 1988). This obscure paper, restarting the field of diatom nanotechnology after over a century, may not be what inspired the many others who later independently entered it. Evelyn E. Gaiser, organizer of the 2003 17th North American Diatom Symposium (NADS), and I decided to arrange the industrial talks from 13 international groups for the NADS Workshop on Diatom Nanotechnology to alternate with biological talks, forcing everyone to stay and listen to one another. This led to great interdisciplinary discussion and collaboration (Gordon *et al.*, 2005b). As scientists are often accused of replacing the truth about events by a neat succession of discovery, I think it is important to record some of what actually transpired.

No survey has been made of the motivations of researchers who entered diatom nanotechnology. For some engineers it is perhaps part of the long history of seeking inspiration from biological structure (Blossfeldt, 1929; Smith, 1981; Bejan, 2000; Jayas, 2002; Goodsell, 2004), including diatoms (Bailey, 1867; Bach and Burkhardt, 1984; Sterrenburg, 2005; Gebeshuber *et al.*, 2009a). For diatomists, many may have been attracted by the surprise that our arcane and underfunded field has practical applications, and that diatomists have essential roles to play in diatom nanotechnology (Gordon and Parkinson, 2005). For both groups there has been a steep learning curve. Summaries of essential diatom background have been written for engineers (Gordon *et al.*, 2005a; Allison *et al.*, 2008).

But the converse has not been done for diatomists, perhaps precisely because the growth of nanotechnology structures (Bäuerlein, 2000; Golovin and Nepomnyashchy, 2006) has not been central to nanotechnology in general, and diatom nanotechnology is leading the whole field of nanotechnology off in a novel direction. So while the history (Feynman, 1960), goals, and methods of classical nanotechnology (Drexler and Peterson, 1991; Brodie and Murray, 1992; Drexler, 1992; Krummenacker and Lewis, 1995; Freitas, 1999; Scherge and Gorb, 2001; Mulhall, 2002; Lu and Zhao, 2004; Ozin, 2005; Sargent, 2005; Liu and Shimohara, 2006; Masayoshi *et al.*, 2006; Rehm, 2006; Renugopalakrishnan and Lewis, 2006; Saliterman, 2006; Visnovsky, 2006; Dupas *et al.*, 2007; Gazit, 2007; Papazoglou and Parthasarathy, 2007; Jotterand, 2008; Reisner, 2008) may provide the diatomist some understanding of engineers, it may prove as relevant as the study of railroad engineering to space travel.

35.2 Diatom taxonomy versus diatom nanotechnology

Exactly what is it that the genome controls that leads to diatom morphogenesis? This is the key question in diatom nanotechnology, because manufacturing is all about control, to achieve the device one wants, and quality control to make sure it works reliably:

> Once the relevant biochemistry and genetic code is understood for the shapes of diatom frustules and, in particular, how such a code may be altered to produce desired (tailored) frustule shapes and features, then shape-tailored diatom frustule templates could be produced at low cost, in large quantities, and in very reproducible shapes and very fine geometries (Sandhage, 2007).

If growing technology, via growing diatoms, is to compete with the classical lithographic approach of MEMS (microelectromechanical systems) or BioMEMS (biological applications of MEMS and/or MEMS incorporating biological components; Saliterman, 2006), it will have to compete in the arena of quality, not just the quantity promised by exponential culturing (Sandhage *et al.*, 2005; Gordon *et al.*, 2009). Here the unknown relationship between genotype and phenotype, between the genome and morphogenesis, may be critical to understand.

Pattern formation during precipitation is a non-equilibrium phenomenon (Nicolis, 1977) that depends on an unstirred solution, so that movement of the precipitating entity (which could

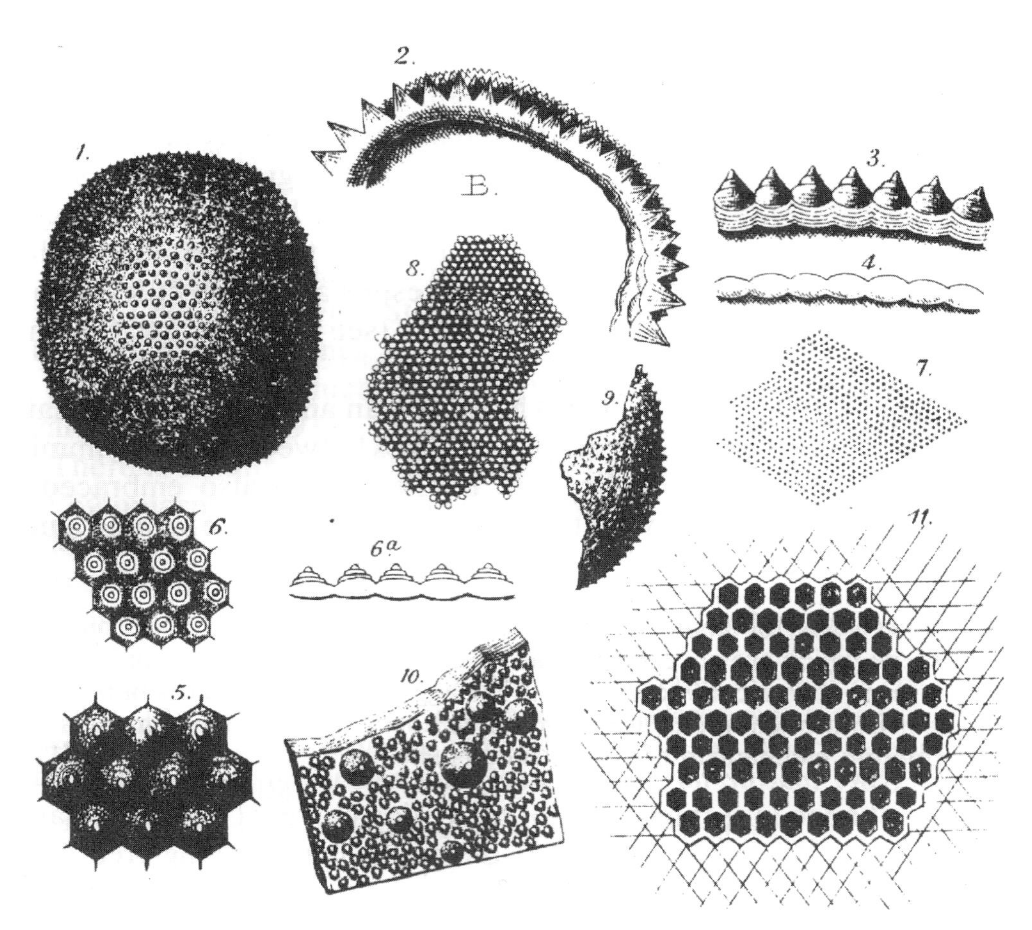

Figure 35.2 Sketches with a microscope of 300× of "artificial diatoms" made of silica precipitated by Max Schultze, except (11), which is *Pleurosigma angulatum* (Schultze, 1863a, b). These were taken at 300×, which meant then the magnification of the microscope, not the magnification of the sketch (Stephen Greenberg, Rare Book Librarian, US National Library of Medicine, personal communication), so that scale bars cannot be provided. The spacing between pores in (11) is 0.4 to 0.6 μm (Patrick and Reimer, 1966).

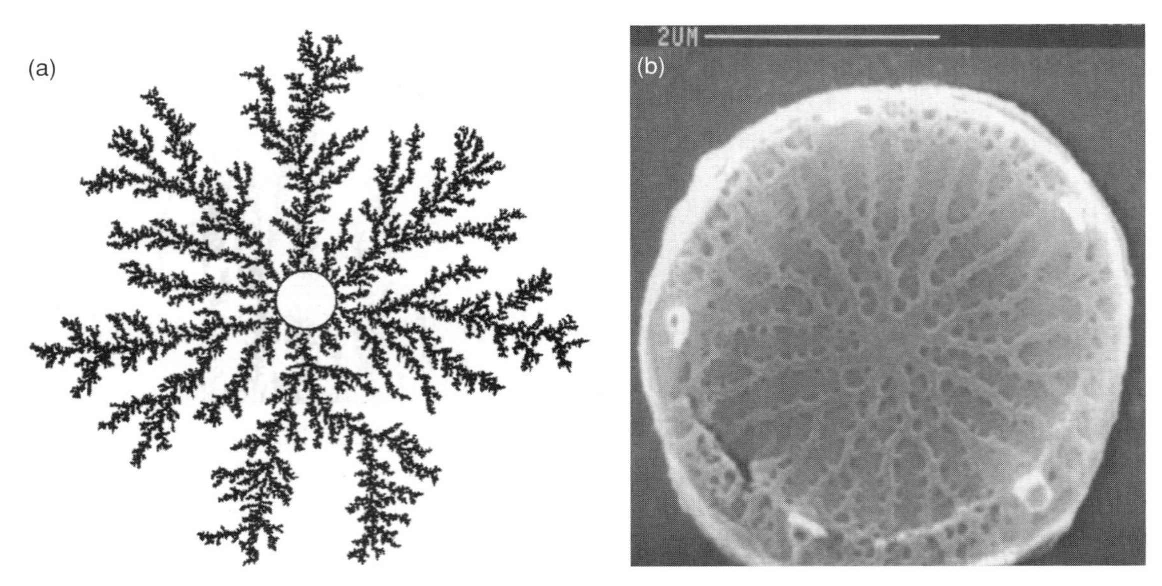

Figure 35.3 (a) A DLA (diffusion limited aggregation) computer simulation of a centric diatom and (b) a centric diatom chosen to match the simulation (courtesy of Hedy J. Kling, bar = 2 μm). (From Gordon and Drum (1994) with permission.)

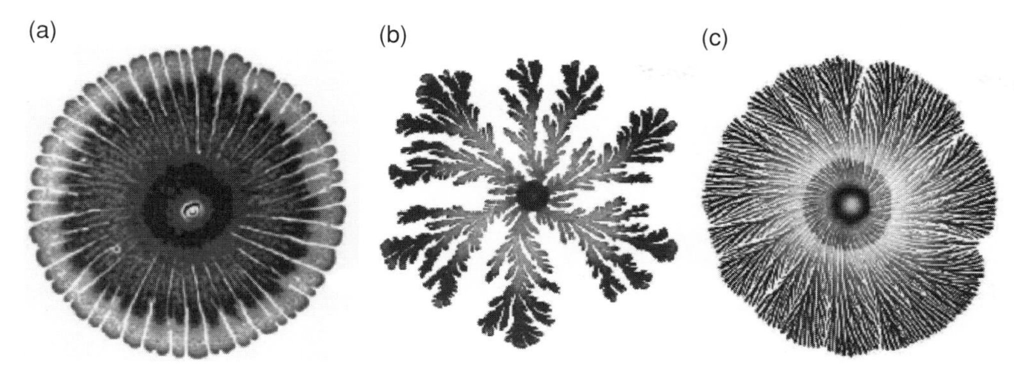

Figure 35.4 Colonies of 5-cm diameter of the bacterium *Paenibacillus dendritiformis*, grown under different nutrient conditions in a thin layer of medium in a Petri dish, roughly resemble centric diatom valves. Note that there is no prepattern in the Petri dish. (a) The pattern at higher food levels when attractive chemotactic signaling is activated. (b) The typical pattern when food chemotaxis dominates the growth at intermediate levels of food depletion. (c) The growth for a very low level of food when repulsive chemotactic signaling is intensified. (From Ben-Jacob (2008) with kind permission of the author and the European Physical Journal (EPJ).)

vary from a silica molecule $Si(OH)_4$ to a 50-nm colloidal silica particle (Iler, 1979)) occurs by diffusion only, at least near the precipitate. Depletion of solute by precipitation then sets up a gradient of the solute in the solvent (Gordon and Drum, 1994), and the rate of diffusion of solute across this gradient, in which the concentration of solute is lower than in bulk solution, limits how fast the precipitate can grow. Disruption of this diffusive boundary layer (DBL) (Rosenhead, 1963; Jorgensen and Des Marais, 1990; Schlichting and Gersten, 2000) may be what is preventing the formation of higher-order structure by current nanotechnology methods (Hildebrand, 2008): the silicalemma, i.e. the flat membrane bag within which a diatom valve is formed, protects the DBL.

Computers had insufficient memory and speed a few decades ago for much simulation of three-dimensional (3D) DLA, so 3D was rarely considered (Sander *et al.*, 1983; Goold *et al.*, 2005; Bourke, 2006). As it happens, a nearly two-dimensional (2D) medium seems critical both to pattern formation by DLA and by related mechanisms in bacteria colonies (Budrene and Berg, 1991, Ohgiwari *et al.*, 1992, Shapiro and Dworkin, 1997, Cohen *et al.*, 2001), whose patterns resemble some diatoms (Figure 35.4). We can take this as the primary reason that the silicalemma is an extremely flat membrane bag (Gordon and Drum, 1994), at least during the first, rapid phase of valve development (Gordon *et al.*, 2009), though the conclusion that 2D is essential still needs mathematical justification.

Bacterial colonies immediately provide the important lesson that, just because a proposed mechanism matches some diatom's pattern, it does not mean that the diatom made it that way. "Ultimately, the proof of a theory of morphogenesis will be its step by step quantitative matching to a computer simulation and/or real-world synthesis, rather than the current approach of shopping for diatoms whose mature valves roughly match a given simulation (Gordon and Drum, 1994, Lenoci and Camp, 2008)" (Gordon *et al.*, 2009), as I did for Figure 35.3. Another important lesson from both bacteria and Schultze's vapor-precipitated silica in making diatom-like patterns is that neither requires previously existing structure, in contrast to the protein templating hypothesis (Hecky *et al.*, 1973; Robinson and Sullivan, 1987; Gautier *et al.*, 2008; Hildebrand, 2008; Brunner *et al.*, 2009; Ehrlich, 2010), which assumes that a prepattern of protein determines where silica is precipitated. There is no prepattern that the bacteria follow that is in any way guided them into position on the agar in the Petri dish.

The presumption that patterns must come from pre-existing patterns is an ancient one, but we are surrounded by phenomena that exhibit the contrary, whether it be ripples in sand, icicles, or the beading of dew drops. These are generally understood as amplifications of perturbations, symmetry-breaking instabilities, or emergent phenomena (Rayleigh, 1892; Locke, 1968; Antonelli, 1985; Crutchfield, 1994; Holland, 1998; Jensen, 1998; Morowitz, 2002; Davies, 2004a; Crommelinck *et al.*, 2006; Turner, 2007). The philosophical conundrum suggesting prepatterns is an ancient one:

> Parmenides (480 BCE) . . . adopted the radical position that change – all change – is a logical impossibility . . . the impossibility of getting something from nothing (Lindberg, 2007).

While much of science is about explaining one level of phenomena on the basis of a "lower" level (Gordon and Chen, 1987), popularized under the term "emergence" (His, 1874;

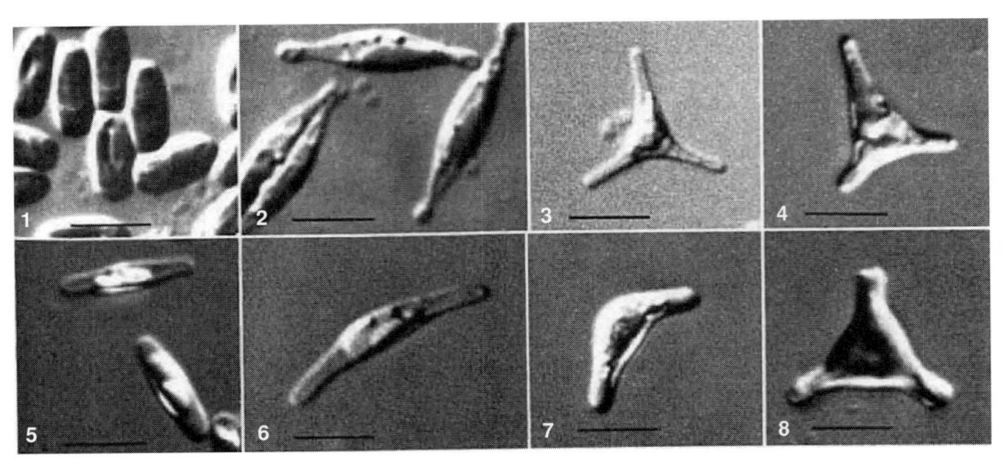

Figure 35.5 Morphotypes of *Phaeodactylum tricornutum* in culture. (1) Oval. (2) Fusiform. (3) "Long-armed" triradiate. (4) Triradiate with unequal arms. (5) "Long" oval. (6) "Bent" (lunate) fusiform. (7) "Boomerang" fusiform. (8) "Short-armed" triradiate. Nomarski optics. Scale bars 10 μm. (From Borowitzka and Volcani (1978) with kind permission of Wiley-Blackwell.)

Locke, 1968; Antonelli, 1985; Yates *et al.*, 1988; Holland, 1998, Jensen, 1998; Johnson, 2001; Morowitz, 2002; Turner, 2007), our understanding of such explanations is rather incomplete (Gordon, 2000; Batterman, 2001; Davies, 2004b; Crommelinck *et al.*, 2006; Bedau and Humphreys, 2008), especially when we consider the need to incorporate the phenomena of non-locality (Aczel, 2002) and consciousness (Tuszynski, 2006). For diatoms, there is now little doubt that proteins catalyze the precipitation of silica (Hecky *et al.*, 1973; Hildebrand, 2008). However, expanding this to the idea that the whole multiscale (Krzhizhanovskaya and Sun, 2007) pattern of a valve is pre-formed by proteins is still an unwarranted projection (Gordon and Drum, 1994; Gordon *et al.*, 2009). Nevertheless, these multiscale steps can all be skipped "by design" (Seckbach and Gordon, 2008) by simply copying diatom valve structure and robotically writing it out in "polyamine-rich ink" onto which silica is precipitated (Xu *et al.*, 2006), short-circuiting both philosophy and explanation.

Some researchers advocate genetic determinism (Keller, 1994), in which every phenotypic detail is specified by the action of genes, and those genes are selected via natural selection over the course of evolution. However, the bacterial colonies in Figure 35.4 are all made by the very same strain of bacteria. Therefore, their pattern is not determined solely by their genetics, and the environmental effects are so great that a casual glance by a diatomist as if they were diatoms might classify them as three different species or genera. In other words, when one examines these 2D growth patterns, one cannot assume a one-to-one relationship between genotype and phenotype. Is this "merely" developmental plasticity (West-Eberhard, 2002, Hall *et al.*, 2004, Jablonka and Lamb, 2005), or does it go beyond reasonable limits to our expectation of a correlation between genotype and phenotype, thereby questioning the very basis of diatom taxonomy (Van de Vijver and Bogaerts, 2010)? The influence of the environment could be so great that our use of diatom valve shapes, patterns, spines, etc. for classification needs to be proven to be robust. One way to test this is to determine if we can deliberately turn one kind of diatom into another using a "compustat," which is a visual analog of the biochemically based chemostat; i.e. selection is based on shape, pattern, and/or color criteria (Gordon, 1996). Diatoms thus become a test bed for these very fundamental biological concepts and presumptions.

Absolute genetic determinism is not the case for diatoms. For example, pennate diatoms are often characterized by a supposedly fixed distance between costae (or number of costae per 10 μm) (Foged, 1986; Van de Vijver and Lange-Bertalot, 2008; Thalera and Kaczmarska, 2009), but individuals in a culture vary in size, so the total number of costae varies from one cell to another. One could guess that the genetics determines the spacing rather than number of costae, but spacing is not exactly constant either (Tropper, 1975; Steinman and Sheath, 1984). For centric diatoms, the circumferential density of repeating structures is an unreliable statistical parameter compared to a simple count (Theriot, 1988). If these seem trivial examples, then consider heterovalvy (Round *et al.*, 1990; Cox, 2006), in which the two valves of a diatom could be mistaken for different species or even separate orders of diatoms (Kociolek and Spaulding, 2002), if we did not see the two valves stuck together. Furthermore, substantial differences between cloned cells have been observed since the nineteenth century (Edwards, 1875), and it is accepted that one of our new "standard" DNA sequenced diatoms (Bowler *et al.*, 2008), *Phaeodactylum tricornutum*, has many alternative "morphotypes" (Borowitzka and Volcani, 1978; Stanley and Callow, 2007) (Figure 35.5). Here is a vividly described event of the initiation of polymorphism via heterovalvy:

> Total conversion from coarsely structured morphologic form to a more finely structured form [of *Mastogloia*] takes place during a single division; hence no intergradations between the 2 forms are evident. Intact frustules with 1 valve of each type leave no reasonable doubt of the ability to make the change (Stoermer, 1967).

The fact that the spacing of the striae themselves change between these *Mastogloia* morphotypes (and for *Cocconeis* (Mizuno, 1982)) makes it unlikely that striae/costal spacing is under direct genetic control in general. Certainly one would expect that the same genome and perhaps the same expressed genes apply to both valves in a diatom with heterovalvy. This is presumed to be the case in other single-celled organisms: "In stentors or in other unicellular forms it is obvious that differences between different regions of the cortex or cytoplasm develop without corresponding nuclear differentiation simply because all parts of the cell draw upon the same nucleus" (Tartar, 1962). Nevertheless, gene products can be targeted in specific directions within a cell (Woods and Bryant, 1993), so perhaps the assumption of spatially isotropic gene expression requires experimental testing in diatoms. Perhaps there is some relationship between heterovalvy and asymmetric cell division (Gordon, 1999; Zhong, 2008).

In trying to understand what produces the phenotype of a diatom, it is important to recognize the contributions that proper mathematical analyses can make, and their implications for experimental probes. For example, the phenomenon of viscous fingering can be shown mathematically to be the "mean field" approximation to DLA (Brener *et al.*, 1991). Viscous fingering is expected in 2D models for diatom morphogenesis that invoke sintering (Gordon and Drum, 1994; Parkinson *et al.*, 1999) or immiscible fluids (Sumper and Brunner, 2006; Lenoci and Camp, 2008; Gordon *et al.*, 2009). If those fluids flow within the silicalemma as it enlarges, the flow itself could alter the pattern of valve thickening (Merks *et al.*, 2003). If the initially precipitated structure is not mechanically rigid (Hildebrand, 2008), compaction may occur (Mendoza and Ramìrez-Santiago, 2005) or buckling of the whole 2D precipitate (Tiffany *et al.*, 2009). If the colloidal silica particles that precipitate (Gordon and Drum, 1994), perhaps because of their embedded zwitterionic silaffins (Sumper and Brunner, 2006), have an electric polarity and an orientation relative to the inner surface of the silicalemma, then phase transitions amongst them are possible (Xu *et al.*, 2008). The number of main costal branches appears to plateau in 2D (Mandelbrot *et al.*, 2002), but not in 3D (Schwarzer *et al.*, 1996), perhaps giving an explanation for the uniform spacing of costae that would be an alternative to sintering (Gordon and Drum, 1994). Variations in pattern with size of the precipitating particle (Tan *et al.*, 1999) or with imposed electric, magnetic (Mansur *et al.*, 2005), or polarized light fields, may help unravel the role of the precipitating entity in the silica deposition vesicles (SDVs: silicalemma plus contents). The effects of light (Brouhard *et al.*, 2003) or magnetic (de Vries *et al.*, 2005) tweezers during valve morphogenesis might be enlightening. If the precipitating particles have long-range interactions, this property would also give altered patterns (Xu *et al.*, 2005). Ostwald ripening (Sugimoto, 1978a, b), the growth of larger "crystals", bumps, or dendritic branches at the expense of smaller ones, may also be involved.

One environmental factor that is generally not considered for microorganisms is the role of gravity, considered to be of potential importance even in the generation of bacterial patterns (Dombrowski *et al.*, 2004). For diatoms, the role of microtubules, whose diffusion-limited polymerization from tubulin depends on gravitationally driven drift through the cytoplasm (Portet *et al.*, 2003), may make their patterning sensitive to microgravity in space (Gordon *et al.*, 2007). Likewise, the immiscible fluid models for diatom morphogenesis (Sumper and Lehmann, 2006; Lenoci and Camp, 2008; Gordon *et al.*, 2009) may prove gravitationally dependent. For starters, we should be testing this on earth first. No one has deliberately grown diatoms yet at different orientations with respect to gravity, or in a centrifuge (Beams and Kessel, 1987) or clinostat (Häder *et al.*, 2005). In general, direct experimental intervention and probing of any sort would seem well worth pursuing. For example, natural salt crystals interfere with valve patterning, giving clues to its mechanism (Gordon and Brodland, 1990).

35.3 The attainable perfection of diatoms for industrial quality control

Quality control is essential to modern industry, including nanotechnology (Gebeshuber *et al.*, 2009b). Diatoms have long been admired for their apparent perfection, for instance as

Figure 35.6 Scanning electron micrograph (SEM) of a nearly perfect circular girdle band (a valvocopula) from the centric diatom *Arachnoidiscus ehrenbergii*. The diameter is about 175 μm (Xu *et al.*, 2006). (From Mary Ann Tiffany, with permission.)

a substitute for (and pre-dating) machined diffraction gratings to test the resolution of microscopes (Redfern, 1853); but clearly, as shown above, much variation exists. For this reason, we have proposed the use of a compustat (Gordon, 1996), FACS (fluorescence-activated cell sorting) (Drum and Gordon, 2003), or field-flow fractionation (Neethirajan *et al.*, 2009) to select for or sort out imperfect from near perfect diatoms or parts. Nevertheless, to the eye some diatoms appear to exhibit remarkably precise patterns and shapes, such as the circular girdle band in Figure 35.6 and the almost perfect symmetries in the Diatomea plate of Haeckel (1904), an observation that may be subjected to quantification. For example, individual centric diatoms can exhibit a high degree of rotational symmetry (Sterrenburg *et al.*, 2007) (Figure 35.7). Raphid pennate diatoms, despite the inherent bilateral asymmetry of their raphes (Drack *et al.*, 2011), can exhibit a high degree of bilateral symmetry. Pores, especially in hexagonal arrays, nearly achieve the regularity of wallpaper patterns (Sterrenburg *et al.*, 2007). Whether the degree of variation or "fluctuating asymmetry" (Møller and Swaddle, 1997) in diatoms is itself inheritable is unknown, let alone the mechanisms by which such near perfection is achieved, nor why cultured diatoms accumulate morphological aberrations (Estes and Dute, 1994), although genomic instability has been implicated (von Dassow *et al.*, 2008). But the first step needs to be a research program of careful quantification (Palmer and Strobeck, 1986) of the precision with which diatoms replicate their structure, at all levels.

35.4 Bionanotechnology and biomimetics of diatoms

We have become accustomed to a rapidly changing technology in which computers, cameras, telephones, and other instruments become smaller, cheaper, and faster. The primary limitation on speed is the speed of light, and the smaller and closer electronic components are, the faster they can work together. Large, regular arrays, such as the pixels in digital cameras, also permit massive parallel data collection (as in a digital photograph), and soon massive parallel computation should be accessible. Many technologies compete in this race, the latest contenders coming from what has been called "bionanotechnology" (Goodsell, 2004; Tan *et al.*, 2004; Guo, 2005; Liu and Shimohara, 2006; Masayoshi *et al.*, 2006; Rehm, 2006; Renugopalakrishnan and Lewis, 2006; Gazit, 2007; Papazoglou and Parthasarathy, 2007; Jotterand, 2008; Reisner, 2008) or "biomimetics" (Dickerson *et al.*, 2008; Tamerler and Sarikaya, 2008; Gebeshuber *et al.*, 2009a). The distinction between these two fields is that in bionanotechnology we use biological organisms or their components directly, whereas in biomimetics we try to emulate what we think organisms are doing, or improve upon them. Both approaches are common in diatom nanotechnology.

It is hard to measure progress towards synthesis of diatom structure, because without knowing how diatoms build themselves, as we have seen above, even replicating the result does not mean that we have replicated the process. For example, honeycomb structures have been attained on spheres (10% yield) of crystalline lamellar aluminophosphates (not amorphous silica) precipitated from an organic solvent (not aqueous, as in diatoms) in two steps followed by three days of autoclaving (in contrast to ambient temperature) (Oliver *et al.*, 1995). While the spheres were "patterned on the submicrometre-to-millimetre scales found in the living world," with selected areas showing pockmarked or hexagonal detail, the claim of "striking similarity" to a chosen diatom is just that, and in the eye of the beholder (Figure 35.8). The situation is analogous to the history of attempts to create a living cell from precursors (Hanczyc, 2008). Some frankly admit " . . . structures resulting from *in vitro* polymerization experiments do not completely mimic structures that are formed *in vivo*, suggesting limitations in our understanding. . . . formation of regularly ordered structures is lacking" (Hildebrand, 2008). As Schultze obtained regular structures, a protected DBL, DLA, and sintering, which function at the mesoscale, may be what is lacking, or perhaps larger-scale structures, such as an MTOC (microtubule organizing center) at the surface of the silicalemma may bias

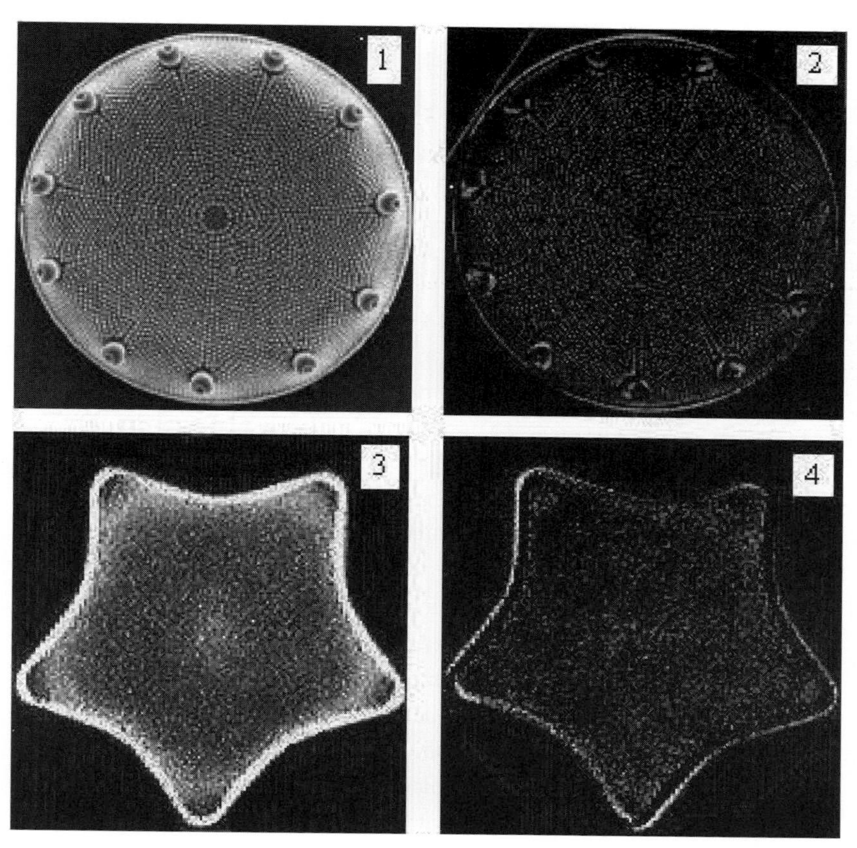

Figure 35.7 A measure of perfection of rotational symmetry of centric diatoms. Frames 1 and 2 show the result of subtracting an SEM image of *Aulacodiscus oregonus* with 11-fold symmetry from its own image rotated by $360/11 \approx 32.7°$. Frame 2 would be all black if symmetry and centering were perfect. Where a near-perfect match obtains, the image is nearly black. Frames 3 and 4 show the same for a light microscope image of *Triceratium formosum* var. *quinquelobata* with five-fold symmetry and rotation by $360/5 = 72°$. (From Sterrenburg *et al.* (2007) with kind permission of Springer Science and Business Media.)

where silica gets deposited within the SDV (Parkinson and Gordon, 1999); cf. Tesson and Hildebrand (2010).

In all of these comparisons of diatoms with analogous processes or computer simulations, to date we are essentially in the position of "The blind men and the elephant" (Saxe, 1873), each working on one seemingly unrelated part of the problem. "These natural assembly processes . . . are usually so complex that at this point in our understanding of biology, we are still unable to reverse engineer them [cf. (Gordon and Melvin, 2003)], in order to deconstruct the various steps in their growth process" (Forbes, 2004). This may require detailed observations of local chemical physical processes by "direct local probing," as has been recognized is needed for DLA patterns in electrodeposition (Léger *et al.*, 1999). The largest known diatom *Ethmodiscus rex* Wiseman & Hendey (Swift, 1973; Villareal *et al.*, 1999; Gordon *et al.*, 2010) perhaps should be considered as the next model diatom for DNA sequencing (Karthick, 2009), with its huge and accessible 2-mm-diameter SDV, to make such research practical.

Five approaches to observing or inferring the time course of valve morphogenesis have been used: electron microscopy of nascent (Borowitzka and Volcani, 1978), or aberrant (Gordon and Drum, 1994) valves, video-microscopy (Pickett-Heaps *et al.*, 1979a, b), fluorescence microscopy (Tesson and Hildebrand, 2010), and by reducing silica availability. The last approach, for instance, shows that the hexagonal structure of some valves may be generated by use of a branching pattern (Fryxell and Hasle, 1977). With adequate silica available, all evidence of the original branching pattern (cf. Figure 35.3) may be obliterated. This case shows the importance of studying the whole process, and not inferring it just from the final result (Gordon *et al.*, 2009). From a general conceptual point of view, considering the possible role of convection inside the silicalemma discussed above, diatom morphogenesis may combine all three aspects of flow, shape, and branching (Ball, 2009a, b, c).

With the understanding that each biomimetic attempt emphasizes a different aspect of diatom structure, I now consider the aqueous assembly of silica colloidal particles of the size found in diatoms (Gordon and Drum, 1994) into a

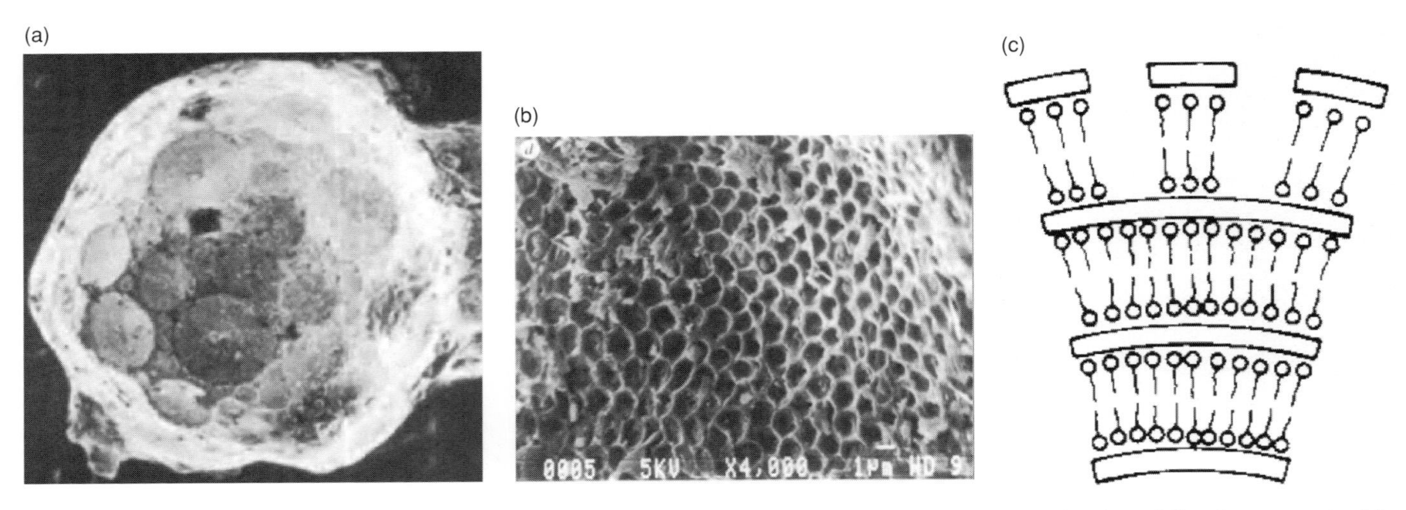

Figure 35.8 (a) Biomimetic crystalline lamellar aluminophosphate structure taken as analogous to a diatom, diameter 1 mm, and (b) a close-up of a hexagonal pattern from another such sphere, with "pores" of 1 μm diameter. The diagram in (c) shows the hypothesized relationship of the inorganic component to the bilayer membranes. (From Oliver *et al.* (1995), with permission.)

"confined" space analogous to the SDV, using multilayered "onion nanovesicles" and a week of "aging" (El Rassy *et al.*, 2005); see also Pinnavala *et al.* (2006). Here the result is a monolayer of colloidal silica particles between each phospholipid bilayer of the onion structure (Figure 35.9). There is no long-range structure, and the colloidal silica particles do not sinter into a solid structure.

In current diatom nanotechnology literature, the distinction between the rapid, 2D growth of a thin silica structure inside a flat silicalemma, and the subsequent slow, 3D thickening (Reimann, 1960; Gordon and Drum, 1994; Hazelaar *et al.*, 2005; Crawford *et al.*, 2009; Gordon *et al.*, 2009; Hildebrand *et al.*, 2009) is generally not considered. A first step towards biomimetic combining of both processes might be the "fractal-mound" of pentacene thin films (Zorba *et al.*, 2006), which could be taken as biomimetic of diatoms.

35.5 Manufacturing goals of diatom nanotechnology

As of the time of writing this chapter (to mid 2009), there were no commercial devices sold that incorporate diatom nanotechnology. Therefore, to assess what might be on the near horizon, I summarize some of the information available from the patent literature, which at least identifies programs that individuals have thus far decided to invest money in. Where there are multiple similar patents or applications, I have just cited one that is representative. Related academic publications are cited in brackets:

1. Silica-binding enzymes from diatoms are "painted" onto electronic chips, in order ". . . to effect the polymerization of silicon dioxide along the interface region of the matrix to form a matrix silicate mesostructure. This method may be used in the semiconductor industry for the preparation of silicate chips that have a variety of end-use applications" (Bryan *et al.*, 2003). [Tomczak *et al.*, 2007]
2. Diatoms provide a porous substrate for batteries (Clough, 2001).
3. Mesoporous materials are prepared "using amphiphilic block copolymers" in a biomimetic method that is presumed to throw light on how diatoms accomplish this, and are used as molecular sieves for reactions and separations (Stucky *et al.*, 2007). Metal oxide/hydroxide-coated diatoms create mesoporous materials that would otherwise have little porosity, for use as catalysts, supercapacitors, electrodes, fuel cells, and water treatment (Harbour and Hartley, 2007) with the advantage that "the actual solid portion of the substrate occupies only from about 10 to 30% of the apparent volume leaving a highly porous material for access to liquid" (Clough, 2005). Metal coatings provide electrically or thermally conducting powders (Clough, 2006). In general, many solid substances may be precipitated onto diatoms to produce 3D structures at ambient temperatures (Dickerson *et al.*, 2007). [This is similar to coating of diatoms with photoluminescent nanoparticles (Cai *et al.*, 2007; Gutu *et al.*, 2009), plastic replicas of diatoms after dissolving the silica in HF (Sandhage *et al.*, 2005), and abiotic silica deposition onto live hot-springs diatoms (Hobbs *et al.*, 2009).]

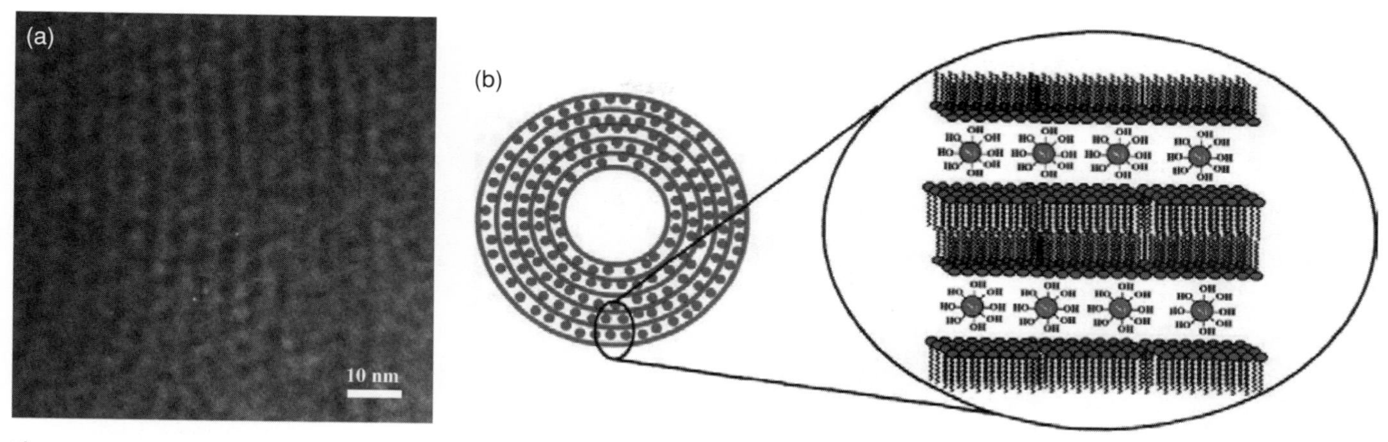

Figure 35.9 (a) Colloidal silica seen in a transmission electron micrograph of an onion nanovesicle, and (b) its interpretation. (From El Rassy *et al.* (2005), with permission.)

4. Silicatein from diatoms is used to catalyze silica precipitation in dental materials (Goldberg *et al.*, 2007) and other patterned materials (Morse *et al.*, 2003; Müller *et al.*, 2007).
5. Silaffin from diatoms is used to immobilize enzymes or even whole cells in precipitating silica (Naik *et al.*, 2005). [Naik *et al.*, 2002, Luckarift *et al.*, 2004]
6. Diatoms coat magnet particles for macromolecular separations (Kusumoto *et al.*, 2006).
7. Use of diatoms as "cages" for delivery of biologically active molecules to tissues (Jung *et al.*, 2007). [Gordon *et al.*, 2009]
8. Diatoms are embedded in the surface of a friction material for brake clutches (Lam *et al.*, 2007). (This is the field of biotribology (Gebeshuber *et al.*, 2005, 2007, 2008).)
9. Holograms are made by two-photon-induced photopolymerization incorporating silaffin. The hologram is hardened by precipitation of silica nanospheres, yielding "nearly fifty-fold in diffraction efficiency over a comparable polymer hologram without silica" (Brott *et al.*, 2007). [(Brott *et al.*, 2001); direct ink writing using polyamine-rich ink is an alternative approach for any structure, including holograms and diatom-like shapes (Xu *et al.*, 2006); see also Deravi *et al.* (2008).]
10. Use of diatoms as prototypes for lightweight structures (Hamm-Dubischar, 2007). [Bailey, 1867; Blossfeldt, 1929; Hamm *et al.*, 2003; Hamm, 2005; Wee *et al.*, 2005]
11. Atom-for-atom replacement of the silica of diatoms by other materials, generally at high temperature by gas/solid displacement reactions, resulting in diatom-shaped ceramics, metal alloys, or Reduction to elemental silicon for possible 3D computer production (Sandhage, 2007, Sandhage and Bao, 2008). [Sandhage *et al.*, 2002; Drum and Gordon, 2003; Bao *et al.*, 2007; Gordon *et al.*, 2009]
12. The optical properties of diatoms include reflectance of infrared light for heat-reflecting cosmetics (Grollier *et al.*, 1991). (This property might help keep cool diatom solar panels for sustainable gasoline (Ramachandra *et al.*, 2009).)
13. A diatom biofilm is "treated with a hydrophobic film-forming agent to form a structured self-cleaning surface" (Axtell III *et al.*, 2006).

While some inventions seem to use only the porosity of diatoms, the fact that this occurs on many size scales (mesoporosity) is what makes diatoms particularly useful in some applications. The fact that we can select diatom species for a given spectrum of mesoporosity, or perhaps even genetically manipulate this characteristic, provides even greater choice than the common use of diatomaceous earth (mentioned in numerous patents not cited here), such as for filter material (Bregar, 1955).

Patent claims are deliberately cast wide, but they can thereby foreshadow future applications of diatom nanotechnology. Here is a micromachining (Parkinson and Gordon, 1999) list from just one patent application (Sandhage, 2007), all prefixed with "micro-" often with "hollow" or "solid":

> Actuator, antenna, bar, bearing, cantilever, capsule, catalyst, cone, cube, cylinder, die, diffraction grating, disk, fiber, filler, filter, funnel, gear, heat exchanger, hinge (cf. (Gebeshuber *et al.*, 2009b)), honeycomb, insulator, lens, lever, light pipe, magnet, membrane, mesh, mirror, mixer, motor, needle, nozzle, piston engine, plate, prism, pulley, pump, reactor, refraction grating, relay, rocket, rotor, sensor, separator, sieve, sphere, spiral, spring, substrate, switch, syringe, tag, tetrahedron, tetrakaidecahedron, transducer, tube, turbine engine, valve, wedge, wheel.

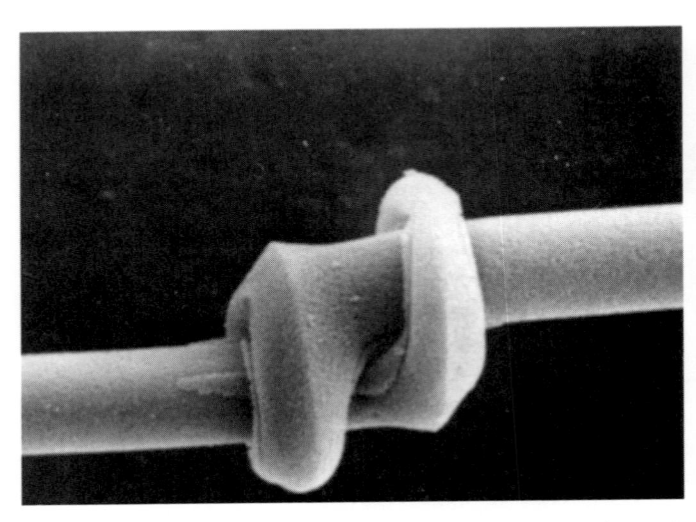

Figure 35.10 Chains of the fossil centric diatom *Syndetocystis barbadensis* were "linked by complex interlocking central processes. . . . From the slight central depression of the valve face there arises a long tubular process (periplekton); this expands at its tip into a ring, which clasps the stem of the process borne in the adjacent valve in the filament". (From Round *et al.* (1990), with permission.)

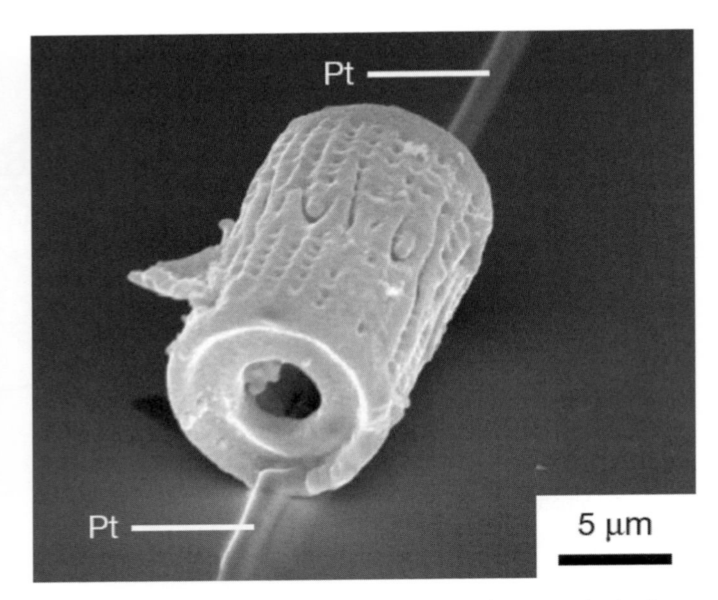

Figure 35.11 A gas sensor with platinum wires attached whose impedance changes measurably in response to parts per million changes in gaseous nitrogen monoxide (NO) concentration. This *Aulacoseira* specimen, a centric diatom, has been converted from silica to silicon. (From Bao *et al.* (2007), with permission.)

Further applications of diatoms for which I did not find any patent literature include:

1. Diatom adhesives (as raphe fluid) work well in wet conditions, whereas most manufactured glues do not (Gebeshuber *et al.*, 2002, 2003). The opposite side of this story is that diatoms are thus masters at biofouling (Molino and Wetherbee, 2008), which is a serious problem for ships and swimming pools.
2. Interlocking devices (Round *et al.*, 1990, Gebeshuber *et al.*, 2009b; Tiffany *et al.*, 2010), such as in Figure 35.10.
3. Diatoms have periodic and layered structure at and below the wavelengths of visible light, and so produce iridescent colors in both transmission and reflection and are photoluminescent (Gordon *et al.*, 2009), or can be used as 3D templates for phosphors (Weatherspoon *et al.*, 2006). Their optical properties (Fuhrmann *et al.*, 2004; Bismuto *et al.*, 2008; Lettieri *et al.*, 2008) can be used for detecting trace amounts of gases (De Stefano *et al.*, 2009), as can a silicon replica (Bao *et al.*, 2007) (Figure 35.11). Atom-substituted diatoms (Sandhage *et al.*, 2005) can have a refractive index above that of silica to create an complete optical band gap of potential use in collimating light beams and optical waveguides (De Stefano *et al.*, 2009). Enhanced photonic or optronics properties, which can also be achieved *in vivo* (Jeffryes *et al.*, 2008), may also lead to improved electricity-producing solar cells.

Leo Szilard encouraged atomic-bomb physicists to go into biology, because here he thought they could do no harm (Aaron Novick, personal communication, 1964; "This radical break, according to one of his friends, signified 'a turning from death to life that seemed to reflect his deep revulsion after Hiroshima'", though he gave other reasons publically (Bess, 1993)). But even diatom nanotechnology may enhance our ability to separate uranium-235 for power or bombs (Indech, 2005). Despite all the positive applications of diatoms and their contributions to fundamental biology, diatomists (like other scientists) cannot escape facing moral dilemmas.

35.6 Outlook

Diatom nanotechnology, although historically older than mainstream nanotechnology, is a new industrial endeavor in which I would estimate $100 million has been spent so far, with no products yet on the market, but with many on the way. In leading us into a manufacturing realm in which we grow rather than lithograph nanocomponents and systems, it is making a major contribution to nanotechnology in general. But because effective control generally requires understanding, diatom nanotechnology is simultaneously drawing us towards solving one of the major outstanding basic problems of biology today, namely the relationship between the genotype and phenotype, i.e. between genome and morphogenesis. The resulting synergy between applied and basic science is refreshing.

35.7 Summary

Diatom nanotechnology is leading to a renaissance in diatom cell biology, because of engineers' need to understand how diatoms build themselves, and because of the money and effort that is thus being put into this perhaps universal basic science problem. Engineers have two motivations: to control and use parts that diatoms build, or to build our own parts by analogous means. Applications that are envisaged, but not yet commercialized, range from the conceptually simple (absorbers, batteries) to sophisticated (trace-gas detectors, computers), and everything in between. To grasp the vast potential impact of diatom nanotechnology I have surveyed the patent literature, where claims are drawn wide, and thus imaginations soar. It would seem that any manufactured item whose miniaturization might be profitable may, at least in part, come to involve diatom nanotechnology.

Acknowledgments

I would like to thank Ille C. Gebeshuber, Karthick Balasubramanian, and the editors for many fine suggestions. This work is dedicated to Frithjof A. S. Sterrenburg and his contagious enthusiasm for diatoms.

References

Aczel, A. D. (2002). *Entanglement: The Greatest Mystery in Physics*, Vancouver: Raincoast Books.

Allison, D. P., Dufrêne, Y. F., Doktycz, M. J., & Hildebrand, M. (2008). Biomineralization at the nanoscale: learning from diatoms. *Methods in Cell Biology*, **90**, 61–86.

Antonelli, P. L. (1985). *Mathematical Essays on Growth and the Emergence of Form*, Edmonton: University of Alberta Press.

Axtell, E. A., III, Sakoske, G. E., Swiler, D. R., *et al.* (2006). Structured self-cleaning surfaces and method of forming same. United States Patent Application Publication US 2006/0246277 A1.

Bach, K. & Burkhardt, B. (1984). *Diatomeen I, Schalen in Natur und Technik/Diatoms I, Shells in Nature and Technics*, Stuttgart: Cramer Verlag.

Bailey, L. W. (1867). Desmids and diatoms. *The American Naturalist*, **1**, 505–517.

Ball, P. (2009a). *Branches: Nature's Patterns: a Tapestry in Three Parts*, Oxford: Oxford University Press.

Ball, P. (2009b). *Flow: Nature's Patterns: a Tapestry in Three Parts*, Oxford: Oxford University Press.

Ball, P. (2009c). *Shapes: Nature's Patterns: a Tapestry in Three Parts*, Oxford: Oxford University Press.

Bao, Z., Weatherspoon, M. R., Shian, S., *et al.* (2007). Chemical reduction of three-dimensional silica micro-assemblies into microporous silicon replicas. *Nature*, **446**, 172–5.

Batterman, R. W. (2001). *The Devil in the Details – Asymptotic Reasoning in Explanation, Reduction, and Emergence*, Oxford: Oxford University Press.

Bäuerlein, E. (2000). *Biomineralization of Nano- and Micro-Structures*, Weinheim: Wiley-VCH.

Beams, H. W. & Kessel, R. G. (1987). Development of centrifuges and their use in the study of living cells. *International Review of Cytolology*, **100**, 15–48.

Bedau, M. A. & Humphreys, P. (eds.) (2008). *Emergence: Contemporary Readings in Philosophy and Science*, Cambridge, MA: MIT Press.

Bejan, A. (2000). *Shape and Structure, from Engineering to Nature*, Cambridge: Cambridge University Press.

Ben-Jacob, E. (2008). Social behavior of bacteria: from physics to complex organization. *European Physical Journal B*, **65**, 315–22.

Bess, M. (1993). *Realism, Utopia, and the Mushroom Cloud: Four Activist Intellectuals and Their Strategies for Peace, 1945–1989 – Louise Weiss (France) Leo Szilard (USA) E.P. Thompson (England) and Danilo Dolci (Italy)*, Chicago, IL: University of Chicago Press.

Bismuto, A., Setaro, A., Maddalena, P., De Stefano, L., & De Stefano, M. (2008). Marine diatoms as optical chemical sensors: a time-resolved study. *Sensors and Actuators B – Chemical*, **130**, 396–9.

Blossfeldt, K. (1929). *Urformen der Kunst, Photographische Pflanzenbilder*, Berlin: Verlag Ernst Wasmuth AG.

Borowitzka, M. A. & Volcani, B. E. (1978). The polymorphic diatom *Phaeodactylum tricornatum*: ultrastructure of its morphotypes. *Journal of Phycology*, **14**, 10–21.

Bourke, P. (2006). Constrained diffusion-limited aggregation in 3 dimensions. *Computers & Graphics-UK*, **30**, 646–9.

Bowler, C., Allen, A. E., Badger, J. H., *et al.* (2008). The *Phaeodactylum* genome reveals the evolutionary history of diatom genomes. *Nature*, **456**, 239–44.

Bregar, G. W. (1955). Diatomaceous earth product and method for its manufacture. United States Patent 2,701,240.

Brener, E., Levine, H., & Tu, Y. H. (1991). Mean-field theory for diffusion-limited aggregation in low dimensions. *Physical Review Letters*, **66**, 1978–1981.

Brewer, D. B. (2006). Max Schultze (1865), G. Bizzozero (1882) and the discovery of the platelet. *British Journal of Haematology*, **133**, 251–8.

Brodie, I. & Murray, J. J. (1992). *The Physics of Micro/Nanofabrication*, New York, NY: Plenum Publishing.

Brott, L. L., Naik, R. R., Pikas, D. J., *et al.* (2001). Ultrafast holographic nanopatterning of biocatalytically formed silica. *Nature*, **413**, 291–3.

Brott, L. L., Naik, R. R., Stone, M. O., & Carter, D. C. (2007). Method and apparatus for use of thermally switching proteins in sensing and detecting devices. United States Patent Application US 2007/0037133 A1.

Brouhard, G. J., Schek, H. T., III, & Hunt, A. J. (2003). Advanced optical tweezers for the study of cellular and molecular biomechanics. *IEEE Transactions on Biomedical Engineering*, **50**, 121–5.

Brunner, E., Richthammer, P., Ehrlich, H., *et al.*, (2009). Chitin-based organic networks: an integral part of cell wall biosilica in the diatom. *Thalassiosira pseudonana. Angewandte Chemie-International Edition*, **48**, 9724–7.

Bryan, B. J., Gaalema, S., & Murphy, R. B. (2003). Apparatus and method for detecting and identifying infectious agents. United States Patent 6,649,356.

Budrene, E. O. & Berg, H. C. (1991). Complex patterns formed by motile cells of *Escherichia coli*. *Nature*, **349**, 630–3.

Cai, Y., Dickerson, M. B., Haluska, *et al.* (2007). Manganese-doped zinc orthosilicate-bearing phosphor microparticles with controlled three-dimensional shapes derived from diatom frustules. *Journal of the American Ceramic Society*, **90**, 1304–8.

Clough, T. J. (2001). Battery element containing efficiency improving additives. United States Patent Application US 2001/0012585 A1.

Clough, T. J. (2005). Metal oxide coated polymer substrates. United States Patent 6,919,035 B1.

Clough, T. J. (2006). Metal non-oxide coated substrates. United States Patent 7,041,370.

Cohen, I., Golding, I., Ron, I. G., & Ben-Jacob, E. (2001). Biofluiddynamics of lubricating bacteria. *Mathematical Methods in the Applied Sciences*, **24**, 1429–68.

Cox, E. J. (2006). Raphe loss and spine formation in *Diadesmis gallica* (Bacillariophyta) – an intriguing example of phenotypic polymorphism in a diatom. *Nova Hedwigia, Beiheft*, **130**, 163–75.

Crawford, S., Chiovitti, T., Pickett-Heaps, J., & Wetherbee, R. (2009). Micromorphogenesis during diatom wall formation produces siliceous nanostructures with different properties. *Journal of Phycology*, **45**, 1353–62.

Crommelinck, M., Feltz, B., & Goujon, P. (eds.) (2006). *Self-Organization and Emergence in Life Sciences*, Dordrecht: Springer.

Crutchfield, J. P. (1994). Is anything ever new? Considering emergence. In *Complexity: Metaphors, Models, and Reality*, ed. G. Cowan, D. Pines, & D. Melzner, Reading, MA: Addison-Wesley.

Davies, P. C. W. (2004a). Emergent biological principles and the computational properties of the universe. *Complexity*, **10**, 11–15.

Davies, P. C. W. (2004b). Emergent complexity, teleology, and the arrow of time. In *Debating Design: From Darwin to DNA*, W. A. Dembski & M. Ruse, Cambridge: Cambridge University Press, pp. 191–209.

De Stefano, L., Maddalena, P., Moretti, L., *et al.* (2009). Nano-biosilica from marine diatoms: a brand new material for photonic applications. *Superlattices and Microstructures*, **46**, 84–9.

De Vries, A. H., Krenn, B. E., Van Driel, R., & Kanger, J. S. (2005). Micro magnetic tweezers for nanomanipulation inside live cells. *Biophysics Journal*, **88**, 2137–44.

Deravi, L. F., Sumerel, J. L., Sewell, S. L., & Wright, D. W. (2008). Piezoelectric inkjet printing of biomimetic inks for reactive surfaces. *Small*, **4**, 2127–30.

Dickerson, M. B., Sandhage, K. H., & Naik, R. R. (2008). Protein- and peptide-directed syntheses of inorganic materials. *Chemical Reviews*, **108**, 4935–78.

Dickerson, M. B., Sandhage, K. H., Nalik, R., & Stone, M. O. (2007). Methods for fabricating micro-to-nanoscale devices via biologically-induced solid formation on biologically-derived templates, and micro-to-nanoscale structures and micro-to-nanoscale devices made thereby. United States Patent Application Publication US 2007/0112548 A1.

Dombrowski, C., Cisneros, L., Chatkaew, S., Goldstein, R. E., & Kessler, J. O. (2004). Self-concentration and large-scale coherence in bacterial dynamics. *Physical Review Letters*, **93** (9), 098103.

Drack, M., Ector, L., Gebeshuber, I. C., & Gordon, R. (2011). A review of the proposed mechanisms for diatom gliding motility: early history to nanofluidics. In *The Diatom World*, ed. J. Seckbach & J. P. Kociolek, Dordrecht: Springer.

Drexler, K. E. (1992). *Nanosystems: Molecular Machinery, Manufacturing, and Computation*, New York: John Wiley & Sons.

Drexler, K. E. & Peterson, C. (1991). *Unbounding the Future: the Nanotechnology Revolution*, New York, NY: Morrow.

Drum, R. W. & Gordon, R. (2003). Star Trek replicators and diatom nanotechnology. *TibTech (Trends in Biotechnology)*, **21**, 325–8.

Dupas, C., Houdy, P., & Lahmani, M. (2007). *Nanoscience: Nanotechnologies and Nanophysics*, Berlin: Springer-Verlag.

Edwards, A. M. (1875). Different diatoms on the same stipes. *Journal of Cell Science*, **15** (new series), 63–64.

Ehrlich, H. (2010). Chitin and collagen as universal and alternative templates in biomineralization. *International Geology Review*, **52**, 661–99.

El Rassy, H., Belamie, E., Livage, J., & Coradin, T. (2005). Onion phases as biomimetic confined media for silica nanoparticle growth. *Langmuir*, **21**, 8584–7.

Estes, A. & Dute, R. R. (1994). Valve abnormalities in diatom clones maintained in long-term culture. *Diatom Resarch*, **9**, 249–58.

Feynman, R. P. (1960). There's plenty of room at the bottom: an invitation to enter a new field of physics. *Engineering and Science*, February. See http://www.zyvex.com/nanotech/feynman.html.

Flake, G. W. (1998). *The Computational Beauty of Nature: Computer Explorations of Fractals, Chaos, Complex Systems, and Adaptation*, Cambridge, MA: MIT Press.

Foged, N. (1986). Diatoms in Gambia; diatoms in the Volo Bay, Greece. *Bibliotheca Diatomologica*, **12**, 1–152.

Forbes, N. (2004). *Imitation of Life: How Biology is Inspiring Computing*, Cambridge, MA: MIT Press.

Freitas, R. A., Jr. (1999). *Nanomedicine*, Georgetown, TX: Landes Bioscience.

Fryxell, G. A. & Hasle, G. R. (1977). The genus *Thalassiosira*: some species with a modified ring of central strutted processes. *Nova Hedwigia*, **54**, 67–98.

Fuhrmann, T., Landwehr, S., El Rharbi-Kucki, M., & Sumper, M. (2004). Diatoms as living photonic crystals. *Applied Physics B – Lasers and Optics*, **78**, 257–60.

Gaul, U., Geissler, U., Henderson, M., Mahoney, R., & Reimer, C. W. (1993). Bibliography on the fine-structure of diatom frustules (Bacillariophyceae). *Proceedings of the Academy of Natural Sciences of Philadelphia*, **144**, 69–238.

Gautier, C., Abdoul-Aribi, N., Roux, C., *et al.* (2008). Biomimetic dual templating of silica by polysaccharide/protein assemblies. *Colloids and Surfaces B-Biointerfaces*, **65**, 140–5.

Gazit, E. (2007). *Plenty of Room for Biology at the Bottom: an Introduction to Bionanotechnology*, Singapore: World Scientific.

Gebeshuber, I. C., Drack, M., & Scherge, M. (2008). Tribology in biology. *Tribology – Materials, Surfaces & Interfaces*, **2**, 200–12.

Gebeshuber, I. C., Gruber, P., & Drack, M. (2009a). A gaze into the crystal ball – biomimetics in the year 2059. *Proceedings of the Institution of Mechanical Engineers, Part C: Journal of Mechanical Engineering Science*, **223**, 2899–918.

Gebeshuber, I. C., Kindt, J. H., Thompson, J. B., *et al.* (2003). Atomic force microscopy study of living diatoms in ambient conditions. *Journal of Microscroscopy*, **212**, 292–9.

Gebeshuber, I. C., Stachelberger, H., & Drack, M. (2005). Diatom bionanotribology – biological surfaces in relative motion: their design, friction, wear and lubrication. *Journal of Nanoscience and Nanotechnology*, **5**, 79–87.

Gebeshuber, I. C., Stachelberger, H., Ganji, B. A., *et al.* (2009b). Exploring the innovational potential of biomimetics for novel 3D MEMS. *Journal of Advanced Materials Research*, **74**, 265–8.

Gebeshuber, I. C., Thompson, J. B., Del Amo, Y., Stachelberger, H., & Kindt, J. H. (2002). *In vivo* nanoscale atomic force microscopy investigation of diatom adhesion properties. *Materials Science and Technology*, **18**, 763–6.

Gebeshuber, L. C. (2007). Biotribology inspires new technologies. *Nano Today*, **2**, 30–37.

Goldberg, A. J., Mather, P. T., & Wood, T. K. (2007). Dental materials, methods of making and using the same, and articles formed thereform. United States Patent Application Publication US 2007/0238808 A1.

Golovin, A. A. & Nepomnyashchy, A. A. (2006). *Self-Assembly, Pattern Formation and Growth Phenomena in Nano-Systems*, Dordrecht: Springer.

Goodale, G. L. (1885). *Physiological Botany. I. Outlines of the Histology of Phaenogamous Plants. II. Vegetable Physiology*, New York, NY: Ivison, Blakeman, Taylor, and Co.

Goodsell, D. S. (2004). *Bionanotechnology, Lessons from Nature*, Hoboken, NJ: Wiley-Liss.

Goold, N. R., Somfai, E., & Ball, R. C. (2005). Anisotropic diffusion limited aggregation in three dimensions: universality and nonuniversality. *Physical Review E*, **72**, 031403 (10 pages).

Gordon, R. (1996). Computer controlled evolution of diatoms: design for a compustat. *Nova Hedwigia*, **112**, 213–16.

Gordon, R. (1999). *The Hierarchical Genome and Differentiation Waves: Novel Unification of Development, Genetics and Evolution*, Singapore & London: World Scientific & Imperial College Press.

Gordon, R. (2000). The emergence of emergence: a critique of "Design, observation, surprise!". *Rivista di Biologia/Biology Forum*, **93**, 349–56.

Gordon, R. & Aguda, B. D. (1988). Diatom morphogenesis: natural fractal fabrication of a complex microstructure. In *Proceedings of the Annual International Conference of the IEEE Engineering in Medicine and Biology Society, Part 1/4: Cardiology and Imaging, 4–7 Nov. 1988, New Orleans, LA, USA*, ed. G. Harris & C. Walker, New York, NY: Institute of Electrical and Electronics Engineers, pp. 273–4.

Gordon, R. & Brodland, G. W. (1990). On square holes in pennate diatoms. *Diatom Resarch*, **5**, 409–13.

Gordon, R. & Chen, Y. (1987). From statistical mechanics to molecular biology: a Festschrift for Terrell L. Hill. *Cell Biophysics*, **11**, i–iii.

Gordon, R. & Drum, R. W. (1994). The chemical basis for diatom morphogenesis. *International Review of Cytolology*, **150**, 243–72, 421–2.

Gordon, R., Drum, R. W., & Thurlbeck, A. (1980). The chemical basis for diatom morphogenesis: instabilities in diffusion-limited amorphous precipitation generate space filling branching patterns. Abstracts, The 39th Annual Symposium of The Society for Developmental Biology, Levels of Genetic Control in Development. Storrs, University of Connecticut, p. 5.

Gordon, R., Hoover, R. B., Tuszynski, J. A., *et al.* (2007). Diatoms in space: testing prospects for reliable diatom nanotechnology in microgravity. *Proceedings of SPIE*, **6694**, V1–V15, DOI:10.1117/12.737051.

Gordon, R., Kling, H. J., & Sterrenburg, F. A. S. (2005a). A guide to the diatom literature for diatom nanotechnologists. *Journal of Nanoscience and Nanotechnology*, **5**, 175–8.

Gordon, R., Losic, D., Tiffany, M. A., Nagy, S. S., & Sterrenburg, F. A. S. (2009). The Glass Menagerie: diatoms for novel applications in nanotechnology. *Trends in Biotechnology*, **27**, 116–27.

Gordon, R. & Melvin, C. A. (2003). Reverse engineering the embryo: a graduate course in developmental biology for engineering students at the University of Manitoba, Canada. *International Journal of Developmental Biology*, **47**, 183–7.

Gordon, R. & Parkinson, J. (2005). Potential roles for diatomists in nanotechnology. *Journal of Nanoscience and Nanotechnology*, **5**, 35–40.

Gordon, R., Sterrenburg, F. A. S., & Sandhage, K. (2005b). A special issue on diatom nanotechnology. *Journal of Nanoscience and Nanotechnology*, **5**, 1–4.

Gordon, R., Witkowski, A., Gebeshuber, I. C., & Allen, C. S. (2010). The diatoms of Antarctica and their potential roles in nanotechnology. In *Antarctica*, ed. M. Masó, M. Masó, & A. Chillida, Barcelona: Editions ACTAR, in press.

Grollier, J. F., Rosenbaum, G., & Cotteret, J. (1991). Transparent cosmetic composition that reflects infrared radiation and its use for protecting the human epidermis against infrared radiation. United States Patent 5,000,937.

Guo, P. X. (2005). A special issue on bionanotechnology – preface. *Journal of Nanoscience and Nanotechnology*, **5**, i–iii.

Gutu, T., Gale, D. K., Jeffryes, C., *et al.* (2009). Electron microscopy and optical characterization of cadmium sulphide nanocrystals deposited on the patterned surface of diatom biosilica. *Journal of Nanomaterials* **2009**, 860536 (7 pages).

Häder, D.-P., Hemmersbach, R., & Lebert, M. (2005). *Gravity and the Behavior of Unicellular Organisms*, New York, NY: Cambridge University Press.

Haeckel, E. (1904). *Kunstformen der Natur (Art Forms of Nature)*, Leipzig: Verlag des Bibliographischen Instituts (in German).

Hall, B. K., Pearson, R. D., & Müller, G. (2004). *Environment, Development and Evolution: Toward a Synthesis*, Cambridge, MA: MIT Press.

Hall, T. S. (1969). *Ideas of Life and Matter. Studies in the History of General Physiology, 600 B.C. to A.D. 1900, Volume 2: From the Enlightenment to the End of the Nineteenth Century*, Chicago, IL: University of Chicago Press.

Hamm, C. E. (2005). The evolution of advanced mechanical defenses and potential technological applications of diatom shells. *Journal of Nanoscience and Nanotechnology*, **5**, 108–19.

Hamm, C. E., Merkel, R., Springer, O., *et al.* (2003). Architecture and material properties of diatom shells provide effective mechanical protection. *Nature*, **421**, 841–3.

Hamm-Dubischar, C. (2007). Method of determining structural prototype data for a technical lightweight structure. United States Patent Application US 2007/0112522 A1.

Hanczyc, M. M. (2008). The early history of protocells: the search for the recipe of life. In *Protocells: Bridging Nonliving and Living Matter*, ed. S. Rasmussen, M. A. Bedau, L. Chen, *et al.*, Cambridge, MA: MIT Press, 3–17.

Harbour, P. J. & Hartley, P. G. (2007). Metal oxide/hydroxide materials. United States Patent Application US 2007/0281854 A1.

Hazelaar, S., van der Strate, H. J., Gieskes, W. W. C., & Vrieling, E. G. (2005). Monitoring rapid valve formation in the pennate diatom *Navicula salinarum* (Bacillariophyceae). *Journal of Phycology*, **41**, 354–8.

Hecky, R. E., Mopper, K., Kilham, P., & Degens, E. T. (1973). The amino acids and sugar composition of diatom cell walls. *Marine Biology*, **19**, 323–31.

Hildebrand, M. (2008). Diatoms, biomineralization processes, and genomics. *Chemical Reviews*, **108**, 4855–74.

Hildebrand, M., Kim, S., Shi, D., Scott, K., & Subramaniam, S. (2009). 3D imaging of diatoms with ion-abrasion scanning electron microscopy. *Journal of Structural Biology*, **166**, 316–28.

His, W. (1874). *Unsere Körperform und das Problem ihrer Entstehung, Briefe an einen befreundeten Naturforscher (Our Body Form and the Problem of its Emergence, Letters to a Friendly Natural Scientist)*, Leipzig: F.C.W. Vogel (in German).

Hobbs, W. O., Wolfe, A. P., Inskeep, W. P., Amskold, L., & Konhauser, K. O. (2009). Epipelic diatoms from an extreme acid environment: Beowulf Spring, Yellowstone National Park, USA. *Nova Hedwigia*, 71–83.

Holland, J. H. (1998). *Emergence: From Chaos to Order*, Reading, MA: Perseus Books.

Iler, R. K. (1979). *The Chemistry of Silica: Solubility, Polymerization, Colloid and Surface Properties, and Biochemistry*, New York, NY: John Wiley & Sons.

Indech, R. (2005). Uranium isotope separation through substitution reactions. United States Patent Application 2005/0287059 A1.

Jablonka, E. & Lamb, M. J. (2005). *Evolution in Four Dimensions: Genetic, Epigenetic, Behavioral, and Symbolic Variation in the History of Life*, Cambridge, MA: MIT Press.

Jayas, D. S. (2002). The role of biology in shaping the engineering profession. In *Canadian Conference on Electrical & Computer Engineering*, ed. W. Kinser, W. Sebak, & K. Ferens, Winnipeg: IEEE, 5.6–5.7. See http://www.ieee.ca/cceceo2/program/pdocs/co2prog1.pdf.

Jeffryes, C., Gutu, T., Jiao, J., & Rorrer, G. L. (2008). Metabolic insertion of nanostructured TiO_2 into the patterned biosilica of the diatom *Pinnularia* sp. by a two-stage bioreactor cultivation process. *ACS Nano*, **2**, 2103–12.

Jensen, H. J. (1998). *Self-Organized Criticality: Emergent Complex Behavior in Physical and Biological Systems*, Cambridge, MA: Cambridge University Press.

Johnson, S. (2001). *Emergence, The Connected Lives of Ants, Brains, Cities, and Software*, New York, NY: Scribner.

Jorgensen, B. B. & Des Marais, D. J. (1990). The diffusive boundary layer of sediments: oxygen microgradients over a microbial mat. *Limnology and Oceanography*, **35**, 1343–55.

Jotterand, F. (ed.) (2008). *Emerging Conceptual, Ethical and Policy Issues in Bionanotechnology*, Berlin: Springer.

Jung, E. K. Y., Langer, R., & Leuthardt, E. C. (2007). Diatom device. U.S. Patent Application Publication number: US 2007/0184088 A1.

Kaandorp, J. A. (1994). *Fractal Modelling: Growth and Form in Biology*, Berlin: Springer-Verlag.

Karthick, B. (2009). Genome sequencing of cells that live inside glass cages reveals their past history. *Current Science*, **96**, 334–7.

Kassner, K. (1996). *Pattern Formation in Diffusion-limited Crystal Growth: Beyond the Single Dendrite*, Singapore: World Scientific.

Keller, E. F. (1994). Rethinking the meaning of genetic determinism. *Tanner Lectures on Human Values*, **15**, 113–39.

Kociolek, J. P. & Spaulding, S. A. (2002). Morphological variation, species concepts, and classification of an unusual fossil centric diatom (Bacillariophyta) from western North America. *Journal of Phycology*, **38**, 821–33.

Kociolek, J. P. & Sullivan, M. J. (1995). *A Century of Diatom Research in North America. A Tribute to the Distinguished Careers of Charles W. Reimer & Ruth Patrick*, Champaign, IL: Koeltz Scientific Books USA.

Krummenacker, M. & Lewis, J. (1995). *Prospects in Nanotechnology: Toward Molecular Manufacturing*, New York: Wiley.

Krzhizhanovskaya, V. V. & Sun, S. (2007). Simulation of multiphysics multiscale systems: introduction to the ICCS'2007 workshop. In *ICCS 2007, Part I, LNCS 4487*, ed. Y. Shi, G. D. van Albada, J. Dongarra, P. M.A. Sloot, Berlin: Springer-Verlag.

Kusumoto, M., Nishiya, Y., Kishimoto, M., & Umebayashi, N. (2006). Device for separation of biological components, and method of separation of biological components using the device. United States Patent Application US 2006/0186055 A1.

Lam, R. C., Chen, Y.-F., & Maruo, K. (2007). Friction material with nanoparticles of friction modifying layer. United States Patent 7,294,388 B2.

Léger, C., Servant, L., Bruneel, J. L., & Argoul, F. (1999). Growth patterns in electrodeposition, *Physica A*, **263**, 305–14.

Lenoci, L. & Camp, P. J. (2008). Diatom structures templated by phase-separated fluids. *Langmuir*, **24**, 217–23.

Lettieri, S., Setaro, A., De Stefano, L., De Stefano, M., & Maddalena, P. (2008). The gas-detection properties of light-emitting diatoms. *Advanced Functional Materials*, **18**, 1257–64.

Lindberg, D. C. (2007). *The Beginnings of Western Science: The European Scientific Tradition in Philosophical, Religious, and Institutional*

Context, Prehistory to A.D. 1450, Chicago, IL: University of Chicago Press.

Liu, J.-Q. & Shimohara, K. (2006). *Biomolecular Computation for Bionanotechnology*, Norwood, MA: Artech House Publishers.

Locke, M. (1968). *The Emergence of Order in Developing Systems*, New York, NY: Academic Press.

Lu, G. Q. & Zhao, X. S. (2004). *Nanoporous Materials: Science and Engineering*, Singapore: World Scientific.

Luckarift, H. R., Spain, J. C., Naik, R. R., & Stone, M. O. (2004). Enzyme immobilization in a biomimetic silica support. *Nature Biotechnology*, **22**, 211–3.

Mandelbrot, B. B., Kol, B., & Aharony, A. (2002). Angular gaps in radial diffusion-limited aggregation: two fractal dimensions and nontransient deviations from linear self-similarity. *Physical Review Letters*, **88**, 055501.

Mansur, J. C., Silva, A. G., Carvalho, A. T. G., & Martins, M. L. (2005). Electrocrystallization under magnetic fields: experiment and model. *Physica A – Statistical Mechanics and Its Applications*, **350**, 393–406.

Masayoshi, E., Keizo, I., Noriaki, O., *et al.* (2006). *Future Medical Engineering Based on Bionanotechnology: Proceedings of the Final Symposium of the Tohoku University 21st Century Center of Excellence Program*, London: Imperial College Press.

Meakin, P. (1998). *Fractals, Scaling and Growth Far From Equilibrium*, Cambridge: Cambridge University Press.

Mendoza, C. I. & Ramìrez-Santiago, G. (2005). Annealing two-dimensional diffusion-limited aggregates. *European Physical Journal B*, **48**, 75–80.

Merks, R. M. H., Hoekstra, A. G., Kaandorp, J. A., & Sloot, P. M. A. (2003). Diffusion-limited aggregation in laminar flows. *International Journal of Modern Physics C*, **14**, 1171–82.

Mills, F. W. & Deby, J. (1893). *An Introduction to the Study of the Diatomacea*, London: Iliffe & Son.

Milsum, J. H. (ed.) (1968). *Positive Feedback, A General Systems Approach to Positive/Negative Feedback and Mutual Causality*, Oxford: Pergamon Press.

Mizuno, M. (1982). Change in striation density and systematics of *Cocconeis scutellum* var. *ornata* (Bacillariophyceae). *Botanical Magazine Tokyo*, **95**, 349–57.

Molino, P. J. & Wetherbee, R. (2008). The biology of biofouling diatoms and their role in the development of microbial slimes. *Biofouling*, **24**, 365–379.

Møller, A. P. & Swaddle, J. P. (1997). *Asymmetry, Developmental Stability, and Evolution*, Oxford: Oxford University Press.

Morehouse, G. W. (1876). Silica films and the structure of diatoms. *Monthly Microscopical Journal*, **15**, 38–40.

Morowitz, H. J. (2002). *The Emergence of Everything, How the World Became Complex*, Oxford: Oxford University Press.

Morse, D. E., Stucky, G. D., & Gaul, J. H. (2003). Method and compositions for binding histidine-containing proteins to substrates. United States Patent Application US 2003/0003223 A1.

Mulhall, D. (2002). *Our Molecular Future: How Nanotechnology, Robotics, Genetics, and Artificial Intelligence will Transform our World*, Amherst, NY: Prometheus Books.

Müller, W. E. G., Schröder, H. C., Lorenz, B., & Krasko, A. (2007). Silicatein-mediated synthesis of amorphous silicates and siloxanes and use thereof. United States Patent 7,169,589 B2.

Mullins, W. W. & Sekerka, R. F. (1963). Morphological stability of a particle growing by diffusion or heat flow. *Journal of Applied Physics*, **34**, 323–9.

Naik, R. R., Stone, M. O., Spain, J. C., & Luckarift, H. R. (2005). Entrapment of biomolecules and inorganic nanoparticles by biosilification. United States Patent Application US 2005/0095690 A1.

Naik, R. R., Stringer, S. J., Agarwal, G., Jones, S. E., & Stone, M. O. (2002). Biomimetic synthesis and patterning of silver nanoparticles. *Nature Materials*, **1**, 169–72.

Neethirajan, S., Gordon, R., & Wang, L. (2009). Potential of silica bodies (phytoliths) for nanotechnology. *Trends in Biotechnology*, **27**, 461–7.

Nicolis, G. P. I. (1977). *Self-Organization in Nonequilibrium Systems, From Dissipative Structures to Order through Fluctuations*, New York, NY: Wiley.

Nordenskiöld, E. (1928). *The History of Biology, a Survey*, New York, NY: Tudor Publishing Co.

Nyhart, L. K. (1995). *Biology Takes Form: Animal Morphology and the German Universities, 1800–1900*, Chicago, IL: University of Chicago Press.

Ohgiwari, M., Matsushita, M., & Matsuyama, T. (1992). Morphological changes in growth phenomena of bacterial colony patterns. *Journal of the Physical Society of Japan*, **61**, 816–22.

Oliver, S., Kuperman, A., Coombs, N., Lough, A., & Ozin, G. A. (1995). Lamellar aluminophosphates with surface patterns that mimic diatom and radiolarian microskeletons. *Nature*, **378**, 47–50.

Ozin, G. A. A. (2005). *Nanochemistry: a Chemical Approach to Nanomaterials*, Cambridge: Royal Society of Chemistry Publishing.

Palmer, A. R. & Strobeck, C. (1986). Fluctuating asymmetry: measurement, analysis, patterns. *Annual Reviews of Ecology and Systematics.*, **17**, 391–421.

Papazoglou, E. S. & Parthasarathy, A. (2007). *BioNanotechnology*, San Rafael, CA: Morgan and Claypool Publishers.

Parkinson, J., Brechet, Y., & Gordon, R. (1999). Centric diatom morphogenesis: a model based on a DLA algorithm investigating the potential role of microtubules. *Biochimica et Biophysica Acta – Molecular Cell Research*, **1452**, 89–102.

Parkinson, J. & Gordon, R. (1999). Beyond micromachining: the potential of diatoms. *Trends in Biotechnology (Tibtech)*, **17**, 190–6.

Patrick, R. & Reimer, C. W. (1966). *The Diatoms of the United States, Exclusive of Alaska and Hawaii, Volume 1*, Philadelphia, PA: Academy of Natural Sciences.

Pickett-Heaps, J. D., Tippit, D. H., & Andreozzi, J. A. (1979a). Cell division in the pennate diatom *Pinnularia*. III – the valve and associated cytoplasmic organelles. *Biologie Cellulaire*, **35**, 195–8.

Pickett-Heaps, J. D., Tippit, D. H., & Andreozzi, J. A. (1979b). Cell division in the pennate diatom *Pinnularia*. IV – valve morphogenesis. *Biologie Cellulaire*, **35**, 199–203.

Pinnavala, T. J., Kim, S.-S., & Zhang, W. (2006). Ultra-stable lamellar mesoporous silica compositions and process for the preparation thereof. United States Patent 7,132,165 B2.

Portet, S., Tuszynski, J. A., Dixon, J. M., & Sataric, M. V. (2003). Models of spatial and orientational self-organization of microtubules under the influence of gravitational fields. *Physical Review E*, **68**, 021903.

Potapova, M. (2008). Charles W. Reimer (1923–2008). See http://www.ansp.org/research/biodiv/diatoms/charles_reimer.php.

Ramachandra, T. V., Mahapatra, D. M., Karthick B., & Gordon, R. (2009). Milking diatoms for sustainable energy: biochemical engineering versus gasoline-secreting diatom solar panels. *Industrial & Engineering Chemistry Research*, **48**, 8769–88.

Rayleigh, L. (1892). On the instability of a cylinder of viscous liquid under capillary force. *Philosophical Magazine*, **34**, 145–54.

Redfern, P. (1853). Mode of isolating Naviculae and other test objects. *Quarterly Journal of Microscopical Science*, **1**, 235–6.

Rehm, B. H. A. (2006). *Microbial Bionanotechnology: Biological Self-Assembly Systems and Biopolymer-Based Nanostructures*, Abingdon: Taylor & Francis.

Reimann, B. E. F. (1960). Bildung, Bau und Zusammenhang der Bacillariophyceenschalen (elektronenmikroskopische Untersuchungen). *Nova Hedwigia*, **2**, 349–73.

Reisner, D. E. (ed.) (2008). *Bionanotechnology: Global Prospects*, Boca Raton, FL: CRC Press.

Renugopalakrishnan, V. & Lewis, R. V. (eds.) (2006). *Bionanotechnology: Proteins to Nanodevices*, Berlin: Springer-Verlag.

Robinson, D. H. & Sullivan, C. W. (1987). How do diatoms make silicon biominerals. *Trends in Biochemical Sciences*, **12**, 151–4.

Rosenhead, L. (ed.) (1963). *Laminar Boundary Layers*, New York, NY: Dover Publications.

Round, F. E., Crawford, R. M., & Mann, D. G. (1990). *The Diatoms, Biology & Morphology of the Genera*, Cambridge: Cambridge University Press.

Saliterman, S. (2006). *Fundamentals of BioMEMS and Medical Microdevices*, Bellingham, Washington: SPIE Press.

Sander, L. M., Cheng, Z. M., & Richter, R. (1983). Diffusion-limited aggregation in three dimensions. *Physical Review B*, **28**, 6394–6.

Sandhage, K. H. (2007). Shaped microcomponent via reactive conversion of biologically-derived microtemplates. United States Patent 7,204,971 B2.

Sandhage, K. H., Allan, S. M., Dickerson, M. B., *et al.* (2005). Merging biological self-assembly with synthetic chemical tailoring: the potential for 3-D genetically engineered micro/nano-devices (3-D GEMS). *International Journal of Applied Ceramic Technology*, **2**, 317–26.

Sandhage, K. H. & Bao, Z. (2008). Methods of fabricating nanoscale-to-microscale structures. United States Patent Application US 2008/0038170 A1.

Sandhage, K. H., Dickerson, M. B., Huseman, P. M., *et al.* (2002). Novel, bioclastic route to self-assembled, 3D, chemically tailored meso/nanostructures: shape-preserving reactive conversion of biosilica (diatom) microshells. *Advanced Materials*, **14**, 429–33.

Sargent, E. H. (2005). *The Dance of Molecules: How Nanotechnology is Changing Our Lives*, Toronto: Viking Canada.

Saxe, J. G. (1873). Blind Men and the Elephant. A Hindoo Fable. *The Poems of John Godfrey Saxe*, Boston, MA: James R. Osgood and Co.

Scherge, M. & Gorb, S. N. (2001). *Biological Micro- and Nanotribology: Nature's Solutions*, Berlin: Springer-Verlag.

Schlichting, H. & Gersten, K. (2000). *Boundary Layer Theory*, Berlin: Springer-Verlag.

Schultze, M. J. S. (1858a). Innere Bewegungserscheinungen bei Diatomeen der Nordsee as den Gattungen *Coscinodiscus*, *Denticella*, *Rhizosolenia*. *Archiv für Anatomie, Physiologie und Wissenschaftliche Medicin*, 330–342 + Table XIII.

Schultze, M. J. S. (1858b). Phenomena of internal movements in Diatomaceae of the North Sea, belonging to the genera *Coscinodiscus*, *Denticella*, and *Rhizosolenia*. With a Plate. *Quarterly Journal of Microscopical Science*, **7**, 13–21 + Plate II.

Schultze, M. J. S. (1863a). Die Structur der Diatomeenschale, verglichen mit gewissen aus Fluorkiesel künstlich darstellbaren Kieselhäuten. *Naturhistorischer Verein der Rheinlande und Westfalens Verhandlungen*, **20**, 1–42 + Table I.

Schultze, M. J. S. (1863b). On the structure of the valve in the Diatomacea, as compared with certain siliceous pellicles produced artificially by the decomposition in moist air of fluo-silicic acid gas (fluoride of silicium). *Quarterly Journal of Microscopical Science, New Series*, **3**, 120–134 + Plate VIII.

Schultze, M. J. S. (1865). Die Bewegung der Diatomeen/The movement of diatoms. *Schultze's Archiv für Mikroskopische Anatomie*, **1**, 376–402 + Table XXIII.

Schwarzer, S., Havlin, S., Ossadnik, P., & Stanley, H. E. (1996). Number of branches in diffusion-limited aggregates: the skeleton. *Physical Review E*, **53**, 1795–804.

Seckbach, J. & Gordon, R. (eds.) (2008). *Divine Action and Natural Selection: Science, Faith and Evolution*, Singapore: World Scientific.

Shapiro, J. A. & Dworkin, M. (eds.) (1997). *Bacteria as Multicellular Organisms*, New York, NY: Oxford University Press.

Slack, H. J. (1870). The patterns of artificial diatoms. *Monthly Microscopical Journal*, **4**, 181–3.

Slack, H. J. (1871). The silicious deposit in Pinnulariae. *Monthly Microscopical Journal*, **6**, 71–4.

Slack, H. J. (1874). On certain beaded silica films artificially formed. *Monthly Microscopical Journal*, **11**, 237–41.

Smith, C. S. (1981). *A Search for Structure: Selected Essays on Science, Art, and History*, Cambridge, MA: MIT Press.

Stanley, H. E. & Ostrosky, N. (1986). *On Growth and Form – Fractal and Non-fractal Patterns in Physics*, Boston, MA: Martinus Nijhoff Publishers.

Stanley, M. S. & Callow, J. A. (2007). Whole cell adhesion strength of morphotypes and isolates of *Phaeodactylum tricornutum* (Bacillariophyceae). *European Journal of Phycology*, **42**, 191–7.

Steinman, A. D. & Sheath, R. G. (1984). Morphological variability of *Eunotia pectinalis* (Bacillariophyceae) in a softwater Rhode Island stream and in culture. *Journal of Phycology*, **20**, 266–76.

Sterrenburg, F. A. S. (2005). Crystal palaces – diatoms for engineers. *Journal of Nanoscience and Nanotechnology*, **5**, 100–7.

Sterrenburg, F. A. S., Gordon, R., Tiffany, M. A., & Nagy, S. S. (2007). Diatoms: living in a constructal environment. In *Algae and Cyanobacteria in Extreme Environments. Cellular Origin, Life in Extreme*

Habitats and Astrobiology, Volume 11, ed. J. Seckback, Dordrecht: Springer, pp. 141–72.
Stoermer, E. F. (1967). Polymorphism in *Mastogloia*. *Journal of Phycolology*, **3**, 73–7.
Stucky, G. D., Chmelka, B. F., Zhao, D., *et al.* (2007). Block copolymer processing for mesostructured inorganic oxide materials. United States Patent US 7,176,245 B2.
Sugimoto, T. (1978a). General kinetics of Ostwald ripening of precipitates. *Journal of Colloid and Interface Science*, **63**, 16–26.
Sugimoto, T. (1978b). Kinetics of reaction controlled Ostwald ripening of precipitates in steady state. *Journal of Colloid and Interface Science*, **63**, 369–77.
Sumper, M. & Brunner, E. (2006). Learning from diatoms: nature's tools for the production of nanostructured silica. *Advanced Functional Materials*, **16**, 17–26.
Sumper, M. & Lehmann, G. (2006). Silica pattern formation in diatoms: species-specific polyamine biosynthesis. *Chembiochem*, **7**, 1419–27.
Swift, E. (1973). Marine diatom *Ethmodiscus rex*: its morphology and occurrence in plankton of Sargasso Sea. *Journal of Phycology*, **9**, 456–60.
Tamerler, C. & Sarikaya, M. (2008). Molecular biomimetics: genetic synthesis, assembly, and formation of materials using peptides. *Materials Research Society Bulletin*, **33**, 504–10.
Tan, W., Wang, K., He, X., *et al.* (2004). Bionanotechnology based on silica nanoparticles. *Medical Research Reviews*, **24**, 621–38.
Tan, Z. J., Zou, X. W., Zhang, W. B., & Jin, Z. Z. (1999). Influence of particle size on diffusion-limited aggregation. *Physical Review E*, **60**, 6202–5.
Tartar, V. (1962). Morphogenesis in *Stentor*. *Advances in Morphogenesis*, **2**, 1–26.
Tesson, B. & Hildebrand, M. (2010). Dynamics of silica cell wall morphogenesis in the diatom *Cyclotella cryptica*: substructure formation and the role of microfilaments. *Journal of Structural Biology*, **169**, 62–74.
Thalera, M. & Kaczmarska, I. (2009). *Gyrosigma orbitum* sp. nov. (Bacillariophyta) from a salt marsh in the Bay of Fundy, eastern Canada. *Botanica Marina*, **52**, 60–8.
Theriot, E. (1988). An empirically based model of variation in rotational elements in centric diatoms with comments on ratios in phycology. *Journal of Phycology*, **24**, 400–7.
Tiffany, M. A., Gebeshuber, I. C., & Gordon, R. (2010). *Hyalodiscopsis plana*, a sublittoral centric marine diatom, and its potential for nanotechnology as a natural zipper-like nanoclasp. *Polish Botanical Journal*, submitted.
Tiffany, M. A., Nagy, S. S., & Gordon, R. (2009). The buckling of diatom valves. In *North American Diatom Symposium (NADS), September 23–27, 2009, Iowa Lakeside Laboratory near Milford, Iowa*, ed. M. B. Edlund & S. A. Spaulding, Milford, IA: Iowa Lakeside Laboratory, University of Iowa, pp. 37–8.
Tomczak, M. M., Slocik, J. M., Stone, M. O., & Naik, R. R. (2007). Bio-based approaches to inorganic material synthesis. *Biochemical Society Transactions*, **35**, 512–5.
Tropper, C. B. (1975). Morphological variation of *Achnanthes hauckiana* (Bacillariophyceae) in the field. *Journal of Phycology*, **11**, 297–302.
Turner, J. S. (2007). *The Tinkerer's Accomplice: How Design Emerges from Life Itself*, Cambridge, MA: Harvard University Press.
Tuszynski, J. A. (ed.) (2006). *The Emerging Physics of Consciousness*, Heidelberg: Springer-Verlag.
Van de Vijver, B. & Bogaerts, A. (2010). Proceedings of Diatom Taxonomy in the 21st Century: Symposium in Honour of H. Van Heurck. *Systematics and Geography of Plants*.
Van de Vijver, B. & Lange-Bertalot, H. (2008). *Cymbella amelieana* sp. nov., a new large *Cymbella* species from Swedish rivers. *Diatom Research*, **23**, 511–18.
Vicsek, T. (1992). *Fractal Growth Phenomena*, Singapore: World Scientific.
Villareal, T. A., Joseph, L., Brzezinski, M. A., *et al.* (1999). Biological and chemical characteristics of the giant diatom *Ethmodiscus* (Bacillariophyceae) in the central North Pacific Gyre. *Journal of Phycology*, **35**, 896–902.
Visnovsky, S. (2006). *Optics in Magnetic Multilayers and Nanosctructures*, Boca Raton, FL: CRC Press.
von Dassow, P., Petersen, T. W., Chepurnov, V. A., & Armbrust, E. V. (2008). Inter- and intraspecific relationships between nuclear DNA content and cell size in selected members of the centric diatom genus *Thalassiosira* (Bacillariophyceae). *Journal of Phycology*, **44**, 335–49.
Weatherspoon, M. R., Haluska, M. S., Cai, Y., *et al.* (2006). Phosphor microparticles of controlled three-dimensional shape from phytoplankton. *Journal of the Electrochemical Society*, **153**, H34–H37.
Wee, K. M., Rogers, T. N., Altan, B. S., Hackney, S. A., & Hamm, C. (2005). Engineering and medical applications of diatoms. *Journal of Nanoscience and Nanotechnology*, **5**, 88–91.
Werner, J. S., Donnelly, S. K., & Kliegl, R. (1987). Aging and human macular pigment density: appended with translations from the work of Max Schultze and Ewald Hering. *Vision Research*, **27**, 257–68.
West-Eberhard, M. J. (2002). *Developmental Plasticity and Evolution*, Oxford: Oxford University Press.
Witten, T. A., Jr. & Sander, L. M. (1981). Diffusion-limited aggregation, a kinetic phenomenon. *Physical Review Letters*, **47**, 1400–3.
Woods, D. F. & Bryant, P. J. (1993). Apical junctions and cell signalling in epithelia. *Journal of Cell Science*, 171–81.
Xu, M. J., Gratson, G. M., Duoss, E. B., Shepherd, R. F., & Lewis, J. A. (2006). Biomimetic silicification of 3D polyamine-rich scaffolds assembled by direct ink writing. *Soft Matter*, **2**, 205–9.
Xu, X. J., Cai, P. G., Ye, Q. L., Xia, A. G., & Ye, G. X. (2005). Effects of long-range magnetic interactions on DLA aggregation. *Physics Letters A*, **338**, 1–7.
Xu, X. J., Wu, Y. Q., & Ye, G. X. (2008). Two-dimensional magnetic cluster growth with a power-law interaction. *Applied Surface Science*, **254**, 3249–54.
Yates, F. E., GarfinkeL, A., Walter, D. O., & Yates, G. B. (1988). *Self-Organizing Systems: The Emergence of Order*, New York, NY: Plenum Press.
Zhong, W. (2008). Timing cell-fate determination during asymmetric cell divisions. *Current Opinion in Neurobiology*, **18**, 472–8.
Zorba, S., Shapir, Y., & Gao, Y. L. (2006). Fractal-mound growth of pentacene thin films. *Physical Review B*, **74**, 245410.

Part VI

Conclusions

36 Epilogue: reflections on the past and a view to the future

JOHN P. SMOL AND EUGENE F. STOERMER

The preceding chapters provide a summary of the major advances (and challenges) related to the application of diatoms to environmental and earth-science issues. A simple comparison between the content and diversity of chapters published in the first edition of this book in 1999 and the current volume should provide considerable satisfaction to practitioners and users of these data. Although many challenges remain, progress has clearly been made on many fronts. Nonetheless, and at the risk of ending the book on a pessimistic note, we reiterate below some of our concerns about the lack of progress and resources dedicated to some of the more basic and fundamental issues related to diatoms.

In the previous edition of this book, we were so bold as to attempt an assessment of future directions in diatom research applications. It seems appropriate, and perhaps humbling, to assess those prognostications and their implications for future directions. Our (1999) statement that "It is very clear that a good deal of effort needs to be devoted to the formalities of taxonomy and nomenclature, which have been sadly neglected for the past century" remains virtually unchanged today. In fact, our enthusiasm for the "great strides (that) have been made very recently in the alpha level taxonomy of diatoms" has proven to be rather naive. Significant increases in taxonomic resolution have certainly been made, partially through application of objective techniques, such as cladistic analysis and numerical shape analysis, but perhaps more importantly through simple, but more thorough, exploration of a much wider range of regions and habitats. This has led to a substantial expansion in the number of regional floristic treatments available, but a more modest increase in the number of professional diatom systematists equipped to refine and order these results. The end result is that the task of those using diatoms in applied studies is not really easier than a decade ago due simply to the mass of un-systematized primary literature now available. Indeed, the most commonly used general freshwater flora reference (Krammer & Lange-Bertalot, 1986–1991) now appears somewhat outdated in both philosophy and content. The early volumes of this series reflected the compressed taxonomic structure of early comprehensive works (Hustedt, 1927–1966). The evolution toward expanded understanding visible in later volumes has increased and continues to accelerate (Krammer, 2000; Lange-Bertalot, 2001; Krammer, 2002, 2003) but has not yet resulted in a comprehensive flora. Given the current rapid and apparently continuing changes in diatom taxonomy and nomenclature, and the technological tools now available, it may be that the days of conventional published floras are at an end. Although not yet fully realized, the Internet offers the possibility of continuously updated, refined, expanded floristic treatments. The possibilities of the Internet as an aid to applied studies are exemplified by, for example, the Internet-based "Catalogue of Diatom Names" (Fourtanier & Kociolek, 2009a,b; http://research.calacademy.org/research/diatoms/names/index.asp), a resource that should make it easier for researchers working on applied problems to sort out the nomenclatural problems that plague many diatom studies.

It remains true that current institutional priorities have not been conducive to placing diatom studies on the same firm scientific base as better-studied groups. Particularly in the United States, several museums that have traditionally served as data and specimen repositories, as well as centers for training, remain on precarious footing. It remains to be seen if the encouraging growth of university laboratories can be maintained, but prospects are not promising. Universities are increasingly faced with a difficult financial situation and the temptation to emphasize research in the better-funded medical and technical fields, at the expense of traditional biology

The Diatoms: Applications for the Environmental and Earth Sciences, 2nd Edition, eds. John P. Smol and Eugene F. Stoermer. Published by Cambridge University Press.

and ecology, increases. Even laboratories successful in applied diatom studies continue to suffer in the increasingly stringent assessment of university financial officers. This problem is recognized by some progressive agencies charged with oversight of the Earth's condition, but they also suffer financial constraints in effectively meeting their mandates.

In the previous volume we envisioned the development of automated enumeration systems or, at least, computer-aided identification systems providing the means of producing more comprehensive population data to support applied studies. Although considerable effort has been devoted to such developments (Kingston & Pappas, 2009), including large and reasonably well-funded studies (e.g. du Buf *et al.*, 1999; du Buf & Bayer, 2002), the general application of these technologies remains elusive.

Given our, at best, modest success in predicting future developments in the previous volume, it is still tempting to make such projections. Future gains in understanding should come from a better comprehension of diatom evolution in the broadest sense. This includes both increased taxonomic resolution and knowledge of the entrained structural and physiological factors involved in influencing species distributions. Based on recent advances in diatom evolutionary history and related fields, the prospects in this area appear bright (Julius & Theriot, this volume). In the marine realm, complete genomes have been published for a centric (Armbrust *et al.*, 2004) and a pennate (Bowler *et al.*, 2008) diatom. Although the evolutionary history of diatoms seems to become ever more complicated (Moustafa *et al.*, 2009), it is clear that we have entered into a new era of understanding. How rapidly this knowledge will be fully translated into applied studies remains to be seen. DNA "barcoding" may eventually provide a means of directly applying genomic techniques to routine identification (Evans *et al.*, 2007).

One of the first areas clarified may be that of issues related to diatom biogeography. It was long thought that diatom species had wide distribution, and this notion was provided with a theoretical basis by Finlay *et al.* (2002). However, some evidence suggests the opposite and tools are now in hand to settle the question. For example, some studies (e.g. Medlin *et al.*, 1991; Amato *et al.*, 2007; Rynearson *et al.*, 2009) have shown that some "common and well known" marine species, that were the basis of suggested pan-global distribution, are actually composite (i.e. the same name applied to taxa that are actually different). There is also evidence (Canter & Jaworski, 1983; Pappas & Stoermer, 2001) that some of the "most common" freshwater species are similarly misconstrued. Application of multiple techniques (Mann *et al.*, 2004) has proven fruitful in resolving relationships in particularly difficult species complexes. The controversial nature of this topic is perhaps evidenced by the fact that there is considerable difference of opinion amongst investigators specializing in various aspects of research involving diatoms, including ourselves. For example, JPS is of the opinion that there are relatively few effective barriers to diatom dispersal, hence diatom community composition and structure is largely controlled by local environmental conditions. EFS, on the other hand, believes that biogeographic factors play larger roles. Having agreed to disagree, we leave the resolution of this topic to the next volume.

Recent advances in genomics and other approaches to decoding taxonomic diversity should also allow researchers to recognize evolutionary changes in their study sets and utilize this information constructively. The most obvious application is in paleoceanography (see, for example, Jordan & Stickley, this volume; Leventer *et al.*, this volume; Romero & Armand, this volume). However, similar evolutionary changes are recognized in ancient lakes (see Mackay *et al.*, this volume), and it has been shown that speciation events can be discernable in post-Pleistocene assemblages (Theriot & Stoermer, 1984; Theriot *et al.*, 2006). We consider it likely that such changes may also be discernable in modern populations under extreme forcing from factors such as radiation, chemical pollution, and climate change. Fully understanding the implications of living in an evolutionary unstable and environmentally stressful world will require long-term studies providing annual-to-decadal resolution, and diatoms appear to be one of the most potent tools for such studies (Smol, 2008).

Clearly, though, despite the concerns we mention above, tremendous progress has been made in applied diatom studies since the publication of the first edition of this book. This progress is evident in the expansion of the number of topics treated in this current volume, and progress and changes evident in revisions of chapters that were included in the first edition. Several new tools and techniques have been introduced, and many tools and techniques have been refined. As summarized by Birks (this volume), much progress has been made in the numerical treatment of diatom data. We think it encouraging that these techniques are now widely accepted and applied in environmental and paleoecological studies. Our general sense is that, as remarkable and useful as many applications have been with developments such as transfer functions and environmental inferences, much of the basic ecological data in, for example, surface-sediment calibration sets remains untapped. We feel that there is doubtless information to be gained in sampling techniques that offer higher spatial and temporal resolution. In a very real sense, we have probably only begun to mine the potential wealth of information diatom studies can provide that are essential to resolving the most pressing problems of the Anthropocene era.

Much has been accomplished for which diatom researchers can be justifiably proud. However, many great works remain to be done.

References

Amato, A., Kooistra, W. H. C. F., Ghiron, J. H. L., *et al.* (2007). Reproductive isolation among sympatric cryptic species in marine diatoms. *Protist*, **158**, 193–207.

Armbrust, E. V., Berges, J. A., Bowler, C., *et al.* (2004). The genome of the diatom *Thalassiosira pseudonana*, ecology, evolution, and metabolism. *Science*, **306**, 79–86.

Bowler, C., Allen, A. E., Badger, J. H., *et al.* (2008). The *Phaeodactylum* genome reveals the evolutionary history of diatom genomes. *Nature*, **456**, 239–44.

Canter, H. M. & Jaworski, G. H. M. (1983). A further study on parasitism of the diatom *Fragilaria crotonensis* Kitton by chytridiaceous fungi in culture. *Annals of Botany*, **52**, 549–63.

du Buf, J. M. H. & Bayer, M. M. (2002). *Automatic Diatom Identification. Machine Perception and Artificial Intelligence, Volume* 51, Singapore: World Scientific Publishing.

du Buf, J. M.H., Bayer, M. M., Droop, S. J. M., *et al.* (1999). Diatom identification: a double challenge called ADIAC. Proceedings of the 10th International Conference on Image Analysis and Processing, Venice, September 27–29, pp. 734–9.

Evans, K. M. A., Wortley, H., & Mann, D. G. (2007). An assessment of potential diatom "barcode" Genes (cox1, rbcL, 18S and ITS rDNA) and their effectiveness in determining relationships in *Sellaphora* (Bacillariophyta), *Protist*, **158**, 349–64.

Finlay, B. J., Monaghan, E. B., & Maberly, S. C. (2002). Hypothesis: the rate and scale of dispersal of freshwater diatom species is a function of their global abundance. *Protist*, **153**, 261–73.

Fourtanier, E. & Kociolek, J. P. (2009a). Catalogue of diatom names: part I: introduction and bibliography. *Occasional Papers of the California Academy of Sciences*, **156** (1).

Fourtanier, E. & Kociolek, J.P. (2009b). Catalogue of diatom names: part II: *Abas* through *Bruniopsis*. *Occasional Papers of the California Academy of Sciences*, **156** (2).

Hustedt, F. (1927–1966). Die Kieselalgen Deutschlands, Österreichs und der Schweiz mit Berücksichtigung der übrigen Länder Europas sowie der angrenzenden Meeresgebeite. In *Rabenhorst's Kryptogramen-Flora von Deutschland, Österreich und der Schweiz*. Leipzig: Akademische Verlagsgesellschaft.

Kingston, J. C. & Pappas, J. L. (2009). Quantitative shape analysis as a diagnostic and prescriptive tool in determining *Fragilariaforma* (Bacillariophyta) taxon status. *Nova Hedwigia, Beiheft*, **135**, 103–19.

Krammer, K. (2000). *Diatoms of Europe. Diatoms of the European Inland Waters and Comparable Habitats. Volume 1. The genus* Pinnularia. Ruggell: A.R.G. Gantner Verlag.

Krammer, K. (2002). *Diatoms of Europe. Diatoms of the European Inland Waters and Comparable Habitats. Volume 3.* Cymbella. Ruggell: A.R.G. Gantner Verlag.

Krammer, K. (2003). *Diatoms of Europe. Diatoms of the European Inland Waters and Comparable Habitats. Volume 4.* Cymbopleura, Delicata, Navicymbula, Gomphocymbellopsis, Afrocymbella. Ruggell: A.R.G. Gantner Verlag.

Krammer, K. & Lange-Bertalot, H. (1986). Bacillariophyceae, 1 Teil, Naviculaceae. In *Süsswasserflora von Mitteleuropa, Band* 2, ed. H. Ettl, J. Gerloff, H. Heynig, & D. Mollenhauer, Stuttgart: Gustav Fischer.

Krammer, K. & Lange-Bertalot, H. (1988). Bacillariophyceae, 2 Teil, Bacillariaceae, Epithemiaceae, Surirellaceae. In *Süsswasserflora von Mitteleuropa, Band* 2, ed. H. Ettl, J. Gerloff, H. Heynig, & D. Mollenhauer, Jena: Gustav Fischer.

Krammer, K. & Lange-Bertalot, H. (1991a). Bacillariophyceae, 3 Teil, Centrales, Fragilariaceae, Eunotiaceae. In *Süsswasserflora von Mitteleuropa, Band* 2, ed. H. Ettl, J. Gerloff, H. Heynig, & D. Mollenhauer, Stuttgart/Jena, Gustav Fischer.

Krammer, K. & Lange-Bertalot, H. (1991b). Bacillariophyceae, 4 Teil, Achnanthaceae, Kritische Erganzungen zu Navicula (Lineolatae) und Gomphonema. In *Süsswasserflora von Mitteleuropa, Band* 2, ed. H. Ettl, J. Gerloff, H. Heynig, & D. Mollenhauer, Stuttgart/Jena: Gustav Fischer.

Lange-Bertalot, H. (2001). *Diatoms of Europe. Diatoms of the European Inland Waters and Comparable Habitats. Volume 2. Navicula* sensu stricto. *10 Genera separated from Navicula* sensu lato. Frustulia. Ruggell: A.R.G. Gantner Verlag.

Mann, D. G., McDonald, S. M., Bayer, M. M., *et al.* (2004). The *Sellaphora pupula* species complex (Bacillariophyceae): morphometric analysis, ultrastructure and mating data provide evidence for five new species. *Phycologia*, **43**, 459–82.

Medlin, L.K., Elwood, H. J., Stickel, S., & Sogin M. L. (1991). Morphological and genetic variation within the diatom *Skeletonema costatum* (Bacillariophyta): evidence for a new species, *Skeletonema pseudocostatum*. *Journal of Phycology*, **27**, 514–24.

Moustafa, A., Beszteri, B., Maier, U. G., *et al.* (2009). Genomic footprints of a cryptic plastid endosymbiosis in diatoms. *Science*, **324**, 1724–6.

Pappas, J. L. & Stoermer, E. F. (2001). Fourier shape analysis and fuzzy measure shape group differentiation of Great Lakes *Asterionella* Hassall (Heterokontophyta, Bacillariophyceae). In *Proceedings of the Sixteenth International Diatom Symposium*, ed. A. Economou-Amilli, University of Athens: Amvrosiou Press, pp. 485–501.

Rynearson, T. A., Lin, E. O., & Armbrust, E.V. (2009). Metapopulation structure in the planktonic diatom *Ditylum brightwellii* (Bacillariophyceae). *Protist*, **160**, 111–21.

Smol, J. P. (2008). *Pollution of Lakes and Rivers: a paleoenvironmental Perspective*, 2nd edition, Oxford: Blackwell-Wiley.

Theriot, E. C., Fritz, S. C., Whitlock, C., & Conley, D. J. (2006). Late Quaternary rapid morphological evolution of an endemic diatom in Yellowstone Lake, Wyoming. *Paleobiology*, **32**, 38–54.

Theriot, E. C. & Stoermer E. F. (1984). Principal component analysis of *Stephanodiscus*: observations on two new species from the *Stephanodiscus niagarae* complex. *Bacillaria*, **7**, 37–58.

Glossary, acronyms, and abbreviations

‰: abbreviation for per mil (or mille, permil); one part per thousand.

δ: the delta notation, which is a relative measure of the difference, typically ratios, of molar concentrations of stable isotopes between a substance and an agreed-upon standard material.

abyssalpelagic zone a subdivision of the water column between the depths of 2000 and 6000 m.

A:C ratio: index of lake trophic status based on the number of frustules of planktonic araphid diatoms divided by the number of frustules of centric diatoms. No longer widely used.

accretion: a gradual increase brought about by natural forces over a period of time; the upward growth of the sedimentary column due to accumulation of matter (= vertical accretion).

ACID: acidity index for diatoms.

acid waters: streams and lakes with a pH value less than 7 or, in surface-water acidification studies, streams and lakes with an acid-neutralizing capacity less than or equal to 0.

acidified waters: streams and lakes that have become increasingly acidic through time.

acidobiontic: according to the pH system of F. Hustedt, acidobiontic diatoms attain their greatest abundances at pH values less than 5.5.

acidophilous: according to the pH ecological classification system of F. Hustedt, acidophilous diatoms are most abundant at pH values below 7.

acytokinesis: cell division limited to multiplication of nuclei but lacking division of cytoplasm.

adnate diatoms: diatoms that grow flat, tightly attached to substrates.

aerial diatoms: diatoms inhabiting periodically to permanently damp substrates which are not aquatic, but are rather exposed to the air. Examples of aerial habitats include soils, plants, bryophytes, rock walls, wood, etc.

aerophilic: "Air loving"; term used for diatoms occurring commonly in subaerial environments, often living exposed to air, and not totally submerged under water (= aerophilous).

aeroterrestrial: adjective describing algae that live both in soils and on prominent, raised objects and receiving moisture only from the atmosphere.

AFDM: ash-free dry mass.

afforestation: the planting or natural regrowth of trees, often of conifer forests on land previously covered by heathland, moorland, grassland, or bog.

agglomerative clustering: methods of cluster analysis that begin with each individual sample in a separate group and then, in a series of steps, combine samples and, later, clusters into new, larger clusters until all samples are members of a single group.

AIC: Akaike Information Criterion.

AIES: Autecological Index of Environmental Stressors.

air roots see pneumorrhizae.

Akaike Information Criterion: numerical criterion introduced by Akaike in 1969 for choosing between competing statistical models.

albedo: the reflectivity of a surface to solar radiation. An ideal white body has an albedo of 100%, while an ideal black body has an albedo of 0%. Typically, albedo ranges from 3% for water at small zenith angles to over 95% for freshly fallen snow.

algal turf: lawn-like growth of filamentous seaweeds.

algorithm: a stated procedure consisting of a series of steps, often repetitive, for solving a problem.

alkalibiontic: diatoms that are most abundant at pH values above 7.0, and assumed to require this condition for growth.

alkalinization: an increase of pH in aquatic systems. This may arise through changes in land use, the natural evolution of lakes or through deliberate treatment (e.g. through liming of lakes which have previously been acidified).

alkaliphilic: diatoms that prefer habitats in which the pH is > 7.0 (= alkaliphilous).

alkaliphilous: diatoms that prefer habitats in which the pH is > 7.0 (= alkaliphilic).

alkenone sea-surface temperature estimates: alkenones are organic compounds. Some organisms, such as some phytoplankton, respond to changes in water temperature by altering the production of long-chain unsaturated alkenones in the structure of their cell. Temperatures can be estimated from alkenones recovered from marine sediments using the ratio of certain alkenones.

allergen: substance that causes an allergic reaction.

allochthonous: material introduced into aquatic environments from external (usually terrestrial) sources. The term may be used in contrast to autochthonous, which refers to material that originates within the aquatic environment.

alluvial: relating to the products of sedimentation by rivers or estuaries.

alpha (α)-mesohaline: salinity between 10 and 18 ppt.

alpha (α)-oligohaline: salinity between 3 and 5 ppt.

amber: fossilized tree resin that may contain small organisms which became trapped while the exudation was still fluid.

amictic: lakes that do not undergo vertical mixing. Includes lakes that possess a perennial ice cover that prevents wind-induced mixing. Such lakes are limited primarily to Antarctica, but are also rarely found in the High Arctic and at high elevations.

AMD: acid mine drainage.

amnesic shellfish poisoning: human illness (often referred to as a syndrome) caused by the consumption of shellfish or other seafood products contaminated with the toxin domoic acid; symptoms appear within 3–5 hours and may include nausea, vomiting, diarrhoea, abdominal cramps, dizziness, hallucinations and, in extreme cases, permanent loss of short-term memory and even death.

amorphous opaline silica: a non-crystalline isotropic mineral ($SiO_2 \cdot nH_2O$). It is often found in the siliceous skeletons of various aquatic organisms, including diatoms.

amphiphilic: characterizes a substance as being both hydrophilic and lipophilic. That means the substance is readily soluble in polar as well as in non-polar solvents.

Amphoroid: a dorsiventral group of marine and freshwater biraphid diatoms characterized by a narrow ventral side and girdle bands and a broader dorsal side so that both valve faces are oriented near the same plane; includes the genera *Amphora*, *Catenula*, *Seminavis*, and *Halamphora*.

anadromous: adjective describing organisms that migrate from the sea to fresh water to spawn.

analcime: a mineral ($NaAlSi_2O_6 \cdot H_2O$); a type of zeolite commonly found in alkaline-rich basalts.

analog matching: a numerical technique in which a similarity or dissimilarity index is used to assess if "fossil" assemblages are represented in a modern "training set" (i.e. are the fossil assemblages represented by modern assemblages?). The analog-matching technique can also be used to estimate limnological conditions for fossil samples by assuming that the "best" matches, or the average of a specified number of best matches (or weighted average, where the percent similarity/dissimilarity is used as the weight) of fossil samples with "modern" samples provides an adequate estimate of the environmental conditions of the fossil sample.

ANC: acid neutralizing capacity; the equivalent capacity of a solution to neutralize strong acids.

ANN: artificial neural networks.

anoxia: the condition of zero oxygen. Anoxic waters contain no dissolved oxygen.

Antarctic Circumpolar Current: an ocean current, sometimes known as the West Wind Drift, that flows eastwards around Antarctica. It is the dominant feature of the Southern Ocean and has a mean temperature of −1 to 5 °C, and a mean salinity of 34.9 to 34.7 psu.

Antarctic Convergence: also termed the Antarctic Polar Front, the phrase refers to a margin around Antarctica where cold, northward-flowing Antarctic waters converge and mix with warmer waters of the Subantarctic. Cold Antarctic waters sink beneath the subantarctic waters, and the associated mixing zones create waters high in productivity. The zone of the Antarctic Convergence is mobile and varies about one half degree of latitude from its mean position.

Antarctic Polar Front: see Antarctic Convergence.

ANTARES: Antarctic Research (French JGOFS program).

Anthropocene: the most recent period in Earth's history associated with significant human impact.

APF: Antarctic Polar Front.

apicomplexans: a parasitic protist group diagnosed by the presence of a unique organelle called an apicoplast.

apomorphic: in phylogenetic systematics, a derived or evolutionary character state.

apparent root-mean-square error: an estimate of the predictive ability of a model based on samples that were used to derive that model, in contrast to error derived from an independent data set. This error is calculated from the sum of the observed minus-inferred values, divided by the total number of samples.

aquifer: a body of rock that contains sufficient saturated permeable material to conduct groundwater.

araphid diatoms: common term for pennate diatoms that lack a raphe.

artificial neural networks: a mathematical structure modeled on the human neural network and designed to solve many numerical problems, particularly in the fields of pattern recognition, multivariate analysis, learning, and memory.

ash-free dry mass: the difference between dry mass and mass of inorganic material that does not burn at high temperatures (ash).

ASP: amnesic shellfish poisoning.

assembly rules: empirically or theoretically derived rules about how organisms group themselves into assemblages.

atoll: barrier reef with islets (motu) and enclosed lagoon developed on top of a sunken volcanic peak.

attributes: characteristics of assemblages that are measured to assess biotic integrity and the environmental stressors affecting biotic integrity (see metrics).

austral: applied or pertaining to the southern biogeographic region.

autecological: adjective referring to the ecological conditions in which a specific species (taxon) occurs.

autecological index: a quantitative inference about environmental conditions in a habitat based on the species composition of organisms in the habitat and the ecological conditions in which those species are usually found.

Autecological Index of Environmental Stressor: an index of water-quality impairment based on the abundance-weighted optima (= weighted average) and tolerances of species along an environmental stressor gradient.

autecology: the study of the relationships between individual organisms or taxa and their environment.

authigenic: a diagenetic mineral or sedimentary rock deposit formed *in situ*.

autochthonous: formed or produced in the place where it is found (as opposed to allochthonous).

autocorrelation coefficient: internal correlation of the observations in a time series, usually expressed as a function of the time lag between observations.

autocorrelogram: a plot of the values of the autocorrelation against the time lag.

automixis: a form of parthenogenesis in which the chromosomes of a haploid gamete divide to form a diploid nucleus without nuclear division.

auxospore: a cell type unique to diatoms that is destined to restore large-size individuals in a diatom population, usually following sexual process. Depending on the species and the stage of development, the auxospore may contain one diploid nucleus (zygote, 2n, a product of syngamy) or two haploid nuclei (n + n, prior to syngamy) in addition to a varying number of degenerating pycnotic nuclei not involved in syngamy.

auxosporulation: the formation of an auxospore, which is a special cell, usually a zygote, produced by diatoms which expands to near the maximum size of a given species, thus compensating for the cell size diminution that occurs during vegetative cell division.

avulsion: the process by which a meandering river creates a new course through erosion of floodplain material, often leaving in its wake floodplain lakes.

bagging tree: a classification or regression tree produced by replicates of the data set in a classification problem and producing an allocation rule for each replicate.

Baikal Drilling Project: a multinational (Russia, Japan, Germany, USA) program (1989–1999) to extract long sedimentary records from Lake Baikal to reconstruct global climate change and tectonic evolution of the Lake Baikal sedimentary basin.

ballast hypothesis: the idea put forward by Klaas and Archer in 2002 that organic matter and ballast mineral fluxes are tightly related below 1000 m.

bar-built estuary: estuaries that are shallow and are separated from the sea by sand spits or barrier islands.

Bartonian: the second geological stage of the middle Eocene (approx. 40–37 Ma), named after the Barton Beds of southern England. It succeeds the first stage of the middle Eocene, the Lutetian.

base cations: the most prevalent, exchangeable cations in soils such as calcium, magnesium, potassium, and sodium that protect soils and surface waters from acidification.

bathypelagic zone: a subdivision of the water column between the depths of 1000 and 2000 m.

baymouth bar: a sandbar formed across the opening of a coastal bay, caused by current transport of sand along the coast. This process eventually results in the formation of a lake as the bay becomes isolated from the larger body of water adjacent to it.

Bayesian inference: an approach to inference based on Bayes' theorem concerned with the consequences of modifying our previous beliefs as a result of acquiring new data. It contrasts with the "classical" approach to inference which begins with a hypothesis test that proposes a specific value for an unknown parameter.

Benguela upwelling system: oceanographic feature, associated with the South Atlantic high pressure cell, located off the southwest coast of Africa. Basically, coastal surface waters are blown offshore by the trade winds, which allow nutrient-rich deeper waters to replace them.

benthic: adjective describing organisms attached to substrates (e.g. rocks, sand, mud, logs, and plants) or otherwise living on the bottoms of aquatic ecosystems (see periphyton).

benthos: organisms living on the bottom of aquatic systems, either attached to substrates or moving along the bottom.

beta-diversity: an aspect of biodiversity obtained by comparing the species diversity between habitats and often involving comparison of the number of taxa that are unique to each of the habitats.

beta (β)-mesohaline: salinity between 5 and 10 ppt.

beta (β)-oligohaline: salinity between 0.5 and 3 ppt.

Bhalme–Mooley Drought Index: an index of drought intensity calculated from monthly precipitation measurements from individual meteorological stations.

bilateral asymmetry: generally refers to an organism that deviates in some aspect from its overall bilateral symmetry.

bilateral symmetry: a basic body plan in which the left and right sides of the organism can be divided into approximate mirror images of each other along the midline.

billabong: Australian term for a stagnant backwater or slough formed by receding floodwater or a branch of a river flowing away from the main stream but leading to no other body of water.

binary divisive techniques: a partitioning procedure that divides a set of observations into two groups, four groups, eight groups, etc., until the groups are too small for further division or their numerical properties suggest further division is not warranted.

binomial error structure: distribution of errors under the assumption of a binomial error distribution where in a series of independent trials the probability of success at each trial is p and the probability of failure q is $1 - p$.

biocoenotic/biocoenosis: a community of organisms inhabiting an area in which the main environmental conditions and organisms adapted to them are uniform.

biofilm: microbial (algal, bacterial, fungal) community in a structured mucilage matrix (produced by the microbial cells) typically on a substrate.

biogenic: formed by organisms.

biogenic silica: also called "biogenic opal." $SiO_2 \cdot nH_2O$, the mineral of which diatom frustules are composed, as well as the siliceous components of radiolarians, chrysophyte algae, sponge spicules, etc.

biogeochemistry: the branch of science that deals with the relation of earth chemicals to biology.

biolog: the stratigraphic distribution of the abundance and diversity of microfossils.

biological carbon pump: a process whereby carbon associated with surface-dwelling plankton is transported to the sediments as sinking particles (e.g. as aggregates and fecal pellets). The transport of organic carbon may be called the "soft tissues pump," while that of inorganic carbon is known as the "hard tissues pump."

biological condition: the similarity of biological attributes (structural and functional characteristics) at sites being assessed, compared to attributes of communities which were historically or currently considered natural for the region.

biological integrity: a biological condition that is close to natural or having little evidence of disturbance by human activities. Specifically, the similarity of community attributes (structural and functional characteristics) at sites being assessed compared to attributes of communities which were historically or currently considered natural for the region. Sometimes defined as "the capability of supporting a balanced, integrated, adaptive community of organisms having a species composition, diversity, and functional organization comparable to that of natural habitats of the region."

biomanipulation: the deliberate alteration of an ecosystem by adding or removing species, especially predators (typically fish), for the purposes of restoration.

biomarker: an original or modified molecule that can be used to indicate the existence or past existence of a living or fossil organism, respectively. The term is also sometimes used for organisms, in addition to biogeochemical indicators.

biomass: mass of biological components of a habitat or ecosystem.

biomass assay: one of many measurements of biomass. Chlorophyll *a* concentration, ash-free dry mass, cell numbers, and biovolume are some of the approaches used to estimate algal biomass.

BioMEMS: a special class of microelectromechanical system (MEMS) where biological matter is manipulated.

biomimetics: the study of the structure and function of living things as models for the creation of materials or products by reverse engineering.

biomineralization: the process by which living organisms produce minerals, often to support existing tissues.

biostratigraphy: the stratigraphic distribution of fossil species.

biotic integrity: see biological integrity.

bioturbation: the mixing of sediment by organisms.

biovolume: the volume occupied by a cell; typically used in relation to cells of phytoplankton and expressed in μm^3.

biozonation: the division of the stratigraphic record and geologic time into intervals characterized by the ranges of key fossil species.

biplot: an ordination diagram of two kinds of entities (e.g. species and samples) that can be interpreted by the biplot rule. This involves projecting points on directions defined by arrows in the biplot.

black ice: transparent ice, which effectively transmits light (as opposed to white ice).

blanket mire: an extensive peatland that forms in wet climates not only in wet depressions but across the landscape.

blooms: large increases in numbers of planktonic organisms, sometimes associated with eutrophication and seasonal temperature changes.

BMDI: Bhalme–Mooley Drought Index.

bog: an ombrotrophic peatland that is hydrologically isolated from the surrounding landscape and therefore receives water and minerals solely from the atmosphere.

bolide: fireball, a brighter than usual meteor.

Bolidomonads: a picoplanktonic, marine, photosynthetic protist group thought to be the closest living ancestor to the diatoms.

Bond cycle: low-frequency, quasi-cycle which occurred during the last ice age and the Holocene *c.* every 3000–7000 years.

boosted tree: a classification or regression tree that attempts accurate prediction and explanation. It combines the strengths of regression trees (models that relate a response to their predictors by repeated binary splits) and boosting (combining many simple models to give improved predictive performance). The final model is an additive regression model in which individual terms are simple trees, fitted in a forward, stage-wise fashion.

bootstrap: a data-based computer-intensive simulation method for statistical inference that can be used to study the variability of estimated characteristics of the probability distribution of a set of observations and provide confidence intervals for parameters in situations where these are difficult or impossible to derive in conventional ways.

bootstrapping: a computer-intensive statistical resampling procedure that randomly generates "new" data sets (e.g. 1000), with replacement, that are the same size as the original data set. The predictive ability of a model is based on estimates derived on samples when they do not form part of the randomly generated data set.

boreal: applied or pertaining to the northern biogeographic regions, such as the northern boreal forest.

bostrychietum: an algal community on mangrove roots that is dominated by the seaweeds *Bostrychia* and *Caloglossa*.

boxplot: graphical display of the important characteristics of groups of numerical data.

box-whisker plot: a graphical tool for displaying the important features of a set of observations in terms of the median, inter-quartile range (box part) and the "whiskers" extending to include all but the outside observations which are displayed separately.

BP: Before Present, typically refers to the radiocarbon standard of years before AD 1950.

brackish water: saline water with a concentration between freshwater and seawater.

Bray–Curtis coefficient: a coefficient that estimates the similarity between quantitative assemblages in two samples that is commonly used in cluster analysis and some ordination methods.

brine composition: the major ions (cations and anions) that characterize water; usually used in reference to saline water.

broken-stick model: a model based on the broken-stick distribution where a set of objects is taken as equivalent to a stick of unit length that is broken randomly into a number of pieces.

Bronze Age: a period of human culture characterized by the use of weapons and implements made of Bronze. In western Europe it lasted from *c.* 2000 to 500 BC.

Brunhes chron: the final normality polarity chron in the magnetostratigraphic timescale, radiometrically dated from *c.* 0.8 Ma to the present.

bryophilous: see bryophytic.

bryophytic: adjective referring to algae associated with mosses (= epibryophytic).

BSEI: back-scattered electron imagery.

BSi: biogenic silica.

C_3 plants: plants whose carbon-fixation products (i.e. 3-phosphoglyceric acid) have three carbon atoms per molecule.

C_4 plants: plants whose intermediate carbon-fixation products (e.g. malate and oxaloacetate) have four carbon atoms per molecule.

C:P ratio: index of trophic status in lakes, or eutrophication and/or habitat availability in estuarine ecosystems, based on the ratio of centric diatom species to pennate diatom species.

CA: correspondence analysis.

cal yr BP: calendar years before AD 1950.

calcareous nannoplankton: normally refers to small plankton (<100 μm in diameter) such as coccolithophorids and dinoflagellates (e.g. thoracospheres) that bear calcified scales (coccoliths) or tests around the cell. The *nanno* is derived from the Greek meaning dwarf. Also spelled nanoplankton.

calcined: refers to a substance, in this case diatomite, heated to a high temperature, usually in excess of 900 °C but below the melting point, causing the loss of moisture, organics, volatiles, and other impurities.

calibration set: a data set of modern biological assemblages, typically from from surface sediments, and associated environmental data.

canonical analysis of principal coordinates: an ordination method where any dissimilarity coefficient between all pairs of objects can be related to external environmental predictors. Its aims are similar to canonical correspondence analysis or redundancy analysis but allows the researcher to use whatever dissimilarity measure she/he wishes to use.

CANOCO: a FORTRAN computer program for CANOnical COmmunity ordination [partial] [detrended] [canonical] correspondence analysis, principal components analysis, and redundancy analysis written by C. J. F. ter Braak. Now version 4.5+.

canonical correspondence analysis: a constrained ordination technique that uses a weighted-average algorithm, in which ordination axes are constrained to be linear combinations of the supplied environmental variables.

CAP: canonical analysis of principal coordinates.

carbonaceous particles: spheroidal fly-ash particles generated by the high-temperature combustion of coal and oil in power stations; can be identified in soils and in lake sediments.

carbonate compensation depth: the depth in the ocean below which calcium carbonate is readily dissolved.

CART: classification and regression trees.

CCA: canonical correspondence analysis.

CDOM: chromophoric (colored) dissolved organic matter.

Cenozoic: the name of the present geological era that followed the Mesozoic era at about 65 Ma. It means "new life" (*kainos* = new and *zoe* = life in Greek). It consists of three geological periods, the Paleogene, Neogene and Quaternary.

Centrales: one of two traditionally recognized orders of diatoms. Members of this order were distinguished by the deposition of silica around a point and the presence of oogamous sexual reproduction. The group is now regarded as paraphyletic.

centric diatom: a diatom belonging to the class Coscinodiscophyceae, commonly with round or polygonal valves.

centric:pennate ratio: index of numbers of centric (primarily planktonic) to pennate (often benthic) diatom species found in a sample. This index is used in estuarine systems to indicate nutrient enrichment and loss of benthic habitat.

CGO: constrained Gaussian ordination.

chemocline: the water depth at which the most rapid change in chemical conditions occurs.

chemostat: a bioreactor to which fresh medium is continuously added, while culture liquid is continuously removed to keep the culture volume constant. By changing the rate with which medium is added to the bioreactor the growth rate of the microorganism can be controlled.

chert: a fine-grained, silica-rich sedimentary rock, that may contain microfossils. Biogenically derived chert may result from the diagenesis of diatomite or radiolarian ooze.

chi-squared distance: a coefficient that estimates the dissimilarity between assemblages (presence/absence, ranks, or quantitative) in two samples that is central to correspondence analysis. It is similar to Euclidean distance but it compensates for different relative frequencies or probabilities of occurrence. There is also a chi-squared metric.

chlorophyll *a*: the most common pigment in all oxygen-evolving photosynthetic organisms such as higher plants, red and green algae. It is best at absorbing wavelengths in the 400–450 nm and 650–700 nm bands of the electromagnetic spectrum.

chlorophyll filament: large plume of nutrient-rich water, spreading offshore into the open ocean in eastern boundary current systems.

chord distance: a coefficient that estimates the dissimilarity between quantitative assemblages in two samples. It is similar to Euclidean distance but it uses square-root-transformed percentage or proportional data.

chronozone: a chronostratigraphic unit; a unit of rock or sediment that serves as reference for material formed during a specific interval of time.

CIE: controlled isotope exchange.

ciliates: a protist group diagnosed by the presence of hair-like cilia that have small microtubles.

circumneutral: diatoms that have their greatest abundance at pH values around pH 7.

clade: a branch in an evolutionary tree.

cladistic analysis: a numerical analysis that analyzes the presence and absence of shared derived characters (e.g. morphological, molecular data) to generate a tree or branching diagram of the evolutionary or phylogenetic relationships among the test organisms.

CLAG: Critical Loads Advisory Group.

classification and regression trees: a robust alternative to regression techniques for determining subsets of explanatory variables most important in the prediction of the response variable. Rather than fitting a model to the data, a tree structure is generated by dividing the data recursively into a number of groups, each division being chosen so as to maximize some measure of the difference in the response variable in the resulting two groups. If the response variables are presence/absence, a classification tree is generated. If the response variables are quantitative, a regression tree is generated.

clastic: sedimentary material derived from the erosion of pre-existing rocks.

clavate: resembling a club, becoming increasingly wide from the base to the distal end.

Clean Water Act: legislation that is designed to protect the uses of waters in the United States of America.

closed-basin lake: a lake with no surface outlet. Water may flow into the lake by a variety of sources (streams, direct precipitation on lake surface, ground-water inflows), but in closed-basin lakes the main way that water leaves is by evaporative processes. Therefore, closed-basin lakes tend to contain higher concentrations of solutes. The oxygen isotopic composition of closed-basin lake waters tends to be higher than open-basin lake waters due to preferential loss of the lighter isotope during evaporation.

closed compositional data: percentage or proportional assemblage compositional data that have a fixed sum (100 or 1) and are thus "closed."

cluster analysis: a set of methods for constructing a sensible and hopefully a useful grouping of an initially ungrouped set of data using the variable values observed on each object.

clustering techniques: techniques that group initially ungrouped objects into groups using, for example, agglomerative clustering, or non-hierarchical cluster analysis such as k-means cluster analysis.

coccolithophorid: a haptophyte alga that possesses calcified scales (coccoliths) around its cell during part of its life cycle.

coccoliths: individual plates of calcium carbonate formed by the single-celled algae coccolithophorids.

coefficient of determination: the square of the correlation coefficient between two variables, giving the proportion of the variation in one variable that is accounted for by the other.

coefficient of variation: a measure of spread for a data set defined as 100 × (standard deviation / mean).

coeval: occurring at the same time.

community structure: properties of biotic communities such as richness, evenness, relative abundances, biomass, spatial patterns, and dominants.

compensation depth: the depth in the ocean where the rate of supply of a particular material equals its rate of dissolution (thus preventing preservation). Normally pertains to calcium carbonate, i.e. the calcium carbonate compensation depth (CCD).

competitive exclusion: when two or more species share the same habitat and resource base, one species may become dominant and exclude other species.

composite diversity: a characterization of species composition that includes both richness and evenness of the species in the habitat or sample.

compositional change: compositional turnover.

compositional turnover: amount of difference in assemblage composition and/or abundance along a known environmental or temporal gradient or along the major direction of variation in a data set.

compustat: a bioreactor in which, using visual criteria via pattern recognition software, a computer selects which individual microorganisms or mutants will die by microbeam ablation or will be allowed to live and reproduce.

conductivity: a measure of water's ability to conduct an electric current. Conductivity depends on the concentration of dissolved salts and increases with increasing concentration of ions.

confidence intervals: a range of values, calculated from the samples, that are believed, with a particular probability, to contain the true but unknown parameter value. A 95% confidence interval implies that when the estimation is repeated many times, then 95% of the calculated intervals would be expected to contain the true parameter value.

conformable: pertaining to an unbroken sequence of strata, characteristic of uninterrupted deposition.

CONISS: constrained incremental sum of squares.

conopeum: siliceous membrane growing off and along the raphe-bearing sternum, which covers a depression on the valve face running along the raphe.

conspecific: belonging to the same species.

constant rate of supply: an assumption that the absolute flux of ^{210}Pb to the sediment–water interface remains constant, regardless of background sedimentation.

constrained Gaussian ordination: a Gaussian ordination in which the ordination axes of assemblage data are also linear combinations of the external environmental variables.

constrained incremental sum of squares: a stratigraphically constrained cluster analysis where groups of adjacent samples or groups of samples are grouped together so as to minimize the within-group sum of squares and thus maximize the between-group sum of squares. Used for zonation of stratigraphical data.

constrained ordination: an ordination of assemblage data but where the ordination axes are constrained to be linear combinations of the external environmental variables. Also known as canonical ordination.

continuous permafrost: permafrost (frozen ground) that occurs when greater than 90% of the exposed land surface is frozen.

copepod: a small crustacean, often the most abundant group found in marine environments.

correlation: a general term for interdependence between pairs of variables usually quantified by a correlation coefficient such as Pearson's product moment correlation coefficient (r) that estimates the linear relationship between pairs of variables.

correlogram: plot of the values of the autocorrelation coefficient. Same as autocorrelogram.

correspondence analysis: an ordination technique that uses a weighted-average algorithm to maximize the dispersion of species or sites in low-dimensional space.

corrie lakes: lakes formed in small, amphitheater-shaped, glacially eroded basins in mountain regions.

coseismic: the coseismic phase of the seismic–interseismic deformation cycle occurs during the earthquake, where the uplifted region quickly subsides, while the offshore region undergoes rapid uplift; this rapid motion generates a tsunami.

cosmopolitan: with global or near-global distribution (i.e. found almost everywhere). For example, those cosmopolitan marine planktonic diatoms found globally, except in the polar regions, are said to indicate warm(er) oceanic waters.

costae: rib-like structures, generally appearing to arise from the central sternum of many pennate diatoms.

covariables: often used as an alternative name for explanatory or predictor variables, but in the context of ordination the term refers more specifically to variables that are not of primary interest but whose effects need to be included and allowed for in the analysis. Also known as concomitant or background variables. They often correspond to incidental or nuisance parameters.

Cretaceous: the last period of the Mesozoic era, spanning from 145 to 65 Ma. During this time there was a high eustatic sea level, both the climate and oceans were warm, and huge deposits of chalk were laid down on the sea floor (*creta* means chalk in Latin). It is also popularly known as the "Age of the Dinosaurs," and the end of the period is marked by a mass extinction event. The period is often abbreviated to K (*kreide* means chalk in German).

Cretaceous–Tertiary (K–T) boundary: the transition between Mesozoic and Cenozoic sediments, and related to a mass extinction event. Often called the K–T boundary (using the abbreviation of the German word for chalk), although with the term Tertiary recently discouraged, researchers are beginning to refer to it as the Cretaceous–Paleogene (K–Pg) boundary.

cristobalite (opal-CT): a crystalline form of silica, blade-bearing microscopic spheres of cristobalite and/or tridymite are packed together to form opal-CT.

critical loads: the values ascribed to the levels of acid deposition (or other pollutants) that soils and surface waters can tolerate before detrimental effects occur.

cross-correlation coefficient: a measure of the correlation between two variables in a time series estimated at different lags.

cross-spectral function: an estimate of the frequencies and periodicities shared between two variables in time-series data. Each time series is decomposed into an infinite number of periodic components and the contributions of these components in the two variables in certain ranges of frequency (spectrum) are estimated to derive a cross-spectral function.

cross-validation: division of data into a training set which is used to estimate the parameters of the model of interest and a test set which is used to assess whether the model with these parameters fits or predicts well. There are several ways of implementing cross-validation including leave-one-out cross-validation.

CRS: *Chaetoceros* resting spores; also refers to the constant rate of supply model used in ^{210}Pb dating of sediments.

crustose: thin algal thallus closely adherent to a living or non-living substratum.

cryoconite holes: depressions on glacier surfaces filled with water and sediment, named for the "cold rock dust," or wind-blown sediment, they contain.

cryophilic: literally cold loving. Applied to organisms which thrive in cold environments.

cryoturbation: the mixing of sediments as a result of freezing and thawing.

cryptic species: species lacking easily discernable morphologically species-specific characters among sexually incompatible individuals.

cryptophytes: a photosynthetic protist group occurring in freshwater and marine habitats, typically possessing two flagella of unequal lengths.

Cveg: *Chaetoceros* vegetative cells.

CWA: Clean Water Act.

cyanobacteria: unicellular, colonial, and filamentous bacteria containing chlorophyll *a* and evolving oxygen during photosynthesis; also called blue-green algae.

Cyanophycean starch: a polysaccharide storage product characteristic of cyanobacteria and similar to Floridean starch. Specifically, it is an alpha 1,4-linked glucan similar to glycogen and amylopectin.

cytokinesis: division of a cell's cytoplasm.

D: Sometimes used as an abbreviation for (lake) depth. Lake depth is also sometimes referred to as Z.

DA: domoic acid.

DAIpo: Diatom Assemblage Index to Organic Pollution.

Dansgaard relationship: at intermediate and high latitudes the oxygen isotope composition of mean annual precipitation correlates directly with changes in temperature with a gradient of approximately $+0.6‰\ ^{\circ}C^{-1}$.

Dansgaard–Oeschger (D–O) events: events related to Heinrich events but they occur more frequently (*c.* every 1.5 ka) and also penetrate into interglacials. First identified in Greenland ice cores, each event consisted of an abrupt warming to near-interglacial conditions, followed by a gradually cooling climate.

DAR: diatom accumulation rate.

DATA: diatom assemblage type analysis.

daughter cell: one of the two cells produced during each mitotic division of a diatom or other unicellular organism.

DBL: diffusive boundary layer.

DCA: detrended correspondence analysis.

DCCA: detrended canonical correspondence analysis.

DCM: deep chlorophyll maximum.

Deccan Traps: volcanic region situated on the Deccan Plateau of west-central India, which was active across the Cretaceous–Tertiary transition (60–68 Ma).

decision tree: a tree representation of a binary key or division of a set of samples, usually developed by a supervised learning algorithm. Includes classification and regression trees.

deep chlorophyll maximum: the peak concentration of chlorophyll in the subsurface water column of an ocean or lake.

Deep Sea Drilling Project: a pioneering international research program for deep-sea drilling (1968–1983). The forerunner to the ODP.

degree of pyritization: the ratio of pyritic iron to pyrtic iron plus acid-soluble iron (total reactive iron). Lower values of degree of pyritization (e.g. <0.42) indicate more aerobic conditions of overlying water, while higher levels may indicate restricted or inhospitable conditions.

deflate: the action of deflation.

deflation: the sorting, lifting, and removal of loose, dry, fine-grained particles by wind.

delta: delta values (i.e. δ) are typically used to express the difference between the stable isotope ratios (e.g. $^{13}C/^{12}C$) of a sample and a widely accepted standard, relative to the standard. Multiplied by 1000, they are reported in per mil.

deltaic: pertaining to a river delta system under the influence of both freshwater and marine waters.

deltaic fan: a depositional feature containing fine-grained sediment that is formed when a river reaches a body of standing water and flow velocity decreases.

demersal: describing fish that live near the sea bed; also, benthic.

dendrite: a branching structure similar to a tree.

dendrochronology: the science dealing with the study of the annual rings of trees in determining the dates and chronological order of past events.

dendroecology: the science that uses annual tree rings to assess past growth rates, from which past climatic and other environmental variables can be inferred.

deployment: as a verb, placing a mooring off a ship and into the water. As a noun, the time a mooring is resident in the water.

desert oasis: as used in this book, a term applied to the McMurdo Dry Valleys, Antarctica, where precipitation is less than 10 cm per year. In these regions, lakes form from glacial meltwaters. Such lakes may be the only liquid water in otherwise dry (or ice-covered) terrain.

deshrinking regression: regression used in weighted averaging and weighted-averaging partial least squares regression to restore estimated or predicted values to the original scales of the training-set samples. It is necessary because of shrinkage that results from taking averages twice.

detrended canonical correspondence analysis: canonical correspondence analysis where the first and later axes are detrended to remove any curvature or "horseshoe" structure that may be an artifact for a particular data set. If a particular scaling is used, the analysis provides a convenient estimate of compositional change or turnover along particular environmental gradients.

detrended correspondence analysis: the detrended form of correspondence analysis. Detrending is a mathematical technique used to remove the "arch" or "horseshoe effect" on the second axis, which is a mathematical artifact.

DF analysis: discriminant function analysis.

DI: diatom-inferred (as in DI-TP = diatom-inferred total phosphorus concentration).

diadinoxanthin: an accessory pigment, or light-absorbing compound, to chlorophyll found in the diatoms. Diadinoxanthin is one of the pigments of the xanthophyll cycle which functions to absorb and dissipate excess energy from light in photosystem II.

diagenesis: postdepositional physical, biological, and chemical alteration of sediments.

diamicton: poorly sorted sediments originating from glacial (till) or marine and terrestrial slope processes.

diastrophism: the process or processes by which the crust of the Earth is deformed, producing continents, ocean basins, plateaus and mountains, flexures and folds of strata, and faults.

Diatom Assemblage Index to Organic Pollution: mathematical index of water quality based on the relative abundances of diatom taxa in a sample.

diatom flux: the accumulation of diatoms over a set period of time.

Diatom Inferred Trophic Index: developed by Agbeti and Dickman in 1989 and used to estimate changes in lake trophic status based on diatom assemblages. The index is based on multiple regression analyses of log-transformed percent abundances of diatom species that were divided into six indicator categories: (i) oligotrophic; (ii) oligomesotrophic; (iii) mesotrophic; (iv) mesoeutrophic; (v) eutrophic; and (vi) eurytopic. Not commonly used.

diatomaceous earth: see diatomite.

Diatomic Index: a mathematical index of water quality based on the relative abundances of diatom taxa in a sample.

diatomite: a porous, lightweight sedimentary rock resulting from accumulation and compaction of diatom remains.

diatoxanthin: an accessory pigment, or light-absorbing compound, characteristic of the diatoms. It is one of the pigments of the xanthophyll cycle which functions to absorb and dissipate excess energy from light in photosystem II.

DIC: dissolved inorganic carbon.

diffraction grating: optical device consisting of a surface with many parallel grooves in it; disperses a beam of light (or other electromagnetic radiation) into its wavelengths to produce its spectrum.

diffuse sources: sources of nutrients, namely nitrogen and phosphorus, from non-point, dispersed sources such as agriculture and storm runoff.

diffusion-limited aggregation: the process whereby particles undergoing a random walk due to Brownian motion cluster together to form aggregates of such particles.

diffusive boundary layer: a thin layer of fluid across which the concentration of a diffusing entity (heat or solute) changes rapidly.

dimictic: two seasonal periods of water-column circulation occurring in a lake. Typical in temperate lakes.

dinoflagellates: a protist group occurring in freshwater and marine habitats, characterized by the presence of two flagella. One flagellum is trailing while the second wraps around the cell like a belt. This group is responsible for "red tides" in coastal marine environments.

direct gradient analysis: the analysis of assemblage data in which samples are positioned not only on the basis of their assemblages but also in relation to their position along one or more known environmental gradients. If one species is considered, regression analysis is used; if two or more species are considered, constrained ordination is used.

discontinuous permafrost: permafrost that covers between 50 and 90% of the exposed land area, where other areas are free of permafrost.

discriminant analysis: a form of multivariate analysis used to determine significant groups.

discriminant function analysis: linear stepwise regression model that attempts to predict group membership based on a linear combination of the interval variables.

DISeen: the diatom index for German lakes developed by Schaumburg and colleagues in 2007 that uses benthic diatoms to provide a measure of lake ecological status.

dissimilarities: a measure of the difference between two samples from a set of multivariate data. For two objects with identical variable values, the dissimilarity (= distance) is usually defined as zero.

dissimilarity index: a metric used to assess how dissimilar two or more sites are; common metrics include Chi-square, chord, and Euclidean distances.

distance-based RDA: distance-based redundancy analysis.

distance-based redundancy analysis: a form of redundancy analysis where any dissimilarity coefficient can be used in place of the Euclidean distance coefficient that is implicit in redundancy analysis.

DITI: Diatom Inferred Trophic Index.

diversity: a characterization of species composition of a habitat or sample that may include richness (number of species) and/or evenness (their relative abundance) of species.

DLA: diffusion limited aggregation.

DM: dry mass.

DNA: deoxyribonucleic acid.

D–O events: Dansgaard–Oeschger events.

DOC: dissolved organic carbon.

domoic acid: water-soluble, acidic amino acid (kainoid group) containing three carboxylic acids; binds to glutamate receptors (kainate subtype) in the central nervous system, stimulating glutamate release and leading to enhanced ion conduction; causes neuroexcitotoxicity and/or neurodegeneration in the brain; responsible for human intoxication syndrome known as amnesic shellfish poisoning (ASP).

domoic acid poisoning: deleterious effects of domoic acid exposure on birds, marine mammals, and other wildlife (generally distinguished from human intoxication); symptoms include abnormal behavior (e.g. ataxia, lethargy) and seizure activity.

DOP: degree of pyritization.

"doubthouse": the transitional period between the "greenhouse" and "icehouse" periods which pertains to the ~16 Ma-long Cenozoic cooling trend between the end of the Early Eocene Climate Optimum (~52–50 Ma) and the Oi-1 event (~34 Ma). The doubthouse has been thus named because global cooling appears to have occurred in a series of steps interrupted by at least one short warming event (the Middle Eocene Climate Optimum). During the doubt-house, isolated ice caps appeared in both hemispheres and seasonal sea ice became established in the central Arctic.

DPDC: Diatom Paleolimnology Data Cooperative.

drowned river valley: the most common type of estuary in temperate climates, found where sea level is rising relative to the land, and seawater progressively penetrates into river valleys.

DSDP: Deep Sea Drilling Project.

DSi: dissolved silicon; occurs in marine and freshwaters predominantly in the form of $Si(OH)_4$ and is also frequently referred to as silicic acid, (dissolved) silicate, and dissolved silica.

dynamic programming algorithm: an algorithm that optimizes a sequence of decisions in which each decision must be made after the outcome of the previous decision becomes known.

dystrophic: waters with a high concentration of humic matter, typically giving them a brownish-yellowish colour.

eastern boundary currents: ocean currents with dynamics determined by the presence of a coastline. Relatively shallow and slow-flowing, found on the eastern side of oceanic basins (adjacent to the western coasts of continents). Subtropical eastern boundary currents flow equatorward, transporting cold water from higher latitudes to lower latitudes. Coastal upwelling often brings nutrient-rich water into eastern boundary current regions, making them productive areas of the ocean.

EC_{50}: concentration of a compound that causes an effect to 50% of the population.

ecogenetics: a combination of the disciplines ecology and genetics.

ecological integrity: the similarity of ecosystem attributes (structural and functional characteristics) at sites being assessed, compared to attributes of communities which were historically or currently considered natural for the region. Ecological integrity is distinguished from biotic integrity because the former includes physical habitat integrity and water-quality characteristics, as well as biotic integrity of assemblages (see also biological integrity).

ecological quality ratio: the ratio between assessed condition and reference condition.

ecotone: a narrow transition zone or region that separates two or more regional biozones.

ecotope: a geographical unit of ecological homogeneity characterized by a specific combination of abiotic conditions.

ecotype: a genetically distinct geographic population within a species, which is adapted to specific environmental conditions such as local selective pressure or physical isolation. Typically, ecotypes exhibit phenotypic differences (such as in morphology or physiology) stemming from environmental heterogeneity and are restricted to one habitat. They can interbreed with other ecotypes of the same species without loss of fertility or vigor.

EDA: exploratory data analysis.

edaphic: in phycological terms, adjective describing algae within or associated with marine intertidal sediments. Also refers to soil conditions.

EDDI: European Diatom Database.

edge effect: refers to the bias introduced into estimates of species optima when lake surveys do not sample the entire range of environmental conditions that a species inhabits. Edge effects refer to truncated species responses at the ends of environmental gradients (e.g. at very low or high values).

EDS: energy dispersive X-ray spectroscopy.

Eemian: the most recent interglaciation; named for marine deposits along a small stream in the eastern Netherlands.

eigenvalue: if A is a square matrix, x is a column vector not equal to 0, and λ is a scalar so that $Ax = \lambda x$, then x is an eigenvector of A and λ is the corresponding eigenvalue. In CA or PCA, the eigenvalue of each axis reflects the proportion of the total variance accounted for by that axis. A measure of the importance of an ordination axis.

El Niño: literally the boy child – refers to the characteristic timing, near Christmas, of the manifestations of El Niño Southern Oscillation (ENSO) events on the South American coast.

electrodeposition: the deposition of a substance on an electrode by the action of electrical current.

EMAP: Environmental Monitoring and Assessment Program (USEPA).

endemic: geographically restricted (i.e. only found in one region). Endemism may be caused by physical and/or oceanographic barriers. Endemic diatom floras exist today, for example, in the Southern Ocean and in the middle-Eocene Arctic. They can similarly occur in freshwater environments. For example, the endemic diatom *Cyclotella baicalensis* is only found in Lake Baikal, Russia.

endosymbiont organism: one which lives within the cells of another organism, often in a mutually beneficial arrangement (for example, algae in coral polyps).

ENSO (El Niño Southern Oscillation): a climate feature of the Pacific Ocean characterized by east–west shifts of the tropical pressure gradient, occurring every four to seven years. During ENSO events, waters of the eastern Pacific are abnormally warm.

Eocene: the second epoch of the Paleogene, spanning from ~56–34 Ma. Its name means "new dawn" (*eos* = dawn and *kainos* = new in Greek) in reference to the emergence of modern mammal groups. The epoch witnessed the transition from a greenhouse world (e.g. culminating in the Early Eocene Climatic Optimum) to an icehouse world (e.g. starting at the Eocene/Oligocene transition), when the climate cooled sufficiently for seasonal sea ice to occur in the Arctic (middle Eocene, around 47 Ma).

Eocene–Oligocene (E–O) transition: the sedimentary transition between two Paleogene epochs. The end of the Eocene (34 Ma) marked the switch towards a cooler global climate, the most significant climate event of the Cenozoic.

eolian: any material or organism that may be carried by the wind. Also spelled aeolian.

EPA: (US) Environmental Protection Agency.

ephemeral: as applied to waterbodies, ephemeral refers to those that flow or are filled with water only for a limited period of time, or season, and are otherwise dry.

epibryophytic: adjective describing algae that live on mosses (= bryophytic).

epigenesis: changes in mineral composition caused by external influences.

epilimnion: the upper stratum of water in a thermally stratified lake that floats over a cold, more dense stratum of water below. The epilimnion is characterized as a layer of warm, turbulent water.

epilithic: adjective describing algae attached to rocks, cement, glass slides, or similar hard surfaces.

epilithon: algae growing attached to rocks and similar hard surfaces.

epipelagic zone: represents ocean water-column depth between 0 and 200 m.

epipelic: adjective describing algae that live on, or in, mud.

epipelon: the highly motile community living on soft substrates, especially mud.

epiphytic: adjective describing algae attached to other algae or aquatic plants.

epiphyton: algae growing attached to other algae or aquatic plants.

epipsammic: adjective describing organisms attached to sand grains.

epipsammon: organisms growing attached to sand grains.

epiterranean: adjective describing organisms that live on the surface of the soil.

epitheca: the larger "upper" half of the diatom frustule.

epixylic: adjective describing organisms that live on wood (= xylophytic).

epizoon: organisms living attached to animals.

epontic: adjective describing organisms occurring in or on the basal layer of sea ice.

EPS: extracellular polymeric substances.

EQR: ecological quality ratio.

error: the deviation of an observed value from an expected or predicted value.

estuarine quality paradox: difficulty in detecting anthropogenic stresses in estuarine and shallow coastal habitats since those mimic the natural (i.e. non-anthropogenic) stresses present in these environments. Discussed by Elliot and Quintino in 2007.

estuary: a semi-enclosed coastal body of water which has a free connection with the open sea and within which seawater is measurably diluted with freshwater derived from land drainage.

exudate: any substance that oozes out (diffuses) from a plant cell or tissue.

EU: European Union.

euaerial: adjective describing algae that live on raised, prominent objects that receive moisture solely from the atmosphere.

Euclidean distance: a measure of dissimilarity between pairs of samples calculated by extending Pythagoras' theorem from two dimensions to the full dimensionality of multivariate data.

euhaline: salinity of 30–40 ppt (definition according to the "Venice System").

euplankton: the "true" plankton; the majority of their life cycle occurring in the water column; opposed to tychoplankton, which occurs intermittently in the water column after being swept up from the bottom.

euphotic zone: the sunlit region of a lake or ocean extending from the air–water interface down to the depth at which the light intensity is only 1% of that at the surface (i.e. the 1% isolume).

euplanktonic: adjective describing truly planktonic organisms that spend their entire life floating in the open water.

European Diatom Database: a data base of diatom training sets and transfer functions from Europe, parts of Africa, and Asia.

euryhaline: referring to an organism with a relatively broad tolerance range for salinity.

eurytopic: adjective describing organisms that live within a broad range of environmental parameters, whether it is temperature, light, nutrient regime, etc.

eustasy: large-scale change in sea level due to the rise and fall of the ocean, not to that of the land.

eustatic: adjective describing a large-scale change in sea level; not a relative change resulting from local coastal subsidence or elevation.

euterrestrial: adjective describing algae that live on soils that periodically dry out.

eutraphentic: adjective referring to algae that consistently occur in eutrophic conditions.

eutrophication: the process of becoming more eutrophic. Eutrophic waters are rich in dissolved nutrients, often from anthropogenic fertilization (for example, sewage outlows or agricultural runoff). Eutrophic waters may experience algal blooms and seasonal oxygen deficiency.

euxinic: adjective describing the condition in which waters are restricted in circulation, stagnant, or anaerobic.

evaporitic lake: see closed-basin lake.

evenness: a measure of the similarity in numbers of organisms of each species in a habitat or sample.

Everglades: a vast marsh, including flooded grasslands and swamp forests, covering much of South Florida, USA.

exopolymers: substances (typically carbohydrates) secreted into the water column by algae.

exotic species: foreign species that are introduced to a region or body of water where they are not native.

exploratory data analysis: an approach to data analysis that emphasizes the use of informal data summarization and graphical procedures not based on prior assumptions about the data structure.

extant: adjective describing organisms that are still living (i.e. not extinct).

extinction: the disappearance of species or larger biological units.

euryhalinity: a species' ability to withstand salinity change.

facies: the physical characteristic of rocks, e.g. lithofacies, lithological characteristics; also used as biofacies, related to the fossil assemblages found in rocks.

facies sequence analysis: analysis of the occurrence of facies changes in order to improve understanding of the relations between different facies and the conditions leading to their formation.

FACS: fluorescence-activated cell sorting.

factor analysis: a form of multivariate analysis in which operations are performed on a correlation or covariance matrix. Eigenvectors are extracted so as to explain as much of the original variance as possible, and each eigenvector is expressed as a factor which can be thought of as representing a theoretical end-member sample; the original samples can be expressed quantitatively as a mixture of factors. Often considered to be synonymous with principal components analysis (PCA) but true factor analysis differs from PCA in several ways.

fall dump scenario: an hypothesis put forward by Kemp and colleagues in 2000 to describe the mass sedimentation of diatoms at the end of summer, when stratification breaks down.

farmer fish: certain Pomacentrid fish that defend territories in which they selectively cultivate algae.

fault plane: a subterranean surface upon which blocks of rock move.

feedback: an event, phenomenon, or process which influences other processes to form a circuit or loop. One that enhances the original signal is called positive feedback, one which reduces the original signal is called negative feedback.

fen: a minerotrophic peatland is one affected by mineral soil waters with the water table at or near the peat surface. Can be further classified into rich and poor fens (see rich fen, poor fen).

field-flow fractionation: a separation technique where a field is applied to a mixture perpendicular to the mixture flow in order to cause separation due to differing mobilities of the various components in the field.

fjord: a long, narrow inlet with steep sides and U-shaped contour, created in a valley carved by glacial activity. Saline water is typically trapped behind a sill near the mouth.

floc: small ($\leq$1 mm to several cm), loosely aggregated mass of living and/or dead organic matter; serves to export material from the photic to aphotic zone of aquatic systems via sinking.

floodplain lake: a lake on a river floodplain, generally created by fluvial activity. Alternate terms include oxbow lakes, meander cutoffs, and billabongs.

Floridean starch: a polysaccharide storage product characteristic of red algae and similar to Cyanophycean starch. Specifically, it is an alpha 1,4-linked glucan similar to glycogen and amylopectin.

fluctuating asymmetry: asymmetry in which deviations from symmetry are randomly distributed about a mean of zero, providing a simple measure of developmental precision or stability.

fluorescence activated cell sorting: a flow cytometry technique for counting and examining microscopic particles suspended in a stream of fluid, in which a laser beam excites one or more fluorescent tags, and the emission of light triggers the cell sorting.

fluorescence microscopy: light microscopy used to study specimens which can be made to fluoresce. The sample can either be fluorescing in its natural form like chlorophyll and some minerals, or treated with fluorescing chemicals.

fluvial: produced by, coming from, or related to a river.

fluvial lake: wider areas of a river where water flow typically slows and sediment may accumulate.

flux: an amount that flows through a unit area per unit time.

Fourier analysis: the analysis of data subject to a fast Fourier transformation based on Fourier's theorem that proposed that any periodic function can be reduced to a series of sine and cosine terms, each represented by an amplitude, a frequency, and a phase.

foraminifera: member of the class Sarcodina; unicellular organisms that secrete an external test of calcium carbonate or of cemented particles, usually consisting of a series of chambers.

forensic science: the use of scientific methods for courts of law.

foreshore: the part of a beach that becomes covered and uncovered by water during the process of tidal rise and fall.

forward selection: a method for selecting a "good" (but not necessarily the "best") subset of explanatory or predictor variables in regression analysis, including constrained ordinations that are, in reality, multivariate regressions. The criterion used for assessing if a variable should be added to an existing model is the change in the residual sum of squares produced by the addition of the variable.

fractionation: partitioning of isotopes between two substances or between two phases that results in a preferential uptake or release of one isotope (i.e. lighter or heavier) relative to another is termed "fractionation."

fragilarioid: describes the freshwater araphid diatoms that are symmetric on the apical and transapical axes, includes the genera *Fragilaria*, *Staurosira*, *Staurosirella*, *Fragilariforma*, and *Stauroforma*, among others.

frequency domain: analysis of observed values in a time series in terms of periodic functions.

front: a boundary where two different (surface) water currents converge or diverge. Convergent fronts are sometimes associated with surface accumulations of giant diatoms. Divergent fronts are often associated with high primary productivity due to upwelled nutrient-rich waters.

frustule: the amorphous siliceous component of the cell wall of a diatom. It is composed of two halves, called valves, that are connected together by a girdle band or cingulum. The distinctive markings and shapes of frustules are used to identify the genera and species of diatoms.

FSS: Forensic Science Service.

fucoxanthin: an accessory pigment, or light-absorbing compound, which is the main light-harvesting pigment of diatoms.

fundamental niche: the entire multidimensional space that represents the total range of conditions within which an organism can function and can occupy in the absence of other species.

GAM: generalized additive model.

gametangium: compartment (in diatoms typically the entire parent-cell) in which gametes develop.

gamete: mature reproductive cell usually possessing a haploid chromosome set, capable of initiating a new diploid individual by fusion with a gamete of the opposite "sex."

gateway: usually refers to an oceanic passage between two landmasses, sometimes connecting separate oceans or seas. The opening or closing of such gateways by plate tectonics may have profound effects on the physical, chemical, and biological composition of the oceans due to the re-routing of surface and deepwater currents.

Gaussian logit regression: mathematical technique that attempts to fit either a Gaussian (bell-shaped) regression curve or an increasing or decreasing monotonic curve to species abundance data generated from a training set.

Gaussian ordination: an ordination procedure based on Gaussian curves that aims to construct one or more latent variables so that these curves optimally fit the species data.

Gaussian response: a normal distribution model.

Gaussian unimodal response model: a species response model in which the species abundance displays a symmetric unimodal relationship with the environmental variable of interest. The resulting curve is defined by the optimum or mode, tolerance, and the curve's maximum.

GCM: general circulation model; also sometimes referred to as global circulation model.

general circulation model: a mathematical model simulating the Earth's atmosphere or/and oceans used for weather forecasting, projecting, and reconstructing climate change.

generalized additive model: a model that uses smoothing techniques to identify and represent possible linear or non-linear relationships between response variables and explanatory variables as an alternative to parametric models.

generalized linear mixed model: generalized linear model that incorporates a random effect vector in the linear predictor.

generalized linear model: a class of regression models that arises from a natural generalization of ordinary models where some function of the expected value of the response variable (link function) is modeled as a linear combination of predictor variables linked by a link function. Different link functions and error structures can be specified, thereby providing a rich array of statistical modeling procedures for a wide range of data types.

generic level indices: indices that use results of identifying genera and usually counting abundances of genera in samples.

generitype: the type specimen of the nominal species typifying a genus.

genome: in diploid organisms, such as diatoms, the genome refers to the two full sets of chromosomes, or genes, that contains the hereditary information encoded in DNA.

genomic instability: the failure to transmit an accurate copy of the entire genome from one cell to its two daughter cells.

geometric mean: a measure of location lying between the arithmetic mean and the harmonic mean, calculated as the nth root of the product of n values.

geomorphology: the study of the classification, description, nature, origin, and development of landforms and their relationships to underlying structures, and the history of geologic changes as recorded by these surface features.

girdle bands: siliceous bands or segments that are usually present between the two valves of a frustule. Also called copulae (singular copula), connecting or intercalary bands, or cingula (singular cingulum).

glacio-isostatic uplift: uplift of the Earth's crust resulting from the release of pressure as glacial ice mass melts away (retreats).

glauconite: a greenish mineral of the mica group which is formed during marine sediment diagenesis.

GLM: generalized linear model.

global circulation model: see general circulation model.

GLR: Gaussian logit regression.

glycerol: an alcohol composed of a backbone of three carbon atoms, each carrying a hydroxyl group; used in the synthesis of lipids.

Gondwana: the name applied to a supercontinent that existed 570–510 Ma in the southern hemisphere. Gondwana included the landmasses of South America, Africa, Madagascar, Australia, New Zealand, Arabia, and the Indian subcontinent.

goodness-of-fit statistic: measure of the agreement between a set of observations and the corresponding values predicted from some model of interest.

Gower coefficient: a similarity coefficient suitable when the variables are mixed, consisting of continuous quantitative variables and categorical variables (e.g. presence/absence variables).

graben: a linear block of the Earth's crust that has dropped down along faults relative to rocks on either side. A depression produced by subsidence between two normal faults.

gradient length: an estimate of the compositional turnover in assemblage data along known environmental gradients or along major ordination axes of variation (latent variables).

greenhouse gases: atmospheric gases that contribute to the "greenhouse effect." These gases are transparent to incoming solar radiation but opaque to outgoing infrared radiation. Example gases include carbon dioxide, water vapor, methane, and chlorofluorocarbons (CFCs).

greenhouse–icehouse transition: the switch from a warm (ice-free) to cool (glaciated) climate, the final phase of which occurred around the Eocene–Oligocene boundary (34 Ma). Sometimes the term "doubthouse" is also used, referring to an intermediate phase between greenhouse and icehouse (e.g. middle Eocene to late Eocene; 49–34 Ma) where global temperatures began to cool and isolated ice caps formed in both hemispheres.

guano: excrement, particularly used in reference to seabirds and bats, sometimes used as fertilizer.

guild: as used here, a group of organisms with similar functions in ecosystems.

gyre: any roughly circular or elliptical region of water with a relatively stagnant stratified central region, surrounded by clockwise or counter-clockwise currents; usually refers to the large subtropical central areas of the oceans.

H': Shannon–Wiener diversity index.

halobion: referring to the biota living in saltwater; a system for classifying organisms according to their salinity tolerances. Typically has five categories: polyhalobous (fully marine), mesohalobous (brackish-water), oligohalobous halophilous (salt-tolerant freshwater), oligohalobous indifferent (tolerate slightly saline water), and halophobous (exclusively freshwater, salt-intolerant).

halocline: depth zone within which salinity changes maximally.

halophilic: adjective describing organisms that prefer, or are most abundant in, habitats that contain a relatively high concentration of salts.

haptophytes: a photosynthetic protist group occurring in freshwater and marine habitats possessing two flagella of slightly unequal length, and an emergent or remnant haptonema during at least part of its life cycle.

h-block cross-validation: a specialized form of cross-validation where blocks of consecutive samples in time series or in spatial data analysis are removed as a test set.

headwater: a body of water that represents the top or uppermost source for a larger system such as a river.

height of curve peak: a parameter of a Gaussian unimodal response curve.

Heinrich events: events linked to ice-rafting pulses from the Laurentide Ice Sheet into the North Atlantic during the last glacial period. They are extreme cold events between D–O events.

hemipelagic: pertaining to or living in half neritic (coastal) and half pelagic (open oceans) waters.

heterokonts: protists with unequal flagella. Heterokonts include both photosynthetic algae and colorless representatives. With respect to nomenclature, heterokont algae are also equivalent to the stramenochromes and roughly equivalent to the older term, chromophytes. The brown seaweeds, diatoms, and chrysophytes are commonly known members.

heteromorphic: existing in two or more morphological conditions.

heterothallic reproduction: sexuality involving parents who are individuals with sex determined (normally for the lifetime of the genotype) as a "male" or "female."

heterotrophic: obtaining nourishment from exogenous organic material, used by organisms unable to synthesize organic compounds from inorganic substrates.

heterovalvy: pertaining to a diatom frustule composed of valves of differing morphologies.

hierarchical Bayesian inference: a general framework for inference where information is heterogeneous or uncertain, processes are complex, and responses depend on scale.

hierarchical cluster analysis: see agglomerative clustering.

hierarchical taxonomic categories: components of the hierarchical system of taxonomic categories arranged in an ascending series of ranks, for example kingdom, division, class, order, family, tribe, genus, section, series, species, variety, and form.

histograms: a graphical representation of a set of observations in which class frequencies are represented by the areas of rectangles centered on the class interval.

histopathological: the post-mortem study (by a pathologist) of the effects of disease or injury on the body.

Holarctic: zoogeographic region that includes the areas of the Earth's northern hemisphere. It is subdivided into the Nearctic (New World) and Palearctic (Old World) regions.

holobiont: both (all) partners in a symbiotic relationship that forms one functional organism (e.g. scleractinian corals, lichens).

Holocene: the name of a geological epoch of the Quaternary period, covering the last ~11,700 years. It means "entirely recent" (*holos* = whole and *kainos* = new in Greek), because it represents modern times.

Holocene Thermal Maximum: warm period between *c.* 9000 and 5000 years BP which is also known as the Hypsithermal, Altithermal, Climatic Optimum, or Holocene Optimum.

holomictic lake: a lake that undergoes complete mixing of its waters during periods of circulation.

holoplankton: organisms completing their entire life cycle in the water column.

homothallic reproduction: sexuality involving parents who are individuals with the same genetic constitution (monoclones); in higher organisms, it is regarded as selfing.

HPLC: high performance liquid chromatography.

humic waters: streams and lakes with high concentrations of colored dissolved organic carbon.

humification: formation of humus, the organic brown or black portion of soils resulting from decomposition of plant and animal material.

hydro-isostasy: crustal adjustment to loading and unloading attributed to water.

hydromorphology: overall shape of the deposit, the underlying ground, hydrological regime, vegetation, and water chemistry. Physical habitat as constituted by the flow regime (hydrology and hydraulics) and the physical template (fluvial geomorphology).

hydrophilous: term used for aerial diatoms requiring permanently moist conditions.

hydroseral succession/hydrosere: an ecological succession of plant communities originating in an aquatic system. Wetland developmental stages of successional transitions from shallow, open water to more terrestrial systems such as fens and bogs.

hydroterrestrial: adjective describing algae that live on permanently wet soils.

hyperalkaline: excessively alkaline.

hyperarid: extremely arid.

hypereutraphentic: adjective referring to algae that occur in hypereutrophic conditions.

hypereutrophic: highly nutrient enriched waters.

hyperhaline: salinity > 40 psu (definition according to the "Venice System").

hypersaline: water with an ionic concentration > 50 g l^{-1}.

hypolimnion: the lower strata of water in a thermally stratified lake that lies beneath a warm, less-dense stratum of water above. The hypolimnion is characterized as a layer of cold, dense water, whose temperature may not change much throughout the stratified period (particularly in deep lakes).

hyporheic zone: the zone beneath and lateral to a stream or river bed, where surface water and groundwater interact. The size, position and extent of the hyporheic zone is defined by hydrology, that is, the porous area through which surface and groundwater are exchanged.

hyposaline: adjective describing water with an ionic concentration between 3 and 20 g l^{-1}.

hypotheca: the smaller "lower" half of the diatom frustule.

hypoxic: adjective describing water deficient in dissolved oxygen (<2 mg O_2 per liter).

Hypsithermal: Holocene climatic optimum.

hysteretic trajectory: relating to a direction of return travel to the origin but by a different route.

IBC: indices of biological condition.

IBI: indices of biotic integrity.

ICDP: Intercontinental Scientific Drilling Program.

ice core: a core of ice taken from the surface of an ice cap or glacier, generally to document past global climatic or environmental change.

ice-wedge polygons: polygon-shaped depressions delimited by ice wedges which form in permafrost. They often fill up with water to form ponds, which result from the expansion of the freeze–thaw cycle. Common in tundra regions of the Arctic.

ICP-AES: inductively coupled plasma atomic emission spectroscopy.

IDP: Practical Diatom Index.

IGCP: International Global Correlation Programme.

iHTR: inductive High-Temperature carbon Reduction.

illite: an aluminosilicate. Originally described from Illinois, hence its name.

impoundment: a body of water confined within an enclosure.

Index alpha (α): an index proposed by G. Nygaard in 1956 to infer lakewater pH from diatom species composition.

index of biotic integrity: quantitative scale for relating effects of human activity to responses of organisms in ecosystems (see biotic integrity).

index of disturbance: a measure of the extent to which lakes have been disturbed by human activities, based on diatoms in sediment cores. The index of disturbance is calculated using species richness, diversity, detrended correspondence analysis scores, and inferred [TP], [Cl], and transparency values.

indifferent: adjective describing organisms that tolerate a wide range of environmental values, such as pH.

indirect gradient analysis: a generic term for ordination methods that only analyze assemblage data and represent the data in a low-dimensional graphical form where the axes are selected to capture the variation in the data as effectively as possible according to an assumed underlying species response model. Interpretation is often aided by a post-hoc regression on external variables.

inertia: as a statistical term, the total variation in assemblage data as measured by the chi-squared statistic of the sample-species table divided by the table's total. It is equal to the sum of all eigenvalues of a correspondence analysis.

infauna: animals living within the sediment substrate.

inference model: a mathematical model, or transfer function, that estimates the value of an environmental variable (e.g. lake-water phosphorus concentration) as a function of biological data (e.g. diatom assemblages).

influence: the effect of each observation in a regression analysis on the estimated regression parameter.

infusorial earth: obsolete synonym of diatomite.

ingression: the entering of the sea on the land at a given place.

insolation: a measure of solar radiation energy received in a given time on a measured area.

Integrated Ocean Drilling Program: an international marine research program utilizing multiple platforms for deep-sea drilling that started operations in 2003 building on the earlier successes of ODP and DSDP.

intercalary valve: valve on the interior of a colony of diatom cells.

interdunal: the valley or trough between two dunes.

interglacial: the time and climatic conditions that separate two glacial or ice-age periods.

internal phosphorus loading: the supply, or return, of phosphorus from the sediments to the zone of primary production.

International Global Correlation Programme: a UNESCO co-operative enterprise that has supported over 400 hundred research programs in the earth sciences since 1972.

interpolation: the process of determining the value of a variable between two known values.

interstadial: a short period of warm climate occurring during overall cooler conditions.

intertidal: between low and high tide.

intertidal zone: region of coastal zone between limits of low and high tide.

Intertropical Convergence Zone: the boundary zone where the northeast trade winds of the northern hemisphere meet the southeast trade winds of the southern hemisphere.

intraformational: within a geologic formation.

IODP: Integrated Ocean Drilling Program.

IPCC: Intergovernmental Panel on Climate Change.

IPS: Indice de Polluo-sensibilité Spécifique.

iron hypothesis: certain areas of the oceans have high levels of plant nutrients, such as nitrates, phosphates, and carbonic acid. However, the phytoplankton do not grow as strongly as may be expected, given the plentiful supply of sunlight and plant food. It was discovered by John Martin that the lack of micronutrients, trace metals, and particularly iron, was a limiting factor for growth of phytoplankton in high-nutrient, low-chlorophyll, ocean surface waters.

IRMS: isotope ratio mass spectrometry.

isogamy: sexual reproduction via the fusion of two morphologically identical gametes.

isoprenoids: biosynthetic compounds with structures based on the isoprene (C_5) unit.

isostasy: the condition of equilibrium, comparable to floating, of the units of the Earth's lithosphere above the asthenosphere. Isostatic compensation and correction occur to maintain this equilibrium, causing relative elevation changes at the Earth's surface.

isostatic rebound: used to refer to the state of gravitational equilibrium where landmasses that were previously depressed by an enormous weight (typically ice sheets or rock) subsequently undergo uplifting after the weight is removed.

isostatic uplift: apparent sea-level change resulting after deglaciation, when landmasses gradually rebounded after being released from the pressure of the huge ice sheets.

isotope: all elements have a fixed number of protons (and hence a stable atomic number), but some elements may have a variable number of neutrons (and thus have forms with different atomic masses). Such forms are called isotopes.

ITCZ: Intertropical Convergence Zone.

ITS rDNA: internal transcribed spacer rDNA: region of ribosomal DNA (rRNA) that is a piece of non-functional RNA located between structural rRNA present on a common precursor transcript of RNA genes.

Jaccard coefficient: a similarity coefficient for use with data consisting of binary (presence/absence) variables. It is estimated for samples i and j as

$$S_{ij} = a/(a + b + c),$$

where a is the number of variables in common to i and j, and b and c are the numbers of variables occurring in sample i and sample j only.

jack-knifing: the simplest form of cross-validation (also known as "leave-one-out" validation) for estimating the root-mean-square error of prediction, where the reconstruction procedure is applied n times using a training set of size $(n - 1)$. In each of the n predictions, one sample is left out, and it is from these samples that the predictive ability of the calibration is evaluated.

JGOFS: Joint Global Ocean Flux Study.

Joint Global Ocean Flux Study: program designed to better understand the controls on the concentrations and fluxes of carbon and associated nutrients in the ocean.

joint plot: an ordination diagram of two kinds of entities (e.g. species and samples) which can be interpreted by the centroid principle. In a correspondence analysis a point representing a species is at the centroid (center of gravity) of the sample points in which the species occurs.

Jurassic: the second geological period (approximately 199–145 Ma) of the Mesozoic era. It was named after the Swiss Jura Mountains, which are characterized by limestones.

K-selected species: populations that produce few offspring and grow more slowly than r-selected species. Typically found in more stable environments.

ka: one thousand years.

karstic wetland: wetlands existing in a limestone landscape, characterized by caves, fissures, and underground streams.

Kazantsevo: eastern Siberian name for the last interglacial.

kernel density estimation plot: a method for the estimation of probability density functions involving a window width or bandwidth and a kernel function. A powerful tool in exploratory data analysis.

kieselguhr: a German word meaning diatomite.

kleptoplasty: a symbiotic phenomenon whereby plastids from photosynthetic protists are sequestered by another organism.

k-means clustering and partitioning: a method of clustering in which from an initial partition or grouping of samples into k clusters, each sample in turn is examined and reassigned if appropriate to a different cluster in an attempt to optimize some predefined numerical criterion that measures in some way the "quality" of the cluster solution.

K_S: half saturation constant: substrate concentration at which the growth rate is half of the maximum.

K-systems analysis: a type of maximum entropy (as opposed to a least-squares) stepwise regression that employs events rather than variables as its fundamental unit.

kurtosis: the extent to which the peak of a unimodal frequency distribution departs from the shape of a normal distribution either by being more pointed (leptokurtic) or flatter (platykurtic).

lagoon: the enclosed body of comparatively shallow saltwater or brackish water separated from the sea by a shallow or exposed sandbank, barrier reef, or barrier islands. Freshwater lagoons also exist.

landmark configuration: the configuration of points that correspond to an anatomical position comparable between objects. There are several types of landmarks used in morphometrics because from such landmarks, distances and angles can be calculated.

La Niña: the name for the cold phase of ENSO, during which the cold pool in the eastern Pacific intensifies and the trade winds strengthen.

Lake Trophic Diatom Index: a mathematical index developed by Kelly and colleagues in 2007 in the UK which uses epilithic and epiphytic diatoms to assess lake ecological status.

lake trophic status: refers to the nutrient and productivity status of a lake; for example lakes with high nutrient content and high productivity are termed eutrophic, lakes of low nutrients and productivity are termed oligotrophic.

last glacial maximum: the time of the maximum advance of the most recent Pleistocene glaciers, commonly dated at about 20,000 years ago.

last interglacial: the interglacial previous to the Holocene, which occurred between c. 128 ka BP and 118 ka BP. It is equivalent to the Eemian period.

latent structure: the underlying structure of a multivariate assemblage data set in terms of unobserved variables (latent variables) that account for major patterns of variation in the data.

latent variable: a variable that cannot be measured directly but is assumed to be related to a number of observed samples. It is selected to "best" explain the data according to an assumed response model, and in ordinations it is the first major ordination axis.

Laurasia: a Mesozoic supercontinent, formerly part of Pangaea, that was composed of North America (Laurentia), Baltica, Siberia, Kazakhstania, and parts of China. The name is a combination of Laurentia and Eurasia.

Law of Large Numbers: a law that describes the stable behavior of the sampling mean of a random variable, such that repeated sampling from the population of a random variable with a finite expected mean will produce an estimated mean that approaches and stays close to the finite expected mean, when the sample size is large.

lead-210 (^{210}Pb) dating: geochronological tool for dating recent (e.g. last 150 years) sediments. ^{210}Pb, with a half-life of approximately 22 years, is part of the radium decay series.

leave-one-out cross-validation: a procedure for reducing bias in estimation and for providing approximate confidence intervals in situations where these are difficult to obtain in the usual way. Each sample is omitted in turn from the data thereby creating a training or modeling set of size $n - 1$ and a test set of size 1. This is repeated $n - 1$ times and the parameter of interest is estimated from the $n - 1$ test-set estimates.

lentic taxa: taxa living in standing water, such as lakes and ponds.

leverage: a term used in regression analysis for those observations that have an extreme value on one or more predictor variable(s). The effect of such observations is to force the fitted model close to the observed value of the response variable leading to a small residual.

LGM: last glacial maximum.

LIA: Little Ice Age.

light tweezers: the use of light to manipulate microscopic objects as small as a single atom. The radiation pressure from a focused laser beam is able to trap small particles. In the biological sciences, these instruments have been used to apply forces in the pN range and to measure displacements in the nm range of objects ranging in size from 10 nm to over 100 μm. Also called optical tweezers.

liming: the addition of limestone, usually in finely powdered form, to the catchments or surfaces of lakes and streams to mitigate the effects of acidification.

limnetic: adjective describing freshwater, salinity <0.5 ppt (definition according to the "Venice System").

linear-logit (sigmoidal) model: a regression model for presence/absence or proportional data in which a monotonic function is fitted in the framework of generalized linear models with a binomial error distribution and a log-link function.

linear regression: a simple linear regression model involving a response variable that is a continuous variable and a single predictor variable that involves the intercept term and the slope term.

linear response model: a species response model in which the abundance of the species increases linearly in relation to the environmental gradient of interest.

lithogenic: sourced or pertaining to rock formation.

lithostratigraphy: the stratigraphic distribution of lithologies.

Little Ice Age: an interval of more frequent cold conditions dated from approximately AD 1300–1850, originally described from western Europe and characterized by the advance of mountain glaciers.

littoral zone: the nearshore area of a lake or ocean, typically defined as the area where rooted aquatic macrophytes can grow. Often delimited by the depth of the photic zone.

LM: light microscopy.

locally weighted scatterplot smoothing: a method of regression analysis in which polynomials of degree one (linear) or two (quadratic) are used to approximate the regression function in particular "neighborhoods" of the space of the predictor variables. It uses weighted least squares with local subsets of the data so as to pay less attention to distant points. It assumes no predetermined model for the entire data set and therefore provides no explicit formula for the fitted curve.

loess: in geological terms, loess consists of wind-blown silt deposits.

LOESS: see LOWESS.

log ratio: an approach useful in the statistical analysis of closed compositional data that involves the use of log-transformed ratios of pairs of variables.

lotic: adjective describing running waters, such as streams and rivers.

Lotic Index: an index, developed by Ludlam and colleagues in 1996, based on the ratio of lotic diatoms to the total number of benthic diatoms in a given habitat.

LOWESS: locally weighted scatterplot smoothing (also called LOESS).

low-pass filter: a procedure that converts one time series into another, e.g. a moving average. A low-pass filter removes short-term random fluctuations.

LSU rDNA: large subunit of the ribosomal DNA.

LTDI: Lake Trophic Diatom Index.

LTMP: long-term monitoring program.

LTSEM: low temperature scanning electron microscopy.

Lutetian: the first stage of the middle Eocene (approximately 49–40 Ma), named after Paris (*Lutetia* in Latin). It precedes the second stage of the middle Eocene, the Bartonian.

Ma: million years ago. Also sometimes abbreviated MYA.

maar: monogenetic volcanic craters, generally less than 2 km in diameter, formed by phreatomagmatic explosions and subsequent collapse.

Maastrichtian: the last stage of the Cretaceous period, lasting from 70–65 Ma. The name refers to the Dutch city of Maastricht where many of the defining fossils were found.

macrophyte: macroscopic forms of aquatic vegetation, commonly referred to as aquatic weeds.

MAGIC: Model of Acidification of Groundwater In Catchments; a dynamic computer model used to estimate acidification.

magnetic susceptibility: a measure of ferromagnetic particles, typically from sedimentary records. Increases in the magnetic signature in a core profile may mark periods of increased erosion.

magnetic tweezers: instruments for exerting and measuring forces on magnetic particles using a magnetic field gradient. Typical applications are single-molecule micromanipulation, rheology of soft matter, and studies of force-regulated processes in living cells. Forces are typically on the order of pico- to nanonewtons.

mangal: a community comprising mangrove trees and associated species.

mangrove swamps: brackish-water coastal wetlands of tropical and subtropical areas that are usually dominated by shrubby halophytes such as mangroves, and are partly inundated by tidal flow.

mariculture: farming of marine organisms.

marine lakes: water bodies of marine origin having essentially the same salt composition as seawater but virtually or entirely land locked and having little or no exchange with the neighboring ocean.

Martin curve: a curve fit through a global data set of mean annual particulate organic carbon (POC) flux into sediment traps at different depths in the ocean. This exponential decline of flux with depth has been the standard description of POC flux to the deep sea for many years.

MAT: modern analog technique.

maximum bias: in a regression model, the residuals (observed – predicted) are divided into equal intervals along the gradient of the predictor variable and the mean bias (observed – predicted) per interval is calculated and the largest absolute value of mean bias for an interval is reported as the maximum bias.

maximum likelihood: an estimation procedure involving maximization of the likelihood or the log-likelihood with respect to the parameters. It is commonly used in generalized linear models.

MC-ICP-MS: multicollector inductively coupled plasma mass spectrometry.

mean: a measure of location or central value for a continuous quantitative variable estimated as the sum of the variable of interest in all samples divided by the number of samples.

mean bias: in a regression model, the mean of the residuals (observed – predicted).

mean field: the main idea is to replace all interactions to any one body with an average or effective interaction. This reduces any multibody problem into an effective one-body problem.

mean higher high water: a tidal datum that equates to the average spring tide water level, determined over a specified period of time.

median: the value in a set of ranked observations that divides the data into two equal parts.

Medieval Warm Period: an interval originally described in Europe of more frequent warm conditions, dated from approximately AD 900 to 1300.

meiofauna: small interstitial animals from 0.1 mm up to 1 mm in size.

MEMS: microelectromechanical systems.

meromictic: adjective describing lakes in which mixing, or circulation, of the entire water body is incomplete on an annual basis. Such lakes possess an upper water mass that circulates, separated from a bottom, non-circulating portion by a strong concentration gradient.

meroplanktonic: literally "partially planktonic," and used to describe organisms that have a staged life strategy where part of their life is spent as part of the benthos and part as plankton. Diatoms described as meroplankton may encyst or form resting cells during their benthic stage and bloom during the plankton stage.

mesic: moist.

mesocosm: an experimental apparatus or enclosure designed to approximate natural conditions, and in which environmental factors can be manipulated.

mesohaline: salinity 5–18 ppt (definition according to the "Venice System").

mesohalobous: diatoms which thrive in salt concentrations of between 0.2 and 30 psu.

mesopelagic zone: a subdivision of the water column between the depths of 200 and 1000 m.

mesoporous: a material containing pores with diameters between 2 and 50 nm.

mesosaline: adjective describing waters with an ionic concentration between 20 and 50 g l^{-1}.

Mesozoic: a geological era (approximately 251–65 Ma), often referred to as the "Age of the Reptiles." The name basically means "middle animals." The era was originally called "Secondary," hence the Tertiary and Quaternary periods which followed are a continuation of the numbering sequence. The era consists of three geological periods: the Triassic, Jurassic, and Cretaceous.

metacommunity: a set of linked local communities connected by the dispersal of at least some of the species present.

metalimnion: a region of rapid temperature decline with depth in a thermally stratified lake. It separates the upper waters (epilimnion) from the lower waters (hypolimnion) and commonly acts as a barrier for nutrients released in the bottom of the lake.

metaphyton: algae found in the littoral (shallow water) region of a lake that are neither attached nor truly planktonic.

metapopulation: a group of partially isolated populations belonging to the same species, namely subpopulations of a natural population, that are partially isolated from one another and are connected by pathways of immigration and emigration.

metrics: literally a measure. As applied here, attributes of assemblages that change in response to human alterations of watersheds.

MHHW: mean higher high water.

Mi-1 glaciation: an early Miocene glaciation characterized by significant expansion of ice sheets on Antarctica, which occurred close to the Oligocene–Miocene boundary at 23 Ma.

microelectromechanical systems: the technology of the very small – merges at the nano-scale into nanoelectromechanical systems (NEMS) and nanotechnology.

microphytobenthos: the assemblage of microalgae (mostly diatoms and cyanobacteria) living in the benthos.

microtidal: coastal ocean or waterway with a mean tidal range <2 meters.

Mid-Pleistocene Revolution (MPR): a transition period *c.* 1 Ma when glacial–interglacial cycles shifted from being forced by obliquity (*c.* 41 ka) to eccentricity (*c.* 100 ka) cycles.

minerotrophic: refers to wetlands that are hydrologically connected to the surrounding landscape via the water table, streams, or groundwater, and are therefore relatively rich in minerals and nutrients and contain water that is circumneutral to alkaline. Nourished by mineral water.

minimal adequate model: a regression model with the smallest number of statistically significant parameters that "explains" the response variable as well or nearly as well as a model with more parameters. An important property of a minimal adequate model is that it has the highest number of degrees of freedom (because of the smallest number of parameters) and hence has maximal predictive power and robustness.

Miocene: the first epoch of the Neogene, spanning from 23 to 5 Ma. The name basically means "less new" (*mio* = less and *kainos* = new in Greek), because it is less modern-looking than the next epoch, the Pliocene. During the Miocene, glaciation intensified on Antarctica and continental climates became drier, with large areas converted from forests to grasslands.

Miocene–Pliocene (M–P) boundary: the sedimentary transition at ~5 Ma between the Neogene epochs, the Miocene and Pliocene.

mire: term used more commonly outside of North America for ecosystems including swamps, bogs, fens, moors, muskeg, and peatlands but is often used synonymously with peatlands. In Europe and Russia, mires are specific to ecosystems that are actively accumulating peat whereas "peatlands" is a more generic term for all peatlands, regardless of whether or not the system is actively accumulating peat.

mitosis: division of a nucleus precisely replicating the genetic constitution of a parental nucleus, usually accompanied by cytokinesis.

ML: maximum likelihood.

MMI: multimetric index.

moat: the portion of open water between a floating ice pan in the middle of a lake and the shore. Common in High Arctic and Antarctic lakes.

modern analog: a surface sediment species assemblage that has similar taxonomic composition to a fossil sample.

modern analog technique: a technique developed in paleoecology in which fossil assemblages are compared, in turn, using an appropriate dissimilarity measure, with each sample in a modern assemblage data set to identify modern analogs as an aid in interpretation of the fossil assemblage in terms of past communities or past environments.

Modified Pantle-Buck Saprobic Index (SI): a mathematical index of water quality based on the relative abundances of diatom taxa in a sample.

MODIS AQUA: moderate resolution imaging spectroradiometer attached to the AQUA (Earth Observing System-PM) satellite.

molecular sequencing: determining the primary structure (or primary sequence) of a large and complex molecule (such as RNA or DNA).

monomictic: one period of annual seasonal circulation occurring in a lake.

monophyletic: a term used by practitioners of cladistic analysis referring to an evolutionary group containing an ancestor and all of its descendants. Creation of such a group is generally regarded as the ultimate goal of taxonomic designations.

monsoon: a wind system that changes direction seasonally, blowing from one direction in summer and the opposite direction in winter; associated with major seasonal shifts in precipitation, bringing heavy rains when blowing onshore.

Monte Carlo permutation test: a procedure for determining statistical significance directly from data without reference to a particular sampling distribution. It is similar to a randomization test except that the permutations are restricted in some way to take account of, or maintain, the sampling design of the observed data (e.g. time series, line transects). For example, in a study involving the comparison of two groups, the data are divided (permuted) repeatedly between groups and for each division (permutation) the relevant test statistic is calculated to estimate the proportion of the data permutations that provide as large a test statistic as that calculated from the observed data.

Monte Carlo simulation: method for finding a solution to mathematical and statistical problems by simulation using random numbers and used when the analytical solution to the problem is either intractable or time consuming.

mooring: typically used to refer to oceanographic equipment; anchored or free-floating lines equipped with various instruments (current meters, sediment traps, etc.) and buoys/drogues for stability to measure oceanographic parameters in the water column.

morpho-genus (morpho-genera): a genus whose description is based solely on morphology (i.e. there are little or no details of its cell ultrastructure and life cycle). Used in a similar context as "form genus" or "organ taxon."

morphospecies: a species recognized on morphological criteria, which may represent more than one molecular identity ("cryptic species").

MS: magnetic susceptibility.

mucilage: polysaccharides secreted by diatoms and other algae to facilitate movement in sediments or on surfaces and/or for attachment; also called mucopolysaccharides.

mucopolysaccharide: any of a group of polysaccharide with high molecular weight that contain amino sugars and often forms complexes with proteins. Also called glycosaminoglycan.

mudrock: a fine-grained, dark gray sedimentary rock, formed from silt- and clay-sized particles; similar to shale but without laminations.

multifrequential periodogram analysis: a statistical method for detecting periodic components in temporal series with unequally spaced samples. The periodic components are detected through the estimation of the corresponding frequencies in a stepwise procedure. All frequencies are re-estimated at each step and a significance test is performed at each step. It can determine if a periodicity is present or absent in portions of the temporal series.

multimetric indices: indices that are composed of two or more metrics.

multinomial distribution: a generalization of the binomial distribution to situations where r outcomes can occur in each trial, e.g. when there are more than two classes.

multinomial logit model: a regression model for proportional data where species parameters are estimated simultaneously with the constraint that the abundances of all species sum to 1. It is difficult to fit and its parameters are difficult to interpret.

multiple linear regression: statistical technique that is used to show the relationship between a response or dependent variable and a number of explanatory or independent variables.

multivariate adaptive regression splines: a regression tree procedure used in the exploration of large data sets where flexible regression models are built by using basic functions to fit separate splines to distinct intervals of the predictor variable.

multivariate analysis: simultaneous analysis of data with high dimensionality.

multivariate analysis of variance: a procedure for testing the equality of the mean vectors of more than two groups for a multivariate response variable. It is directly analogous to analysis of variance of univariate data except that the groups are compared on all response variables simultaneously.

multivariate regression: a regression analysis with two or more response variables and one or more predictor variables.

MYA: million years ago. Also abbreviated as Ma.

NADW: North Atlantic Deep Water.

nannoplankton: see nanoplankton.

nanoplankton: microorganisms drifting in the sea whose size range between 2 and 20 μm. Also spelled nannoplankton.

nanotechnology: a field of applied science and technology covering a very broad range of topics. The main unifying theme is the control of matter on a scale smaller than one micrometer. Nanotechnology cuts across many disciplines, including colloidal science, chemistry, applied physics, and biology.

NAO: North Atlantic Oscillation.

National Wetlands Working Group: a national wetland science committee consisting of experts from government, non-government, university, and private-sector agencies across Canada.

natron: a mineral ($Na_2CO_3 \cdot 10H_2O$) that is very soluble in water and is common in soda lakes.

naviculoid: biraphid diatoms having a boat-like outline in valve view, used to describe numerous genera similar in outline to *Navicula*, many of which were formerly circumscribed as groups of species within that genus.

negative multinomial distribution: a generalization of the negative binomial distribution in which >2 outcomes are possible on each trial and sampling is continued until a predefined number of outcomes of a particular type are obtained.

Neogene: the period of Earth history spanning the interval from ~23 to ~2.6 Ma, encompassing the Miocene and Pliocene epochs.

neogenesis: the formation of new minerals, as by diagenesis or metamorphism.

Neolithic Age: the new Stone Age; historical period of human activity that occurred *c.* 5000–3000 BC in Europe.

Neoglacial: a time of glacial readvance during the Holocene.

nepheloid layer: a layer rich in particles overlying the sea floor, sourced from particle fallout from overlying waters or stripped from sediments by bottom currents.

neritic: occurring in shallow marine waters; usually refers to waters overlying the continental shelves.

neurotoxin: toxin causing damage to and/or impaired functioning of nerve cells or tissue.

NGP: Northern Great Plains.

NHG: Northern Hemisphere Glaciation.

niche: the ecological role of a species in an assemblage, conceptualized as a multidimensional space of which the coordinates are the various parameters representing the conditions of existence of the species.

niche dimensions: the range of environmental conditions that determine the place in a community occupied by an organism.

nitrogen fixing: a process by which nitrogen is taken from its relatively inert molecular form (N_2) in the atmosphere and converted into nitrogen compounds.

NMDS: non-metric multidimensional scaling.

non-indigenous: a species is considered to be "non-indigenous" if it is located in a region outside of its native geographic range. Non-indigenous species are also referred to as introduced, non-native, exotic, alien, or invasive.

non-linear CAP: non-linear canonical analysis of principal coordinates.

non-linear canonical analysis of principal coordinates: canonical ordination procedure in which the relationship between assemblage data and environmental variables are non-linear, for example assemblage composition at different distance from a pollution source.

non-metric multidimensional scaling: an ordination or indirect gradient analysis method in which only the ranks of the dissimilarity or similarity coefficients are used to produce a low-dimensional representation of the data.

non-parametric: making no distributional assumptions about the population under study. Often non-parametric methods involve only the ranks of the observations rather than the observations themselves.

non-parametric additive model: a model in which the predictor variables have an additive effect on the response variable.

North Atlantic Deep Water: a highly saline water mass of the North Atlantic that is formed in the Greenland and Labrador seas and usually found at 2000–4000-m water depth.

North Atlantic Oscillation: a large-scale fluctuation in atmospheric pressure in the Atlantic Ocean between the high-pressure system near the Azores and the low-pressure system near Iceland, quantified in the North Atlantic Oscillation (NAO) Index.

Northern Hemisphere Glaciation: continental-scale glaciation which started in the north about 3 Ma.

null model: a statistical model where there is no statistically significant relationship between variables, usually a response variable and one or more predictor variables.

nursery lake: lake in which some species of salmon (such as some sockeye salmon) spend the first one to three years of their life. Sockeye salmon often return to their nursery lake system to spawn.

NWWG: National Wetlands Working Group.

obliquity: tilt of the Earth's axis which varies between 22.1° and 24.5° with *c.* 41 ka periodicity.

Ocean Drilling Program: an international research program (1983–2003) for deep-sea drilling that followed the DSDP and preceded the IODP.

Ockham's principle of simplicity: an expression of the principle of parsimony presented by the English philosopher William of Ockham (or Occam) (1280–1349), namely "*entia non sunt multiplicanda praeter necessitatem*" (a plurality (of reasons) should not be posited without necessity) and "*frustra fit per plura, quad fieri potest per pauciora*" (it is vain to do with more what can be done with less). Also referred to as Ockham's razor.

ODP: Ocean Drilling Program.

Oi-1 glaciation: the major early-Oligocene glaciation, marking the start of the current "icehouse," which occurred close to the Eocene–Oligocene boundary at 34 Ma and involved major and rapid ice-sheet expansion on Antarctica.

Oligocene: the youngest epoch of the Paleogene, spanning from 34 to 23 Ma. Its name basically means "few new" (*oligos* = few, *kainos* = new in Greek), in reference to the fact that few new mammalian faunas evolved during this epoch. The earliest Oligocene marks the most dramatic cooling event in the entire Cenozoic era when permanent ice sheets grew rapidly on Antarctica and the Earth transited from a greenhouse–doubthouse into an icehouse. The Oligocene is also marked by regression of tropical broad-leaf forests and expansion of grasslands globally.

oligohaline: salinity 0.5–5 ppt (definition according to the "Venice System").

oligohalobous: adjective describing organisms that generally occur in salt concentrations less than 0.2 psu.

oligonucleotide: short nucleic acid polymer, typically with twenty or fewer bases.
oligotraphentic: adjective describing algae that obligately occur in oligotrophic conditions.
oligotrophic: a term generally applied in aquatic science as indicating low nutrient availability and hence low primary production.
ombrotrophic: adjctive describing soil or vegetation that is very low in minerals and nutrients and is acidic as these environments are hydrologically isolated from their surrounding landscape and receive nutrients/minerals/water exclusively from the atmosphere through precipitation.
Ontario Trophic Status Model: an empirical mass-balance model for estimating past, present, and future lake-water phosphorus concentration, presented by N. Hutchinson and colleagues in 1991.
ontogenetic: adjective of ontogenesis, which describes the origin and development of an individual organism from embryo to adult; also used to describe pathways of successional development in landscapes.
oogamy: sexual reproduction via fusion of a relatively large non-motile egg and small motile sperm.
ooze: fine-grained sediment characterized by a single type of microfossil (for example, a diatom ooze).
opal: an amorphous silicate mineral. See also biogenic silica.
open-basin lake: lake with a surface outlet. Water may flow in and out of the lake by a variety of routes (streams, direct precipitation on lake surface, groundwaters); therefore evaporation within the lake is usually limited and the lake water tends to contain lower concentrations of solutes (see closed-basin lake). The oxygen isotopic composition of open-basin lake waters tend to be lower than closed-basin lake waters and are often similar to groundwaters.
optimal divisive partitioning: a stratigraphically constrained partitioning procedure that finds the optimal partitioning for 2, 3, 4, . . . , 20 groups that is not hierarchically nested. The partitions are found so that the divisions yield the smallest residual variation (sum-of-squares or information content). Used in zonation of stratigraphical data.
optimum: a measure of the “preferred” environmental conditions of a taxon, usually with respect to a particular environmental variable; the value of that environmental variable at which the taxon is most abundant.
ordination: a collective term for statistical techniques that attempt to arrange sites in low-dimensional space based on their species composition (e.g. a data set consisting of many sites and species can effectively be summarized by one or more ordination axes).
ordination axis: latent variable, theoretical explanatory variable, eigenvector (see eigenvalue).
ordination diagram: a scatter plot of the sample, species, or environmental variable scores. Can be a biplot or a joint plot.
organic acids: weakly dissociated acids that include humic acids generated by the natural decomposition of organic matter.
orthophosphate: the form of phosphorus (PO_4^{3-}) which can be directly assimilated by algae.
osmoregulation: the set of processes that enable a cell or an organism to maintain the osmotic concentration of internal fluids within some narrow range, despite fluctuations in the external medium.
Ostwald ripening: the tendency for a population of particles of diverse sizes to grow in diameter over time, by a process in which the smaller particles dissolve preferentially, because of their higher solubility, with subsequent crystallization onto larger particles, making them even larger.
outlier: an observation that appears to deviate markedly from other observations.
over-fitting: the fitting of statistical models that contain more unknown parameters than can be justified by the data.
oxbow lake: an often crescent-shaped lake that is formed when a meander of a river is cut off from the main channel.
oxygen isotope stratigraphy: a stratigraphy based on the ratio of stable isotopes of oxygen (^{16}O and ^{18}O). The isotopic ratios are used, for example, to infer the amount of ice on land (as glaciers) and so document the waxing and waning of glacial periods.
Pacific Decadal Oscillation: a pattern of Pacific climate variability that shifts phases on at least interdecadal timescale, usually about 20 to 30 years. The PDO is detected as warm or cool surface waters in the Pacific Ocean, north of 20° N.
Pacific Northwest Index: a terrestrial climate index that is a composite of air temperature, total precipitation, and snowpack depth at specific sites in the Pacific northwest USA.
paleobathymetry: study of water depths of the past.
Paleocene: the first period of the Cenozoic era, spanning from ~65 to ~56 Ma. During this time the climate changed from a greenhouse world to an icehouse world, and mammal faunas underwent rapid radiation and dominated the land areas.

Paleocene–Eocene Thermal Maximum: the Paleocene–Eocene boundary (56 Ma) was marked by a rapid and significant rise in temperature (about 6 °C). Consequently there were corresponding rises in sea level and carbon dioxide concentrations.

paleoenvironmental proxy: an indicator that can be used to infer past environmental conditions. These may include microfossils, such as diatoms, as well as indicators such as pollen, sediment grain size, etc.

Paleogene: the period of Earth history spanning the interval from ~65 to 23 Ma. During this time mammal faunas underwent rapid radiation and dominated the land areas.

paleogeography: the configuration of past landmasses, oceans, and seaways.

paleolimnology: the sciences that uses the physical, chemical, and biological information stored in lake and river sediments to infer past environments.

paleomagnetic chronology: a stratigraphy or chronology based on the ages of the reversals of the Earth's magnetic field.

paleomarsh-surface elevation: when a fossil diatom assemblage is calibrated by an appropriate transfer function, an estimate of the former elevation of the host sediment sample is generated. Multiple samples collected from a sediment core will produce a sequence of paleomarsh-surface elevation change. The resulting diagrams, plotting elevation against depth, show changes in the balance between sediment accumulation and relative sea level.

paleotempestology: the study of past tropical cyclone activity by means of geological proxies as well as historical documentary records.

paleotidal: refers to former tidal conditions.

paleotsunamis: tsunamis that occurred prior to the twentieth century. They are studied by means of geological proxies as well as historical documentary records.

Palmer cell: a microscope slide adapted to hold 0.1 ml of water in a shallow well under a coverglass. The depth of the well is shallow enough that the short working distance of a 40× objective will focus on the bottom of the well where algae settle.

paludification: the waterlogging and swamping or outward spread of wet, peat-forming environments over adjacent terrestrial areas.

palynology: the study of pollen and spores. A micropaleontological method based on the identification and counting of fossil pollen and spore types that have been preserved in lake and mire deposits. It allows the reconstruction of past vegetation. Palynology also refers to the study of other palynomorphs such as dinoflagellate cysts and acritarchs.

palynomorph: any organic particle (normally of size 5–500 μm) composed of chitin, pseudochitin, or sporopollenin, often found in marine and freshwater sediments. They originate from a wide spectrum of organisms, including some green algae, dinoflagellates, acritarchs, pollen, spores, fungi, and insect mouth parts.

PAM: pulse amplitude modulated.

Pangaea: a huge supercontinent, including all landmasses, that existed in the Paleozoic and Mesozoic eras before splitting up into several smaller supercontinents (e.g. Laurasia and Gondwana), and then finally into today's continents. The name means the "entire Earth" after *pan* (entire) and *Gaia* (Earth). The ocean that surrounded Pangaea is called Panthalassa (of which the Pacific is the last remnant).

panmixis: a population of organisms that is characterized by random mating, resulting in an interchange of genetic material within the group.

parallel computation: a form of computation in which many calculations are carried out simultaneously, operating on the principle that large problems can often be divided into smaller ones, which are then solved concurrently ("in parallel") on separate processors or computers.

parametric: a statistical procedure that makes distributional assumptions about the population under study (e.g. normal distribution, binomial distribution).

paraphyletic: a term used by practitioners of cladistic analysis referring to an evolutionary group containing an ancestor and some of its descendants. These groups are considered undesirable by most taxonomists in taxonomic schemes.

parautochthonous: intermediate between autochthonous and allochthonous; used to refer to a mixed assemblage of which the autochthonous and allochthonous parts cannot be discerned.

PARFLUX: the abbreviated name for the McLane Research Laboratories line of particle flux sediment traps.

partial CA: partial correspondence analysis.

partial canonical correspondence analysis: a canonical correspondence analysis where the effects of one or more covariables are partialed out prior to the canonical correspondence analysis.

partial CCA: partial canonical correspondence analysis.

partial correspondence analysis: a correspondence analysis where the effects of one or more covariables are partialed out prior to the correspondence analysis.

partial DCA: partial detrended correspondence analysis.

partial DCCA: partial detrended canonical correspondence analysis.

partial detrended canonical correspondence analysis: a detrended canonical correspondence analysis where the effects of one or more covariables are partialed out prior to the detrended canonical correspondence analysis.

partial detrended correspondence analysis: a detrended correspondence analysis where the effects of one or more covariables are partialed out prior to the detrended correspondence analysis.

partial least-squares regression: an alternative to multiple regression in which, instead of using the original predictor variables directly, a new set of regression variables are constructed as linear combinations of the predictor variables, so as to have maximal covariance with the response variable subject to the constraint of being uncorrelated with the other regressors.

partial PCA: partial principal components analysis.

partial principal components analysis: a principal components analysis where the effects of one or more covariables are partialed out prior to the principal components analysis.

partial RDA: partial redundancy analysis.

partial redundancy analysis: a redundancy analysis where the effects of one or more covariables are partialed out prior to the redundancy analysis.

partial techniques: statistical modeling techniques where the effects of one or more covariables are allowed for (partialing out) prior to the analysis.

partitioning: dividing a set of samples into a series of non-overlapping groups; starting with all samples and progressively dividing them according to some specified mathematical criterion.

passive samples: samples included in ordinations that have no influence on the position of other samples (i.e. active samples) in the analysis. For example, fossil samples with no corresponding environmental data.

P:B (P/B): abbreviation for the ratio of planktonic to benthic diatoms. Also referred to as P/L.

PCA: principal components analysis.

PCoA: principal coordinates analysis.

PCR: polymerase chain reaction.

PDO: Pacific Decadal Oscillation.

P – E: precipitation minus evaporation.

Pearson's correlation coefficient (r): A measure of the strength of the linear relationship between two variables. It can take on the values from −1.0 to 1.0, where −1.0 is a perfect negative (inverse) correlation, 0.0 is no correlation, and 1.0 is a perfect positive correlation.

peat: substance composed of partially decomposed plant remains with >65% organic matter (dry weight) and <20–35% inorganic content. This is the material that accumulates in peatlands.

peatland: a specialized type of wetland (organic wetland often called mire in Europe) whereby the rate of plant decomposition is much lower than plant production resulting in accumulations of partially decayed plant detritus from previous years' growth. This term does not differentiate between systems that are actively accumulating peat and those that are not.

pelagic: occurring in the deep ocean beyond the continental shelves, or in the deep parts of a lake.

pelagomonads: a picoplanktonic, marine, photosynthetic protist group related to other chlorophyll-*a*- and -*c*- bearing protists.

Pennales: a diatom order with members diagnosed by the presence of bilateral symmetry.

pennate diatom: a diatom belonging to the class Fragilariophyceae (araphid pennate diatoms) or to the class Bacillariophyceae (raphid pennate diatoms), commonly with bipolar, elongate valves.

per mille (also per mil, permil, and ‰): literally, parts per thousand; the conventional unit for expressing isotope delta (δ) values.

perched lake: a lake that is isolated above the groundwater table by a layer of rock or organic material.

periglacial lake: a lake originating on, or in immediate contact with, a glacier.

periodicities: periods or rhythmic occurrences of an event or process.

periodogram: a graphical representation of the results of a harmonic or spectral analysis that determines the period of the cyclical terms in a time series.

periphyton: microorganisms attached to substrata (e.g. rocks, sand, mud, logs, and plants).

periplekton: a modified rimoportula consisting of an outwardly projecting hollow stem bearing two arms at its distal ends that clasp the stem of a similar structure arising from an adjacent frustule.

perizonium: secondary covering of an auxospore consisting of closed and/or open rings or strips laid in specific orientation relative to the direction of auxospore expansion.

permafrost: ground remaining at or below 0 °C continuously for at least two years.

permanently open ocean zone: the region of the Southern Ocean between the winter sea-ice edge and the Antarctic Polar Front (APF). As implied by the name, these waters are never covered by sea ice.

persistence: in the context of diatom ecology, the survival time of the ecological system or part of the system (e.g. populations).

PETM: Paleocene–Eocene Thermal Maximum.

phosphor: chemical compound that emits light when excited by particles such as electrons.

phosphatase: an enzyme that cleaves phosphate molecules from organic molecules that have phosphate groups attached.

photoautotrophic: adjective describing organisms that use carbon dioxide as a source of carbon and light as a source of energy.

photoluminescence: a process in which a substance absorbs photons (electromagnetic radiation) and then re-radiates photons. Quantum mechanically, this can be described as an excitation to a higher energy state and then a return to a lower energy state accompanied by the emission of a photon.

photopolymerization: any polymerization reaction that requires light for the propagation step.

photosystem I and II: protein complexes involved in photosynthesis. Photosystem I is most reactive at the wavelength of 700 nanometers, while photosystem II is at 680 nanometers.

photic zone: the volume of water in a water body that is delineated by the atmosphere–water interface downwards to a depth where light intensity falls to 1% of that at the surface. The thickness of this layer corresponds with the depth in the water column, where light penetration is sufficient to support photosynthesis.

phreatic: adjective describing water in the saturated zone below the static water table.

phylogenetic analysis: a method of classification based on the study of evolutionary relationships between species. The criterion of common ancestry is fundamental and is assessed primarily by recognition of shared derived character states. See cladistic analysis.

phytobenthos: the photosynthetic organisms associated with submerged surfaces in aquatic ecosystems. Taxonomically, these organisms can include algae, bryophytes, and vascular angiosperms.

phytodetritus: detritus ultimately of phytoplanktonic origin, including intact or fragmented cells and also fecal matter from zooplankton, that sinks rapidly to the sea floor. Pulses of phytodetritus are episodic and important as a source of food to the deep-sea benthos.

phytolith: a siliceous structure that is formed inside the plant cell, and released into the surroundings after the cell's death. Sometimes referred to as "silicon phytolith" or "opal phytolith." The name means "plant stone" and these structures are known in the fossil record from late Devonian onwards. Grass phytoliths are known from the Cretaceous onwards.

phytoplankton: photosynthetic organisms suspended in the water and not capable of determining their own position in the water column, particularly their horizontal position. Some of these organisms can migrate vertically in the water column, but their position is largely determined by currents.

picoplankton: microorganisms drifting in the seas and freshwaters whose size ranges between 0.2– and 2 μm.

PICT: pollution-induced community tolerance.

pinnipeds: semi-aquatic marine mammals (or fin-footed mammals), including seals, sea lions, and walruses.

PIRLA project: Paleoecological Investigation of Recent Lake Acidification project.

P:L (P/L): the ratio of planktonic to littoral diatoms.

planktivore: an organism that feeds on plankton.

plankton: bacteria, algae, protozoa, fungi, small animals, and other organisms suspended in the water and not capable of determining their own position in the water column, particularly their horizontal position. Some of these organisms can migrate vertically in the water column, but their position in largely determined by currents.

plankton:benthos: the ratio of plankonic forms to benthic forms in a diatom assemblage.

plateau response model: a species response model where the modeled response along an environmental gradient rises and then forms a plateau. Such a model is very rare in diatom response models.

playa lake: a shallow intermittent lake in an arid or semi-arid region, which is filled only during wet periods and then subsequently dries via evaporation, usually leaving deposits of soluble salts.

Pleistocene: the first of two geological epochs within the Quaternary period, for the time from approximately 2.6 Ma to 11,700 years ago.

pleniglacial: full glacial; maximum of a glaciation.

Pliocene: the second and youngest epoch of the Neogene spanning from ~5 to 2.6 Ma.

Plio-Pleistocene boundary: the sedimentary transition between the Pliocene and Pleistocene epochs at ~2.6 Ma.

PLS: partial least squares.

pluvial lake: a lake formed in a period of exceptionally heavy rainfall; in the Pleistocene epoch the term refers to a lake, formed during a moist interval, that is now either extinct or exists as a remnant.

pneumorrhizae: the roots of some mangrove trees that emerge from the trunk and branches (also called prop roots in *Rhizophora*) or grow back up into the air from buried root branches.

POC: particulate organic carbon.

point sources: nutrients, especially nitrogen and phosphorus, from a well-defined single point such as a pipe (e.g. sewage effluent, industrial wastewater).

Polar Front: sharp gradient in temperature that forms the boundary between the polar cell and the Ferrel cell on each hemisphere. The Polar Front arises as a result of cold polar air meeting warm tropical air. See also Antarctic Convergence.

pollution-induced community tolerance: a measure for indicating damage to an algal community grown on artificial substrate, which uses the selective pressure of a toxicant on sensitive species, so that the community as a whole shows an increase in tolerance for the toxicant.

polyamine: organic compounds having two or more primary amino groups.

polyhaline: salinity 18–30 ppt (definition according to the "Venice System").

polyhalobous: adjective describing organisms that thrive in fully marine conditions, with a salt concentration exceeding 30 practical salinity units (psu).

polymictic: adjective describing lakes that experience frequent or continuous periods of water-column mixing throughout their annual cycle, including (i) cold polymictic lakes that warm after ice-out to mix at 4 °C (usually located in polar regions), and (ii) warm polymictic lakes that cool to temperatures above 4°C and mix (usually located in tropical regions).

polynya: area of open water surrounded by ice.

polyphyletic: systematic term referring to a taxon that does not share a common ancestor that gave rise exclusively to individuals or taxa considered to be members of the taxon. Polyphyletic taxa are not monophyletic, and require taxonomic revision.

polysaccharide: a carbohydrate which consists of a number of linked sugar molecules, such as starch or cellulose.

poor fen: a fen that is weakly connected to the water table, is low in base cations, has pH <5.0, has a lower number of species than rich fens, and in which *Sphagnum* mosses dominate.

POOZ: permanently open ocean zone.

postseismic: the postseismic phase of the seismic–interseismic deformation cycle is the rapid recovery after the earthquake. Deformation is recovered at first rapidly (the postseismic phase), and then more slowly (the interseismic phase). This cycle, typically repeating over some hundreds of years, may be superimposed on longer-term patterns of net uplift or subsidence.

power-law relationship: an empirical relationship between two variables that has a power component (e.g. power of 2) in the model equation.

power spectrum: a function for a stationary time series that has the properties that define the "power" or the contribution to the total variation of the time series made by the frequencies in the particular band.

ppt: parts per thousand.

practical diatom index: a mathematical index of water quality based on the relative abundances of certain diatom taxa in a sample.

Praetiglian: a European time-stratigraphic term that refers to cold climate environments and sediments deposited during the late Pliocene (*c.* 2.4 Ma) and preceding the Tiglian warm conditions.

PRC: principal response curve.

precession: relates to the wobble of the Earth's axis during a precessional cycle with a period of *c.* 21 ka.

prepattern: an assumed direct relationship between the spatial distribution of a purported morphogen and the visible pattern that develops in an organism.

preseismic: the preseismic phase of the seismic–interseismic deformation cycle occurs between earthquakes, when the shallow part of the subduction zone is locked, strain accumulates on the fault, and the coastline rises (with subsidence offshore).

principal components analysis: an ordination method for indirect gradient analysis that finds the best-fitting linear combination of variables (latent variables) to minimize the total residual sum of squares and finds subsequent linear combinations that are uncorrelated to previous combinations. It assumes a linear species response model to these latent variables. In the process it transforms original variables in the multivariate data into new composite variables that are uncorrelated and account for decreasing proportions of the variance of the data. The main aim of the method is to reduce the dimensionality of the data and to find the latent variables.

principal coordinates analysis: an ordination method for indirect gradient analysis in which the required coordinate values for the axes are found from the eigenvectors of a matrix of inner products of any dissimilarity or similarity coefficient. The object points are mapped onto the resulting low-dimensional ordination space so that the distances between the objects are as close as possible to the original dissimilarity (= distance) or transformed similarity. Also known as classical or metric scaling.

principal response curves: a multivariate method for the analysis of repeated measurement designs that tests and displays treatment effects that change across time. It is based on redundancy analysis (= reduced rank regression) that is adjusted for changes across time in the control treatment. It focuses on the time-dependent treatment effects. In some instances, space can be substituted for time in ecological studies.

probe: natural or synthetic molecule used to detect a specific biological target via recognition of a genetic sequence (e.g. oligonucleotide probe) or chemical structure (e.g. antibody probe); normally requires reporter (e.g. fluorescence, luminescence, etc.) to visualize probe binding to target.

productivity: rate of production.

proglacial lake: a lake formed either by the damming action of a moraine or an ice dam during the retreat of a melting glacier, or one formed by meltwater trapped against an ice sheet due to isostatic depression of the crust around the ice.

proline: a non-polar amino acid.

protein templating hypothesis: the assumption that diatom valve patterns are the result of a prepattern in which the morphogen is a protein.

proteoglycans: proteins conjugated to several units of polysaccharides.

protist: eukaryotic organisms lacking differenetiation into complex organs and tissues.

pseudoaerial: adjective describing algae that live on rocks moistened by a fairly steady source of water, such as waterfall spray, springs, or seeps.

psu: practical salinity unit = the total concentration of ionic components in g per kg water (= parts per thousand).

pycnocline: a rapid change in water density with depth caused by decreasing temperature (thermocline) and/or increasing salinity (halocline); this layer becomes a barrier for water mixing from above and below resulting in water-column stratification.

pyrite: an iron sulfide compound that is found in soils and sediments, and is formed in the absence of oxygen.

pyritization: a natural process involving the replacement of the original material with the mineral pyrite.

quadratic: a function of predictor variables in a regression model where the variable x_i is combined with x_i^2.

quantile–quantile plot: a plot for comparing two probability distributions, where the coordinates are the quantiles for different values of the cumulative probabilities. Quantiles are values that divide a frequency distribution or probability distribution into equal ordered subgroups such as quartiles or percentiles.

quantile regression: a technique to estimate the conditional quantiles of a response variable distribution in a linear model that provides a more complete view of possible causal relationships between variables.

quartiles: the values that divide a frequency distribution or a probability distribution into four equal parts.

quasibiennial oscillation: a self-generating, dynamic relation in the atmosphere that appears as an oscillation in temperature, pressure, and other climatic and oceanographic parameters, occurring roughly every two years.

quasi-likelihood: a function that can be used in place of a conventional log-likelihood in maximum-likelihood estimation when it is not possible and/or desirable to make a particular distributional assumption about the observations.

Quaternary: consists of the Pleistocene and the Holocene epochs, spanning the last *c.* 2.6 Ma.

r-selected species: populations that are adapted for rapid colonization and growth. Offspring produced are numerous and mature rapidly. Consequently, such populations grow and reproduce quickly, and tend to be good colonizers of new or disturbed habitats. Opposite to K-selected species.

r^2: coefficient of determination.

RACER: Research on Antarctic Coastal Ecosystem Rates.

radiocarbon (^{14}C) dating: a dating technique that utilizes the ratio between the unstable (^{14}C) and stable (^{12}C) isotopes of carbon preserved in once-living organic materials. The technique can produce reliable dates back to about 40 ka.

raised bog: an ombrotrophic peatland (bog) with a distinctly dome-shaped profile elevated above the surrounding land.

Ramsar: wetlands of international importance designated under the convention of wetlands of international importance. The Ramsar convention was adopted in Ramsar, Iran, in 1971.

random forest: a classification or regression tree designed to produce reliable predictions and that avoids being over-fitted. Bootstrap samples are drawn to construct multiple trees and each tree is grown with a randomized set of predictors. A large number of trees (500–2000) are grown, hence a "forest" of trees.

random settling method: a method described by Moore in 1973 that assists in the random dispersion of diatoms onto glass coverslips, placed in a petri dish, through the slow removal of the suspension via capillary action.

range: the numerical difference between the largest and the smallest values for a variable in a data set.

rank dissimilarities: dissimilarities between all pairs of samples reduced to ranks, in either ascending or descending order.

raphe: a slit through the valve face of monoraphid and diraphid diatoms, usually situated along the apical axis. This is the structure which enables a diatom cell to move over substrates.

raphid: describing a diatom that has a raphe on one or both valves.

rarefaction analysis: a means of standardizing assemblage samples for estimating taxonomic richness to a common sample size and for estimating the number of taxa that would be expected in samples if all the samples had the same count size.

rate-of-change analysis: a method for estimating the rate of compositional change in stratigraphical assemblage data. Compositional dissimilarity between adjacent pairs of samples is calculated and this is standardized to units of dissimilarity or difference per unit time.

RDA: redundancy analysis.

realized niche: that part of the fundamental niche actually occupied by a species in the presence of other species.

realm: a term often used by paleontologists and paleoceanographers to describe the extent of a paleo-ocean or climatic zone.

Redfield ratio: the approximate molar ratios of carbon, nitrogen, and phosphorus found in living phytoplankton (C:N:P of 106:16:1).

redox-sensitive metals: metals that undergo phase changes with oxidation–reduction reactions under hypoxic or anoxic conditions.

redundancy analysis: a constrained ordination technique based on principal components analysis in which the ordination axes are constrained to be linear combinations of the environmental variables. Being based on principal components analysis, it assumes a linear species–environment response model. Also known as reduced-rank regression.

reference lake: a lake that has been minimally impacted by anthropogenic activity that can be used as a benchmark against which to judge the status of impacted lakes.

regime shift: a rapid reorganization of an ecosystem from one relatively stable state to another.

regional occupancy: species occurrence, reproduction, and survival within a specified geographical or environmental region.

regression: (i) a statistical technique that describes the dependence of one variable on another (cf. correlation, which assesses the relationship between two variables); (ii) when used in the context of training sets, the "regression" step refers to the estimation of species parameters (e.g. optima, tolerances) from the species abundances in the training set; (iii) retreat of the sea from a land area.

regression analysis: a frequently applied statistical modeling technique that serves as a basis for studying and characterizing a system of interest by formulating a mathematical model of the relationship between a response variable, y, and set of x explanatory variables. The choice of model may be based on previous knowledge or on considerations derived from an exploratory data analysis.

regression diagnostics: procedures designed to investigate the assumptions underlying particular forms of regression analysis such as normality and distribution of residuals, or to examine the influence of particular data points or small groups of data points on the estimated regression coefficients.

regressive processes: see transgressive and regressive processes.

relative sea level: the height of the sea measured relative to a tide gauge on land. Changes in RSL will be a combination of ocean- and land-level changes.

Remane's brackish-water rule: a model of the generalized penetration of marine, freshwater, and brackish-water animals into an estuary in relation to salinity. The diversity of marine and freshwater animals is shown as a percentage of total species diversity in each source habitat. Diversity of brackish-water animals is shown as a subdivision of marine animals. Around salinity of 5 psu, there is a pronounced minimum in the total animal diversity, with c. 20% of the species number of freshwater habitats. Named after A. Remane.

repeated measures: an experimental design in which the measurements are taken at two or more points in time on the same set of experimental units.

residence time: the average time water spends within a specified region, such as an estuary or reservoir.

resting spores: diatoms form cells with reduced metabolic activity called resting spores in times of nutrient or environmental stress. Generally the spores are heavily silicified frustules and contain significant levels of photosynthetic products that allow them to return to normal cell functioning if returned to improved environmental conditions.

retting: a stage in the manufacturing of vegetable fibers which involves submerging plant stems such as flax, hemp, or jute to separate the fibers from the woody part of the plant.

reverse engineering: the process of analyzing the construction and operation of a product or organism in order to manufacture or computer-simulate a similar one.

rhizosolenid: any one of several planktonic genera characterized by large, elongate, spindle-shaped cells.

rich fen: a fen that is high in base cations, has pH > 5.0, is more species-rich than poor fens, and contains brown mosses, grasses, and sedges.

richness: the number of species (or higher specified taxonomic level, e.g. number of genera) in a habitat or sample.

rift lake: a lake that develops in a long and narrow continental trough formed by normal faults.

RMSE: root-mean-square error.

RMSEP: root-mean-square error of prediction.

rookery: a breeding ground or colony of gregarious animals, such as some seabirds.

root-mean-square error: the square root of the sum of the differences between the observed and estimated values for a variable squared and divided by the number of objects. The square root of the mean squared error.

root-mean-square-error of prediction: similar to root-mean-square error except that the estimated values are based on some form of cross-validation such as leave-one-out. This has a lower bias and is a more realistic guide to the performance of a model than the root-mean-square error.

RSM: redox-sensitive metal.

running mean: a method of smoothing a time series to reduce the effects of random variation and to reveal any underlying trend or seasonality. Weights can be used in the averaging.

S: the number of algal taxa (species and their varieties) in a specified count; species richness.

SAF: Subantarctic Front.

salinization: the process of increasing salt concentration in water bodies. Salinization arises from a variety of mechanisms of which decreased effective precipitation, land-use change, and upstream abstraction are the most common.

salinity: the total quantity of dissolved salts, usually expressed as the sum of the ionic concentration of the four major cations (sodium, magnesium, calcium, and potassium) and four major anions (carbonate, bicarbonate, sulfate, and chloride) in mass or milliequivalents per liter.

salt marshes: salt marshes are intertidal environments that are distinguished from neighboring tidal mud- or sandflats by a covering of vascular halophytic vegetation.

sample: sampling unit, individual, object, site.

sample score: the coordinates of a sample along an ordination axis. Sample scores can represent the position of biological communities along the coenocline.

sample-specific error: prediction errors for individual samples estimated by bootstrapping or Monte Carlo simulation in association with an environmental-inference model. The magnitude of such errors can depend on several factors, including taxonomic composition and model robustness.

Sangamon: stratigraphic name referring to the last interglacial stage in North America. The term is a name for a paleosol (buried soil) widespread in central USA. The term generally refers to the period between 125 and 70 ka that encompasses both warm and cool periods.

saprobic: a term associated with the amount of organic waste in an aquatic ecosystem.

sapropel: a fine-grained aquatic sediment rich in organic material, often of a greenish or black color.

SAV: submerged aquatic vegetation.

saxitoxin: tricyclic perhydropurine alkaloid that functions as a neurotoxin by blocking voltage-gated sodium channels and impairing signal transduction in excitable cells; over 30 naturally occurring analogs of differing toxic potencies are known; responsible for human intoxication syndrome known as paralytic shellfish poisoning (PSP).

SD: standard deviation.

SDV: silica deposition vesicle.

sea-ice algae: algae living in brine pockets within sea ice, frozen in when the ice forms in fall, and released to produce a large pulse of primary production when ice melts the following spring.

seagrass: flowering plant rooted in marine sediments belonging to several families, but not members of the grass family.

sea-level index point: a sea-level index point is a sample of known age from a known geographical position (horizontally and vertically) that possesses an indicative meaning. Index points collected from an area can be used to fix the former altitude of relative sea level in that locality.

seasonal sea-ice zone: or otherwise known as the seasonal ice zone (SIZ): the oceanographic zone affected by seasonal sea-ice cover. The zone varies in surface-water mixing and stratification through the annual season. The zone maintains mobile productivity levels as a function of sea-ice melting and retreat.

sediment trap: an instrument used in oceanography and limnology to measure the quantity of sinking particles, consisting generally of a funnel or tube and a collection recipient. Multiple collection cups can be programmed for rotation to assist in long-term particle capture and analysis.

seismite: a structure present in soft sediments, formed by tectonic activity, which can be used to determine past earthquakes.

SEM: scanning electron microscope.

semi-cryptic species: species characterized by very subtle differences in diagnostic characters among sexually incompatible individuals; in diatoms usually morphological characters.

semi-terrestrial: transitional between terrestrial and aquatic; only seasonally/periodically submersed.

senescent: term defining aging or damaged cells that have decreased rates of physiological processes, such as photosynthesis and respiration, compared to cells of actively growing populations.

seston: the particulate matter suspended in bodies of water such as lakes and seas. It applies to all particulates, including plankton, organic detritus, and inorganic material.

seta (pl. setae): in the case of some centric diatoms (e.g. *Bacteriastrum* and *Chaetoceros*), it is a long tubular extension of the valve. The name means "bristle" or "stiff hair" in Latin.

SEWT: summer epilimnetic water temperature.

shale: a fissile rock composed of layers of fine-grained silt and clay-sized sediments; similar to mudrock, but breaking into thin, platy laminations.

shallow lake: lakes typically with a mean depth of less than 3 m where, under a favorable light climate, productivity would typically be dominated by benthic communities of algae or macrophytes rather than by phytoplankton.

Shannon–Wiener diversity index (H'): informational index used to measure diversity of algal assemblages; expressed as "bits" ($\log_2$) or "nits" ($\log_e$) per individual.

SI: (Modified Pantle-Buck) Saprobic Index.

significance of zero crossings of the derivative: a graphical tool for use in association with smoothing methods in the analysis of temporal series that helps to answer which observed features are "real" as opposed to being spurious sampling artifacts.

silaffins: peptides that have been implicated in the biogenesis of diatom biosilica.

silica deposition vesicle: a structure involved in the formation of the siliceous valves of diatoms responsible for the sequestering and deposition of silica.

silica transporter genes: a gene family responsible for encoding portions of the mechanisms for silica uptake in diatoms.

silicalemma: a membrane in which diatoms precipitate silica during valve formation. New valves of the diatom frustule form within the structure.

silicate pump hypothesis: an explanation put forward by Dugdale and colleagues in 1995, which describes the preferential recycling of nutrients, rather than opal, in surface waters resulting in enhanced export of opal to the deep ocean.

silicatein: an enzyme used to capture silicate and build silica nanostructures.

SIMI: Stander's Similarity Index.

similarity index: a coefficient expressing the similarity of two multivariate samples, usually a ratio based on cumulative differences calculated between abundance of each variable in the two samples.

SIMS: secondary ion mass spectrometry.

sintering: a process in metallurgy of welding together small particles of metal by applying heat below the melting point; also the bonding of powdered or colloidal materials by solid-state reactions at a temperature lower than that required for the formation of a liquid phase.

SIP: sea-ice presence.

Sites A to L: alphabetically identified mooring location sites in McMurdo Sound.

Site ALB-1F: Alboran Sea sediment-trap site 1F.

Site ALOHA: A Long-Term Oligotrophic Habitat Assessment sediment-trap site.

Site BG: Barnes Glacier sediment-trap site in McMurdo Sound.

Site C: one of the stations of the survey line extending from the western North American coast into the north Pacific Ocean.
Site CB: Cape Blanc sediment-trap site.
Site CH: Chile sediment-trap site.
Site CV: Cape Verde sediment-trap site.
Site CWR: Central Walvis Ridge sediment-trap site.
Site DG1–2: Debenham glacier sediment-trap sites 1 and 2 in McMurdo Sound.
Site EA 1–3: Eastern Atlantic sediment-trap sites 1–3.
Site EBC: Eastern Boundary Current trap site.
Site EIT: Erubus Ice Tongue sediment-trap site.
Site ESTOC: European Station for Time-series in the Ocean sediment-trap site.
Site GBN: Guinea Basin north sediment-trap site.
Site GBZ: Guinea Basin south sediment-trap site.
Site GH: Granite Harbor sediment-trap site.
Site HP1, 2: Hut Point sediment-trap sites 1, 2.
Site JAM: Java Mooring sediment-trap site.
Site JV7: Jervis Inlet position 7, British Columbia fjords, Vancouver, sediment-trap site.
Site KG1–3: King George Island sediment-trap sites 1–3.
Site KH94–4: The Japanese Hakuhou-maru cruise code and station number for the sediment-trap site off Adélie Land.
Site M2–3: Mooring sites 2 and 3 of the French Antares/JGOFS program.
Site MS: McMurdo Sound ice-shelf sediment-trap site.
Site MST-8, -9: Moored sediment trap sites 8 and 9 off Somalia.
Site NH1, 3: New Harbor sediment trap sites 1, 3.
Site NU: Namaqua upwelling sediment-trap site.
Site PAPA: also called Station P (Pacific), one of the stations of the survey line extending from the western North American coast into the north Pacific Ocean.
Site PF1: Polar Front sediment-trap site 1.
Site PZB-1: Prydz Bay sediment-trap site 1.
Site R13–48: RACER sediment-trap sites 13–48.
Site SBB: Santa Barbara Basin sediment-trap site.
Site SP: Scripps Pier sediment-trap site.
Site TI: Tent Island sediment-trap site in McMurdo Sound.
Site WA: Western Atlantic sediment-trap site.
Site WR2: Walvis Ridge 2 sediment-trap site.
Site WS1–4: Weddell Sea sediment trap sites 1 to 4.
SITG: sea-ice taxa group.
SIZ: seasonal sea-ice zone.
SIZer: significance of zero crossings of the derivative.
skewed unimodal response model: a unimodal species response model that is not normally distributed and has a long thin tail on one side of the mode.
skewness: the lack of symmetry in a probability or frequency distribution. A distribution is said to have positive skewness when it has a long tail to the right and to have negative skewness when it has a long tail to the left.
smectites: dioctahedral (montmorillonite) and trioctahedral (saponite) clay minerals that possess high cation-exchange capacities.
smolt: a juvenile salmonid of one or more years old that emigrates from fresh- to saltwater. It is during this stage the the juvenile salmon undergoes the physiological changes needed to cope with a marine environment.
smoothing: a term that can be applied to almost any statistical technique that involves fitting a model to a set of observations. It is usually used for methods that use computing power to highlight structure very effectively by taking advantage of people's ability to draw conclusions from well-designed graphics. Examples include LOWESS (= LOESS) and kernel density functions.
sodium polytungstate: a water-soluble, pH-neutral heavy liquid ($3Na_2WO_4\ 9WO_3 \cdot H_2O$) with a maximum attainable solution density of 3.1 g cm^{-3} at 25 °C used to separate diatoms from other siliceous particles by floatation.
sound: a long and/or broad inlet of the ocean, generally with its larger part extending roughly parallel to the shore.
Southern Westerlies: as the circumpolar wind belt of southern high to mid latitudes, the Southern Westerlies have considerable influence on ocean circulation and global climate.
span: the proportion of a set of observations used in LOWESS (= LOESS) in which a locally weighted regression is fitted as part of deriving LOWESS curves.
spatial autocorrelation: the correlation between samples that are different distances apart in spatial data. Because of it, one can often predict the values of a variable in samples in space from the values of the variable in other samples whose spatial positions are known.
spawners: fish in their reproductive stage, ready to deposit eggs or sperm.
species level indices: indices that are based on identifying taxa to the species level and usually counting abundances of species in samples.
species response model: a model of the response of a species (presence/absence or abundance) in relation to one or more environmental variables.

spectral analysis: a statistical procedure which decomposes a time series into a spectrum of cycles of different lengths.

spectral density functions: the population counterpart of a periodogram which is a graphical tool in time-series analysis where the time series is decomposed into an infinite number of periodic components. Estimates of the contributions of these components in certain ranges of frequency are termed the spectrum of the series.

Spermatophyta: vascular plants that reproduce by seeds; most commonly used to refer to flowering plants or angiosperms.

split-plot: an experimental or sampling design used when comparisons of different types of treatments or environmental variables can be made at different scales. A split-plot design is a hierarchical design with two levels of units – whole-plots containing split-plots. Split-plots are the lowest level of sampling unit, namely the samples in a data set. Examples include samples-within-lakes, plots-along-transects, samples-within-time series (in a study of permanent plots).

square-root transformed data: data that have been transformed to the form $y = \sqrt{x}$; often used to stabilize variances and to make random variables more suitable for techniques such as regression.

squared chord distance coefficient: a simple measure of dissimilarity used, for example, to assess floristic dissimilarity between diatom samples.

squared residual distance: the distance squared between an observed value and the predicted value of a variable or sample in a statistical model. Least-squares estimation attempts to minimize these distances.

s.s: (Latin) *sensu stricto*, in the strict sense.

SST: sea-surface temperature.

stalked diatoms: diatoms that are attached to substrates by mucilaginous stalks.

standard deviation: the most commonly used measure of the spread of a set of observations. It is equal to the square root of the variance.

standard deviation units of compositional turnover: the length of a CA, DCA, CCA, or DCCA ordination axis (range of sample scores) expressed in standard-deviation units of compositional turnover. The tolerance of the species' curves along the axis is close to 1 after rescaling, and each curve therefore rises and falls over about four standard deviations. A gradient of more than four standard deviations can thus be expected to have no species in common.

standard error of the mean: the standard deviation of the sampling distribution of a statistic. The standard error of the mean of n observations is $\sigma/\sqrt{n}$ where σ^2 is the variance of the original observation.

Stander's Similarity Index: index used to compare the statistical similarity of two samples based on the species present and their relative abundances.

Station A: an alphanumeric station in Lützow–Holm Bay.

statistical independence: two events are said to be independent if knowing the outcome of one tells us nothing about the other.

statistical principle of parsimony: the general principle that among competing statistical models, all of which provide an adequate fit for a set of data, the one with the fewest parameters is to be preferred.

steinkern: the fossilized outline of a hollow organic structure formed when sediment or mineral is consolidated within the structure and the structure itself disintegrates or dissolves; an internal cast.

stenohaline: referring to an organism with a relatively narrow tolerance range for salinity.

Stephanodiscoid: refers to diatoms belonging to or related to the genus *Stephanodiscus*.

stratocladistics: a cladistic analysis that uses stratigraphic position to support or refute a phylogenetic hypothesis.

striae: linear rows of puncta/areolae, usually oriented along the transapical axis, between the costae in pennate diatoms.

subaerial: adjective describing algae that live in soils and other non-aquatic habitats.

Subantarctic Front: the front, in the Southern Ocean, to the south of the Antarctic Polar Front (APF). To the north of the SAF is the Polar Frontal Zone (PFZ) and to its north is the Subantarctic Zone (SAZ).

subglacial lakes: lakes that lie between overlying glacial ice and the underlying bedrock.

sublittoral: in marine systems, the shore area between low tide and a depth of *c.* 100 m. The term is also used in limnology for the region below the littoral zone.

submerged aquatic vegetation: a diverse assembly of rooted macrophytes that grow below the water surface and are found in shallow water areas of variable salinity; often termed "seagrasses" in marine systems, although they are not true grasses.

subsaline: water with an ionic concentration between 0.5 to $3\ g\ l^{-1}$.

subsidence: loss of land elevation by compaction of sediments.

subterranean: adjective describing algae that live at depth in the soil.

subtidal: below the low tide level.

sun-spot cycles: a cycle in the number of spots (faculae) on the surface of the Sun.

supplementary samples: see passive samples.

supratidal: the shore area just above high tide level.

surface-sediment calibration: a technique whereby contemporary limnological characteristics of a suite of lakes are related to the species composition of surface sediments (upper c. 0.5–3 cm) in order to estimate the range and optima of taxa relative to some environmental gradient(s). Also called "training sets."

SWAP: Surface Water Acidification Project.

SWF: stepwise fluorination.

sylvic peat: peat composed of woody material or is derived from trees.

symbiosis: two or more organisms living together, often to their mutual benefit (mutualism), but also may be parasitic (one benefits at the expense of the other), or commensal (one benefits while the other is neither harmed nor benefited).

symbiotic: living in symbiosis.

symmetry breaking: a phenomenon where small fluctuations acting on a system crossing a critical point decide a system's fate by determining which branch of a bifurcation is taken.

synecology: the study of the relationships between biotic communities and their environment.

syngamy: union of the nuclei from two gametes, following fertilization, to produce a zygotic nucleus.

synrift: occurring during the active rifting of a rift basin.

taiga: the world's largest terrestrial biome stretching over Eurasia and North America, situated below the tundra biome and characterized by boreal forest.

tailwater: water located immediately downstream from a hydraulic structure, such as a dam.

taphocoenosis: the fossil assemblage as it results from all taphonomic processes (death, burial, diagenesis, discovery); the stage following the thanatocoenosis (death assemblage).

taphonomy: the branch of paleoecology concerned with the processes that affect material before its permanent burial; in the case of fossil organisms, the processes that alter the composition of the death assemblage relative to the living assemblage.

TBT: tri-*n*-butyl tin.

TDI: Trophic Diatom Index.

TDIL: Trophic Diatom Index for Lakes.

TD Index: a ratio based on the percentage of "warm" diatom specimens relative to the total percentage of "warm" and "cold" specimens in a sample; used as a semi-quantitative indicator of relative temperature differences between samples.

tectonism: synonym of diastrophism, which is defined as a general term for all movement of the Earth's crust produced by tectonic processes, including the formation of ocean basins, continents, plateaus, and mountain ranges.

teleconnection: relationship between climate variations or anomalies that occurs in regions of the world physically far apart (usually thousands of kilometers).

temporal autocorrelation: the internal correlation of observations in a time series. It measures the extent of the linear relation between values at time points that are a fixed interval (or lag) apart.

temporal series: series of time-ordered samples common in paleoecology and paleolimnology where the time intervals between samples are not equal.

temporal-series analysis: analysis of temporal series to detect trends, periodicities, short-term variation, and irregular variations.

tenfold cross-validation: the division of data into 90% and 10% ten times. The 90% set is used to estimate the parameters in some model of interest ("training set") and the 10% set is used to assess if the model with these parameter values fits the data adequately ("test set").

tephra: volcanic ash. When found in a sediment core, and the tephra layer can be linked to a known volcanic eruption, the layer can be used to help date the sediment core.

tephrochronology: a chronological technique that utilizes the presence of volcanic ashes of known age in deposits to ascribe a date to the deposit.

teratological: adjective describing visible aberrations in diatom valves caused by interruption or alteration of normal development, often associated with exposure to unnatural concentrations of metals or chemical toxins.

terminal valve: valve on the end of a colony of diatom cells.

Termination 1: period of rapid deglaciation at the end of the last ice age to the start of the Holocene.

terrestrialization: a pathway for peat formation whereby a lake or shallow water body is gradually infilled with accumulated debris from organic and inorganic sources to become more terrestrial in nature.

terrigenous: sediments derived from the land, especially by erosive action.

test set: a data set deliberately excluded from statistical modeling based on a training or calibration set to be use as a test where parameters from the statistical model are used to predict values in the test set and to assess the predictive power of the model.

Tg: teragram; equivalent to 10^{12} grams or one megatonne (one million metric tonnes).

thanatocoenotic: pertaining to a community/assemblage of deceased organisms.

thermocline: the location in a water body where temperature changes at a maximal rate with depth in the water column.

thermohaline circulation: large-scale, asymmetric ocean circulation currents that are driven by density gradients (linked to temperature and salinity), and mechanical forces such as tides and winds.

thermokarst lakes: periglacial features formed in areas of permafrost, where ground ice has thawed and filled with water. These depressions can result from climate change, fire, vegetation change, etc.

thin-plate splines: a technique for the analysis of shape of organisms using landmark configurations. Landmarks for two organisms, each regarded as being fixed to a thin metal plate, will necessarily be bent in the fitting of one form to the other. The degree of distortion involved can be expressed as energy generated by bending one plate so as to make its points conform with those of the second plate.

tidal datum: chart datum based on a phase of the tide.

tidal flats: tracts of wet, low-lying, level land that are inundated regularly by ocean tides.

tidal head: farthest point upstream where a river is affected by tidal fluctuations.

tidal prism: volume of water that flows in and out of an estuary with the flood and ebb of the tide, excluding any contribution from river inflow.

till: unsorted glacial sediment, which may contain clays, sands, gravels, and boulders.

time-domain approach: the direct analysis of the observed values in a time series involves the time-domain approach, whereas analysis in terms of periodic functions is the frequency-domain approach. Common techniques in the time-domain approach are autocorrelation coefficients, autocorrelograms, cross-correlation coefficients, and cross-correlograms.

time series: a series of observations over a long period of time, usually at regular intervals, of a random variable. The observed movement and fluctuations of many such series consist of four components – secular trend, seasonal variation, cyclical variation, and irregular variation.

TN: total nitrogen.

TOC: total organic carbon.

tolerance: in diatom research, the range of an environmental variable which an organism or population can survive or one standard deviation of a Gaussian unimodal response model estimated by Gaussian logit regression or as a weighted standard deviation estimated by weighted-averaging regression.

toxigenic: adjective describing organisms intrinsically capable of toxin(s) biosynthesis.

toxin: substance that is a specific product of the metabolism of living cells or organisms, which can be poisonous on contact with or absorption by body tissues of other organisms by interacting with macromolecules such as enzymes or cellular receptors; term biotoxin is also often used as a qualifier to denote a biological source of the toxin.

TP: total phosphorus.

training set: a survey of biotic assemblages preserved in the surface sediments of lakes with an associated set of contemporaneous environmental measurements. Used to generate transfer functions or inference models. Synonymous with "surface-sediment calibration sets."

transfer function: a mathematical function that describes the relationships between biological species and environmental variables that allow the past values of an environmental variable (e.g. pH, salinity) to be inferred from the composition of a fossil assemblage.

transgression: flooding of a land area by a rise in relative sea level resulting in an onshore migration of the high-water mark.

transgressive and regressive processes: periods of marine transgression are indicated by an upward transition from salt-marsh peats to estuarine clays and silts whilst a regression may lead to a change from estuarine silts and clays to salt-marsh peats. The precise nature and characteristics of transgressive or regressive processes are local and site dependent. Stratigraphical descriptions can allow basic mapping of the varying paleogeographies of the coastline, bathymetries, and tidal dispositions, and changing sediment fluxes.

transitional waters: bodies of surface water in the vicinity of river mouths which are partially saline in character as a result of their proximity to coastal waters but which are substantially influenced by freshwater flows (Directive 2000/60/EC of the European Parliament).

Triassic: the first geological period (approximately 251–199 Ma) of the Mesozoic era, named after the three layers of sediments (red beds, chalk, and black shales) which characterize parts of northern Europe; *trias* (Latin) means triad.

tri-*n*-butyl tin: a chemical substance capable of killing living organisms (a biocide) used in marine paints to prevent fouling. Abbreviation: TBT.

trimmed mean: a method of estimating the mean that is less affected by the presence of outliers than the usual estimator, the mean. It involves dropping a proportion of observations from both ends of the sample before calculating the mean of the remainder.

triplot: an ordination diagram with three kinds of entities of which all pairs form biplots. Examples are triplots of species, samples, and environmental variables in redundancy analysis or CCA.

tripoli: a light-colored powdery sedimentary rock that forms as a result of weathering of chert. This term was originally applied to diatomite deposits, but this is now considered to be an incorrect usage.

trona: a white sodium-based mineral, which occurs in columnar layers or masses in saline deposits: $Na_2(CO_3)$, $Na(HCO_3){\cdot}2H_2O$.

TROPH 1: a diatom-based trophic index developed by T. Whitmore to estimate changes in lake trophic status. The index is based on the ratio of diatoms indicating high trophic status to those indicating low trophic conditions.

Trophic Diatom Index for Lakes: a mathematical index developed by Stenger-Kovacs and colleagues in 2007 in Hungary using epiphytic diatom samples from shallow lakes to assess trophic status.

troposphere: layer of atmosphere (air) extending about 12 to 15 km upwards in the tropics and ~8 km near the poles from the Earth's surface.

TSI: total solar irradiance, which represents the amount of incident radiative energy arriving at the top of the Earth's atmosphere.

TSM: (Ontario) Trophic Status Model.

tsunami: large ocean wave usually caused by an underwater earthquake or mass movement of sediment.

tundra: treeless subpolar biome characterized by permafrost, low temperatures, and a short growing season.

turbidity: a visual property of water that changes as a function of the quantity of suspended and dissolved matter present. The sources can be organic (biological origin) or inorganic (geologic origin).

turnover: the fraction of an assemblage which is exchanged or lost per unit time or per unit of an environmental variable.

TWINSPAN: two-way indicator species analysis.

two-way indicator species analysis: a method for partitioning large data sets of assemblages into groups using the first correspondence analysis axis as a basis for division. The algorithm continues to produce 2, 4, 8, 16, etc. groups unless a resulting group is too small to justify further division. Species are then grouped on the basis of their indicator value in relation to the groups of samples.

tychoplankton: literally "accidentally plankton," and used to describe organisms that are often accidentally entrained in the water column and, as a result, can live as either benthic organisms or part of the plankton. For example, benthic diatoms which lie on the sea-floor substrate may be uplifted by currents into the surface waters amongst the plankton.

UKAWMN: United Kingdom Acid Waters Monitoring Network.

uniformitarian: adjective describing a scientific philosophy that assumes that the natural processes that operated in the past are the same as those that can be observed operating in the present.

unimodal response: the expected non-linear response of a biological species to an environmental variable along an environmental gradient. The abundance of a species is expected to be at its maximum at the center of its range.

upwelling: phenomenon that involves wind-driven motion of dense, cooler, and usually nutrient-rich water towards the ocean or lake surface, replacing the warmer, usually nutrient-depleted surface water.

USEPA: United States Environmental Protection Agency.

UV: ultraviolet (radiation).

UV-B: The B component of the spectral ultraviolet radiation band that has a wavelength of 280 to 320 nm. This wavelength is the most damaging to biological forms.

vagal: relating to the cranial nerve that is concerned in regulating heart beat and rhythm of breathing.

valves: the siliceous cell wall of a diatom (frustule) is composed of two halves, which are called valves. These valves often separate when the protoplasm is oxidized out of cells.

valvocopula: the girdle band adjacent to a valve.

variance: a measure of the variability in the values of a random variable. It is estimated by the squared difference between the variable and its mean.

variance inflation factor: an indicator of the effect that other predictor variables have on the variance of a regression coefficient of a particular predictor variable.

variation partitioning: the partitioning of the variation in a data set into the variation uniquely explained, in a statistical sense, by two or more sets of predictor variables, into the covariance between variable sets, and into the unexplained component. It most commonly uses (partial) CCA or (partial) RA.

varve: sediment deposited during the course of one year composed of two or more visible seasonal layers (e.g. a light summer layer and a dark winter one).

Venice System: a final resolution from a symposium held in Venice, Italy, 1958, on the Classification of Brackish Waters, and published in 1959.

Vienna PeeDee Belemnite: a global standard derived from the bulk geochemical signature of Cretaceous belemnite guards from the Peedee Formation of southeast USA, against which the carbon and oxygen isotopes in geochemical samples are measured. Although originally called PDB, a Vienna-based laboratory calibrated a new reference sample when the original sample was exhausted.

Vienna Standard Mean Ocean Water: the isotopic water standard defined by the International Atomic Energy Agency which serves as a reference standard for comparing hydrogen and oxygen isotope ratios, mostly in water samples.

viscous fingering: the formation of patterns in a morphologically unstable interface between two fluids.

VPDB: Vienna PeeDee Belemnite.

VSMOW: Vienna Standard Mean Ocean Water.

WA: simple two-way weighted averaging.

WACALIB: a computer program developed by Line and colleagues in 1994 that is used to develop transfer functions (e.g. weighted-averaging regression and calibration, and other techniques).

WA-PLS: weighted-averaging partial least squares regression and calibration.

wadi: a stream bed, channel, or ravine that is usually dry but may fill with water in the rainy season; a term used in the desert regions of Africa and southwest Asia.

Walther's law of facies: only those facies and facies areas can be superimposed primarily which can be observed beside each other at the present time; consequently, vertical facies successions reflect horizontal spatial relations.

Water Framework Directive (2000/60/E): legislation (WFD) that came into force in 2000. This is the most substantial piece of water legislation ever produced by the European Commission, and provides the major driver for achieving sustainable management of water across Europe.

wavelength: pertaining to sea state, the horizontal distance between the crests of adjacent waves.

wavelet power-spectral analysis: an approach to representing signals in time series that is analogous to the Fourier approximation of trigonometric sine and cosine functions but which uses a linear combination of wavelet functions. Its strength is its ability to describe local phenomena more accurately than an expansion of sines and cosines can. There are two main types of wavelet functions – mother wavelets and father wavelets. Both oscillate about zero but they damp down to zero so the function is localized in time or space. Mother wavelets integrate to 0, father wavelets integrate to 1. Mother wavelets are thus good at representing the detail and high-frequency parts of a signal, whereas father wavelets are good at representing the smooth and low-frequency parts of a signal.

weathering: the mechanical (physical), chemical, and biological breakdown of rocks and soils through contact with the planet's atmosphere, but without movement (= erosion).

weighted averaging: a technique used to estimate either: (i) the optimum of a taxon (weighted-averaging regression) based on measured values of environmental variables from the lakes in a training set, where the weight is proportional to the species abundance; or (ii) an environmental variable from the species composition of a sample, based on estimates of species parameters from a training set, where species are weighted relative to their abundance (weighted-averaging calibration).

weighted-averaging calibration: the estimation of environmental values at a site from optima of the species present by an average of the optima weighted by the relative abundance of the species present.

weighted-averaging method: an estimation procedure based on the unimodal response model where the optimum (mode, ideal point) of a species is estimated by weighted-averaging regression, for example in correspondence analysis.

weighted-averaging partial least squares: an extension of weighted-averaging regression where partial least squares regression is used to find components within the modern assemblage data that will maximize the covariance between the species' weighted averages and the environmental variable of interest. It uses residual structure in the species data to improve the estimates of the species' parameters in the final prediction model. A WA-PLS first component is the same as a two-way weighted-averaging model that uses an inverse deshrinking regression. Prediction or reconstruction (calibration) of an environmental variable from a fossil assemblage is done by weighted averaging of the species' parameters, as in weighted-averaging calibration.

weighted-averaging regression: the estimation of species' optima and tolerances with respect to known environmental variables. The optimum is estimated as an average of the environmental variable weighted by the relative abundance of the species presence. The tolerance is estimated as a weighted standard deviation.

wet mount: entrapment of a suspension of microorganisms under a coverglass on a microscope slide.

wetlands: habitats with characteristics of both dry land and bodies of water, typically occurring in low-lying areas (edges of lakes, ponds, streams, and rivers or isolated depressions) that receive freshwater or in tidal areas protected by waves. Surface water is typically at, above, or just below the land surface for enough time to restrict growth of plants to those adapted to wet conditions and promote development of soils characteristic of wet environments.

WFD: Water Framework Directive.

white ice: ice that appears white in color due to entrapment of air or other materials during melting and refreezing. This type of ice is far less transparent to light than black ice.

WHOI: Woods Hole Oceanographic Institution.

whole-plot: the term is derived from an experimental design called the split-split design (see split-plot). An example is samples taken from within a lake where the lake is the whole-plot.

Wm^{-2}: watts per square meter.

WTN: water total nitrogen; mass of organic and inorganic nitrogen compounds per liter of water.

WTOC: water total organic carbon; mass of carbon bound in different organic fractions per liter of water.

WTP: water total phosphorus; mass of organic and inorganic phosphorus compounds per liter of water.

Würm: the last glacial stage of the Pleistocene in the European Alps, equivalent in age to the Weichselian in northern Europe, the Devensian in Britain, and the Wisconsinian of North America.

xeric: see xeritic.

xeritic: adjective describing aerial algae and other organisms that are exposed to long periods of desiccation between brief periods of moisture.

Younger Dryas: a period of abrupt cool climate (stadial) which occurred just prior to the Holocene between c. 12.9 and 11.7 ka BP. Its cause is most commonly related to a massive influx of freshwater from Lake Agassiz into the North Atlantic ocean, slowing down thermohaline circulation.

yr BP: years before present.

XRF: X-ray fluorescence.

zeolite: a generic term for a large group of hydrous aluminosilicates that have sodium, calcium, and potassium as their chief metals.

zooxanthellae: photosynthetic autotrophs living as intracellular mutualistic symbionts in coral reef animals and protists.

zwitterion: a chemical compound that carries a total net charge of zero and is thus electrically neutral, but carries positive and negative charges on different atoms.

Index